IONIC INTERACTIONS IN NATURAL AND SYNTHETIC MACROMOLECULES

IONIC INTERACTIONS IN NATURAL AND SYNTHETIC MACROMOLECULES

EDITED BY

ALBERTO CIFERRI

ANGELO PERICO

WILEY

A JOHN WILEY & SONS, INC., PUBLICATION

Published by John Wiley & Sons, Inc., Hoboken, New Jersey
Published simultaneously in Canada

Library of Congress Cataloging-in-Publication Data:

Ionic interactions in natural and synthetic macromolecules / edited by
Alberto Ciferri, Angelo Perico.
p. cm.
Includes bibliographical references.
ISBN 978-0-470-52927-0
1. Macromolecules. 2. Ion-ion collisions. 3. Supramolecular chemistry.
I. Ciferri, A. II. Perico, Angelo.
QD381.I64 2012
547'.7045723–dc23

2011041448

Printed in Singapore

10 9 8 7 6 5 4 3 2 1

CONTENTS

PREFACE

Two editions of *Supramolecular Polymers* (Dekker 2000 and CRC Press 2005) have dealt with the self-assembly of structures in the nano- and up-dimensional range in terms of molecular (chemical and shape) recognition and growth mechanisms. This third book is devoted to a more detailed description of how the components of chemical recognition are modulated to produce the variety of properties that characterize the supramolecular organization of functional systems and adaptive polymers. Fundamental approaches are highlighted, along with descriptions of a selected number of sophisticated applications.

Ionic interactions definitively play a primary role in the self-assembly of functional biological and synthetic systems. Theoretical features of purely ionic interactions have been extensively described in the past, and structural features of several real systems in which ionic interactions play a prevailing role have been experimentally characterized. However, in the majority of real cases, ionic interactions are modulated (reinforced or antagonized) by the occurrence of other, charge-independent interactions that produce important alterations of the structure and properties of the systems.

The first section of the book includes chapters focused on fundamental aspects of purely ionic interactions. The properties of simple salts described by *Marcus* are essential for the analysis of specific details of their interaction with organic substrates. Polyelectrolyte theories are presented by *Perico*, who places particular emphasis on the role of ion condensation and like-charge attraction on the stabilization of biological assemblies. Systems with opposite fixed charges include host–guest complexes, described by *Schneider*, which will facilitate quantitative comparisons with the polymeric ion pair systems such as polyampholytes, described next by *Thirumalai* and coworkers, and layered polyelectrolytes, described by *Holm* and coworkers.

The second section of the book includes chapters focused on fundamental aspects of mixed interactions. *Ciferri* discusses correlations of mixed systems and the role of mixed interactions in the Hofmeister effects. Polyelectrolytes with a significant number of hydrophobic groups, described by *Olea*, and polyelectrolyte–surfactant/protein assemblies, described by *Ciferri and Perico*, offer vivid evidence for the variety of supramolecular structures that are promoted by the microsegregation of apolar segments or associated proteins. Also in the second section, *Nikonenko* and coworkers describe the structure and properties of charged membranes resulting from the microsegregation of hydrophilic channels within a hydrophobic matrix. Metal–ion coordination for reversible polymers is discussed by *Marcelis* and coworkers, whereas a broader range of the ligand interaction is described by *Lombardi* and coworkers in terms of protein thermodynamics, and the chapter by *Hermans* highlights the role of charges on the conformation of helical polypeptides and proteins

The third section of the book focuses on functions and applications of various systems, and is opened by the chapter by *Crumbliss and Parker Siburt* on ion transport across living cells. The biological transport mechanisms are brilliant examples of the way nature has engineered a complex sequence of ionic mixed interactions for a specific function. The chapter by *Chan and Wang* describes DNA–lipid assemblies that generate pseudoliposomic structures used for targeted gene and drug delivery. The chapters by *Chiarelli and De Rossi* and by *Tiwari and Kim* feature stimulus-responsive intelligent gels, with emphasis on biomedical applications and muscle-like actuators. The electromechanochemical functioning of the gels entails consideration of readjustment kinetics of forces and related fluxes, which are handled by continuum mechanics approaches.

The next three chapters deal with functional layered assemblies: *Knoll* and coworkers describe a variety of driving interactions exploited for the formation of nanostructured films; *Seantier and Deratani* emphasize the sequential polycation–polyanion approach and the formation of multilayered membranes; and *Cavallo and Comoretto* discuss the application as photonic crystals of ordered periodic structures formed by materials having alternating refractive indices. Next, the chapter by *Pourcelly* and coworkers describes complex separation processes (particularly water treatments) using charged membranes, and also the membrane-based energy conversion in fuel cells technology. Efficient energy sources, such as lithium batteries, are characterized by cationic conductivity even in the absence of water. Polymer gel electrolytes, described in the final chapter by *Ward and Hubbard*, are intermediate between typical liquid and solid polymeric electrolytes, and allow a molecular understanding of ionic conductivity.

The goal of this book is to provide a coordinated and comprehensive representation of ionic mixed interactions, and actual or potential applications. Proper design of the molecular structure of self-assembling building blocks requires the strategic inclusion of groups that will modulate their contributions so that tailored properties and even complex functions may be produced. The unified presentation of the systems described here, characterized by different blends of ionic and ionic mixed interactions, allows the identification of general mechanisms and a correlation between, at first sight, unrelated phenomena.

The result was made possible by the dedication and patience of the contributors. Extensive discussions followed the preliminary draft, and attempts were made to reach a consensus over controversial issues. The Editors express their appreciation to all colleagues who have cooperated in the preparation of the book, to the Chemistry Department at Duke University, and to the Institute for Macromolecular Studies (ISMAC) of the National Research Council.

The editors wish to honor the memory of their late colleague Eligio Patrone for his lifelong commitment to cultural and scientific values.

Alberto Ciferri
Angelo Perico

CONTRIBUTORS

Rigoberto Advincula
Department of Chemistry
University of Houston
Houston, TX 77204-50031

Dario Cavallo
Department of Chemistry and Industrial Chemistry
University of Genoa
Via Dodecaneso 31,
16146 Genova
Italy

Juan J. Cerdà
Instituto de Fisica Interdisciplinar y Sistemas Complejos
IFISC (CSIC-UIB)
Universitat de les Illes Balears
E-07122 Palma de Mallorca
Spain

Peggy P.Y. Chan
Department of Chemical Engineering
Monash University
Clayton, Vic. 3800
Australia

Piero Chiarelli
Institute of Clinical Physiology CNR
Via Moruzzi 1
56124 Ghezzano (Pi)
Italy

Alberto Ciferri
Department of Chemistry
Duke University
Durham, NC 27078

Davide Comoretto
Department of Chemistry and Industrial Chemistry
University of Genoa
Via Dodecaneso 31
16146 Genova
Italy

Alvin L. Crumbliss
Department of Chemistry
Duke University
Durham, NC 27078

Danilo De Rossi
Faculty of Engineering
University of Pisa
56100 Pisa
Italy

André Deratani
Institut Européen des Membranes
Université Montpellier II
CC 047 Place Eugène Bataillon
34095 Montpellier cédex 5
France

Nicel Estillore
Department of Chemistry
University of Houston
Houston, TX 77204-50031

Bae-Yeun Ha
Department of Physics and Astronomy
University of Waterloo
Waterloo, ON
Canada N2L 3G1

Jan Hermans
Department of Biochemistry and Biophysics
University of North Carolina
Chapel Hill, NC 27514

CHRISTIAN HOLM
Institute for Computational Physics
Universitat Stuttgart
70569 Stuttgart
Germany

HUGH V. ST. A. HUBBARD
School of Physics and Astronomy
University of Leeds
Leeds LS2 9J
UK

KWANG J. KIM
Active Materials and Processing Laboratory
Department of Mechanical Engineering
University of Nevada
Reno, NV 89557

WOLFGANG KNOLL
AIT Austrian Institute of Technology
Donau City, Vienna
Austria

KIM DE LANGE
Laboratory of Organic Chemistry
Wageningen University
Dreijnplein 8
6703 HB Wageningen
The Netherlands

ANGELA LOMBARDI
Department of Chemistry
University of Naples Federico II
Via Cintia
80126 Naples
Italy

ORNELLA MAGLIO
Department of Chemistry
University of Naples Federico II
Via Cintia
80126 Naples
Italy

Antonius T.M. Marcelis
Laboratory of Organic Chemistry
Wageningen University
Dreijnplein 8
6703 HB Wageningen
The Netherlands

Yizhak Marcus
Institute of Chemistry
Edmond Safra Campus
The Hebrew University of Jerusalem
Jerusalem 91904
Israel

Flavia Nastri
Department of Chemistry
University of Naples Federico II
Via Cintia
80126 Naples
Italy

Victor V. Nikonenko
Department of Physical Chemistry
Kuban State University
149 Stavropolskaya St.
350040 Krasnodar
Russia

Andrés F. Olea
Departamento de Ciencias Químicas
Facultad de Ecología y Recursos Naturales
Universidad Andrés Bello
Santiago, Chile

Jos M.J. Paulusse
Laboratory of Organic Chemistry
Wageningen University
Dreijnplein 8
6703 HB Wageningen
The Netherlands

Angelo Perico
Istituto per lo Studio delle Macromolecole (ISMAC)
Consiglio Nazionale delle Ricerche (CNR)
Via De Marini 6
16149 Genova
Italy

Natalia D. Pismenskaya
Department of Physical Chemistry
Kuban State University
149 Stavropolskaya St.
350040 Krasnodar
Russia

Gérald Pourcelly
Institut Européen des Membranes
Université Montpellier II
CC 047 Place Eugène Bataillon
34095 Montpellier cedex 5
France

Baofu Qiao
Institute for Computational Physics
Universitat Stuttgart
70569 Stuttgart
Germany

Hans-Jörg Schneider
FR Organische Chemie
Universität des Saarlandes
D 66041 Saarbrücken
Germany

Bastien Seantier
Institut Européen des Membranes
Université Montpellier II
CC 047 Place Eugène Bataillon
34095, Montpellier cédex 5
France

Claire J. Parker Siburt
Department of Chemistry
Duke University
Durham, NC 27078

Dave Thirumalai
Institute for Physical Sciences
University of Maryland
College Park, MD 20742

Rashi Tiwari
Sibley School of Mechanical and Aerospace Engineering
Cornell University
Ithaca , NY 14850

Ngo Minh Toan
Institute for Physical Sciences
University of Maryland
College Park, MD 20742

LiShan Wang
Institute of Bioengineering and Nanotechnology
The Nanos 136889
Singapore

Ian M. Ward
School of Physics and Astronomy
University of Leeds
Leeds, LS2 9J
UK

Andrey B. Yaroslavtsev
Kurnakov Institute of General and Inorganic Chemistry
RAS, Leninsky Prospect 31a
119991 Moscow
Russia

Han Zuilhof
Laboratory of Organic Chemistry
Wageningen University
Dreijnplein 8
6703 HB Wageningen
The Netherlands

PART I

FUNDAMENTALS

CHAPTER 1

ION PROPERTIES

YIZHAK MARCUS

1.1 IONS AS CHARGED PARTICLES

Ions are defined as particles that carry electrical charges. Condensed phases, solids and liquids, are electrically neutral; that is, ions exist in them in combinations of positively and negatively charged particles—cations and anions, respectively—that may be bound or relatively free to migrate. An electrolyte is a neutral combination of cations and anions that can exist as a chemical substance capable of dissociating into its constituent ions in a suitable environment, for example, in an aqueous solution.

Ions may be monatomic (such as Na^+ or Br^-), may consist of a few atoms (such as NH_4^+ or SO_4^{2-}) or even considerably more than a few (such as $HOC_6H_4CO_2^-$ or $(C_4H_9)_4N^+$), or be much larger, consisting of very many atoms. In some of such cases, they may be referred to as polyions, constituting the dissociated part of polyelectrolytes. Some polypeptides, proteins, nucleic acids, and similar biological moieties, but also suitable synthetic molecules, are examples of polyelectrolytes. Throughout this chapter, a generalized ion is designated by $I^{z\pm}$, but when used as a subscript the charge is dropped, and $_I$ symbolizes a quantity pertaining to an ion.

Some substances that are not dissociated in solution, but are capable of donating a hydrogen ion to a basic environment, thus turn into an anion—these are weak acids, such as acetic acid, CH_3COOH forming $CH_3CO_2^-$. Other substances are able to add on a hydrogen ion in an acidic medium and turn into a cation—these are weak bases, such as aniline, $C_6H_5NH_2$ forming $C_6H_5NH_3^+$. A special category consists of zwitterions: These turn into anions or cations depending on the pH of the medium (hydrogen ion deficiency or excess), an example being glycine, $^+H_3NCH_2COO^-$, turning into $^+H_3NCH_2COOH$ in acidic media and into $H_2NCH_2CO_2^-$ in basic ones. The properties discussed

Ionic Interactions in Natural and Synthetic Macromolecules, First Edition. Edited by Alberto Ciferri and Angelo Perico.

in this chapter pertain to ions that have been formed either by strong electrolytes directly on their dissolution or by weak electrolytes in suitable media.

1.1.1 Isolated Ions

Isolated ions may be regarded as existing in an ideal gaseous state, that is, devoid of interactions with other particles or their surroundings in general. Some quite large ions are produced in mass spectrometers, but commonly isolated ions consist of relatively few atoms. They may, however, be the centers of clusters consisting of the ion proper surrounded by a small number of solvent molecules.

The primary characteristics of isolated ions are the amount of electrical charge they carry, their mass, their shape, and their size. The amount of charge is given in terms of a multiple, z_I, of the elementary units of the charge of a proton (positive) or an electron (negative), namely $e = 1.60218 \times 10^{-19}$ C. Within the scope of this book, the absolute values of z_I for isolated ions range from 1 to 4 for monatomic ones and possibly somewhat larges for some complex ions. Highly ionized atoms that may be produced artificially or result from nuclear reactions are not considered here.

The mass of isolated ions is generally specified per mole of ions; a mole consists of a very large number of individual particles, Avogadro's number: $N_A = 6.02214 \times 10^{23}$ mol^{-1}. The unit of molar mass, M_I, is kg mol^{-1}, but generally, M_I is given in g mol^{-1}.

The shape of monatomic ions is, of course, strictly spherical when isolated, but they may be deformed slightly by external forces (strong electrical fields). Ions that consist of several atoms may have any shape, but common ones are planar (NO_3^-, CO_3^{2-}), tetrahedral (NH_4^+, SO_4^{2-}), octahedral ($Fe(CN)_6^{4-}$), elongated (SCN^-), or more irregular ($CH_3CO_2^-$, HCO_3^-). The sizes of ions in the isolated state, however, are difficult to specify because the electrons in their periphery extend indefinitely around the inner electronic shells and the nuclei of the atoms.

An attribute of a generalized isolated ion I with charge z_+ or z_- ($I^{z\pm}$,g) is its self-energy that is due to its charge. Per mole of isolated ions, this self-energy is

$$E_{\text{self}}(I^{z\pm}, g) = N_A z^2 e^2 / 4\pi\varepsilon_0 r_I, \tag{1.1}$$

where $\varepsilon_0 = 8.85419 \times 10^{-12}$ C^2 J^{-1} m^{-1} is the permittivity of free space, and r_I is the radius of the ion. Since, however, as stated above the size of an isolated ion is an ill-defined quantity, so must be its radius, hence the self-energy. Still, this concept is employed in the discussion of the hydration of ions in Section 1.3.4.

Other thermodynamic quantities that pertain to isolated ions, on the other hand, are well defined. The standard molar Gibbs energy and the enthalpy of formation, $\Delta_f G°(I^{z\pm},g)$ and $\Delta_f H°(I^{z\pm},g)$, of many ions from the elements in their standard states, and the standard molar entropy and constant-pressure heat capacity, $S°(I^{z\pm},g)$ and $C_P°(I^{z\pm},g)$, all at temperature $T = 298.15$ K have been reported in the National Bureau of Standards (NBS) tables,[1] with values for additional ions being included in the book by Marcus.[2] The standard molar volume of an isolated ion is a trivial quantity, the same for all ions: $V°(I^{z\pm},g) = RT/P° = 0.02479\ m^3\ mol^{-1}$ at 298.15 K, where $R = 8.31451\ J\ K^{-1}\ mol^{-1}$ is the gas constant and $P°$ is the standard pressure of 100 kPa.

The ionization process leading from an atom, a radical, or a molecule to a cation requires the investment of energy, expressed by the ionization potential, ΣI_p. The sum sign, Σ, is used for the ionization potential I_p because this process of ionization can proceed in several stages up to the finally produced positive ion, the cation. The electron capture by an atom, a radical, or a molecule to form an anion releases energy that is expressed as the electron affinity, *EA*, of such a moiety. On the other hand, the capture of an electron by a negative ion, an anion, is a very unlikely event, so that only a single electron may generally be added to a neutral species in the *EA* process. These energies, in electron volt units (1 eV/particle = 96.483 kJ mol^{-1}), have been reported for many ions in the aforementioned book[2] and are shown in Table 1.1 for some ions.

There are two further relevant properties of ions that depend little, if at all, on whether the ion is isolated in the ideal gas phase or is present in a condensed phase, a solid or a liquid. These are magnetic susceptibility, χ_I, and polarizability α_I (or the corresponding molar refractivity, $R_{I\infty}$). Most ions are diamagnetic; that is, they are repulsed out from a magnetic field. Exceptions are ions that have an unpaired electron in their electronic shells: These are paramagnetic. The molar magnetic susceptibilities, χ_{Im}, range from a few to several tens of $-10^{-12}\ m^3\ mol^{-1}$; that is, they have the dimensions of molar volumes. For a paramagnetic ion having n unpaired electrons, $\chi_{Im} = +1.676n(n+2) \times 10^{-9}\ m^3\ mol^{-1}$ at $T = 298.15$ K. Values of the diamagnetic χ_{Im} for many ions have been reported.[2]

The polarizability of an ion also has the dimension of a volume, of the order of $10^{-30}\ m^3$ per ion. The molar refractivity is

$$R_{I\infty} = (4\pi N_A/3)\alpha_I = 2.5227 \times 10^{24}\alpha_I. \quad (1.2)$$

It is the latter quantity that is estimated experimentally for neutral species, and is additive for the constituting ions if the ions are isolated (in a gas or infinitely dilute solution). The reported[2] individual ionic values are based on the arbitrary but reasonable value of $R_D^\infty(Na^+) = 0.65 \times 10^{-6}\ m^3\ mol^{-1}$, where R_D^∞, the infinite dilution value obtained from the refractive index at the sodium D line (589 nm), is used in lieu of the infinite wavelength value R_∞.

The values of these molar ionic properties for several selected ions in the ideal gas state are shown in Table 1.1.

1.1.2 Ions in Aqueous Environments

As a thought process, a single ion $I^{z\pm}$ may be transferred from the ideal gas phase into water, but this process involves the passage through the water–gas

TABLE 1.1. Properties of Some Isolated Ions[2]

Ion	z	M_I (G mol^{-1})	$\Delta_f H_I^\circ$ (kJ mol^{-1})	$\Delta_f G_I^\circ$ (kJ mol^{-1})	S_I° (J K^{-1} mol^{-1})	C_{pI}° (J K^{-1} mol^{-1})	$\sum I_p/EA$ (eV)	χ_m (-10^{12} m^3 mol^{-1})	R_D^∞ (10^{-6} m^3 mol^{-1})
H^+	1	1.008	1536.2	1523.2	108.9	20.8	1318	−6.6	−0.1
Li^+	1	6.94	685.8	685.8	113.0	20.8	526	−3	0.08
Na^+	1	22.94	609.4	580.5	148.0	20.8	502	2.3	0.65
K^+	1	39.10	514.3	487.3	154.6	20.8	425	11.2	2.71
NH_4^+	1	18.04	630	681	186.3	34.9	458	11.5	4.7
Mg^{2+}	2	24.31	2348.5	2300.3	148.7	20.8	2201		−0.7
Ca^2	2	40.08	1925.9	1892.1	154.9	20.8	1748	−1	1.59
Fe^{2+}	2	55.85	2749.9	2689.6	180.3	25.9	2334	19.6*	2.1
Fe^{3+}	3	55.85	5712.8	5669.1	174.0	20.8	5296	15.6*	3.2
Cl^-	−1	35.45	−233.1	−241.4	154.4	20.8	−349	28	8.63
Br^-	−1	79.91	−219.1	−245.1	163.6	20.8	−324	39	12.24
I^-	−1	126.91	−197	−230.2	169.4	20.8	−295	57	18.95
SCN^-	−1	58.08			166.5	43.2	−207	35	17
NO_3^-	−1	62.01	−320	−272.8	245.2	44.7	−378	23	10.43
ClO_4^-	−1	99.45	−344	−266.8	263.0	62.0	−454	34	12.77
$CH_3CO_2^-$	−1	59.04	−504.2	−464.1	278.7	61.4	326	32.4	13.87
HCO_3^-	−1	61.02	−738	−702	257.9	50.6			10.9
$H_2PO_4^-$	−1	96.99	−1280	−1190	286.7	62.5			14.6
CO_3^{2-}	−2	60.01	−321	−300.9	246.1	44.4	226	34	11.45
SO_4^{2-}	−2	96.07	−758	−704.8	363.6	62.4	1660	40	13.79
HPO_4^-	−2	95.98			283.0	67.8			

* Paramagnetic ion.

interface with not well-defined consequences. There are, therefore, very few experimental determinations that can be applied unambiguously to individual ions in aqueous solution.

The more common process that can be carried out experimentally is to dissolve in water an entire electrolyte (consisting of a matched number of cations and anions to produce a neutral species). Conditions can be chosen for approximating infinite dilution, that is, the dissolution of an infinitesimal amount of electrolyte in a finite amount of water or a mole of electrolyte in a very large amount of water. It is then possible to deal with the molar quantities pertaining to the aqueous electrolyte at infinite dilution and devise some means to deduce from the measured quantities those pertaining to the individual ions.

At infinite dilution, the aqueous ions are remote from each other, and hence are surrounded by water only, with which each ion interacts. Therefore, the individual ionic quantities are additive, weighted by their stoichiometric coefficients in the electrolyte: ν_+ cations C^{z+} and ν_- anions A^{z-}. For a 1:1 electrolyte such as NaCl $\nu_+ = \nu_- = z_+ = |z_-| = 1$, for a 2:1 electrolyte such as K_2SO_4 $\nu_+ = |z_-| = 2$ and $z_+ = \nu_- = 1$. The quantitative aspects of such assignments of individual ionic properties are detailed in the appropriate sections below.

At this place, the interactions of the ions with the surrounding water are dealt with in a qualitative manner. Cations have the water molecules oriented toward them, with one of the lone pairs of electrons, carrying a fractional negative charge, pointing at them. In some cases, this results in a coordinate bond, the fractional charge penetrating unoccupied electronic orbitals of the cations. Transition metal cations, such as Fe^{2+}, and also small multivalent cations such as Mg^{2+}, tend to form such bonds with water. A definite coordination geometry and number of the nearest water molecules, those in the first hydration shell, result: 6 in a regular octahedron for Mg^{2+}. These water molecules are polarized by the charge of the cation and tend to form such hydrogen bonds with water molecules in a second hydration shell that are somewhat stronger than those prevailing in pure water, so that this second shell remains with the cation as it moves in the solution.

This does not mean that the water molecules in either shell are permanently bound to the cation: On the contrary, there is a dynamic exchange of water molecules with the bulk water outside the hydration shells. For some cations (e.g., Mg^{2+}), though, this exchange is relatively slow—only $10^{5.2}$ exchanges occur per second—compared with $10^{8.5}$ exchanges per second for a cation such as Ca^{2+}, also doubly charged, but that is not as strongly bound to the water. The latter cation, as well as the alkali metal ones (except, perhaps, Li^+), does not have a definite number of water molecules coordinated to them in their hydration shell. Instead, they have a distribution with a fractional average number, and there is no evidence for a stable second hydration shell. Such ions still affect the hydrogen-bonded structure of the surrounding water and in the cases of large univalent cations (K^+, Rb^+, and Cs^+) break this structure up to some extent. This is manifested by the faster flowing ability of the water (lower viscosity) in solutions containing such ions (Section 1.4.3).

Anions in aqueous solutions have the water molecules pointing one (or both in some cases) of their hydrogen atoms toward them, resulting in hydrogen bonds. Anions tend to be larger and have relatively smaller electric fields than common cations and have no definite coordination number of water molecules hydrating them but rather a distribution of values with an average that may be rather small for singly charged anions (<2). Multivalent anions, especially oxyanions, such as CO_3^{2-} or SO_4^{2-} bind more water molecules by accepting hydrogen bonds from them, and anions such as HSO_4^- or $H_2PO_4^-$ also donate hydrogen bonds to adjacent water molecules. Large univalent anions, such as Cl^- and ClO_4^- break the water structure around them and lower the viscosity of the solutions.

Ions with hydrophobic groups in their periphery around a buried charge, such as $(C_6H_5)_4B^-$ or $(C_4H_9)_4N^+$, are generally only poorly hydrated, but they enhance the water structure, in the manner characteristic of pure water (tetrahedrally oriented hydrogen bonds), as do uncharged organic molecules with similar hydrophobic groups. Ions containing hydrophobic parts, such as $C_6H_5CO_2^-$ and $CH_3NH_3^+$, arrange water molecules differently around such parts, $-C_6H_5$ and $-CH_3$, compared with the arrangement near the hydrophilic part, $-CO_2^-$ or $-NH_3^+$. This behavior is even more characteristic of ionic surfactants, such as $C_{12}H_{25}SO_3^-$ or $(C_8H_{17})_2NH_2^+$. The long organic chains tend to associate in aqueous solutions forming micelles or other structures, the hydrophobic "tails" being removed from immediate contact with the water into which the hydrophilic "heads" stick out.

1.1.3 Ions in Other Environments

Although water is the most important solvent, in general and within the context of this chapter, ions do exist also in other environments—in nonaqueous and mixed solvents, and in condensed phases without any solvents—as molten salts. Some of the properties of ions in nonaqueous and mixed solvents are discussed in Section 1.5.

Molten salts are conventionally classed as high-melting ones (examples being molten NaCl and NH_4NO_3) and room temperature ionic liquids (RTILs; examples being pyridinium tetrafluoroborate and 1-methyl-3-butylimidazolium bis(trifluoromethane-sulfonyl)imide). The high-melting salts (the melting temperatures are 1074 K for NaCl and 443 K for NH_4NO_3) are probably not relevant in the context of this chapter. RTILs, or more generally ionic liquids (ILs) that have melting temperatures ≤373 K, may have some relevance. They are characterized by negligible vapor pressures, relatively high viscosities, but adequate electrical conductivities, and are in vogue in recent years as being friendly to the environment contrary to some solvents that have been in use in industry. Their properties are not discussed further here.

1.1.4 Charged Macromolecules

This introduction to charged macromolecules is only cursory, the subject being fully treated in Chapters 3, 4, and 6. Polymers that carry ionically dissociable groups are called polyelectrolytes, ionizing partly or completely to polyions. Synthetic polyelectrolytes generally have a linear polymeric chain, the hydrophobic backbone, to which side groups with hydrophilic ionizable groups are attached. When these are ionized, these covalently attached ions are the "fixed ions" to which correspond mobile "counterions" of opposite charge. An example is polystyrene sulfonic acid $-CH_2CH(C_6H_5SO_3H)-$, a strong, practically completely ionized polyelectrolyte. On the other hand, a weak polyelectrolyte, polymethacrylic acid $-CH_2C(CH_3)(CO_2H)-$, is only partly ionized, depending on the pH of the solution. A polyelectrolyte that has the charged sites as parts of the polymer chain itself is polyethyleneimine (-immonium) $(-C_2H_4NH_2^+C_2H_4-)$ with its counterions. Such synthetic linear polyelectrolytes have a regularly repeating structure. They may be cross-linked to form water-insoluble materials that can exchange the counterions with other ones in the imbibed or adjacent solution: ion exchange resins, gels, and membranes. Biological examples of polyelectrolytes are proteins and nucleic acids that share many features with synthetic polyelectrolytes, but they do not have the regular repeating structures of the latter.

The special features exhibited by polyelectrolytes are due to the proximity along the polymeric chain (the backbone) of charges of the same kind that affect (repulse) one another. Even if the polyelectrolyte itself is at high dilution in the solution, the fixed ions attached to it can never be said to be at infinite dilution. The pH and ionic strength of the solution in which the polyelectrolyte is situated may affect its expansion or contraction, evidenced by changes in the viscosity of the solution. In the case of strong flexible polyelectrolytes, the end-to-end distance of the chain is shortened on increasing concentration due to self-screening of the charges. Furthermore, counterions with charges opposite to those of the fixed ions tend to associate with them to some extent (see Chapters 3 and 6). If ions or molecules with hydrophobic groups are present in the solution, these tend to associate with the hydrophobic backbone or hydrophobic parts of the side groups carrying the charges. In particular, such association becomes important when surfactants are present in the solution. The solubility of polyelectrolytes in water is affected by cosolvents that have poor solvation power for the charges; this may lead to micellization or precipitation of the polyelectyrolyte. Such effects are fully discussed in Chapters 3, 4, and 6.

Polyelectrolytes are characterized by being linear or branched (unless cross-linked), by their degree of polymerization, that is, their average molar mass (MW) and the dispersion of this average, the lengths, l, along the backbone of the segments that carry the charged groups, and the degree of ionization, $0 \le \alpha \le 1$.

Rigid polyelectrolytes are characterized by large values of the intrinsic persistent length, q_{in}, measuring their conformational rigidity (for instance, in DNA). This quantity should be many-fold times that of l, the distance between charges. Flexible polyelectrolytes have instead low values of q_{in} and are extended in order to reduce the coulombic repulsion between the charges. Moreover, conformational changes occur by the screening actions of counterions and co-ions, if present.

A highly ionized ($\alpha \rightarrow 1$) linear polyion in aqueous solution may be approximated by a charged rod with a cylindrical sheath of water of hydration. As α is decreased, the hydration water envelopes overlap only partly, down to low degrees of ionization ($\alpha \rightarrow 0$) where the charged sites are hydrated individually. However, the hydrophobic part of the polyion causes the arrangement of water molecules near the polymer backbone to differ from that near the ionic groups. In addition, counterions attracted electrostatically to the fixed ions bring their own water of hydration with them, and if ion pairs are formed, the hydration shells of fixed and counterions overlap. Thus, a complicated picture of the hydration of polyelectrolytes emerges, depending on the flexibility of the polymeric backbone, the segment length l, the extent of ionization α, the nature of the fixed ions and of the counterions, and on the presence or absence of additional electrolytes in the solution containing or in contact with the polyelectrolyte.

1.2 SIZES OF IONS

1.2.1 Ionic Radii in Crystals

Diffraction of X-rays and neutrons in crystals has been used for the determination of the interatomic distances, or in the case of ionic crystals, the interionic distances. The interionic distances could be measured with modern technology with an accuracy of ±0.0001 nm. They are fairly regularly additive in the sizes of the ions, so that radii may be assigned to monatomic ions and to polyatomic ones having a globular shape: tetrahedral, octahedral, or cubic. The manner of splitting the interionic distances into ionic radii is not straightforward, however.

For monatomic ions, over many years, the suggested Goldschmidt radii (based on the refractivity of ions in crystals leading to $r(F^-) = 0.133$ nm and $r(O^{2-}) = 0.132$ nm) were in substantial agreement with the Pauling radii (based on the screening of effective nuclear charges by the electronic shells, leading to $r(F^-) = 0.136$ nm and $r(O^{2-}) = 0.140$ nm). Subsequently, a completely different set was suggested by Gourary and Adrian, based on the loci of minima in the electron density between the ions, leading to $r(F^-) = 0.116$ nm, hence much larger cation radii than those of Goldschmidt and Pauling. Ionic radii r_I depend somewhat on the temperature,[3] but this dependence is of no consequence for temperatures in aqueous solutions under ambient pressure and may play a role only in high-melting crystalline or molten salts. The dependence on the coordination number and geometry in the crystals was established in detail by Shannon and Prewitt,[4] modifying slightly the Pauling set of radii. Marcus[2] summarized the reported values and provided a "selected" set of ionic radii, r_I, for the most commonly observed coordination numbers that correspond with the hydration numbers of ions in aqueous solutions.

Kapustinskii obtained interionic distances for crystals containing polyatomic ions from the dependence of thermochemically derived lattice energies of the crystals on these distances. A set of thermochemical radii was established for polyatomic ions by Jenkins and Thakur[5] from these interionic distances in crystals with monatomic counterions using the Goldschmidt radii of the latter. As mentioned above, such radii make sense for globular polyatomic ions, and a "selected" set was provided by Marcus.[2] For polyatomic ions deviating strongly from a globular shape, the concept of a radius is ambiguous. When oblate (disklike) ions such as NO_3^- or prolate (rodlike) ions such as SCN^- are considered, then the half-axes a and b of the ellipsoids of revolution may be used for estimation of the volume: $(4\pi/3)a^2b$ and $(4\pi/3)ab^2$ with $a > b$, respectively. Caution should be used with these values when considering the distance to which other particles (solvent molecules or other ions) can approach such nonglobular ions.

1.2.2 Ionic Radii in Aqueous Solutions

Diffraction methods have been used also on aqueous salt solutions for establishing the distances $d_{\text{I-O}}$ between the centers of ions and those of the oxygen atoms of adjacent water molecules, as summarized by Marcus[6] and by Ohtaki and Radnai.[7] It was found that if the radius of a water molecule, $r_W = 0.138$ nm, is deducted from the $d_{\text{I-O}}$ values, the results correspond quite well with the set of Pauling radii of the ions.[8] The values of $d_{\text{I-O}}$ reported by various authors and pertaining to different concentrations have a mean uncertainty of ±0.002 nm, much worse than the values in crystals. Still, the "selected" ionic radii[2] r_I add up with r_W (= 0.138 nm) within this uncertainty to the $d_{\text{I-O}}$ values, suggesting their use for the radii of aqueous ions, as listed in Table 1.2. The electrostatic forces acting on ions in condensed phases, crystals, and solutions are of similar magnitudes, as appear from the relevant energetics: lattice energies and energies of solvation. Hence, the use of the r_I derived from measurements on crystals for the radii of ions in solution is reasonable.

The concept of a constant radius for the water molecule, r_W, was challenged, however, by David and Fourest.[9] They claimed that the electric field of the ions polarizes the water molecules adjacent to them and, for multiply charged ions, squeezes these molecules somewhat in the direction of the ions. They suggested that r_W decreases from 0.143 nm for the alkali metal cations down to 0.133 nm for the trivalent lanthanide cations. The use of the mean value, $r_W = 0.138$ nm, increases the uncertainty of the ionic radii in solution from ±0.002 to ±0.005 nm according to these authors.

1.2.3 Ionic Volumes in Solution

Henceforth, the solvent (water) is designated by subscript $_1$ (or by subscript $_W$ as above for the radius) and the solute electrolyte by subscript $_2$, as is generally the practice. The individual ionic contributions to a (molar) property of the electrolyte are designated by subscript $_I$. Properties that pertain to infinite

TABLE 1.2. Properties of Some Ions in Aqueous Solutions at 298.15 K[2]

Ion	r^a (nm)	V_I^∞ (cm^3 mol^{-1})	V_{Ielec}^{∞} ^{b} (cm^3 mol^{-1})	h_{Ielec}^{∞} ^{b}	h_{Icomp}^{∞} ^{b}	$\Delta_{hydr}H_I^\infty$ (kJ mol^{-1})	S_I^∞ (J K^{-1} mol^{-1})	$\Delta_{hydr}G_I^\infty$ (kJ mol^{-1})	C_{PI}^∞ (J K^{-1} mol^{-1})
H^+	0	–5.4				–1103	–22.2	–1056	–71
Li^+	0.069	–6.4	–12.9	4.4	4.0	–531	–8.8	–481	–9
Na^+	0.102	–6.7	–8.6	2.9	4.5	–416	36.8	–375	–28
K^+	0.138	3.5	–5.9	2.0	3.5	–334	80.3	–304	–58
NH_4^+	0.148	12.4				–329	74.7	–292	–1
Mg^{2+}	0.072	–32.2	–52.5	11.4^c	10.0	–1949	–182.5	–1838	–158
Ca^2	0.100	–28.9	–38.5	10.0^c	9.0	–1602	–97.5	–1515	–169
Fe^{2+}	0.078	–34.4		(11.8)^c		–1972	–182.1	–1848	–188
Fe^{3+}	0.065	–60.2		(21.1)^c		–4462	–382.5	–4271	–204
Cl^-	0.181	23.3	–4.0	1.4	2.0	–367	78.7	–347	–56
Br^-	0.196	30.2	–3.5	1.2	1.8	–336	104.6	–321	–60
I^-	0.220	41.7	–2.8	1.0	1.5	–291	133.5	–283	–50
SCN^-	(0.213)	41.2		0.5^c	0.1^b	–311	166.5	–287	42
NO_3^-	(0.200)	34.5		1.4^c	0.7^b	–312	168.8	–306	–1
ClO_4^-	0.240	49.6	–2.4	0.8		–246	206.2	–214	46
$CH_3CO_2^-$	(0.232)	46.2			4.0^b	–425	108.8	–373	97
HCO_3^-	(0.156)	28.9			1.7^b	–384	120.6	–368	18
$H_2PO_4^-$	(0.200)	34.6				–522	114.7	–473	37
CO_3^{2-}	(0.178)	6.7		(10.9)^c	14.4^b	–1397	0.9	–1315^d	–159
SO_4^{2-}	0.230	25.0	–13.8	4.6	10.4^b	–1138^d	63.2	–1090	–138
HPO_4^-	(0.200)	18.7					10.9		–102

[a] Values in parentheses are approximate since ions are not nearly spherical; see text.
[b] The values are from Marcus.[12]
[c] The values are from Marcus[26]; those in parentheses are the mean values for ions of the given charge.
[d] Corrected values.

dilution of the solute are designated by superscript $^{\infty}$ and those of the pure solvent by superscript *.

The volumes that are to be assigned to ions in aqueous solutions are related to the concentration-dependent densities, ρ, of the solutions at constant temperature and pressure. Consider a solution made up from n_1 moles of water and n_2 moles of electrolyte. The apparent molar volume of the latter, ${}^{\varphi}V_2 = (V - n_1V_1^*)/n_2$, is, per mole of electrolyte, that part out of V, the total volume of the solution, remaining for the electrolyte once the volume $n_1V_1^*$ (that the water would have had if there were no effect due to the ions) is subtracted. In a solution of density ρ made up from $n_1 = 55.51$ moles of water (1 kg) and $n_2 = m_2$ moles of electrolyte (i.e., at a molality m_2), the apparent molar volume is obtained from the densities as ${}^{\varphi}V_2 = M_2/\rho + 1000(\rho - \rho_1^*)/\rho\rho_1^*m_2$, M_2 being the molar mass of the solute and ρ_1^* the density of pure water.

This apparent molar volume of the solute is not necessarily the actual volume that should be assigned to the electrolyte because the water near ions does not have the same molar volume as pure water has. The water near the ions is compressed, electrostricted, by the electrical fields of the ions. The volume to be assigned to the ions is the partial molar volume, which for a solution of molality m_2 is

$$V_2 = {}^{\varphi}V_2 + m_2(\partial^{\varphi}V_2/\partial m_2)_T. \tag{1.3}$$

As the molality diminishes on approaching infinite dilution, the second term on the rhs diminishes too, so that on extrapolation to infinite dilution ${}^{\varphi}V_2$ becomes equal to the standard partial molar volume of the electrolyte: ${}^{\varphi}V_2^{\infty} = V_2^{\infty}$.

At infinite dilution, the contributions of the cations and anions are additive as mentioned in Section 1.1.2, but the way to split the measured V_2^{∞} to the contributions from the individual ions, $V^{\infty}(I^{z\pm},aq)$, requires the knowledge of the value for just one ion, from which those of other ions are then derived. The value $V^{\infty}(H^+,aq) = -5.4\ cm^3\ mol^{-1}$ at 298.15 K has been accepted as a reasonable assumption, and the derived values for some other ions are shown in Table 1.2. In view of the steps that have led to this assumption, the ionic values have uncertainties of at least $\pm 0.2z_I\ cm^3\ mol^{-1}$, increasing with the ionic charges. Note that for some cations, and in particular for multivalent ones, the values of $V^{\infty}(I^{z\pm},aq)$ are negative. Such cations produce a large diminution of the volume of the water surrounding them, or in other words cause a large electrostriction.

Most authors have dealt with the standard partial molar volumes of ions, V_I^{∞}, that is, the infinite dilution values. The values of the ionic V_I at finite concentrations are not known accurately because the additivity of the individual volumes breaks down due to interionic interactions. The apparent molar volumes of electrolytes can be expressed as ${}^{\varphi}V^2 = {}^{\varphi}V_2^{\infty} + S_V c^{1/2} + bc$, where c is the molar concentration (in mol dm^{-3}), S_V is the theoretical slope of the square root term (Debye–Hückel theory; 1.85 $dm^{3/2} \cdot mol^{-1/2}$ at 298.15 K), and b is an empirical parameter specific for each electrolyte.

The standard partial molar volume of an ion in aqueous solution, V_{I}^{∞}, can be construed as being the sum of its intrinsic volume, $V_{\mathrm{Iintr}}^{\infty}$, and the electrostriction that the ion has caused in the water around it, $V_{\mathrm{Ielec}}^{\infty}$, the latter being a negative quantity. Attempts have been made to estimate each of these quantities independently, but so far the results are not very conclusive.

The intrinsic volume of an ion in a solution is a rather nebulous concept, but it should be independent of the concentration and, substantially, also of the temperature. If the molar volume of the "bare" unhydrated ion, $(4\pi N_{\mathrm{A}}/3)r_{\mathrm{I}}^{3}$, is considered to represent $V_{\mathrm{Iintr}}^{\infty}$, then a much too small negative or even a positive quantity must be assigned to $V_{\mathrm{Ielec}}^{\infty}$ in order to make up the measured V_{I}^{∞}. Thus, the radius of the "bare" ion should be modified (enlarged) for it to be useful in this respect. This enlargement has been supposed to represent the void spaces between the water molecules and the ion and among themselves. A factor of $k = 1.23$ was proposed[10] for the alkali metal and the halide ions, producing

$$V_{\mathrm{Iintr}}^{\infty} = (4\pi N_{\mathrm{A}}/3)(kr_{\mathrm{I}})^{3}, \tag{1.4}$$

a more reasonable estimate, in fair agreement with other ones. These, for V_{2intr}, are the sums of the cation and anion values, such as the extrapolated molar volumes of molten alkali metal halides down to room temperature or the limits of V_2 at very high concentrations, where all the water present is already completely compressed, so that $V_{\mathrm{2elec}} \to 0$.[11]

The electrostriction caused by an ion has been estimated on the basis of the electrostatic effects the very high electric field of an ion has on the dielectric medium, the water with which the ion is surrounded. One consequence of this field is a large pressure exerted on the water; the other is the sharp decrease of the dielectric permittivity of the latter, down to dielectric saturation. Such calculations are complicated due to the mutual dependences of the pressure and the compressibility on the one hand, and of the field strength and the permittivity on the other. Values of $V_{\mathrm{Ielec}}^{\infty}$ have recently[12] been calculated for the alkali metal and alkaline earth metal cations and the halide, perchlorate, and sulfate anions for aqueous solutions at five temperatures between 273.15 and 373.15 K, with those for 298.15 K being shown in Table 1.2. The corresponding intrinsic volumes are $V_{\mathrm{Iintr}}^{\infty} = V_{\mathrm{I}}^{\infty} - V_{\mathrm{Ielec}}^{\infty}$, and of course, all are positive.

1.3 THERMODYNAMICS OF AQUEOUS IONS

Ions in aqueous solutions are characterized by several thermodynamic quantities, in addition to the volumes discussed above. Some of these quantities are the molar heat capacities (at constant pressure) and entropies; others are the molar changes of enthalpy or Gibbs energy on the transfer of an ion from its isolated state in the ideal gas to the aqueous solution. The latter quantities pertain also to the dissolution of an electrolyte in water, being parts in a thermodynamic cycle describing the transfer of the electrolyte itself to the gas phase, its dissociation there into its constituent ions that are then transferred into the solution. In all these measures and processes, it is not possible to deal experimentally (contrary to thought processes) with individual ions but always only with entire electrolytes or with such combinations (sums or differences) of ions that are neutral. The assignment of values to individual ions requires the splitting of the electrolyte values by some extrathermodynamic assumption that cannot be proved or disproved within the framework of thermodynamics.[2]

1.3.1 Molar Heat Capacities of Aqueous Ions

When the heat of solution of an electrolyte in water to form a dilute solution is measured calorimetrically at several temperatures, the standard partial molar (constant pressure) heat capacity of the electrolyte, C^{∞}_{P2}, is obtained from the temperature coefficient of heats of dilution, extrapolated to infinite dilution. Alternatively, the difference between the specific heat of a dilute solution of the electrolyte and that of water is obtained by flow microcalorimetry to yield the same quantity. Such determinations are accurate to ±1 to ±3 J K^{-1} mol^{-1}.

As mentioned above, it is necessary to assume a value for one ion in order to obtain the so-called "absolute" standard molar ionic heat capacities, C^{∞}_{PI}, these values being additive at infinite dilution. The value $C^{\infty}_{P}(H^+, aq) = -71 \pm 14$ J K^{-1} mol^{-1} at 298.15 K has been suggested[13] on equating the standard molar heat capacities of aqueous tetraphenylphosphonium and tetraphenylborate ions (the TPTB assumption). These ions should have nearly the same value due to their chemical similarity and similar sizes and the charges of opposite sign being buried well inside the tetraphenyl structure.[14] Unfortunately, the C^{∞}_{PI} of these bulky ions are large, hence the uncertainty involved in equating them, due to slight differences in the sizes and induced partial charges in the phenyl rings, is also large. However, no more satisfactory method for splitting C^{∞}_{P2} into the C^{∞}_{PI} of the constituent ions was found. Values of the latter are shown in Table 1.2. The values of multicharged ions are seen to be large and negative, and those of polyatomic ions are more positive (or less negative) than those of monatomic ions of the same charge class.

1.3.2 Molar Entropies of Aqueous Ions and Their Entropies of Hydration

The standard molar entropies of aqueous electrolytes, S^{∞}_2, are obtained from the temperature coefficients of the electromotive forces of galvanic cells or of the solubilities of sparingly soluble salts. The values for individual ions need an assumption concerning the value of one ion, as in the cases of the standard molar volumes and heat capacities. The value chosen is $S^{\infty}(H^+,aq) = -22.2 \pm 1.4$ J K^{-1} mol^{-1} at 298.15 K, based on data for thermocells.[15] The derived values for other ions, so-called "absolute" standard molar ionic entropies, are shown in Table 1.2. The values are seen to increase with the masses of the ions but to be small or negative for multicharged ions.

The standard molar entropy of hydration of an ion is $\Delta_{hydr}S_I^{\infty} = S_I^{\infty} - S_I^{\circ}$, the difference between its standard molar entropy in the aqueous solution (Table 1.2) and the standard molar entropy of the isolated ion in the ideal gas phase (Table 1.1). These standard molar entropies of hydration of ions are related to the effect that ions have on the structure of water. Various schemes of expressing this relationship have been proposed,[16] the main point being the subtraction from $\Delta_{hydr}S_I^{\infty}$ of the entropic effect of a neutral molecule of a similar size as the ion has and of the electrostatic effects of the electric field of the ion in terms of the entropy. The latter are obtained from the temperature coefficient of the Born expression for the Gibbs energy of hydration (Section 1.3.4). The remainder then expresses the effect of the ion on the water structure: If positive, the ion is said to be structure-breaking; if negative, it is structure-making. To the former category belong large univalent ions (e.g., K^+, I^-, ClO_4^-) and to the latter small or multivalent ions (e.g., Li^+, Mg^{2+}, CO_3^{2-}), while others are borderline in this respect (Section 1.4.3).

1.3.3 Enthalpies of Hydration of Ions

When an ion is transferred, in a thought process, from its isolated state in the ideal gas phase into water at infinite dilution, a large amount of energy is released due to the interaction of the ion with the surrounding water. Its molar volume is compressed in this process from RT/P° to the standard partial molar volume in the solution. The net relevant energetic amount is the change in the enthalpy, $\Delta_{hydr}H_I^{\infty}$, but this cannot be determined experimentally for individual ions. The values for complete electrolytes, $\Delta_{hydr}H_2^{\infty}$, are obtained from their heats of solution and lattice energies, yielding the differences between the standard molar enthalpy of formation of the infinitely dilute aqueous solute[1] and the sum of the standard molar enthalpies of formation of the ideal gaseous ions (Table 1.1).

A value must be estimated for one ion in order to split the experimentally available values dealing with entire electrolytes into the ionic contribution. The value $\Delta_{hydr}H^{\infty}(H^+,aq) = -1103 \pm 7$ kJ mol^{-1} results from equating the standard enthalpies of hydration of the tetraphenyphosphonium and tetraphenylborate ions (the TPTB assumption). The choice of these ions has been briefly discussed in Section 1.3.1, but contrary to the case of the heat capacities, the values of $\Delta_{hydr}H_I^{\infty}$ for these bulky ions are small compared with those of small ions, so that the uncertainty involved in equating the values for these reference ions is also small. This estimate for the hydrogen ion is compatible with several other reliable values suggested on the basis of other considerations, ranging from -1091 ± 10 to -1104 ± 17 kJ mol^{-1}. The values for a number of ions are shown in Table 1.2 and are expected to be accurate to within $\pm 7z_I$ kJ mol^{-1}.[2] The values are all negative, as expected (heat is released), of similar magnitude for singly charged ions, whether cations or anions (though becoming less negative with increasing sizes) but becoming considerably more negative for multicharged ions, by a factor of the order of z^2.

On departing from infinite dilution, molar enthalpies of hydration of electrolytes may be estimated by adding to the sum of the cation and anion values of $\Delta_{\text{hydr}}H_{\text{I}}^{\infty}$ the relative partial molar heat content of the solute, L_2, equal and of opposite sign to the experimentally measurable[17,18] enthalpy of dilution of the electrolyte, $\Delta_{\text{dil}}H_2$. At finite concentrations, the heat content and the enthalpy of hydration may therefore be smaller or larger than at infinite dilution, depending on the enthalpies involved in the interactions between neighboring ions. These are obtainable from the temperature derivatives of the activity coefficients

$$L_2 = -\nu RT^2(\partial \ln \gamma_{\pm}/\partial T)_{P,m}, \tag{1.5}$$

where ν is the number of ions in a formula of the solute electrolyte.

1.3.4 Gibbs Energies of Hydration of Ions

The standard molar ionic Gibbs energy of hydration $\Delta_{\text{hydr}}G_{\text{I}}^{\infty}$ has traditionally been estimated from the Born expression, resulting from the following idealized process. Consider an isolated ion in the gaseous phase, $\text{I}^{z\pm}$(g), that is discharged, producing a neutral particle. The electric self energy $E_{\text{self}}(\text{I}^{z\pm},\text{g})$ (Section 1.1.1) must be provided for this process. The neutral particle is then transferred into the bulk of liquid water, there being no electric energetic component for crossing the gas/liquid boundary. The neutral particle is then charged up to the original value, producing the infinitely dilute aqueous ion, $\text{I}^{z\pm}$,aq$^{\infty}$. The energy of interaction with the surrounding water is thereby released. This depends on the permittivity of the water $\varepsilon^* = 4\pi\varepsilon_0\varepsilon_{\text{r}}^*$, where ε_{r}^* is the temperature- and pressure-dependent relative permittivity ($\varepsilon_{\text{r}}^* = 78.4$ at 298.15 K and ambient pressure). The net effect of this idealized process representing the hydration of the ion is

$$\Delta_{\text{hydr}}G_{\text{I}}^{\infty} = (N_{\text{A}}e^2/4\pi\varepsilon_0)z_{\text{I}}^2 r_{\text{I}}^{-1}(1-1/\varepsilon_{\text{r}}^*). \tag{1.6}$$

The problem with this mode of calculation is the use of the same value of the radius r_{I} for the ion in the aqueous solution and the isolated state and the use of the relative permittivity of pure water for the description of the interaction of the ion with its immediate surroundings, where dielectric saturation, due to the high electric field of the ion, occurs. Various schemes have been proposed to counter this problem, such as adding a quantity Δr to the ionic radius and/or splitting the process into two spatial regions: one adjacent to the ion, where dielectric saturation occurs and $\varepsilon_{\text{r}}^* \approx n_{\text{D}}^2$ (the square of the refractive index), and the other beyond this, where the bulk value ε_{r}^* prevails. The use of such devices permits the estimation of reasonably correct $\Delta_{\text{hydr}}G_{\text{I}}^{\infty}$ values.[19]

Such values of the standard molar Gibbs energy of hydration, $\Delta_{\text{hydr}}G_{\text{I}}^{\infty}$, must be thermodynamically consistent with the combination of the standard molar enthalpy and entropy of hydration:

$$\Delta_{\text{hydr}}G_{\text{I}}^{\infty} = \Delta_{\text{hydr}}H_{\text{I}}^{\infty} - T\Delta_{\text{hydr}}S_{\text{I}}^{\infty}. \quad (1.7)$$

The value for the hydrogen ion, $\Delta_{\text{hydr}}G^{\infty}(\text{H}^+,\text{aq}) = -1056 \pm 6\ \text{kJ mol}^{-1}$ is thus consistent[19] with the values of $\Delta_{\text{hydr}}H^{\infty}(\text{H}^+,\text{aq})$ and $S^{\infty}(\text{H}^+,\text{aq})$ given above and $S°(\text{H}^+,\text{g})$ in Table 1.1 as required. It is also compatible with the estimate[15] $-1066 \pm 17\ \text{kJ mol}^{-1}$ but not with $-1113 \pm 8\ \text{kJ mol}^{-1}$ obtained from the cluster pair approximation used by Kelly et al.[20] As discussed by Marcus,[21] the assumptions involved in the latter value lead to a surface potential of water of $\Delta\chi = 0.34 \pm 0.08$ V, which in turn, is not consistent with the recent estimate[22] of $\Delta\chi = 0.1$ V deemed to be the most nearly correct one.

So-called "real" standard molar Gibbs energies of hydration are obtained from the electromotive force of cells consisting of a downward flowing jet of aqueous solution and a solution along the surface of a tube concentric with the jet, with a small air (vapor) gap between them. The "real" $\Delta_{\text{hydr}}G_{\text{I}}^{\infty\text{R}}$ differ from the thermodynamic ones $\Delta_{\text{hydr}}G_{\text{I}}^{\infty}$ by $z_{\text{I}}F\Delta\chi$, where the algebraic value of the ionic charge z_{I} is to be used and where $F = 96485.3\ \text{C mol}^{-1}$ is Faraday's constant. The uncertainties connected with the value of $\Delta\chi$ make the use of the measurable "real" standard molar Gibbs energies of hydration unattractive for obtaining individual ionic values for the desired quantity, $\Delta_{\text{hydr}}G_{\text{I}}^{\infty}$.

1.4 ION TRANSPORT

Ions in solution move around spontaneously due to their thermal energy. The speed of their movements is a quantity that can be determined experimentally for individual ions, contrary to the thermodynamic quantities dealt with in Section 1.3. The ions may carry with them some of their hydration shells, depending on how strong the bonding between the ion and the water of hydration is. The movement of ions depends on the presence or absence of fields, that is, gradients in the forces that cause the ions to migrate. An external field could be a pressure gradient, causing the flow of the solution as a whole. It could be an electrical field, causing ions of opposite charges to move in opposite directions. A directional concentration gradient at finite concentration causes directional diffusion of ions. The inherent movement of ions in the absence of a field is their self-diffusion and can occur at infinite dilution or at finite ones in a homogeneous solution.

1.4.1 Self-Diffusion of Ions

The rate of self-diffusion of ions is commonly obtained from other transport quantities, such as the conductivity, but can as well be determined by the use of isotopic labeling of the ions. For this purpose, a diaphragm cell with equal concentrations of the electrolyte in the two stirred compartments is employed, in one of which ions of one kind are labeled by a radioactive tracer. Other methods of measuring diffusion of ions, for example, by NMR with nonradioactive isotopes, have also been used. The rates of migration of the labeled ion truly measures self-diffusion at the nominal concentrations employed. When these are extrapolated to infinite dilution, the corresponding value of the limiting diffusion coefficient, D_I^∞, is obtained. The values are of the order of 10^{-9} m^2 s^{-1}.

Values of the limiting self-diffusion coefficients of the ions dealt with in this chapter at 298.15 K are shown in Table 1.3. It is seen that the more strongly hydrated an ion is, the lower is its rate of self-diffusion, an exception being the hydrogen ion. This ion does not diffuse in water by massive movement of the ion carrying its hydration shell but by the Grotthuss mechanism of the positive charge hopping from one water molecule to the next, and hence is much faster.

1.4.2 Ionic Conductivities

The most characteristic properties of ions are their abilities to move in the direction of an electrical field gradient imposed externally, cations toward the

TABLE 1.3. Transport Properties of Some Ions at 298.15 K[2]

Ion	D_I^∞ (10^{-9} m^2 s^{-1})	λ_I^∞ (cm^2 Ω^{-1} mol^{-1})	r_{ISI} (nm)	B_I (dm^3 mol^{-1})
H^+	9.311	349.8	0.026	0.068
Li^+	1.029	38.7	0.238	0.146
Na^+	1.334	50.1	0.184	0.085
K^+	1.957	73.5	0.125	−0.009
NH_4^+	1.958[a]	73.6	0.125	−0.008
Mg^{2+}	0.706	106.1	0.174	0.385
Ca^{2}	0.792	119.0	0.155	0.298
Fe^{2+}	0.719	107	0.172	0.42
Fe^{3+}	0.604	204	0.136	0.69
Cl^-	2.032	76.4	0.121	−0.005
Br^-	2.08	78.1	0.118	−0.033
I^-	2.045	76.8	0.120	−0.073
SCN^-	1.758	66	0.142	−0.022
NO_3^-	1.902	71.5	0.129	−0.045
ClO_4^-	1.792	67.4	0.137	−0.058
$CH_3CO_2^-$	1.089	40.9	0.225	0.246
HCO_3^-	1.185	44.5	0.207	0.13
$H_2PO_4^-$	0.879	33	0.279	0.34
CO_3^{2-}	0.923	138.6	0.133	0.294
SO_4^{2-}	1.065	160	0.115	0.206
HPO_4^-	0.439	66	0.279	0.382

[a] Calculated from the molar conductivity.

negatively charged cathode, anions in the opposite direction. The rate of movement of ions in an electric field is governed by their mobilities u_I^∞, measuring their speed in m s^{-1} at unit field, 1 V m^{-1}.

When an external electric field is imposed on an electrolyte solution by electrodes dipped into the solution, the solution conducts a current. The intensity of the current is proportional to the potential difference between the electrodes, the proportionality coefficient being the resistance of the solution. This resistance, or its reciprocal, the conductivity of an electrolyte solution, is readily measured accurately with an alternating potential at a rate of ~1 kHz in a virtually open circuit in order to avoid electrolysis at the electrodes. The molar conductivity Λ_2 can be extrapolated to infinite dilution by an appropriate theoretical expression (see below), yielding Λ_2^∞. This quantity can be split into the ionic contributions, the limiting molar ionic conductivities λ_I^∞, by using experimentally measured (and extrapolated to infinite dilution) transport numbers, t_+^∞ and $t_-^\infty = 1 - t_+^\infty$. For a binary electrolyte, $\Lambda_2^\infty = \lambda_+^\infty + \lambda_-^\infty$ and $\lambda_+^\infty = t_+^\infty \cdot \Lambda_2^\infty$. Values of the limiting ionic molar conductivities in water at 298.15 K are shown in Table 1.3. Molar ionic conductivities, λ_I^∞, are $|z|$ time the formerly widely used equivalent conductivities, and their commonly used units are S cm^2 mol^{-1} (S = Ω^{-1}). For many ions, the λ_I^∞ are accurate to ±0.01 S cm^2 mol^{-1}.

The mobilities, u_I^∞, of ions at unit electric field gradient are directly proportional to the limiting ionic molar conductivities:

$$u_I^\infty = \lambda_I^\infty / |z| F, \tag{1.8}$$

as are also the self-diffusion coefficients:

$$D_I^\infty = RT\lambda_I^\infty / z^2 F^2. \tag{1.9}$$

In fact, the latter have been obtained for most ions from the conductivities rather than from isotope labeling. Ion mobilities (hence molar conductivities and self-diffusion coefficients) increase with increasing temperatures and a fivefold increase in Λ_2^∞, the limiting molar conductivity of many electrolytes, has been reported between 273 and 373 K. This is mainly due of the corresponding decrease of the viscosity of the solvent. However, the transference numbers are also temperature-sensitive, though only mildly.

The mobility of an ion and hence its electric conductivity depends on its size and on the viscosity of the solvent, η_W^*, for aqueous solutions. According to Nernst, Stokes, and Einstein, a quantity called the Stokes radius may be assigned to an ion:

$$r_{ISt} = (F^2/6\pi N_A)|z_I|/\eta_W^* \lambda_I^\infty. \tag{1.10}$$

The parameter 6 in the denominator arises from the assumption of perfect slipping of the hydrated ion in the aqueous environment; otherwise, for perfect sticking, the parameter would be 4. Ionic Stokes radii, shown in Table 1.3, are of the same order of magnitude as the ionic radii r_I measuring the sizes of ions in crystals and solutions, shown in Table 1.2, but are not directly related to them, except for large tetraalkylammonium ions. In fact, in many cases, the Stokes radii are smaller than the crystal ionic radii, although they are supposed to pertain to the hydrated ions and ought to be larger than the latter. It is interesting to note that the Stokes radii of ions are not sensitive to the solvents in which they are dissolved, as the products $\eta_1^*\lambda_I^\infty$ of the viscosities of the solvents and the ionic molar conductivities (the Walden products) are approximately constant.[23]

At finite concentrations, ion–ion interactions cause the conductivities of electrolytes to decrease, not only if ion pairs are formed (Section 1.6.2) but also due to indirect causes. The modern theory, for example, that of Fernandez-Prini,[24] takes into account the electrophoretic and ionic atmosphere relaxation effects. The molar conductance of a completely dissociated electrolyte is

$$\Lambda_2 = \Lambda_2^\infty - Sc_2^{1/2} + Ec_2 \ln c_2 + J'(R')c_2 - J''(R'')c_2^{3/2}. \qquad (1.11)$$

Here S, E, J', and J'' are explicit expressions, containing contributions from relaxation and electrophoretic effects, the latter two depending also on ion-distance parameters R.

1.4.3 Ionic Effects on the Viscosity

The dynamic viscosity of water, η_W^*, although rather small compared with other liquids, is caused by the extensive network of hydrogen bonds existing in it that must be partly broken for the water to flow.[16] Ions affect the dynamic viscosity of the solution, η, some electrolytes enhancing it, whereas others diminishing it. The effect is described up to fairly concentrated solutions by the Jones–Dole expression:

$$[(\eta/\eta_W^*) - 1] = Ac_2^{1/2} + Bc_2. \qquad (1.12)$$

The A coefficients can be calculated theoretically from the conductivities, but the B coefficients are empirical and are obtained as the limiting slopes of plots of $[(\eta/\eta_W^*) - 1]c_2^{-1/2}$ versus $c_2^{1/2}$. In order to have the individual ionic effects on the viscosity, the B coefficients must be split according to some reasonable assumption. The generally accepted one relates to the mobilities of the ions: $B_+/B_- \approx u_+/u_-$. Over a fairly wide temperature range in water, this leads to $B(Rb^+,aq) = B(Br^-,aq)$, although $B(K^+,aq) = B(Cl^-,aq)$, valid over a narrow temperature range, has often been used, differing by $\pm 0.002\ dm^3\ mol^{-1}$.[25] Selected viscosity B coefficients of ions at 298.15 K are listed in Table 1.3. They are seen to be positive for small and multivalent ions but negative for univalent large ions.

These algebraic signs have led to the classification of ions into water structure makers ($B > 0$, increasing the viscosity) and water structure breakers

($B < 0$, diminishing it and increasing the fluidity of the solution). The absolute magnitudes of the B coefficients indicate the extents of the effects of the ions on the structure of water in the solution. They agree with other measures of such effects, for example, those derived from the relaxation of NMR signals and from the entropies of hydration of ions (Section 1.3.2).[16] As the temperature is increased, the negative B values become less negative and may change sign at a characteristic temperature. This is explained by the diminishing extent of hydrogen-bonded structure in the water as the temperature is raised, so that structure-breaking ions have less structure to break.

1.5 ION–SOLVENT INTERACTIONS

Water is the most important solvent for electrolytes dissociating into ions. Therefore, the properties of ions in aqueous solutions have been described above in detail. There are some aspects of the interaction of ions with water that have not been touched upon in the previous discussion. One of them is the number of water molecules in the hydration shells. This aspect requires, of course, departure from the strict primitive model described below since it recognizes the molecular nature of the aqueous solvent. This number must be defined operationally since diverse methods are sensitive in different ways to this number.

However, ions can be found also in nonaqueous solvents and in mixed solvent consisting of water and another solvent. A brief account of the relevant ion properties in such environments is presented here. The number of solvent molecules surrounding an ion, its solvation number, is of interest, as mentioned in Section 1.2.3 concerning hydration numbers. More definite are the thermodynamic quantities related to the transfer of ions from a source solvent, say water, to a target solvent or solvent mixture. As for the thermodynamic quantities characterizing ions in aqueous solutions dealt with above, also the transfer quantities can be measured for entire electrolytes only, and an extrathermodynamic assumption is required in order to describe the transfer properties of individual ions.

If an ion exists in a mixture of solvents, its solvation by the components of the mixture depends not only on its affinity to each of these but also on the mutual interactions of the solvents. The composition of the solvation shell of the ion will generally differ from that of the bulk solvent. This can be determined from thermodynamic data, including the standard molar Gibbs energy of transfer of the ion from the source solvent (say water) to the target aqueous solvent mixture.

1.5.1 Solvation Numbers

What is said in the following regarding hydration numbers generally applies also to solvation numbers. Water molecules around ions are arranged in concentric shells: The nearest ones, in the first hydration shell, are relatively strongly bound to the ion and move together with it, as do at least some of the water molecules in a second hydration shell, if present. Large ions with a single charge (e.g., $(C_2H_5)_4N^+$ and larger tetraalkylammonium ions) may not have a hydration shell altogether, but have the water in enhanced icelike tetrahedral structures around them as for nonionic hydrophobic solutes, but in any case large ions (e.g., K^+ and Cl^-) lack a well-formed second hydration shell. A second shell characterizes multiply charged small ions (e.g., Mg^{2+}), the water molecules in it being hydrogen-bonded to those in the first shell more strongly than the hydrogen bonds in pure water. Beyond the hydration shells, the water molecules are still affected by the electric field of the ionic charge and the possibility of being hydrogen-bonded to the inner water molecules, but the hydrogen-bonded structure of this region is less ordered than in pure water (see above). Only further out from the ion does the water become bulk water, having the properties of pure water, unless, of course, at finite concentrations of the electrolyte, counterions show the effects of their presence.

Hydration numbers are the time-average numbers of water molecules residing in the first (and second, if formed) hydration shell of ions. Coordinate bonds may be formed between the water molecules in the first hydration shell and the ion, the bonds being then strongly directional, and a definite hydration number results, equaling the coordination number (e.g., 4 for Be^{2+}, 6 for Mg^{2+} and Fe^{2+}). If only nondirectional electrostatic association takes place, then geometric constraints may occur, smaller ions having smaller hydration numbers than larger ions, although the water molecules are bonded more energetically to the former. Over time, water molecules depart from the hydration shells and others come in, so that the time-average is a dispersion of numbers with a noninteger average value.

There are several methods for the determination of the hydration numbers, but there is unfortunately no strong agreement between the results. If a primitive model of the ions (hard spheres) and the solvent (a compressible dielectric) is employed, it may be assumed that the electric field of the ion causes a certain molar compression of the electrostricted solvent in its solvation shell independent of the nature of the ion.[26] The molar compression of electrostricted water is $\Delta V_{\mathrm{Welec}} = -2.9\ \mathrm{cm^3\ mol^{-1}}$ at 298.15 K[12,26] obtained from the compressibility and the pressure derivative of the relative permittivity of pure water. The ratio

$$h^{\infty}_{\mathrm{Ielec}} = V^{\infty}_{\mathrm{Ielec}}/\Delta V_{\mathrm{Welec}} \tag{1.13}$$

between the molar ionic electrostriction, $V^{\infty}_{\mathrm{Ielec}}$, and $\Delta V_{\mathrm{Welec}}$ can be construed to represent the time-average hydration number of the ion.

Alternatively, the ion and the water in its first hydration shell may be considered to be uncompressible by an external pressure, the electric field having produced the maximal possible compression. Then the hydration number is defined by the compressibility as

$$h_{\mathrm{Icomp}}^{\infty} = 1 - (\partial V_{\mathrm{I}}^{\infty}/\partial P)_T / \kappa_{\mathrm{W}}^{*} V_{\mathrm{W}}^{*}. \quad (1.14)$$

Here $(\partial V_{\mathrm{I}}^{\infty}/\partial P)_T$ is a negative quantity, the standard partial molar compressibility of the ion, and κ_{W}^{*} is the isothermal compressibility of water at ambient pressures. To obtain individual ionic values of $(\partial V_{\mathrm{I}}^{\infty}/\partial P)_T$ from experimental values for electrolytes requires the assumption of a value for one ion. That for chloride, $(\partial V_{\mathrm{I}}^{\infty}/\partial P)_T(\mathrm{Cl}^-, \mathrm{aq}) = -16.5 \pm 1.5\ \mathrm{cm}^3\ \mathrm{GPa}^{-1}\ \mathrm{mol}^{-1}$ at 298.15 K, was suggested.[27] The hydration numbers from these two methods are shown in Table 1.2 and are in only fair mutual agreement. Other values were obtained by other methods and for ions for which no other value of the hydration number is available, the approximation $h_{\mathrm{I}}^{\infty} = 0.360|z|/(r_{\mathrm{I}}/\mathrm{nm})$ can be used.[2]

Hydration numbers are expected to diminish as the concentration of the electrolyte increases, mildly at low concentrations but strongly when the hydration shells of oppositely charged ions start to overlap. The average distance between ions in a solution is inversely proportional to the cube root of the concentration:

$$d^{\mathrm{av}} = N_{\mathrm{A}}^{-1/3} \cdot \left(\sum \nu_{\mathrm{I}} c_{\mathrm{I}}\right)^{-1/3} = 1.1844 \cdot \left(\sum \nu_{\mathrm{I}} c_{\mathrm{I}}\right)^{-1/3}\ \mathrm{nm}, \quad (1.15)$$

the summation extending over all the ions present with their stoichiometric coefficients ν_{I} at concentrations c_{I} in mol dm^{-3}.[28] The size of a hydrated ion may be taken as $d_{\mathrm{I\text{-}O}} + r_{\mathrm{W}} = r_{\mathrm{I}} + 2r_{\mathrm{W}}$ (Section 1.2.2), so that it is possible to estimate the concentration at which the hydration shells start to overlap. For aqueous NaCl with ionic radii r_{I} of 0.102 and 0.181 nm for the cation and anion and with $r_{\mathrm{W}} = 0.138$ nm, the sum of the radii of the hydrated ions is 0.835 nm, so that at a concentration of 1.43 mol dm^{-3} the hydration shells start to overlap. Below the overlap limit, experimental values of $(\partial V_{\mathrm{I}}/\partial P)_T$ may be used for the estimation of the hydration numbers at finite concentrations from the expression for h_{Icomp} given above.

The molar compression of solvents other than water in the electric field of an ion, $\Delta V_{1\mathrm{elec}}$, has been determined from their compressibility and the pressure derivative of their relative permittivity as it was for water. Typical values are considerably larger than for water, but relative to the molar volume of the pure solvent, V_1^{*}, some are of the same magnitude as for water (0.161).[26] Solvation numbers have been deduced from the ratios of the ionic electrostriction volume, $V_{\mathrm{Ielec}}^{\infty}$ (when available for various ion/solvent systems), and the molar compression of the solvent, $\Delta V_{1\mathrm{elec}}$. They are commensurate with the hydration numbers of univalent ions but smaller than the hydration numbers (about one-half) for multivalent ions, probably because of geometric constraints for the bulkier solvents.

1.5.2 Salting Out and Salting In

As the ions are solvated, there remains less "free" solvent to accommodate other solutes, so that a general result of the presence of ions in a solution is a diminished solubility of nonelectrolyte solutes, marked by subscript $_3$. (The mutual interactions of ions are treated in Section 1.6.) This phenomenon is called "salting out" and is expressed up to fairly high concentrations, c_2, of the electrolyte by $\log(s_3^*/s_3) = kc_2$, where s_3^* is the solubility of the nonelectrolyte in the absence and s_3 that in the presence of the electrolyte. The coefficient k is called the Setchenov salting-out constant. There are some systems (see below) where $k < 0$, and salting in then occurs, the solubility being enhanced by the presence of the electrolyte, but as a rule $k > 0$ and it is diminished. The magnitude of k increases with the molar volume of the nonelectrolyte, V_3, and with the intensity of solvation of the electrolyte. The latter can be described by the electrostriction that the ion causes, and this leads to the McDevit and Long[29] formulation for the Setchenov constant:

$$k = -V_3 V_{2\text{elec}}/(\ln 10)RT\kappa_T. \tag{1.16}$$

Here $V_{2\text{elec}}$ is the molar electrostriction by the electrolyte, the (negative) difference between the partial molar volume of the electrolyte and the sum of the intrinsic volumes of its ions and κ_T is the isothermal compressibility of the solvent.

This discussion pertains to electrolytes consisting of "small" ions, such as Na^+ and Cl^-, to solvents such as water, and to nonpolar solutes, such as benzene. In systems like these, the coulombic forces between the ions and the solvent leading to their solvation cause the salting out. This does not exclude the possibility of direct interactions between the solute and the ions by hydrogen bonding or donor–acceptor interactions, which may overcompensate the coulombic forces and cause salting in ($s_3 > s_3^*$) (see Chapter 6). Other cases of salting in occur with very poorly solvated ions (tetraalkylammonium ones in water) and nonelectrolytes that are more polar that the solvent (have a higher permittivity).

1.5.3 Ion Transfer Thermodynamics

The transfer of an ion from water, as the source solvent, to a target solvent or solvent mixture has thermodynamic consequences in terms of energies, entropies, volumes, and so on. It is, in principle, possible to evaluate these from a comparison of the solvation energetics of the ions, that is, transfer from the ideal gas into the source solvent on the one hand and into the target solvent or mixture on the other. However, this procedure would involve rather small differences between large numbers and be inaccurate. It is, therefore, preferable to determine the transfer energetics directly, and derive from them, if desirable, the solvation quantities. This preference arises from the fact that the properties of hydrated ions and the hydration properties of ions have been most extensively and accurately determined.

The sign of the standard molar Gibbs energy of transfer of an ion $I^\pm$ from water to the target solvent (or mixture) S, $\Delta_t G_I^\infty(I^\pm, aq \rightarrow S)$, determines whether this ion prefers to be in an aqueous environment ($\Delta_t G_I^{\infty} > 0$) or have S as its near environment ($\Delta_t G_I^{\infty} < 0$). Values of the standard molar Gibbs energy of transfer electrolytes, $\Delta_t G_2^{\infty}$, can be determined for sparingly soluble electrolytes from their relative solubilities in water and in S, or for well-soluble electrolytes electrochemically from the electromotive force of cells, the two back-to-back half-cells of which have water and S as the solvent but being otherwise identical, or from polarographic half-wave potentials in water and in S. The standard molar enthalpy of transfer of electrolytes, $\Delta_t H_2^{\infty}$, can be determined from their relative heats of solution in water and in S and the standard molar entropy of transfer electrolytes, $\Delta_t S_2^{\infty}$, can be determined from the temperature coefficients of the solubilities, electromotive forces, or half-wave potentials establishing the $\Delta_t G_2^{\infty}$ values. The standard molar volumes of transfer $\Delta_t V_2^{\infty}$ are obtained from the apparent molar volumes, and these, in turn, from the densities of the solutions in water and in S.

In attempts to split the electrolyte values into the individual ionic contribution, it is expedient to employ the same extrathermodynamic assumption for all these three thermodynamic quantities in order to arrive at consistent values, obeying

$$\Delta_t G_2^{\infty} = \Delta_t H_2^{\infty} - T\Delta_t S_2^{\infty}. \tag{1.17}$$

A variant of the TPTB assumption (Section 1.3.1), namely that the transfer quantities of tetraphenylarsonium cations and tetraphenylborate anions are equal, has been widely used (the TATB assumption). Although Ph_4As^+ is slightly larger than Ph_4P^+, which in turn, is slightly larger than BPh_4^-, these small differences are deemed to lead to results within experimental errors of the transfer data and be of little consequence for the individual ionic transfer energetics.[30] The results for transfer of ions into many pure organic solvents have been summarized by Marcus[2] and those for transfer into mixed aqueous-organic solvents by him with coworkers.[31–33] Some results for the standard molar ionic volumes of transfer, $\Delta_t V_I^{\infty}$, and heat capacity of transfer, $\Delta_t C_{PI}^{\infty}$, into pure organic solvents have also been reported.[34]

The transfer of many ions from water into most organic solvents is an unfavorable process that would not occur spontaneously: $\Delta_t G_I^{\infty}(I^{\pm}, aq \rightarrow S) > 0$. This is generally the case for transfer of cations into alcohols, ketones, esters, nitriles, nitro compounds, and halogenated alkanes. Exceptions are the transfer of Ag^+ and Cu^+ into nitriles and of "soft cations" (such as Ag^+, Cd^{2+}, and Hg^{2+}) into solvents such as pyridine and those in which the oxygen donor atom is thio-substituted. Furthermore, solvents that are strong electron pair donors, such as ethane-1,2-diol, propylene carbonate, amides, sulfolane, dimethyl sulfoxide, and trimethyl phosphate, have $\Delta_t G_I^{\infty}(I^{z+}, aq \rightarrow S) < 0$, and cations are preferentially solvated by them. Anions, in general, prefer the aqueous environment that provides hydrogen bonds with them, an exception being the strong hydrogen bond donor solvent 2,2,2-trifluoroethanol. Bulky ions, whether cations or anions, do prefer the less structured (less hydrogen-bonded) environment of most organic solvents. This pertains to tetraalkylammonium cations, the tetraphenyl moieties mentioned above, triiodide, picrate, and so on.

1.5.4 Preferential Solvation of Ions in Mixed Solvents

When an ion is located in a solvent mixture, $S_A + S_B$, its environment generally has a composition differing from that of the bulk mixture due to preferential solvation of the ion by one of the components of the solvent mixture. It stands to reason that if an ion $I^{\pm}$ has a favorable Gibbs energy of transfer into a pure component of the mixture, say S_A, $\Delta_t G_I^{\infty}(I^{\pm}, aq \rightarrow S_A) < 0$, but an unfavorable one into the other component, $\Delta_t G_I^{\infty}(I^{\pm}, aq \rightarrow S_B) > 0$, then it will be preferentially surrounded by molecules of S_A. This statement must be modified in view of the mutual interactions of the solvent components of the mixture. In some cases, the preferential solvation in the mixture is practically complete, and then selective solvation takes place. The quantitative aspect of these preferences is of interest, that is, what fraction of the solvation shell of the ion is occupied by each solvent component, and there are two approaches that have been used in order to deal with it.[35]

In the quasi-lattice quasi-chemical (QLQC) approach,[36] the ion $I^{\pm}$ and the molecules of the two solvents, S_A and S_B, are distributed on sites of a quasi-lattice characterized by a lattice parameter Z. It specifies the number of neighbors each particle has, independently of the nature of the particles. The pair-wise interaction energies e_{IA}, e_{IB}, e_{AA}, e_{AB}, and e_{BB}, weighted according to the numbers of the corresponding nearest neighbors, determine the total configurational energy of the system. Ideal entropy of mixing of the particles on the quasi-lattice sites is assumed. The quasi-chemical aspect relates to the relative strength of the mutual interactions of the solvent molecules and those with the ion. A set of equations is provided by this approach to determine the local (i.e., around the ion) mole fraction of component S_A: $x_{IA}^{L}(= 1 - x_{IB}^{L})$, hence the solvation number of the ion by S_A: Zx_{IA}^{L}, and eventually the equilibrium constant for the replacement of S_B by S_A: $K_B^A = x_{IA}^{L} x_B / x_A x_{IB}^{L}$.[37] These expressions require the excess Gibbs energy of mixing of the equimolar solvent mixture ($x_A = 0.5$) in the absence of the solute and the standard molar Gibbs energy of transfer of the ion $I^{\pm}$ from solvent S_A to the mixture as a function of the composition up to $x_B = 1$. The latter data establish the lattice parameter Z within ±2 units.

The inverse Kirkwood–Buff integral method (IKBI) does not involve a model such as the QLQC method does, hence is rigorous, but requires derivatives of the Gibbs energy of transfer and the excess Gibbs energy of mixing[38] with respect to the solvent composition. These functions, however, may often not be known sufficiently accurately for obtaining meaningful derivatives. The Kirkwood–Buff integrals need for their evaluation, in addition to these derivatives, also the isothermal compressibility of the mixture and the partial molar volumes of the ion and the two solvent components in it as a function of the solvent composition. The expressions yielding the local mole fraction x_{IA}^{L}

require, furthermore, an estimate of the correlation volume, that is, the volume around the ion in which it affects the composition of the local solvent mixture. Finally, there is the problem in obtaining the required information from experimental data, in that the latter pertain to entire electrolytes, and their application to single ions has some bearing on the meaning of the Kirkwood–Buff integrals. Application of the TATB assumption (Section 1.5.3) concerning the splitting of the standard molar Gibbs energy of transfer of an electrolyte into the ionic contributions has been employed to circumvent this theoretical problem.[35]

Both methods have been applied to several ions in aqueous and completely nonaqueous solvent mixtures, provided that ionic dissociation is complete, with fair agreement when they have been applied to a given system.[35]

1.6 ION–ION INTERACTIONS

At practical concentration, that is, beyond infinite dilution, ions in solution interact with each other electrostatically due to their charges: Ions of like charge sign repel each other and those of unlike charge sign attract each other. These interactions compete with the thermal movement of all the particles in the solution and are screened by the dielectric medium of the solvent in which the ions are located. Water, having a high dielectric permittivity, is very effective in this screening. The overall interactions, involving ion solvation in addition to ion–ion interactions are quite complicated, and approximations have to be applied in order to handle the resulting behavior of the ions theoretically.

The simplest approximation is called the "restricted primitive model" and considers the ions as charged conducting spheres dispersed uniformly in a continuum made up of a compressible dielectric. The ions are characterized by their charges (sign and magnitude) and sizes (radii), the solvent, whether single or a mixture, by its permittivity, compressibility, and thermal expansibility. Within this model and in dilute solutions of electrolytes, the well-known Debye–Hückel theory describes the chemical potentials of the electrolyte, μ_2, and that of the solvent, μ_1, sufficiently well. The former is directly related to the activity of the electrolyte and to its mean ionic activity coefficient, $\gamma_{2\pm}$ (on the molal scale) or $y_{2\pm}$ (on the molar scale). The latter, μ_1, is directly related to the activity of the solvent, a_1, and to its osmotic coefficient,

$$\varphi_1 = -(1000/\nu_2 m_2 M_1)\ln a_1, \tag{1.18}$$

where ν_2 is the number of ions per formula of the electrolyte of molality m_2, and M_1 is the molar mass of the solvent (in g mol^{-1}; for water $1000/M_1 = 55.51$). The solvent activity is roughly equal to the ratio of the vapor pressure of the solvent in the solution to that of the pure solvent: p_1/p_1^* (when the vapor pressures are small and the vapors can be considered to approximate ideal gases). It is also related to the depression of the freezing point and elevation of the boiling point of the solvent.

1.6.1 Activity and Osmotic Coefficients

The limiting (Debye–Hückel) expression for the mean molal activity coefficient of an electrolyte is $\log(\gamma_{2\pm}) = -Az_+z_-I^{1/2}$, where A depends on the solvent and the temperature ($A = 0.511$ in water at 298.15 K). The logarithm of $\gamma_{2\pm}$ is thus proportional to the square root of the ionic strength $I = 0.5\sum_I m_I z_I^2$, the summation extending over all the ions in the solution, which may contain a mixture of electrolytes. This limiting expression is valid only up to, say, $m = 0.01$ m (~0.01 mol dm^{-3}) for uni-univalent electrolytes.

Beyond this concentration, the *extended* Debye–Hückel expression

$$\log(\gamma_{2\pm}) = -Az_+z_-I^{1/2}/(1 + BaI^{1/2}) \tag{1.19}$$

should be used, valid, in turn, up to, say, $m = 0.2$ m (~0.15 mol dm^{-3}). Beyond this concentration again, or for higher valent electrolytes, a linear term in the ionic strength, bI, has to be added, up to quite high concentrations. In these extended expressions, B is a solvent- and temperature-dependent constant and a is the mean distance of closest approach of the ions (their diameters, if considered the same for cations and anions), whereas the coefficient b is a completely empirical fitting parameter. For many purposes the product $Ba = 1.5$ kg$^{1/2}$ mol$^{-1/2}$ at any temperature and for any solvent may be used, the onus of fitting the experimental values of $\log(\gamma_{2\pm})$ is then placed on the value of the parameter b.

A key quantity in the Debye–Hückel theory is the screening length, κ^{-1}, the average reciprocal of the radius of the "ionic atmosphere" surrounding an ion in the solution. The square of screening length is proportional to the ionic strength of the solution:

$$\kappa^2 = (2N_A^2e^2/1000\varepsilon_0\varepsilon_r RT)I, \tag{1.20}$$

this being the source of the dependence of $\log(\gamma_{2\pm})$ on $I^{1/2}$. The numerical value of this screening length is thus $\kappa = 502.9\varepsilon_r^{-1/2}(T/\mathrm{K})^{-1/2}(I/\mathrm{mol\ dm^{-3}})^{1/2}$ nm^{-1} and depends on the reciprocal square roots of the relative permittivity and of the temperature, becoming $\kappa = 3.26(I/\mathrm{mol\ dm^{-3}})^{1/2}$ nm^{-1} for water at 298.15 K. The dimensionless product κa features in the denominator of the extended Debye–Hückel expression for $\log(\gamma_{2\pm})$ and in the expression for the osmotic coefficient. The logarithm of the solvent activity is given by the Debye–Hückel theory as

$$\ln a_1 = -\ln(1 + \nu_2 m_2 M_1/1000) + (V_1/24\pi N_A)\kappa^3\sigma(\kappa a). \tag{1.21}$$

The $\sigma(\kappa a)$ function is

$$\sigma(\kappa a) = 3(\kappa a)^{-3}[(1+\kappa a)-(1+\kappa a)^{-1}-2\ln(1+\kappa a)] \approx 1-1.5(\kappa a)+1.8(\kappa a)^2, \quad (1.22)$$

where the approximation is valid for $\kappa a \leq 1$.

An alternative formulation for the activity and osmotic coefficients is that of Pitzer[39] that includes the Debye–Hückel limiting law but treats differently its extension to practical concentrations for fully dissociated electrolytes, in fact up to several moles per kilogram or per cubic decimeter. For 1:1 electrolytes, the resulting expressions are $\ln(\gamma_{2\pm}) = \mathrm{f}_\gamma + B_\gamma m + C_\gamma m^2$ and

$$\varphi_1 = 1 + \mathrm{f}_\varphi + B_\varphi m + C_\varphi m^2, \quad (1.23)$$

the functions f are the electrostatic Debye–Hückel terms and the B and C coefficients are electrolyte-specific fitting parameters. For electrolyte types other than 1:1, factors in the charge numbers and stoichiometric coefficients have to be included. Two universal constants, $b = 1.2$ and $\alpha = 2.0$, are employed in the full expressions as well as the solvent- and temperature-dependent A_φ arising from the Debye–Hückel theory. For details, the series of papers by Pitzer and coworkers should be consulted.[40]

1.6.2 Ion Pairing

In electrolyte solutions consisting of relatively poorly solvated ions and of solvents of relatively low permittivity, the screening of the charges by the solvent is inadequate to prevent ionic association at sufficiently high concentrations. In most cases, the association stops at ion pairing: one cation with one anion, but in a solvent of low permittivity, triple ions may be formed or even larger associates. The treatment of ion pairing offered by Bjerrum, using the restricted primitive model (Section 1.5), has been validated over time, and can be formulated as follows.[41] Ions that are nearer each other than a certain distance

$$q = z_+ z_- N_A e^2/8\pi\varepsilon_0\varepsilon_r RT = (z_+ z_-/2) l_B, \quad (1.24)$$

where l_B is the Bjerrum length (but necessarily more remote than the distance of closest approach a), are considered to be paired, those at larger distances from each other are free. Due to the electrostatic forces, only ions of opposite charge signs are likely to approach each other to a distance $\leq q$. An equilibrium constant, K_A, may be formulated for the equilibrium $C^{z+} + A^{z-} \rightleftarrows C^{z+}A^{z-}$. If the fraction α of the c_2 molar electrolyte is dissociated and $1 - \alpha$ is paired, then

$$K_A = (1-\alpha)/\alpha^2 c_2. \quad (1.25)$$

A dimensionless parameter $b = q/a$ is employed and the resulting association constant is

$$K_A = (4\pi N_A/1000)b^3 Q(b) \tag{1.26}$$

using the function $Q(b) = \int_2^b x^{-4}\exp(x)\mathrm{d}x$, with x as an auxiliary variable, and the integral is to be solved numerically. At 298.15 K, the value of the key variable b for any solvent is given by

$$\log(b) = 1.448 + \log|z_i z_j| - \log(\varepsilon_r) - \log(a/\mathrm{nm}), \tag{1.27}$$

log(298.15 K/T) should be added to the numerical constant at other temperatures. The distance of closest approach a is taken as the mean diameter of the ions that should not be smaller than the sum of their crystal ionic radii: $a \geq r_+ + r_-$. The values of log $Q(b)$ have been tabulated[42] and range from −1.358 at $b = 2.1$ (the lowest practical value) through zero for $b = 5.9$ to positive values at large b for which $Q(b) \approx \exp(b)/b^4$. The parameters b and $Q(b)$ depend on the temperature and the solvent (through its relative permittivity ε_r) and on the electrolyte through the mean diameter of its ions, a, and are readily calculated. From them, the association constant K_A is obtained and the fraction associated, $1 - \alpha$, as a function of the concentration c_2.

Experimentally, the dissociated fraction of the electrolyte, α, is commonly obtained from conductivity data, although other techniques have also been widely employed. In the case of the conductivities, the expression

$$\Lambda_2 = \Lambda_2^\infty - S(\alpha c_2)^{1/2} + E(\alpha c_2)\ln(\alpha c_2) + J'(R')(\alpha c_2) - J''(R'')(\alpha c_2)^{3/2} \tag{1.28}$$

has to be used (Section 1.4.2), recognizing that the actual concentration of the ions is αc_2 instead of c_2.

In aqueous solutions, with its high relative permittivity, ion pairing between univalent ions is rare, unless they are only poorly hydrated and can approach each other to within $q = 0.357$ nm (at 298.15 K). However, for more highly charged ions, appreciable ion pairing does occur in aqueous solutions at concentrations of the order of 1 mol dm^{-3} for 1:2 or 2:1 type salts and even at concentrations of 0.1 mol dm^{-3} for 2:2 or higher types of salts. The equilibrium constants K_A for the latter (e.g., $MgSO_4$) are of the order of 100 to 200 dm^3 mol^{-1}.

In this formal theory, based on the restricted primitive model of the electrolyte solution, there is nothing that indicates how intimately the cation and anion are bound together in the ion pair. In fact, for sufficiently low ε_r values or large charges of the ions, the cutoff distance q beyond which ions are considered to be free is manifold larger than the mean ionic diameters a. It is clear that solvent molecules penetrate the intervening distance, and if the "restricted" is removed from the model, it is possible to specify how many solvent molecules separate the cation and the anion. If this number is zero, this means that a "contact ion pair" (CIP) is formed, if it is 1 then a "solvent-shared ion pair" (SIP) is formed, if it is 2, then a "solvent-separated ion pair" (SSIP or S2IP) is formed. Several forms of such ion pairs can exist at equilibrium with each other, depending on the solvent permittivity and the electrolyte concentration. Methods for ascertaining the situation with regard to these forms of ion pairs have been reviewed.[41]

REFERENCES

1. D. D. Wagman, W. H. Evans, V. B. Parker, R. H. Schumm, I. Halow, S. M. Bailey, K. L. Churney, R. L. Nutall. *J Phys Chem Ref Data* **1972**, 11 (Suppl. 2), 1–391 and more recent supplementary data.
2. Y. Marcus. *Ion Properties*. Dekker, New York, 1997.
3. G. A. Krestov. *Thermodynamics of Solvation*. Ellis-Horwood, Chichester, 1991.
4. R. D. Shannon and C. T. Prewitt. *Acta Cryst B* **1969**, 25, 925; **1970**, 26, 1046.
5. H. B. D. Jenkins and K. P. Thakur. *J Chem Educ* **1979**, 56, 576.
6. Y. Marcus. *Chem Rev* **1988**, 88, 1475.
7. H. Ohtaki and T. Radnai. *Chem Rev* **1993**, 93, 1157.
8. Y. Marcus. *J Solution Chem* **1993**, 12, 271.
9. F. H. David and B. Fourest. *New J Chem* **1990**, 21, 167.
10. P. Mukerjee. *J Phys Chem* **1961**, 65, 740.
11. Y. Marcus. *J Solution Chem* **2010**, 39, 1031.
12. Y. Marcus. *J Phys Chem B* **2009**, 113, 10285.
13. M. H. Abraham and Y. Marcus. *J Chem Soc Faraday Trans 1* **1988**, 82, 3255.
14. Y. Marcus. *J Chem Soc Faraday Trans* **1987**, 83, 339.
15. B. E. J. Conway. *Solution Chem* **1978**, 7, 721.
16. Y. Marcus. *Chem Rev* **2009**, 109, 1346.
17. H. S. Harned and B. B. Owen. *The Physical Chemistry of Electrolyte Solutions*, 3rd ed. Reinhold, New York, 1958.
18. R. A. Robinson and R. H. Stokes. *Electrolyte Solutions*, 2nd ed. Butterworths, London, 1965.
19. Y. Marcus. *J Chem Soc Faraday Trans* **1991**, 87, 2995.
20. C. P. Kelly, C. J. Cramer, D. J. Truhlar. *J Phys Chem B* **2006**, 110, 16066.
21. Y. Marcus, in D. V. Bostrelli, eds. *Solution Chemistry Research Progress*. Nova Science, Hauppauge, NY, 2008; pp. 51–68.
22. V. I. Parfenyuk. *Coll J* **2002**, 64, 588.
23. Y. Marcus. *Russ J Electrochem* **2008**, 44, 16.
24. R. Fernandez-Prini. *Trans Faraday Soc* **1969**, 65, 3311.
25. H. B. D. Jenkins and Y. Marcus. *Chem Rev* **1995**, 95, 2695.
26. Y. Marcus. *J Phys Chem B* **2005**, 109, 18541.
27. J. G. Mathieson and B. E. Conway. *J Solution Chem* **1974**, 3, 455.
28. Y. Marcus. *J Solution Chem* **2009**, 38, 513.

29. W. F. McDevit and F. A. Long. *J Am Chem Soc* **1952**, 74, 1773.
30. Y. Marcus. *Pure Appl Chem* **1952**, 58, 1721.
31. C. Kalidas, G. T. Hefter, Y. Marcus. *Chem Rev* **2000**, 100, 819.
32. G. T. Hefter, Y. Marcus, W. E. Waghorne. *Chem Rev* **2002**, 102, 2773.
33. Y. Marcus. *Chem Rev* **2002**, 107, 3880.
34. Y. Marcus. *Pure Appl Chem* **1996**, 68, 1495.
35. Y. Marcus. *Solvent Mixtures*. Dekker, New York, **2002**.
36. Y. Marcus. *Austr J Chem* **1983**, 36, 1718.
37. Y. Marcus. *J Chem Soc Faraday Trans 1* **1988**, 84, 1465.
38. A. Ben-Naim. *Cell Biophys* **1988**, 12, 255.
39. K. S. Pitzer. *Activity Coefficients in Electrolyte Solutions*, R. M. Pytkovicz, ed. Vol. 1. CRC Press, Boca Raton, FL, 1979; Chapter 7.
40. K. S. Pitzer and G. Mayorga. *J Phys Chem* **1973**, 77, 2300 and following papers by Pitzer et al.
41. Y. Marcus and G. Hefter. *Chem Rev* **2006**, 106, 4585.
42. Y. Marcus. *Introduction to Liquid-State Chemistry*. Wiley, Chichester, UK, 1977; pp. 241–245, 267–279.

CHAPTER 2

IONIC INTERACTIONS IN SUPRAMOLECULAR COMPLEXES

HANS-JÖRG SCHNEIDER

2.1 INTRODUCTION/APPLICATIONS

Ion pairing plays a major role in the design of supramolecular complexes, especially in an aqueous environment, where the presence of charges also secures sufficient solubility. Related complexes have been discussed already in books and reviews on supramolecular chemistry,[1] also with an emphasis on the underlying binding mechanism.[2] Salt bridges between host-and-guest compounds bearing complementary charges allow reversible formation of associations with many targets of biological importance, such as nucleotides, saccharic acids, peptides, biogenic amines, and so on. The analyses of synthetic host–guest complexes contribute to a better understanding of interactions also in biological systems, where distinction between the many noncovalent forces within large associations often is difficult. In contrast, synthetic complexes can be so designed, that single interactions can be well studied; at the same time the underlying geometry in solution of lower molecular weight associations is easier to characterize than with, for example, biopolymers. Also, the use of macrocyclic host compounds leads to a 1:1 **stoichiometry** in most complexes. Aggregation with a **composition** beyond 1:1 can occur particularly at higher concentration, and their interference can be studied by dilution experiments.[3]

The most important applications of synthetic host–guest complexes are still in the analytical field, and they also play an increasingly important role in technologies such as separations, in the design and delivery of drugs, of polymers with smart functions, and so on. In principle, suitable host compounds can be designed for any desired target and can be applied in any medium including the gas phase. Particularly, the hosts can be equipped for optical or

Ionic Interactions in Natural and Synthetic Macromolecules, First Edition. Edited by Alberto Ciferri and Angelo Perico.

electrochemical detection and may be implemented in a large variety of sensors. The optimization of affinity, leading to low detection limits, can be achieved by increasing the number of ionic sites within a host structure. The equally important issue of selectivity can be met by steric constraints, particularly with macrocyclic hosts (see Section 2.7). Promising new applications such as selectively responsive chemomechanical polymers for sensors and actuators emerge with intelligent materials based, for instance, on polyelectrolyte gels described by Chiarellii and De Rossi in Chapter 15 of this book. An interesting feature of correspondingly designed host–guest complexes is the possibility to control their behavior by external effectors, for instance, by pH changes. This opens on a molecular basis the way to new intelligent materials.[4]

2.2 SYSTEMS DOMINATED ENTIRELY BY ION PAIRING: IONOPHORES AND SO ON

Salt bridges between anions and cations can be used either way, with cationic or anionic host molecules and guest molecules bearing complementary charges. Most of such artificial host–guest complexes were designed for anion recognition,[5] association with metal ions rely more often on donation from electron pairs in neutral carriers such as crown ethers, bearing oxygen, nitrogen, or sulfur atoms. Structures **1** and **2** shows classical examples of how salt bridging in confined spaces leads to selective binding of anions, such as azide[6] (Fig. 2.1); the macrocycle **2** bears permanent charges and binds therefore independent of the pH anions, for example, Cl^- with $K = 270$, Br^- with 2150, and I^- with 6500 (K as $[M^{-1}]$, with ($X = (CH_2)_n$, $n = 6$), in water).[5e,d] Guanidinium hosts such have the advantage to work also at higher pH than amines, as the corresponding pK is 13.5. The guanidinium function can be fixed within a bicyclic framework such as **3**, which allows to attach substituents R securing discrimination of chiral anions.[7]

The binding strength for complexes between anions and polyamines increases linearly with the degree of protonation (Fig. 2.2), and thus can be controlled by the pH of the medium. The affinities between the anions, which

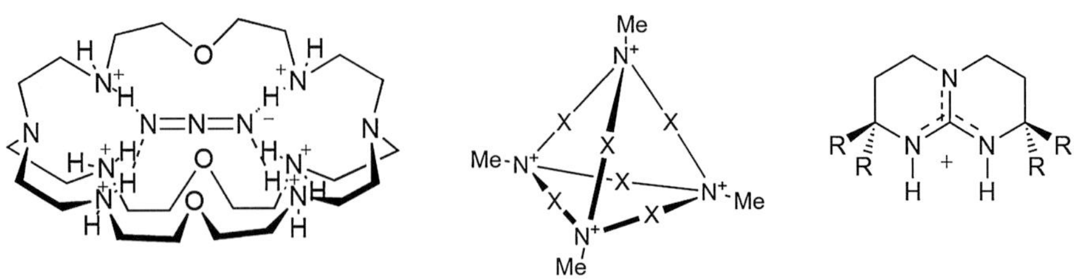

1 **2** **3**

FIGURE 2.1. Examples of cationic hosts for ion pairing with anionic guests.

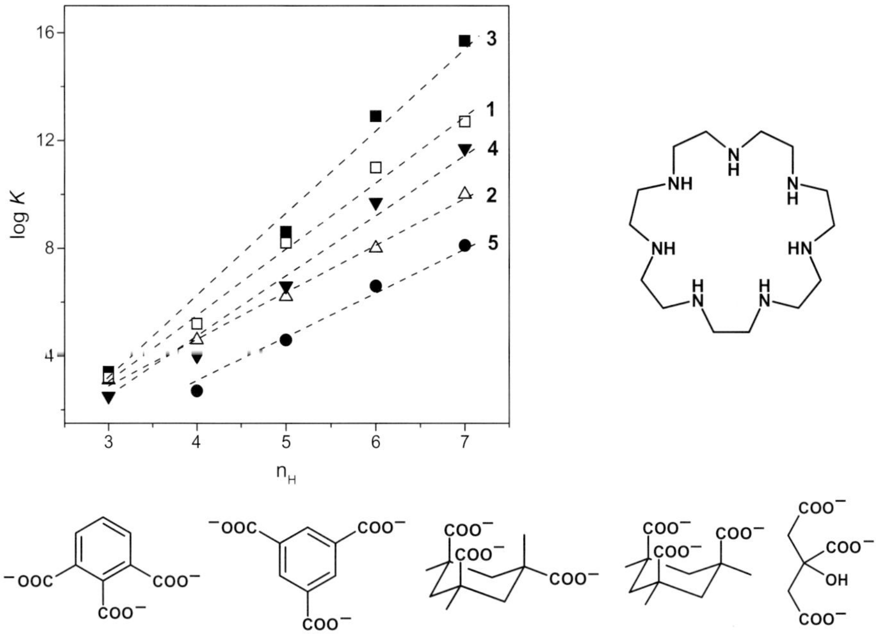

1 2 3 4 5

FIGURE 2.2. Binding strength (as log *K*) of anions **1–5** with a macrocyclic polyamine as function of the amine protonation degree (n_H). Reprinted with permission from H.-J. Schneider, A.Yatsimirsky. *Chem Soc Rev* **2008**, 37, 263. Copyright 2008 The Royal Society (London).[8]

all bear three carboxylate groups, differ as consequence of different geometric matching with the nitrogen centers, with a selectivity increase with increasing total binding strength (see Section 2.7).[8]

Structures such as **6** (Fig. 2.3) represent hosts bearing negative charges, which effectively bind positively charged guest molecules, including ammonium ions. The sulfanato groups in calixarenes[9] **6** ($n = 4$, 6, or 8) form salt bridges over a large range of pH; the aryl moieties in the cavity allow simultaneously cation–π interactions , which then leads to simultaneous formation of **6b**, being in rapid equilibrium with **6a**. The tetraphenol units in the resorcarene host **7** are fully deprotonated at higher pH, and then form the open conformer **7a**; only the partially protonated form **7b** allows ion paring with guest molecules such as the neurotransmitter acetylcholine Addition of such ammonium guest molecules to the host leads with uptake of two protons to formation of form **7b**, which is stabilized by hydrogen bonds between the phenolic groups. Thus, the system represents a simple element of a proton pump, where protons are released or taken up by additional effector molecules such as acetylcholine.[10]

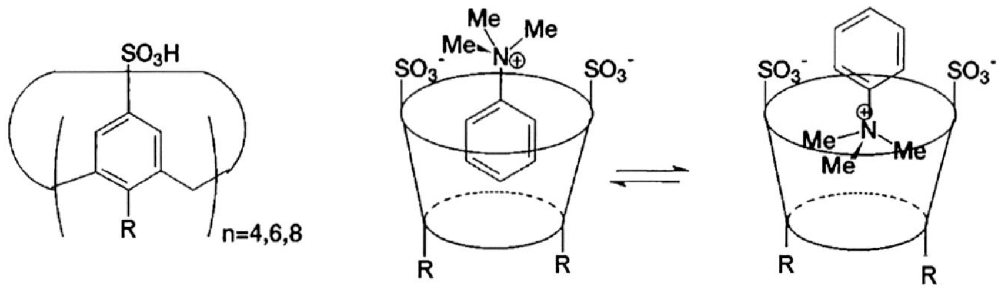

6 6a 6b

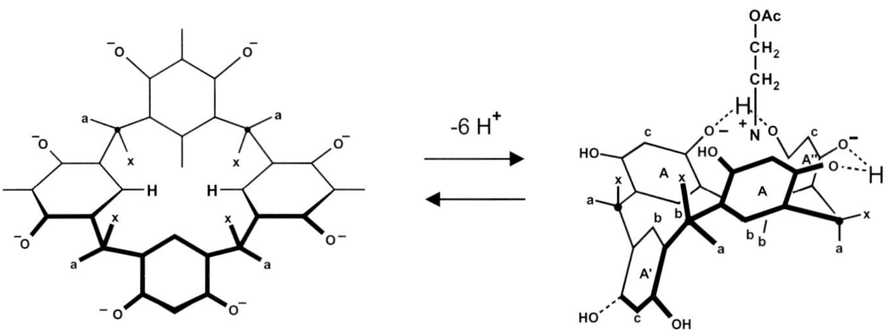

7a 7b

FIGURE 2.3. Examples of ion pairing with anionic hosts and cationic guests.

TABLE 2.1. Complexation Thermodynamics for Reactions of Fully Protonated Amines with the Citrate Trianion in Water

Amine	n	ΔH	$T\Delta S$	ΔG
$H_2N\text{-}(CH_2)_2\text{-}NH_2$	2	0.9	21	20.4
$H_2N\text{-}(CH_2)_3\text{-}NH\text{-}(CH_2)_3NH_2$	3	−0.9	23	23.6
$H_2N\text{-}(CH_2)_3\text{-}NH(CH_2)_4NH(CH_2)_3NH_2$	4	5.5	39	33.7

All values in kJ mol^{-1}; extrapolated to $I = 0$.
n, number of charges in the amine.
Source: E. García-España, P. Díaz, J. M. Llinares, A. Bianchi, *Coord Chem Rev* **2006**, 250, 2952.[11]

2.3 ENTHALPIC VERSUS ENTROPIC CONTRIBUTIONS/HYDROGEN BOND CONTRIBUTIONS

Ion pairing in aqueous media is, as discussed in many electrochemistry textbooks, almost entirely entropy-driven, as formally deduced from the temperature dependence of the equilibrium constant K in the Bjerrum or the Fuoss equation. The dominance of the $T\Delta S$ term in the often-even endothermic ion pair formation has been corroborated for many salt bridge complexes (Table 2.1),[11] and also for many inorganic associations.[12]

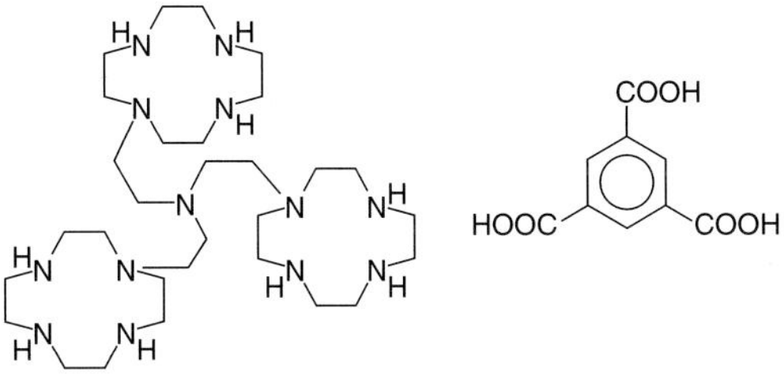

pH 3 : $\Delta H = -50$, $T\Delta S = -10$ kJ/mol^{-1}

pH 7 : $\Delta H = 0$, $T\Delta S = 40$ kJ mol^{-1}

FIGURE 2.4. Change of binding mechanisms with pH: Only at higher pH does the entropy-driven ion pairing dominate.[13]

Complexes between amines and carboxylic acids can form via salt bridges, and also via hydrogen bonds, although hydrogen bonding is weak in aqueous media. In consequence, the binding mechanism may change with the pH value of the medium, as exemplified in Figure 2.4.[13] Here, ion paring with the characteristic entropy-driving force occurs only at higher pH; at lower pH, hydrogen bonding with the electroneutral free acid takes over.

2.4 QUANTIFICATION OF SALT BRIDGE CONTRIBUTIONS/ ADDITIVITY OF BINDING INCREMENTS

The classical quantification of free energy of ion pair formation in aqueous media is that by the Bjerrum equation

$$K = (4\pi \mathrm{N}/1000)(z_A z_B e^2/\varepsilon \mathrm{k} T)^3 \mathrm{Q(b)} \tag{2.1}$$

or by the simpler Fuoss equation

$$\log K = -2.6 + 3\log a + 243\, z_A z_B/\varepsilon a, \tag{2.2}$$

which both describe the dependence of K on the dielectricity constant ε and the permanent ion charges z_A and z_B; the parameters Q and a characterize the distance between the ions. Indeed, one finds with a multitude of experimental data a fairly linear dependence between log K and $z_A z_B$ (Fig. 2.5).

The data in Figure 2.2 already suggest that in a most simple description each ion pair contributes—within the same geometrically matching complex—the same binding increment. If one analyses a large number of supramolecular complexes in which an optimal contact between the positive and negative charge centers is secured—which is common practice in well-designed synthetic complexes—one indeed finds a linear dependence between the total binding free energy ΔG_t and the number n of salt bridges that can build up within the supramolecular structure. The increments $\Delta\Delta G_s$ for each salt bridge are simply added up for all contacts that are possible at a given time, counting, for example, 4x$\Delta\Delta G_s$ if a cation can simultaneously approach four anionic

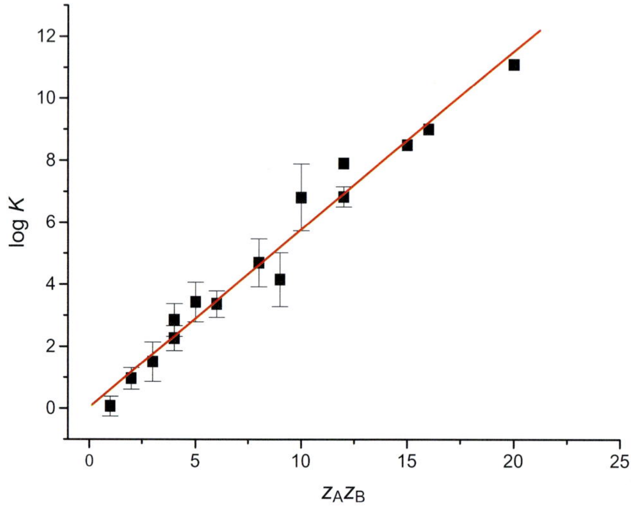

FIGURE 2.5. Ion pair association constants of 203 ion pairs as function of charge product z_Az_B. The points show mean values of log K; vertical bars the respective standard deviations. Slope a = 5.3 Å of the Fuoss equation. Reprinted with permission from H.-J. Schneider, A. Yatsimirsky. *Principles and Methods in Supramolecular Chemistry*, Wiley, Chichester, 2000.[1c].

centers, a situation occurring, for example, in complexes discussed below. The resulting correlation is surprisingly linear for a manifold of experimental data, including anions and cations of very different polarizability and charge delocalization, such as metal, sulfonium, or ammonium cations, and halide, sulfonate, phosphate, carboxylate, or phenolate anions. Many of the underlying data stem from supramolecular complexes in which not only salt bridges but also other binding forces contribute. As will be shown in Section 2.6, one can dissect in these cases the different contributions, and then incorporate the salt bridge increments within the correlation in Figure 2.6.

2.5 SALT EFFECTS/INFLUENCE OF CONFORMATIONAL FLEXIBILITY

From the slope of the correlation line in Figure 2.6, one can deduce an average increment for a single salt bridge in the range of $\Delta\Delta G_s = 5 \pm 1$ kJ mol^{-1}. A similar value has been found from the analyses of many inorganic ion pairs.[12] The value is rather constant as most of the underlying measurements have been done in media of similar ionic strength, often with salt/buffer concentrations around 10 mM. The dependence of ΔG on the ionic strength has been evaluated with inorganic salts,[12] and with several supramolecular complexes.[1c] Figure 2.7 shows, with two examples,[14] a very linear correlation in line with the Debye–Hückel equation (2.3),

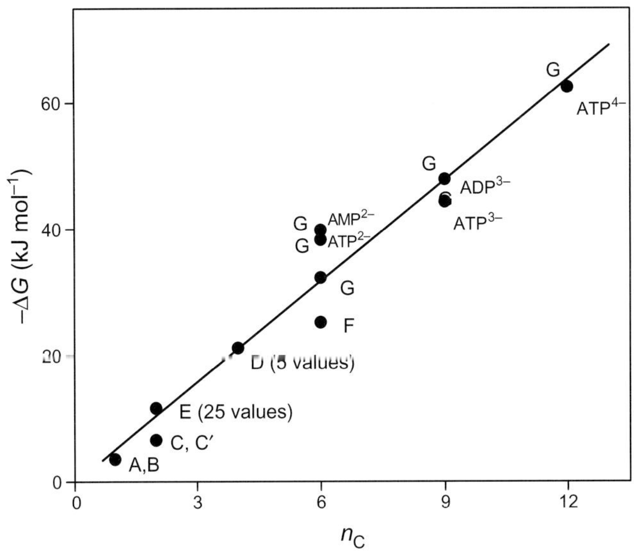

FIGURE 2.6. Experimental complexation free energies ($-\Delta G$; kJ mol^{-1}) versus number n of salt bridges in ion pairs, in water. A, B and C, C′ = complexes of a tetraphenolate cyclophane (4−) with Me_4N^+ and an azoniacyclophane (4+) with mono- and dianionic naphthalene derivatives; D = anionic (sulfonate or carboxylate) with cationic (ammonio) triphenylmethane derivatives; E = organic dianions with organic dications; F = cationic azamacrocyle (6+ charges) with aliphatic dicarboxylates; G = cationic azacrowns with adenosine mono-, di-, and triphosphates. From H.-J. Schneider, A. Yatsimirksi. *Principles and Methods in Supramolecular Chemistry*, Wiley, Chichester, 2000, p. 9, Copyright 2000 by Wiley.[1c]

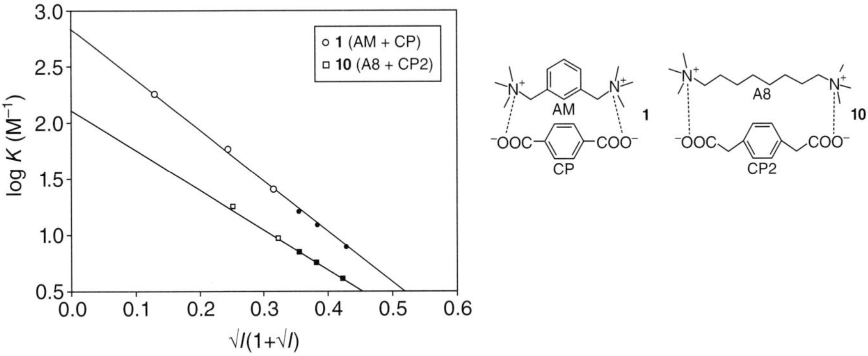

FIGURE 2.7. Debye–Hückel correlation of log K and ionic strength I of the medium for two ion pairs with differing distance between the charges. Reprinted with permission from A. Md. Hossain, H.-J. Schneider. *Chem Eur J* **1999**, 5, 1284. Copyright 1999 Wiley VCH.[14]

$$\log K = \log K_0 + m\sqrt{I}/(1+\sqrt{I}). \qquad (2.3)$$

In this case, the ionic strength I contribution of the reaction partners, which are ions themselves, could be taken care of by dilution experiments, leading at $I = 0$ with $\Delta\Delta G_s = 8 \pm 1$ kJ mol^{-1} for a salt bridge increment, which is, as expected, much higher than that observed under usual measuring conditions, where buffers are used to maintain a constant I as well as pH. The slope m of the correlation line is close to that predicted by Debye–Hückel; for one association one observes log $K_0 = 2.85$ and m = 4.6, for the other log $K_0 = 2.21$ and m = 3.9. The observed slopes are not too far from the theoretical value of $m = 4.7$ for such a 2 + 2 combination with $z = 2$. The linear correlations are surprising for complexes that represent ions of very anisotropic shape and charge distribution.[13–15] Also, the correlations hold with such supramolecular structures even if forces other than ion pairing contribute significantly (examples in Section 2.6).

Supramolecular complexes such as the ones in Figure 2.7 offer the opportunity to analyze also the effect of conformational preorganization. The presence of single bonds that upon complex formation can feel restriction of free rotation is expected to lower association energies due to entropy losses. Corresponding losses have been estimated to amount to $T\Delta S = 1.5$ to up to 6 kJ mol^{-1}.[16,17] Comparison of experimental free energies of many complexes that exhibit a variable number n of rotatable single bonds that separate the charges leads to a fairly linear correlation (Fig. 2.8), indicating that the disadvantage due to the presence of a single bond here is only 0.5 kJ mol^{-1}. The

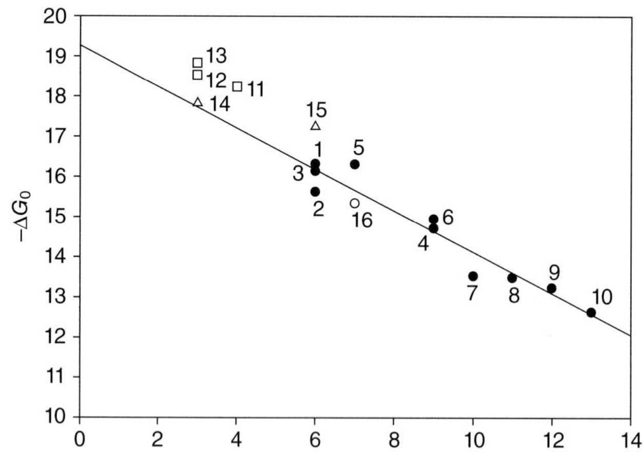

FIGURE 2.8. Dependence of free energy of the ion pairs with the number of rotatable single bonds. For assignment of the data points to different complexes, see Hossain and Schneider.[14] Regression analysis leads to the equation $\Delta G_{calcd} = 0.52n - 19.3$. Reprinted with permission from A. Md. Hossain, H.-J. Schneider. *Chem Eur J* **1999**, 5, 1284. Copyright 1999 Wiley VCH.[14]

disadvantage may, in these complexes, where the charges are separated by longer alkyl chains, be diminished by a hydrophobic attraction between the chains. However, related studies with hydrogen bonding as binding functionalities lead with 1.3 kJ/(mol and single bond) also to a smaller disadvantage than usually anticipated; here the measurement medium chloroform would not allow compensation by solvophobic attractions. In consequence, preorganization is a condition for optimal localization of the interacting and matching functions, but depends to a lesser degree on pre-fixing within more rigid conformations.

2.6 SYSTEMS WITH ADDITIONAL INTERACTIONS/ARTIFICIAL RECEPTORS FOR NUCLEOTIDES AND NUCLEOSIDES

Most supramolecular complexes, particularly those for the use in aqueous media, provide, in addition to ion pairing, other noncovalent interactions. Complexes of cyclodextrins (a cyclic heptaamylose with $n = 7$) equipped with a different number m of positively charged amino groups, and nucleotides with an increasing number of negative charges show the large increase of association constants with the number of possible salt bridges, with K values going up stepwise from 90 to over 3 million (Fig. 2.9).[18]

We will illustrate with one example how the different binding contributions can be dissected. The cyclophane in Figure 2.10 binds electroneutral nucleosides as well as the charged nucleotides; the difference $\Delta\Delta G$ of thc binding free energy ΔG between these is almost independent of the nucleobase in the different nucleotides, and amounts to 10 ± 1 kJ mol^{-1}. This value agrees exactly with ion pair increments deduced form the analysis of many other salt bridges (Fig. 2.6), indicating the presence of two salt bridges with 5 kJ mol^{-1} each. Indeed, one phosphate, bearing two charges, can at a given time be in contact with one cationic site of the cyclophane. Similar additive salt bridge contributions are observed with resorcarene complexes similar to **7**, here also with the expected decrease of ΔG_{cplx} as function of the charge separation distance d.[10] In line with this, the association constant K of an ion pair with α,ω-dianions decreases from $K = 170$ (M^{-1}) for $r = 6.0$ Å only to $K = 72$ (M^{-1}) for $d = 9.3$ Å.[19] From all noncovalent interactions, ion pairing as well as coulombic forces are those with the smallest distance dependence.

2.7 SELECTIVITY IN COMPLEXES WITH ION PAIR CONTRIBUTION

As evident from Figure 2.6, and corroborated by many examples, the difference in binding free energy is usually very small for simple ion pairs XY in aqueous solution, as long as the charges remains the same. Obviously, selectivity is seen between ions of different charge due to the increased number of salt bridges; Figure 2.3 also illustrates such an increase. It should

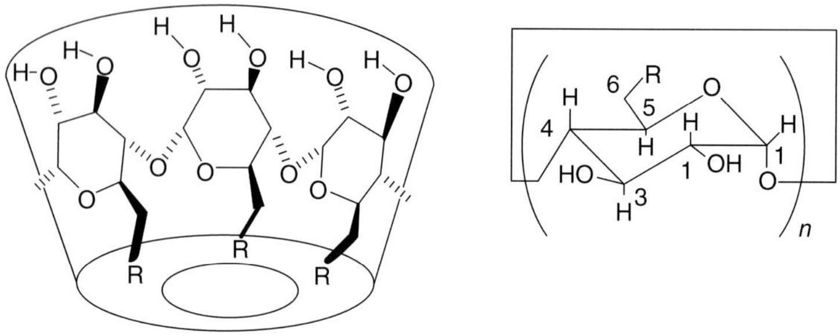

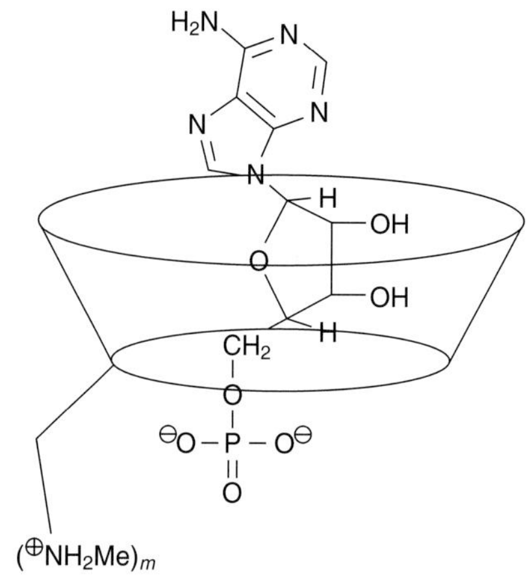

	β-CD, $m = 0$	β-CD, $m = 1$	β-CD, $m = 7$
AMP^{2-}	90	14.100	126.000
ATP^{4-}	–	98.000	3.240.000

FIGURE 2.9. Binding constants K (in $[M^{-1}]$) of amino-β-cyclodextrin with adenosin-phosphates.

be remembered that selectivity in terms of the binding constant ratio K_A/K_B for two different complexes A and B necessarily increases with the amount of the total ΔG_{cplx}.[8] Only if one partner exerts steric constraints, typically in macrocycles, can one expect selectivity as a function of the shape of the guest ion. An example is the azide complex **1**; another one is the host **2** that can discriminate between different halide anions. Ligand **8** (Fig. 2.11) contains three pyridine units within a narrows space which upon protonation shows in acetonitrile a pronounced selectivity for Cl over the Br or I anion.[20] Organic ions, however, where the charges are located at different locations, can exhibit quite significant selectivity by different geometric matching (Fig. 2.3). Often higher selectivity toward organic ions is gained by providing additional binding mechanisms besides ion pairing within a supramolecular complexes.[2,5]

Another way to reach larger selectivity is the implementation of, for example, Cu ions within a rigid framework such as **9**, which then binds, for instance, azide about 1000 times stronger than, for example, acetate or even

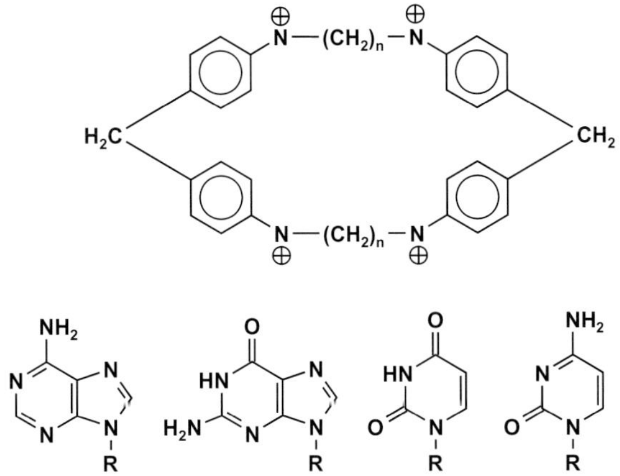

	A	G	U	C
Ribose	10	6.5	6	7
Ribose-$OPO_2^{2\ominus}$	19.3	15.9	17.3	18.3
$\Delta\Delta G$	9.3	9.4	11.3	11.3

FIGURE 2.10. Energies of complexation ΔG_{HG} (kJ mol^{-1}) with a cyclophane for nucleotides and nucleoside, and difference $\Delta\Delta G$ between nucleoside and nucleotide.

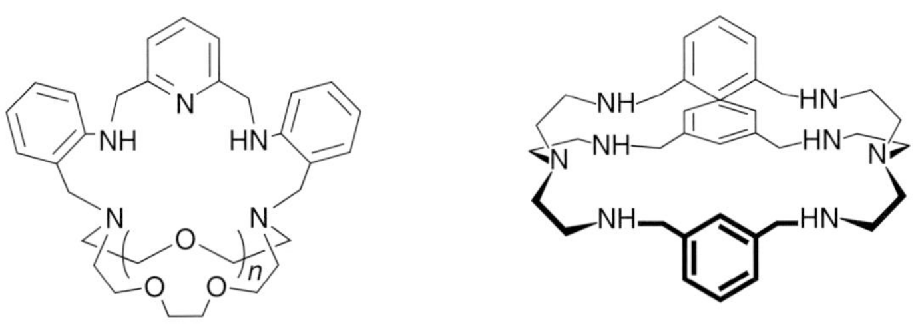

8 **9**

FIGURE 2.11. Structures of ligands **8** and **9**.

sulfate, although the latter bears two charges.[21] Metal ions not only provide the positive charge but also can act as a scaffold that holds the ligands containing, for example, the hydrogen bond donor groups in the appropriate positions.[22] Implementation of metal ions such as Cu or Zn into polymer chains has also been shown to extend the recognition of many anionic effector molecules, leading to shape changes of such intelligent materials.[23]

2.8 CONCLUSIONS

Ion pairing constitutes the most important binding contribution in a manifold of supramolecular complexes, including those with biopolymers. The analysis of well-designed synthetic complexes allows to quantify the different binding forces, also by systematic comparison of reaction partners differing by their charges. The free energy of complexation ΔG can be described for a multitude of ion pairs just by the charge products z_A and z_B of the ions, independent of their different polarizability and diameters. Measurements at typical intermediate ionic strength show even rather constant binding increments of $\Delta\Delta G = 5 \pm 1$ kJ mol^{-1} for each salt bridge in water, as long as contact between the charge centers is secured. The observed affinities drop only a little with increasing distance between the charges. Salt effects on the binding energies can be described by surprisingly linear Debye–Hückel correlations, in spite of the different polarizability and diameters, and also including highly anisotropic complexes in which forces other than ion pairing contribute significantly. The correlations shows for the ionic strength of zero a binding increment increase to $\Delta\Delta G = 8 \pm 1$ kJ mol^{-1} for each salt bridge. The presence of single bonds separating the charges within host and guest ions is expected to lead to significant entropic disadvantage, but it leads to only small losses of ΔG at around 1 kJ mol^{-1} per single bond. Supramolecular binding sites in synthetic polymers function also by ion pairing with ionic effector molecules; the recognition possibilities can be enhanced by the introduction of transition metal ions, and lead to new intelligent materials.

REFERENCES

1. Selected monographs: (a) J.-M. Lehn. *Supramolecular Chemistry. Concepts and Perspectives*. Wiley-VCH, Weinheim, 1995; (b) F. Vögtle. *Supramolecular Chemistry: An Introduction*. Wiley, Chichester, 1993; (c) H.-J. Schneider and A. Yatsimirsky. *Principles and Methods in Supramolecular Chemistry*. Wiley, Chichester, 2000; (d) J. W. Steed and J. L. Atwood. *Supramolecular Chemistry*. Wiley, New York, 2000; P. Cragg. *A Practical Guide to Supramolecular Chemistry*. Wiley, New York etc, 2005.
2. H.-J. Schneider. *Angew Chem Int Ed Engl* **2009**, 48, 3924.
3. See e.g. M. Rekharsky, Y. Inoue, S. Tobey, A. Metzger, E. Anslyn. *J Am Chem Soc* **2002**, 124, 14959.
4. H.-J. Schneider. *Intelligent Material*. M. Shahinpoor and H.-J. Schneider, eds. Royal Society of Chemistry, Cambridge, UK, 2007; pp. 100–120.
5. (a) *The Supramolecular Chemistry of Anions*. A. Bianchi, K. Bowman-James, and E. García-España, eds. Wiley-VCH, Weinheim, Germany, 1997; (b) V. Amendola and L. Fabbrizzi. *Acc Chem Res* **2006**, 39, 343; (c) K. Bowman-James. *Acc Chem Res* **2005**, 38, 671; (d) F. P. Schmidtchen and M. Berger. *Chem Rev* **1609**, 1997, 97; (e) F. P. Schmidtchen. *Top Curr Chem* **2005**, 255, 1.

6. B. Dietrich, J. Guilhem, J. M. Lehn, C. Pascard, E. Sonveaux. *Helv Chim Acta* **1984**, 67, 91.
7. K. A. Schug and W. Lindner. *Chem Rev* **2005**, 105, 67; A. Echavarren et al. *J Am Chem Soc* **1988**, 111, 4994; G. Müller, J. Riede, F.P. Schmidtchen. *Angew Chem Int Ed Engl* **1988**, 27, 1516.
8. H.-J. Schneider and A. Yatsimirsky. *Chem Soc Rev* **2008**, 37, 263.
9. J. M. Lehn, R. Meric, J. P. Vigneron, M. Cesario, J. Guilhem, C. Pascard, Z. Asfari, J. Vicens. *Supramol Chem* **1995**, 5, 97; P. Lhotak, S. Shinkai. *J Phys Org Chem* **1997**, 10, 273.
10. H. J. Schneider, D. Güttes, U. Schneider. *J Am Chem Soc* **1988**, 110, 6449.
11. E. García-España, P. Díaz, J. M. Llinares, A. Bianchi. *Coord Chem Rev* **2952**, 2006, 250.
12. (a) A. de Robertis, C. De Stefano, C. Foti, O. Giuffre, S. Sammartano. *Talanta* **2001**, 54, 1135; (b). C. De Stefano, O. Giuffre, S. Sammartano. *Faraday Trans* **1998**, 94, 2395.
13. C. Bazzicalupi, A. Bencini, A. Bianchi, L. Borsari, C. Giorgi, B. Valtancoli, C. Anda, A. Llobet. *J Org Chem* **2005**, 70, 4257.
14. A. Md. Hossain and H.-J. Schneider. *Chem Eur J* **1999**, 5, 1284.
15. H.-J. Schneider, R. Kramer, S. Simova, U. Schneider. *J Am Chem Soc* **1988**, 110, 6442.
16. P. Groves, M. S. Searle, M. S. Westwell, D. H. Williams. *Chem Commun* **1994**, 1519, zit. Lit.
17. J. P. Carver. *Pure Appl Chem* **1993**, 65, 763; M. S. Searle, D. H. Williams. *J Am Chem Soc* **1992**, 114, 10690; J. Gomez, E. J. Friere. *J Mol Biol* **1995**, 252, 337.
18. A. V. Eliseev and H. J. Schneider. *J Am Chem Soc* **1994**, 116, 6081; see also P. Schwinté, R. Darcy, F. O'Keeffe. *Perkin Trans 2* 1998, 805.
19. See also M. I. Nelen, A. V. Eliseev. *Perkin Trans 2* **1997**, 7, 1359.
20. D. Esteban-Gomez, C. Platas-Iglesias, A. de Blas, L. Fabbrizzi, T. Rodriguez-Blas. *Chem Eur J* **2008**, 14, 5829.
21. V. Amendola, M. Bonizzoni, D. Esteban-Gómez, L. Fabbrizzi, M. Licchelli, F. Sancenón, A. Taglietti. *Coord Chem Rev* **2006**, 250, 1451; V. Amendola, L. Fabbrizzi. *Chem Comm* **2009**, 513.
22. J. Perez and L. Riera. *Chem Soc Rev* **2008**, 37, 2658.
23. H.-J. Schneider and R. M. Strongin. *Acc Chem Res* **2009**, 42, 1489.

CHAPTER 3

POLYELECTROLYTE FUNDAMENTALS

ANGELO PERICO

3.1 INTRODUCTION

Polyelectrolytes are polymers bearing ionizable groups along the polymer chain, which in polar solvent (usually water) can dissociate and yield charged chains, macroions, and small counterions. The importance of polyelectrolytes is widely recognized due to their fundamental occurrence in living systems and their use in many industrial applications.[1–3] Well-known examples of such charged systems are many natural polymers, as well as synthetic polymers such as sulfonated polystyrene, polyvinylamine, and polyacrylic acid. Many biopolymers are charged cationic or anionic polyelectrolytes or anphiphilic polymers, and electrostatic interactions are found to be relevant for virtually all biomolecular systems and processes. Proteins are polyampholytes with both acidic and basic groups; nucleic acids, DNA, and RNA are among the strongest natural polyelectrolytes, and some of the naturally occurring polysaccharides are also polyelectrolytes (for instance, hyaluronic acid). Biological evolution points to large macromolecules rich in hydrophobic groups responsible for specific structures and balanced by a lot of charges to maintain water solubility and prevent aggregation. Each charge makes a molecule more hydrophilic and it contributes a counterion that can explore the solution on its own and gives the solution a great deal of entropy.[4] The general behavior of polyelectrolyte systems is critically dependent on electrostatic interactions, of which description is difficult due to the long-ranged nature of these forces. At the same time, the long-ranged electrostatic interactions may be controlled via screening lengths and counterion condensation, contributing to a wide range of possibilities of organization. Electrostatic interactions mediated by salt between DNA and countermacroions, proteins, or protein complexes (e.g., DNA–Lac repressor and DNA–octamer in nucleosome) are fundamental for DNA regulation and for the organization of chromatin.[5]

Ionic Interactions in Natural and Synthetic Macromolecules, First Edition. Edited by Alberto Ciferri and Angelo Perico.

In this chapter, we will focus on selected fundamental aspects of polyelectrolyte theory of importance in this book on ionic interactions. As a first step, we derive and discuss the Poisson–Boltzmann (PB) equation relating the electrostatic potential to the charge density due to the all-ion species in the polyelectrolyte solution, together with the important Debye–Hückel (DH) approximation, introducing the important topics of screened ionic interactions (Section 3.2). The condensation theory is described in Section 3.3 as a common approach to deal with the nonlinearity of the PB equation while retaining the simplicity of the DH approximation. Giving the importance of counterion condensation as a general framework for polyelectrolyte solutions, condensation theory is here first derived in its extended form, characterized by the dependence of the condensation volume from salt concentration. The classic Manning condensation theory is obtained in the limit of low ionic strength. In Section 3.4, a recent application of the extended condensation approach to like-charge attraction mediated by counterions between strong polyelectrolytes is reported. This like-charge attraction leads to the formation of polyelectrolyte clusters physically due to the increase of condensation volume with decreasing of polyelectrolyte separation distances and with increasing of the number of chains in the assembly. In Section 3.5, the effect of counterion correlation on strong polyelectrolyte solutions is introduced, demonstrating that attraction between like-charged polyelectrolytes is obtained only in the presence of multivalent counterions, in the absence of the entropic effect due to the expansion of condensation volume. Finally, in Section 3.6, the assembly of polyelectrolytes with oppositely charged macroions is analytically treated at a fundamental level with applications to emblematic relevant biological cases. Polyelectrolyte multilayers will be treated in Chapter 5 of the book.

3.2 PB EQUATION AND THE DH APPROXIMATION

The classic topic of the PB equation and of its DH linear approximation, widely described in standard books and reviews, will be shortly summarized, following recent excellent reviews.[3,4]

The electrostatic potential inside a given macroscopic system is given by the Poisson equation that relates the potential to the spatial distribution of the charges responsible for this potential:

$$\nabla^2\psi(\mathbf{r}) = -(4\pi/\varepsilon)\rho(\mathbf{r}). \tag{3.1}$$

ρ is the electric charge density at position $\mathbf{r}$ in the system, which for a discrete distribution of M point charges may be written as

$$\rho(\mathbf{r}) = e\sum_{i=1}^{M} z_i\delta(\mathbf{r}_i - \mathbf{r}), \tag{3.2}$$

with e being the unit charge, z_i being the valence of the point charge (negative for anions and positive for cations), and δ being the Dirac function. The Poisson equation is exact, provided the charges do not interact significantly with the uniform medium, here taken as water of dielectric constant ε (=80). By assuming that m of the M charges are fixed in space the potential should satisfy the equation

$$\nabla^2\psi(\mathbf{r}) = -(4\pi/\varepsilon)e\sum_{i=1}^{m} z_i\delta(\mathbf{r}_i-\mathbf{r}) - (4\pi/\varepsilon)e\sum_{j=m+1}^{M} n_j z_j \exp(-w_j/k_BT), \tag{3.3}$$

with n_j and w_j, respectively, the number density and the potential of mean force of the mobile charge (j). Introducing into Equation 3.3 the further approximation that the potential of mean force w_j on the mobile charge j is equivalent to the electrostatic energy that this charge would have in position $\mathbf{r}$ due to the constrained electrostatic potential

$$w_j = ez_j\psi(\mathbf{r}), \tag{3.4}$$

the PB equation in its general form is eventually obtained.

When Equation 3.3 is applied to the case of a single polyelectrolyte with N fixed unit negative charges, the PB equation in the region outside the polyelectrolyte becomes

$$\nabla^2\psi(\mathbf{r}) = -(4\pi/\varepsilon)\rho_{\text{out}}(\mathbf{r}) = -(4\pi/\varepsilon)e\sum_{i} z_i n_i \exp(-ez_i\psi(\mathbf{r})/k_BT), \tag{3.5}$$

with $\rho_{\text{out}}(\mathbf{r})$ being the density of mobile charges in the region outside the polyelectrolyte and n_i being the number density of charges of valence z_i at a point where $\psi = 0$.

Note that the PB equation is a mean-field theory since the mobile ions in the region outside the polyelectrolyte affect each other only through their average contribution to the mean-field potential: Ion–ion correlations between the mobile ions are completely ignored. The problem of counterion correlations will be approached in Section 3.5.

The solution of the PB equation is complicated due to its high degree of nonlinearity, associated with the exponential term containing the electrostatic potential. Two boundary conditions are required to fix the values of the integration constants for this second-order equation. As the charges of the macroion are located on its surface, the potential on the macroion surface should obey Gauss's law:

$$\nabla\psi(\mathbf{r}) = -4\pi\sigma(\mathbf{r})/\varepsilon, \tag{3.6}$$

with σ being the surface charge density of the macroion. Equation 3.6 may be considered as the first boundary condition to Equation 3.5. For a cylindric polyelectrolyte of radius a, a cell model may be adopted where the counterions and salt ions are contained between a first cylinder corresponding to the polyelectrolyte and a second cylinder of greater radius R. When there is no salt, or in the presence of only counterions, the cell should be electroneutral and the radial electric field near the cell periphery must vanish by virtue of the Gauss's law. This defines the second boundary condition:

$$\nabla\psi(\mathbf{r})\big|_{r\to R} = 0. \tag{3.7}$$

In this model, the PB equation may be solved exactly.[6] In the presence of an excess of added salt, the second boundary condition becomes

$$\lim_{R\to\infty}\psi(R) = 0, \tag{3.8}$$

and this Gouy–Chapman model[7,8] is solved numerically, with n_i representing the concentration of the bulk salt.

The electrostatic potential for this model as a function of the radial distance, r, is reported in Figure 3.1a for two polyelectrolytes having a linear charge density of one charge in 0.17 and 0.715 nm, respectively. Lowering the linear charge density lowers the magnitude of the potential, but increasing the salt concentration also reduces the potential outside the cylinder.

Going back to Equation 3.5, it can be seen that if the electrostatic potential is weak, the PB equation may be linearized by simply expanding the exponential term on the right side of the equation. The first term in the expansion is identically zero due to electroneutrality in the bulk solution, which implies

$$\sum_i z_i n_i = 0, \tag{3.9}$$

while the second term leads to the DH equation[9]

$$\nabla^2\psi(\mathbf{r}) = 4\pi\left(e^2/\varepsilon k_B T\right)\left(\sum_i z_i^2 n_i\right)\psi(\mathbf{r}), \tag{3.10}$$

It is useful to introduce two important constants characterizing the ion solution:

$$l_B = e^2/\varepsilon k_B T, \tag{3.11}$$

the Bjerrum length, or length at which the electrostatic interaction between two unit charges equals the thermal energy, which depends only on temperature and dielectric constant of the solution; and

$$I = (1/2)\sum_i z_i^2 n_i, \tag{3.12}$$

the ionic strength of the solution.

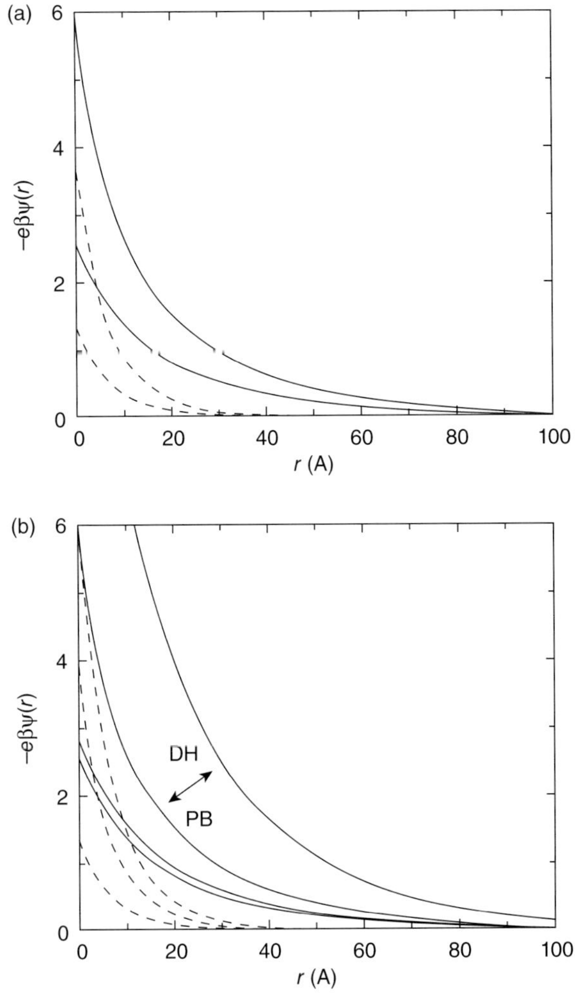

FIGURE 3.1. Dimensionless electrostatic potential at a distance r from a negatively charged cylindrical surface with radius 1 nm in a bulk 1:1 salt solution. Salt concentration: 0.01 M (solid lines); 0.1 M (dashed lines). The upper curve in each pair corresponds to a linear charge density of one unit charge per 0.17 nm (e.g., DNA), and the lower curve corresponds to one charge per Bjerrum length, $l_B = 0.715$ nm. (a) Poisson–Boltzmann (PB) results; (b) comparison with the Debye–Hückel (DH) results (upper curve in the related pairs). Reproduced from *DNA Interactions with Polymers and Surfactants*. R. S. Diaz and B. Lindman, eds. Wiley, Hoboken, NJ, 2008; figures 1.2 and 1.6)

Then the DH equation 3.10 may be written in the standard compact form of a Helmholtz equation:

$$\nabla^2 \psi(\mathbf{r}) = \kappa^2 \psi(\mathbf{r}), \tag{3.13}$$

$$\kappa^2 = 8\pi l_B I, \tag{3.14}$$

with κ being the Debye screening constant. The inverse of the Debye screening constant, κ^{-1}, is the Debye screening length, which is a measure of the range of electrostatic interactions in the ion solution: The smaller the screening length, the greater the ion concentration. It should be remembered that throughout the book CGS units have been assumed; therefore, if it is desired to give salt and ion concentrations in molarity units (moles per liter), ceteris paribus, Equation 3.14 should be written as

$$\kappa^2 = 8\pi (N_A / 10^3) l_B I, \tag{3.15}$$

with N_A being the Avogadro number.

In the following part of this chapter, an Z:1 salt will be assumed, with Z being the positive valence of counterions and 1 being the valence of the negative co-ions, for instance, NaCl ($Z = 1$) or $CaCl_2$ ($Z = 2$). In this case, Equation 3.12 becomes

$$I = (1/2) c_s (Z^2 + Z)), \tag{3.16}$$

with c_s being the salt concentration.

The DH potential outside an infinitely long, negatively charged cylinder of radius a and charge separation b, in the cell model with boundary conditions Equations 3.6 and 3.8, is[10]

$$\psi = -2(e/\varepsilon b) K_0(\kappa r)/[\kappa a K_1(\kappa a)], \tag{3.17}$$

with $K_0(x)$ and $K_1(x)$ the modified Bessel functions of order zero and one, respectively. The adimensional interaction energy, $e\psi/k_B T$, taking into account Equation 3.11, turns out to be proportional to the dimensionless Manning linear charge density of the polyelectrolyte[11]:

$$\zeta = l_B / b, \tag{3.18}$$

to be discussed later in the context of the polyelectrolyte critical behavior.

The electrostatic potential as a function of the radial distance r, obtained in the DH approximation is compared with the exact PB potential in Figure 3.1b. For high charge densities and low salt concentrations, the DH approximation is poor due to the importance of the nonlinear contributions. The linearization of the PB equation becomes more realistic, lowering the electrostatic potential. Note, however, that increasing the salt concentration increases ion correlations, and therefore the performance of the mean-field PB theory itself decreases.

For a charged sphere of radius a and surface charge eq surrounded by an excess salt solution, the DH equation may be solved using as the first boundary condition the expression derived from Equation 3.6,

$$\nabla\psi(r)|_{r=a} = -eq/a^2\varepsilon \tag{3.19}$$

and Equation 3.8 as the second boundary condition to give for the potential outside the sphere,[10]

$$\psi(r) = (eq/r\varepsilon)\exp[-\kappa(r-a)]/(1+\kappa a) \rightarrow_{a\to 0} (eq/\varepsilon)\exp[-\kappa r]/r. \tag{3.20}$$

The pair interaction u_{ij} between a charge q_ie in i and a charge q_je in j is derived taking the limit $a \to 0$ of Equation 3.20:

$$\begin{aligned} u_{ij}(r) &= k_BTq_iq_jl_B\exp(-\kappa r)/r \\ &\rightarrow_{\kappa\to 0} k_BTq_iq_jl_B/r. \end{aligned} \tag{3.21}$$

This is the screened coulomb interaction in an ionic solution characterized by a screening constant κ. In the limit of zero ion concentration or $\kappa = 0$, the coulomb interaction is obtained. The limits of validity of the screened coulomb interaction are again related to low charge densities to avoid nonlinear accumulation of counterions. It can be anticipated that in the context of the condensation approaches, the high linear charge density on the polyelectrolytes is strongly reduced by critical counterion condensation, allowing the use of the short-range DH screened interaction to take into account the intercharge interactions between polyelectrolytes by the short-range DH screened interaction.

3.3 EXTENDED CONDENSATION THEORY

The condensation theory in its extended form includes the effects of salt concentration and of the structure of the "interacting" polyelectrolytes. It is presented here following the derivation of Schurr and Fujimoto[12,13] that gives, in the limit of low ionic strength, the classic Manning's theory.[11,14] This general approach allows one to discuss the approximations involved in condensation theory and to prepare the tools to treat interactions between polyelectrolytes, especially like-charged polyelectrolytes.

3.3.1 Two-Phase Counterion Binding Theory

It is worthwhile to give a full derivation of the extended condensation theory, given originally[12] in a very compact way, with the main objective of clarifying limits and differences with the classic approach. Moreover, the latter is so often used and criticized that it seems useful to give all the elements needed for a full understanding.

The condensation theory is a two-phase theory where the polymer domain and the bulk solution are treated separately with the ions distributed between the two phases.[4] Consider the reaction to condense n_C counterions near a polyelectrolyte constituted of N equivalent negative unit charges:

$$P(0)+n_C C \rightleftarrows P_j(n_C), \tag{3.22}$$

with C, $P(0)$, and $P_j(n_C)$ being the concentrations of counterions, of naked polyelectrolytes, and of polyelectrolytes carrying n_C condensed counterions in the particular configuration j between the possible N polyelectrolyte sites. If each species obeys Henry's law, for the corresponding chemical potential we have

$$\begin{aligned} \mu_P &= \mu_P^0 + k_B T \log(c_P / c_0) \\ \mu_C &= \mu_C^0 + k_B T \log(c_s / c_0) \\ \mu_j &= \mu_j^0 + k_B T \log(c_j / c_0), \end{aligned} \tag{3.23}$$

with c_P, c_s, c_j, and c_0 the molar concentrations of naked polyelectrolytes, counterions, polyelectrolyte–counterion complex in configuration j, and pure water. The chemical potential of each species is referred to the standard state of the species in pure water (c_0 = 55.6 M in standard conditions).

At equilibrium, the chemical potentials of the two sides of reaction (Eq. 3.22) should be identical, to yield after exponentiation

$$c_j(n_C)/c_P = (c_s/c_0)^{n_C} \exp\{-[\mu_j^0 - \mu_P^0 - n_C \mu_C^0]\}. \tag{3.24}$$

The square brackets in Equation 3.24 represent the difference of the standard chemical potential in pure water of complex j, with respect to naked polyelectrolyte and counterion. It is evaluated as

$$\Delta\mu^0 = [\mu_j^0 - \mu_P^0 - n_C \mu_C^0] = F_j^{el}(n_C) - F_P^{el}(0) + n_C \Delta F_C. \tag{3.25}$$

$F_j^{el}(n_C)$ is the sum of all the electrostatic interactions in the configuration j of the counterion–polyelectrolyte complex; $F_P^{el}(0)$ is the sum of all the electrostatic interactions between the charged sites of the naked polyelectrolyte; and ΔF_C is the free energy change upon transferring a counterion in its standard state to a single binding site in its standard state, corresponding to an equilibrium constant of[1]

[1] Note that the equilibrium constant introduced here represents ion binding. It is different from the equilibrium constant discussed in Chapter 7 representing salt binding to a site on a polymer with no specific reference to ionic interaction.

$$\beta = \exp[-\Delta F_C / k_B T]. \tag{3.26}$$

Inserting Equation 3.25 into 3.24 and rearranging we get

$$c_j(n_C)/c_P = (c_s\beta/c_0)^{nC} \exp\{-[F_j^{el}(n_C) - F_P^{el}(0)]/k_B T\}. \tag{3.27}$$

Summing Equation 3.27 over all configurations and all possible values of n_C, the grand partition function, χ, is obtained:

$$c/c_P = \chi(c_s, T, N) = \sum_{n_C=1}^{N} (c_s\beta/c_0)^{nC} \sum_j \exp\{-[F_j^{el}(n_C) - F_P^{el}(0)]/k_B T\}, \tag{3.28}$$

with c being the total concentration of counterion–polyelectrolyte complexes.

Taking into account that the concentration of polyelectrolytes in the standard state, c_P^0, should be conserved in the complex state

$$c_P^0 = c + c_P, \tag{3.29}$$

it is found that in the limit of large N and of nonzero condensed unit charges per site, defined as

$$r = n_C Z / N, \tag{3.30}$$

the naked polyelectrolytes are very few and therefore $c/c_P = \chi \gg 1$ and $c/c_P^0 \approx 1$. In this approximation, the average number of bound counterions is identical to the average number of bound counterions per complex, or

$$\langle n_C \rangle = c_s \frac{\partial \log \chi(c_s, T, N)}{\partial c_s} \tag{3.31}$$

(the partial derivative is taken at constant $F_j^{el}(n_C)$ and β). After derivation of Equation 3.31, it is assumed that nearly uniform counterion configurations on the polyelectrolyte dominate the partition function χ and that they have approximately the same electrostatic energy:

$$\chi(c_s, T, N) = \sum_{n_C=1}^{N} (c_s\beta/c_0)^{nC} \exp\{-[F^{el}(n_C) - F^{el}(0)]/k_B T\}(N/n_C)^{nC}, \tag{3.32}$$

the last power term in Equation 3.32 being the total number of uniform configurations of the polyelectrolyte complexed to n_C counterions (each counterion having N/n_C binding possibilities independently from the other counterions).

The usual approximation of the maximum term is assumed to avoid the full calculation of the sum over n_C in the partition function χ: This amounts to ignore the n_C summation in Equation 3.32 while n_C is substituted with $<n_C>$. The extremum condition applied to the logarithm of the resulting equation gives upon exponentiation and rearranging, using the definition Equation 3.30, with $n_C = <n_C>$:

$$r = vZE^{-1}c_s \exp\left[-\frac{\partial(F^{el}/k_BT)}{\partial\langle n_C\rangle}\right], \tag{3.33}$$

with E, the base of natural logarithms.

Here, the normalized equilibrium constant β/c_0 is identified with the Manning's free volume:

$$v = \beta/c_0, \tag{3.34}$$

which is the object of the following discussion.

Within the counterion condensation model, each unit charge is estimated to take the effective value

$$q_{\text{eff}} = e(1-r), \tag{3.35}$$

and the electrostatic interaction between the effective charges on the polyelectrolyte is calculated as a screened coulomb interaction in the DH approximation to get

$$F^{el}(\langle n_C\rangle)/k_BT = (1-r)^2 l_B NS, \tag{3.36}$$

with S being the interaction sum on the polyelectrolyte,

$$S = (1/2N)\sum_{i\neq l} \exp[-\kappa r_{il}]/r_{il}. \tag{3.37}$$

Note that since the electrostatic interaction is calculated only between the strongly reduced effective charges due to counterion condensation, the DH interaction potential turns out to be a suitable approximation to the PB equation to calculate the electrostatic interaction sum of Equation 3.37. In addition, the structure of the polyelectrolyte appears as completely general under the unique condition that the sites should be equivalent. This allows one to treat a single polyelectrolyte as a linear rod as well as clusters of parallel rods.

Then the final equation of the counterion binding theory is derived from Equation 3.33, using Equations 3.36 and 3.37, to eventually yield

$$\Delta(r, v, c_s) = r - vZE^{-1}c_s \exp[2(1-r)l_B ZS] = 0. \tag{3.38}$$

Knowing the number of condensed unit charges per polyelectrolyte site r, the number of condensed counterion of valence Z, θ, is obtained as

$$\theta = r / Z. \tag{3.39}$$

The binding relation of Equation 3.38 contains two unknown variables, the fraction of condensed unit charges per site, r, and the condensation volume per site, v, to be determined as a function of the salt concentration, c_s. In the original Manning approach to the condensation theory, the additional condition, necessary to solve Equation 3.38, implies that the volume v does not depend on the salt concentration c_s, or $dv/dc_s = 0$ for all c_s and also that $dr/dc_s \cong 0$ for small c_s, which in turn means that $(\partial\Delta/\partial c_s)_{r,v} \cong 0$ for low c_s. Schurr and Fujimoto[12] demonstrated that the assumption $dv/dc_s = 0$ is incorrect and yields invalid results at finite c_s, and therefore proposed the alternative equation

$$(\partial\Delta / \partial c_s)_{r,v} = 0, \tag{3.40}$$

which is simply an extension to all the c_s values of the classic condition limited to small c_s. Equation 3.40 gives, as a nontrivial solution,

$$r = 1 - [\kappa l_B Z(-\partial S / \partial\kappa)]^{-1}. \tag{3.41}$$

When this value of r, independent of v, is substituted into Equation 3.38, the appropriate equation for the volume v as a function of c_s is obtained:

$$v = r\{c_s Z \exp[2(1-r) Z l_B S - 1]\}^{-1}. \tag{3.42}$$

A deep motivation of condition Equation 3.40 in terms of compatibility with the classic condensation assumptions and of physical consequences of Equations 3.41 and 3.42 on the extended theory, was recently given.[15] For polyelectrolyte clusters, such as those discussed in the following section, the assumption of Equation 3.40 is the best possible choice for the quantity Δ between all the explicit functions of the salt concentration still compatible with Manning's boundary conditions for $c_s \to 0$. Due to Equation 3.40, the extended condensation theory gives properties dependent on salt concentration and on the structure of the polyelectrolyte array as required by physical prescriptions: The condensed unit charges per site r increase in comparison with the case of a single polyelectrolyte, the free volume per site v becomes sensitive to the structure of the polyelectrolyte array even in the zero salt limit, and r and v decrease with salt concentration increase.[12,15] In addition, the extended theory gives, for a uniformly charged cylinder, r-values in agreement with those obtained by the nonlinear PB equation.[12]

3.3.2 Classic Condensation Theory: Single Polyelectrolyte in the Limit of Low Salt Concentration

For a single polyelectrolyte, assumed to be a charged rod carrying N unit charges with a spacing b and a Manning linear charge density given by Equation 3.18, Equation 3.37 becomes

$$S=(1/b)\sum_{n=1}^{\infty}\exp(-\kappa bn)/n=-(1/b)\log[1-\exp(-\kappa b)]. \tag{3.43}$$

With this interaction sum, Equations 3.41 and 3.42 become

$$r=1-[\exp(\kappa b)-1]/\kappa l_B Z, \tag{3.44}$$

and

$$v=rE/\{c_s Z\exp[-2(1-r)Z\zeta\log(1-\exp(-\kappa b))]\}, \tag{3.45}$$

which are the results for the condensed unit charges and condensation volume per site in the extended condensation theory of a single rod polyelectrolyte. In the limit of low salt concentration, the following results for the Manning classic condensation theory are recovered:

$$r_M=\lim_{c_s\to 0} r=1-1/\zeta Z \tag{3.46}$$

$$v_M=\lim_{c_s\to 0} v=4\pi Eb^3(N_A/10^3)(1+Z)(\zeta-1/Z). \tag{3.47}$$

The free energy for the single polyelectrolyte rod may, in general, be given as the sum of the DH electrostatic interactions between all the charges on the polyelectrolyte after condensation, plus the ideal free energy of mixing the free and bound counterions and solvent molecules. The first term, from Equations 3.36 and 3.43, is

$$F^{el}/k_BT=-N\zeta(1-r)^2\log[1-\exp(-\kappa b)]. \tag{3.48}$$

The second term may be, in general, given in the entropic form,

$$F^{en}/k_BT=n_C\log(c_{\text{cond}}/c_s), \tag{3.49}$$

with n_C and c_{cond} being the total number of counterions per polyelectrolyte (see Eq. 3.30) and the concentration of counterions in the condensed layer around the polyelectrolyte:

$$n_C=Nr/Z \tag{3.50}$$

$$c_{\text{cond}}=n_C/Nv=r/vZ \tag{3.51}$$

Equation 3.49 as a function of r and v turns out to be

$$F^{en}/k_BT=(Nr/Z)\log(r/vZc_s). \tag{3.52}$$

The total free energy for the single polyelectrolyte is finally given by the sum of Equations 3.48 and 3.52:

$$\begin{aligned}F/k_BT&=F^{el}/k_BT+F^{en}/k_BT\\&=-N\zeta(1-r)^2\log[1-\exp(-\kappa b)]+(Nr/Z)\log(r/vZc_s).\end{aligned} \tag{3.53}$$

In the extended condensation theory, the free energy (Equation 3.53) should be calculated by directly substituting for r and v the optimized expressions given in Equations 3.44 and 3.45, while in the low ionic strength limit it should be calculated using the classic Equations 3.46 and 3.47. When the latter calculation is done, one obtains the well-known Manning classic result for a single polyelectrolyte rod:

$$F/k_BT = -(N/Z)(2-1/\zeta Z)\log[1-\exp(-\kappa b)]-(N/Z)(1-1/\zeta Z). \quad (3.54)$$

The free energy of the classic condensation theory was originally obtained simply by maximizing the sum of the DH electrostatic interaction free energy plus the mixing free energy in respect of the condensed counterion per site.

The condition for condensation is[16]

$$r \geq 0. \quad (3.55)$$

This introduces a critical linear charge density ζ_c, over which there is condensation,

$$\zeta \geq \zeta_c, \quad (3.56)$$

with

$$\begin{aligned}\zeta_c &= \kappa l_B/\log(1+\kappa l_B Z)\\ &= 1/Z \quad c_s \to 0.\end{aligned} \quad (3.57)$$

In the classic condensation theory ($c_s \to 0$), the critical linear charge density is independent of salt concentration and simply equal to the inverse of the counterion valence. In the extended condensation theory, it increases with salt concentration starting from the zero ionic strength limit: At 0.1 M, the increase is 33%, 99%, and 187% for $Z = 1, 2, 3$, respectively (see Fig. 3.2a).

The differences between the extended and the classical theory are, in the case of a single polyelectrolyte rod, very small for the free energy (Fig. 3.2b) and number of condensed unit charges (Fig. 3.2c) and more relevant for the condensation volume (Fig. 3.2d), with the extended theory always yielding smaller values.

3.4 ATTRACTION AND CLUSTERS OF LIKE-CHARGED STRONG POLYELECTROLYTES IN THE CONDENSATION APPROACH: MONOVALENT COUNTERIONS AND EFFECT OF CONDENSATION VOLUME

The electrostatic interactions, mediated by counterions, between like-charged polyelectrolytes may lead to attraction and aggregation in clusters. It is well known that genomic DNA, both in eukaryotes and prokaryotes, is densely

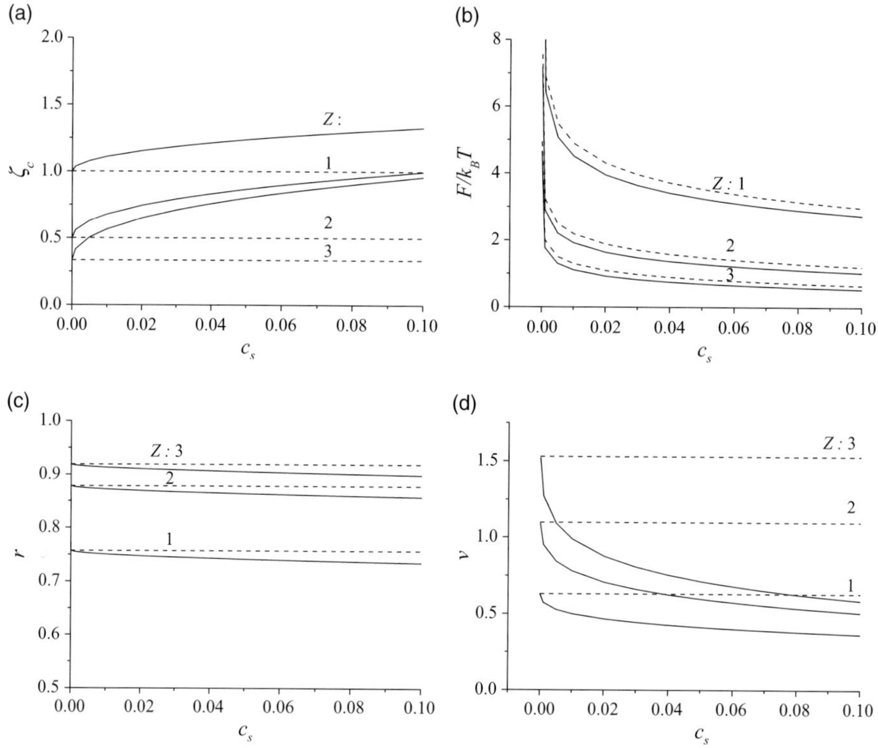

FIGURE 3.2. Condensation on a single polyelectrolyte rod of linear charge density $\zeta = 4.118$ (DNA in water at room temperature), as a function of salt concentration c_s, for counterion valence $Z = 1, 2, 3$. Dashed curves = limit theory; full curves = extended theory. (a) Critical linear charge density ζ_c; (b) normalized free energy; (c) number of condensed unit charges, r; (d) condensation volume per site v (M^{-1} units).

packed without posing repulsive problems due to the balance of the strong negative charges of the phosphates by counterion-induced attraction.[17,18] Light and neutron scattering provided extensive evidence for the existence of aggregates or clusters of strong polyelectrolytes, both in monovalent and multivalent counterion solutions. For short DNA of the order of 160 bp, it was shown by static and dynamic light scattering (S/D-LS) that, while at low DNA concentration, repulsion prevails, at a certain critical concentration, increasing with monovalent salt concentration, aggregates are formed as demonstrated by a several hundredfold scattering increase and by electron cryomicroscopy visualization.[19] This clustering was confirmed by small angle neutron scattering (SANS) experiments and again by S/D-LS on a 480 bp DNA as a function of NaCl concentration.[20] In addition, aggregation states of strong polyelectrolytes such as sodium poly(styrene sulfonate),[21–23] and poly(*N*-methyl-2-vinylpyridinium chloride)[24] were seen with similar techniques. It is worthwhile to observe that the dimensions of multichain domain structures were found to decrease with increasing counterion valence that has a large influence on the structure and dynamics according to the Manning linear charge density.[23] In addition, these studies indicate that polyelectrolyte structure and symmetry dictate the geometry of the cluster.[23] This concept will be applied in the following sections to short, stiff DNA displaying a cylindrical symmetry that should favor a parallel organization in hexagonal clusters.

Coulomb interactions and screened coulomb interactions, such as those emerging from the linearized DH approximation and even the same mean-field PB approach, can predict only repulsion between like-charged polyelectrolytes in a solution of small salts of any valence (see Section 3.2).[18,25] Several computer simulations of interacting charged rods and DNA molecules show that attraction arises from counterion correlations not included in mean-field approaches.[18,26–29] In a recent molecular dynamics (MD) simulation,[30] attraction between short dsDNA in monovalent (NaCl) and divalent ($MgCl_2$) salts was demonstrated: The attraction is stronger in divalent salt because DNA specifically interlocks minor-groove-to-minor-groove bridged by $MgCl_2$.[31] Theoretical approaches taking into account counterion correlations, obtain like-charge attraction in the presence of multivalent counterions (see Section 3.5).[32–35] The model of polyelectrolyte interaction[32] in counterion correlation theory localizes counterions around the polyelectrolytes, thus favoring energetically driven counterion correlations leading to attraction.

Alternatively, attraction between like-charged rodlike polyelectrolytes has been obtained qualitatively in the context of the classic condensation theory,[25,36] and quantitatively[16] in the framework of the extended condensation theory.[12,13] These condensation approaches do not explicitly consider counterion correlation but only the entropic effect of the expansion of the condensation volume near the polyions because of the superposition of the condensation volumes of the two polyelectrolytes. The possibility that this type of attraction may lead to the formation of loose clusters of strong rodlike polyelectrolytes has been considered in the framework of the classic condensation theory.[37] The above-mentioned quantitative condensation theory of like-charged polyelectrolyte attraction[16] has been extended to describe the interaction free energy of assembly of polyelectrolytes, such as rod hexagonal clusters,[15] and is reported below.

3.4.1 Electrostatic Free Energy for Polyelectrolyte Clusters

The electrostatic free energy of two parallel rodlike polyelectrolytes and of more complex parallel clusters of polyelectrolytes is derived[15,16] within the framework of the extended condensation theory of Section 3.3. The m polyelectrolytes in the cluster are modeled as parallel long rods of in-phase uniformly spaced unit charges. Each single polyelectrolyte has N unit charges, or charge sites, with site spacing b.

The total free energy of the cluster is given in mNk_BT units, or adimensional free energy per site, and is reported as the difference of the actual free energy in the polyelectrolyte assembly to that at infinite separation ($x \to \infty$). The assemblies are completely defined by only one internal cluster parameter, x, chosen as the inter-rod separation distance in a two-rod model or in a triangular cluster, and as the radius of the circumscribed circle to the first hexagon in a hexagonal cluster.

$$\begin{aligned} \Delta F(x)/mNk_BT &= [F(x) - F(\infty)]/mNk_BT \\ F(x) &= F_1 + F_{1-2}(x) + F_e(x). \end{aligned} \tag{3.58}$$

The free energy takes into account the free energy cost to condense the counterions on the polyelectrolyte assembly and all the electrostatic interactions between the sites after counterion condensation. The first term, F_1, is m times the total DH electrostatic free energy associated with the ionic repulsions among sites along the same line; the second term, $F_{1-2}(x)$, is the total DH electrostatic free energy associated with the ionic repulsion between sites, belonging to different lines for a total of $\binom{m}{2}$ line couples, each one characterized by a specific (x-dependent) separation distance; the third term, $F_e(x)$, describes the entropic free energy cost to condense θ counterions of valence Z per polyelectrolyte site from the bulk solution. Note that the electrostatic interaction free energies, F_1 and F_{1-2}, are calculated between sites of reduced charge, $e(1 - \theta Z)$, due to counterion condensation. This motivates the use of the DH approximation to the PB equation in their calculation.

In the following, the above electrostatic free energy will be calculated for the polyelectrolyte cluster models illustrated in Figure 3.3: two-rod model, triangular model, and hexagonal model.

Triangular Clusters The simple case of a triangular cluster is used to illustrate the approach: Here $m = 3$ and x is the common unique separation distance between the three couples of rods—1-2, 1-3, and 2-3.

The first term, F_1/mNk_BT, amounts to the normalized DH repulsions of one site with the remaining equal sites on a rod (see Eq. 3.36):

$$F_1/mNk_BT = l_B(1-r)^2 S_1, \tag{3.59}$$

with S_1 the interaction sum of one charge on a site with all the charges on the same rod:

$$S_1 = N^{-1}\sum_{i<j}^{N} \exp(-\kappa r_{ij})/r_{ij} = b^{-1}\sum_{n=1}^{\infty} \exp(-\kappa bn)/n, \tag{3.60}$$

where r_{ij} is the distance between two sites on an uniformly spaced straight line. In the limit of an infinite rod, Equation 3.60 may be summed, eventually giving for Equation 3.59:

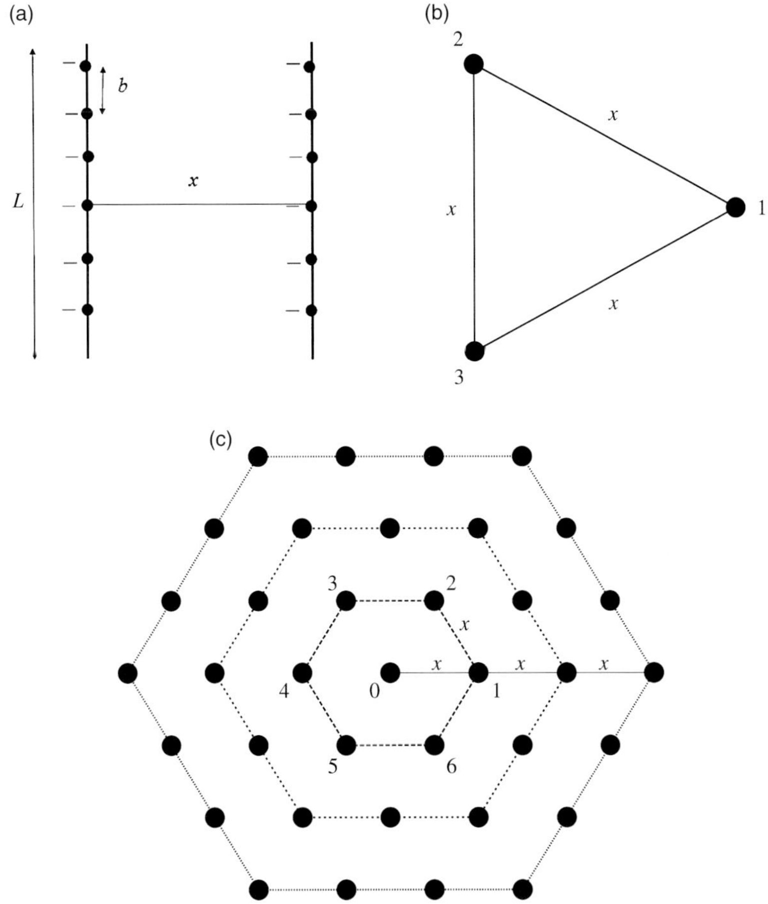

FIGURE 3.3. (a) Two-rod model. The two-rod polyelectrolytes, located at a distance x, are parallel with in-phase uniformly spaced unit charges, and charge separation b. (b) Triangular cluster: a section normal to the three in-phase parallel polyelectrolytes. (c) Hexagonal cluster: a section normal to the parallel polyelectrolytes; the model includes three polyelectrolyte layers and the distance parameter x of the cluster, coincident with the radius of the circumscribed circle to the first hexagon.

$$F_1 / mNk_BT = -\zeta(1-r)^2 \log[1-\exp(-\kappa b)]. \tag{3.61}$$

The second term $F_{1\text{-}2}(x)/mNk_BT$ amounts to the normalized DH repulsions between one site on a rod and all sites on a different rod (because all the inter-rod distances are identical) and is given by

$$F_{1-2}(x)/mNk_BT = -l_B(1-r)^2 S_{1-2}, \tag{3.62}$$

with $S_{1\text{-}2}$ being the interaction sum of one charge on a rod with all the charges on a different rod:

$$S_{1-2} = N^{-1}\sum_{i<j}^{N} \exp(-\kappa r_{ij})/r_{ij} + \exp(-\kappa x)/(2x)$$
$$= b^{-1}\sum_{n=1}^{\infty} \exp\left[-\tilde{\kappa}\left(n^2+\tilde{x}^2\right)^{1/2}\right]\Big/\left(n^2+\tilde{x}^2\right)^{1/2} + \exp(-\kappa x)/(2x), \tag{3.63}$$

where r_{ij} now reads as the distance between two sites on two different parallel lines of in-phase uniformly spaced charges, separated by a distance x. In the second equation in Equation 3.63, the dimensionless quantities

$$\tilde{x} = x/b; \quad \tilde{\kappa} = \kappa \mathrm{b} \tag{3.64}$$

are introduced.

In the limit of lines of infinite length, Equation 3.63 may be summed, eventually giving for Equation 3.62:

$$F_{1-2}(l)/mNk_BT = \zeta(1-r)^2 K_0(\kappa x), \tag{3.65}$$

with $K_0(\kappa x)$ being the modified Bessel function of zero order.

Similarly, the entropic contribution $F_e(x)$ to condense $\theta = r/Z$ counterions per each one of the mN sites in the cluster is proportional to the total number of condensed counterions times the logarithm of the ratio of the counterion concentration in the condensation volume to the counterion concentration in the bulk, or

$$F_e(x)/mNk_BT = (r/Z)\log(c_{\text{cond}}/c_s). \tag{3.66}$$

Finally, using Equation 3.51,

$$c_{\text{cond}}/c_s = r/vZc_s, \tag{3.67}$$

Equation 3.66 becomes

$$F_e(x)/mNk_BT = (r/Z)\log[r/(Zvc_s)]. \tag{3.68}$$

Hexagonal Clusters In Figure 3.3c, a section perpendicular to the cluster axis of a hexagonal cluster is represented, with x being the radius of the circle circumscribed to the first hexagon. The cluster includes a rod in the center, the first layer of 6 rods, the second and third layers of 12 and 18 rods, respectively, and so on. As a consequence, the cluster model, both including the central rod and the first layer of rods, gives $m = 7$, while $m = 19$ or 37 are obtained by sequentially including the second and third layers. Equations 3.59 and 3.68, for the single rod electrostatic interaction free energy and for the entropic free energy terms, respectively, are found explicitly independent of the structure of the cluster model, while the free energy for electrostatic interactions between different rods depends explicitly on the structure of the cluster because the rods are at different distances in the hexagonal array. Thus, Equation 3.62 for the model including up to the first hexagonal layer becomes

$$F_{1-2}(x)/Nk_BT = l_B(1-r)^2\left[12S_{1-2}(\kappa x)+6S_{1-2}\left(\kappa x\sqrt{3}\right)+3S_{1-2}(2\kappa x)\right]. \quad (3.69)$$

Taking into account that

$$K_0(y)/b = \lim_{N\to\infty} S_{1-2}(y), \quad (3.70)$$

Equation 3.69 in the limit of rods of infinite length becomes

$$F_{1-2}(x)/mNk_BT = (\zeta/m)(1-r)^2\left[12K_0(\kappa x)+6K_0\left(\kappa x\sqrt{3}\right)+3K_0(2\kappa x)\right]. \quad (3.71)$$

Note that, for hexagonal clusters, $F_{1-2}(x)$ is not simply proportional to the number of rods in the cluster.

For the models including further layers, the free energy term F_{1-2}, calculated taking into account all the interactions between the sites on the m different rods, is a straightforward extension of Equations 3.69 and 3.71.

The total interaction sum per site S of the array of equivalent polyion charge sites is calculated as follows:

$$S = S_1 + S_{1-2}(x), \quad (3.72)$$

for two parallel lines of in-phase uniformly spaced charges at a separation distance x (see Eq. 3.60 and 3.63);

$$S = S_1 + S_{1-2}(x) + S_{1-3}(x), \quad (3.73)$$

for a triangular cluster, and

$$S = S_1 + 3S_{1-2}(x) + 2S_{1-2}\left(x\sqrt{3}\right) + S_{1-2}(2x), \quad (3.74)$$

for a site on a rod on the first layer of a hexagonal cluster. Similar equations are obtained for a site on a rod on the next layer, and so on.

These definitions of the interaction sum should be inserted into the general expressions for r and v of Equations 3.41 and 3.42, which, of course, become functions of the structure of the cluster model.

By inserting these charge and volume fraction per site into Equations 3.59, 3.62, and 3.68, the free energy difference of 3.58 is eventually obtained for the interacting cluster as a function of the cluster characteristic distance, at different values of counterion valence, ionic strength, and polyelectrolyte linear charge density.

Note that r and θ should be nonnegative in the whole range of the physical variables:

$$r \geq 0; \quad \theta \geq 0. \quad (3.75)$$

This inequality dictates the general condition for counterion condensation for a given cluster characterized by a distance x and a structural parameter $S(x)$

$$(-\partial S/\partial \kappa) \geq (\kappa l_B Z)^{-1}. \tag{3.76}$$

In the limiting case of infinite distances, Equation 3.76 reduces to the condition for condensation on a single polyelectrolyte of Equations 3.56 and 3.57 (see discussion there) and to the definition of the critical charge density ζ_c.

It can be anticipated here that if $\zeta \geq \zeta_c$ (supercritical polyelectrolyte cluster), Equation 3.76 is satisfied for any cluster for any x and therefore there is condensation in the whole separation range. For undercritical polyelectrolytes, Equation 3.76 predicts the possible existence of a critical distance, x_c: Above this distance there is no counterion condensation because the polyelectrolyte rods are sufficiently separate to remain below critical conditions, while below x_c condensation starts occurring. On the extreme case of very low linear charge densities, $\zeta \ll \zeta_c$, or weakly charged polyelectrolytes (this happens at $\zeta \leq \zeta_c/2$ in the case of two-rods), the polyelectrolytes remain below critical conditions in the whole separation range: There is no condensation, and clusters cannot be formed.

3.4.2 DNA Clusters

In this subsection, we consider DNA in water ($\varepsilon = 80$) at room temperature ($T = 298$ K), with $b = 0.17$ nm, $l_B = 0.701$ nm, and linear charge density $\zeta = 4.123$. Figure 3.4 reports the free energy difference per site $\Delta F(x)/mNk_BT$ (panel A), the condensed charge number per site r (panel B), and the condensation volume per site v (panel C), calculated from Equations 3.58, 3.41, and 3.42, respectively, as a function of the characteristic distance x (equal to the inter-DNA distance for two parallel DNA rods; to the triangle side length in the triangular cluster and to the radius of the circumscribed circle to the first hexagonal layer in the hexagonal cluster), for counterion valence $Z = 1$ and salt concentration $c_s = 0.1$ M. In each figure, five models of DNA assembly are compared: two-rod model—representing the interaction of two in phase DNA rods; triangular cluster model—representing three DNA rods at a separation distance x; hexagonal cluster model hex1—including the central rod and the first hexagonal layer; model hex2—including the central rod and the first two layers; and model hex7—including the central rod and the first seven layers. Note that after reaching the seventh layer the central rod becomes numerically equivalent to any other rod of the first hexagonal layer.

As x approaches zero, all the free energy curves should approach infinity because of the modified Bessel function of zero order describing direct electrostatic interactions between DNA rods. As x approaches infinity, all the curves should approach zero as we have subtracted the free energy of noninteracting DNA lines. Figure 3.4a shows that the free energy difference has a

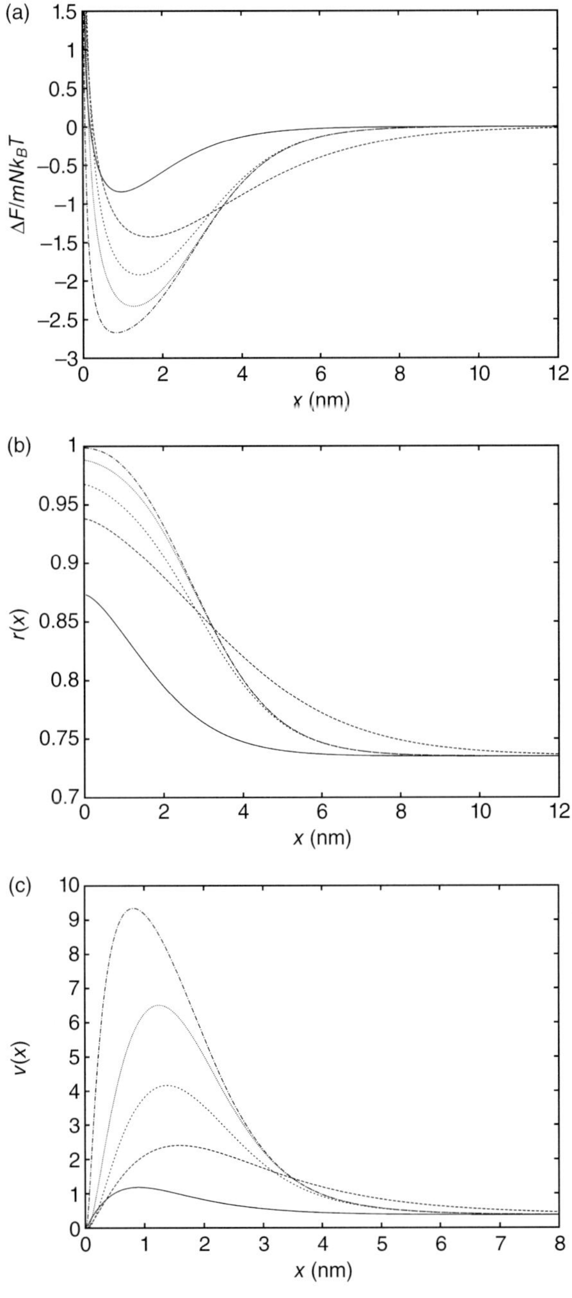

FIGURE 3.4. DNA cluster models as a function of x at ionic strength $I = 0.1$ M and counterion valence $Z = 1$: two-rod model (full curve); triangular model (dashed curve); and hexagonal cluster—model hex1, including only the first layer (short-dashed curve); model hex2, including the first two layers (dotted curve); model hex7, including the first seven layers, asymptotic case (dash-dotted curve). (a) Normalized free energy difference per site; (b) condensed charge number per site r; (c) condensation volume per site, v (M^{-1} units). Reproduced from Perico and Rapallo.[15]

minimum for all models indicating that DNA prefer to form clusters instead of repelling. The free energy minimum becomes deeper when the number of DNA participating in the assembly is increased. The absolute minimum is reached for the full hexagonal cluster, including seven layers. Note also that the x value corresponding to the minimum decreases with increasing the number of DNA, smoothly falling below 1 nm. When the number of layers in the hexagonal cluster increases, the DNA come closer together and the cluster becomes more compact and stable.

Figure 3.4b shows that the condensed charge number per site increases for all the models as the parameter x decreases, showing that condensation increases with the assembly of DNA, the increase being greater the larger the number of DNA participating to the assembly.

To rationalize the free energy behavior of the clusters of Figure 3.4a, we should take into account the condensation volume per site, which is illustrated in Figure 3.4c. The curves of all models show a maximum at an intermediate characteristic distance, corresponding to the superposition of the condensation volumes of the individual DNA participating to the cluster. This maximum, which increases with the number of DNA participating in the assembly, is responsible for the minimum in the free energy of the clusters on account of the entropy contribution of Equation 3.68.

The above conclusion is supported by consideration of the three components of the free energy per site. This is done in Figure 3.5 for the hexagonal model including one layer. Here, for a better comparison, the reduced free energy at infinite is not subtracted. The figure shows that the term F_1, the self

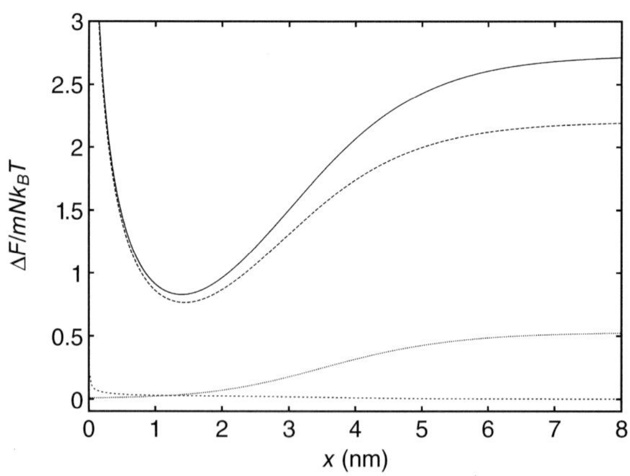

FIGURE 3.5. The normalized free energy components per site as a function of x at ionic strength $I = 0.1$ M, counterion valence $Z = 1$, and at the DNA linear charge density $\zeta = 4.123$ for model hex1: total free energy (full curve); single chain component F_1/mNk_BT (dotted curve); interacting chain component F_{1-2}/mNk_BT (short-dashed component); and entropic contribution F_c/mNk_BT (dashed curve). Reproduced from Perico and Rapallo.[15]

electrostatic energy of the individual polyelectrolytes, is always low and decreases regularly with x. Therefore, it cannot be responsible of the minimum. Also, the term F_{1-2}, corresponding to the electrostatic interaction of different DNA, is very low and only at very small distances begins to increase. Therefore, its effect on the free energy is almost negligible with the exclusion of the region of very small x separation, where it contributes to the strong electrostatic repulsion. The behavior of the free energy is dominated by the entropic contribution, F_e, which is particularly sensitive to the charge and condensation volume per site. Ultimately, it is this entropic contribution that determines the existence of a minimum in the free energy. Similar behavior is shown when more layers are involved, with the entropic term becoming more and more important with respect to the other two terms. At higher counterion valence and for decreasing ionic strength, the behavior is similar to that shown in Figure 3.5.

What is the role of the counterion valence Z on the formation of the clusters? Since the condensation model may break down with multivalent counterions due to their possible complex interaction with the polyelectrolytes, the calculations were limited to $Z = 1, 2$ just to indicate what condensation theory dictates in its point charge approximation for counterions.

Figure 3.6a represents the free energy difference $\Delta F(x)/mNk_BT$ for different Z, at ionic strength $I = 0.1$ M for a DNA cluster including seven layers. The minimum in the free energy is deeper in the case of monovalent counterions, consistently with the entropic origin of the cluster. Figure 3.6b reports the same curves at an ionic strength $I = 0.001$ M: Again the minimum is deeper for monovalent counterions, but the equilibrium distances are greater and the minima deeper, and therefore the cluster is larger and more stable than at $I = 0.1$ M.

3.4.3 Loose Clusters at Lower Linear Charge Densities

According to condensation theory, a critical charge density occurs for a single polyelectrolyte (essentially dependent on the inverse of counterion valence) below which condensation does not occur. For two parallel polyelectrolytes, condensation may occur at short distance even though the single polyelectrolyte is subcritical. This situation may generate interesting cases of attraction/repulsion for moderately charged polyelectrolytes in contrast to highly charged polyelectrolytes (where there is always attraction) and to weakly charged polyelectrolytes (where there is always repulsion).[16] Equation 3.76 dictates the conditions for the three above cases for any cluster which, of course, depend on the organization of the cluster.

Figure 3.7 illustrates $\Delta F(x)/mNk_BT$, at $I = 0.1$ M, as a function of x, for different values of the linear charge densities from highly to weakly charged polyelectrolytes ($\zeta = 3.0; 2.0; 1.0; 0.5$). It can be seen that for strong polyelectrolytes ($\zeta > 1.0$), the cluster is well formed both for model hex1 (Fig. 3.7a) and model hex7 (Fig. 3.7b). Interestingly, a loose cluster is also formed for

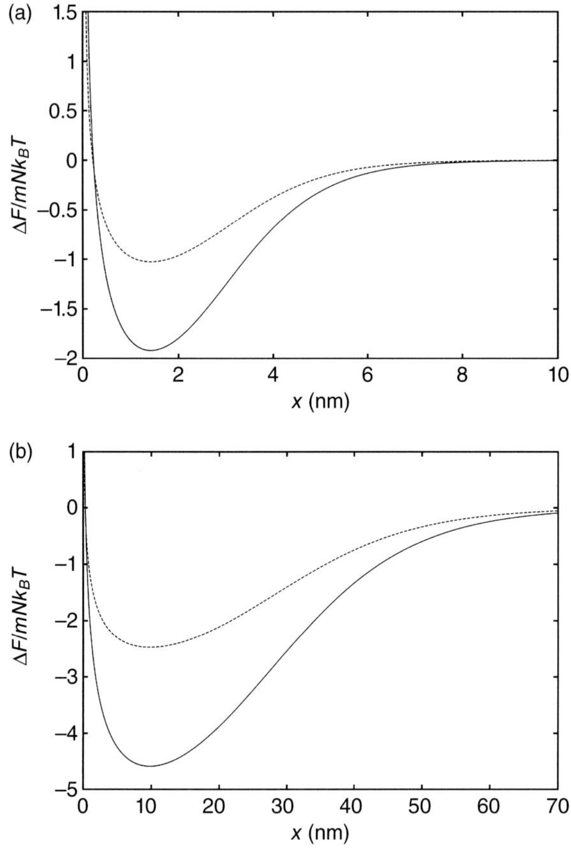

FIGURE 3.6. The normalized free energy difference per site for the DNA cluster hex7 as a function of x at counterion valence $Z = 1, 2$ (the curves increase with Z). (a) Ionic strength $I = 0.1$ M; (b) ionic strength $I = 0.001$ M. Reproduced from Perico and Rapallo.[15]

$\zeta = 1.0$ for both the above cluster models, while even for $\zeta = 0.5$ a loose cluster is formed only in the case of the more dense cluster model, hex7, where the number of DNA participating to the assembly is larger. Figure 3.7 shows that, for any cluster model, the polyelectrolyte assembly is stronger (deeper minimum at a smaller cluster radius), the higher the charge density ζ and, at a given charge density, the greater the number of polyelectrolytes in the cluster. To better understand the formation of the different clusters as a function of the linear charge density, we may take advantage of Equation 3.76, dictating the conditions required to produce condensation within a given cluster.

Figure 3.8 illustrates the critical distance x_c as a function of the linear charge density at $I = 0.1$ M. For undercritical polyelectrolytes, upon decreasing the

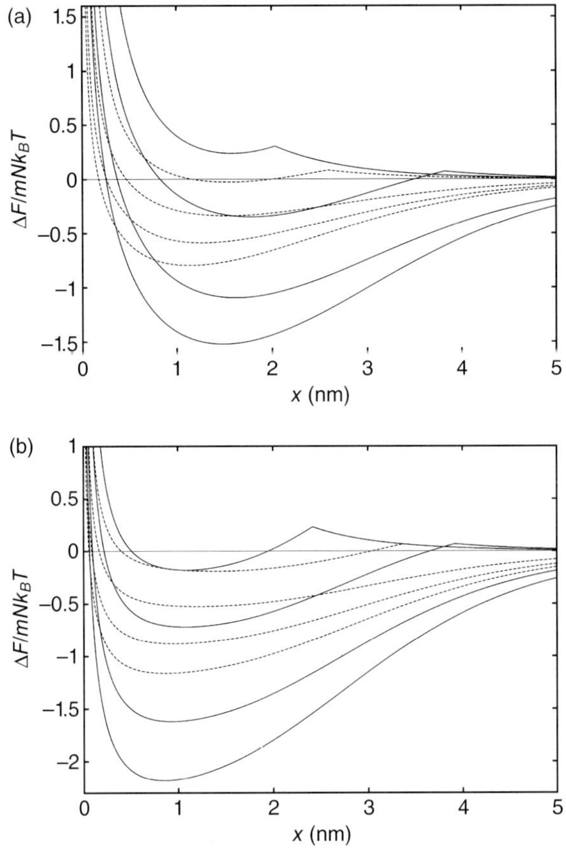

FIGURE 3.7. The normalized free energy difference per site as a function of x at ionic strength $I = 0.1$ M, for different values of linear charge densities ($\zeta = 3.0, 2.0, 1.0, 0.5$ from bottom to top), and Z = 1 (full curves) and Z = 2 (dashed curves). (a) Model hex1; (b) model hex7. Reproduced from Perico and Rapallo.[15]

cluster distance counterion condensation starts at x_c and condensation may cause the free energy of the cluster to reach a minimum at short distances. The figure shows that, given a fixed ionic strength and counterion valence, for $\zeta \geq \zeta_c$, condensation occurs at any distance for all the clusters, while for $\zeta < \zeta_c$, the onset of condensation begins at a distance decreasing with the polyelectrolyte charge strength and increasing with the number of chains involved in the cluster. The latter consideration is the motivation for the formation of a loose cluster even for $\zeta = 0.5$ for the denser cluster model (hex7). At higher valence, condensation, and eventually cluster formation, occurs at lower linear charge density due to the dependence of the critical charge density on the inverse of the counterion valence (Equations 3.56 and 3.57). The results at lower ionic strengths are qualitatively similar.

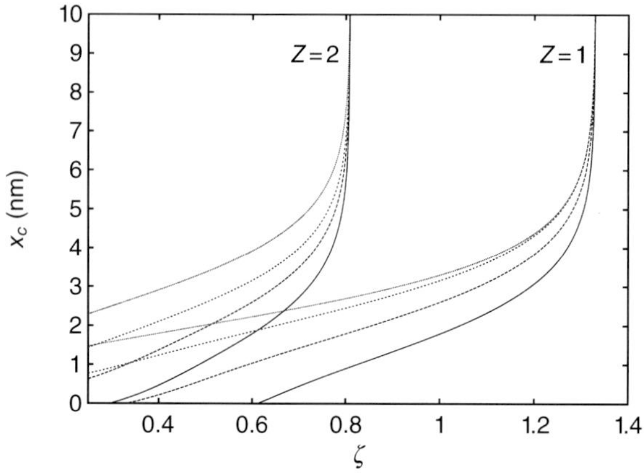

FIGURE 3.8. The critical distance x_c as a function of the polyelectrolyte linear charge density ζ for $I = 0.1$ M and counterion valence $Z = 1, 2$. Note that $\zeta_c = 1.33$ and 0.808 at $Z = 1, 2$, respectively. Two-rod model (full curve); triangular model (dashed curve); hexagonal cluster—model hex1 (short-dashed curve) and model hex7 (dotted curve). Reproduced from Perico and Rapallo.[15]

3.4.4 Concluding Remarks for Section 3.4

A consideration of the reliability of the above quantitative results, obtained in the extended condensation theory approach, is in order.

In condensation theory, counterions are considered as mobile point charges that are dynamically confined in a condensation volume around the polyelectrolytes, an approximation that can break down as counterions become multivalent and displaying more complex interactions with polyelectrolytes.[3] Counterion correlation theory[32] does not suffer from this approximation because it localizes the counterions near the polyelectrolytes ignoring the entropic cost to condense counterions from the bulk, thus favoring energetically driven attraction, mediated by multivalent counterions. For this reason, the latter is probably the proper theory to describe polyelectrolyte attraction in the presence of multivalent counterions (see Section 3.5). The most important result of the extended condensation approach is that electrostatic clusters between strong polyelectrolytes may also be formed in the presence of monovalent counterions, in agreement with several experiments in aqueous NaCl solutions in the absence of multivalent counterions,[19–24] and some recent MD simulations.[30] On the contrary, counterion correlation is small for monovalent counterions, and the correlation theory predicts only repulsion.[32] For multivalent counterions, the two mechanisms of expanded condensation volume and counterion correlation should be in competition and it is possible that the latter mechanism prevails, leading to cluster formation with increasing counterion valence, in agreement with experiments.[17,18] A possible unified theory, including the enhancement of condensation volume with the cluster compaction and counterion correlation for multivalent counterions, is expected to clarify the roles of the different physical contributions and possibly correlate with the whole extended experimental findings both in biological and synthetic systems. Electrostatic clustering may be an important component in the real assembly of short DNA and portions of the order of the persistence length of highly charged polyelectrolytes. The contributions of polyelectrolyte interaction mediated by counterion condensation, has been recently stressed in important biological supramolecular assemblies such as double helix formation, the DNA clustering, and the parallel packing arrangements in RNA folding.[38,39]

3.5 ATTRACTION OF LIKE-CHARGED POLYELECTROLYTES FOR MULTIVALENT COUNTERIONS: COUNTERION CORRELATION APPROACH

Polyelectrolyte solutions frequently contain monovalent salt, such as NaCl. Since correlations between monovalent ions are negligible,[32,40] their effect is mainly to provide screened interactions via the Debye screening constant and a strong contribution to the free energy of the polyelectrolyte–counterion clusters due to counterion condensation in the condensation volume near the polyelectrolytes. These effects are mainly described in Sections 3.3, 3.4, and 3.6 in the context of the extended condensation theory, though results are also presented for higher counterion valence to describe the entropic contribution of counterion condensation. In a solution of multivalent salt, multivalent counterions do not condense randomly around the polyelectrolytes but the position of a counterion takes into account the positions of near counterions building a three-dimensional order dictated by counterion correlation.[32,40]

In this section, a simple model for the attraction between like-charged polyelectrolytes, induced by counterion correlation, is described.[32,33] Consider two parallel rodlike polyelectrolytes, as described in Figure 3.3a of Section 3.4, at a separation distance of d corresponding to an adimensional quantity x:

$$x = d / b \tag{3.77}$$

when normalized with the charge spacing, b, on the polyelectrolyte. Assuming that around each polyelectrolyte there is the condensation of n mobile counterions of valence Z, each charge site along the polyelectrolytes is characterized by an occupation variable:

$$\begin{aligned} \sigma_{ij} &= 1, \text{ if a counterion is condensed at the site} \\ &= 0, \text{ otherwise} \\ i &= 1, N;\ j = 1, 2, \end{aligned} \tag{3.78}$$

with i describing the site along the polyelectrolyte j. The occupation variables should obey the constraint

$$\sum_{i=1}^{N} \sigma_{i1} = \sum_{i=1}^{N} \sigma_{i2} = n. \tag{3.79}$$

The coulombic electrostatic interaction energy between all the sites on the two polyelectrolytes is

$$H / k_B T = \frac{\zeta}{2} \sum_{i,i'=1}^{N} \sum_{\substack{j,j'=1; \\ (i,j)\neq(i',j')}}^{2} (1 - Z\sigma_{ij})(1 - Z\sigma_{i'j'}) b / r(i, j; i', j'), \tag{3.80}$$

with

$$r(i, j; i', j') / b = \sqrt{|i - i'|^2 + (1 - \delta_{jj'}) x^2}. \tag{3.81}$$

The partition function is then calculated as

$$\chi = \sum_{\{\sigma_{ij}\}}{}' \exp(-H / k_B T), \tag{3.82}$$

The sum is, overall, the possible configurations of the occupation numbers, and the prime implies that the sum is done under the constraint of Equation 3.79. The force between the two polyelectrolytes is obtained as

$$F = (k_B T / b) \frac{\partial \log(\chi)}{\partial x}. \tag{3.83}$$

For values of N that are not too great, an exact calculation of the force is possible, otherwise a Monte Carlo simulation is required. The results of Figure 3.9 demonstrate that for monovalent counterions ($Z = 1$), the force is always repulsive, while for $Z \geq 2$ the force becomes attractive for sufficiently high number of condensed counterions.

The above results were obtained for interacting charge lines. The theory was extended to finite polyelectrolyte diameter confirming the results of the simple model.[34] For macromolecules of finite diameter, fewer multivalent counterions are needed to induce attraction and the charge-charge correlations along the polyelectrolyte are of very short range.

These results demonstrate the occurrence of attraction between like-charged polyelectrolytes for multivalent counterions, where correlations become important, but cannot be considered valid for monovalent counterions where correlations are negligible. The above correlation model does not take into account the free energy cost to condense counterions (of any valence)

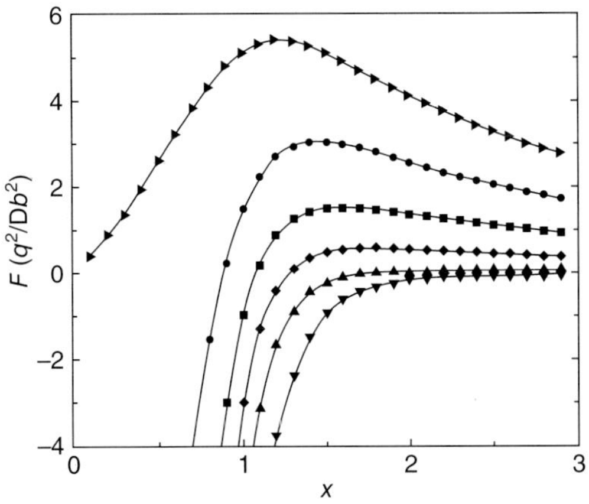

FIGURE 3.9. Force versus distance between the polyelectrolytes for $N = 20$, $Z = 2$, $\zeta = 2.283$ (polymethacrylate), and $n = 5, \ldots, 10$ from top to bottom in the Monte Carlo simulation. Positive force indicates repulsion, while negative force indicates attraction. Reproduced from figure 1 of Arenzon et al.[33a]

near the polyelectrolytes to balance the energetic interactions between condensed counterions; this may be the reason for its failure in predicting attraction between like-charged polyelectrolytes in the presence of monovalent counterions. A generalized theory, valid for any counterion valence, should include the entropic effect to condense the counterions in a condensation volume dependent on the polyelectrolyte assembly, together with the energetic effect of counterion correlations in the condensation volume. The latter effect is negligible for monovalent counterions where the condensation volume effect prevails. For multivalent counterions, the two effects are in competition and the correlation effect favoring attraction could prevail.

3.6 INTERACTIONS BETWEEN POLYELECTROLYTES AND OPPOSITELY CHARGED MACROIONS

Electrostatic interactions mediated by salt counterion are responsible for the association of oppositely charged macroions, which play a crucial role in biology, for instance, in chromatin organization and function.[5,41–43] Similarly, DNA and RNA wrap around oppositely charged protein complexes and nanoparticles in many near turns.[44,45] In this section, the interactions between polyelectrolytes and oppositely charged macroions are considered and treated according to the counterion condensation/release mechanism in the condensation approach.[44,46]

Many attempts have been made, in recent years, to model electrostatic interactions of linear polyelectrolytes with countermacroions of different three-dimensional structures such as spheres or cylinders.[14,39,42,44,46–54] The large size of macromolecules and aggregates involved in such complexes reduces the possibility of application of atomic-scale models, so that coarse-grained simplified models are currently developed and applied to explain experimental data or give hints to the answers of structural questions of great biological interest.

A general analytic description of the electrostatic interactions mediated by salt between DNA and oppositely charged spherical proteins or protein complexes, ultimately responsible for the wrapping of DNA around the surface of the protein, is given here following recently proposed electrostatic models (Section 3.6.1).[42,44,48,49]

The wrapping of DNA around the octamer in nucleosome, the action of the chromatin remodeling factor Cockayne syndrome group B (CSB), and the condensation of chromosomes in archaea by the histone-like protein methanogen chromosomal protein 1 (MC1), together with a wrapping-around model of Lac operator/Lac repressor paradigmatic interaction, will be approached (Section 3.6.2).[44,46]

3.6.1 Strong Anionic Polyelectrolytes and Oppositely Charged Spheres

Figure 3.10 illustrates the model and relative parameters of the interactions between DNA and oppositely charged proteins. DNA is represented as a negatively charged elastic cylinder of contour length L and radius a (= 1 nm) and the protein as a positively charged sphere of radius R_0 and surface charge number, Q_0. Describing a charged globular protein or protein complex as a uniformly charged sphere is a drastic simplification, useful in treating the important role played by the counterion condensation/release mechanism.

When the positive sphere approaches the negatively charged DNA, DNA and sphere try to minimize the electrostatic interaction in such a way that

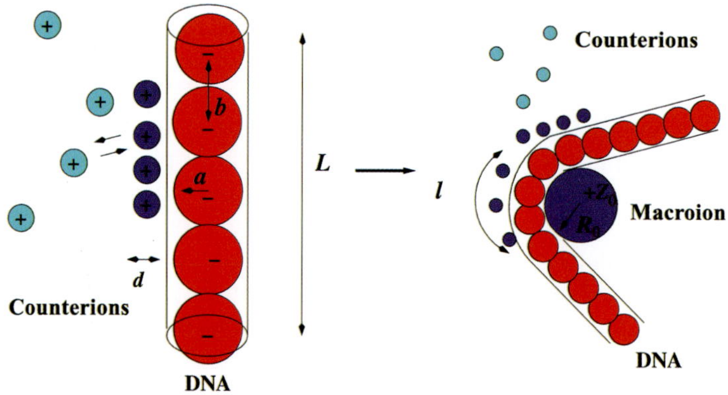

FIGURE 3.10. DNA and protein interact as oppositely charged cylinder and sphere, respectively. Reproduced from figure 1 of Arcesi et al.[44]

roughly Q_0 DNA negative charges will be as near as possible to the positive Q_0 charges of the sphere.

The binding free energy of this DNA sphere model is calculated as the difference of the actual free energy of the sphere with a length l of DNA wrapped around, to that of isolated straight DNA and sphere[44,55]:

$$\Delta F(l) = F_s(L-l) + F_c(l) + F_{c-s}(l) + F_b(l) - F_s(L) - F_c(0). \tag{3.84}$$

The first term, $F_s(L-l)$, is the electrostatic free energy in the condensation regime of the piece of straight DNA of length $L-l$; the second term, $F_c(l)$, is the electrostatic free energy of a spherical macroion after it has been bent around a length l of DNA; the third term, $F_{c\text{-}s}$, describes the electrostatic interaction of straight DNA and the complex, sphere-wrapped DNA; and finally, the last term, $F_b(l)$, is the elastic energy necessary to bend a piece of DNA of length l with a radius of curvature $R_0 + a$, corrected by the additional electrostatic repulsion due to DNA bending. The contributions to Equation 3.84 will be calculated in this section following Arcesi and coworkers.[44]

The first term, $F_s(L-l)$, is calculated according to Equation 3.54, taking into account that the total charge of the straight DNA of length $L-l$ is simply $(L-l)/b$:

$$F_s(l)/k_BT = -[(L-l)/b](2-\zeta^{-1})\log[1-\exp(-\kappa b)] - [(L-l)/b](1-\zeta^{-1}) \tag{3.85}$$

The spherical DNA–protein complex has a radius given by the sum of the radius of the protein sphere and the radius a of DNA cylinder wrapped on it:

$$R = R_0 + a, \tag{3.86}$$

and a total surface charge number Q given by the sum of Q_0 positive charges of the protein sphere and of l/b negative DNA charges:

$$Q = Q(l) = Q_0 - l/b. \tag{3.87}$$

The total charge Q is assumed uniformly distributed on the surface of the spherical complex: certainly a rough approximation that could be overcome taking into account the geometry of the wrapping of DNA on the sphere, the electrostatic interactions between the charges on the sphere, and the electrostatic repulsions between DNA turns when the strengths of the involved charges dictate a wrapping of many turns.[49–51] Note, however, that the screening due to salt condensation at physiological conditions should reduce the latter effect (see Sections 3.4 and 3.5). If the length of the bent DNA equals the isoelectric bending length,

$$l_{\text{iso}} = bQ_0, \tag{3.88}$$

then the complex has a zero charge.

The second term, $F_c(l)$, in the absence of counterion condensation on the sphere, is simply the usual electrostatic charging free energy of a sphere of global charge $Q(l)$ in a dilute electrolyte solution of dielectric constant ε and Debye screening length κ, that is obtained from the sphere potential of Equation 3.20 for $a = R$ and $q = Q(l)$[10,44]:

$$F_c^{\text{charging}}(l) = eQ(l)\psi(R)$$
$$= k_BT(l_B/2R)(Q_0 - l/b)^2 A(R);\ A(R) = 1 - \kappa R/(1+\kappa R). \quad (3.89)$$

In the case where the charge $Q(l)$ is strongly positive or negative (highly charged regime), the entropic contribution of condensation should be added to the charging free energy of Equation 3.89[44]:

$$F_c^{\text{cond}} = k_BT|Z(l)|(1-\zeta_{\text{sphere}}^{-1})\Omega \quad |Q(l)| > Q_1. \quad (3.90)$$

Note that the approximation of uniform distribution of the charge $Q(l)$ on the spherical surface enables the simple evaluation of this important entropic contribution.

Here, $|Q(l)|(1-\zeta_{\text{sphere}}^{-1})$ is the number of condensed counterions on the sphere and Q_1 is the minimum charge value at which we can have condensation on the sphere (see below). $k_BT\Omega$ is the entropic free energy cost to confine one counterion near the sphere from the bulk. It can be obtained from the logarithm of the ratio of the counterions concentration in the condensation layer near the sphere, c_{cond}, and in the bulk, c_s:

$$k_BT\Omega = k_BT\log(c_{\text{cond}}/c_s). \quad (3.91)$$

This expression is valid only if the concentration of condensed counterions, c_{cond}, is greater than the bulk salt concentration, c_s, which in turns means that $|Q(l)|$ should be greater than Q_1, the value for which $c_{\text{cond}} = c_s$. The counterion concentration c_{cond} is simply given by the number of condensed counterions divided the condensation volume v. Here the volume of the condensation layer is estimated from the volume of the Gouy–Chapman region.[42,44] Alternatively, the condensation volume v in the classic or extended condensation theory could be used with qualitatively similar results (see Section 3.3). Considering that near the surface of the sphere there is the uniform radial electrostatic field,

$$E = eQ(l)/\varepsilon R^2, \quad (3.92)$$

then the thickness d_{GC} of the Gouy–Chapman region is estimated from the condition to get an electrostatic energy equal to k_BT for a monovalent ion at a distance d_{GC} from the sphere surface in the field E: $k_BT = ed_{\text{GC}}E$. Thus, the number of charges per unit volume in the condensed layer is

$$\begin{aligned} c_{\text{cond}} &= |Q(l)|\left(1-\zeta_{\text{sphere}}^{-1}\right)/\left(4\pi R^2 d_{\text{GC}}\right) \\ &= |Q(l)|^2\left(1-\zeta_{\text{sphere}}^{-1}\right) l_B/\left(4\pi R^4\right), \end{aligned} \tag{3.93}$$

and from Equations 3.90 and 3.91, we get for $|Q(l)| > Q_1$ an extra contribution:

$$F_c^{\text{cond}}(l) = k_B T|Q(l)|\left(1-\zeta_{\text{sphere}}^{-1}\right)\log\left[2|Q(l)|^2\left(1-\zeta_{\text{sphere}}^{-1}\right) l_B^2/\kappa^2 R^4\right] |Q(l)| > Q_1. \tag{3.94}$$

Taking into account Equations 3.87 and 3.88, it is found that Equation 3.94 gives a nonzero contribution only in the ranges $l < l_{\min}$ and $l > l_{\max}$ with

$$\begin{aligned} l_{\min} &= l_{\text{iso}} - bQ_1 \\ l_{\max} &= l_{\text{iso}} + bQ_1. \end{aligned} \tag{3.95}$$

The total contribution due to the free energy of the complex thus becomes

$$F_c(l) = F_c^{\text{charging}}(l) + F_c^{\text{cond}}(l). \tag{3.96}$$

It remains to estimate ζ_{sphere}, that is, the Manning parameter for the sphere (equivalent to ζ for a cylinder). Following the calculation of the Manning parameter $\zeta = l_B/b$ for a straight linear array of charges, the Manning parameter for a sphere ζ_{sphere} was estimated as[44]

$$\zeta_{\text{sphere}} = |Q(l)| l_B A(R)/2R. \tag{3.97}$$

Finally, Q_1 is obtained as the solution for Q of the equation $c_{\text{cond}} = c_s$[44]:

$$Q_1 = Q_{\max}(2R/(l_B A(R)))\left\{1+\sqrt{1+\kappa^2 R^2 A^2(R)/2}\right\}/2; \quad Q_{\max} = 2R/(l_B A(R)). \tag{3.98}$$

Here $Q_{\max}$ is the absolute value of the charge of the sphere, after condensation.

The third term of Equation 3.84, $F_{c\text{-}s}$, describes the interaction of DNA and the complex. This is calculated as the free energy of a point of charge Q^*e in the electrostatic potential $\psi(r)$ of straight DNA. Note that Q^* is a linearly decreasing function of l equal to $Q(l)$ for $l_{\min} < l < l_{\max}$; a positive constant equal to $Q_{\max}$ for $l < l_{\min}$; and a negative constant equal to $-Q_{\max}$ for $l > l_{\max}$.[44]

The potential $\psi(r)$ is calculated solving the DH equation exactly for a cylinder of radius a and renormalizing the DNA linear charge due to Manning condensation (see Eq. 3.17):

$$\psi(r) = -(2e/\varepsilon l_B \kappa a K_1(\kappa a)) K_0(\kappa r). \tag{3.99}$$

By using this potential $\psi(r)$, bearing in mind that the distance between the axis of DNA cylinder and the complex point charge is R, the free energy $F_{c\text{-}s}$ is obtained as

$$F_b(l) = -2k_BTQ^*(l)K_0(\kappa R)/[\kappa a K_1(\kappa a)]. \tag{3.100}$$

Finally, the last term of Equation 3.87, $F_b(l)$, is the elastic energy necessary to bend a piece of DNA of persistence length l_p and contour length l with a radius of curvature R, corrected by the additional electrostatic repulsion due to DNA bending[44,56]:

$$F_b(l) = k_BTl_pl/(2R^2) + k_BTl\zeta/\left[8b(\kappa R)^2\right]. \tag{3.101}$$

The second term in Equation 3.101 corresponds to the Odijk–Skolnik–Fixman (OSF)[57,58] work against electrostatic repulsion required to bend DNA with a curvature R, in which validity is restricted to the conditions $\kappa R > 1$. This term amounts to a very low contribution.

The optimal length l^* of the wrapped DNA around the sphere is obtained by searching the minimum of $F(l)$ or $\Delta F(l)$. Alternatively to l^*, the wrapping may be described by a bending angle around the sphere:

$$\theta^* = l^*/R, \tag{3.102}$$

or as numbers of turns (or fraction) of DNA around the protein:

$$n_t^* = l^*/2\pi R \tag{3.103}$$

The typical behavior of the normalized binding free energy, $\Delta F(l)/k_BT$ of Equation 3.84, of the system DNA–protein (for a sphere of radius R_0 equal to 3.0 nm and surface charge Q_0 = 10, 20, and 100, at a salt concentration of 0.1 M) is shown in Figure 3.11 as a function of the bent DNA length l.

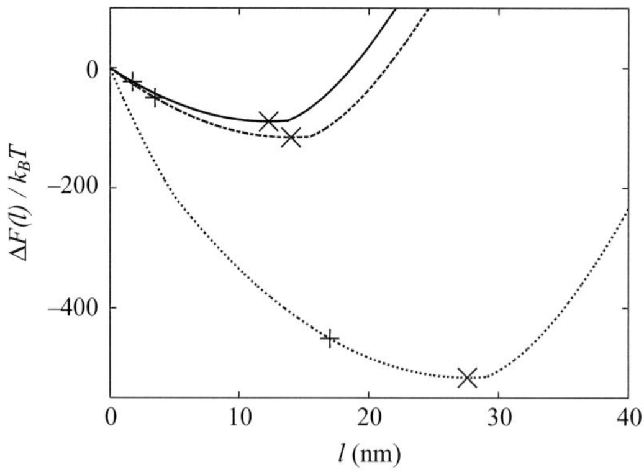

FIGURE 3.11. The total binding free energy of Equation 3.84 as a function of the bent DNA length l for R_0 = 3 nm and Q_0 = 10 (full line), 20 (dashed line), 100 (dotted line). Salt concentration c_s = 0.1 M. The binding free energy values at $l = l_{iso}$ and at $l = l^*$ are indicated on the curves by "+" and "×," respectively. Reproduced from figure 3 of Arcesi et al.[44]

The free energy functions always display a minimum at values $l^* > l_{iso}$, indicating an important wrapping of DNA around the protein, increasing with R_0 and Q_0 (l^* and l_{iso} are indicated on the curves by "×" and "+," respectively). The difference $l^* - l_{iso}$ is a measure of the length of DNA, negatively overcharging the sphere. It is the release of positive counterions from DNA, as DNA bends, that is responsible of the counterintuitive phenomenon of the overcharging. The overcharging of the sphere demonstrated in Figure 3.11 is similar, using the proper parameters, to that previously theoretically found for nucleosome.[44,59,60]

The value of l^* is very near to, but lower than that of, l_{max} (see Eq. 3.95) due to the matching of the dependence on l of two main terms in the free energy: the decreasing of electrostatic interactions and counterion condensation with the reducing of the length of straight DNA; the increasing of the electrostatic free energy of the sphere both in the charging part and in the condensation part. While the charging free energy of the sphere has a very low increase with l, near l^* there is a strong entropic loss generated by the sudden increase of counterion condensation on the sphere as l becomes greater than l_{max}. The contribution to l^* of the DNA elastic energy and of DNA–sphere interaction, although small, is significant.

In conclusion, we may physically summarize that DNA initially bends around the sphere mainly by electrostatic interactions between opposite charges of order Q_0 on the sphere and on DNA, favored by entropic release of counterions on the straight DNA and by reduction of the charging free energy of the sphere due to compensation of the positive charges of the sphere with negative charges of wrapped DNA, with the only opposition being the DNA elastic energy. The latter energy is always very low for a normal DNA of 50 nm persistence length and for enough large radius of the sphere.

The qualitative behavior of the free energy of the complex does not change, moving from DNA to other strong polyelectrolytes and changing reasonably the radius and surface charge of the spherical oppositely charged macroion. Increasing the polyelectrolyte persistence length, the length of wrapped polyelectrolyte around the sphere decreases linearly. As the negative charge of the complex increases, a condensation of positive counterions on the sphere begins, and the wrapping stops.

3.6.2 Calculation of DNA Wrapping around Proteins or Protein Complexes in Eukaryotes and Prokaryotes

We report here the wrapping of DNA around protein complexes, calculated searching the minimum of the interaction free energy, Equation 3.84, for particular cases of great biological interest.[44,46]

Nucleosome The first mechanism of condensation of the genetic information in the nucleus of eukaryotes is the wrapping of DNA around the histone octamer to form a nucleosome core particle, a complex of a histone protein

TABLE 3.1. Number of DNA Turns n_t* around the Octamer in Nucleosome, CSB, MC1, and Lac-Repressor Proteins as Determined by the Present Model and as Resulting from Experiments

Nucleosome Q_0: ($R_0 = 3.4$)	100	134	Experiment:
	1.08	1.29	1.58[64]–1.59[62]
CSB($Q_0 = 14$) R_0:	4	6	Experiment:
	0.60	0.80	0.95[68]
MC1 Q_0:	10	12	Experiment:
R_0 (nm): 1.5	0.32	0.34	0.32:[70]
1.7	0.34	0.36	
Lac Q_0: ($R_0 = 4.5$ nm)	20	25	Experiment:[82]
	0.69	0.71	0.90–1.0

octamer of radius around 3.4 nm with 147 bp of DNA wrapped around it in 1.75 turns. The histone fold domain, driven by electrostatic interactions, organizes the central 129 of the 147 bp in 1.59 superhelical turns, leaving the remaining 9 bp on the two DNA ends almost straight.[61–63] The parameters describing the electrostatic wrapping of DNA around the octamer are simply the radius of the histone octamer ($R_0 = 3.4$ nm) and the octamer positive charge number located on the surface of the octamer ($Z_0 = 134$). With these values, at physiological conditions ($c_s = 0.1$ M), a wrapping of 1.29 turns (see Table 3.1) is calculated,[44] which must be compared with the known experimental values around 1.58–1.59.[62,64]

The bending of linker DNA in chromatin may be explained, in a similar fashion, by electrostatic interactions with the positively charged N terminal H3 tail.[49] Some bending of linker DNA was confirmed by a recent 9 Å resolution X-ray structure of a tetranucleosome by Richmond and coworkers,[65] describing the chromatin structure as a two-start superhelix. More recently, a compact interdigitated structure model of the 30-nm chromatin fiber proposed by Robinson et al.[66,67] on the basis of extensive electron microscopy (EM) experiments also suggests a bent linker.

CSB Remodeling Factor The CSB protein is a member of a family of ATP-dependent chromatin remodeling factors. CSB remodels nucleosomes by wrapping/unwrapping DNA depending on ATP binding and hydrolysis.[68] While the CSB three-dimensional structure has not yet been obtained, there is evidence that CSB is a functional dimer of approximately 360 kg mol^{-1} both *in vivo* and *in vitro*. Its homodimerization, occurring through interactions with ATPase domain, is essential for ATP hydrolysis of CSB.[69]

The CSB dimer was modeled as a sphere with a radius around 6 nm with surface charge number estimated by the difference of positive and negative charged amino acids corresponding to a balance of +14.[44] The electrostatic contribution, at 0.1 M 1:1 salt, to this wrapping was evaluated in 0.80 turns

(Table 3.1). Beerens et al.[68] reported (by scanning force microscopy [SFM]) a DNA wrapping on the CSB (dimer) around 42 nm, or 0.95 turns of DNA around its surface (Table 3.1). This suggests that the large CSB dimer, carrying a net positive charge +14, interacts sufficiently with the highly negative charged DNA to motivate almost quantitatively the observed wrapping of almost one turn. The experimental wrapping is of 42 nm in the case of no hydrolysis for the bound ATP. When ATP hydrolysis occurs, a reduction of the wrapping is observed by SFM due to an equilibrium between hydrolyzed and not hydrolyzed species.[68] This unwrapping could be partially explained by the electrostatic model assuming that ATP hydrolysis causes either a contraction or a dissociation of CSB, reducing both the protein radius and surface charge. This model shows that a simple localized chemical reaction induces, through salt-mediated electrostatic forces, a collective rearrangement by switching on/off the DNA wrapping.[44]

Prokaryotes/Archaea Proteins: MC1 In prokaryotes and archaea, chromosomal proteins induce compaction of DNA, thus structuring the chromosome in a manner similar to the histone octamer in eukaryotes, and for these reasons are generally referred to as histone-like proteins.[70] Methanogen chromosomal protein 1, MC1, is the most abundant chromosomal protein in various species of the family of methanogenic archaea: It is found as a basic polypeptide of 93 amino acids with a net positive charge of +12, associated with DNA *in vivo* at the ratio of 1 protein for 100–170 bp.[71,72] Its solution nuclear magnetic resonance (NMR)-imaged three-dimensional structure shows a globular protein with an overall radius near 1.7 nm with an electrostatic surface map displaying a prevailing global positive charge that is expected to strongly interact with DNA.[72] The bending of DNA by interaction with MC1 has been measured by dark-field EM and electrophoresis, and studied by molecular modeling.[70,73,74] Visualization of DNA molecules by EM shows sharp kinks in DNA. The use of short DNA of 176 bp containing a preferential binding site in the center of the DNA sequence, coupled to simple flexible polymer models, allowed a precise determination of a bending angle of 116° or 0.32 turns.

The electrostatic model of Equation 3.84 with a radius around the value above determined by NMR and a positive charge near 12 was used to calculate the possible DNA bending. The results are reported in Table 3.1: The electrostatic model predicts, for $R_0 = 1.7$ and $Z_0 = 12$, a value of 130°, very near to the experimental one. Thus, the electrostatic model shows that even such a small protein may wrap the DNA around its surface in a measure of 0.36 turns corresponding to a bending angle of 130°. The effect of many of these proteins acting on long DNA filaments can produce a significant condensation.

A possible three-dimensional structure of a condensed DNA-MC1 complex is schematized in Figure 3.12, assuming a 130° DNA bending around MC1, random orientation of the bending angles, and straight DNA elsewhere.[44] The length of straight linker was randomly chosen in the range of 82–152 bp, while a constant number of 18 bp were bent around the spherical protein (the blue

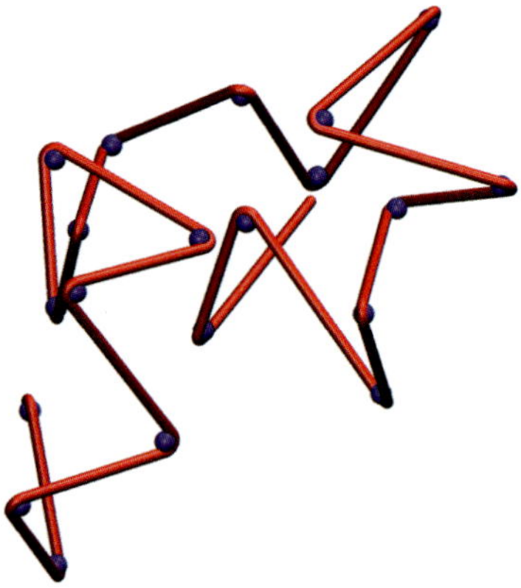

FIGURE 3.12. Three-dimensional model of DNA in the archaea cell, condensed by interactions with histone-like protein MC1. Average DNA length per MC1 protein: 135 bp, with 18 bp (corresponding to a bending angle of 130°) over 135 bent around MC1. The orientation of each bending plane is randomly chosen. DNA = red; protein = blue. Reproduced from figure 6 of Arcesi et al.[44]

sphere). A single-site Lennard-Jones potential was assigned to beads centered on each DNA bp (the red spheres) in order to include long-range interactions between DNA segments.

The Lac Operon Complex Recent results in the modeling of the Lac operon complex are here reported.[46] Gene regulation is often achieved by interaction of DNA with proteins both in eukaryotes and in prokaryotes: a paradigm of it is the Lac repressor–Lac operator interaction.[75,76] The Lac repressor is a protein of 360 amino acids that associates into a homotetramer of 154,520 Da molecular mass organized in a dimer of dimers linked in by a four-helix bundle at the C-terminus of each monomer subunit. The tetramer has a V shape with two different contacts with DNA operators at the tips of the V.[77,78] The binding of the Lac repressor tetramer (LacI) to the Lac operon inhibits transcription of the genes necessary for cell metabolism of lactose as an energy source. In the presence of binding to the repressor of an inducer (natural as allolactose or gratuitous as isopropyl-β-D-1-thiogalactoside [IPTG]), the conformation of the repressor is specifically changed to reduce the affinity for the operator while the expression of the genes is activated again.[76,79,80] Less is known about the actual configuration of the complex LacI–DNA, particularly in the case when one of the LacI–DNA binding domains is related to the primary operator O1 and the other to the weaker auxiliary operator O3. One of the main questions related to the organization of the 93 bp DNA O1-O3 segment is the occurrence of a wrapping-around or a wrapping-away model.[81–83] Qualitative arguments have been reported supporting one or the other model. Thermodynamic data by Tsodikov et al.[82] suggest a fine DNA wrapping path

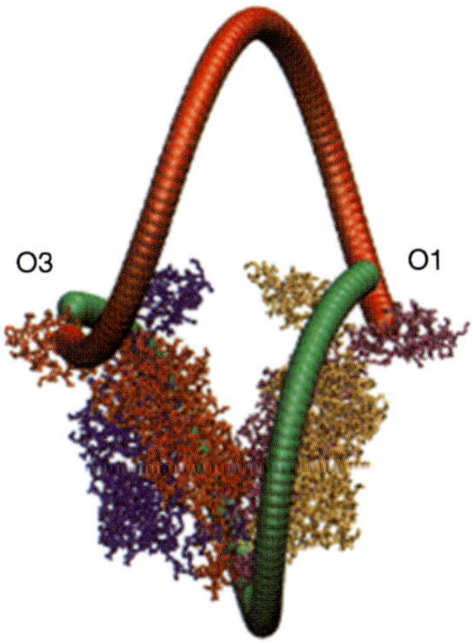

FIGURE 3.13. Schematic representation of the two proposed configurations for the DNA in the Lac operon complex: the left-handed wrapped-away (red) and wrapped-around (green) configurations. Each DNA base pair is represented as a sphere with radius of 0.5 nm, the nonhydrogen atoms in the protein are represented as sticks with different colors for each of the four monomers in the LacI tetramer (P1 is blue, P2 is red, P3 is orange, and P4 is purple). Reproduced from figure 1 of La Penna and Perico.[46]

around LacI and a possible wrapping-around model due to large, nonspecific electrostatic interactions between DNA and LacI (Fig. 3.13).

Using the latter suggestion, the LacI–DNA electrostatic interactions were calculated quantitatively using the general analytical approach to protein–DNA electrostatic interactions embodied in Equation 3.84.[46] The calculation gives a quantitative estimate of the nonspecific electrostatic free energy originally suggested as responsible of the wrapping-around model.[82] Under physiological conditions, a generic electrostatic contribution to the wrapping around of 0.70 turns is calculated, given for LacI a radius of $R_0 = 4.5$ nm and a number of 20–25 positive surface charges (see Table 3.1). Additional local specific contributions, such as the local binding of the two DNA operators (O3 and O1) to the two tetramer binding domains at the tips of the V-shaped LacI tetramer contribute the remaining 0.2 turns to reach the 0.9 turns corresponding to the contour length of the O3-O1 segment. Thus, the analytical model supports DNA wrapping around the tetramer according to the model originally proposed by Tsodikov et al.[82]

The above results were further supported by MD simulations. A fluctuating coarse-grained DNA model interacting with a fixed-in-space atomistic tetramer showed that wrapped-away configurations of DNA rarely occur. Atomistic MD simulations of the relaxation of the all-atom model showed that the DNA can be accommodated in a wrapping-around fashion, with only slight distortions in both DNA and the protein assembly.[46] Even though the wrapped-away configuration cannot be excluded, there are several evidences for the wrapped-around configuration: extended thermodynamics experiments,[82] the electrostatic analytical calculations, the MD simulations of coarse-grained models, and the MD relaxation of the all-atom model.[46]

All the above examples demonstrate that the attitude of DNA to wrap around protein complexes is, to a great extent, due to electrostatic interactions mediated by salt counterions. On this basis, the statement that most binding proteins may organize flanking DNA by wrapping is reinforced. Advanced models of electrostatic interactions between polyelectrolytes and countermacroions may therefore greatly help in explaining important problems in biology.

REFERENCES

1. F. Oosawa. *Polyelectrolytes*. Marcel Dekker, New York, 1971.
2. J. L. Barrat and J. F. Joanny. Theory of polyelectrolyte solutions. *Adv Chem Phys* **1996**, 94, 1.
3. M. Mandel. *Physical Properties of Polyelectrolyte Solutions*. AIM Editions, Pacini, Pisa, 1999.
4. M. Ullner. Polyelectrolytes. Physicochemical Aspects and Biological Significance. In *DNA Interactions with Polymers and Surfactants*. R. S. Diaz and B. Lindman, eds. Wiley, Hoboken, 2008.
5. R. F. Bruinsma. *Physics of Protein-DNA Interaction*. Les Houches Courses. Springer-Verlag, Berlin, 75, **2002**.
6. R. M. Fuoss, A. Katchalski, S. Lifson. *Proc Natl Acad Sci USA* **1951**, 37, 579.
7. G. L. Gouy. *J de Phys* **1910**, 9, 457.
8. D. L. Chapman. *Philos Mag* **1913**, 25, 475.
9. P. Debye. *E Huckel Phys Z* **1923**, 24, 185.
10. T. H. Hill. *Arch Biochem Biophys* **1955**, 57, 229. **1962**, 97, 505.
11. G. S. Manning. *Q Rev Biophys* **1978**, 11, 179.
12. J. M. Schurr and B. S. Fujimoto. *Biophys Chem* **2002**, 101-102, 425.
13. J. M. Schurr and B. S. Fujimoto. *J Phys Chem B* **2003**, 107, 4451.
14. G. S. Manning. *J Am Chem Soc* **2003**, 125, 15087.
15. A. Perico and A. Rapallo. *J Chem Phys* **2011**, 134, 055108.
16. S. Pietronave, L. Arcesi, C. D'Arrigo, A. Perico. *J Phys Chem B* **2008**, 112, 15991.
17. V. A. Bloomfield. *Biopolymers* **1997**, 44, 269. and references therein.
18. W. M. Gelbart, R. F. Bruinsma, P. A. Pincus, V. A. Parsegian. *Physics Today* **2000**, 53, 38.
19. P. Wissenburg, T. Odijk, P. Cirkel, M. Mandel. *Macromolecules* **1995**, 28, 2315.
20. R. Borsali, H. Nguyen, R. Pecora. *Macromolecules* **1998**, 31, 1548.
21. M. Sedlak and E. J. Amis. *J Chem Phys* **1992**, 96, 817, 826.
22. J. J. Tanahatoe and M. E. Kuil. *J Phys Chem B* **1997**, 101, 5905.
23. Y. Zhang, J. F. Douglas, B. D. Ermi, E. J. Amis. *J Chem Phys* **2001**, 114, 3299.
24. B. D. Ermi and E. J. Amis. *Macromolecules* **1998**, 31, 7378.

25. J. Ray and G. S. Manning. *Langmuir* **1994**, 10, 2450.
26. N. Gronbech-Jensen, R. J. Mashl, R. F. Bruinsma, W. M. Gelbart. *Phys Rev Lett* **1997**, 78, 2477.
27. M. Stevens. *J Phys Rev Lett* **1999**, 82, 101.
28. M. Deserno, A. Arnold, C. Holm. *Macromolecules* **2003**, 36, 249.
29. E. Allahyarov, G. Gompper, H. Lowen. *Phys Rev E* **2004**, 69, 041904.
30. B. Luan and A. Aksimentiev. *J Am Chem Soc* **2008**, 130, 15754.
31. X. Qiu, K. Andresen, L. W. Kwok, J. S. Lamb, H. Y. Park, L. Pollack. *Phys Rev Lett* **2007**, 99, 38104.
32. Y. Levin. *Rep Progr Phys* **2002**, 65, 1577.
33. (a) J. J. Arenzon, J. F. Stilk, Y. Levin. *Eur Phys J B* **1999**, 12, 79; (b) A. Diehl, H. A. Carmona, Y. Levin. *Phys Rev E* **2001**, 64, 011804.
34. A. Naji and R. R. Netz. *Eur Phys J E* **2004**, 13, 43.
35. M. L. Henle and P. A. Pincus. *Phys Rev E* **2005**, 71, 060801.
36. J. Ray and G. S. Manning. *Macromolecules* **1997**, 30, 5739.
37. J. Ray and G. S. Manning. *Macromolecules* **2000**, 33, 2901.
38. G. S. Manning. *Biopolymers* **2003**, 69, 137.
39. V. L. Murthy and G. D. Rose. *Biochemistry* **2000**, 39, 14365.
40. A. Y. Grosberg, T. T. Nguyen, B. Shklowskii. *Rev Mod Phys* **2002**, 74, 329.
41. J. Widom. *Annu Rev Biophys Biomol Struct* **1998**, 27, 285.
42. H. Schiessel. *J Phys: Cond Matter* **2003**, 15, R699.
43. B. Jayaram, K. McConnell, S. B. Dixit, A. Das, D. L. Beveridge. *J Comp Chem* **2002**, 23, 1.
44. L. Arcesi, G. La Penna, A. Perico. *Biopolymers* **2007**, 86, 127.
45. A. A. Zinchenko, T. Sakaue, S. Araki, K. Yoshikawa, D. Baigl. *J Phys Chem B* **2007**, 111, 3019.
46. G. La Penna and A. Perico. *Biophys J* **2010**, 98, 2964.
47. K. K. Kunze and R. R. Netz. *Phys Rev Lett* **2000**, 85, 4389.
48. K. K. Kunze and R. R. Netz. *Europhys Lett* **2002**, 58, 299.
49. A. Perico, G. L. Penna, L. Arcesi. *Biopolymers* **2006**, 81, 20.
50. A. G. Cherstvy and R. G. Winkler. *J Chem Phys* **2004**, 120, 9394.
51. A. G. Cherstvy and R. G. Winkler. *J Phys Chem B* **2005**, 109, 2962.
52. R. G. Winkler and A. G. Cherstvy. *Phys Rev Lett* **2006**, 96, 066103.
53. A. G. Cherstvy and R. G. Winkler. *J Chem Phys* **2006**, 125, 064904.
54. A. A. Zinchenko and N. Chen. *J Phys Condens Matter* **2006**, 18, R453.
55. A. Perico and A. Ciferri. *Chem Eur J* **2009**, 15, 6312.
56. G. S. Manning. *Macromolecules* **2001**, 34, 4650.
57. T. Odijk. *J Polym Sci* **1977**, 15, 477.
58. J. Skolnik and M. Fixman. *Macromolecules* **1977**, 10, 94.
59. K. K. Kunze and R. R. Netz. *Phys Rev E* **2002**, 66, 011918.
60. E. Gurovitch and P. Sens. *Phys Rev Lett* **1999**, 82, 339.
61. K. Luger, A. W. Maeder, L. K. Richmond, D. F. Sargent, T. J. Richmond. *Nature* **1997**, 389, 251.

62. C. A. Davey, D. F. Sargent, K. Luger, A. W. Maeder, T. J. Richmond. *J Mol Biol* **2002**, 319, 1097.
63. T. J. Richmond and C. A. Davey. *Nature* **2003**, 423, 145.
64. A. Hamiche, P. Shultz, V. Ramachrishnam, P. Oudet, A. J. Prunell. *J Mol Biol* **1996**, 257, 30.
65. T. Schalch, S. Duda, D. F. Sargent, T. J. Richmond. *Nature* **2005**, 436, 138.
66. P. J. J. Robinson, L. Fairall, V. A. Huynh, D. Rhodes. *Proc Natl Acad Sci USA* **2006**, 103, 6503.
67. P. J. J. Robinson and D. Rhodes. *Curr Op Struct Biol* **2006**, 16, 336.
68. N. Beerens, J. H. J. Hoeijmakers, R. Kanaar, W. Vermeulen, C. Wyman. *J Biol Chem* **2005**, 280, 4722.
69. M. Christiansen, T. Thorslund, B. Jochimsen, V. A. Bohr, T. Stevnsner. *Febs J* **2005**, 272, 4306.
70. E. Le Cam, F. Culard, E. Larquet, E. Delain, J. A. Cognet. *J Mol Biol* **1999**, 285, 1011.
71. M. Imbert, B. Laine, N. Helbecque, J. P. Mornon, P. J. Henichart, P. Sautiere. *Biochim Biophys Acta* **1990**, 1038, 346.
72. F. Paquet, F. Culard, F. Barbault, J. C. Maurizot, G. Lancelot. *Biochemistry* **2004**, 43, 14971.
73. J. A. H. Cognet, C. Pakleza, D. Cherny, E. Delain, E. Le Cam. *J Mol Biol* **1999**, 285, 997.
74. G. De Vuyst, S. Aci, D. Genest, F. Culard. *Biochemistry* **2005**, 44, 10369.
75. F. Jacob and J. Monod. *J Mol Biol* **1961**, 3, 318.
76. M. Lewis. *C R Biologies* **2005**, 328, 521.
77. A. M. Friedman, T. O. Fishmann, T. A. Steitz. *Science* **1995**, 268, 1721.
78. M. Lewis, G. Chang, N. C. Horton, M. A. Kercher, H. C. Pace, M. A. Shumacher, R. G. Brennan, P. Lu. *Science* **1996**, 271, 1247.
79. R. Daber, S. Stayrook, A. Rosenberg, M. Lewis. *J Mol Biol* **2007**, 370, 609.
80. C. E. Bell and M. Lewis. *Curr Opin Struc Biol* **2001**, 11, 19.
81. S. E. Halford, D. M. Gowers, R. B. Sessions. *Nature Struct Biol* **2000**, 7, 705.
82. O. G. Tsodikov, R. M. Saecker, S. E. Melcher, M. M. Levandoski, D. E. Frank, M. W. Capp, M. T. J. Record. *J Mol Biol* **1999**, 294, 639.
83. D. Swigon, B. D. Coleman, W. K. Olson. *Proc Natl Acad Sci USA* **2006**, 103, 9879.

CHAPTER 4

POLYELECTROLYTE AND POLYAMPHOLYTE EFFECTS IN SYNTHETIC AND BIOLOGICAL MACROMOLECULES

NGO MINH TOAN, BAE-YEUN HA, AND DAVE THIRUMALAI

4.1 INTRODUCTION

The charged nature of biological macromolecules (not limited to DNA, RNA, and proteins) and synthetic polymers makes the study of polyelectrolyte (PE) properties important. Shape fluctuations in charged biomolecules control a number of functions such as transcription and packaging of DNA and RNA into phage heads, as well as interactions involving nucleic acids with other biological macromolecules (higher order structural organization of DNA upon interaction with histones, for example). An example of functional significance is the spectacular phenomenon of reversible condensation of DNA into toroidal structure in the presence of multivalent cations. Upon condensation, the volume occupied by DNA decreases by nearly four or more orders of magnitude. Description of counterion-induced condensation of DNA and synthetic PEs requires accounting for density fluctuations associated with counterions and their coupling to conformations of the macromolecule. Similarly, the phase diagram of charged synthetic polymers, especially in aqueous solution, is complicated and is critically dependent on ion concentration as well their size and shapes. The behavior of PEs is governed by a number of factors such as the intrinsic properties of the backbone (extent of flexibility), length scale associated with ion fluctuations, and even the chemical nature of the counterions. As a result, it has been difficult to develop a coherent theory that can capture all the properties of PEs.

Ionic Interactions in Natural and Synthetic Macromolecules, First Edition. Edited by Alberto Ciferri and Angelo Perico.

In this chapter, we confine ourselves predominantly to the study of the properties of an isolated PE and the not so unrelated problem of polyampholytes (PAs) and develop concepts that have been particularly useful in understanding a number of aspects of RNA folding and a few problems associated with DNA. One focus, described in Section 4.2, is to elucidate the current understanding of how the persistence of charged chains, a basic property of polymers, depends on salt concentration. Although considerable work has been done in this area (see an excellent review[1] by Barrat and Joanny that still remains relevant), a clean solution to this problem has not been obtained. This is probably because there are multiple scenarios that depend on the molecular weight of the polymer and the concentration of chemical characteristics of the counterions. In Sections 4.3 and 4.4, we provide a simple picture of shapes of PAs, which are polymers with random distribution of positively and negatively charged monomers. In Section 4.5, we describe the responses of PEs to mechanical force, which not only have provided insights into the problem of electrostatic contribution to persistence length of charged polymers but also have been instrumental in describing length scale-dependent elasticity of single- and double-stranded DNA. We describe simulations (Section 4.6) and theory (Section 4.7) of collapse of PAs and PEs, which prove to be important in understanding folding of RNA. The chapter ends with concluding remarks in Section 4.8.

4.2 PERSISTENCE LENGTH OF PEs

The problem of electrostatic persistence length of PEs has been studied for more than three decades using a variety of methods, including analytical treatment, experiments, and simulations. Due to an interplay of a number of length scales, compared with the case of neutral polymers, a complete theoretical understanding of persistence length changes in PE as a function of salt concentration is difficult. The electrostatic persistence length is defined as the contribution to the chain total persistence length in addition to the "bare" (intrinsic) value by electrostatic interactions among the charged monomers. Theoretical description of the characteristics of an isolated PE chain began with two pioneering and independent studies by Odijk and by Skolnick and Fixman (hereafter referred to as OSF theory), which addressed the problem of electrostatic contribution to persistence length of an intrinsically stiff chain. Besides the usual chain intrinsic stiffness of a neutral chain, which is characterized by bending rigidity (persistence length, Kuhn length, monomer length, and interaction strength), electrostatic interactions among monomers can further stiffen the chain. The interplay of the chain intrinsic stiffness and the electrostatic interactions, which are usually modeled by screened coulomb potential using the Debye–Hückel (DH) theory, potentially gives rise to a number of scenarios, which require a number of approaches.

4.2.1 Intrinsically Stiff Chains: The OSF Theory

Odijk[2] and, independently, Skolnick and Fixman[3] were the first to develop a theory for persistence length of PEs. Following Barrat and Joanny[1], we model a PE as a wormlike chain (WLC) with a contour length L and "bare" persistence length l_0, and that carries charges separated by a distance A along its contour. The electrostatic interaction between two charges separated at a spatial distance r is modeled by the DH potential:

$$v_{\rm DH}(r) = k_B T l_B \frac{\exp\left(-\dfrac{r}{\lambda_D}\right)}{r}, \quad (4.1)$$

where $l_B = e^2/(\varepsilon k_B T)$ (in Gaussian units as in Higgs and Joanny[4] and Dobrynin et al.[5]) is the Bjerrum length and is about 0.7 nm at room temperature, and λ_D (or, interchangeably, κ^{-1}) is the Debye screening length, which is given by

$$\kappa^{-1} = \lambda_D = \frac{1}{\sqrt{4\pi l_B I}}, \quad (4.2)$$

with I being the ionic strength of the solution.

For simplicity, we confine the chain in two dimensions so it is a planar curve. The results in three dimensions are similar. The chain configuration is characterized by the angle function (of arc length s) $\theta(s) = \mathbf{t}(s)\dot{\mathbf{t}}(0)$, where $-L/2 \le s \le L/2$ and $\mathbf{t}(s)$ is the tangent vector at s. We assume that the chain is almost linear (strong charge limit) or $\theta(s)$ is small, so that the electrostatic interaction between two charges at s_1 and s_2 can be expanded around the rodlike conformation, $v_{\rm DH}(|s_2 - s_1|)$. With these conditions, the chain Hamiltonian $H(\theta)$ can be written for a given configuration $\{\theta(s)\}$ as

$$H[\theta] = H_0 + \frac{1}{2}k_B T \int_{-L/2}^{L/2} ds \int_{-L/2}^{L/2} ds' \frac{d\theta(s)}{ds}[l_0\delta(s-s') + K(s,s')]\frac{d\theta(s')}{ds}. \quad (4.3)$$

Here, H_0 is the electrostatic energy of a rod, the term proportional to l_0 is the usual bending energy of the WLC that is proportional to the square of radius of curvature, and the electrostatic interactions described using the kernel $K(s, s')$ which, for $s > s'$, is

$$K(s,s') = \frac{1}{A^2}\int_{-L/2}^{s'} ds_1 \int_{s}^{L/2} ds_2 v'_{\rm DH}(s_2 - s_1)\frac{(s_2 - s)(s' - s_1)}{(s_2 - s_1)}, \quad (4.4)$$

with $v'_{\rm DH}(s) = dv_{\rm DH}/ds$. The result has been obtained with the only assumption that within λ_D the chain remains in an almost rodlike configuration. For very long chains, $L/\lambda_D \gg 1$, we can extend the limits in the integrals of Equation 4.4 to infinity and $K(s, s')$ becomes a function of $s - s'$,

$$K(s)=\frac{1}{6A^2}\int_0^\infty dx\frac{x^3}{x+s}v'_{\mathrm{DH}}(x+s), \tag{4.5}$$

which has the Fourier transform

$$\begin{aligned}\tilde{K}(q)&=\int_0^\infty ds\exp(iqs)K(s)\\&=l_{\mathrm{OSF}}\frac{2\kappa^2}{q^2}\left(\frac{\kappa^2+q^2}{q^2}\ln\left(\frac{\kappa^2+q^2}{\kappa^2}\right)-1\right),\end{aligned} \tag{4.6}$$

where

$$l_{\mathrm{OSF}}=l_B/(4A^2\kappa^2) \tag{4.7}$$

is called the celebrated OSF length. The statistical properties of the chain are obtained by integrating the Boltzmann factor exp($H([\theta]/k_BT)$ over all possible configurations, that is, over all functions $\theta(s)$. Since the energy is a quadratic function of θ, the integration can be carried out analytically. For instance, we can calculate the mean squared angle $\langle\theta(s)^2\rangle$ between the chain directions at the origin and at arc length s. This quantity characterizes the local flexibility of the chain. With the approximate Hamiltonian (Eq. 4.3), we have

$$\langle\theta(s)^2\rangle=\frac{4}{\pi}\int_0^\infty dq\frac{\sin^2(qs/2)}{q^2}\frac{1}{l_0+\tilde{K}(q)}. \tag{4.8}$$

For a neutral chain, $\tilde{K}(q)=0$ in Equation 4.8 and we have

$$\langle\theta(s)^2\rangle=\frac{s}{l_0}, \tag{4.9}$$

which varies linearly with s, and the proportional constant is the inverse of the persistence length.

For a charged chain, $\langle\theta(s)^2\rangle$ cannot be analytically calculated. Thus, to describe the expected scaling behavior, we work in the limits of small and large s. At large s,

$$\langle\theta(s)^2\rangle=\frac{s}{l_0+l_{\mathrm{OSF}}}, \tag{4.10}$$

which expresses the fact that at long length scales, the chain conformation can be described by an effective persistence length l_0+l_{OSF}, which is the sum of a "bare" and of an electrostatic contribution. It indicates that the influence of the screened electrostatic interactions can extend far beyond their range λ_D since in weakly screened solutions, l_{OSF} is much larger than λ_D. The electrostatic persistence length also decreases with salt concentration as $l_{\mathrm{OSF}}\simeq I^{-1}$, whereas the Debye screening length has a slower decay $\lambda_D\simeq I^{-1/2}$.

On the other hand, for small s, Equation 4.8 reduces exactly to that of a neutral chain in Equation 4.9. This means that the chain statistics at short scales are not modified by electrostatic interactions. The crossover between the "intrinsic" regime described by Equation 4.9 and the "electrostatic" regime described by Equation 4.10 takes place when the electrostatic interactions become strong enough to perturb the statistics of the neutral flexible chain. The crossover length s_c can be obtained qualitatively from the following argument. If a small chain section, of length $s < \kappa^{-1}$, is bent to form an angle θ, the cost in "bare" curvature energy is $k_B T l_0 \theta^2/s$, while the electrostatic energy is $k_B T l_B (s/A)^2(\theta^2/s)$. The two energies are comparable for $s \simeq s_c$, which gives $s_c \sim A(l_0/l_B)^{1/2}$.

The picture that emerges from this calculation is that the chain flexibility depends on the length scale. At short scales, $s < s_c$, the chain structure is determined by its bare rigidity l_0, while the electrostatic rigidity (Eq. 4.7) dominates at large scales. Large length scale properties, such as the radius of gyration, can be determined by applying the standard formula for semiflexible chains of persistence length l_{OSF}.

The only approximation required to obtain the OSF length is the expansion that yields Equation 4.3. The calculation is therefore consistent if the angle $<\theta(\kappa^{-1})^2>$ is small compared with unity, which holds good if the bare persistence length is much larger than l_{OSF}. In other words, the requirement $<\theta(\kappa^{-1})^2> \ll 1$ is then equivalent to $s_c l_{OSF}/(l_0(l_0 + l_{OSF})) \simeq s_c/l_0 \ll 1$; that is, the angular deflection of the chain must be small when the crossover region is reached. In other words, the angular fluctuations that take place before the electrostatic interactions can come into play and rigidify the chain should not be too strong. If these fluctuations are too large, that is, if the chain is too flexible, the perturbation expansion that underlies the OSF calculation can break down. The criterion for the validity of the calculation can be simply written, in the limit of weak screening, as $l_0 > A^2/l_B$. This implies that the OSF calculation should be directly applicable to stiff chains such as DNA ($l_0 \sim 50$ nm), but has to be reconsidered for flexible chains such as polystyrene sulfonate ($l_0 \sim 1$ nm).

Multiple lines of inquiry have questioned the validity of the OSF theory (see below). All of them hinge on the assumption that when chain fluctuations are taken into account, the persistence length is renormalized and the dependence on λ_D deviates from the predictions of OSF theory. Starting with the work of Barrat and Joanny,[6] several variational calculations have shown that, when the chains are intrinsically flexible, the electrostatic persistence length $l_e \sim \lambda_D^1$. This dependence is valid when the controlling parameter $(l_0 l_B/A^2) \sim O(1)$, that is, when the chain is in the so-called nonasymptotic regime. Li and Witten[7] argued that, unlike the variational theories, approximate inclusion of fluctuations still leads to the result predicted by OSF even if the backbone is flexible. In other words, the OSF result is always valid at least for long chains regardless of the backbone stiffness. In the above referenced variational theories, it is generally believed that the electrostatic interactions generally stiffen the chain significantly so that the trial Hamiltonian should consist of terms that account for the electrostatic-mediated interaction rigidity.

The controversies associated with the validity of the OSF theory were further addressed using two related theoretical studies that used different variational methods.[8,9] Using the most general variational approach these studies showed that the OSF result is recovered whenever $(l_0 l_B/A^2) \ll 1$. Thus, in both the stiff and intrinsically flexible limit, the classical result is obtained, provided the chain is long. Indeed, these theories supported a scaling-type argument due to Khokhlov and Khachaturian[10] (KK), who argued that chain fluctuations merely renormalize A by a blob with dimension D with $(D/A)e$ charge. The physical that emerges is that fluctuations do not invalidate the qualitative aspect of the OSF result. Rather, they effectively reduce the direct distances between charges so as to renormalize the bare parameters l_0, A, charge per blob, as was first recognized by KK. In this limit, local fluctuations inside the blobs are strong enough to reduce significantly the direct distance between two consecutive charges by the factor $(l_0 l_B/A^2)^{1/3}$, which is much less than one. This results in much stronger effective coulomb repulsion between two consecutive blobs. Since the length scale of l_p is much larger than the blob size D and ordering as implied in these cases does not refer to the local structure of blobs, we can expect that the local fluctuations inside a blob not to affect significantly the property of the chain much beyond the length scale D. As a result, we obtain qualitatively similar dependence of l_p on λ_D as that obtained for stiff polyions. Note that this is relevant only for the asymptotic case where each blob contains a large number of segments that roughly obey Gaussian statistics. The variational theories also predict interesting crossover behavior in which the electrostatic persistence length $\sim\lambda_D^y$ where the exponent $y \leq 1$.

4.2.2 The Linear and Sublinear Dependence of l_e on λ_D

The OSF theory has formed the basis for interpreting both experiments and simulations. Indeed, a number of experiments have confirmed the quadratic dependence of the electrostatic persistence length l_e on the Debye length λ_D.[3,11–13] However, many other experiments[14–22] and simulations[21,22] have shown that the dependence of l_e on λ_D is linear, or even sublinear. The reason for this discrepancy seems to be that the OSF theory was originally derived for intrinsically stiff PEs, those that have their intrinsic persistence length much larger than separation of charges on the backbone $l_{OSF} \gg A$, such as dsDNA and carboxymethylcellulose (CMC).[3] In the latter experiments, intrinsically more flexible PEs, such as ssDNA and RNA, were examined. Thus, chain fluctuations or finite size effects can indeed produce deviations from the OSF theory. The deviations from OSF behavior have lead to other approaches to study flexible PEs.

Barrat and Joanny,[23] on the other hand, used a variational method in which they replaced the actual Hamiltonian with a "trial" one, H_t, and minimized the resulting free energy $F_{var} = \langle H_t \rangle - TS_t$, where the average is taking with the Boltzmann weight $\exp(-\beta H_t)$, with β being the inverse of the thermal energy

k_BT. Here, H_t is that of a neutral chain under uniform tension. For flexible chains, they also found the total persistence length scales as λ_D. Dobrynin[24] also modeled a PE as a WLC with electrostatic interactions and evaluated the bending angle fluctuations in the framework of the Gaussian variational principle. For semiflexible chain and strongly charged flexible chain, $l_e \sim \lambda_D$, whereas the dependence is sublinear for weakly charged chains. Ha and Thirumalai (HT)[25] used a self-consistent variational theory to calculate the total persistence length of a polyion. The theory is general for both flexible and stiff chains. They found that

$$l_p \sim \begin{cases} l_o + l_{\mathrm{OSF}}, & \text{if } l_{\mathrm{OSF}} \ll l_o \\ (l_o \omega_c)^{1/2} \lambda_D, & \text{if } l_{\mathrm{OSF}} \gg l_o \end{cases}. \tag{4.11}$$

Therefore, for stiff chains ($l_o \gg l_{\mathrm{OSF}}$), the electrostatic persistence length l_e is $l_{\mathrm{OSF}} \sim \lambda_D^2$, whereas for flexible chains ($l_{\mathrm{OSF}} \gg l_o$), l_e is not much different from l_p and scales as λ_D^1.

The effect of valence in driving collapse of RNA was vividly illustrated using small-angle X-ray scattering (SAXS) experiments on a ~200 nucleotide *Azoarcus* ribozyme.[13] From the measured scattering intensity they obtained the distance distribution, $P(r)$. The dependence of the square of the radius of gyration, which corresponds to the second moment of $P(r)$, depends sensitively on the valence of the counterion (see Fig. 4.1). The ribozyme is extended at

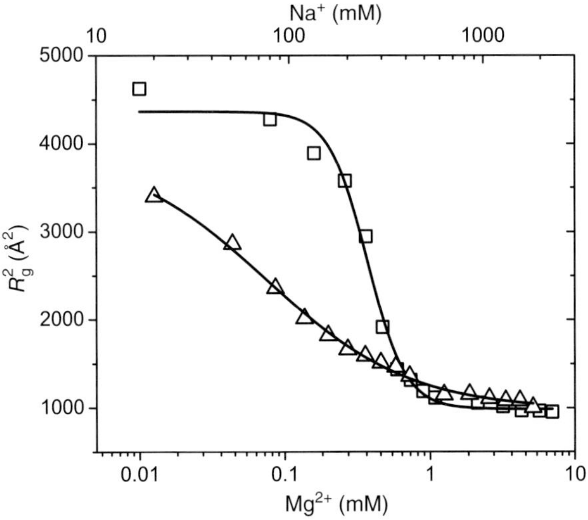

FIGURE 4.1. The dependence of R_g^2 on Mg^{2+} (squares) and Na^+ (triangles) concentration. Solid lines are the fits using the Hill equation in the form $A\left\{1-[Mg^{2+}]^n / \left(C_m^n + [Mg^{2+}]^n\right)\right\} + y_0$, where A, n, and y_0 are adjustable parameters. The best fit values for n are 3.33 and 1.20 for Mg^{2+} and Na^+, respectively. Clearly, cooperativity of the folding transition increases as the valence of ions increases. (Reproduced from Caliskan et al.[13])

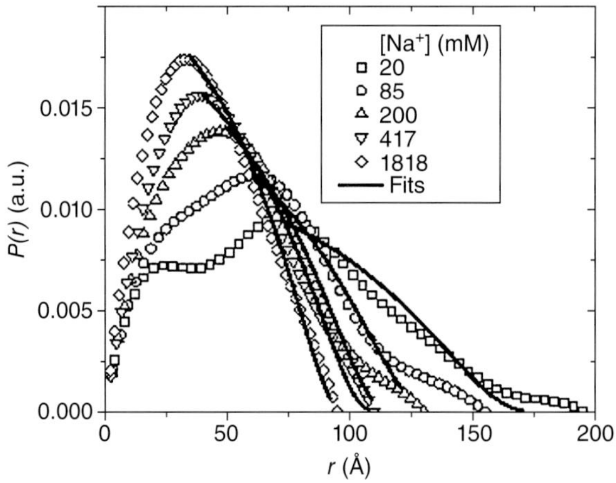

FIGURE 4.2. Distance distribution functions $P(r)$ at various Na^+ concentrations at 32°C are obtained by inverting the measured scattering intensity by a Fourier transform. The solid lines are fits of $P(r) \sim \exp(-(1/(1 - x^2)))$, which is the expected asymptotic result for WLC. The concentrations of counterions are given in the insets. (Reproduced from Caliskan et al.[13])

low cation concentrations and is compact at elevated values of the counterion concentration (Fig. 4.1). The collapse transition is highly cooperative in Mg^{2+} and is much less so in Na^+ (Fig. 4.1). It was found that $P(r)$ (Fig. 4.2) could be well fit using the asymptotic form for the end-to-end distribution for WLCs.[26] The experiments showed that the persistence length of this RNA changes dramatically from about 3 nm at low ionic concentration to about 1 nm at high-salt concentrations. More importantly, for both Na^+ and Mg^{2+}, the persistence length scales as λ_D^2, which is consistent with the OSF theory. From this finding, we find that the intrinsic persistence length of RNA is ~1 nm (Fig. 4.3).

The evidence of the linear dependency of the electrostatic persistence length on λ_D is provided in a number of experimental studies. Tricot[14] analyzed the intrinsic viscosity-molecular weight dependence of a number of PEs such as CMC in solutions with I ranging from 0.005 to 1.0 M. The linear dependency is found for other types of charged polymers, whose total persistence length is on the order of 10 nm or less (see Fig. 4.4). Tinland et al.[16] measured the self-diffusion coefficient of ssDNA fragments using fluorescence recovery after photobleaching to infer the persistence length. The fitted total persistence length measured in angstroms is found to be $l_p = 6.4210^{-8} + 4I^{-1/2}$, with I measured in molars. Perhaps, the most convincing evidence for the linear variation is found in recent single-molecule stretching experiments. Saleh et al.[17] combined single-molecule stretching data of ssDNA at very low forces with scaling arguments to obtained $l_e \sim I^{-0.51\pm0.04}$ or $l_e \sim I^{-0.40\pm0.04}$. The similar behavior is once again reproduced using a general property of the force–extension curve (FEC) at point where the relative extension is about 1/2 by

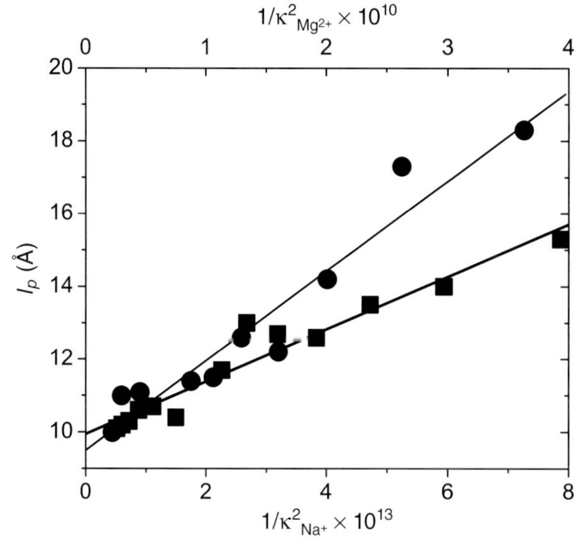

FIGURE 4.3. Dependence of l_p on $1/\kappa^2$ in Mg^{2+} (solid circles) and Na^+ (solid squares). Lines represent fits to the data. Note that the $1/\kappa^2$ for Mg^{2+} is given on top. (Reproduced from Caliskan et al.[13])

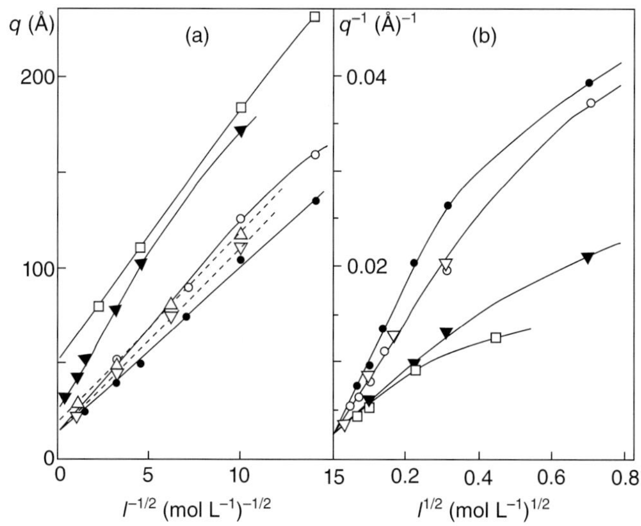

FIGURE 4.4. Variations of the experimental persistence length with the reciprocal of the square root of the ionic strength for sodium poly[((acrylamido)methyl) propanesulfonate] (▼), sodium (carboxymethyl)cellulose (□), sodium polyacrylate (○), sodium poly(styrenesulfonate) (●), and the sodium salt of an alternating copolymer of isobutyl vinyl ether and maleic anhydride at two degrees of neutralization $\alpha = 0.5$ (△) and at $\alpha = 1.0$ (▽). (Reproduced from Tricot.[14])

Toan and Thirumalai.[20] Toan and Micheletti[19] also found the linear dependence for synthetic RNA by fitting the FEC data by Seol et al.[12] using the Thick chain model.[19,27]

4.3 PAs

In this section, we present a simple physical model for describing a PA chain, especially the weakly charged case, with a random distribution of opposite charges, in an electrolyte solution (e.g., NaCl dissolved in water). Here, we do not attempt to elaborate on the known results in the literature.[4,5] There has been much progress in understanding the physical properties of a PA chain under various conditions. For instance, the shape of a PA chain has been studied extensively for varying solvent quality (good vs. poor solvent) and for different charge distributions (e.g., a varying degree of excess charge) (see Higgs and Joanny[4] and Dobrynin et al.[5] and references therein). However, the earlier effort has been focused on the low salt and high salt limits.[4,5] Our main motivation here is to offer a unified picture in which both limits are integrated coherently. In appropriate limits, our description reproduces known results. Also, in our approach, the crossover between the two limits is captured in a more transparent manner. This effort will be beneficial for further illustrating the various competing effects in shaping a PA chain and for offering a more coherent picture of such a system.

4.3.1 Charge Correlations and Screening

Much of our discussion will rely on the concept of "screening," which is well understood for a simple electrolyte (e.g., NaCl dissolved in water).[28] Select any charge and place it at the origin. Since the entire system is electrically neutral, the charge at the origin will be surrounded and shielded by the ionic cloud of the opposite charge. If the energy of this system tends to shrink the ionic cloud,[1] the entropy opposes this tendency. This means that the thickness of the ionic cloud depends on the ion concentration and the temperature.[28] This idea can be extended to PA charges. While both species (i.e., PA charges and salt ions) contribute to screening, their roles in shaping the PA chain are opposite. Along this line, it proves useful to distinguish between "self-screening" (screening of a PA charge by other PA charges) and salt screening (i.e., screening of PA charges by salts). Here, self-screening refers to the tendency of opposite charges on the chain to be spatially correlated. This effect is responsible for the electrostatic compaction of a PA chain (under the right conditions) and thus competes with the latter effect since the surrounding salt ions tend to diminish this tendency[4,5]; there is a competition for the spatial correlation of

[1] Despite the positive energy stored in the ionic cloud, the total energy is negative and favors the formation of a denser cloud.

a PA charge with salt ions. Despite the difference between self-screening and salt screening, the electrostatic free energy of a PA chain can be obtained by treating both kinds of charges as forming a DH solution (or an electrolyte at the low electrostatic coupling limit[2]). This is most evident within the theoretical framework known as random phase approximation.[29,30] The level of approximations assumed for this is similar to that for the DH approach. At this level, chain connectivity does not influence the electrostatic free energy of such a (randomly charged) PA chain (for a given chain size).[29,30] Obviously, nonelectrostatic terms are different for the PA chain and for the surrounding electrolyte.

Consider a PA chain consisting of N monomers of size b each, which is inside an imaginary volume V of radius R. The PA chain is assumed to be weakly charged. Let $f_\pm$ be the fraction of positively and negatively charged groups. Then $f_\pm \ll 1$. For a neutral chain, $f_+ = f_-(\equiv f)$. Let $n_\pm$ be the total density of positive and negative ions whether they are free or on the chain; let ε be dielectric constant of the solvent and k_BT the thermal energy. The Bjerrum length, at which the coulomb energy of two electronic charges is equal to the thermal energy k_BT, is then expressed as $l_B = e^2/\varepsilon k_BT$. (Here, we adopt Gaussian units as in Higgs and Joanny.[4] and Dobrynin et al.[5]) Finally, the Debye screening length, κ^{-1}, is given by the relation $\kappa^2 = 4\pi l_B(n_+ + n_-)$. Outside V, $n_\pm$ tends to a constant, n_0 ($2n_0$ is the salt concentration at infinity). Inside V, however, the salt ion concentration can be perturbed by the PA charges. Nevertheless, we write as $\kappa^2 = \kappa_{\mathrm{PA}}^2 + \kappa_0^2$ inside V, where $\kappa_{\mathrm{PA}}^2 \sim Nfl_B/R^3$ and $\kappa_0^2 = 8\pi l_B n_0$. Our DH approach will reproduce known results based on a more elaborated approach, in which the salt concentration inside V is determined more systematically.[4]

The electric potential due to a point charge e is given by

$$\Psi(\mathbf{r}) = \frac{e}{\varepsilon}\frac{e^{-\kappa r}}{r}. \tag{4.12}$$

This is a potential screened by both PA charges and salt ions. To isolate the contribution of PA charges to screening, one has to subtract the salt contribution as follows:

$$\begin{aligned} \Psi_{\mathrm{PA}} &= \lim_{r\to 0}\left[\Psi(r) - \frac{e}{\varepsilon}\frac{e^{-\kappa_0 r}}{r}\right] \\ &= -\frac{e}{\varepsilon}(\kappa - \kappa_0). \end{aligned} \tag{4.13}$$

[2] Crudely speaking, the low electrostatic coupling means that the interaction between charges is weaker than the thermal energy k_BT. More precisely, the electric energy of a PA chain per charge in Equation 4.14 (and the energy of the surrounding salt ions per ion) should be smaller in magnitude than k_BT. In the low salt limit, one can find $\kappa_{\mathrm{PA}}\ell_B < 1$. This is the condition required for a weakly charged PA chain.

The electrostatic energy gain per charge due to self-screening or the polarization free energy of the PA chain is

$$\frac{E_{\mathrm{PA}}}{k_B T} = \frac{1}{2}\frac{e\Psi_{\mathrm{PA}}}{k_B T} = -\frac{1}{2} l_B \left(\sqrt{\kappa_{\mathrm{PA}}^2 + \kappa_0^2} - \kappa_0\right). \tag{4.14}$$

The factor 1/2 is to avoid double counting of interaction pairs. The total polarization free energy of the PA chain can be written as

$$\frac{F_{\mathrm{PA}}}{k_B T} \sim -Nfl_B \left(\sqrt{\frac{Nfl_B}{R^3} + \kappa_0^2} - \kappa_0\right), \tag{4.15}$$

where κ_{PA} is expressed explicitly in terms of PA parameters. Note here that the electrostatic free energy differs from the energy contribution by a numerical prefactor. Clearly, this free energy favors chain collapse.

4.3.2 Flory Theory: A Θ Solvent Case

In Flory theory, the nonelectrostatic free energy can be expanded in powers of the monomer density $\rho = N/R^3$ as

$$\frac{F}{k_B T} \sim \left(vb^3\rho^2 + wb^6\rho^3\right) R^3, \tag{4.16}$$

where v and w are the second and third virial coefficient, respectively.[31] The total free energy is the sum of F_{PA} and F:

$$\frac{F_{\mathrm{total}}}{k_B T} \sim \left(vb^3\rho^2 + wb^6\rho^3\right) R^3 - Nfl_B \left(\sqrt{\frac{Nfl_B}{R^3} + \kappa_0^2} - \kappa_0\right). \tag{4.17}$$

Here, we mainly focus on a θ solvent in which $v = 0$, even though our general approach can readily be extended to other cases (see Higgs and Joanny[4] and Dobrynin et al.[5]). For $v = 0$, we show that our approach reduces to existing ones in low salt and high salt limits. In this case, the two-body term, that is, the first term in Equation 4.16, vanishes. The equilibrium size of a PA chain is then determined by the balance between F_{PA} and the three-body term, that is, the second term in Equation 4.16.[4] The equilibrium R satisfies

$$\left(\frac{Nfl_B}{R^3}\right)^2 \left(\frac{Nfl_B}{R^3} + \kappa_0^2\right)^{-1/2} - wb^6 \left(\frac{N}{R^3}\right)^3 = 0. \tag{4.18}$$

Because of the functional form of this relation, one can easily find that $R \sim N^{1/3}$. The dependence of R on other parameters is more involved. After analyzing the relation in Equation 4.18 in the two limiting cases, $\kappa_0 \to 0$ and $\kappa_0 \to \infty$, we will present our numerical solution for the intermediate range of κ_0. For simplicity, we set $w = 1$.

Low Salt Limit $\kappa \to 0$. In this case, Equation 4.18 reproduces the known result[4]

$$R \sim N^{1/3} b \left(\frac{b}{f l_B} \right)^{1/3}. \tag{4.19}$$

The chain size decreases as f increases, as expected based on the following picture. For larger f, the effect of self-screening or the PA effect is more pronounced and shrinks the chain size more effectively.

High Salt Limit $\kappa \to \infty$. In this case, Equation 4.18 reduces to

$$\left(\frac{N f l_B}{R^3} \right)^2 \frac{1}{\kappa_0} - b^6 \left(\frac{N}{R^3} \right)^3 = 0. \tag{4.20}$$

This reproduces the known result for the high salt limit[4]

$$R \sim N^{1/3} b \left(\frac{\kappa_0 b^3}{f^2 l_B^2} \right)^{\frac{1}{3}}. \tag{4.21}$$

The chain is swollen by added salts. This finding is consistent with our earlier assertion that self-screening competes with salt screening.

Crossover At the crossover between the two limits, the PA size expressed in Equation 4.19 becomes comparable with that in Equation 4.21. This leads to the following crossover condition[4]:

$$\frac{f l_B}{b^2} \sim \kappa_0. \tag{4.22}$$

In fact, the term on the left-hand side is the reciprocal of the self-screening length, κ_{PA}, for a PA in the low salt limit:

$$\kappa_{PA} \sim \sqrt{\frac{N f l_B}{R^3}} \sim \frac{f l_B}{b^2}. \tag{4.23}$$

The crossover condition thus reads $\kappa_{PA} \sim \kappa_0$.

For the intermediate salt range, we have solved Equation 4.18 numerically and plotted our results in Figure 4.5. Our results in Figure 4.5 show how added salts expand the PA chain. A key to understanding this behavior is that the tendency for PA charges to be correlated is weakened by the surrounding salt

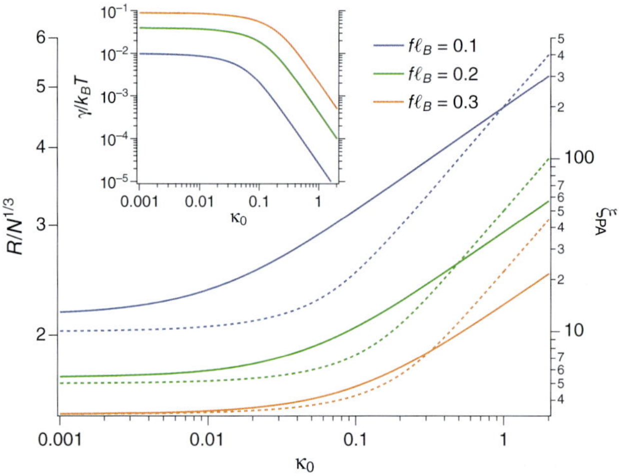

FIGURE 4.5. The equilibrium size of a polyampholyte chain as a function of κ_0 for a few choices of f: $fl_B = 0.1, 0.2, 0.3$. Here, all lengths are given in units of b. Our results show how the PA chain is swollen by salt screening, which tends to weaken the polyampholyte effect: The competition for the screening of PA charges by salt ions is more pronounced at high salt concentrations. Also plotted is the size of polyampholyte blobs, ξ_{PA} (the same color scheme used for R-κ_0 graphs is adopted). Our result for R and ξ_{PA} are useful for illustrating the crossover from the low salt to the high salt limit. Finally, we have plotted the surface tension of a PA chain, $\gamma \sim 1/\xi_{PA}^2$ as a function of κ_0 (see the relevant discussion around Eq. 4.26).

ions since they are equally correlated with the salt ions. This competition underlies the κ_0 dependence of R shown in Figure 4.5.

In summary, our DH approach combined with the Flory theory offers a unified framework for describing the equilibrium properties of a PA solution in the presence or absence of added salts. It not only reproduces known results in the low salt and high salt limits but also shows the crossover between the two limits. In the next section, we complement this approach using the blob picture of a PA chain.

4.3.3 Blob Picture of a PA Chain

The notion of blobs has been proven to be useful for illustrating various aspects of polymer systems.[31] The section is devoted to presenting a blob picture of a PA solution, offering a "visual" guide to the approach presented in earlier subsections. For a weakly charged PA chain, which we consider here, the chain statistics at short length scales will not be perturbed by the PA effect. Within this length scale, the PA chain resembles an ideal chain or a random walk (in a θ solvent). Beyond this, the PA effect will collapse the chain. This means that there exists a special length, denoted as ξ_{PA}, that separates these two regimes. To obtain this length, consider a section of the chain consisting of g monomers. For sufficiently small g (>1), the size of this section is given by $r \sim bg^{1/2}$. The free energy of this section is dominated by the entropic term r^2/gb^2. As g increases, the PA term becomes more important. To set up the crossover condition, find g at which the two terms are equally important:

$$\left(\frac{r^2}{gb^2}\right)_{r \sim bg^{1/2}} \sim gfl_B \left(\sqrt{\frac{gfl_B}{r^3} + \kappa_0^2} - \kappa_0\right)_{r \sim bg^{1/2}}. \tag{4.24}$$

This leads to

$$gfl_B \left(\sqrt{\frac{fl_B}{g^{1/2}b^3} + \kappa_0^2} - \kappa_0\right) \sim 1 \tag{4.25}$$

The g value that satisfies Equation 4.25 is the minimum number of steps the random walk has to take until it starts to feel the PA effect. The blob size is then $\xi_{\text{PA}} \sim bg^{1/2}$. For $N \gg g$, the chain can be viewed as a compact stack of many such blobs. This means that $R \sim \xi_{\text{PA}}(N/g)^{1/3} \sim bg^{1/6}$.

Low Salt Limit As $\kappa_0 \to 0$, Equation 4.25 results in $g \sim b^2/(fl_B)^2$ and $\xi_{\text{PA}} \sim b^2/fl_B \sim \kappa_{PA}^{-1}$, consistent with Higgs and Joanny.[4]

High Salt Limit For large κ_0, Equation 4.25 reduces to $g \sim \kappa_0^2 b^6/(fl_B)^4$. The length equivalent is then $\xi_{\text{PA}} \sim \kappa_0 b^4/(fl_B)^2 \sim \kappa_0/\kappa_{\text{PA}}^2$, in agreement with Higgs and Joanny[4].

In both limits, this blob picture reproduces the results for R in Equations 4.19 and 4.21 obtained by free energy minimization. Also, one can show that $F_{\text{PA}} \sim k_BT \times (N/g)$. This is simply the free energy cost for "redirecting" the random walk so as to fill the space compactly; hence, k_BT per blob.

Intermediate Salt Concentration For the intermediate range of κ_0, g has been obtained numerically and plotted in Figure 4.5 (see the dotted lines) for a few choices of f; we have used the same color scheme as for the R-κ_0 plot. As shown in the figure, ξ_{PA} grows in magnitude as κ_0 increases. This is paralleled with the earlier finding that the PA chain is swollen by salt.

Surface Tension Our DH approach to a PA chain is based on the assumption that N is arbitrarily large. In this case, the "surface effect" is minimal, as long as the PA chain is overall spherical. Under different conditions, however, this picture may break down.[5] For instance, if stretched by an external force or a net charge repulsion, a PA chain will break into many smaller subunits. An important contribution to the PA free energy arises from the fact that PA charges on the periphery (the surface of the chain) are less effectively screened by other PA charges than those inside. This unfavorable free energy cost per unit area is called the surface tension, denoted as γ. Following the scaling approach adopted in Dobrynin et al.[5], γ is the number of blobs on the surface per unit area and is readily given by

$$\frac{\gamma}{k_B T} \sim \frac{1}{\xi_{\mathrm{PA}}^2} \sim \begin{cases} (fl_B)^2/b^2, & \text{low salt} \\ (fl_B)^4/(\kappa_0 b^4)^2, & \text{high salt} \end{cases}. \tag{4.26}$$

Finally, we have plotted the surface tension γ as a function of κ_0 in Figure 4.5 (see the inset). Since γ is a direct result of PA effects, its magnitude is diminished as κ_0 increases, as shown in the figure.

While the notion of the surface tension will be useful in some contexts (e.g., the formation of a necklace globule[5]), we only focus on overall spherical PAs here. An excess charge on a PA chain can drive a conformational transition to a pearl necklace structure from a more spherical globule (referred to as a PE effect). This may be realized at a low salt limit. It is worth noting that the low salt limit here is different from that for the PA effect since the PE effect is longer ranged. Salt screening is not felt sensitively by the PA effect, as long as $\kappa_{\mathrm{PA}} \sim fl_B/b^2 \gtrsim \kappa_0$, which is N-independent. On the other hand, the PE effect will be screened unless $R\kappa_0 \lesssim 1$ or $\kappa_0 \lesssim 1/R$. For the long-chain case, this imposes a very strict condition on salt concentrations. In conclusion, the low salt limit for the PA effect can be realized in a wide range of salt concentrations, while that for the PE effect is prohibitively narrow, as long as $N \gg 1$. We thus focus on overall "spherical" PAs, whether they are collapsed or swollen (e.g., self-avoiding walk chains).

4.4 PAS WITH EXCESS CHARGES

For several reasons, PA chains can carry an unbalanced, excess charge. Recall $f_\pm$ is the fraction of positive and negative charges on the PA chain. Then the chain will carry a net charge unless $f_+ = f_-$. The net charge repulsion now enters into the picture and competes with the PA effect. To focus on this PE effect, smear out all the charges on a PA chain, which is assumed to be inside an imaginary volume V. The excess electrostatic energy stored in this volume is expressed as

$$\begin{aligned} \frac{E_{\mathrm{PE}}}{k_B T} &= \frac{l_B}{2}(f_+ - f_-)^2 \rho^2 \iint_{\mathbf{r},\mathbf{r}' \in V} \frac{e^{-\kappa|\mathbf{r}-\mathbf{r}'|}}{|\mathbf{r}-\mathbf{r}'|} d\mathbf{r} d\mathbf{r}' \\ &\sim V l_B (f_+ - f_-)^2 \rho^2 \left(\frac{1}{\mathbf{k}^2 + \kappa^2} \right)_{\mathbf{k}=0} \\ &\sim \frac{l_B (f_+ - f_-)^2}{\kappa^2} \rho^2 R^3. \end{aligned} \tag{4.27}$$

Here, the subscript of κ_0 was dropped for simplicity; $\mathbf{k}$ is the Fourier conjugate to $\mathbf{r} - \mathbf{r}'$. Only $\mathbf{k} = 0$ contributes to E_{PE} since the monomer density is assumed to be uniform inside V. The PE free energy thus scales as

$$\frac{F_{\mathrm{PE}}}{k_B T} \sim \frac{l_B (f_+ - f_-)^2}{\kappa^2} \rho^2 R^3. \tag{4.28}$$

This estimate is relevant for the case $R \gg \kappa^{-1}$. It is tempting to relate this to the two-body term in Equation 4.16 and interpret the PE term as renormalizing the second virial coefficient v, as in Higgs and Joanny.[4] Strictly speaking, this reasoning is valid only if each "chain segment" remains spherical, in the sense that its width is equal to or comparable with its length. A crucial concept that describes the PE effect on chain shape is the electrostatic persistence length, denoted here as l_e, which was originally introduced over three decades ago.[2,3] While this concept has been widely used in the literature, ironically, there has been a controversy over whether $l_e \sim \kappa^{-2}$, as originally predicted[8,32,33] (also see references therein).

With this subtlety in mind, let us insist on using the PE term in Equation 4.28 and compare it with the PA term in Equation 4.15. These two effects are comparable if $F_{\mathrm{PE}} \sim |F_{\mathrm{PA}}|$. For the case $\kappa_0 > \kappa_{\mathrm{PA}}$, this implies that

$$(f_+ + f_-)^2 l_B \sim (f_+ - f_-)^2 \kappa^{-1}. \tag{4.29}$$

(One can arrive at this by considering Equation 4.11 in Higgs and Joanny.[4]) For a random PA chain, $f_+ = f_-$ and $|f_+ - f_-| \sim 1/\sqrt{N}$. Despite some uncertainty in the κ dependence of Equation 4.29, it is clear that the PE effect can be easily dominated by the PA effect. For different solvent chemistry or polymerization processes (i.e., $f_+ \neq f_-$), however, the PE term can be dominant and will expand the chain. As a result, $R \sim N^{\nu}$, where $\nu \approx 3/5$ if $R\kappa \gg 1$ or $\nu = 1$ if $R\kappa \lesssim 1$. We will not attempt to refine earlier results for PA chains with excess charges, especially in regard to the κ dependence of their size. As more results become available, they can be incorporated into our approach.

4.5 ELASTIC RESPONSE OF FLEXIBLE PEs

The response of flexible PEs such as ssDNA and RNA to mechanical force has further given a fundamental understanding of elasticity of PEs and has further clarified the changes in persistence length as the ionic concentration is varied. In fact, double-stranded DNA molecules, which as we have seen are intrinsically stiff PEs, have long been the major polymers of interest in single-molecule stretching experiments.[34–37] For dsDNA and related stiff PEs, a WLC model readily explains the measured force–extension data[35–37] as well as the distribution of the end-to-end distance.[38] For flexible PEs, however, it has been difficult to account for the data using polymer models alone, thus raising the possibility that microscopic structures might matter. For these systems, it is probably the strong interplay between the electrostatic effects and the small intrinsic stiffness of the molecules that complicates the physics of stretching. For ssDNA and RNA in particular, base pairing between pairs of nucleotides along the chain backbone could be an additional complicating factor. Here, we will just focus on the situations in which the base pairing interactions are negligible.

We begin with a survey of experiments to indicate the diversity of responses of charged flexible PEs. Dessinges et al.[39] used magnetic tweezers to obtain FECs for 11-kb ssDNA molecules, in the range of force from 0.05 pN to about 50 pN. In order to extract the structural information of the molecules, the authors used an extensible freely jointed chain with electrostatic interactions. The electrostatics is modeled using the DH potential with an effective charge density, ν, along the backbone first described by Zhang et al.[40]:

$$E_{\text{elec.}} = \frac{\nu^2}{\varepsilon}\int ds_i ds_j \frac{\exp\left(-\frac{r_{ij}}{\lambda_D}\right)}{r_{ij}}, \tag{4.30}$$

where $\varepsilon = 80$ is the dielectric constant of water, r_{ij} is the spatial distance between two points i and j, and the double integral is taken along the chain contour. The force–extension data for this model is determined through Monte Carlo simulations. It appears that the model reproduces well stretching data in several different ionic conditions. In particular, in solution of 10 mM phosphate buffer with $\lambda_D = 1.87$ nm, where the base pairing interactions are suppressed, the FEC of the ssDNA appears to be almost straight for at least two decades of force in the log-linear scales (see fig. 4 of Dessinges et al.[39]). At the values of $\nu = 1.28e/\text{nm}$ and Kuhn length of 1.6 nm, the theoretical FEC follows very well the experimental data. The extensible WLC and the extensible freely jointed chain models without electrostatic interactions can only fit the FEC for forces at least 20 pN or even higher. Even so, the persistence length obtained with the extensible WLC model is too small, $l_p = 0.21$ nm. (It should be noted that although the FEC portion for $f \geq 20$ pN was plotted in the same figure, the data had actually been obtained previously for overstretched dsDNA in a solution at a much higher ionic concentration of 150 mM Na^+ by Rief et al.[41])

In 2004, Seol et al.[12] built synthetic RNA constructs made of only uracils, or poly(U) (no possibility for base pair interactions) and examined their elastic properties using optical tweezers. The solutions considered have the ionic strength ranging from 5 to 500 mM, and the force range is from about 0.5 to 50 pN. The authors used, instead, a modified extensible WLC model with electrostatic interactions where the relative extension is given by[6,36]

$$x = 1 - \int \frac{dq}{2\pi} \frac{1}{l_p q^2 + f/K_B T} + \frac{f}{S}, \tag{4.31}$$

where S is the stretch modulus in unit of force (pN) and the scale-dependent persistence length is $l_p = l_0 + l_{OSF}\tilde{\kappa}(q)$. Thus, it is assumed a priori that the electrostatic persistence length is given by the OSF theory. The modified freely joined chain (FJC) model with electrostatic interactions could not fit the data well at high ionic concentrations. However, the WLC model appears to reproduce the experimental FEC at high concentration of Na^+ down to about 10 mM. The resulting persistence length does not depend on the force scale for concentration up to 500 mM, whereas it decreases by twofold from $f = 1$ pN to about 50 pN.

More recently, Saleh et al.[17,18] used magnetic tweezers to stretch ssDNA molecules that are specially treated to avoid base pairing even at the lowest forces. The ionic concentrations were varied over a broad range (20–5000 mM of Na^+), and the minimum force is about 0.08 pN, whereas the maximum force is similar to those in the aforementioned works. Instead of modeling electrostatic through the DH potential, they use the blob picture and scaling arguments[31,42] for self-avoiding chains. Besides, demonstrating for the first time the Pincus regime $x \sim f^{2/3}$ at salt concentrations up to 2000 mM, the data also shows the logarithmic dependence of the extension on the force, as observed earlier.[12,39,40] The two regimes are demarcated by a crossover force f_c at a characteristic extension x_c. When all the FECs at different salt concentrations are normalized by f_c and x_c, they collapse fairly well onto a single master curve. The scaling arguments show that $f_c \sim k_BT/l$ and $x_c \sim (v/l^3)^{1/3}$, where l is the Kuhn length, which is twice the total persistence length, and v is the excluded volume between the Kuhn segments. Since f_c and x_c can be extracted from the FECs, v and l then can be estimated without any assumption on their relation. As a result, $l \sim I^{-0.51\pm0.04}$, which is almost close to $I^{-0.5}$ or the Debye length λ_D. In addition, at $I = 300$ M, there is effectively no excluded volume effect due to the counterion condensation (see next section) that renders the excluded volume v negligibly small (θ condition). Under this condition, the FEC can then be fitted very well using the standard WLC model[36] but not with the FJC model. The extracted persistence length l_p is 0.62 ± 0.01 nm, which can be treated as the "bare" persistence length of ssDNA molecules. The seemingly diverse responses of a single chain can be explained using a unified theory (see below).

In 2006, Toan and Micheletti[19] used their thick chain model, which views any polymer as a tube with uniform thickness, to fit the FEC data in Seol et al.[12] The excluded volume effects along the chain backbone lead to an effective persistence length. The extracted parameters are the effective monomer length and the polymer thickness, both of which can be combined in a simple formula to produce the persistence length. Besides producing a good fit to the data for almost all salt conditions, the model shows that the effective thickness of RNA varies linearly with the Debye length. More importantly, it has been shown that the resulting persistence length can be fitted with $l_p = l_0 + c\lambda_D$, where $l_0 \approx 0.66$ nm and $c = 0.43$ is a constant. Thus, in contrast to the results by Seol et al.,[12] which used the WLC model with electrostatic interactions that assumed an OSF-type persistence length, the dependence here is linear or somewhat sublinear.

In a recent paper published in 2010, Toan and Thirumalai[20] used a combination of geometrical arguments and scaling theory to derive a model-free unified theory for semiflexible polymer stretching at high force. The theory can be used for PEs in moderate to high salt concentrations. One of the main results of the theory is that the apparent elasticity of a polymer is inherently force-dependent. The FEC will appear to be that of a WLC model in the force range of $\mathcal{F}_l = k_BT/l_p$ to $\mathcal{F}_h \approx 4k_BTl_p/b^2$, with b being the monomer length, and the polymer behaves as an FJC for forces higher than $\mathcal{F}_h$. Although similar results for WLC-like models were obtained previously,[43–45] the results are general and hold good for any semiflexible chain regardless of the details of the monomer–monomer interactions. The numerical values of the two crossover forces for typical ssDNA and RNA molecules are $\mathcal{F}_h \approx 4$ pN and $\mathcal{F}_h \approx 50$ pN, which turn out to be relevant to the reported discrepancy in the elastic behaviors of structurally similar molecules, ssDNA and RNA. The higher crossover force is about the maximum force than can be achieved in most of the experiments,[12,17,39] which means that the WLC behavior is more likely to be observed. This is also true for the data cited in Dessinges et al.,[39] where the FJC fits well the data above 50 pN (in fact, the data is from another work[41] done at a different ionic concentration [see above]). In fact, fitting the same high force data using a generalized formula for both the WLC and FJC regimes, Toan and Thirumalai[20] show that the fit is reasonably good and produce a physically meaningful persistence length of 0.72 nm. Moreover, the theory naturally leads to a simple rule, called the 1/2 rule, which states that the extension x at force $f = \mathcal{F}_l = k_BT/l_p$ is 1/2. That means the persistence length could be quickly estimated from the FEC without doing any fit by calculating the force at $x = 1/2$. Indeed, the rule does lead to the estimate of the persistence lengths of RNA[12] and ssDNA[17] in different ionic solutions that are similar to each other and also scale linearly in the Debye length (see Fig. 4.6).

4.6 SIMULATIONS OF COLLAPSE OF AN ISOLATED PE CHAIN

A theoretical description of PE collapse is relevant in a number of applications including RNA folding and conformational fluctuations of disordered proteins. The initial event in the folding of RNA is counterion-driven collapse. The kinetics of collapse of this process is driven by a number of factors such as solvent quality, valence, shape, and size of counterions. Although the final structure is unaltered, the pathways in the coil-globule transition is dependent on the details of interactions between ions and RNA. In addition, DNA collapse into toroidal structures can also mediated by counterions. These considerations make the study of collapse of PEs (and the related PAs) important.

Lee and Thirumalai[46] analyzed the collapse of flexible PEs in poor solvents using simulations. In their model, each monomer is approximated as a van der Waals sphere with radius $b/2$ that carries a charge of $-e$. Two successive monomers are connected by a spring with spring constant $3k_BT/b^2$. The PE chain is

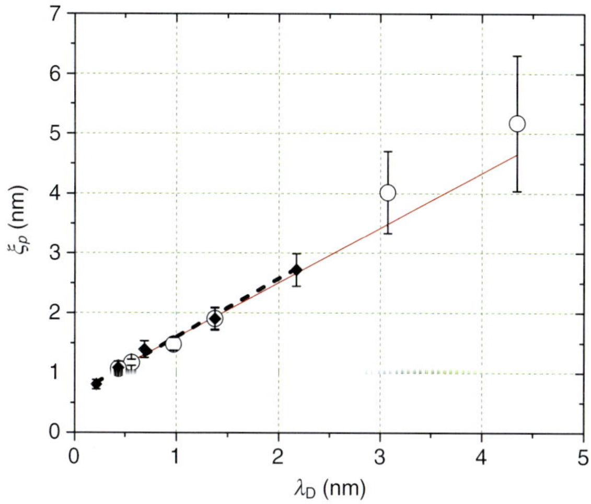

FIGURE 4.6. Analysis of the experimental results. Estimates of $l_p(\lambda_D)$ for poly(U) (circles) and ssDNA (filled symbols) using the 1/2 rule. Linear fits to the data (solid and dashed lines) yielded the bare persistence lengths $l_p \approx 0.67$ nm for poly(U) and $l_p \approx 0.63$ nm for ssDNA. (Reproduced from Toan and Thirumalai.[20])

in a box with counterions, whose valence varies from +1 to +4. The number of ions is such that the total system is neutral. The nonelectrostatic interaction between the particles (monomers or counterions) i and j with radii σ_i and σ_j, respectively, that are at a spatial distance r_{ij} is modeled using the Lennard-Jones potential

$$H_{\mathrm{LJ}}(r_{ij}) = \varepsilon_{\mathrm{LJ}}\left[\left(\frac{r_0}{r_{ij}}\right)^{12} - 2\left(\frac{r_0}{r_{ij}}\right)^{6}\right], \tag{4.32}$$

where $r_0 = (\sigma_i + \sigma_j)/2$. The parameter $\varepsilon_{\mathrm{LJ}}$ is used to control the quality of the solvent, which is expressed in terms of the second virial coefficient υ_2 roughly equal to the excluded volume

$$\upsilon_2 = \int_{\upsilon} \mathrm{d}^3\mathbf{r}\left(1 - e^{-\beta H_{\mathrm{LJ}}}\right) \tag{4.33}$$

For poor solvents or hydrophobic $\upsilon_2 < 0$, and the strength of the hydrophobicity depends on the value of υ_2 (see Fig. 4.7). The number of monomers considered was $N = 240$, for which the effective θ temperature corresponds to $\varepsilon_{\mathrm{LJ}} = 0.5$ and $\upsilon_2 = -1.9b^3$. The coulomb potential is used to model interactions between the charged particles i and j with valences z_i and z_j, respectively,

$$\frac{H_C(r_{ij})}{k_B T} = \frac{l_B z_i z_j}{r}. \tag{4.34}$$

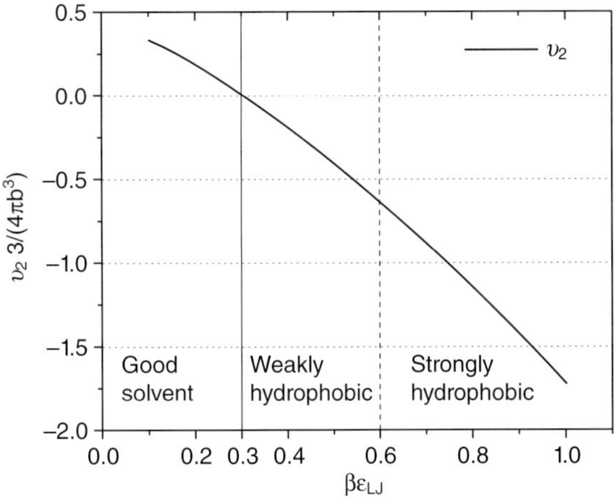

FIGURE 4.7. Second virial coefficient as a function of ε_{LJ} from Equation 4.33. The classification of solvent quality based on the values of υ_2 are shown.

The chain dynamics is followed using Brownian dynamics simulations. The time unit is $\tau = b^2\zeta/2k_BT$, where ζ is the friction coefficient of a monomer.

4.6.1 Effect of Valence and Solvent Quality on Collapse Dynamics

The dynamics of collapse of the PE chain is determined by a balance between hydrophobic interactions and charge renormalized electrostatic potentials. The chain would be in extended conformation if the electrostatic repulsion is dominant. On the other hand, the valence-dependence effective electrostatic attractions between monomers due to the Manning condensation (of counterions onto the chain) could make the chain collapse. In particular, let us consider the case of near θ and hydrophobic conditions.

In near-θ solvent ($\varepsilon_{LJ} = 0.3$ and $\upsilon_2 = -0.06b^3$), the collapse kinetics is monitored through the normalized radius of gyration $R_g(t)/R_c$, where $R_c = 4.08b$ is the size of compact globule (Fig. 4.8). The collapse is strongly dependent on z, and occurs readily for $z = 3$ and 4 within the time of 150τ from initial condition. For $z = 2$, the collapse only happens after 600τ, whereas it is not seen for $z = 1$ even up to the time limit of the simulation. There are two effects of counterion condensation: (1) The overall charge of the polyanion is greatly reduced, and (2) there is an effective attraction due to the excess charge that effectively makes the solvent poor and gives rise to the collapse. In the hydrophobic condition $\upsilon_2 \leq -10.3b^3$, on the other hand, the neutral chain would adopt a compact conformation. Thus, the counterion condensation would further accelerate the collapse. Indeed, with $\varepsilon_{LJ} = 2.0$ and $\upsilon_2 = -25.8b^3$, the collapse already occurs at $t = 200\tau$ for $z = 2$ and at about 600τ for $z = 1$.

The collapse is also found to happen in three stages. The first stage corresponds to the condensation of the counterions to the charged backbone that

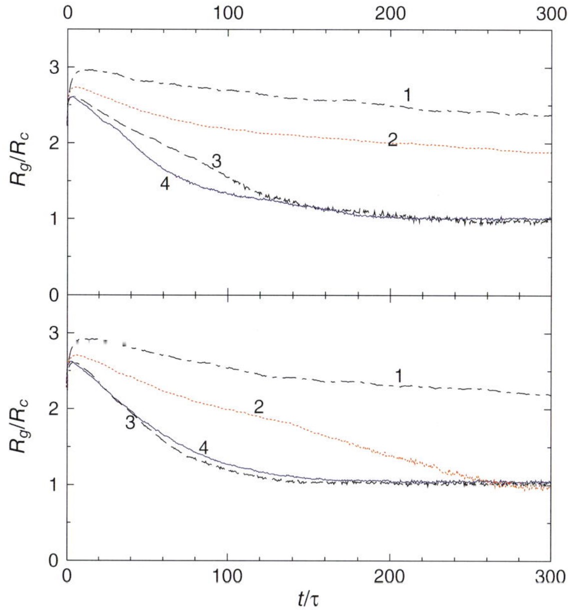

FIGURE 4.8. The time dependence of the radius of gyration following a quench from θ solvent to poor solvent condition. The value of $N = 240$ and $l_B = 5.3b$. Top panel is for $\upsilon_2 = -0.06b^3$, whereas bottom panel is for $\upsilon_2 = -3.69b^3$. The numbers on the curves denote the valence of the counterions.

is driven mostly by diffusion and occurs on the order of 25τ independently of the valance z. The second stage is the formation of pearl necklace structures of globular clusters containing both monomers and counterions. The clusters are mainly local, namely the monomers in a cluster are predominantly neighbors, and are connected by strings. In the third stage, all the clusters merge and the largest cluster grows at the expense of smaller ones (see below for a theoretical explanation).

The collapse mechanisms depend on the valency of the counterions. For $z = 1$, the monovalent counterions when condensed combine with the charges on the backbone to form random dipoles of magnitude $p \approx eb$. When the attraction between two dipole exceeds k_BT, that is, the spatial distance is less than l_B, contacts between distinct segments of chain can form. Because this is a very short distance, the globular clusters are predominantly local and their sizes are very small, while the number of them is large. Thus, it takes a very long time to reach the global compact structure. When the counterions are multivalent, they can both neutralize the negative charge on the monomer and provide an excess charge of $|(z - 1)e|$. Bare monomer charges that are separated by a large distance along the contour can be attracted to the positive charge in a process called "ion bridging." The range over which such attractive interactions are effective increases with z. Thus, the size of the initial clusters and the efficiency of the collapse also increase with z.

4.6.2 Phase Diagram in the (υ_2, z) Plane

The observations concerning the structure of the collapsed globules together with simple scaling arguments can be used to construct a valence-dependent phase diagram for strongly charged PEs in poor solvents ($\upsilon_2 < 0$). When counterion condensation takes place, the decrease in the effective charge of the polyanion can be computed by equating the chemical potential of the free and condensed counterions (two-phase approximation). Generalizing the arguments of Schiessel and Pincus[47] to arbitrary z, the authors find that the total charge of the PE decreases from Nf_e (f_e is the fraction of charge, which is 1 here) to $N\tilde{f}_e \approx k(L/l_B)(1/z)$, where $k \approx -\ln\phi$ and ϕ is the volume fraction of the free counterions. The size of the PE is given by $L \approx k^2b^2N/l_Bz^2$, provided $l_B > k^2z^{-2}bN^{1/2}$. As the quality of the solvent decreases to a level such that thermal blob size $\xi_T \approx b^4/|\upsilon_2| < \xi_{el} \approx l_Bz^2/k^2$ (size of the electrostatic blob), then the chain condenses to a globule. The boundary dividing the stretched and collapsed conformation is obtained by equating ξ_{el} and ξ_T and is given by $|\upsilon_2| \approx b^4k^2/l_Bz^2$.

In the globular phase, two regimes are found, one corresponding to the Wigner crystal and the other a Wigner glass. The boundary between the two is obtained by equating the gain in the energy upon condensation ($\approx ze^2/d\varepsilon$) to the attractive interaction due to the poor solvent quality ($kT\upsilon^2b^{-6}$). This leads to the condition $|\upsilon_2| \approx z^{1/2}b^{5/2}l_B^{1/2}$. To determine the boundary more precisely we have to account for the induced attraction between monomers after counterion condensation. Thus, the actual determination of the boundary separating the Wigner crystalline regime and the glassy regime requires equating $ze^2/d\varepsilon$ and $kT\upsilon_{2R}^2b^{-6}$, where υ_{2R} is the renormalized second virial coefficient and d is approximately the distance between the condensed counterion and the backbone charge. This argument shows that the shape of the PE chain requires, in a nontrivial way, the coupling of electrostatic interactions and effects coming from solvent quality. In light of this, the boundary indicated in Figure 4.7 should be regarded as qualitative.

The boundary between the two collapsed phases is difficult to determine from the simulation because of the extreme slow dynamics in the glassy phase. For the purpose of illustrating the phase diagram, we assume that the ionic glass is an equilibrium phase. To validate the phase diagram, extensive simulations were performed for all z (= 1, 2, 3, 4) at values of l_B^{-1} and $|\upsilon_2|$ indicated by asterisks in Figure 4.9. The structural conformations are in qualitative accord with the predicted phase diagram.

4.7 THEORY OF COLLAPSE DYNAMICS

Although there are a number of theoretical models that describe the collapse kinetics of homopolymers with applications to protein folding, there is only one study that has considered the rate of PA and PE collapse. The theory proposed by Lee and Thirumalai[48] (LT) adopts an approach developed by

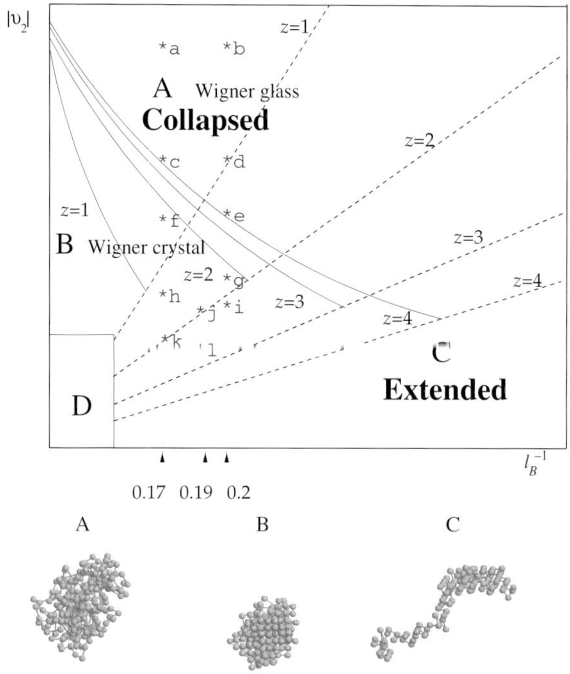

FIGURE 4.9. Valence-dependent diagram of states in the ($|\upsilon_2|$ and l_B^{-1}) plane for strongly charged PE. The dashed lines represent the boundary between stretched and collapsed states and depend on z. The z-dependent solid lines in the collapsed region separate Wigner crystalline region from the Wigner "glassy" region. The asterisks are the simulation results with each data point corresponding to four z values ($z = 1$–4). Pictures of the conformation of the chain in regions A, B, and C are also shown. The l_B^{-1} values are indicated by arrows. The letters (a–l) near the asterisks corresponds to values of $|\upsilon_2|/b^3 = 62.34, 62.34, 25.82, 25.82, 15.20, 12.36, 7.51, 4.47, 3.91, 3.69, 0.34, 0.06$, respectively.

Pitard and Orland[49] to describe uncharged homopolymer collapse in poor solvents. For simplicity, we consider the collapse of PAs, which as we will argue bears a strong resemblance to the behavior expected for PEs as well.

A model system, which may captures some generic aspects of biomolecules, is a PA, which is a linear polymer chain that has both positive and negative charges along the backbone. The very large muscle protein titin associated with sarcomere contains regions that have PA characteristics as do many intrinsically disordered proteins. Because of the simultaneous presence of positive and negative charges, the shape of PAs is determined by competition between the two conflicting interactions. The repulsion between the charges tends to swell the chain (the PE effect), while the attraction (PA effect) tends to collapse the chain. It is known that if the total charge, Q, of PA is less than $Q_c \sim \pm Ne$ (N is the number of monomers), the chain adopts a globular conformation, and is extended otherwise. If $Q > Q_c$, then it can be shown, using an analogy to Rayleigh instability of charged droplets, that the chain can be visualized in terms of pearl necklace structures, which is similar to that found in weakly charged PEs. The LT theory showed that collapse to a global globular structure, which can only take place when $Q < Q_c$, occurs in two major steps. In the first stage, the chain forms a metastable pearl necklace structure reminiscent of the equilibrium structures for $Q > Q_c$. In the second stage, the pearls (domains) merge, leading to the compact globular conformation. The theory for the intermediate time regime (when pearl necklace structures form) was developed by adopting the procedure suggested by Pitard and Orland.[49] The LT theory showed that the square of the radius of gyration decays as

$$R_g^2(t) \approx R_g^2(0)\left(1-\left(\frac{t}{\tau_{PA}}\right)\right)^{\alpha}, \tag{4.35}$$

where $\alpha = 5/4$ and the characteristic time $\tau_{PA} \sim N^{\frac{1}{\alpha}}$. On the timescale τ_{PA}, metastable pearl necklace structures (local structures connected by strings) form. Interestingly, formation of such structures have also been observed in the coil-globule transition in uncharged homopolymers.

4.7.1 Dynamics of PE Collapse

The theory described for PA can qualitatively predict the collapse dynamics of strongly charged PEs. Let a strongly charged PE (fraction of charged monomers is close to unity) chain be initially in the θ solvent with respect to the neutral polymer. At $t = 0$, we imagine a quench to low enough temperatures so that counterions start to condense. Upon condensation of counterions, the conformation of the chain approximately resembles that of a PA. The relevant length scales for PE are the electrostatic blob length $D \approx l_B z^2/k^2$ and $\xi_T \approx a_0(\theta/(\theta - T))$, where z is the valence of counterions, $k = -ln\phi$ (ϕ is the volume fraction of free counterions), and θ is the collapse temperature of uncharged polymer. If ξ_T is not too small, the PE chain undergoes a sequence of structural changes en route to the collapsed conformation. After counterion condensation PE evolves toward a metastable pearl necklace structure. For weakly charged PEs, such structures are the equilibrium conformations in poor solvents. The dynamics of this process can be described using the theory developed to describe collapse of PA. We assume that shortly following the quench to low temperature, the counterions condense onto the PE chain. The timescale for this process is diffusion limited, with values that are much smaller than time in which the macromolecule relaxes. Upon condensation of a multivalent cation ($z \geq 2$), the effective charge around the monomer becomes $((z - 1)e)$. If the locations of the divalent cation are random and if the correlations between the counterions are negligible, then in the early stages, a PE chain with condensed counterions may be mapped onto an evolving random PA. With this analogy, we suggest that the pearl necklace-like structures that are found in the collapse of strongly charged PEs should also be governed by Equation 4.35.

4.7.2 Late Stages of Collapse

At long times, the pearl necklace structures merge to form compact collapsed structures. This occurs by the largest cluster growing at the expense of smaller ones, which is reminiscent of the Lifshitz–Slyozov (LS) mechanism.[50] If this analogy is correct, then we expect the size of the largest cluster $S(t)$ to grow as $S(t) \sim t^{\frac{1}{3}}$. The collapse is complete when $S(t) \sim R_g(t \to \infty) \approx N^{\frac{1}{3}}$, which implies that the characteristic collapse time is ~N.

Langevin simulations for strongly charged flexible PEs where the collapse is induced by multivalent counterions have shown that late-stage coarsening indeed occurs by the LS growth mechanism. Figure 1 in Reference 48 shows a representative example of the conformations that are sampled in the dynamics of approach to the globular state under θ solvent conditions. Both panels show that in the later stages of collapse, the largest clusters in the necklace globule grow and the smaller ones evaporate. This lends support to the proposed LS[50] mechanism. To estimate, we have used simulations to calculate the number of particles that belong to the largest cluster $N_S(t)$ as a function of time. Figure 2 in Reference 48 shows the linear increase of the number of particles $N_S(t)$ at long times from which the growth time increases linearly with N as expected from the LS mechanism. The change of slope for long times is due to the finite size effects and indicates the completion of the globule formation.

The mechanism of approach to the globular state for PA and PE should be similar. The collapsed conformations are reached via metastable pearl necklace structures. For PE, the driving force for forming such structures is the counterion-mediated attractions. Charge fluctuations in PA lead to pearl necklace structures. The lifetime of pearl necklace structures depends on energy barriers separating merged and unmerged clusters. The necklace globule conformation consists of n globules with nearly vanishing net charge that are in local equilibrium. The free energy of the ith globule is $F_i \sim (4\pi/3)\Delta F R_i^3 + 4\pi\sigma R_i^2$, where R_i is the radius of the ith globule and σ is the surface tension. Note that Δf is the same before and after the merger of clusters. The free energy difference between a conformation consisting of two clusters and one in which they are merged is $\Delta F \sim \left(8\pi\sigma - 4(2)^{2/3}\pi\sigma\right)(N/n)^{2/3}$. Charge fluctuation in each globule is $q_0(N/n)^{1/2}$. If the coulomb energy fluctuation of each globule ($\delta E \sim (N/n)q_0^2 / a_0(N/n)^{2/3}$) is less than the free energy difference between the conformation with two separated clusters and the one where they are merged, then the system spontaneously grows to a large cluster.

4.8 CONCLUSIONS

We have only described a few topics dealing with properties of isolated PEs and PAs with some applications to biological systems. There are a variety of issues especially such as the viscosity of PEs that are puzzling and warrant explanation. In addition, bundling of charged systems (actin and microtubule, for example) is another topic with applications in biology that have not been adequately described. Only recently, PE concepts have been applied to understand conformational fluctuations in intrinsically disordered proteins.[51] Detailed theoretical and experimental studies are needed to clarify the role that PE effects and ion–PE interactions play both in the synthetic and biological world.

ACKNOWLEDGMENTS

This work was supported by a grant from the Natural Sciences and Engineering Research Council of Canada to B.Y.H. and the National Science Foundation to N.M.T. and D.T.

REFERENCES

1. J.-L. Barrat and J.-F. Joanny. *Adv Chem Phys* **1996**, 84, 1.
2. T. Odijk. *Polym Sci B Polym Phys* **1977**, 15, 477–483.
3. J. Skolnick and M. Fixman. *Macromolecules* **1977**, 10, 944–948.
4. P. G. Higgs and J.-F. Joanny. *J Chem Phys* **1991**, 94, 1543–1554.
5. A. V. Dobrynin, R. H. Colby, M. Rubinstein. *J Polym Sci B Polym Phy* **2004**, 42, 3513–3538.
6. J.-L. Barrat and J.-F. Joanny. *Europhys Lett* **1993**, 24, 333–338.
7. H. Li and T. A. Witten. *Macromolecules* **1995**, 28, 5921–5927.
8. B.-Y. Ha and D. Thirumalai. *J Chem Phys* **1999**, 110, 7533–7541.
9. R. R. Netz and H. Orland. *Europhys J B* **1999**, 8, 81–98.
10. A. R. Khokhlov and K. A. Khachaturian. *Polymer* **1982**, 23, 1742–1750.
11. C. G. Baumann, S. B. Smith, V. A. Bloomfield, C. Bustamante. *Proc Natl Acad Sci U S A* **1997**, 94, 6185–6190.
12. Y. Seol, G. M. Skinner, K. Visscher. *Phys Rev Lett* **2004**, 93, 118102.
13. G. Caliskan, C. Hyeon, U. Perez-Salas, R. M. Briber, S. A. Woodson, D. Thirumalai. *Phys Rev Lett* **2005**, 95, 268303.
14. M. Tricot. *Macromolecules* **2010**, 17, 1698–1704.
15. W. F. Reed, S. Ghosh, G. Medjahdi, J. Francois. *Macromolecules* **1991**, 24, 6189–6198.
16. B. Tinland, A. Pluen, J. Sturm, G. Weill. *Macromolecules* **1997**, 30, 5763–5765.
17. O. A. Saleh, D. McIntosh, P. Pincus, N. Ribeck. *Phys Rev Lett* **2009**, 102, 068301.
18. D. McIntosh, N. Ribeck, O. A. Saleh. *Phys Rev E* **2009**, 80, 041803.
19. N. M. Toan and C. Micheletti. *J Phys Condens Matter* **1997**, 18, S269–S281.
20. N. M. Toan and D. Thirumalai. *Macromolecules* **2010**, 43, 4394–4400.
21. U. Micka and K. Kremer. *J Phys Condens Matter* **1996**, 8, 9463–9470.
22. U. Micka and K. Kremer. *Europhys Lett* **1997**, 38, 279–284.
23. J.-L. Barrat and J. E. Joanny. *Europhys Lett* **1993**, 24, 333.

24. A. V. Dobrynin. *Macromolecules* **2005**, 38, 9304–9314.
25. B.-Y. Ha and D. Thirumalai. *Macromolecules* **1995**, 28, 577–581.
26. C. Hyeon and D. Thirumalai. *J Chem Phys* **2006**, 124, 104905.
27. N. M. Toan, D. Marenduzzo, C. Micheletti. *Biophys J* **2005**, 89, 80–86.
28. L. D. Landau and E. M. Lifshitz. *Statistical Physics*, 3rd edn. Butterworth Heinemann, Oxford, 1980.
29. J. Wittmer, A. Johner, J. F. Jonnay. *Europhys Lett* **1993**, 24, 263–268.
30. A. V. Dobrynin and M. Rubinstein. *J Phys II (France)* **1995**, 5, 677–695.
31. P.-G. de Gennes. *Scaling Concepts in Polymer Physics*. Cornell University Press, Ithaca, 1979.
32. A. Gubarev, J. M. Y. Carrillo, A. V. Dobrynin. *Macromolecules* **2009**, 42, 5851–5860.
33. M. Fixman. *J Phys Chem B* **2010**, 114, 3185–3196.
34. S. B. Smith, L. Finzi, C. Bustamante. *Science* **1992**, 258, 1122–1126.
35. M. D. Wang, H. Yin, R. Landick, J. Gelles, S. M. Block. *Biophys J* **1997**, 72, 1335–1346.
36. J. F. Marko and E. D. Siggia. *Macromolecules* **1995**, 28, 8759–8770.
37. C. Bouchiat, M. D. Wang, J. F. Allemand, T. Strick, S. M. Block, V. Croquette. *Biophys J* **1999**, 76, 409–413.
38. F. Valle, M. Favre, P. De Los Rios, A. Rosa, G. Dietler. *Phys Rev Lett* **2005**, 95, 158105.
39. M.-N. Dessinges, B. Maier, Y. Zhang, M. Peliti, D. Bensimon, V. Croquette. *Phys Rev Lett* **2002**, 89, 248102.
40. Y. Zhang, H. Zhou, Z.-C. Ou-Yang. *Biophys J* **2001**, 81, 1133–1143.
41. M. Rief, H. Clausen-Schaumann, H. E. Gaub. *Nat Struct Biol* **1999**, 6, 346–349.
42. P. Pincus. *Macromolecules* **1976**, 9, 386–388.
43. J. Kierfeld, O. Niamploy, V. Sa-yakanit, R. Lipowsky. *Eur Phys J E* **2004**, 14, 17–34.
44. L. Livadaru, R. R. Netz, H. J. Kreuzer. *Macromolecules* **2003**, 36, 3732–3744.
45. A. Rosa, T. X. Hoang, D. Marenduzzo, A. Maritan. *Macromolecules* **2003**, 36, 10095–10102.
46. N. Lee and D. Thirumalai. *Macromolecules* **2001**, 34, 3446–3457.
47. H. Schiessel and P. Pincus. *Macromolecules* **1998**, 31, 7953–7959.
48. N. Lee and D. Thirumalai. *J Chem Phys* **2000**, 133, 5126–5129.
49. E. Pitard and H. Orland. *Europhys Lett* **2010**, 41, 467.
50. I. M. Lifshitz and V. V. Slyozov. *J Phys Chem Sol* **1961**, 19, 35–50.
51. S. Mueller-Spaeth, A. Soranno, V. Hirschfeld, H. Hofmann, S. Rueegger, L. Reymond, D. Nettels, B. Schuler. *Proc Natl Acad Sci U S A* **2010**, 107, 14609–14614.

CHAPTER 5

MODELING THE STRUCTURE AND DYNAMICS OF POLYELECTROLYTE MULTILAYERS

JUAN J. CERDÀ, CHRISTIAN HOLM, and BAOFU QIAO

5.1 INTRODUCTION

Polyelectrolyte multilayers (PEMs) are complex structures in which the property of polyelectrolytes (PEs) to complex with oppositely charged PEs is exploited to form a sequence of polymeric layers made of alternating types of PEs on top of a charged surface (see Fig. 5.1). PEMs are usually obtained using the so-called layer-by-layer (LbL) technique,[1–4] which has several variants, among them are as follows:

1. The conventional dipping process where the substrate is dipped into a solution of PEs of opposite charge. In the LbL technique, one sequentially exposes a negative (respectively positive) substrate to a positively (respectively negative) charged PE solution followed by a rinsing step in order to dispose of the supernatant ions and PEs. Then the substrate is dipped into a solution of PE of opposite charge. Repeating this simple procedure can result in a buildup of hundreds of charged layers. A great majority of the current PEMs via the LbL technique are built up using an aqueous solution as the solvent, although the use of other solvents have been explored in some works.[5]
2. An alternative to the traditional method is known as the hydrodynamic LbL.[6] This method has the advantage that, by spinning the substrate, the enhanced mass transport can speed up the deposition process by roughly one order of magnitude.

Ionic Interactions in Natural and Synthetic Macromolecules, First Edition. Edited by Alberto Ciferri and Angelo Perico.

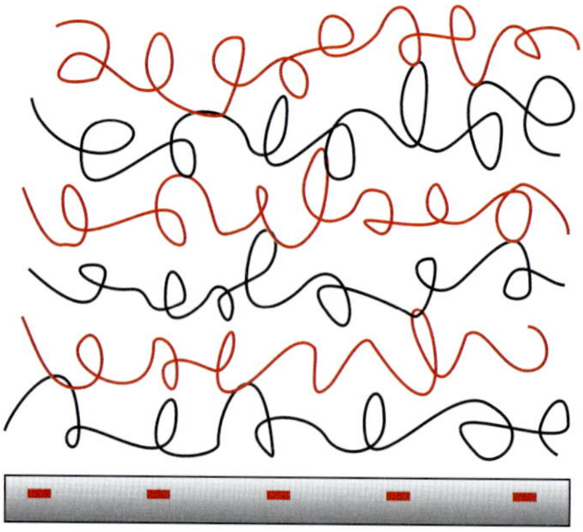

FIGURE 5.1. Schematic representation of a PEM, where the substrate is negatively charged. Polycations and polyanions are in black and red, respectively.

3. In the spin-coating LbL variant,[7,8] drops of PE solution fall on top of a spinning disk that allows a fast removal of the loosely attached material due to the centrifugal force acting onto the material on the surface of the disk.
4. A fourth method to create a PEM is to use a combination of LbL and the spraying technique,[9,10] in which dipping steps have been replaced by a spraying deposition of the surface with an alternating succession of positively and negatively charged solutions.
5. Recently, covalent assembly LbL technology has been developed, which is capable of preparing more chemically and mechanically stable PEMs. See Chapter 17 for a detailed discussion.

From a historical perspective, it was Decher et al.[1,2] in the early 1990s who demonstrated the feasibility of the dipping LbL technique to form stable PEMs. Nonetheless, it should be mentioned that the LbL technique is not only applied to the production of PEMs but also has a much more widespread use (for a review of the technique and its scope, see Ariga et al.[3] and Wang et al.[11]). Those building-up techniques are a vivid example of how one can create complex structures with a degree of control at the nanometer scale using simple macroscopic procedures. The versatility of such processes to create PEMs has allowed the fabrication of thin multilayer films made of synthetic PEs, DNA, lipids, and proteins, which has resulted in a boost of novel applications in recent years. For instance, PEMs have a potential as matrix materials for enzymes and proteins in sensor applications[12] (see also Chapter 17), and also as a matrix for active components in solar cells. PEMs are used as a coating for protecting and controlling the healing process of damaged arteries.[13] In addition, PEMs are used as membranes in fuel cells,[14] for nanofiltration, gas separation, protein purification, and catalysis.[15–17] PEMs are also found in the fabrication of nonlinear optical materials[18] and colored electrochromic electrodes (future display devices), as well as in tailoring the properties of photonic crystals[19] (see Chapter 19 for a detailed review) and photovoltaic cells.[20,21] Other uses of PEMs include analyte separation processes (chromatography)[22] and the fabrication of thin-walled hollow micro- and nanocapsules (see Chapter 17, Khopade et al.,[23] and other references therein). These capsules have a great potential to be drug carriers and nanoreactors. Other uses of PEMs include, for instance, their ability to control the distance between fluorophores and plasmonic nanostructures.[24] Other examples of the huge potential of PEMs for applications include the possibility of doing electrochemical nanopatterning of conducting PEMs, the synthesis of dual intelligent surfaces where PEMs are combined with polymer brushes (see Chapter 17), the fabrication of PEMs designed to be contact-killers of bacteria that do not need aid from other biocidal species,[25–27] and also the creation of liposome-containing PEMs designed to work as biocompatible coatings for local drug delivery.[28] It is also worth mentioning the synthesis and use of PEMs with enantiospecific properties that has received some attention, though its development is still in its infancy.[29,30]

5.2 STATE OF THE ART IN THE STUDY OF PEMs

The current knowledge about PEMs has been summarized in a few reviews.[2,31–37] There exist also some other recent reviews on PEs that discuss their own works on PEMs (see, e.g., Dobrynin[38] and Messina[39]). Glinel et al.[40] and Sukhishvili[41] have covered the advances on responsive PEMs, while Schönhoff et al.[42] have dealt with the hydration and internal properties of PEMs. The development of biomimetic films has been recently assessed by Picart.[43] On the other hand, Sukhishvili et al.[44] have revised the intricate relation between PEMs and complexes that can hamper the formation of PEMs in certain situations. The correlation between the internal structure and dynamics of PE multilayers, on the one hand, and their functional properties, on the other, has been reviewed by von Klitzing.[45] The advances on mechanical properties of PEMs and films with extreme wetting properties have been summarized by Jaber and Schlennoff.[46] Polypeptide multilayer films have been covered by Haynie et al.[47,48] During the last two decades, many works have been devoted to the study of the fundamental properties of PEMs. From the experimental point of view, the current state of the art in the study of the properties of PEMs, which constitute a key reference for those interested in the study of PEMs using models, can be roughly summarized according to the following key topics:

1. *The Buildup of PEMs*: Many studies have been devoted to the elucidation of the factors that lead to a linear or an exponential growth of PEMs.[43,45,49–56] According to Hoda and Larson,[55] a rough idea of the differences in the thickness one can attain depending on the growth mode

followed by the PEM can be obtained considering a PEM made of 20 bilayers: Such a PEM will have a typical thickness $H \in$ (50 nm, 100 nm) if it grows linearly, while its thickness will be typically in the range of $H \in$ (0.1 μm,10 μm) when it follows the exponential growth mode.[55] In our opinion, the most probable mechanism that accounts for the different growth modes involves the chain interpenetration mechanism[50,55,57]: The linear mode is assumed to occur when no PE chains can diffuse within the multilayer, and the exponential mode is expected when chains can diffuse through the multilayers. The level of interpenetration of the chains depends on many factors, and thus, for instance, recent studies[58] have shown that by changing the type of anion used during the poly(styrene sulfonate) (PSS)/poly(diallyldimethylammonium chloride) (PDDA) PEM growth, the growth mode can change from linear to exponential, and in some cases, the resulting growth mode can be in between those. It should be mentioned that an alternative model to explain the exponential growth based on the surface roughness has also been proposed.[59–61] This second model is based on the assumption that if the ratio of the net charge left on the PEM after the adsorption of one layer δQ to the total charge carried by the adsorbed chains during the last deposition step Q is larger than 50%, then the PEM will grow exponentially.[61] Other aspects related to the buildup of PEMs have been also considered during the last years. Interestingly, numerical simulations[36,37] have shown that under certain circumstances a PE bilayer could be a thermodynamically unstable system and only via a fast deposition of a third layer of PE would the system be able to stabilize and lead to the formation of PEMs, being thus a kinetically trapped state. Unfortunately, so far, the aging effects in the first stages of the PEM formation have experimentally received little attention. In addition, several studies have dealt with the effect in the buildup of PEMs of the charge density of the PE[62–70]: Some of them have shown that there seems to be a minimum charge density that is necessary for the formation of stable films. Thus, for instance, Schoeler et al.[64] have shown that the addition of oppositely charged polyions, when the charge density is below a certain critical value they have found, can lead to the removal of almost the whole previously deposited layer and therefore PEMs cannot grow. Nonetheless, some experiments have reported the opposite, that is, the possibility to build up PEMs made with low-charge-density PEs, and Bertrand et al.[66] have challenged the critical charge threshold idea and have suggested that what is required is a proper matching of the charge density of both types of adsorbing PEs. Another aspect that has been considered to affect the buildup is the effect of the type of salt in the PE dipping solution and its ionic strength[62,71–74] as well as the temperature preparation.[73–76] Other issues related to the buildup of PEMs that have also received considerable attention are the way PEs intermix over several layers,[2,74,75] the ion and water distribution inside PEMs,[77–79] the influence of the substrate on the properties of the first layers,[80,81] and the role of nonelectrostatic short-range interactions on the stability and properties of PEMs.[82] Those aspects have been specially addressed for the case of PEMs made of weak PEs[40,44,83–90] in which the crucial influence of pH during the buildup has been thoroughly studied[74,86,91] also including the case of PEMs containing pH-sensitive star polymers.[79,92]

2. *The Stability of the Already Formed PEMs*: An important point in order to understand the mechanical properties of already formed PEMs is to know if they are equilibrium structures or, on the other hand, they are kind of frozen states. From the studies done so far, it seems that the answer is strongly dependent on the type of PEs under consideration and on the environmental conditions affecting the PEM. Thus, poly(ethylene oxide) (PEO)-poly(acrylic acid) (PAA) multilayers have been observed to possess a glass temperature transition that is strongly sensitive to the assembly pH and the ionic strength.

   On the other hand, studies on poly(allylamine hydrochloride) PAH/PAA systems show that such PEMs do not exhibit a glass temperature, and they remain in a glassy state even at elevated temperatures due to some cross-linking reactions that take place when temperature is increased.[93] In turn, Kovacevic et al.[83] have measured the effect of exposing the PEMs to solutions containing both salts and PEs. They have found that below a certain salt concentration the PEM stays in a glassy state and dissolution is not possible. Instead, above the critical salt concentration, solutions composed of salt plus free PEs can manage to dissolve the PEMs though the minimum requested PE concentration for that mixture, depending on the details (length, degree of charge, etc.) of the free PE chains and the details of those PEs that form part of the PEM. It is worth mentioning that the interaction of already formed PEMs with surfactants is a topic that has received marginal attention so far (see Iturri Ramos et al.[94] and references therein). Another topic of research is related to those PEMs made of weak PEs. The pH of the solution in contact with the PEM can largely change the properties of the already built up films.[95] Thus, for instance, for PAH/PSS films, there exists a critical pH below which the PEM breaks and releases a large part of the polymers, roughly retaining just those PEs bound electrostatically to the substrate.

3. *The Mechanical Properties of PEMs*: The stretching properties of several types of planar PEMs have been studied in detail by Mertz et al.[96] In addition, some studies have already dealt with the mechanical stimuli on PEM microcapsules. Studies about the elastic modulus for thermoresponsive PEMs have been performed by Wang et al.[97] The compression ratio of formed PEMs as a function of the type of salt and concentration during assembly has been studied by Gong and Gao[98] In a recent work, Fruh et al.[99] have focused on the changes of orientation of the PEs when

a PEM is under lateral mechanical stress: In such studies, PEMs are observed to suffer a plastic deformation that leads to an irreversible transition from a coiled to an uncoiled state.

4. *Hollow Multilayer Nanocapsules*: The synthesis and study of hollow PEM nanocapsules and their properties has received also considerable attention during the last years (see Chapter 17 and Wang et al.[11], Cui et al.[27], Glinel et al.[40], Schönhoff et al.,[42] and Peyratout and Dähne[100]), including their study via numerical simulations[101] and theoretical approaches,[102] as well as discussions about the pros and cons of their use in drug delivery.[103]
5. *Free Molecules within PEMs*: There are two main areas of study of free particles within PEMs, one of which deals with the study of the uptake and release of those particles,[104,105] and the other of which deals with the immobilization and diffusion dynamics of the molecules, including lateral transport, inside PEMs.[106–109] The mechanisms of diffusion of molecules do depend on the properties of the surrounding and the interactions of the molecules; thus, particles can diffuse through water pockets and channels via hopping to the neighborhood of a charged PE to another neighborhood of similar characteristics, via the following of some PE segmental motion, or via a combination of several of those mechanisms. A related topic is the study of the transport of species through PEMs that act as selective membranes: Interesting progress in this area has been made by Rmaile et al. (see Rmaile et al.[110] and references therein), and other studies dealing with the permeability of PEMs for diverse molecules exist (see, e.g., Chapter 17 as well as Mathew et al.[111]).

All these works have contributed to unraveling many puzzling questions and, in some cases, have put forward new, very interesting questions. Nonetheless, despite the progress gained during these two decades of research, still many unresolved issues remain to be addressed. For instance, further work is needed to identify the factors that contribute to stabilizing the multilayers, especially in the case of weak PEs that exhibit a pH-dependent charge distribution. There is also a need to obtain a better understanding of the growth modes and the factors that may cause a switch from one mode to the other, as well as a need for a more detailed study of the conditions under which stable, free-standing PEMs can be obtained. Another issue almost unexplored is the effect of the chain stiffness and the length of the PE: To our best knowledge, only preliminary numerical studies have been performed on the effects of chain stiffness on the buildup of PEMs.[37] In numerical simulations, the stiffness of the chains can be introduced via angular potentials, and in more refined approaches using torsional potentials. In our coarse-grained (CG) simulations, we restrict ourselves, for simplicity, to the use of angular potentials of the form:

$$V(\phi) = \frac{\kappa}{2}(\phi - \phi_0)^2, \tag{5.1}$$

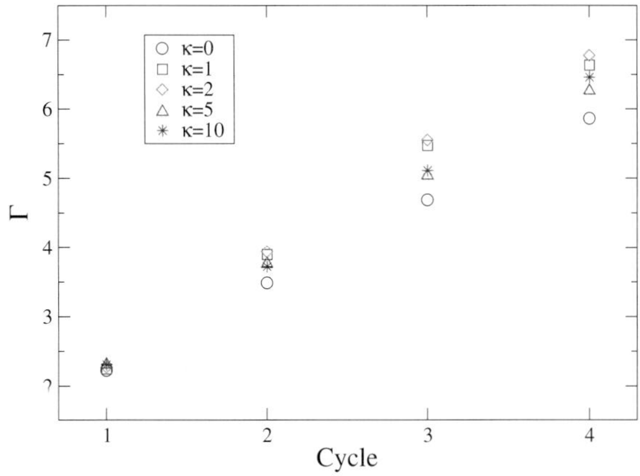

FIGURE 5.2. The total monomer surface coverage Γ as a function of the deposition cycle is shown for several degrees of stiffness of the chain (κ). Adapted from Cerdà et al.[37]

where ϕ is the angle between two vectors having as origin a non-end monomer and pointing to the previous and the next monomer in the PE sequence, respectively. $\phi_0 = \pi$ is the equilibrium angle, and κ is known as the bending or stiffness constant. Figure 5.2 shows the surface coverage Γ (equal to the number of monomers adsorbed per unit of area, where areas and distances are measured in reduced units using the diameter of the monomers as the unit length) as a function of the deposition cycle (number of deposited PE layers) for several values of the stiffness constant $\kappa \in (0, 10)$ in numerical simulations with a bare substrate charge density set to $\Sigma_S = 0.5$. PE monomers are modeled following a primitive model with a number of beads per PE equal to $N = 20$ and a charge density $f = 1/3$, that is, one in each three monomers bears a net charge (see Section 5.4.2 for more details about the model). The stiffness in the PE chains is introduced by applying the angle potential in Equation 5.1 to all the nonending monomers of the PE chains. Figure 5.2 shows that according to those numerical simulations,[37] the effect of the chain stiffness on the adsorption of the the first layer is minimal. Nonetheless, as the number of layers (the cycle) increases, the effect of the stiffness becomes more noticeable. It can be observed that completely flexible chains exhibit a smaller surface coverage, while a larger surface coverage corresponds to chains with a degree of stiffness $\kappa = 2$. Thus, the introduction of small amounts of stiffness in the chain seems to improve the adsorption, while an excess of stiffness seems to deteriorate the adsorption. The behavior of PEMs as a function of the degree of stiffness of the chains is a topic that, in our opinion, deserves to be studied in more detail.

Another topic in the field of PEMs that has received little attention so far is the effects induced by external electrical fields during the deposition stage as well as on already formed PEMs, which remains almost unexplored (see Section 5.6 for a brief summary of the state of the art).

The modeling of PEMs can help to unravel some of these unknowns and provide valuable knowledge; yet, as we will see, the current models and techniques probably need to be improved for us to be able to tackle such challenges. In what remains of this chapter, we explain the current state of the art in PEM modeling and propose issues we believe should be addressed in order to push the modeling of PEMs one step forward.

5.3 THEORETICAL AND ANALYTICAL APPROACHES TO THE MODELING OF PEMs

The first method of trying to describe the formation and properties of PEMs is the use of simple models for theoretical and analytical description. First, one should point out former works dealing with the adsorption of a single layer of PE. An early review of the results obtained from mean-field calculation was performed by Andelman and Joanny.[112] More recently, Chang and Yethiraj[113] have presented an integral equation theory for the behavior of salt-free PEs at charged surfaces. Dobrynin and Rubinstein have studied the adsorption of PE via scaling theory[114] as well as the effect of short-range interactions on the PE adsorption[115] via a strongly correlated Wigner liquid approach for two-dimensional (2D) dilute and semidilute adsorbed layers, and a mean-field approach in the case of dense adsorbed layers. Shafir and Andelman[116] have reexamined the mean-field equations for PE adsorption and have focused on the adsorption-depletion transition. Subsequently, Shafir and Andelman[117] and Wang[118] have improved mean-field and self-consistent field frameworks to study the adsorption of one layer of PE onto charged surfaces, mostly focusing on the overcharging and charge inversion phenomena that are thought to be an important initial step for PEM formation. In the case of Wang,[118] the influence of salt concentrations on the adsorption was also studied. Ubbink and Khokhlov[119] developed a Poisson–Boltzmann theory to characterize the charge adsorption of semiflexible PE, which allowed them to predict the equilibrium density of the adsorbed phase, and the interstrand separation for a wide range of ionic conditions. Recently, Manghi and Aubouy[120] have used a mean-field Poisson–Boltzmann–Edwards theory and a scaling approach to describe the variations of the monomer concentration, the electrostatic potential, and the local grafting density with the distance to the surface. They have found that the issue of overcharging–supercharging relies more on the chain length than on the bulk volume fraction of PE and salt concentration. Another approach has been the use of density functional theories to model the adsorption of PE near oppositely charged surfaces.[121–124] Recently, a nonlocal density functional theory (NLDFT) has been developed to study the effect of the ion valence, salt concentration, and the chain length of the polyions on the PE adsorption onto surfaces governed by pure electrostatic interactions.[125]

In the case of PEMs, there have been very few attempts to describe theoretically the electrostatic self-assembly; the first theories were built up on several assumptions that were difficult to test experimentally. We should point out an important assumption that is inherent in all descriptions, namely that PEMs are in thermodynamic equilibrium. This should be kept in mind when one compares them with experimental results.

Netz and Joanny[126] considered the formation of multilayers in a system of semiflexible PEs, assuming that the deposited layer structure was fixed, providing a solid charged substrate for the next layer. Park et al.[127] applied a similar idea to flexible PEs. Both of these models neglect interpenetration (interdigitation) and chain complexation between the layers, which are commonly found in experimental works about PEMs. The opposite limit of strong intermixing of PE between neighboring layers was considered by Castelnovo and Joanny[128] by incorporating the complex formation between oppositely charged PEs into self-consistent field equations. Unfortunately, these equations are limited to solutions of high ionic strength, where electrostatic interactions can effectively be treated as short-range interactions of the Debye–Hückel type. Despite these huge efforts, the strong correlations that exist between oppositely charged polyions provide a formidable challenge to their theoretical description.

More recently, Lavalle et al.,[50] based on their experimental findings, have suggested a heuristic model to explain why exponential growth occurs. The authors argue that the distinction between linearly and exponentially growing films is dictated by the PE chain's ability to diffuse freely in and out of the PEM. After PE chains are adsorbed to form the topmost layer, the adsorbed PE chains not only cover the surface but also diffuse into the PEM. In the second adsorption step of such a method, even more PEs can be adsorbed, the amount of which is roughly proportional to the existing film thickness, thus leading to an exponentially growing film. In the case where the PEM is glass-like, no fast chain diffusion on the timescale of the adsorption process can occur; the PEs can only be adsorbed on the surface (with some interdigitation), and hence one observes a linearly growing film.

Other explanations for film growth are due to Shafir and Andelman,[129] who use a mean-field theory. They have found that a strong short-range attraction between the two types of polymers is essential for the formation of PEMs and that the adsorbed amount per layer always decreases in the first layers, and increases later and saturates, if the short-range interactions between PE are large enough leading to a steady growth of the PEM. Shafir and Andelman have also reported the need for increasing the strength of short-range interactions if one increases the degree of charge of PE in order to keep the PEM stable. On the other hand, the addition of salt lowers the threshold value of the short-range interactions needed to keep the PEM stable. In the last years, the exponential growth has also been addressed by Wang[130] and Hoda and Larson.[55] Wang has modeled the LbL deposition of flexible PEs via a self-consistent field (SCF) approach, and the findings are in agreement with molecular dynamics (MD) simulations and most of the experimental work. They have observed PEMs to build up under poor solvent conditions but not under good solvent conditions. Hoda and Larson[55] have developed a one-dimensional model that relaxes the quasi-steady state assumption of the previous models[50] and they are able to describe nicely the buildup of exponentially growing PEMs. The model developed by Hoda and Larson has the advantage that it can be compared with experimental results, and the initial comparisons between the model and the experiments agree nicely. The results of Hoda and Larson support the idea that the transition from exponential to linear growth happens when the dipping time is not long enough for unbound polycations to uniformly distribute inside the film. To reach exponential growth, it is essential that very few unbound polycations move out of the film during the rinsing steps. Nonetheless, no general agreement about the rationale behind the linear and the exponential growth has been reached. One fact that can contribute to mask the picture is the fact that the type of growing regime can depend on the allowed adsorption time: If one gives the PEs enough time to reach adsorption equilibrium, one would observe an exponentially growing PEM. If they are not given enough deposition time, the PEM can grow linearly.

Among the theoretical works, one should also remark on the recent work by Wang,[131] who has studied the structure and charge compensation of PEMs via an SCF that avoids many of the assumptions employed previously in similar works. The study focuses mainly on the poor solvent regime, and provides hints toward the fact that PEMs can be difficult to build up in good and theta solvent conditions. The study also shows that within such a model it is not possible to form PEMs from PE bulk solution mixtures, which provides further support in the direction that the formation of PEMs via LbL can be a nonequilibrium process.

Finally, for nonplanar substrates, one could mention the micromechanical scaling theory developed by Biesheuvel et al.[102] for multilayer capsules containing ionizable segments. Such capsules are reported to have a wide plateau in the range of pH and salt concentrations where the equilibrium size is independent of such parameters, whereas out of such plateau, the size of the capsule increases whatever the value of the pH, and also when the ionic strength is increased. Remarkably, their theoretical approach seems to describe the influence of pH on capsule size almost quantitatively without the use of fitting parameters.

5.4 MODELING PEMs VIA NUMERICAL SIMULATIONS

To deal with the adsorption of PE and the buildup of PEMs, the ideal would be to use atomistic simulations where all the interactions between the different particles are taken into account in an explicit way and the solvent (mostly water) is explicitly included. Unfortunately, this kind of simulation to mimic PEM growth has a prohibitive computational cost due to the large number of particles that would have to be simulated and the slow relaxation times of the polymers, and only small systems of PE close to a surface designed to infer and extrapolate results to PEMs are available (see Section 5.4.1). To speed up the calculations, one could use CG molecules while still maintaining the molecules of solvent explicitly. This has been the approach taken by Reddy et al.[132] to study the adsorption and dynamics near a flat charged surface, but the computational time is still prohibitively large and it has been possible only to study single short CG PE chains consisting of 16 monomers. Due to the limitations of the current attempts in performing atomistic and/or explicit solvent simulations for PE adsorption, it is not surprising that the first attempts to study PE adsorption, as well as the buildup of PEMs, have been performed using even more coarse grain models. Those attempts are discussed in Section 5.4.2.

5.4.1 Atomistic MD

The number of works using an atomistic approach to study PE adsorption is comparatively scarce, and limited to monolayer adsorption: Chialvo and Simonson[133] have reported a united atom (UA)-level study on the adsorption of short PSS chains (10-mers) onto a graphene surface. In the UA-level simulations, the hydrogen atoms chemically bonded to carbons are represented implicitly by adjusting the force field parameters of such carbons to include the interactions originating from the implicit hydrogens. On the other hand, Horinek et al.[134] have performed all-atom (AA) MD simulations mimicking atomic force microscopy (AFM) pulling experiments of fragments of spider silk peptides up to 16 amino acids long. The excellent agreement between the experimental and numerical desorption forces showed that classical simulations with explicit water can capture the essence of hydrophobic attraction.

Those attempts clearly show that even though atomistic MD simulations are desirable to the aim of quantitatively understanding the inner structures and dynamics of PEMs, it is beyond the ability of current computational resources to build a real multilayer structure via this kind of simulation. In recent years, we have taken a complementary approach in which two regions of PEMs, that is, the substrate–PEM interfacial region and the central region of PEMs, are separately investigated via MD simulations at an AA level in order to infer properties that we can relate to the behavior of PEMs with an accuracy that only atomistic simulations can yield. In our work, PSS and poly(diallyldimethylammonium) (PDADMA) are chosen as the polyanion and polycation, respectively (see Fig. 5.3). The PSS/PDADMA PEM is one of the most experimentally investigated PEM systems, which facilitates the comparison of our simulation results with the available experimental data.

PSS Adsorption Monolayer Due to the significance in applications, such as colloidal stabilization and coating, PE adsorption has been extensively studied for several decades. Reviews of experimental, theoretical, and computational studies have been collected in Dobrynin,[38] Netz and Andelman,[135] Claesson et al.,[136] Dobrynin and Rubinstein.[137] However, the understanding of the adsorption mechanism and of the microscopic structure of PEs after adsorption is still limited: The experimental instruments can only probe the

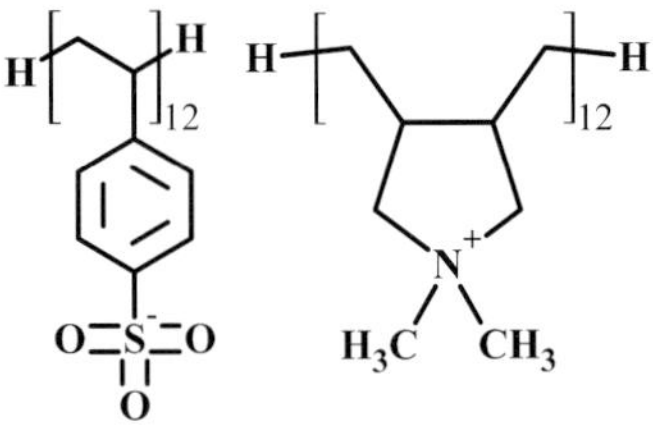

FIGURE 5.3. Schematic representation of (a) PSS and (b) PDADMA under study. The degree of polymerization is 12 for both PSS and PDADMA. Adapted from Qiao et al.[147]

structure up to the length scale of tens of nanometers (see, e.g., Block and Helm[138] and Roiter et al.[139] and Chapter 17), and the theoretical and computational works done so far always involve some approximations.

Due to the constant boost in computing resources, it is nowadays feasible to investigate the adsorption of (oligomeric) PEs via atomistic-level simulations. At such levels, atoms of both PEs and solvents are represented explicitly, generating a higher resolution than CG models in both the PE and the solvent. Actually, there already exists work on PE adsorption at an atomistic level: In Chialvo and Simonson,[133] the adsorption of PSS-Li^+ onto a graphite wall has been investigated at different surface charge densities, namely, $\sigma_s = 0$, 0.101, and −0.101 C m^{-2}. The interfacial structure of waters and PSS, and in particular the charge density profiles, were investigated in the presence of $BaCl_2$ or $LaCl_3$ multivalent salts. In Chialvo and Simonson,[133] simulations of PSS oligomers (degree of polymerization [DP] = 8) were performed at the UA level.

To provide further insight into this topic, AA MD simulations were performed.[140] The aim was to understand the adsorption ability of PSS under different substrate features: hydrophilicity versus hydrophobicity, and neutral versus charged surfaces. This is expected to be useful in elucidating the buildup of PEMs.

The substrates can be principally divided into two categories: hydrophobic surface and hydrophilic surface. The hydrophobic surface is modeled by an implicit structureless graphite wall using the Steele 10-4-3 wall potential[141–143]:

$$U_{sf}(z) = 2\pi\rho_s \Delta\varepsilon_{sf}\sigma_{sf}^2 \left\{ \frac{2}{5}\left(\frac{\sigma_{sf}}{z}\right)^{10} - \left(\frac{\sigma_{sf}}{z}\right)^4 - \frac{\sigma_{sf}^4}{3\Delta(0.61\Delta + z)^3} \right\}, \tag{5.2}$$

where the subscripts s and f stand for surface particles (i.e., carbon) and the other particles, respectively, z is the distance from the adsorbing surface, $\Delta = 0.335$ nm is the separation between the graphene sheets, and $\rho_s = 114$ nm^{-3} is the number density of carbon atoms in the graphite wall. The Lennard-Jones (LJ) interaction parameters can be found in Chialvo and Simonson.[133] It is worth noting that the Steele 10-4-3 wall potential arises from the integration of the LJ 12-6 interaction in three dimensions when the graphene sheet structure is taken into consideration. The effective cutoff distance of the Steele 10-4-3 potential is about 5.8 nm, where the error from the Steele 10-4-3 potential is about equal to that of LJ 12-6 potential at a cutoff distance of 1.2 nm (the cutoff distance of 1.2 nm is used for the other LJ interactions). In Figure 5.4, the Steele 10-4-3 potential of the sulfur atoms from PSS is plotted.

To model the hydrophilic substrate, polar hydroxyl groups (–OH) are fixed on top of the implicit structureless graphite wall at a number density of 19.1 nm^{-2}. The partial charge of hydroxyl hydrogens, 0.408 e, is taken from Horinek et al.,[134] while the one of the hydroxyl oxygens is set to be –0.408 e to neutralize the hydroxyl group.

For each of the hydrophobic and hydrophilic substrates, three types of surface charge conditions were applied with the surface charge density $\sigma_s = 0$, 0.073, and 0.164 C m^{-2}, respectively. That is, six systems were investigated in total. A summary of the adsorbing substrate is reported in Table 5.1.

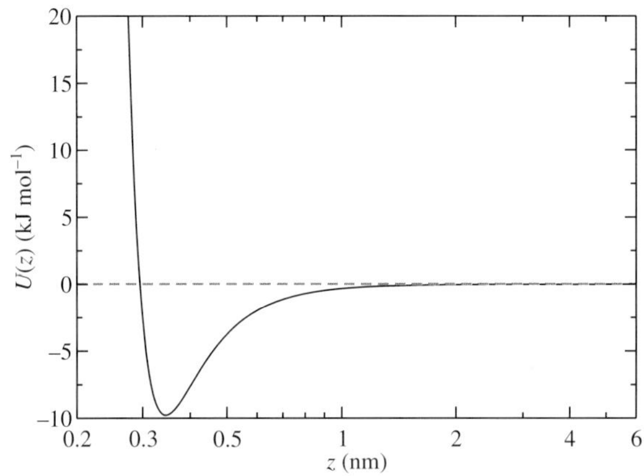

FIGURE 5.4. Steele 10-4-3 wall potential of sulfonate sulfur from PSS. Adapted from Qiao et al.[140]

TABLE 5.1. Features of the Adsorbing Substrate

System[a]	ρ_{OH}^b	σ_s^c	Cl^-
A.1	19.1	—	—
A.2	19.1	0.073	16
A.3	19.1	0.164	36
B.1	—	—	—
B.2	—	0.073	16
B.3	—	0.164	36

[a] Systems labeled with A (B) represent those with hydrophilic (hydrophobic) adsorbing surface.
[b] Number density of surface hydroxyl groups in unit of nm^{-2}.
[c] Surface charge density in unit of C m^{-2}. The counterions of the surface charges are Cl^-.

We used a PSS monomer concentration of 0.4 M, which is composed of 21 PSS chains with DP of 12, 252 Na^+ and 32,760 water molecules. The system size is built as $5.964 \times 5.9027 \times 29.0\ nm^3$ in $x \times y \times z$ dimensions. A hydrophobic neutral wall is added at $z = 29$ nm to avoid the escape of particles.

The extended simple point charge (SPC/E) water model[144] is employed with the geometry constrained using the SETTLE algorithm.[145] The force field parameters of PSS are based on the OPLS-AA force field.[146] For the missing force field parameters (i.e., the partial charge) of the sulfonate group, some modifications have been performed[147]: The partial charges of the sulfonate sulfur (1.48 e) and the oxygen (−0.68 e) are based on that of a related neutral group, namely, sulfonamide ($-S(=O)_2OR$). To keep the overall charge of −1 *e*, the partial charge of the benzene carbon chemically bonded to the sulfur atom has been modified to be −0.44 e. To verify the modified force field for PSS, radial distribution functions (RDFs) from an aqueous PSS-K solution were obtained, and very good agreements were achieved for both of the calculated RDFs (between sulfonate sulfur and K^+, and between sulfonate oxygen and water oxygen) when compared with those reported in Vishnyakov and Neimark.[148] Note that a very precise force field needs the comparison with experimental data, which is unfortunately lacking for PSS.

Classical MD simulations have been performed using the package GROMACS.[149] In all the simulations, 2D periodic boundary conditions (PBCs) were applied in the x–y plane. Neighbor searching is performed up to 1.2 nm. LJ interactions and forces are cut off at 1.2 nm, and long-range dispersion corrections to potential energy and pressure are applied.

In the production stage (200 ns), the PME algorithm[150,151] is used to perform the electrostatic interaction calculation with a direct space cutoff of 1.2 nm and a Fourier grid spacing of 0.12 nm. The reciprocal summation of the Ewald formula is performed in three dimensions (3D), but a force and potential correction via the dipole term[152] is performed in the z direction to produce a pseudo-2D summation. Furthermore, to reduce the errors of the cyclic electrostatic interactions of the unwanted z-direction, a void space of $z_{\text{void}} = 58$ nm is added in the z-direction, which results in an overall z box length of 87 nm. A canonical ensemble (NVT) is employed with the temperature fixed via a Nose–Hoover thermostat ($T = 298$ K). The time step of integration of 2.5 fs is applied and all covalent bonds lengths are constrained using the Linear Constraint Solver (LINCS) algorithm.[153]

It is noteworthy that to reach better statistics and remove the effect of the initial structure, three independent parallel runs, which were built with different random seeds, are performed for each system. Based on the calculations of the density profile as a function of time, we have found that a production run of 100 ns is long enough to accomplish the PSS adsorption in all the investigated systems. Therefore, only the last 100 ns simulation runs were used for calculating the equilibrium structure.

The number density profile $\rho(z)$ of the PSS sulfur atoms is computed to study the structure of adsorbed PSS sulfonate groups, as shown in Figure 5.5.

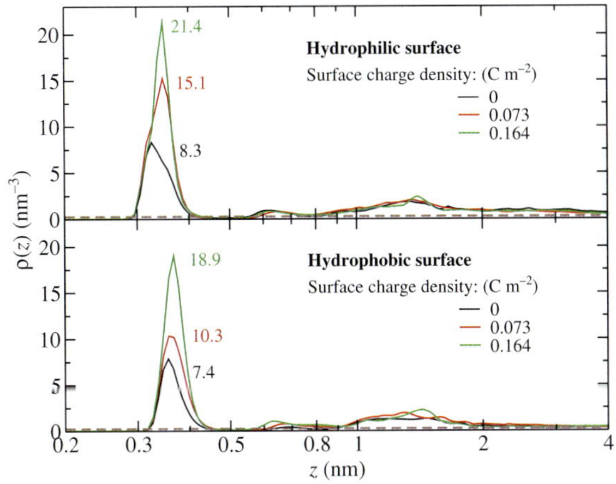

FIGURE 5.5. Number density profile of sulfur atoms from PSS. The values of number densities of the first peaks are given in the corresponding color. The horizontal dashed lines represent the average density of 0.25 nm^{-3}. Adapted from Qiao et al.[140]

We define the term "sulfonate group" to stand for the group $[CSO_3]^-$, which has an overall charge of –1 e and its geometric center located close to the position of the sulfur atom.[147] Our results indicate that in all investigated systems the first adsorption layer occurs in the region up to about 0.5 nm away from the surface. Therefore, the region of the first adsorption layer of PSS sulfonate groups is defined to be $z \leq 0.5$ nm.

In Figure 5.5, we can see that the first adsorption peak is located at $z \approx 0.35$ nm from the adsorbing surface, which approximately corresponds to the position of the strongest attractive interaction of the Steele 10-4-3 wall potential, as shown in Figure 5.4. Moreover, the fluctuations of the density profiles in Figure 5.5 extend up to about 2 nm away from the adsorbing surface, which is also consistent with the Steele 10-4-3 wall potential. These findings indicate that the Steele 10-4-3 wall potential plays a significant role in the adsorption process of the PSS sulfonate groups.

Figure 5.5 also shows that in the system with the hydrophobic neutral surface, the mere presence of the Steele 10-4-3 wall potential is strong enough to form a stable PSS adsorption layer, which is consistent with some CG-level Monte Carlo (MC) works,[154,155] where short-range attractive interaction of nonelectrostatic origin was also found to be necessary to produce stable PEMs.

The effect of the strength of the surface charge density is demonstrated by the increase in height for the first adsorption peak. By increasing the surface charge density from 0 to 0.073 to 0.164 C m^{-2}, the peak density in the first adsorption layer increases from 8.3 to 15.1 to 21.4 nm^{-3} in systems with hydrophilic surfaces. Correspondingly, the increase occurs also in systems with hydrophobic surfaces, but smaller density values are obtained, namely, 7.4, 10.3, and 18.9 nm^{-3}. We thus concluded that the presence of surface charges increases the adsorption of PSS sulfonate groups in the first adsorption layer regardless of whether the surface is of hydrophilic or hydrophobic nature.

On the other hand, by comparing this information with data in systems with a neutral surface, we note that the presence of hydrophilic surface groups also promotes the adsorption of PSS in the first adsorption layer.

The increased adsorptions of PSS in the presence of polar surface groups and surface charges indicate that PSS behaves like a hydrophilic PE, which suggests that the hydrophobic nature of the PSS backbones is screened by the outward extending sulfonate groups.

Experimentally, Block and Helm[138,156] have found that even though the flat PSS adsorption layer dominates, some adsorbed PSS chains dangle into the above PSS solution. In particular, at the very low NaCl concentration of 1 mM, the adsorbed PSS chains (DP ≈ 1910) can dangle into the PSS solution up to about 120 nm, which is about 30% of the contour length.[156] Moreover, in a CG-level work,[157] the partially standing polyion chains (up to ≈5 out of DP = 10) were also observed under good solvent condition (see fig. 1 of Reddy and Yethiraj[157]).

In the present work, we also observed two kinds of conformations of adsorbed PSS chains (see Fig. 5.6). The predominant conformation is that the PSS backbones stand roughly parallel to the adsorbing surface, while in the less probable conformation, the PSS backbones are standing approximately vertical to the surface. This vertical PSS conformation is evidently ascribed to the hydrophilic nature of PSS chains. For hydrophobic polymers, only the flat chain conformation can be reasonably expected.

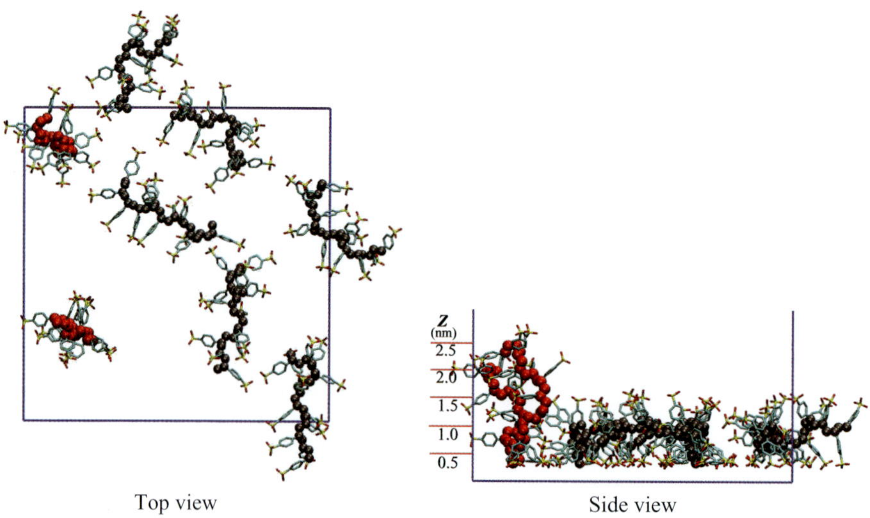

FIGURE 5.6. Two types of conformations of adsorbed PSS chains. Adapted from Qiao et al.[140]

Central Region of PSS/PDADMA PEM First we should note that it is impossible to build a physically meaningful layer structure manually due to the lack of the thickness (or, alternatively, concentration of PEs) of each layer. To the aim of understanding the inner structures of PEMs, polyelectrolyte complexes (PECs) are commonly investigated as a rough approximation. Experimentally, PEMs and PECs composed of PSS and PDADMA have been characterized using FTIR[158] and ^{1}H[159] and ^{13}C[107] solid-state nuclear magnetic resonance (NMR). On the basis of such pioneering works, it has been found that the microscopic structure of PSS/PDADMA PEMs far from the substrate is very similar to that observed in PSS/PDADMA PECs. Moreover, some works using atomistic simulations of PECs have already been performed to attain insight on PEM systems.[160–162]

Even though it would be desirable to perform systematic studies of the influence of water and salt concentrations on PSS/PDADMA PECs by varying both concentrations independently, such works are not possible due to the giant computational requirement of atomistic MD simulations. Given the fact that Schlenoff and coworkers[163] have experimentally performed quantitative studies of salt and water concentrations in PSS/PDADMA PEMs using the doping equilibrium method, we have been using similar concentrations of water and NaCl in our recent computational study[147] (see Table 5.2).

We were using a DP = 12 for the oligomers of PSS and PDADMA. Some preliminary simulations using different DPs (1, 3, 5, 10, and 12) have found a weak dependence of the chain length when DP ≥ 3 based on the calculation of dielectric constant of water. Similar findings were reported by Spruijt and coworkers[164] very recently, where four different DPs (20, 50, 150, and 510) of both polycation and polyanion are investigated in PAA/poly(*N*,*N*-dimethylaminoethyl methacrylate) (PDMAEMA) PECs. Both PAA and PDMAEMA are strong PEs, which is also true for the PSS and PDADMA investigated by us.

In our recent work,[147] classical MD simulations are performed using the package GROMACS.[149] The SPC/E water model[144] was employed, with the geometry fixed by the SETTLE algorithm.[145] The other force field parameters were based on the OPLS-AA force field[146] (see also the previous section for those of PSS). The partial charges of PDADMA are taken from Thomas and

TABLE 5.2. System Compositions and Concentrations of PE Monomer and Salt

System	Number of Components[a]				Concentration (mol L^{-1})	
	PSS	PDADMA	H_2O	NaCl	c_{PE}	c_{NaCl}
i	35	35	2940	25	2.82	0.168
ii	35	35	3738	82	2.56	0.500
iii	35	35	4536	180	2.33	1.000

[a] Degree of polymerization of 12 for both PSS and PDADMA.

Elcock[165]: nitrogen −0.32 e; carbon of the methyl groups 0.15 e; carbons in the ring bonded to the nitrogen 0.21 e; and hydrogen 0.06 e.

In the production stage, a time step of integration of 2 fs was applied with all covalent bonds involving hydrogen atoms constrained using the LINCS algorithm.[153] It lasted for 100 ns, where the isobaric-isothermal ensemble was used with the temperature scaled via a Nose–Hoover thermostat ($T = 298$ K) in combination with a Parrinello–Rahman barostat algorithm ($P = 1$ bar). The neighbor searching was done up to 1.2 nm and LJ interactions and forces were also cut off at 1.2 nm. The long-range interactions were handled via PME method[150,151] with a direct space cutoff of 1.2 nm and Fourier grid spacing 0.12 nm. Long-range dispersion corrections for the energy and pressure were applied.

In the following, S refers to the sulfur atoms from PSS, and N stands for the nitrogen atoms from PDADMA. They are used to represent the charge groups on PSS and PDADMA (i.e., $[CSO_3]^-$ and $[(CH_2CH_3CH_3CH_2)N]^+$, respectively) (see Fig. 5.3). Such selection is done because those atoms are close to the geometrical center of the groups.

Since PEs are charged, is critical to understand the structures of PSS/PDADMA PEMs by investigating the ion pair formation that is due to the electrostatic interactions. Two types of charge compensations have already been proposed for PEMs: the intrinsic mechanism (polyanions pair with polycations) and the extrinsic mechanism (polyions pair with salt ions). To evaluate these two charge compensation mechanisms inside PSS/PDADMA PEC, the following criteria were defined on the basis of the assumption that every charged PE group can be charge compensated by a single oppositely charged PE group or salt ion. Certain sulfur (nitrogen) atoms are not charge compensated if the closest distance to oppositely charged nitrogen and Na^+ (sulfur and Cl^-) is larger than 0.64 nm, which is the first minimum on S-N radial distribution function. Otherwise, it is considered to be charge compensated: An intrinsic charge compensation refers to the situation that the sulfur (nitrogen) atom has a nitrogen (sulfur) atom that is located closer to it than any Na^+ (Cl^-); otherwise, it is called extrinsically compensated.

The probabilities of the intrinsic and extrinsic charge compensations of PSS and PDADMA are given in Figure 5.7. The contribution from noncompensation is always below 1%, and therefore is neglected in what follows. For system (i) ($c_{NaCl} = 0.168$ M), the intrinsic charge compensation accounts for about 90%, whereas only about 10% for the extrinsic one, which evidently shows that the intrinsic charge compensation mechanism dominates over the extrinsic mechanism in PSS/PDADMA systems. As expected, with the increase of the salt concentration, the intrinsic charge compensation decreases (from ~90% to ~60–70%).

The slow dynamics of water in PEMs has been experimetally addressed by Kulcsar et al.,[166] but further studies are needed. A way to obtain information about the dynamics of water molecules is through the study of the behavior of the dielectric constant. Some recent simulation studies[167–169] have provided

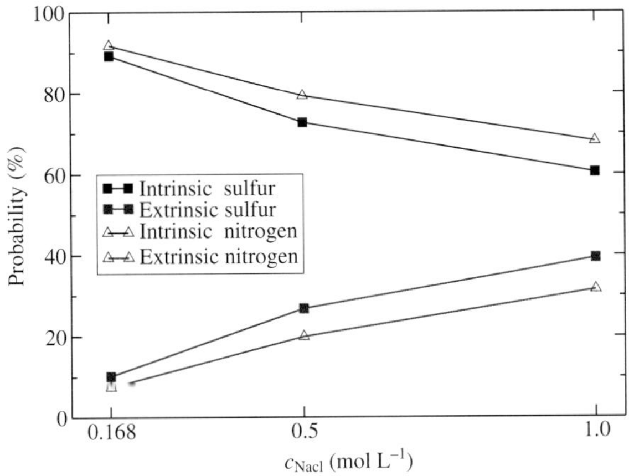

FIGURE 5.7. Probabilities of sulfur on PSS and nitrogen on PDADMA that are intrinsically and extrinsically charge compensated. Adapted from Qiao et al.[147]

TABLE 5.3. Slow Dynamics of Water

System	ε^b	Diffusion coefficient^c
i^a	11.7 ± 1.4	3.61 ± 0.01
	11.1 ± 1.2	3.49 ± 0.03
	9.9 ± 0.6	3.65 ± 0.01
ii^a	14.0 ± 0.6	6.02 ± 0.01
	14.0 ± 0.4	6.51 ± 0.01
iii^a	16.2 ± 0.7	9.77 ± 0.03
	16.5 ± 0.4	9.00 ± 0.07
Exp.^{d}	<19[42,170]	~2[171]
SPC/E (pure)	71 ± 6[172]	260 (our work)

^{a} Three parallel simulations are performed for system (i), which has the slowest dynamics of water, while two for systems (ii) and (iii). In the parallel runs, only the random seeds in building the initial structures are different.
^{b} Dielectric constant of water. Error estimates are determined by block averages of 20 ns in length. For a pure water system, the experimental value of ε is 78.3.[173]
^{c} Unit: 10^{-7} cm^2 s^{-1}. For pure SPC/E water system, the diffusion coefficient of water is obtained at ambient condition in an extra simulation of 4000 SPC/E water.
^{d} Experimental data based on PSS/PDADMA PEMs.

an important insight about the dielectric constant of water in concentrated electrolyte solutions where electrostatic interactions can be almost twice as strong as in pure water. A similar scenario is expected inside PEMs, and in order to obtain a rough idea of the dynamics of water inside those structures we have focused on the behavior of the static dielectric constant (i.e., static relative permittivity) and the diffusion coefficient of the water molecules inside a PSS/PDADMA mixture. The results are depicted in Table 5.3.

Our results show that the dielectric constants for all the studied mixtures lie within the range of 9–17, and therefore are coherent with the upper boundary value of 19 predicted in experiments for PSS/PDADMA PEMs.[42,170] The fact that the dielectric constant of water in the PSS/PDADMA mixtures under study is much lower than for pure water implies that the interactions among charged groups are much stronger (by a factor of 4–5), which is one of the causes of the slow dynamics observed in those systems.

The simulated diffusion coefficients of water were in the range of $3.5–9.8 \times 10^{-7}$ cm^2 s^{-1}, which are in the same order of magnitude as the experimental value of about 2×10^{-7} cm^2 s^{-1} measured by pulsed field gradient diffusion NMR.[171] Comparing with the diffusion coefficient of 2.6×10^{-5} cm^2 s^{-1} in a pure SPC/E water system, one can conclude that in the PSS/PDADMA PECs under study the water molecules diffuse about two orders of magnitude slower than in a pure water system. Therefore, the obtained results of the water diffusion lead to the same conclusion as our dielectric measurements: Water molecules inside PE mixtures exhibit very slow dynamics when compared with water molecules in a bulk environment.

5.4.2 CG Numerical Simulations

There are several approach levels to coarse-graining polymers. The closest to the AA description is the approach known as the united level where the hydrogen atoms linked to a same carbon and the carbon atom are reduced to a single site. The model preserves the carbon-carbon angle and bond length but the hydrogens have been coarsened. A subsequent coarsening approach is to replace several of such united sites by a single site. Models using this last approach are commonly referred to as bead spring models, where the connectivity between the sites of the polymer is maintained by potentials mimicking spring behavior: either harmonic, finitely extensible nonlinear elastic (FENE) (see next section, ca. Eq. 5.4), or more sophisticated ones. Thus, the polymer chain looks like a pearl necklace. As explained at the end of Section 5.2, the stiffness of the chains can be introduced in the CG models via angular potentials, and when more refined approaches are needed via adding torsional potentials. Such sites typically interact via LJ potentials, which can be shifted if only repulsive short-range interactions are desired among sites (commonly Weeks–Chandler–Anderson [WCA]-like potentials[174]). The interactions among beads, aside of reproducing the steric repulsion between the groups of atoms, must also be set in such a way that mimics the effects of the solvent on the chemical units (good and poor solvents). For PEs, one faces the additional complication that the solvent is not inert, but is strongly polarizable. The extension of the bead spring models to PEs goes by simply adding point charges to the centers of some beads. Leaving some beads neutral, one can regulate the charge density. The system is made charge-neutral by modeling the released counterions by hard spheres of opposite size. The ion size is usually taken in the range of 4–5 Å, which is the typical size of an ion including its first hydration shell.[175] The water is modeled as a dielectric continuum with a fixed dielectric constant representative of the solvent and of the temperature. In the case of a simple electrolyte solution and just hard sphere potentials, this model is known as the restricted primitive model (RPM) of electrolytes. Some more details on modeling aspects can be found in various reviews on dealing with the simulations of charged systems.[34,38,39,176]

The first simulational models aimed to understand PE multilayering were developed by Messina et al.[154,155] where the multilayering onto a charged sphere and rod were performed via MC simulations. With only the electrostatic interaction, the formed PEMs were always found to be unstable, indicating the necessity of having additional attractive short-range forces of nonelectrostatic origin to obtain stable PEMs. Later, Messina[177] studied the multilayering on charged planar surfaces, confirming also for this surface geometry the need of additional nonelectrostatic forces to keep the PEMs stable. The main findings obtained until 2004 were reviewed in Messina et al.[34] Later, Panchagnula et al.[178] used the same model with Langevin simulations to revisit PEM growth on spherical substrates. The results of Panchagnula et al. suggested that stratified multilayers around adsorbing particles where only temporal structures whose intermixing could be slowed down via the formation of strong ionic pairs between oppositely charged monomers. Subsequently, Messina[179] studied the effect of the image forces (arising from the dielectric mismatch between surfaces and solution) on the adsorption of PE in planar surfaces. It was shown that image forces can considerably reduce the degree of adsorption of PE and can even inhibit the charge inversion process in the topmost layer.

The possibility of forming PEMs from salt-free solutions containing both polycations and polyanions together was studied via MD by Abu-Sharkh.[180] According to the results of Abu-Sharkh, it is possible to obtain thermodynamic stable PEM purely from mixed solutions. This is, however, a very controversial issue that is in contradiction to most experimental reports, and as far as we know no further numerical study has been performed to confirm or refute the findings of Abu-Sharkh. It should be remarked that Abu-Sharkh used a CG model where chains were modeled not as a bead spring or freely jointed chain but using bending and dihedral potentials, which make chains semiflexible. The use of flexible chains enhances the entropic contribution arising from the differences in entropy between the adsorbed chain and the released ions from the surface. If this gain in free energy due to the entropic contribution of the released counterions (see, e.g., von Klitzing[45]) is the main driving force for complexion, this could partially explain the observed behavior. To our best knowledge, no other numerical studies have focused on the study of semiflexible chains and their effects on PEM growth and stability until now to verify these results. Abu-Sharkh[180] has also found the stability as well as the microstructure of the PE layers to be especially sensitive to the strength of the nonelectrostatic short range attraction between the PE and the charged substrate that was in line with the earlier observations.

The effects of the length and fraction degree of charge of PE on PEM formation of planar substrates were studied by Patel et al.[181] using Langevin simulations, confirming the importance of short-range interactions for all fraction degrees of charge, the considerable intermixing between chains deposited on subsequent steps, and the validity of the three-zone model for PEMs. Shortly after, Panchagnula et al.[182] extended their previous work on charged particles[178] and reinforced the idea that the presence of ionic pairs between monomers of opposite PE types has an important role on the stabilization of PEMs, and that multilayers are trapped nonequilibrium states. At that time, the multilayer assemblies of PE and nanoparticles were studied by Jeon et al.[183] via Langevin simulations showing that films assembled from nanoparticles show a better stratification but have rougher surfaces.

In a subsequent study by Abu-Sharkh,[184] the minimum energy configuration of PEMs was studied using a similar model to that used previously.[180] It was observed that PE in the PEMs were kinetically trapped due to the strong electrostatic interactions; that is, the PE were not in the lowest energy structure that could be attained, although the trapped structure was indeed layered. Shortly after, Patel et al.[61] presented the most exhaustive numerical study on the growth of PEMs on planar substrates up to the present, where multiple parameters such as degrees of charge, Bjerrum length, and monomer–monomer and monomer–surface interactions were investigated. The hypothesis that surface overcharging is crucial for stable film growth was confirmed, as well as the important role that both electrostatic and short-range interactions play in the formation of PEMs. A linear growth was observed in the studies conducted by Patel et al. Another outcome of that work was the confirmation of the fuzzy nature of PEMs (molecules in one layer interpenetrate other layers), although almost perfect periodic oscillations of the density differences between monomers belonging to positively and negatively charged PEs have been found. The authors believed that their simulations were in equilibrium; however, our own investigations showed that many of the PEMs formed by Patel et al.[61] would probably not form if enough time is allowed to properly equilibrate the two initial bilayers due to the formation of complexes that tend to dissolve the adsorbed PEs. This is due to the very slow equilibration times involved, and the slow dynamics of the whole system.

Although the following works do not directly address PEM formation, some recent simulational works studying the adsorption of PE chains on top of charged planar substrates should be mentioned: Carrillo and Dobrynin[185] have performed a very detailed study of the properties of the first adsorbed layer using the model previously used by Patel et al.,[61] mainly focusing on the effects of the different parameters on the surface coverage and thickness of the adsorbed layer. Grand canonical MC simulations have been used by Narambuena et al.[186] to study the adsorption properties of the first layer. Charge reversal was obtained in that study without the need to use other additional short-range attractions of the PE to the surface other than the electrostatic one. From this, they concluded that the entropic contribution of the counterion release from the double layers of the PE and the substrate was the key factor in the driving force for the adsorption of PE. This was suggested long ago as the main factor for single PE adsorption, but, as has been found before, is probably not a sufficient driving force to produce stable PEMs.

Despite the fact that the RPM of PEMs does not describe well any short range structures of PEMs due to the lack of detailed atomistic potentials and the lack of a proper description of the water, they are currently (and probably will be for some more time) the only feasible approaches to numerically study PEMs beyond the mean-field levels. Simulational studies based on the RPM have already demonstrated their potential to describe PEMs, but as will be shown in the next section, one should improve them in order to obtain a more accurate physical description of the structure and dynamics of PEMs. The necessary improvements should still be, of course, on the level of a CG description so that they will still be able to simulate PEMs in computational times that are orders of magnitude faster than a fully featured atomistic simulation.

5.4.3 Refinements in the CG Simulations

One important topic that must be addressed in the simulation of PEMs is the way the chains are allowed to complex with each other. The protocol followed by Patel et al.[61] is to place the polyanion or polycation solution into a parallelepiped box containing a charged surface in the plane x–y with $L_x \sim L_y$ and L_z that is about four times larger than L_x in order to ensure there is enough material to be adsorbed onto the surface, and let the chains adsorb within a certain simulation time. After some time, one declares the adsorption process to be finished, and removes the excess PEs. The dipping will be repeated with a solution of oppositely charged PEs. That approach is taken because the system to mimic has a $2D + h$ geometry, which means it has PBCs along two directions (x and y), but it is finite along the z-direction. To compute long-range electrostatic interactions under period boundary conditions, Patel et al. use the P^3M algorithm[187] with a gap between successive replicas in the z-direction equal two $2L_z$ in order to reduce the interaction between the unwanted replicas that P^3M generates along that direction. Nonetheless, the previous approach has several efficiency drawbacks from the computational point of view: The electrostatics are not fully $2D + h$; most of the computer power is devoted to moving chains that are far from surface and they will take a lot of time to reach it; and, most severely, it is difficult to know if one really has reached the saturation limit of the surface for a given concentration of PE, because in small boxes with a fixed amount of PE, the concentration in the "bulk" diminishes as the adsorption proceeds. Furthermore, the depletion zone near the surface that forms due to the differences between the speed of diffusion of chains and the adsorption rate can create the false impression that a saturation regime has been reached in a short running time interval. In a work on the adsorption of a monolayer, Carrillo and Dobrynin[185] have introduced the so-called "stirring step" where after a certain time, chains that remain unadsorbed are randomly repositioned in order to improve adsorption rates. Carrillo and Dobrynin[185] showed that without stirring steps,

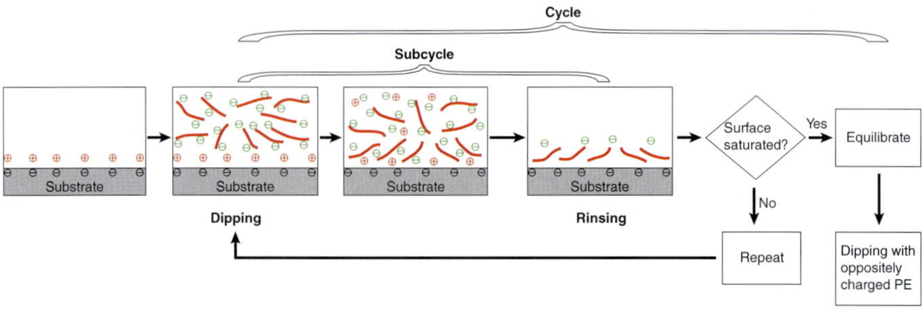

FIGURE 5.8. Diagrammatic scheme of the new proposed protocol that ensures saturation of the surface at equilibrium with a given density of monomers ϕ, and allows to substantially reduce the computer time needed to deal with the study of PEMs. Adapted from Cerdà et al.[36]

the adsorption might not be in the real saturation regime in the simulations performed by Patel et al.[61] The introduction of stirring steps helps to palliate the problems related to adsorption, but still the previous drawbacks persist.

In order to try to improve the current situation, we have proposed a new protocol that serves a threefold purpose: to improve computational efficiency, which will allow us to get calculations faster and be able to simulate larger systems; to ensure the equilibrium adsorption; and to get closer to the usual experiment conditions where the PE concentration remains almost constant during the adsorption process. Figure 5.8 depicts schematically the proposed protocol which uses smaller boxes along the z-direction in order to focus the computational power in the surroundings of the substrate. For a given experimental deposition cycle, the new protocol performs several dipping-rinsing subcycles using the same type of PE. In this way, the system has the chance to reach an equilibrium with the correct bulk concentration of PE. Thus, there is no need of performing from time to time stirring steps. Furthermore, in order to account exactly for the long-range electrostatic forces in this $2D + h$ geometry, without the need to spend a large amount of computer power on empty space regions (gaps between replicas in the z-direction), we have used a P^3M together with an electrostatic layer correction (ELC) term.[188,189] This algorithm gives in principle the exact forces which are present for $2D + h$ geometry. The use of ELC allows us to reduce the empty gap along the z-direction, which is no longer the region $z \in [L_z, 3\,L_z]$ but just $z \in [0.8\,L_z, L_z]$, saving a substantial amount of computer time because grids with fewer nodes along the z-direction can be used in P^3M for a given accuracy. Further details about the Langevin dynamics simulations performed, the model, and the protocol can be found for the interested reader in Cerdà et al.[36,37] We should comment that there exists also the possibility of using adapted Ewald sums for the $2D + h$ geometry, but they turn out to be much less efficient than our method of choice (P^3M combined with ELC).

The improved method has been applied to the study of the stability of the PE bilayers made of PEs with opposite charge. Due to the gain in computer efficiency, systems larger than those previously used have been studied: Simulated substrates have an area of size $A = 40 \times 40$ (in units of the square of the particle diameter). Our CG simulations[36,37] have been able to reproduce the trends observed in the numerical simulations of Carrillo and Dobrynin[185] for monolayers. Nonetheless, for bilayers, our results show that the behavior can be much more complex than previously reported in numerical simulations.[61] To study the stability of the first layers, it is convenient to test different combinations of the strength of the attractive interaction between the monomers and the surface (represented by a well depth ε_{m-s}), and the strength of the interactions among the monomers themselves (represented by a well depth ε_{m-m}). In our CG studies,[36,37] we have addressed these issues using Langevin dynamics simulations in which the translational Langevin equation of motion for a particle i is given by[190]

$$M_i \frac{d\mathbf{v}_i}{dt} = \boldsymbol{F}_i - \Gamma_T \boldsymbol{v}_i + \xi_i^T, \tag{5.3}$$

where $\boldsymbol{F}_i$ is the total force acting on the particle i, $M_i = M$ is the mass, and Γ_T is the translational friction constant. ξ_i^T are Gaussian random forces, each of zero mean and satisfying usual fluctuation–dissipation relations. As is usual in CG simulations, in order to avoid working with extreme numbers, we use a reduced system of units in which the reduced mass M is given in units of the characteristic experimental mass of the monomers M_e, and the lengths are given in units of the experimental characteristic monomer diameter σ_e. The reduced time t is defined as $t = t_e\sqrt{\varepsilon_e/(M_e\sigma_e^2)}$, where t_e is the experimental characteristic time of the dynamics of the monomers, and ε_e is the characteristic strength of the nonelectrostatic interacting potential among real monomers. The interaction potentials among the different particles in the system are given in units of ε_e. The reduced temperature T is defined as $T \equiv k_B T_e/\varepsilon_e$, where T_e is the experimental temperature. It is worthwhile to comment that for equilibrium simulations, the values of the mass and the friction constant Γ_T are irrelevant. The particle mass is chosen to be $M = 1$ and we take $\Gamma_T = 1$ because we observed that these values produced a conveniently fast relaxation to equilibrium. In our simulations, the polymers are modeled using a bead spring chain model, each chain containing $N = 20$ beads of diameter σ_m. Consecutive beads inside the chain interact via a FENE potential:

$$V_{\text{FENE}}(r) = \frac{-K_f\, r_{\max}^2}{2} \ln\left(1 - \left(\frac{r}{r_{\max}}\right)^2\right), \tag{5.4}$$

where $K_f = 30/\sigma_m^2$ and $r_{\max} = 1.5\sigma_m$. In our model, all the chain beads (monomers) in the system interact via a truncated-shifted LJ potential, sometimes known as the WCA[174] potential:

$$V_{\mathrm{LJ}}(r,\varepsilon,\sigma,r_{\mathrm{cut}})=\begin{cases}4\varepsilon\left(\left(\frac{\sigma}{r}\right)^{12}-\left(\frac{\sigma}{r}\right)^{6}-\left(\frac{\sigma}{r_{\mathrm{cut}}}\right)^{12}+\left(\frac{\sigma}{r_{\mathrm{cut}}}\right)^{6}\right), & \text{for } r<r_{\mathrm{cut}}\\ 0, & \text{for } r\geq r_{\mathrm{cut}}\end{cases}, \tag{5.5}$$

where $\sigma = \sigma_m$, $\varepsilon = \varepsilon_{m-m}$, $r_{\mathrm{cut}} = r_{\mathrm{cut},m-m}$, and $r = |\mathbf{r}_i - \mathbf{r}_j|$ is the relative distance between the centers of two beads. For simplicity, we take $\sigma_m = 1$, $r_{\mathrm{cut},m-m} = 2.5\sigma_m$. The fraction of charged monomers on each chain is denoted by f, meaning that one in each f consecutive monomers carries a charge q located in its center. Thus, for instance, $f = 1$ corresponds to all beads bearing a charge q, whereas $f = 1/3$ implies a sequence of charge for the beads $q, 0, 0, q, 0, 0, q, 0, 0, \ldots$.

The steric forces among counterions are also modeled via a WCA potential (Eq. 5.5), with parameters $\sigma = \sigma_m$, $\varepsilon = \varepsilon_{m-m}$, and $r_{\mathrm{cut}} = 2^{1/6}\sigma_m$ that mimics a repulsive soft sphere. Counterions interact with the chain beads via a similar repulsive WCA potential. The adsorbing surface (located at $z = 0$) is modeled as a squared lattice array of fixed sites, each site bearing a negative unit charge. To keep charge neutrality, the same amount of ions of opposite charge is added to the solution. Ions interact with the surface and the remaining particles with the same WCA repulsive potential as the counterions of the polymer chains. All beads interact with the adsorbing surface via a LJ potential where r now represents the distance between the particle and the surface. The energy parameters for the particle–surface interaction are $\varepsilon = \varepsilon_{m-s}$. In all studied cases, $\sigma = \sigma_m$, and $r_{\mathrm{cut}} = r_{\mathrm{cut},m-m} = 2.5\sigma_m$.

Electrostatic interactions between any two charged particles with charge q_i and q_j and separated by a distance r_{ij} are given by the coulomb potential:

$$\frac{U_{\mathrm{coul}}(r)}{k_B T_e} = l_B \frac{q_i q_j}{r_{ij}}, \tag{5.6}$$

where l_B is the Bjerrum length $l_B \equiv e^2/(\varepsilon_r\, k_B T_e)$ which defines the length scale at which the thermal energy $k_B T_e$ is equal to the coulomb interaction between two elementary charges e in a dielectric medium of dielectric constant ε_r. In our simulations, we have set the Bjerrum length to the value $l_B = 3$. Unless otherwise specified, charged particles are always monovalent ions with charge $q_i = \pm 1$ in our system of reduced units. The solvent is modeled in these CG simulations as a continuum of dielectric constant ε_r. Electrostatic interactions are observed to be essential in order to maintain the PEM layering structure: In our preliminary simulations, we have observed that if we take a formed PEM and we lower systematically the value of all the charges, a progressive remix of the multilayers occurs. For values of the charges lower than one-tenth of their original values, the PE layers completely remix, forming a thick monolayer composed of a mixture of positive and negative PEs, and any trace of the former PEM structure is completely lost. On the other hand, precedent CG models as well as our own show that the formation of PEMs using only electrostatic interactions seems impossible when one disregards completely the short-ranged noncoulombic forces. Thus, at least from what is infered from numerical simulations performed so far, both coulomb and noncoulombic forces are essential, and as we will see later, their combined effect on the PEM formation is highly nontrivial.

To map our simulation units to real experimental units, we note that the Bjerrum length in water at room temperature is about 7.14 Å, giving us the mapping that $1\sigma = 7.14$ Å, and a value of $\Sigma = 0.5$ for the charge surface density (number of net charges per substrate area) will correspond to a physical system where the surface charge density has a value of 0.18 C m^{-2}, and each bead in the chain will roughly correspond to three real chemical monomers of sodium poly(stryrensulfonate) (NaPSS), each of them with a size of 2.5 Å, which are values reasonably close to those used in experiments.

The previous model constitutes a first approach to the study of the stability of the first layers and has been used in the results presented henceforth. Figure 5.9 shows the temporal evolution of three different systems studied using the previous model (A: $\varepsilon_{m-m} = 0.2$, $\varepsilon_{m-s} = 1.4$; B: $\varepsilon_{m-m} = 1.0$, $\varepsilon_{m-s} = 1.4$; C: $\varepsilon_{m-m} = 1.0$, $\varepsilon_{m-s} = 10$) during the deposition of PE^- on top of an initial PE^+ monolayer previously deposited on top of a bare negatively charged substrate with a surface charge density set to $\Sigma_S = Q_{\text{total}}/\text{Area} = 0.5$. In these simulations, the degree of charge of the PE is set to $f = 1/3$, that is, one in each three monomers is charged. The evolution in Figure 5.9 is depicted via snapshots of the systems taken at subcycles 5 and 20, and a last snapshot taken after the deposition is complete and the scheduled equilibration has been performed.

From Figure 5.9, we can see that complexation that destroys the bilayer and thus the possibility of building up a PEM can occur if the monomer–surface interaction is of the order of the thermal energy, $\varepsilon_{m-s} \approx 1$. It is interesting to note that two different modes of complexation are observed in our model: At low values of the monomer–monomer interaction $\varepsilon_{m-m} \sim 0.2$ (system A), the complexes that tend to escape from the bilayer involve just a few chains. If, however, the monomer–monomer attraction is of the order of the thermal energy, $\varepsilon_{m-m} \sim 1$ (system B), complexation occurs by, first, a mixing of the two types of PE followed by a dewetting of the surface in order to enhance the number of contacts among monomers. The cylindrical shape of the formed complex in the late stages in system B is simply due to the PBC.

For larger values of the monomer–surface interaction $\varepsilon_{m-s} = 10$ (system C), an apparent stable bilayer can be obtained as observed by Patel et al.[61] Nonetheless, Figure 5.10 shows that if enough running time is allowed to the simulation, the bilayer formed in system C suffers a complexation process roughly similar to that observed in system B (which has the same monomer–monomer interaction strength). The increase of the strength of the monomer–surface interaction ε_{m-s} seems only to slow down the kinetics of the dewetting process. We are currently studying whether there is a critical value of the monomer–surface interaction ε_{m-s} above which the dewetting is prevented and how the stiffness of the chain can influence such dewetting process. Furthermore,

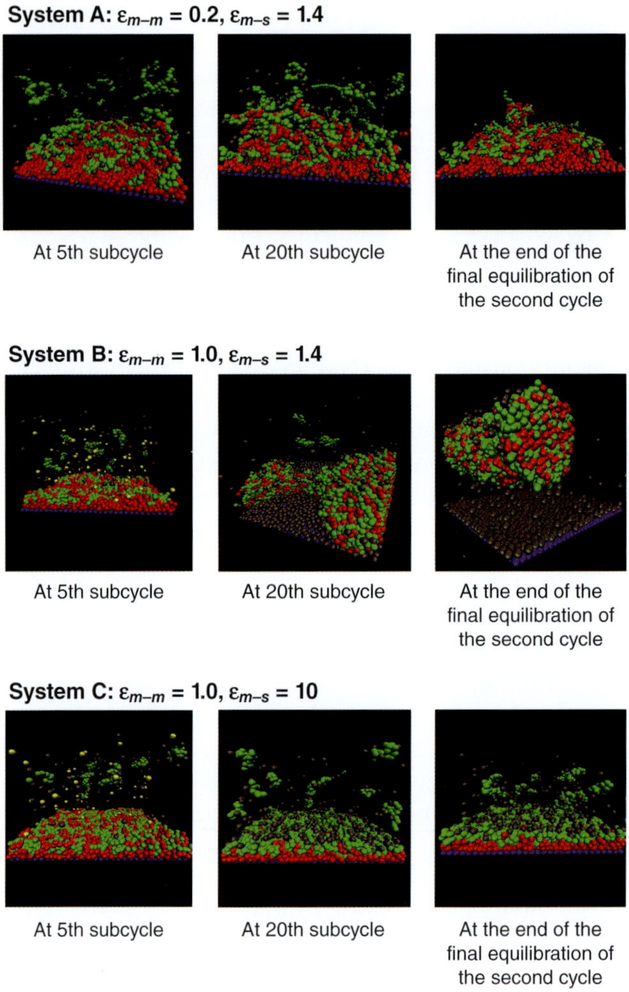

FIGURE 5.9. In order to compare the stability of bilayers, the temporal evolution of three different systems (A, B, and C) during the gradual deposition of PE^- (green chains), while building up the second layer (second cycle) is shown via snapshots taken at subcycles 2, 20, and the last one taken at the end of the equilibration process done after the deposition is complete (left, center, and right columns, respectively). The three systems share all parameters except the monomer–monomer ε_{m-m} and the monomer–surface ε_{m-s} interactions. In the three cases, the surface charge density is set to $\Sigma_S = 0.5$, the number of beads to $N = 20$, the charge degree is set to $f = 1/3$ (see text), and the chains are fully flexible. System A has ($\varepsilon_{m-m} = 0.2$, $\varepsilon_{m-s} = 1.4$), system B has ($\varepsilon_{m-m} = 1.0$, $\varepsilon_{m-s} = 1.4$), system C has ($\varepsilon_{m-m} = 1.0$, $\varepsilon_{m-s} = 10$). The color code is as follows: Blue particles represent the negative charged substrate sites and white particles are their co-ions; red particles are the monomers of the PE^+ chains and yellow particles are their counterions; green particles represent the monomers of the PE^- chains and gray particles are their counterions. Adapted from Cerdà et al.[36,37]

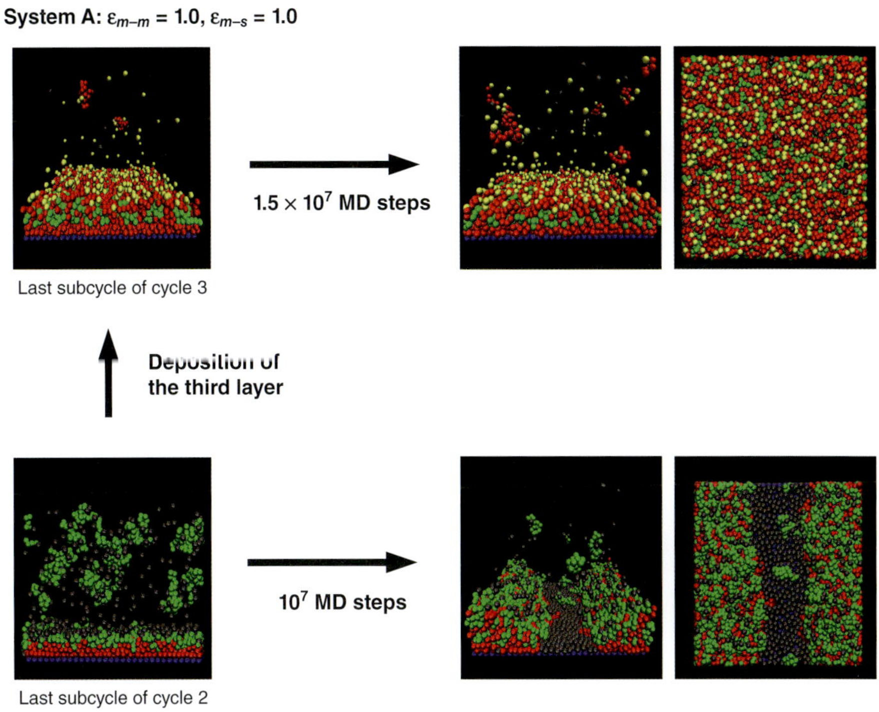

FIGURE 5.10. The plot shows a comparison of the different evolutions that system C (see Fig. 5.9) may suffer once the deposition of PE^- in the second cycle is completed. In the lower path, the bilayer has been equilibrated for 10^7 MD steps and the lateral and top views of the system after the run are shown. A clearly ongoing dewetting process is observed in that case. In the upper path the PE^+ chains that constitute the third layer are added right after the last subcycle of cycle 2; that is, we add the third layer without allowing the bilayer to equilibrate. Once the third layer has been completely deposited (after the end of the third cycle), an extra equilibration period of 1.5×10^7 MD steps is applied. Following this last path, no signs of dewetting are observed. Adapted from Cerdà et al.[37]

Figure 5.10 shows that if the third deposition (now of PE^+, cycle 3) is deposited on top of a bilayer (system C) that has not been allowed to equilibrate after concluding the last deposition of PE^- in the second cycle, then a stable three-layer seems to occur. Our long simulations of this three-layer system do not show any apparent change or modification in the three-layer that could hint at the beginning of a complexation process. Figure 5.10 indicates that the fast deposition of a third layer on top of a bilayer can help to stabilize a system that otherwise will lead to dewetting and therefore to the impossibility of building up a PEM. The simulations performed by Patel et al.[61,181] (which use a model roughly similar to our model) are possibly close to this kind of systems. It would be worthwhile to gain a deeper understanding of why a

$[PE^+,PE^-,PE^+]$ system will remain stable, while a corresponding $[PE^+,PE^-]$ system suffers from dewetting. This could be related to the experimental findings that structures with an odd number of layers are more stable than those with an even number of layers. Nonetheless, in our CG simulations, the addition of a new deposition of PE^- to system C, that leads to a final system $[PE^+,PE^-,PE^+,PE^-]$ (four cycles), does not lead to the destabilization the system. The long simulations we have performed for this last system do not show any onset toward a complexation or dewetting process. Henceforth, we will focus on the characterization of this four-layer system. As observed previously by Patel et al.[181] for four-layer structures, the majority of newly PE^+ adsorbed monomers are added to a distance about 2σ corresponding to the formation of a second layer of positively charged chains, while a small amount of monomers remain on the top of the third (negatively) charged layer.

5.5 TOWARD A BETTER PHYSICAL DESCRIPTION OF PEMs USING NUMERICAL MODELS

5.5.1 The Influence of Salt Ions

Complexation phenomena in PEMs has long been known to hamper the formation of PEMs. Although theoretical stability diagrams can look very simple,[83] experiments show that in fact stability diagrams can be quite complex.[44] Thus, it could be possible that the addition of small amounts of salt could broaden the range of stability of PEMs in the CG models. The rationale of this is that the formation of free complexes in solution, due to its reduced size, may have less tolerance to the presence of salt ions that weakens the ion pair interactions than PEMs. Thus, moderate amounts of salt could bias the equilibrium between free complexes and PEMs to the side of PEMs, promoting and enhancing of the stability of multilayers. Therefore, a basic issue to be studied is the determination of the stability diagram of bilayers and trilayers as a function of the main parameters that could affect the equilibrium between free complexes and PEMs.

To our best knowledge, no simulation has dealt with the study of the salt effects in the buildup of PEMs. In fact, even a more basic issue that consists of getting insight into the effects that salt ions have on the properties of the first layer of adsorbed PE onto a charged substrate has received little attention so far from the simulational point of view. Wang et al.[125] have studied both, numerically via MC simulations and theoretically via an NLDFT, the PE adsorption onto a charged hard wall, where $l_B = \sigma$, and the degree of charge of the PE if $f = 1$. The van der Waals interactions as well as whatever other kind of short-range nonelectrostatic attractive interactions have been neglected in that study. Aside from the electrostatic interactions, all the remaining interactions are modeled as hard sphere potentials. The solvent is modeled as a dielectric continuum. Thus, the approach taken by Wang et al. can be considered to be valid in the limit of good solvent conditions in a considerable electrostatic screened solution. In such limits, they have found NLDFT to predict quite reasonably well the results obtained through numerical simulations. Under that approach, they have found that the PE attraction to an oppositely charged surface just amounts to a neutralization effect, and that it is little influenced by the length of the PE and the valence of the counterions. No significant sign of charge inversion has been observed in their simulations, and adsorption onto a surface with the same charge sign is observed only when multivalent counterions are present.

On the other hand, when attractive short-range attractions are included in a PEM model, the behavior becomes more complex.[36] Furthermore, if salt ions are added to that system, the behavior greatly increases in complexity. Thus, for instance, Figure 5.11 shows some preliminary results about the behavior of the surface coverage Γ (equal to the number of monomers adsorbed per unit of area) as a function of the salt concentration ϕ_{salt} for monolayers simulated using a similar model to those used in Figures 5.9 and 5.10. The introduction of salt ions with valence $v = 1$ does not substantially modify the amount of adsorbed PE when $\phi_{salt} \leqq 0.03$; just a slight increase can be observed. Instead, a huge apparent improvement of the adsorption is observed for salt ions with valence $v = 2$. Figure 5.12 shows what happens. As can be seen in panels A and B, the salt ions with valence $v = 1$ scarcely distort the layer, they mostly just top the layer as counterions of PE would do. On the other hand, salt ions with valence $v = 2$ (notice orange ions within the red zone) are able to substantially

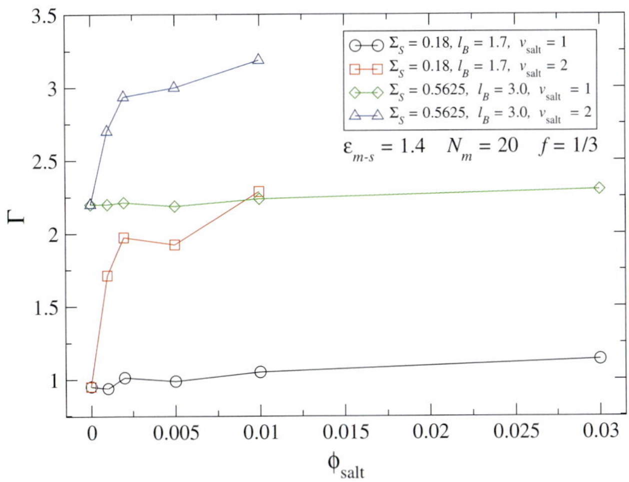

FIGURE 5.11. The plot depicts the preliminary results using our coarse-grained model for the surface coverage Γ of a monolayer of PE^{+} as a function of the salt concentration ϕ_{salt} for different types of salts, monovalent $v_{salt} = 1$, and divalent $v_{salt} = 2$ ions. The remaining parameters of the system are the same as in Figure 5.9 except for the surface charge density Σ_S (see the plot legend). Adapted from Cerdà et al.[36]

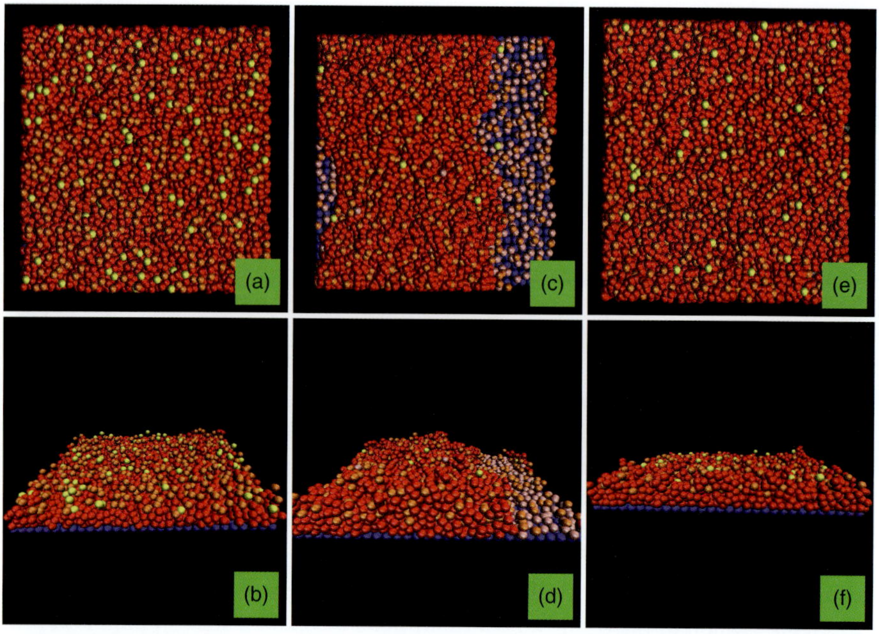

FIGURE 5.12. In this figure, three different systems with a single deposited layer of PE^+ chains containing salt ions are depicted after rinsing. Parameters are similar to those of systems depicted in Figure 5.9 except the salt concentration ϕ_{salt}. Panels A and B represent the top and lateral views for $\phi_{salt} = 0.03$ and monovalent salt ions, valence $\nu = 1$. Panels C and D represent the top and lateral views for $\phi_{salt} = 0.03$ and divalent salt ions, valence $\nu = 2$. Panels E and F represent the same as in panels C and D but $\phi_{salt} = 0.005$. The code of colors is similar to Figure 5.9; in addition, pink particles are positive salt ions and orange particles negative salt ions with valence ν. Adapted from Cerdà et al.[36]

enter the layer and intercalate between PEs (see panels D and F), reducing the repulsion between PE chains, which allows more PEs to be adsorbed. A side effect is that at large salt concentrations (panels C and D) the PE no longer wets the whole substrate. At this point, a series of fundamental questions about the suitability of the CG model must be formulated: How much do the observed salt effects depend on the specific coarse-graining model one assumes? If monovalent salt ions were modeled to be of a different size from monomers, would these lead to a different behavior? Another important question is related to the fact that an implicit solvent is assumed, but it is clear that inside the monolayer the amount of solvent can be different than in bulk. Therefore, the effective dielectric constant may be different and coulomb interactions among charged objects within the layer should then be stronger than in the bulk. How important are those effects on the stability of PEMs, in which extent do they modify the equilibrium between free complexes (pair ions interacting softly due the bulk dielectric constant) and PEMs (with pair ions strongly interacting due to the drop of the dielectric constant)?

5.5.2 The Effect of Solvent within PEMs

The dielectric continuum approximation in the CG models using an implicit solvent is under usual conditions actually a very good approximation.[167,168] Basically, for distances larger than 1 nm, the potential of mean force between an ion pair in an implicit solvent and in an explicit water surrounding becomes basically indistinguishable. However, the presence of many ions can actually change the dielectric properties of water, as was shown via atomistic simulations by Hess et al.[167,168] and Gavryushov and Linse.[169] The effective relative dielectric constant of the water is decreased when the amount of added salt is increased, and so the effective interactions between nearby ions can increase as much as a factor of 4–5. This effect could be simply taken into account by lowering the dielectric constant in the PEM, which in fact would lead to electrostatic interactions four to five times larger in the PEM than in the bulk. This effect could lead to the observed glassification of multilayer films when the water content is lowered upon drying. Particularly dry films are then glued together by strong electrostatic forces. In this way, the amount of bound water within the PEM plays an important role. Special care must be taken, however, due to the interfacial regions. A solid substrate has a very low dielectric constant of around 5, whereas the bulk water has a relative dielectric constant of around 80. These differences will lead to additional electrostatic interface effects.

The determination of the amount of water within PEMs has received considerable attention from the experimental side.[42,73,191–193] Tanchak et al.,[191] for example, have studied experimentally the water distribution inside PEMs and found that water did not completely penetrate to the substrate but that there was an exclusion zone near the substrate. Jaber and Schlenoff,[192] working at zero salt concentration, have reported a substantial reduction of the number of water molecules surrounding the charged sites inside the PEMs. Thus, for instance, *PAH/PSS* PEMs have about six water molecules per ion pair, *PDADMA/PSS* PEMs have around seven molecule waters per ion pair, and NaCl ion pairs have around 15 molecules of water associated to them. Schönhoff et al.[42] have estimated the effective dielectric constant inside PEMs to be around $\varepsilon \sim 30$ for *PAH/PSS* systems, and $\varepsilon < 19$ for *PDADMA/PSS* systems, which indeed leads to a dramatic change of the dielectric constant of water in bulk $\varepsilon \sim 80$. In our opinion, all these studies clearly show that this point needs to be explored, namely to take into account the change in the dielectric constant among the different regions within the system. Of course, an AA simulation model with explicit solvent improving the atomistic simulations shown in the section "Central Region of PSS/PDADMA PEM" would be ideal since these effects would be automatically included. However, as has been shown before, with that approach, it is currently not possible to tackle real PEM systems due to the large size and long timescales involved. The most simple approach to try to circumvent the problem of the solvent could be to start with a very simple three-region approach of different homogeneous dielectrics. In that approach, the substrate, layer region, and bulky water region would have different dielectric constants. Due to the dielectric boundaries, the calculation of the electrostatic interactions is necessarily more complex than in the case of just one homogeneous dielectric background. For planar geometries, the electrostatic boundary conditions can be fulfilled through the inclusion of virtual image charges. Based on this method, we have developed two very efficient and accurate algorithms, which can in one case even reach almost linear scaling[194,195] and which easily allow for the presence of three dielectric slabs. An even more refined model would be to allow the dielectric constant in the multilayer region to be a variable function $\varepsilon = \varepsilon(x,y,z)$ of position. In this way one could also incorporate the local salt concentration, which would affect the dielectric constant. This case is much more difficult to handle efficiently such that it can be used within a standard MD or MC calculation. Some possible candidates for such algorithms could be extensions of standard multigrid algorithms[196,197] and Maggswellian lattice algorithms.[198–204]

An alternative approach to improving the description of the solvent involves inferring the effective potentials between CG monomers from atomistic simulations where several short PE chains with explicit solvent are simulated in an environment similar to that one would expect to find in experiments inside PEMs (reduced amount of water and a fixed PE concentration). Similar approaches have been already used in the modeling of heterogeneous membranes.[205] They are just a particular case of multiscale modeling techniques[206] that are explored by many groups.[207–211] The application of other multiscaling strategies including adaptive resolution algorithms[212] to the study of PEMs is a worthwhile topic of research. Without doubt, PEMs pose a real challenge for the development of numerical methods that give an accurate physical description but are still efficient in CPU time.

5.5.3 Study of the Diffusive Behavior of PE in PEMs

In the section "Central Region of PSS/PDADMA PEM," we have shown the first steps toward the numerical calculation of the diffusion properties for PEMs using atomistic simulations. Despite being more exact than any CG approach, such diffusion coefficients are still a little bit off from the experimental[213] because the PEM conditions are not fully captured in such models. As the level of coarse-graining increases, the larger is the mismatch with experimental values. In fact, CG models, by their intrinsic nature, cannot aspire to give diffusion values that match with experimental ones, but they can be used to study more general aspects of the behavior of the diffusion coefficients in PEMs. Thus, for instance, Figure 5.13 displays the dependence of the diffusion coefficient with temperature for a four-layer PEM with values for the parameters such as those used in the previous system C. The four-layer

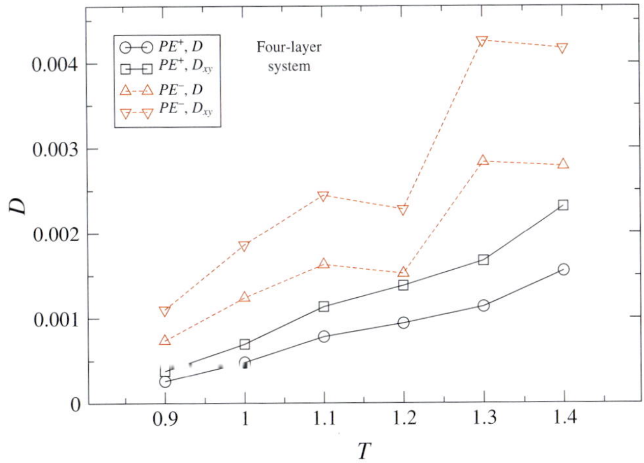

FIGURE 5.13. The total and lateral diffusion coefficients (D and D_{xy}, respectively) for system C (see Fig. 5.9) after the fourth cycle are shown as a function of temperature for both types of PE: PE^+ are depicted by black continuous lines (circles for D and squares for D_{xy}, and PE^- are depicted by red dashed lines (up triangles for D and down triangles for D_{xy}).

PEM has been built up according to our proposed model (Section 5.4.3; see also Cerdà et al.[36,37]). Two different diffusion coefficients are measured for monomers of PE within the multilayers: the total 3D diffusion coefficient D given by

$$\langle (\boldsymbol{r}(\Delta t) - \boldsymbol{r}(0))^2 \rangle \equiv 6D\Delta t, \tag{5.7}$$

where $\boldsymbol{r} = (x,y,z)$ represents the position vector of monomers of PE^+ or PE^- depending on for which type of PE the diffusion coefficient is measured, and the lateral diffusion coefficient for monomers D_{xy}, which is given by

$$\langle (\rho(\Delta t) - \rho(0))^2 \rangle \equiv 4D_{xy}\Delta t, \tag{5.8}$$

where $\rho = (x,y)$ represents the components of the position vector parallel to the substrate. Given the strong intermixing of the several layers and the short thickness of the PEM (around $9\sigma_m$), we did not attempt to measure the different values of D and D_{xy} as a function of the distance to the substrate: Rather, a single value of the diffusion coefficients D and D_{xy} will be reported for the whole multilayer. The diffusion coefficients are measured by recording the positions of particles (frames) at intervals of 2500 MD time steps. Approximately 500 frames were recorded for each case in order to obtain reasonable statistics.

In Figure 5.13, we use circles and squares to denote the 3D diffusion coefficient D and the lateral diffusion coefficient D_{xy}, respectively, for PE^+. For the case of PE^-, red triangles up and down are used, respectively. As expected, it can be seen that both diffusion coefficients, D and D_{xy}, increase with temperature. The values of the coefficients of diffusion for PE^- are much larger than the values of those for PE^+. This behavior is compatible with the fact that the first PE layer in direct contact with the fixed negatively charged surface is made of PE^+ and has a stronger friction and therefore less mobility than subsequent layers, while the last PE layer in those four-layer PEMs is made of PE^- and it is more mobile because no other layers are deposited on top of it. It is also remarkable that the values of D and D_{xy} are approximately related via the expression $D \approx 2/3D_{xy}$ which implies that the diffusion coefficient along the z-direction D_z is almost negligible when compared with the value of D_{xy}. This last fact strongly suggests that chains tend to move much faster in the direction parallel to the substrate than in the direction perpendicular to it.

5.5.4 Study of the Thermodynamical Stability of PEMs

Using our proposed model, a simple test can be performed to assess the hypothesis of Abu-Sharkh,[180] which states that PEMs deposited from salt-free solutions on charged planar surfaces are thermodynamically stable structures that form spontaneously regardless of the method of deposition. We choose a PEM created taking after our system C (after the completion of the fourth cycle) at temperature $T = 1.0$. Now we turn the monomer–monomer interaction from attractive to purely repulsive by setting a new cutoff for the monomer–monomer interaction $r_{\text{cut},m-m} = 2^{1/6}\sigma_m$, and we run the new parameters for some millions of MD steps. Our simulations show that the redissolution of the PEM is unavoidable, and just a tiny layer of PE^+ adsorbed directly onto the negative substrate remains. Clearly, the mixture we have got comes from a formed PEM and therefore, if we reset the value of $r_{\text{cut},m-m}$ of the monomer–monomer interaction to its original value and allow for enough time, one should observe hints of the rebuilding of the PEM if the suggestion made by Abu-Sharkh is true. Nonetheless, our results clearly show that the system tends to remain in the unadsorbed disordered state, and no sign of PEM rebuildup is observed. Rather, we observe the formation of a complex of PE^+ and PE^- plus a monolayer of PE^+ adsorbed onto the substrate with a few chains of PE^- also adsorbed. This result clearly works against the hypothesis of Abu-Sharkh[180] and provides a further hint in the direction that complexation in the bulk favorably forms layers on top of a charged substrate.

5.6 CHALLENGING TOPICS IN THE NUMERICAL SIMULATIONS OF PEMs

As we explained before, a refined primitive model would probably be helpful in understanding the structure and dynamics of PEMs. A first challenge is to have an improved CG model that is able to predict more accurately the interactions of PEs in a PEM. A stringent test for the new models would be, for example, the ability to reproduce numerically the force profiles obtained from AFM experiments.[134,156,214] So far, experiments have only dealt with a single adsorbed chain that is pulled away from a substrate. It would be desirable to have experimental AFM force profiles of chains pull out from PEMs since in this way one could also test the entanglements within PEM layers. Another challenging topic will be the investigation of the kinetics of PEMs. Can one detect and measure the kinetic barriers associated with the proposed metastability of PEMs, as well as perform a sound test of the hypothesis of the glassy and the liquid PEM states?[83] This topic will be even more challenging than the study of polymeric glass transition, a theme whose complexity has accompanied us throughout the last three decades without a final result. Also, surprisingly, no numerical work so far has dealt with the issue of the effects of the ionic strength and type of salt ions on the growth and properties of PEMs. Experimentally, the effects of salt on the stability of PEMs have been recently found to be much more complex than formerly thought, including reentrant transitions to the stable regime for moderate concentrations of salt at low and high fraction degrees of charge.[44]

A related topic that can also bring valuable insight into the glass–liquid PEMs hypothesis is the numerical study of the self-diffusion of PE within PEMs. It should be noted that the diffusion of particles and macromolecules inside a PEM has not been considered in numerical studies so far. To our best knowledge, only very few experimental works have already started to address this fundamental issue. For instance, Nazaran et al.[213] have studied the lateral mobility of PEMs via fluorescence recovery after photobleaching (FRAP). The study shows that the density of charged sites along the polymer is the most important parameter, and it has the effect that when the charge density increases, the diffusion coefficient decreases. Added stiffness reduces the interpenetration of PEs in the PEM and increases the lateral mobility. The diffusion coefficients are found to be the largest on the surfaces, and decrease when further layers adsorb on the top. After two double layers of PE, one reaches conditions that are very similar to the "bulk" conditions, and the value of the diffusion coefficient remains constant. Sagidullin et al.[215] have studied the diffusion of probe molecules in polymer systems. Pores in PEMs are not usually static but are so-called breathing pores; that is, they form due to the interaction with the particles and close afterward. More recently, Yoo et al.[216] have measured the interdiffusion of PEs during electrostatic LbL assembly via adsorbing viruses onto the surface of the PEM. The viral movement is analyzed to probe the mobility of the PE of the PEM. Crouzier and Picart[217] have as well reported the diffusion of poly-L-lysine (PLL) inside different PEMs.

Another challenging topic is the modeling of weak PEs and the numerical study of PEMs made of such types of PEs. The dependence of the degree of charge of weak PE on the pH adds an extra degree of complexity to the CG modeling of such types of charged polymer chains. For MC simulations, several approaches have been developed,[218–221] but to our knowledge no good MD procedure has been developed so far, and none has been applied to the study of PEMs.

Numerical studies have focused up to now on the growth of flexible PE chains on top of simple substrate geometries since it is the most simple case. It would definitely be worthwhile to also investigate the influence of semiflexible PEs on PEM growth. The growth of stable PEMs on top of porous microgels where the "effective force" between the PEM and the substrate seems to be the entanglement of the PEs and the microgel[222] is another issue worth studying, as well as the growth process of PEMs onto patterned geometries. Experiments by Pallandre et al.[223] have shown that it is possible to control laterally the size of PEMs via nanostripped surfaces with alternating attracting and repulsive patterned stripes. Numerical simulations could help to provide a better understanding of which are the parameters to have stable stripped PEMs with optimal properties.

The effects of external fields on the building up of PEMs and the modifications those fields can induce on the properties of already formed PEMs are issues that could be worth investigating. Ngankam and Tassel[224] and Olsen and Tassel[225] have reported very interesting experimental studies on the effects of electrical fields on the adsorption of a single PE layer. Ngankam and Tassel[224] have recently shown than the application of a modest electrical field during the adsorption of a charged polymer leads to a continuous adsorption almost linearly in time that does not seem to reach saturation. Later, Olsen and Tassel[225] experimentally investigated if deprotonation is a necessary condition for having such a type of endless adsorption under the action of a constant electrical field. They have found that continuous adsorption is possible not only for weakly charged PEs but also for strongly charged PEs via a counterion condensation mechanism. These previous findings suggest that the application of electrical fields during the buildup phase of PEMs could dramatically change their properties, allowing an improved tailoring of their functions. External fields are able to induce a mechanical stretching of PEMs. This could be also of interest in the formation of pores of controllable size that close after the external field is switched off.[96]

Despite the 8 years that have passed since the first numerical simulations of PEMs,[154] the amount of unknowns waiting to be solved is substantial. It is our belief that many of them can be properly addressed by using refined primitive models without needing to wait a few decades until computers are fast enough to make atomistic simulations of PEM systems feasible.

5.7 SUMMARY

In this chapter, we have reviewed several topics related to PEMs: its current state, how to build up such multilayers, their applications, the current existing approaches to modeling the PEMs, and the perspectives in the modeling of such complex multilayers. To model PEMs in the area of the current existing approaches, we have first reviewed the theoretical models and the few atomistic attempts to obtain results of relevance for PEMs. Subsequently, a detailed review on the CG attempts used so far to model PEMs was presented. We have tried to show that current primitive models need to be further refined in order to reach a better physical description of PEMs. Some ideas about what should be the first features to be improved have been suggested, and those include primarily the dipping protocol and the treatment of electrostatic interactions in the layers. We also discussed the main challenging topics that numerical studies should address using refined primitive models, at least from our perspective.

PEMs are not only relevant to all the current technological applications and the large potential for new ones, as we have discussed in Section 5.1, but they also offer us a superb opportunity to extend and test our knowledge about the modeling approaches of charged systems that can encompass all aspects of science, from statistical mechanics and material science to computational physics.

ACKNOWLEDGMENTS

J.J. Cerdà acknowledges the financial support of the Spanish Ministerio de Educación y Ciencia, postdoctoral grant no. EXP2006-0931, and C. Holm and B. Qiao acknowledge support through the SPP 1369 and a DFG grant HO 1108/17-2.

REFERENCES

1. J. Schmitt, G. Decher, G. Hong. *Thin Solid Films* **1992**, 210/211, 831.
2. G. Decher. *Science* **1997**, 277, 1232–1237.
3. K. Ariga, J. P. Hill, Q. Ji. *Phys Chem Chem Phys* **2007**, 9, 2319–2340.
4. J. B. Schlenoff. *Langmuir* **2009**, 25, 14007–14010.
5. V. K. Kamineni, Y. M. Lvov, T. A. Dobbins. *Langmuir* **2007**, 23, 7423–7427.
6. S. T. Dubas and J. B. Schlenoff. *Macromolecules* **1999**, 32, 8153–8160.
7. J. Cho, K. Char, J.-D. Hong, K.-B. Lee. *Adv Mater* **2001**, 13, 1076–1078.
8. P. A. Chiarelli, M. S. Johal, D. J. Holmes, J. L. Casson, J. M. Robinson, H.-L. Wang. *Langmuir* **2002**, 18, 168–173.
9. J. B. Schlenoff, S. T. Dubas, T. Farhat. *Langmuir* **2000**, 16, 9968–9969.
10. M. Kolasinska, R. Krastev, T. Gutberlet, P. Warszynski. *Langmuir* **2009**, 25, 1224–1232.
11. Y. Wang, A. S. Angelatos, F. Caruso. *Chem Mater* **2008**, 20, 848–858.
12. D. Trau and R. Renneberg. *Biosens Bioelectron* **2003**, 18, 1491–1499.
13. B. Thierry, F. Winnik, Y. Merhi, M. Tabrizian. *J Am Chem Soc* **2003**, 125, 7494–7495.

14. J. Xi, Z. Wu, X. Teng, Y. Zhao, L. Chen, X. Qiu. *J Mater Chem* **2008**, 18, 1232–1238.
15. R. Malaisamy and M. Bruening. *Langmuir* **2005**, 21, 10587–10592.
16. M. L. Bruening, D. M. Dotzauer, P. Jain, L. Ouyang, G. L. Baker. *Langmuir* **2008**, 24, 7663–7673.
17. S. Datta, C. Cecil, D. Bhattacharyya. *Ind Eng Chem Res* **2008**, 47, 4586–4597.
18. L. Jiang, F. Lu, Q. Chang, Y. Liu, H. Liu, Y. Li, W. Xu, G. Cui, J. Zhuang, X. Li. *Chem Phys Chem* **2005**, 6, 481–486.
19. A. Arsenault, J. Halfyard, Z. Wang, V. Kitaev, G. Ozin, I. Manners, A. Mihi, H. Miguez. *Langmuir* **2005**, 21, 499–503.
20. S. Wang, C. Li, F. Chen, G. Shi. *Nanotechnology* **2007**, 18, 185707.
21. L. Zhao, T. Ma, H. Bai, G. Lu, C. Li, G. Shi. *Langmuir* **2008**, 24, 4380–4387.
22. M. Kamande, K. Fletcher, M. Lowry, I. Warner. *J Sep Sci* **2005**, 28, 710–718.
23. A. Khopade, N. Arulsudar, S. Khopade, J. Hartmann. *Biomacromolecules* **2005**, 6, 229–234.
24. K. Ray, R. Badugu, J. R. Lakowicz. *Chem Mater* **2007**, 19, 5902–5909.
25. J. A. Lichter and M. F. Rubner. *Langmuir* **2009**, 25, 7686–7694.
26. J. A. Lichter, K. J. Van Vliet, M. F. Rubner. *Macromolecules* **2009**, 42, 8573–8586.
27. D. Cui, A. Szarpak, I. Pignot-Paintrand, A. Varrot, T. Boudou, C. Detrembleur, C. Jrme, C. Picart, R. Auzly-Velty. *Adv Funct Mater* **2010**, 20, 3303–3312.
28. D. V. Volodkin, P. Schaaf, H. Mohwald, J. Voegel, V. Ball. *Soft Matter* **2009**, 5, 1394–1405.
29. W. Ouyang, M. Muller, D. Appelhans, B. Voit. *ACS Appl Mater Interfaces* **2009**, 1, 2878–2885.
30. H. H. Rmaile and J. B. Schlenoff. *J Am Chem Soc* **2003**, 125, 6602–6603.
31. P. Hammond. *Curr Opin Colloid Interface Sci* **2000**, 4, 430–442.
32. M. Schönhoff. *Curr Opin Colloid Interface Sci* **2003**, 8, 86–95.
33. M. Schönhoff. *J Phys Condens Matter* **2003**, 15, R1781–R1808.
34. R. Messina, C. Holm, K. Kremer. *J Polym Scit B Polym Phys* **2004**, 42, 3557.
35. R. von Klitzing, J. Wong, W. Jäger, R. Steitz. *Curr Opin Colloid Interface Sci* **2004**, 9, 158–162.
36. J. J. Cerdà, B. Qiao, C. Holm. *Soft Matter* **2009**, 5, 4412–4425.
37. J. J. Cerdà, B. Qiao, C. Holm. *Eur Phys J Spec Top* **2009**, 177, 129–148.
38. A. V. Dobrynin. *Curr Opin Colloid Interface Sci* **2008**, 13, 376–388.
39. R. Messina. *J Phys Condens Matter* **2009**, 21, 113102.
40. K. Glinel, C. Déjugnat, M. Prevot, B. Schöler, M. Schönhoff, R. V. Klitzing. *Colloids Surf A Physicochem Eng Aspects* **2007**, 303, 3–13.
41. S. A. Sukhishvili. *Curr Opin Colloid Interface Sci* **2005**, 10, 37–44.
42. M. Schönhoff, V. Ball, A. R. Bausch, C. Dejugnat, N. Delorme, K. Glinel, R. von Klitzing, R. Steitz. *Colloids Surf A* **2007**, 303, 14–29.
43. C. Picart. *Curr Med Chem* **2008**, 15, 685–697.
44. S. A. Sukhishvili, E. Kharlampieva, V. Izumrudov. *Macromolecules* **2006**, 39, 8873–8881.
45. R. von Klitzing. *Phys Chem Chem Phys* **2006**, 8, 5012–5033.

46. J. A. Jaber and J. B. Schlenoff. *Curr Opin Colloid Interface Sci* **2006**, 11, 324–329.
47. D. T. Haynie, L. Zhang, J. S. Rudra, W. Zhao, Y. Zhong, N. Palath. *Biomacromolecules* **2005**, 6, 2895–2913.
48. D. T. Haynie and W. Zhao. *J Nanosci Nanotechnol* **2009**, 9, 3562–3567.
49. J. B. Schlenoff and S. T. Dubas. *Macromolecules* **2001**, 34, 592–598.
50. P. Lavalle, C. Picart, J. Mutterer, C. Gergely, H. Reiss, J.-C. Voegel, B. Senger, P. Schaaf. *J Phys Chem B* **2004**, 108, 635–648.
51. P. Kujawa, P. Moraille, J. Sanchez, A. Badia, F. Winnik. *J Am Chem Soc* **2005**, 127, 9224–9234.
52. M. Salomäki, I. A. Vinokurov, J. Kankare. *Langmuir* **2005**, 21, 11232–11240.
53. C. Porcel, P. Lavalle, V. Ball, G. Decher, B. Senger, J.-C. Voegel, P. Schaaf. *Langmuir* **2006**, 22, 4376–4383.
54. N. Laugel, C. Betscha, M. Winterhalter, J.-C. Voegel, P. Schaaf, V. Ball. *J Phys Chem B* **2006**, 110, 19443–19449.
55. N. Hoda and R. G. Larson. *J Phys Chem B* **2009**, 113, 4232–4241.
56. G. Liu, Y. Hou, X. Xiao, G. Zhang. *J Phys Chem B* **2010**, 114, 9987–9993.
57. C. Picart, J. Mutterer, L. Richert, Y. Luo, G. D. Prestwich, P. Schaaf, J.-C. Voegel, P. Lavalle. *Proc Natl Acad Sci U S A* **2002**, 99, 12531–12535.
58. G. Liu, Y. Hou, X. Xiao, G. Zhang. *J Phys Chem B* **2010**, 114, 9987–9993.
59. B. Schoeler, E. Poptoshev, F. Caruso. *Macromolecules* **2003**, 36, 5258–5264.
60. R. A. McAloney, M. Sinyor, V. Dudnik, M. C. Goh. *Langmuir* **2001**, 17, 6655–6663.
61. P. A. Patel, J. Jeon, P. T. Mather, A. V. Dobrynin. *Langmuir* **2006**, 22, 9994–10002.
62. J. E. Wong, H. Zastrow, W. Jaeger, R. von Klitzing. *Langmuir* **2009**, 25, 14061–14070.
63. R. Steitz, W. Jaeger, R. von Klitzing. *Langmuir* **2001**, 17, 4471–4474.
64. B. Schoeler, G. Kumaraswamy, F. Caruso. *Macromolecules* **2002**, 35, 889–897.
65. K. Glinel, A. Moussa, A. M. Jonas, A. Laschewsky. *Langmuir* **2002**, 18, 1408–1412.
66. P. Bertrand, A. Jonas, A. Laschewsky, R. Legras. *Macromol Rapid Commun* **2000**, 21, 319–348.
67. K. Kamburova, V. Milkova, I. Petkanchin, T. Radeva. *Biomacromolecules* **2008**, 9, 1242–1247.
68. N. Fujii, K. Fujimoto, T. Michinobu, M. Akada, J. P. Hill, S. Shiratori, K. Ariga, K. Shigehara. *Macromolecules* **2010**, 43, 3947–3955.
69. E. Guzman, V. San Miguel, C. Peinado, F. Ortega, R. G. Rubio. *Langmuir* **2010**, 26, 11494–11502.
70. L. Lundstrom-Hamala, E. Johansson, L. Wagberg. *Starch/Starke* **2010**, 62, 102–114.
71. M. Salomäki and J. Kankare. *Macromolecules* **2008**, 41, 4423–4428.
72. A. E. E. Haitami, D. Martel, V. Ball, H. C. Nguyen, E. Gonthier, P. Labbe, J. Voegel, P. Schaaf, B. Senger, F. Boulmedais. *Langmuir* **2009**, 25, 2282–2289.
73. M. Gopinadhan, O. Ivanova, H. Ahrens, J.-U. Günther, R. Steitz, C. A. Helm. *J Phys Chem B* **2007**, 111, 8426–8434.
74. G. Liu, J. Zhao, Q. Sun, G. Zhang. *J Phys Chem B* **2008**, 112, 3333–3338.

75. M. Gopinadhan, H. Ahrens, J.-U. Gunther, R. Steitz, C. A. Helm. *Macromolecules* **2005**, 38, 5228–5235.
76. M. Cornelsen, C. A. Helm, S. Block. *Macromolecules* **2010**, 43, 4300–4309.
77. H. Riegler and F. Essler. *Langmuir* **2002**, 18, 6694–6698.
78. O. M. Tanchak, K. G. Yager, H. Fritzsche, T. Harroun, J. Katsaras, C. J. Barrett. *J Chem Phys* **2008**, 129, 084901.
79. A. Garg, J. R. Heflin, H. W. Gibson, R. M. Davis. *Langmuir* **2008**, 24, 10887–10894.
80. M. Kolasinska and P. Warszynski. *Bioelectrochemistry* **2005**, 66, 65–70.
81. C. Buron, C. Filiatre, F. Membrey, C. Bainier, D. Charraut, A. Foissy. *Colloids Surf A* **2007**, 305, 105–111.
82. A. Guyomard, G. Muller, K. Glinel. *Macromolecules* **2005**, 38, 5737–5742.
83. D. Kovacevic, S. van der Burgh, A. de Keizer, M. Stuart. *Langmuir* **2002**, 18, 5607–5612.
84. E. Kharlampieva and S. Sukhishvili. *Langmuir* **2003**, 19, 1235–1243.
85. D. Kovacevic, S. van der Burgh, A. de Keizer, M. A. Cohen Stuart. *J Phys Chem B* **2003**, 107, 7998–8002.
86. M. Muller, J. Meier-Haack, S. Schwarz, H. Buchhammer, E. Eichhorn, A. Janke, B. Kessler, J. Nagel, M. Oelmann, T. Reihs, K. Lunkwitz. *J Adhes* **2004**, 80, 521–547.
87. C. Buron, C. Filiatre, F. Membrey, C. Bainier, D. Charraut, A. Foissy. *J Colloid Interface Sci* **2007**, 314, 358–366.
88. E. Kharlampieva, J. F. Ankner, M. Rubinstein, S. A. Sukhishvili. *Phys Rev Lett* **2008**, 100, 128303.
89. M. Elzbieciak, M. Kolasinska, P. Warszynski. *Colloids Surf A* **2008**, 321, 258–261.
90. J. Fu, J. Ji, L. Shen, A. Kuller, A. Rosenhahn, J. Shen, M. Grunze. *Langmuir* **2009**, 25, 672–675.
91. K. Itano, J. Choi, M. F. Rubner. *Macromolecules* **2005**, 38, 3450–3460.
92. L. A. Connal, Q. Li, J. F. Quinn, E. Tjipto, F. Caruso, G. G. Qiao. *Macromolecules* **2008**, 41, 2620–2626.
93. L. Shao and J. L. Lutkenhaus. *Soft Matter* **2010**, 6, 3363–3369.
94. J. Iturri Ramos, I. Llarena, S. Moya. *J Mater Sci* **2010**, 45, 4970–4976.
95. J. Dejeu, S. Diziain, C. Dange, F. Membrey, D. Charraut, A. Foissy. *Langmuir* **2008**, 24, 3090–3098.
96. D. Mertz, J. Hemmerle, F. Boulmedais, J.-C. Voegel, P. Lavalle, P. Schaaf. *Soft Matter* **2007**, 3, 1413–1420.
97. F. Wang, L. Li, C. Gao. *Macromol Chem Phys* **2009**, 210, 2022–2028.
98. X. Gong and C. Gao. *Phys Chem Chem Phys* **2009**, 11, 11577–11586.
99. J. Fruh, R. Kohler, H. Mohwald, R. Krastev. *Langmuir* **2010**, 26, 15516–15522.
100. C. S. Peyratout and L. Dähne. *Angew Chem Int Ed* **2004**, 43, 3762–3783.
101. M. Stukan, V. Lobaskin, C. Holm, O. Vinogradova. *Phys Rev E* **2006**, 73, 021801.
102. P. M. Biesheuvel, T. Mauser, G. B. Sukhorukov, H. Mohwald. *Macromolecules* **2006**, 39, 8480–8486.
103. B. G. De Geest, G. B. Sukhorukov, H. Möhwald. *Expert Opin Drug Deliv* **2009**, 6, 613–624.
104. A. Guyomard, B. Nysten, G. Muller, K. Glinel. *Langmuir* **2006**, 22, 2281–2287.

105. B. Wang, C. Gao, L. Liu. *J Phys Chem B* **2005**, 109, 4887–4892.
106. J. K. Carr, R. D. Himes, C. H. Keung, D. L. Burden, P. K. Walhout. *Langmuir* **2009**, 25, 8330–8339.
107. R. N. Smith, L. Reven, C. J. Barrett. *Macromolecules* **2003**, 36, 1876–1881.
108. M. McCormick, R. N. Smith, R. Graf, C. J. Barrett, L. Reven, H. W. Spiess. *Macromolecules* **2003**, 36, 3616–3625.
109. R. V. Klitzing and H. Mohwald. *Macromolecules* **1996**, 29, 6901–6906.
110. H. H. Rmaile, T. R. Farhat, J. B. Schlenoff. *J Phys Chem B* **2003**, 107, 14401–14406.
111. J. Mathew, C. Aravindakumar, U. K. Aravind. *J Memb Sci* **2008**, 325, 625–632.
112. D. Andelman and J. Joanny. *C R Acad Sci IV* **2000**, 1, 1153–1162.
113. R. Chang and A. Yethiraj. *Prog Org Coat* **2003**, 47, 331–336.
114. A. V. Dobrynin and M. Rubinstein. *Macromolecules* **2001**, 34, 1964–1972.
115. A. V. Dobrynin and M. Rubinstein. *J Phys Chem B* **2003**, 107, 8260–8269.
116. A. Shafir and D. Andelman. *J Chem Phys* **2003**, 119, 2355.
117. A. Shafir and D. Andelman. *Phys Rev E Stat Nonlin Soft Matter Phys* **2004**, 70, 061804.
118. Q. Wang. *Macromolecules* **2005**, 38, 8911–8922.
119. J. Ubbink and A. R. Khokhlov. *J Chem Phys* **2004**, 120, 5353–5365.
120. M. Manghi and M. Aubouy. *Phys Chem Chem Phys* **2008**, 10, 1697–1706.
121. C. N. Patra, R. Chang, A. Yethiraj. *J Phys Chem B* **2004**, 108, 9126–9132.
122. O. Pizio, K. Bucior, A. Patrykiejew, S. Sokolowski. *J Chem Phys* **2005**, 123, 214902.
123. Z. Li and J. Wu. *Phys Rev Lett* **2006**, 96, 048302.
124. K. Bucior, J. Fischer, A. Patrykiejew, R. Tscheliessnig, S. Sokolowski. *J Chem Phys* **2007**, 126, 094704.
125. L. Wang, H. Liang, J. Wu. *J Chem Phys* **2010**, 133, 044906–044913.
126. R. Netz and J. Joanny. *Macromolecules* **1999**, 32, 9013.
127. S. Park, M. Rubner, A. Mayes. *Langmuir* **2002**, 18, 9600–9604.
128. M. Castelnovo and J. F. Joanny. *Langmuir* **2000**, 16, 7524–7532.
129. A. Shafir and D. Andelman. *Eur Phys J E* **2006**, 19, 155–162.
130. Q. Wang. *J Phys Chem B* **2006**, 110, 5825–5828.
131. Q. Wang. *Soft Matter* **2009**, 5, 413–424.
132. G. Reddy, R. Chang, A. Yethiraj. , *J Chem Theory Comput* **2006**, 2, 630–636.
133. A. A. Chialvo and J. M. Simonson. *J Phys Chem C* **2008**, 112, 19521–19529.
134. D. Horinek, A. Serr, M. Geisler, T. Pirzer, U. Slotta, S. Q. Lud, J. A. Garrido, T. Scheibel, T. Hugel, R. R. Netz. *Proc Natl Acad Sci U S A* **2008**, 105, 2842–2847.
135. R. R. Netz and D. Andelman. *Phys Rep* **2003**, 380, 1–95.
136. P. M. Claesson, E. Poptoshev, E. Blomberg, A. Dedinaite. *Adv Colloid Interface Sci* **2005**, 114–115, 173–187.
137. A. V. Dobrynin and M. Rubinstein. *Prog Polym Sci* **2005**, 30, 1049–1118.
138. S. Block and C. A. Helm. *Macromolecules* **2009**, 42, 6733–6740.
139. Y. Roiter, O. Trotsenko, V. Tokarev, S. Minko. *J Am Chem Soc* **2010**, 132, 13660–13662.
140. B. Qiao, J. J. Cerdà, C. Holm. *Macromolecules* **2011**, 44(6), 1707–1718.

141. W. A. Steele. *J Phys Chem* **1978**, 82, 817–821.
142. R. F. Cracknell, D. Nicholson, N. Quirke. *Phys Rev Lett* **1995**, 74, 2463.
143. J. Stecki. *Langmuir* **1997**, 13, 597–598.
144. H. J. C. Berendsen, J. R. Grigera, T. P. Straatsma. *J Phys Chem* **1987**, 91, 6269–6271.
145. S. Miyamoto and P. A. Kollman. *J Comput Chem* **1992**, 13, 952–962.
146. W. Jorgensen, D. Maxwell, J. Tirado-Rives. *J Am Chem Soc* **1996**, 118, 11225–11236.
147. B. Qiao, J. J. Cerdà, C. Holm. *Macromolecules* **2010**, 43, 7828–7838.
148. A. Vishnyakov and A. V. Neimark. *J Chem Phys* **2008**, 128, 164902–164911.
149. B. Hess, C. Kutzner, D. van der Spoel, E. Lindahl. *J Chem Theory Comput* **2008**, 4, 435–447.
150. T. Darden, D. York, L. Pedersen. *J Chem Phys* **1993**, 98, 10089–10092.
151. U. Essmann, L. Perera, M. L. Berkowitz, T. Darden, H. Lee, L. Pedersen. *J Chem Phys* **1995**, 103, 8577.
152. I.-C. Yeh and M. L. Berkowitz. *J Chem Phys* **1999**, 111, 3155–3162.
153. B. Hess, H. Bekker, H. J. C. Berendsen, J. G. E. M. Fraaije. *J Comput Chem* **1997**, 18, 1463–1472.
154. R. Messina, C. Holm, K. Kremer. *Langmuir* **2003**, 19, 4473–4482.
155. R. Messina. *J Chem Phys* **2003**, 119, 8133–8139.
156. S. Block and C. A. Helm. *J Phys Chem B* **2008**, 112, 9318–9327.
157. G. Reddy and A. Yethiraj. *J Chem Phys* **2010**, 132, 074903.
158. T. Farhat, G. Yassin, S. T. Dubas, J. B. Schlenoff. *Langmuir* **1999**, 15, 6621–6623.
159. L. N. J. Rodriguez, S. M. D. Paul, C. J. Barrett, L. Reven, H. W. Spiess. *Adv Mater* **2000**, 12, 1934–1938.
160. W. Zhao, B. Zheng, D. T. Haynie. *Langmuir* **2006**, 22, 6668–6675.
161. L. Zhang, W. Zhao, J. S. Rudra, D. T. Haynie. *ACS Nano* **2007**, 1, 476–486.
162. N. Hoda and R. G. Larson. *Macromolecules* **2009**, 42, 8851–8863.
163. J. B. Schlenoff, A. H. Rmaile, C. B. Bucur. *J Am Chem Soc* **2008**, 130, 13589–13597.
164. E. Spruijt, A. H. Westphal, J. W. Borst, M. A. Cohen Stuart, J. van der Gucht. *Macromolecules* **2010**, 43, 6476–6484.
165. A. S. Thomas and A. H. Elcock. *J Am Chem Soc* **2007**, 129, 14887–14898.
166. A. Kulcsar, J.-C. Voegel, P. Schaaf, P. Kekicheff. *Langmuir* **2005**, 21, 1166–1170.
167. B. Hess, C. Holm, N. van der Vegt. *Phys Rev Lett* **2006**, 96, 147801.
168. B. Hess, C. Holm, N. van der Vegt. *J Chem Phys* **2006**, 124, 164509.
169. S. Gavryushov and P. Linse. *J Phys Chem B* **2006**, 110, 10878–10887.
170. P. A. Neff, B. K. Wunderlich, R. von Klitzing, A. R. Bausch. *Langmuir* **2007**, 23, 4048–4052.
171. C. Wende and M. Schönhoff. *Langmuir* **2010**, 26, 8352–8357.
172. P. G. Kusalik, M. E. Mandy, I. M. Svishchev. *J Chem Phys* **1994**, 100, 7654–7664.
173. M. W. Mahoney and W. L. Jorgensen. *J Chem Phys* **2000**, 112, 8910–8922.
174. J. D. Weeks, D. Chandler, H. C. Andersen. *J Chem Phys* **1971**, 54, 5237.
175. J. Israelachvili. *Intermolecular and Surface Forces*. Academic, London, 1992.

176. C. Holm and K. Kremer. *Electrostatic Effects in Soft Matter and Biophysics*, vol. 46. C. Holm, P. Kékicheff, R. Podgornik, eds. Kluwer Academic Publishers, Dordrecht, NL, 2001.
177. R. Messina. *Macromolecules* **2004**, 37, 621–629.
178. V. Panchagnula, J. Jeon, A. V. Dobrynin. *Phys Rev Lett* **2004**, 93, 037801.
179. R. Messina. *Phys Rev E* **2004**, 70, 051802.
180. B. Abu-Sharkh. *J Chem Phys* **2005**, 123, 114907.
181. P. Patel, J. Jeon, P. Mather, A. Dobrynin. *Langmuir* **2005**, 21, 6113–6122.
182. V. Panchagnula, J. Jeon, J. Rusling, A. Dobrynin. *Langmuir* **2005**, 21, 1118–1125.
183. J. Jeon, V. Panchagnula, J. Pan, A. V. Dobrynin. *Langmuir* **2006**, 22, 4629–4637.
184. B. F. Abu-Sharkh. *Polymer* **2006**, 47, 3674–3680.
185. J.-M. Y. Carrillo and A. V. Dobrynin. *Langmuir* **2007**, 23, 2472–2482.
186. C. F. Narambuena, D. M. Beltramo, E. P. M. Leiva. *Macromolecules* **2007**, 40, 7336–7342.
187. R. W. Hockney and J. W. Eastwood. *Computer Simulation Using Particles*. IOP, London, 1988.
188. A. Arnold, J. de Joannis, C. Holm. *J Chem Phys* **2002**, 117, 2496–2502.
189. A. Arnold, J. de Joannis, C. Holm. *J Chem Phys* **2002**, 117, 2503–2512.
190. M. P. Allen and D. J. Tildesley. *Computer Simulation of Liquids*, 1st edn. Clarendon Press, Oxford, 1987.
191. O. M. Tanchak, K. G. Yager, H. Fritzsche, T. Harroun, J. Katsaras, C. J. Barrett. *Langmuir* **2006**, 22, 5137–5143.
192. J. A. Jaber and J. B. Schlenoff. *Langmuir* **2007**, 23, 896–901.
193. O. Ivanova, O. Soltwedel, M. Gopinadhan, R. Ko ihler, R. Steitz, C. A. Helm. *Macromolecules* **2008**, 41, 7179–7185.
194. S. Tyagi, A. Arnold, C. Holm. *J Chem Phys* **2007**, 127, 154723.
195. S. Tyagi, A. Arnold, C. Holm. *J Chem Phys* **2008**, 129, 204102.
196. C. Sagui and T. Darden. *J Chem Phys* **2001**, 114, 6578.
197. S. Banerjee and J. A. Board, Jr. *J Comput Chem* **2005**, 26, 957–967.
198. A. Maggs. *J Chem Phys* **2002**, 117, 1975.
199. A. Maggs. *J Chem Phys* **2004**, 120, 3108.
200. J. Rottler and A. C. Maggs. *Phys Rev Lett* **2004**, 93, 170201.
201. J. Rottler and A. C. Maggs. *J Chem Phys* **2004**, 120, 3119.
202. I. Pasichnyk and B. Dünweg. *J Phys Condens Mat* **2004**, 16, 3999–4020.
203. A. C. Maggs. *Phys Rev Lett* **2006**, 97, 197802–197804.
204. J. Rottler. *J Chem Phys* **2007**, 127, 134104–134109.
205. S. A. Pandit and H. L. Scott. *Biochim Biophys Acta Biomembr* **2009**, 1788, 136–148.
206. T. Murtola, A. Bunker, I. Vattulainen, M. Deserno, M. Karttunen. *Phys Chem Chem Phys* **2009**, 11, 1869–1892.
207. B. Ensing, S. O. Nielsen, P. B. Moore, M. L. Klein, M. Parrinello. *J Chem Theory Comput* **2007**, 3, 1100–1105.
208. H. Wang, C. Junghans, K. Kremer. *Eur Phys J E* **2009**, 28, 221.
209. M. Praprotnik, C. Junghans, L. D. Site, K. Kremer. *Comp Phys Comm* **2008**, 179, 51.

210. M. Praprotnik, K. Kremer, L. D. Site. *Phys Rev E* **2007**, 75, 017701.
211. L. D. Site and K. Kremer. *Int J Quantum Chem* **2005**, 101, 733.
212. M. Praprotnik, L. D. Site, K. Kremer. *J Chem Phys* **2005**, 123, 224106.
213. P. Nazaran, V. Bosio, W. Jaeger, D. Anghel, R. von Klitzing. *J Phys Chem B* **2007**, 111, 8572–8581.
214. T. Pirzer, M. Geisler, T. Scheibel, T. Hugel. *Phys Biol* **2009**, 6, 025004.
215. A. Sagidullin, J. Meier-Haack, U. Scheler. *Magn Reson Imaging* **2007**, 25, 541–543.
216. P. J. Yoo, N. S. Zacharia, J. Doh, K. T. Nam, A. M. Belcher, P. T. Hammond. *ACS Nano* **2008**, 2, 561–571.
217. T. Crouzier and C. Picart. *Biomacromolecules* **2009**, 10, 433–442.
218. G. Berghold, P. van der Schoot, C. Seidel. *J Chem Phys* **1997**, 107, 8083.
219. S. Ulrich, A. Laguecir, S. Stoll. *Macromolecules* **2005**, 38, 8939–8949.
220. S. Ulrich, M. Seijo, A. Laguecir, S. Stoll. *J Phys Chem B* **2006**, 110, 20954–20964.
221. S. Ulrich, M. Seijo, S. Stoll. *J Phys Chem B* **2007**, 111, 8459–8467.
222. J. E. Wong, C. B. Müller, A. Laschewsky, W. Richtering. *J Phys Chem B* **2007**, 111, 8527–8531.
223. A. Pallandre, A. Moussa, B. Nysten, A. Jonas. *Adv Mater* **2006**, 18, 481–486.
224. A. P. Ngankam and P. R. V. Tassel. *Proc Natl Acad Sci U S A* **2007**, 104, 1140–1145.
225. C. Olsen and P. R. V. Tassel. *J Colloid Interface Sci* **2009**, 329, 222–227.

PART II

MIXED INTERACTIONS

CHAPTER 6

IONIC MIXED INTERACTIONS AND HOFMEISTER EFFECTS

ALBERTO CIFERRI

6.1 INTRODUCTION

A wide variety of electrostatic interactions control the structure and the properties of natural and synthetic macromolecules. This book focuses on ionic interactions, most of which have often been theoretically modeled, avoiding consideration of other electrostatic contributions. The properties of some real systems may indeed exhibit a prevailing role of ionic interaction. In general, it is however necessary to distinguish between several types of ionic interactions and recognize the occurrence of other electrostatic contributions that mutually reinforce or antagonize each other. This situation greatly increases the difficulty of a detailed analysis of these systems but, on the other hand, it is the origin of an outstanding variety of structures, properties, and functions.

This chapter presents a description of the structure–property relation, and underlying molecular mechanisms, for systems that exhibit prevalent ionic interactions or include a substantial contribution of other electrostatic components. These systems are grouped, in the ensuing six sections, according to the ionic or ionic mixed interaction highlighted below.

Purely ionic interactions involve full, permanent charges, covalently or supramolecularly bound to macromolecules, and mobile ions. The definition includes the mutual interaction, intra- and intermolecular, among all fixed charges on the macromolecule as well as the mutual interaction among all mobile ions. In polyelectrolytes (see Section 6.2), all fixed charges have similar signs causing *repulsive* interactions responsible, for instance, for the well-known chain expansion and Donnan effects. On the other hand, in polyampholytes (see Section 6.3), the fixed charges have complementary signs, causing *attractive* interactions (ion pairing). Under neutral or isoelectric

Ionic Interactions in Natural and Synthetic Macromolecules, First Edition. Edited by Alberto Ciferri and Angelo Perico.

conditions, these attractive ionic interactions are responsible for the well-known chain globularization and insolubilization. Ion pairing is also relevant to the interactions controlling the occurrence of salt bridges in proteins, polyelectrolyte multilayers, and classical host–guest complexes (see Section 6.3).

Small monovalent ions (counter- or co-ions) mitigate (screen) the interaction among the fixed charges according to the Debye–Hückel model, causing seemingly opposite effects on flexible polyelectrolytes (reduced chain expansion and swelling) and on polyampholytes (increased globular size and swelling).

A special feature is the condensation of mobile ions that occurs particularly for conformationally rigid polyelectrolytes. It is manifested by *attractive* interactions among like-charged macromolecules.

Coordination occurs when a central metal ion is surrounded by multifunctional ligands that may or may not have charged anionic substituents. The interaction is relevant to *reversible* coordination polymers and supramolecular networks. Stronger, coordinated covalent bonds may occur with multivalent ions. The ligand donates at least a pair of electrons to a shared orbital. This interaction is relevant to coordination polymers and metalloproteins (see Section 6.4 and Chapters 10, 11, and 13).

Ionic mixed hydrophobic interactions arise from the simultaneous presence of hydrophobic and charged groups within the macrolmolecule. The hydrophobic groups do not partake in the water H-bonded structure and tend to separate into microdomains, as in the cases of micelles or globular proteins. The driving force for the formation of the (improperly named) hydrophobic bond has indeed an entropic origin that has been related to the structurization of water.[1] In hydrophobic polyelectrolytes, proper macromolecular design modulates the competition between the self-association of hydrophobic segments and the ionic interaction. Novel structures and tailored properties are thus produced (see Section 6.5). Complexes of polyelectrolytes with surfactants or proteins are extensively discussed in Chapter 8.

Ion–polymer interaction in the absence of water is manifested by the association of ionic liquids (ILs) with polyelectrolytes or polar polymers, and leads to the formation of solid polymeric electrolytes used for efficient energy sources (see Section 6.6 and Chapter 21 for gel polymeric electrolytes). Cationic conductivity is controlled by the polymer/salt composition, by the occurrence of ion pairs within the IL, and, possibly, by interaction of the latter with the polymeric substrate. Strong polymer–salt interaction in the absence of water is also exhibited by high melting salts and polyamides.

Selective ion binding is an important mixed interaction, having a prevailing charge-independent nature, which is manifested by a large increase of the solubility of charged or uncharged polar polymers in the presence of selective ions of the Hofmeister series. This interaction is relevant to the stability of biopolymers and is superimposed to the ionic interactions related to charge screening or ion condensation. The analysis of the Hofmeister effects presented in Section 6.7 allows the identification of charge-dependent and charge-independent contributions.

Several excellent reviews and books that deal with individual topics of the above classification are referred to in the corresponding sections of this chapter, or in other chapters of this book.

Ionic mixed interactions are important in fundamental areas such as the science of supramolecular organization and the functioning of biological systems. It has been pointed out that most manifestations of life are based on the control of ionic strength in the narrow physiological range.

Possible applications, emerging or well established, are briefly highlighted in the individual sections. These include biomedical and pharmaceutical applications (e.g., drugs and gene delivery; see Chapter 14), functionalized surfaces, membranes, and networks (for nanoreactors, sensors, separation processes, see Chapters 16–20), energy devices (lithium batteries and fuel cells; see Chapters 20 and 21), nanostructured materials (for optical and electronic devices, see Chapter 19), coating, drug reduction, and solubilization described in earlier reviews.

6.2 POLYELECTROLYTES

Polyelectrolytes (PE) are natural or synthetic polyanions or polycations in which the charge is due to pH-dependent ionization of groups covalently linked to main or side chains. Ionic strength or pH alterations cause chain expansion or contraction, evidenced by corresponding viscosity changes, as extensively documented for flexible polymers.[2]

The definition of strong or weak polyelectrolytes refers to the extent of fixed charges, controlled by synthesis and by the dissociation pK. For instance, poly(sodium styrenesulfonate) pK ~2 is fully dissociated for reasonable pH values, in contrast to poly(acrylic acid) pK ~5. (Scheme 6.1)

The increase of polymer concentration causes a decrease of the end-to-end distance of strong, flexible polyelectrolytes as evidenced by the simulation data illustrated in Figure 6.1. It is attributed to screening of electrostatic interactions, or reduced charge repulsion, in the more concentrated solutions.[3]

6.2.1 Mobile Ions Distribution

Counterions and co-ions (i.e., ions of added monovalent salts) screen the repulsion of fixed charges and the Debye screening length[4] is reduced to

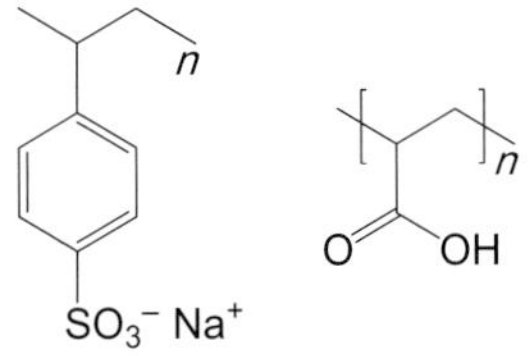

SCHEME 6.1

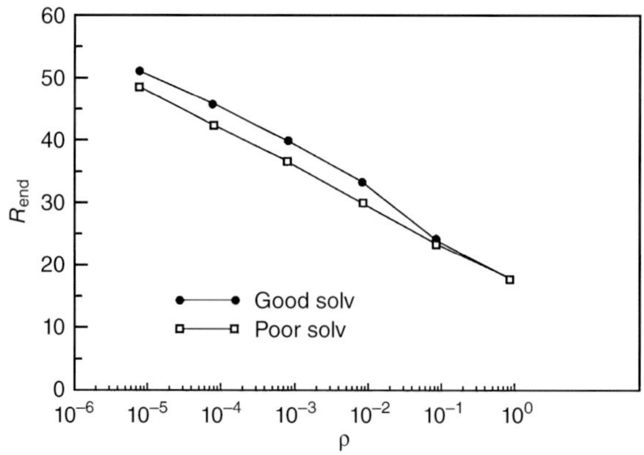

FIGURE 6.1. Variation of the end-to-end distance (R_{end}) for a flexible polyelectrolyte over a concentration range (ρ) from extremely diluted to dense solutions in good solvents and in θ solvents. Simulation data from Micka and Kremer,[3] with permission.

values below 10 Å at ionic strength ~0.1 M (see Chapter 3). Correspondingly, a flexible polyelectrolyte has an extended conformation at very low salt concentration due to electrostatic *repulsion* of fixed charges. It acquires a more compact conformation when the fixed charges are screened by clouds of counter- and co-ions having a prevalent complementary charge.

In the classical strong polyelectrolyte regime, the prevailing electrostatic interaction leads, for flexible and conformationally rigid chains, to an alternative type of interaction between fixed and mobile ions. According to the Oosawa–Manning model of ion condensation, a finite fraction of free counterions becomes dynamically confined in a thin condensation layer surrounding extended segments of charged chains.[4–7] Within the confinement volume, each fixed charge is balanced 80% on average by tightly held complementary monovalent charge, independently of the ionic strength of the bulk solution. The interaction between mobile and fixed charges is considerably stronger than in the case of screening, and can be described by a binding constant (see Chapter 3). The condensation volume per fixed charge can be theoretically calculated and from it the thickness of the condensation layer may be estimated.[8,9] For a rigid DNA chain having a molecular diameter of ~2 nm (Fig. 6.2a), the thickness of the condensation layer turns out to be ~0.12 nm at 0.1 M.[5] It is now well established that counterion condensation is manifested in several systems.[10]

Delocalization within the confinement volume produces the surprising result that two like-charged, rigid DNA molecules attract and find equilibrium within a condensation cylinder having a calculated diameter of ~5.24 nm[8,9] (Fig. 6.2b). The condensation layer now contributes 1.12 nm to the condensation cylinder. This conclusion shows that the merging of two like-charged molecules is associated with a considerable increase in the condensation

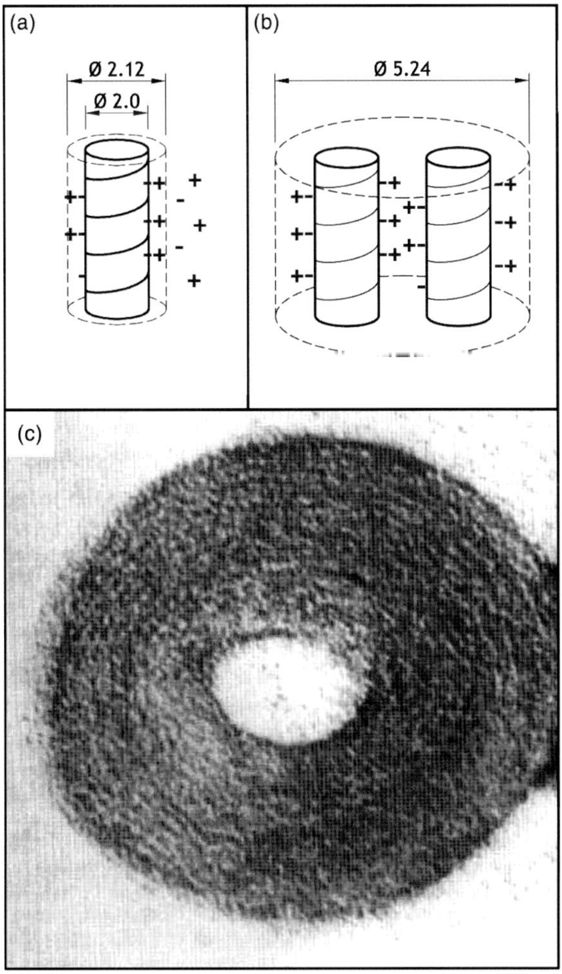

FIGURE 6.2. Schematic representation of: (a) a strong rigid polyelectrolyte (2 nm molecular diameter) within a condensation layer (ionic strength 0.1 M); (b) attraction of two like-charged rigid polyelectrolytes within the enlarged condensation layer (data from Perico and Ciferri[9]); (c) the toroidal DNA condensate of the genome of λ bacteriophage from Gelbart et al.[21] Eelectron micrograph courtesy of J.-L. Sikorav.

volume. Recent calculations reveal that larger clusters of rigid molecules can likewise be stabilized by a further increase of the condensation volume.[11]

6.2.2 Rigid and Flexible Chains

As indicated above, charge repulsion induces chain expansion for polymers with a flexible backbone.[4] However, the polyelectrolyte behavior of conformationally rigid polymers, such as DNA (charge spacing = 0.17 nm), is significantly different from that of the flexible ones.[12]

The intrinsic persistent length ($l_{p\,\mathrm{in}}$), measured in the presence of counterions able to screen the electrostatic repulsion, is a measure of the conformational rigidity. $l_{p\,\mathrm{in}}$ is defined by the relationship[13]:

$$l_{p\,\mathrm{T}} = l_{p\,\mathrm{in}} + l_{p\,\mathrm{el}}, \tag{6.1}$$

where $l_{p\,\mathrm{el}}$ is the electrostatic contribution to the total persistence length $l_{p\mathrm{T}}$. For instance, $l_{p\mathrm{T}}$ determined for double-helical DNA (~150 bp) attains the value of 540 Å in the presence of 0.2 M NaCl in contrast to much lower values reported for synthetic flexible polyelectrolytes at comparable salt concentration[14] (see Chapter 4 for a more detailed analysis).

The intrinsic persistence length is actually a measure of the *local* structural rigidity of a polymer chain. A rigid worm-like chain will nevertheless acquire a *global* flexibility when its contour chain length (L) is several times $l_{p\,\mathrm{in}}$. In fact, DNA is often described as a rigid, rod-like molecule for molecular length up to the persistence length, and as a Kuhn chain at larger molecular length. Differences in chain length may greatly alter the mode of aggregation of a given polymer in the presence of simple ions, surfactants, and proteins (see below).

6.2.3 Conformational and Phase Changes

The helix-random coil transition, which may be triggered by alterations of pH and ionic strength for polypeptides, DNA, and polysaccharides, has been extensively investigated (see Chapter 12).[15–17] These systems offer examples of the competition between ionic and H-bonding interactions. The effect of pH may reflect the ionization equilibria of different residues along the chain. For instance, in the case of DNA the protonization of the bases competes with the charges on the phosphates, destabilizing the double helix and lowering the transition temperature. Ionic strength effects were not very pronounced for the helix-coil transition in polypepdides, likely because both conformations are exposed to the solvent.

The isotropic-nematic transition of polyelectrolytes at low ionic strength could have relevance for the possible formation of mesophases of charged, rod-like biopolymers *in vivo*. The occurrence of a twisting effect that disfavors parallel alignment and reduces the field of stability of the mesophase was pointed out, but only tentative phase diagrams were elaborated.[18,19] Results for the DNA/0.3 M NaCl system evidence a further complication due to the encroachment of the regions of stability of liquid crystalline and crystalline phases.[20]

The formation of soluble aggregates or insoluble precipitates has been extensively investigated, particularly in the case of DNA in its rod- and worm-like conformation.[17] The toroidal assembly of the genome of the λ bacteriophage is circumferentially wound with interaxial spacings only 30% larger than the diameter of DNA. Like-charge attraction may control the structure in Figure 6.2c.[21] The prevailing opinion is that like-charge attraction may contribute to aggregation in water solutions of monovalent cations, but may not always be sufficient to overcome the residual negative charge (about 20%) on the double-helical B-form.[11]

Particularly interesting is the effect of ionic strength, pH, and temperature on the stability of folded DNA within a protein capside, as in the N4 virus. On increasing temperature, ejection of extended DNA, degradation of the empty capside, and melting of DNA were sequentially detected.[22]

Also of great interest is the interaction of polyelectrolytes with complementary charged surfactants or lipids.[23–29] In this case the hydrophobic tails of the surfactant may be regarded as side chains supramolecularly attached to the polyelectrolyte backbone. A variety of structures may be produced depending upon properties of the polyelectrolyte and the surfactant. These systems are highlighted in Section 6.5, and an extensive discussion is presented in Chapter 8.

6.2.4 Networks

The occurrence of cross-linkages prevents the free diffusion of individual chains included in a covalent or in a supramolecular polyelectrolyte network. The polymer acts as its own membrane and swells as a result of the equilibrium between elastic and osmotic forces. The swelling of a polyelectrolyte gel by a saltwater solution is strongly enhanced by a contribution due to the disproportion of mobile ions, similar to that occurring across a semipermeable membrane in classical Donnan equilibria[30] (see Fig. 6.3). The treatment of Donnan equilibria predicts that for a polycationic network, a larger concentration of

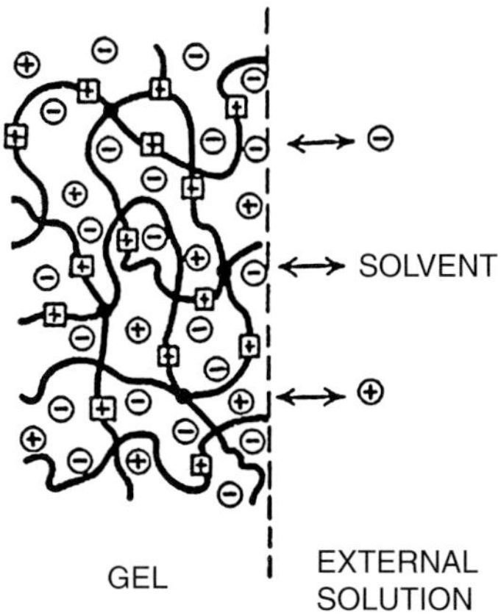

FIGURE 6.3. Schematic representation of a polycationic gel in equilibrium with a salt solution. From Flory.[30a]

anions occurs within the solution admitted into the network than in the external solution. The treatment shows that the sum of the concentration of mobile anions and cations is larger within the former solution, generating the osmotic contribution that enhances network swelling. The disproportion and the Donnan effect vanish when there is a large excess of a simple electrolyte (above 0.2 M), corresponding to an efficient screening of the fixed charge (see also Chapters 9 and 15).

Extensive theoretical investigation and experimental verifications of the swelling of polyelectrolyte networks were provided in early literature.[30] The theoretical expression for the degree of swelling of a charged network (the well-known Flory–Huggins equation; see Chapter 15) reverts to the conventional expression for an uncharged network when the ionic term vanishes at large salt concentration or low ionization degree.

6.3 ION PAIRING

6.3.1 Polyampholytes

Polyampholytes (PA) carry fixed charged groups from anionic and cationic residues. The number of complementary charges (N^+ and N^-) is controlled by synthesis and by solution pH.[1,31–36] Synthetic polyampholytes may be highly charged (e.g., one charge per repeating unit) and be obtained as random, block, alternating copolymers. On the other hand, in natural polyampholytes such as proteins, only about 20% of residues may carry a charge. At pH values between the pKs of acid and basic residues a polyampholyte carries both positive and negative charges. The net charge will be equal to zero at the isoelectric pH (pI) at which pseudoneutral behavior due to the formation of ion pairs is exhibited. pI is determined by the ratio R of acidic to basic groups and their respective pKs. A suitable expression is[1]:

$$(1+10^{\mathrm{pK}_a-\mathrm{pI}})/(1+10^{\mathrm{pI}-\mathrm{pK}_b})=R, \qquad (6.2)$$

which yields $\mathrm{pI} = (\mathrm{pK}_a + \mathrm{pK}_b)/2$ in the case of equimolar polyampholytes ($R = 1$). When the net charge is large, or at extreme pH values, the polyampholyte may be regarded as a cationic or anionic polyelectrolyte (*polyelectrolyte regime*).

Intramolecular charge *attraction* promotes compact conformations and a decreasing solubility in water when pI is approached (*polyampholyte regime*). Added ions selectively screen the attraction of fixed charges. Consequently, coil dimension, swelling, and solubility may increase in a discrete salt concentration range (up to ~0.2 M). This behavior is typical of polyampholytes and opposite to that observed with polyelectrolytes and is often described as an *inverse polyelectrolyte effect*. The salt concentration at which the net excluded volume is zero represents a theta condition.[31,32] Superimposition of the attrac-

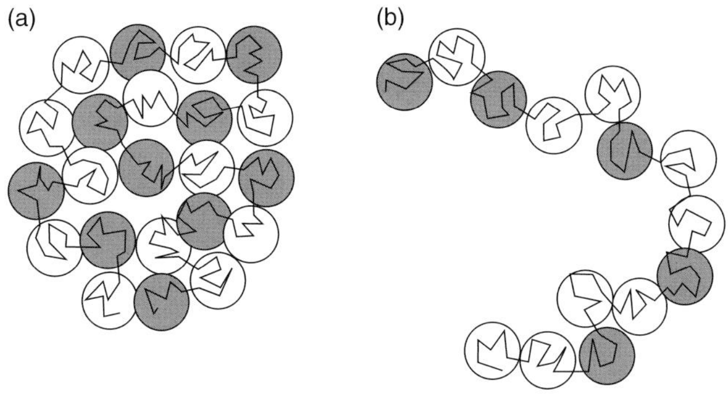

FIGURE 6.4. Globular (a) and linear (b) assembly of blobs in a polyampholyte chain. Light or dark blobs represent prevailing positive or negative charge. Interactions are screened over a long distance in the globular assembly. From Kantor and Kardar,[31] with permission.

tive interactions due to the hydrophobic interaction of nonpolar groups along the chain enhances the conformational and solubility characteristics of polyampholytes when pI is approached.

A well-established theoretical treatment analyzes the competion between the electrostatic free energy (based on the Debye–Hückel theory in the limit of ~0.2 M ionic strength) and the interfacial free energy of the polymer (see Chapter 4). The theory elucidates the main structural organization of the polyampholyte chain.[31,32] A neutral collapsed chain is regarded as a liquid drop undergoing a Rayleigh-type instability when the net charge increases. The Rayleigh model predicts that a liquid droplet splits into two smaller drops when the electrostatic free energy reaches a critical value comparable to the interfacial energy. The Kantor and Kardar model,[31] illustrated in Figure 6.4, shows a *connected* series of pearls (or blobs) with prevailing positive or negative charge. In the globular structure (a) the excess charge interaction of individual blobs is screened out over long distances. When the overall excess charge exceeds a critical value, the expanded conformation (b) is favored. Neutral spacers between the charged residues are considered, but their hydrophobic interaction is not accounted for.[33] The analysis of hydrophobic polyelectrolytes reported in Section 6.6 suggests the occurrence of an additional stabilization of the pearling conformation due to hydrophobic attraction.

The basic predictions are consistent with experimental findinings.[1,34–37] The specific viscosity–pH dependence for random polyampholytes exhibits a minimum at pH ~ pI, consistent with the occurrence of a compact conformation in the polyampholyte regime. A minimum coil size was also detected when charge balance was achieved by alteration of the comonomers fraction. Polyampholytes with a significant proportion of hydrophobic groups are actually insoluble. Solubility and network swelling are consistent with the trend of viscosity in dilute solutions. The specific viscosity was found to increase at pH larger or smaller than pI, supporting the occurrence of chain expansion due to repulsive interactions in the polyelectrolyte regime.[1,34–37] The reduction of the attraction of complementary charges due to mobile ion screening at pH ~ pI and low ionic strength (inverted PE effect) is supported by swelling and melting data for collagen I,[38] and by specific viscosity data of Zheng et al.[39] They used an amphoteric cellulose derivative containing anionic carboxymethyl groups and cationic 2-hydroxy-3-(trimethylammonio) propyl substituents.

Block polyampholytes having equal number of cationic and anionic charges were also investigated. Phase separation reflected complexation of the two blocks, in analogy to the behavior of mixtures of oppositely charged polyelectrolytes.[40] Soluble aggregates, occurring above a critical charge unbalance, include sections of the balanced complex and branches carrying the uncompensated charges. Micellar structures stabilized by a charged corona were therefore suggested.[41] A double helical conformation of the balanced complex ($N^+ = N^-$) was evidenced by Monte Carlo (MC) simulation.[42]

6.3.2 Ion Pairing in Proteins

Proteins necessarily include a large number of less polar or apolar groups (up to 80%). One might expect that strong hydrophobic interaction could alter the behavior expected from ion pair formation by fixed anionic and cationic residues (these ion pairing attractions are often somewhat inappropriately described as *salt bridges*). Nevertheless, it is remarkable that the main experimental features of polyampholytes described above are manifested also by proteins such as gelatin or bovine serum albumine.[1,37] A possible interpretation is that, in the case of polyampholytes, both ionic and hydrophobic interaction are favoring compact structures and reduced solubility.

The contribution of intramolecular salt bridges, occurring within sequences of groups also displaying hydrophobic and hydrogen bond interaction, hinges on the complex problem of the intra- and intersupramolecular stabilization of biomacromolecules. Current results suggest that salt bridges may have a stabilizing or destabilizing role depending upon their geometrical orientation and location in the context of the overall sequence and structure[43,44] (see also Chapter 12). For instance, Kumar and Nussinov discuss short-range stabilizing (separation distances within 4 Å) and long-range destabilizing ion pairs.[44] We may therefore expect that the strength of salt bridges will depend upon details of the distribution of polar and apolar groups. In extreme cases, the formation of the ion pair may be inhibited.

6.3.3 Polyelectrolyte Multilayers

The electrostatic attraction between complementary fixed charges on polycations and polyanions is the main driving force for the assembly of polyelectrolyte multilayers.[45,46] The latter are engineered by a layer-by-layer (LbL) deposition. The process is performed by first deposing a layer of a polyelectrolyte over a compatible substrate. The latter is then alternatively dipped and extracted (or sprayed) into solutions of polyelectrolytes having charges of opposite sign, allowing intermediate rinsing steps and final drying. The alternating deposition produces odd–even effects on the last deposited layer (e.g., water contact angle). A final hydrophobic capping layer may be added to reduce the water affinity and readsorption by the assembly. The film thickness has been found to increase either linearly or exponentially with the number of deposited bilayers and hundreds of complementary layers can be assembled. Polyelectrolyte multilayers have a great interest for a wide range of applications, particularly matrices and functionalized surfaces (nanoreactors, sensors), membranes (fuel cells, nanofiltration, protein purification), and nanocapsules for drug delivery, coating, and photonic crystals.[47,48] An impressive amount of research has been carried out; a 2005 review lists over 1000 references.[46] Details on the fundamental aspects of LbL deposition are given in Chapter 5, and selected applications are given in Chapters 17–19.

In spite of the great interest in layered polyelectrolytes, their fundamental description is still incomplete. Short-range, charge-independent contribution and the entropic contribution of mobile ion release accompanying deposition have been considered. It is still an open question whether the layered deposition corresponds to an equilibrium or a frozen state.[49]

Simulation studies are hampered by the large number of particles involved and by the slow relaxation of polymers. Nevertheless, results for systems with short chains and a limited number of layers offer valid hints to the complex parameters affecting the multilayers (Chapter 5). The occurrence of exponential versus linear growth was related to the ability of polyelectrolyte chains to diffuse into and out of the multilayer.[50] Simulation of diffusion properties suggested that chains in contact with the supporting surface have more restricted motion than chains in the subsequent layers. Moreover, faster movement occurs in the plane of the multilayer than in the perpendicular direction. Simulation studies for systems with a limited number of layers (i.e., negative surface, 1st positive layer, 2nd negative layer, 3rd positive layer, etc.) offer partial support to the experimental observation that the odd or even number of layers affects the stability or the dewetting of the system (Chapter 5). Assessments of the roles of direct ion pairing between the fixed charges, mediated by salt ions and water, were also presented. Ion pairing is amplified in dried films due to the reduction of the effective dielectric constant.[51,52] Increasing the amount of salt, particularly divalent ions, caused a significant increase of surface coverage.[47] Substitution of Cl^- with Br^- counterions enhanced the swelling and reduced the mechanical properties of the films, an effect consistent with the analysis of Hofmeister effects presented in Section 6.7.[53,54]

6.3.4 Ion Pairing in Host–Guest Complexes and Inorganic Salts

Ion pairing in low molecular weight (1.1) host–guest compounds, as well as in inorganic salts, has been extensively investigated[55,56] (see also Chapters 1 and

2). The results often allow an assessment of the multiple additive interactions between binding sites in reversible complexes that exhibit geometric complementarity (chelate effect), including possible reinforcement by cooperative effects. These results ought to be considered for the understanding of ion pairing in polymers, and for the design of novel supramolecular structures.

The Bjerrum equation quantifies the association constant in terms of the equilibrium distance between the pairing ions (q) referred to their mean diameter (a)[55,56]:

$$K = (4\pi N_A/1000)(q/a)^3 Q(q/a). \tag{6.3}$$

Association constants have been extensively determined for simple salts and for organic complexes.[55,56] Ion pairing is rare for inorganic salts but it is favored by concentration, poor hydration, and valence. For $MgSO_4$, K is in the order of 100–200 dm^3 mol^{-1}.[56] The cutoff distance, q, at which no ion pairing is considered to occur may be significantly larger than the distance of closest approach (the mean diameter a). The role of hydration on ion pair formation has also been considered.[56] A discussion of ion pairing occurring in the absence of water is presented in Section 6.6. The dissociation of a salt is commonly assessed from conductivity data.

The decrease of the association constant K with ionic strength for typical organic complexes in water is illustrated in Figure 6.5a. Remarkable correlation is shown with the Debye–Hückel prediction for ion screening[55]:

$$\lg K = \lg K_0 + m\sqrt{I/(1+\sqrt{I})}. \tag{6.4}$$

Therefore, absolute values of K for organic compounds may be larger than for inorganic salts due to nonionic and shape-dependent interactions,[55,57] but the screening of ion pairs follows theoretical expectations. Other noncovalent interactions occurring for the organic complexes may sometimes be dissected from the ion pairing contribution (see Chapter 2). The linear dependence of K upon the product of fixed ion charges ($z_A z_B$, predicted by Eq. 6.4) was verified for a large number of organic complexes differing in shape and charge distribution.[55] The implication is that an increase of the number of salt bridges in a given complex produces a constant increment of the binding free energy, a concept useful for the design of stronger assemblies.

Figure 6.5b illustrates host–guest complexes assembled at two different pH values at which either H bonds or electrostatic ion pairs are the stabilizing forces. It is seen that in the latter case (pH = 7), the stabilization of the complex is entirely entropy driven. This surprising result is confirmed by a variety of thermodynamic data for the formation of salt bridges in water solutions, and appears in contrast to the expectation that pair-wise associations should be driven by enthalpy effects.[55] Salt bridges in protic solvents are indeed stabilized by enthalpy effects.[55] Positive $T\Delta S$ contributions to a variety

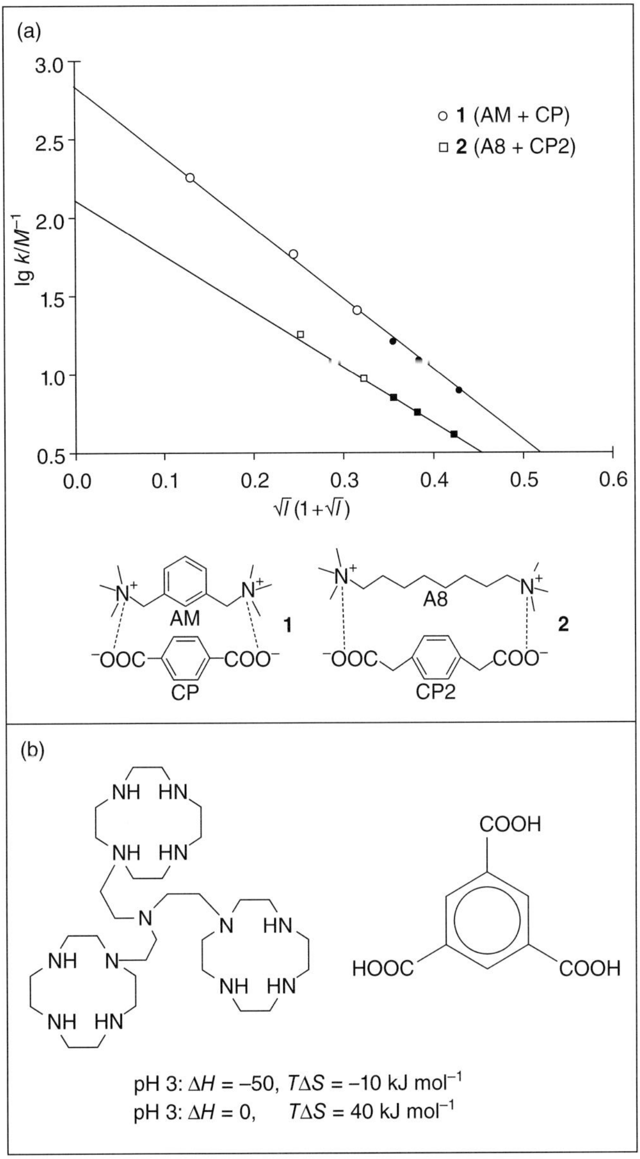

FIGURE 6.5. (a) Debye–Hückel relation (Eq. 6.4) between association constants and ionic strength for two ion pairs 1 and 2 having differing spacings between the charged groups; (b) change in binding mechanism for host–guest complexes assembled at different pH. From Schneider,[55] with permission.

of assembly processes have often been interpreted in terms of a disorganization of the water structure.[56] Schneider suggested a contribution due to a decreasing electric shielding of the solvent when the ions forming a pair are getting closer.[55] It would be interesting to assess if the release of the Debye–Hückel screening cloud, which must accompany ion pair formation, could contribute to the observed positive entropy of formation.

6.4 ASSOCIATIONS MEDIATED BY MULTIVALENT MOBILE IONS

6.4.1 Supramolecular (Reversible) Coordination Polymers

Supamolecular polymers are characterized by a continuous scrambling of bound and free unimers. Depending upon the strength of the supramolecular bond, and its complexation/decomplexation rates, these polymers may be regarded as reversible (equilibrium) or irreversible systems (see Chapters 10, 11, and 13).

Scheme 6.2 includes an example of bifunctional ligands able to form water-soluble, reversible metallosupramolecular polymers by coordination with multivalent metal ions such as Zn^{2+}, Co^{2+}, and Fe^{2+} (see Chapter 10). The bifunctional ligand based on pyridine-2, 6-dicarboxyilic acid connected by tetra- or hexa-ethylene oxide spacers forms reversible polymers (and rings) in which a single Zn^{2+}metal ion is coordinated to two terdendate groups (1:2 complexes: ML_2).[57] The stability constants $K_1 = (ML)/(M)(L)$ corresponds to the nonbonding 1:1 complex, and $K_2 = (ML_2)/(ML)(L)$ corresponds to the polymerization process. The estimated value of K_2, ~105.5 M^{-1}, was somewhat smaller than K_1 on account of a small repulsion of the second by the first ligand group.

The theory of isodesmic supramolecular polymerization predicts a degree of polymerization roughly corresponding to the square root of the binding constant (DP ~ $2(\nu K)^{1/2}$).[58–60] These expectations are in line with the above experimental data.[57] The assembly mechanism of coordination polymers appears therefore to conform to the general scheme of isodesmic supramolecular polymerization.

Han et al. investigated bifunctional bis-terpyridines connected by rigid spacers of different lengths and polymerized with Fe^{2+} (Scheme 6.3).[61] A significant effect of the pyridine substituents on the electrochromic properties of

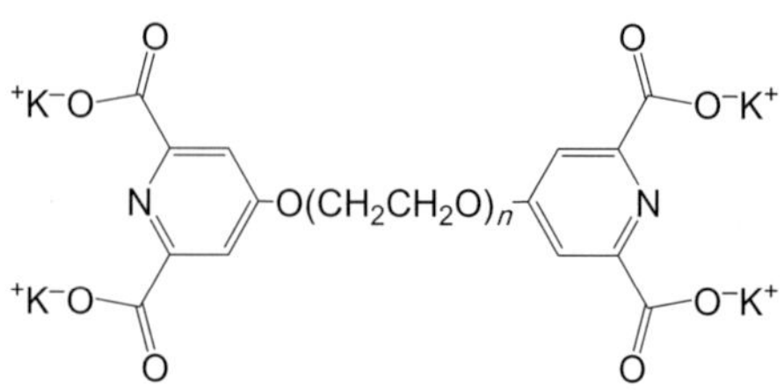

SCHEME 6.2

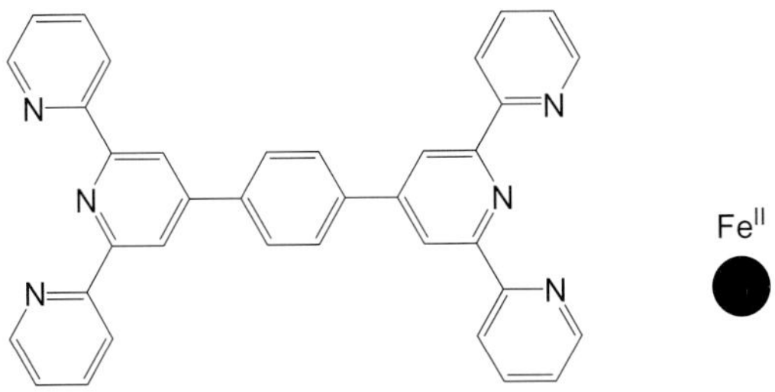

SCHEME 6.3

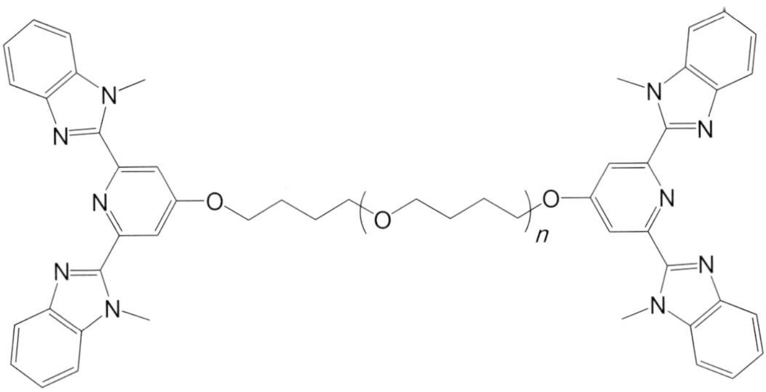

SCHEME 6.4

the supramolecular polymers was observed. The effect has possible applications in molecular switches and optical and electronic devices.[61]

Beck et al.[62] investigated ditopic monomers based on short poly(tetrahydrofuran) segments terminated at either ends by the tridentate ligand 2, 6-bis(1′-methylbenzimidazolyl)-4-oxypyridine). These compounds formed linear polymers by 2:1 complexation with single transition metals ions from $Co(Cl0_4)_2$ or $Zn(Cl0_4)_2$ (Scheme 6.4).[62] The nature and size of the core segment, and the type of the metal ligand, influenced the properties. The increase of intrinsic viscosity from 18.2 to 30.1 mL g^{-1} and the enhancement of mechanical properties upon complexation suggested a sizeable degree of polymerization.[62]

In the above cases a linear assembly is produced by a metal ion that bridges two ligands that bear a complementary charge or a strong polarity. The case of a columnar supramolecular assembly based on stacks of the tapered amphiphile12-ABG-B15C5 is instead illustrated in Figure 6.6. The disk-like repeating unit includes crown ether *endo*receptors in the center, surrounded by aromatic moieties, in turn surrounded by melted alkyl tails of the *exo*receptors. The variety of supramolecular interactions perpendicular to the disk surface is enhanced by complexation with 0.4 M triflate salt per mole of the crown ether (B15C5).[63] Schneider has discussed related types of complex interactions involving, for example, ions–π attraction.[55]

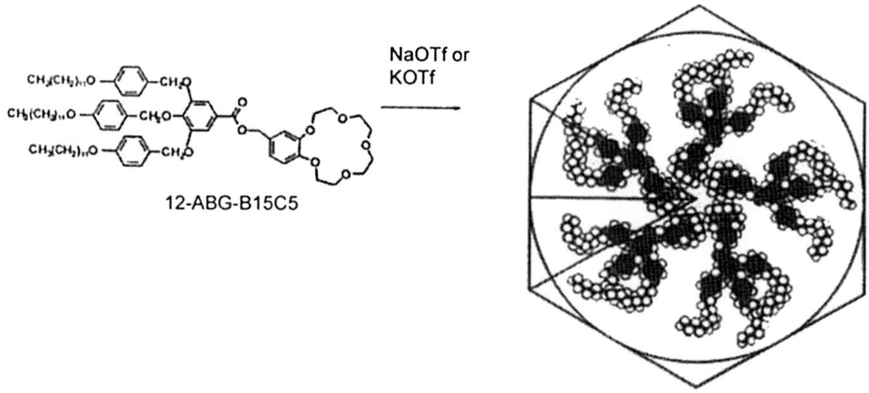

FIGURE 6.6. The assembly of the complex between 12-ABG-B15C5 and trifilate salts produces undiluted hexagonal columnar mesophases based on stacks of disks containing five molecular units. From Percek et al.[63]

6.4.2 Intermolecular Bridges and Networks

It has been often suggested that a divalent cation (e.g., Ca^{2+}) may favor a bridge, or a correlation, between two fixed anionic charges along different polyelectrolyte chains. Attraction between two like-charged polyelectrolytes may also be enhanced (see Chapter 3).

Three-dimensional networks in which multivalent ions act as cross-linkages preventing the free diffusion of individual chains have also been reported. Three-dimensional networks were prepared using the same bifunctional ligands illustrated in Scheme 6.2 with lanthanide ions replacing transition metal ions. Lanthanide ions have the capability to complex three pyridine-2,6-dicarboxylic acid groups. Gels formed using La^{3+} and Nd^{3+} ions exhibited viscoelastic and thermoreversible properties that were unprecedented with respect to the formation of other strong materials obtained by self-assembly of low molecular weight compounds in aqueous solution.[57]

The 3:1 complexes with the larger lantanium ions[64] (from $La(NO)_3$ or $Eu(NO)_3$) were also formed with the tridentate ligand bis(2, 6-bis(1′-methylbenzimidazolyl)-4-hydroxypyridine). Metallosupramolecular gels were obtained using mixtures such as Co^{2+} and La^{3+} (~95/5%). The gel included linear chains based on sequences involving Co^{2+} ions and cross-linkages involving La^{3+} ions. Thermochemo- and mechanoresponses and light-emitting properties were exhibited, depending upon the nature of the metal ion and the swelling agent. A reversible gel-sol transition was exhibited at ~100°C at which the orange color typical of Co^{2+} linear polymers persisted in solution, suggesting degradation of the La^{3+} cross-linkages. Preliminary stress recovery experiments were reported.[64]

The mechanical properties of networks formed by labile chain bonds and cross-linkages have recently been discussed.[65–67] Dynamic scrambling of bonds and cross-linkages results in long-range elasticity characterized by compliance under stress and enhanced recovery when stress is released. It should be interesting to investigate the occurrence of scrambling processes in the coordination networks.

The cross-linkages alleged to form on collagen under the action of chromium(III) salts (~5%), as done in the tanning industry, might also be described in terms of a network. Earlier data on the properties of collagen I tendons swollen in $KCr(SO_4)_2$ water solutions suggested the permanence of a network structure resisting extensive washing.[68] More recent results on soluble collagen reveal that partial denaturation of the native triple helix is necessary before soluble complexes with bound chromium are evidenced.[69a] A detailed mechanism for the unusually strong binding of chromium salts to collagen remains elusive, although carboxylic acid groups of aspartate and glutamate side chains are involved. These conjectures appear consistent with data reported by Nijenhuis[69b] on the cross-linking of hydrolyzed poly(acrilamide) by chromium(III) salts (see also Chapter 10).

6.4.3 Metalloproteins

Metal ions play an important role on the stabilization of the structure of metalloproteins.[70a–f] The metal ion is frequently coordinated to groups containing oxygen, nitrogen, and sulfur along the sequence of a given or partner proteins. A rather strong, coordinated covalent bond is involved (cf. Chapters 10, 13, Messerschmidt et al.[70h]) The so-called zinc finger motif is based on the coordination of a zinc ion to histidine and cysteine residues along α-helical and antiparallel β-sheet sequences (e.g., Scheme 6.5 from Messerschmidt et al.[70b]). In hemoglobin, four subunits are coordinated by the iron(II) ion through the planar, macrocyclic ligand protoporphyrin (IX) and the imidazole nitrogen atom of a histidine residue (see also Chapters 11 and 12).

In addition to the structural roles highlighted by the examples above, functions such as transport, enzyme, and signal transducers are typical of metalloproteins. Metals participate directly in the chemistry performed by

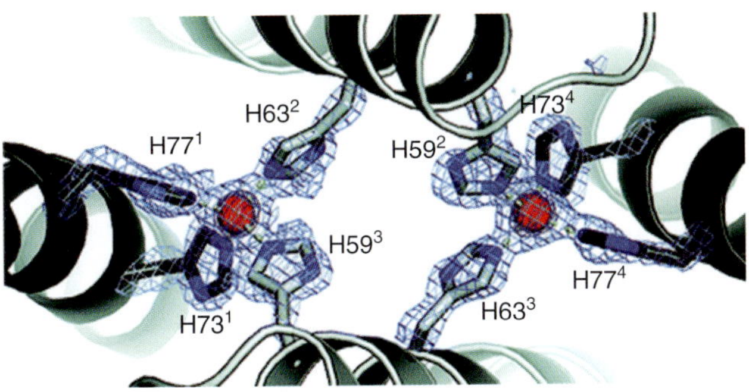

SCHEME 6.5

metalloenzymes and expand the chemical tool box available to a biological system beyond the possibilities of carbon chemistry.[70d] In this sense, hemoglobin can also be viewed as a metalloenzyme which utilizes the Fe(II) center to reduce nitric oxide.[70e] A classic metalloenzyme example, carboxypeptidase, binds zinc with two histidines and a glutamate. The substrate peptide binds directly to the zinc and the first coordination sphere is completed with a water molecule. This direct metal binding mechanism is required for the peptide cleavage to occur, and the identity of the metal can be varied[70f] (see Chapters 10–13).

Metal transport proteins directly interact with the respective metal cargo in order to control the reactivity of the metal ion as it is moved through the biological system. For example, the human transport protein transferrin tightly sequesters iron to prevent it from precipitating or producing reactive oxygen species during transport. This tight sequestration is made possible by binding Fe(III) with two tyrosines, a histidine, and an aspartate residue. Two additional oxygen ligands are provided by a carbonate anion that is required for tight iron binding.[70g]

6.5 HYDROPHOBIC POLYELECTROLYTES

All synthetic and natural polyelectrolytes carry nonpolar or hydrophobic groups (hydrophobes) that have a tendency to self-assemble in water solutions (see Chapter 7). The overall conformation represents a balance of the repulsive interactions of the fixed charges, favoring extended sequences, and the attractive interactions of the hydrophobes, favoring compact structures. The competition between these interactions can be controlled by reducing the electrostatic interaction (e.g., at pH < pK or high C_s), or by intentionally increasing the amount of nonpolar groups on the main or side chains by proper synthesis or side-chain modification.[71–76] In hydrophobically modified polyelectrolytes (HMPE), sometimes called amphiphilic polyelectrolytes, the amount of hydrophobes may be increased by copolymerization (e.g., maleic acid and styrene), and random, block, and alternating copolymers have been investigated. The objective is to control the competition between electrostatic repulsion and hydrophobic self-association so that materials with novel solubility, adsorption, rheological, and biomedical properties may be designed. Association structures promoted by the hydrophobic microdomains may show analogies with conventional surfactant micelles and be relevant to delivery mechanisms. HMPEs may drastically influence the rheological properties of aqueous solution, and are used in several industrial processes.[71–76]

Chitosan is a polyelectrolyte produced by partial deacetylation (DA) of chitin. The amount of fixed charge is controlled by the DA value and pH (the pK of its amino group is ~6). The modification of chitosan by conjugation of amino groups with hydrophobic moieties, having carboxy or aldhehyde functionalities, is illustrated in Scheme 6.6. Inter- and intramolecular aggregation

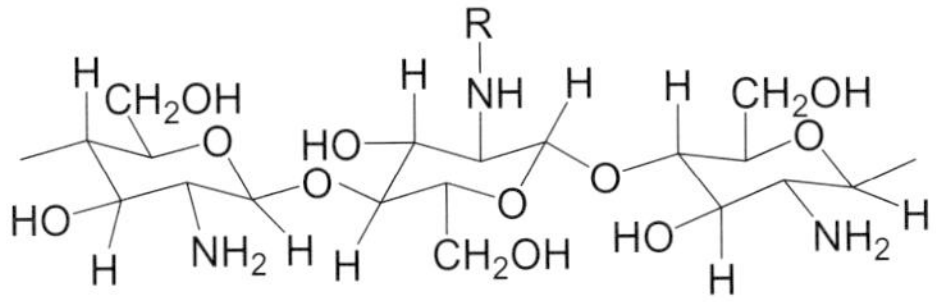

Chitosan: R = H or $COCH_3$
Hydrophobically modified chitosan: R = CO-alkyl (C_6–C_{18})

SCHEME 6.6

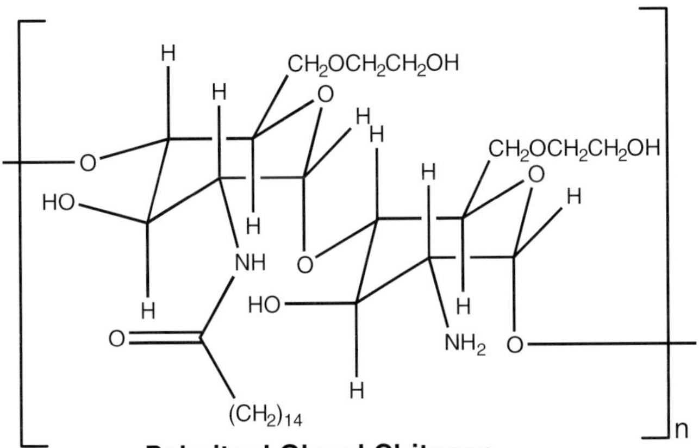

Palmitoyl Glycol Chitosan

SCHEME 6.7

enhance the formation of clusters (size in the 50–150 nm range) stabilized by physical cross-linkages centered on hydrophobic residues.[71] The aggregates form complexes with hydrophobic drugs, surfactants, dyes, and DNA that have potential interest for drug and gene therapy. The attachment of hydrophobic fatty acid groups to glycol chitosan promotes the formation of vesicles capable of entrapping and slow releasing water-soluble drugs (Scheme 6.7).[72d]

The interaction between dextran sulfate (DxS), of different charge density, and three cationic amphiphiles varying with respect to hydrophobicity, was investigated by means of equilibrium dialysis and surface tension. The results evidence a cooperativity in the polyelectrolyte–amphiphile interaction that starts at a certain amphiphile concentration and depends on both the hydrophobicity of the amphiphile and the charge density of DxS.[72a] Hydrophobically modified cationic dextrans have biomedical interest as permeation enhancer of heparin, and sequestration of bile acids.[72b,c] A more detailed account of these and other HMPE biomedical applications is presented by Bromberg.[73,74]

Khokhlov and Khatchaturian suggested that the competition between hydrophobic attraction and electrostatic repulsion leads to the stabilization of a cylindrical globule for flexible polyelectrolytes in the absence of added salt.[77]

Subsequent theoretical analysis by Dobrynin et al. suggests, however, that the cylindrical assembly becomes unstable and a more extended pearl (necklace) conformation allows a reduction of electrostatic repulsion and a decreased free energy of the polyelectrolyte.[78] In the pearling conformation, charged beads are connected by charged linear strands along a single polyelectrolyte chain. The pearls-on-a-string conformation is favored over the extended one in the semidilute concentration range in which the correlation length of the solution becomes comparable to the length of the strings between neighboring pearls.[79]

Several simulation studies for charged, flexible polymers in poor solvents support the occurrence of the pearling conformation for hydrophobic polyelectrolytes.[80a,b] An analysis of experimental data for HMPEs (e.g., polymaleic acid-co-alkyl vinyl ethers[80c–f]) is presented in Chapter 7.

The conformation is the result of a Rayleigh instability experienced by a charged cylinder, in analogy to the model originally adopted by Kantor and Kardar and described in Section 6.3[31] (see also Chapter 4). The effective charge fraction experimentally determined for highly charged poly(styrene-co-styrene sulfonate) was significantly smaller than expected on the basis of the condensation theory.[81] The results are consistent with a theoretical analysis of the counterion distribution around a pearl.[82]

One must therefore conclude that the pearling conformation is a general manifestation of the competition between repulsive and attractive interactions. In the case of polyelectrolytes, the attractive component is enhanced by hydrophobic groups, poor solvents, increasing concentration of polymer, salt, and surfactant. Indeed, even the counterion-mediated attraction due to ion condensation can produce a necklace structure, as evidenced by the comparison of simulation data reported by Lee and Thirumalai for polyelectrolytes and polyampholytes (Fig. 6.7).[83] In the case of polyampholites, attractive electrostatic interactions of salt bridges between opposite charges reinforce hydrophobic attractive contribution.

The interaction between polyelectrolytes and surfactants, or lipids, carrying a complementary charge connected to a hydrophobic tail is extensively described in Chapter 8. The structure of the assembly is affected by the conformational rigidity and the contour length of the polyelectrolyte. For flexible chains, the compensation of the complementary charges on the polymer and on the surfactant produces a structure in which the polymer chains is wrapped up around pseudo-micellar aggregates of the surfactant.[9,23]

However, chains having a large persistence length, comparable to the contour length, will not be induced to fold around a pseudo-micellar structure stabilized by weak cohesion forces.[9,23] A coating of the polymer by the individual surfactant molecules will avoid the cost of bending the rigid chain.[23–26] The molecular recognition of the hydrophobic tails and their self-assembly will be manifested by their interdigitation, promoting the growth of a hexagonal structure, as schematized in Figure 6.8 for a short, rigid DNA (150 bp) (see also Chapter 8). The release of the ions confined in the condensation layer provides an additional (entropic) stabilization.[9] These aggregates have poten-

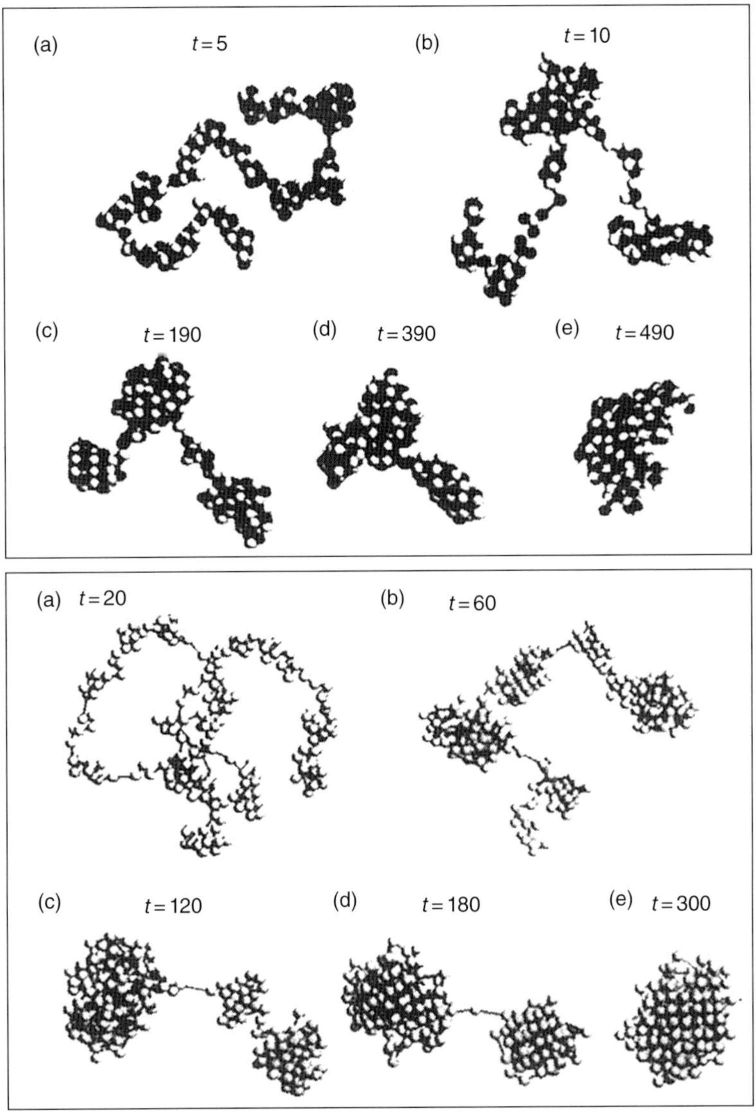

FIGURE 6.7. Time evolution (t) of conformations during the approach to a globular state under θ conditions for polyampholytes (above) and polyelectrolytes (below) having 240 charged residues. Counterions are included only in the simulation of polyelectrolytes. Dark and bright shades indicate complementary charges. From Lee and Thirumalai,[83] with permission.

tial applications as delivery vehicles for genetic material in gene and drug therapy.[27]

In the case of giant DNA (bp > 50 kbp), a variety of condensed structures have been documented that appear more complex than the extended ones illustrated in Figure 6.8 for short DNA segments coated by a cationic

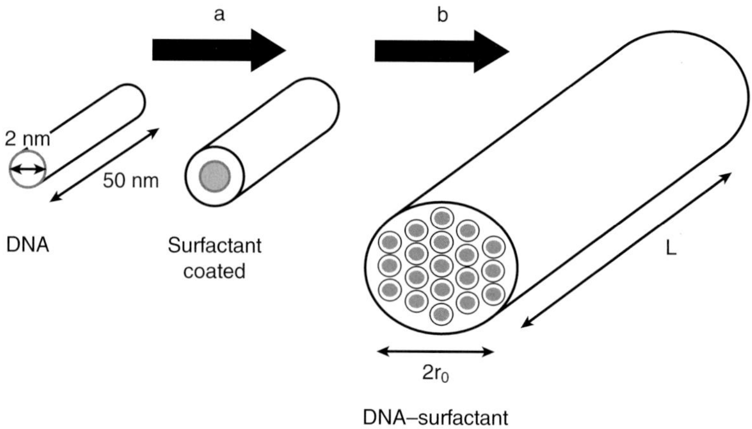

FIGURE 6.8. Sketch for an assembly of a monovalent cationic surfactant with a short, rigid (150 bp) DNA segment. From McLoughlin et al.,[25] with permission.

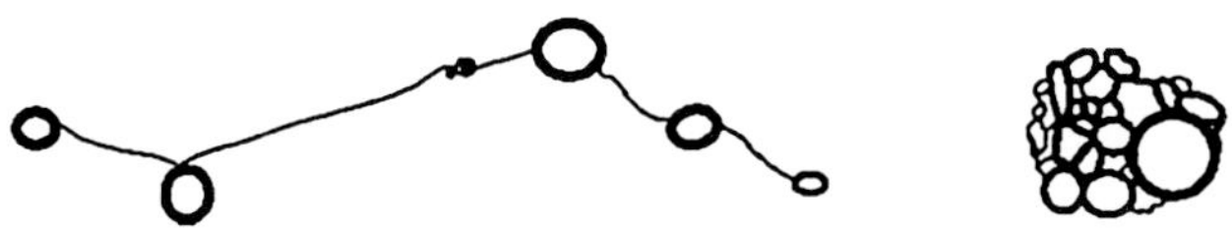

FIGURE 6.9. Schematic representations of atomic force microscopy (AFM) images of the structures shown by a giant DNA (165 kbp) in 0.2 μM (left) and 1.0 μM (right) propanediyl-1,3-bis(dimethyldodecylammonium bromide). The ring-on-a-string structure evolves into a compact network of many fused rings. From Miyazawa et al.,[29] with permission.

surfactant.[28,29] A pearls-on-a-string structure was induced by a Gemini surfactant on a 48.5 kbp DNA.[28] The authors suggested that the beads (diameters in the order of 10 nm) are formed by the wrapping of DNA around micellar-like structures of the surfactant, and suggest a possible similarity to the chromatin assembly. However, as evidenced by the analysis of Perico and Ciferri,[9] it is difficult to justify the bending of DNA segment around structures, such as the micellar ones, that are not stabilized by strong cohesion forces (an extensive discussion is presented in Chapter 8).

Miyazawa et al.,[29] working with a 165 kbp DNA and a Gemini surfactant, reported the formation of rings-on-a-string structure (Fig. 6.9) where the diameter of the rings (70–160 nm) was in the order of that observed for the toroids (see Fig. 6.2c and Chapter 8).[24] Upon increasing surfactant concentration, the structure evolved into a more compact conformation due to a gradual increase of the number of rings. The authors justified the occurrence of such structure in terms of a balance between the bending energy and the energy associated with distributing DNA segments into multiple rings.[29]

6.6 POLYMERIC IONIC LIQUIDS, POLYMERS-IN-SALT SYSTEMS

The association of an IL to a polymer in the undiluted state produces systems referred to as solid polymeric electrolytes. These systems have great interest for applications, particularly in the area of highly efficient energy sources. Their study facilitates the identification of the ionic species that contribute to the mechanisms of ion–polymer interaction.

ILs are salts melting at low melting temperature (even below room temperature). They may include organic cations (e.g., imidazolium, pyridinium) paired with inorganic (e.g., halides) or organic (e.g., acetate, bis(trifluoromethanesulfonyl)imide) anions.[84–87]

Most important properties of IL are the ion conductivity and the bulk viscosity. The correlation between these parameters is evidenced by Walden plots such as the one reported in Figure 6.10. The plot includes the ideal conductivity versus viscosity dependence exhibited by a true IL (e.g., a dilute KCl solution).[85–87] Neutral ion pairs should behave as molecular solvents, not contributing to conductivity. Data for several systems based on tetraalkylphosphonium cations and several anions (some included in Scheme 6.7) are seen to fall below the ideal line, suggesting a behavior intermediate between that of a true IL and a true neutral ion pair. The behavior reflects the occurrence of a correlation (degree of association) between the ions in a pair.[87]

Solid polymeric electrolytes are based either on ILs polymerized along the main chain or mechanical blends of ILs and polymers. In either case the aim

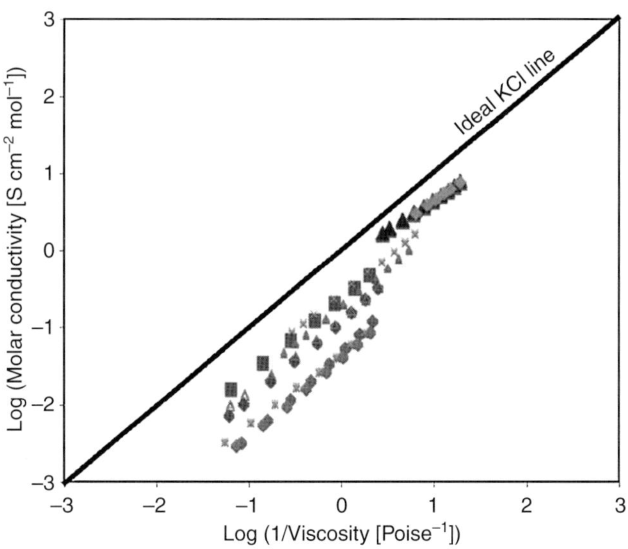

FIGURE 6.10. Conductivity–viscosity Walden plot for several phosphonium salts over the temperature range 30–100°C. Ethylammonium nitrate and α-picolinium triflate are included. Adapted from Green et al.[86]

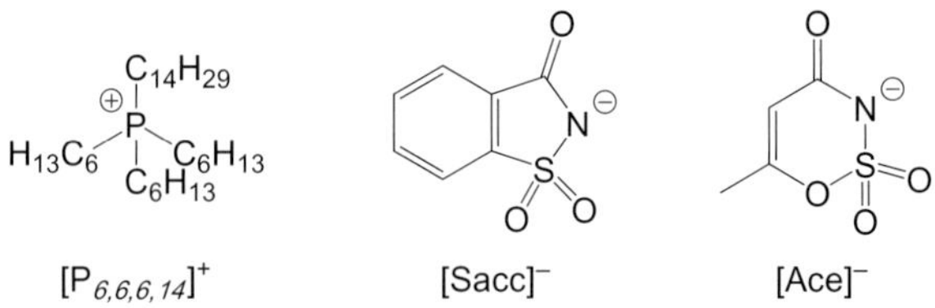

SCHEME 6.8

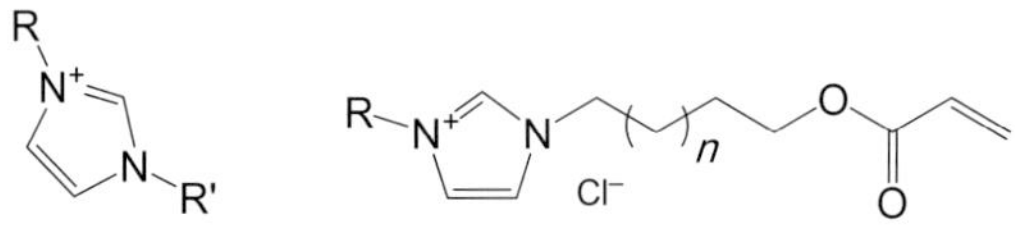

SCHEME 6.9

is to couple properties of ILs, such as ion conductivity, stability, and nonflammability to the mechanical properties of polymers.

Polymerizable versions of ILs, tailored for specific applications, were obtained by incorporation of acryloyl or vinyl groups, usually on an IL cation. Scheme 6.8 illustrates the structure of polycationic 1-(8-(acryloyloxy)alkyl)-3-methylimidazolium chloride.[86] Polymeric ILs (PILs) exhibit polyelectrolyte behavior (Scheme 6.9). Tunable structures having hydrophobic, hydrophilic, and amphyphylic character have been reported (for recent reviews see Green et al. [86] and Ueki and Watanabe.[88]) Applications in the areas of solid polymer electrolytes (SPEs), CO_2 sorption, nanotube dispersion, ionic electroactive polymers, and nanostructured materials have been described. Gel polymer electrolytes have also been considered (see the following and Chapter 21).[89,90]

Mechanical blends in which there is not a chemical connectivity between the components are defined by two classes of systems according to components content. In the polymer-in-salt systems, the salt content (>50wt %) prevails. These systems have stimulated recent interest particularly for applications in lithium and lithium-ionic batteries (e.g., cell phones, portable computers, vehicles with electrical propulsion). The polymer-in-salt systems are regarded as particularly efficient energy sources due to a reduction of the share of anions in the electric charge transfer. Moreover, greater ionic mobility is expected, substantially decoupled with the segmental relaxation of the polymeric chain.

A definite model for the conductivity and for the ion–polymer interaction in SPEs is still a controversial matter. In the case of a polymer-in-salt electrolyte (e.g., polyacrylonitrile with lithium bis(trifluoromethane sulfone imide $LiN(CF_3SO_2)_2$), ionic conductivity is not reduced by the segmental mobility of the polymer matrix, provided a high degree of ionic aggregation occurs. Bushkova et al. suggested that above a critical concentration all clusters are in close contact forming an infinite cluster that promotes fast cationic transport.[91] Forsyth et al. elaborated a dynamic percolation model and a broad percolation threshold in which the connectivity percolation of ionic clusters is decoupled from the segmental motion of the bulk system.[92]

The possibility of a direct interaction of lithium ions with a polyacrylonitrile substrate was considered by Ferri et al.[93] They specifically suggested a salt-induced reduction of the crystallinity of the polymer. However, this possibility may be inconsistent with the lack of a depression of the melting temperature caused by LiCl in the absence of water for polymers such as polyethylene terephalate or polyoxyethylene reported by Valenti et al.[94] The latter results suggest that solid polymeric electrolytes that included polar groups able to bind lithium ions would show decreased crystallinity and thus increased segmental mobility of the matrix that could possibly enhance cationic conductivity.

Gel polymeric electrolytes represent a situation intermediate between the solid PIL and typical liquid electrolytes. Lithium cells include a network of an uncharged polar polymer, swollen by a compatible diluent with a dissolved lithium salt, sandwiched between a lithium metal anode and a porous carbon cathode that receives electrons from the external circuit. As in the case of PILs, the interplay between ionic conductivity and bulk viscosity (see Fig. 6.10), and the direct interaction between the polymer and the ions, appear to be the most relevant issues.[89] The analysis of these systems, detailed in Chapter 21, showed cases in which the degree of association (ion pairs, charged triplets, etc.) increased with salt concentration, but was occasionally reduced (using polyethylene oxide [PEO] matrices) with respect to the corresponding liquid electrolyte. The latter effect substantiates a direct interaction of lithium ions with the polymer network.

6.7 HOFMEISTER EFFECTS

6.7.1 Specific Ion Binding

The pioneering work of Hofmeister gave evidence of how salts affect the solubility properties of proteins such as globulin and albumin. Later work analyzed the detailed role of salts on the conformational stability of proteins and other polar macromolecules. The salt interaction with charged or polar groups on the macromolecule appeared to involve features that could not be justified by considering only charge-dependent effects such as ion screening or condensation over fixed charges. Early literature confirmed the following series for the ability of ions to increase the solubility and the swelling of polar polymers and isoelectric proteins, or depressing their melting temperature[94–97]:

$$SO_4^{2-} < F^- < Cl^- < Br^- < NO_3^- < I^- < SCN^-,$$

$$K^+ < Mg^{2+} < Na^+ < Cs^+ < Li^+ < Ca^{2+}.$$

Related Hofmeister series were shown to represent a variety of physicochemical phenomena including ion hydration, the surface tension and structure of water.[33,96] The effects of single ions are addictive ones, and salts formed by anions and/or cations on the extreme right of the series are strong *salting-in* agents. Early reports showed that a large solubility of isoelectric proteins in LiCl or $Ca(SCN)_2$ solutions occurs well above the concentration (~0.2 M) at which charge-dependent effects (i.e., chain expansion and Donnan effects) ought to be minimized. A specific, charge-independent effect prevailing with the salting-in agents was therefore postulated.[95]

Various molecular interpretations of this effect have been proposed, but none has gained definite acceptance.[96] A recent interpretation, based on dispersion forces, supports a charge-independent contribution to Hofmeister effects.[96a] A satisfactory thermodynamic description considers a preferential association (binding) of the salting-in agents to the polymeric substrate.[38] Early experimental data supported the occurrence of such a preferential association. The analysis was based on the determination of the initial (C_s^i) and final (C_s^f) concentration of salt solutions before and after equilibration with amorphous[95b] or crystalline collagen.[97] Data in Figure 6.11a evidence a close

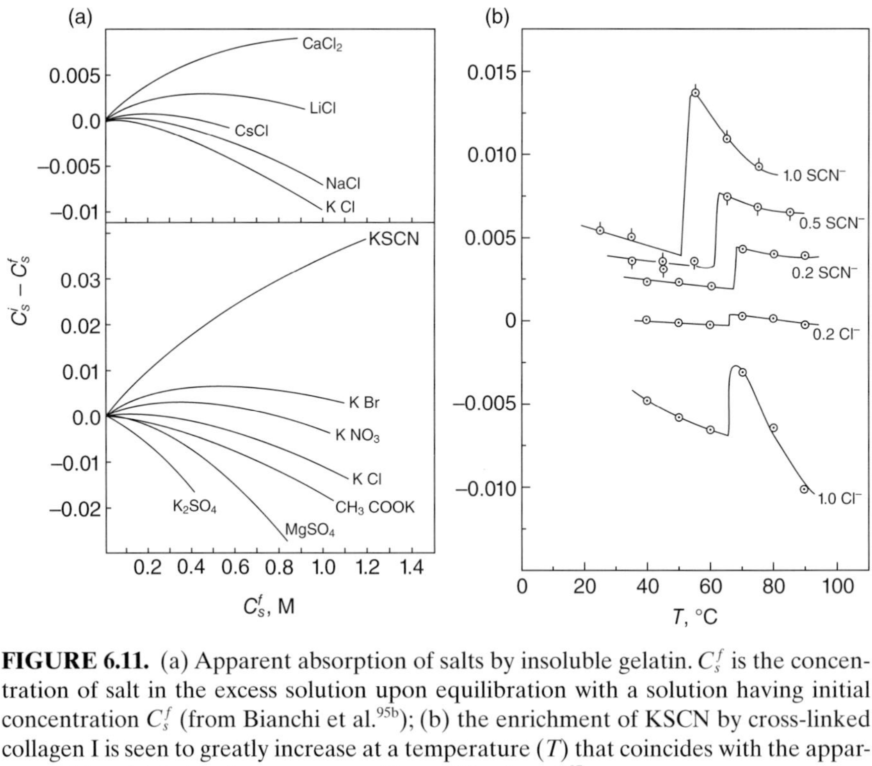

FIGURE 6.11. (a) Apparent absorption of salts by insoluble gelatin. C_s^f is the concentration of salt in the excess solution upon equilibration with a solution having initial concentration C_s^f (from Bianchi et al.[95b]); (b) the enrichment of KSCN by cross-linked collagen I is seen to greatly increase at a temperature (T) that coincides with the apparent melting (shrinkage) temperature. From Ciferri et al.[97]

TABLE 6.1. Binding Free Energy Obtained from Melting and from Enrichment Data (Ciferri et al.[97])

Salt	Temperature (°C)	$-\Delta F_0 = p\ RT \ln(1 + K_s\ a_s)$, cal mole^{-1}		
		Melting (Eq. 6.5)	Analytical (Bianchi et al.[95b])	Analytical (Ciferri et al.[97])
KSCN (1 M)	75	115.7	55.2	48.0
KCl (1 M)	80	23.2	11.5	6.9

$\Delta H_u = 2$ Kcal mol^{-1} (Flory and Garrett[105]). Hydration (mole water/mole residue): $0 \rightarrow 2$ (Ciferri et al.[97])

correspondence of the apparent ion adsorption to the protein to the ranking of ions in the Hofmeister series. The binding free energy calculated from the data in Figure 6.11a is reported on Table 6.1.

Salt enrichment is temperature dependent and a particularly large increase of the salt associated to the polymer is observed during the melting transition of collagen fibers (Fig. 6.11b). Moreover, the melting transition is strongly lowered by increasing KSCN concentration. Similar effects were noticed in connection with helix-coil transitions.[95a] Preferential association in solution of the disorganized polymer, when a larger number of binding sites becomes available, is typical of *allosteric* transitions that are thermodynamically described by adding a binding term to the conventional free energy of solution, as embodied by the relationship[98,99]:

$$1/T_m - 1/T_m^0 = (R/\Delta H_u)(V_u/V_1)(v_1 - \chi v_1^2) + (R/\Delta H_u) p \ln(1 + K_s C_s). \quad (6.5)$$

The first term on the right-hand side (RHS) of Equation 6.5 accounts for the melting temperature depression due to a nonbinding diluent having volume fraction v_1 and thermodynamic quality expressed by the Flory–Huggins parameter χ (classical diluent effect). The second term on the RHS of Equation 6.5 accounts for an additional depression due to specific binding characterized by the number of independent binding sites p, the salt activity C_s, and the salt binding constant K_s. The latter will include the contributions of constituent ions. Moreover, the competition between salts and water molecules for binding to the same polar sites can be theoretically described[58] by introducing the water binding constant K_w in Equation 6.5. Equation 6.5 does not specifically account for charge-dependent effects. Therefore, the diluents and the binding effects are expected to be particularly evident above ~0.1 M when ionic interaction are screened out, but must necessarily occur in the whole concentration range (it is to be noticed that the charge-dependent interaction in the condensation regime is described by a different binding constant between fixed charges and mobile ions; see Chapter 3).

The experimental variation of the melting temperature of collagen fibers as a function of salt concentration is compared with theoretical expectations

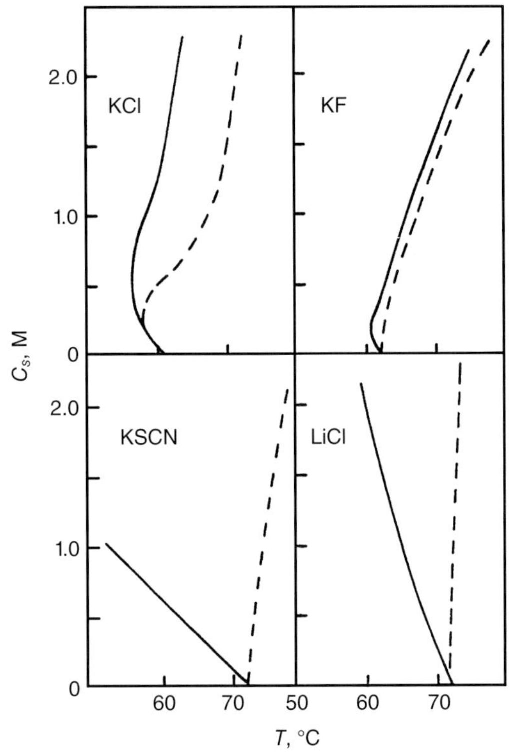

FIGURE 6.12. Full lines: experimental melting curves for cross-linked collagen I in solutions of two salting-out agents (KCl and KF) and two salting-in agents (KSCN and LiCl). Broken lines: theoretical curves calculated from Equation 6.5 using $K_s = 0$, $\Delta H_u = 2$ K cal mol^{-1} (Flory and Garrett[105]) and χ parameters assessed from intrinsic viscosity and swelling data (Ciferri[38]). It is seen that Equation 6.5 with $K_s = 0$ satisfactorily describes the experimental behavior only for the salting-out agents. From Ciferri.[38]

in Figure 6.12. The curves allow resolution of the contributions of specific binding and of the diluent effect. Using χ values independently assessed from viscosity and swelling equilibrium data (see Table 6.2), the binding term in Equation 6.5 was evaluated for the salting-in agents KSCN and LICl.[33,97] Results are included in Table 6.1. Comparison with the values independently obtained from the analytical data lends plausible support to the operational validity of Equation 6.5.

6.7.2 Diluent Effect

Omitting the binding free energy term, Equation 6.5 represents the classical expression of polymer solution theory for the variation of the solubility or melting temperature with the amount and thermodynamic quality of the diluenty. The experimental data for the melting of collagen I (see Fig. 6.12) for

TABLE 6.2. Viscosity Results for Gelatin (Ciferri and Kudaibergenov[33])

C_s (M)	$[\eta]_{40}$ (dl g^{-1})	χ	θ (K)	$-\psi$	$-\kappa$ (55°C)
KCl					
0	0.440	0.4995	394	0.028	0.033
0.05	0.442	0.4994	412	0.023	0.029
0.15	0.470	0.4991	442	0.022	0.031
0.40	0.430	0.4994	410	0.022	0.028
1.0	0.382	0.4997	385	0.019	0.022
2.0	0.332	0.4999	355	0.014	0.015
3.0	0.243	0.5001	276	0.010	0.009
KSCN					
0.3	0.438	0.4994	413	0.022	0.028
0.7	0.478	0.4992	444	0.023	0.031
1.5	0.502	0.4900	529	0.016	0.026
2.5	0.498	0.4899	820	0.007	0.018
4.0	0.484	0.4899	(−102)	−0.009	0.002
6.0	0.414	0.4993	239	−0.024	−0.017
8.0	0.284	0.4998	311	−0.040	−0.038

The entropy (ψ) and enthalpy (κ) of dilution parameters are related to the excess free energy of dilution by the relationship $\Delta F_1/RTv_1^2 = 1/2 - \chi = (\psi - \kappa) = \psi(1-\theta/T)$, where θ is the theta temperature.[30a]

two salting-out agents (KCl and KF) are well represented by the theoretical curves using the χ parameter determined for salt solutions of different salt concentration (see Table 6.2[33]). Therefore, the melting behavior in the presence of KCl and KF conforms to the conventional treatment of polymer solutions.[30a]

Data in Table 6.2 reveal that the χ parameter was close to 0.5 over a salt concentration range 0.05–8 M. Moreover, its entropy (ψ) and enthalpy (κ) components were both negative and their absolute values decreased with ionic strength (see footnote on Table 6.2). The thermodynamic quality of the solvent appears therefore improved by addition of KCl from the standpoint of dilution entropy, and reduced from the standpoint of the exothermicity of dilution. The small salting-in effect observed up to about 0.2 M concentration of KCl and KF was attributed to the reduction of the attraction of complementary charges predicted by the theory of polyampholytes (Section 6.3). The re-entrant effect (reduced solubility or salting out) visible at concentrations >~0.3 M can therefore be unambiguously attributed to the reduced exothermicity of the salt solution toward the screened polyampholyte.[33] This interpretation appears more general and far superior to that based on charge screening effects. Multimeric associations that form soluble aggregates are also favored.[100]

The diluent effect is a thermodynamic rendition that includes the salting-out behavior exhibited by nonpolar solutes. The latter behavior was the objective of earlier theories of the Hofmeister series (e.g., the hydration theory[96e] or the electrostriction theory[56,96f]) that were not fit to describe the behavior of polar and charged solutes, nor the differences between the roles of anions and cations.[58]

The separation into weak (electrostatic screening + solvophobic) and strong (specific binding) interactions embodied in Equation 6.5 might be regarded as an oversimplification. However, a similar separation is made for binary systems between bound (hydration) and bulk water. The separation permits an operational assessment of dilution and binding parameters. For the latter, additional molecular features are highlighted below.

6.7.3 pH Shifts and Location of Specific Interaction

The molecular description of the binding process is still an elusive one.[96,101,102] The binding constant of KSCN to collagen is rather small (~2.5 M^{-1}) consistently with the additive contribution of various ions and with the observed lack of saturation of the available peptide bonds even at high molarity.[33,97] Only a small fraction (up to 25%) of the available residues appear to be potential binding sites for any given salt.[97]

At salt concentration below 0.1 M, pH shift have been observed (e.g., Fig. 6.13). The occurrence of a pH shift suggests an alteration of the total

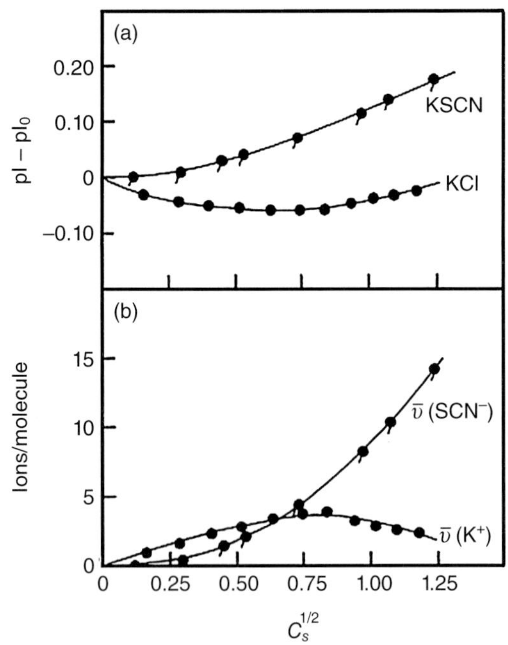

FIGURE 6.13. (a) Shift of the isoelectric pH for a $pI_0 = 4.8$ gelatin at 40°C versus the square root of molarity for KSCN and KI; (b) ion per molecule selectively bound according to $pI = pI_0 - C(N^+ - N^-)$. C = constant[1] and the bracket represents the inequality of the number of cationic and anionic charges. From Ciferri et al.[97]

charge on the polymer due to salt binding.[38,101] Note, however, that considerable difficulties have often been reported in interpreting pH shifts.[97,102].In any event, the Debye screening length falls below 3 Å at $C_s > 0.1$ M when all polyelectrolyte and Donnan effects should be screened out. The continuing depression of the stability of collagen cannot, therefore, be attributed to polyelectrolyte effects induced by selective ion binding.

Moreover, where does specific ion binding occur? To the relatively few fixed charges, to partial charges such as peptide bonds occurring with every residue, or to the numerous apolar (hydrophobic) sites along the polymer chains? Direct pairing between fixed and mobile opposing charges does not appear to be essential to the binding process. In fact, destabilization and increased solubility are promoted by salting-in cations (i.e., LiCl), even for polymers containing amide links (i.e., polycaproamide) but no titrable amino-acid residues. Moreover, the largest destabilization is promoted by salts for which both the anion and the cation are strongly bound[38] (i.e., $Ca(SCN)_2$). On decreasing the number of amide bonds on the nylon series, the depression of the melting temperature of the polymers decreased and vanished altogether for a polyethylene chain,[94] also ruling out a direct binding to apolar groups.

The polar peptide bond, rather than charged or apolar groups, has therefore been regarded as the binding site. Stoichiometric complexes involving model peptides and LiBr, or KI, with amide bonds in solution, were in fact reported.[102] Siegel and Martin have discussed metal ion complexes and charge distribution over the peptide bond.[103] The peptide bond is neutral through most of the pH range but can form adducts at the oxygen atom. Schneider has described several supramolecular complexes stabilized by electrostatic interactions different from the ion pair interaction, particularly interactions between a permanent charge and an opposing *partial* charge. For instance, complexes between crown ethers and alkali metal cations are stabilized by the attraction between the ion and the free electron pairs on a heteroatom of the ligand.[55]

Still unsettled is the molecular basis for the specificity of the large, weakly hydrated anions, and of the small, strongly hydrated, net structure-forming cations. Some relevant suggestions emerge from the occurrence of a melting temperature depression caused by LiCl for uncharged polyamides in the absence of water.[94] In the absence of water, there should be neither competition of ions and water molecules for binding to peptide bonds, nor ionization of mobile or fixed ions, nor ion hydration. LiCl is not an IL; its melting temperature is in the order of 600°C, larger than that of NaCl reflecting a shorter distance between Li and Cl ions. A reduction of its melting temperature due to binding to the amide bond and a strong correlation of the two ions in the absence of solvation could be expected.[56] Therefore, a complete dissociation of LICl into the constituent ions, and the accompanying hydration occurring in water solution, may actually be unessential. Moreover, no depression was caused by KBr, containing the salting-in anion Br^-. Therefore, properties specific to the *naked* ions are suggested to play a primary role in Hofmeister effects (see Section 6.8).

TABLE 6.3. Influence of Monovalent Ion Binding on Polymer Net Charge and Solubility

Polymer	C_s (M)	Mobile Ion	Net Charge	Solubility
PE +	<0.1	Cation	↑	↑
PE +	<0.1	Anion[a]	↓	↓
PE –	<0.1	Anion	↑	↑
PE –	<0.1	Cation[a]	↓	↓
PA (pI) +,–	<0.1	Anion	↑	↑
PA (pI) +,–	<0.1	Cation	↑	↑
PE, PA	>0.1	Anion, cation	—	↑
Polar	>0.1	Anion, cation	—	↑

[a] Divalent ions → intermolecular aggregation.

6.7.4 Hofmeister Series Reversals

Reversals of the order of the Hofmeister series have been recently discussed in connection with synthetic anionic or cationic colloids carrying polar or nonpolar groups.[96g] Reversals were reproduced by molecular dynamics simulation, and the interplay between charge and polarity was discussed in terms of alteration of surface tension. Actually, reversals of the Hofmeister series were reported and interpreted several years ago in connection with phase diagrams for isoelectric and cationic proteins.[95a] For instance, the denaturation temperature of collagen measured under isoelectric condition (pH ~ 7) is known to be lowered by salting-in agents (KSCN) and increased by salting-out agents (KCl). However, at pH ~ 2 and low salt concentration, when the protein carries a net cationic charge, the order of effectiveness of the anion series was reversed. This trend is not just consistent but actually reinforces the interpretation of the Hofmeister series in terms of binding and diluent effects. In fact, at low ionic strength the selective binding of SCN^- is expected to cause a decrease of the electrostatic repulsion along the cationic chain, and thus increased protein stabilization or decreased solubility. Table 6.3 presents a general schematization of the way ion binding affects the solubility characteristics of polyelectrolytes, polyampholytes, and uncharged polar polymers as a function of ionic strength. Occurrence of reversals on the Hofmeister series appear to be a consequence of charge-dependent effects for charged polymers at low ionic strength.

6.7.5 Encroachment of Charge-Dependent and Charge-Independent Interactions

The occurrence of pH shifts and the reversal of the Hofmeister series discussed above offer compelling evidence that charge-independent and charge-dependent effects are intertwined, but might be dissected by proper analysis of the role of the extent of ionization and ionic strength. A reanalysis of earlier

TABLE 6.4. Apparent Melting Temperatures for Collagen I (Ciferri[38])

C_s (M, pH)	$T_m - T_m^*$(°C)			
	KCl	KSCN	$CaCl_2$	$Ca(SCN)_2$
0.1, 7.4	−3.0	−4.0	−6.0	−9.0
0.3, 7.4	−1.5	−9.0	−14.0	−22.0
0.6, 7.4	−0.5	−20.0	−20.0	
1.0, 7.4	0	−26.0		
10^{-4}, 6.0	0.5		1.5	
10^{-3}, 6.0	2.0		2.0	
10^{-2}, 6.0	5.0		3.0	
0.2, 6.0		−5.0	−6.0	
0.6, 6.0		−9.0	−8.0	
1.0, 6.0		−14.0	−18.0	
0.3, 2.3		1.0	−1.0	
0.6, 2.3	3.0	4.0	5.0	
1.0, 2.3	10.0	1.0	8.0	
1.6, 2.3	17.0	−5.0	−1.0	

T_m* = T_m at $C_s = 0$ (48°C at pH = 7.4; 42°C at pH = 6.0; 34°C at pH = 2.3).
Low C_s ($10^{-4} - 10^{-2}$ M): data from Bianchi et al.[95a]

and reccent data covering a wide range of pH and ionic strengths allows a clear operational distinction between charge-dependent and charge-independent contributions for a given protein. Table 6.4[38] summarizes data on the thermal stability of collagen I fibrils reported by two groups of investigators.[95a,101] The difference between the apparent melting temperature of collagen I measured in water (T_m^*) and in salt solutions (T_m) is reported as a function of salt type, salt concentration, and pH of the equilibrium solution. Results and conclusions are highlighted for nonbinding (KCl) and binding (KSCN, $CaCl_2$, and $Ca(SCN)_2$) salts.

KCl At the isoelectric pH (7.4) and low salt concentration, the data show a significant depression of the melting temperature which is attenuated as C_s increases. As discussed above (the diluent effect and Fig. 6.12), the trend is attributed to the screening of the attraction of complementary fixed charges that is gradually balanced by the competing reduction of the exothermicity of dilution.

Under nonisoelectric conditions (pH 6 and 2.3), the data of both groups of investigators reveal a significant and continuous increase of the melting temperature up to rather large C_s values. This trend is attributed to screening of the repulsion of fixed cationic charges at pH = 2.3 reinforced, as C_s increases, by the worsening exothermicity of dilution. In particular, T_m^* in pure water was some 12°C lower at pH = 2.3 than at pH = 7.4. This difference, which can be unambiguously attributed to the repulsion of the unscreened cationic charges, appears overcompensated at 1 M KCl (pH = 2.3) by the encroachment of the worsening diluents' power.

KSCN At the isoelectric pH, the depression of T_m is much larger than observed with KCl. This finding is attributed to the specific binding of SCN^-, supported by the data in Figure 6.11. The encroachment with a salting-out effect due to a decreasing diluent power is observed for KSCN only at salt concentration much larger than those reported in Table 6.4, when more binding sites are saturated.[97] The suggestion, often found even in recent literature, that such a large depression reflects the electrostatic repulsion of selectively bound SCN^- ions should be definitively rejected for several reasons. At the salt molarity considered in Table 6.4, the available peptide bonds are far from being fully saturated.[38] Analysis of enrichment data (Fig. 6.11a) reveals that at $C_s^f = 0.5$ M, only 0.02 M represents the net KSCN enrichment within the protein. Consideration of pH shifts[97] (Fig. 6.13) further reveals that only a small fraction of salt bound (~8%) corresponds to selectively bound SCN^- ions. Under these conditions, any additional charge on the polymer induced by binding of negatively charged anions would be completely screened out by the excess of free ions in solutions.[38] In fact, no individual ion disproportion was detected in the solution within and outside collagen networks, demonstrating the lack of Donnan effects above 0.1 M concentration.[38,97] Moreover, the depression due to $Ca(SCN)_2$ when both ions are strongly bound is seen to be much larger than that due to either KSCN or $CaCl_2$ when only one ion is selectively bound.

Under near isoelectric conditions at pH = 6, the depression of T_m is less pronounced than at pH = 7.3. This is attributed to a competing screening effect due to a small, net cationic charge associated with Lys and Arg residues (evidenced by a 6% reduction of T_m^* in pure water with respect to T_m^* at pH = 7.4; see also the footnote on Table 6.4). On the other hand, the screening of the net cationic charge at pH = 2.3 causes an increase of T_m up to 0.6 M KSCN. However, a reversal is seen to occur at still larger concentrations, suggesting that once the repulsion of the net cationic charges is mitigated, the ongoing depression due to charge-independent salt binding again becomes the overriding effect.

$CaCl_2$ At the isoelectric pH, similar effects to those observed with KSCN are evidenced. Under nonisolectric conditions, the increase of T_m with salt concentration in the range from 10^{-4} to 10^{-2} M reported at pH = 6 by Freudenberg et al.[101] was interpreted in terms of a perplexing superimposition of two charge-dependent effects: a stabilizing effect due to screening, and a destabilizing effect resulting from binding of the divalent Ca^{2+} ions to the negatively charged Glu and Asp residues (suggested by pH shifts), thus enhancing the net cationic charge. However, at lower pH (2.3) and in a larger salt concentration range (0.3–1.3 M) a re-entrant effect is evidenced by the data of Bianchi et al.[95a] The increase of the melting temperature up to ~1 M was attributed to prevailing screening, and the ensuing reduction of T_m was attributed to the prevailing charge-independent salt binding. Consistent with this interpretation, the increase of T_m observed with $CaCl_2$ is larger than observed with KSCN at pH = 2.3. In fact, if binding of the SCN^- anion resulted in a compensation of the net cationic charge on the protein, it would have caused a larger increase of T_m with respect to the cation Ca^{2+}, contrary to what is observed.

6.8 MAIN CONCLUSIONS

The conformational features promoted by ionic mixed interaction are of great relevance to the assembly of complex biostrucures. An elongated, fully charged chain has been the accepted model for the conformation of strong polyelectrolytes. However, it appears that a wide variety of attractive interactions favor a pearling conformation. Deviations from the fully stretched model may arise from hydrophobic groups, as evidenced by the analysis of hydrophobic polyelectrolytes, or from the occurrence of complementary fixed charges along a given chain, as evidenced by the analysis of polyampholytes. Moreover, these deviations were shown to occur also from alteration in the concentration of the polyelectrolyte, ionic strength, solvent quality, and interactions with complementary charges carried by other molecules or molecular assemblies (e.g., polymers, surfactants, micelles, histones). Pearls or blobs predicted by theory, demonstrated by simulation, and actually observed, include folds of section of a given polyelectrolyte, or folds surrounding clusters of other molecules. In either case, driving forces for thc assembly of rather complex structures of biological relevance can now be suggested.

In the case of folds within a given polyelectrolyte, proper account of chain rigidity and like-charge attraction ought to describe complex conformations such as that of the DNA toroids (see Fig. 6.2c).[21] The structures including folds of a polyelectrolyte around clusters of other molecules ought to be predicted by the interplay between the flexural rigidity of the polyelectrolyte and the association constant of the complementary assembly, as discussed in Perico and Ciferri.[9] In the aforementioned work, the wrapping of a short and rigid DNA molecules (150 bp) around the complementary histone octamer (nucleosome) was forced by the strong cohesion of the histone octamer.[9] Extension of the same mechanism to the case of giant DNA that has a global semirigid conformation should offer a justification for the superstructure of the chromatin fiber, regarded as a helical sequence of sections of rigid DNA wrapped around the histone octamer.[9,104]

The analysis of the Hofmeister series, including the role of salts in the absence of water, invites some conclusions that shed considerable new light on this still incompletely understood effect. It is too often implied that only charge-dependent ionic interactions (namely screening and condensation) control the interactions between salts and charged polymers. However, the reported results evidence that this oversimplified approach ought to be finally rejected and replaced by consideration of ionic mixed interactions such as the diluent effect and the specific binding. Both effects include charge-dependent and charge-independent contributions prevailing, respectively, in the low (<~0.1 M) and in the higher ionic strength range.

The diluent effect describes the relatively weak interactions due to screening of the fixed ions (prevailing at low ionic strength) and the reduced compatibility between the screened/bound polymer and the bulk of the salt solution (prevailing at higher strength).

The relatively stronger binding interaction includes a charge-dependent contribution related to the alteration of the fixed charge on the polymer, evidenced by pH shifts at $C_s < 0.1$ M. At a larger ionic strength, any charge-dependent contribution is inconsequential, and the overwhelming contribution is the decrease of free energy of the solution due to the spontaneous association of the ions to peptide bonds, in analogy to allosteric phenomena. The observation that the contribution of salting-in ions persists when there are no titrable groups on the polymer, and even in the absence of water, admits no other interpretation. The complex encroachment of charge-dependent and charge-independent interactions could be evidenced and dissected using data on the thermodynamic stability of a given protein in a wide pH, salt type, and concentration range.

The roles of ion hydration, of ion pairing of the salts, and the origin of the specificity of the weakly hydrated anions or the strongly hydrated cations, are still unsettled points. Long ago, Robinson and Jenks suggested that anion binding reflects more the poor interaction of the ions with water than a strong preference for the polar groups of the solute.[96b] The salting in promoted by cations was attributed to more definite classes of compounds. Consistently with these earlier suggestions, cations coordinate water molecules by interaction with a lone electron pair, whereas hydration of anions is promoted by weaker H-bonds. Moreover, negative free energies for the transfer of large anions or cations from water to an organic solvent were reported.[56]

Additional support to Robinson and Jenks' suggestion is offered by the more recent determination of melting temperature depression of polyamides in absence of water. The depression was found to be a strong one with a salt containing a salting-in cation, but a negligible one with a salt containing a salting-in anion.[94] Therefore, under anhydrous conditions, the naked cations, at variance with the naked anions, still exhibit a strong interaction with the substrate that is evidently not due to hydration. Dispersion forces, which depend upon specific ion polarizability and first ionization potential,[96a] are likely a primary driving force for the competing effects of hydration and binding to polar sites.

ACKNOWLEDGMENTS

The author expresses his great appreciation to the contributors of the book for stimulating discussions and relevant information. Great appreciation is also manifested to Dr. J.L. Sikorav for making available the image in Figure 6.2.

REFERENCES

1. C. Tanford. *Physical Chemistry of Macromolecules*. Wiley, New York, 1961.
2. (a) H. Zhou, G.-Q. Song, Y.-X. Zhang, J. C. M. Jiang, T. E. Hogen-Esch, R. Dieing, L. Ma, L. Haeussling. *Macromol Chem Phys* **2001**, 202, 3057; (b) A. Ciferri. *Chem Eur J* **2010**, 16, 10930.
3. U. Micka and K. Kremer. *Europhys Lett* **2000**, 49, 189.
4. F. Oosawa. *Polyelectrolytes*. Marcell Dekker, New York, 1971.
5. G. S. Manning. *Q Rev Biophys* **1978**, 11, 179; *J Am Chem Soc* **2003**, 125, 15087.
6. B. O'Shaughnessy and Q. Yang. *Phys Rev Lett* **2005**, 94, 048302.
7. C. Holm, J. F. Joanny, K. Kremer, R. R. Netz, P. Reineker, C. Seidel, T. A. Vilgis, R. G. Winkler. *Adv Polym Sci* **2004**, 166, 111.
8. S. Pietronave, L. Arcesi, C. D'Arrigo, A. Perico. *J Phys Chem B* **2008**, 112, 15991.
9. A. Perico and A. Ciferri. *Chem Eur J* **2009**, 15, 6312.
10. V. A. Bloomfield, D. M. Crothers, I. Tinoco. *Nucleic Acids Structures, Properties, Functions*. University Science Books, Sausalito, CA, 2000.
11. A. Perico and A. Rapallo. *J Chem Phys* **2011**, 134, 055108.
12. C. Holm, M. Rehanan, W. Oppermann, M. Ballauff. *Adv Polym Sci* **2004**, 166, 1.
13. T. Odijk. *J Polym Sci Polym Phys* **1977**, 15, 477.
14. M. Mandel and J. Schouten. *Macromolecules* **1980**, 13, 1247.
15. D. Poland and A. Scheraga. *Theory of Helix-Coil Transitions in Biopolymers*. Academic Press, New York, **1970**.
16. W. Zhang and S. Nilsson. *Macromolecules* **1993**, 26, 2866.
17. R. S. Dias. In *DNA Interactions with Polymers and Surfactants*, R. S. Diaz and B. Lindman, eds. Wiley, Hoboken, NJ, 2009.
18. A. Stroobants, H. Lekkerkerker, T. Odjik. *Macromolecules* **1986**, 19, 2232.
19. A. R. Khokhlov. In *Liquid Crystallinity in Polymers*, A. Ciferri, ed. VCH Publishers, New York, 1991.
20. T. E. Strzelecka, M. W. Davidson, R. L. Hill. *Nature* **1988**, 331, 457.
21. W. M. Gelbart, R. F. Bruinsma, P. A. Pincus, V. Parsegian. *Phys Today* **2000**, 53, 38.
22. G. Rialdi, P. Profumo, A. Ciferri. *Biopolymers* **1969**, 8, 701. *J Biol Chem* **1973**, 248, 2963.
23. A. Ciferri. *Macromol Phys Chem* **1994**, 195, 457.
24. V. A. Bloomfield. *Curr Opin Struct Biol* **1996**, 6, 334.
25. D. McLoughlin, M. Delsanti, C. Tribet, D. Langevin. *Europhys Lett* **2005**, 69, 461.
26. I. M. Kulic and H. Schiessel. In *DNA Interactions with Polymers and Surfactants*. R. S. Dias and B. Lindmann, eds. Wiley, Hoboken, NJ, 2008.
27. L. Koltover, T. Salditt, J. O. Radler, C. Safinya. *Science* **1998**, 281, 76.
28. X.-L. Wang, X.-H. Zhang, M. Cao, H.-Z. Zheng, B. Xiao, Y. Wang, M. Li. *J Phys Chem B* **2009**, 113, 2328.
29. N. Miyazawa, T. Sakaue, K. Yoshikawa, R. Zana. *J Chem Phys* **2005**, 122, 44902.
30. (a) P. J. Flory. *Principles of Polymer Chemistry*. Cornell Univ. Press, Ithaca, NY, 1953; (b) A. Kathchalsky, S. Lifson, H. Eisenberg. *J Polym Sci* **1952**, 8, 476.

31. Y. Kantor and M. Kardar. *Phys Rev E* **1995**, 51, 1299.
32. A. V. Dobrynin, R. H. Colby, M. Rubinstein. *J Polym Sci Polym Phys B* **2004**, 42, 3513.
33. A. Ciferri and S. Kudaibergenov. *Macromol Rapid Commun* **2007**, 28, 1953.
34. (a) A. Katchalsky and I. R. Miller. *J Polym Sci* **1954**, 13, 57; (b) J. Mazur, A. Silberberg, A. Katchalsky. *J Polym Sci* **1959**, 35, 43.
35. H. Dautzemberg, W. Jarger, J. Koetz, B. Philipp, C. Seidel, D. Stecherbina. *Polyelectrolyte Formation, Characterization, Application*. Hanser, Munich, 1994.
36. S. E. Kudaibergenov. *Adv Polym Sci* **1999**, 144, 115.
37. A. B. Lowe and C. I. McCormick. *Chem Rev* **2002**, 102, 4177.
38. A. Ciferri. *Biopolymers* **2008**, 89, 700.
39. G. Z. Zheng, G. Meshitsuka, A. Ishizu. *J Polym Sci Polym Phys* **1995**, 33, 867.
40. B. Phyllip, H. Dautzemberg, K. J. Linow, J. Klotz, W. Davidoff. *Prog Polym Sci* **1998**, 14, 823.
41. M. Castelnovo and J. F. Joanny. *Macromolecules* **2002**, 31, 4531.
42. A. Baumketner, H. Shimizu, M. Isobe, Y. Hiwatari. *J Phys Condens Matter* **2001**, 13, 10279.
43. G. L. Makhatadze, V. L. Loladze, D. N. Ermolenko, X. C. Chen, S. T. Thomas. *J Mol Biol* **2003**, 327, 1135.
44. S. Kumar and R. Nussinov. *Chembio Chem* **2002**, 3, 604.
45. G. Decher. *Science* **1997**, 277, 1232.
46. X. Arys, A. M. Jonas, A. Laschewsky, R. Legras, F. Malwitz. In *Supramolecular Polymers*, 2nd ed. A. Ciferri, ed. CRC Press, Boca Raton, FL, 2005.
47. J. J. Cerda, B. Quiao, C. Holm. *Soft Matter* **2009**, 5, 4412.
48. A. C. Arsenault, J. Halfyard, Z. Wang, V. Kitaev, G. A. Ozin, I. Manners, A. Mihi, H. Miguez. *Langmuir* **2005**, 21, 499.
49. (a) B. Abu-Sharkh. *J Chem Phys* **2005**, 123, 114907; (b) R. R. Netz, J. F. Joanny. *Macromolecules* **1999**, 71, 4007; (c) D. Kovacevic, S. van der Burgh, A. de Kleizer, M. A. Cohen Stuart. *Langmuir* **2002**, 18, 5607.
50. P. Lavalle, C. Picart, C. Mutterer, J. Gergely, H. Reiss, J. C. Voegel, B. Senger, P. Schaaf. *J Phys Chem B* **2004**, 108, 635.
51. M. Schönhoff, V. Ball, A. R. Bausch, C. Dejugnat, N. Delorme, K. Glindel, R. von Klitzing, R. Steitz. *Colloid Surf* **2007**, 303, 14.
52. N. Hoda and R. G. Larson. *Macromolecules* **2009**, 42, 8851.
53. M. Salomäki and J. Kankare. *Macromolecules* **2008**, 41, 4423.
54. O. M. Tanchak, K. G. Yager, H. Fritzsche, T. Harroun, J. Katsaraas, C. H. Barret. *J Chem Phys* **2008**, 129, 084901.
55. H.-J. Schneider. *Angew Chem Int Ed* **2009**, 48, 3924; **1999**, 30, 1471, 1991.
56. Y. Marcus and G. Hefter. *Chem Rev* **2006**, 106, 4585; Y. Marcus. *Ion Properties*. Dekker, New York 1997.
57. (a) T. Vermonden, J. van der Gucht, P. de Waard, A. T. M. Marcelis, N. A. M. Besseling, E. J. R. Sudholter, G. J. Fleer, M. A. Cohen Stuart. *Macromolecules* **2003**, 36, 7035; (b) T. Vermonden, W. M. de Vos, A. T. M. Marcelis, E. J. R. Sudhölter. *Eur J Inorg Chem* **2004**, 11, 2847; (c) T. Vermonden, M. J. van Steenbergen, N. A. M.

Besseling, A. T. M. Marcelis, W. E. Hennink, E. J. R. Sudhölter, M. A. Cohen Stuart. *J Am Chem Soc* **2004**, 126, 15802.

58. D. Puett and A. V. Rajagh. *J Macromol Sci A* **1968**, 2, 111.
59. A. Ciferri. In *Supramolecular Polymers*, 2nd ed. A. Ciferri, ed. CRC Press, Boca Raton, FL, 2005.
60. P. van der Schoot, A. P. Schenning, E. W. Meijer. *Science* **2006**, 313, 80.
61. F. S. Han, M. Higushi, D. G. Kurth. *Adv Mater* **2007**, 19, 3928.
62. J. B. Beck, J. M. Ineman, S. J. Rowan. *Macromolecules* **2005**, 38, 5060.
63. V. Percek, J. Heck, G. Johansen, G. Tomazos, M. Kawagumi. *J Macromol Sci A* **1994**, 111, 1031.
64. J. B. Beck and S. J. Rowan. *J Am Chem Soc* **2003**, 125, 13922.
65. P. Cordier, F. Tournihac, C. Souliè-Ziakovic, L. Lieber. *Nature* **2008**, 451, 977.
66. R. Reuteanauer, F. Buhler, P. J. Boul, S. J. Candau, J.-M. Lehn. *Chem Eur J* **2009**, 15, 1893.
67. A. Ciferri. *Chem Eur J* **2009**, 15, 6920.
68. A. Ciferri, L. V. Rajagh, D. Puett. *Biopolymers* **1965**, 3, 461.
69. (a) E. M. Brown. *J Am Leather Chem Assoc* **1997**, 92, 225; (b) K. te Nijenhuis. *Macromol Symp* **2001**, 171, 189.9.
70. (a) A. Messerschmidt, R. Huber, K. Weighardt, T. Poulos. *Handbook of Metalloproteins*. Wiley, New York, 2001; (b) E. N. Salgado, R. J. Radford, F.A. Tezcan. *Acc Chem Res* **2010**, 43, 661; (c) M. Faiella, C. Andreozzi, R. T. M. de Rosales, V. Pavone, O. Maglio, F. Nastri, W. F. DeGrado, A. Lombardi. *Nat Chem Biol* **2009**, 5, 882; (d) J. J. R. Frausto da Silva, R. J. P. Williams. *The Biological Chemistry of the Elements: The Inorganic Chemistry of Life*, 2nd ed. Oxford University Press, Oxford, 2001; (e) C. Bonaventura, R. Henkens, K. D. Weaver, A. I. Alayash, A. L. Crumbliss. *Protein Reviews*, M. Bolognesi, G. di Prisco, C. Verde, eds. Springer Verlag, 2008; (f) K. Burger. *Biocoordination Chemistry: Coordination Equilibrium in Biologically Active Systems*. Ellis Horwood, New York, 1990; (g) A. L. Crumbliss. Personal communication; (h) S. R. Baten, M. S. Heville, D. R. Turner, eds. *Coordination Polymers*. R.S.C., Cambridge, 2009.
71. K. Y. Lee, W. H. Jo, I. C. Kwon, S. Y. Jeong. *Langmuir* **1998**, 14, 2329.
72. (a) B. Persson, A. Hugerth, N. Caram-Lelham, L.-O. Sundelöf. *Langmuir* **2000**, 16, 313; (b) M. Nichifor, D. Cristea, A. Carpov. *J Biomater Sci* **2000**, 28, 15; (c) M. Nichirof, X. X. Zhou, W. Beille, D. Cristea, A. Carpov. *J Pharm Sci* **2001**, 90, 681; (d) W. Wang, A. M. McConaghy, L. Tetley, I. Uchegbu. *Langmuir* **2001**, 17, 631.
73. L. Bromberg. *Curr Pharm Biotechnol* **2003**, 4, 339.
74. L. Bromberg. In *Handbook of Polyectrolytes and their Applications*, S. K. Tripathy, J. Kumar, H. S. Nalwa, eds. American Scientific Publishers, Valencia, CA, 2002.
75. A. Hashidzume, Y. Morishima, K. Szczubialk. In *Handbook of Polyectrolytes and their Applications*, S. K. Tripathy, J. Kumar, H. S. Nalwa, eds. American Scientific Publishers, Valencia, CA, 2002.
76. I. Ghimici, M. Bercea, E.-S. Dragan. *J Macromol Sci B* **2009**, 48, 1011.
77. A. R. Khokhlov and K. A. Khatchaturian. *Polymer* **1982**, 23, 1742.
78. A. V. Dobrynin, M. Rubinstein, S. P. Obukhov. *Macromolecules* **1996**, 29, 2974.

79. (a) A. V. Dobrynin, M. Rubinstein. *Macromolecules* **1999**, 32, 915; (b) A. V. Dobrynin, M. Rubinstein. *Macromolecules* **2000**, 33, 8097; (c) P. G. de Gennes, P. Pincus, R. M. Velasco, F. Brochard. *J Phys (Paris)* **1976**, 37, 1461.
80. (a) Q. Liau, A. V. Dobrynin, M. Rubinstein. *Macromolecules* **2006**, 39, 1920; (b) S. Ulrich, A. Laguecir, S. Stoll. *J Chem Phys* **2005**, 122, 094911; (c) J.-L. Hsu, U. P. Strauss. *J Phys Chem* **1987**, 91, 6238; (d) V. S. Zdanowicz, U. P. Strauss. *Macromolecules* **1993**, 26, 4770; (e) A. F. Olea, R. G. Barraza, I. Fuentes, B. Acedevo, F. Martinez. *Macromolecules* **2002**, 35, 1049; (f) Q. Qiu, A. Lou, P. Somasundaran, B. A. Pethica. *Langmuir* **2002**, 18, 5921.
81. W. Essafyi, F. Lafuma, D. Baigt, C. E. Williams. *Europhys Lett* **2005**, 71, 938.
82. A. Chepelanskii, F. Mohammad-Rafiee, E. Trizac, E. Raphael. *J Phys Chem B* **2009**, 113, 3743.
83. N. Lee and D. Thirumalai. *J Chem Phys* **2000**, 113, 5126.
84. K. Ohno. *Electrochemical Aspects of Ionic Liquids*, 2nd ed. Wiley, Hoboken, NJ, 2011.
85. W. Hu, E. I. Cooper, C. A. Angell. *J Phys Chem B* **2003**, 107, 6170.
86. O. Green, O. Grubjesic, S. Lee, M. L. Firestone. *Polym Rev* **2009**, 49, 339.
87. K. J. Fraser, E. L. Izgordina, M. Forsyth, J. L. Scott, D. R. MacFarlane. *Chem Commun* **2007**, 37, 3817.
88. T. Ueki and M. Watanabe. *Macromolecules* **2008**, 41, 3739.
89. J. P. Southall, H. V. Hubbard, S. F. Johnston, V. Rogers, G. R. Davies, J. E. McIntyre, I. M. Ward. *Solid State Ionics* **1996**, 85, 51.
90. J. Y. Song, Y. Y. Wang, C. C. Wan. *J Power Sources* **1999**, 77, 183.
91. O. V. Bushkova, V. M. Zhukowsky, B. L. Lirova, A. L. Kruglyashov. *Solid State Ionics* **1999**, 119, 217.
92. M. Forsyth, J. Z. Sun, D. R. MacFarlane, J. Hill. *J Polym Sci A* **2000**, 38, 341.
93. A. Ferry, L. Edman, D. R. MacFarlane, J. Z. Sun. *J Appl Phys* **1999**, 6, 2346.
94. (a) B. Valenti, E. Bianchi, A. Tealdi, S. Russo, A. Ciferri. *Macromolecules* **1979**, 9, 117; (b) E. Bianchi, A. Ciferri, A. Tealdi, R. Torre, B. Valenti. *Macromolecules* **1974**, 7, 495; (c) F. Hofmeister. *Archiv fuer Experimentelle Phatologie und Pharmakologie* **1887**, 24, 247 (translated by W. Kunz, J. Henle, B. W. Ninham. *Curr Opin Coll Interface Sci* **2004**, 9, 19).
95. (a) E. Bianchi, A. Ciferri, C. Conio, D. Puett, L. V. Rajagh. *J Biol Chem* **1967**, 242, 1361; (b) A. R. Docking, E. Heyman. *J Phys Chem* **1939**, 43, 513.
96. (a) P. LoNostro, B. W. Ninham, S. Milani, A. LoNostro, G. Pavesanto, G. Baglioni. *Biophys Chem* **2006**, 124, 208; (b) D. R. Robinson, P. Jenks. *J Am Chem Soc* **1965**, 87, 2462; (c) L. Vrbka, P. Jungwirth, P. Baduin, D. Touraud, W. Kunz. *J Phys Chem B* **2006**, 110, 7036; (d) X. Chen, T. Yang, S. Kataoka, P. S. Cremer. *J Am Chem Soc* **2007**, 129, 12272; (e) J. H. McBain. *Colloid Science*. Heath, London, 1950; (f) F. A. Long, W. F. McDevitt. *Chem Rev* **1952**, 51, 119; (g) N. Schwierz, D. Horinek, R. R. Netz. *Langmuir* **2010**, 26, 7370; (h) K. D. Collins. *Methods* **2004**, 34, 300.
97. A. Ciferri, R. Garmon, D. Puett. *Biopolymers* **1967**, 5, 439.
98. T. A. Orofino, A. Ciferri, J. J. Hermans. *Biopolymers* **1967**, 5, 773.
99. J. Peller. *J Phys Chem* **1959**, 63, 1194.
100. W. Lin, L. Yan, C. Mu, W. Li, M. Zhang, Q. Zhu. *Polym Int* **2002**, 51, 233.

101. U. Freudenberg, S. H. Beherens, P. B. Welzel, M. Muller, M. Grimmer, K. Salchert, T. Taeger, K. Schmidt, W. Pompe, C. Werner. *Biophys J* **2007**, 92, 2108.

102. J. Kurtz and W. F. Harrington. *J Mol Biol* **1966**, 17, 440; J. Abu-Isa. *J Polym Sci A* **1971**, 9, 199.

103. H. Siegel and R. B. Martin. *Chem Rev* **1982**, 82, 385.

104. H. Schiessel. *J Phys Condens Matter* **2003**, 15, R699.

105. P. J. Flory and R. R. Garrett. *J Am Chem Soc* **1958**, 80, 4832.

CHAPTER 7

HYDROPHOBIC POLYELECTROLYTES

ANDRÉS F. OLEA

7.1 INTRODUCTION

For many years, much effort has been dedicated to the study of polyelectrolytes carrying nonpolar side groups attached to the main chain. Such amphiphilic polyelectrolytes have a variety of solution properties that make them important in many industrial applications and also as simplified models of natural polyelectrolytes. First, they exhibit the ability to produce highly viscous solutions above a critical polymer concentration. This property has been used in many industrial applications such as enhanced oil recovery.[1,2] Second, these polyelectrolytes may form hydrophobic microdomains in aqueous solution, where nonpolar molecules can be dissolved. In this sense, the amphiphilic macromolecule behaves as a surfactant and the formed microdomain has a micelle-like structure. This feature has been applied in the development of drug delivery systems.

In the following, we review the formation and properties, in aqueous solution, of some common amphiphilic polyelectrolytes and others that have been functionalized with hydrophobic groups. Special emphasis is given to the aggregates formed by self-association and the transition that takes the polymer chain from an extended to a compact form. The main results of the necklace model are reviewed and compared with experimental results obtained for hydrophobic (HPE) and hydrophobically modified (HMPE) polyelectrolytes in the low concentration regime. Finally, the solubilization of nonpolar molecules in these hydrophobic microdomains will be discussed using the pseudophase approach.

Ionic Interactions in Natural and Synthetic Macromolecules, First Edition. Edited by Alberto Ciferri and Angelo Perico.

7.2 HYDROPHOBIC AND HYDROPHOBICALLY MODIFIED POLYELECTROLYTES (HPE, HMPE)

The amphiphilic character of a polyelectrolyte is generated by creation of a formal charge on the side chain, and/or by increasing the number of hydrophobic groups on the main or side chain. In a pioneering work, Strauss et al. synthesized amphiphilic polyelectrolytes by quaternization of poly(vinylpiridine) with dodecylbromide.[3,4] In this case, all repeat units with a reacted pyridine are amphiphilic, and therefore their structure is similar to that of common surfactant molecules. For this reason these HPE have been called "polysoaps." In addition, the authors showed that these HPE form intramolecular aggregates when the content of hydrophobic groups reaches a critical value. Although the first example of HPE was a polycation, most of the HPE that have been studied lately are anionic HPE. Engberts et al. synthesized a series of cationic HPE by cyclopolymerization of methyl-n-alkylmethyldiallylammonium halides, with n = 1–12.[5–9]

Examples of anionic HPE are synthetic weak polyelectrolytes such as polycarboxylic acids, which adopt a relatively compact globular structure in water when the hydrophobic effect is larger than the repulsion of ionized groups. However, the electrostatic repulsion can be increased by increasing pH, and the polymer chain stretches out.

Maleic anhydride (MA) form alternating copolymers with a number of hydrophobic monomers such as styrene, vinylpyrrolidone, n-alkylvinylethers, and olefins.[10] These copolymers can be transformed in HPE by hydrolysis of the anhydride in basic medium (see Scheme 7.1). The most common examples of this class of HPE are alternating copolymers of maleic acid and n-alkylvinyl ethers (MA-nVE)[11] or 1-olefins (MA-n),[12,13] in which their hydrophobicity depends on the length of the side alky chain (see Chart 7.1). In these HPE all repeat units are amphiphilic and therefore they have also been termed polysoaps.

On the other hand, the hydrophobicity of a polyelectrolyte can be increased by incorporating hydrophobic groups on the main or side polymer chain.[14] This can be accomplished by direct copolymerization of hydrophobic and hydro-

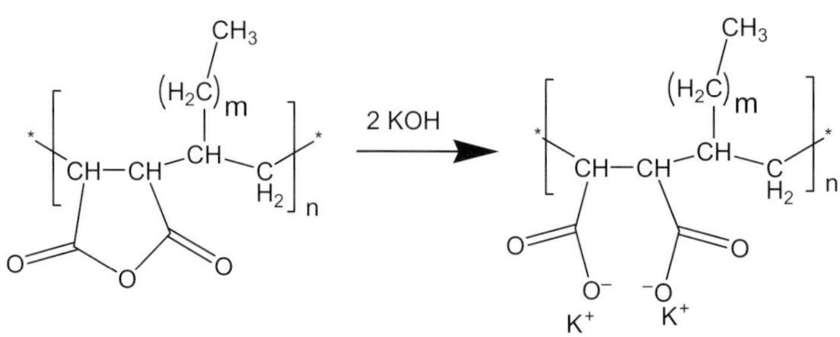

SCHEME 7.1

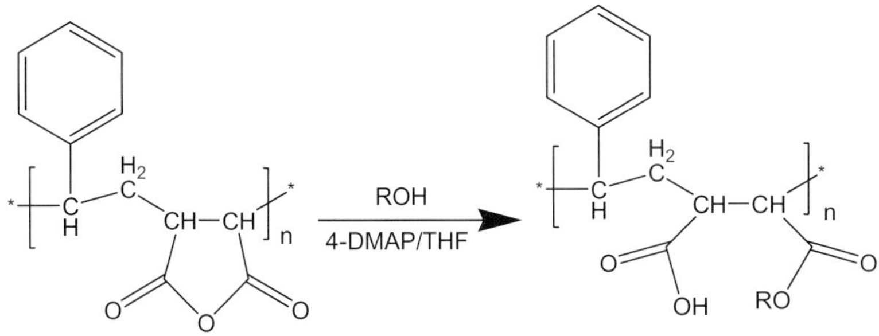

SCHEME 7.2

philic monomers, by grafting of hydrophilic polymers on hydrophobic polymers, and by functionalization of reactive polyelectrolytes. The reaction of hydrophobic molecules with reactive groups on the side polymer chain generates HMPE, where the hydrophobic groups are randomly distributed along the polymer chain. For instance, hydrophobically modified poly(acrylic acid) (PAA) have been obtained by reaction of the carboxylic group with alkylamines.[15–17] In this way, alkyl chains with n carbon atoms (n = 8, 12, 14, 18) have been attached to the polymer chain. Chitosan, partially deacetylated chitin, has been modified by reaction with dodecyl aldehyde.[18,19] A derivative of poly(aspartic acid), PASP, was obtained by reaction of poly(succinimide) with dodecylamine, followed by hydrolysis.[20] Copolymers of MA have been modified by hydrolysis of the anhydride with alkyl amines[21] or alkyl alcohols,[22,23] forming monoamides and monoesters, respectively, with different alkyl chain lengths (Chart 7.1, Scheme 7.2).

A novel class of HMPE has been obtained by random copolymerization of a hydrophobic monomer with a hydrophilic monomer.[5–8,24–28] Morishima et al. have studied a series of copolymers of sodium 2-(acrylamido)-2-methylpropanosulfonate (AMPS) with methacrylamides substituted with hydrophobic groups[24–28] (Chart 7.1).Cationic copolymers were obtained by random cyclopolymerization of dimethyldiallylammonium and methyl-n-alkyldiallylammonium salts, with n = 6, 8, 10, 12.[8,9]

A different class of HMPE has emerged whereby PAA segments are grafted onto Pluronic poly(ethylene oxide)-*b*-poly(propylene oxide)-*b*-poly(ethylene oxide) (PEO-PPO-PEO) block copolymers via C–C bonding.[29–31]

7.3 SELF-ASSEMBLY OF HYDROPHOBIC POLYELECTROLYTES

In aqueous solution, normal amphiphilic molecules self-aggregate, forming micelles. Micelles can have different forms and sizes but they are mostly spherical with a hydrophobic core and a hydrophilic corona (see Chapter 8). The aggregation number denotes the number of molecules forming the aggregate and for small micelles this number is in the order of 100. The formation

Hydrophobic polyelectrolytes

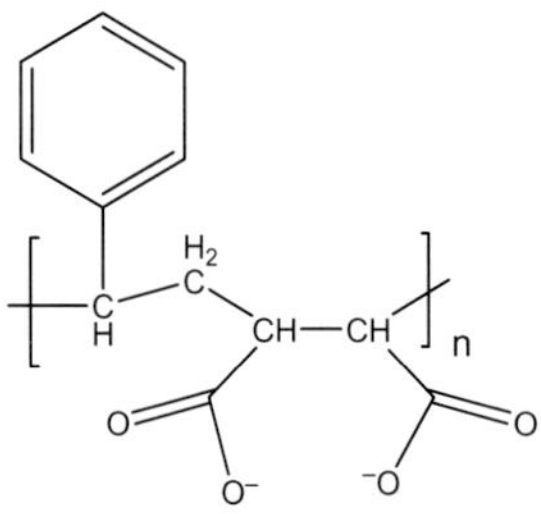

MA-S

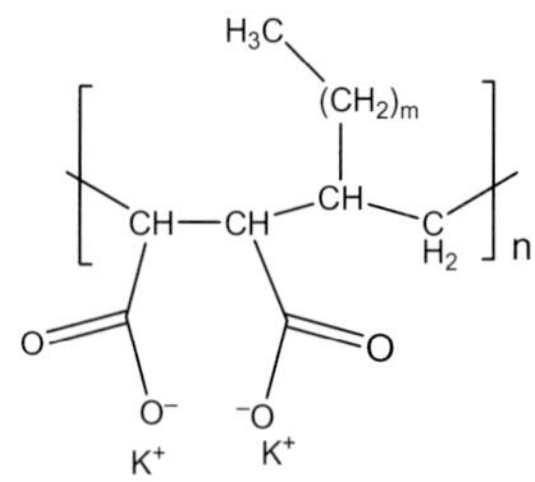

MA-n

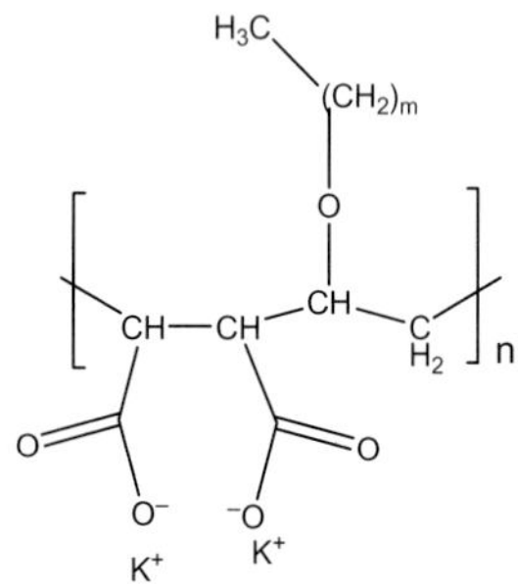

MA-nVE

Hydrophobically modified polyelectrolytes

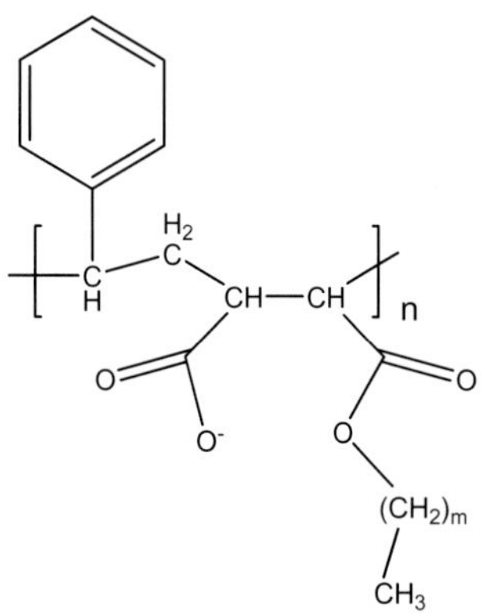

MA-S-n

HMPE obtained by random copolymerization

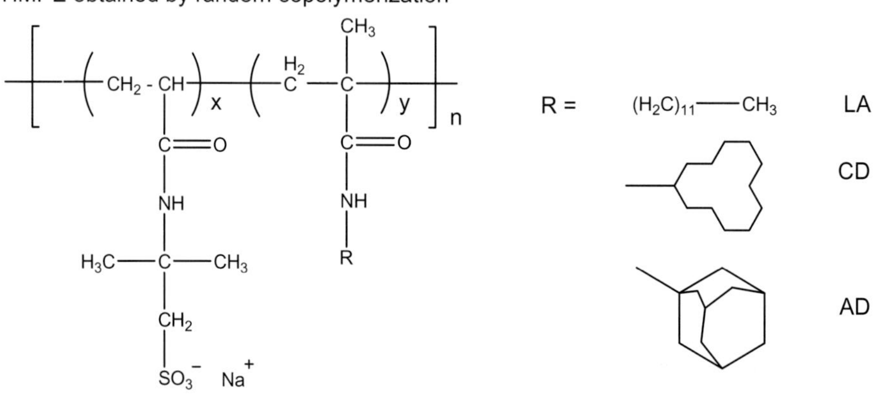

AMPS-R

CHART 7.1

of micelles is the result of a balance between the hydrophobic bonding and the electrostatic repulsion of ionic groups. The hydrophobic effect originates from the entropic change required to locate a nonpolar group in a water cavity.[32,33] A positive entropy change is attributed to the reorganization of water after the hydrophobic group has been removed from it. Thus, micelle formation requires two opposing forces: the hydrophobic effect that is the driven force for aggregation and the electrostatic repulsion that prevents unlimited growth of micelles. Consequently, the micelle formation is a cooperative process where a number of amphiphilic molecules participate simultaneously. The concentration at which the aggregation occurs has been termed critical micelle concentration, cmc, and is related to the change of free energy per monomer added to a micelle by $\Delta G^0 = RT \ln \text{cmc}$.

In the case of hydrophobic polyelectrolytes, the presence of nonpolar groups attached to the polyion chain has two main effects: hydrophobic bonding and long-range attractive interactions. These attractive forces are opposed to the chain expansion due to the electrostatic repulsion of ionic groups. Thus, the conformation adopted by the macromolecule is the result of a subtle balance between these opposing forces. Therefore, in aqueous solution, the self-aggregation of HPE can lead to intrapolymer or interpolymer aggregates. Intrapolymer aggregates are formed by HPE and HMPE at low polymer concentration, and they have been termed polymeric micelles and unimolecular micelles. Interpolymer micelles are formed by amphiphilic block copolymers and by HPE and HMPE at high concentration. For these two kinds of aggregates the meaning of thc aggregation number and of a critical aggregation concentration (cac*) need to be specified. For unimolecular micelles the aggregation number is the number of amphiphilic units forming part of the aggregate, and as the aggregation is independent of the polymer concentration, formally there is no cac*. For interpolymer micelles the aggregation number is the number of chains forming the micelle and the cac* is the concentration of polymer chains at which the micelles are formed. For block copolymers the values of the cac* are two orders of magnitude lower than those measured for normal surfactants.

The transition between the compact and the stretched out forms has been observed in aqueous solution of various HPE such as poly(methacrylic acid) (PMA),[34,35] alternating copolymers of maleic acid with alkyl vinyl ethers,[11,36] 1-olefins,[12,37] or styrene,[38–40] and with HMPE derived from alternating copolymers of MA and styrene MAS-n.[41]

This process has been studied by different techniques: potentiometric titration,[42–44] viscometry titration, calorimetry,[45,46] Raman spectroscopy,[47] pH jump,[35,48–50] and fluorescence probing.[51] The results indicate that the transition is highly cooperative, taking place within a narrow range of pH, and can be viewed as a two-state transition analogous to the helix-coil transition of polypeptides.

Fluorescence probing methods have been used to investigate the microenvironment and conformational transitions of HPE in aqueous solutions and to study the dynamics of the transition. The concept is to introduce fluorescent

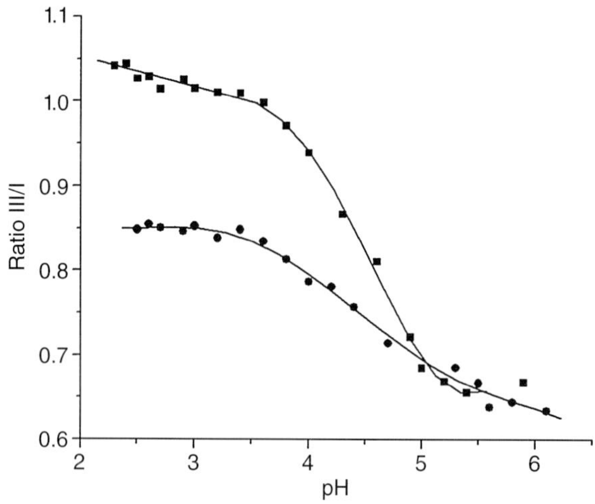

FIGURE 7.1. Fluorescence intensity ratio III/I of pyrene (2 μM) as a function of pH in PMA solution (1 g L^{-1}) containing [NaCl] = 100 mM. (■) PMA540K, (●) PMA8K. A transition from a compact to an open conformation is suggested when pH increases. Reproduced from Olea et al.[35]

probes into the hydrophobic microdomain, which exhibit photophysical properties that are strongly dependent on the medium.[52–54] With pyrene, for example, spectral characteristics such as fluorescence intensity, ratio III/I of the intensities of the third and first vibronic bands, ratio monomer/excimer, and fluorescence lifetime have been used to monitor the conformational transition and to report on the local properties of the site where pyrene is located. Figure 7.1 shows the changes of the ratio III/I of the pyrene fluorescence emission as a function of pH for PMA of different molecular weights. It can be seen that at low pH the ratio is close to 1 and fairly constant, indicating that pyrene is completely solubilized into a hydrophobic environment provided by the polymer. This result suggests that at low pH the macromolecule adopts a hypercoiled form in order to minimize the hydrophobic effect and to maximize long-range interactions between the methyl groups. Initial neutralization of the chain changes its conformation slightly, up to a value between pH 5 and 6, where the process is drastically accelerated. At this pH, the HPE chain stretches to the extended form because of the electrostatic repulsion of the ionized COOH groups. As a consequence, the probe is expelled into the aqueous phase, and the ratio I/III increases. The conformational transition between these two forms takes place at pH 4–6.

The curves have the same pattern found by different groups using different photophysical properties of a wide variety of probes.

Although many HPE and HMPE exhibit this transition, a structure–property relationship of their associative properties in aqueous solution has not been well established. The effect of hydrophobic bonding on the macromolecular conformation and the pH at which the transition occurs, was first addressed by Strauss and his collaborators,[11,44,55] and later by others.[12,41,51,56,57] In these studies a series of alternating copolymers of maleic acid and alkyl vinyl ethers, MA-nVE, with n ranging from 1 to 16, were studied (Chart 7.1). The results indicate that the copolymers with n = 1 and 2 adopt a random coil conformation over the whole range of pH, while copolymers with n = 4, 6, and 8 undergo a transition from compact to extended forms upon increase of the neutralization degree. Polyelectrolytes with longer hydrocarbon side chains maintain a compact structure even at high pH.[11,44,51,55,56] Similar results were obtained for poly(maleic acid-co-1-olefins) MA-n, with n ranging from 6 to 18 (Chart 7.1). The data indicate that just MA-6 exhibits a pH-induced conformational transition. All other copolymers adopt a compact configuration over the entire range of pH.[12,57]

The nature of the driven force for the hydrophobic aggregation of HPE has been a controversial matter. The enthalpy and entropy changes associated to the transition from the extended to the compact form of PMA, determined by potentiometric titration and calorimetric measurements, were both negative.[43] This result is exactly the opposite of what is to be expected if hydrophobic bonding is responsible of this transition. In addition, in partially esterified PMA, the transition occurred at the same pH as in PMA, although the ester groups should have larger hydrophobic bonding than COOH groups. From these results it was concluded that the compact form of PMA is mainly stabilized by van der Waals interactions between methyl groups. However, it was also shown that the enthalpy of the transition have a positive temperature coefficient, which is characteristic of phenomena driven by hydrophobic interaction. From these data it is concluded that the pH-induced conformational transition of PMA it is not only due to the hydrophobic effect, but long-range interactions also play an important role.

The self-assembly of HMPE, obtained by random copolymerization, is quite interesting because these polymers can form intra- or inter-pseudo-micelles, depending on the polymer structure. McCormick et al.[58] have shown that copolymers with a block sequence of hydrophobic units exhibit a tendency for interchain association, whereas copolymers with random distribution of these units tend to aggregate within a single polymer chain. In these copolymers, the conformation adopted by the intrapolymer aggregate is mainly determined by the number of hydrophobic monomer units. At low content of hydrophobic units the polymer chain will adopt an extended conformation, as does PMA at high pH. With increase in the hydrophobic content, the polymer collapses into a highly compact structure.

Copolymers of AMPS and methacrylamides substituted with bulky hydrophobic groups, such as lauryl (LA), cyclododecyl (CD), and adamantyl (AD) (see Chart 7.1), were studied in a wide range of concentration.[24] The results show that polymers having the CD and AD groups form unimolecular aggregates even at high concentration (up to ca. 7 wt %), whereas the polymer having LA groups form intrapolymer pseudo-micelles only at low concentration (below 0.2 wt %), due to the tendency of the LA group to associate intermolecularly. The aggregation of AMPS-LA copolymers begins when the fraction of LA units is ca. 20 mol %.[25]

7.4 STRUCTURE AND PROPERTIES OF MICELLE-LIKE AGGREGATES

7.4.1 Aggregation Numbers of Intramolecular Micelles

The shape of HPE and HMPE in dilute salt-free aqueous solution is determined by the fine balance between electrostatic repulsion of the PE charges and the hydrophobic interactions of nonpolar groups and water. For neutral and weakly charged PE the hydrophobic effect causes the polymer chain collapse into a compact globular structure. By increasing charge, the globule becomes unstable due to the increased strength of electrostatic repulsion, and the polymer chain may form smaller charged globules connected by linear strings. This behavior is suggested by consideration of the Rayleigh instability of a charging droplet, which leads to a split once a critical charge is reached. In the case of a polymer, the chain reduces its energy by adopting a necklace structure where compact beads are joined by linear strings. The necklace conformation was first proposed by Kantor and Kardar[59] in the case of polyampholytes, which are treated in Chapters 4 and 6. For HPE and HMPE the necklace conformation has been predicted by a scaling theory developed by Dobrynin and Rubinstein[60–64] (see also Chapter 4). In this section we summarize the basic predictions of the theory and compare with simulations results and experimental data.

In the scaling approach, the shape of a charged HPE is determined by the balance of electrostatic repulsion and the surface tension of the globule.[60,63] The beads are smaller charged globules, each having m_b monomers, connected by strings of length l_s. The number of beads N_b, the bead size m_b, and the string length l_s are scaled to the degree of polymerization (DP) N, bond length b, fraction of charged monomers f, Bjerrum length l_B, and the reduced temperature τ by the following relations:

number of monomers in a bead:

$$m_b \approx b\tau l_B^{-1} f^{-2},$$

length of the string:

$$l_s \approx b^{3/2}\tau^{1/2} l_B^{-1/2} f^{-1} \approx b m_b^{1/2},$$

and number of beads N_b:

$$N_b \approx N b^{-1} \tau^{-1} l_B f^2.$$

A diagram of states was obtained as a function of charge f and quality solvent τ, where a cascade of transitions between states with different numbers of beads per polymer chain is predicted.[60,63]

By increasing the fraction of charged monomers (i.e., titration of PMA) the number of beads increases, while the number of monomers in a bead decreases. This theory has been extended to dilute and semidilute solutions where the effects of counterion condensation were considered.[61,64] HMPE and the equilibrium conformations have been characterized using the ratio of block sizes and the ratio of block energies of hydrophobic and charged blocks.[62] The diagram of states obtained, in dilute salt-free solution, shows that HMPE chain aggregates into a necklace or cylindrical structure. In this case, the strings that connect intrachain globules are formed at the expense of the hydrophobic energy of unassociated repeat units. Thus, the number of monomers connecting two neighboring beads is determined by the balance between electrostatic repulsion and hydrophobic energy of unassociated monomer units.

Therefore, the model predicts that a single charged polymer chain of HPE or HMPE can form a necklace containing several intrapolymer beads joined by charged strands along the polymer backbone. This theory suggests that several pseudo-micelles are formed by one polymer chain, and not all repeat units are involved in the hydrophobic microdomain.

The necklace conformation has also been obtained by simulations studies.[60,64–67] Monte Carlo simulations of titration of HPE have shown that during the titration process the HPE chain goes through a number of conformations such as collapsed spherical globule, cylinder, necklace, and fully extended chain.[60,66]

In summary, the predictions of the theory and simulations are the following:

1. Depending on the chain charge density, HPE and HMPE adopt a number of necklace conformations with different number of beads.
2. The number of beads in the necklace conformation increases with increasing chain size.
3. The number of beads increases with increasing degree of neutralization.
4. The increase in the number of beads results in the decrease of their size, and the decrease of the number of monomers inside a bead.
5. In the strong hydrophobic regime the necklace conformation is not formed at any pH for polymer with DP = 40 and the spherical globule conformation is observed.
6. HPE with DP > 40 form dense spherical or cylindrical conformations at low degree of neutralization. The increase of charge density in the polymer chain induces transitions between different conformations, including the necklace.

There have been few studies intending to estimate the number of repeat units involved in the intramolecular hydrophobic domains formed by HPE and

HMPE.[12,34,56,57,68–71] Most of these studies have used the alternating copolymers of maleic acid with alkyl vinyl ethers and 1-olefins, and have been carried out by potentiometric titration and time-resolved fluorescence quenching (TRFQ). The latter method was developed to determine the aggregation number of common surfactant micelles,[72–77] and a rigorous analysis of the experimental conditions required for the application of this method to polysoaps was given by Binana-Limbele and Zana.[56] Basically, the TRFQ method requires the measurement of the fluorescence intensity, as a function of time, of a probe that is dissolved completely into the micelle, in the presence of a quencher that is totally micellized, that is, both probe and quencher are immobile. Under conditions such that $[P]/[M] \ll 1$ and $[Q]_M/[M] \approx 1$, where $[P]$, $[Q]_M$, and $[M]$ are the molar concentrations of probe, quencher, and micelles, respectively, the time-dependent quenching data fit the equation:

$$I(t) = I(0)\exp\{-A_2 t - A_3[1-\exp(A_4 t)]\}, \tag{7.1}$$

where A_2, A_3, and A_4 are adjustable parameters and $I(0)$ is the intensity at time zero. In the case that both probe and quencher do not migrate on the fluorescence lifetime, the parameters have the following physical meaning:

$$A_2 = k_0 \quad A_3 = \frac{[Q]_M}{[M]} = \bar{n} \quad A_4 = k_Q,$$

where $\bar{n}$ denotes the mean number of quenchers in a micelle, and k_0 and k_Q are the rate constant of decay in the absence and presence of quencher, respectively.

The aggregation number N is calculated from $\bar{n}$, $[Q]_M$, and $[M]$ according to:

$$N = \frac{[\text{HPE}]-[\text{HPE}]_F}{[\text{M}]} = \frac{[\text{HPE}]-[\text{HPE}]_F}{[\text{Q}]_M}\bar{n}, \tag{7.2}$$

where [HPE] is the concentration of polymer in mole of repeat unit and $[\text{HPE}]_F$ is the concentration of monomer units that are not forming part of the polymer micelle. In this equation, there are two sources of error in the determination of N, namely $[Q]_M$ and $[\text{HPE}]_F$. In the case that the quencher is not completely incorporated into the polymeric micelle, $[Q]_M < [Q]_T$, and therefore N would be underestimated.[70] On the other hand, the determination of $[\text{HPE}]_F$ raises problems because it is not possible to measure this quantity directly. For this reason, in most of the studies it has been assumed that $[\text{HPE}]_F = 0$, that is, all repeat units are involved in the hydrophobic microdomain. According to theory and simulation results, this assumption seems to be valid at low pH (in the hypercoiled state) and for polymers with repeat units carrying large hydrophobic groups. This method has been used to study the size of intramolecular aggregates formed by poly(maleic acid-co-alkyl vinyl

TABLE 7.1. Aggregation Numbers of HPE at Different Degrees of Neutralization of Maleic Acid, Determined by TRFQ. The DP Is Given for Comparison

Copolymer	DP	N	Copolymer[a]	DP	N
MA-4VE[106]	500	19	MA-8	33 (21)	36 (18)
MA-5VE[106]		13	MA-10	35 (21)	35 (29)
MA-6VE[69,70]	1100	24; 45	MA-12	29 (16)	27 (62)
MA-8VE[71]	660	45	MA-14	30 (19)	35 (87)
MA-10VE[56] $\alpha = 0.1$	1000	90	MA-16	26 (19)	20 (160)
MA-10VE[56] $\alpha = 0.5$	1000	52	MA-18 (1%)	24 (20)	20 (214)
MA-10VE[56] $\alpha = 1$	1000	15	MA-18 (10%)	(20)	(115)
MA-10VE[56] $\alpha = 1$	100	16	MA-18 (20%)	(20)	(86)
MA-16VE[56] $\alpha = 1$	140	60			
MA-16VE[56] $\alpha = 1$	4000	50			

[a] Values of N were measured at $\alpha = 1$; from Olea et al.[57] and Claracq et al.[80]

ethers), MA-nVE, and poly(maleic acid-co-1-olefins), MA-n, and the results are summarized in Table 7.1.

The experimental results obtained for MA-nVE and MA-n (see Table 7.1) agree well with the theory, namely: (1) The aggregation numbers have been found to be independent of the concentration and ionic strength, which indicate that pseudo-micelle formation involves only individual polymer chains.[56,69] (2) For MA-nVE with high degrees of polymerization (DP > 100), N is lower than DP, suggesting that each polymer chain gives rise to a number of hydrophobic domains, connected by polymer segments where the nonpolar moiety of the repeat unit is exposed to water.[56,69,70] (3) In the case of MA-n copolymers, with low degrees of polymerization (DP < 40), the aggregation numbers have been found to be equal to the DP, within the experimental error.[12,57] This means that just one polymer aggregate arises from one polymer chain. The different behavior observed for these two series of copolymers has been attributed to the large difference in molecular weight. Although in both series of copolymers the intramolecular aggregate are of similar size, that is, the aggregation numbers in both systems are of the same order of magnitude, just in alkyl vinyl ethers (with DP in the range of 1700–4000), it is possible to form more than one pseudo-micelle by macromolecule. (4) For strong hydrophobic polyelectrolytes, that is, long side alkyl chains, the dense spherical globule conformation is observed even at high degrees of neutralization.[12,57]

However, there are some discrepancies between theory and experiment that need to be clarified:

1. For MA-nVE with DP > 100, the necklace conformation is adopted even at low degree of neutralization.[56,69,70]
2. In addition, Binana-Limbele and Zana[56] have found that the aggregation number of pseudo-micelles formed by MA-10VE is nearly independent

of the DP in fully neutralized samples. The simulation predicts that this number should decrease as the number of beads increases with the chain length.

3. Attempts to evaluate the effect of neutralization degree on the size of polymeric aggregates have been carried out by two different groups, but the results are contradictory.[56,70] The theory predicts that an increase of charge density on the polymer chain induces a transition from spherical to necklace conformation, an increase on the number of beads, and consequently a decrease on the aggregation number.[61] In the pearling conformation, where polymeric aggregates are connected by no associated repeating units, the assumption that $[HPE]_F = 0$ is obviously not valid under any circumstance. Therefore, the experimental values of N are overestimated and need to be corrected. Currently, there is no way to measure $[HPE]_F$ directly and an approximate method, based on the ability of HPE solutions to dissolve water-insoluble compounds, has been applied.[56,70] In this method it is assumed that the solubility S of a nonpolar compound is proportional to the number of hydrophobic repeat units involved in the polymeric aggregate.[56,69] Thus, at all pH the aggregation number is corrected by the ratio S_1/S_0, where S_1 and S_0 denote the solubility in the extended and compact states of the polymer chain, respectively. Once again, it is assumed that in the compact form (low pH) all repeat units are forming the hydrophobic microdomain. With this method, Binana-Limbele and Zana have shown that, according to the theory, the aggregation number of micelles formed by MA-10VE decreases from 90 to 15 with increasing neutralization degree.[56] A similar result was obtained for MA-8VE by measuring the excimer-to-monomer fluorescence intensity ratio, I_e/I_m, as a function of polymer concentration.[71] At constant concentration of pyrene, this ratio is considered a measure of the local concentration of pyrene in the aggregate.[78] The value of I_e/I_m increases with increasing concentration of polymer reaching a maximum at 50–100 ppm of polymer. This concentration corresponds to the polymer concentration needed to solubilize the given amount of pyrene. From these data it is found that one pyrene molecule is bound to 40–50 repeat units, and that this number decreases with increasing pH. These results are opposed to that reported by Zdanowicz and Strauss for MA-6VE, who reported that N is independent of pH and equal to 45.[70]

4. In the high concentration regime the theory predicts that the size of the beads increases with polymer concentration while their number per chain decreases.[64] However, in a study of the effect of concentration on the aggregation of MA-18 by using small-angle neutron-scattering (SNS), Shih et al. have shown that the hydrophobic domains are formed by an intermolecular association of several polymer chains.[79] In addition, the value of N decreases with increasing polymer concentration, that is, in

the low regime concentration (0.25%), the aggregation number is 234 (12 polymer chains), whereas this number is reduced to 86 (5–6 polymer chains) in the high regime concentration (20%).

A similar result was found for the aggregates formed by copolymers of styrene and maleic acid in the concentration range 1–10%. A hydrodynamic radius equal to 1.7 nm was obtained by fluorescence anisotropy measurements.[80] Assuming that MA-S aggregates have the same density as bulk polystyrene, a sphere with this radius must be formed by five copolymer chains.

5. Shih et al. have also shown that at high polymer concentration the size of the aggregates formed by fully neutralized MA-n depends strongly on the side chain length,[81] namely the number of repeating units per aggregate decreases from 214 to 18 by decreasing n from 18 to 8. Concomitant to this change the nature of the aggregates also varies. For MA-n with n = 18 to 12, the hydrophobic domains are formed by an intermolecular association of several polymer chains, while MA-8 and MA-10 form an intramolecular pseudo-micelle.

7.5 SOLUBILIZATION OF NONPOLAR MOLECULES INTO PSEUDO-MICELLES

In the last decades much effort has been devoted to the study of amphiphilic polymers that form aggregates in aqueous solution, and have therefore proved to have potential applications in biology and pharmacy.[82–86] The most important feature of these polymer aggregates is their capability of solubilizing water-insoluble compounds. This property is of practical importance for drug delivery systems. For this type application the goal is to use a micellar-like system where the drugs, generally nonpolar molecules, can be dissolved in aqueous solution. The substrate solubilized into the aggregate is then released by a simple thermodynamic equilibrium with the medium, or by a drastic change of the external conditions, such as pH.[82] The efficiency of this approach is dependent upon the stability of the aggregate, and upon the partitioning of the drugs between the aqueous phase and the hydrophobic microdomains.[82,87,88] The most extensively studied systems are block copolymers composed of hydrophilic and hydrophobic segments.

In aqueous solution, HPE provides a hydrophobic environment where nonpolar molecules can be solubilized, and some systems have been proposed as drug containers.[89–92] In the regime of low concentration, where only intramolecular aggregates are formed, there is no multimolecular association and hence no cac*. Thus, these aggregates remain stable even at high dilution. It is interesting to evaluate the distribution coefficient of different substrates in order to establish a relationship of this property with the structure of both polymer and solute.

7.5.1 Measurement of Distribution Coefficients

Different models have been applied for analyzing solute incorporation to micelles formed by common surfactants.[72,93,94] In the pseudophase model the incorporation of a substrate into micelles is treated as a distribution between the aqueous phase and the micellar phase. As a consequence, the distribution constant can be expressed on a molar concentration basis, K_s, as:

$$K_S = \frac{[\mathrm{S}]_M}{[\mathrm{S}]_W[\mathrm{D}]_M}\frac{f_M}{f_W}, \tag{7.3}$$

where $[\mathrm{S}]_W$ and $[\mathrm{S}]_M$ denote molar concentrations of substrate in aqueous and micellar pseudophase, respectively, and $[\mathrm{D}]_M$ is the concentration of micellized surfactant, and f_W and f_M are the activity coefficient of the substrate in the respective phase. In the case of intramolecular aggregates the concentration of micellized surfactant is given by the concentration of monomer units forming the pseudo-micelle, $[\mathrm{HPE}]_M$. Alternatively, this equilibrium constant can be defined on a mole fraction basis, K_X, by:

$$K_X = \frac{X_M}{X_W}\frac{f_M}{f_W}, \tag{7.4}$$

where X_M and X_W are the substrate mole fractions in the aggregate and aqueous phases.

Both quantities are related by:

$$K_X = \frac{55.5K_S}{1 + K_S[\mathrm{S}]_W}. \tag{7.5}$$

At low substrate occupation numbers $K_S[\mathrm{S}]_W = [\mathrm{S}]_M/[\mathrm{HPE}]_M \ll 1$, and Equation 7.3 can be approximated by $K_X = 55.5K_S$.

The experimental solute partitioning coefficients are related to the standard free energy of transfer by:

$$\Delta\mu_t^0 = \mu_M^0 - \mu_W^0 = -RT\ln K_X. \tag{7.6}$$

To get a better understanding of the forces which contribute to solute incorporation into the pseudo-micelle, $\Delta\mu_t^0$ is assumed to include additive contributions from hydrophilic and hydrophobic groups:

$$\Delta\mu_t^0 = \Delta\mu_{Hy}^0 + n_C\Delta\mu_C^0, \tag{7.7}$$

where $\Delta\mu_{Hy}^0$ denotes the contribution of the hydrophilic group, $\Delta\mu_C^0$ is the incremental free energy per methylene group, and n_C is the number of these groups attached to the hydrophilic parent group. Within this approach the thermodynamic parameters obtained are the partition coefficient K_S, and the standard free energy of transfer, $\Delta\mu_t^0$.

The equilibrium constants, K_S, can be experimentally determined by solubilization, separation, and spectroscopic methods. For the sake of comparison it is important to use the same expression for the distribution constant.

There have been relatively few studies enabling an assessment of the solubilization capacity of organic compounds by intramolecular pseudo-micelles.[12,37,57,91,95] Studies of solubilization of p-alkyl-substituted phenols, p-nitrophenol, and N-alkyl-nitroimidazoles into the hydrophobic microdomains formed by MA-n copolymers have been carried out using the pseudophase model. The equilibrium constants, K_S, were obtained from the initial slope of plots of $[S]_W$ against $[S]_M$ according to Equation 7.1. The concentration of substrate in water was measured by UV after the separation of a small part of the aqueous phase by ultrafiltration. As discussed above, in the case of MA-n of low DP all monomer units are included in the pseudo-polymer micelle (see Table 7.1). Therefore, $[HPE]_M$ is assumed to be equal to the concentration of monomer units. To assess the effect of the substrate and polymer structure on this equilibrium, two series of experiments were carried out: (1) a homologous series of substrates were partitioned between water and the microdomain provided by MA-18 and (2) one substrate was distributed between water and the polymer pseudo-micelles formed by the MA-n copolymers. The values of K_S obtained in (1) and (2) are collected in Tables 7.2 and 7.3, respectively. The data show that the equilibrium constant increases

TABLE 7.2. Distribution Constants for Transfer of Organic Substrate between the Aqueous Phase and Intramolecular Aggregates Formed by MA-18

HPE	Substrate	K_S	Substrate	K_S
MA-18	Phenol	216	1-Methyl-2-nitroimidazole	340
	p-Methylphenol	337	1-Ethyl-2-nitroimidazole	464
	p-Ethylphenol	489	1-Propyl-2-nitroimidazole	573
	p-Propylphenol	706	1-Butyl-2-nitroimidazole	651
	p-Nitrophenol (pH 5.0)	191	1-Pentyl-2-nitroimidazole	767
	p-Nitrophenol (pH 8.0)	91	1-Hexyl-2-nitroimidazole	871

TABLE 7.3. Distribution Constants for Transfer of Phenol and p-Nitrophenol between the Aqueous Phase and Intramolecular Aggregates Formed by MA-n

HPE	Phenol	p-Nitrophenol (pH 5.0)	p-Nitrophenol (pH 8.0)
MA-10	38		
MA-12	96	86	18
MA-14	131	118	31
MA-16	177	148	60
MA-18	216	191	91

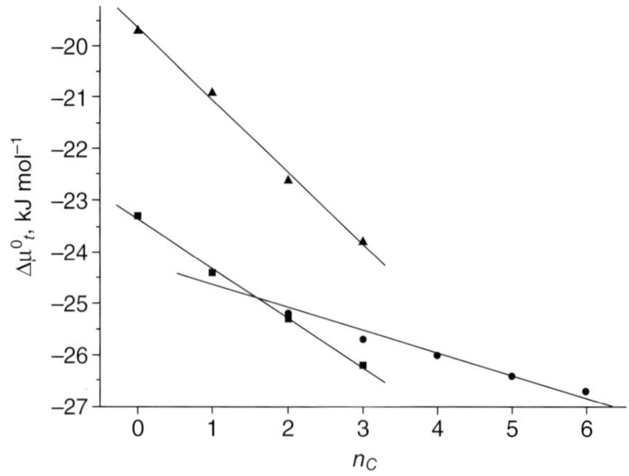

FIGURE 7.2. Free energies of transfer, ($\Delta\mu_t^0$), of *p*-alkylphenols from water to PA-18 (■) and to SDS (▲); and of N-alkyl-2-methylnitroimidazole from water to PA-18 (●) as a function of the number of carbon atoms of the alkyl chain of the organic molecule. Reproduced from Salamanca et al.[91]

with increasing size of both the side alkyl chain and the alkyl group of the organic substrate.

The standard free energy for the transfer of substrates from the aqueous phase to the pseudo-micellar phase, calculated from Equation 7.4, is plotted in Figure 7.2 against the number of carbon atom in the alkyl chain. For comparison, the results reported for transfer to sodium dodecyl sulfate (SDS) micelles have been included. The intercepts obtained from these plots are −23.4 and −24.2 for p-alkylphenol and N-alkyl-2-methylnitroimidazole in MA-18, respectively, and −19.6 kJ mol^{-1} in SDS micelles. This result suggests that the phenol and nitroimidazole groups interact more strongly with the polymer chain, and consequently dissolves more in the MA-18 aggregates than in SDS micelles. However, the values of $\Delta\mu_C^0$ obtained from the slopes (0.97, 0.44, and 1.42 kJ mol^{-1}) are much lower than determined for transfer to heptane (3.27 kJ mol^{-1}). The low values of $\Delta\mu_C^0$, measured in ionic micelles, have been explained in terms of penetration of water molecules into the aggregate, which reduces the micelle hydrophobicity. However, a comparison of the ratio III/I values obtained in ionic micelles and in MA-18 reveals that the hydrophobicity is higher in the polymer micelles. In addition, quenching studies suggest that the penetration of neutral molecules is more restricted in the polymer than in simple micellar systems. Thus, a possible explanation for the low value of $\Delta\mu_C^0$ for the transfer of a methylene group from water to the pseudo-micelle formed by MA-18 is the need of a conformational reorganization of the polymer chains to accommodate the alkyl chain of the organic substrate.

In the distribution of phenol and p-nitrophenol between water and the microdomains exhibited by MA-n, the dependence of $\Delta\mu_t^0$ upon the length of the side alkyl chain reflects the change in μ_M^0 due to a change in the polymer structure. It is interesting to note that a linear relation exists between the values of $\Delta\mu_t^0$ and the number of methylene groups in the side alkyl chain. However, in this case the slope is just 0.34 kJ mol^{-1}, which is much lower than the contribution of one methylene group to the hydrophobic stabilization of the pseudo-micelle (1.67 kJ mol^{-1}), measured in copolymers of maleic acid and alkyl vinyl ethers.[11] Probably, this difference can be due to the pearling conformation adopted by MA-nVE copolymers. In any case, an increase of aggregate size, due to an increasing number of carbon atoms in the alkyl side chain, will result in more negative values of μ_M^0. Consequently, largest changes of free energies of transfer (and distribution coefficient) will be observed for polymers with the longest side alkyl chain, that is, MA-18. This conclusion is in line with the results indicating that the environment provided by polymer aggregates becomes more nonpolar with increasing size of the side chain.[57] Interestingly, this effect is even more important at pH 8.0 where the slope is almost twice the value measured at pH 5.0. If we assume that only the neutral form dissolves into the hydrophobic aggregates, then a possible explanation is that at pH 8.0 the change of $\Delta\mu_t^0$ includes the polymer effect on the displacement of the acid–base equilibrium of p-nitrophenol.

Solubilization studies for HPE that aggregate in a necklace structure are scarce. Luminescence measurements have been used to determine the partition constants of pyrene between the microdomains provided by MA-nVE (n = 4, 6, 10, 12, 16) and the aqueous phase.[95] With this method, K_S is obtained as the inverse of the concentration of monomer unit forming part of the polymer aggregate, when the contribution of pyrene in water and in the aggregate to the ratio I/III are equal.[95,96] The results indicate that the partition constant increases with decreasing degree of neutralization of maleic acid and with increasing length of the alkyl side chain. The effect of pH on K was attributed to an increase of polymer hydrophobicity, while the variation of K with the number of carbon atoms in the alkyl side chain was explained in terms of volume changes of the polymer aggregates. The volume of pseudophase is proportional to the fraction β of monomer units forming part of the aggregate. Assuming that $\beta = 1.0$ for MA-16VE in the whole pH range, the values of β for the other systems were calculated from the ratio of the respective partition constants. The values of K and β are presented in Table 7.4.

The numbers of monomer units forming the aggregates were calculated using the values of β and the DP. The aggregation numbers were deduced from previous work of the same group. The results indicate that for MA-16VE and MA-12VE, the pearls are close and separated by short strings. On the other hand, the pearls formed by MA-nVE with n = 4, 6, and 10 are separated by long strings. In this approach the increase in the equilibrium constant with increasing polymer hydrophobicity, that is, increasing length of the alkyl side chain, has not been considered and the whole change in K has been ascribed to changes in β.

TABLE 7.4. Values of the Distribution Constant *K* and Fraction of Monomer Units Involved in the Aggregates Formed by MA-nVE

HPE	K_S	β	Nm	N	DP
MA-16VE	1.0×10^5 ($\alpha = 0.5$)	1.0	4000	50	4000
	7.6×10^4 ($\alpha = 1.0$)	1.0	4000	50	4000
MA-12VE	5.6×10^4 ($\alpha = 0.5$)	0.66	462	—	700
	4.1×10^4 ($\alpha = 1.0$)	0.64	448	—	700
MA-10VE	1.4×10^4 ($\alpha = 0.5$)	0.20	200	52	1000
	6.3×10^3 ($\alpha = 1.0$)	0.10	100	15	1000
MA-6VE	5.4×10^3 ($\alpha = 0.5$)	0.10		24	
MA-4VE	4.7×10^3 ($\alpha = 0.0$)	0.11	26	19	260
MA-18	4.5×10^4 ($\alpha = 1.0$)	1.0	25	25	25

7.5.2 Solubilization of Drugs by Polymer Micelles

The ability of true polymer micelles to solubilize a wide variety of agents has led to their use as drug containers. Most of the studies have been carried out with amphiphilic block copolymers, and just a few examples of solubilization in HPE have been reported. The location of a drug into the aggregate will be determined by hydrophobic, electrostatic, and specific interactions between the drug and the monomer units. Thus, drugs can be incorporated to the polymer aggregate by physical entrapment or by reaction to form a conjugate.

Copolymers of MA have been used in medicine and pharmacy as drugs per se, drug carriers,[91,92] support for enzymes,[97,98] and protein modifiers.[99] The main reasons for their potential use are: biocompatibility, pH-dependent structure, and the hydrophobicity that can be modulated by changing the comonomer or by reaction of the anhydride group with alkylalcohols or alkylamines. In addition, the reactivity of the anhydride group provides a route to the formation of polymer conjugates by reaction with drugs[90,100] or proteins.[101,102] Finally, it has been shown that copolymers of vinylpyrrolidine and dimethyl MA are potential carriers for renal drug targeting.[98,103]

A series of drugs with a nitroimidazole parent ring (ornidazole, metronidazole, and tinidazole), which are antiparasitic and antibacterial agents, were solubilized in polymer aggregates formed by MA-n at different pH.[91] The values of K_S in aqueous solution of MA-14 (253, 332, and 337 for tinidazole, ornidazole, and metronidazole, respectively) are of the same order of magnitude that those measured for N-alkyl-2-methylnitroimidazole, and slightly dependent on the polymer structure. Interestingly, the effect of pH on K_S is important only for MA-8, that is, the values of K_S decrease with increasing pH. This change is attributed to a pH-induced transition from a compact to a random form. In other words, increasing pH reduces the drug solubility, and consequently most of the drug is released to the aqueous phase when the pH is increased.

Other example is the solubilization of clofazimine, an antibiotic used for the treatment of leprosy, in the aggregates formed by MA-1VE. This copolymer was hydrophobically modified by partial esterification with dodecanol, and then mannosyl residues were attached to the resulting copolymer PMA-1VE-C12 with the aim to enhance the affinity to mannose receptors. The amount of clofazimine solubilized in the presence of both copolymers is about 1 g of drug by 1 g of polymer, that is, one molecule for two monomer units.

Studies of polymer conjugates outnumber the direct dissolution studies. A series of bioactive molecules with disinfectant and/or odorant properties, such as tyhmol, eugenol, phenylethyl alcohol, and citronellol, have been attached to copolymers of MA and styrene or vinyl acetate. The release of active moieties was studied by infrared spectroscopy.[104] Copolymers of MA and methyl vinyl ether have been used as reactive polymers to link nucleic acids and proteins.[99,105] An example of polymer conjugates is the conjugate of neocarzinostatin, NCS, an antitumor protein, with a copolymer of styrene and MA MA-S designated as SMANCS, which has been developed as a potential anticancer agent.[101] In order to increase the conjugate hydrophobicity, the copolymer was esterified with 30–50% butylalcohol. The nature of the alkyl group and the extent of esterification are important factors in determining the binding affinity of the conjugate to albumin, its lipid solubility, and the biological function of the attached drug. In addition, by increasing the molecular weight of NCS, its plasma half life was increased over a 10-fold range.

REFERENCES

1. J. E. Glass *Polymers in Aqueous Media: Performance Through Association. Advances in Chemistry Series, 223*. American Chemical Society, Washington, DC, 1989.
2. C. Taylor and A. Nasr-El-Din. *J Petrol Sci Eng* **1998**, 19, 265–280.
3. U. P. Strauss and E. G. Jackson. *J Polym Sci* **1951**, 6, 649.
4. U. P. Strauss and N. L. Gersheld. *J Phys Chem* **1954**, 58, 747.
5. J. Kevelam and J. B. F. N. Engberts. *J Colloid Interf Sci* **1996**, 178, 87–92.
6. J. Kevelam and J. B. F. N. Engberts. *Langmuir* **1995**, 11, 793–797.
7. G. J. Wang and J. B. F. N. Engberts. *Langmuir* **1996**, 12, 652–656.
8. G. J. Wang and J. B. F. N. Engberts. *J Phys Org Chem* **1998**, 11, 305–320.
9. Y. J. Yang and J. B. F. N. Engberts. *J Org Chem* **1991**, 56, 4300–4304.
10. B. C. Trivedi and B. M. Culbertson. *Maleic Anhydride*. Plenum Press, New York, 1982.
11. P. L. Dubin and U. P. Strauss. *J Phys Chem* **1970**, 74, 2842–2847.
12. D. Y. Chu and J. K. Thomas. *Macromolecules* **1987**, 20, 2133–2138.
13. F. Martinez, E. Uribe, A. F. Olea. *J Macromol Sci-Pure Appl Chem* **2005**, A42, 1063–1072.

14. D. N. Schulz, J. Bock, P. L. Jr Valint. Synthesis and characterization of hydrophobically associating water-soluble polymers. In *Macromolecular Complexes in Chemistry and Biology*. Springer-Verlag, Heidelberg, 1994; pp. 3–13.
15. K. T. Wang, I. Iliopoulos, R. Audebert. *Polym Bull* **1988**, 20, 577–582.
16. I. Iliopoulos. *Curr Opin Colloid Interface Sci* **1998**, 3, 493–498.
17. B. Magny, F. Lafuma, I. Iliopoulos. *Polymer* **1992**, 33, 3151–3154.
18. A. L. Kjøniksen, B. Nyström, C. Iversen, T. Nakken, O. Palmgren, T. Tande. *Langmuir* **1997**, 13, 4948–4952.
19. A. L. Kjøniksen, C. Iversen, B. Nystrom, T. Nakken, O. Palmgren. *Macromolecules* **1998**, 31, 8142–8148.
20. M. Suwa, A. Hashidzume, Y. Morishima, T. Nakato, M. Tomida. *Macromolecules* **2000**, 33, 7884–7892.
21. M. C. Kramer, J. R. Steger, Y. X. Hu, C. L. McCormick. *Polymer* **1996**, 37, 4539–4546.
22. G. H. Hu and J. T. Lindt. *J Polym Sci Part A Polym Chem* **1993**, 31, 691–700.
23. F. Martinez, G. Neculqueo, M. Torres, A. F. Olea. *Bol Soc Chil Quim* **2001**, 46, 137–141.
24. Y. Morishima, S. Nomura, T. Ikeda, M. Seki, M. Kamachi. *Macromolecules* **1995**, 28, 2874–2881.
25. H. Yamamoto, M. Mizusaki, K. Yoda, Y. Morishima. *Macromolecules* **1998**, 31, 3588–3594.
26. S. Yusa, M. Kamachi, Y. Morishima. *Langmuir* **1998**, 14, 6059–6067.
27. X. Wu, Y. J. Qiao, H. Yang, J. B. Wang. *J Colloid Interf Sci* **2010**, 349, 560–564.
28. H. Yamamoto, A. Hashidzume, Y. Morishima. *Polym J* **2000**, 32, 745–752.
29. L. Bromberg. *J Phys Chem B* **1998**, 102, 1956–1963.
30. L. Bromberg and M. Temchenko. *Langmuir* **1999**, 15, 8627–8632.
31. L. E. Bromberg and D. P. Barr. *Macromolecules* **1999**, 32, 3649–3657.
32. C. Tanford. *The Hydrophobic Effect*. John Wiley, New York, 1988.
33. W. Blokzijl and J. B. F. N. Engberts. *Angew Chem Int Ed* **1993**, 32, 1545–1579.
34. D. Y. Chu and J. K. Thomas. *J Phys Chem* **1985**, 89, 4065–4070.
35. A. F. Olea, H. Rosenbluth, J. K. Thomas. *Macromolecules* **1999**, 32, 8077–8083.
36. U. P. Strauss, B. W. Barbieri, G. Wong. *J Phys Chem* **1979**, 83, 2840–2843.
37. R. G. Barraza, A. F. Olea, C. E. Valdebenito, V. Dougnac, I. Fuentes. *J Colloid Interf Sci* **2004**, 275, 434–438.
38. N. Ohno, K. Nitta, S. Makino, S. Sugai. *J Polym Sci Polym Phys Ed* **1973**, 11, 413–425.
39. S. Sugai and N. Ohno. *Biophys Chem* **1980**, 11, 387–395.
40. S. Sugai, K. Nitta, N. Ohno. Hydrophobic domains of maleic acid copolymers. In *Microdomains in Polymer Solutions*. Plenum Press, New York, 1985; pp. 13–32.
41. A. F. Olea, B. Acevedo, F. Martinez. *J Phys Chem B* **1999**, 103, 9306–9313.
42. A. Katchalsky and P. Spitnik. *J Polym Sci* **1947**, 2, 432–448.
43. M. Mandel, J. C. Leyte, M. G. Stadhouder. *J Phys Chem* **1967**, 71, 640–649.
44. U. P. Strauss and M. S. Schlesinger. *J Phys Chem* **1978**, 82, 571–574.

45. V. Crescenzi, F. Quadrifoglio, F. Delben. *J Polym Sci A-2* **1972**, 10, 357–368.
46. F. Delben, V. Crescenzi, F. Quadrifoglio. *Eur Polym J* **1972**, 8, 933–935.
47. J. L. Koenig, A. C. Angoud, J. Semen, J. B. Lando. *J Am Chem Soc* **1969**, 91, 7250.
48. M. Irie and W. Schnabel. *Makromol Chem Rapid Commun* **1984**, 5, 413–417.
49. B. Bednar, H. Morawetz, J. A. Shafer. *Macromolecules* **1985**, 18, 1940.
50. S. Sugai, N. Ohno, K. Nitta. *Macromolecules* **1974**, 7, 961–963.
51. W. Binana-Limbele and R. Zana. *Macromolecules* **1987**, 20, 1331–1335.
52. J. K. Thomas. *Chemistry of Excitation at Interfaces*. ACS Monograph, 191. American Chemical Society, Washington, DC, 1983.
53. K. Kalyanasundaram. *Photochemistry in Microheterogeneous Systems*. Academic Press, Orlando, FL, 1987.
54. M. A. Winnik and F. M. Winnik. Fluorescence studies of polymer association in water. In *Structure–Property Relations in Polymers*. American Chemical Society, Washington, DC, 1991; pp. 485–505.
55. U. P. Strauss, V. S. Zdanowicz, Y. Z. Zhong. Determination of microdomain size of hydrophobic polyelectrolytes by luminescence quenching. In *Macro-ion Characterization: From Dilute Solutions to Complex Fluids*. ACS, Washington, DC, 1994; pp. 154–161.
56. W. Binana-Limbele and R. Zana. *Macromolecules* **1990**, 23, 2731–2739.
57. A. F. Olea, R. G. Barraza, I. Fuentes, B. Acevedo, F. Martinez. *Macromolecules* **2002**, 35, 1049–1053.
58. Y. H. Chang and C. L. McCormick. *Macromolecules* **1993**, 26, 6121–6126.
59. Y. Kantor and M. Kardar. *Europhys Lett* **1994**, 27, 643–648.
60. A. V. Dobrynin, M. Rubinstein, S. P. Obukhov. *Macromolecules* **1996**, 29, 2974–2979.
61. A. V. Dobrynin and M. Rubinstein. *Macromolecules* **1999**, 32, 915–922.
62. A. V. Dobrynin and M. Rubinstein. *Macromolecules* **2000**, 33, 8097–8105.
63. A. V. Dobrynin and M. Rubinstein. *Prog Polym Sci* **2005**, 30, 1049–1118.
64. Q. Liao, A. V. Dobrynin, M. Rubinstein. *Macromolecules* **2006**, 39, 1920–1938.
65. P. Chodanowski and S. Stoll. *J Chem Phys* **1999**, 111, 6069–6081.
66. S. Ulrich, A. Laguecir, S. Stoll. *J Chem Phys* **2005**, 122, 094911.
67. H. J. Limbach and C. Holm. *J Phys Chem B* **2003**, 107, 8041–8055.
68. U. P. Strauss and B. W. Barbieri. *Macromolecules* **1982**, 15, 1347–1349.
69. J.-L. Hsu and U. P. Strauss. *J Phys Chem* **1987**, 91, 6238–6241.
70. V. S. Zdanowicz and U. P. Strauss. *Macromolecules* **1993**, 26, 4770–4773.
71. Q. Qiu, A. Lou, P. Somasundaran, B. A. Pethica. *Langmuir* **2002**, 18, 5921–5926.
72. M. Tachiya. *J Chem Phys* **1982**, 76, 340.
73. N. J. Turro and L. A. Yekta. *J Am Chem Soc* **1978**, 100, 5951.
74. M. Tachiya. Stochastic and diffusion models of reactions in micelles and vesicles. In *Kinetics of Nonhomogeneous Processes*. John Wiley & Sons, New York, 1987; pp. 575–650.
75. M. H. Gehlen and F. C. De Schryver. *Chem Rev* **1993**, 93, 199–221.
76. M. Almgren. *Adv Colloid Interf Sci* **1992**, 41, 9–32.

77. R. G. Alargova, I. I. Kochijashky, M. L. Sierra, R. Zana. *Langmuir* **1998**, 14, 5412–5418.
78. J. H. Kim, M. M. Domach, R. D. Tilton. *Langmuir* **2000**, 16, 10037–10043.
79. L. B. Shih, E. Y. Sheu, S. H. Chen. *Macromolecules* **1988**, 21, 1387–1391.
80. J. Claracq, S. F. C. R. Santos, J. Duhamel, C. Dumousseaux, J. M. Corpart. *Langmuir* **2002**, 18, 3829–3835.
81. L. B. Shih, D. H. Mauer, C. J. Verbugge, C. F. Wu, S. L. Chang, S. H. Chen. *Macromolecules* **1988**, 21, 3235–3240.
82. N. Nishiyama and K. Kataoka. *Adv Polym Sci* **2006**, 193, 67–101.
83. K. Kataoka, A. Harada, Y. Nagasaki. *Adv Drug Deliv Rev* **2001**, 47, 113–131.
84. K. Kataoka and A. V. Kabanov. *Colloids Surf B* **1999**, 16, 1–2.
85. L. Bromberg. *J Control Release* **2008**, 128, 99–112.
86. A. Mahmud, X. B. Xiong, H. M. Aliabadi, A. Lavasanifar. *J Drug Target* **2007**, 15, 553–584.
87. A. Lavasanifar, J. Samuel, G. S. Kwon. *Adv Drug Deliv Rev* **2002**, 54, 169–190.
88. M. Y. Kozlov, N. S. Melik-Nubarov, E. V. Batrakova, A. V. Kabanov. *Macromolecules* **2000**, 33, 3305–3313.
89. A. M. Sundell and H. J. G. Luttikhedde. *Pure Appl Chem* **2008**, A36, 1045–1059.
90. E. Bacu, G. C. Chitanu, A. Couture, P. Grandclaudon, G. Singurel, A. Carpov. *Eur Polym J* **2002**, 38, 1509–1513.
91. C. H. Salamanca, R. G. Barraza, B. Acevedo, A. F. Olea. *J Chil Chem Soc* **2007**, 52, 1025–1029.
92. I. Hernandez-Valdepena, M. Domurado, J. Coudane, C. Braud, J. F. Baussard, M. Vert, D. Domurado. *Eur J Pharm Sci* **2009**, 36, 345–351.
93. M. Almgren, F. Grieser, J. K. Thomas. *J Am Chem Soc* **1979**, 101, 279–291.
94. L. Sepulveda, E. A. Lissi, F. Quina. *Adv Colloid Interf Sci* **1986**, 25, 1–27.
95. O. Anthony and R. Zana. *Macromolecules* **1994**, 27, 3885–3891.
96. M. Wilhelm, C. L. Zhao, Y. C. Wang, R. L. Xu, M. A. Winnik, J. L. Mura, G. Riess, M. D. Croucher. *Macromolecules* **1991**, 24, 1033–1040.
97. T. Hirano, T. Todoroki, S. Kato, H. Yamamoto, P. Caliceti, F. Veronese, H. Maeda, S. Ohashi *J Control Release* **1994**, 28, 203–209.
98. Y. Yamamoto, Y. Tsutsumi, Y. Yoshioka, H. Kamada, K. Sato-Kamada, T. Okamoto, Y. Mukai, H. Shibata, S. Nakagawa, T. Mayumi. *J Control Release* **2004**, 95, 229–237.
99. C. Ladaviere, T. Delair, A. Domard, A. Novelli-Rousseau, B. Mandrand, F. Mallet. *Bioconjug Chem* **1998**, 9, 655–661.
100. H. Kamada, Y. Tsutsumi, Y. Yoshioka, Y. Yamamoto, H. Kodaira, S. Tsunoda, T. Okamoto, Y. Mukai, H. Shibata, S. Nakagawa, T. Mayumi. *Clin Cancer Res* **2004**, 10, 2545–2550.
101. H. Maeda. *Adv Drug Deliv Rev* **1991**, 6, 181–202.
102. H. Maeda, L. W. Seymour, Y. Miyamoto. *Bioconjug Chem* **1992**, 3, 351–362.
103. H. Kamada, Y. Tsutsumi, K. Sato-Kamada, Y. Yamamoto, Y. Yoshioka, T. Okamoto, S. Nakagawa, S. Nagata, T. Mayumi. *Nat Biotechnol* **2003**, 21, 399–404.

104. G. C. Chitanu, G. G. Bumbu, A. Stoleriu, A. Carpov, C. Vasile, A. Anghelescu-Dogaru, M. Rinaudo. *Polym Degrad Stabil* **1999**, 65, 75–85.

105. C. Ladaviere, L. Veron, T. Delair, A. Domard, C. Pichot, B. Mandrand. *J Appl Polym Sci* **1997**, 65, 2567–2577.

106. B. W. Barbieri and U. P. Strauss. *Macromolecules* **1985**, 18, 411–414.

CHAPTER 8

ASSOCIATION OF POLYELECTROLYTES TO SURFACTANTS AND SUPRAMOLECULAR ASSEMBLIES: COMPETITIVE ROLE OF CHAIN RIGIDITY AND ASSEMBLY STABILITY

ALBERTO CIFERRI and ANGELO PERICO

8.1 INTRODUCTION

Ionic and ionic mixed interactions involving polyelectrolytes and complementary charged polymers or amphiphiles are of great relevance within the context of supramolecular self-assembling processes, their applications, and the functioning of active and adaptive biological systems.[1–3] For instance, DNA–cationic lipid complexes are being considered for gene and drug delivery[1,4] (cf. Chapter 14). DNA–histone octamers in nucleosome modulate supramolecular structures that enable transcription, replication, recombination, and repairs.[5] The complexation of polyelectrolytes having opposite charge leads to the formation of layered films considered in an impressive number of possible applications[6] (cf. Chapters 5, 17–19). Associations based on synthetic polyelectrolytes and surfactants with 8–18 carbon atoms have been considered in applications such as sensors, encapsulation, coating, solubilization, catalysis, food technology, drag reduction, and water purification[7–9] (cf. Chapter 20).

The main driving forces for the self-assembly of these structures are the ionic interactions (mediated by mobile ions) between charged groups on the polyelectrolyte and on the complementary moiety, as well as the hydrophobic interaction between apolar sections of the components. Structural details are modulated by molecular parameters of the components, primarily the extent

Ionic Interactions in Natural and Synthetic Macromolecules, First Edition. Edited by Alberto Ciferri and Angelo Perico.

and distribution of charged and hydrophobic groups, ionic strength, and chain flexibility. Extensive experimental and theoretical investigation of the role of such parameters has been reported in reviews and books.[1–3,9–16]

This chapter focuses on systems in which the role of the conformational rigidity of the polyelectrolyte is analyzed for systems in which the complementary moiety is able to form supramolecular assemblies in the absence of the charged polymer. The systems considered here are based on polyelectrolyte–surfactant and on polyelectrolyte–protein complexes. For these systems, the shape and the stability of the complementary assembly are additional structural parameters. Our aim is the identification of a general framework allowing the prediction of these structures in terms of the interplay between the stability of the assembly and the persistence length of the complementary polyelectrolyte.[17]

Section 8.2 includes details clarifying concepts that will be used in the following sections: polyelectrolyte flexibility and the shape and the stability of the micellar and of the protein components. Section 8.3 deals with assemblies of flexible polyelectrolytes with surfactants. It will be shown that these assemblies exhibit surfactant clusters within loops of the polymer chain, sharing some of the features exhibited by hydrophobic polyelectrolytes (cf. Chapters 6 and 7). Section 8.4 deals with assemblies of rigid polyelectrolytes (i.e., short DNA segments) with surfactants. It will be shown that the rigidity of the polymer prevents the formation of spherical clusters, but favor the formation of hexagonal composite structures. Section 8.5 deals with the complexes of rigid polyelectrolytes with protein assemblies. It will be shown that the cohesion of the protein assembly induces folding even in a rod-like polymer. The approach is thereafter generalized to the case of worm-like polyelectrolytes exhibiting local rigidity and large contour length (i.e., giant DNA), thus providing a plausible model for the supramolecular organization of complex biostructures such as that of the chromatin fiber.[18,19]

8.2 CHAIN FLEXIBILITY, SURFACTANTS, AND PROTEIN ASSEMBLIES

8.2.1 Polyelectrolyte Conformation

A fully charged strong polyelectrolyte has an extended conformation. When the charge is screened by counterions, a coiled conformation will be assumed by flexible polyelectrolytes. However, an extended conformation will be retained even by screened rigid polyelectrolytes. The total persistence length is in fact defined by the sum of its electrostatic and intrinsic components[20]:

$$l_{p\,\mathrm{T}} = l_{p\,\mathrm{in}} + l_{p\,\mathrm{el}} \tag{8.1}$$

The electrostatic contribution is related to the Debye screening length ($l_{p\,\mathrm{el}} \sim 4\kappa^{-1}$, cf. Chapter 3) and is negligible above 0.1 M ionic strength. Some

TABLE 8.1. Intrinsic Persistence Length of Molecular and Supramolecular Polymers

System	Shape	l_p, Å	References
Polyethylene/p-xylene	Linear	6.5	Flory,[21] Ramachandran et al.[22]
Isotactic poly(hydroxyl butyrate)	Linear	29.1	Beaucage et al.[23]
Cellulose acetate/DMAC	Linear	70	Bianchi et al.[24]
Cellulose/DMAc + 5%LiCl	Linear	110	Kamide et al.[25]
DNA/0.2 M NaCl	Helical	540	Mandel and Schouten[26]
Poly(p-benzamide)/ DMAc + 3%LiCl	Linear	750	Ying and Chu[27]
Actin/phalloidin stabilized + buffer	Helical	600×10^2	Gittess et al.[28]
Microtubules/taxol stabilized + buffer	Tubular	7900×10^2	Gittess et al.[28]

measured persistence lengths are collected in Table 8.1.[21–28] Persistence lengths in 0.2 M NaCl solutions for the flexible polyelectrolytes considered in the following section are in the order of 10 Å. The persistence length measured in the presence of 0.2 M NaCl for DNA chains considered in Section 8.4 is in the order of 500 Å (roughly corresponding to 150 base pairs [bp]). Much larger values of persistence lengths have been reported for screened supramolecular polymers (e.g., up to the millimeter range for microtubules).

The persistence length is a measure of the conformational rigidity of a polymer chain. If the contour length L of a chain is comparable to the value of its persistence length, the chain is regarded as a rod. However, when $L \gg l_p$, for instance in the case for giant DNA (number of bp greater than 50 kbp), the conformation is that of a semirigid worm-like chain, endowed with considerable flexibility.[21]

In connection with the ability of flexible or rigid polymers to wrap around complementary spherical assemblies (cf. seq.), it is to be noted that folding a flexible polymer requires a minimal amount of conformational energy (e.g., a trans vinyl chain can make a full turn with ~4 cis-trans isomerizations requiring ~500 cal bond^{-1}).

Furthermore a large fraction of mobile counterions (about 0.8 per fixed charge) are strongly held by strong polyelectrolytes and their release (e.g., following substitution by protein macroions) is accompanied by a considerable entropy gain per ion (Oosawa–Manning condensation; cf. Chapters 3 and 6).

For conformationally rigid polymers and biopolymers, as well as for large discotic compounds or supramolecular complexes, the geometrical asymmetry is responsible for excluded volume effects (hard interactions) favoring the formation of mesophases that bears similarity to the mesophases observed with small amphiphilic molecules in virtue of soft interactions.[29] Treatments specific to polyelectrolytes have been developed.[29c] Liquid crystallinity due to hard interactions is manifested above critical values of the molecular asymmetry (length-to-diameter or diameter-to-thickness ratios). In the case of random, flexible polymers the distribution of polar and apolar groups does not usually favors effects of microsegregation. In fact, for linear polymers the critical asymmetry restricts the stabilization of mesophases to values of persistence length larger that about 50 Å.[29c]

8.2.2 Surfactant Micelles

While a hydrophilic substance, such as sugar, is thermodynamically compatible and soluble in water, a hydrophobic substance, such as oil, will phase separate from water. Simple amphiphilic molecules, such as the anionic sodium dodecyl sulfate (SDS) or the cationic dodecil trimethyl ammonium bromide (DTAB), having a hydrophilic head and a hydrophobic tail, will solubilize due to the head and at the same time try to get out of solution due to the hydrophobic tail. As a result, with increasing amphiphile concentration a surface layer (having heads in the water and tails out of it) or a spherical coating around a drop of oil may occur. Eventually, at a critical micelle concentration (cmc), which decreases with the concentration (C_s) of added salt, micelles appear where the tails are hidden in the interior of a sphere and hydrophilic heads are concentrated on the rim. Values of micellar diameter, aggregation number, and cmc may greatly depend upon surfactant type, and are frequently in the order of 5 nm, 100, and 10 mM, respectively.[29–35] Time-resolved fluorescence quenching (TRFQ) is the elite method for assessing aggregation numbers.[33] The free energy of formation of globular micelles and micellar parameters can be deduced, according to current treatments, from the balance of the transfer of the hydrocarbon tails from water to the micellar core (favorable contribution and main driving force for micellization), the formation of the interface (unfavorable contribution, mitigated by increasing aggregation number thus favoring cooperative micellar growth), and the repulsion of the charges at the rim (unfavorable contribution decreasing with micellar size).[36a] Typical values of the formation free energy of a micelle in binary surfactant solutions are in the order of 10 Kcal mol^{-1}.[36b]

The detailed aggregation process, schematized in Figure 8.1, reveals that further addition of surfactant above the cmc cause a rather step increase of the number and size of the micelles without a significant increase of the free amphiphile concentration. The behavior is typical of nucleation-induced unidimensional growth processes, including nonisodesmic (cooperative, helical) supramolecular polymerization and the assembly of some biological systems.[29,31,37] Following Israelachvili, finite-size distributions are produced by nucleated unidimensional growth, in contrast to two- and three-dimensional growth processes that have features of phase transitions.

Neutron small-angle scattering data for SDS and other ionic surfactants have confirmed the model of a spherical micelle. Nuclear magnetic resonance (NMR) studies revealed that the first segment of the tail assume a nearly trans

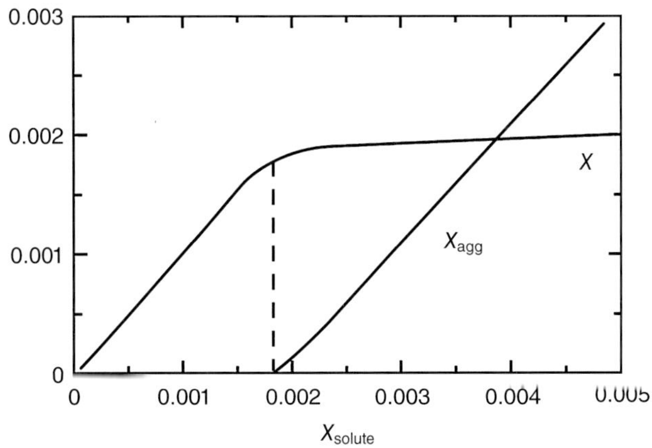

FIGURE 8.1. Schematic representation of the variation of the mole fraction of free surfactant (x) and of the mole fraction of aggregated surfactant (x_{agg}) with the total solute mole fraction. The sudden break in the increase of x corresponds to the cmc.

conformation while the end segments have conformation similar to that of a random hydrocarbon.[32] As the surfactant concentration is further increased, there is an evolution toward an elongated shape (that allows reduction of repulsive ionic interactions and may be marked by a second cmc[35]), often evolving into hexagonal-lamellar sequences[30–32] (Fig. 8.2). The elongated micelle may undergo a linear growth process eventually attaining dimensions in the micrometer range. This process has been described as growth coupled with the development of supramoleculat liquid crystallinity,[29a,b] although some linear growth may occur even in the absence of a nematic phase.[35b] With even fewer water molecules, inverted spherical and cylindrical micelles may be observed. Surfactants and phospholipids (amphiphilic molecules with a hydrophilic head and one or more hydrophobic tails) may form vesicles (liposomes): closed bilayers with hydrophobic tails inside and the charged heads exposed to water, the typical organization of cellular membranes. Liposomes are regarded as delivery vehicles in drug and gene therapy (cf. Chapter 14). The prediction of the organization of complex molecular architectures is hampered by the need for a detailed description of the distribution of the amphiphile tails.

The clustering of hydrophobic segments is not necessarily associated with the occurrence of charged groups. Micellar structures are formed by nonionic detergents such as ethylene oxide condensates of n-dodecanol or n-octadecanol. Mixed micelles formed by ionic and/or nonionic amphiphiles are of great interest (e.g., for separation processes) and have been extensively investigated. In such cases, the hydrophobic attraction of the tails is balanced by the hydration and the steric interaction of the ethylene oxide segments at the rim. The cmc is much lower than for ionic detergents, and large aggregation numbers, increasing with the length of the hydrophobic segment, were reported.[38a]

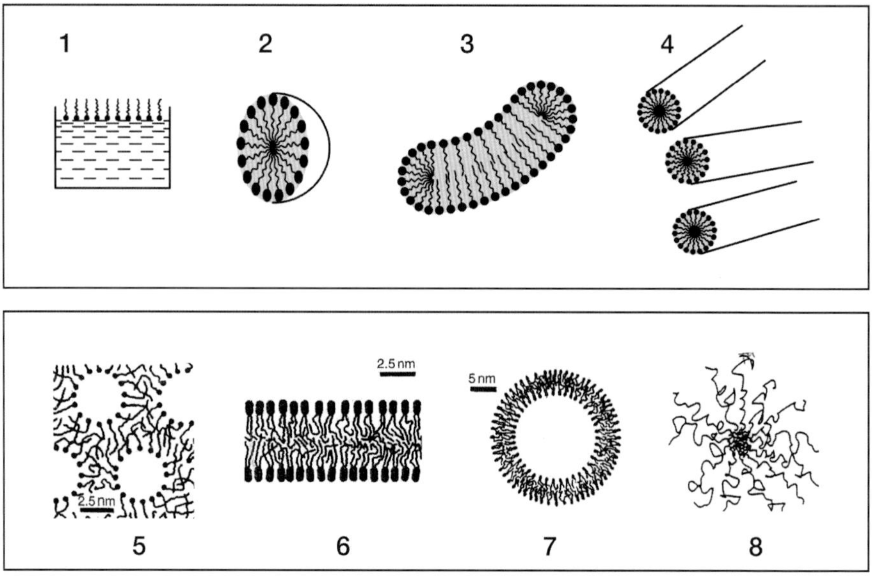

FIGURE 8.2. Schematic organization of various amphiphiles. *Surfactants*: 1, segregation at water interface below the cmc; 2, spherical micelle; 3, elongated micelle; 4, hexagonal, lamellar assemblies; 5, inverted micelle. *Double chained phospholipids*: 6, bilayer; 7, vesicle. *Coil-coil diblock copolymers*: 8, core-corona micelle.

Theoretical and experimental analysis of mixed nonionic detergents (e.g., dodecyl octaoxyethylene glycol monoether and dodecyl hexaoxyethylene glycol monoether) in water supported the formation of mixed micelles having elongated shapes.[38b] Mixed micelles formed by a ionic, rod-like surfactant and a nonionic long chain alcohol cosurfactant were also investigated.[38c] The alcohol, regarded as a spacer, was preferentially partitioned in regions of lower curvature, where it was particularly efficient in reducing the charge density and repulsion at the rim. Micelles formed by amphiphilic block copolymers have also been investigated.[34] The core–corona assembly of a copolymer composed by a solvophobic and a solvophilic block is schematized in Figure 8.2.

Spherical micelles are formed in isotropic solutions, but liquid crystalline mesophases (nematic cylindrical or disk-like shapes, hexagonal, smectic, biaxial, and complex cubic phases) are eventually formed when concentration is increased.[31,32,39]

8.2.3 Histone Octamers

The protein assembly that is considered here is the histone octamer, a strongly held association of eight globular proteins, wound up and interconnected by a long DNA chain to form the superstructure of the 10 and 30 nm chromatin fiber. The latter is actually a polynucleosome assembly stabilized by both

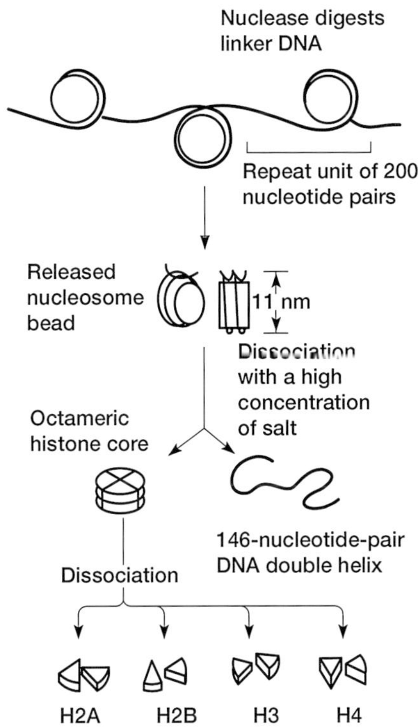

FIGURE 8.3. Schematic representation of the components of the chromatin assembly. From Alberts et al.[40]

histone–histone and histone–DNA interactions and may be regarded as a chain of beads on a string (Fig. 8.3). Nucleosomes can be isolated by enzymatic digestion of the DNA linker (~10–90 bp), and further dissociated into short (146 bp) DNA segments and histone octamers. The latter can in turn be dissociated into the constituent proteins only in concentrated salt solutions.[18,19,40–44] The reassociation is not completely reversible, supporting the occurrence of a large formation constant for the octamer complex.[45]

The surprising feature deduced from X-ray studies of nucleosomes is the ability of rigid DNA segment of ~150 bp (comparable to the persistence length) to fold around the histone octamer that has a radius of 3.4 nm[19] (cf. seq.).

8.3 FLEXIBLE POLYELECTROLYTES AND SURFACTANTS

When a complementary charged, flexible polymer is added to a surfactant/water solution, self-assembly of polyelectrolyte–surfactant aggregates occurs. Binding of the surfactant to the polymer is conveniently investigated using

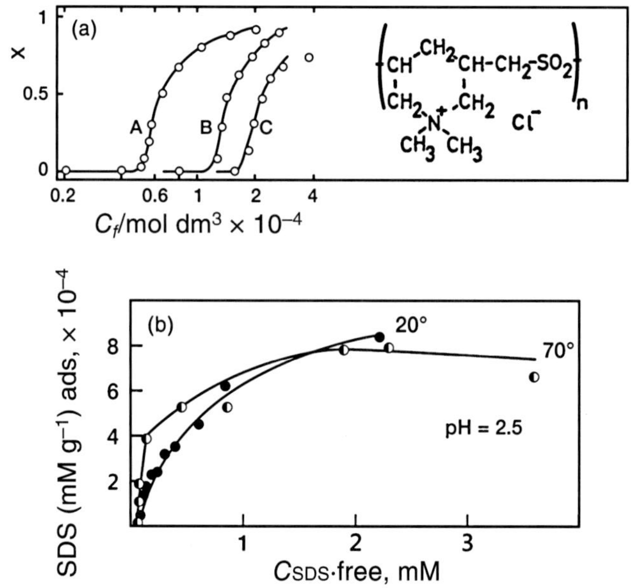

FIGURE 8.4. Binding isotherms. (a) X = Number of SDS bound/number of cationic sites on the PAS polymer poly(dimethyldiallylammonium chloride-co-sulfur dioxide) (formula on the right-hand side [RHS]) plotted versus the concentration of free SDS. NaCl concentration: A = 20, B = 50, and C = 100 mM/dm^3; (b) adsorbed SDS (mM/g) of denatured collagen at the indicated temperatures and pH, plotted versus the concentration of free SDS. Adapted from Shirahama et al.[46a] and Henriquez et al.[46b]

surfactant-selective electrodes that measure the difference between the analytical and the free surfactant concentration at a given polymer concentration. Typical binding isotherms are illustrated in Figure 8.4. The first case (a) is that of a polycation (PAS: formed by repeating units based on dimethyldialliyl-ammonium chloride and sulfur dioxide) interacting with SDS.[46a] The second case (b) is that of gelatin (a weak polycation at pH = 2.5) interacting with SDS.[46b] Both systems include a significant number of hydrophobic units. The coulombic attraction of the complementary charged groups is reinforced by the hydrophobic attraction of the surfactant tails. In the case of the gelatin–SDS systems, direct hydrophobic bonding (tails of the surfactant to the polymer backbone) could be a contribution since the potential number of cationic binding sites (ε-amino residues) is small, and sulfate anions do not bind to peptide bonds.

The onset of the deviation of the electrode response from the reading for the pure surfactant solution allows the assessment of a critical aggregate concentration (cac). For systems displaying strong ionic interactions, the cac can be several orders of magnitude smaller than the cmc for the surfactant (polymer-free) solution. In Figure 8.4a, the cac is seen to increase with ionic strength. From other studies it is known that the cac increases also with

TABLE 8.2. Binding Parameters for the PAS–SDS System (cf. Fig. 8.4a)

Cs, mol/dm^3	0.02	0.05	0.10
u	25.6	26.3	26.1
K, mol^{-1} dm^3	601	271	180

polymer concentration and decreases with surfactant tail length and charge density on the polyelectrolyte. Systems in which one or both components are neutral exhibit weaker effects. Corresponding effects are observed in the case of the cmc of the pure surfactant.[7,14]

For strong polyelectrolytes and complementary ionic surfactant, the sigmoid shape of the isotherms reveals a cooperative binding process. Saturation (when all sites are bound to one complementary surfactant molecule) is not always observed. Analysis of the shape of the isotherms allows an evaluation of the isodesmic binding constant K and a parameter of interaction between adjacent binding sites (cooperativity parameter u).[10,46,47] The pioneering results of Shrihama and coworkers, reported in Table 8.2,[46a] reveal that u changes little with ionic strength in comparison to K. They argued that the strength of primary electrostatic binding is reduced by the salt and that, once anchored to the charged groups, cooperativity is induced by subsequent interaction among the hydrophobic tails.

The above effects are consistent with a long advocated model based on the assembly of micelle-like clusters within polymer loops. The fixed charge on the polyelectrolyte cause a reduction of the ionic repulsion among the head groups of the surfactant, in a similar way to the effect of added salts.

While binding of surfactants to strong polyelectrolytes is usually cooperative,[48] for the system gelatin–SDS (Fig. 8.4b) values of $K \sim 1800$ M^{-1} and $u \sim 1$ were determined, revealing strong binding but no cooperativity. The latter effect may be attributed to the low charge density of gelatin. In fact, Wei and Hudson reported that the cooperativity parameter for SDS binding to cationic chitosan decreased with decreasing charge density (deacetylation).[49]

Zana reported an interesting analysis of the way the aggregation number of the surfactant relates to the hydrophobicity of the polyelectrolyte.[50,51] He finds that the aggregation number can be smaller or larger than observed for the binary solutions. Moreover, hydrophilic polyelectrolytes bind cooperatively only opposite charged surfactants, whereas proteins such as bovine serum albumine or gelatin may exhibit a noncooperative followed by a cooperative binding step. In the latter step, the aggregation number (from fluorescence of pyrene data) can be smaller than for pure SDS solutions. A necklace (beads along a string) model, schematized in Figure 8.5, was suggested for a weakly hydrophobic polyelectrolyte that includes preexisting microdomains of the polymer and new microdomains formed by the surfactant. A decrease of the aggregation number is expected if hydrophobic segments of the polymer are included in a pseudo-micelle.[52]

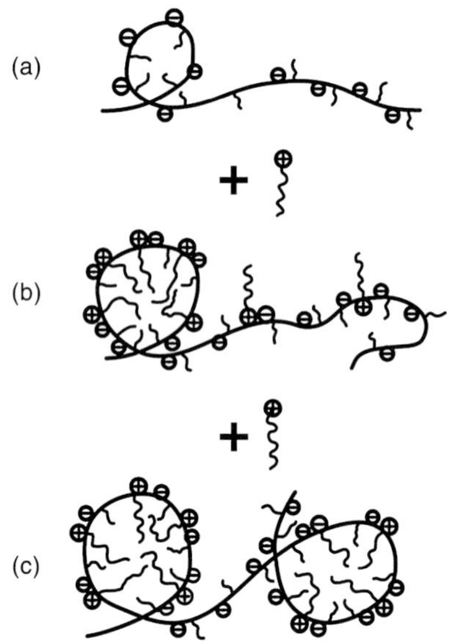

FIGURE 8.5. Model for the assembly of a flexible polyanion and a cationic surfactant. (a) Preexisting microdomains are swollen by the surfactant. (b) and (c) Additional microdomains upon further association of surfactant. Adapted from Anthony and Zana.[51]

Similar occurrences were suggested by other authors. Guo et al.,[53] also working with the BSA/SDS system, suggested that the hydrophobic segments of the protein act as a nucleus for small micellar-like clusters that showed an aggregation number of ~35, smaller than the value for SDS micelles in water. Henriquez et al.[46] reported that SDS causes a depression of the denaturation temperature of collagen that is much larger than observed with sodium methyl sulfate and other strong salting-in agents (cf. Fig. 8.6). They inferred that the prevailing interaction involves binding of the tails of the surfactant to hydrophobic sites on the protein exposed during the melting transition. Since the cmc of SDS occurs at the low range of concentration considered in Figure 8.6, they concluded that an unusual organization of the surfactant clusters is involved, with segments of the protein able to form loops and partake in the hydrophobic core.

Bekturov et al.[54] considered the competition of the hydrophilic and hydrophobic segments of random copolymers (2,5-trimethyl-4-vinylpiperidol-4 and acrylic acid) in the presence of both SDS and a nonaqueous solvent. The low viscosity observed in the absence of methanol (Fig. 8.7) was attributed to the folding of the polymer stabilized by SDS. The increase of viscosity above 50% methanol was attributed to the opening up of the folded structure, allowing a greater exposure of hydrophobic segments. The subsequent decrease was attributed to an inverse clustering, with hydrophilic segments in the core of the structure.

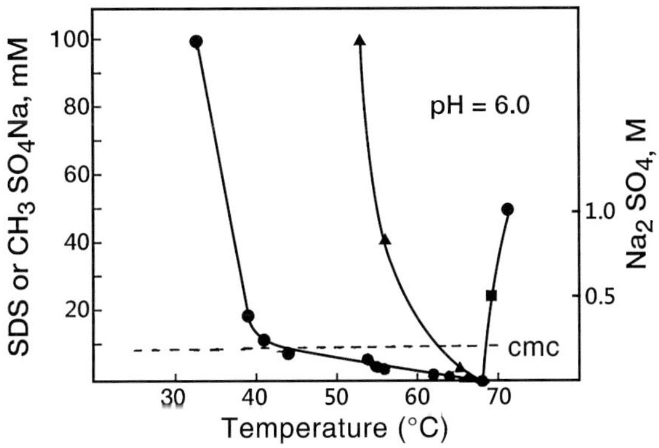

FIGURE 8.6. Variation of the apparent melting temperature of collagen with the concentration of either SDS, sodium methylsulfate, or sodium sulfate (squares) at pH = 6.0.The broken line indicates the cmc for the binary water/SDS system. Adapted from Henriquez et al.[46b]

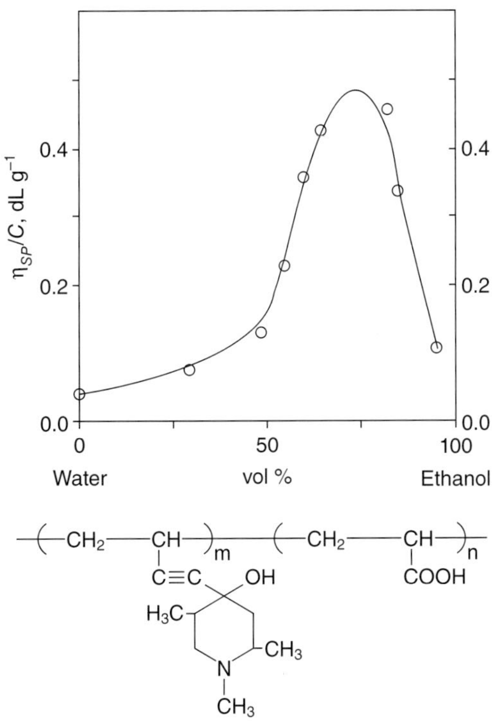

FIGURE 8.7. Specific viscosity of a cationic polyampholyte having 60% 2,5-trimethyl-4-vinylpiperidol-4 and 40% acrylic acid bound to excess SDS (molar ratio > 4) versus the water/methanol composition of the solvent. From Bekturov et al.[54]

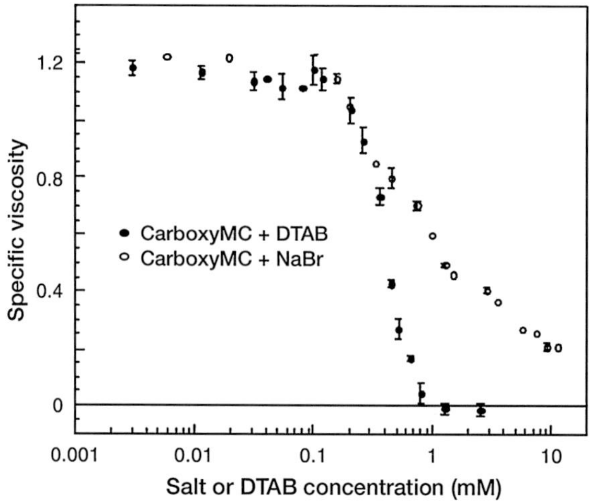

FIGURE 8.8. Specific viscosity (referred to water) of water solutions containing carboxymethyl cellulose and either the DTAB surfactant or the salt NaBr. Polymer concentration is 127 mg L^{-1}. Adapted from Langevin.[10]

Viscosity measurement allows a detailed analysis of the collapse of the polyelectrolyte due to complexation with the surfactant. Figure 8.8 illustrates the variation of the specific viscosity of water solutions containing a fixed amount of carboxymethyl cellulose in the presence of either a simple salt (NaBr) or a complementary surfactant (DTAB).[10] Below the cac, the viscosity slightly decreases with the concentration of salt or surfactant. The effect was attributed to the screening of the polymer charges by the counterions, reducing the electrostatic contribution to the persistent length. Above the cac, the chain collapse induced by the surfactant appears much larger than that induced by the salt. The hydrophobic tails of the surfactant magnify the screening effect by a reduction of the local dielectric constant, thus enhancing stronger, cooperative binding to the fixed charges on the polymer.

Several theoretical studies have been reported (cf. e.g., Ober and Wegner[9] and Diamant and Andelman[14]). Nagarajan presented a thermodynamic analysis of uncharged polymers wrapped around a spherical surfactant micelle.[36] The free energy of formation of the polymer-micelle complex included two terms that decreased the free energy of the micelle and favored its binding to the polymer (decrease of the hydrophobic surface area exposed to water for both the micelle and the polymer). A third term (the steric repulsion between polymer segments and surfactant head groups) disfavored instead the formation of the polymer-bound micelle. Although Nagarajan did not consider polyelectrolytes, his analysis includes interesting details regarding the alteration of micellar parameters induced by the polymer. In particular, the cmc and the size of the polymer-free systems were significantly decreased, and shape changes were occasionally exhibited (i.e., cylindrical to globular). The phenomenological analysis of Diamant and Andelman[14a] focused instead on the effect of small associating molecules on chain statistics, rather than on structural details of the micelles. They associated the cac to a local instability (partial collapse) of a flexible polymer chain occurring when intrinsic intrachain repulsions are overcome by attractive correlations resulting from association with smaller molecules. They predict that the cac should be lower than the cmc, increasing with ionic strength and (slightly) with polymer charge, in line with the experimental observations described above. Nylander et al. interpreted the reduction of the cmc in terms of a decreased Debye screening length.[14c]

Wallin and Linse reported interesting Monte Carlo (MC) simulations representing the effects of chain flexibility, linear charge density, micellar size, and polymer hydrophobicity.[13] Figure 8.9 illustrates the interplay between chain flexibility and complexation of the ionic charges. The more flexible polymer in Figure 8.9a is wound around a spherical micelle without a significant cost in terms of conformational energy. A reduced flexibility allows optimal electrostatic interaction at the expense of an increased internal stress (Fig. 8.9b). When the persistence length of the polymer is larger than half the circumference of the micelle, the electrostatic interaction is reduced (Fig. 8.9c). The release of small counterion by the polymer contributes to its folding. The cac was also influenced by flexibility, being lower for the more flexible polymer. The role of the micellar size and hydrophobicity of the polymer were also considered. Larger reduction of the cac corresponded to larger micellar diameters.[13c] For more hydrophobic polymers, a substantial penetration of the micelle core was observed, accompanied by a decrease of the cac.[13d]

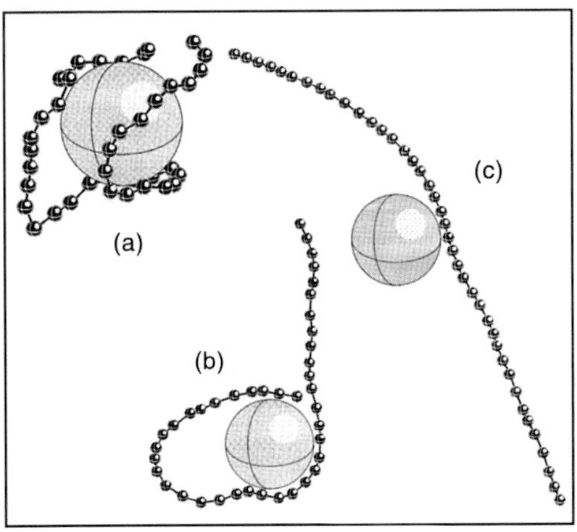

FIGURE 8.9. MC simulation of complexes between micelle polyelectrolytes. The flexibility of the polymer decreases with the increase of average angles between consecutive segments 135° (a), 165° (b), and 175° (c). Adapted from Wallin and Linse.[13a]

8.3.1 Conclusions

The body of the experimental and theoretical results outlined above allows some general conclusions regarding the nature of the association between flexible polyelectrolytes and surfactants. One conclusion is that several characteristics of the complexes are similar to those of surfactant micelles formed in the absence of polymer. In fact, both micelles and polyelectrolyte–surfactant complexes may exhibit nucleation steps, cooperative growth, and sizeable aggregation numbers. While the cac is smaller than the cmc, similar dependences upon ionic strength, polymer concentration, charge density, and tail length were observed. As discussed in Section 8.2, surfactant micelles undergo alterations of their shape and parameters following the addition of nonpolymeric charged or uncharged components.[38] However, a distinctive feature of the association of surfactants to polyelectrolytes is the possible inclusion of polymer segments within the hydrophobic core formed by the surfactant (cf. above discussion of data in Wallin and Linse,[13d] Shirahama et al.,[46] Hansson and Almgen,[52] Guo et al.,[53] and Fig. 8.5).

Hydrophobic polyelectrolytes, in which sizeable hydrophobic segments are covalently linked to the polymer chain (e.g., alternating copolymers of maleic acid and alkyl vinyl ethers; cf. Chapter 7 and Binana-Limbele and Zana[51b]), have also been shown to undergo micellar-like structurization in the absence of surfactant. Depending upon the degree of polymerization (DP) and the length of the alkyl segment, necklace (or pearling) conformations were observed with charged beads connected by charged linear strands along a single polyelectrolyte chain. The micellar-like beads exhibited aggregation numbers up to 100 (determined by TRFQ). The theoretical basis for the formation of such necklace conformation, described in Chapters 4 and 6, is based on a Raleigh-type instability resulting from imbalance of electrostatic and interfacial free energies. Deviations from the fully stretched conformation of polyelectrolytes may be induced by a variety of attractive interactions, including the occurrence of hydrophobic sequences, complementary fixed charges either along a polyampholyte chain or carried by other molecules, alteration of the concentration of the polyelectrolyte, ionic strength, and solvent quality (cf. Chapter 6). *It thus appears that polyelectrolyte–surfactant interaction can be reconduced to a general scheme involving the competition between attractive and repulsive interaction valid for either covalently or supramolecularly linked components.* The detailed structure of the complexes would have to be determined for each specific case. If the term “micelle” is used for the spherical structures described in Section 8.2, characterized by a detailed distribution of the amphiphile tails, terms such as “pseudo-micelle” or “micellar-like” ought to be consistently used for hydrophobic clusters wrapped by polymer chains.

The balance of forces on hand for the flexible polyelectrolyte–surfactant systems (the ionic repulsion of the charged sites including the counterion condensation/release mechanism, and the hydrophobic attraction of apolar segments) is necessarily modulated by the conformational energy of the polyelectrolyte. The latter contribution was evidenced by the simulation illustrated in Figure 8.9, where it was necessarily implied that the sum of the contributions stabilizing the surfactant–polymer complex is larger than the energy required to fold the polymer chain. This condition appears to be fulfilled by conventional micelle due to their considerable free energy of formation (cf. Section 8.2, expected to be further increased by binding of the polyelectrolyte) compared to the negligible conformational energy loss for bending a flexible polymer (cf. Section 8.2). However, situations involving rather rigid polymers (cf. Fig. 8.9c) do require an alternative approach, as discussed in the following section.

8.4 RIGID POLYELECTROLYTES AND SURFACTANTS

In the interaction of a polyelectrolyte and an oppositely charged ionic micelle (Fig. 8.9), there is the possibility of a destabilization of the spherical micelle in favor of a cylindrical structure in which compensation of the charges on the polyelectrolyte by the hydrophilic heads of single surfactant molecules still occurs. In this conformation, the hydrophobic interaction of the surfactant tails could also be preserved by the growth of an interdigitated structure having hexagonal symmetry. Bending of the rigid polyelectrolyte would not therefore be required unless conformational or morphological changes are triggered, or the contour length greatly exceeds the persistence length (cf. Section 8.6).

This mode of interaction between rigid polymers and complementary surfactants was first suggested by one of the present authors,[55] who proposed the assembly scheme in Figure 8.10, which is most effectively illustrated by the splendid images in Figure 8.11. The model was suggested by the anomalous behavior of adsorption isotherms reported by Shirahama et al. for DNA–dodecylpiridinium (or dodecyl trimethyammonium) systems.[11,56] These isotherms evidenced two cooperative binding steps upon increasing surfactant concentration. The fist step was suggested to involve binding of the ligand

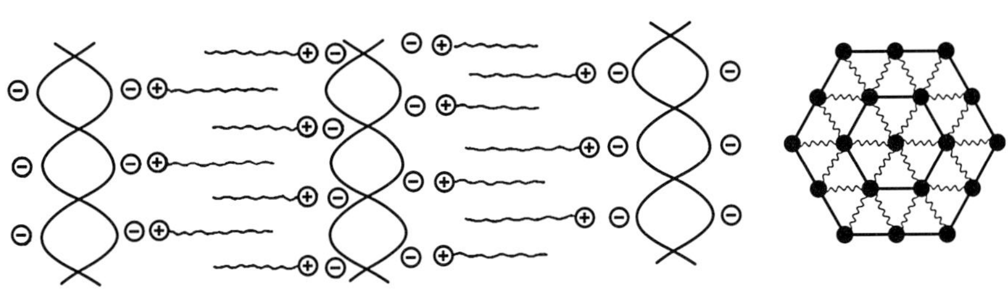

FIGURE 8.10. Scheme for the assembly of a rigid polyanion and a cationic surfactant. The basic hexagonal unit has a central molecule, here represented as a double helical polyelectrolyte, and an outer layer of polyanions having residual charges toward the aqueous solution. Adapted from Ciferri.[55]

FIGURE 8.11. Pictorial representation of the hexagonal assembly of rod-like DNA segments (~150 bp) with a cationic surfactant. The edge on view with only tree DNA chains evidences a more detailed interdigitation of the hydrophobic tails of the surfactant. Images produced with POVRAY (Vision Raytracer, Williamston, Australia) and kindly contributed by Dr. Cecilia Leal.

(surfactant) to the charged sites on the polymer (e.g., the single, central double helix in Fig. 8.10). The second step was suggested to involve the assembly of additional ligand to bound ligand, resulting in the growth of a first layer of a hexagonal, interdigitated structure stabilized by both ionic and hydrophobic interactions.

Support to the model in Figure 8.10 was produced by several authors.[16,57–61] Lamellar and hexagonal structures were reported by Koltover et al. from small angle X-ray scattering (SAXS) studies for DNA decorated by a cationic liposome.[57] Definite support was recently offered by Langevin and coworkers, who presented an excellent analysis of the DNA (147 bp)–dodecyltrimethylammonium bromide–H_2O system, which they characterized by light and neutron scattering studies.[58] They discussed the hexagonal structure formed by two concentric layers of surfactant-decorated DNA around a central DNA. They reported that the assemblies have remarkably well-defined diameters, lengths, and internal organization. Additional support to the model was offered by the calculation of relevant parameters for DNA–surfactant hexagonal assemblies, as indicated below.[17]

In the presence of a high concentration of cationic surfactants, an almost stoichiometric association of DNA to the surfactant would cause neutralization of the complex. The separation distance between the decorated DNAs may be evaluated, neglecting the contribution of like-charge attraction (cf. Chapter 3). The hydrophobic alkyl tails are spaced 0.17 nm along the polymer axis, corresponding to the axial separation between unit fixed charges. Two parallel neighboring DNA–surfactant complexes will find equilibrium at a distance d', where the free energy per phosphate group reaches a minimum. The hydrophobic attraction will reach a maximum when the two tails, bound to corresponding phosphates, superimpose.

The equilibrium distance may be estimated as $d' = h_{\text{tail}} + 2a'$, where h_{tail} is the length of the hydrophobic tail (~1.8 nm for a fully extended, 12 methylene units chain), and a' is the distance between the axis of the polyelectrolyte cylinder and the center of the surfactant head (~1.2 nm). The result is $d' \approx 4.2$ nm. Note that the latter value is greater than the value (2.92) calculated for the separation distance due to like-charge attraction in the absence of surfactant[58] (cf. Chapter 3). Assuming that the radius r of the assembly is on the order of the equilibrium separation distance between two parallel DNA-bound surfactants, we have that the shortest distances in the hexagonal cluster are given by $r \approx d' \approx 4.2$ nm. Hexagonal clusters of surfactant-decorated DNAs, including l layers, will have diameters $d_l = 2lr + 2a'$, yielding 10.8 nm for a single-layer cluster including 7 DNAs, and 19.2 nm for a double-layer cluster including 19 DNAs. McLaughin et al.[58] found experimentally that a cluster of decorated DNA including two layers had estimated values of $r = 4$ and $d_2 = 24$ nm. The overall satisfactory agreement of calculations, scattering data,[58] and binding isotherms[11,56] support the stabilization of the hexagonal cluster by interdigitation of the hydrophobic tails of the bound surfactant.[55]

As mentioned earlier, saturation (i.e., binding occurring to all available sites) is not always revealed by binding isotherms. Nevertheless, the occurrence of situations in which more than one surfactant molecule is bound to each site has been suggested, eventually predicting a charge inversion on the polyelectrolyte.[62] An estimate of the average number of surfactants per charged phosphate site $\langle y \rangle$ may be made on the basis of the free energy per site $E(y)$:

$$\langle y \rangle = \sum_{l=1}^{N_{\max}} y \exp(-\beta E(y)) \Big/ \sum_{l=1}^{N_{\max}} \exp(-\beta E(y)). \tag{8.2}$$

$E(y)$ may be calculated, following Kuhn et al.,[62] as the sum of the following terms: (1) 1/2 of the hydrophobic interaction between y surfactants on a charged site, and the surfactants on the two nearest neighbor sites; (2) the electrostatic repulsions between the y surfactant heads (charge +1) plus the attraction between all the hydrophobic surfactant tails at the given phosphate site; and (3) the electrostatic repulsion per charged site along the rod-like polyelectrolyte following the association of y surfactants per charged site. The result (for equal distances between the surfactant heads on the phosphate sites) is:

$$\beta E(y) = \beta\chi \quad y = 1$$

$$\beta E(y) = \beta\chi[y + y(y-1)/2] + (l_B/a)\sum_{i=1}^{N_{max}} v_i(y)a/d_i(y) \tag{8.3}$$

$$-(l_B/b)(-1+y)^2\log(1-\kappa b) \quad y > 1,$$

where l_B is the Bjerrum length, b is the unit charge separation along the polyelectrolyte (equal to 0.17 nm for DNA), $\beta\chi$ is the hydrophobic interaction between surfactants, assumed equal to –3.5 for a typical cationic surfactant with 12 methylene groups,[62] and κ^{-1} is the Debye screening length (cf. Chapter 3). With this geometry we calculate $\langle y \rangle \sim 1.31$ suggesting some overcharging in the case of DNA. Note that the last term in the second equality in Equation 8.3, representing the repulsion between the effective charges along the polyelectrolyte, is positive and dominant as *y increases*, thus preventing too much overcharging. Note also that if $\beta\chi$ decreases to –7.0, that is, the hydrophobic attraction increases, the average number of surfactants per phosphate site $\langle y \rangle$ increases to 2.10, demonstrating that the strength of the hydrophobic interaction enhances the binding of more surfactants to the polyelectrolyte, eventually causing overcharging at very high strength. Larger values of $\langle y \rangle$ may be obtained also by arranging the surfactants molecules in different geometries, favoring hydrophobic attraction relative to head repulsions.

It was earlier suggested[55] that growth of the basic hexagonal structure along the lateral and longitudinal dimensions could be a model for the formation of the macroscopic organization of some biological and composite materials, characterized by a regular distribution of rigid and flexible macromolecular units.[52,57] Lateral growth (D) should favor an increase of interdigitating hexagonal layers (disk-like growth), while longitudinal or rod-like growth (L) should favor stacking of individual hexagons having a length corresponding to the length of the polyelectrolyte. Simple D growth, and simultaneous D and L growth, have features of all-or-none phase transitions (systems with dimensionality indexes of 1/2 and 1/3, respectively, according to Israelachvili's formalism[31]). Lateral growth may be controlled by the polyelectrolyte/surfactant composition. For instance, an excess DNA could terminate lateral growth, leaving uncompensated charges on the external hexagonal layer. It is to be noted that whereas extensive growth is expected for a phase transition, experimental data for DNA–surfactant systems evidenced only a limited lateral and longitudinal growth. Secondary effects have been discussed.[58]

In cases in which lateral growth is effectively prevented, longitudinal growth may be treated as an isodesmic supramolecular polymerization (dimensionality index = 1) and is expected to attain a DP simply related to an association constant K[29,37]:

$$\mathrm{DP} \sim 2(\mathrm{v}K^{1/2}). \tag{8.4}$$

For the case at hand, K should reflect the hydrophobicity of the aliphatic tails distributed along the cross-section of the hexagons. The experimental data quoted above for DNA–surfactant systems[58] suggested a stacking in the order of four basic hexagonal units. If the longitudinal association of DNA–surfactant hexagons is regarded as an isodesmic supramolecular polymerization,[29,37] such a low DP is consistent with a small but not vanishing value of K obtained from Equation 8.4. It should be possible to obtain much larger values of longitudinal aggregation under nonisodesmic conditions when supramolecular liquid crystallinity in concentrated polyelectrolyte–surfactant solutions occurs (cf. Section 8.2).[1,2,9,63,64] According to the current theory of liquid crystallinity in polymers, the critical volume fraction v* at which nematic order appears is related to the persistence length l_p and to the diameter d of the polymer by the relationship[29c]:

$$v^* \sim \text{const}(d/l_p). \tag{8.5}$$

Using the value of l_p reported for DNA in Table 8.1, Equation 8.5 predicts a critical concentration in the order of 0.1 that has not been simple to verify due to the encroachment with the field of stability of the crystalline phase of DNA.[65] However, in the case of the growing DNA–surfactant assembly, an increase of the diameter of the assembly and, particularly, of the persistence length is anticipated (cf. e.g., the large values of l_p reported in Table 8.1 for tubular assemblies). The ratio d/l_p is therefore expected to greatly decrease. This fact, coupled with the occurrence of the small association constant for longitudinal stacking, should promote the simultaneous occurrence of enhanced growth and liquid crystallinity at low DNA concentration. More detailed description of the development of liquid crystallinity in supramolecular assemblies is described in Ciferri.[29b] The underlying concepts were applied to the fibrillogenesis of keratin and collagen II.[63] In the latter case, lateral growth is biochemically inhibited by the decoration of growing assemblies with protoglycans.[63a]

8.4.1 Conclusions

The body of results presented evidence that polyelectolyes having a large persistent length, comparable to the contour length, are unable to wrap around a spherical micelle. The situation illustrated in Figure 8.9c of the preceding section, where the shapes of a spherical micelle and of a rod-like polymer are retained, is not realistically attained since it does not allow for the compensation of complementary charges. The high energy cost for folding the rigid chain can be avoided by the formation of a cylindrical structure, where a balance of the charges on the polyelectrolyte and on the surfactant is achieved, simultaneously with the preservation of the hydrophobic interaction in the core of the assembly. The distribution of the aliphatic tail in the cylindrical structure bears analogy to the organization of tails in elongated micelles in the absence of polymer (cf. Section 8.2 and Fig. 8.2). The basic hexagonal unit, having seven polyelectrolyte molecules decorated by interdigitated tails of the surfactant, is a model for the regular organization of rigid and flexible components for synthetic and biological composite materials.[63]

The cases discussed in the two preceding sections illustrate the behavior of either flexible or extremely rigid (L ~ l_p) polyelectrolytes vis-à-vis a complementary surfactant able to form various micelle-like structures in which electrostatic and hydrophobic balance are achieved without a large change in stabilization energy (cf. Section 8.2). In the following section, we shall consider instead the behavior of the same rigid polyelectrolyte considered here vis-à-vis a complementary assembly having a rigid structure, unable of morphological reorganization driven by ionic interaction with the polymer.

8.5 RIGID POLYLECTROLYTES AND PROTEIN ASSEMBLIES

The third model considered here is exemplified by nucleosomes, the complexes known to form between DNA and the oppositely charged histone octamers, a strongly held association of eight globular proteins (cf. Section 8.2 and Fig. 8.3). The nucleosome structure is well established by X-ray diffraction studies,[18,19,41,42] and a noteworthy feature is the ability of a rather rigid DNA segment of ~130 bp, shorter than the DNA persistence length, to wrap around the undeformed octamer (radius 3.4 nm) in 1.58 helical turns.[41]

A recent theoretical study by Arcesi et al.[43] extends previous calculations presented by Shiessel[19] and Manning,[44] and explains the apparent contradiction that DNA segments with a contour length comparable to the persistence length retain sufficient flexibility to wrap around a quasi-spherical complementary assembly. They consider the combined role of electrostatic interaction between the components and the entropic gain associated with the release of the counterions condensed over the highly charged polyelectrolyte. The histone octamer was tacitly assumed to be an absolutely stable assembly under the adopted conditions.[43] Their approach is summarized below in a general form valid for rigid, strong polyelectrolytes and undeformable, complementary spheres (cf. also Fig. 8.12 and Chapter 3, Section 3.6).

The binding free energy of a rigid polyelectrolyte–sphere complex is calculated as the free energy difference when a length l of polyelectrolyte is wrapped around the sphere, and when an isolated, rod-like polyelectrolyte of contour length L coexists with a noninteracting sphere of radius R_0:

$$\Delta F(l) = F_s(L-l) + F_c(l) + F_{c-s}(l) + F_b(l) - F_s(L) - F_c(0). \tag{8.6}$$

The first term, $F_s(L - l)$, is the electrostatic free energy in the condensation regime of a straight segment of polyelectrolyte of length $L - l$; the second term, $F_c(l)$, is the electrostatic free energy of a spherical macroion that has, around it, a length l of polyelectrolyte, including the free energy cost for coun-

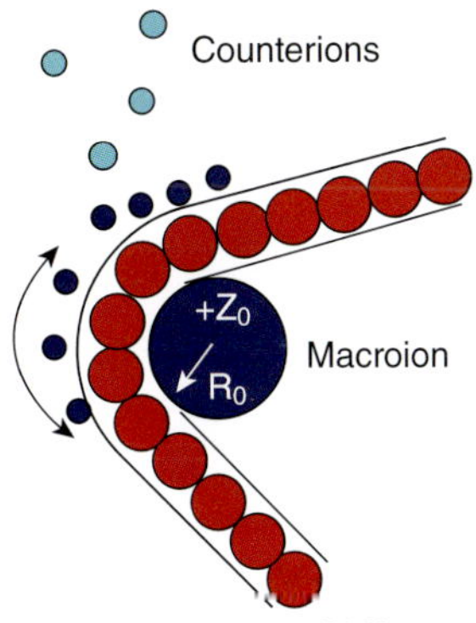

FIGURE 8.12. Model for the association of a rigid polyelectrolyte (DNA) with a complementary charged protein complex (histone octamer). From Arcesi et al.,[43] with permission.

terion condensation at large surface charge; the third term, F_{c-s}, describes the electrostatic interaction between the straight polylectrolyte and the complex of the sphere-wrapped polyelectrolyte; and the last term, $F_b(l)$, is the elastic energy necessary to bend a segment of polyelectrolyte of length l. It is assumed that the spherical polyelectrolyte–protein complex has a radius R given by the sum of the radius of the sphere R_0 and the radius a of the polyelectrolyte segment wrapped on it ($R = R_0 + a$), and that its total number of unit charges Q is the sum of Q_0 positive charges of the sphere and l/b complementary polyelectrolyte charges $Q = Q(l) = Q_0 = l/b$. The optimal length l^* of rigid polyelectrolyte wrapped around the sphere can be obtained by searching the minimum of $F(l)$ or $\Delta F(l)$. Alternatively to l^*, the wrapping may be described by a bending angle around the sphere ($\theta^* = l^* / R$) or by the number of polyelectrolyte turns around the sphere ($n_t^* = l^* / 2\pi R$).

The normalized binding free energy, $\Delta F(l)/k_BT$ of Equation 8.6, for the polyelectrolyte–protein complex is given in Figure 8.13 as a function of the bent polymer length l (for simplicity here we use the DNA parameters, but the theory equally applies to strong polyelectrolytes). Spheres of charge $Q_0 = 10, 20, 100$ and radius $R_0 = 1.5, 5.0$ nm are considered at 0.1 M salt concentration. The minimum in the free energy indicates a wrapping of the polyelectrolyte around the protein at values l^* that increase with R_0 and Q_0. The value of l_{iso} indicated in Figure 8.13 corresponds to the isoelectric bending length:

$$l_{iso} = bQ_0, \tag{8.7}$$

or length of wrapped DNA for which the complex has a zero net charge. The difference $l^* - l_{iso}$ is a measure of the length of wrapped polyelectrolyte negatively overcharging the positive sphere. It is the release of positive counterions

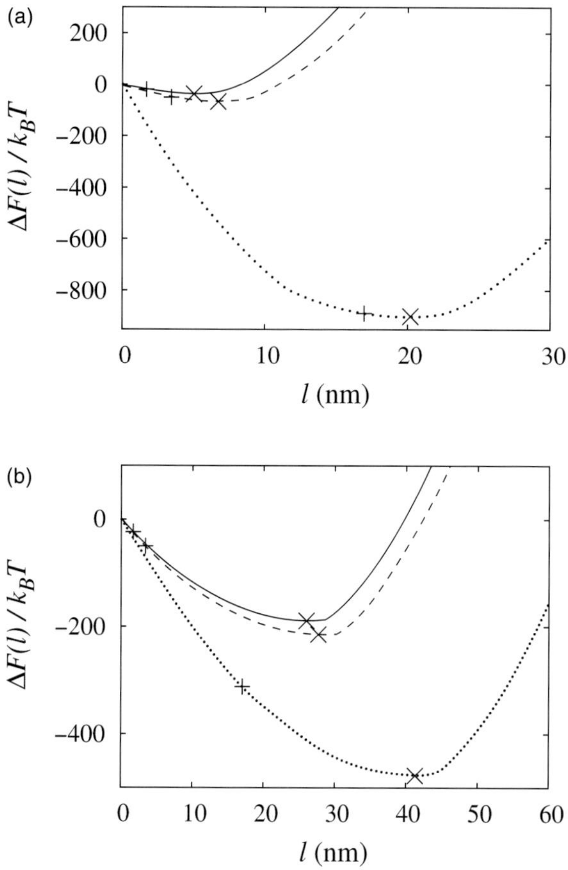

FIGURE 8.13. Binding free energy between a rigid polyelectrolyte (DNA) and a charged spherical particle having different values of the radius R_0 (panel a: $R_0 = 1.5$ nm; panel b: $R_0 = 5$ nm) and charge $Q_0 = 10$ (solid lines), 20 (dashed lines), and 100 (dotted lines) plotted versus the bent length of the rigid polymer. Salt concentration 0.1 M. The minima in the free energy at $l = l^*$ are indicated by X. The + sign indicates the free energy at $l = l_{iso}$ representing the length of the wrapped segment for the electrostatically balanced complex with zero net charge. From Arcesi et al.,[43] with permission.

from the polyelectrolyte as it bends around the sphere that is responsible for overcharging.

The length of the wrapped polyelectrolyte around the protein at equilibrium l^* is obtained by simply setting the derivative of the free energy equal to zero (cf. Chapter 3, Section 3.6.1, and definitions therein):

$$l^* = l_{iso} - (bR/l_B A)\{(2-\zeta^{-1})\log[1-\exp(-\kappa b)] + (1-\zeta^{-1})\}$$
$$- (bR/l_B A)[2K_0(\kappa R)/\kappa a K_1(\kappa a)] - l_p[b^2/2l_B A(R)R] - [b\zeta/8\kappa^2 R l_B A(R)]. \tag{8.8}$$

The value of l^* turns out to be due to the matching of the dependence upon l of two main terms in the free energy: (1) the decrease of electrostatic interactions and counterion condensation upon reducing the length of the straight polyelectrolyte and (2) the increase of the electrostatic free energy of the sphere (both in the charging and in the condensation regimes). The charging free energy of the sphere causes a very low free energy increase with l until the value of l^* is attained. As the sphere is greatly overcharged, counterion condensation on the sphere begins with a sudden increase of the free energy. The dependence of l^* on Q_0 in Equation 8.8 reflects the contribution of l_{iso} in Equation 8.7, and is therefore linear. On the other hand, the contribution of the bent polyelectrolyte over the value of l_{iso}, responsible for the overcharging of the sphere, is independent of Q_0, but strongly increases with R_0. The contributions to l^* of the polyelectrolyte elastic energy and of the direct polyelectrolyte–sphere interaction are small, but significant. From Equation 8.8, the role of the polyelectrolyte persistence length in the wrapping up of the sphere may also be appreciated. An increase of the persistence length corresponds to a linear decrease of the wrapped polyelectrolyte length.

In Figure 8.14, the overcharging degree $(l^* - l_{iso})/l_{iso}$ as a function of the polyelectrolyte persistence length l_p is reported for a sphere of radius 3 nm at $C_s = 0.1$ M and for a polyelectrolyte of linear charge density equal to that of DNA ($\zeta = 4.123$). The isoelectric point is at $l_p = 448$ nm, while there is no wrapping ($l^* = 0$) for l_p greater than 512, 575, and 1086 nm for $Q_0 = 10, 20$, and 100, respectively, because the polyelectrolyte is too rigid to wrap around the protein. It is noteworthy that these values are much larger than the DNA persistence length, confirming the DNA propensity to wrap around oppositely charged complexes, which is of such great importance in biology. In the limit of flexible polyelectrolyte (zero persistence length), the overcharging degree is maximum:

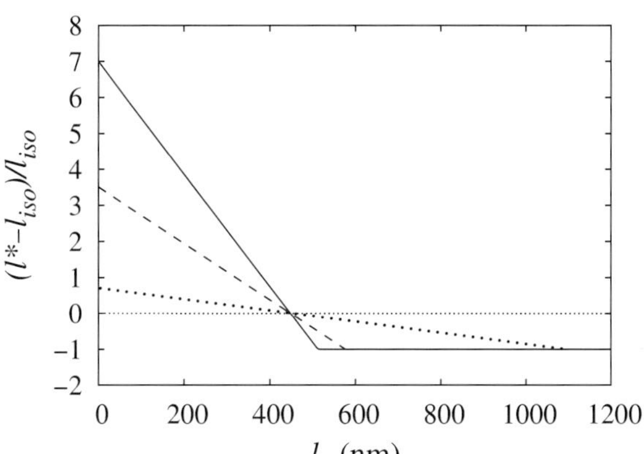

FIGURE 8.14. Degree of overcharging of a sphere of radius $R_0 = 3$ nm and charge $Q_0 = 10$ (solid lines), 20 (dashed lines), and 100 (dotted lines) as a function of the persistence length l_p of the polyelectrolyte (linear charge density $\zeta = 4.123$). Salt concentration 0.1 M. From Arcesi et al.,[43] reproduced with permission.

7.05, 3.52, and 0.71 nm for $Q = 10$, 20, and 100, respectively to be compared with the smaller values (6.23, 3.11, and 0.62 nm) in the DNA case.

8.5.1 Conclusions

The latter approach applied to the case of the DNA–histone octamer complex ($R_0 = 3.4$ nm, $Q_0 = 134$) yields $l^* = 36$ nm corresponding to 1.29 turns of the rigid DNA molecule around the protein complex. The figure favorably compares with experimental data indicating that DNA makes 1.58 turns around the protein complex. Considering the reported persistence length of DNA (Table 8.1), it is evident that a segment length about half the persistence length can make a full turn around the octamer. Therefore, a polymer segment with a length even smaller that its persistence length cannot be unambiguously described as a rigid rod. It will retain a residual flexibility that will be manifested under the action of the combined effect of charge compensation with the spherical protein and the associated release of mobile counterions. However, the essential role of the strength and stability of the complementary spherical complex is to be stressed. In fact, the case described in the preceding section (the same rigid DNA segment interacting with surfactants) evidences that, rather than inducing DNA folding, a spherical micelle (characterized by a weak cohesion and possible morphological rearrangements) was destabilized in favor of a cylindrical structure compatible with the rigid polyelectrolyte.

In the following section we shall consider the case of polyelectrolytes having a contour length much larger that the persistence length. Their worm-like conformation allows the possibility of mimicking the beads on a string assembly of the chromatin fiber.

8.6 WORM-LIKE CHAINS, SURFACTANTS, AND PROTEIN ASSEMBLIES

The so-called giant DNA is a macromolecule having a contour length order of magnitude larger than its persistence length. The prevailing conformation will be that of a worm-like chain, encroaching Gaussian statistic and the Kuhn chain.[29a] The condensed structures formed by interaction of cationic surfactants or protein assemblies with giant DNA may therefore be expected to differ from the structure formed by short, rod-like DNA having a contour length comparable to the persistence length, considered in the two preceding sections (cf. Figs. 8.11 and 8.12). These expectations are confirmed by the following experimental and theoretical analyses.

8.6.1 Giant DNA and Surfactants

Wang et al. investigated the association of a 48 kbp λ-DNA (contour length 16 μm) to the Gemini surfactant hexyl-α, ω-bisdodecyldimethylammonium

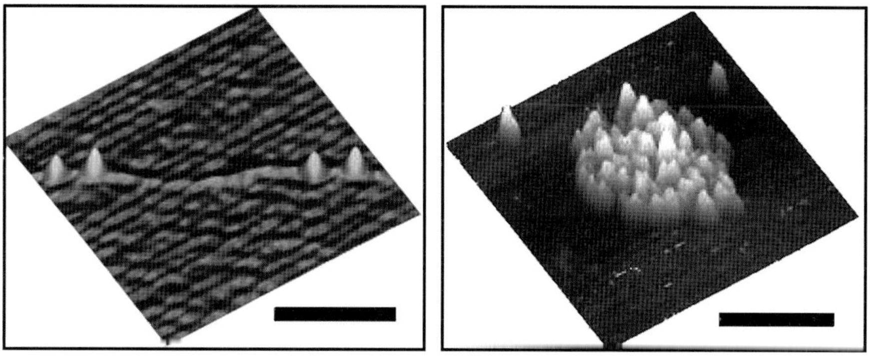

FIGURE 8.15. Morphologies obtained from AMF following 1 and 8 minutes interaction of giant DNA (48.5 bp) with 10 μm hexyl-α, ω-bisdodecyldimethylammonium bromide. Scale bar = 200 nm. Adapted from Wang et al.[66]

bromide having two aliphatic tails and two headgroups linked by a spacer.[66] Using atomic force microscopy (AFM) they followed the time evolution of the shape of the complexes. Figure 8.15 shows three-dimensional morphologies detected 1 and 8 minutes following the mixing of DNA and surfactant. A beads-on-a-string organization is observed. As time elapsed, the beads grow in number but not in size (their diameter was in order of 30 nm and the length of the DNA segment involved in one bead was in the order of 100 nm). Moreover, the beads appeared to approach each other, without coalescing, by stepwise inclusion of more DNA segments. The authors suggested that the driving force for the formation of the beads was the hydrophobic association of the tails of the surfactant electrostatically bound to DNA. The specific structure of the beads was not elucidated, but a directing role of the molecular structure of the surfactant was suggested, based on the observation that beads were not formed by Gemini surfactants with short spacer length, or by single-chained surfactants. With the latter, random globules and rings (toroids, cf. seq.) were instead noticed. The authors recognized that accommodating a 50 nm persistence length into a 30 nm bead implied a large curvature of the rigid DNA, and suggested a specific role of the double head of the Gemini surfactant in locally bending the polyelectrolyte.

Miyazawa et al. investigated the even larger giant T4 DNA (156 kbp, contour length 57 μm) in the presence of a similar dimeric (Gemini) surfactant.[67] AFM images reported in Figure 8.16 show partially folded and folded structures corresponding to increasing amounts of surfactant. The image on the left clearly reveals a ring-on-a-string structure with a ring diameter up to ~160 nm, corresponding to ~16 turns. The image on the right shows a network structure composed of many fused rings. Differences between the ring-on-a-string structure and the beads-on-a-string structure reported by Wang et al. were pointed out.[66] One important difference between these structures is the

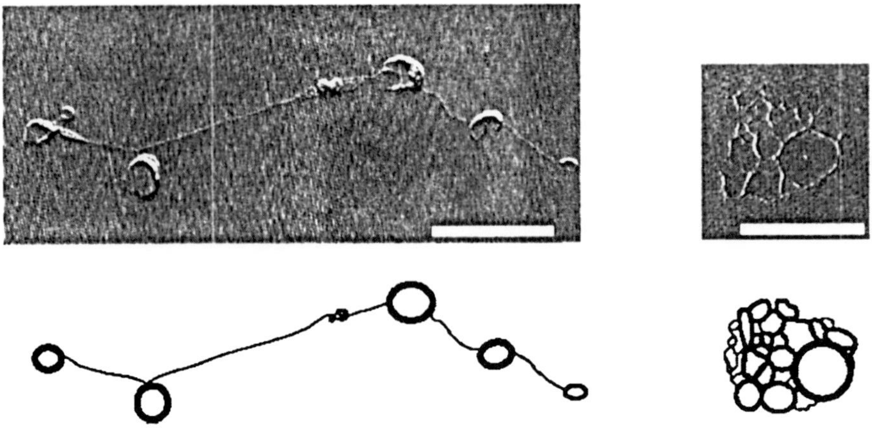

FIGURE 8.16. AMF images and their schematic representation for giant DNA (157 bp) and the Gemini surfactant propanediyl-1,3-bis(dimethyldodecylammonium bromide) 0.2 μm (left, partially condensed) and 1.0 μm (right, condensed). Scale bar = 1 μm. Adapted from Miyazawa et al.[67]

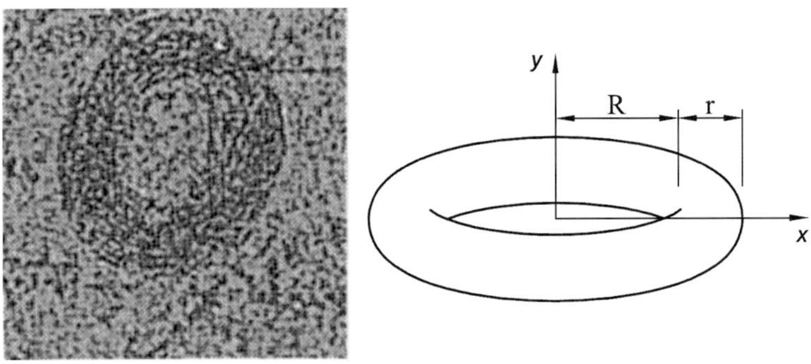

FIGURE 8.17. The toroidal DNA condensate of the genome of the λ bacteriophage. Electron micrograph by kind permission of Dr. J.-L. Sikorav.

toroidal appearance of the rings illustrated in Figure 8.16. Another difference is the size of the beads in Figure 8.15, which is much smaller than the size of the rings in Figure 8.16.

8.6.2 Toroids

In toroidal assemblies, the polyelectrolyte is circumferentially wound, making multiple rings even in the absence of surfactants (cf. Chapter 6 and Miyazawa et al.[67–69]). Figure 8.17 includes an electron micrograph and a schematization of the toroidal parameters: the major radius R and the thickness r.[69] The occurrence of toroids was observed even for single uncharged chains of suitable contour length and rigidity in poor solvents. Their formation has been described as a phase transition from a disordered to an ordered state.[69c] The detailed morphology of toroids is sensitive to variation in persistence length and surface energy. Their formation in polyelectrolytes has been attributed to a balance between the electrostatic repulsion and the surface energy, thus controlling the thickness of the ring. There is an analogy with the Raleigh-type instability that has been advocated to interpret the formation and the size of spherical mini globules along a single charged chain observed with polyampholytes and hydrophobic polyelectrolytes even in the absence of surfactants (cf. Chapter 6).[70] Electrostatic repulsion limits the size of the mini globule in the latter case and the thickness of the toroid in the present case. When a surfactant is present, the thickness is almost constant and independent of surfactant concentration. Effects similar to those limiting the growth of the hexagonal structures described in Section 8.4 are possibly operative.

The size of rings was calculated by a minimization of the energy loss associated with bending (bending energy is part of the total energy of a ring that also includes surface and cohesion energies). Theoretical studies suggest diameters in the order of 70–140 nm for DNA toroids (much larger than the diameters of 30 nm reported for the beads in Fig. 8.15).[66,68] The preference for the conformation of several rings spaced by suitable strings (as opposed to a single ring) observed with DNA was attributed to a favorable contribution of the entropy for distributing rings and strings along a one-dimensional array.[67] A theory including like-charge attraction (in the absence of surfactants), and the associated entropy gain due to the release of mobile counterions (cf. Chapters 3 and 6), was outlined by Gelbart et al.[68a]

8.6.3 Giant DNA and Protein Assemblies

The complex between the histone protein and a short DNA segment described in Section 8.5 can be regarded as a model for the nucleosome, the repeating unit of polynucleosomes and chromatin fibers (cf. Fig. 8.3). The former may be described as a chain of beads on a string where the nucleosomes are beads connected by DNA linkers having a length between 10 and 90 bp. The latter is based on the super-helical arrangement of the polynucleosome with partly stacked octamers. Due to the cohesion of the histone octamer, its association to giant DNA is to some extent simpler to handle than the interaction of the latter with surfactants. In fact, the forces controlling the folding of DNA around the octamer are well established (cf. Figs. 8.12 and 8.13), and the structure of the latter is not fundamentally altered by the polyelectrolyte, as is the case with surfactants. At the *in vivo* level, the condensation of genetic information in the suprastructure of chromatin does, however, require a variety of remodeling factors that locally unwrap DNA to allow for the occurrence of transcription, replication, and repair mechanisms.[5]

The assessment of the local bending of the DNA chain is based on the calculations embodied in Equation 8.6, corresponding to the schematization given in Figure 8.12. In general, given a polyelectrolyte interacting with an oppositely charged sphere, there is a critical value of the persistence length $(l_p)_{iso}$ (isoelectric persistence length, function of the linear charge density on the polyelectrolyte and of the radius of the sphere, but independent on the surface charge of the latter), at which the length of the polyelectrolyte segment wrapped around the sphere equals the segment length necessary to compensate the surface charge of the sphere. For instance, $(l_p)_{iso}$ turns out to be 448 nm (see Fig. 8.14) for a linear charge density equal to that of DNA ($\zeta = 4.123$) and a sphere of radius $R_0 = 3$ nm. The latter $(l_p)_{iso}$ value is much larger than the actual DNA persistence length (50 nm). Note that for $l_p > (l_p)_{iso}$, the wrapped length is always lower than l_{iso}: the polyelectrolyte stiffness prevents the overcharging of the sphere. Upon further increase of stiffness, a limit value is reached the higher the surface charge of the sphere, above which the polyelectrolyte can no longer wrap around the sphere. For the above example, the largest values of the persistence length allowing wrapping are 512, 575, and 1086 at $Q_0 = 10$, 20, and 100, respectively. This demonstrates the effect of varying the rigidity of the polyelectrolyte and the role of the shape and charge of the complementary macroion in Equation 8.8 and Figure 8.14.

8.6.4 Conclusions

The reciprocal stresses posed on the conformation of a worm-like polyelectrolyte and on the structural organization of the oppositely charged component are less severe than observed for rigid polyelectrolytes interacting with complementary surfactants or protein assemblies. Thus, the occurrence of considerable flexibility allows worm-like chains to fold around oppositely charged micellar-like aggregates, a possibility that was manifested by flexible chains (Section 8.3) but not by the rigid ones (Section 8.4). Furthermore, worm-like polyelectrolytes bound to undeformable macroions allow the possibility of additional structurization due to the direct interaction of protein complexes along the polymer chain.

In the case of the beads-on-a-string organization formed by Gemini surfactants, it has not been simple to elucidate the detailed organization of the surfactant within the beads. The basic electrostatic attraction of the complementary charges and the basic hydrophobic attraction of the apolar tails are always operative, but the detailed arrangement of the components appears sensitive to details of the molecular structure of the particular surfactant.[66]

Very long polyelectrolyte chains were considered, and therefore the occurrence of rings-on-a-string conformations is not surprising. In fact, toroids occur even in the absence of the surfactant, in which case concentric rings are stabilized by like-charge attractive interactions (cf. e.g., Chapters 3, 6, and Gelbart et al.[68]). In the presence of surfactant, like-charge interaction is likely replaced by the hydrophobic interaction (i.e., interdigitation) of the apolar surfactant tails coating the polyelectrolyte chain. In either case (with or without surfactant), the formation of ring is favored by minimization of the bending energy of the polyelectrolyte and by entropy consideration.[67]

In Chapter 3 (Section 3.6), the wrapping mechanism of DNA around protein is extended to complexes in cases of great biological interest such as the nucleosome in the organization of chromatin in eukaryotes, the remodeling factors, the Lac operon in the regulation of the genetic information, and the chromosomal or histone-like proteins in DNA condensation in prokaryotes. All these examples show that the DNA wrapping length l^* is consistently increasing with the radius and with the surface charge of spherical macroions. These examples of the functioning of the incredible genetic machinery confirm the great attitude of DNA to interact (wrap/unwrap) extensively with proteins to manage the enormous library of genetic information.

8.7 SUMMARY AND CONCLUDING REMARKS

We reported data for systems displaying the association of polyelectrolytes to complementary charged surfactants, micelles, and protein assemblies. In all cases, electrostatic and hydrophobic interactions are the main driving forces for the formation of complexes. An instability, involving the competition of electrostatic and surface energies, favors conformations based on mini globules or rings along a single charged chain. However, the detailed structure of the complexes is controlled by molecular parameters of the components, such as the conformational rigidity of the polyelectrolyte, and the shape and cohesion of the complementary moiety. Furthermore, a mutual interaction between these parameters occurs. For instance, a rigid polyelectrolyte induces morphological rearrangement in complementarily charged assembly held by weak cohesion forces (i.e., spherical micelles), but an assembly with a strong cohesive energy (i.e., the nucleosome) evidences a latent flexibility in a rigid polyelectrolyte.

To better evidence and classify the role and the interrelation of these variables, we organized the data in different sections dealing with flexible polyelectrolytes and surfactants, rigid polyelectrolytes and surfactants, rigid polyelectrolytes and protein assemblies, and worm-like polyelectrolytes with surfactant or proteins assemblies. A great variety of structures is described, and theoretical approaches to their interpretation are highlighted. We emphasize the relevance of the underlying mechanisms to the science of supramolecular organization, and to the functioning and mimicking of biological systems.

ACKNOWLEDGMENTS

We express our great appreciation to the contributors of the book for stimulating discussions, and to Dr. Cecilia Leal for preparing the images in Figure 8.11.

REFERENCES

1. S. Zhou and B. Chu. *Adv Mater* **2000**, 12, 545.
2. P. S. Cheluskin, E. A. Lysenko, T. K. Bronich, A. Eisenbergh, A. V. Kabanov, V. A. Kabanov. *Docklady Phys Chem* **2004**, 395, 72–75.
3. J. C. T. Kwak, ed. *Polymer-Surfactant Systems*. Marcel Dekker, New York, 1998.
4. (a) S. V. Vinogradov, E. V. Batrakova, A. V. Kabanov. *J Drug Target* **2004**, 12, 517–526; (b) P. L. Felgner and G. M. Ringold. *Nature* **1989**, 337, 387; (c) Y. Wang, S. Gao, W. H. Ye, H. S. Yoon, Y. Y. Yang. *Nat Mater* **2006**, 5, 791.
5. J. Beerens, J. H. J. Hoeijmakers, R. Kanaar, W. Vermuelen, C. J. Wyman. *J Biol Chem* **2005**, 280, 4772.
6. X. Arys, A. M. Jonas, A. Laschewsky, R. Legras, F. Malwitz. *Supramolecular Polymers*, 2nd ed., A. Ciferri, ed. CRC Press, Boca Raton, FL, 2005.
7. *Interactions of Surfactants with Polymers and Proteins*, CRC Press, London, 1993.
8. A. Ciferri and S. Kudaibergenov. *Macromol Rapid Commun* **2007**, 29, 1953, 1962.
9. K. Ober and G. Wegner. *Adv Mater* **1997**, 9, 17.
10. D. Langevin. *Adv Colloid Interface Sci* **2009**, 147–148, 170.
11. K. Shrihama. *Polymer-Surfactant Systems*, J. C. T. Kwak, ed. Marcel Dekker, New York, 1998.
12. P. Linse, L. Piculell, P. Hanson. *Polymer-Surfactant Systems*, J. C. T. Kwak, ed. Marcel Dekker, New York, 1998.
13. T. Wallin and P. Linse. (a) *Langmuir* **1995**, 12, 305; (b) *J Phys Chem* **1996**, 100, 17873; (c) *J Phys Chem B* **1967**, 101, 5506; (d) *Langmuir* **1998**, 14, 2940; (e) *J Chem Phys* **1998**, 109, 5089.
14. (a) H. Diamant and D. Andelman. *Macromolecules* **2000**, 33, 8050; (b) R. Zana. *Polymer-Surfactant Systems*, J. C. T. Kwak, ed. Marcel Dekker, New York, 1998; (c) T. Nylander, P. Kicheff, B. W. Ninha. *J Colloid Interface Sci* **1994**, 164, 136.
15. R. Nagarajan. *J Chem Phys* **1989**, 90, 1980.
16. (a) R. S. Diaz and B. Lindman. *DNA Interactions with Polymers and Surfactants*. Wiley, New York, 2008; (b) S. Q. Zhou, D. H. Liang, C. Burger, F. J. Yeh, B. Chu. *Biomacromolecules* **2004**, 5, 1256.
17. A. Perico and A. Ciferri. *Chem Eur J* **2009**, 15, 6312.
18. J. Widom. *Annu Rev Biophys Biomol Struct* **1988**, 27, 285.
19. H. Schiessel. *J Phys Condens Matter* **2003**, 15, R699.
20. T. Odijk. *J Polym Sci* **1977**, 15, 477.
21. P. J. Flory. *Statistical Mechanics of Chain Molecules*. Wiley, New York, 1969.
22. R. Ramachandran, G. Beucage, A. S. Kulkarni. *Macromolecules* **2008**, 41, 9802.
23. G. Beaucage, S. Rane, S. Sukumaran. *Macromolecules* **1997**, 30, 4158.
24. E. Bianchi, A. Ciferri, G. Conio, L. Lanzavecchia, M. Terbojevich. *Macromolecules* **1986**, 19, 430.
25. M. Kamide, M. Saito, T. Norisuye. *Polym J* **1987**, 19, 1173.
26. M. Mandel and J. Schouten. *Macromolecules* **1980**, 13, 1247.
27. Q. Ying and B. Chu. *Macromolecules* **1987**, 20, 871.
28. F. Gittess, B. Mickey, J. Nettleton, J. Howard. *J Cell Biol* **1993**, 120, 923.

29. (a) A. Ciferri. *Supramolecular Polymers*, 2nd ed., A. Ciferri, ed. CRC Press, Boca Raton, FL, 2005; (b) A. Ciferri. *Liq Cryst* **2004**, 31, 1487; (c) A. R. Khokhlov. *Liquid Crystallinty in Polymers*, A. Ciferri, ed. VCH Publishers, New York, 1991.
30. A. Y. Grosberg and A. R. Khokhlov. *Giant Molecules*. Academic Press, San Diego, CA, 1977.
31. J. N. Israelachvili. *Intermolecolar and Surface Forces*. Academic Press, London, 1992.
32. (a) Y. Moroy. *Micelles: Theoretical and Applied Aspects*. Plenum, New York, 1992; (b) E. D. Goddart. *Colloid Surf* **1986**, 19, 301.
33. R. G. Allargova, I. I. Kochijashky, M. L. Sierra, R. Zana. *Langmuir* **1998**, 14, 5412.
34. A. Halperin, M. Tirrel, T. P. Lodge. *Adv Polym Sci* **1991**, 100, 31.
35. (a) F. Mackintosh. *Europhys Lett* **1990**, 12, 697; (b) P. Linse. *J Phys Chem* **1993**, 97, 13896.
36. (a) R. Nagarajan. *J Chem Phys* **1989**, 90, 1980; (b) R. Nagarajan and R. Ganesh. *J Chem Phys* **1989**, 90, 5843; (c) L. Mailbaum, A. R. Dinner, D. Chandler. *J Phys Chem B* **2004**, 108, 677.
37. P. van der Schoot. *Supramolecular Polymers*, 2nd ed., A. Ciferri, ed. CRC Press, Boca Raton, FL, 2005; P. van der Schoot, A. P. Schenning, E. W. Meiyer. *Science* **2006**, 313, 80.
38. (a) B. M. Schick, S. M. Atlas, F. R. Eirich. *J Phys Chem* **1962**, 66, 1326; (b) H. G. Thomas, A. Lomakin, D. Blankshtein, G. B. Benedeck. *Langmuir* **1997**, 13, 209; (c) A. Ben-Shaul, D. H. Rorman, G. V. Hartland, W. M. Gelbart. *J Phys Chem* **1986**, 90, 5286.
39. A. Ciferri, ed. *Liquid Crystallinity in Polymers*. VCH, New York, 1991.
40. B. Alberts, D. Bray, M. Raff, K. Roberts, J. D. Watson. *Molecular Biology of the Cell*, 3rd ed. Garland Publishing, New York, 1955; p. 343.
41. (a) A. Davey, D. F. Sargent, K. Luger, A. W. Maeder, T. J. Richmond. *J Mol Biol* **2002**, 319, 1097; (b) K. Luger, A. W. Maeder, L. K. Richmond, D. F. Sargent, T. J. Richmond. *Nature* **1997**, 389, 251.
42. J. Richmond and C. A. Davey. *Nature* **2002**, 415, 145.
43. L. Arcesi, G. La Penna, A. Perico. *Biopolymers* **2007**, 86, 127.
44. G. S. Manning. *J Am Chem Soc* **2003**, 125, 15087.
45. A. Ruiz Carrillo, J. L. Jorcano, G. Elder, R. Lurz. *Proc Natl Acad Sci U S A* **1979**, 76, 32840.
46. (a) K. Shirahama, H. Fuasa, S. Sugimoto. *Bull Chem Soc Jpn* **1981**, 54, 375; (b) M. Henriquez, E. Lissi, E. Abuin, A. Ciferri. *Macromolecules* **1994**, 27, 6834.
47. B. Zimm and J. K. Bragg. *J Chem Phys* **1959**, 31, 526; G. Schwarz. *Eur J Biochem* **1970**, 12, 442.
48. K. Tailberg and B. Lindman. *J Phys Chem* **1989**, 93, 1478.
49. Y. C. Wei and S. M. Hudson. *Macromolecules* **1993**, 26, 4151.
50. R. Zana. *Polymer-Surfactant Systems*. J. C. T. Kwak, ed. Marcel Dekker, New York, 1998.
51. (a) O. Anthony and R. Zana. *Langmuir* **1996**, 12, 3590; (b) W. Binana-Limbele and R. Zana. *Macromolecules* **1990**, 23, 2731.
52. P. Hansson and M. Almgen. *J Phys Chem* **1995**, 99, 16684.

53. X. H. Guo, M. Zhao, S. H. Chen, J. Teixeira. *Biopolymers* **1990**, 29, 335.
54. E. A. Bekturov, S. E. Kudaibergenov, G. S. Kanapyonova. *Polym Bull* **1984**, 11, 551.
55. A. Ciferri. *Macromol Phys Chem* **1994**, 195, 457.
56. K. Shirahama, K. Taknashima, N. Tasikawa. *Bull Chem Soc Jpn* **1987**, 60, 43.
57. I. Koltover, T. Talditt, J. O. Radler, C. R. Safinya. *Science* **1998**, 281, 78.
58. D. M. McLoughin, M. Delsanti, C. Tribet, D. Langevin. *Europhys Lett* **2005**, 69, 461.
59. D. F. Evans and H. Wennerstrom. *The Colloidal Domain*, 2nd ed. Wiley-VCH, New York, 1999.
60. K. K. Evert, H. M. Evans, A. Zidovska, F. N. Bouxein, A. Amad, C. R. Safinya. *J Am Chem Soc* **2006**, 128, 3998.
61. K. Thalberg and B. Lindman. *Surfactants in Solution*, Vol. 11, K. L. Mittal, D. Shash, eds. Plenum Press, New York, 1991.
62. P. S. Khun, Y. Levin, M. C. Barbosa. *Physica A* **1999**, 274, 8.
63. (a) A. Ciferri. *Liq Cryst* **2007**, 34, 693; (b) A. Ciferri. *Prog Polym Sci* **1995**, 20, 1081.
64. M. Antonietti, C. Burger, A. Thunemann. *Trends Polym Sci* **1997**, 5, 262.
65. T. E. Strzelecka, M. W. Davidson, R. L. Rill. *Nature* **1988**, 331, 457.
66. X.-L. Wang, X.-H. Zhang, M. Cao, H.-Z. Zheng, B. Xiao, Y. Wang. *J Phys Chem B* **2009**, 113, 2328.
67. N. Miyazawa, T. Sakaue, K. Yoshikawa, R. Zana. *J Chem Phys* **2005**, 122, 044902.
68. (a) W. M. Gelbart, R. F. Bruinsma, P. A. Pincus, V. Parsegian. *Phys Today* **2000**, 53, 38; (b) V. A. Bloomfield. *Curr Opin Struct Biol* **1996**, 6, 334; (c) N. V. Hud and K. H. Downing. *Proc Natl Acad Sci U S A* **2001**, 98, 14925; (d) W. B. Fu, X. L. Wang, X. H. Zhang, S. Y. Ran, M. Li. *J Am Chem Soc* **2006**, 128, 15040.
69. (a) T. Sakaue. *J Chem Phys* **2004**, 120, 4004, 6299; (b) M. R. Stukan, V. A. Ivanov, A. Y. Grosberg, W. Paul, K. Binder. *J Chem Phys* **2003**, 118, 3392; (c) K. Yoshinawa, M. Takahashi, V. V. Vasilevskaya, A. R. Khokhlov. *Phys Rev Lett* **1996**, 76, 3029.
70. (a) Y. Kantor and M. Kardar. *Phys Rev E* **1995**, 51, 1299; (b) A. V. Dobrynin and M. Rubinstein. *Macromolecules* **1999**, 32, 915; ibid. **2000**, 33, 8097.

CHAPTER 9

ION TRANSFER IN AND THROUGH CHARGED MEMBRANES: STRUCTURE, PROPERTIES, AND THEORY

VICTOR V. NIKONENKO, ANDREY B. YAROSLAVTSEV, AND GÉRALD POURCELLY

9.1 INTRODUCTION

The main aim of this chapter is to give a state-of-the-art view on the structure and transport phenomena in charged artificial membranes. This class of materials includes ion exchange, reverse osmosis (RO), and nanofiltration (NF) membranes. All these membranes are largely used in separation processes; the main applications are described in Chapter 20 of this book.

Over the past few years, a remarkable progress in the preparation of membrane materials and in the understanding of their functioning has been made. The significant extent of this progress is due to a relatively novel approach to membranes, regarded as nanomaterials constructed from macromolecules having properties determined by their structure at the nanometer scale. A great number of membranes are obtained by self-assembly strategies allowing the formation of supramolecular structure.[1–4] It is remarkable that the structure and, hence, the properties of biological and artificial membranes have considerable similarities.[1] In both cases, the main elements of the structure are amphiphilic compounds having both hydrophilic and hydrophobic components. The mutual interaction of hydrophobic segments and the interaction of hydrophilic polar groups with aqueous medium result in self-organized stable structures. In particular, nanoscale hydrophilic channels selectively permeable to water and ions are formed within a hydrophobic matrix. In the case of perfluorinated sulfonated ion exchange membranes (IEM) of the Nafion® type

Ionic Interactions in Natural and Synthetic Macromolecules, First Edition. Edited by Alberto Ciferri and Angelo Perico.

(DuPont, Wilmington, DE), the basic structural element has similarities with an inverse micellar morphology. The negative charges of sulfonated groups fixed on the perfluorinated matrix favor cation enrichment within the membrane inner pore solution, so that mainly cations play the role of carriers within the channels transpiercing the membrane body.

In this chapter we try to describe how the structure of charged membranes conditions their properties. Basic approaches to the description of ion and water transfer in such membranes are given. Additionally, we consider surface phenomena accompanying the electric current transfer across IEMs. Concentration polarization and related surface phenomena such as "water splitting" and current-induced convection are described. The mechanisms of these phenomena are analyzed in relation with the properties of a thin surface layer of the membrane. Novel methods of surface modification and the effect of such modification on the overall membrane properties and electrochemical behavior are presented.

9.2 STRUCTURE OF CHARGED MEMBRANES

9.2.1 Biological Membranes: Self-Organizing Systems

A cell membrane is an enclosing film that acts as a selective barrier, within or around the cell.[2,3] Cell membranes control the movement of substances in and out of cells.

According to the fluid mosaic model of Singer and Nicolson,[4] the cell membrane is a complex structure made up of many different components, such as proteins, phospholipids, and cholesterol. The relative amounts of these components vary from membrane to membrane, and the type of lipids can also vary. Amphiphilic phospholipids constitute the major component of the cell membrane. They consist of a phosphate "head" (circles in Fig. 9.1) and a lipid "tail" (lines) that are, respectively, compatible and incompatible with water.

A lipid bilayer is spontaneously formed when the lipid is exposed to water: the hydrophobic "tails" are pointed toward the center of the sheet while the hydrophilic "heads" are exposed to water.

The plane geometry is not the only possible one. In aqueous solution, phospholipids can form two spherical structures as well: the liposome (a closed bilayer), and the micelle (Fig. 9.2).[5,6]

The interior of the bilayer contains almost no water and also excludes nonelectrolyte molecules or salts that dissolve in water but not in oil. The passive transport of water, small noncharged molecules, or ions is facilitated by transmembrane protein channels, which are watery pathways embedded into the lipid bilayer[7,8] and are permanently attached to the membrane. More typically, a cylindrical arrangement of identical or homologous proteins closely packed around a water-filled pore through the plane of the lipid bilayer is observed.[9] Active transport requires additional transmembrane proteins, the transporters, that by complex biochemical reactions and the energy of

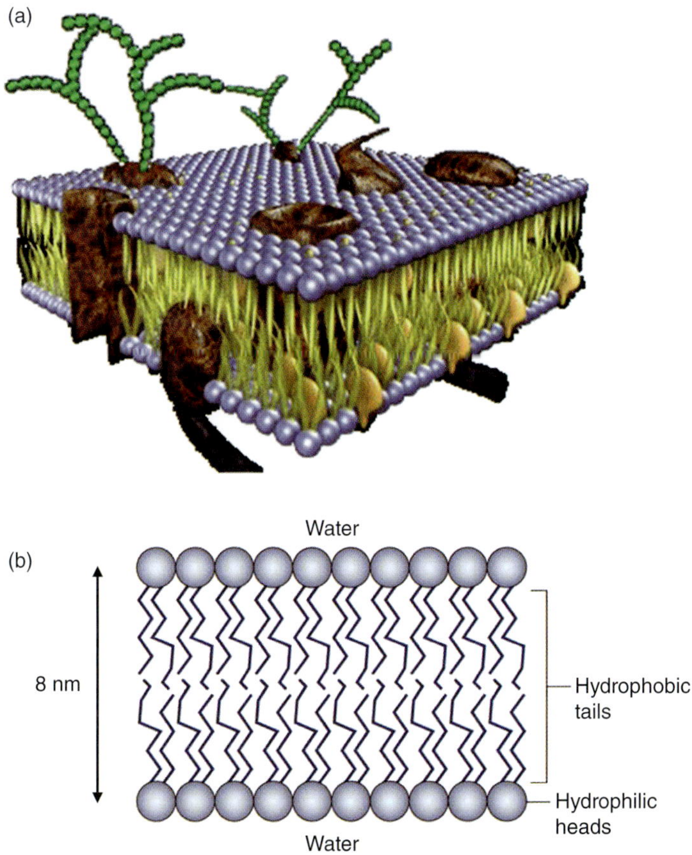

FIGURE 9.1. The structure of a cell membrane (a) and of its main fragment, a phospholipid bilayer (b).

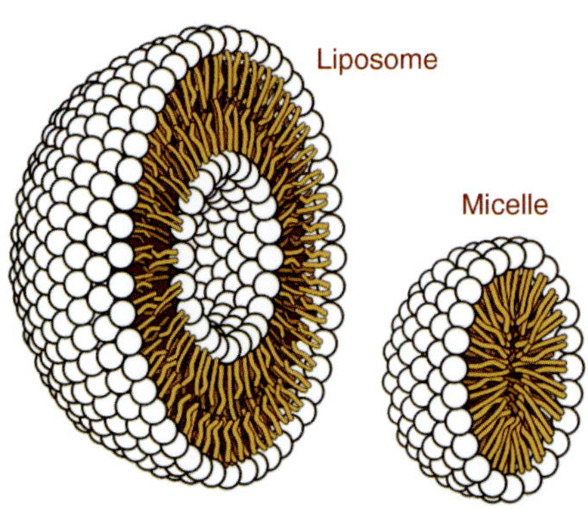

FIGURE 9.2. Spherical structures, which can be formed by phospholipids in solution: the liposome and the micelle. The picture is adapted from Garrett and Grisham.[280]

adenosine triphosphate (ATP) are able to force ions or small molecules through the membrane against their concentration gradient.

This brief description of the cell membranes structure and transport serves to point out similarities and differences between the structure and functionality of biological and synthetic membranes.[1,6] This book contains other chapters treating the interactions and transport in living systems: the protein structure is considered in Chapters 8 and 12, and the iron transport in living cells is described in Chapter 13.

9.2.2 IEMs

Types of Membranes Synthetic IEMs are constructed using flexible polymer chains having aliphatic, aromatic, or perfluorinated residues containing functional groups such as $-COOH$, $-SO_3H$, $-NH_3OH$, and others. Protons or OH^- ions of these groups may be replaced by cations or anions, respectively, present in solutions contacting the membrane. A large number of membrane materials is currently available.[10–12]

The diversity of practical applications of membranes determines a variety of requirements expected from them and, ultimately, promotes the development of a wide range of membrane materials. Generally, IEMs are classified following the charge of the exchanged ions.[13] Thus, the cation exchange membrane can exchange cations associated with the functional groups. For example, in the case of sulfonic acid group, the following reaction occurs:

$$\text{R-SO}_3\text{H} + \text{Na}^+(\text{solution}) \Leftrightarrow \text{R-SO}_3\text{Na} + \text{H}^+(\text{solution}), \tag{9.1}$$

where R is the matrix of membrane. In turn, anion exchange membranes can exchange anions, for example:

$$\text{R-NH}_3\text{OH} + \text{Cl}^-(\text{solution}) \Leftrightarrow \text{R-NH}_3\text{Cl} + \text{OH}^-(\text{solution}). \tag{9.2}$$

Depending upon material structure and the method of preparation, IEMs can generally be divided into two groups: homogeneous and heterogeneous membranes. Homogeneous membranes are obtained by copolymerization of monomers that produces homogeneous material (Fig. 9.3a). Heterogeneous membranes include macroparticles (size from 1 to 50 μm) of different polymer materials; for example, Russian MA-40 (SchekinoAzote, Schekino, Russia) anion exchange membranes include particles of polystyrene–divinylbenzene anion exchange resin and polyethylene (Fig. 9.3b). A more detailed classification, proposed by Molau, includes, in addition to homogeneous and heterogeneous types, interpolymer membranes, microheterogeneous graft and block polymer membranes, and snake-in-the-cage membranes.[14]

A membrane includes one or several layered films. Monopolar membranes have one-film geometry. Bipolar membranes are composed of two layers, a cation and an anion exchange one.[10,11,15–18]

(a)
(b)
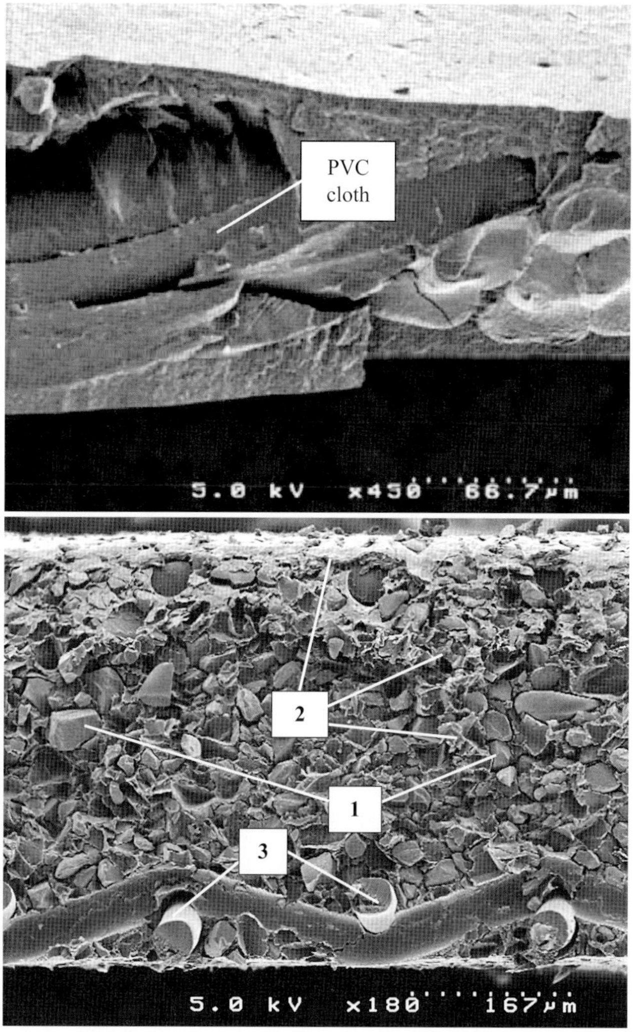

FIGURE 9.3. Electronic microphotographies of IEM. (a) Surface and cross-section of a homogeneous AMX (Tokuyama Soda Co.) anion exchange membrane, a copolymer of styrene, and divinylbenzene[281]; (b) cross-section of a heterogeneous MA-40 anion exchange membrane. 1, ion exchange resin particles; 2, polyethylene; 3, reinforcing cloth.

Supramolecular Interactions and Structuring in Dry State: Phase Separation The main structural feature of IEMs is the hydrophobic backbone of hydrocarbon or perfluorinated chains that antagonizes the hydrophilic character of the functional groups. Together with the flexibility of the polymer chains, this combination leads to self-organization during the formation of membranes.[19,20] With a proper distribution of the hydrophobic and hydrophilic

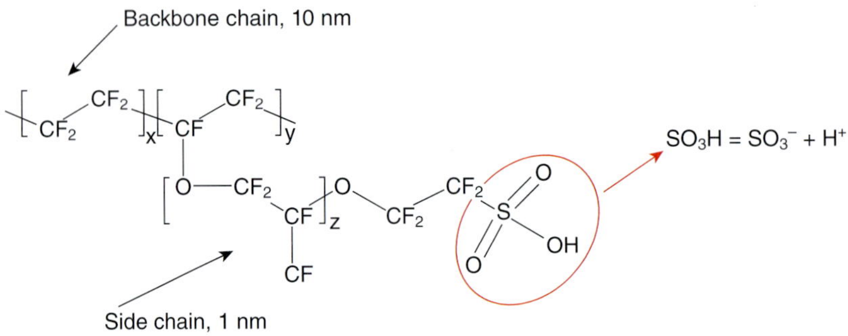

FIGURE 9.4. A perfluorinated sulfonated polymer (supra)molecule. The $-SO_3H$ group is susceptible to dissociate in hydrated state.

components, hydrocarbon or perfluorinated chains act as the matrix of the membrane, while the functional residues are aggregated in small pockets or clusters. The size of these clusters may be in the order of a few nanometers. The hydrophobic component thus contributes to the morphological stability of the membrane, while the hydrophilic domains may provide a system of conducting channels.[20,21]

Perfluorosulfonated acid (PFSA) membranes, in comparison to other types of IEMs, have been studied in more detail due to their applications in *chlor-alkali* production and in fuel cells. Hence, we will focus mainly on this type of membranes, using them as example in order to elucidate their structure–property relation. The chemical structure of the perfluorinated polymer of Nafion membranes is shown in Figure 9.4. The polymer backbone has side chains that carry the sulfonic (or carboxylic) cation exchange groups.[20] The most important feature of the structure in Figure 9.4 is the occurrence of rather long side chains that decouple the motion of the backbone and the charged groups. It is this feature that will facilitate the formation of hydrophilic conducting channels.

Eisenberg[22] and subsequently other authors[20,23–25] suggested that due to dipole–dipole interaction, ion pairs (formed by a fixed ion and a counterion, initially H^+) combine in multiplets able to exclude segments of the hydrophobic backbone (Fig. 9.5). Steric hindrance limits the number of ion pairs in a multiplet. If the number of fixed charges is sufficiently large, and the polymer matrix is sufficiently hydrophobic and elastic, the occurrence of multiplets in larger clusters is favored. The sum of the energies of hydrophobic interactions between backbone segments and of electrostatic attraction between dipoles prevails over the conformational energy of the polymer matrix[25] (Fig. 9.5).

Structure Evolution with Swelling The pores of IEMs containing hydrophilic functional groups are prone to hydration. Even in the dry state, equilibrated with the atmospheric air, a sulfonic acid group in Nafion links two water

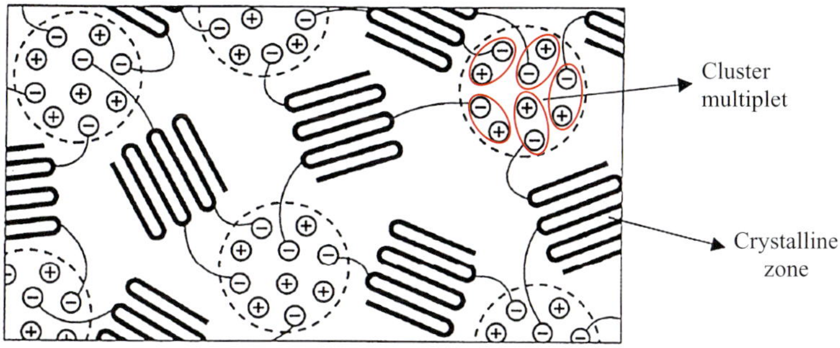

FIGURE 9.5 Structuring of perfluorinated membranes in dry state. Adapted from Shiryaeva and Victorov.[25]

molecules.[21] When an IEM is placed into a water solution, water is adsorbed into the hydrophilic clusters seeking to form ion hydration shells. As a result, the size of clusters increases, and at a certain water content, they form a percolation system.[26–28]

The details of morphology of ion exchange, in particular that of perfluorosulfonic membranes, has been the subject of a great number of publications.[1,20,24,29–32] The main experimental technique applied is small angle scattering with neutrons (SANS) or X-rays (SAXS).[26,29,32,33] The basic model of Nafion morphology, first proposed by Gierke,[23,29] adequately describes its main features and explains the transport properties. The membrane is regarded as a system of hydrated clusters with a diameter of about 4 nm connected by channels with 1 nm in diameter and length. Following Gierke, the side chain charges are located on the periphery of the cluster, which has spherical form in first approximation. This form minimizes the surface energy by reducing the contact of hydrophobic polymer chains with aqueous solution. Thus, the reverse micellar-like structure can be attributed to the membrane: more or less connected spherical aqueous domains are embedded in the hydrophobic polymer matrix (see Fig. 9.6 for low water contents).

The solution filling the inner part of the clusters is charged; it contains mainly counterions produced by the dissociation of functional groups. They neutralize the charge of the side chain charges, forming with them an electrical double layer (EDL). A pore contains a small quantity of co-ions as well, which are ions carrying a charge of the same sign as the fixed ions. The co-ions enter the membrane and are present there, mainly in the pore center zone, under the action of their concentration gradient created by higher concentration of these ions in the outer solution. However, their concentration in the pore solution is generally much lower than that of counterions due to the presence of fixed ions. This effect of ionic disproportion (of uneven ionic distribution) occurring in the solution outside and inside the membrane is called the Donnan effect (see Chapters 9 and 17). Just that causes selective uptake and selective transport (permselectivity) of counterions in IEMs. Figure 9.7 shows

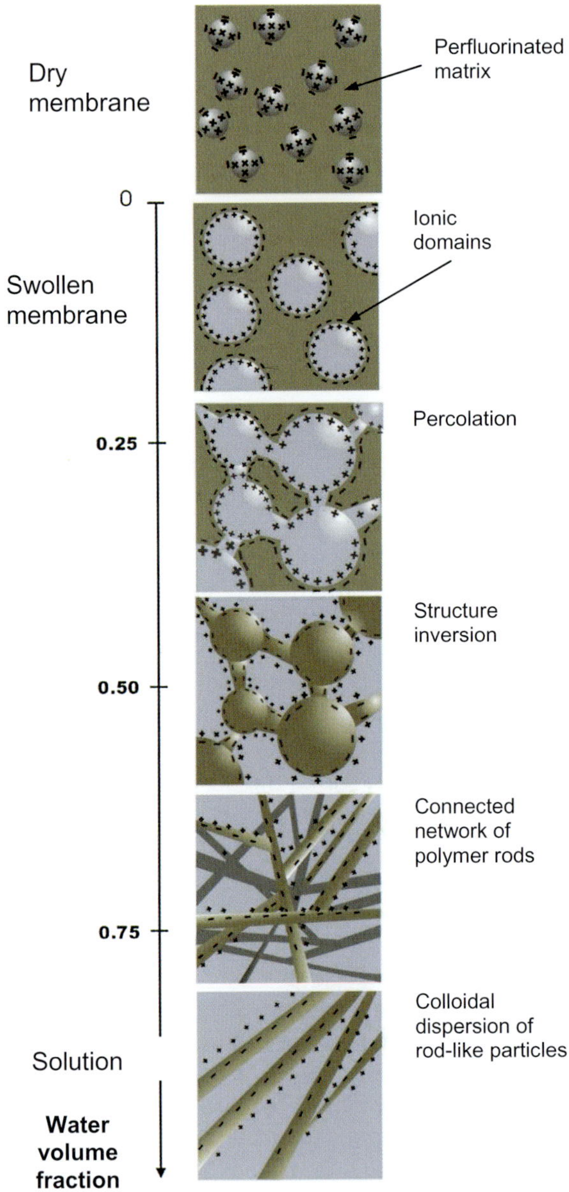

FIGURE 9.6. Evolution of perfluorinated IEM nanostructure with increasing water content. Adapted from Gebel.[26]

schematically the distribution of ions near a pore wall, presented for simplicity as a charged plane.

The structure of EDL at inner interfaces, as well as at outer interfaces, is generally determined by electrostatic interactions and thermal motion. Electrostatic forces result in separation of cations and anions near the charged

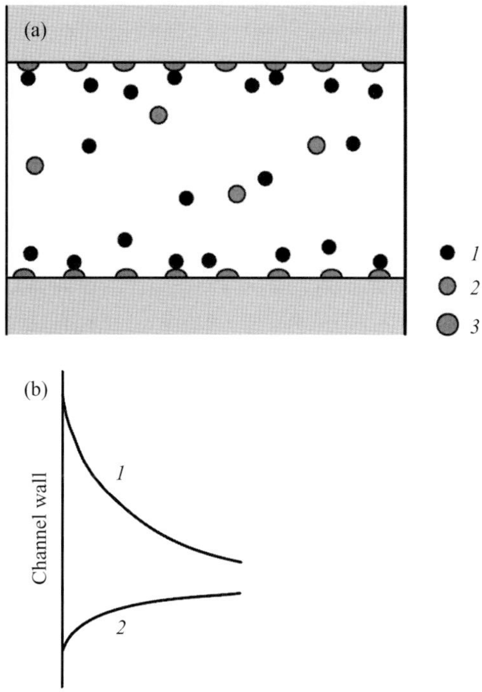

FIGURE 9.7. Counterion (1) and co-ion (2) distribution within a pore near a wall bearing fixed charges (3). The concentration of counterion near the pore wall is essentially higher than that of co-ion. Both concentrations approach that of the external electrolyte with increasing distance from the wall. Adapted from Yaroslavtsev et al.[64]

wall, while the thermal motion tends to equalize the distribution of ions. The distribution of ions within an EDL is described by the Gouy–Chapman theory, taking into account both electrostatic and thermal interaction. This theory gives ion distribution schematically shown in Figure 9.7. Note that the concentration of co-ions in the pore solution decreases with decreasing ratio of the pore radius to the Debye length, the latter characterizing the EDL thickness. Traditionally, when applying this theory, a continuous variation of dielectric permittivity is assumed.[34,35] However, it should be borne in mind that these continuum theories neglect structural inhomogeneity in the vicinity of the inner interface. In particular, Brownian dynamics simulation[36] and other methods[21,34] show that the application of Poisson–Boltzman and Poisson–Nernst–Planck continuum equations leads to an overestimate of the shielding effects (i.e., too high thickness of the diffuse EDL) when the pore radius is less than two Debye lengths.

A number of other investigations of the structure and related properties of Nafion and other perfluorinated membranes were successively reported. Generally, these publications have confirmed the main features of Gierke's model, namely the cluster-channel morphology of these membranes, as well as the characteristic sizes. However, not all the authors agree with the spherical

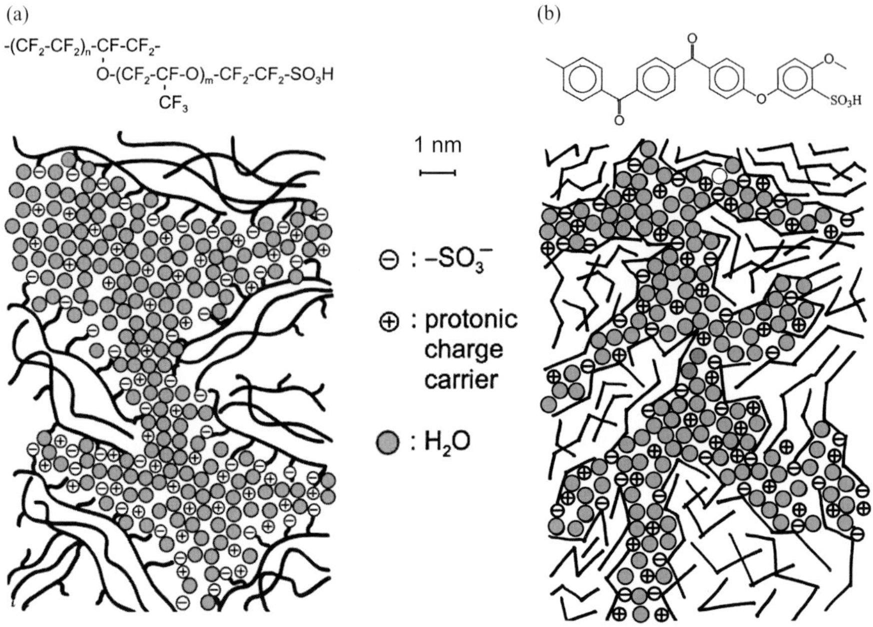

FIGURE 9.8. Two-dimensional structure of a Nafion (a) and a sulfonated polyetherketone (b) membranes on nanometer scale. Adapted from Kreuer.[38]

form of clusters and cylindrical channels argued by Gierke. A review of the papers considering different geometric forms of clusters and channels is made by Rollet et al.[33] In particular, some authors argued for lamellar organization of planar clusters with short connecting channels.[31,37] A rather realistic presentation of two-dimensional structure of Nafion is made by Kreuer[38] (Fig. 9.8a).

At high water contents (which can be attained in commercial perfluorinated membranes at high temperatures in pressure vessels),[26] the membrane structure undergoes essential evolution (Fig. 9.6). Gebel[26] has examined the structure of the swollen PFSA membranes depending on the water content using SAXS and SANS techniques. He has found that at 50% water content, the reverse micellar-like structure transforms into a direct micellar-like one: hydrophobic chains form domains intercalated into the aqueous phase. The side chain charges are located at the polymer/solvent interface and oriented into the aqueous phase. At very high water content, the membrane represents a network of connected polymer rod-like particles: the experimental SANS spectra[26] corresponded to cylindrical particles with 2.5 nm as radius for both the shape and the level of scattered intensity. According to Gebel,[26] the formation of connected cylindrical particles is the best compromise between the minimization of the interfacial energy (which is attained at spherical morphology) and the packing constraints blocking the formation of polymer spheres. Besides, PFSI solutions obtained by using appropriate solvents were studied by small-angle scattering[39] and magnetic resonance techniques.[40] A colloidal structure formed of polymer rod-like shaped particles was evidenced. The value of the particle radius was found to be between 2 and 2.5 nm, and the length was estimated to be larger than 30 nm.[39]

The size of hydrophilic clusters and their distribution within the membrane depend on such factors as the nature of polymer, concentration of fixed ions, and the degree of polymer cross-linking. Kreuer[38] has carried out a cognitive analysis of the difference in structure and properties of two sulfonic membranes—one with a perfluorinated matrix (Nafion) and another with a polyetherketone one—by applying the SAXS techniques. The polyetherketone backbone is less hydrophobic and less flexible, and the sulfonic acid functional group is less acidic and, therefore, also less polar. As a consequence, the hydrophobic/hydrophilic separation of sulfonated polyetherketones is less pronounced, and the hydrophilic channels are narrower and less connected compared to Nafion (Fig. 9.7). At the same time, the hydrophilic/hydrophobic interface is larger and, therefore, the average separation of neighboring sulfonic acid functional groups is also larger. The water is strongly confined in the narrow channels of the polyetherketone polymers, leading to a significantly lower dielectric constant of the water of hydration (about 20 compared to almost 64 in fully hydrated Nafion).[41] As for the transport properties, the narrower channels in sulfonated polyetherketones reduce the electroosmotic drag and water permeation in comparison to Nafion, while maintaining high proton conductivity, at high water contents.[38]

Role of Side Chains: Swelling in Alcohol Solvents Important information on the side-chain architecture of Nafion was found by Haubold et al.[32] via a SAXS study (Fig. 9.9).

The side chains form a transition region between the aqueous domain and the hydrophobic polymer backbone. The hydration effect of the hydrophilic sulfonic acid groups covalently bounded to the hydrophobic polymer extends within the transition region. Hence, the side chains are partially hydrated. With increasing degree of hydration/solvation, the swelling of the membrane occurs via this region: the side chains unfold in a way that the thickness of the transition region ($s/2$, Fig. 9.9) increases, while that of the purely aqueous region decreases.[32] In aqueous solutions, the $s/2$ value for a Nafion membrane is about 1.7 nm. When methanol (or another alcohol solvent) is present in solution, the membrane swells additionally.[42] For example, in LiCl methanol solution, the thickness of Nafion 117 grows up to 25% and its volume up to 70%.[43] While $s/2$ increases with swelling, the distance between the centers of opposite sulfonic groups (c in Fig. 9.9) decreases.[32] Apparently, when an alcohol is added in an aqueous solution contacting a Nafion membrane, its molecules are initially sorbed by the zones adjoining hydrophobic fragments of the perfluorinated matrix devoid of fixed charges and not available for water. This is in accordance with the lower polarity of alcohol in comparison to water (1.68 and 1.84 Debye for methanol and water, respectively).[13] Hence, there is a

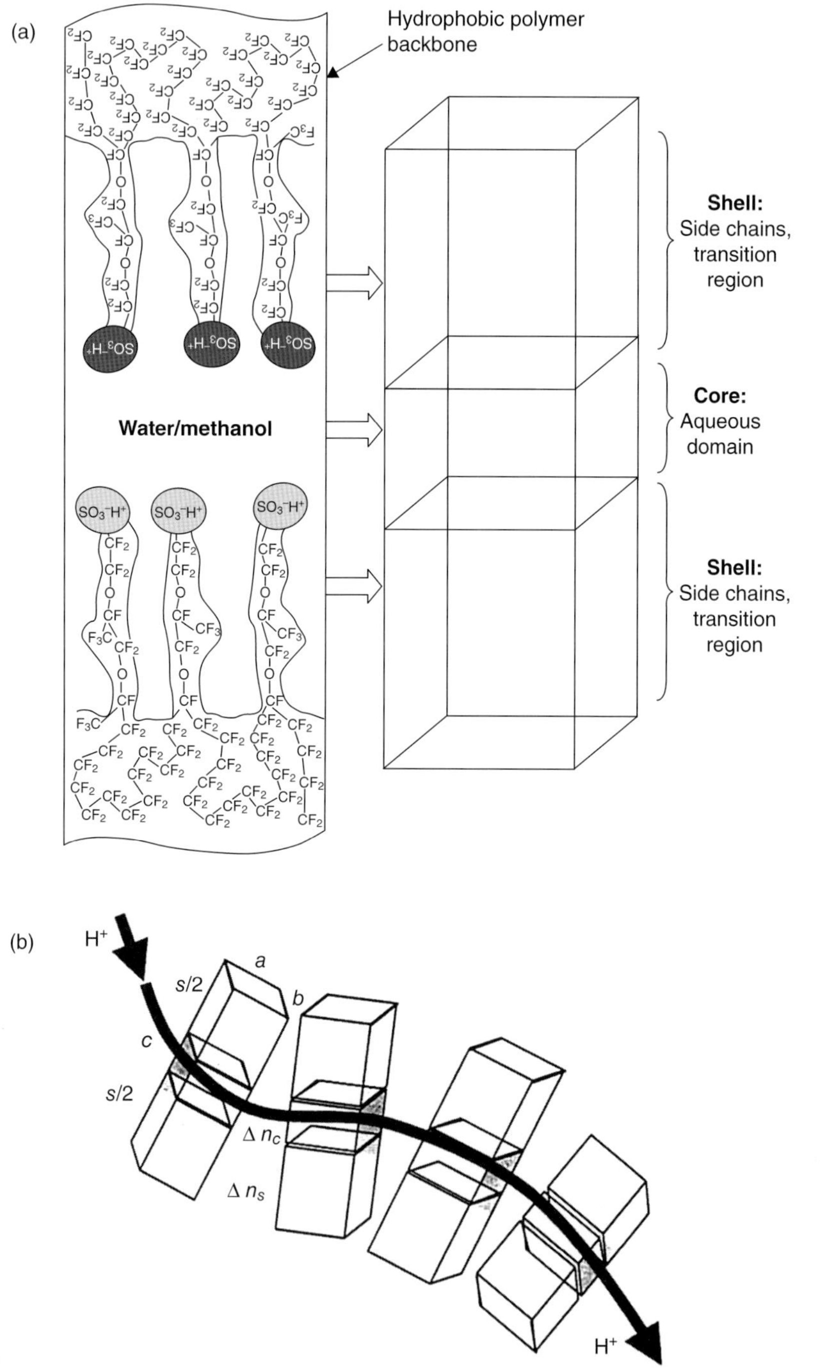

FIGURE 9.9. Schematization of the Nafion structure taking into account the transition region formed by side chains. (a) The main elements of the structure; (b) the passage of a counterion through the membrane. The sense of geometric parameters *a*, *b*, *c*, and *s*/2 is explained in the text. Adapted from Haubold et al.[32]

certain solubility of alcohol in the hydrophobic perfluorinated domains.[44] However, the dielectric constant of water and alcohol is rather different: 81 for water and 32 for methanol. Hence, a noticeable increase in the repulsion between neighboring charged sites should be expected when transferring from water to alcohol solutions. That should lead to an increase of the cross-section of the structure (the product $a \times b$ [Fig. 9.9]),[32] leading to growth of the length of side chains and increase in membrane volume.

The property of side chains to unfold is supported by additional data, for example by electronic structure calculations,[45] showing that the unfolding requires 18 kJ mol^{-1}. Treatment of the Nafion conductivity data led Chaabane et al.[43] to conclude that the space between the charged sites belonging to opposite pore walls (parameter c in the Houbold's model) decreases with increasing methanol content.

Structure of Heterogeneous Membranes Heterogeneous membranes (Fig. 9.3b) are characterized by a more complicated system of pores. Often these membranes are fabricated from ion exchange resin particles (forming ion-conducting phase) and polyethylene (or other bonding agent), strengthening the membrane. The structure of ion exchange resin is similar to that of homogeneous IEMs; it contains micro- and mesopores. Moreover, there are large macropores (cavities, fissures) confined between ion exchange and inert bonding agent.[13] The existence of these two types of pores is evidenced by contact etalon porometry.[46] This method allowed Berezina et al.[47] to obtain pore size distribution for Russian MK-40 (SchekinoAzote) and MA-40 membranes showing two maximums, the first one at about 10 nm and corresponding to micro- and mesopores, and the second one related to macropores having size about 1000 nm. The pores of the first type are located within the ion exchange particles, which are sulfonated (MK-40) or aminated (MA-40) styrene divinylbenzene copolymer. The pores of the second type are the spaces between different particles in a membrane.

Note that sometimes homogeneous membranes contain also inert reinforcing material. For example, Neosepta membranes prepared by the paste method contain a poly(vinylchloride) (PVC) cloth and PVC fine powder constituted of particles much smaller than polyethylene ones in heterogeneous membranes.[48,49] The monomers forming the paste are copolymerized and subsequently sulfonated or aminated to produce IEMs with sulfonic acid or quaternary ammonium groups. Hence, in this case the ion exchange resin produced in the copolymerization presents a continuous phase, which includes PVC particles separated in space and a PVC cloth.

9.3 ION AND WATER TRANSFER IN IEMs

9.3.1 Transfer Mechanisms at Microscale

Ion transport in solutions and organic IEMs is based on two types of transport mechanism at microscale: diffusion in general sense and convection. It is understood that the movement of ions under the influence of external electric field can also be seen as a diffusion process: migration of an ion in the electric field results from the fact that the frequency of ion jumps in the field direction is higher than in the opposite one.[50–52] Generally, the rate of diffusion processes is determined by the product of the ion mobility and concentration. The first factor is controlled by the nature of the mobile ion and the structure of conducting channels. The ion concentration depends on the charge density of fixed ions and the pore radius. Convective transport is determined by the product of the velocity of the liquid center of mass and the concentration of the given particle.

By analogy with crystals, three diffusion mechanisms may be distinguished in charged membranes: *diffusion by vacancies*, *relay*, *and solvation diffusion*.[50] The first one occurs as a series of jumps from one potential well located near a fixed ion to another one near a neighboring fixed ion, under the condition that the receiving potential well is vacant. In the relay diffusion, the ion can push out a neighboring counterion from its well; the latter passes in interstitial state or pushes out another ion.[50] The solvation mechanism is similar to the ion transport in interstitial space of ionic crystals or in solutions: it is a jump movement from one environment to another.

The height of potential barrier for ion jumps is determined by the energy of electrostatic interactions of hydrated ions, by the chemical interaction of the ion and a fixed group, and by the energy necessary for moving the polymer chains apart to form a channel needed for the passage of ion. The distance between two neighboring wells, λ, should be equal to the distance between two fixed ions and depends on the exchange capacity. For usual ion exchangers, λ is in the range 0.5–1.0 nm[1]; in perfluorosulfonic membranes, this distance within a hydrated cluster is about ~0.8 nm.[1,21]

In the literature considering the transport of protons in PFSA membranes, two mechanisms are discussed[21]: vehicular motion, which is movement of the center of mass of proton in an aqueous environment (solvation diffusion as mentioned above), and structural diffusion or Grotthuss mechanism, which is proton shuttling through hydrogen bonded network.[53] Recently Elliot and Paddison[54] have reported the results of atomistic computational simulations according to which both the vehicular and Grotthuss (relay or shuttling) mechanisms contribute to the mobility of the protons. Zundel and Eigen cations play an important role in charge transport. The $H_5O_2^+$ ions appear to be the dominant charge carriers at relatively low water content, while at higher water contents, $H_9O_4^+$ cations form as solvent-separated ions. The contribution of Grotthuss shuttling increases with increasing water content. Similar results are found in other papers.[55,56]

9.3.2 Heterogeneity on Nanometer Scale and the Role of Dimensions

We have seen in the previous section that ion exchange materials are heterogeneous and there are several scales of heterogeneity. On the nanometer scale, the homogeneous aqueous media are confined within a hydrophobic phase. This morphology results in relatively narrow transport pores and a very large internal interface. The physicochemical behavior of the conducting domains differs essentially from that of free solutions. First, the internal solution in nanometer pores is charged, as the EDL thickness at the pore walls is of the same order as, if not higher than, the pore radius. The electric current is mainly carried by the ions with the charge of opposite sign vis-à-vis the fixed ions, and that assures the charge permselectivity of IEMs. Second, new transport features, such as electroosmotic drag, occur within the conducting domains.

The transport properties of central parts of the hydrophilic domains differ from that near the internal interface. The very high proton conductivity of hydrated polymers relies on the presence of liquid water as a thermodynamically distinct phase in these domains.[21,45] The water in the center of sufficiently large pores behaves like in free solution.[21,57] The elementary proton transport mechanism there is the same as in the bulk solution: the Grotthuss shuttling is dominant; proton passes from one H_3O^+ (or $H_5O_2^+$ or even larger) complex to a neighboring one.[57] Besides, the dielectric permeability, ε, approaches 81, the value corresponding to free solution (Fig. 9.10). The relative volume of bulk-like water within the pore depends on the water content. According to electronic structure calculations,[21,58] two to three water molecules per sulfonic acid group (n) in perfluorinated membranes are necessary for proton dissociation; when six water molecules are added, separation of the dissociated proton from the sulfonate anion is observed. However, when $n = 6$, all the water within a pore show decreased value of the dielectric permeability (Fig. 9.10). Following Kreuer et al.,[21] only at $n > 14$ it is possible to distinguish the bulk-like water within a pore and to talk about two-phase system there.

The water molecules near the functional groups are structured and largely lose their mobility.[1,21] There is a decrease in polarity and the rate of relaxation, as well as an increase in spatial and orientation order when compared to bulk water; as a result, ε decreases with approaching a functional group (Fig. 9.10). The counterions are concentrated near the functional groups; their mobility there is lower as well; the side chains contribute by increasing the encumbering of the space (Fig. 9.9). At the distance lower than about 0.5 nm from a fixed group, the concentration of counterions decreases due to finite size of hydrated ion. When approaching the pore center, the counterion concentration decreases as well, following the Gouy–Chapman law of ion distribution in EDL (Fig. 9.7).

9.3.3 Functional Properties and Their Relation to the Nanostructure

The main membrane properties, such as permselectivity and conductivity, depend on the ratio of the central and peripheral liquid volumes within the pores. Too high volume of the central part, which occurs in large pores, leads to loss in permselectivity: with increasing pore radius, the co-ion and counter-ion concentrations become closer, hence, the contribution of co-ions to the

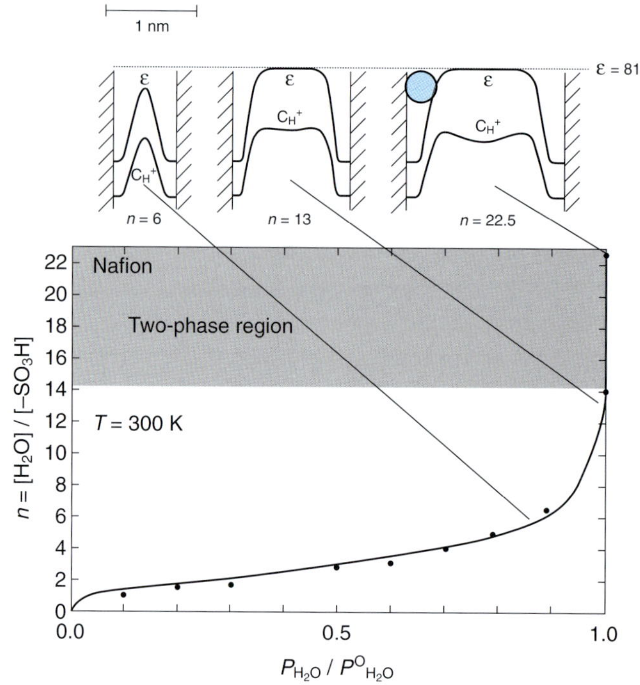

FIGURE 9.10. Hydration isotherm (water content, n, as a function of relative water vapor pressure) for Nafion 117,[162] and the distribution of the dielectric constant[58] and protonic charge carrier concentration across the hydrated hydrophilic pores for three different values of n (top). A hydrated counterion is shown near the pore wall. Adapted from Kreuer et al.[21] T, temperature.

charge transfer increases. The decrease in the amount and the radius of pores results in decreasing conductivity. The membrane morphology can be controlled by changing polymer nature[38] and the concentration of functional groups. In particular, an increase in the length of backbone and side chains leads to increasing the hydrophobic constituent of the membrane and reducing the number and the size of hydrophilic pores.

When the water content increases, the membrane conductivity increases.[26,59] There are several reasons for this. First, the increase in water content enhances percolation between the clusters (see Fig. 9.6), which gives rise to an exponential increase in conductivity according to the common percolation model.[60] Second, when the space available for ion transfer enlarges while the hydrophobic domains remain the same, the average concentration (per unit volume of membrane) and the mobility of ions in the membrane increase.

Note also that the picture of membrane morphology presented above is averaged in time. There is a rapid exchange of molecules between the bulk and interface regions within a pore.[61] Moreover, the pore is not a stiff formation. Its morphology varies with time within certain limits and has a flickering character.[1]

9.3.4 Irreversible Thermodynamics Approach to Transport Description

There are a number of different approaches to describe ion and water transport in membranes. Some books[52,62,63] and reviews[35,64–66] are dedicated to this subject. Phenomenological approach resulting from the irreversible thermodynamics provides a relatively simple mathematical description, which is however not directly linked to the membrane structure. The structure–properties relations can be described within other approaches, which take into account knowledge on membrane structure, consider the membrane as a multiphase system, and take into account or not the nonuniform distribution of concentrations and electric potential within one phase. In this section, we will consider the principals of mathematical description following these both general axes.

The Onsager Phenomenological Equations The strength and weakness of the description of solute and solvent transfer through the membranes within the framework of irreversible thermodynamics is that no information about the membrane structure and transport mechanism is needed. This approach gives general relations between fluxes and driving forces verified for any type of membrane. The individuality of a given membrane is described through a number of phenomenological coefficients characterizing the system. These coefficients cannot be found within irreversible thermodynamics. Often, their experimental evaluation is sufficient for some engineering purposes. Otherwise, the transport equations given by irreversible thermodynamics should be completed with other equations/boundary conditions, which take into account the details of membrane structure.

At equilibrium, electrochemical potential, μ_i, of any mobile species i (a solvent, an ion, or a molecule) is the same everywhere in the membrane (which can generally have a multiphase structure). If nonzero gradients $d\mu_i / dx$ (for the sake of simplicity we consider one-dimensional [1D] transfer normal to the membrane surface) are present, fluxes of different species appear. When the system is not too far from equilibrium, linear relations between all the fluxes and all the forces can be written as[63,67,68]:

$$J_i = -\sum_j L_{ij} d\mu_j / dx, \tag{9.3}$$

where phenomenological coefficients L_{ij} reflect the effect of thermodynamic force $F_j = -d\mu_i / dx$ on the flux density of species i, J_i. Besides material properties, L_{ij} can depend on the temperature, pressure, and concentrations, but not on fluxes and forces.

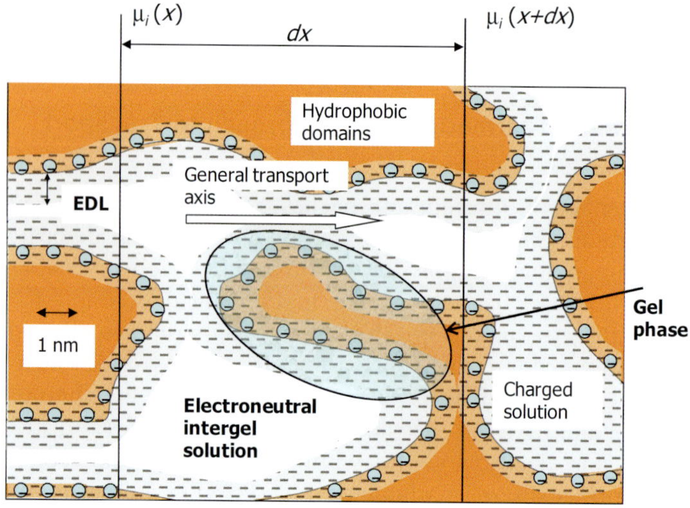

FIGURE 9.11. Schematic presentation of IEM structure including its main elements, in particular fixed ions (shown as circles with –) and electric double layer at the internal interfaces. Gel phase includes fixed ions with polymer matrix and EDL balancing the charge of fixed ions. Adapted from Larchet et al.[150]

To make more physics in the sense of Equation 9.3, note that there is no net transfer in the direction parallel to the membrane surface, and, as a consequence, the electrochemical potential $\mu_i(x)$ is the same at each point belonging to the plane drawn in this direction independently of the phase through which the plane passes (Fig. 9.11). dx in Equation 9.3 may be interpreted as a distance between two planes normal to the transport axis: the first one corresponds to electrochemical potential value $\mu_i(x)$, the second one to $\mu_i(x + dx)$.[52,69,70] dx should be much smaller than the membrane thickness d (to justify the use of differential equation), but noticeably higher than the characteristic size of a phase element, in order that the slice between x and $x + dx$ can serve as a "representative" membrane volume possessing all the structure features. Thus, L_{ij} depends on the properties of all phases present within this slice and varies continuously with x.

Equation 9.3 shows that J_i depends not only on the force applied to species i, F_i, but on the forces applied to other species. The mechanisms of interaction between fluxes of different species (when, e.g., ionic fluxes produced by an electric force give rise to water flow, electroosmosis) are ignored in irreversible thermodynamics.

It is convenient to present L_{ij} coefficients as functions of the concentration of "virtual" electroneutral solution, which is thought to be in local equilibrium with a small volume/thin slice of membrane. The virtual solution was first introduced by Kedem and Katchalsky[71]; it may actually be present in the central part of large pores (if the pore radius is higher than the Debye length) (Fig. 9.11) or considered as a hypothetical solution (if the double layers at opposite walls of the pores are overlapped). As there is usually local equilibrium at external membrane interfaces, the virtual solution at these interfaces is identical to the solution contacting the corresponding side of the membrane.

The generalized electrochemical potential μ_i in Equation 9.3 can be expressed as a function of concentration c_i (or activity a_i) of species i, electrical potential φ, and pressure p in the virtual solution:

$$\mu_i = \mu_i^0 + RT \ln a_i + z_i F\varphi + \bar{V}_i p, \tag{9.4}$$

where z_i is the charge and $\bar{V}_i$ the partial molar volume of species i; and R, T, and F are the gas constant, the temperature, and the Faraday constant, respectively.

The choice of fluxes (J_i) and conjugated thermodynamic forces (F_i) in transport equations may be different, but not arbitrary: with the right choice of J_i and F_i, the sum of products J_iF_i should give the dissipation function.[63,67] Transport coefficients depend on the choice of fluxes and forces. In the case of the use of Onsager's form, Equation 9.3, coefficients L_{ij} were found experimentally by Narebska et al.[72] for the Nafion 120/NaCl membrane system as functions of NaCl solution concentration. In other forms, with other set of the forces and fluxes, these coefficients are also reported in Spiegler and Kedem[73] and Auclair et al.[74]

Along with Onsager's Equation 9.3, other equations (Stefan–Maxwell, Spiegler, Kedem–Katchalsky, and others)[52,75–78] are used; their review is given in Zabolotsky and Nikonenko[52] and Konturri et al.[63] These systems differ by the choice of fluxes and forces, and consequently, by the transport coefficients; each of them has certain advantages and disadvantages. Generally speaking, they are mathematically equivalent[52] in the sense that it is possible to pass from one system to another by a simple transformation of variables.

The famous Onsager's reciprocal theorem:

$$L_{ij} = L_{ji}, \tag{9.5}$$

allows one to reduce the number of independent phenomenological coefficients. In the case of three different species (a counterion, a co-ion, and a solvent present in a membrane), the number of independent coefficients is equal to 6.

The Kedem–Katchalsky Equations Perhaps it is the system of Kedem–Katchalsky equations[75] that is of the greatest interest for the practical description of transfer through membranes. In differential form, these equations are written as[74]:

$$J_v = -L_p\left(\frac{dp}{dx} - \sigma RTc_s\nu\frac{d\ln a_\pm}{dx}\right) + \beta i = -L_p\left(\frac{dp}{dx} - \sigma\frac{d\pi}{dx}\right) + \beta i, \tag{9.6}$$

$$J_i = -P\frac{dc_i}{dx} + \frac{it_i}{z_i F} - c_i L_p (1-\sigma)\frac{dp}{dx}, \tag{9.7}$$

$$\frac{d\varphi}{dx} = -\frac{i}{\kappa} - \frac{RT}{F}\left(\frac{t_+}{z_+}\cdot\frac{d\ln a_+}{dx} + \frac{t_-}{z_-}\cdot\frac{d\ln a_-}{dx} - \beta c_s \nu F \frac{d\ln a_\pm}{dx}\right) - \beta\frac{dp}{dx}, \tag{9.8}$$

where J_v, J_i, and i are the thermodynamic fluxes (the densities of volume and solute species i fluxes, and the current density, respectively); c_i and c_s are the molar concentration of ion i and that of the salt in the virtual solution of the membrane; π и p are the osmotic and hydrostatic pressures, and φ is the electric potential in the virtual solution, respectively; z_i is the charge number of ion i; $\nu = \nu_+ + \nu_-$ is the stoichiometric number; subscripts $_+$ and $_-$ relate to cation and anion, respectively; and d is the membrane thickness. The form of Equations 9.6–9.8 is so that they are applicable to a cation or an anion exchange membrane. However, let us keep in mind a cation exchange membrane with negatively charged fixed ions; hence, cations will be counterions, and anions, co-ions.

Three thermodynamic forces are considered: mechanical (dp/dx), electric ($d\varphi/dx$), and chemical. The latter is expressed through the gradient of concentration or that of osmotic pressure, which are linked between them[52,75]:

$$\nabla\pi = RTc_s\nu\nabla\ln a_\pm, \tag{9.9}$$

where $a_\pm$ is the mean ionic activity of the electrolyte. Note that the gradient of osmotic pressure can be also expressed through the gradient of water (solvent) activity (a_w)[13]:

$$\nabla\pi = -\frac{RT}{\bar{V}_w}\nabla\ln a_w. \tag{9.10}$$

The equivalence of Equations 9.9 and 9.10 follows from the Gibbs–Duhem equation:

$$c_w\nabla\mu_w + c_s\nabla\mu_s = \nabla p, \tag{9.11}$$

and assumption $c_w\bar{V}_w \approx 1$. It can be verified by substituting Equations 9.12 and 9.13 for water and electrolyte chemical potentials into Equation 9.11:

$$\nabla\mu_w = RT\nabla\ln a_w + \bar{V}_w\nabla p, \tag{9.12}$$

$$\nabla\mu_s = \nu_+\nabla\mu_+ + \nu_-\nabla\mu_- = \nu RT\nabla\ln a_\pm + \bar{V}_s\nabla p. \tag{9.13}$$

Here, $\bar{V}_s$ and $\bar{V}_w$ are the partial molar volume of the solvent and electrolyte, respectively.

The well-known van't Hoff's law for a nonelectrolyte solution ($\pi = c_s RT$) is obtained from Equation 9.10 when using the solvent mole fraction, X_w, as the solvent activity, and applying the following approximations: $\ln X_w = \ln(1 - X_s) \approx -X_s \approx -n_s / n_w$ and $V \approx n_w \bar{V}_w$. Here n_s and n_w are the mole numbers of the solute and the solvent, respectively, and V is the volume of the system.

The main interest of the Kedem–Katchalsky equations is that the transport coefficients in these equations are just those usually used in characterization of membrane transport properties. For this reason these six transport coefficients are called practical: electrical conductivity (κ); diffusion permeability (P); hydraulic permeability (L_p); transport number of the counterion (t_+); electroosmotic permeability (β); and Staverman reflection coefficient (σ). The sense of the latter may be understood from the following limit cases: $\sigma = 1$, if the membrane reflects completely the solute when it transfers with convective flow through the membrane, and $\sigma = 0$, if there is no retention. The physical sense of coefficients κ, P, and L_p coincides with that of coefficients used in linear equations of Ohm, Fick, and Darcy, respectively, generalized by Equations 9.6–9.8. These linear equations follow directly from Equations 9.6–9.8, if only one driving force is kept nonzero (e.g., the Darcy law results from Eq. 9.6 where $\nabla\pi$ and i are set zero). t_+ shows the fraction of electric charge carried by counterion in conditions where gradients of concentration and pressure are zero; and β is the proportionality coefficient between the volume flux and the current, also under zero gradients of concentration and pressure.

Note that a different form of Kedem–Katchalsky equations is applied as well[75]: instead of the pressure gradient, the volume flux is used in Equations 9.7 and 9.8. This slightly changes the sense of coefficients κ, P, and t_+.[52,75]

Equations 9.6–9.8 are on the basis of membrane characterization.[47,72,78–83] Each of the coefficients terms one of the membrane transport properties, which can be measured under conditions when only one driving force is applied. Thus, diffusion permeability P should be measured when $\nabla p = i = 0$; more details about measuring procedure and the passage from the integral permeability found from the experimentally measured diffusion flux to local coefficient P can be found in Berezina et al.,[47] Zabolotsky and Nikonenko,[52] and Auclair et al.[74]

As we have mentioned above, it is possible to pass from the Onsager to the Kedem–Katchalsky equations by mathematical transformation. Hence, the coefficients present in these two sets of equations are linked.[74,84] Here we present the expressions for κ, t_i, and P,[74] which are the most commonly used parameters:

$$\kappa = (z_+^2 L_+ + z_-^2 L_-)F^2, \tag{9.14}$$

$$t_i = \frac{z_i^2 L_i}{z_+^2 L_+ + z_-^2 L_-} = \frac{z_i^2 L_i F^2}{\kappa}, \quad i = +,-, \tag{9.15}$$

$$P = \frac{2RTg}{c}\left[\frac{\kappa t_- t_{+app}}{F^2} + (L_{+-} - m_s M_w L_{-w})\right] \approx \frac{(z_+ L_+ t_- + |z_-| L_- t_+)RT}{c}, \tag{9.16}$$

where:

$$L_+ = L_{++} + (z_- / z_+)L_{+-}, \quad L_- = L_{--} + (z_+ / z_-)L_{+-} \tag{9.17}$$

are the conductance coefficients of the individual ions $_+$ and $_-$, respectively. The frequently used approximation in Equation 9.16 (the right-hand side after sign ≈) is obtained when assuming $L_{+-} - m_s M_w L_{-w} = 0$, $g = 1$, and $t_{+app} = t_+$, where m_s is the molal salt concentration in the membrane and M_w is the water molar mass. Note that the central part of this equation (after sign =) is written for 1:1 electrolyte, while the right-hand side for any single electrolyte, $c = |z_i|c_i$ is the electrolyte concentration in the virtual solution in Eq dm^{-3}. Similar expressions for L_p, β, and σ can be found in Auclair et al.[74]

The experimental determination of Onsager or Kedem–Katchalsky coefficients is laborious and delicate. This work can be facilitated if one restricts oneself with an incomplete set of coefficients: for example, Staverman coefficient σ is rarely used in ion exchange membrane characterization as it is very close to 1. Another possibility is the use of additional relations between the coefficients.[74,78,84] Such a relation can be obtained[74,84] under the condition that the difference between the cross coefficients reflecting the interaction of coions with counterions (L_{-+}) and water ($m_s M_w L_{-w}$) is small compared with the diagonal coefficient L_{--}:

$$P = \frac{2RT\kappa t_- t_{+app} g}{F^2 c}, \tag{9.18}$$

where t_{+app} is the apparent transport number of the counterion, $t_{+app} = t_+ - mM_w t_w$, $m = z_+ \nu_+ m_s$, and t_w is the water transport number in membrane; g is the activity factor for the virtual solution[74,85]: $g = 1 + d\ln y_\pm / d\ln c$; and $y_\pm$ is the mean molar activity coefficient.

Equation 9.18 is a generalization of the Nernst–Einstein equation (see the next section) for the case where the osmotic and electroosmotic water transfer are implicitly taken into account. This or similar equations were used and verified[74,79] in different conditions. In the case of perfluorinated Nafion 120 and MF-4SK membranes, a deviation from Equation 9.18 was found,[74] which can be interpreted, following Kedem and Perry,[79] as a consequence of relative hydrophobicity of these membranes.

Equation 9.18 and similar relations[78] are successfully used in the characterization of ion exchange membranes.[74,78,80,82,86] Normally, the transport number t_i^* is easier to calculate from Equation 9.18 or a similar relation when knowing the values of κ and P, in comparison with the evaluation from direct measurements of ionic fluxes.[87]

The Nernst–Planck Equation Together with the laws of Ohm, Fick, and Darcy, it is as well possible to deduce the Nernst–Planck equation from the Kedem–Katchalsky equations. However, that is simpler to do starting from the Onsager Equation 9.3.[51,52,88] If the cross-phenomenological coefficients are neglected in Equation 9.3 and Equation 9.4 is used with p = const, one immediately finds:

$$J_i = -L_i \frac{d\tilde{\mu}_i}{dx} = -L_i \left(\frac{d \ln a_i}{dx} + z_i F \frac{d\varphi}{dx} \right) \tag{9.19}$$

where L_i is the diagonal conductance coefficient.

The gradient of the electrochemical potential can be expressed through the ion activity (a_i) and electrical potential (φ) in any phase constituting the membrane. In some cases, it is more convenient to use the virtual solution because in this case a_i and φ have not jumped at the external boundaries of the membrane (if only the condition of quasi-equilibrium double electric layers at these boundaries holds true, that is verified at underlimiting current densities).[89–91] L_i characterizes the transport properties of a slice located between planes x and $x + dx$.

When applying Equation 9.19 for a homogeneous phase, the diffusion coefficient of individual ion, D_i, is usually used instead of L_i:

$$D_i = \frac{L_i RT}{c_i}, \tag{9.20}$$

where c_i is the ion concentration in the considered phase. Substituting Equation 9.20 into Equation 9.19 yields:

$$J_i = -D_i \left(\frac{dc_i}{dx} + z_i c_i \frac{F}{RT} \frac{d\varphi}{dx} \right), \tag{9.21}$$

where the variation of the activity coefficient with concentration is neglected.[85,88]

Equation 9.21 is known as the Nernst–Planck equation. It shows that the ion transport occurs under the influence of concentration (the first term in the right-hand side) and electric potential (the second term) gradients.

If not considering the electrochemical gradient as a single driving force, there is no reason to think that the proportionality coefficient before the concentration and electric gradients should be the same. Hence, the equation relating the ionic flux with these two forces is often written in the form[88]:

$$J_i = -D_i \frac{dc_i}{dx} + \frac{z_i}{|z_i|} u_i c_i \frac{d\varphi}{dx}, \tag{9.22}$$

using ionic mobility u_i together with the diffusion coefficient.

Equation 9.22 generalizes formally two linear laws: the Fick and the Ohm laws. When comparing Equations 9.21 and 9.22, we obtain the relation:

$$u_i = D_i \frac{|z_i| F}{RT}, \tag{9.23}$$

known as the Nernst–Einstein relation.

As we have seen, neglecting the cross coefficients in the Onsager equations leads automatically to the Nernst–Einstein relation. This relation is implicitly used each time when Equation 9.21 is applied. It is approximately verified in a number of cases. However, there are deviations related to electroosmotic[92] and/or electrophoretic[93] contributions into electrical conductance, as well as to correlation effects in diffusion[50] and some other effects.[94] Taking into account these deviations is rather complicated. Equation 9.18 gives such an example of taking into account the interactions between counterions and water. It is easy to show that when assuming $g = 1$ and $t_{+app} = t_+$, Equation 9.18 written for a homogeneous phase transforms to Equation 9.23. In this case P is replaced by the electrolyte diffusion coefficient presented (for a single symmetric electrolyte) as $D = 2D_+D_-/(D_+ + D_-)$, where D_i is the diffusion coefficient of the ith *ion*; the transport number of the ith ion is also expressed through D_i: $t_i = D_i/(D_+ + D_-)$. As well, when applied to solution, Equation 9.7 is written with D instead of P, c_i and t_i are referred to the solution, and σ is set to zero (as there is no reflection of any species); product $L_p dp/dx$ may be replaced by J_v, according to Equation 9.6.

The Nernst–Planck equation is widely used in the theory of ionic transfer in membrane systems.[52,63] It is notably simpler than the equations of irreversible thermodynamics, but takes into account two main contributions into ionic transport: diffusion and migration. This equation may be also deduced from the microscopic consideration, which allows one to determine the limits of applicability of this equation[51] when treating the transport at nanometer scale.

There are several generalizations of the Nernst–Planck equation, including the Onsager and Kedem–Katchalsky equations described above. To take into account convective flow, the so called extended Nernst–Planck equation[13,88] is used:

$$J_i = -D_i\left(g\frac{dc_i}{dx} + z_i c_i \frac{F}{RT}\frac{d\varphi}{dx}\right) + c_i V, \tag{9.24}$$

where g is the activity factor (defined above) and V is the center of mass velocity of the fluid flow. Often V is taken as the solvent velocity, which is justified in diluted solutions. Besides, g is assumed to be equal to 1 (as in Eq. 9.21), which is true when the gradient of the activity coefficient is neglected. Instead, g can be calculated using different approaches.[82,83] In the case of 1D flow, V is numerically equal to the volume flux density, J_v. V is calculated as proportional to the gradient of hydrostatic pressure and/or electric potential (electroosmosis)[92] (e.g., with help of Eq. 9.6) or found from experimental volume flux through the membrane. When solution near a membrane is considered, Equation 9.24 is written in two-dimensional (2D) form, and the Navier–Stokes equation is used to find $\vec{V}$.

Convective transport is very important in pressure-driven processes,[95,96] in dialysis across biological or imitating membranes.[97] In the case of IEMs, the convective contribution in the overall transport through the membrane, expressed by the c_iV term, is rather low. However, this transport is important in fuel cells (water management), in electrodialysis (ED) of concentrated solutions, and in some other cases considered in Section 2.6.

9.3.5 Modeling the Structure–Property Relations

There are a number of models specially developed for linking the structure and transport properties of membranes. Generally, they may be divided in two great groups: *heterogeneous models*, which consider a membrane as a disperse multiphase system; and *continuum models*, describing the transfer within one phase. During the last few years, significant progress was achieved in applying molecular dynamics (MD) for modeling macroscopic behavior of ions and water within a membrane pore.[21] In this method, an ensemble of individual particles is considered, and a system of the Newton equations is solved as time evolving. In this section, these approaches will be briefly described, including some results obtained with the help of MD simulations.

Continuum Models

Teorell–Meyer–Sievers (TMS) Model Historically, the first model successfully applied to explain the main properties of charged membranes was that developed independently in 30 years of the last century by Teorell[98] and by Meyer and Sievers.[99] It is often called the TMS model. This model provides the main ideas of mathematical description of membrane transport and makes the basis of a number of more recent and more sophisticated models.

As we mentioned above, the Nernst–Planck equation is easier to be applied in comparison with complete irreversible thermodynamics approach, as it needs only one diffusion coefficient to be known when describing the transport of an ion. The TMS model is based on this equation; it considers a membrane as a single homogeneous phase (charged gel), which is an aqueous solution of matrix polymer chains, fixed, and mobile ions. Together with Nernst–Planck's Equation 9.21, the local electroneutrality assumption in the membrane:

$$z_+\bar{c}_+ + z_-\bar{c}_- = Q, \tag{9.25}$$

(Q is the concentration of fixed ions, the ion exchange capacity) and surrounding solutions:

$$z_+c_+ + z_-c_- = 0, \tag{9.26}$$

as well as the equation expressing current flowing through the individual ionic flux densities:

$$i = (z_+J_+ + z_-J_-)F, \tag{9.27}$$

and Donnan equilibrium at the interfaces (as boundary conditions):

$$\frac{\bar{c}_+^{1/z_+}}{\bar{c}_-^{1/z_-}} = K_D \frac{c_+^{1/z_+}}{c_-^{1/z_-}}, \tag{9.28}$$

are used to obtain a boundary value problem. The bar under a magnitude signifies that it belongs to the membrane phase.

The Donnan equilibrium relation is obtained when assuming the equality of electrochemical potential for each ion species i present in the membrane/solution system:

$$RT \ln a_i + z_i F\varphi = RT \ln \bar{a}_i + z_i F\bar{\varphi}. \tag{9.29}$$

The ion activities implicitly entering into Equation 9.29 are generally functions of the pressure in each phase. Two relations follow from Equation 9.29: one for activities:

$$\bar{a}_+^{1/z_+} / \bar{a}_-^{1/z_-} = a_+^{1/z_+} / a_-^{1/z_-}, \tag{9.30}$$

and the other for the potential difference (pd) between two phases:

$$\Delta\varphi_D = \bar{\varphi} - \varphi = -\frac{RT}{z_+F} \ln \frac{\bar{a}_+}{a_+} = -\frac{RT}{z_-F} \ln \frac{\bar{a}_-}{a_-}. \tag{9.31}$$

When introducing ionic concentrations (molar) in Equation 9.30 instead of activities ($a_i = c_i y_i$ and $\bar{a}_i = \bar{c}_i \bar{y}_i$), we arrive at Equation 9.28, where the Donnan equilibrium coefficient, K_D, is expressed through the ratio of mean ionic activity coefficients:

$$K_D = (y_\pm / \bar{y}_\pm)^{1/z_+ - 1/z_-}, \tag{9.32}$$

where $y_\pm = \left(y_+^{\nu_+} y_-^{\nu_-}\right)^{1/\nu} = \left(y_+^{1/z_+} y_-^{-1/z_-}\right)^{(1/z_+ - 1/z_-)^{-1}}$; a similar expression can be written for the mean ionic activity coefficient in the membrane, $\bar{y}_\pm$.

It is assumed (the osmotic theory of adsorption)[13] that the activity coefficients take into account osmotic pressure. Often the $\exp(\Delta\pi\bar{V}/RT)$ term ($\Delta\pi$ is the difference in osmotic pressure between the phases, $\bar{V}$ is the electrolyte molar volume) is introduced separately in the expression for K_D,[13,100] though it is believed that its variation is negligible when analyzing the dependence of $\bar{c}_i$ on c_i.

Usually ion exchange materials have a high concentration of fixed ions (high ion exchange capacity), close to 1 M or higher.[13] This results in a strong exclusion of co-ions, so that their concentration is small in comparison with $\bar{Q}$. When using approximations $\bar{c}_- \ll \bar{Q}$ and $\bar{c}_+ \approx \bar{Q}$, we can simplify Equation 9.28. In the case of a symmetrical electrolyte ($z_+ = -z_- = z$) we obtain:

$$\bar{c}_A = \frac{(K_D)^z}{\bar{Q}} c_A^2, \quad \bar{c}_1 = \bar{Q} + \bar{c}_A, \tag{9.33}$$

where subscript $_A$ is used for the co-ion and $_1$ for the counterion.

The Donnan equation, in the form of Equation 9.28 or 9.33, allows the description of selective adsorption of counterions. Together with this property, the TMS models gives a qualitatively adequate description of other fundamental membrane properties: electric conductivity, transport number, and membrane potential.[101] When introducing another counterion species, 2, and applying Equations 9.28 and 9.31 for each pair 1-A and 2-A, it is possible to describe counterion competitive transfer,[52] bionic potential,[13,102] and so on. This model is widely used for biological as well for artificial membranes.

More details about the TMS model can be found in References 13, 62, and 101.

More recent continuum models may be divided into three groups:

1. Models treating the membrane as a single phase (charged gel), where static and kinetic parameters (concentration of fixed charges, diffusion coefficients, etc.) can continuously change along the coordinate without explicit relation to the structure.[103,104] The option of variation in space of the fixed charge concentration is the feature of these models in comparison with the original TMS theory. Generally, they show that heterogeneity in fixed charge distribution leads to increasing permselectivity in comparison to a membrane having homogeneously distributed charges with the same average concentration. The reason is that the permselectivity is controlled by a layer with the higher local concentration of fixed charge; the layers with low fixed charge concentration have minor impact on global membrane behavior.
2. Capillary models,[105–107] more known as "space-charge models,"[30,108–111] which consider ion and water transfer as occurring within a capillary pore with charged walls.
3. Models applying the percolation theory[112] and simulating the ion transfer by flow of a liquid through a system of voids connected by narrow channels.[1]

Space Charge Capillary Models *Capillary models* are based on the knowledge on membrane nanostructure; they provide detailed information on the transport parameters. First, these models were developed for the description of electrokinetic phenomena within microcapillaries with charged walls. Later on they were improved and adopted for transport phenomena in membranes. These models allow the inclusion of effects specific for interfaces: variation of the dielectric constant with the distance from the pore wall,[108] adsorption,[110] ion hydration effects, finite ion sizes,[113] and others.[110] The role of some of these effects in formation of streaming potential,[114] pore conductivity,[107,115] and permselectivity[116] is studied. This type of model has been shown as very useful in describing the transport of ions and fluid in and around nanometer-sized objects.[117] The research field where the object of study is the fluid transfer in such nanostructures was only recently identified as nanofluidics, but it has a rather rich history, in particular, in membrane science.[118] Nanofluidics has

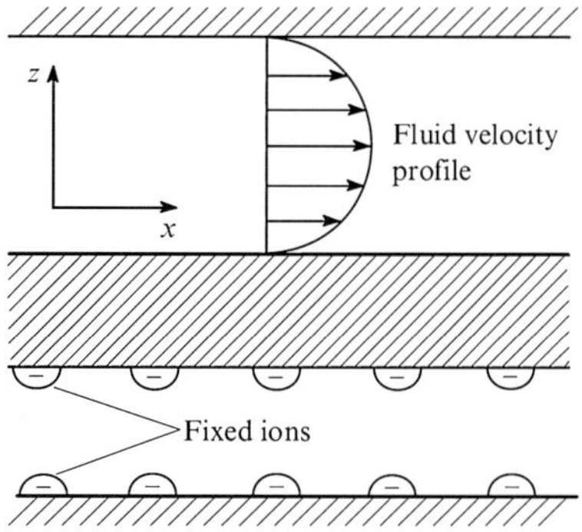

FIGURE 9.12. Schematic representation of the structure of a membrane containing parallel pores with charged walls. The hatched area represents the polymer matrix. Adapted from Zabolotsky and Nikonenko.[52]

experienced considerable growth in recent years, as is confirmed by significant scientific and practical achievements. One hundred nanometers enable the occurrence of phenomena that are impossible at bigger length scales.

Let us consider the main features of space-charge models.

The governing equation describing the ion transport in a pore schematically shown in Figure 9.12 is written in 2D geometry as the extended Nernst–Planck equation with the convective term (Eq. 9.24). The diffusion coefficient is usually assumed to be the same as in free solution. The cross-sectional distribution of local concentration, $\tilde{c}_i$, is described by the Poisson–Boltzman equation.[103] The surface density of the electric charge on the pore wall is $\sigma =$ 110–440 mC m^{-2}; and the pore radius ranges from 0.5 to 5 nm.[29] The fluid flow (assumed occurring only in axial direction) is described by the Navier–Stokes equation when taking into account the effect of axial pressure gradient and the body force produced by the action of the axial electric potential gradient on the space charge.[105,108]

Calculation of phenomenological coefficients within space-charge models allowed Cwirko and Carbonell[119] and later on other authors[120] to bridge the gap between two different approaches, the irreversible thermodynamics and microscopic model description. Thus, it becomes possible to evaluate macroscopic Onsager's L_{ij} coefficients as functions of membrane nanostructure parameters: the pore radius, charge density of the walls, and others. The authors[119] have compared and calculated experimental (determined by Narebska et al.)[72] L_{ij} coefficients for a Nafion membrane, and found a rather good agreement.

We have mentioned that the use of the continuum Poisson–Boltzman or Nernst–Planck and Poisson equations supposes the ions considered as points; the counterions are permitted to approach too closely the fixed groups that results in overestimation of the shielding effect. Eikerling and Kornyshev et al.[34,113] apply the modified Poisson–Boltzman equation, which takes into account the size of proton complexes and the real distribution of the charge of the fixed sulfonate groups. In this case, the calculation of the distribution of protons and coulomb energetic barriers is in good agreement with the results of MD simulations.[121] Paul and Paddison[122] have used the Poisson–Boltzman equation in calculations of the hydrated proton self-diffusion coefficient in a Nafion pore through a statistical mechanical model as a function of the distance from the $-SO_3^-$ group. They considered a rather real geometry of cylindrical pores with hydrophobic walls where the sulfonate groups protrude into the pore at some distance (0.4 nm in the calculations) similar to Haubold's model[32] (Fig. 9.9). Following their calculations, the proton mobility is the maximum in the center region of the pore and decreases when approaching the $-SO_3^-$ group; the mobility reaches zero at the distance of about 1 Å to the group. In the vicinity of the $-SO_3^-$ ions the electric field is so strong that the permittivity of water undergoes dielectric saturation, and any positive charge falls are linked to an ion pair that makes negligible its contribution to the ionic current.[122]

Recently Bazant et al.[123,124] have paid attention to the fact that there are steric effects near a wall nonpermeable for ions at high voltages: solvated counterions are crowded there; due to steric repulsion and increased viscosity of the condensed layer, the double-layer capacitance decreases and the electroosmotic mobility saturates, respectively. The former can explain observed high frequency flow reversal in AC electroosmosis.

The *percolation theory*[112] is well adapted for the description of transfer in systems with a developed network of channels, where a part of channels exhibits a good ionic conductivity (such as the clusters in perfluorinated membranes) while the other part shows a low conductivity (the interclusters channels). This theory allows explanation for the dependence of the membrane conductivity on the water content (w) (e.g., in volume percent).[1,59] At low water content, a part of intercluster channels is not conducting; hence, there is no percolation and the clusters present on the left-hand and on the right-hand sides of the membrane are found isolated one from the other. As the water content increases, a more and more increasing number of clusters become connected by channels, thus forming a network penetrating throughout the whole membrane (Fig. 9.6). There is a percolation threshold water content, w_0, corresponding to insulator–conductor transition. At $w < w_0$ the membrane conductivity is nearly zero, while at $w > w_0$ it increases exponentially.[1,59,125]

Multiphase Models Analysis of the structure of charged membranes (Section 9.2) showed that they are porous materials involving in general case micro-, meso-, and macropores. While the definition of these kinds of pores mentions several characteristics (including the length of action of adsorption forces),[126] we will focus only on one parameter: the ratio of the pore radius, r, to the Debye length, λ, characterizing the EDL thickness at the inner pore wall. If $r < \lambda$, we deal with a micropore (normally, $r < 1$ nm); if $r > \lambda$, but r and λ are of the same order of magnitude, we have a mesopore; and if $r \gg \lambda$

($r > 50$ nm), it is a macropore. It is important that in spite of complicated structure of charged membranes, it is possible to distinguish more or less homogeneous regions, which are conventionally called phases.[19] One of these phases comprises electroneutral solution, which fills the central regions of meso- and macropores where the dielectric permittivity and other properties are close to those in free solution[21,45,57] (Fig. 9.10). Homogeneous PFSA membranes such as Nafion contain clusters and channels, which are mesopores and micropores, respectively; there are no macropores; hence, the volume fraction of electroneutral solution is rather small. In the case of heterogeneous membranes, this parameter is several times greater: a relatively large volume of solution is confined in macropores, which are cavities, cracks between ion exchange resin and polyethylene particles. As soon as the first electroneutral phase is distinguished, the remained volume may be referred to the second phase. It is called the "gel phase,"[13,52,70,127] and formed by an ensemble of micropores and the electric double layers in the internal solution of large pores. Besides, this phase includes the fixed charged groups together with the polymer matrix (Fig. 9.11). Sometimes, within the gel phase, the third phase, which is the aggregate of interlaced hydrophobic polymer chains not permeable for ions and solvent, is considered.[70] The third phase may also include (nonconducting) inert binder used to improve mechanical strength, such as polyethylene in heterogeneous membranes.

It is clear that there is no distinct boundary between the intergel electroneutral solution and the EDL that belongs to the gel phase. However this separation represents a useful means, allowing one to simplify the mathematical description. Mafé et al.[128] carried out such a separation within a continuous space charge model when calculating membrane electric conductivity.

The main idea of modeling the transport within a membrane presented as a multiphase system is in attributing to each phase some physicochemical properties, and then to describe the properties characterizing the whole membrane as functions of the single phase properties. Let us consider a macroscopic volume in the form of a layer containing all phases of the membrane (Fig. 9.11). The thickness of this layer, dx, is small as compared to the membrane thickness, and we assume that the phases are in equilibrium with one another. If the matter transfer across this layer is described, in accordance with irreversible thermodynamics, by Equation 9.3 or Equation 9.19 (which is a simplification of the former), the problem is to find the effective transport coefficient L_i, characterizing a layer of thickness dx of the membrane, as a function of coefficients $L_i^{(k)}$, which characterize individual phases k, and the structural and geometric parameters, which describe the shape and mutual position of the phases. This formulation addresses the effective medium approach[129] intensively developed in relation to a great variety of systems and transfer phenomena: electrical conduction,[130] diffusion,[131] heat transfer,[132] optics, and others.[133,134] This approach originated in the work by Maxwell who, in as early as 1873,[135] reported a solution to the problem of conductivity of a system constituting a conductive continuum medium with dispersed spheres whose specific conductivity differed from that of the medium. A number of researchers, including Rayleigh, Lichtenecker, Bruggeman, Landau and Lifshitz, and others[136–138] contributed to the development of the effective medium approach. The obtained results are often generalized by a hypothesis called "principle of generalized conductivity,"[138] according to which the function relating L_i to $L_i^{(k)}$ is independent of both the nature of applied force and the substance transferred.

As for the application of this approach to ion exchange membranes, several forms of functions relating effective medium conductivity to that of individual phases were proposed. Some of them concern the diffusion,[136,137] the others electrical conductance.[127,136,139,140] Gnusin et al.[70,81,138] have developed a comprehensive model named "microheterogeneous model," which treats conductance coefficients L_i and $L_i^{(k)}$ instead of particular coefficients. This approach allows one to find not only efficient electrical conductivity,[141–144] but diffusion,[145] permselectivity,[146,147] and other[81,148,149] properties starting from one set of membrane parameters.

In the microheterogeneous model, two phases are considered: the gel phase, which is a nanopore medium, globally electroneutral, and including the polymer matrix; and the electroneutral solution filling the inner parts of meso- and macropores, as well as fissures and cavities. The properties of the latter are assumed to be the same as those of the outer equilibrium solution. The relation between the conductance coefficients of individual phases and the effective membrane conductance coefficient, L_i, is expressed in the form[70,138]:

$$L_i = \left[f_1 \left(L_i^g \right)^{\alpha} + f_2 \left(L_i^s \right)^{\alpha} \right]^{1/\alpha}, \tag{9.34}$$

where L_i^g refers to the gel phase, and L_i^s to the interstitial electroneutral solution; f_1 and f_2 are the volume fractions of these phases, respectively; α is the structural parameter characterizing the position of the phases in relation to the transport axis: when the phases are disposed in parallel to the transport axis, $\alpha = 1$, in the case of serial disposition $\alpha = -1$, in other cases $-1 < \alpha < 1$.

The gel phase responses perfectly to the assumptions made in the TMS model, hence, this model is used to describe the transport there. Magnitudes L_i^s and L_i^g are expressed by the ionic diffusion coefficients, D_i^s and D_i^g, and the concentrations, c_i^s and c_i^g in the corresponding phases, in accordance with Equation 9.8: $L_i^s = D_i^s c_i / RT$, $L_i^g = D_i^g c_i^g / RT$. Concentrations c_i and c_i^g are linked by the Donnan relations, Equation 9.28 or 9.33, as local equilibrium is assumed between the both phases.

Taking into account that the ions are present in both phases and using Equation 9.33, their concentration in the membrane is written as (for co-ion A)[70]:

$$c_A^* = f_1 \frac{K_D}{Q^g} c_A^2 + f_2 c_A, \tag{9.35}$$

where the first term shows the contribution of the gel phase, and the second that of the interstitial solution; and K_D and Q^g are the Donnan coefficient and the exchange capacity of the gel phase, respectively. In spite of small value of f_2 (typically less then 0.1 in homogeneous membranes and about 0.2 in heterogeneous ones), the electrolyte sorption by the "intergel" phase is dominant, especially in diluted solutions, due to co-ion exclusion from the EDL in micro- and mesopores. For this reason, the amount of electrolyte sorbed by the gel is small, and Equation 9.33, which is an approximation of the Donnan relation presumed to be true for diluting solutions of strong electrolytes, is verified for concentrations up to 1–2 M, for conventional IEMs.

Thus, the membrane equilibrated with a binary solution is characterized by six parameters: two static, K_D and Q^g; two structural, f_1 and α; and two kinetic ones, diffusion coefficients of counterion, D_1^g, and co-ion, D_A^g, in the gel phase. It is assumed that the diffusion coefficients in the interstitial solution, D_i^s, are the same as in free solution. When knowing these parameters, L_i coefficients can be calculated as functions of the interstitial solution concentration, as was explained above. Then the electrical conductivity, κ, the ion transport number, t_i, and the diffusion permeability, P, can be calculated by using Equations 9.14–9.16, giving the links between the coefficients of transport equations written in Onsager's and Kedem–Katchalsky's forms. Equations 9.14–9.16 can be applied at any coordinate in a membrane considering c as the local concentration. Thus, the microheterogeneous model can be incorporated into a number of boundary value problems modeling the ion and solvent transfer in membrane systems.[150,151]

To find the parameters of the microheterogeneous model, some experiments should be made and the data obtained treated. To determine all six parameters, ion exchange capacity, electrolyte uptake, conductivity, and diffusion permeability—all properties as functions of electrolyte concentration—should be found. The algorithm of data treatment is described in Zabolotsky and Nikonenko.[52] However, often only some parameters between the six are of interest. For example, the volume fraction of the gel (or intergel phase) is a quite important characteristic of membrane. This parameter may be relatively easily found from the concentration dependence of membrane conductivity (κ_a) under alternative current. Equation 9.28a allows a simple approximation at $\alpha \rightarrow 0$[70]:

$$\kappa_a = \left(\kappa^g\right)^{f_1} \left(\kappa^s\right)^{f_2}. \tag{9.36}$$

Since the gel conductivity, κ^g, only slightly depends on electrolyte concentration (as the co-ion sorption by the gel is very low), the $\ln \kappa_a - \ln \kappa^s$ correlation, according to Equation 9.36, should be linear with f_2 as the slope. Numerical calculations show that near the "isoconductance point" (where $\kappa_a = \kappa^g = \kappa^s$), κ_a is almost independent on α,[70] and in the range of $0.1\ c_{iso} < c < 10\ c_{iso}$ the $\ln \kappa_a - \ln \kappa^s$ dependence may be approximated by a straight line up to $|\alpha| = 0.2$. Taking into account that for most ion exchange membranes α lays in the range

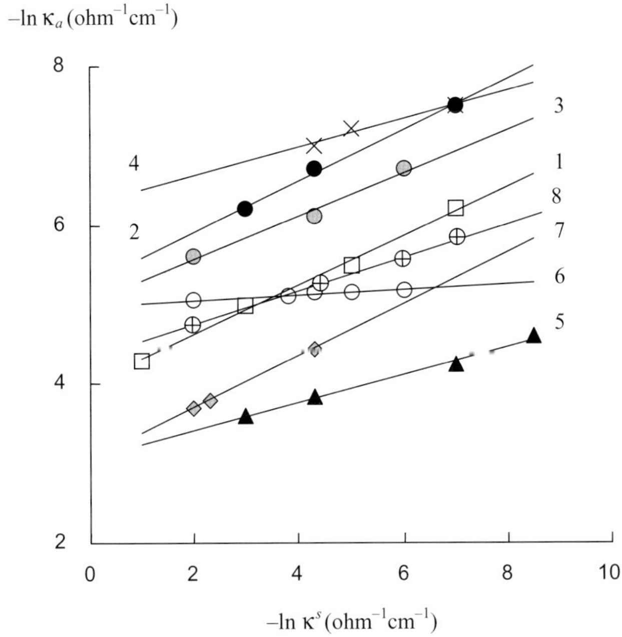

FIGURE 9.13. Specific conductivity of IEMs under AC (κ_a) as a function of solution-specific conductivity (κ^s) in ln–ln coordinates for different membranes. 1, C-60/65 in HCl; 2 and 3, Amberplex C-1 in NaCl and KCl, respectively; 4, Amberplex A-1 in NaCl; 5, AMF C-103 in NaCl; 6, MF-4SK in NaCl; 7, Nepton CR-61 in KCl; 8, MK-40 in NaCl. Adapted from Zabolotsky and Nikonenko.[52]

0.1–0.3, Equation 9.36 may be confidently applied, if the concentration is not too far from the isoconductance point c_{iso}. As Figure 9.13 shows, the slope for the homogeneous membranes (MF-4SK, AMF, CM2) is essentially lower than that for the heterogeneous ones. Thus, f_2 is equal to 0.01 for a perfluorinated MF-4SK, and about 0.2 for heterogeneous MK-40; this quantity reaches 0.45 for a special macroporous membrane.[141]

Zabolotsky and coworkers[141,152] have compared the values of f_2 found from conductivity measurements and applying Equation 9.36, and by using other methods: sorption of electrolyte, standard contact porosimetry,[46] and differential scanning calorimetry (DSC). The close values of f_2 obtained (e.g., for a MK-40 membrane: 0.17 ± 0.02 from conductivity[141]; 0.1 ± 0.02[152] and 0.26 ± 0.02[141] from sorption; 0.23 ± 0.05 from porosimetry[141]; and 0.21 from DSC)[141] testifies that the microheterogeneous model adequately presents the membrane structure and the transport laws.

The facility of use of Equation 9.36 for treatment of experimental conductivity data, which are often available, the possibility to describe other important properties, the diffusion permeability, and the ion transport numbers, result in the fact that this model has been shown as a quite suitable for ion exchange membrane characterization.[12,47,143,144,147,153–157]

There is interference of the microheterogeneous model with other, more detailed structural models, namely with the Haubold model,[32] built on the basis of SAXS experimental data. From the measurements of electric conductivity of a Nafion 117 membrane in mixed water–methanol LiCl solutions and the treatment of the data with the microheterogeneous model, it was found[43] that the volume fraction of electroneutral interstitial solution (f_2) in the membrane decreases with increasing methanol content in solution (X_{MeOH}), at least at $X_{MeOH} < 60$ vol %. The same result was obtained by evaluating f_2 with the help of the Haubold model and parameters resulted from SAXS measurements. The decrease in f_2 with growing methanol content (and growing overall membrane swelling) is explained by unfolding of side chains during membrane swelling in methanol (see Section 1.2 and Fig. 9.9). The latter results in a decrease in the distance between the sulfonic ions bonding to the opposite walls of the pore.

9.3.6 Water Transport

Thermodynamic Consideration: Total Driving Force While the considered above models deal with the membrane transport of ions as well as water (or other solvent), they are mainly directed at the ion transfer description. In literature, water transfer is given special attention. The understanding of this phenomenon is important for improving dialysis and ED as well as fuel cells. In fuel cells, the so called water management[158–160] is one of the major problems in low-temperature technology as partial dehydration of proton-conducting membrane results in sharp increase of its resistance.[21,42] In ED, high water flux does not allow concentrated brine to be obtained.[10] On the other hand, water transfer contributes greatly in effective work of bipolar membranes.[161]

The mechanisms of water transport in membranes, similarly to the ion transport, include self-diffusion; chemical or Fickian diffusion produced by gradient of water chemical activity; and permeation produced by a hydrostatic pressure gradient, which is convection transfer. The water transport occurring under the action of electric field and caused by the interaction between counterions and water molecules (electroosmotic drug)[21,35] is often considered apart as no external force is applied to water molecules.

Generally, the thermodynamic force applied to water is expressed as the gradient of water chemical potential and includes, in accordance with Equation 9.4, the hydrostatic pressure gradient and the gradient of Napierian logarithm of water activity:

$$\nabla\mu_w = RT\nabla \ln a_w + \bar{V}_w \nabla p, \tag{9.37}$$

where $\bar{V}_w$ is the partial molar volume of water. Note that Equation 9.37 may be applied also to other solvents; thus, the subscript $_w$ denotes any solvent in general case. As the osmotic swelling pressure (π) inside the membrane is linked with the solvent activity (a_w)[13]:

$$\pi = -\frac{RT}{\bar{V}_w}\ln a_w, \tag{9.38}$$

in any case where a gradient of water chemical potential occurs, a swelling pressure gradient is induced. In particular, when a membrane separates two solutions of different concentration, the water transfer occurs as chemical diffusion and permeation.[21,62]

Substituting Equation 9.38 into Equation 9.37 yields:

$$\nabla\mu_w = \bar{V}_w\nabla(p-\pi). \tag{9.39}$$

This equation enables linking the water chemical potential gradient with $\nabla(p-\pi)$, that is, the difference between the hydrostatic and osmotic pressures, which was called[21] "total pressure" gradient. Thus, the "total pressure" gradient can be considered as a unique driving force of solvent transfer. Due to Equation 9.39, a correlation between the "chemical" (found from chemical diffusion data) and "permeation" (found from solvent permeation experiments) diffusion coefficients can be established.[21] As follows from papers by Kreuer et al.[54,162] and other authors[163,164] reviewed in Kreuer et al.,[21] the "permeation" diffusion coefficient is always greater than the "chemical" one. This result is not expected if the elementary mechanisms of water transport are the same in chemical diffusion and permeation water transfer.[21] The reason of permeation transport "priority" is that in addition to quasi-random walk process there is another transport component. That is, following Kreuer et al.,[21] viscous flow accompanied by a certain "slip" at the interface with the hydrophobic domain. It is known that in macroscopic systems this slip introduced by Navier is negligible.[88] However, it becomes important in micro- and nanometer channels.[165,166] The situation with the correlation between two types of water transport is similar to that occurring when counterion diffusion coefficient found from diffusion measurements is compared with that obtained from conductivity data by applying the Nernst–Einstein relation, Equation 9.23. In this case the second value is often higher mainly because of additional transport mechanism, electroosmosis, generating solvent flow within membrane pore, and facilitating counterion transport in the same direction (convection conductivity).[13,94]

On the basis of Equation 9.39, the water flux density in the membrane may be written as (in 1D geometry):

$$J_w = J_w^o + J_w^{eo} = -A\frac{d(p-\pi)}{dx} + \frac{t_w i}{F}, \tag{9.40}$$

where coefficient A is proportional to the membrane hydraulic permeability and inversely proportional to the water viscosity[167] as it takes place in classical Darcy's law originally formulated for the hydrostatic pressure gradient only.

Equation 9.40 contains two terms. The first one represents the Darcy law written for the "total pressure" gradient and describing the solvent flux under the action of the external pressure gradient (hydraulic filtration) and the capillary (or swelling) pressure gradient (osmotic flux)[168]; the latter being proportional to $d\pi/dx$. When an electric current of density i is applied, the second term, the electroosmotic flux, proportional to the product of i and the water transport number, t_w, is added.[167] Instead of t_w, often the water drug coefficient is used in Equation 9.40. The latter shows the number of water molecules transported per proton,[169] while t_w is the number of water moles transferred through the membrane with one Faraday of electricity.[62,170]

Note that the term proportional to $d\pi/dx$ may be written also as $-D_w dc_w/dx$ with diffusion coefficient, D_w, and concentration, c_w, of water in the membrane,[171] when taking into account Equation 9.10. In this case, the Fickian mechanism of water transport is emphasized.

The use of the "total pressure" in the Darcy equation is justified by clear correlation between the swelling and the microscopic growth of the clusters in Nafion.[172] The involving of pressure gradient in water transfer phenomenon through an ion exchange membrane is well explicated by a scheme proposed by Meier and Eigenberger[167] (Fig. 9.14). This scheme is in conformity with the membrane structure models reviewed above (Section 9.1), which take their origin from the cluster channel model of Gierke.[23] On the side of high water activity (diluted solution or pure water), the swelling of hydrophilic domains (clusters) produces an increase in their size and an increase in the tension of polymer chains; the elastic matrix of the membrane is more strongly stretched at this side. On the side of low water activity (concentrated solution), the size of clusters and the tension of elastic matrix is lower.

When starting from irreversible thermodynamic approach, one arrives at a similar phenomenological description of the solvent transfer, which is the

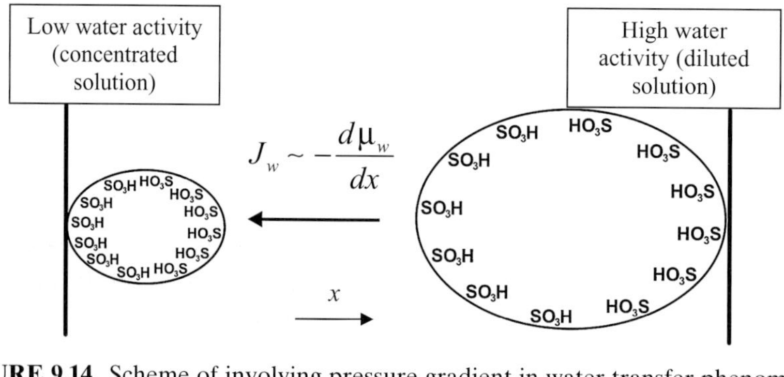

FIGURE 9.14. Scheme of involving pressure gradient in water transfer phenomenon. Adapted from Meier and Eigenberger.[167]

Kedem–Katchalsky equation in the form where the gradient of osmotic pressure enters explicitly as a driving force, Equation 9.6:

$$J_v = -L_p\left(\frac{dp}{dx} - \sigma\frac{d\pi}{dx}\right) + \beta i. \tag{9.41}$$

The difference between Equations 9.40 and 9.41 is in the fact that Equation 9.40 is applied to the transfer of water while Equation 9.41 to that of the volume, $J_v = \bar{V}_w J_w + \bar{V}_s J_s$, where $\bar{V}_s$ and J_s are the partial molar volume and the flux density of the solute, respectively. As a consequence, the reflection coefficient σ appears in Equation 9.41. For better understanding, consider the situation where a membrane separates pure water and a solution. The water flux through the membrane is zero, if, according to Equation 9.40, the difference in hydrostatic and osmotic pressures across the membrane is zero: $\Delta p = \Delta\pi$. However, the volume flux is zero if $\Delta p = \sigma\Delta\pi$, Equation 9.41. In the last case, the water flux is not zero, but the volume transferred by water is compensated by that transferred by the solute in the opposite direction. The water flux becomes zero only if the solute reflection is ideal: $\sigma = 1$.

The importance of involving σ becomes more evident when considering the so-called anomalous osmosis through a membrane.

Anomalous Osmosis The Darcy law under the form of Equation 9.40 as applied to osmotic transfer (when $\nabla p = i = 0$) assumes that the solvent flux is proportional to the difference in osmotic pressures between two solutions separated by the membrane, and the direction of this flux is from the more diluted solution into more concentrated one. However, in several cases the solvent flow deviates from this law. Anomalous osmosis was observed in numerous ion exchange membranes,[106,173,174] including Nafion.[174] In the case of positive anomalous osmosis, the solvent flux is higher than it can be expected from the Darcy law; and in negative anomalous osmosis the situation is presented as inverse: the solvent flux is low and even may be directed from the more concentrated into the more diluted solution.[13,62] The theory of anomalous osmosis was developed by Schlögl[175] and later by Kedem and Katchalsky,[75] and others.[106,176] The reason of anomalous osmosis is that the osmotic pressure is not the only driving force for solvent transfer.[13] The other driving force is the gradient of electric potential induced due to the difference in diffusion mobilities of co- and counterions in membrane pores.[13,62] As follows from the theory of Kedem and Katchalsky,[75] based on the irreversible thermodynamics and frictional forces formalism, the negative anomalous osmosis may be expected in the case where the electrolyte flow across the membrane is elevated (the membrane permeability and electrolyte concentration are high), and the counterion transport number is low; the reflection coefficient σ is then negative. In this case, the co-ions diffuse within a membrane pore faster than the counterions at the beginning that gives rise to an electric field accelerating the flow of counterions. Co- and counterions diffusing from the more concentrated to the less concentrated solution transport water in the direction, which is opposite to that of the water osmotic flux. If the flux of water transported with ions is higher than the osmotic flux, the overall water flux becomes directed toward the diluted solution. Koter[106] has modeled the osmosis transfer in the framework of a space charge capillary model, and found that the negative osmosis for an anion exchange membrane (AM1) with a HCl solution can be explained if one assumes a relatively low transport number of Cl^- counter-ion and a high mobility of H^+ co-ion in the membrane.

Ion Transfer by Convection in IEMs The contribution of different terms in Equations 9.40 and 9.41 depends on the membrane and solution under study. As well, the mechanism of ion transfer becomes dependent on the peculiarities of the solvent transfer. Bobreshova et al.[177] and Shel'deshov et al.[178] have evaluated the contribution of the ion convection transport within the pores of heterogeneous membranes when varying the hydrostatic pressure difference, the electrolyte concentration, and the current density. They have applied the extended Nernst–Planck equation, Equation 9.24, taking into account that the convection is caused by both electroosmotic and pressure-driven water transfer. By comparing their simulation with experimental data, the authors[178] have found that the convective transfer through the MA-40 heterogeneous membrane is controlled by an array of mesopores connecting two neighboring macropores. This result is conformed with the hydrodynamic model of fluid flow in biporous media developed by Filippov et al.[134,179] The electroosmotic contribution decreases with increasing electrolyte concentration (due to reduction of the EDL), and at 2 M NaOH the pressure-driven contribution to the convection caused by a small pressure difference in 30 kPa becomes about 10 times higher than the electroosmotic one. Hence, when applying a pressure difference across a membrane together with an electric current, it becomes possible to operate the convection flux. For example, if a sufficiently elevated pressure is applied in the concentration compartments of an ED stack, the solvent is driven out of these compartments, and a more concentrated solution is obtained. This is especially important when using hydrodynamically closed concentration compartments without solution circulation (Fig. 9.15; see also Chapter 19 of this book). Generally, combining electric and pressure gradients across a membrane opens up large possibilities in operating ED separation and concentration.

Water Transfer in an ED Cell Equations 9.40 and 9.41 are applicable to one membrane. In an ED stack (Fig. 9.16a), cation and anion exchange membranes are alternatively arranged to form desalting and concentrating compartments[10–12] (see also Chapter 21). For describing an ED process, at least two membranes making an elementary cell with one desalting and one concentrating compartments should be involved. Zabolotsky et al.[52,180] have developed a model for ion and solvent transfer for an ED cell with a cation and an anion exchange membranes where there is no circulation through the concen-

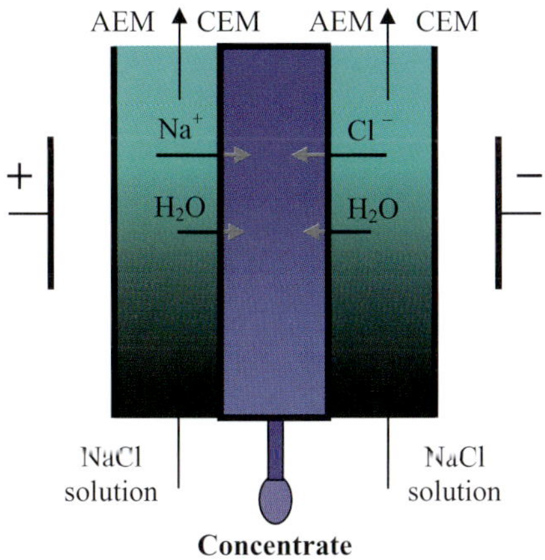

FIGURE 9.15. Scheme of an ED cell involving a concentration compartment with closed entrance for feed solution. AEM, anion exchange membrane; CEM, cation exchange membrane.

tration compartment: the entrance is closed (Fig. 9.15). Water enters this compartment by electroosmosis, mainly within the hydration shell of counterions. The concentrate flows out via a capillary.

The transport equations applied in the model result from integration of the Kedem–Katchalsky equations. The total fluxes of salt (s) and water (w) through both membranes into the concentration compartment are written as follows[52,180]:

$$J_{s\,tot} = -P_s(c_b - c_d) + \eta\frac{i}{F}, \tag{9.42}$$

$$J_{w\,tot} P_w(c_b - c_d) + t_w\frac{i}{F}, \tag{9.43}$$

where $P_s = P_s^a + P_s^c$ and $P_w = P_w^a + P_w^c$ are the electrolyte diffusion and water osmotic permeabilities related to the cell pair, respectively; P_s^a, P_s^c, P_w^a, and P_w^c relate to anion exchange (a) and cation exchange (c) membranes, respectively; $t_w = t_w^a + t_w^c$ is the water transport number per cell pair; $\eta = 1 - t_+^a - t_-^c$ is the current efficiency of electromigration transfer; and c_b and c_d are the electrolyte concentration in concentration (brine) and desalination compartments, respectively.

The brine concentration expressed in mole fractions is found as:

$$N_s = \frac{j_{s\,tot}}{j_{s\,tot} + j_{w\,tot}}. \tag{9.44}$$

The results of calculations of N_s as a function of current density with the use of Equations 9.42–9.44 are in a good agreement with the experiment.[180] The

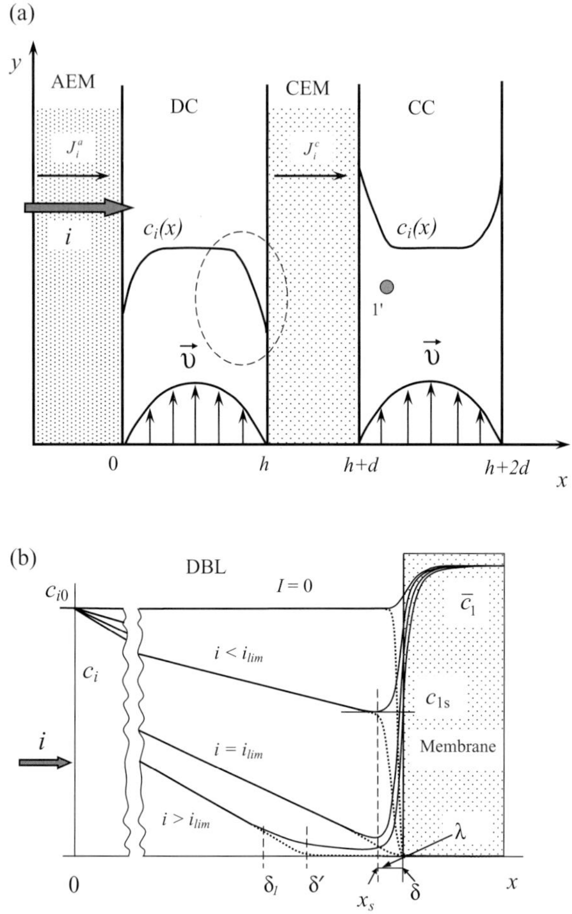

FIGURE 9.16. (a) Scheme of an ED cell with anion exchange (AEM) and cation exchange (CEM) membranes; DC and CC are desalting and concentrating compartments, respectively. The points 1 and 1′ show where the tips of Luggin's capillaries are placed for measuring potential difference. (b) Schematic view of salt counterion (solid line) and co-ion (dashed line) concentration profiles in the diffusion boundary layer (DBL) of thickness δ near a CEM at different current densities. The DBL is divided in an electroneutral ($0 \le x \le \delta_1$) and a space charge ($\delta_1 < x \le 1$) regions. The intersection of the straight line extending the linear concentration profile in the electroneutral region with the x-axis gives the effective thickness of DBL (δ); λ is the thickness of quasi-equilibrium electric double layer. Adapted from Urtenov et al.[91]

fitted parameters for a system involving MK-40 and MA-40 membranes and concentrated LiCl aqueous solutions are gathered in Table 9.1.

The numerical values of parameters are well agreed with other data. Thus, the t_w/η value shows how many water moles are transported into the brine compartment with 1 mole of salt. This value should be close to the hydration

TABLE 9.1. Transport Characteristics of the MK-40/MA-40 Membrane Pair in LiCl Solutions[180]

η, %	$P_s \times 10^8$, m s^{-1}	$P_w \times 10^5$, m s^{-1}	$t_w, \dfrac{\text{mole } H_2O}{F}$
91 ± 6	2 ± 1	0.66 ± 0.13	10.5 ± 0.9

number of LiCl in concentrated water solutions, which conform to the membrane pore solution with the concentration close to 3–4 M. The hydration number of LiCl in 3 M solution is approximately equal to 12.[85]

It is shown[180] that the main membrane characteristics controlling ED con centration process are the permselectivity determining the value of η, and the electroosmotic permeability determining t_w. In a less degree, the process is affected by the diffusion and osmotic permeabilities.

9.4 CONCENTRATION POLARIZATION IN ED

9.4.1 Current-Induced Concentration Gradients

As noted above, the main functional property of IEMs is permselectivity toward the ions of a certain sign of charge (counterions). Namely, this feature determines the applications of this type of membranes in ED, enabling electrolyte concentration to decrease and increase in different ED compartments (Fig. 9.16a). However, the same property is the reason of restrictions in the mass transfer rate across IEMs in ED, which is referred to the current induced concentration polarization phenomenon. Thus, the understanding of concentration polarization is important for the successful application of IEMs and their improvement. In this section, we briefly consider this phenomenon.

When a direct current of density i is switched on across a membrane, during the first moments the current in the membrane is carried mainly by the counterions, while in the solution the contribution of both cations and anions is comparable. Hence, immediately after the switching, the migration flux of counterions through the membrane will be higher than their flux from the solution bulk to the interface. As a result, the concentration of counterions (it can be shown that the same concerns the co-ions) will decrease at one membrane side and increase at the other. The changes in concentrations will continue up to the moment when increasing with time diffusion transfer compensates completely the difference of migration fluxes in the solution and the membrane.

At any time, the flux density ingoing to the membrane/solution interface from the solution is equal to the flux density outgoing into the membrane bulk. This condition can be written as:

$$(J_i)_s = \left(-D\frac{\partial c_i}{\partial x} + \frac{it_i}{z_iF}\right)_s = \frac{iT_i}{z_iF}, \tag{9.45}$$

where $(J_i)_s$ is the flux density of ion i through the interface; the middle part of the Equation 9.45 refers to the solution: D is the electrolyte diffusion coefficient in solution, C_i and t_i are the concentration and transport number of ion i in solution; the right-hand part refers to the membrane: T_i is called the integral[13,181] or effective[52,182] transport number of this ion in the membrane. T_i is defined as the current fraction carried by ion i through the interface or the membrane in steady state under all forces applied:

$$T_i = \frac{z_i F (J_i)_s}{i}. \tag{9.46}$$

T_i can differ from the electromigration transport number in the membrane, t_i^{mb} (t_i in Eq. 9.7 written for a membrane) owing to the different migration contributions into the transfer: diffusion and convection. Nevertheless, if the external solution concentration is not too high, T_i is rather close to t_i^{mb}. Normally, commercial membranes are highly permselective to counterions[83]: the concentration of co-ions is much lower than that of counterions, hence, T_1 and t_i^{mb} verge toward 1.

Equation 9.45 expresses the condition of flux continuity at the interface, and is verified in unsteady and steady state. In steady state, the total flux density of any ion is the same at any point in the solution and the membrane.

The emergence of concentration gradients at membrane/solution interface under effect of external driving forces is called, following the International Union of Pure and Applied Chemistry (IUPAC) recommendations,[183] "concentration polarization." This term is used in electro- as well as in pressure-driven membrane processes.[10] In all cases, the reason of concentration polarization is that the membrane has the ability to transport some species more readily than the other(s): the retained species are concentrated at the interface while the concentration of transported species decreases. Thus, concentration polarization phenomenon is inherent to membrane separation processes. Note that sometimes concentration polarization is understood[184] as a complex of effects related to the formation of current-induced concentration gradients near a membrane (electrode) surface, including, in particular, the overlimiting current phenomenon.

According to classical electrochemistry, formation of concentration gradients near the surface of a membrane (or an electrode) results in the limitation of the current density, i. With increasing i, the electrolyte concentration at the interface, c_s, decreases (Fig. 9.16b). When c_s approaches zero (becomes much smaller than the bulk concentration, c^0), the current density reaches its limiting value (i_{lim}). The expression for i_{lim} is readily obtained from Equations 9.45 and 9.46, when writing the concentration gradient of the counterion (1) at the membrane surface as:

$$\left(\frac{\partial c_1}{\partial x}\right)_s = -\frac{c_1^0 - c_{1s}}{\delta}, \tag{9.47}$$

where δ is the thickness of the Nernst diffusion boundary layer (DBL) in the depleting solution. According to Equation 9.47, δ is the distance from the membrane to the intersection point of the tangent drawn to the concentration profile at the interface with the line corresponding to the bulk solution concentration. Substituting Equation 9.47 into Equation 9.45 and setting $c_{1s} = 0$ yields:

$$i_{\text{lim}} = \frac{FDc^0}{\delta(T_1 - t_1)}, \tag{9.48}$$

where $c^0 = z_1 c_1^0 = -z_2 c_2^0$ is the bulk electrolyte concentration in Eq $L.^{-1}$

Equation 9.48 was first obtained by Peers in 1956.[185]

Equation 9.48 assumes that the local electroneutrality condition is hold and no other charge carries besides salt anion and cation present in the solution. Within this approach, when i tends to its limiting value (i_{lim}), the potential drop over a membrane surrounded by two DBLs tends to infinity. However, in real membrane or electrode systems, the limiting current density can be exceeded in several times (Fig. 9.17) due to a complex of effects arising at the membrane (electrode) surface under the combined action of electric current

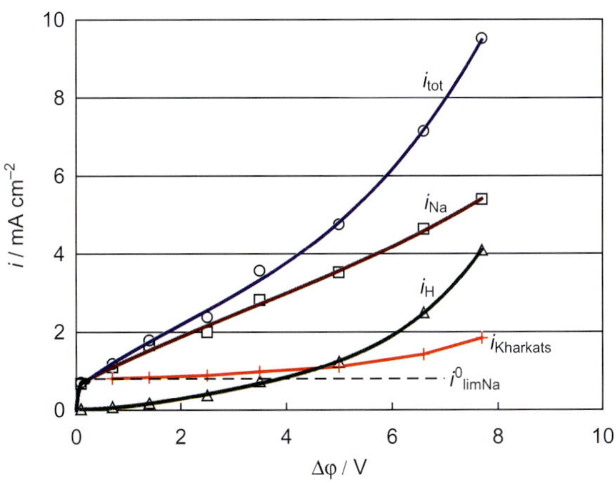

FIGURE 9.17. Current–voltage curves for a desalting compartment formed of a cation exchange membrane MK-40 and an anion exchange membrane MA-40 with 3 × 3 cm^2 of active membrane area and 1 mm of intermembrane space; the positions of measuring electrodes are shown in Figure 9.1a. A 0.002 M NaCl is flowing between the membranes with a velocity of 3.2 $cm\ s^{-1}$. The total current (i_{tot}) and partial currents of Na^+ (i_{Na}) and H^+ (i_{H}) ions through the MK-40 membrane, as well as the Kharkats current (i_{Kharkats}) and the limiting current density ($i^0_{\text{lim}\,Na}$), are shown; i_{tot}, i_{Na}, and i_{H} are experimental data,[188] i_{Kharkats} is calculated using Equation 9.36, and $i^0_{\text{lim}\,Na}$ is found by the intersection of tangents drawn to i_{tot} at $i = 0$ and at the inclined plateau region. Adapted from Urtenov et al.[91]

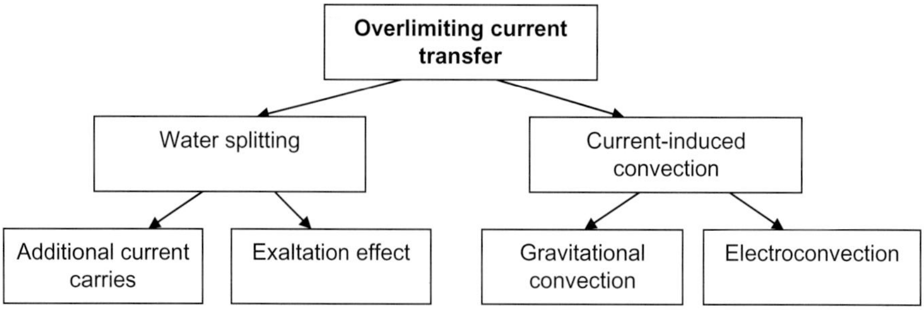

FIGURE 9.18. Scheme of overlimiting current transfer mechanisms.[203]

and concentration variations (produced by the same current) at the interface. In early period of ED, the use of underlimiting current was prescribed.[182,186] However, now the overlimiting current regimes are widely used in ED and especially in electrodeionization of diluted electrolyte solutions.[187–190] We will see below that the main effect responsible for the intensive overlimiting transfer is electroosmosis of the second kind,[191–193] also named induced-charge electroosmosis.[194,195] This effect is used in a number of nano- and microfluidic devices[118,124,194,196] such as electrokinetic micropumps.[197,198] Besides, intensive current regimes are applied in electrophoresis,[196] electrodeposition,[199] layering of colloid crystals on electrode surfaces,[200] and other domains related to separation science.[64,115,193]

9.4.2 Mechanisms of Overlimiting Current

In literature,[89,188,189,201–203] four effects explaining the overlimiting current transfer are discussed (Fig. 9.18). The first two are related to generation of H^+ (OH^-) ions (usually named "water splitting")[15,186] at the membrane/solution interface. The water splitting in membrane systems was observed by a number of researchers,[15,17,18,186,204,205] and for a long time the transfer of current by additional carriers, H^+ and OH^- ions, was considered as the main and often the only reason for the overlimiting conductance.[186] However, the generation of H^+ and OH^- ions causes another, less evident mechanism of overlimiting transfer, the exaltation effect, first studied by Kharkats.[206] The emergence of H^+ and OH^- ions near the interface disturbs the electric field that can increase (exalt) the salt counterion transfer: for example, the OH^- ions generated into the depleted diffusion layer adjacent to a cation exchange membrane attract the salt cations from the solution bulk toward the interface.

There are two other mechanisms of overlimiting conductance that contribute to salt counterion transfer. These are two types of current-induced convection, gravitational convection and electroconvection. They both provide additional mixing of depleted solution, in comparison with forced convection. This mixing is produced by local vortices resulting from the action of two volume forces different in nature.

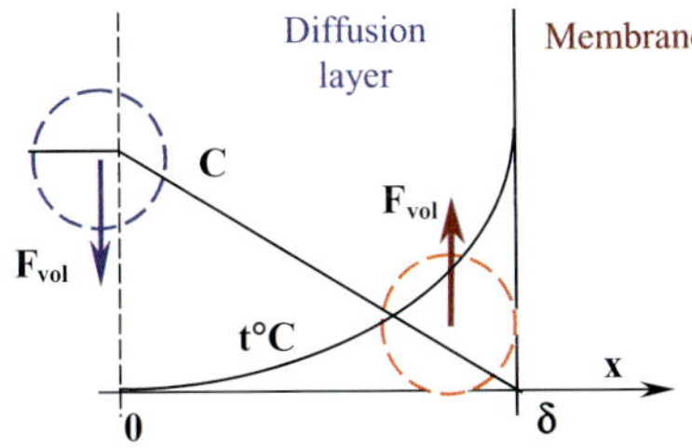

FIGURE 9.19. Scheme of gravitational convection.

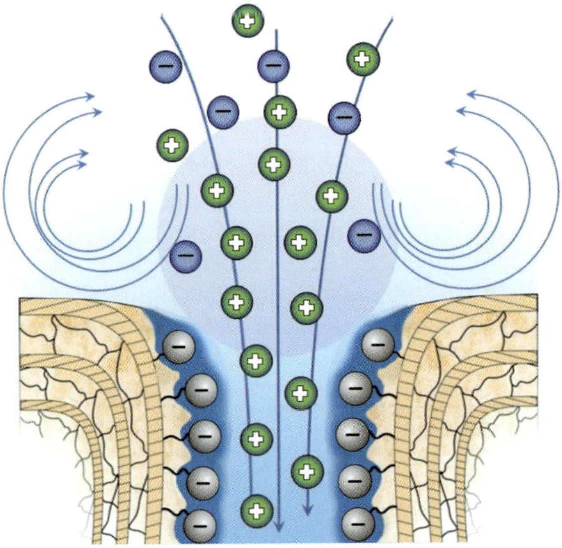

FIGURE 9.20. Scheme of electroconvection near the entrance in a membrane nanopore: the action of electric current on the space charge produces an excess in local pressure, which generates a pair of vortices. Adapted from Nikonenko et al.[203]

Gravitational convection develops due to the nonuniform distribution of solution density, which causes Archimedean volume force, $\mathbf{F}_{vol}$, bringing the liquid in motion (Fig. 9.19). Near the membrane, solution is more diluted then in the bulk. Moreover, its temperature is elevated due to Joule heat within a layer with high electric resistance. As a result, near the surface, the Archimedean buoyant force acts upon a parcel of fluid vertically upward, while in the bulk, the body force applied to a parcel of fluid acts in opposite direction. This couple of forces produces vortex motion of the fluid in the space near the membrane.

Electroconvection arises due to the action of the electric field on the electric space charge in the boundary depleted solution.[89,184,191,192] The mechanism of generation of a pair of vortices is shown in Figure 9.20. For this, a local increase in space charge density is needed. As the membrane surface is not electrically homogeneous, this local increase is possible in different scales, from nanometer (shown in Fig. 9.20) to several tens of micrometer, in the case of heterogeneous membranes (Fig. 9.3b). Rubinstein and Zaltzman[184] have theoretically shown that electroconvective mixing develops even if the membrane surface is homogeneous, due to fluctuations at the wall. A slight distortion of membrane surface (making the surface waved) results in precipitation of the onset of overlimiting conductance and a respective increase of the current compared to those for a flat membrane.[184]

The structure of depleted DBL near an ion exchange membrane at different currents is shown in Figure 9.16b. Numerical and analytical solutions[89–91,184,207] have shown that the following three regions may be distinguished within the DBL (Fig. 9.16b). The quasi-electroneutral region of thickness δ_1 is next to the solution bulk. It is separated from the membrane surface by a space charge region (SCR), which adjoins the membrane surface and where the salt co-ion concentration is very small. The SCR may be divided into a migration zone (of thickness δ_2) and a quasi-equilibrium boundary layer (of thickness δ_3, denoted as λ in Fig. 9.16b). In the migration zone (which is absent at $i < i_{\text{lim}}$), the diffusion contribution to the counterion flux is negligible in comparison with the electromigration one. In the quasi-equilibrium zone, the diffusion and migration terms in Nernst–Planck's Equation 9.21 are high and have opposite signs, so that the J_1 term may be neglected, then c_1 and E are linked by the Boltzmann relation.

The fact that the DBL is distinctly divided into electroneutral and SCRs allows integration of flux equations in the region $0 \le x \le \delta'$ at $i \ge i_{\text{lim}}$, under the condition that the concentrations of all species are negligible at $x \le \delta$ in comparison with c_{10}. This integration conducts to an equation relating the flux density of counterion, J_1, with that of H^+ (OH^-) ions produced in the water splitting, J_w[52]:

$$J_1 = \frac{2D_1 c_{10}}{\delta'} + \frac{D_1}{D_w} J_w, \tag{9.49}$$

where subscript $_1$ refers to the salt counterion and $_w$ to the ion generated into the depleted solution in the course of water splitting (OH^- in the case of cation exchange membrane). Equation 9.49 is a generalization of Peers' Equation 9.48 and Kharkats' equation[206] deduced under the LEN assumption. δ' enters Equation 9.49 instead of δ in Peers' and Kharkats' equations. Thus δ' acts the role of the effective thickness of the DBL; geometrically, it is the distance from $x = 0$ to the point where the straight line following the counterion concentration profile in the electroneutral region cuts the x-axis (Fig. 9.16b). If we assume $\delta' = \delta_0$, where δ_0 is the DBL thickness formed near an "ideal" membrane not involving coupled convection (δ_0 may be calculated by the Leveque equation[88] for lamillar flow between two plates), the first term in Equation 9.49 gives a constant value, the limiting current density. In this case, the current density found by Equation 9.49 is called Kharkats' current.[91]

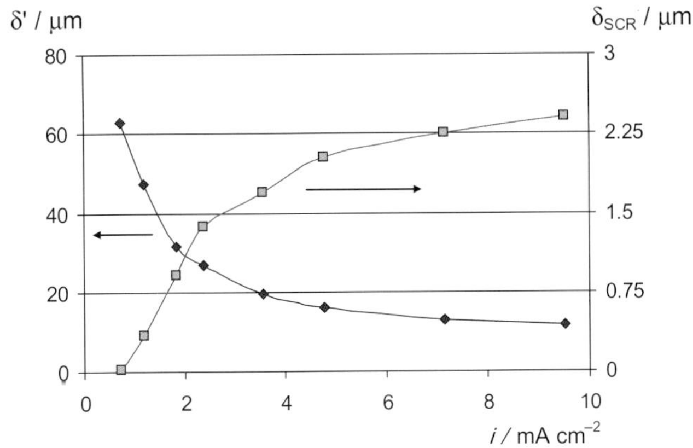

FIGURE 9.21. Effective thickness of the DBL (δ') and the thickness of SCR (δ_{SCR}) as functions of the current density applied to the membrane system described in Figure 9.17. Adapted from Urtenov et al.[91]

Equation 9.49 shows that the overlimiting increase of the salt counterion flux density may be due to only two causes: a decrease in the effective thickness of the DBL, δ', or an increase in the water splitting rate, J_w, which gives rise to the exaltation effect. The exaltation effect is rather low due to low value of the D_1/D_w ratio: it is about 0.25 in the case where ion $_1$ is Na^+, and $_w$ is OH^-.

When knowing the experimental values of J_1 and J_w , it is possible to evaluate δ' from Equation 9.49 (Fig. 9.21).

The decrease in δ' with increasing current density may be produced by current-induced convection mixing the boundary solution and shifting the outer edge of the DBL toward the membrane. As we mentioned above, this partial destruction of the DBL may be produced by gravitational convection or by electroconvection. The contribution of the gravitational convection in overlimiting current transfer is important when the variation in fluid density near the membrane and the spacing between the membranes (h) are high while the solution flow velocity (V) is low. Otherwise, when h, solution bulk concentration (c^0), and current density (i) are low, and V is high, gravitational convection is negligible. Theoretically, its contribution may be evaluated[208] via Rayleigh and Richardson numbers. In experiments, gravitation convection was negligible when $h < 1$ mm, $c^0 < 0.02$ M NaCl, and $V > 0.07$ cm s^{-1} [209] or $h < 6$ mm, $c^0 < 0.05$ M NaCl, and $V > 0.4$ cm s^{-1}.[203]

As opposed to gravitational convection, electroconvection is more intensive with diluting solution, since the EDL thickness as well as extended SCR[184] increases in this case. In the case of 0.002 M NaCl circulating through a short (3 cm) ED channel, the effective thickness of DBL can decrease to as low as 15 μm, under a rather strong, 8 V, potential difference over a pair cell (Fig. 9.21). The numerical calculation of the extended SCR thickness (by fitting the current–voltage curve[91] presented in Fig. 9.17) shows that this value may attain

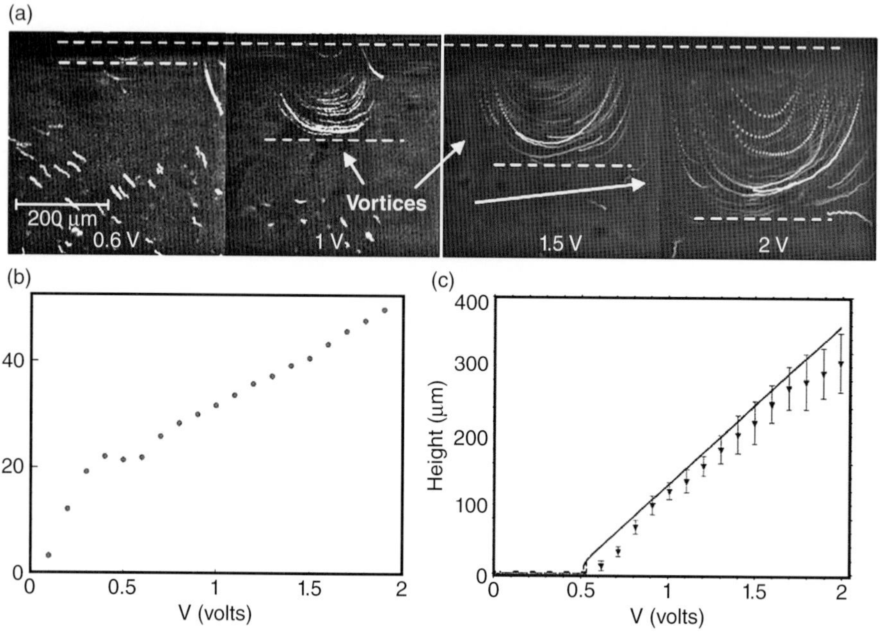

FIGURE 9.22. (a) Time-lapse snapshots of the experimental cell seeded with tracer particles showing "quasi-steady-state" streamlines. The membrane is situated at the top boundary of each image; the applied voltage (volts) is indicated. (b) Corresponding current–voltage curve. (c) The measured size of the vortical structures appearing at the limiting current density. Adapted from Rubinstein et al.[210]

few micrometers and become much higher than the equilibrium Debye length, λ (several nanometers).

The direct experimental observation of the nonequilibrium electroosmotic vortices together with simultaneous registration of current–voltage curves at overlimiting currents in an ED cell was realized by Rubinstein et al.[210] (Fig. 9.22). For the visualization, the authors have used tracers and a transparent cell. The membrane was positioned horizontally so that no gravitational convection appears: when a direct current is applied, a depleted DBL forms under a cation exchange membrane separated from a copper anode by a 0.005 M $CuSO_4$ solution. The heavier concentrated DBL is over the membrane under study.

9.5 RELATIONSHIPS BETWEEN ELECTROCHEMICAL BEHAVIOR OF ION EXCHANGE MEMBRANES AND THEIR BULK AND SURFACE STRUCTURE: MEMBRANE MODIFICATION

In previous sections, we have become acquainted with the structure and main properties of ion exchange membranes. A number of different membranes (about 50 kinds) are commercially available.[11,12] In spite of this variety, they cannot satisfy increasing requirements of science and industry. For different separation processes, various properties are needed. Thus, in the case of deep water desalination by ED, the ability of the membrane to generate electroconvection is of great interest, while in the case of electrolyte concentration, the main desirable property is the ability to restrain the water transfer. For membranes used in fuel cells, the major characteristics are high electric conductivity, low gas permeability, high thermal stability, and low dependence of the conductivity on humidity.[21,211–213] In all cases, the membranes must have high selectivity toward counterion transport.

One of the most evident approaches for mass transfer improvement in membranes is their bulk and/or surface modification. Modification opens wide opportunities for preparation of membranes with different properties on the basis of a rather small number of commercially available membranes. A small amount of additive materials only slightly increases the costs of membrane production, hence this method is believed economically effective. We shall consider some examples of IEM modification aimed at improving their characteristics, necessary for different process.

9.5.1 Membrane Surface Modification

Membrane Profiling One of the first ideas of surface modification was membrane profiling in order to obtain relief (waved, undulated, or corrugated) surface.[189,214–217] The gain in mass transfer rate is determined by several effects. First, the active membrane area available for mass transfer increases. Second, there is an increase in fluid mixing by better hydrodynamic situation. Third, electroconvective mixing becomes more effective as the tangential component of driving force increases.[184,193] Fourth, the water splitting rate decreases, as the concentration polarization of the profiled membrane is lower. All these four effects are synergic in increasing mass transfer. As a result, the mass transfer coefficient, growing with diluting feed solution, can be up to 10 times higher than that in the same ED stack with conventional spacer (at 0.05 mM NaCl in the feed solution).[218]

Permselectivity among Different Ion Species Permselectivity of charged membranes toward ions of certain sign is their main property. It is possible to prepare membranes permselective to specific ion species. Permselectivity among ions of the same sign of charge in ED is governed by the difference in affinity of different ionic species with the membrane (ion exchange equilibrium constant) and the difference in the ionic mobility.[219] This specific permselectivity can be achieved by surface treatment of the membrane. One of the methods aiming at increasing membrane permselectivity toward monocharged counterions is surface coating with a thin ion exchange layer having a low concentration of fixed sites with the charge sign opposite to that of the fixed sites of the membrane bulk. Thus, a cation exchange membrane was coated with a (slightly charged) anion exchange layer, for example containing amines.[220,221] This thin layer serves as a barrier toward multicharged ions while only slightly restraining the transfer of monocharged ones. This coating can be made, for example, by immersing a cation exchange membrane in a cationic weakly basic polyelectrolyte (such as polyethyleneimine) solution.[222] A relative transport number of the calcium to sodium ions dropped about 10 times during ED of a sodium chloride–calcium chloride solution. Moreover, this layer being a cationic surface-active agent can prevent the increase in the electric resistance of the membrane due to organic fouling.[222]

In Hu et al.,[223] quaternized chitosan was used as a cationic layer for surface modification of heterogeneous and homogeneous commercial cation exchange membranes to enhance their specific permselectivity. The transport number of the metallic multicharged cations was decreased significantly in binary systems (H^+/Zn^{2+} and H^+/Al^{3+}) after surface modification due to the electrostatic and steric barrier created by a thin chitosan layer.

Sata[219] has classified and reviewed methods resulting in obtaining specific permselectivity: (1) to vary cross-linkage of ion exchange membranes, (2) to form tight surface layers, (3) to decrease/increase hydrophilicity of the membrane surface by introducing specific ion exchange groups or to impregnate hydrophilic compounds, (4) to control permselectivity of anions by photoirradiation using membranes with a photoresponsive group, and (5) to control permselectivity through thermally responsive anion exchange membranes with temperature.

There is a large number of applications of IEMs with specific permselectivity.[219–226] Let us mention nitrate–chloride (drinking water denitrification),[224] fluoride–other anions (drinking water defluoridation),[225] chloride–sulfate (potable salt production from sea water)[226] separations, and others.[219]

Increasing Fouling Resistance Increasing fouling resistance can be obtained by the use of heterogeneous polyethylene-based anion exchange membranes modified by a hydrophilic anion exchange coating. The modified membranes showed a smaller fouling tendency than commercial ones in the desalination of recycled water.[227] Membrane modification with high molecular mass surfactants has mitigated fouling by organic compounds during ED of aqueous solutions.[228] Antifouling effect was also achieved by coating ion exchange membranes with a 12-μm-thick urethane acrylate layer.[229]

Lowering Water Splitting It is known that the rate of water splitting at the interface of anion exchange membranes generally is higher than that at the interface of cation exchange ones.[10,204,230] The reason is that secondary and tertiary amino groups often serving as ionogenic fixed sites in anion exchange membranes are strongly catalytically active in relation to water dissociation reaction.[15,17] Even when the ionogenic fixed sites are quaternary amino groups, sometimes these groups are not stable and transform into tertiary amino groups under the action of intensive electric current. An example is the case of AMX anion exchange membrane described by Choi and Moon.[231] High water splitting rate at anion exchange membrane in ED treatment of natural waters leads not only to lower current efficiency, but some other nondesirable effects. In particular, in this case, the desalinating stream becomes acid that hampers the removal of weak acid anions, since they pass in nondissociated form. Moreover, alkalinization of the concentrating stream increases the risk of the hardness salts precipitation on the concentrating membrane surface.

It is possible to lower the water splitting rate by surface modification of anion exchange membrane. The case of modification of MA-40 with a strong polyelectrolyte solution is described in Pismenskaya et al.[232] The idea is to convert secondary and tertiary amino groups initially present in MA-40 membrane into quaternary ones, which are much less catalytically active toward water dissociation reaction.[15,17] Figure 9.23 shows the results of measurements of pH of the solution outgoing from ED desalination compartments formed by modified, MA-40M, or unmodified, MA-40, anion exchange membranes together with different cation exchange membranes, MK-40 or CMX (Neosepta®, Tokuyama Corporation, Shibuya, Japan). The rate of water splitting at the MA-40 membrane is higher than that at the MK-40 one. Hence, when the current density becomes equal or higher than i_{lim} (where the potential drop over a pair of the membranes with desalting solution between them, $\Delta\varphi$, is higher than 1 V), the pH of the outlet desalted solution becomes acidified: the amount of H^+ ions generated at the MA-40 membrane interface into the desalination compartment is higher than the amount of OH^- ions generated at the MK-40 interface. If MA-40 is replaced with the MA-40M membrane, there is no pH variation up to 3 i_{lim} (up to $\Delta\varphi \approx 9$ V), that is, the rate of water splitting at MA-40M is as low as that at MK-40. When MA-40M is used together with a CMX membrane, the desalted solution becomes alkalinized (in the interval between i_{lim} and 3 i_{lim}) because the rate of water splitting at CMX is higher than that at MA-40M. Note that the properties of modified MA-40M membrane was stable at list during 300 hours.[232]

Increasing Electroconvective Mixing According to classical physicochemical hydrodynamics,[233] the thickness of DBL and, hence, the limiting current density, Equation 9.48, are not functions of the surface properties. This is a consequence of the no-slip condition for the velocity at a solid wall. However, a strong correlation between the limiting (and overlimiting) current density and the contact angle on the membrane surface was established[203] (Fig. 9.24, Table 9.2). The contact angle, which is a measure of the surface hydrophobicity (determined in the swollen state 20 s after water drop application while the bottom of the membrane remains in contact with the equilibrium solution), increases in the range MK-40 < CMX < MK-40 + F < Nafion-117 (Table 9.2). The overlimiting mass transfer increases in the same range. Hence, the more the surface is hydrophobic, the higher the overlimiting mass transfer rate.

Note that the slip boundary condition was introduced by Navier in the form[166]:

$$u_{\text{slip}} = b\left(\partial V_y / \partial n\right)_{x=\delta}, \tag{9.50}$$

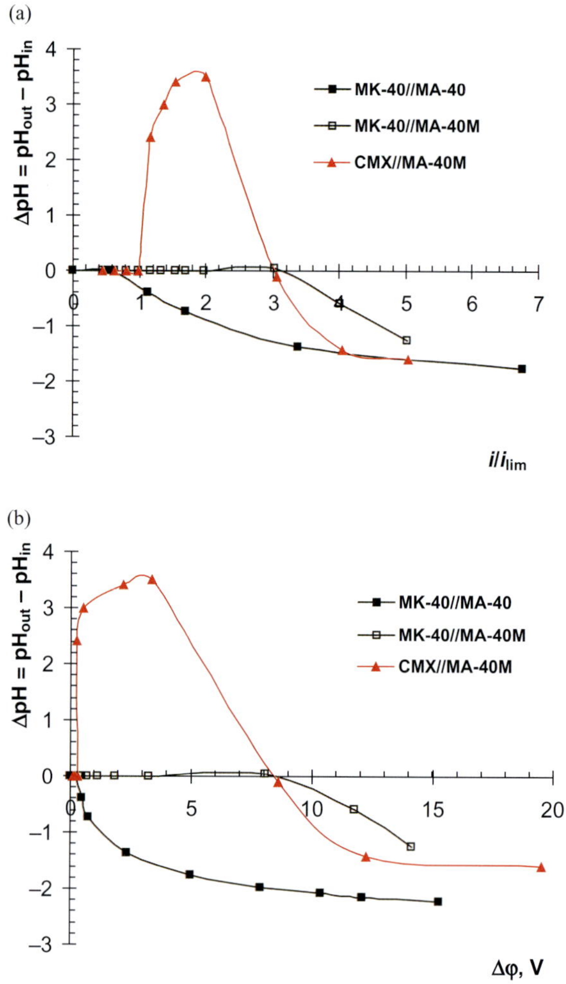

FIGURE 9.23. Difference of pH in the outlet and inlet solutions of the desalination compartment versus the current to limiting current ratio (a) and potential drop over a cell pair (b). $C = 0.005$ M; $h = 1$ mm; $S = 3 \times 3$ cm^2; $V = 1.6$ cm s^{-1}. Adapted from Pismenskaya et al.[232]

which relates the fluid tangential velocity u_{slip} at the surface and the shear strain rate normal to the surface, $\partial V_y/\partial n$, via the slip length b; the velocity of the surface is set to zero. $b = 0$ relates to the no-slip condition; u_{slip} increases with increasing b.

The nonslip condition is supported by macroscopic experiments, in particular, in electrochemistry when nonintensive current regimes are applied.[88] However, when fluid volumes are reduced, the impact of surface phenomena increases and the slip of fluid at the interface becomes important.[166,234] The slip

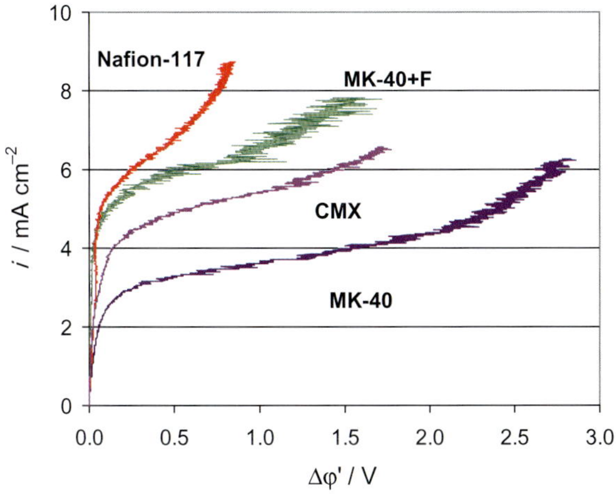

FIGURE 9.24. Current–voltage curves of MK-40, CMX, Nafion-117, and modified MK-40+F membranes in horizontal position with the depleted DBL under the membrane. $\Delta\varphi'$ is the corrected pd reduced by the ohmic contribution.[201] The limiting current density $i_{\lim} = 2.7$ mA cm^{-2} is calculated using the Lévêque equation[88] deduced when applying the no-slip condition. Adapted from Nikonenko et al.[203]

TABLE 9.2. Contact Angles for Different Cation-Exchange Membranes[203]

Membrane	MK-40	CMX	MK-40+F	Nafion-117
Contact angle, degrees	39 ± 5	49 ± 3	53 ± 4	64 ± 4

velocity depends strongly on the degree of surface hydrophobicity. On the hydrophobic surface, there is a repulsion of water molecules, which causes a slip of the fluid over the surface.[235,236]

Slip velocities on both hydrophilic and hydrophobic surfaces have been studied using a variety of experimental techniques.[235–237] The values of slip length reported vary from several nanometers[236] to as high as several tens of micrometers.[237,238] Majumder et al[237] have found $b = 25$ μm from the pressure-driven flow rate measured through aligned multiwall carbon nanotubes incorporated in a membrane. The flow velocity normalized at 1 bar was found to be 25 cm s^{-1}, that is, four or five orders of magnitude faster than the evaluation by the Hagen–Poiseuille law. Hydrophobicity and, hence, the slip length can be significantly amplified by roughness and can reduce friction due to trapped nanobubbles.[239]

Obviously, the slip of fluid at hydrophobic surface takes place in the case of ion exchange membranes. This effect increases with increasing hydrophobicity, which explains the results presented in Figure 9.24.

The properties of MK-40 + F membrane present a particular interest. This membrane is prepared by casting a thin (of several micrometers thickness) film of a sulfonated perfluorocarbon polymer on the surface of heterogeneous MK-40 membrane. The increase in the contact angle on the surface of the modified membrane is due to hydrophobic properties of the caste film. Note that besides increasing overlimiting current density, the modified membrane is characterized by lower water splitting[240,241] due to homogenization of the current lines distribution near the surface.[203]

9.5.2 Membrane Bulk Modification

Preparation and Properties of Organic/Inorganic Materials The membrane surface properties control mainly separation processes. However, in other cases, such as fuel cells, it is desired to modify all the membrane volumes, since when the membrane thickness is minimized, all the membrane volumes should be active.[212] Hybrid organic/inorganic materials are the main subject of investigations in this field.[213,242,243] Generally, these composite materials present high interest due to their extraordinary properties, electrical, mechanical, thermal, and others, which arise from the synergism between the properties of the components.[11,244]

First works in the field of organic/inorganic materials were devoted to the synthesis of a large class of sulfonate membranes on the base of layered zirconium phosphate containing organic chains with $-SO_3H$ group. Alberti et al.[245–247] have synthesized a large number of new sulfonate materials with high proton conductivity, which can be promising for fuel cells.[248,249] However, their mechanical properties are not very attractive for membrane applications.

The most simple and common way of polymer membrane modification is the use of small inorganic particles (mainly nanoparticles). They can be presented both by substances capable of generating mobile ions during dissociation or by inert compounds.[212,242] Among various types of inorganic components, which can be used for membrane modification, oxides of polyvalent elements (silicon, aluminum, zirconium) are used most frequently.

The modification of membrane materials by inorganic nanoparticles can be performed by two ways: first, by membrane casting from solutions containing finely dispersed additives[212,213] and second, by nanoparticle synthesis in the membrane matrix (*in situ*).[213,250–252] The first way is the most simple, but it does not always get successful results because finely dispersed particles can easily form aggregates. It reduces significantly the efficiency of membrane doping and sometimes results in the loss of mechanical strength and in macroporosity increase. Hence, high gas permeability is a typical defect of these membranes. Thus, the stabilization of their surface by various surfactants can be very attractive.[253,254]

On the other hand, the synthesis of nanoparticles may occur *in situ*, within the membrane matrix. The nanopores can uptake one of the initial reagents, for example, cations of polyvalent elements. After that, it is possible to carry out the further synthesis of nanoparticles in the same pores, using them for restriction of reaction volume and the size of formed particles. In this case the pores play a role of nanoreactors. The pore walls can isolate the formed particles from each other and reduce the surface tension, providing thermodynamic stability of nanoparticles.

A review of the methods of organic/inorganic materials preparation is made by Xu.[11]

Note that modification of homogeneous ion exchange membranes (such as perfluorinated Nafion and MF-4SK membranes) by inorganic particles results in noticeable improvement of these membranes (both ion conductivity and selectivity are improved),[250,255–257] while it is not the case of heterogeneous membranes such as MK-40.[251,252] The reason is that heterogeneous membranes contain both micro- and macropores. Therefore, the size of acid zirconium phosphate particles (from X-ray data) in MK-40 is much higher than that in MF-4SK membrane (2–5 nm).

Nanocomposites on the basis of ion exchange membranes, containing nanoparticles of organic substances, were recently investigated. The most important additive of this type is polyaniline. Such composites can be obtained by aniline polymerization in membrane matrix (e.g., in Nafion),[258] as well as in their solutions with the subsequent casting.[259] In the latter case the size of polyaniline (or polyaniline/Nafion interpolymer) particles is not limited by the size of membrane pores and changes from several up to tens of nanometers depending on the concentration of solutions used for synthesis and way of membrane preparation.[259]

Composite membranes on the basis of MF-4SK and polyaniline possess mixed ion and electronic conductivity in the case of the bulk modification. Contribution of electronic conductivity can reach 60–70%, if polyaniline was obtained *in situ* in membrane matrix.[258] These composites have the mixed proton–electron conductivity, which keeps the high value, but the composite diffusion permeability in acid solution decreases after modification.[257] At the same time, proton conductivity of membranes obtained by casting after aniline polymerization in MF-4SK solution passes through the maximal value for low polyaniline concentration.[259] The increase in polyaniline content results in both decrease of proton conductivity and mechanical properties of membranes. The addition of small amounts of polyaniline results in the linkage of a part of protons, in an increase of defect concentration, and probably in improving of membrane microstructure. Further increase in polyaniline concentration results in the decrease in current carrier concentration (the protons are bonded with nitrogen by strong hydrogen bonds SO_3-H–N-). Large polyaniline particles can exclude some proton conductivity channels from transport process.[259] The morphology transitions inside of the nanostructure of the basic membrane lead to a decrease in the conductivity, diffusion, and electroosmotical permeability of surface modified by polyaniline composite materials.[257,259] The properties of the obtained samples depends on the way of the polyaniline intercalation. As a result, the samples of MF-4SK/polyaniline after surface modification proved themselves as quite efficient for ED concentration of salt solutions[260] and for fuel cells.[258]

Composite MF-4SK/polyaniline membranes have essentially changed diffusive permeability in comparison with the starting material. A reduction of water transport number is also noted. It can be explained by decrease in membrane hydrophilic property.[259]

Special attention should be given to the problem of asymmetry of transport processes in membrane materials. Membranes with asymmetric permeability have been obtained by oxide nanoparticles introduction with a gradient distribution. The difference in the diffusion permeability in opposite directions can achieve 40%.[255,256] Anisotropic composite membranes on the basis of MF-4SK and polyaniline[145,259,261] were also obtained. The asymmetry of diffusion permeability and ion conductivity was found for these membranes. Similar properties were obtained for MK-40 and MA-41 membranes modified by tetrabutylammonium or dodecyl sulfate.[262,263]

Between other approaches to form composite membranes, nonconducting polymers impregnation by molecules or nanoparticle of substances containing mobile ions can be mentioned.[212,213] In order to have the possibility to dissolve such additives, the polymer must contain electronegative atoms. Membranes on poly(ethylene oxide) $(-CH_2-CH_2-O-)_n$ basis are often used for these purposes. Initial material has very low conductivity; however, the presence of a large number of oxygen atoms determines good solubility of various salts or acids in it. Their dissociation in membrane matrix provides high concentration of mobile ions and results in high ionic conductivity of such materials. Ion transfer in these materials takes place by means of cations motion from one group of electronegative atoms to another on a background of intensive segmental mobility of the polymer matrix providing reorganization of coordination polyhedrons. Such membranes with high lithium conductivity are often used in lithium batteries.[264–266]

Membrane materials can be doped by acids in order to increase proton conductivity as in the case of introduction of phosphoric acid into polybenzimidazole.[267,268] New membranes with high proton conductivity on the base of poly(vinyl alcohol) doped by phenol-sulfonic acids were also recently synthesized.[88,269]

Mechanism of the Increase in Ionic Conductivity of Polymer Membranes by Their Modification with Inorganic Nanoparticles The increase in ionic conductivity of composite materials was discovered by Liang.[270] This phenomenon can be attributed to the formation of additional defects, caused by the sorption processes at the phase boundary.[271,272] The optimal result should be achieved in the case of weak base (such as polyaniline) incorporation into Nafion membrane. The incorporation of small polyaniline amounts results in membrane conductivity increase. But if its concentration is more than 2 vol %, the membrane conductivity decreases with increasing polyaniline content. It is caused by the exception of too much carriers (protons) from the transfer process due to the sorption processes.

An increase in proton conductivity of hybrid membranes can be also achieved by incorporation of nanoparticles containing acid groups (acid zirconium phosphate).[252,256,273,274] In the last cases we can suppose that acid zirconium phosphate dissociation can increase carrier concentration. Similar results were obtained in the case of hydrophobic (metal) nanoparticles incorporation.[275] Moreover, Nafion-type membranes are strong acids[21] and the degree of dissociation of sulfonate groups is close to 100% in the hydrated state. Hence, the defect concentration (the oxonium ion in the frame of water molecules) is very high. Another suggestion, which is discussed in the literature, is the additional water molecules sorption by hydrophilic particles.[212] This can assist the proton transfer in membrane matrix according to the Grotthuss mechanism. But in the case of hydrophobic silver and polyaniline nanoparticles, the increase in proton conductivity cannot be explained by additional water sorption.

Proton diffusivities calculated from ^1H nuclear magnetic resonance (NMR) relaxation are mainly controlled by proton and water molecule migration in wide pores where most of the protons and water molecules are located.[276] These values are much higher than those calculated from membrane conductivity, since the latter are controlled by proton and water transfer through the narrow channels. The membrane modification by different dopants leads to the increase in ionic conductivity. At the same time, ^1H NMR relaxation rate in modified membranes can be both higher or lower in comparison with that of the unmodified samples.[276] This shows that the increase in proton conductivity can be attributed to the influence of incorporated nanoparticles on the size of narrow channels controlling ionic conductivity.

The formation of dopant nanoparticles takes place in the membrane pores. These particles occupy a part of the pore volume. The driving force of the pore expansion is osmotic pressure, which is determined mostly by the concentration of acid protons. The total amount of protons remains the same after modification, in the most cases. The amount of water molecules within the membrane matrix remains the same or even slightly increases after incorporation of hydrophilic particles. Incorporation of hydrophobic particles results in a decrease in water uptake by the modified membrane, but in this case, a noticeable pore volume increase can also be observed.[277]

Figure 9.25a,b shows the mechanism of the effect of nanoparticle growth within a large pore on the size of narrow channel linking two neighboring pores.[277] Expansion in the channel radius results in an increase in membrane ionic conductivity, which is controlled by proton transfer through such channels with low proton mobility. The activation energy of conductivity should decrease and approaches the activation energy for aqueous acidic solutions (5 kJ mol^{-1}).

Figure 9.25b,c shows that when the size of nanoparticles is not too high, they do not encumber all the space within a pore, however they allow an increase in channel radius. Further increase in nanoparticles volume (>3–4%

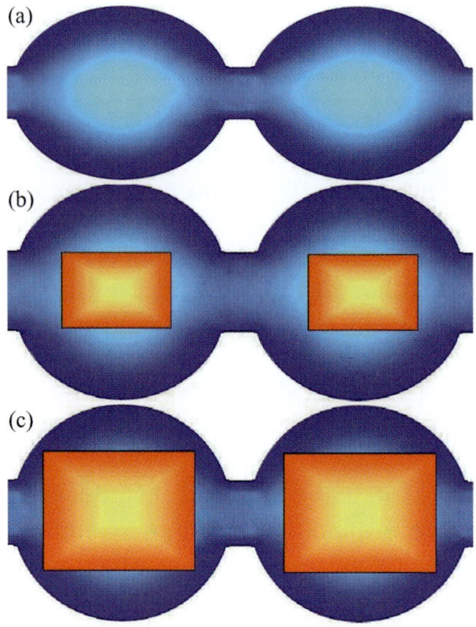

FIGURE 9.25. Scheme of pore structure for the (a) initial and (b, c) modified by nanoparticles membranes.

of total membrane volume) leads to essential encumbering of pore volume resulting in a decrease of conductivity and a growth of activation energy. The increase in nanoparticle volume up to 3–4 vol % leads also to a decrease in water uptake.[277]

Since a nanoparticle occupies the central part of a pore, the space available for free water molecules is found reduced. As a result, the volume fraction of bound water molecules making part of hydration shells increases. As well, the water activity increases, which leads to growing osmotic pressure, Equation 9.10, which is the cause of the increase in the pore size of modified membranes.

The dopant nature significantly affects the membrane conductivity. Thus, membranes doped with silica and zirconia are more conductive in comparison with the membranes doped with hydrophobic metallic particles. This effect is due to additional water sorption on the hydrophilic particle surface. Further conductivity increase can be achieved by simultaneous membrane modification by silica and phosphotungstic acid (PWA).[272,278] Most likely, the additional growth of proton conductivity can be explained by the increase in charge carrier (protons) concentration, which is accompanied by water uptake increase.

High proton conductivity at low humidity is mentioned among the main advantages of hybrid membranes.[212,279] A decrease in humidity results in partial membrane dehydration and in a dramatic decrease of pore volume.

The first advantage of the hybrid membranes in comparison with unmodified ones is that the pore and channel sizes in the first ones retains higher. But the main reason of low proton conductivity of membranes at low humidity is high distance between oxygen atoms participating in proton transfer. In hybrid membranes additional groups containing oxygen ions are incorporated into the membrane matrix. They also participate in the proton transfer process. These two facts results in the proton conductivity increase in hybrid materials at low humidity, which is very important for their application in fuel cells. And in the case of simultaneous membrane modification by silica and PWA, the effect of proton conductivity increase at low humidity became much more marked.[279] It means that the change of pores and channels size is important, but the nature of dopant particles is very important too. This fact makes the search of new approaches to membrane modification very promising.

ACKNOWLEDGMENTS

The chapter is prepared in the framework of the French–Russian laboratory "Ion-Exchange Membranes and Related Processes." We are grateful to CNRS, France, and RFBR, Russia, for financial support of this work.

ABBREVIATIONS AND SYMBOLS

Abbreviations

DBL	diffusion boundary layer
ED	electrodialysis
EDL	electrical double layer
IEM	ion exchange membrane
NPP	Nernst–Planck and Poisson (equations)
pd	potential difference
PFSA	perfluorosulfonated acid (ionomer)
TMS	Teorell–Meyer–Sievers (model)

Symbols

a_i, $a\pm$	ion and electrolyte molar activity, respectively
c_i	molar concentration of ion i
c^0, c_1^0	electrolyte and counterion concentration in the solution bulk, respectively
c_{1s}	counterion concentration at the membrane surface
d	thickness of the membrane

D electrolyte diffusion coefficient
D_i diffusion coefficient of ion i
F Faraday constant
f_1, f_2 volume fraction of gel and intergel regions in the membrane
i current density
i_{lim} limiting current density
J_i ionic flux density
J_v volume flux density
K equilibrium constant
L_D Debye length
L_p hydraulic permeability coefficient
p hydrostatic pressure
P membrane diffusion permeability
Q membrane ion exchange capacity
R universal gas constant
t time
t_i transport number of ion i
t_w water transport number
T absolute temperature
T_i effective transport number of ion i in the membrane
V velocity
$\bar{V}_s$ partial molar volume of electrolyte
$\bar{V}_w$ partial molar volume of water
x normal to membrane coordinate
$y\pm$ molar activity coefficient
z_i charge number of ion i

Greek Symbols

δ Nernst's diffusion layer thickness
δ' effective thickness of the DBL
κ electric conductivity, S m^{-1}
μ_i electrochemical potential
ν solution viscosity
π osmotic pressure
σ Staverman reflection coefficient
φ electric potential
Δ difference in a quantity
∇ gradient operator

Indices

A co-ion
g superscript denoting that the quantity relates to the gel phase
i ionic species
s superscript denoting that the quantity relates to the interstitial solution
w water, species produced by water splitting
1 counterion

REFERENCES

1. S. F. Timashev. *Physical Chemistry of Membrane Process*. Ellis Horwood Series in Physical Chemistry. Ellis Horwood, New York, 1991.
2. B. Alberts, A. Johnson, J. Lewis, et al. *Molecular Biology of the Cell*. Garland Science, New York, 2002.
3. H. Lodish, A. Berk, L. S. Zipursky, et al. *Molecular Cell Biology*. Scientific American Books, New York, 2004.
4. S. J. Singer and G. L. Nicolson. *Science* **1972**, 175, 720.
5. D. Evans and H. Wennerstrom. *The Colloidal Domain: Where Physics, Chemistry, Biology, and Technology Meet*. Wiley, New York, 1999.
6. A. Pohorille and M. A. Wilson. *Cell Mol Biol Lett* **2001**, 6, 369.
7. E. Gouaux and R. Mackinnon. *Science* **2005**, 310 (5753), 1461.
8. M. H. Cordes, A. R. Davidson, R. T. Sauer. *Curr Opin Struct Biol* **1996**, 6 (1), 3.
9. B. Hille. *Ion Channels of Excitable Membranes*. Sinauer Associates, Sunderland, MA, 2001.
10. H. Strathmann. *Ion-exchange Membrane Separation Processes*. Membrane Science and Technology Ser. 9. Elsevier, Amsterdam, 2004.
11. T. Xu. *J Membr Sci* **2005**, 263, 1.
12. R. K. Nagarale, G. S. Gohil, V. K. Shahi. *Adv Colloid Interface Sci* **2006**, 119, 97.
13. F. Hellferich. *Ion Exchange*. McGraw-Hill, New York, 1962.
14. G. E. Molau. *J Membr Sci* **1981**, 8, 309.
15. R. Simons. *Electrochim Acta* **1984**, 29, 151.
16. R. Simons. *J Membr Sci* **1993**, 82, 45.
17. V. I. Zabolotskii, N. V. Shel'deshov, N. P. Gnusin. *Russ Chem Rev* **1988**, 57, 801.
18. H. Strathmann, J. J. Krol, H. J. Rapp, et al. *J Membr Sci* **1997**, 125, 123.
19. K. A. Mauritz, D. A. Mountz, D. A. Reuschle, et al. *Electrochim Acta* **2004**, 50, 565.
20. K. A. Mauritz and R. B. Moore. *Chem Rev* **2004**, 104, 4535.
21. K. D. Kreuer, S. Paddison, E. Spohr, et al. *Chem Rev* **2004**, 104, 4637.
22. A. Eisenberg. *Macromolecules* **1970**, 3, 147.
23. T. D. Gierke and W. Y. Hsu. The cluster-network model of ion clustering in perfluorosulfonated membranes. A. Eisenberg, H.L. Yeager. Proceedings of the ACS Symposium on Perfluorinated Ionomer Membranes. Washington, 1982.

24. E. J. Roche, M. Pinéri, R. Duplessix. *J Polym Sci Polym Phys Ed* **1982**, 20, 107.
25. I. M. Shiryaeva and A. I. Victorov. *Fluid Phase Equilibria* **2001**, 180, 115.
26. G. Gebel. *Polymer* **2000**, 41, 5829.
27. L. Rubatat, A. L. Rollet, G. Gebel, et al. Evidence of Elongated Polymeric Aggregates in Nafion. *Macromolecules* **2002**, 35, 4050.
28. D. Y. Galperin and A. R. Khokhlov. *Macromol Theory Simul* **2006**, 15, 137.
29. T. D. Gierke, G. E. Munn, F. C. Wilson. *J Polym Sci , Polym Phys Ed* **1981**, 19, 1687.
30. M. W. Verbrugge and R. F. Hill. *J Electrochem Soc* **1990**, 137 (3), 886.
31. Y. K. Tovbin and N. F. Vasutkin. *Colloids Surf A Physicochem Eng Aspects* **1999**, 158, 385.
32. H. G. Haubold, T. Vad, H. Jungbluth, et al. *Electrochim Acta* **2001**, 46, 1559.
33. A. L. Rollet, O. Diat, G. Gebel. *J Phys Chem* **2002**, 106, 3033.
34. M. V. Fedorov and A. A. Kornyshev. *Electrochim Acta* **2008**, 53 (23), 6835.
35. J. Fimrite, H. Struchtrup, N. J. Djilali. *J Electrochem Soc* **2005**, 152, A1804.
36. B. Corry, S. Kuyucak, S. H. Chung. *Chem Phys Lett* **2000**, 320, 35.
37. M. H. Litt. *Polym Prepr* **1997**, 38, 80.
38. K. D. Kreuer. *J Membr Sci* **2001**, 185, 29.
39. B. Loppinet, G. Gebel, C. E. Williams. *J Phys Chem* **1997**, 101, 1884.
40. S. Schlick, G. Gebel, M. Pineri, F. Volino. *Macromolecules* **1991**, 24, 3517.
41. S. J. Paddison, G. Bender, K. D. Kreuer, et al. *J New Mater Electrochem Syst* **2000**, 3, 293.
42. V. Saarinen, K. D. Kreuer, M. Schuster, et al. *Solid State Ionics* **2007**, 178, 533.
43. L. Chaabane, G. Bulvestre, C. Larchet, et al. *J Membr Sci* **2008**, 323, 167.
44. V. M. Barragan, J. P. G. Villaluenga, M. P. Godino, et al. *J Power Sources* **2008**, 185, 822.
45. S. J. Paddison, L. R. Pratt, T. A. Zawodzinski Jr. *J New Mater Electrochem Syst* **1999**, 2, 183.
46. Y. M. Volfkovich, V. S. Bagotzky, V. E. Sosenkin, et al. *Colloids Surf A Physicochem Eng Asp* **2001**, 187–188, 349.
47. N. P. Berezina, N. A. Kononenko, O. A. Dyomina, et al. *Adv Colloid Interface Sci* **2008**, 139 (1–2), 3.
48. Y. Mizutani. *J Membr Sci* **1990**, 49, 121–144.
49. J. J. Krol, M. Wessling, H. Strathmann. *J Membr Sci* **1999**, 162, 155.
50. P. Kofstad. *Nonstoichiometry. Diffusion and Electrical Conductivity in Binary Metal Oxides*. Wiley-Interscience, New York, 1972.
51. R. P. Buck. *J Membr Sci* **1984**, 17, 1.
52. V. I. Zabolotsky and V. V. Nikonenko. *Ion Transport in Membranes (in Russian)*. Nauka, Moscow, 1996.
53. N. Agmon. *Chem Phys Lett* **1995**, 244, 456.
54. J. A. Elliott and S. J. Paddison. *Phys Chem Chem Phys* **2007**, 9 (21), 2602.
55. M. Esai Selvan, D. J. Keffer, S. Cui, et al. *Mol Simul* **2010**, 36 (7–8), 568.
56. C. K. Knox and G. A. Voth. *J Phys Chem B* **2010**, 114, 3205.
57. A. A. Kornyshev, A. M. Kuznetsov, E. Spohr, et al. *J Phys Chem* **2003**, 107, 3351.

58. S. Paddison and J. Annu. *Rev Mater Res* **2003**, 33, 289.
59. G. Pourcelly, A. Oikonomou, H. D. Hurwitz, et al. *J Electroanal Chem* **1990**, 287, 43.
60. R. Zallen. *The Physics of Amorphous Solids*. Wiley Interscience, New York, 1998.
61. V. I. Volkov, Y. A. Dobrovolsky, M. S. Nurmiev, et al. *Solid State Ionics* **2008**, 179 (1–6), 148.
62. N. Lakshminarayanaiah. *Transport Phenomena in Membranes*. Academic Press, New York, 1969.
63. K. Konturri, L. Murtomäki, J. A. Manzanares. *Ionic Transport Processes in Electrochemistry and Membrane Science*. Oxford University Press, New York, 2008.
64. A. B. Yaroslavtsev, V. V. Nikonenko, V. I. Zabolotsky. *Russ Chem Rev* **2003**, 72, 393.
65. A. Z. Weber and J. Newman. *Top Appl Phys* **2009**, 113, 157.
66. C.-Y. Wang. *Chem Rev* **2004**, 104, 4727.
67. D. Kondepudi and I. Prigogine. *Modern Thermodynamics—From Heat Engines to Dissipative Structures*. John Willey & Sons, New York, 1999.
68. S. R. Caplan and A. Essig. *Bioenergetics and Linear Nonequilibrium Thermodynamics*. Harvard University Press, Cambridge, 1983.
69. K. S. Spiegler. *Trans Faraday Soc* **1958**, 54, 1408.
70. V. I. Zabolotsky and V. V. Nikonenko. *J Membr Sci* **1993**, 79, 181.
71. O. Kedem and A. Katchalsky. *J Gen Physiol* **1961**, 45, 143.
72. A. Narebska, S. Koter, W. Kujawski. *Desalination* **1984**, 51, 3.
73. K. S. Spiegler and O. Kedem. *Desalination* **1966**, 1, 311.
74. B. Auclair, V. Nikonenko, C. Larchet, et al. *J Membr Sci* **2002**, 195, 89.
75. O. Kedem and A. Katchalsky. *Trans Faraday Soc* **1963**, 59, 1918.
76. P. Meares. *J Membr Sci* **1981**, 8, 295.
77. E. A. Mason and H. K. Lonsdale. *J Membr Sci* **1990**, 51, 1.
78. S. Koter, W. Kujawski, I. Koter. *J Membr Sci* **2007**, 297, 226.
79. O. Kedem and M. Perry. *J Membr Sci* **1983**, 14, 249.
80. A. Narebska and S. Koter. *Electrochim Acta* **1993**, 38, 815.
81. N. P. Gnusin, N. P. Berezina, N. A. Kononenko, et al. *J Membr Sci* **2004**, 243, 301.
82. C. Larchet, B. Auclair, V. Nikonenko. *Electrochim Acta* **2004**, 49, 1711.
83. C. Larchet, L. Dammak, B. Auclair, et al. *New J Chem* **2004**, 28, 1260.
84. C. R. Paterson and C. R. Gardner. *J Chem Soc* **1971**, 2254.
85. R. A. Robinson and R. H. Stokes. *Electrolyte Solutions*. Butterworths, London, 1959.
86. L. Lebrun, E. Da Silva, G. Pourcelly, et al. *J Membr Sci* **2003**, 227, 95.
87. R. Lteif, L. Dammak, C. Larchet, et al. *Eur Polym J* **2001**, 37, 627.
88. J. S. Newman. *Electrochemical Systems*. Prentice Hall, Englewood Cliffs, NJ, 1973.
89. B. Zaltzman and I. Rubinstein. *J Fluid Mech* **2007**, 579, 173.
90. K. T. Chu and M. Z. Bazant. *SIAM J Appl Math* **2005**, 65, 1485.
91. M. A.-K. Urtenov, E. V. Kirillova, N. M. Seidova, et al. *J Phys Chem B* **2007**, 111, 14208.

92. R. Schlögl. *Stofftransport Durch Membranen*. Steinkopff-Verlag, Darmstadt, 1964.
93. P. Turq, J. Barthel, M. Chemla. *Transport, Relaxation, and Kinetic Processes in Electrolyte Solutions*. Springer-Verlag, Berlin, 1992.
94. G. Pourcelly, P. Sistat, A. Chapotot, et al. *J Membr Sci* **1996**, 110, 69.
95. S. Bandini and L. Bruni. Transport phenomena in nanofiltration membranes. In *Comprehensive Membrane Science and Engineering*, E. Drioli, L. Giorno, eds. Elsevier Science, 2010; Vol. 2, p. 67.
96. G. Jonsson and F. Macedonio. Fundamentals in reverse osmosis. In *Comprehensive Membrane Science and Engineerining*, E. Drioli, L. Giorno, eds. Elsevier Science, 2010; Vol. 2, p. 1.
97. M. Caputo and C. Cametti. *J Theor Biol* **2008**, 254, 697.
98. T. Teorell. *Proc Soc Exp Biol Med* **1935**, 33, 282.
99. K. H. Meyer and J. F. Sievers. *Helv Chim Acta* **1936**, 19, 649.
100. E. Glueckauf and R. E. Watts. *Proc R Soc Lond* **1962**, A 268, 339.
101. Y. Tanaka. Theory of Teorell, Meyer and Sievers (TMS Theory). Chapter 4 in: *Membr Sci Technol* **2007**, 12, 59.
102. V. V. Nikonenko, K. A. Lebedev, V. I. Zabolotsky, et al. *Eur Polym J* **1997**, 33, 1057.
103. A. V. Sokirko, J. A. Manzanares, J. Pellicer. *J Colloid Interface Sci* **1994**, 168, 32.
104. A. A. Moya and J. A. Moleón. *J Electroanal Chem* **2010**, 647, 53.
105. S. Levine, J. R. Marriott, G. Neale, et al. *J Colloid Interface Sci* **1975**, 52, 136.
106. S. Koter. *J Membr Sci* **2002**, 206 (1–2), 201.
107. V. Garcia-Morales, J. Cervera, J. A. Manzanares. *J Electroanal Chem* **2007**, 599, 203.
108. A. G. Guzmán-Garcia, P. N. Pintauro, M. V. Verbrugge. *AIChE J* **1990**, 36, 1061.
109. X. Lefebvre and J. Palmeri. *J Phys Chem* **2005**, 109, 5525.
110. A. Yaroshchuk, Y. Boiko, A. Makovetskiy. *Langmuir* **2009**, 25, 9605.
111. A. Szymczyk, H. C. Zhu, B. Balannec. *J Phys Chem B* **2010**, 114, 10143.
112. R. Zallen. *The Physics of Amorphous Solids*. John Wiley & Sons, New York, 1983.
113. M. Eikerling and A. A. Kornyshev. *J Electroanal Chem* **2001**, 501, 196.
114. A. Mansouri, C. Scheuerman, S. Bhattacharjee, et al. *J Colloid Interface Sci* **2005**, 292, 567.
115. A. Höltzel and U. Tallarek. *J Sep Sci* **2007**, 30, 1398.
116. Y. Yang and P. N. Pintauro. *Ind Eng Chem Res* **2004**, 43, 2957.
117. R. B. Schoch, J. Han, P. Renaud. *Rev Mod Phys* **2008**, 80, 839.
118. J. de Jong, R. G. H. Lammertink, M. Wessling. *Lab Chip* **2006**, 6, 1125.
119. E. H. Cwirko and R. G. Carbonell. *J Membr Sci* **1992**, 67, 227.
120. A. Szymczyk, P. Fievet, B. Aoubiza, et al. *J Membr Sci* **1999**, 161, 275.
121. P. Commer, A. Cherstvy, E. Spohr, et al. *Fuel Cells* **2003**, 2, 127.
122. R. Paul and S. J. Paddison. *J Chem Phys* **2005**, 123, 224704.
123. M. Z. Bazant, M. S. Kilic, B. Storey, et al. *Adv Colloid Interface Sci* **2009**, 152, 48.
124. M. Z. Bazant and T. M. Squires. *Curr Opin Colloid Interface Sci* **2010**, 15, 203.
125. N. P. Berezina and L. V. Karpenko. *Rus Colloid J* **2000**, 62, 676.
126. M. M. Dubinin. *Russ Chem Rev* **1982**, 51, 605.

127. N. P. Gnusin, V. D. Grebenyuk, M. V. Pevnitskaya. *Electrochemistry of Ion Exchangers*. Nauka, Novosibirsk, 1972.
128. S. Mafé, J. A. Manzanares, P. Ramirez. *Phys Chem Chem Phys* **2003**, 5, 376.
129. A. G. Rojo and H. E. Roman. *Phys Rev* **1988**, 37, 3696.
130. H. L. Duan, B. L. Karihaloo, J. Wang, et al. *Phys Rev B* **2006**, 73 (17), 174203.
131. V. M. Starov and V. G. Zhdanov. *Adv Colloid Interface Sci* **2008**, 137 (1), 2.
132. R. Pal. *Mater Sci Eng A* **2008**, 498 (1–2), 135.
133. M. Zhdanov. *Geophysics* **2008**, 73 (5), F197.
134. S. I. Vasin, A. N. Filippov, V. M. Starov. *Adv Colloid Interface Sci* **2008**, 139, 83.
135. J. C. Maxwell. *Treatise on Electricity and Magnetism*. Clarendon, Oxford, 1873.
136. R. Wodzki and A. Narebska. *Angew Macromol Chem* **1980**, 88, 149.
137. J. S. Mackie and P. Meares. *Proc R Soc Lond A* **1955**, 232, 498A.
138. N. P. Gnusin, V. I. Zabolotskii, V. V. Nikonenko, et al. *Zh Fiz Khim* **1980**, 54, 1518.
139. W. Y. Hsu, T. D. Gierke, C. J. Molnar. *Macromolecules* **1983**, 16, 1945.
140. N. N. Belaid, B. Ngom, L. Dammak, et al. *Eur Polym J* **1999**, 35, 879.
141. M. T. Bryk, V. I. Zabolotskii, I. D. Atamanenko, et al. *Sov J Water Chem Technol* **1989**, 11, 14.
142. N. Pismenskaya, E. Laktionov, V. Nikonenko, et al. *J Membr Sci* **2001**, 181, 185.
143. P. V. Vyas, P. Ray, S. K. Adhikary, et al. *J Colloid Interface Sci* **2003**, 257, 127.
144. X. T. Le, T. H. Bui, P. Viel, et al. *J Membr Sci* **2009**, 340, 133.
145. A. N. Filippov, V. M. Starov, N. A. Kononenko, et al. *Adv Colloid Interface Sci* **2008**, 139 (1–2), 29.
146. K. A. Lebedev, V. V. Nikonenko, et al. *Sov Electrochem* **1987**, 23, 459.
147. X. T. Le. *Adv Colloid Interface Sci* **2008**, 325, 215.
148. N. P. Berezina, N. Gnusin, O. Dyomina, et al. *J Membr Sci* **1994**, 86, 207.
149. L. X. Tuan and C. Buess-Herman. *Chem Phys Lett* **2007**, 434, 49.
150. C. Larchet, S. Nouri, B. Auclair, et al. *Adv Colloid Interface Sci* **2008**, 139, 45.
151. P. Huguet, T. Kiva, O. Noguera, et al. *New J Chem* **2005**, 29, 955.
152. V. I. Zabolotskii, V. V. Nikonenko, O. N. Kostenko, et al. *Zh Fiz Khim* **1993**, 67, 2423.
153. A. Elattar, A. Elmidaoui, N. Pismenskaia, et al. *J Membr Sci* **1998**, 143, 249.
154. J.-H. Choi, S.-H. Kim, S.-H. Moon. *Adv Colloid Interface Sci* **2001**, 241, 120.
155. L. X. Tuan, M. Verbanck, C. Buess-Herman, et al. *J Membr Sci* **2006**, 284, 67.
156. J. P. G. Villaluenga, V. M. Barragán, M. A. Izquierdo-Gil, et al. *J Membr Sci* **2008**, 323, 421.
157. M. Kumar, S. Singh, V. K. Shahi. *J Phys Chem B* **2010**, 114, 198.
158. A. Z. Weber and J. Newman. *J Electrochem Soc* **2006**, 153, A2205.
159. M. Eikerling. *J Electrochem Soc* **2006**, 153, E58–E70.
160. S. Litster, C. R. Buie, T. Fabian, et al. *J Electrochem Soc* **2007**, 154 (10), B1049–B1058.
161. A. J. B. Kemperman, ed. *Handbook on Bipolar Membrane Technology*. Twente University Press, Enschede, 2000.

162. K. D. Kreuer. *Solid State Ionics* **1997**, 97, 1.
163. T. A. Zawodzinski, T. E. Springer, J. Davey, et al. *J Electrochem Soc* **1993**, 140, 1981.
164. S. Hietala, S. L. Maunu, F. Sundholm. *J Polym Sci B Polym Phys* **2000**, 38, 3277.
165. P. Thompson and S. Troian. *A General Boundary Condition for Liquid Flow at Solid Surfaces*. Nature, London, 1997.
166. M. Z. Bazant and O. I. Vinogradova. *J Fluid Mech* **2008**, 613, 125.
167. F. Meier and G. Eigenberger. *Electrochim Acta* **2004**, 49, 1731.
168. M. Eikerling, Y. I. Kharkats, A. A. Kornyshev, et al. *J Electrochem Soc* **1998**, 145, 2684.
169. T. A. Zawodzinski, J. Davey, J. Valerio, et al. *Electrochim Acta* **1995**, 40, 297.
170. G. Xie and T. Okada. *Electrochim Acta* **1996**, 41, 1569.
171. T. Okada, G. Xie, O. Gorseth, et al. *Electrochim Acta* **1998**, 43, 3741.
172. A.-L. Rollet, G. Gebel, J.-P. Simonin, et al. *J Polym Sci* **2001**, 39 (2), 548.
173. M. Tasaka, Y. Kondo, M. Nagasawa. *J Phys Chem* **1969**, 73, 3181.
174. J. P. García-Villaluenga, B. Seoane, V. M. Barragán. *J Colloid Interface Sci* **2003**, 263, 217.
175. R. Schlögl. *Z Phyzik Chem (Frankfurt)* **1955**, 3, 73.
176. H. Fujita and Y. Kobatake. *J Colloid Interface Sci* **1968**, 27, 609.
177. O. Bobreshova, I. Aristov, P. Kulintsov, et al. *J Membr Sci* **2000**, 177, 201.
178. N. V. Shel'deshov, V. V. Chaika, V. I. Zabolotskii. *Russ J Electrochem* **2008**, 44, 1036.
179. A. N. Filippov, S. I. Vasin, V. M. Starov. *Colloids Surf A* **2006**, 282–283, 272.
180. V. I. Zabolotskii, A. A. Shudrenko, N. P. Gnusin. *Sov Electrochem* **1988**, 24, 689.
181. J. Manzanares and K. Kontturi. Diffusion and migration. In *Encyclopedia of Electrochemistry. V. 2. Interfacial Kinetics and Mass Transport*, A. J. Bard, M. Stratmann, E. J. Calvo, eds. VCH-Wiley, Weinheim, 2003.
182. S.-T. Hwang and K. Kammermeyer. *Membranes in Separation*. Wiley, New York, 1975.
183. W. J. Koros, Y. H. Ma, T. Shimidzu. *Pure Appl Chem* **1996**, 68, 1479.
184. I. Rubinstein and B. Zaltzman. *Phys Rev E* **2000**, 62, 2238.
185. A. M. Peers. *Discuss Faraday Soc* **1956**, 21, 124.
186. C. Forgacs, N. Ishibashi, J. Leibovitz, et al. *Desalination* **1972**, 10, 181.
187. T. A. Davis, V. Grebenyuk, O. Grebenyuk. *Membrane Technology in Chemical Industry*, S. P. Nunes, K.-V. Peinemann, eds. Wiley-VCH, Weinheim, 2001.
188. V. I. Zabolotsky, V. V. Nikonenko, N. D. Pismenskaya, et al. *Sep Purif Technol* **1998**, 14, 255.
189. J. Balster, M. H. Yildirim, D. F. Stamatialis, et al. *J Phys Chem B* **2007**, 111, 2152.
190. J. Lu, Y.-X. Wang, J. Zhu. *Electrochim Acta* **2010**, 55, 2673.
191. S. S. Dukhin. *Adv Colloid Interface Sci* **1991**, 35, 173.
192. N. A. Mishchuk and S. S. Dukhin. *Interfacial Electrokinetics and Electrophoresis*, A. Delgado, ed. Marcel Dekker, New York, 2002.
193. N. A. Mishchuk. *Adv Colloid Interface Sci* **2010**, 160 (1–2), 16.
194. T. M. Squires and M. Z. Bazant. *J Fluid Mech* **2004**, 509, 217.
195. F. C. Leinweber and U. Tallarek. *J Phys Chem B* **2005**, 109, 21481.

196. N. A. Mishchuk, T. Heldal, T. Volden, et al. *Electrophoresis* **2009**, 30, 3499.
197. A. Ajdari. *Phys Rev E* **2000**, 6, 45.
198. D. Kim, J. D. Posner, J. G. Santiago. *Sens Actuators A* **2008**, 141, 201.
199. V. M. Volgin and A. D. Davydov. *Electrochim Acta* **2004**, 49, 365.
200. M. Trau, D. A. Saville, I. A. Aksay. *Science* **1996**, 272, 706.
201. E. I. Belova, G. Y. Lopatkova, N. D. Pismenskaya, et al. *J Phys Chem B* **2006**, 110, 13458.
202. V. A. Shaposhnik, V. I. Vasil'eva, O. V. Grigorchuk. *Adv Colloid Interface Sci* **2008**, 139, 74.
203. V. V. Nikonenko, N. D. Pismenskaya, E. I. Belova, et al. *Adv Colloid Interface Sci* **2010**, 160, 101–123
204. J. J. Krol, M. Wessling, H. Strathmann. *J Membr Sci* **1999**, 162, 145.
205. Y. Tanaka. *J Membr Sci* **2010**, 350, 347.
206. Y. I. Kharkats. *Sov Electrochem* **1985**, 21, 917.
207. J. A. Manzanares, W. D. Murphy, S. Mafé, et al. *J Phys Chem* **1993**, 97, 8524.
208. I. Rubinstein, B. Zaltzman, O. Kedem. *J Membr Sci* **1997**, 125 (1), 17.
209. V. I. Zabolotsky, V. V. Nikonenko, N. D. Pismenskaya. *J Membr Sci* **1996**, 119 (2), 171–181.
210. S. M. Rubinstein, G. Manukyan, A. Staicu, et al. *Phys Rev Lett* **2008**, 101, 236101.
211. M. A. Hickner, H. Ghassemi, Y. S. Kim, et al. *Chem Rev* **2004**, 104, 4587.
212. D. J. Jones and J. Roziere. Chapter In *Handbook of Fuel Cells—Fundamentals, Technology and Applications*, Vol. 3, W. Weslich, A. Lamm, H. A. Gasteiger, eds. John Wiley & Sons, Ltd., New York, 2003; p. 447.
213. A. B. Yaroslavtsev. *Solid State Ionics* **2005**, 176, 2935.
214. N. P. Gnusin, M. V. Pevnitskaya, V. K. Varentsov, et al. USSR patent No 216622. Publ. 21.10.72, 1972.
215. V. I. Zabolotsky, V. V. Nikonenko, N. D. Pismenskaya, et al. Russ. Patent No 2033850, B01 D 13/02, 1995.
216. G. Eigenberger, H. Strathmann, A. Grabovskiy. Germany Patent WO 2005/009596 A1, B01D 61/44, 2005.
217. V. I. Zabolotskii, S. A. Loza, M. V. Sharafan. *Russ J Electrochem* **2005**, 41, 1053.
218. C. Larchet, V. I. Zabolotsky, N. Pismenskaya, et al. *Desalination* **2008**, 222 (1), 489.
219. T. Sata. *J Membr Sci* **2000**, 167, 1.
220. K. Takata, Y. Yamamoto, T. Sata. *J Membr Sci* **2000**, 179, 101.
221. K. Takata, H. Ihara, T. Sata. *Angew Macromol Chem* **2003**, 236, 67.
222. T. Sata and Y. Mizutani. *J Polym Sci* **2003**, 17, 1199.
223. Y. Hu, M. Wang, D. Wang, et al. *J Membr Sci* **2008**, 319, 5.
224. K. Kerose, F. Janowski, V. A. Shaposhnik. *J Membr Sci* **1997**, 127, 17.
225. Z. Amor, S. Malki, M. Taky, et al. *Desalination* **1998**, 120, 263.
226. R. Yamane, Y. Mizutani, H. Motomura, et al. *Denki Kagaku* **1964**, 32, 277.
227. Y. Oren, I. Rubinstein, C. Linder, et al. *Environ Eng Sci* **2002**, 19, 512.
228. V. D. Grebenyuk, R. D. Chebotareva, S. Peters, et al. *Desalination* **1998**, 115, 313.

229. R. K. Nagarale, V. K. Shahi, R. Schubert, et al. *J Colloid Interface Sci* **2004**, 270, 446.
230. V. V. Nikonenko, N. D. Pis'menskaya, E. I. Volodina. *Russ J Electrochem* **2005**, 41, 1205.
231. J.-H. Choi and S.-H. Moon. *J Colloid Interface Sci* **2003**, 265, 93.
232. N. D. Pismenskaya, E. I. Belova, V. V. Nikonenko, et al. *Desalin Water Treat* **2010**, 21, 109.
233. V. G. Levich. *Physicochemical Hydrodynamics*. Prentice-Hall, New York, 1962.
234. H. A. Stone, A. D. Stroock, A. Ajdari. *Annu Rev Fluid Mech* **2004**, 36, 381.
235. N. Churaev, V. Sobolev, A. Somov. *J Colloid Interface Sci* **1984**, 97, 574.
236. C.-H. Choi, K. J. A. Westin, K. S. Breuer. *Phys Fluids* **2003**, 15 (10), 2897.
237. M. Majumder, N. Chopra, R. Andrews, et al. *Nature* **2005**, 438, 44.
238. O. I. Vinogradova and G. E. Yakubov. *Langmuir* **2003**, 19, 1227.
239. O. I. Vinogradova, N. F. Bunkin, N. V. Churaev, et al. *J Colloid Interface Sci* **1995**, 173, 443.
240. V. V. Nikonenko, N. D. Pismenskaya, E. I. Belova, et al. Proceedings of XVIII Mendeleev Congress on General and Applied Chemistry. Moscow, 2007.
241. M. V. Sharafan, V. I. Zabolotskii, V. V. Bugakov. *Russ J Electrochem* **2009**, 45 (100), 1162.
242. A. B. Yaroslavtsev. *Russ J Inorg Chem* **2000**, 45, 249.
243. B. Bonnet, D. J. Jones, J. Roziere, et al. *J New Mater Electrochem Syst* **2000**, 3, 87.
244. G. Kickelbick. *Prog Polym Sci* **2003**, 28, 83.
245. G. Alberti, U. Costantino, J. Kornyei, et al. *Reactive Polym* **1985**, 4, 1.
246. M. Casciola, U. Costantino, A. Peraio, et al. *Solid State Ionics* **1995**, 77, 229.
247. G. Alberti and M. Casciola. *Solid State Ionics* **1997**, 97, 177.
248. E. W. Stein, A. Clearfield, M. A. Subramanian. *Solid State Ionics* **1996**, 83, 113.
249. C. Y. Ortiz-Avila, C. Bharwaj, A. Clearfield. *Inorg Chem* **1994**, 33, 2499.
250. E. Y. Voropaeva, I. A. Stenina, A. B. Yaroslavtsev. *Russ J Inorg Chem* **2008**, 53, 1531.
251. S. A. Novikova, E. I. Volodina, N. D. Pis'menskaya, et al. *Russ J Electrochem* **2005**, 41, 1070.
252. A. S. Shalimov, S. A. Novikova, I. A. Stenina, et al. *Russ J Inorg Chem* **2006**, 51, 700.
253. V. A. Gerasin, T. A. Zubova, F. N. Bahov, et al. *Nanotechnol Russ* **2007**, 2, 90.
254. B. S. Bokshtein and A. B. Yaroslavtsev. *Diffusion of Atoms and Ions in Solids* [in Russian]. MISIS ed. MISIS, Moscow, 2005.
255. E. Y. Voropaeva, I. A. Stenina, A. B. Yaroslatsev. *Russ J Inorg Chem* **2008**, 53, 1677.
256. A. S. Shalimov, A. I. Perepelkina, I. A. Stenina, et al. *Russ J Inorg Chem* **2009**, 54, 356.
257. H. Tang, Z. Wan, M. Pan, et al. *Electrochem Commun* **2007**, 9, 2003.
258. I. A. Stenina, A. A. Lisova, I. Y. Pinus, et al. *Russ Chem Bull* **2008**, 57 (11), 2261.
259. N. P. Berezina, N. A. Kononenko, A. A.-R. Sytcheva, et al. *Electrochim Acta* **2009**, 54 (8), 2342.

260. N. P. Berezina, A. A.-R. Kubaisy, S. V. Timofeev, et al. *J Solid State Electrochem* **2006**, 11, 378.
261. M. Hibino, H. Itoh, K. Kinosita. *J Biophys* **1993**, 64, 1789.
262. N. P. Gnusin, N. P. Berezina, N. A. Kononenko. *Sov Electrochem* **1987**, 23, 127.
263. M. Winter and R. J. Brodd. *Chem Rev* **2004**, 104, 4245.
264. A. M. Bernardes, D. C. R. Espinosa, J. A. S. Tenório. *J Power Sources* **2004**, 130, 291.
265. O. V. Bushkova, I. P. Koryakova, B. I. Lirova, et al. *Dokl Akad Nauk* **2006**, 407, 634.
266. A. L. Rusanov, D. Y. Lihatchev, K. Myullen. *Russ Chem Rev* **2002**, 71 (9), 862.
267. Y. A. Dobrovolsky, P. Jannasch, B. Lafitte, et al. *Russ J Electrochem* **2007**, 43, 489.
268. Y. A. Dobrovolsky, A. V. Pisareva, L. S. Leonova, et al. *Int Sci J Altern Energy Ecol* **2004**, 12, 36.
269. V. V. Starkov, Y. A. Dobrovolsky, N. V. Liskov, et al. *Int Sci J Altern Energy Ecol* **2007**, 6, 24.
270. C. C. Liang. *J Electrochem Soc* **1973**, 120, 1289.
271. J. Maier. *Prog Solid State Chem* **1995**, 23, 171.
272. A. B. Yaroslavtsev. *Russ Chem Rev* **2009**, 78 (11), 1013.
273. R. Jiang, H. R. Kunz, J. M. Fenton. *J Membr Sci* **2006**, 272, 116.
274. G. Alberti, M. Casciola, D. Capitani, et al. *Electrochim Acta* **2007**, 52, 8125.
275. S. A. Novikova, G. Y. Yurkov, A. B. Yaroslavtsev. *Mendeleev Commun* **2010**, 20, 89.
276. E. Y. Voropaeva, E. A. Sanginov, V. I. Volkov, et al. *Russ J Inorg Chem* **2008**, 53, 1536.
277. S. A. Novikova, E. Y. Safronova, A. A. Lysova, et al. *Mendeleev Commun* **2010**, 20, 156.
278. E. Y. Safronova, I. A. Stenina, A. B. Yaroslavtsev. *Russ J Inorg Chem* **2010**, 55, 13.
279. A. B. Yaroslavtsev. *Russ J Inorg Chem* **2010**, 55, 13.
280. R. H. Garrett and C. M. Grisham. *Biochemistry*. Brooks-Cole, Belmont, CA, 2005.
281. N. Pismenskaia, P. Sistat, P. Huguet, et al. *J Membr Sci* **2004**, 228 (1), 65.

CHAPTER 10

REVERSIBLE COORDINATION POLYMERS

KIM DE LANGE, JOS M.J. PAULUSSE, ANTONIUS T.M. MARCELIS, and HAN ZUILHOF

10.1 INTRODUCTION

In conventional polymers, the monomeric units are connected through covalent bonds. However, when noncovalent bonds are used to create polymers, materials with interesting new properties are obtained. These "supramolecular" polymers consist of self-assembled chains of small molecules held together by reversible noncovalent interactions.[1] These interactions have to be highly directional, because otherwise only randomly shaped aggregates will be formed. Supramolecular polymers have been prepared based on many types of noncovalent interactions, such as hydrogen bonding, hydrophilic and hydrophobic interactions, π–π stacking, as well as metal–ligand coordination. Depending on the strength of these interactions, the noncovalent bonds may break and recombine relatively easy. Consequently, continuous exchange of bound and free monomers occurs in supramolecular polymers. The reversible nature of these bonds imparts these polymers with several unique properties such as stimuli responsiveness and self-healing abilities.[2] Furthermore, the synthesis of supramolecular polymers does not require an initiator, the obtained polymers are dynamic, and termination processes during the polymerization are limited to chain stoppers only. Processing of covalent polymers typically has to be carried out at high temperatures and pressures, hence severely limiting their applicability. The bonds in supramolecular polymers, however, are readily broken and heating of these polymers is therefore accompanied by a much larger decrease in viscosity. In addition, supramolecular polymers behave very differently under stress as compared to conventional polymers. Supramolecular polymers may thus have the potential to combine

Ionic Interactions in Natural and Synthetic Macromolecules, First Edition. Edited by Alberto Ciferri and Angelo Perico.

the material properties of conventional polymers with the ease of processability of small molecules.[3]

Coordination polymers are a special class of supramolecular polymers in which the reversible bond is caused by metal–ligand interactions. The coordinative bond is a strong bond compared to hydrogen bonds and a weak bond compared to covalent bonds. Furthermore, it is highly specific and directional. Extensive literature is available on coordination polymers in the solid state. Metal-organic frameworks (MOFs), for example are one- (1D), two- (2D), or three-dimensional (3D) coordination networks that are built up of ligands and metal ions in the solid state. They have promising applications as gas storage materials, in gas purification, and in heterogeneous catalysis.[4,5] These coordination polymers only exist in the solid state and dissolution typically leads to disassembly into small molecules. Since these polymers cannot be classified as reversible, they fall outside the scope of this chapter. For information on the synthesis and properties of these coordination polymers the reader is referred to some excellent reviews on this topic.[6–8]

In this chapter we will give an overview of the history and recent advances in the field of reversible coordination polymers. We highlight the effects of environmental influences and physical parameters on reversible coordination polymers, discuss useful techniques for their characterization, and give insight in applications and prospects for reversible coordination polymers.

10.2 ANATOMY OF A REVERSIBLE COORDINATION POLYMER

Reversible coordination polymers are dynamic structures and a fixed chain length can therefore not be assigned (Fig. 10.1). Their length changes continuously around a well-defined average, because the bonds between the monomers are continuously broken and reformed. This also entails that the degree of polymerization (DP) changes when the environmental conditions change. Multiple physical parameters can be distinguished that influence the DP, such as strength of the metal–ligand bond, presence of chain stoppers, concentration, temperature, pH, and solvent. To obtain reversible coordination polymers, the coordinative bond has to be strong enough to maintain a chain-like struc-

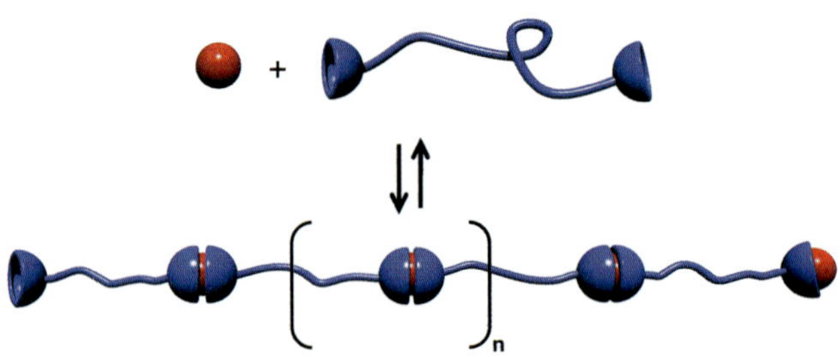

FIGURE 10.1. Formation of a linear coordination polymer.

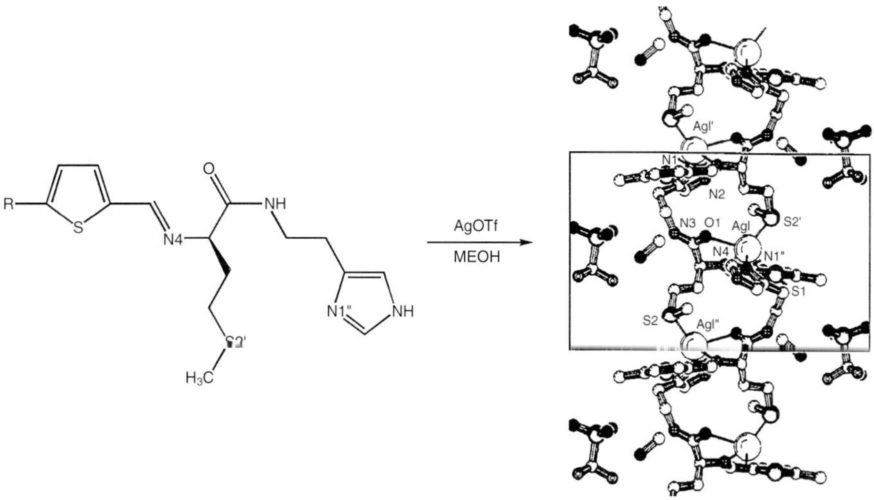

FIGURE 10.2. Complexation of a peptide-based ligand with silver(I) triflate leading to silver(I) stereoregular coordination polymers.[14]

ture in solution. On the other hand, kinetic inertness of the coordinative bond results in loss of the dynamic character of the system. Loss of dynamic character is, for example, observed in ferrocene-based polymers.[9] Ample use has been made of the robustness of ruthenium-terpyridine complexes in the preparation of, for example, block copolymer architectures.[10–12] Solutions of these polymers do also not display ligand exchange, although ligand exchange could be achieved by chemical means, for example, by the addition of acids.[13]

The first reversible coordination polymers actually recognized as such were stereoregular coordination polymers developed by van Koten and coworkers.[14] They investigated the complexation of peptide-based multidentate ligands with silver(I) to give kinetically labile coordination complexes in methanol (Fig. 10.2). The average chain length of these polymeric complexes was shown to be dependent on concentration and temperature. The crystal structure of the silver(I)-ligand complex showed that each ligand molecule contained three donor atoms (imine N4, methionine S2′ and imidazole N1″). The Ag(I) centers coordinated to three ligand groups, which resulted in a stereoregular helical structure. Vapor pressure osmometry of concentrated solutions helped to establish that the coordination polymer consisted of at least 11 units. In dilute solution, however, only monomeric species were present.

Rehahn and coworkers used proton nuclear magnetic resonance (NMR) spectroscopy to study coordination polymers prepared from kinetically labile coordination complexes (Fig. 10.3).[15] The monomers consisted of two *o*-phenanthroline residues connected by a rigid spacer, which prevented the formation of small rings. Hexyl side chains were attached to the spacer to make the monomers soluble in apolar, noncoordinating solvents such as

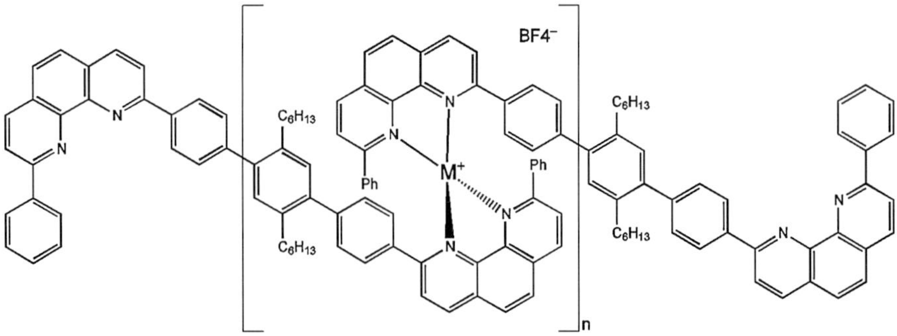

FIGURE 10.3. Kinetically labile coordination polymer formed in apolar solvents, M = Cu(I) or Ag(I).[15]

tetrachloroethane, toluene, and chloroform. Complexation of the obtained monomers with Cu(I) and Ag(I) resulted in the formation of reversible coordination polymers. The chemical shift of the protons of the ligands was dependent on whether the monomer functioned as a free ligand, a chain end group, or an inner chain repeat unit. At equimolar stoichiometry, all peaks in the NMR spectrum corresponding to the end groups disappeared, which is indicative of the formation of high molecular mass polymers. The reversibility of these coordination polymers became clear when extra ligand was added to the formed polymers. The addition of excess ligand resulted in chain degradation, as was confirmed by the reappearance of the end group absorptions in the NMR spectrum as well as by the decrease in viscosity. Coordination polymers were only formed when an apolar solvent like 1,1,2,2-tetrachloroethane was used. When more polar solvents were used, for example acetonitrile or pyridine, no high molecular mass polymers were observed due to the competition of those solvents for the metal ions.

Michelsen and Hunter prepared porphyrin ligands with two pyridine moieties. Complexation with cobalt(II) resulted in the formation of self-assembled polymers in chloroform (Fig. 10.4).[16] Size-exclusion chromatography (SEC) was used to estimate the DP and it was found that a 7 mM solution consisted of polymers with 100 monomeric units. After addition of porphyrins with only one pyridine group, the average length of the polymer decreased, showing the reversibility of the system.

10.3 METAL–LIGAND BOND STRENGTH

The strength of the metal–ligand coordinative bond has a strong influence on the DP and thus on the physical properties of a coordination polymer. The stronger the bond, the less likely the ligand is to dissociate from the metal ion, resulting in higher degrees of polymerization (Fig. 10.5). The average DP for

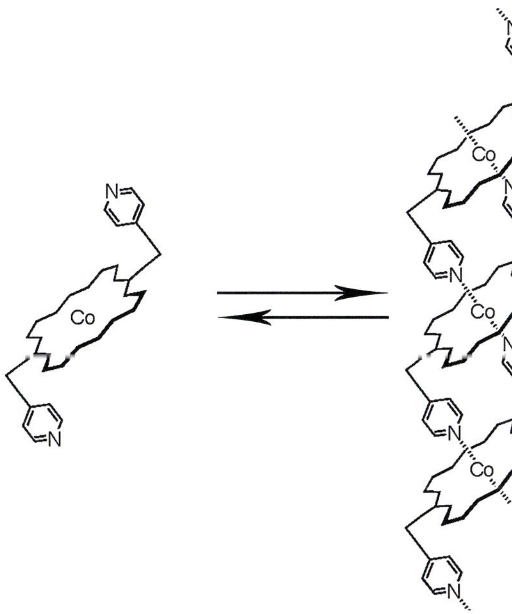

FIGURE 10.4. Self-assembly of a cobalt porphyrin dimer via coordination of a covalently attached pyridine ligand.[16]

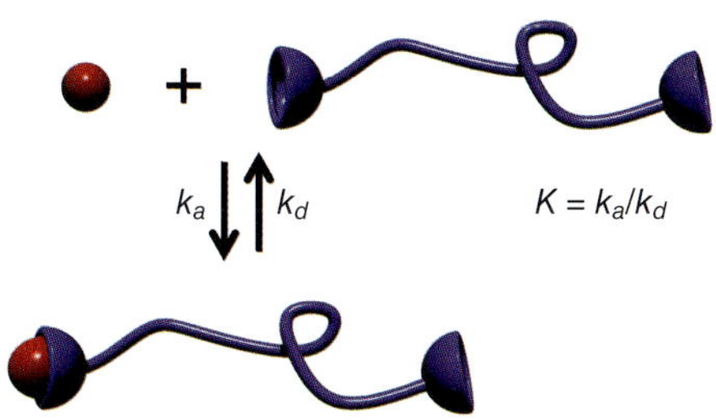

FIGURE 10.5. Schematic representation of the formation of a coordination bond with its thermodynamic and kinetic parameters.

coordination polymers is dependent on the growth mechanism. In a recently published review, the polymerization of supramolecular polymers is outlined.[17] For linear coordination polymers growth can occur through two association mechanisms, namely isodesmic and ring-chain polymerization. Isodesmic polymerizations are characterized by the fact that the formation of a single noncovalent bond is identical at all steps of the polymerization process and no cyclic intermediates are present.[1] The DP of such polymers is strongly related to the association constant (K) and monomer concentration ($[M_1]$):

$$[M_i] = K^{-1}(K[M_1])^i \qquad i = 1, 2, \ldots, \infty.$$

The theory of isodesmic polymerization predicts a DP roughly corresponding to the square root of the binding constant (Chapter 6). Another growth mechanism in coordination polymers is the ring-chain polymerization, which is characterized by the fact that linear oligomers and polymers are in equilibrium with their cyclic counterparts. We will discuss this mechanism later on in this chapter.

In general, high degrees of polymerization are only obtained with sufficiently high association constants and concentrations. However, if the association constant is too high, reversibility is often lost because the reversible bonds in the polymers become kinetically inert. In this situation, the composition of rings and chains reflects the conditions during their formation and these coordination polymers behave like covalent polymers. Changing the topology of the ligand or the metal ion can alter the association constant. When multiple binding sites are present in the monomer, the bond will become stronger due to the chelating effect. The use of multiple binding sites to increase the binding constant is an interesting strategy to be used with coordination polymers and has already successfully been used with hydrogen bonds. In Chapter 2 of this book, Schneider describes the additivity of binding increments corresponding to multiple ion pairs. Furthermore, use of different metal ions results in different binding constants. Nickel ions, for instance, bind weaker to ligands than palladium ions.[18] Platinum ions bind even stronger, so choosing the right metal ion is crucial to achieve the desired properties. Temperature also influences the strength of the metal–ligand bond. Increasing temperature results in lower association constants, causing the coordination polymers to break more easily, thereby decreasing the average DP.

10.4 LIGAND EXCHANGE KINETICS

In addition to association constants, ligand exchange kinetics markedly influences the properties of reversible coordination polymers. In general, high association constants lead to high molecular weights and slow ligand exchange dynamics. However, a polymer with slow rates of complexation (k_a) and decomplexation (k_d) behaves differently from a polymer with fast exchange kinetics. Craig and coworkers showed that it was possible to separate the contributions of the association constant and exchange rate through careful ligand design and metal ion choice.[19] They prepared pincer metal–ligand coordination motifs and by incorporating bulky substituents they were able to tune ligand exchange while the thermodynamics of association remained constant. Model studies confirmed that significantly higher exchange rates were achieved for ligands with methyl substituents instead of ethyl substituents, because the ethyl substituents on the pincer ligand shield the metal complex from incoming ligands.

To study the effect of ligand exchange on material properties, they prepared networks from poly(4-vinylpyridine) (PVP) and bifunctional Pd(II) and Pt(II) pincer-based cross-linkers in DMSO (Fig. 10.6).[20]

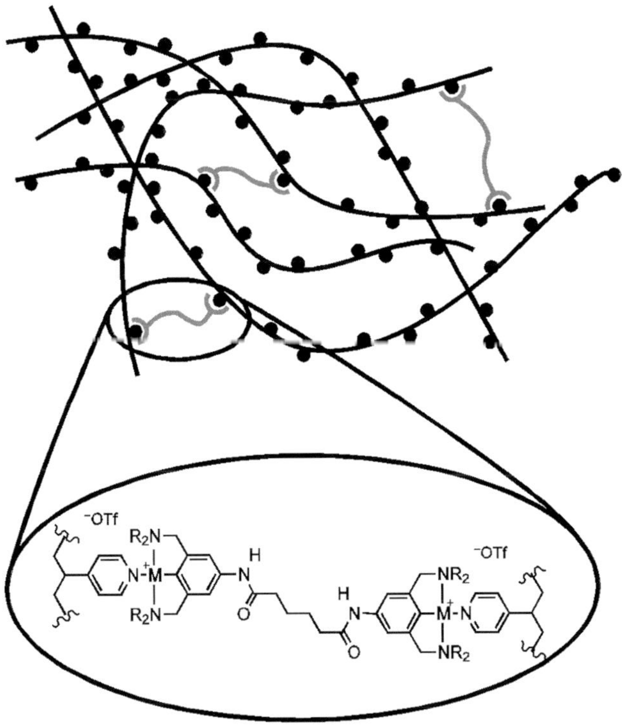

FIGURE 10.6. Schematic representation of reversible cross-links between PVP through coordination with bis(M^{II}-pincer) complexes; $OTf^- = [CF_3SO_3]^-$; R = Me or Et; M = Pd or Pt.[34]

The effect of ligand exchange dynamics on viscosity was dramatic; an 80-fold increase in viscosity was observed for the pincer complex containing ethyl groups. Changing the metal center from Pd(II) to Pt(II) resulted in even higher viscosities. This was attributed to the higher association constant of Pt(II) as compared to Pd(II).[21] Furthermore, the effect of solvent nucleophilicity on viscosity was studied to discern the dominant pathway for ligand exchange in networks. The influence of the different solvents (CH_2Cl_2, DMF, and DMSO) on the viscosity suggests that a solvent-assisted pathway dominates the ligand exchange process. In this way it was shown that material properties can be tuned by ligand design. In supramolecular polymers the structure of the monomeric units greatly determines the properties of the bulk material and being able to tune material properties by ligand design opens up new routes toward tailored materials.

Sijbesma and coworkers studied the effects of ligand exchange kinetics on polymer properties as well.[22] Phosphorus ligands were prepared with different steric and electronic properties. Complexation with palladium(II) dichloride resulted in high-molecular-weight polymers in the melt (Fig. 10.7). Three

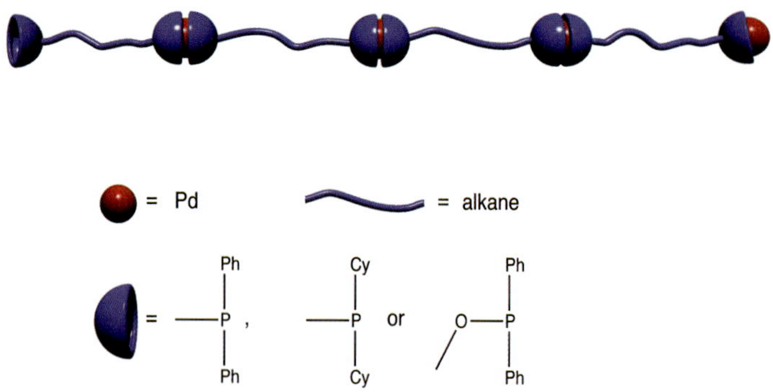

FIGURE 10.7. Linear coordination polymers from alkyldiphenylphosphine, alkyldicyclohexylphosphine, and alkyldiphenyl-phosphinite ligands.[22]

different palladium-based polymers were dissolved in chloroform, and the formation of low-molecular-weight (cyclic) material was monitored by the appearance of new signals in the ^{31}P NMR spectra. Dissolution of the coordination polymers based on diphenylphosphine and diphenylphosphonite ligands readily resulted in the formation of new species of considerably lower molecular masses. Half-lives of 13 and 11 hours were observed, respectively. The equilibration process of the dicyclohexylphosphine-based coordination polymer, however, was so slow that after 11 days only minute formation of monomeric cyclic material was observed. These ligands are known to be excellent σ-donors resulting in strong bonds. Moreover, the steric hindrance of the cyclohexyl group results in slow ligand exchange.

In another study the same group prepared coordination polymer networks by complexing rhodium(I) and iridium(I) with bifunctional diphenylphosphinite ligands.[23] Rhodium(I) and iridium(I) are both able to coordinate four phosphinite ligands, but the ligand exchange kinetics of iridium(I) phosphinite complexes are known to be much slower compared to the corresponding rhodium(I)-based systems. In order to assess ligand exchange kinetics, studies were performed in which metal ion precursors were added to the corresponding triphosphinite and tetrakisphosphinite complexes. Because formation of new complexes can only occur by dissociation of the phosphinite ligands, the rate of formation of new complexes is representative for the ligand exchange rate. The combination of iridium(I) chloride with alkyldiphenylphosphinite led to a system in which the ligand exchange processes were much slower compared to the rhodium(I)-based system. In both cases, stable gels were formed and sonication caused liquefaction within minutes. The rhodium(I)-based system reformed a gel almost immediately after sonication was stopped, whereas the iridium(I)-based system required standing for 1.5 hours at room temperature to reconstitute the gel. The fact that at −20°C re-gelation of the iridium(I)-based gel took several days points to ligand exchange processes as

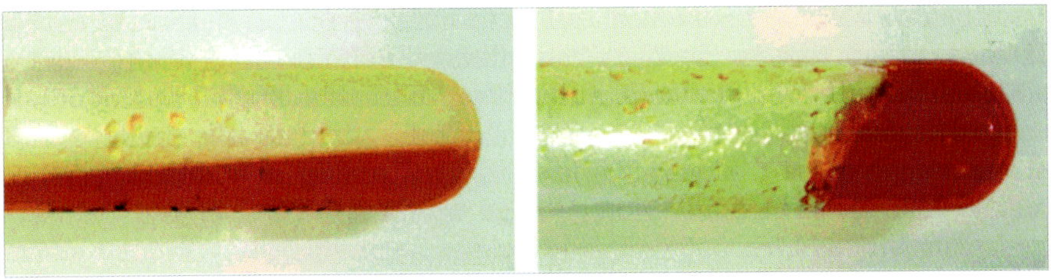

FIGURE 10.8. Iridium(I)-based gel with diphenylphosphinite telechelic polytetrahydrofuran 3 days after sonication, kept at –20°C (left) and after 1 hour at 20°C.[23]

the limiting factor for reconstitution of the gel. After placing the sample at room temperature, gelation occurred within 1 hour (Fig. 10.8).

10.5 CHAIN STOPPERS

In reversible coordination polymers, monomers with only one ligand group act as chain stoppers and decrease the average DP. Even a small amount can significantly affect the average chain length and consequently the material properties. The presence of chain stoppers is a drawback in the preparation of high-molecular-weight reversible coordination polymers. As is the case in any step polymerization, the presence of an excess of either bifunctional ligands or metal ions can limit the DP. Therefore, in order to obtain polymers with a high DP, pure bifunctional material is essential, but achieving the correct stoichiometry is equally important.

Sudhölter and coworkers discovered that stoichiometry was very important in the preparation of their reversible coordination polymers. They studied a set of water-soluble bifunctional ligands based on pyridine-2,6-dicarboxylic acid groups connected at the 4-position of the pyridine ring by ethylene oxide spacers (Fig. 10.9).[24] Complexation of the bifunctional ligands with zinc(II) ions resulted in water-soluble reversible coordination polymers. Increasing the metal-to-ligand ratio until 1.0 resulted in an increase in viscosity. Beyond a ratio of 1.0, viscosity decreased again. So an excess of either ligand or metal ion results in chain stoppers, which results in extra chain ends and a decrease in viscosity. The experiments also nicely show the reversibility of the system.

Rowan and coworkers showed the reversibility of their coordination polymers by titration experiments as well. They prepared bifunctional ligands consisting of 2,6-bis(1'-methylbenzimidazolyl)pyridine (BIP) as the metal coordinating group (Fig. 10.10).[25] Two BIP ligands were connected by a 1,4-diethynylphenylene spacer and together with zinc ions coordination polymers were formed. The ultraviolet-visible (UV-Vis) absorption spectrum of the monomer showed an absorption band associated with the BIP moiety at 321 nm. Upon addition of zinc ions this band was red-shifted to 339 nm until

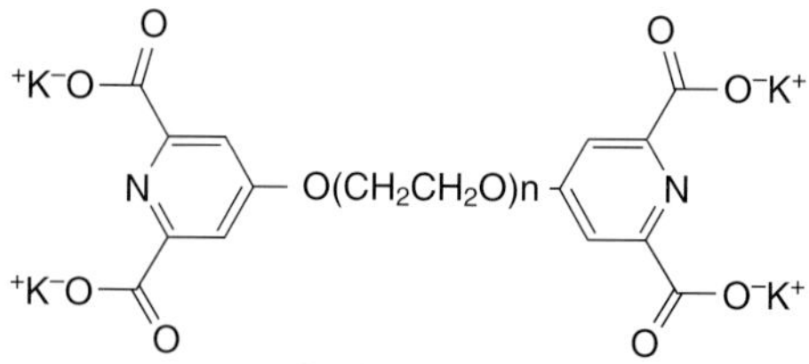

C4 (n = 4)
C6 (n = 6)

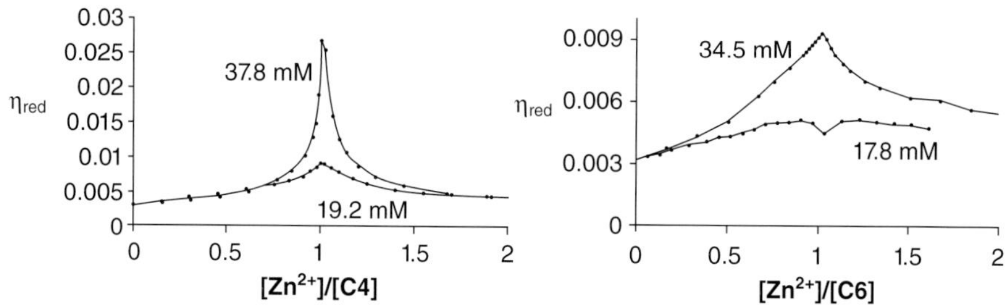

FIGURE 10.9. Water-soluble bifunctional ligand **C4** and **C6** (top) and reduced viscosity of ligands as a function of metal-to-ligand ratio for ligand **C4** (left) and **C6** (right).[24]

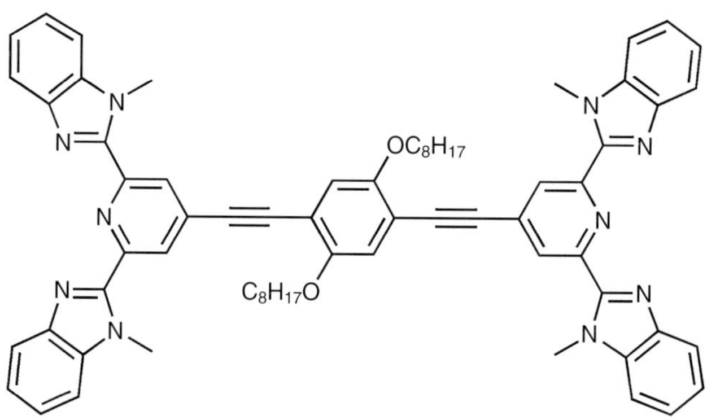

FIGURE 10.10. Chemical structure of the ditopic monomer based on the 2,6-bis(1'-methylbenzimidazolyl)pyridine (BIP) ligand.

a metal-to-monomer ratio of 1:1 was reached and the polymer was formed. Beyond this point, the subsequent addition of zinc ions caused new spectral changes. The absorption band at 339 nm was split into two bands and the spectral changes leveled off when a metal-to-ligand ratio of 2:1 was reached, indicating the breakdown of the polymer. Viscosity measurements were in agreement with the UV–Vis results, since a steady increase of the reduced viscosity was observed until a Zn^{2+}/monomer ratio of 1. Beyond this point a

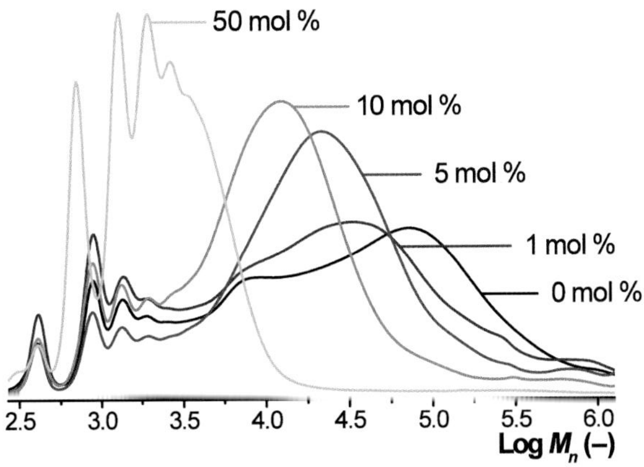

FIGURE 10.11. SEC traces of a palladium-based diphenylphosphine coordination polymer in the presence of different amounts of chainstoppers.[37]

decrease in viscosity was observed, confirming the depolymerization of the coordination polymer by the formation of chain ends.

In some cases, chain stoppers are deliberately added to a solution of reversible coordination polymers to demonstrate the linear nature and reversibility of coordination polymers (Fig. 10.11). Like in condensation polymerization reactions, where the molar mass of a condensation polymer is limited by the presence of monofunctional compounds, the addition of specific amounts of chain stopper will result in a linear decrease of the DP.[22,26,27]

10.6 RING–CHAIN EQUILIBRIUM

Intrinsic to many polymerization reactions is the formation of cyclic polymers (Fig. 10.12). The formation of rings in reversible coordination polymers has many similarities with ring formation observed during condensation polymerization. As early as 1950, Jacobson and Stockmayer studied the presence of rings in polycondensation reactions.[28] They determined the ring fraction as a function of monomer concentration. By carrying out polymerizations under dilute conditions they observed an increased degree of cyclization that was attributed to an increased probability to form intramolecular bonds. For strainless rings, composed of chains that obey Gaussian statistics, the probability that a chain is in a ring configuration follows $K_i = EM_i = Bi^{-5/2}$, where K_i is the molar cyclization equilibrium constant, EM_i the equilibrium effective molarity, i the DP, and B the effective molarity of the cyclic monomer if it were strainless. Ercolani and coworkers studied the distribution of cyclic monomers in dilute solution and included the possible occurrence of strained rings in their theory.[29] For strained rings, the effective molarity is lower than predicted

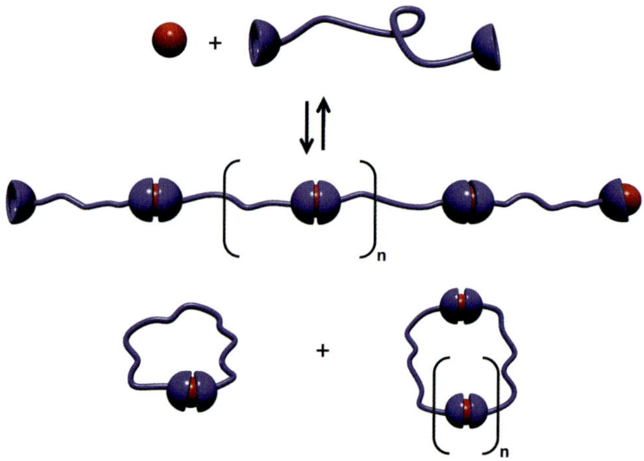

FIGURE 10.12. Schematic representation of the formation of linear polymers and rings in reversible coordination polymers.

because the formation of strained rings is less favored, resulting in a smaller ring fraction.

Inherent to ring formation in reversible systems is the critical cycle concentration (C^{CR}). The critical cycle concentration is the concentration below which all monomers are present as rings and above which the concentration of rings remains constant and the excess of monomer leads to the formation of linear chains.

Dormidontova and coworkers analyzed the reversible self-assembly of linear oligomers end-functionalized by ligands capable of forming 2:1 complexes with metal ions by means of combined Monte Carlo (MC) simulations and analytical modeling.[30] They assumed the free oligomers to be fully flexible, without an energetic penalty for bending any of the chemical bonds connecting the monomer units of an oligomer. For this system, ring formation increased with decreasing concentration. Furthermore, the fraction of rings decreased considerably for metal–ligand pairs with a high cooperativity of complexation. A large enthalpy gain for the second ligand–metal bond formation promotes the formation of 2:1 ligand–metal complexes at the expense of 1:1 ring complexes. So, cooperativity can be used in such a way that longer polymers will be obtained.

Sudhölter and coworkers investigated the ring–chain equilibrium of a set of water-soluble bifunctional ligands based on pyridine-2,6-dicarboxylic acid groups connected at the 4-position of the pyridine ring by ethylene oxide spacers (Fig. 10.9).[24] The bifunctional ligands were complexed with zinc ions. At low concentrations and equimolar stoichiometry a large fraction of the ligand monomers was incorporated in rings The fact that rings and chains gave different proton NMR signals for the protons in the spacer made it possible to determine the amount of rings and chains present. These measurements showed that increasing concentration of the 1:1 complexes led to a decrease in ring fraction and higher viscosities. In addition to concentration, the length of the spacer had a strong influence on the ring fraction as well. In general, the longer the spacer between the ligand groups, the smaller the probability that both ligand groups will bind to the same metal ion, and thus the smaller the ring fraction will be. Interestingly, NMR and viscosity measurements showed that for the longer spacer (**C6**) a larger amount of rings was present at all concentrations. This was explained by the fact that the bifunctional ligand with the shortest spacer (**C4**) could not form monomeric rings due to ring strain. This result was confirmed by viscosity measurements, because **C4**-Zn(II) complexes exhibited higher viscosities than **C6**-Zn(II) complexes. Furthermore, the viscosity measurements showed a dip in viscosity for **C6**-Zn(II). This dip can be explained by the fact that at low concentrations **C6** forms monofunctional rings with Zn(II) that have a smaller contribution to the viscosity. Increasing the temperature also affected the ring–chain equilibrium. For almost all polymers, covalent and supramolecular, the viscosity decreases with increasing temperature due to the higher mobility of the molecules. However, for the **C4** and **C6** zinc complexes, viscosity increased upon increasing temperature due to ring opening.[24]

The ring–chain equilibrium in high-molecular-weight coordination polymers was also studied by complexation of bifunctional phosphorus ligands with palladium(II) dichloride in chloroform. Ligand exchange in the complexes established a ring–chain equilibrium, which was studied by SEC and ^{31}P NMR. The concentration dependence of the equilibria was shown to conform to theory and a critical cycle (or ring) concentration was found.[22] The dimeric ring was significantly more abundant than the monomeric ring, implying that a monomeric ring was not strainless. According to theory, the equilibrium concentration of strainless rings is a monotonically decreasing function of ring size caused by the decreasing statistical probability that chain ends meet for larger spacer lengths.

10.7 NETWORKS

In linear polymers, the degree of functionality is 2. This means that all monomers, when polymerized, can join with only two other monomers. Increasing functionality of (at least some) monomers leads to nonlinear polymers, and network structures are often obtained.[26] In coordination polymers the functionality of the monomers can be increased in two ways: by increasing the number of donor groups (e.g., multifunctional ligands) or by increasing the number of donor atoms a metal ion can accommodate (e.g., lanthanide ions as compared to first-row transition metal ions).[31,32]

Reversible coordination polymer networks in water were prepared by Sudhölter and coworkers.[32] Bifunctional ligands based on pyridine-2,6-dicarboxylic acid groups with two different spacer lengths (6 and 7.4 ethylene

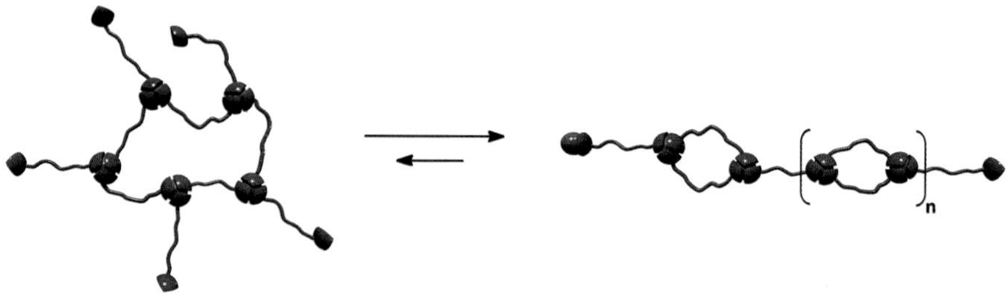

FIGURE 10.13. Branching leads to more chain ends. This shifts the equilibrium toward linear chains, which include rings.

oxide units) were mixed with lanthanide ions (Nd^{3+} and La^{3+}). The properties of the obtained gels were studied by rheology. At very high concentrations (500 g L^{-1}), a viscoelastic material was obtained. The scaling of the elastic modulus, relaxation time, and zero-shear viscosity with concentration were in good agreement with the predictions of Cates' model.[33] This model describes the dynamics of linear equilibrium polymers and therefore it was concluded that only a few cross-links were present in the gel and that mainly linear polymers were formed (Fig. 10.13). The formation of mostly linear polymers is probably due to the fact that branching inherently leads to more chain ends, which is energetically unfavorable.

Rowan and coworkers prepared stimuli-responsive reversible coordination polymer networks by the use of both transition and lanthanide metal ions in acetonitrile.[31] They prepared bifunctional ligands from two tridentate BIP ligands (2,6-bis(1'-methylbenzimidazolyl)-4-hydroxypyridine) connected by an ethylene oxide spacer (Fig. 10.14). Addition of zinc ions led to precipitation and no gel formation was observed. After the addition of only 3% of lanthanide ions (La^{3+} or Eu^{3+}) a gel-like material was obtained. Different combinations of transition metal and lanthanide ions led to gel-like materials that all showed thermoresponsive behavior. A reversible gel-sol transition was observed around 100°C and caused cleavage/re-formation of the lanthanide–ligand bond. The obtained gels exhibited thixotropic (shear-thinning) behavior; shaking the gel resulted in a free-flowing liquid and after standing for about 20 seconds the gel reformed again. Rheological studies confirmed this behavior and showed that at shear stresses up to 155 Pa the gel responded mostly elastically. Shear stresses above this value caused bond breaking and resulted in a large drop of the storage modulus corresponding with the behavior of a free flowing liquid.

As discussed earlier, Craig and coworkers prepared cross-linked polymer networks and studied the influence of the dynamics of the cross-linker on the viscoelastic properties of the network.[34] They cross-linked poly(4-vinylpyridine), PVP, with different bis(M^{II}-pincer) alkylamine ligands with Pd(II) and Pt(II) as the metal ions (Fig. 10.6). The addition of only 2% (by functional group) of

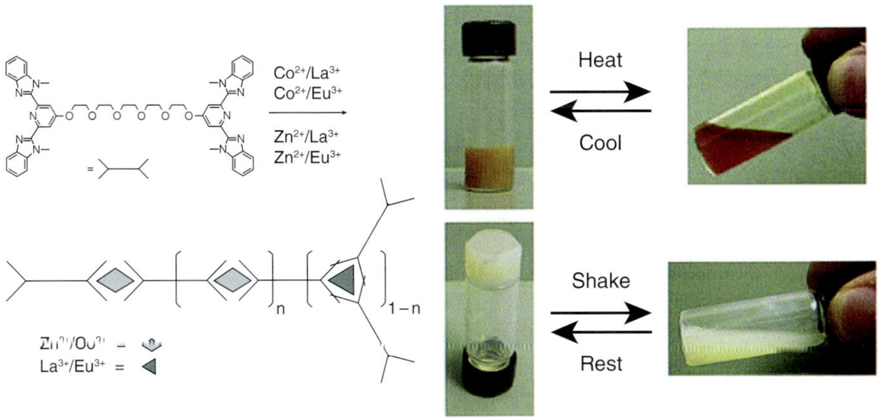

FIGURE 10.14. Schematic representation of a metallo-supramolecular gel-like material using a combination of lanthanide and first-row transition metal ions mixed with a bifunctional BIP ligand (left) and the thermoresponsive (right, top) and mechanoresponsive (right, bottom) nature of the obtained gels.[31]

Pd^{II} pincer molecules to PVP in DMSO led to a 2000 times higher viscosity than for PVP alone. The increase in viscosity was not due to complexation alone, but actual cross-links were formed, since complexation with monomeric ligands did not result in an increase in viscosity.

10.8 SOLVENT INFLUENCE

While most reversible coordination polymers are prepared in apolar and noncoordinating solvents, preparing reversible coordination polymers in polar and coordinating solvents is challenging. Apolar solvents hardly compete with ligand monomers and therefore a kinetically inert metal–ligand bond is formed with higher degrees of polymerization. When coordinating solvents are employed, the coordinative bonds break more readily due to the competition of the solvent with the ligands for the metal ions and a more dynamic system is obtained. The preparation of reversible coordination polymers in coordinating solvents requires strong coordinative interactions, but still a dynamic system is obtained. An environmentally friendly and cheap coordinating solvent is, of course, water, but research on reversible coordination polymers in water or other coordinating solvents is still limited.

One of the few water-soluble coordination polymers known up to now is the class of reversible coordination polymers prepared from bifunctional ligands based on pyridine-2,6-dicarboxylic acid groups. Complexation with lanthanide ions resulted in the formation of gels (Fig. 10.15).[35] Furthermore, gels were also obtained by the cross-linking of high molecular weight, partly hydrolyzed poly(acrylamide) (HPAAm) with trivalent chromium ions in

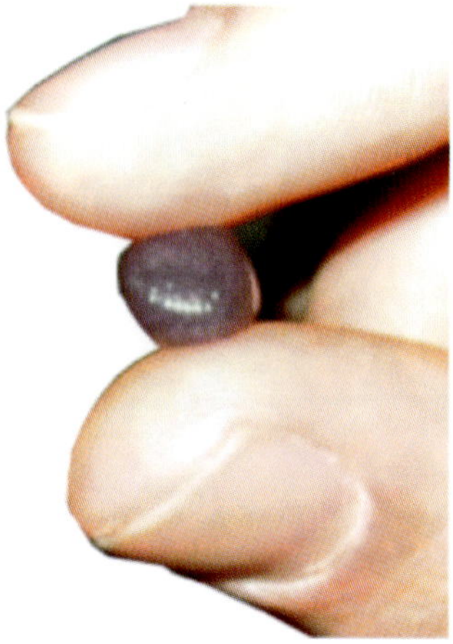

FIGURE 10.15. Picture of a sample of 500 g L^{-1} of bifunctional ligand **C6** and 0.7 equivalent of $Nd(NO_3)_3$ in water.[35]

aqueous solution.[36] Carboxylic acid groups were involved in the complexation of chromium(III) and HPAAm, because cross-linking did not occur in unhydrolyzed poly(acrylamide). The formed cross-links consisted of complexes of chromium(III) dimers with two COO^- groups. This reversible coordinative interaction was shown to be strongly pH dependent, since a sufficiently low pH was required to ensure the presence of free chromium(III) ions, while a sufficiently high pH was required to deprotonate carboxylic acid groups so they could act as ligands. Network formation proceeded slowly and an equilibrium state was not reached within 12 hours. The addition of citric acid slowed down the gelation process even more and the cross-link density decreased. This was due to the fact that citric acid withdraws chromium ions from the solution by complexation as a chelate.

10.9 CHARACTERIZATION OF REVERSIBLE COORDINATION POLYMERS

The kinetically labile bond between metal and ligands in reversible coordination polymers often requires a different approach in the characterization of these polymers than what can be used for covalent, irreversible polymers. Analytical techniques that are often used for the characterization of polymers are matrix-assisted laser desorption/ionization–time-of-flight (MALDI-TOF) mass spectrometry and SEC. These techniques, however, are not always successful for the characterization of reversible coordination polymers. The equilibrium bond between the monomers is often broken within the experimental timescale, thereby obscuring the measurements. On top of that, metal ions can complicate characterization even more by interacting, for example, with the column material.

10.9.1 NMR

NMR spectroscopy is often used for the characterization of reversible coordination polymers. The experimental timescale of the NMR experiment can be as short as a few milliseconds and in most cases this is shorter than the lifetime of the coordinative bond. Upon complexation, a change in chemical shift is often observed for nuclei close to the metal ion. Proton NMR is used most frequently,[15] but phosphorus[37] and fluorine[14] are nuclei that also often display marked shift changes upon complexation.

The ring–chain equilibrium of coordination polymers has been successfully characterized with diffusion-ordered spectroscopy (DOSY) NMR.[22,24] With DOSY NMR it is possible to determine the diffusion coefficient of a molecule. Because, in most cases, rings are smaller than chains, they move faster and have higher diffusion coefficients as compared to chains.

10.9.2 Viscometry and Rheology

Polymers in solution greatly enhance the viscosity of that solution, even at very low concentrations. Furthermore, the viscosity of polymer solutions increases with the average chain length of the polymers and for reversible coordination polymers, viscosity measurements have been successfully applied to study the effects of temperature, concentration, and metal-to-ligand ratio on the average chain length.[24,25,38]

In addition to the increase in viscosity, polymer solutions and melts can be viscoelastic as well. This means they both have liquid-like (viscous) and solid-like (elastic) behavior, depending on the experimental timescale. Reversible coordination polymers also possess viscoelastic properties, and these properties are typically studied by rheology. Rheology is the study of the flow and deformation of materials in which the response of the material to an oscillating stress or strain is measured. For the characterization of viscoelastic materials the ratio between the amplitude of the maximum stress (σ_0) and maximum strain (γ_0) as a function of the frequency is measured.

The rheological behavior of linear coordination polymers can be compared with that of wormlike micelles, because these micelles break and recombine reversibly as well.[39,40] Rheology studies on linear reversible coordination polymers are scarce. For reversible coordination networks, considerably more rheological data have been reported.[35,41,42]

10.9.3 SEC

SEC is the most frequently applied technique to obtain detailed information about the molecular mass distribution of polymers. It separates molecules on the basis of their hydrodynamic volume or size, and calibration with similar polymers with well-defined molecular weights is often necessary to obtain the exact molar mass distribution. This can be a disadvantage for newly designed materials with functional moieties that are not yet well described in the polymer literature. Furthermore, interactions with the SEC column may complicate the situation in the case of ionic polymers, resulting in incorrect values for the molar masses. SEC has been used for the characterization of reversible coordination polymers, although in most cases this technique is not useful for reversible coordination polymers. Important for the applicability of SEC in reversible coordination polymers is the kinetic stability of the complexes. If dissociation is too fast, the polymers will dissociate during the measurement, resulting in lower molar masses than expected, and the actual molecular weight distribution before chromatography is only correctly reflected by the SEC trace in the limit of kinetically inert coordination complexes. Phosphine-based coordination polymers were successfully characterized by SEC, because equilibration was slow enough.[22] High-molecular-weight polymers were observed, corresponding to a DP of up to 100 monomers. The elution of the coordination complexes at high hydrodynamic volumes on SEC demonstrated the polymeric nature of the materials. It also indicated that the kinetics of ligand exchange were slow under the chromatographic conditions, because equilibration during chromatography is expected to lead to dissociation and low-molecular-weight and/or cyclic material.

10.10 APPLICATIONS AND POTENTIALS

The application potential of reversible coordination polymers in material science is rapidly increasing. The reversibility of the bonds in supramolecular polymers gives these polymers some unique properties in comparison to their covalent counterparts. Since the viscosity of supramolecular polymers strongly depends on temperature, processing these polymers at high temperatures is easy due to the low viscosity, while the polymer properties remain intact at room temperature. Furthermore, stimuli-responsive materials can be obtained because the strength of the reversible bond is highly dependent on external factors such as temperature, pH, mechanical forces, and so on. The incorporation of metal ions in reversible coordination polymers can give the polymers novel properties, like light-emitting capabilities and catalytic properties.[43]

10.10.1 Gel Formation

Recently Rowan and coworkers studied the gelation mechanism of a ditopic ligand, consisting of a 2,6-bis(1'-methylbenzimidazolyl)-4-oxypyridine (BIP) moiety attached to either end of an oligo(ethylene glycol) core, in the presence of zinc ions (Fig. 10.16).[44] Gelation occurred via the flocculation of semicrystalline colloidal particles due to phase separation and crystallization. By the

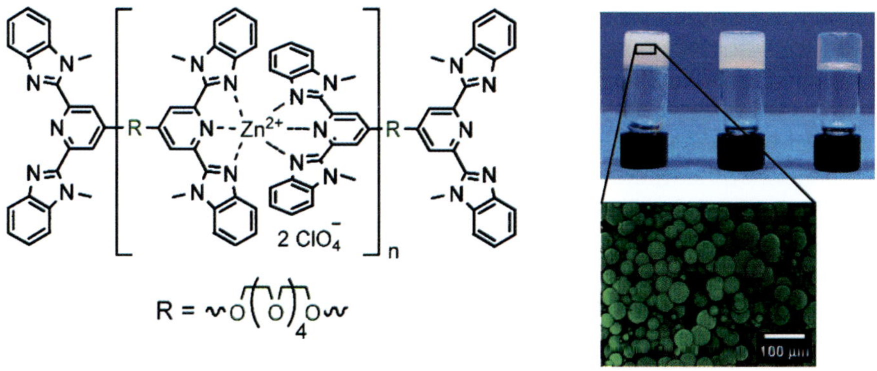

FIGURE 10.16. Schematic representation of the formation of metallosupramolecular polymeric aggregates.[45]

addition of a nonsolvent (water) or poor solvent (ethylene glycol) to the good solvent (DMSO), they were able to tune the gelation.[45] In this system the coordination polymers are not directly involved in gel formation, but the formed colloidal particles are responsible for the gel-like behavior. The obtained gels varied from highly opaque (high water/ethylene glycol content) to fully transparent (low water/ethylene glycol content). As the fraction of good solvent increased, particle size reduced, which was accompanied by an increase in sol fraction, a decrease in opacity, and an increase in gelation time.

To further study the gelation mechanism of these coordination polymers, the same ligands were prepared, but with three different ethylene oxide spacer lengths (4, 5, and 6 ethylene oxide units).[46] These different spacer lengths resulted in differences in solubility, and hence, in different gel characteristics. The shear modulus of the different ligand–metal solutions was strongly dependent on ligand solubility. The lower solubility of the ligand with the shortest spacer was responsible for the formation of dense colloidal particles and a lower particle volume fraction, which resulted in a lower shear modulus. On the other hand, the increased solubility of the ligand with the longest ethylene oxide spacer also resulted in a lower shear modulus. This was due to the larger sol fraction under equivalent conditions.

10.10.2 Catalysis

Ultrasonic scission of the weak coordinative palladium–phosphorus bond in palladium(II) coordination polymers proved to be completely selective and fully reversible.[47,48] This reversible and selective breaking of coordinative bonds was used to develop a method for the activation of homogeneous catalysts. Activation of latent catalysts is usually achieved by thermally or chemically induced dissociation of ligands to give catalytically active species from coordinatively saturated metal complexes. Sijbesma and coworkers were the first to use mechanical forces to selectively dissociate ligands from a metal centre, resulting in a catalytically active species.[49,50] They designed latent catalysts that combined polymer-functionalized ligands with a metal center. Polymeric ligands of sufficient size are a requirement for ultrasound-induced ligand dissociation, because enough force has to be transduced via this polymer to induce bond scission. Hence, small molecules are hardly affected by mechanical forces. The metal–ligand bonds in silver(I) complexes with polymeric N-heterocyclic carbene (NHC) ligands were shown to break rapidly when exposed to ultrasound, and now these complexes were able to catalyze transesterification. A 65% conversion of vinyl acetate and benzyl alcohol into benzyl acetate was achieved after 1 hour of sonication. Without sonication, only 3% conversion was obtained. The mechanical origin of scission was evident from the dependence of the catalytic activity on molecular weight, since the thermal scission of the Ag–C bond is not expected to be molecular weight dependent.

10.10.3 Complex Coacervate Core Micelles

Complex coacervate core micelles (C3Ms) are micelles composed of a diblock copolymer with a polyelectrolyte block and another oppositely charged species, wherein the associative phase separation is arrested by at least one neutral water-soluble component.[51] Generally, the resulting micelles are spherical in shape and show a rather narrow size distribution. These micelles are sensitive to solution parameters that affect the ionic complex, such as salt concentration and pH. Cohen Stuart and coworkers prepared C3Ms from a positive diblock copolymer (poly(2-vinyl-N-methylpyridinium iodide)-*b*-poly(ethylene oxide), $PMVP_{41}$-*b*-PEO_{205}) and an oppositely charged reversible coordination polymer (pyridine-2,6-carboxylic acid based bifunctional ligands complexed with zinc nitrate,[24] L_2EO_4).[52] Mixing dilute solutions of $PMVP_{41}$-*b*-PEO_{205}, L_2EO_4, and Zn^{2+} resulted in self-assembled structures (Fig. 10.17). No self-assembly took

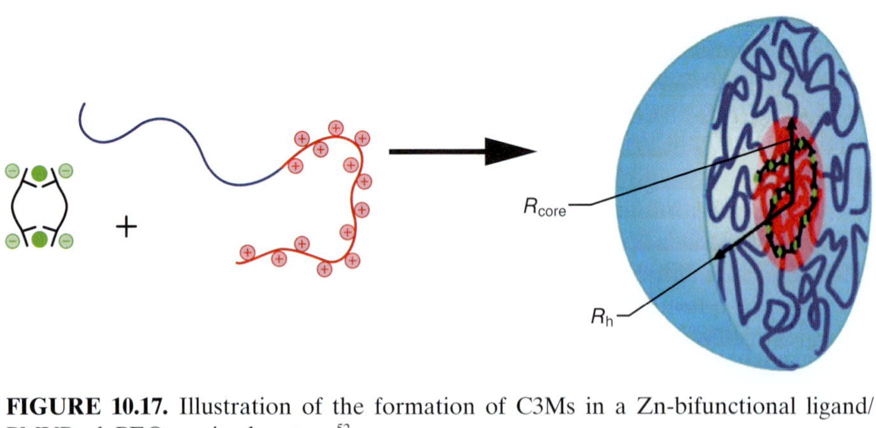

FIGURE 10.17. Illustration of the formation of C3Ms in a Zn-bifunctional ligand/ $PMVP_{41}$-*b*-PEO_{205} mixed system.[52]

place when either zinc ions or bifunctional ligands were absent. The largest structures were obtained at equimolar concentrations and when the positive charges and negative charges were compensated. Surprisingly, C3Ms were formed with very dilute solutions of Zn-L_2EO_4, where normally only oligomeric rings are present. The reason that indeed longer polymers are formed is because of the high local concentration inside the complex coacervate core of the micelle, and C3M formation is further promoted by the positively charged PMVP blocks. Furthermore, the addition of neodymium ions instead of zinc ions to the diblock copolymer resulted in a faster formation of C3Ms. On the other hand, excess of Nd^{3+} ions led to a breakdown of the C3Ms, while an excess of Zn^{2+} ions had little effect on the micelles.[53] This phenomenon was attributed to the charge inversion that can occur in this system upon addition of Nd^{3+} ions.

10.11 CONCLUSIONS

In this chapter an overview is given of the research on reversible coordination polymers during the start of the 21st century. The development of reversible coordination polymers has become a rapidly increasing field of interest, and more and more research groups focus on the development of "small" structures that are able to self-assemble into polymeric materials. In this way, new materials will be obtained that can possess stimuli-responsive and self-healing abilities on a macroscopic level. The introduction of metal ions into these reversible systems gives them even more new features, such as magnetic, optical, and electrical properties, but also applications in catalysis are within reach. In the development of reversible coordination polymers, a crucial goal remains obtaining reliable and large degrees of polymerization. Information on these DPs is scarce and often only viscosity changes are reported. Depending on the polymerization mechanism the DP can be deduced from the equilibrium constant (Chapter 6).

One of the main challenges in the field of reversible coordination polymers is the preparation of hydrogels. Hydrogels are gels in which the aqueous solvent is immobilized by the presence of small amounts of gelators.[54] Generally, these gelators are large covalent polymers, but the use of low-molecular-weight organic molecules will offer several advantages, especially for biological systems. The search for biodegradable polymers for the use in, for example, controlled drug release, tissue engineering, and capture of pollutants is ongoing and the use of reversible coordination polymers can be a promising tool in the preparation of these materials. To develop biocompatible and water-soluble reversible coordination polymers that result in the formation of hydrogels upon complexation with metal ions will be a major challenge. Most of the reversible coordination polymers developed so far are insoluble in water or consist of a dynamically inert bond, resulting in the loss of their reversible character. To develop reversible coordination polymers in water, a relatively strong coordination bond is needed because water competes with the ligand groups for the metal ions. It is expected that specifically in this area big steps will be taken with reversible coordination polymers.

REFERENCES

1. A. Ciferri. *Supramolecular Polymers*. CRC Press, Boca Raton, FL, 2005.
2. M. Burnworth, D. Knapton, S. J. Rowan, C. Weder. *J Inorg Organomet Polym Mater* **2007**, 17, 91.
3. B. J. B. Folmer, R. P. Sijbesma, R. M. Versteegen, J. A. J. van der Rijt, E. W. Meijer. *Adv Mater* **2000**, 12, 874.
4. A. U. Czaja, N. Trukhan, U. Müller. *Chem Soc Rev* **2009**, 38, 1284.
5. C. Janiak. *Dalton Trans* **2003**, 14, 2781.
6. G. Ferey. *Chem Soc Rev* **2008**, 37, 191.
7. J. L. C. Rowsell and O. M. Yaghi. *Micropor Mesopor Mater* **2004**, 73, 3.
8. W. L. Leong and J. J. Vittal. *Chem Rev* **2010**, 111, 688.
9. D. E. Herbert, J. B. Gilroy, W. Y. Chan, L. Chabanne, A. Staubitz, A. J. Lough, I. Manners. *J Am Chem Soc* **2009**, 131, 14958.
10. C. A. Fustin, P. Guillet, M. J. Misner, T. P. Russell, U. S. Schubert, J. F. Gohy. *J Polym Sci Polym Chem* **2008**, 46, 4719.
11. E. C. Constable. *Chem Soc Rev* **2007**, 36, 246.
12. R. D. Archer. *Inorganic and Organometallic Polymers*. Wiley-VCH, Chichester, 2001.
13. J. F. Gohy, B. G. G. Lohmeijer, U. S. Schubert. *Chem Eur J* **2003**, 9, 3472.
14. J. F. Modder, K. Vrieze, A. L. Spek, G. Challa, G. Van Koten. *Inorg Chem* **1992**, 31, 1238.
15. U. Velten and M. Rehahn. *Chem Commun* **1996**, 2639.
16. U. Michelsen and C. A. Hunter. *Angew Chem Int Ed* **2000**, 39, 764.
17. T. F. A. De Greef, M. M. J. Smulders, M. Wolffs, A. P. H. J. Schenning, R. P. Sijbesma, E. W. Meijer. *Chem Rev* **2009**, 109, 5687.
18. D. T. Richens. *Chem Rev* **2005**, 105, 1961.
19. W. C. Yount, H. Juwarker, S. L. Craig. *J Am Chem Soc* **2003**, 125, 15302.
20. W. C. Yount, D. M. Loveless, S. L. Craig. *J Am Chem Soc* **2005**, 127, 14488.
21. J. D. Atwood. *Inorganic and Organometallic Reaction Mechanisms*, 2nd ed. VCH, Weinheim, 1996.
22. J. M. J. Paulusse, J. P. J. Huijbers, R. P. Sijbesma. *Macromolecules* **2005**, 38, 6290.
23. J. M. J. Paulusse, D. J. M. van Beek, R. P. Sijbesma. *J Am Chem Soc* **2007**, 129, 2392.
24. T. Vermonden, J. van der Gucht, P. de Waard, A. T. M. Marcelis, N. A. M. Besseling, E. J. R. Sudhölter, G. J. Fleer, M. A. Cohen Stuart. *Macromolecules* **2003**, 36, 7035.
25. P. K. Iyer, J. B. Beck, C. Weder, S. J. Rowan. *Chem Commun* **2005**, 319.
26. P. J. Flory. *Principles of Polymer Chemistry*. Cornell University Press, Ithaca, NY, 1964.
27. R. P. Sijbesma, F. H. Beijer, L. Brunsveld, B. J. B. Folmer, J. H. K. K. Hirschberg, R. F. M. Lange, J. K. L. Lowe, E. W. Meijer. *Science* **1997**, 278, 1601.

28. H. Jacobson and W. H. Stockmayer. *J Chem Phys* **1950**, 18, 1600.
29. G. Ercolani, L. Mandolini, P. Mencarelli, S. Roelens. *J Am Chem Soc* **1993**, 115, 3901.
30. C. C. Chen and E. E. Dormidontova. *J Am Chem Soc* **2004**, 126, 14972.
31. J. B. Beck and S. J. Rowan. *J Am Chem Soc* **2003**, 125, 13922.
32. T. Vermonden, W. M. de Vos, A. T. M. Marcelis, E. J. R. Sudhölter. *Eur J Inorg Chem* **2004**, 2847.
33. M. E. Cates. *Macromolecules* **1987**, 20, 2289.
34. W. C. Yount, D. M. Loveless, S. L. Craig. *Angew Chem Int Ed* **2005**, 44, 2746.
35. T. Vermonden, M. J. van Steenbergen, N. A. M. Besseling, A. T. M. Marcelis, W. E. Hennink, E. J. R. Sudhölter, M. A. Cohen Stuart. *J Am Chem Soc* **2004**, 126, 15802.
36. K. te Nijenhuis. *Macromol Symp* **2001**, 171, 189.
37. J. M. J. Paulusse and R. P. Sijbesma. *Chem Commun* **2003**, 1494.
38. U. Velten, B. Lahn, M. Rehahn. *Macromol Chem Phys* **1997**, 198, 2789.
39. M. E. Cates and S. J. Candau. *J Phys Condens Matter* **1990**, 2, 6869.
40. M. E. Cates. *J Phys Chem* **1990**, 94, 371.
41. Y. Q. Zhao, J. B. Beck, S. J. Rowan, A. M. Jamieson. *Macromolecules* **2004**, 37, 3529.
42. D. Xu, J. L. Hawk, D. M. Loveless, S. L. Jeon, S. L. Craig. *Macromolecules* **2010**, 43, 3556.
43. R. Dobrawa and F. Würthner. *J Polym Sci Polym Chem* **2005**, 43, 4981.
44. W. G. Weng, A. M. Jamieson, S. J. Rowan. *Tetrahedron* **2007**, 63, 7419.
45. W. G. Weng, Z. Li, A. M. Jamieson, S. J. Rowan. *Macromolecules* **2009**, 42, 236.
46. W. Weng, Z. Li, A. M. Jamieson, S. J. Rowan. *Soft Matter* **2009**, 5, 4647.
47. J. M. J. Paulusse and R. P. Sijbesma. *Angew Chem Int Ed* **2004**, 43, 4460.
48. J. M. J. Paulusse and R. P. Sijbesma. *Chem Commun* **2008**, 4416.
49. S. Karthikeyan, S. L. Potisek, A. Piermattei, R. P. Sijbesma. *J Am Chem Soc* **2008**, 130, 14968.
50. A. Piermattei, S. Karthikeyan, R. P. Sijbesma. *Nat Chem* **2009**, 1, 133.
51. M. A. Cohen Stuart, B. Hofs, I. K. Voets, A. de Keizer. *Curr Opin Colloid Interface Sci* **2005**, 10, 30.
52. Y. Yan, N. A. M. Besseling, A. de Keizer, A. T. M. Marcelis, M. Drechsler, M. A. Cohen Stuart. *Angew Chem Int Ed* **2007**, 46, 1807.
53. Y. Yan, N. A. M. Besseling, A. de Keizer, M. A. Cohen Stuart. *J Phys Chem B* **2007**, 111, 5811.
54. L. A. Estroff and A. D. Hamilton. *Chem Rev* **2004**, 104, 1201.

CHAPTER 11

STRUCTURAL AND FUNCTIONAL ASPECTS OF METAL BINDING SITES IN NATURAL AND DESIGNED METALLOPROTEINS

ORNELLA MAGLIO, FLAVIA NASTRI, and ANGELA LOMBARDI

11.1 INTRODUCTION

Metalloproteins take part in a variety of life-sustaining processes in bacterial, plant, and animal kingdoms and catalyze difficult reactions with efficiency and selectivity that few other natural or artificial molecules can achieve.[1,2] For this rcason, structural and functional studies on metalloproteins have been the focus of many years of research. These studies require a simultaneous and accurate analysis of both the polypeptide chain and the metal cofactor herein embedded. In fact, the plethora of mixed interactions that occur between the metal cofactor and the protein environment mutually affects the properties of each other, thus enhancing, diversifying, or tuning their individual functions.

Metalloprotein functions benefit from the unique physicochemical properties of metal cofactors, such as their small volume and simple structure, positive charge, Lewis acidity, rigid/flexible coordination sphere, specific ligand affinity, different stable oxidation states, varying electron spin configuration, and high mobility.[3,4] Despite the widespread distribution of metal ions in proteins, the number of different metal cofactors is almost limited. Therefore, often the same metal cofactor is able to serve a number of roles, either structural or functional. Understanding the molecular basis of protein-metal recognition is a fundamental issue in order to shed light on the detailed mechanisms of important processes in which metal ions are involved and to engineer novel metalloproteins with programmed properties.

Ionic Interactions in Natural and Synthetic Macromolecules, First Edition. Edited by Alberto Ciferri and Angelo Perico.

This chapter discusses the main properties of metal ions in proteins. First, it describes the amino acids that act as ligands and their possible binding modes. The most representative metal ions in biological systems are briefly outlined, mainly regarding their preferred geometry and their functions. Finally, the chapter focuses on two classes of iron-containing metalloproteins (heme proteins and carboxylate-bridged diiron proteins) in order to illustrate how the same metal cofactor can be engaged in a number of different roles. The various first and second ligation sphere interactions, which finely tune the cofactor properties, thus effecting such different functions, will be highlighted. The models developed by us for these two classes of metalloproteins will be also described, summarizing principles and methods for designing artificial metalloproteins.

11.2 CHEMISTRY OF COORDINATION COMPOUNDS: AN OVERVIEW

This section contains an overview of the main properties and bonding in coordination compounds. Coordination chemistry is of paramount importance in biochemistry; in fact, the biologically relevant compounds, in which metal ions play an important role, may be considered, as are coordination compounds, in which the metal ions are bound to complicated molecular architectures.

In particular, metalloprotein structure and function are strictly related to the plethora of interactions that occur between the protein chain and the metal prosthetic group, first of all the coordinate bond. For a more comprehensive and rigorous treatment, the readers may refer to several reviews and textbooks.[5–9]

Numerous metal ions in proteins are transition elements. They possess energetically accessible *d* orbitals, which are partially filled with electrons, in at least one oxidation state. The number, the shape, and the symmetry properties of *d* orbitals all contribute to the extraordinary rich chemistry (including spectroscopic, magnetic, and catalytic properties) exhibited by transition metal complexes, either classical metallorganic or bioinorganic.

The typical transition metal atom has one *s*, three *p*, and five *d* atomic orbitals that possess geometrical and energetic properties suitable for bonding. Transition metal ions may form with ligands different types of bonds, from highly covalent to highly electrostatic. Much of their remarkable structural and functional behavior resides in this versatility.

11.2.1 Coordination Chemistry: Some Definitions and Terminology

In this section, definitions of some concepts that will be used throughout this chapter are given.

Transition elements exhibit a number of peculiar properties, which distinguish them from other elements: (1) they are all metals; (2) most of them display numerous oxidation states; (3) they are generally positively charged and act as Lewis acid; and (4) they have an unparalleled propensity for forming coordination compounds with Lewis bases.

A **Lewis acid** is a molecular entity that is an electron pair acceptor and therefore able to react with a Lewis base to form a Lewis adduct by sharing the electron pair furnished by the Lewis base. A **Lewis base** is a molecular entity able to provide a pair of electrons and thus capable of coordination to a Lewis acid, thereby producing a Lewis adduct.

According to the International Union of Pure and Applied Chemistry (IUPAC) nomenclature, **coordination** implies "the formation of a covalent bond, the two shared electrons of which have come from only one of the two parts of the molecular entity linked by it, as in the reaction of a Lewis acid and a Lewis base to form a Lewis adduct." This definition relates also to the bond formed in coordination compounds (also called complexes), the so-called "coordinate bond"; the terms "coordinate covalence," "coordinate link," and "dative bond" are obsolete synonyms.[10]

Ligands are the atoms or groups of atoms that are bonded to the metal ion. They are electron pair donors (Lewis base) and are classified as monodentate or polydentate, according to the number of electron pairs (one or more) donated to the metal ion. In metalloproteins, the ligands are often derived by the polypeptide chain (**endogenous ligands**). As described later on in this chapter, these comprise side chain groups, carbonyl and amide main chains, N-terminal amino groups, and C-terminal carboxylate groups. **Exogenous ligands** are ligands not derived from the polypeptide chain; they comprise small inorganic entities (e.g., oxide, hydroxide, water) as well organic compounds, such as porphyrins.[11–13]

The **coordination number** is the number of lone pairs that ligands donate to the metal.

The inner sphere containing those atoms directly interacting with the metal ion is called **first coordination sphere**; the **second coordination sphere** contains those groups interacting with the inner sphere ligands. The number of atoms in these spheres will depend on the size and charge of the metal ion and on the sizes of the ligand groups.

The **coordination geometry** refers to the arrangement of the ligand around the metal ion.

The distribution of coordination numbers and the coordination geometry differ significantly among the metal ions. However, for the majority of metals, the most frequent coordination numbers are four and six. Coordination numbers and geometry in metalloproteins are related to the function that the metal ion achieves. For example, structural sites, which generally stabilize or direct the folding of the protein, exhibit common, coordinately saturated geometries that are well precedented in small-molecule/metal ion complexes. By contrast, functional metal binding sites often show more unusual ligation geometries, which are largely preorganized in the folded apo-protein.

The **prosthetic group** is defined as the nonamino acid portion of a conjugated protein. Examples include cofactors such as flavines or cytochromes, as well as lipids and polysaccharides, which are the prosthetic groups of lipoproteins and glycoproteins, respectively.[14]

Cofactors are organic molecules or ions (usually metal ions) that are required by an enzyme for its activity. They may be attached either loosely or tightly to the enzyme. A cofactor binds with its associated protein (**apo-enzymes**), which is functionally inactive, to form the active enzyme (**holo-enzyme**).[14]

11.2.2 The Coordinate Bond

The coordinate bond is defined as the interaction between a cation and an anion or neutral molecule possessing a lone pair of electrons. In different complexes, it can span from an electrostatic to a covalent character. Different theories have been developed to study the bond in coordination compounds; the most widely used are the **crystal field theory** (CFT) and the **ligand field theory** (LFT).[5–9] They are based on different assumptions: The first considers the bond as entirely electrostatic; the latter, which arose from an application of molecular orbital theory, is built on the CFT and includes the possibility of some covalency.

The CFT (developed by Hans Beth in 1929)[15] considers the ligand lone pair as a simple point negative charge or point dipole, which repels the d electrons of the central metal ion. Beth showed that each of the various states, arising from a particular electron configuration of an ion, may be split into two or more groups with different energies (terms) when the ion is introduced into a crystalline environment of definite symmetry. Beth also described how to calculate the splitting magnitudes, assuming the interactions between the ligands and the electrons in the unhybridized *d* orbitals as purely electrostatic. Based on this assumption, all electrons, which are in metal ion orbitals in the free ion, are considered remaining in orbitals that are 100% metal ion orbitals. This model is very simple; nevertheless, it captures the essence of the electronic structure of coordination compounds. The splitting of the orbitals can be conveniently used to understand, interpret, and predict the magnetic behavior, optical spectra, thermodynamic stability, and structures of coordination compounds.

Using the methods of group theory, it is possible to see that in the presence of an octahedral field (point group O_h), the *d* orbitals are split into a lower energy triple degenerated group (t_{2g}; d_{xy}, d_{yz}, and d_{xz} orbitals) and a higher energy double degenerated group (e_g; d_{z^2} and $d_{x^2-y^2}$ orbitals). The separation between the two sets of orbitals is called the ligand field splitting parameter (Δ_o) (see Fig. 11.1). One of the most important aspects of CFT is that not all ligands are identical in determining the Δ_o value. For transition metal compounds, it is clear from the multitude of colors available for a given metal ion when the ligands or stereochemistry are varied. It has been established that the ability of ligands to cause a different energy splitting of the *d* orbitals is essentially independent of the metal ion and it varies systematically with the identity of the ligand. Based on these considerations, ligands may be arranged in a **spectrochemical series,** a list of ligands ranked in order of their ability to

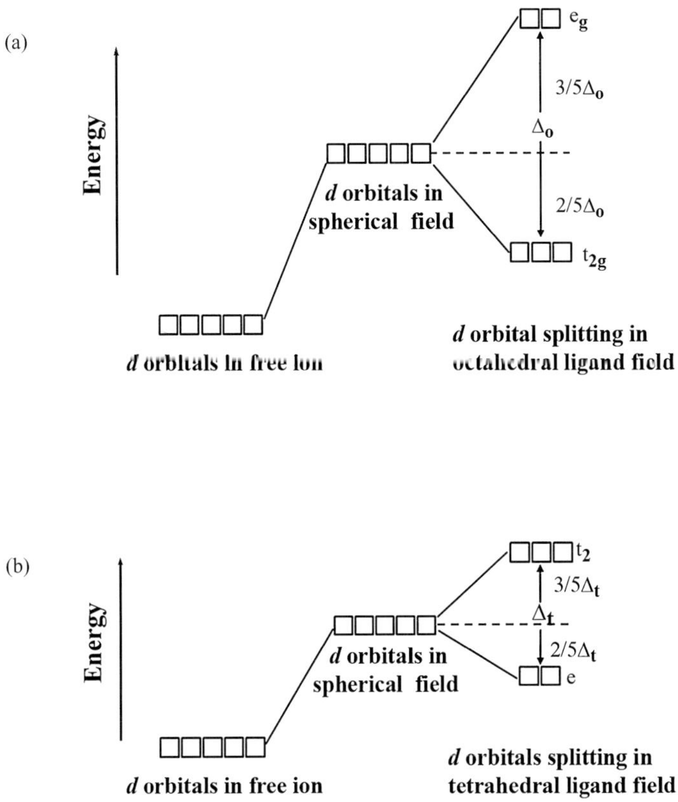

FIGURE 11.1. Splitting of d orbital energies by (a) octahedral and (b) tetrahedral field of ligands. Energies are not in scale.

cause large Δ_o separations (the donor atom in bidentate ligand is underlined; EDTA, ethylenediaminetetraacetic acid; py, pyridine; en, ethylenediamine; bipy, 2,2′-bipyridine; phen, phenanthroline):

$$I^- < Br^- < S^{2-} < \underline{S}CN^- \sim Cl^- < N\underline{O}_2^- < N_3^- < F^- < OH^- < C_2O_4^{2-} < H_2O < \underline{N}CS^-$$
$$< EDTA^{4-} < NH_3 \sim py \sim en < bipy < phen < \underline{N}O_2^- < \underline{C}N \sim CO.$$

In the presence of a tetrahedral field (point group T_d), the *d* orbitals are split into two sets, but the *e* orbitals lie below the t_2 orbitals (see Fig. 11.1). The CFT approach can be easily extended to other geometries, such as the square-planar (point group D_{4h}), square-pyramidal (point group C_{4v}), and trigonal-bipyramidal (point group D_{3h}). Lowering the symmetry, more splitting occur; therefore, more parameters are needed to decipher the energy of the orbital groups.

The CFT, although simple and useful in interpreting physical properties for numerous coordiantion compounds, is not sufficient to explain some properties and experimental evidences, which need a covalent contribution to be claimed. For example, electron paramagnetic resonance (EPR) and nuclear magnetic resonance (NMR) measurements show that appreciable unpaired electron density lies on the ligand and not on the metal. Further, the order of ligands in the spectrochemical series and the infrared (IR) spectra of simple carbonyl compounds (e.g., $M(CO)_6$) cannot be explained on the basis of purely electrostatic interactions.

The LFT, first proposed by Van Vleck, can be considered an extension of the CFT in that it moves from a purely electrostatic approach to one that considers a covalent contribution in the metal–ligand bond. Thus, the orbitals involved in the bond are no longer pure *d* metal orbitals, but there is overlap between the *d* metal orbitals and the ligand orbitals. Even in the case of a very highly covalent bond, which is best treated using the molecular orbital theory, the symmetry properties remains the same as for the crystal field model. In order to introduce covalency into the basic electrostatic model of CFT, several adjustments were introduced, such as the Racah interelectron repulsion parameter (B), introduced in the analysis of absorption spectra. It should be pointed out that the B value of the free ion usually decreases upon complex formation. The decreased entity reflects the covalent contribution to the coordinate bond. This appears to be a consequence of the metal electrons being delocalized over a larger volume of space in the complex than in the free ion. Ligands have been arranged in a series such that, for a given metal ion, the B value decreases down the series (the covalent character of the bond increases):

$$F^- > H_2O > NH_3 > en > \text{-}NCS^- > Cl^- \sim CN^- > Br^- > S^{2-} \sim I^-.$$

This series is called a nephelauxetic series (from the Greek nephelē [cloud] and auxēsis [growth]; hence, cloud-expanding).[9] It should be noted that the most polarizable ligands give the lowest B values and vice versa; also, hard bases are on the left and soft bases are on the right (see also Section 11.3.2).

In summary, depending on the nature of the metal ion and the ligand, the total bond strength has prevalently electrostatic (ion pair) or prevalently covalent contributions. The metal ion–ligand association will correspondingly exhibit supramolecular (reversible) or covalent (irreversible) features. The case of reversible coordination polymers is discussed in Chapter 10. The class of irreversible coordination polymers has recently stimulated attention in the context of synthetic materials displaying unusual properties.[16] The present chapter will show that extensive investigations of the interactions involved in metalloproteins allow a great opportunity for mastering the design and driving the function of artificial model compounds.

11.3 METAL COFACTORS AND METALLOPROTEIN FUNCTIONS

Of the 70,695 protein crystal structures contained in the Protein Data Bank (PDB) as of January 2011, 29,270 (approximately 41% of the known proteins)

TABLE 11.1. Metals in the Protein Data Bank (January 2011)

Metal	Hits	Metal	Hits	Metal	Hits
Na	2797	V	70	Pd	18
Mg	5626	Cr	9	Ag	11
K	942	Mn	1490	Cd	546
Ca	5264	Fe	3837	Ir	4
		Co	471	Pt	68
		Ni	551	Au	39
		Cu	806	Hg	424
		Zn	6297		
		Total: 29,270			

contain metal ions (including weakly bound metal as sodium), with an approximate breakdown reported in Table 11.1. Numerous different metal ions are found in association with proteins in living systems. Na^{+}, K^{+}, Mg^{2+}, Ca^{2+}, Zn^{2+}, Mn^{2+}, Ni^{2+}, $Cu^{+/2+}$, $Fe^{2+/3+}$, and $Co^{2+/3+}$ are most frequently found to bind proteins under physiological conditions.[17–28] These metal ions are coordinated to various donor atoms from protein residues, mainly nitrogen, oxygen, and sulfur, which are provided by amino acid side chains (see next section). Certain metal ions are also found in proteins coordinated by organic macrocyclic ligands, such as porphyrin, corrin, and chlorin groups. Notable examples are Fe, Mg, and Co ions, which are the components of heme, chlorophyll, and cobalamins, respectively.[20,29] Metal ions have been selected over an evolutionary period of 2–3 billion years on the basis of their unique physicochemical properties and bioavailability.[30] They serve a variety of functions in proteins, spanning from protein structure stabilization to catalysis, signal transduction, nitrogen fixation, photosynthesis, and respiration,[24,31] as reported in Figure 11.2. Metal ions can activate chemical bonds and make them more amenable to reactions. Certain metals are very useful for their rich redox chemistry. On the other hand, metal ions may bind various domains of the protein together, thus acting as a template, bringing reactive groups in the correct relative orientation for activity, and making the protein functional. Sometime, only one specific metal ion, in a specific oxidation state, can be employed to aid in catalysis, and its removal or its replacement by a different metal ion is often accompanied by a loss or reduction in the biological activity of the protein.

Figure 11.2 evidences that the same metal ion can serve numerous and different functions. As it will be detailed later in this chapter, heme or carboxylate-bridged diiron cofactors participate in many processes essential to life, such as electron transfer, oxygen transport, storage, and activation. As a consequence, the biological function of a metalloprotein is subtly related to the protein architecture and, most important, to the chemical and coordination properties, which the protein is able to impart to the metal. If the knowledge of the protein

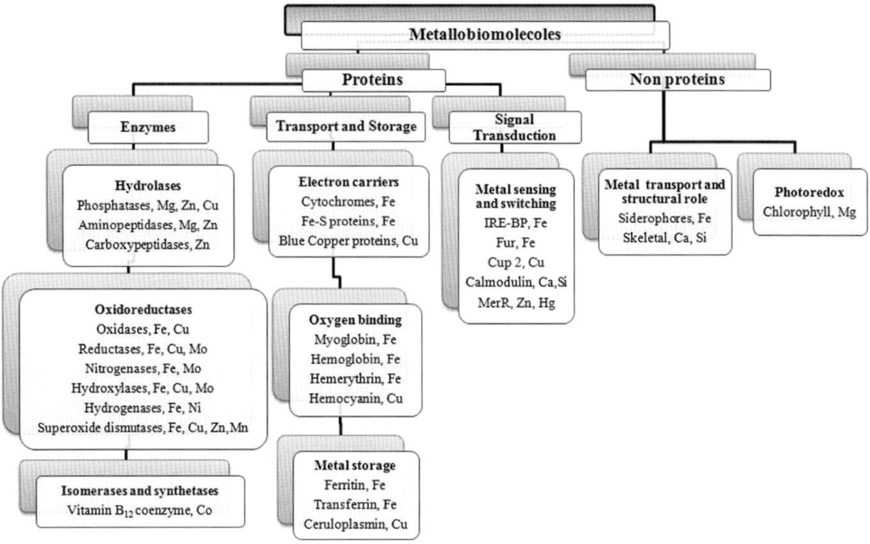

FIGURE 11.2. Classification of metalloproteins into different groups according to their functions.[28]

structure plays a key role in understanding the minimal requirements for function, in a similar way, information about the properties of functional sites in metalloproteins, geometry, and nature of ligands can clarify the relationship between active site geometry and chemistry, shed light on the evolution of these functional sites, and allow the development of methods for predicting metalloprotein function from structure and for constructing new metal binding sites into a given protein scaffold or modifying an existing metal site.

Several computational methods have been used for identifying and characterizing metal binding sites in proteins.[3,32] The growing number of three-dimensional structures, both in solution and in solid state, has allowed more detailed statistical analyses[33–36] on metal binding sites in protein and led to the development of online databases relevant to metal coordination chemistry in proteins.[37–40] All the collected data are valuable for understanding how a protein selects a specific metal ion, the role of ions in protein function, and some fundamental properties of metal ions in metalloproteins, such as the relative abundance of different metal ions in proteins, the preferred coordination geometries, as well as the type and frequency of amino acid residues that participate in their coordination and the role of second shell ligands in stabilizing the metal binding site geometry.

11.3.1 Amino Acid Residues as Ligands for Metal Ions

In metalloproteins, the amino acid residues that most often coordinate the metal ions are histidine (His), cysteine (Cys), aspartic (Asp), and glutamic

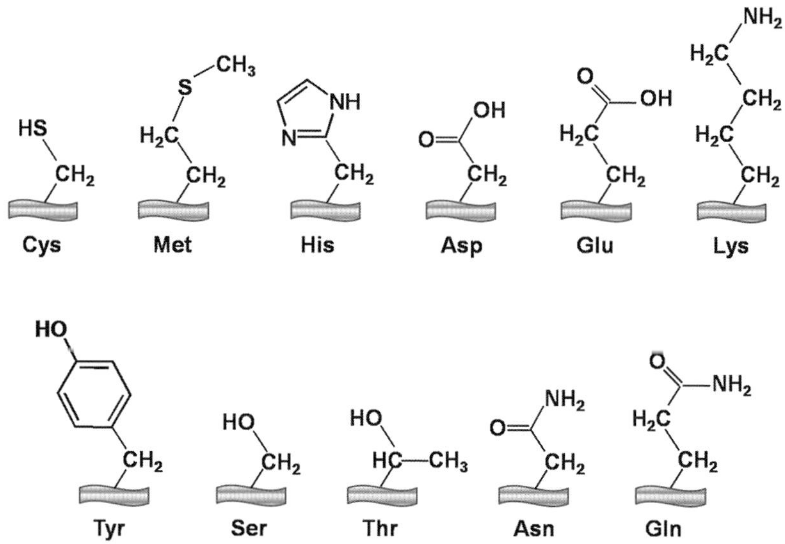

FIGURE 11.3. Amino acids containing donor atom(s) in their side chains. (Adapted from Gray, Stiefel, Valentine and Bertini.[2])

(Glu) acid, followed by tyrosine (Tyr), methionine (Met), serine (Ser), lysine (Lys), glutamine (Gln), asparagine (Asn), threonine (Thr), and glycine (Gly). With the exception of Gly, which coordinates a metal ion through the backbone carbonyl oxygen, the other amino acids participate in coordination almost exclusively through their side chains containing sulfur, oxygen, or nitrogen as donor atoms (Fig. 11.3).

His, Glu, and Asp are very frequently encountered in the metal coordination sphere. They may display more than one ligation mode, as described below.

Histidine may coordinate the metal ions either through N^{ε} or through N^{δ}. In protic solvent, at neutral/basic pHs, the NH proton of free His participates in a tautomeric equilibrium, the N^{ε} protonation form (tautomer II) being energetically more stable than the N^{δ} protonation form (tautomer I) (Fig. 11.4a).[41,42]

Electronic and/or chemical differences of N^{ε} versus N^{δ} metal interactions (e.g., the ability to deprotonate the N-H group), as well as steric requirements for binding, may reverse the stability of the two tautomers and induce preferential binding for either one of the two nitrogen. Indeed, coordination through N^{δ} or N^{ε} imidazole nitrogens is almost equally frequent in metalloproteins.[43] The N^{δ}-metal interaction locates the His main chain closer to the metal, with respect to the N^{ε}-metal interaction. This arrangement may provide a more stable metal–ligand coordination, which may result in a favorable condition for an electron transfer center. On the other hand, the N^{ε} coordination leaves more space available in the active site for binding exogenous molecules and for catalysis. Indeed, in the electron transfer type 1 copper site, the His binds

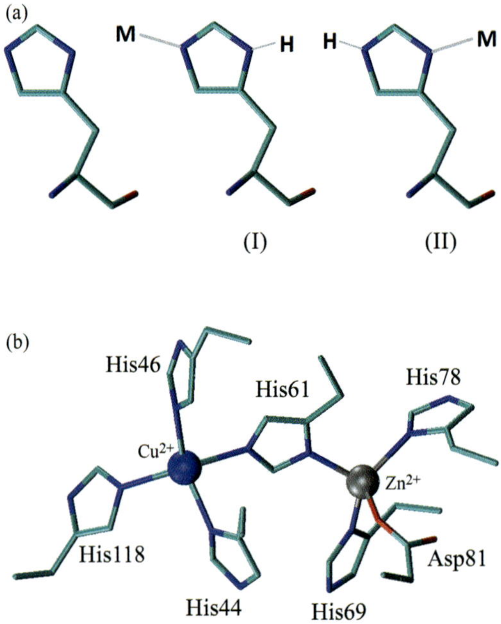

FIGURE 11.4. Histidine as ligand. (a) The two tautomeric forms, (I) and (II), of its neutral imidazole side chain; (b) Cu,Zn-SOD active site (PDB code 1SXA)[47]—an example of His acting as bridging ligand.

the metal ion through its N^{δ} nitrogen, whereas in almost all type 2 copper sites, the His is coordinated through the N^{ε} nitrogen.[43] This histidine coordination mode is also found in trinuclear and binuclear copper sites.[44,45]

In mononuclear iron-containing proteins, the histidine ligands adopt the N^{ε} tautomeric conformation. By contrast, diiron proteins ligate predominantly His via N^{δ}. The only exception is hemerythrin (Hr), in which all five His ligands are coordinated through their N^{ε}. Iron heme protein also preferentially coordinate His through the N^{ε} atom, as it will be discussed in more detail in Section 11.4.3.2.

The His binding modes for mononuclear zinc ion are variable: The majority of His ligands found in zinc protein structures coordinate the metal ion through the N^{ε} atom; however, coordination with N^{δ} atoms has also been observed. The oxidized form of the Cu,Zn-superoxide dismutase (SOD)[46,47] represents an example of the ability of His to act as a bridging ligand between two metal ions (Cu^{2+} and Zn^{2+}) (Fig. 11.4b); this bridge is broken upon copper reduction.

Glutamic and aspartic acids coordinate through their carboxylate side chains, which differ only in one bond length, thus displaying essentially identical properties (pK_a, ligation modes). As depicted in Figure 11.5, they may coordinate the metal ion in either a bidentate or *syn/anti* monodentate mode, the *syn* stereochemistry being the most common.[13,48,49]

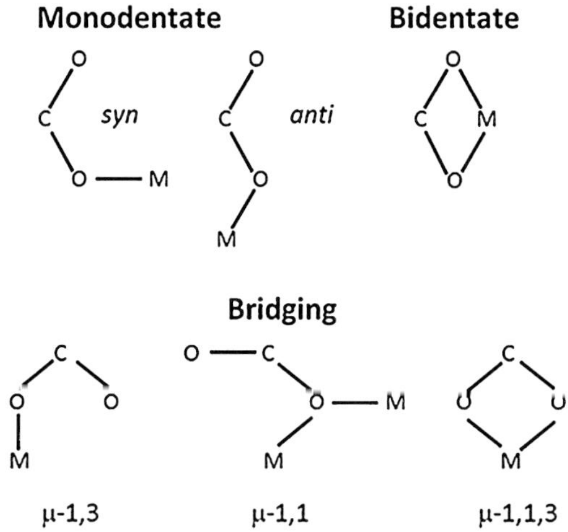

FIGURE 11.5. Carboxylate binding modes.

The Glu and Asp carboxylate groups can also act as a bridging ligand in dinuclear metal centers, using one or both oxygen atoms (Fig. 11.5, bottom).

Recently, Dudev and Lim accurately delineated the factors governing the carboxylate binding mode (monodentate vs. bidentate) in metalloproteins.[50] They investigated several factors that may affect metal–ligand interactions and, consequently, the monodentate/bidentate equilibrium: the immediate neighbors of the metal-bound carboxylate, the nature and coordination number of the metal, the total charge of the metal complex, and the relative solvent accessibility and rigidity of the metal binding pocket. Their results suggest that the carboxylate binding mode is determined mainly by the competition between the metal cation, on one hand, and the neighboring nonacidic ligands (water, backbone carbonyls) from the metal binding pocket, on the other, for binding the second oxygen of the COO^- group.[50,51] Bidentate carboxylate binding is preferred over the monodentate mode when the second carboxylate oxygen interacts more favorably with the metal cation with respect to the first or second shell ligand(s). This occurs under the following conditions: (1) the metal cation has a high coordination number, giving rise to a crowded coordination environment; (2) the metal cation is a good Lewis acid that can accept charge from the second carboxylate oxygen; (3) the metal positive charge is not neutralized by charge transfer from negatively charged ligands in the metal complex; (4) the metal cation is relatively *large* and can accommodate *bulky* protein main chain/side chain dipoles that do not stabilize the second carboxylate oxygen as well as the metal cation; and (5) the first or second shell ligands lack hydrogen bond donors or provide only poor hydrogen bond donors, which do not stabilize the second carboxylate oxygen as well as the metal cation. The relative rigidity of the metal binding site structure can also affect the monodentate/bidentate equilibrium.

The same authors also showed that the carboxylate binding mode modulates the affinity and/or selectivity for a given metal cofactor, as well as the protein function. By shifting the equilibrium toward bidentate carboxylate binding, selectivity toward trivalent lanthanide cations over the naturally occurring divalent cations (Mg^{2+} and Ca^{2+}) is observed. Further, the carboxylate denticity tunes the number of Asp/Glu that can be accommodated in the metal first coordination sphere.

The carboxylate binding mode is crucial in determining functions, for example, in signal transduction. In EF-hand regulatory proteins, substitution of Ca^{2+} with Mg^{2+} causes a switch from bidentate to monodentate binding mode in a highly conserved Glu residue, with concomitant loss of signal transduction.[52] The interconversion between mono- and bidentate Asp/Glu also occurs in *polynuclear* metal binding sites; the so-called "carboxylate shift" has been recognized as an important phenomenon in several catalytic processes.[53] The diiron component R2 of ribonucleotide reductase (RNR) is probably the most representative example of a dinuclear enzyme where a carboxylate shift mechanism has been suggested. Changing the mode of carboxylate binding allows this protein to create an accessible coordination site for the incoming O_2, as discussed in Section 11.4.4.1.

In summary, the ligand binding mode is a significant determinant of the metal ligand interactions. It may modulate the metal binding site affinity/selectivity for a given metal cofactor, and may play a key role in the accomplishment metalloprotein functions. It may represent an interesting design tool in engineering new metal binding sites with desired properties.

11.3.2 Metal Ion and Ligand Preferences

Metal ion preferences for a certain amino acid side chain follow the general rules of coordination chemistry and often can be interpreted using the *theory of hard and soft acids and bases* (HSAB theory). The terms *hard* and *soft* were coined by Pearson to describe the coordination properties of metal ions and ligands.[54] As a general rule, "hard" metal ions favor binding to "hard" donor atom, while "soft" metal ions favor "soft" donor atom. Table 11.2 reports a classification of acids and bases by hard/soft criteria. The idea underlying in

TABLE 11.2. Hard and Soft Acids and Bases

	Acids	Bases
Hard (class a)	H^+, Li^+,Na^+, K^+, Mg^{2+}, Ca^{2+}, Al^{3+}, Cr^{3+}, Fe^{3+}, Co^{3+}	NH_3, RNH_2, N_2H_4, H_2O, OH^-, RS^-, O^{2-}, SO_4^{2-}, RCO_2^-, F^-
Borderline	Fe^{2+}, Co^{2+}, Ni^{2+}, Cu^{2+}, Zn^{2+}, Pb^{2+}, Sn^{2+}	Imidazole, pyridine
Soft (class b)	Cu^+, Ag^+, Au^+, Cd^{2+}, Hg^{2+}, Pt^{2+}, Pt^{4+}, Pd^{2+}	H^-, R^-, C_2H_4, CN^-, CO, SCN^-, R_3P, RSH, I^-

this scheme is fairly simple. All metal ions have acid properties and are divided into hard (class a) and soft (class b). Hard metal ions, as a consequence of their small radii and their high positive charge, have a low polarizability and a small electron density to share with a ligand. They hold tightly to their electrons and therefore the electron cloud is not readily distorted; their unshared valence electrons are not easily excited. As a consequence, hard metal ions form most stable complexes with donor atoms that have a high electronegativity, a low polarizability, and small radii, that is, with hard bases. The resulting hard acid/base interaction will be mainly electrostatic.

Soft metal ions have a low positive charge, are readily polarized (an electron cloud readily deformed) and are more prone to sharing electron density. They prefer to bind soft bases, and the soft metal ion/ligands interaction will have a greater degree of covalence.

In the interaction of a hard metal ion with a soft ligand, the metal does not readily accept the ligand electron density, and the resulting complex will be less stable since both partners are "incompatible." As always in nature, there is not a definite division between hard and soft features. In fact, certain metal ions and ligands display intermediate behavior and may interact reasonably well with both hard and soft species.

In metalloproteins, only three elements are predominantly involved in metal ion coordination. Ligands that bond through oxygen (carboxylates, phenolates, carbonyls, and hydroxyls) are hard. The only common soft ligands are thiolates and thioethers, while most nitrogen-containing ligands are classified as borderline, being capable of ligating both hard and soft metals (Table 11.2).

The stabilities of complexes formed by different types of metal ions follow sequences that are, to a large degree, independent of the ligand when steric selectivity is minimal.[8,18,55–57]

The stabilities of monovalent ion complexes are $Cu^+ > Na^+ > K^+$; for bivalent ions, the stabilities usually follow the Irving–Williams order, irrespective of the particular ligand involved[58]:

$$(Mg^{2+}, Ca^{2+})^+ < Mn^{2+} < Fe^{2+} < Co^{2+} < Ni^{2+} < Cu^{2+} > Zn^{2+}.$$

Few stability constants are available for trivalent cations, but generally, the order observed is Co^{3+}(low spin) $> Mn^{3+} > Fe^{3+} > Cr^{3+} > Al^{3+}$.

The observed trends are largely determined by electrostatic effects, and can be related to the ionic potential z/r (where z is the ionic charge and r the ionic radius) and to the ionization potential I_n from M to M^{n+}. Figure 11.6 reports a plot of z/r against I_2 for bivalent ions. The separation between *hard* and *soft* metal ions is clearly evident; the plot also generates the Irving–Williams's stability order.[18] This behavior was a major factor in the acceptance of the CFT, briefly discussed in Section 11.2.

Even though the hard/soft theory gives a reasonable view of the metal ligand preferences, other factors drive metal ion selectivity in both chemistry

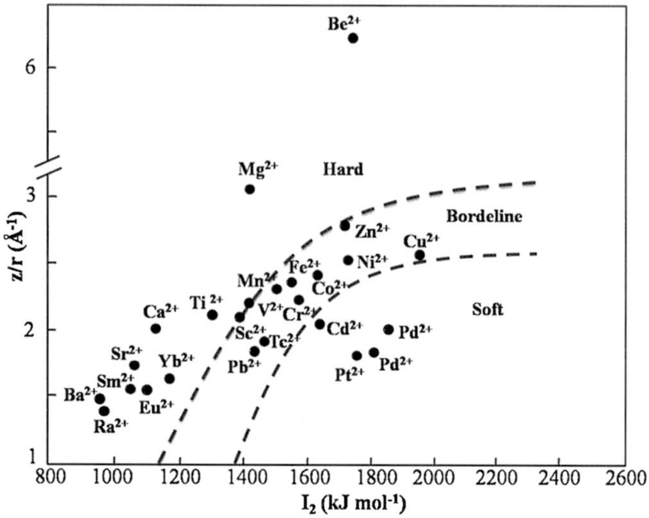

FIGURE 11.6. Plot of z/r (where z is the ionic charge and r is the ionic radius) against I_2 (ionization potential) for bivalent cations. (Adapted from Frausto da Silva and Williams.[18])

and biology, such as stereochemical properties of the ligands, preferred coordination geometry of the metal ion, and spin pairing stabilization.

Based on all these issues, it is possible to delineate for each metal ion the residue coordination preferences: Less easily polarized ions, such as Ca^{2+}, should preferentially bind less easily polarized residue groups, such as the carboxyl group in glutamate and aspartate, whereas more easily polarized ions, such as Zn^{2+}, should prefer easily polarized residue groups, such as the sulfhydryl group in cysteines.[59] In agreement with these evaluations, analysis of the amino acid donor atoms in metalloprotein active sites reveals that Ca^{2+} and Mg^{2+} generally bind oxygen atoms, rather than nitrogen or sulfur. Calcium ions coordinate, through largely electrostatic interactions, the side chains of negatively charged residues (Asp and Glu), noncharged residues (Asn and Gln), and the main chain carbonyl oxygen and water. In particular, Asp and Glu are the amino acids most frequently observed in the Ca^{2+} coordination sphere. This is not surprising; the carboxylate groups of these residues provide charge balance for Ca^{2+} ion at most sites. Furthermore, the aspartate and glutamate side chains are able to participate in the network of hydrogen bond interactions that stabilize most Ca^{2+} binding sites.

There are striking similarities as well as some interesting differences between binding sites containing divalent magnesium and manganese. Similarly, both ions preferentially coordinate oxygen. In agreement with its borderline behavior in the hard/soft classification (see Fig. 11.6), Mn^{2+} more frequently binds nitrogen than Mg^{2+} does. Some Mn^{2+} enzymes contain up to three imidazole nitrogens bound to the metal and all structurally characterized Mn^{3+} proteins contain at least one His in the coordination environment.

Cu^{2+}, Fe^{2+}, and Ni^{2+} are most often coordinated by His, followed by Cys and less often by Met, Asp, and Glu. In the case of Cu^{2+}, which is a "softer" acid with respect to Fe^{2+} and Ni^{2+}, Met is more abundant than Glu and Asp; the opposite occurs for Fe^{2+} and Ni^{2+}.

Co^{2+} shows a strong degree of preference for His, but Asp and Glu are also very often present in its coordination sphere.

Zn^{2+} and Cd^{2+} (soft acids) display a large preference for Cys sulfur atom (soft base); the next most abundant amino acids in their coordination spheres are His and Glu/Asp, respectively. This observed trend follows the decrease in the "soft character" of the donor atom.

11.3.3 Ligands in the First Coordination Sphere Specify the Role of the Metal Ion: The Case of Zn^{2+}

The ligand types in the first coordination sphere are very effective in specifying the role of the metal ion within the protein. This finding is clear evident in zinc-containing proteins.

Based on the primary coordination sphere, four distinct types of zinc binding sites in metalloproteins have thus far been identified, as schematically depicted in Figure 11.7: catalytic, structural, cocatalytic, and protein interface sites.[60–67]

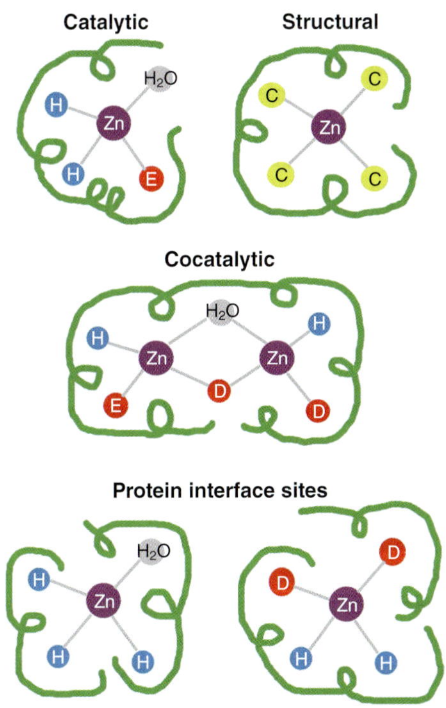

FIGURE 11.7. Schematic representation of zinc binding sites, with the different roles played by the metal ion. One-letter code is used to indicate amino acids.

Catalytic sites are most common motif in zinc enzymes.[64] Generally, they are composed of a bound water molecule and three protein ligands, such as histidine, aspartate/glutamate, and cysteine. Water is always a ligand to the catalytic zinc and it is important for the catalytic mechanism. In catalytic zinc sites, ligands can be separated by short or long amino acid spacers (from one to three residues to six to seven residues), with the length of the spacer often dictated by the ligand environment. The zinc ion (*borderline hard/soft acid*) polarizes the water molecule, making it a potent nucleophile for many hydrolytic reactions (e.g., in esterase, phosphatases, peptidases).

The second common motif in zinc proteins are *structural sites*.[65] In contrast to catalytic sites, in structural sites the zinc ion is coordinated by four protein ligands. The most recurring amino acid in such sites is Cys; however, His is also found, usually together with Cys. Structural sites may be found in addition to catalytic sites in the same protein (e.g., alcohol dehydrogenase).

Cocatalytic zinc sites are found in enzymes containing two or more zinc ions, and/or other transition metal ions in close proximity to each other, which operate in concert as a catalytic unit.[67,68] Thus, the metal ions in cocatalytic sites are important for the overall fold of the protein, as well as for its catalytic function.

A structural characteristic of these centers is the bridging of two metal ions by a single amino acid side chain, usually the carboxylated oxygens of Asp, and sometimes a water molecule, as well the imidazole ring of His and the carboxylate oxygens of Glu or carbamylated Lys. Apart from the bridging ligand, aspartate and histidine predominate as ligands in cocatalytic zinc sites; however, they also contain unusual zinc ligands, such as amide carbonyls provided by Asn, Gln, and peptide backbone, hydroxyl groups from Ser, Thr, and Tyr, and the nitrogen of Lys or N-terminal amino acid. Cys ligands are very rare.[68]

Protein interface zinc sites were first identified based on the observations that zinc can have an impact on the quaternary structure of a protein (formation of dimers, trimers, etc.).[61] They are composed of amino acid ligands that are located on the binding surface between two protein subunits or interacting proteins, and generally have the coordination properties of catalytic or structural zinc binding sites. Examples of zinc binding sites in metalloproteins are reported in Figure 11.8.[69–71]

The properties of the Zn^{2+} ion may account for the different roles played by this ion in metalloproteins. According to the HSAB theory, Zn^{2+} ion can be classified as borderline, thus capable of binding both soft and hard donor atoms (see previous section), as indeed found in Zn-containing metalloproteins. The hard/soft behavior of the ligand is responsible for the different electrostatic/covalent contributions to the coordinate bond, which in turn specify the metal site toward different functions. Soft donor atom are mainly found in the coordination sphere of the structural Zn^{2+} site, where a higher degree of covalence makes the coordinate bond stronger, as required by a structural site. The presence of hard donor atoms in the Zn^{2+} site increases the

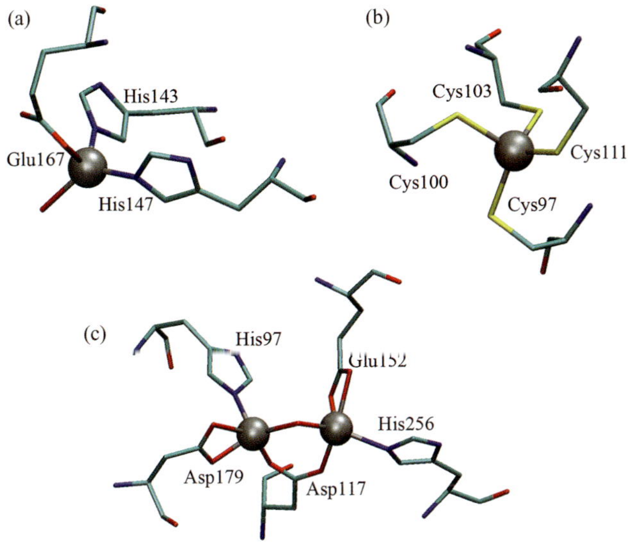

FIGURE 11.8. Zinc binding sites in enzymes. (a) Thermolysin (PDB code 1ESP)[69]; (b) alchohol dehydrogenase (PDB code 8ADH)[70]; (c) *Aeromonas proteolytica* aminopeptidase (PDB code 1AMP).[71]

electrostatic character of the coordinate bond, making the site more likely to accommodate structural changes required by catalysis.

More structural details of Zn^{2+}-containing metalloproteins will be given in Section 11.3.4.6.

11.3.4 Metal Ions in Biological Systems

In this section, the coordination properties of the most representative metal ions that play an important role in biological systems are illustrated. Iron is not discussed in this section because it will entirely be the object of the next two sections.

Calcium Ca^{2+} ion has a variety of functional roles in proteins, including (1) behaving as a structure-forming switching control, as in calmodulins (CaMs) and many other Ca^{2+}-sensor proteins[18,19,72]; (2) enhancing protein stability[73]; and (3) acting as an electrostatic control on the redox potential of free radical species in heme-containing redox proteins, such as peroxidases.[74] In particular, Ca^{2+} binding sites are often classified as continuous or discontinuous types.[75–77] Discontinuous Ca^{2+} binding sites are formed by residues that are spatially proximate in the folded structure but distant in the primary sequence, whereas continuous sites are composed of amino acids adjacent in the sequence, usually in flexible loop regions, such as the helix-loop-helix in the EF-hand motif.[78–80]

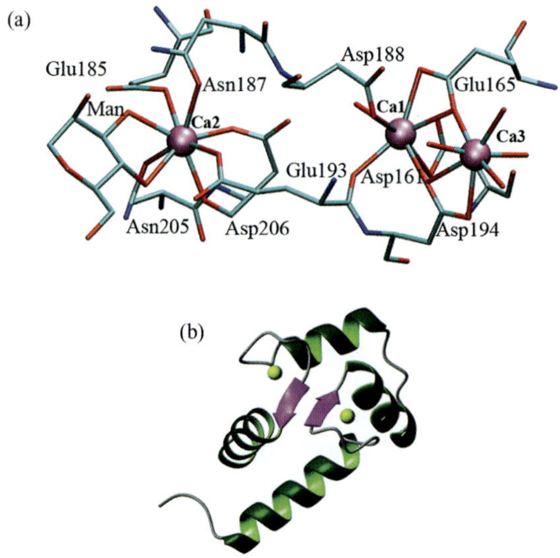

FIGURE 11.9. Calcium binding sites in proteins. (a) Active sites of C-type mannose-binding protein (PDB code 2MSB)[81]; (b) the EF-hand motif in the N-terminal domain of CaM (PDB code 1EXR). (Reprinted with permission from Gifford et al.[80] Copyright 2007 Biochemical Society.)

The calcium preferred coordination numbers range from six to eight, and the calcium binding sites, observed in natural proteins, frequently exhibit pentagonal-bipyramidal or distorted-octahedral geometries, with a large range of deviation in bond lengths and angles. Oxygen donor atoms from glutamates and aspartates are mainly found in the coordination sphere, which well correlates with the hard acid behavior on the Ca^{2+} ion.

Some representative examples of Ca^{2+}-containing metalloproteins are reported in the following.

The *C-type mannose-binding protein*, which mediates numerous cell surface carbohydrate recognition events, contains a calcium binding site. The crystal structure of the C-type mannose-binding protein, complexed with an oligosaccharide, revealed the presence of three calcium sites: calcium 1, 2, and 3, as depicted in Figure 11.9a.[81]

The ligands of calcium 3 are carboxylate groups of Glu165 and Asp194, along with four water molecules. Both carboxylate oxygen atoms of Glu165 also bound calcium ion 1, whereas only one carboxylate oxygen atoms of Asp194 is a ligand for this calcium site. Asp188, Asp161, Glu193 (through its main chain carbonyl oxygen), and a water molecule complete the coordination sphere of the calcium 1, providing an epta coordination for both calcium ions. The calcium 2 site has been found to directly bind to the terminal mannose unit of the oligosaccharide ligand. The Ca^{2+} coordination number in this site is eight, and the geometry is pentagonal bipyramidal. Oxygen atoms from the side chain of Glu185, Asn187, Glu193, and Asn205, and the main chain carbonyl oxygen of Asp206, constitute the base of the bipyramid. A carboxylate oxygen of Asp206 forms one apex of the bipyramid, and two mannose OH groups bisect the other apex.

Another important Ca-containing protein is CaM, a regulatory protein, involved in a variety of cellular calcium-dependent signaling. The calcium binding site in CaM involves seven oxygen atoms at the vertices of a pentagonal bipyramid, each about 2.4 Å from the central Ca^{2+} ion. This particular structural motif is the so-called "EF hand,"[78–80] in which the seven oxygen donor atoms come not only from protein side chains but also from main chain carbonyl oxygens and water molecules. In Figure 11.9b, the EF-hand motif of the N-terminal domain of CaM is reported.[80]

Manganese Like magnesium, manganese forms mainly six-coordinated, octahedral complexes, but four- and five-coordinated Mn^{2+} binding sites have been also reported. Mn^{2+} occurs in metalloproteins as mononuclear and dinuclear centers. An example of mononuclear and binuclear Mn^{2+} enzymes is reported in the following.

Manganese peroxidase (MnP) is a heme-containing glycoprotein.[82] It contains a ferric iron protoporphyrin IX heme cofactor and requires hydrogen peroxide as well as Mn^{2+} for its activity.[83] MnP can oxidize a wide range of substrates, from small phenolic compounds to high molecular weight chlorolignins. This enzyme contains a mononuclear manganese center, in which Mn^{2+} ion binds two glutamates, one aspartate, one heme propionate, and two water molecules in an octahedral geometry.

The dinuclear manganese centers share structural similarity with diiron centers, which will be discussed in Section 11.4.4.

Manganese catalase uses a dimanganese center to catalyze the disproportionation of hydrogen peroxide into water and molecular oxygen. Both manganese ions are six-coordinated, with an approximately octahedral geometry. The crystal structure of manganese catalase from *Lactobacillus plantarum* reveals that each individual Mn^{2+} ion is coordinated by two glutamates and one histidine.[84] One of the glutamate carboxylate (Glu66) forms a μ-1,3 bridging interaction with the metal ions, while the other two, Glu148 and Glu35, interact with the individual ions in a different manner: The Glu35 carboxylate acts as monodentate ligand in the Mn1 subsite; Glu148 interacts with the metal ion in a bidentate chelating manner in the Mn2 subsite. Two single-atom solvent bridges (oxo, hydroxo, or aquo) lie between the manganese ions and provide electronic coupling between the two metal centers. A water molecule completes the coordination sphere of the Mn1 subsite, as depicted in Figure 11.10.

Cobalt Cobalt receives great attention in bioinorganic chemistry studies, owing to its geometrically sensitive spectroscopic signatures as well as its rich ligand substitution and redox chemistry.[2]

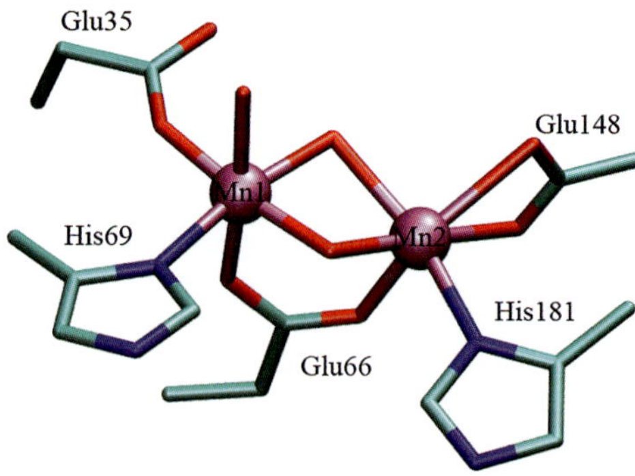

FIGURE 11.10. Active site of manganese catalase from *Lactobacillus plantarum* (PDB code 1JKU).[84]

Although cobalt is less frequently encountered in metalloenzymes than the other first-row transition metals (e.g., manganese, iron, copper, and zinc), it is nevertheless part of vitamin B_{12}, an important cofactor in several vitamin B_{12}-dependent enzymes.[85] This cofactor is involved in numerous biological processes, the most important of which is the formation of erythrocytes. Numerous protein crystal structures have appeared in the literature, with various forms of cobalamin, which all share a similar coordination geometry around the cobalt ion.[33] Similar to iron in the heme, cobalt in cobalamins is coordinated by four almost coplanar nitrogens, part of the corrin ring. One axial position is occupied by a nitrogen donor atom from 5,6-dimethylbenzimidazole (Fig. 11.11a). The sixth coordination position is occupied by a different group: CN- in vitamin B_{12}; deoxyadenosine in adenosylcobalamin (also called coenzyme B_{12}), and a methyl group in methylcobalamin. This coordination site is labile, which is an important property for activity.[85] Incorporation of cobalt into the corrin ring modifies the redox potentials of cobalt, making accessible three consecutive oxidation states (+3, +2, and +1).[8]

A noncorrin cobalt-containing protein is *methionine aminopeptidase*, an important catalyst for N-terminal protein modification, thus being involved in functional regulation, intracellular targeting, and protein turnover.[86,87] This enzyme exhibits a dinuclear Co^{2+} metal binding site; the cobalt ions are coordinated by the side chain atom of aspartic, glutamic, and histidine residues that are organized in an approximately octahedral geometry (Fig. 11.11b).

Nickel The coordination number of Ni^{2+} ion rarely exceeds six. In small-molecule complexes, the octahedral (six-coordinated) and square planar (four-coordinated) geometries are preferred, even though rather fewer examples of trigonal-bipyramidal (five-coordinated), square-pyramidal (five-coordinated), and tetrahedral (four-coordinated) geometries have been reported.[88]

In metalloproteins, octahedral geometry should be considered as favorable for Ni^{2+}.[88]

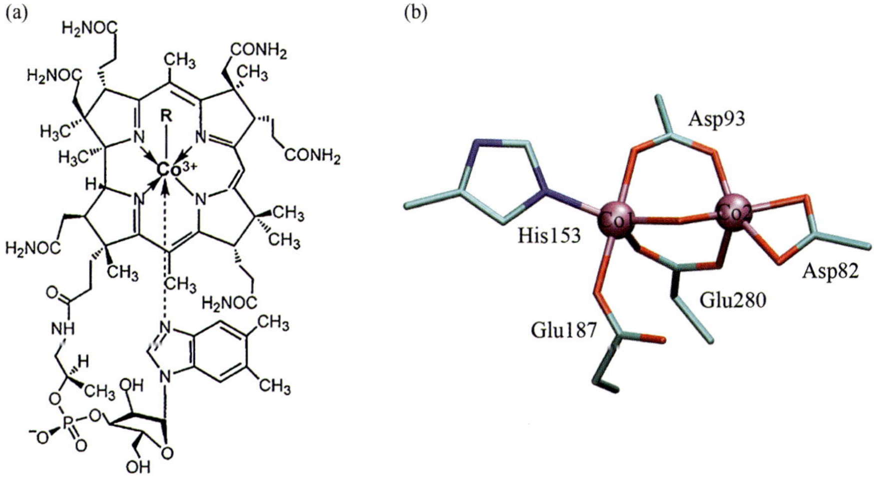

FIGURE 11.11. Corrin and noncorrin cobalt binding sites. (a) Structure of cobalamin in B12-dependent enzymes. R is deoxyadenosine in AdoCbl, methyl in MeCbl, -OH in hydroxocobalamin, and -CN in vitamin B12. (b) Active site of methionine aminopeptidase (PDB code 1XGS).[86]

Ni-containing enzymes catalyze five distinct biological activities, including urea hydrolysis, reversible hydrogen oxidation, interconversion of carbon monoxide and carbon dioxide (often associated with acetate metabolism), methane generation, and superoxide dismutation.[89] In proteins, Ni^{2+} can be present in mononuclear and dinuclear centers. Representative nickel-containing enzymes are glyoxylase, aci-reductone dioxygenase (ARD), nickel SOD (Ni-SOD), urease, NiFe hydrogenase, and CO dehydrogenase. Almost all these enzymes show nickel centers with similar coordination geometry.

Glyoxylase catalyzes the conversion of methylglyoxal, a toxic species that forms covalent adducts with DNA, to lactate.[90] A single nickel ion in an octahedral coordination environment acts as a Lewis acid catalyst and remains in the +2 state throughout the isomerization reaction. The nickel is coordinated by two His residues and two glutamate. Two water molecules complete the primary coordination sphere.

Ni-SOD, like other SODs (the Cu/Zn-SODs, the Fe-SODs, the Mn-SODs), catalyzes the conversion of superoxide to O_2 and H_2O_2. Ni-SOD switches between a square planar N2S2 (Ni^{2+}) and a square pyramidal N3S2 (Ni^{3+}) coordination environment, made up by two Cys residues (sulfur donor atoms) and two/three nitrogen donor atoms, two of them from a histidine residue.[91] The most striking feature of the active site in the oxidized and reduced states is the dual conformation of His1. In the reduced form, His1 side chain is not a Ni ligand, but forms hydrogen bonds to glutamic residue and the backbone carbonyl of a valine residue, suggesting a positively charged side chain. In this form, Ni^{2+} ion binds to the N-terminal amine of His1, the backbone amide of

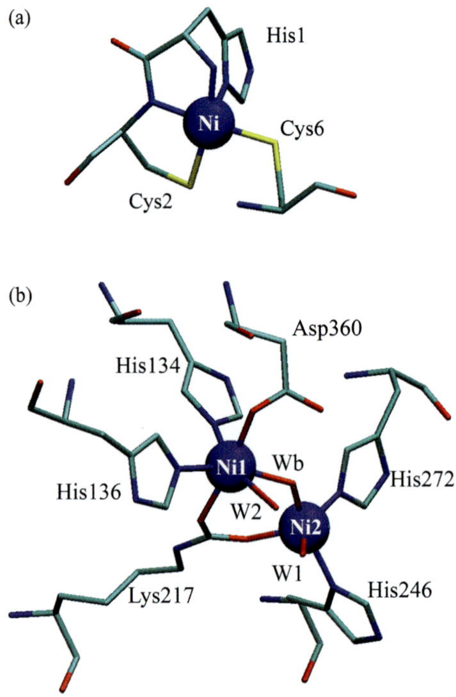

FIGURE 11.12. (a) Ni-SOD active site structure (PDB code 1T6U).[91] (b) Urease dinickel active site structure (PDB code 1EJX).

Cys2, and the thiolates of Cys2 and Cys6. In the oxidized form, the His1 imidazole becomes an axial Ni ligand, making a five-coordinate square pyramidal site (see Fig. 11.12a). The Cys sulfur ligands appear to poise the $Ni^{3+/2+}$ redox couple in the appropriate range for catalyzing both the reduction and oxidation of superoxide and to serve as a proton donor during catalysis.[91]

A dinuclear nickel active site has been found in *urease*. This enzyme plays a central role in the global nitrogen cycle because it catalyzes the hydrolysis of urea, which is excreted by vertebrates, into ammonia and bicarbonate. Several structures of urease and site-directed variants in the presence and absence of substrates and inhibitors are available.[92,93] In the urease dinickel center, the two nickel ions (Ni1 and Ni2) reside at a distance of 3.7 Å. They are bridged by a carbamylated lysine and a water/hydroxyl group. Two histidines and a water molecule coordinate Ni1 in a square-pyramidal geometry. Ni2 is hexacoordinated and shows a pseudo-octahedral geometry: two histidines, a water molecule, and one aspartate coordinate this metal ion (see Fig. 11.12b). A fourth water molecule is present in the active site, interacting with the others through hydrogen bonds, and completes a tetrahedral cluster of water/hydroxide molecules.

Copper Two oxidation states are usual for copper, Cu^{+} and Cu^{2+}, and different coordination numbers are associated with these oxidation states.

Cu^{+}, with a d^{10} electronic configuration and a soft acid behavior, prefers to bind soft sulfur ligands, with a coordination number of two, three, or four (tetrahedral geometry). A few examples of pentacoordination are also known (square pyramid). By contrast, Cu^{2+}, with a d^{9} electronic configuration and a borderline behavior, favors nitrogen ligands and higher coordination numbers, between four and six, with very plastic and distorted geometries.[8] With coordination number six, the Jahn–Teller effect excludes the regular octahedral. Regular geometries are rare and the distinction between square-planar and tetragonally distorted octahedral coordinations is generally hard to achieve.

Based on the coordination and the types of ligands, as well as on the spectroscopic features and the function of the metal site, mononuclear copper sites in proteins have traditionally been classified into three types: type 1 "blue" (T1), type 2 (T2) "normal," and type 3 (T3) "coupled-binuclear" sites. While the "type 1" centers are normally implicated in electron transfer, the "type 2" and "type 3" centers are involved in substrate binding and activation.[94,95]

Type 1 Cu sites are present in blue copper proteins, such as plastocyanin, azurin, and stellacyanin.[96] They are called blue proteins for their intense blue color at ~ 600 nm, arising from S(Cys) → Cu^{2+} charge transfer. In all these proteins, a central Cu^{2+} is coordinated by three equatorial residues (two histidine and one cysteine), with a geometry that is more or less trigonal. Additional more weakly bound axial ligands are found: A methionine sulfur atom and a glutamine side chain amide oxygen atom represent the fourth ligand in plastocyanin and in stellacyanin, respectively, whereas Cu^{2+} in azurin is coordinated by a glycine backbone carbonyl oxygen. In azurin, a Met residue is also in the Cu^{2+} coordination environment, but at a longer distance (~3.1 Å) (see Fig. 11.13a).[97]

Blue copper proteins function in electron transfer processes and their role involves the reduction of the Cu^{2+} into Cu^{+}, which prefers the tetrahedral coordination (when four-coordinate). From the functional point of view, the trigonal coordination observed for the Cu^{2+} site is a compromise between the stereochemical and electronic requirements of Cu^{+} and Cu^{2+}.[98] The trigonal Cu^{2+} environment in the blue copper proteins is much closer to a tetrahedron than to a tetragonal environment; thus, the geometry changes accompanying reduction are minimal: The copper ligand distances increase by 0–0.15 Å, while the angles do not change significantly. The blue copper proteins represent a remarkable example in which the protein matrix dictates the metal ion coordination geometry, forcing it toward unusual geometries. It is an entatic state that enhances the proteins' reactivity. The "entatic" (or "strained") state hypothesis was first proposed by Vallee and Williams.[99,100] They analyzed numerous metalloproteins, focusing on redox-active copper, and suggested that the active site in an enzyme is constrained by the surrounding protein matrix to adopt a geometry similar to that of the transition state that would

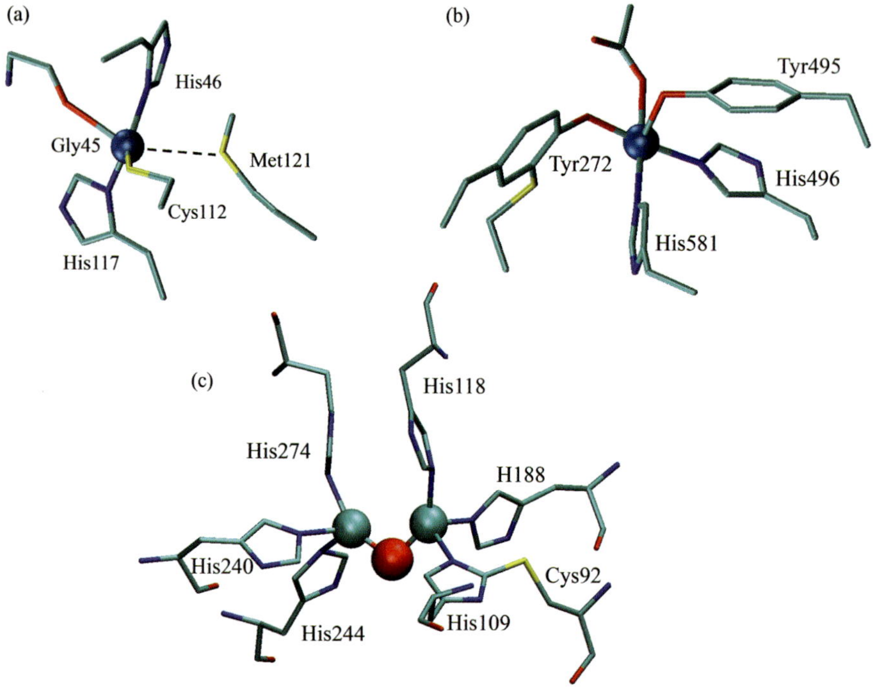

FIGURE 11.13. (a) Type 1 copper site in azurin (PDB code 1AZU)[97]; (b) type 2 copper site in galactose oxidase (PDB code 1GOF)[101]; (c) type 3 copper site in catechol oxidase (PDB code 1BT1).[102]

exist in a similar unconstrained system, thereby reducing the reorganizational energy.

The type 2 copper sites, found, for example, in Cu,Zn-SOD and in galactose oxidase, do not share similar geometries.

The active site geometry of *galactose oxidase*[101] is nearly square pyramidal, as depicted in Figure 11.13b. Two histidine (His496 and His581), one tyrosinate (Tyr272), and a weakly coordinated water molecule (pH 7) or an acetate ion (pH 4.5) occupy equatorial positions. A fifth tyrosinate ligand (Tyr495) is involved in an elongated axial bond.

Unlike galactose oxidase, the redox-active Cu^{2+} ion in *Cu,Zn-SOD*[47] is coordinated by four histidine residues in a distorted square planar geometry (see Fig. 11.4b).

"Type 3" copper sites are dinuclear sites, with each copper ion bound to three histidines; in the oxidized form, the two copper ions are bridged by an OH^- or a water molecule. The simplest of the metalloproteins containing type 3 sites are hemocyanin (Hc), tyrosinase (TY), and catechol oxidase (CatO). The type 3 sites are also found in multicopper oxidases (MCOs), which include tree and fungal laccase (Lc), ceruloplasmin, and ascorbase oxidase. MCOs use a minimum of four Cu centers: a "blue," type 1 Cu site and a trinuclear Cu cluster composed of a "normal" type 2 Cu, and a binuclear type 3 Cu site. All together, these sites catalyze the four-electron reduction of O_2 to water, with concomitant oxidation of substrates.

All proteins containing the type 3 copper centers bind dioxygen, and their dicopper centers have similar geometric and electronic structures. Despite these similarities, these proteins carry out different biological functions: Hc is the oxygen transport and storage protein in arthropods and mollusks; tyrosinase catalyzes the orthohydroxylation of phenols, with further oxidation of the product catechol to *o*-quinone 5; CatO catalyzes a two-electron transfer reaction during the oxidation by molecular oxygen of a broad range of *o*-diphenols (such as caffeic acid) to the corresponding *o*-quinones.

Analysis of the CatO crystal structure gave information regarding the ligands and the active site geometry.[102] In the oxidized form of CatO, both Cu^{2+} metal ions are four-coordinate, in an approximately trigonal pyramidal geometry, by three histidine residues, and a bridging ligand (a solvent molecule or, most likely a hydroxide ion; see Fig. 11.13c). The coordination geometry changes upon reduction of the enzyme. In particular, the metal-metal separation increases significantly to 4.4 Å, while the histidines move only slightly and no significant conformational change is observed for other residues of the protein. In the X-ray structure of the reduced state, one copper ion is four-coordinated by three histidine residues and a water molecule, and its coordination sphere matches a distorted trigonal pyramid. On the contrary, based on the angle values His-Cu-His, the coordination of the other Cu^{+} ion can be described as square planar, with one missing coordination site.

Zinc The zinc coordination number ranges between two and eight. In biological macromolecules, four, five, and six are most frequently observed, with tetracoordination in tetrahedral geometry being the most prevalent. Trigonal-bipyramidal and square-pyramidal pentacoordinated geometries, as well octahedral hexacoordinated geometry has been reported (see Fig. 11.14).[8]

As already outlined in Section 11.3.3, Zn^{2} ion plays different roles in proteins[103]: catalytic,[64] structural,[65] cocatalytic,[66,67] and protein interface.[61]

Structural Zn^{2+} sites are generally tetrahedral with all the four ligands from protein.

The first zinc metalloproteins recognized to have a **structural site** were horse *alcohol dehydrogenase*[104] and the regulatory subunit of *aspartate carbamoyltransferase*.[105]

The first coordination sphere is similar in both metalloproteins: The zinc ion is tetrahedrally coordinated to the four cysteines, preventing access of water or substrate to its coordination sphere. The role of zinc apparently is to maintain the structure of the protein in its immediate vicinity.

A tetrahedral coordination is also found in *zinc finger proteins*. This protein family plays a critical role in many cellular functions, including transcriptional

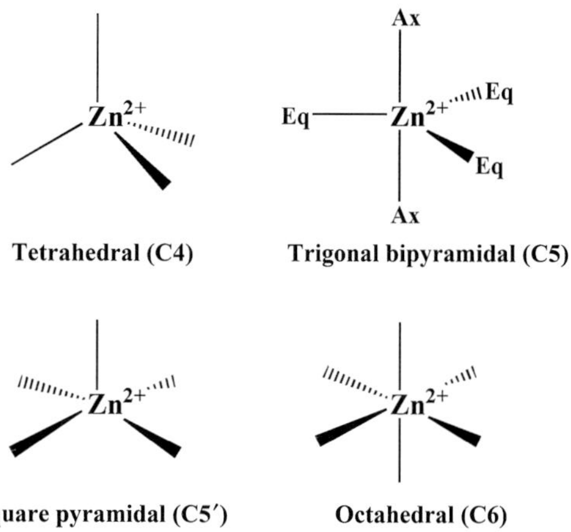

FIGURE 11.14. Ideal geometries for four-, five-, and six-coordinated zinc. Ax = axial; Eq = equatorial.

regulation, RNA binding, regulation of apoptosis, and protein–protein interactions. They are classified into several different types based on the number and order of the Cys and His residues that bind the zinc ion in the secondary structure of the finger.[106] Among the different zinc finger types, the C2H2 type (Cys2His2) is known as the classical zinc finger and it is one of the best studied.[107–111] In these proteins, the side chains of the conserved cysteine and histidine residues tetrahedrally coordinate a Zn^{2+} ion, and these metal–ligand interactions promote the formation of well-defined secondary structures and stabilize the global fold of the protein.

In **catalytic sites**, Zn^{2+} could adopt a coordination number of four, five, or six, probably reflecting the need for Zn^{2+} to adopt different coordination numbers during the enzymatic reaction.[112] The geometry in the free state is frequently distorted-tetrahedral or trigonal-bipyramidal. One more feature for a catalytic zinc site is the existence of an open coordination sphere; that is, the zinc binding polyhedron contains at least one water molecule in addition to three or four protein ligands. Carboxypeptidase, termolysin, and carbonic anhydrase (CA) are enzymes belonging to this class of zinc-containing proteins. The carboxypeptidases catalyze the degradation of food proteins, leading to the formation of amino acids. Three-dimensional structures are available for several members of the carboxypeptidase family. The catalytic zinc site of carboxypeptidase A[113] is composed of three amino acid residues, namely, His69 (N^{δ}), Glu72 ($O^{\varepsilon 1}$ and $O^{\varepsilon 2}$), and His196 (N^{δ}), and a water molecule, arranged in a distorted tetrahedral coordination. This site is highly conserved throughout the extended carboxypeptidase family.

There are also several crystalline-derived structures of the thermolysin family. In this family, the coordination geometry of the active site is approximately tetrahedral with three protein ligands, two histidines, and a glutamic residue, and a bound water molecule, although a distorted trigonal bipyramidal coordination is observed in some inhibitor complexes.[114]

The CA family, involved in the physiology of CO_2 transport, has been classified in three gene families α, β, and γ.[115] These three classes have no significant sequence identity, and have structurally distinct overall folds. Yet, despite their structural differences, the active sites of all three classes contain a single zinc ion that is essential for catalysis. In particular, the α-class is one of the best studied from a mechanistic point of view. This class is characterized by high affinity for zinc, where the geometry of the conserved histidine residues favors zinc binding, and is destabilized by the binding of other metals. The X-ray structures of α-CAs[116,117] from several mammals have revealed a catalytic zinc, tetrahedrally coordinated by three histidine residues (two of them bind the metal ion through their N^{ε}), and a hydroxide or a water molecule as a fourth ligand.

The **cocatalytic zinc sites** occur only in multimetal zinc. This group of enzymes includes alkaline phosphatase (with two zinc ions and one magnesium ion), phospholipase C (three zinc ions), nuclease P1 (three zinc ions), and leucine aminopeptidase (two zinc ions). In phospholipase C[118] (see Fig. 11.15), an aspartic acid forms a carboxylate bridge between the two ions (Zn1 and Zn3), and a second bridge is formed by a water molecule or an OH^-. The last zinc ion (Zn2), designated as the catalytic zinc, contains a bound water molecule, essential for catalysis, and a His2Glu metal polyhedron, similar to those found in other single catalytic zinc sites. Metal coordination is completed by

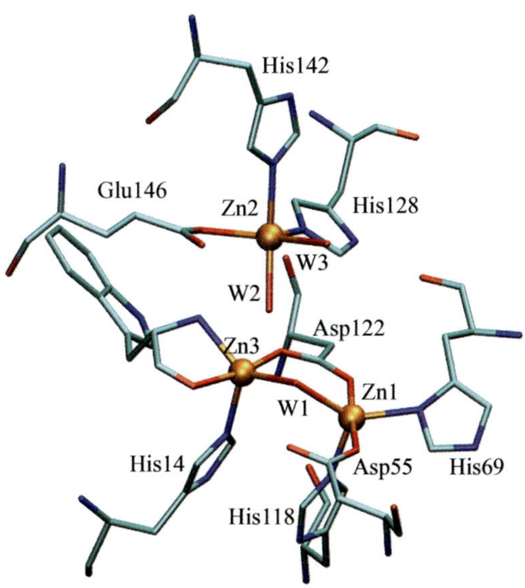

FIGURE 11.15. The cocatalytic zinc site in phospholipase C from *Bacillus cereus* (PDB code 1AH7).[118]

two further water molecules so that all three metal ions are approximately trigonal bipyramidal.

11.4 COORDINATION SITE SPECIFICITY AND METALLOPROTEIN FUNCTIONS: A LESSON FROM IRON

It is difficult to imagine life without the iron ion. With rare exceptions, virtually all studied organisms from *Archaea* to man are dependent on iron for survival, and numerous cellular function depend on the presence of iron coordinated by a protein.[119–122] Iron is the second most abundant metal on the earth's surface, falling closely behind aluminum and in near equivalent concentration to calcium and sodium.

Despite the ubiquitous distribution and abundance of iron in the biosphere, the biological availability of iron is extremely limited by its insolubility.[123,124] Iron in simple aqueous solution may access two principal oxidation states, the ferrous (Fe^{2+}) and the ferric (Fe^{3+}). The one-electron transfer between these states is easily accomplished: Common reducing agents will usually convert aqueous ferric iron to ferrous, while molecular dioxygen will promote the reverse reaction. Even though these properties render iron attractive for biological redox processes, they can become a problem in terms of low bioavailability and toxicity. The description of the properties of iron in biology is preceded by a brief description of the inorganic chemistry of iron, as reported in the following.

11.4.1 The Inorganic Chemistry of Iron

The bioavailability of iron strictly depends on the relative solubility of ferric and ferrous salts in the biological fluids. Whereas Fe^{2+} is extremely water soluble, Fe^{3+} is quite insoluble due to the occurrence of hydrolysis phenomenon.[8] Ferrous salts when dissolved in water, in the absence of oxygen, give solutions containing the hexaquo cation $[Fe(H_2O)_6)]^{2+}$. On the opposite, ferric salts produce highly acidic solution due to the hydrolysis phenomenon, and the hexaquo cation $[Fe(H_2O)_6)]^{3+}$ predominates in solution only at very acidic pH (pH $\approx$ 0). The hydrolytic processes are complicated, and in the presence of anions with appreciable coordinating tendencies, they are further confused by the displacement of water molecules from the iron coordination sphere. In aqueous solution of salts such as perchlorate, the following equilibria occur:

$$[Fe(H_2O)_6)]^{3+} + H_2O \leftrightarrow [Fe(H_2O)_5(OH)]^{2+} + H_3O^+ \quad K = 10^{-3.05}$$

$$[Fe(H_2O)_5(OH)]^{2+} + H_2O \leftrightarrow [Fe(H_2O)_4(OH)_2]^+ + H_3O^+ \quad K = 10^{-3.26}$$

and also

$$2[Fe(H_2O)_6)]^{3+} + H_2O \leftrightarrow [Fe(H_2O)_4(OH)_2Fe(H_2O)_4]^{4+} + H_3O^+ \quad K = 10^{-2.91}.$$

Hydrolysis of ferric solutions involves first deprotonation of the the hexaquo cation $[Fe(H_2O)_6)]^{3+}$ to produce hydroxo- and oxo- species, thus leading initially to mono- and dinuclear species, which subsequently interact to produce further species of higher nuclearity. Raising pH above 2–3 induces further condensation with the formation of colloidal gels, and eventually of hydrous iron (III) oxide precipitation.

Fe^{3+} forms a variety of cationic, neutral, and anionic complexes, but an interesting feature of its coordination chemistry is a marked preference for oxygen donor ligands as opposed to nitrogen donor ligands, as it can be expected from its character of hard Lewis acid (see also Section 11.3.2). Fe^{2+} is a borderline soft Lewis acid and prefers ligation to soft Lewis basis, such as nitrogen donor ligands. In fact, whereas amines of Fe^{3+} are unstable and dissociate in water, the complex $[Fe(NH_3)_6]^{2+}$ is known and stable; also, there are fewer Fe^{2+} complexes with oxygen donor ligands, such as acetylacetonate and oxalate, as opposed to the high stability of Fe^{3+} complexes with phenols and enols.

Fe^{3+} complexes may be high or low spin, and even though the octahedral stereochemistry is the most common, a number of different geometries is known. Most Fe^{2+} complexes are octahedral and ligands such as 2,2′-bipyridine (bipy), 1,10-phenanthroline (phen), and CN^- form low-spin complexes. The chemistry of Fe^{2+} is strongly influenced by the possibility of oxidation to Fe^{3+}, and most of its salts are unstable with respect to air oxidation. However, the standard reduction potential (E°) for the Fe^{3+}/Fe^{2+} couple strongly depends on the nature of the ligands and on the pH. Table 11.3 reports a selection of the E° values for some iron complexes.[8]

The Fe^{3+}/Fe^{2+} system provides an excellent example of the effect of ligands on the relative stabilities of oxidation states since replacement of one ligand by another results in redox potential changes over the range of +1.12 to −0.30 V. The nature of the ligands affects the position of the equilibrium between the oxidized and reduced form of the iron ion. If the reduced form is more stabilized by the coordination than the oxidized form, the redox potential will increase; if the oxidized form is more stabilized than the reduced form, the opposite is found. Fe^{3+}, being more highly charged than Fe^{2+}, is stabilized

TABLE 11.3. E° Values at 25°C for Some Fe^{3+}/Fe^{2+} Couples in Acid Solutions

Fe^{3+}			Fe^{2+}	E° (V)
$[Fe(phen)_3)]^{3+}$	$+e^-$	↔	$[Fe(phen)_3)]^{2+}$	1.12
$[Fe(bipy)_3)]^{3+}$	$+e^-$	↔	$[Fe(bipy)_3)]^{2+}$	0.96
$[Fe(H_2O)_6)]^{3+}$	$+e^-$	↔	$[Fe(H_2O)_6)]^{2+}$	0.77
$[Fe(CN)_6)]^{3+}$	$+e^-$	↔	$[Fe(CN)_6)]^{2+}$	0.36
$[Fe(C_2O_4)_3)]^{3-}$	$+e^-$	↔	$[Fe(C_2O_4)_2)]^{2} + (C_2O_4)^{2-}$	0.02
$[Fe(EDTA)]^-$	$+e^-$	↔	$[Fe(EDTA)]^{2}$	−0.12
$[Fe(quin)_3)]$	$+e^-$	↔	$[Fe(quin)_2)] + quin^-$ [a]	−0.30

[a] $quin^-$ = 5-methyl-8-hydroxyquinolinate.

by negatively charged ligands such as the anions EDTA and the hydroxyquinoline. On the opposite, neutral ligands, such as phen- and bipy-, relatively stabilize the Fe^{2+} form. The influence of the pH on the redox potential is evident from the different air oxidation susceptibility showed by the iron aquo complexes. The value of E° for the couple involving the aqua-ion $[Fe(H_2O)_6)]^{3+}/[Fe(H_2O)_6)]^{2+}$ is 0.77 V: In atmospheric condition, the oxygen is able to oxidize Fe^{2+}. This oxidation is slow at acidic pH, but if the solution became alkaline, precipitation of the iron(III) hydroxide causes a dramatic change in the redox potential (from positive value to negative one), thus rendering, at alkaline pH, Fe^{2+} air oxidation very fast.

The effect of the ligands in modulating the redox potential of the Fe^{3+}/Fe^{2+} couple is particularly evident in natural iron-containing proteins. As will be outlined in Section 11.4.3.4, cytochromes are examples of how first and second shell interactions render the Fe^{3+}/Fe^{2+} couple an efficient electron transfer system.

11.4.2 The Biochemistry of Iron

Proteins containing iron catalyze several important reactions, ranging from electron transfer, to oxygenation, to reversible O_2 binding.[125] Iron in proteins is coordinated by ligands ranging from soft donors such as sulfide, as in the in iron-sulfur clusters, to hard donors such as carboxylate in diiron oxoproteins.[18] In the cells, very little amount of iron is contained as Fe^{2+}, and extremely low amount as Fe^{3+}. As outlined above, under aqueous aerobic conditions, the Fe^{3+} is the thermodynamically favored species over Fe^{2+}; therefore, iron biochemistry is largely determined by the low solubility of the ferric hydroxide species.[126] To overcome the poor availability and potential toxicity of iron, and achieve effective homeostasis, organisms have evolved several mechanisms in order to strictly regulate their iron balance.[127–129]

Iron Assimilation and Homeostasis by Living Systems The adult human body contains ≈4 g of iron, of which about 3 g are present as hemoglobin in erythrocytes. These cells circulate for about 100–120 days in the body before their components are recycled by macrophages. The heme constituent of hemoglobin is broken down into Fe^{3+} and biliverdin. The biliverdin is reduced to bilirubin, which is released into the plasma and recirculated to the liver bound to albumin. The iron is released into the plasma, and is recirculated by a carrier protein called transferrin. This recycling process is very efficient since it is necessary to daily absorb, from diet sources, only very low amount of iron, about 1 mg day^{-1}, to compensate for some iron loss.[123]

A diversity of metabolic pathways has evolved to accommodate organisms' need for iron, which regulate the absorption, transport, storage, and mobilization of cellular iron.[121,123,129] A detailed description of these mechanisms can be found in Chapter 13. To solve the problem of taking up iron into cells, organisms have developed numerous iron solubilization technologies.

Duodenal absorption of dietary iron requires the presence of several enzymes and transporters, which are involved in the following: reduction of dietary Fe^{3+} to Fe^{2+}, and subsequent binding to ferrous iron transporter; binding of heme dietary iron and internalization into the cells; transport of Fe^{2+} into the enterocyte cytosol; and release into the blood circulation.

In mammals, birds, fish, and amphibians, a highly conserved protein called transferrin is needed to move the major load of extracellular iron into cells.[123,124,129,130] Bacteria and fungi evolved different tools to scavenge iron from the environment. These organisms produce low molecular weight ferric ion specific chelating agents, known collectively as siderophores.[126,131] These molecules can be classified as phenolates–catecholates or hydroxamates, or mixtures of the two forms.

In dealing with iron, biological systems face the interesting problem that, whereas iron metabolism requires the metal in both the ferrous and ferric states, the presence of free ferrous iron is potentially very dangerous. In fact, the one-electron reduction of dioxygen by Fe^{2+} results in superoxide formation, which in turn leads to the well-known Haber–Weiss–Fenton sequence generating the hydroxyl radical, $OH^{\bullet}$[119]:

$$Fe^{2+} + O_2 \rightarrow Fe^{3+} + O_2^{-\bullet}$$

$$2O_2^{-\bullet} + 2H^{+} \rightarrow H_2O_2 + O_2$$

$$Fe^{2+} + H_2O_2 \rightarrow OH^{\bullet} + OH^{-} + Fe^{3+}.$$

The hydroxyl radical, perhaps the most powerful oxidant encountered in biological systems, will attack proteins, nucleic acids, and carbohydrates, and initiate chain-propagating lipid peroxidation. Biological systems have evolved complex mechanisms to control the levels of superoxide and hydroxyl radicals. Two classes of metalloenzymes, namely SODs and catalases, are involved in these mechanisms: SOD enzymes catalyze the disproportion of superoxide anion, while catalases catalyze the disproportion of hydrogen peroxide. These enzymes work in concert to protect the organism, which utilizes dioxygen, from the potentially toxic by-products of oxygen metabolism.[19]

Iron Proteins: A Plethora of Different Interactions Modulate Reactivity

The bulk of iron present in the human body is bound to the oxygen transporting hemoglobin in the red blood cells, to the muscle oxygen storage protein myoglobin, or is stored by ferritin and hemosiderin. Less than 1% of this iron is bound to the various iron enzymes and redox proteins or is being transported through the blood by transferrin.[125]

Iron sites in proteins are usually divided into two categories depending on the different coordination environments: iron sites containing the metal ion inserted into a porphyrin ligand, and sites composed of iron ions coordinated by only protein residues.[2,18,19] The former sites are contained in the well-known heme protein family, whereas nonheme iron sites occur in several classes of

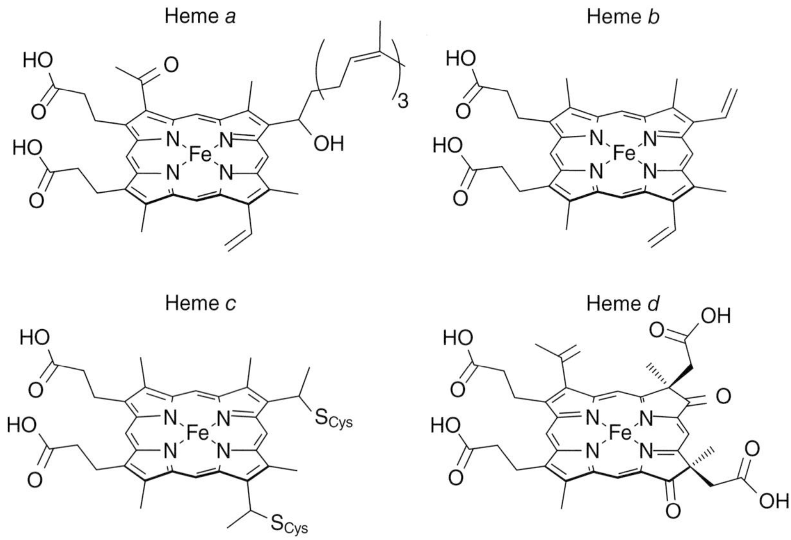

FIGURE 11.16. Chemical structures of commonly occurring natural hemes *a*, *b*, *c*, and *d*. (Reprinted with permission from Reedy and Gibney.[132] Copyright 2004 American Chemical Society.)

proteins, such as the mononuclear iron proteins, the diiron carboxylate-bridged proteins, and the iron-sulfur proteins.[2,18,19] In the following sections, the most important interactions involved in the modulation of heme functions and reactivity will be discussed. A similar analysis will be described on the diiron carboxylate protein family.

11.4.3 Heme Proteins: Mixed Interactions in Modulation of Their Functions

The name heme refers to a family of related iron–porphyrin complexes, all of which share the same central skeleton but differ in their side chain substituents and in their mode of attachment and coordination to the protein chain. Figure 11.16 reports the chemical structures of the commonly occurring natural hemes.[132,133] The most commonly found heme in proteins is the *b*-type, an iron protoporphyrin IX noncovalently bound to the protein. It is thought that all heme derivates use *b*-type hemes as template.[134,135] A common modification is attaching a *b*-type heme covalently to the protein, thereby forming a *c*-type heme.[136] The covalent bonds, made between the heme vinyls at positions 8 and 13 of the porphyrin macrocycle and two specific cysteine residues, form a classical Cys-Xaa-Xaa-Cys-His (CXXCH) *c*-type heme binding motif. Here, the histidine acts as one of the heme ligands and Xaa are two arbitrary amino acids other than Cys. Other modifications of *b*-type hemes lead to *a*-, *o*-, and *d*-type hemes, which are found, for example, in the terminal oxidases of bacteria and eukaryotes.[134]

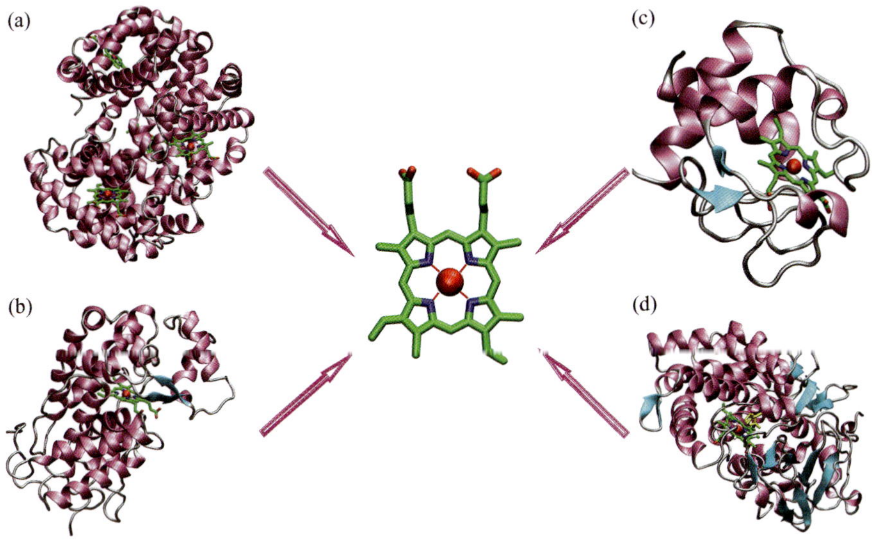

FIGURE 11.17. Different protein environments specify the heme toward different functions. Crystal structure of (a) deoxyhemoglobin (PDB code 1HGA),[139] (b) HRP (PDB code 1ATJ),[140] (c) horse hearth cytochrome *c* (PDB code 1AKK),[141] (d) Cyt-P450cam (PDB code 5CP4).[142]

Variations in the shape, volume, and chemical composition of the binding site, in the mode of heme binding and in the number and nature of heme protein interactions, result in significantly different heme environments in proteins with different functions in biology.[137,138] Examples are reported in Figure 11.17.[139–142] Heme proteins are involved in dioxygen storage and transport, electron transfer, hydroxylation and oxidation of organic substrates, and hydrogen peroxide disproportion (Table 11.4).[138] Therefore, the protein matrix controls the intrinsic reactivity of a similar prosthetic group, selecting one reaction as the only or predominant one.

High-resolution X-ray crystallographic data as well NMR analysis available for an increasing number of heme proteins have deeply contributed to the development of systematic studies aimed at the understanding of the reactivity and structural properties of heme proteins at the molecular level. Some of the results from these survey can be accessed through several useful databases, such as the Heme Protein Database (HPD),[133] the ProMISe database,[143] and the MDB database.[37] Moreover, site-directed mutagenesis studies have allowed to address the role of individual residues on the modulation of heme properties.

It is well ascertained that the fine-tuning of heme functions is achieved through an intricate set of different interactions between the cofactor and the protein matrix. Coordinate interactions dominate in the first coordination shell: the number, type, and donor properties of the axial ligands directly contribute to a first imprinting of heme properties.[144–151]

TABLE 11.4. Iron Coordination Sites in Biological Systems: The Most Common Heme Proteins[a]

Protein Classes	Heme–Iron Coordination	Axial Iron Ligands	
		Endogenous	Exogenous[b]
Cytochrome c oxidase heme *a* ('cyt *a*') Cytochrome b/b_6 Cytochrome b_5 Cytochrome *c* (classes III and IV) Cytochrome c_{554} (hemes 3 and 4) Cytochrome cd_1 nitrite reductase (*c*-domain) Hydroxylamine oxidoreductase (heme *c*)	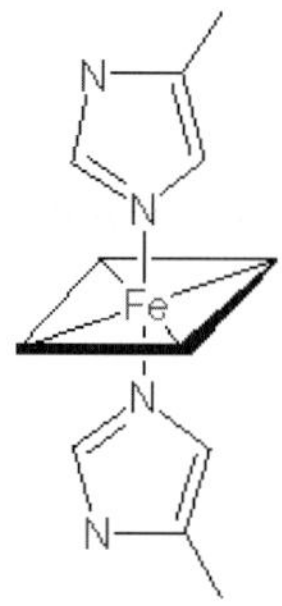	N^{ε}_{His}; N^{ε}_{His}	
Soluble cytochrome b_{562} Cytochrome *c* (cyt c_1, classes I, IIb, and IV)	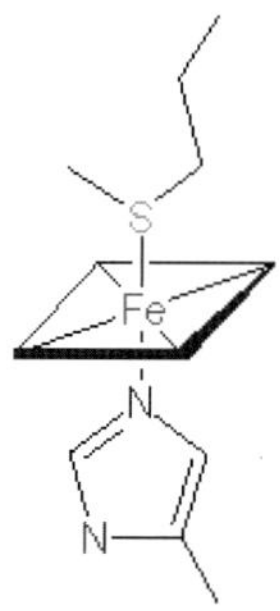	N^{ε}_{His}; S^{δ}_{Met}	
Heme thiolate proteins Heme chloroperoxidase Nitric oxide synthases P450 proteins	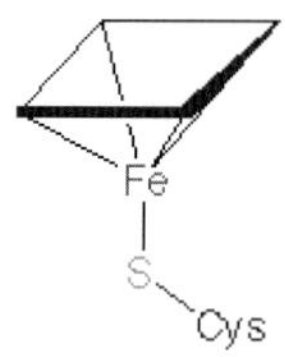	S^{γ}_{Cys}	O_2, H_2O_2, CO, NO
Cytochrome c oxidase heme a_3 Cytochrome c_{554} Cytochrome *c′* Globins Guanylate cyclase (soluble) Hydroxylamine oxidoreductase Nitrophorins Peroxidases	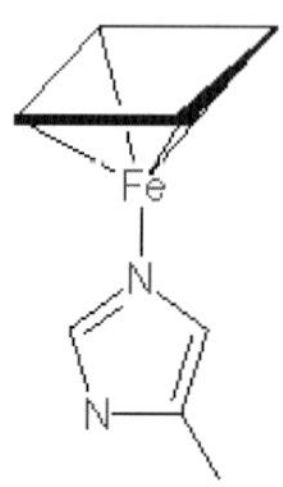	N^{ε}_{His}	O_2, CO, NO, CN^-
Catalases	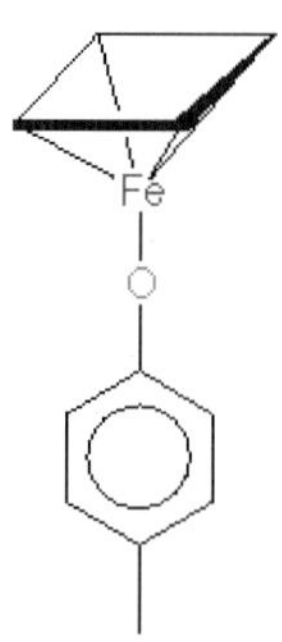	O^{η}_{Tyr}	H_2O_2, O_2

[a] Adapted from the ProMISe Database.[143]

[b] Most frequently observed.

The second coordination sphere is dominated by hydrophobic, ion pairing, and hydrogen bonding interactions, which control several features of the heme environment, such as the local dielectric constant, the heme exposure, and the ligand orientation. Further, the protein directs long-range interactions and regulates the accessibility of solvent and substrates into the active site, thus allowing to selectively discriminate between different ligands and/or substrates. These factors all contribute to the functional versatility of the heme. In the following sections, a comparative discussion of some representative heme proteins will be given in order to dissect the contribution of interactions in tuning heme properties.

First Coordination Sphere Interactions: Role of the Macrocycle and of the Axial Ligand(s) The porphyrin macrocycle consists of four pyrrolic units linked by four methine bridges. It contains an 18-electron π system and exhibits aromaticity. The resonance energy is estimated to be 1670–2500 kJ mol^{-1}.[152] The porphyrin ring is a dianionic chelating agent, which provides a relatively rigid hole for a cation, close to but smaller than the diameter of high-spin Fe^{2+}or Fe^{3+}, with the four pyrrole nitrogen as its ligands.[18] While the porphyrin ligand is a good four-centered σ donor, it is also an effective π acceptor. The energy level diagram of the iron porphyrin unit makes it clear that the metal ion and the ligand are a strongly cooperative unit. The high-spin iron in the ring, whether in the Fe^{2+} or Fe^{3+} states, is quite close in energy to the low-spin state, and indeed to intermediate spin state. This spin state equilibrium balance is one of the delicate controls, which the porphyrin ring, together with the axial ligands, exhibits over the metal ion. In the low-spin state, Fe^{2+} is made into a good π donor, which strongly assists binding of several exogenous ligands (O_2, CO, NO).[18]

The iron in the porphyrin ring adopts a square pyramidal or octahedral geometry. Depending on the oxidation and spin state of the iron ion, several different configurations are found. In particular, iron is coplanar or nearly coplanar with the porphyrin plane in almost all low-spin ferric and low-spin ferrous complexes.[153] On the contrary, it is displaced from this plane in high-spin ferric and high-spin ferrous complexes. Further, in high-spin complexes, the displacement of the iron is accompanied by significant doming of the porphyrin ring.

Dealing with the first coordination sphere, the next valuable descriptive feature of heme proteins is the nature of the protein ligands. With respect to the axial ligands, heme proteins are generally classified depending on either they contain a five-coordinate or six-coordinate heme complex. Table 11.4 reports a list of the principal heme proteins found in nature, grouped according to heme iron coordination, taken from the ProMISe database.[143] In the globins, the protein provides only one axial ligand, histidine, whereas the opposite side is open for O_2 binding. Likewise, in the heme enzymes (cytochrome P450 [Cyt-P450], peroxidases, catalases), the surrounding protein provides only one axial ligand (Cys, His, or Tyr), whereas the sixth site is open for the binding of substrates. In the electron carriers, the protein provides two axial ligands, typically histidine or methionine, which ensures a low-spin state of the iron ion and a low reorganization energy of the site. In the following sections, several examples will be described in order to highlight the contribution of the different ligands and different interactions, which direct the heme toward a selected function.

Histidine as Iron Axial Ligand: Interactions Driving Ligand Orientations

The first coordination sphere of iron in heme proteins is characterized by a high content of nitrogen as donor atoms. Four of them are invariably provided by the porphyrin macrocycle and lie in an equatorial plane with respect to the iron ion. The two axial positions, referred usually as the fifth and sixth coordination position in the octahedral geometry, can be occupied by one or two ligands; in this last case, they can be equal or different. In the globin family, the protein supplies the imidazole ring of a histidine as an axial ligand. The mono-histidine five-coordinated heme is the most frequently observed coordination motif in heme proteins.[132] In the six-coordinated heme motif, the protein chains supply a second ligand, which can be methionine, giving rise to a His/Met six-coordinated heme, as found in the cytochrome c family,[136] or histidine, giving rise to a bis-His six-coordinated heme, as found in the cytochrome b_5.[153] Coordination motifs bearing histidine in combination with ligands different from His and Met are relatively rare, and only cases of lysine, carboxylate, or phenolate binding are known.

As outlined in Section 11.3.1, His can coordinate a metal ion through the N^{δ} or N^{ε} imidazole nitrogens. Even though the propensities of His N^{δ} versus N^{ε} coordination are variable among metalloprotein active sites, in heme proteins invariably, with only one exception,[154] the His is bonded to the heme iron through the N^{ε} nitrogen. The orientation of the axial ligand, with respect to the porphyrin plane, has been suggested to be one of the mechanisms by which the protein matrix fine-tunes the heme properties. This hypothesis originated from the evidence that, even within a common set of axial ligands and heme substituents, the physicochemical properties of heme proteins vary widely. It has been demonstrated that different histidine conformations can shift the redox potential of heme proteins,[155] and that the spectroscopic, electronic, and magnetic properties of ferriheme model complexes can depend on the orientation of axially coordinated planar ligands.[156,157] Ligand orientation can also control the coordination of substrates to heme proteins.[158]

Starting from early 1980s, numerous studies, by Walker, have systematically analyzed the relationships between metalloporphyrin properties and axial ligand orientation through theoretical calculations and structural analysis on simple models.[157,159] More recently, the growing number of heme protein structures available in the PDB has stimulated statistical analysis of natural proteins in order to give a more complete view of the problem.[160,161]

One of the main aspects that is still a matter of debate relates with the type of interactions that render one orientation preferred respect to another. By

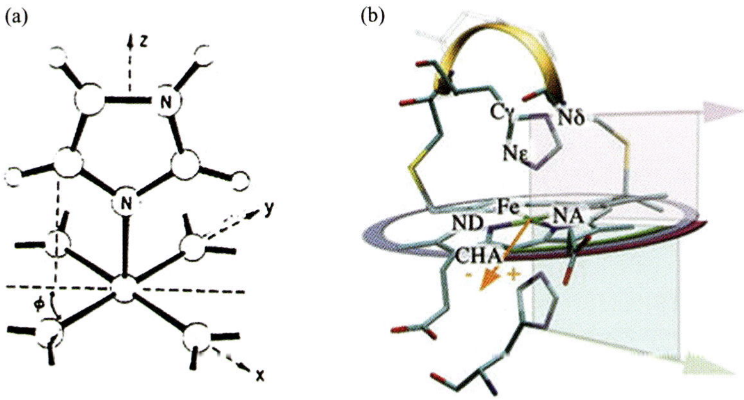

FIGURE 11.18. (a) The orientation of imidazole plane with respect to the porphyrin, as defined by Scheidt and Chipman. (Reprinted with permission from Scheidt and Chipman.[162] Copyright 1986 American Chemical Society.) (b) The orientation of the histidine ligand as defined by Gunner and coworkers. Alpha is defined after the histidine vector (C^{γ}-N^{δ}) is projected onto the heme plane. Alpha of +45° and −45° is pointing toward the heme nitrogen NA and ND, respectively. (Reprinted with permission from Fufezan et al.[161] Copyright 2008 Wiley.)

using extended Huchel calculations, Scheidt and Chipman[162] provided an explanation of the finding that, in many structurally characterized metalloporphyrin models, the axial ligands tend to prefer sterically unfavorable orientations. Calculations were interpreted by analyzing the extent of π interactions between the imidazole and the metal ion, as a function of the rotation angle ϕ. The ϕ dihedral angle was defined as the angle between the imidazole plane and a plane perpendicular to the porphyrin plane and passing through a porphyrinato nitrogen atom (see Fig. 11.18a).

Maximum steric repulsion between imidazole and the porphyrin atoms is expected to be at imidazole orientation eclipsing the metal equatorial nitrogen bonds, that is, at ϕ values near 0°, and minimum repulsion at ϕ values near 45°. On the contrary, experimental evidence demonstrated that ϕ values smaller than 45° were preferred. Calculations on a set of representative complexes, such as five-coordinate metalloporphyrin imidazole complexes and hexacoordinated metalloporphyrin complexes, in which one ligand is imidazole, and the second is either imidazole or a different ligand, demonstrated that the orientation of the axial ligands is not determined by nonbonded interactions, which should favor ϕ values near 45°. On the contrary, the most significant variation in bonding parameters, as a function of the rotation angle, was found for the π bond between the imidazole and the metal ion. Metal pπ and imidazole pπ interactions favor ϕ values near 0°. Therefore, the orientation of the imidazole ligands is the result of the competition between steric repulsion and π bond interactions. The orientation was also found to be independent from the metal ion d^n configuration, and from the presence or nature of the second axial ligand. In fact, in hexacoordinated complexes, the effect of coplanar or perpendicular imidazole planes orientation is negligible with respect to the preferences toward smaller ϕ values. However, in hexacoordinated low-spin complexes, which are characterized by smaller nonbonded distances, steric repulsion between imidazole and porphyrin are more pronounced. This observation was consistent with the finding that hexacoordinated species showed wider range of accessible ϕ values. The only exception was found for the low-spin iron(III) hexacoordinated complexes, where the magnitude of orbital overlaps depends on the relative imidazole orientation, being mostly favored at parallel imidazole rings orientation and smaller ϕ values.

The effect of axial ligand orientations on the spectroscopic properties were analyzed by combined studies with EPR and Mossbauer spectroscopies, and structural and theoretical analysis on a series of well-defined model compounds.[156,162–164] It was demonstrated that the EPR signals is correlated with the mutual orientation of the axial ligands, with the parallel or approximately parallel orientations giving rise to the well-known low-spin rhombic EPR spectrum. The perpendicular orientation leads to an unusual EPR spectrum called "large g_{max}," which consists of a single observed line (at very low temperatures) with a relatively large g value ($g > 3.3$). Finally, the observed EPR spectra, and therefore the mutual ligand orientation, were also correlated with the reduction potential of the iron heme couple. Perpendicular ligand orientation can determine positive shifts of the redox potential of up to 50 mV over that observed for parallel alignments.[163] All these studies revealed a good correlation between the effects of mixed interactions in modulating the properties of axial ligands and, as a consequence, the spectroscopic and redox potential of heme complexes.

The results obtained for the heme protein models have been basic to the understanding of similar behaviors observed in natural proteins. Heme model systems are usually simpler than their natural counterparts and therefore the relationships between type of interactions, ligand orientation and functional and spectroscopic properties have been quite clearly rationalized.

In natural heme proteins, the presence of numerous interactions in their folded structure render the dissection of the problem more complicated, and this argument is still a matter of discussion. Among the factors that can influence the orientation of coordinated imidazoles, the following interactions seem to be dominant in natural systems: hydrogen bonds between the imidazole $N^{\delta}H$ group and H bond acceptors of the protein; nonbonded interactions of the imidazole ring with the protein backbone and side chains; nonbonded interactions of the imidazole with the porphyrin atom skeleton and the side chains of cysteines covalently bound to heme (in cytochrome *c* heme proteins); and electrostatic interactions and the presence of charged amino acid side chains.[159,160,165]

By using density functional theory (DFT) calculations and PDB statistical analysis, Knapp and coworkers notice that in all the examined structures the $N^{\delta}H$ group of the imidazole ligated to heme is oriented toward the propionate groups of the heme, thus suggesting a possible role of electrostatic interactions between the propionates and the imidazole in driving the orientation of the axial ligands.[160,166] Force field and DFT computations supported this hypothesis.[166]

More recently, a similar search in the PDB, supported by molecular mechanics calculations was done by Gunner and coworkers.[161] They investigated ligand orientation in different classes of heme proteins, which they classified as (1) ligating *b*-type hemes (51 hemes), (2) ligating *c*-type hemes where the histidine ligand is not part of the CXXCH motif (42 hemes), and (3) ligating *c*-type hemes where the histidine ligand is part of the CXXCH binding motif (58 hemes). Ligand orientation was described by the α angle defined as the projection of the His ligand vectors onto the heme plane (Fig. 11.18b). Analysis of the PDB showed that no preferred orientation is found in the class of ligating *b*-type hemes and *c*-type heme where His is not part of the CXXCH motif; in fact, values of α angle span over the whole range. On the contrary, restriction to only small values of the α angles are observed for the *c*-type ligation in which His is part of the CXXCH motif. The majority of the structures showed α angles close to $0 \pm 25°$. Independent of heme type, those orientations that eclipse heme nitrogens ($\pm45°$, $\pm135°$) are less frequently observed. Calculations confirmed that the major restriction imposed on the ligand orientation is the covalent attachment of the heme to the cysteine of the protein chain. This constraint is more severe when His is part of the CXXCH motif (>40 kcal mol^{-1}). Other interactions are also responsible of the ligand orientation, in the following order: hydrogen bonds with the imidazole ligands ($\approx$5 kcal mol^{-1}), quantum effects due to orbital interactions and ligand field distortions (1–3 kcal mol^{-1}), and finally, electrostatic interactions ($\approx$1 kcal mol^{-1}).[161]

Histidine in Five-Coordinate Heme Proteins: Modulation of Activity by Second Shell Interactions in Globins and Peroxidases One of the most striking evidence of the contributions of different interactions to the functional specificity of heme proteins appears from a comparative analysis of globins and peroxidases. Figure 11.19 reports a comparison of the metal binding site in myoglobin[167] and horseradish peroxidase (HRP).[140] Both families of proteins are examples of five-coordinated heme with a histidine axial ligand, namely the proximal histidine. In both proteins, exogenous ligands bind at the distal side of the heme plane, opposite to the proximal His, giving rise to a distorted octahedral coordination of the iron site. Even though these two protein families contain the same axial ligand, an intricate set of interactions at the proximal and distal site dramatically change the properties of the axial histidine, and specify the distal face toward a particular function. In order to have a comprehensive view of the molecular interactions involved in the fine-tuning of the properties of a five-coordinated mono-histidine heme, a very brief description of the globins and the peroxidases structures will be given in the following. Then, the role of these interactions will be analyzed by comparing the two different heme environments.

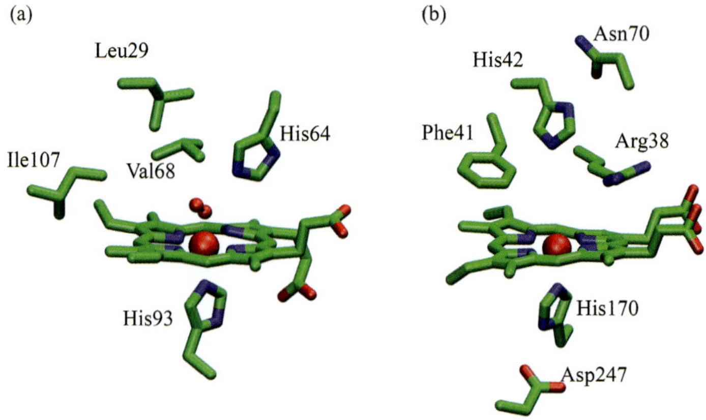

FIGURE 11.19. Five-coordinated heme proteins with His as axial ligand. Details of the heme binding site in (a) oxymyoglobin (PDB code 1MBO)[167] and (b) in HRP (PDB code 1ATJ).[140]

The huge amounts of structural, biochemical, and physiological information on hemoglobins and myoglobins far exceed that for any other known protein family. High-resolution X-ray crystallographic data as well NMR analysis available for an increasing number of globins in all the coordination states, such as the oxy, deoxy and carbon-monoxy forms, and the met-oxidized state, have deeply contributed to the understating of all the mechanism by which dioxygen is transferred and stored in living organisms.[168–172] The globins contains the *b*-type heme (Fig. 11.19), with the iron ion stabilized in the Fe^{2+} state. They are excellent examples of how nature has devised to manage and control the chemistry of the iron ion in the Fe^{2+} oxidation state by encapsulating it into complex macromolecules. In this form, heme is able to bind gaseous ligands, such as O_2, CO, and NO.[172,173] The heme in the globins can be oxidized to the Fe^{3+} state by reaction with ferricyanide or by standing in air. The proteins in this form, also referred as the Met- state, bind at the sixth coordination position different anions such as OH^-, CN^-, N_3^-, and F^-, or a water molecule.[138] The heme group in the globins is embedded in the protein molecule in such a way that there is access for O_2 to a small cavity, where it coordinates to the Fe^{2+} in the sixth position of the octahedron. Myoglobin is a protein that stores O_2 and is monomeric, while hemoglobin is a tetramer that carries O_2 from lungs to tissues. Hemoglobin contains two distinct type of globins, namely the α- and β-chains, differing in length and composition.[138] The most common combination is the $\alpha_2\beta_2$ type, known as the HbA, even though several other combinations such as HbA_2 or $\alpha_2\delta_2$, HbF or $\alpha_2\gamma_2$, are found. HbA has a molecular weight of 64,500 Da, consisting of two identical α-chains of 141 amino acids each, and two identical β-chains of 146 amino acids each. Despite some differences in their sequences, each of the monomers within the hemoglobin tetramer is very similar to a myoglobin molecule, and has one heme group and one globin chain. The three-dimensional arrangement in the globins is very highly conserved. They all contain fragments in α-helical conformations: the α-chain of HbA includes eight helical segments, designated A–H, whereas the β-chain is very similar but lacks a D helix. In a nomenclature common to the globin proteins, each amino acid residue is designated by its position in a helix or interhelical segment. For example, α87His and β92His are both His F8, the eighth residue in the F helix in their respective subunits.

Tight binding of the heme cofactor in the globin pocket is accomplished through two types of interactions: the coordinate bond between imidazole and the heme iron, and numerous contacts between the protein residues and the heme ring. These interactions are mostly with hydrophobic amino acids, which predominate at the heme protein interface. As an example, Table 11.5 lists the

TABLE 11.5. Heme Contacts[a] in the Crystal Structure of Human Deoxyhemoglobin[b]

Residue[c]			α Subunit			β Subunit
B13 (D)	Met	>	4.0	Leu		3.9 (1) II: V
C7 (E)	Tyr		3.4 (1) III: M	Phe		3.7 (4) II: V, III: M
CD1 (D)	Phe		3.5 (8) III: R, M, II/III	Phe		3.6 (5) III: M, IV: M, II/III
CD3 (E)	His		2.8 (4) III: P*	Ser	>	4.0
CD4 (D)	Phe		3.4 (4) III: P 3.8 (4) III: P	Phe	>	4.0
E7 (D)	His		3.2 (9) I: R	His		3.2 (14) I: R, IV: R, III/IV 3.2 (11) I: R, IV: R, III/IV
E10 (D)	Lys		3.4 (5) IV: M 3.6 (1) IV: M	Lys		3.8 (4) IV: P 3.0 (5) IV: P*
E11 (D)	Val		4.0 (1) I/IV	Val		3.6 (8) I: R, V, M
E14 (D)	Ala	>	4.0	Ala		3.8 (2) I: M, IV: M
F4 (P)	Leu		3.8 (2) IV: M 3.7 (5) IV: M, P, I/IV	Phe		3.9 (3) IV: M
F7 (P)	Leu		3.4 (2) IV: P	Leu		3.8 (1) IV: P 3.5 (2) IV: P
F8 (P)	His		3.1 (15) I–IV: R, I/IV	His		3.0 (18) I–IV: R, I/IV
FG3 (P)	Leu		3.3 (7) III: R, P, III/IV	Leu		3.5 (5) III: R, P, III/IV
FG5 (P)	Val		3.5 (6) II: R, V, II/III	Val		3.9 (2) II: V, II/III
G4 (P)	Asn		3.4 (4) II: V, M	Asn		3.5 (3) II: M
G5 (P)	Phe		3.5 (5) II: M, I: R,V, I/II	Phe		3.6 (5) II: R, M, I/II
G8 (D)	Leu		3.5 (2) I: R, I/II 3.8 (1) I/II	Leu		3.4 (3) I: R, V
H19 (P)	Leu		3.5 (4) I: R, V, M	Leu		3.4 (3) I: R, V

[a] Interatomic distances ≤ 4.0 Å counted as contacts.
[b] From Fermi et al.[174] The minimum contact distances are indicated followed by the number of contacting pairs in parentheses. The number of the pyrrole(s) with which contact is made are also indicated as following: R, ring; M, V, P, methyl, vinyl, propionic side chains; I/II methane bridge between rings I and IIm, and so on; asterisks (*) denote hydrogen bonds.
[c] Letters in parentheses denote as follows: D, distal; P, proximal; E, edge on contact.

heme contact found in the deoxyhemoglobin structure at 1.74 Å resolution, as reported by Fermi et al.[174] Only few contacts are mediated by polar interactions: Lys E10 in the β subunit and His CD3 make hydrogen bonds with the macrocycle propionate group. Majority of the contacts are found with aliphatic residues, such as Leu, Ile, Met, Val, and Ala. Finally, Phe and Tyr aromatic residues interact with the heme macrocycle through π stacking or edge-to-face interactions.[132]

One of the distinctive features of the globins is their ability to discriminate between small molecule with similar size, shape, and charge distribution, such as O_2, CO, and NO. Site-directed mutagenesis studies coupled to structural and spectroscopic characterizations on myoglobins have allowed to define that electrostatic interactions and steric constraints are predominantly involved in the mechanism of ligand discrimination.[169,172] The composition of the distal site in myoglobin and hemoglobin are highly conserved across species line. Key residues in the heme distal site, which faces the E helix fragment, are Phe CD1, Val E11, Leu B10, and finally, His E7 (see Fig. 11.19a). As evident from Table 11.5, Val E11 and Phe CDl are mainly involved in hydrophobic contacts with the heme ring, thus contributing to the stability of the globin–heme complex. Myoglobin binds CO and NO with higher affinity with respect to O_2.[172] This may appear incongruent for an O_2-binding protein; however, the relative affinities of heme for CO and O_2 drastically change when the heme is embedded in the globin protein matrix. In fact, the partition constant (M), defined as the ratio of the equilibrium association constants for CO and O_2 (K_{CO}/K_{O2}), is ≈20,000 for free protoheme, whereas that for sperm whale myoglobin is only 25.[175–177] The major factors that have been postulated to explain these altered properties are steric hindrance within the heme pocket and electrostatic interactions between the bound ligand and distal histidine.[178] The effects of steric hindrance on the relative affinities of myglobin for CO and O_2 have been often referred to the different binding mode of these two ligands. In particular, since CO binds linearly to the heme iron with respect to the binding mode of O_2, which binds at an angle normal to the heme,[179] it was proposed that the presence of His and Val at the distal side provides unfavorable interactions with the CO. Therefore, a slight bending in the binding mode of this ligand was suggested, which in turn should lower the affinity. More recently, the X-ray crystal structure of the carbomonoxy- form of myoglobin, solved at high resolution, demonstrated that the binding of CO to the heme iron is linear, and that the protein chain undergoes a slight conformational change in order to accommodate CO in its almost perpendicular binding mode to the heme.[180,181] Site-directed mutagenesis studies supported the hypothesis that the distal His contributes to ligand discrimination mainly through electrostatic interactions. Distal His interacts with a water molecule, which must be displaced from the site upon ligand binding. This displacement is compensated by the strong hydrogen bond interaction between His and the polar FeO_2 complex. On the contrary, little electrostatic stabilization effect can be envisaged for the formation of the neutral FeCO complex.

The role of the distal and proximal site residues in fine-tuning the properties of the heme site is also well represented in the peroxidase family. Heme-containing peroxidases are enzymes that catalyze a variety of oxidation reactions using hydrogen peroxide.[182,183] Several enzymes have been isolated, sequenced, and structurally characterized. Usually, they are classified, as defined by Welinder,[184] according to the organism from which they have been isolated. To class I belong intracellular prokaryotic peroxidases, to class II extracellular fungal peroxidases, and to class III secretory plant peroxidases.[182–184] The structure of yeast cytochrome *c* peroxidase (CCP) was the first peroxidase structure solved in the early 1980s.[185,186] Today, there is at least one representative structure from each of the main three classes.[187] However, the majority of the information on the structure–activity relationship have been obtained from the analysis of CCP and of HRP.[187] The catalytic process of heme peroxidases occurs through several steps, which first involve the reaction of hydrogen peroxide with the heme to give a two-electron oxidized protein and water.[188] The oxidized protein is called compound I, and contains an oxoferryl center (Fe(IV)=O), and a porphyrin or protein radical cation, depending on the enzymes. Two single-electron transfers from substrates reduce compound I to compound II and then to the ferric resting state. This sequential one-electron pathway can be replaced by a two-electron transfer since some substrates directly reduce the protein to the resting state, without the formation of compound II.[182,183,187]

The distal cavity of heme peroxidases is the site of the interaction with hydrogen peroxide and it is characterized by two completely invariant amino acids, the distal histidine and the distal arginine (see Fig. 11.19b). These two residues play a major role in the formation and stabilization of compound I. The distal His (His42 in HRP, His52 in CCP) mainly acts as a general acid–base catalyst in facilitating the peroxide anion binding to the heme and in assisting the subsequent heterolytic cleavage of the O–O bond during the compound I formation.[186] The conserved distal Arg (Arg38 in HRP, Arg48 in CCP) is involved in the charge stabilization, mediated by its positively charged guanidinium group, of the developing OH- leaving group of hydrogen peroxide. Further, once the O–O bond is cleaved, Arg contributes to stabilize the oxoferryl species via hydrogen bond formation. Substitution of these key residues dramatically influences the peroxidase activity of the enzymes. Mutation of the His42 to Ala or Leu decrease the rate of compound I formation by a factor of $\approx 10^6$, replacement of Arg38 by Leu gives a protein for which the rate of compound I formation is decreased by a factor of $\approx 10^3$ with respect to the wild-type enzyme.[189] Thus, the proximal and distal environments of the heme in peroxidases account for the high activity of these enzymes, thanks to a global mechanism for peroxide activation, known as the push-pull concept.[190] To effectively mediate the activation of hydrogen peroxide and to function as oxidative catalysts, the proteins in the resting state are stabilized in the ferric heme form. As a result, peroxidases show unusually negative redox potentials when compared with the positive values of globins.[133,191]

One of the key factors in the modulation of the redox potential of five-coordinated His heme complexes is the presence of hydrogen bond interactions between the N^{δ} of the axially coordinated imidazole and another group in the surrounding protein proximal environment. In particular, the presence in HRP of a strong hydrogen bond with an aspartate side chain (Asp247) determines an increase in the basicity, relative to the globins, of the coordinating His170, thus helping to stabilize high–oxidation-state intermediates, such as the high-valent oxoferryl center (Fe(IV) = O), by a charge relay system. The Fe^{3+}/Fe^{2+} reduction potential of HRP (isozyme C) was determined to be −0.270 V.[191] In contrast, the globins, in which the N^{δ} of the axially coordinated imidazole forms an hydrogen bond with a main chain carbonyl oxygen, show positive values of the redox potential, 0.170 V for hemoglobin and 0.046 V for myoglobin.[181]

In conclusion, the analysis of peroxidases and globins reveals that second shell interactions are key factors in the modulation of the properties of a mono-histidine heme coordination motif. Hydrogen bond networks and electrostatic interactions all contribute in specifying the properties of the site by tuning the heme redox potential and the distal site machinery.

Interactions Involved in the Modulation of the Redox Potential: The Cytochromes Cytochromes are crucial for life since they constitute key components of the electron transfer systems in the respiratory chain and in photosynthesis.[192] As such, they are usually associated with membrane systems. The term cytochromes refer to both an individual protein or to a subunit of a larger enzyme. Cytochromes can contain either heme *a*, *b*, or *c*. They are also characterized by different axial coordinations, as listed in Table 11.4.

All cytochromes transfer electrons via a reversible change in the oxidation state of heme iron between the Fe^{3+}/Fe^{2+} state. They are generally six-coordinated, with at least one His as axial ligand, whereas the sixth ligand varies among them. Usually, the iron ion is in the low-spin configuration in both the oxidation states. The capability of cytochromes to efficiently transfer the electrons is strictly related to the high covalent character of the iron–porphyrin coordinate bond (see Section 11.2.2). An efficient overlap between the t_{2g} orbitals of the iron and the orbitals of the porphyrin allows the electrons to enter or leave an orbital having π overlap with the π^* antibonding molecular orbitals of the ring.[7] In this way, the iron orbitals are extended out to the periphery of the macrocycle and efficiently transfer the electrons.

Since huge amounts of X-ray structures are available in the literature, only a brief description of the most representative cytochrome folds will be reported. The readers can refer to the HPD[133] for a punctual analysis of the structures of various cytochrome classes and their relationship with the redox potential.

Regarding their structural organization, cytochromes are usually classified as all-α, $\alpha + \beta$ and all-β classes.[2] Among the all-α, *c*-type cytochromes are characterized by a well-conserved fold, usually referred to as cytochrome *c* fold.[192] Although there is some variability in the mode of attachment of the heme group to the polypeptide chain, in most cases the CXXCH pentapeptide segment constitutes the attachment site. Further, the heme group 4-vinyl substituent is exposed to the solvent. An intriguing feature of the cytochrome *c*-fold is the finding that the two heme propionates are deeply buried. The thermodynamic cost of fitting polar residues into a hydrophobic cage seems to be largely compensated by a strong hydrogen bond network of interactions between the propionate groups and three almost invariantly protein residues, that is, Arg38, Tyr48, and Trp 59, found in all the eukaryotic cytochrome *c*. The protein manages to stabilize an unnatural arrangement, that is, polar groups into a hydrophobic environment, through a complex set of strong bonding interactions: Propionates form two hydrogen bonds with the ε-nitrogen of Trp and the hydroxyl group of Tyr, and ion pair interactions with Arg.

Another type of all-α cytochrome is constituted by the four-helix bundle fold, as appears in the cytochrome b_{562}.[192] The structure is composed of four nearly parallel α-helices that closely pack together. The *b*-type heme group is found parallel to the bundle axis, with the two axial ligands being the ε-nitrogen of His102 and the δ-sulfur of Met7. The heme environment is asymmetrical: The heme face pointing toward the His ligand is rather exposed to the solvent, while the side facing the Met is in a very hydrophobic environment.

Finally, the cytochrome b_5 fold is characterized by the $\alpha + \beta$ motif. Cytochromes with the b_5 fold are small proteins, constituted by two hydrophobic cores of aromatic and aliphatic side chains, divided by a β-sheet. The protein is organized into two domains: The larger domain contains the heme group, and it is highly structured, with four helices; the smaller domain has a structural role. The heme group is buried within the hydrophobic core, with only one propionate group fully exposed to the solvent, and the other bended back to the molecule and stabilized by a hydrogen bond with a Ser residue.[193,194]

Cytochrome redox potentials span over the range of –0.30 to +0.40 V (see Fig. 11.20). In the biological electron transfer chains, the midpoint redox potential of each constituent controls the direction of favorable transport between the individual cofactors from food to oxidant. Therefore, the protein matrix controls the heme redox potential very efficiently.

Several additive factors tune the redox potential of the heme group: the nature of the axial ligands, the inductive effects of the porphyrin substituent, and the composition of the protein environment. Several theoretical and experimental approaches have been undertaken in order to dissect the contributions of these distinct effects in the modulation of the heme redox potential.[195–203] First, the nature of the axial ligand is evident from the observation that the high-potential cytochromes display His/Met coordination, with respect to the low-potential ones, which use bis-His coordination. This is in agreement with the electron-withdrawing power of Met with respect to the donor properties of the His ligand. According to the HSAB theory (see Section 11.3.2), Met ligand, a soft base, favors the ferrous heme, whereas His, with a borderline hard/soft character, may preferentially stabilize the ferric heme.

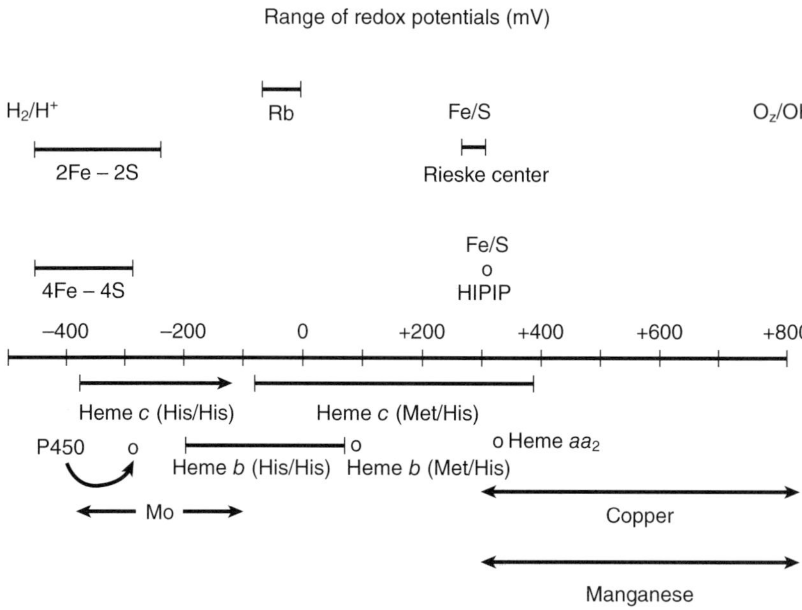

FIGURE 11.20. The redox range of heme proteins in comparison with other metal sites. (Adapted from Frausto da Silva and Williams.[18])

Therefore, a shift in the redox potential toward more negative values is expected by replacing the Met ligand with His (see Fig. 11.20). For example, analysis of the E° redox potential of several Met80 mutants of the horse heart cytochrome *c* showed that a shift of the E° values from 262 mV of the native form, to 41 mV in the Met80His and −390 mV in the Met80Cys mutants, respectively, occur.[202] This perfectly agrees with the increase in the hard character of the ligand, moving from methionine to histidine and cysteinate (see Table 11.2).

Electrostatic interactions, mediated by the protein matrix, appear to be the other dominant factor in tuning the heme redox potential. The early theoretical work of Kassner demonstrated the role of the dielectric constant of the protein matrix.[195] By studying the bis-pyridine mesoheme complex in nonaqueous environment, Kassner's theory concluded that when the heme is embedded in a nonpolar environment, such as the one present in high-potential cytochromes, it experiences a local environment of low dielectric constant. Considering only the charge of the iron ion and of the dianionic porphyrin ring, the reduced form can be formally considered as a neutral species. Therefore, a hydrophobic environment stabilizes the reduced state of the heme with respect to the +1 formally charged ferric porphyrin form, and shifts the redox potential toward positive values. Similar analyses were performed by studying the correlation between the heme solvent exposure and redox potential values. More recently, Gunner and coworkers used multiconformation continuum electrostatic (MCCE) calculations in order to correlate the experimentally observed electrochemical midpoint potential values (E_m) of the different heme groups with the composition and folding of the surrounding proteins.[198] In particular, the quantitative analysis of the factors that modulate heme electrochemistry included the interactions of the heme with its ligands, the solvent, the protein backbone, and the side chains. The redox potentials calculated by MCCE were found to be in good agreement with the experimental values, and demonstrated that no individual contribution, but the global combination of each single effect, provides an explanation of the observed experimental redox potential values. Based on the analysis of the potentials of eight structurally characterized *c*-type cytochromes, Gray and coworkers demonstrated that the reduction potentials of heme proteins can be tuned through variations in the heme exposure to solvent.[197] In conclusion, the contribution of the axial iron ligation and the protein encapsulation of the heme cofactor fine-tune the heme redox potential.

Cysteine as Iron Axial Ligand: Cyt-P450 Five-coordinate heme proteins with cysteine ligation are observed in Cyt-P450, nitric oxide synthase, and chloroperoxidases.[204–208] In fact, all these enzymes are characterized by an axial cysteinate sulfur donor ligand trans to the dioxygen or peroxide binding site.

Cyt-P450s catalyze a wide variety of reactions, including epoxidation, N-dealkylation, O-dealkylation, S-oxidation, and hydroxylation of aromatic or aliphatic compounds by the activation of molecular dioxygen.[204] They have been found in every mammalian tissue and organ, and are also highly distributed in plants, bacteria, and yeasts. Since numerous Cyt-P450s have been discovered, they are classified in different families, on the basis of their sequence similarity, when appropriately aligned. A good link for the information on the Cyt-P450s can be found at "The Cytochrome P450 Homepage."[209] Apart from the microbial enzymes, the majority of Cyt-P450s are membrane-bound enzymes, which catalyze the conversion of highly hydrophobic substrates, such as steroids, fatty acids, and xenobiotics, such as drugs and environmental pollutants. Since these enzymes are entrapped in the membranes, their extractions, solubilizations, and purifications are extremely difficult processes. Therefore, the main structural and functional data have been obtained on a soluble form, extracted from *Pseudomonas putida*,[204] a bacterium able to catalyze camphor hydroxylation.

The main reaction they catalyze is a monoxygenase reaction:

$$RH + O_2 + NAD(P)H + H^+ \rightarrow ROH + H_2O + NAD(P)^+.$$

The two electrons derived from NAD(P)H are shuttled to the heme center via electron transfer proteins (flavins and iron-sulfur cofactors).

The structural fold of the $CytP450_{cam}$ is highly conserved among the other Cyt-P450s.[204] The molecule can be described as flat-triangular shaped, with an

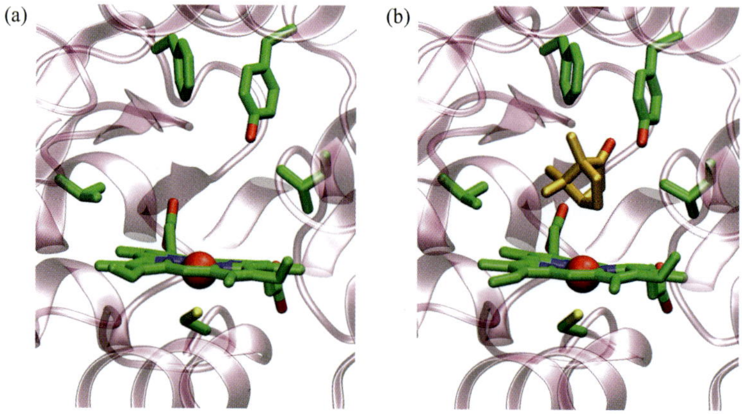

FIGURE 11.21. A five-coordinated heme protein with Cys as axial ligands. Details of the heme binding site of Cyt-P450$_{cam}$ (a) in the substrate free form (PDB code 1PHC)[210] and (b) with the camphor substrate bound at the distal site (PDB code 5CP4).[142]

α-helix-rich domain and a β-sheet-rich domain.[204] The typical shape derives from a particular folding pattern, arising from the C- and N-terminal fragment arrangement around the heme. The C-terminal half forms the inner core and two helices bracket either side of the heme. The C-terminal β-structure is at the surface. The N-terminal half is composed primary of helices, which surround the C-terminal fragment.

The heme group in Cyt-P450$_{cam}$ is buried in the interior of the enzyme, unlike peroxidases or cytochromes, where at least one heme edge in exposed to the surface, and it is linked to the protein through a few salt bridges between basic residues and heme propionates (see Fig. 11.21a).[210] The proximal side of the heme is closer to the protein surface and thought to be the docking site for Cyt-P450's redox partner. The distal side forms the substrates and dioxygen binding site (see Fig. 11.21b).[142]

Most P450 enzymes share similar chemistry, which can be described with a general catalytic reaction cycle.[204] The resting state of Cyt-P450 has a ferric heme with water (or hydroxyl) as the sixth ligand. The first key step in the cycle is the binding of the substrate, which is accompanied by a change in the spin state of the heme iron and a shift of the redox potential from −300 mV to about −170 mV.[204] The substrate-triggered redox potential shift ensures the efficient transfer of an electron from the redox partner, NADH or NADPH, to the enzyme, which causes the reduction of the Fe^{3+} ion. Oxygen binds the ferrous ion to give a ferric–superoxide complex, which is converted to a ferric peroxy anion, via a second electron transfer. The ferric hydroperoxo intermediate is unstable and, upon protonation, fragments to give a ferryl intermediate that can be formulated as a porphyrin radical cation Fe(IV) species, resembling the peroxidase compound I. The following step is believed to be the extraction of a hydrogen atom from the substrate by this high-valent iron species, forming a substrate radical, which leads to product via an oxygen rebound mechanism.[204]

One of the main aspects of the Cyt-P450s that is still a matter of debate is the role of the thiolate ligand in driving the chemistry of this class of enzyme. Differently from peroxidases, the distal site does not contain acid–base catalytic groups, which can assist in the cleavage of the O–O bond. Spectroscopic studies, together with crystal structure and catalytic investigations, suggested that the main support in providing an electron "push" to facilitate the cleavage of the O–O bond is provided by the thiolate ligand, which is a better electron-donating group with respect to the peroxidase His axial ligand. Nevertheless, a proton shuttling mechanism must be involved in the global mechanism in order to drive proton to the activated dioxygen. The main route seems to be correlated to the presence of highly conserved residues, such as Thr252 and Asp251.[211]

The binding pocket of the camphor substrate is mostly composed of hydrophobic residues. Only the hydroxyl group of Tyr96 directly contacts the camphor substrate. Other residues, such as Val295 and Val247, Phe87, and Thr185 impose several constraints by orienting the substrate relatively to the iron-bound oxidizing species and by restricting its mobility.

Cyt-P450s are a good example of how the noncovalent tuning of the secondary coordination sphere, mainly by hydrophobic and hydrogen bond interactions, drive the heme toward a particular chemistry. This is particularly evident from site-directed mutagenesis studies: In order to convert peroxidases into Cyt-P450s, or vice versa, simply swapping the axial ligands is not sufficient. First, steric differences between the two ligands and the local network of H bonds impose restriction to the design. Further, in Cyt-P450s, the Cys ligand acts as H bond acceptor from a main chain NH, while in peroxidase the His ligand acts as an H bond donor. Therefore, the local electrostatic environments of the two enzymes are totally different, and a plethora of interactions should be constructed in order to change one function into the other.

11.4.4 Carboxylate-Bridged Diiron Metalloproteins

Carboxylate-bridged diiron metalloproteins represent a further example of how a similar iron prosthetic group can accomplish different roles, depending on the protein environment (see Fig. 11.22).

Diiron sites are found in a functionally diverse class of proteins involved in oxygen binding and activation.[212–218] These proteins, referred to as diiron carboxylate proteins, share a number of structural and mechanistic commonalities. In the majority of these enzymes, the dimetal sites are bridged by a combination of oxo, hydroxo, or carboxylate donors, and are housed within a very simple four α-helix bundles; two histidine and four carboxylate ligands represent a mostly conserved protein-derived ligand set. The only exception is Hr, which contains in its active site a high number of His ligands. The function of diiron carboxylate proteins range from dioxygen transport and

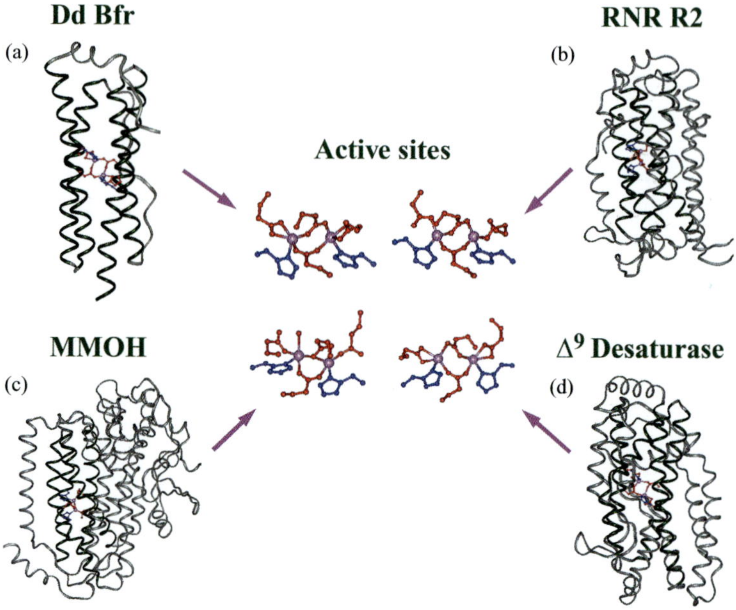

FIGURE 11.22. Four helix bundle motif housing the diiron site and details of the metal binding site in (a) bacterioferritin (PDB code 1NF4),[224] (b) RNR-R2 (PDB code 1XIK),[238] (c) MMOH (PDB code 1FYZ),[249] and (d) Δ^9-desaturase (PDB code 1AFR),[264] as derived from X-ray crystallography.

activation, to phosphoryl transfer, to iron storage.[212–216] Deeply characterized members of this class of proteins include **soluble methane monooxygenase** (MMO)[218] and the growing subclass of bacterial multicomponent monooxygenases (including toluene monooxygenases, phenol hydroxylase, alkene monooxygenase), which hydroxylate a variety of organic substrates[219–221]; the **ribonucleotide reductase R2 subunit** (RNR-R2), which generates a tyrosyl radical essential for the reduction of ribonucleotides to deoxyribonucleotides in DNA biosynthesis[214]; the stearoyl-acyl carrier protein (ACP) Δ^9-desaturase (Δ^9-desaturase), which introduces a double bond into saturated fatty acids[222]; **ferritins** and **bacterioferritins**, which use iron as a substrate for ferroxidation and iron storage[223,224]; and **Hr** and **myohemerythrin**, which reversibly bind and transport oxygen.[225] More recently, the diiron carboxylate protein family broadened to include four membrane-associated enzymes, identified on the basis of six conserved amino acids, which constitute the iron binding motif. These proteins are **alternative oxidase** (AOX), **plastid terminal oxidase** (PTOX), **5-demethoxyquinone hydroxylase** (DMQ hydroxylase), and **Mg-protoporphyrin IX monomethylester hydroxylase** (MME hydroxylase).[226–228]

The functional differences between all these proteins can be ascribed to a different mechanism in their interaction with dioxygen. In fact, most of these enzymes catalyze redox reaction by the use of dioxygen, binding it initially in the diferrous (bis-Fe^{2+}) state. The next step is the formation of a diferric peroxide intermediate, whose fate depends on the specific environment of the diiron site.[213,215,216] For example, in Hrs, the cycling between the diferrous and diferric/peroxo states is the basic mechanism for reversible oxygen binding. In some enzymes, such as MMOH (soluble methane monooxygenase hydroxylase subunit) and RNR-R2, the diferric peroxide intermediate evolves toward the formation of highly oxidative species.

How the diiron carboxylate proteins tune the properties of a single inorganic cofactor to obtain such a diversity of function is a question that is currently being addressed through spectroscopic and mechanistic studies.[229] Our knowledge about the diiron carboxylate protein family has expanded considerably in the last years as a consequence of the increasing availability of both sequence and structural data. All the available data allow to evaluate, at an atomic level, the structural features required for the accomplishment of specific catalytic functions. Similarly to the behaviors of the heme center in heme proteins, the main factors that specify the functions of diiron protein are an intricate set of interactions between the cofactor and the protein matrix. In the following section, a brief description of the most representative members of the diiron carboxylate protein family is reported, with a detailed analysis of the interactions that modulate their active site structure and function.

Dioxygen Activation at the Diiron Carboxylate Proteins: MMO, RNR, and Stearoyl-ACP Δ^9-Desaturase Among diiron carboxylate proteins, MMO, RNR-R2, and Δ^9-desaturase are very valuable for a comparative analysis. Even though a superimposition of their crystal structures shows similarity in their overall fold, their active sites show significant differences. In addition, their reaction with dioxygen gives rise to different intermediates, thus providing different chemistries.

MMO activates dioxygen to primary catalyze mono-oxygenation reactions.[213] They are found as multienzymatic complex in methanotrophic bacteria, where they catalyze the selective oxidation of methane to methanol, an essential step in the metabolic cycle of these organisms. MMOs are made up of three components: the hydroxylase subunit (MMOH), which houses the diiron site; the reductase component (MMOR), which accepts electrons from NADH and transfers them to the hydroxylase (for the reduction of the diiron site); and the regulatory protein (MMOB), which couples electron transfer to substrate oxidation.[213] The catalytic chemistry of this system is extraordinary in that it can catalyze the cleavage of the stable C–H bond in methane, the oxidation of a remarkable variety of substrates (saturated, nonsaturated, aromatic, heterocyclic halogenated hydrocarbons, etc.), and promote the formation of numerous chemically different products.[230]

RNRs are enzymes responsible for the conversion of ribonucleotides to their deoxyribonucleotide counterparts, thereby providing the precursors needed for both the synthesis and the repair of DNA. RNRs have been divided into four classes on the basis of their cofactors. Class I RNRs consist of a homodimer of two subunits, R1 and R2. The RNR-R2 subunit contains the diiron site and activates dioxygen for the generation of a stable tyrosyl radical.[231] This radical is essential to the overall mechanism of reduction of ribonucleotides, which occurs at the R1 subunit active site. This site contains the substrate binding and redox-active cysteine residues that are involved in the reduction of nucleoside diphosphates (NDPs).[214,217,232–238]

Finally, Δ^9-desaturase performs dehydrogenation reactions by activating molecular oxygen at the diiron site for introducing double bonds into fatty acids. Δ^9-Desaturase is a key enzyme of fatty acid synthetic metabolism in higher plants. Located in plastid stroma, it catalyzes desaturation of stearoyl-ACP, introducing a double bond into the fatty acid chain between *C*9 and *C*10 to form oleoyl-ACP.[239]

Despite their functional diversity, MMOH, RNR-R2, and Δ^9-desaturase proteins share a common mechanism, which occurs during their cycling through the different diiron oxidation states. In all these proteins, the ligating modes of the active site carboxylate ligands have been shown to be flexible, giving rise to a carboxylate shift phenomenon, occurring upon changes in the oxidation state of the diiron site. The catalytic relevance of the carboxylate shifts, and the interactions involved in this phenomenon, will be highlighted in the following.

The MMOH dimetal site, rich in carboxylates, tends to consume oxygen rather than binding it reversibly. During the catalytic cycle, a variety of intermediates has been proposed and numerous spectroscopic techniques and theoretical calculations have been used to identify these compounds.[240–246] MMOH binds O_2 in the diferrous state, leading to a diferric intermediate with a symmetrical bridging peroxo group.[240–242] Then, the site evolves through several reactions, giving rise to a reactive high-valent iron-oxo intermediate diferryl (bis-Fe^{4+}), named compound Q.[243]

The crystal structure of MMOH has been obtained in different crystal forms and oxidation states and in the presence of various substrates and products.[247–255] In all the structures of MMOH oxidized diiron(III) form, both irons have a six-coordinate distorted octahedral environment (Fig. 11.23a),[254] and reside in the center of a four-helix bundle, which provides two EXXH motifs. Fe1 is coordinated to the δ1-nitrogen atom of a histidine residue, His147, a monodentate carboxylate, Glu114, and a terminal water molecule. Fe2 is ligated by His246, two monodentate carboxylates, Glu209, and Glu243. The terminal water molecule coordinated to Fe1 forms a hydrogen bond with the uncoordinated oxygen atoms of both Glu114 and Glu243. The two iron ions are bridged by Glu144, a hydroxide ion and a third ligand that is variable in the different crystal structures. Depending on the crystal form, the species and the temperature of data collection, bridging water,[242] hydroxide,[243] $H_3O_2^-$,[249] acetate,[241] and formate[249] have been reported as the third ligand. Changes in the structure occur upon reduction of the binuclear iron cluster to the diferrous [Fe^{2+}-Fe^{2+}] form (Fig. 11.23b).[252] One specific ligand, Glu243, modifies its coordination mode, giving rise to the so-called "carboxylate shift." Glu243

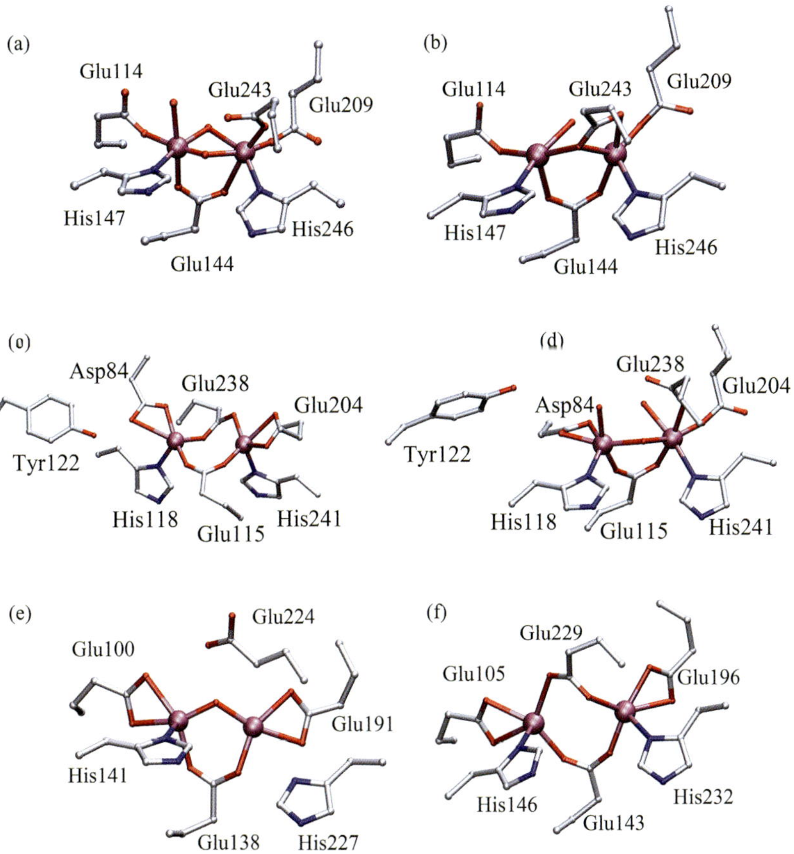

FIGURE 11.23. Active site structures of carboxylate-bridged diiron proteins. (a) Diferric (PDB code 1MTY)[251] and (b) diferrous (PDB code 1FYZ)[249] form of MMOH from *Methylococcus capsulatus* (Bath). (c) Diferric (PDB code 1XIK)[238] and (d) diferrous (PDB code 1RIB)[237] form of RNR-R2. (e) Diferric form of Δ^9-desaturase from *Hedera helix* (ivy) (PDB code 2UW1)[266] and (f) diferrous form of Δ^9-desaturase from *Ricinus communis* (castor) (PDB code 1AFR).[264]

displaces the bridging hydroxide ion and becomes a bidentate chelating ligand to Fe2, while it forms a monodentate interaction with Fe1. As a consequence, both iron atoms becomes five-coordinated. The ligand flexibility allows ligands to coordinate the Fe centers differently, depending on the enzymatic needs: as bidentate ligands (bridging or chelating), when saturation of the first coordination sphere of Fe ions is required, and as monodentate (terminal) ligands, when one (or more) vacant coordination site(s) in the Fe center are needed (e.g., upon dioxygen or substrate coordination) to enable a certain reaction step to take place. The consequence of these structural changes is that both five-coordinated iron(II) ions react with O_2.

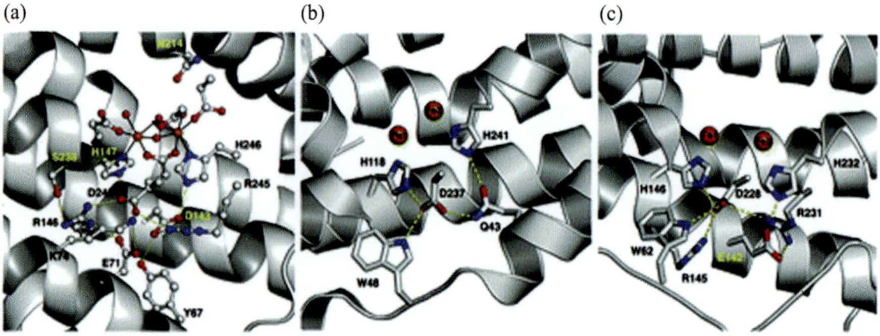

FIGURE 11.24. Hydrogen bonding network behind the diiron centers of (a) MMOH, (b) RNR-R2, and (c) Δ^9-desaturase. (Reprinted with permission from Sazinsky and Lippard.[261] Copyright 2006 American Chemical Society.)

The diiron center is located on the surface of a hydrophobic cavity, which mainly contains hydrophobic side chains, with the exception of Cys151 and Thr213. The occurrence of Cys151 at this location suggests a possible role for the sulfhydryl moiety as a one-electron source and/or as a proton donor. The threonine may also participate in the oxidation reaction by functioning as a proton source. Such a role for Thr213 would be mechanistically analogous to that of Thr302 in the related heme oxygenase, cytochrome P450-Δ2B4.[256] Furthermore, a hydrogen bonding network between residues belonging to the second coordination sphere, is observed (Fig. 11.24a). The network extends from the two iron-coordinating histidines, His147 and His246, to the surface of the active site. The involved residues are Asp143 and Asp242, each of which is hydrogen bonded to the $N^{\varepsilon}H$ of the histidine ligands,[247] and to surface residues, Tyr67 and Lys74.

The iron center in the R2 subunit of class I RNR is employed specifically for the activation of the Tyr122 radical and does not play a role in the subsequent catalytic chemistry that leads to nucleotide reduction. Tyr122 functions to initiate multiple-turnover, one-electron oxidation of a remote cysteine in the R1 subunit. The tyrosyl radical and the binuclear iron center are deeply embedded within the protein, far away from the active site of R1 (~35 Å). Therefore, long-range electron transfer is thought to occur to form a cysteinyl radical on a cysteine residue (Cys439) at the R1 substrate binding site.

The coordination geometry at the diiron core of R2 resembles that of MMOH in that each R2 domain houses a dinuclear iron cluster, buried in a four-helix bundle. As in the case of MMOH, the iron site is coordinated by two histidines and four carboxylate side chains in the two (D/E) XXH motifs, the same motifs observed in MMOH. In addition, in R2 the diferric ions are coordinated by two water molecules and a bridging oxo group, whereas they are μ-hydroxo bridged in MMOH. The coordination of one iron ion is close to a regular octahedron, whereas the coordination of the other iron is more distorted due to the bidentate binding of Asp84 (Fig. 11.23c).[238] The two His ligands bind to the iron ions through their δ-nitrogen atoms and are involved in a hydrogen bonding network through their N^{ε} atoms. His118 forms a hydrogen bond with one carboxylate oxygen atom of Asp237, which is also involved in a hydrogen bond with the ε-nitrogen of Trp48. The other side chain oxygen of Asp237 forms a hydrogen bond to Gln43, which is itself hydrogen-bonded to His241.[257]

A striking similarity between the binuclear iron clusters of R2 and MMOH is the change in ligand environment occurring upon reduction of the diiron site. As observed for MMOH, the structure of the reduced R2 form has a very different coordination environment with respect to the oxidized form (Fig. 11.23d),[237] thanks to the "carboxylate shift." The reduction of the diiron site causes an increase in the Fe1-Fe2 distance, a loss of the μ-oxo bridge and a rearrangement of one of the bridging carboxylates Glu238. This residue moves from terminal in Fe2 to a bridging position for both iron ions. In addition, the Asp84 undergoes a shift from bidentate to monodentate terminal position to Fe1, a shift that has no analog in the counterpart reduction of MMOH. The different coordination mode of Asp84 brings it closer to the catalytically functional Tyr122, thus facilitating electron transfer between them. The formation of the Tyr122 radical is the first step in the overall reaction of RNR. This residue is about 5.3 Å away from Fe1, and is positioned into a hydrophobic pocket. Several investigations on mutant proteins demonstrate that this hydrophobic environment stabilizes the tyrosyl radical by shielding it from potential electron donors not involved in the normal electron transfer pathway.[258]

The carboxylate shift occurring in R2 generates a highly symmetric reduced diiron site, with an unusually low coordination number: two carboxylates bridge two four-coordinate ferrous centers; the coordination sphere about each metal ion is completed with a single monodentate carboxylate and one imidazole donor.[259] The iron coordination geometry can be described as distorted tetrahedrons, whereas in the reduced form of MMOH the ferrous irons are five-coordinated.

There are also some noticeable differences in the active site environment of MMOH and R2.

The RNR binuclear active site is tightly packed and situated in a deep cavity far away from the protein surface. This feature prevents the formation of possible binding pocket for substrates, and this may account for the lack of monooxygenase activity for R2 with respect to MMO, despite the similarity of their iron centers. Regarding the RNR catalytic mechanism, it is believed that MMOH and RNR share a common path of intermediates in the first steps.[259,260] The presence of open coordination positions on both iron ions in the reduced form of R2 allows the binding of dioxygen in a bridging mode. Subsequently, in analogy with the MMOH cycle, the formation of a diferric peroxide intermediate has been suggested, even though this intermediate has only been detected in some R2 mutants (R2-D84E and R2-W48F/D84E) from *Escherichia coli*, which have exactly the same iron ligands as MMOH.[241] The diferric peroxide intermediate evolves to generate intermediate X, which has been identified to contains a μ-oxo/μ-hydroxo [FeIII(μ-O)(μ-OH)FeIV] core structure, similar to that proposed for the intermediate Q of MMOH. To complete the catalytic cycle, the presence of another intermediate, labeled U, has been hypothesized. The identity of this intermediate has not been determined unambiguously, but the available data are consistent with the presence of a cationic tryptophan radical,[257] which would supply the fourth electron needed for the complete reduction of dioxygen. The best candidate for this function is Trp48, which is close to the surface of the protein. Furthermore, Trp48 is also coupled to the iron ligand His118 through hydrogen bonds with Asp237. This finding demonstrates how the second and also the third coordination shells "communicate" with the diiron site, thus assisting in catalytic function (Fig. 11.24b).

The hydrogen bonding network in RNR-R2 is similar to that observed in Δ^9-desaturase, even though no striking evidences are until now available for the presence of a radical in the site of Δ^9-desaturase (Fig. 11.24c).[261] Crystal structures are available for the diferrous Δ^9-desaturase from *Ricinus communis* (castor)[262–264] and for the diferric form from *Hedera helix* (ivy).[265,266] Each monomer of the castor enzyme homodimer contains a boomerang-shaped cavity, adjacent to the diiron site, capable of binding a stearoyl moiety in a gauche conformation. The Fe^{2+}-Fe^{2+} active state of the enzyme resembles that of MMOH (Fig. 11.23f).[264] The site shows a highly symmetrical structure: Each iron ions is five-coordinate, with a distorted square pyramidal geometry. Fe1 is coordinated by Glu105 in a bidentate manner and by His146. Two carboxylates, Glu143 and Glu229, bridge the two iron ions, whereas Glu196 and His232 are bidentate and monodentate ligands to Fe2, respectively. A water molecule is observed in the vicinity, but not at coordinating distance, of the diiron site (at 3.0 Å from Fe1 and at 3.3 Å from Fe2). The crystal structure of castor desaturase has been obtained only in the reduced Fe^{2+}-Fe^{2+} form, presumably as a result of photochemical reduction under the X-ray beam. A detailed knowledge of the Fe^{3+}-Fe^{3+} resting state of the enzyme has been derived from the X-ray structure of the ivy desaturase.[265,266] The diferric form of ivy desaturase is characterized by a shorter distance between the two irons, with respect to the reduced form (~3.2 Å vs. ~4.2 Å) (see Fig. 11.23e).[266] Further, a μ-oxo bridge is observed between the two ions. The decreased iron–iron distance is strictly related to changes in the coordination sphere of the diiron site. Fe1 coordinates His141 and makes a bidentate interaction with the Glu100 side chain. Glu138, which is equivalent to the Glu143 in the castor structure, forms a bridge across the two irons. Fe2 interacts directly with only two residues, Glu138 and Glu191, which act as monodentate and bidentate ligand, respectively. Differently from the His232 in the castor structure, His227 is not a ligand in the ivy structure, and instead weakly interacts with the iron via a water molecule located 2 Å from Fe2. The remaining coordination position in the coordination sphere of each iron is occupied by a bidentate small ligand, derived from the expression medium. The most significant change in residue position between the oxidized and reduced active sites is a shift in the side chain of Glu224, corresponding to Glu229 in castor desaturase. As already observed for MMOH and R2, Glu224 undergoes a large "carboxylate shift" in the oxidized structure to leave the side chain pointing away from the diiron center, and making no interaction with either iron. Glu224 has pulled completely away from the diiron center to form hydrogen bonding interaction with outer-sphere His198. The conformational rearrangement of Glu224 side chain creates space for the formation of the μ-oxo bridge. One side chain oxygen atom of Glu224 interacts with Thr194, a residue just outside of the active site. A conserved threonine at this position is also observed in the structure of MMOH, and it has been suggested to play a role in proton transfer. The role of Thr194 in desaturase catalysis is unclear, but mutation of the corresponding Thr199, in the castor enzyme, to an aspartic and glutamic acid resulted in a switch from desaturase to oxidase chemistry.[267] Therefore, the conformational change in Thr194 between the oxidized and reduced desaturase structures and its potential interactions with the oxidized active site suggest that Thr194 is potentially involved in the catalytic cycle of the desaturases.

The analysis of these three enzymes highlights how similar sites may play different roles. However, subtle differences are needed in order to account for different reactivity and reaction mechanism.

Dioxygen Reversible Binding at Diiron Site: Hr One of the first described diiron carboxylate protein, Hr,[225] reveals some unique features not observed in the other proteins of this class later discovered. Hr can bind and release O_2 reversibly and functions as an oxygen carrier protein similar to hemoglobin in vertebrates; it is the only O_2 carrier in the diiron carboxylate protein family. Hr typically occurs in marine organisms, but bacterial fusion-domain homologues are also known.[225,268–270] Recently, Hr-like domains located in the bacteria have been identified and characterized.[271,272] The main differences between Hr and the other diiron proteins is that the Hr metal binding site is relatively rigid and rich in His ligands. This environment specifically stabilizes both the diferrous and diferric/peroxide states and constitutes the molecular basis for reversible oxygen binding.

The Hr diiron site comprises five histidine and two carboxylate residues (aspartate and glutamate) (Fig. 11.25). Three histidine residues bind to one iron, and two bind to the other. The glutamate and aspartate residues bridge the metal ions, in a μ1,1,3 mode, and a water molecule is bound to both iron ions.[225] There are also interesting differences in the His binding mode: In Hr, all of the histidine residues bind through their ε-nitrogen atoms, while in other diiron proteins, the δ-nitrogen is bound to the metal.

Diferrous and diferric forms of Hr have been characterized both spectroscopically and crystallographically.[225,271,273–276] The fully reduced deoxy state (deoxyHr) maintains one coordinatively saturated six-coordinate ferrous ion (Fe1) and one five-coordinate center (Fe2) (Fig. 11.25). Binding and concomitant two-electron reduction of O_2 occur at Fe2, the open coordination site.

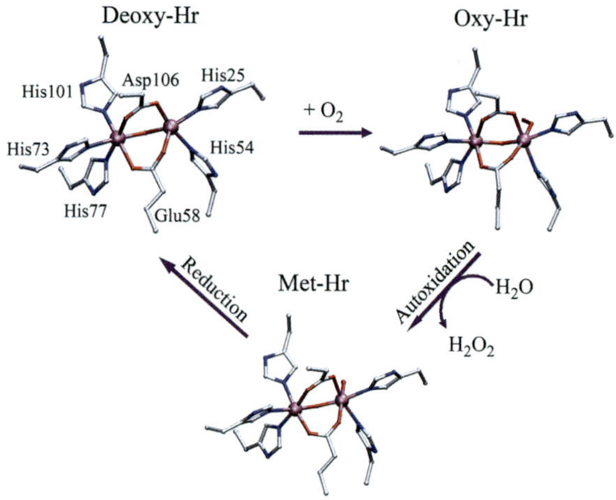

FIGURE 11.25. Three different states of hemerythrin as derived from crystal structures: deoxyhemerythrin (PDB code 2AWC),[271] oxyhemerythrin (PDB code 1HMO),[276] and methemerythrin (PDB code 2AWY).[271]

Dioxygen reacts at the iron site and generates oxyhemerythrin, a species that is best described as a peroxo adduct, with the iron ions in the ferric oxidation state.[275] Thus, the Hr diiron site allows only terminal O_2 binding, whereas, as described in the previous section, the O_2-activating diiron sites allow O_2 bridging of the irons. Hr is the only diiron carboxylate protein for which the nature of the O_2 adduct is well established. This is due mainly to the high stability of the O_2 adduct at room temperature.

Another structural difference between Hr and RNR-R2/MMOH diiron sites lays in their different coordination flexibility. The flexibility of RNR-R2 and MMOH sites, occurring through the carboxylate shift phenomenon, determines significant conformational changes, when these proteins interact with dioxygen, and constitute an important functional adaptation of the site for oxygen activation.

In Hr, the oxygen binding does not lead to any significant conformational change: A very small rearrangement of the diiron coordination sphere occurs during the interconversion of the deoxy-, oxy-, and met- forms of the protein.[225]

11.5 METALLOPROTEIN DESIGN AND ENGINEERING

As highlighted in the previous sections, almost half of all known proteins contain metal cofactor(s) that are able to serve numerous roles, either structural or functional, such as nitrogen fixation, molecular oxygen reduction, and water oxidation.[2,18,19,277–280] It is remarkable that, despite the variety of functions in which metalloproteins are involved, the number of utilized metal-based prosthetic groups is relatively small. It appears that both catalytic activity and stability of the metal cofactor are regulated by different mixed and specific interactions with the protein matrix, ranging from coordinate and hydrogen bond, to hydrophobic and ionic interactions.[13,19,281–287]

As described in the previous section, the protein matrix shapes the primary coordination shell of the metal ion dictating the composition, number, and geometry of the ligands. In addition, residues in the second shell, the immediate surroundings of the primary coordination sphere, influence a variety of structural and chemical facets, including hydrogen bonding interactions to the ligands, pK_a values of the ligand, metal center oxidation state, and redox potential. Finally, neighboring side chains also exert steric and chemical controls over the ability of the metal ion(s) to bind or discriminate the substrates and to accommodate conformational changes. Many of these interactions have been described in other chapters of this book. In particular, sequences involving complementarily charges residues (Chapters 4 and 6) or apolar substituents (Chapters 6 and 7) promote compact conformation that affect the local dielectric constant, and thus the ionization and the strength of ionic interaction (Chapter 12), metal center oxidation state, and redox potential. Suitable sequences may thus become involved in hydrogen bonds (Chapter 12). The ion pair formation between two fixed charges on the protein along the same or different protein chains promotes inter- or intramolecular associations that might compete with the coordination by a metal ion. The occurrence of salt bridges may have a significant influence on the stabilization of specific conformations, as discussed in Chapter 12.

Understanding at a molecular level the mechanism by which the protein matrix finely tunes the environment of the metal cofactor, thus producing such different chemistries, is of fundamental importance in both basic and applied science. In fact, this would shed light on the detailed mechanisms of fundamental biological processes, and would allow researchers to construct new metalloproteins with desired properties and functions.

Today, several fundamental aspects of metal cofactor assembly and metalloprotein functions have been clarified, mainly through structural-functional studies on natural metalloprotein and related mutants, generated by site-directed mutagenesis.[278,288–293] For example, mutation studies have shown that small changes to an enzyme structure can have large effects on its catalytic properties.[288,289] Chemical modification of a single amino acid residue can profoundly affect the activity of the enzyme, and a single-site mutation can alter its enantioselectivity.[288,289]

Despite these progresses, several questions are still open. While the control of the functional specificity by the primary coordination sphere is almost well understood, the contribution of medium- and long-range interactions is still not completely rationalized, and their molecular description remains to be determined. Thus, it is not surprising that many efforts are being devoted to the development of metalloprotein mimics aimed to (1) provide further insights for structure-activity relationships; (2) understand the minimal requirements for function; (3) reproduce the properties of the parent natural proteins in smaller molecules, with improved properties with respect to the natural counterpart, such as higher stability and greater efficiency; and (4) most importantly, construct new, tailor-made molecules useful for biomedical, pharmaceutical, biological, and environmental applications.

Over the years, a large number of low–molecular-weight metal complexes, containing designed organic ligands, has been developed as metalloenzyme mimics. They have been basic in elucidating structure and function of metalloproteins and metalloenzymes. Some of them are key components for chemical industries in that they are able to catalyze the same reactions catalyzed by biological enzymes, such as conversion of nitrogen to ammonia, methane hydroxylation, and CO oxidation or insertion.[277,294–296] Synthetic catalysts and biological enzymes operate under different conditions, the first generally at high temperatures and pressures in organic solvent, whereas biological systems work in water, at ambient temperatures and pressures, as schematically depicted in Figure 11.26. As clearly outlined by Lu in a recent review,[277] chemical catalysts and biocatalysts have their own advantages and disadvantages. Chemical catalysts are generally smaller, can be produced in high quantity and at low cost through chemical synthesis, and possess high stability and resistance to severe conditions of temperatures and pressures. On the other hand, chemical catalysts often fail in reproducing several features of biocatalysts, such as high turnover number under mild conditions and high selectivity. Further, biocatalysts are synthesized and operate under environmentally

	NITROGEN FIXATION
CHEMICAL SYSTEM	*(α-Fe catalyst, 400-550 °C, 100 atm)* $N_{2(g)} + 3H_{2(g)} \rightarrow 2NH_{3(g)}$
BIOLOGICAL SYSTEM	*(nitrogenase)* $N_2 + 8H^+ + 8e^- + 16MgATP \rightarrow 2NH_3 + H_2 + 16MgADP + 16Pi$
	METHANE HYDROXYLATION
CHEMICAL SYSTEM	*(Ni catalyst, 700-900 °C, 1-25 bar)* $CH_4 + H_2O \rightarrow CO + 3H_2$ *(Cu/Zn catalyst, 250-280 °C, 70-110 bar)* $CO + 2H_2 \rightarrow CH_3OH$
BIOLOGICAL SYSTEM	*(methane monooxygenase)* $CH_4 + O_2 + NADH + H^+ \rightarrow CH_3OH + H_2O + NAD^+$
	CO OXIDATION
CHEMICAL SYSTEM	*(Fe/Cu catalyst, > 200 °C)* $CO + H_2O \rightarrow CO_2 + H_2$
BIOLOGICAL SYSTEM	*(CO dehydrogenase)* $CO + H_2O \rightarrow CO_2 + 2H^+ + 2e^-$

FIGURE 11.26. Comparison between chemical and biological catalysts in selected reactions. (Adapted from Lu.[277])

benign conditions. Combining the most desirable features of synthetic and biological catalysts would represent a formidable goal for chemists. Despite the tremendous progress made in this field, the functional specificity of the natural systems has not been completely achieved. Therefore, it still is a daunting challenge to reproduce in artificial molecules the intricate control mechanisms orchestrated by natural systems, and hopefully their efficient regio-, stereo-, and enantioselectivity.

11.5.1 Strategies in Metalloprotein Design

Due to the inherent complexity of metalloprotein structures, tailoring synthetic models requires the development of sophisticated molecular architectures that distill the quintessential elements believed to be responsible for the activities. Thus, peptide- and protein-based models seem better candidates to mimic both the structural characteristics and reactivity of the natural systems. Metalloprotein design is a formidable task; in fact, it involves two issues to be addressed simultaneously: (1) the correct fold of the protein chain and (2) the coordination requirements of the metal ion.[278,297] Concerning the first point, it is important to outline that numerous interactions contribute to protein folding: Hydrophobic interactions, with a perfect shape complementarity, provide a powerful driving force for folding, whereas polar interactions are essential for conformational specificity.[298–301]

For the construction of the metal binding site, the following demanding aspects should be considered.

- An appropriate number of coordinating residues should be placed around the metal center, in the proper position and with the correct side chain conformation, in order to satisfy the coordination geometry dictated by the metal.
- Secondary shell interactions should be carefully considered. They refer to a network of interactions between the coordinating residues and the surrounding protein matrix. Secondary shell interactions occur, for example, if the coordinating residues have more than one functional group capable of interacting with the metal ion (such as His, Asp, Glu; see Fig. 11.3). One of them will bind the metal ion, leaving the other free to form other interactions, usually H bonds, with the protein backbone, with side chains, and occasionally, with water molecules. They are very important in positioning the ligand in the correct orientation for binding and in modulating the properties of the metal site.

A further level of complexity for the successful design of metalloproteins implies focusing the stability/function trade-off, which is essential for structure and function. Thus, many other key interactions, including short- and long-range interactions, should accurately be considered.

- Structural metal binding sites in proteins frequently achieve stability by binding metal ions in coordinately saturated ligand environments, with idealized ligand–metal bond geometries (see Section 11.3.2).[13,19,302]
- Functional sites in metalloenzymes frequently contain coordinately unsaturated metal ions, sometimes with unusual geometries, and properly positioned for binding substrates and undergoing dynamic changes in configuration during catalysis.[13,19,302] Thus, the construction of metalloenzyme active sites reflects a delicate interplay between opposing requirements, for tight binding of the metal cofactor versus function, as it will be also delineated in the next sections by analysis of artificial systems.
- The active sites of metalloproteins are frequently preorganized in the absence of metal ions, which requires the burial of polar ligands at the expense of folding free energy.[303] The preorganization imparts tight and geometrically specific metal binding by assuring that the protein imparts its own structural preferences on the metal rather than vice versa.

Considering the above mentioned aspects, it appears that metal ion coordination by amino acid side chains imposes strong structural constraints on the surrounding residues due to the well-defined bond distances and angles required to fulfill the metal coordination geometry. As a consequence, the introduction of a metal binding site into folded scaffold may disrupt its well-defined tertiary structure. The simultaneous optimization of the hydrophobic interactions, which constitute the main driving force for folding, and of the metal ligand interactions critically test our understanding of the principles of protein folding; further, it will allow to succeed with the design of any desired metalloprotein.

The increasing number of structural data nowadays available for metalloproteins and synthetic analogs, together with advances in biophysics and computational biology, has deeply contributed to the significant progress in the field of metalloprotein design, which is successfully moving toward the development of functional metalloprotein mimics. Numerous excellent reviews on metalloprotein design and engineering has recently appeared in the literature, which also discuss the different strategies applied to designing mimics of metal binding site.[277,278,297,299,304–317]

Metalloprotein models can be designed rationally, on the basis of available information on protein structure, in particular at the metal binding site. This *structure-based* strategy basically includes three different approaches.

The first approach involves a miniaturization process, suitable to build the minimal peptide sequence, which contains sufficient information for proper folding and for an accurate reconstruction of the active site structure.[305,318] This approach holds the advantage that the designed systems are generally simple enough, and therefore they can be easily synthesized and characterized, and simultaneously the polypeptide sequences are of sufficient size and chemical diversity to accommodate the metal binding site. The miniaturization process can be rationally organized once a structural knowledge of the parent natural

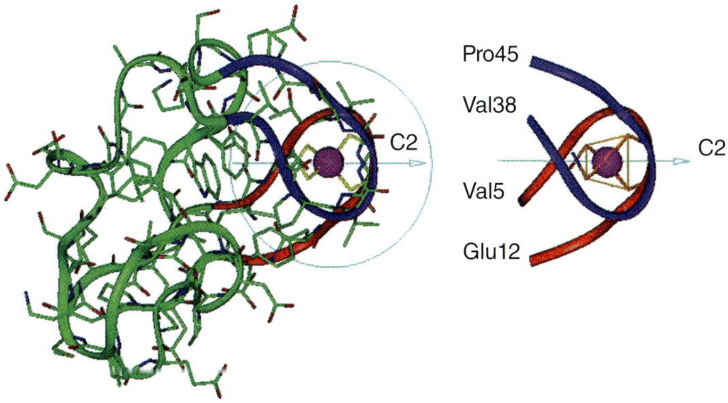

FIGURE 11.27. The full structure of rubredoxin (left) (PDB code 8RXN),[319] whose active site can be dissected in two symmetry-related idealized β-hairpins, residues 5–12, and 38–45 (right).

system to be miniaturized is available. It is necessary to define (1) the type and number of constituents to be assembled, (2) the structure to be reconstructed, and (3) the function to be reproduced. These aspects are strictly related. Metalloproteins are well suited to be miniaturized. As illustrated in Figure 11.27, the metal center represents a pivot point where spheres of variable diameters that circumscribe part of the protein are centered.[318,319] The larger the diameter of the sphere, the larger the number of constituents included in the model. For a sphere of a given diameter around the metal center, the constituents therein circumscribed are, in general, unconnected parts. Several strategies can be conceived to obtain a miniaturized model. Two or more parts could spontaneously associate to give folded noncovalent self-assembled oligomers, or two or more parts could be covalently connected to give folded monomers.

An alternative approach uses native protein scaffolds, in which a metal binding site can be engineered.[277,278,288] This approach entails a careful choice of the scaffold, which should tolerate the mutations required for the construction of the metal site. In some cases, the exchange of a single residues have been shown to disrupt the well-defined tertiary structure of folded proteins, forming molten globule-like folds.[300] Thus, the replacement of several residues, required for the construction of a metal binding site, may be detrimental to the structure. The use of native scaffolds with a stable secondary and tertiary structure allows less concern about the decreased stability that may derive from the introduction of the liganding residues. This may have the great advantage of a simpler design, as it mainly focuses on reproducing the metal binding site architecture into the template protein.[277,278]

The *de novo* design strategy has been successfully applied to the development of metalloprotein models.[298,299,301] In its purest and most challenging form, *de novo* metalloprotein design involves the construction of a polypeptide

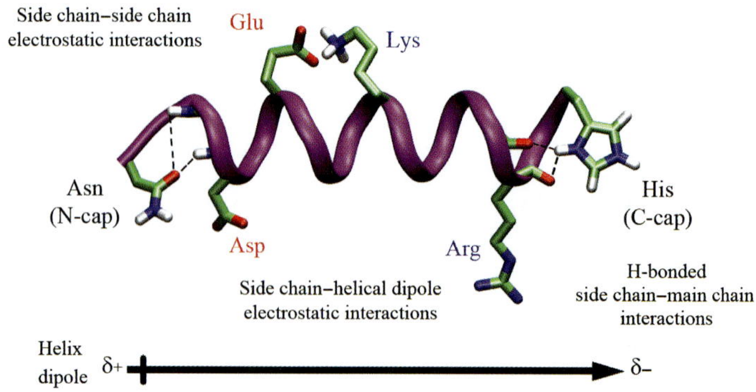

FIGURE 11.28. H bonds and electrostatic interactions for the α-helical conformation. (Adapted from Bryson et al.[321])

sequence, which is not directly related to any natural protein, intended to fold into a well-defined three-dimensional structure and to incorporate a metal cofactor into a precise geometry. The success in obtaining uniquely folded proteins through *de novo* design imply the introduction and optimization of the large number of short- and long-range noncovalent interactions, which all together contribute to the stability of a protein.[298,299,301,320,321] These include the correct placement of hydrophobic and hydrophilic amino acids into the sequence, H bonds, and electrostatic interactions, the intrinsic conformational propensity of each amino acid. These interactions are schematically depicted in Figure 11.28 for the α-helical conformation.[321]

The relative contribution of each noncovalent interaction should be carefully considered, and many efforts in *de novo* design have been devoted to measuring the energetic contribution of each interaction.[299] For example, relative to Ala, a buried Leu or Phe residue contributes ~2–5 kcal mol^{-1} to stability, and buried H bonds or salt bridges further increase stability by as much as 3 kcal mol^{-1}, expressed on a per-residue basis.[322] Short-range (in sequence) H bonds between small polar side chains and backbone amides can be stabilizing by 1–2 kcal mol^{-1}, as in helix N-caps.[323] Exposed H bonds and salt bridges are less stabilizing; nevertheless, they have been demonstrated to be important for conformational specificity.[299] Finally, each residue in a protein contributes its intrinsic conformational preference for adopting a given set of ϕ, ψ torsional angles.[324–326] With the exception of Pro, the energetic contribution of this effect spans a range of ~1 kcal mol^{-1} for the commonly occurring amino acids. Thus, the information required for folding is distributed throughout the chain in a network of diverse interactions, each with different energetic contributions. All types of interactions need to be included and optimized for a successful design, and no single residue or interaction type may be sufficient "to determine the fold."

Another important aspect of a successful design strategy is the concept of negative design, through which a uniquely folded structure can be favored by destabilizing alternative folds, thus providing conformational specificity. For example, inclusion of a large number of hydrophobic residues in a designed protein can lead to multiple, stable conformations; therefore, a correct balance between apolar and polar interactions are demanded for folding and function.[298,299,327,328] Examples of the importance of the use of polar residues in negative design include (1) the placement of hydrophilic groups that are solvent exposed in the desired fold but buried in alternatively folded structures[329]; (2) inclusion of buried H bonds and salt bridges that are less stabilizing than hydrophobic interactions, but can be formed only in the desired structure[330–333]; and (3) solvent-accessible electrostatic interactions that are very weakly stabilizing of the desired fold but strongly destabilize alternatives.[334–337] Other elements of negative design include conformational restraints from residues such as Gly and Pro that are often used to interrupt elements of secondary structure. Shape complementarity and, more important, a lack thereof provide a third medium for negative design.

Because of the complexity of protein folding, a trial-and-error approach is often necessary for *de novo* designing a metalloprotein with a native-like structure. Through several cycles of design, synthesis, characterization, and redesign, it is possible to fine-tune the structural properties of the initial model, and to tailor a functional metal site into the interior.[298,338,339] Some illustrative examples of models developed by us will clarify this concept in the next sections.

All the above illustrated strategies take advantage of the growing number of computational procedures for structure and sequence prediction, which are enormously contributing to design uniquely packed structures and metal binding sites within either *de novo* or native protein scaffolds. Computer algorithms have been developed to redesign the hydrophobic core of either native or *de novo* designed proteins, and also to design interfacial and fully solvent-accessible regions of protein structure. Finally, computational algorithms have been developed for designing a metal binding site by searching protein structures for backbone geometries suitable to accommodate the liganding side chains in the appropriate conformation for metal binding.[340,341]

In contrast to rational design, design through combinatorial and evolutionary methods does not require a prior knowledge of the protein structure. Protein with the desired structure and function can be selected from large pools, generated by binary patterning of hydrophobic/hydrophilic amino acids,[342–344] cassette mutagenesis, error-prone polymerase chain reaction, or gene shuffling techniques on the DNA for redesign.[288,289,345,346] Combinatorial approaches are powerful when combined with a selection scheme for an appropriate function.

Outstanding examples of designed and engineered metalloproteins, developed by using the different strategies herein described, have appeared in the literature over the years. Among these, worth mentioning are the results reported by Dutton and coworkers on the construction of an O_2 transport protein, made up of a de novo designed four helix bundle encompassing

(a) (b)

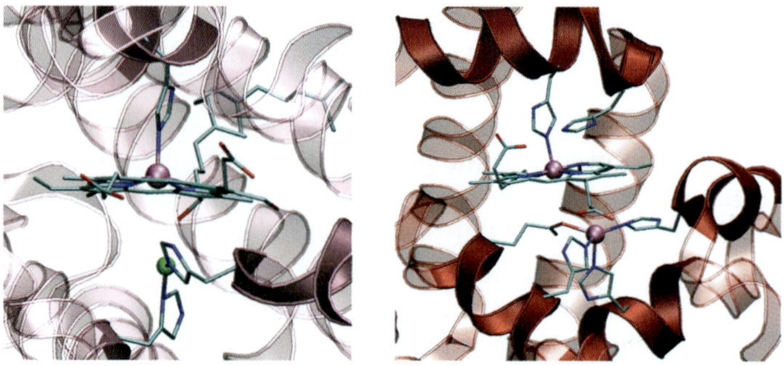

FIGURE 11.29. Designed metalloproteins using native scaffolds. (a) Catalytic heme CuB center in COX (PDB code 1AR1) and (b) the designed heme–copper model in sperm whale myoglobin (PDB code 3X9Z).[348]

bis-His ligated hemes.[347] Using native protein scaffolds, Lu and coworkers described, in an elegant contribution, the rational design of a structural and functional model of nitric oxide reductase (NOR).[348] This artificial protein was engineered by introducing three histidines and one glutamate, predicted as ligands in the active site of NOR, into the distal pocket of myoglobin, as illustrated in Figure 11.29.

We have approached the challenge of reproducing by design metalloprotein active sites. In particular, we centered our attention on iron-containing proteins, and we developed models for heme proteins (mimochromes),[349,350] iron-sulfur proteins[318] and DFs.[351,352] All these models have a common characteristic: They are fully or partially C_2 symmetric systems, intended to reproduce the quasi-symmetrical structure of a metalloprotein. The use of C_2 symmetry is particularly advantageous: It simplifies the design, synthesis, and structural characterization of the models, even though it leaves open the problem of possible diastereomeric forms, as it will be shown in the following description.

In this section, we will summarize our recent results on heme protein and DF models.

11.5.2 Mimochromes: Heme Protein Models

Over the years, a large number of peptide- and protein-based heme protein mimetics have been developed, using the above described strategies, in order to determine how the protein matrix tunes the properties of the heme to evoke the wide variety of activities. They differ in structural complexity, starting with simple microperoxidases,[353] chelated deuterohemes,[354] and peptide–porphyrin conjugates,[355,356] and progressing to more elaborate architectures, based on native[278,357] and *de novo* designed protein scaffolds.[347,358–363]

In developing peptide–porphyrin conjugates as heme protein mimetics (mimochromes), we applied a miniaturization process.[364] The miniaturization process, as already mentioned in the previous section, is a *structure-based* approach that can be applied once a structural knowledge of the parent natural system to be miniaturized is available. A detailed analysis of heme protein structures showed that the prosthetic group in natural heme proteins is strongly kept inside the protein structure by a large number of interactions, as illustrated in Table 11.5 for the deoxyhemoglobin. These interactions were replaced in mimochromes by a few strong local constraints. The prototype molecule, mimochrome I,[364] was patterned after the F helix of hemoglobin β-chain.[174] The smallest sequence, required for a complete coating of one face of the heme, was identified in a nine-residue peptide, which contains a central His residue to ligate the heme iron. Leu residues were placed at position $i - 4$ and $i + 4$ relative to the His (N- and C-termini) because of their high propensity to be accommodated in α-helical conformation[322] and to hydrophobically interact with the heme macrocycle. Two copies of the peptide were covalently linked to the porphyrin propionic groups through the ε-amino function of Lys,[8] obtaining a *pseudo* C_2 symmetric dimer (Fig. 11.30).[364] The spectroscopic and structural features of mimochrome I cobalt and iron complexes indicate that they bind heme iron in a low-spin bis-His-ligated complex as envisioned in the design, even though some unexpected features were observed.[364–366] The iron complex showed quite low solubility in water (in the μM range) in a wide range of experimental conditions.[364,366] Its detailed characterization was henceforth strongly limited. On the contrary, the cobalt complex was soluble in water (up to mM concentration), but two equally abundant isomers were obtained. Because of the flexibility of the linker between the peptide and the

(a)

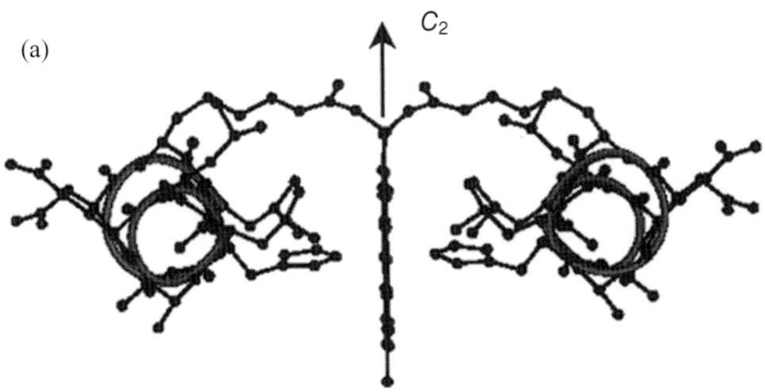

(b)

Mimochrome I	Ac-L-A-Q-L-H-A-N-K-L-NH_2
Hemoglobin Leu[88]-Leu[96]	L-S-E-L-H-C-D-K-L

FIGURE 11.30. (a) Molecular model of mimochrome I, showing the C_2 symmetry axis. (b) Comparison of the amino acid sequences of mimochrome I and of the Leu[88]-Leu[96] hemoglobin fragment. (Reprinted with permission from Lombardi et al.[349] Copyright 2001 American Chemical Society.)

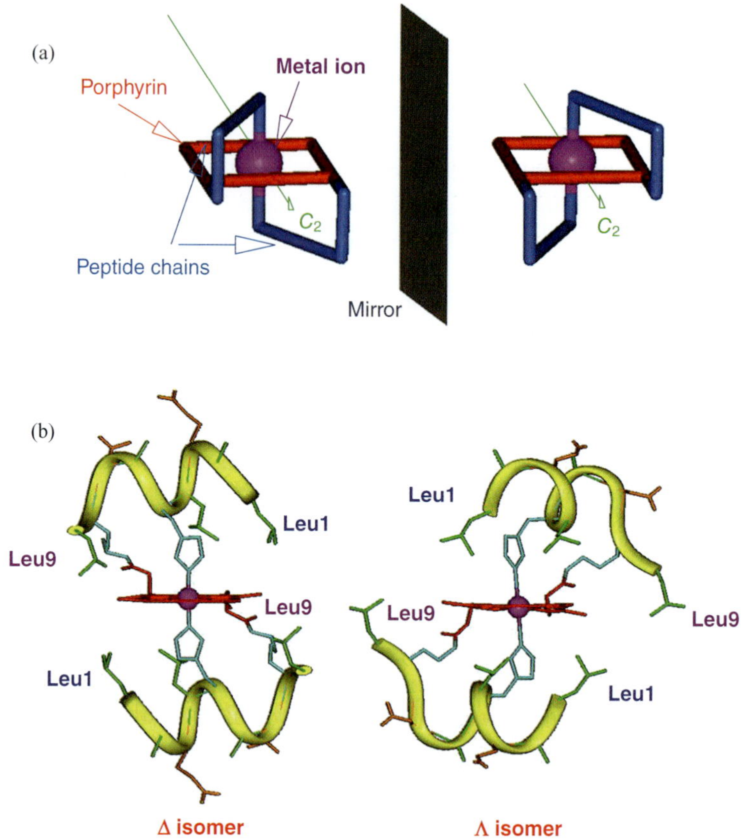

FIGURE 11.31. Mimochrome Δ and Λ isomers.[365] (a) Schematic representation of the two possible orientations of the peptide chains around the metal center, which give rise to the two Δ and Λ diastereomers. (b) Average molecular structures of Co(III)-mimochrome I Δ and Λ isomers, as obtained from NMR experimental data and restrained molecular dynamics (RMD) calculations.

deuteroporphyrin ring, each peptide chain can be positioned either above or below the porphyrin plane, giving rise to Δ and Λ diastereomers (see Fig. 11.31a), which were structurally characterized in solution by NMR (Fig. 11.31b).[365] As an important result, both diastereomers are stable and clearly distinguishable, even at very acidic conditions ($pH < 1$), because the cobalt–histidine coordination is strong enough to preserve the sandwich structure. On the contrary, the iron complex is partly unstable because a fast interconversion between the two diastereomeric forms occurs. During the interconversion process both peptide chains must move from one face of the porphyrin to the other, and the porphyrin ring becomes exposed to the solvent. Henceforth, aggregation by stacking of the porphyrin ring may occur, and low water solubility of Fe(III)-mimochrome I is observed.

The unpredicted, but experimentally observed Λ and Δ isomers of Co(III)-mimochrome I clarify the concept of diastereomerism that may occur when polypeptides coordinate to metal ions. Therefore, great attention should be paid in the design in order to overcome the problem of diastereomer formation. To drive the folding of mimochromes into a unique diastereomeric form, or into stable, noninterconvertible diastereomers, two strategies were used. They were aimed to reduce the flexibility of the linker by inserting further interactions between the two peptide chains and/or between peptide chains and the porphyrin.

A first redesign strategy focused on molecules with longer (14 residues) sequences, modeled in both α-helical (residues 1–10) and extended (C-terminal part) conformations. This approach led to a mimochrome II molecule, where the stabilization of the Δ isomer was achieved (see Fig. 11.32a).[367]

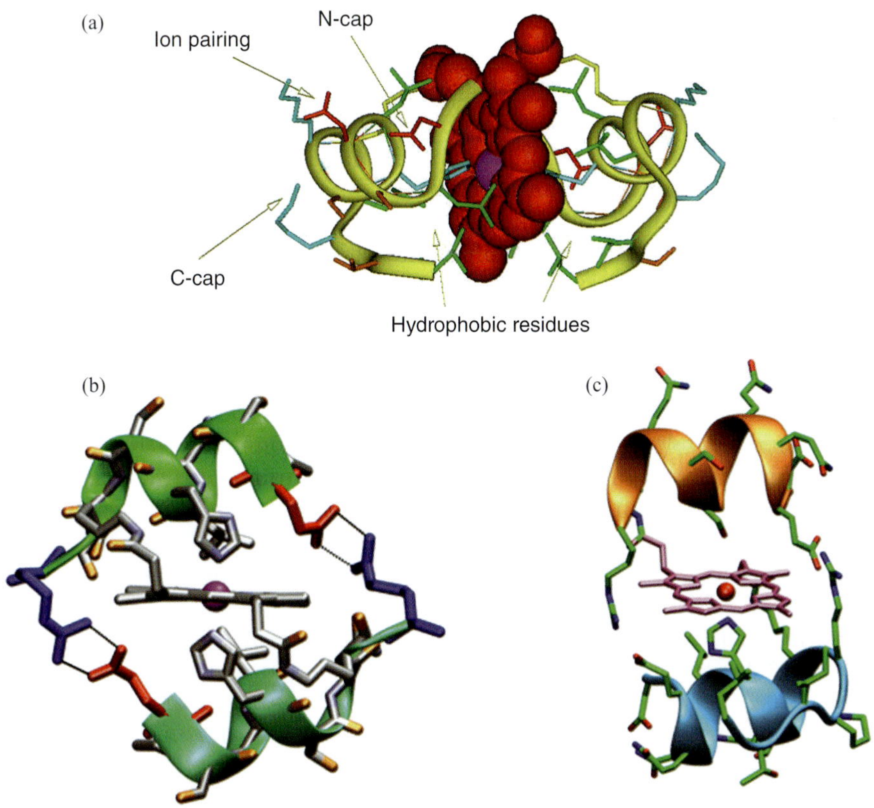

FIGURE 11.32. (a) Molecular model of Fe(III)-mimochrome II.[367] (b) Average molecular structure of Co(III)-mimochrome IV as obtained from NMR experimental data and RMD calculations (PDB code 1VL3).[369] (c) Molecular model of Fe(III)-mimochrome VI.[370]

A second strategy consisted in the introduction, into the mimochrome I sequence, of amino acid substitutions that may provide *intra*molecular, *inter*-helical interactions, leading to mimochrome IV.[368,369] A careful analysis of the NMR structure of Co(III)-mimochrome I Δ and Λ isomers showed that the Cα atoms of Leu1 of one helix and of Leu9 of the other helix are about 12 Å apart in both isomers (see Fig. 11.31b). At this distance, the carboxylate side chain of a Glu residue at position 1 of one peptide chain could ion-pair with the guanidine group of an Arg residue at position 9 of the other peptide chain (both side chains being modeled in an appropriate extended conformation). This condition may occur for both Λ and Δ isomers. It is expected that these substitutions may perturb the helical conformation marginally because of the high propensity of Arg and Glu for helical structure.[299,322] Actually, a positively charged residue (Arg) at the C-terminus, and a negatively charged residue (Glu) at the N-terminus, reduces the intensity of the helix dipole and the helix would be more stable.[299] Thus, mimochrome IV differs from mimochrome I for Glu1 and Arg9, which replace Leu1 and Leu9, respectively. Further, the solvent exposed Ala2 and Ala6 in mimochrome I were replaced by Ser residues in mimochrome IV in order to further increase the water solubility of the new molecule. The spectroscopic characterization in solution of the iron and cobalt derivatives suggested the presence of the Λ isomer as unique species. The NMR structure of the diamagnetic Co(III)-mimochrome IV confirmed the ability of the molecule to adopt a unique topology and revealed the peptide chains to be in helical conformation, as designed, as reported in Figure 11.32b. Thus, the insertion of *intra*molecular *inter*chain interactions was successful in favoring the formation of one of the two possible diastereomers. Interestingly, the heme redox potential of mimochrome IV (−80 mV, vs. standard hydrogen electrode (SHE), at pH 7) falls within the range observed for natural cytochrome *b* and bacterial multiheme cytochrome *c*, which present a bis-His coordination.[133] Future studies will focus on mimochrome IV analogs containing differently charged residues in their sequences in order to correlate charge distribution with heme redox potential.

The last achievement was the design of molecules capable of stabilizing a five-coordinate heme complex, providing an empty site available for catalyzing oxidation reactions.[370,371] In particular, mimochrome VI embodies some of the key elements for functioning as peroxidase-like catalyst. The peptide moieties were designed to create a proximal and a distal site environment of the heme, similarly to heme proteins: (1) a 14-residue peptide bears a His residue at position 6 as the axial ligand to the heme (proximal face); and (2) a 10-residues peptide is devoid of heme coordinating residue, and may create a cavity around the metal ion (distal face). Figure 11.32c depicts the hypothetical model structure. The spectroscopic and functional characterization indicated that Fe(III)-mimochrome VI is an efficient heme protein model, which displays a peroxidase-like catalytic activity.[370] With the use of hydrogen peroxide, it efficiently catalyzes the oxidation of several substrates, with a typical Michaelis–Menten mechanism and with several multiple turnovers. Fe(III)-

mimochrome VI peptide framework, despite its small structure (a total of 24 amino acid residues), confers higher efficiency to the porphyrin cofactor. Three important outcomes deserve highlighting: (1) Fe(III)-mimochrome VI efficiently catalyzes the oxidation of different substrates, such as 2,2′-azino-bis(3-ethylbenzothiazoline-6-sulfonic acid) (ABTS) and guaiacol, by activating H_2O_2, and efficiently catalyzes the nitration of phenols; (2) Fe(III)-mimochrome VI displays a very high specific activity (104 mol mg^{-1} s^{-1} for ABTS oxidation), with respect to highly purified HRPs (91 mol mg^{-1} s^{-1} for ABTS oxidation at pH 4.6): this value highlights both its high catalytic efficiency and small molecular mass, compared with natural peroxidases that contain more than 300 amino acid residues; and (3) Fe(III)-mimochrome VI exhibits multiple turnover kinetics: more than 4000 turnovers within 10 minutes were observed in the ABTS oxidation. All these features indicate that Fe(III)-mimochrome VI is an attractive, low-molecular-weight heme enzyme, which will serve as an excellent scaffold to further develop designed heme-based biocatalysts.

The results herein reported for the mimochromes demonstrated that small molecules are able to reproduce some of the features of their natural counterparts. They also highlight that in small models, which lack the complex architecture of the natural protein, the careful engineering of numerous and diverse interactions, such as hydrophobic, ion pair, and coordination interactions, is crucial for the development of molecules with unique structures and functions.

11.5.3 DFs: Carboxylate-Bridged Diiron Metalloprotein Models

In the field of diiron carboxylate proteins, small synthetic models have been successfully developed in order to reproduce the structures and spectroscopic properties of the various redox sates of the cofactor, including a high-valent diferryl intermediate.[372–375] These models share similarities with the intermediates in the reaction cycle of the natural proteins, thus allowing a better interpretation of the chemistry supported by the natural systems. A current important goal is to design more complex models to allow the evaluation of environmental variables, such as the role of second shell ligands, and also to provide sites for binding substrates as well as the dimetal cofactor. Our group, in collaboration with DeGrado's group at the University of Pennsylvania, used a *de novo* design approach for the development of the DF (Due Ferri) family of artificial proteins, as models of diiron metalloproteins. The DF proteins were developed through an iterative process of design and rigorous characterization, which allowed us to diagnose the problems associated with the initial models, and to improve the models in subsequent designs. Thus, the problem of designing the DF family was approached through several steps. The first involved the design of stable, uniquely folded proteins that contain the metal binding site, DF1.[302,376,377] To improve the functional properties of the initial model, DF1 was subjected to several changes in the sequence, as well in the loop structure (DF1 and DF2 subsets).[303,378–384] The design further evolved with

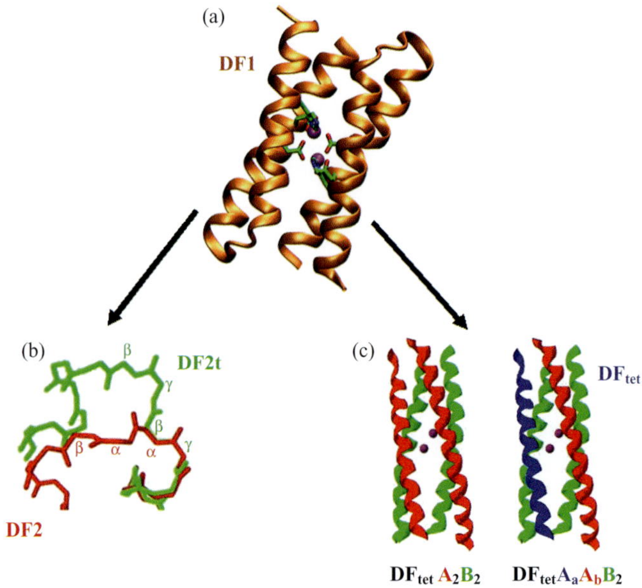

FIGURE 11.33. (a) Crystal structure of di-Zn(II)-DF1 (PDB code 1EC5).[368] (b) Backbone structure of the turns in DF2 (red) and DF2t (green).[381] (c) Models of DF_{tet}.[387]

the construction of versions in which the four helices are distinct chains that come together by noncovalent self-assembly (DF_{tet}) (see Fig. 11.33).[385–387] Each class has different advantages. For example, the symmetric nature of the dimeric derivatives simplifies interpretation of data, whereas analogs of the four-chain DF_{tet} constructs can be mixed and matched to allow easy generation of combinatorial diversity. The last developed protein, the functional DF3, integrates all the results derived from previously characterized models.[388,389]

DF protein design evolved from a retrostructurally analysis of carboxylate-bridged diiron proteins, which feature a helical Glu-Xxx-Xxx-His motif, discussed in Section 11.4.4.[377] Although the folds of diiron proteins are frequently complex, their active sites are housed within a very simple pseudo-222-symmetric four-helix bundle. The helices have a slight left-handed tilt as in classical four-helix bundles and antiparallel four-stranded coiled coils. Thus, their geometries may be discussed in terms of the heptad nomenclature of coiled coils.[390,391] Almost invariably, each helix contains a single coordinating Glu residue in an "a" position of the heptad motif, which projects toward the center of the bundle (Fig. 11.34a). His residues, located four residues downstream at a "d" position in two helices contribute the remaining side chain ligands to the diiron site. Small side chains generally occur at the analogous "d" positions of the other two helices. Importantly, these side chains control the access of oxygen and substrates to the diiron center.

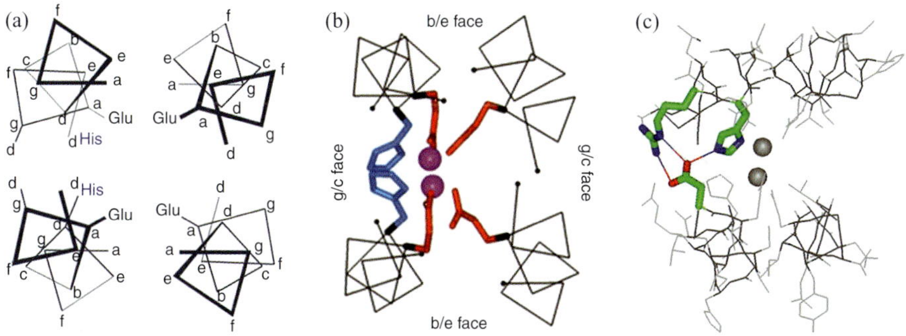

FIGURE 11.34. (a) Schematic diagram of the four-helix bundle defining the active site of a typical diiron protein. (b) Top view of the four-helix bundle harboring the active site of bacterioferritin.[392] (c) The Asp-Glu-Xxx-Arg-His motif: The hydrogen bonding between the Asp side chain from one helix and the His and Arg of a second helix is illustrated.

Other residues also help shape the active sites of diiron proteins. The top and bottom of the active site is defined by the side chains projecting from "d" and "a" of neighboring heptads. The sides of the active site are defined by residues at positions "g" and "c" on two faces of the four-helix bundle, and "b" and "e" on the other two faces, as reported in Figure 11.34b for bacterioferritin from *E. coli*.[392] The residues at the b/e faces of the helical bundle tend to be tightly packed and appear to serve a structural role. In contrast, the c/g faces appear to be important for function; one g/c face presents side chains that help define the entry and binding of substrates. The opposite c/g face may help tune the properties of the coordination site by hydrogen bonding to the coordinating His residues. For example, the active sites of MMO and desaturase contain two copies of the sequence Asp_g-Glu_a-Xxx_b-Arg_c-His_d. The Asp at a "g" position engages in an extensive hydrogen-bonded network with a His_d and Arg_c in a neighboring helix, as illustrated in Figure 11.34c.[261] The contribution of these interactions to functional properties of the diiron site has been discussed in Section 11.4.4.

Based on these structural observations, the design of diiron protein models represents a very significant structural challenge, requiring the burial of six ionizable side chains (four Glu and two His) and several side chain/side chain hydrogen bond interactions. Thus, it tests one's ability to design structurally fine-tuned networks of interacting groups with sub-angstrom accuracy.

DF1 is made up of two 48-residue helix-loop-helix (α_2) motifs, able to specifically self-assemble into an antiparallel four-helix bundle.[376,377] To provide a Glu_4His_2 liganding environment for the diiron center, each α_2 subunit contains a Glu_a-Xxx_b-Xxx_c-His_d sequence in helix 2, which donates a His side chain (His39/His39′) as well as a bridging Glu carboxylate (Glu36/Glu36′) to the site. A second Glu carboxylate ligand on helix 1 (Glu10/Glu10′) provides a fourth

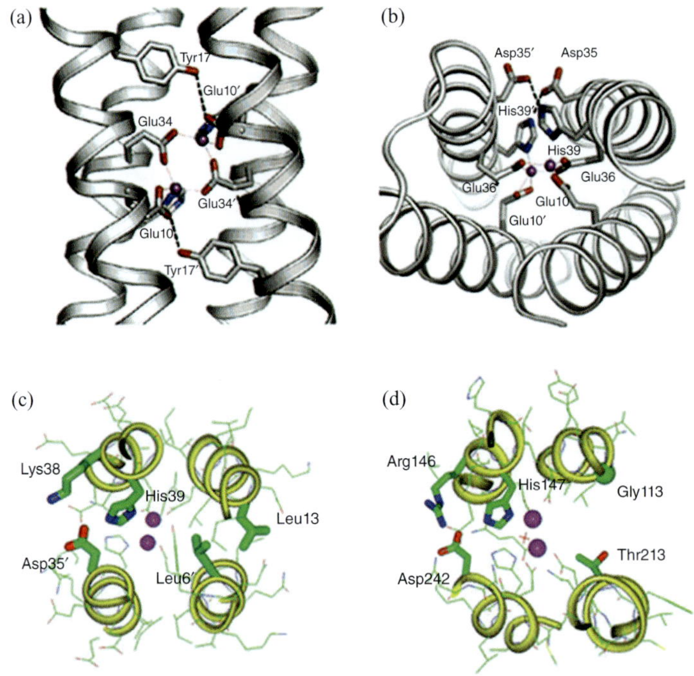

FIGURE 11.35. DF1 second shell hydrogen bonds between (a) Tyr17 and Glu10′, and between (b) Asp35 and His39′ (prime numbers refer to residues belonging to symmetry-related chains). (c) Hydrogen-bonded network consisting of Lys/Asp/His in DF1 versus (d) Arg/Asp/His, as observed in MMO.

protein ligand per metal ion (Fig. 11.35a,b). Liganding side chains were placed in the appropriate rotamers to allow interaction with a diiron center. The final positions of the helices were dictated by three different requirements: (1) the geometry of the liganding site was restrained to bind diiron with two bridging Glu carboxylates, two nonbridging Glu side chains, and the N^{δ} of two His side chains; (2) the helical packing angles and distances were constrained to match those typically observed in the active sites of diiron proteins; and (3) precise twofold symmetry between the two pairs of helices was enforced.

Each metal ion is five-coordinate, and a sixth vacant site lies on adjoining faces of the two metal ions, providing a potential site for binding of exogenous ligands. Satisfaction of side chain packing requirements was ensured by positioning hydrophobic side chains to fill the core as dictated by the steric environment of the backbone structure. In addition, Glu and Lys residues were chosen for helix-favoring, solvent-accessible sites to provide water solubility and to drive the assembly into the desired antiparallel topology. Finally, an idealized γ-α_L-β interhelical loop was included between the two pairs of helices. The DF1 sequence was also carefully engineered to include second shell interactions, which are crucial for defining structural and functional properties of metal binding sites. Thus, in DF1 a Tyr residue at position 17 donates a second shell hydrogen bond to the nonbridging E10′ of the other monomer in each of the α_2 subunits (the same interaction exists between the symmetrically related pair, Y17′ and E10). Similarly, an Asp residue at position 35′ forms a hydrogen bond with the imidazole N^{ε} of the His ligand in the neighboring helix of the dimer. This Asp is further involved in a salt bridge interaction with a Lys at position 38. This hydrogen bond network consisting of Lys/Asp/His in DF1 is similar to that observed in the active sites of natural proteins, such as MMO, where the lysine residue is replaced by an arginine, to form an Arg/Asp/His cluster (see Fig. 11.35c,d).

DF1 was designed for maximal stability.[376,377] Thus, the interior of the protein was efficiently packed with a large number of hydrophobic side chains, resulting in a high degree of conformational stability. The structural characterization proved DF1 to be an excellent structural model of diiron proteins. It adopts a folded, native-like conformation in the apo- and metal-bound forms and was able to bind metal ions such as Zn^{2+}, Co^{2+}, and Mn^{2+} in the correct geometry and stoichiometry.[302,376] Unfortunately, DF1 was not able to support any function because the access to its active site was hampered by the compact hydrophobic core around the metal center.

To improve the functional properties of the initial model, subsequent work focused on opening the cleft for binding small ligands to the dimetal center. DF1 was thus subjected to several changes in the sequence, as well as in the loop structure. In particular, substitution of Leu13 and Leu13′ in DF1 with smaller side chain amino acids, such as Ala and Gly, afforded a cavity large enough to allow the access of small molecules to the metal center. The L13A-DF1[378] and L13G-DF1[379] variants, as well as the DF2 subset,[303,381–383] bind exogenous ligands, such as phenol and acetate, and display ferroxidase activity. However, optimization of the catalytic activity required further changes in the second shell ligands in order to provide the protein with stability and, at the same time, with the flexibility required for function. Modeling suggested that residues at both positions 13 and 9 play a critical role for the accessibility to the metal site. The next design evolved into DF3,[388,389] aimed to introduce a phenol binding site into the DF1 scaffold, converting it in a phenol oxidase, similar to the natural enzymes AOX and PTOX.[226–228] The elaboration of phenol oxidase activity in DF3 required expansion of the active site cleft to accommodate phenols. Previously, possible solutions to this problem were explored in a combinatorial manner through the design of DF_{tet}, which is composed of four unconnected helical peptides.[384–387] Mixing different DF_{tet} peptides produced catalysts for the oxidation of 4-aminophenol.[387] Unfortunately, the complex stoichiometry, marginal stability, and tendency to undergo ligand exchange reactions hampered attempts to fully characterize the structure and properties of the DF_{tet} assemblies. This led us to introduce the corresponding mutations into the more well-characterized DF1 framework. We first introduced Gly residues at both positions 9 and 13 of both α_2 monomer, with the aim to open up the active site, and create a cavity large

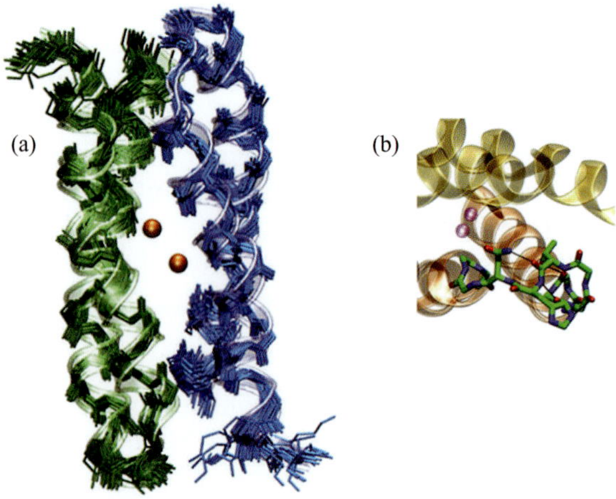

FIGURE 11.36. (a) di-Zn(II)-DF3 solution structure. (b) Details of the loop structure showing the network of hydrogen bonds (PDB code 2KIK).[388]

enough to accommodate exogenous ligands. However, the introduction of helix-destabilizing glycine residues and the loss of the hydrophobic driving force strongly destabilized the fold of the protein. Indeed, a single mutation of Leu13,13′ to glycine destabilizes DF1 by 10.8 kcal mol^{-1} $dimer^{-1}$, precluding the introduction of the second glycine residue.[302] To compensate for the thermodynamic cost of carving an active site access channel onto the protein, it was necessary to optimize the interhelical loop conformation (Fig. 11.36). To increase the conformational stability of the DF scaffold, we modified the sequence of the interhelical turn, which adopts a "rose-like" α_R-α_L-β conformation.[381] The original Val24-Lys25-Leu26 of DF1 was changed to Thr24-His25-Asn26. In models, His25 appeared capable of forming stabilizing hydrogen bonded C-capping interactions of helix 1, while Asn26 could either form N-capping interactions with helix 2 or with the carbonyl group of Thr24, depending on its rotamer. The newly designed protein showed improved water solubility (up to 3 mM) and active site accessibility while retaining the unique native-like structure, as assessed by NMR structural characterization.[388] Most importantly, DF3, with its well-defined active site, displays ferroxidase and oxidase activity. In fact, it catalyzes the two-electron oxidation of 3,5-ditert-butyl-catechol to the corresponding 3,5-ditert-butyl-quinone. The catalytic efficiency toward structurally related substrates illustrates the selectivity of the active site.

The development of the DF family of artificial proteins, based on the four-helix bundle fold, clearly demonstrates the trade-off between conformational stability and function. Catalytic activity in *de novo* designed scaffold required engineering with angstrom-level precision: (1) the primary and second shell ligands; (2) the residues that define the shape and accessibility of the active site cavity; and (3) thermodynamic stability, achieved by careful design of the helical packing and the interhelical turn. Our initial design focused on the construction of a thermodynamically stable framework, able to bind the dimetal cofactor, and to react with molecular oxygen. In subsequent studies, the basic framework was reengineered, with the aim of expanding the active site pocket, to allow binding of organic substrates. Although the first two objectives were achieved, the expansion of the active site—with concomitant loss of packing and dehydration of apolar side chains—proved to be unfeasible within the initially designed sequence. When the active site cavity was expanded, the desired fold became energetically unfavorable, leading to either unfolded protein or off-pathway formation of misfolded aggregated species. In DF3, the stability and conformational specificity was successfully increased through the redesign of a turn quite distant from the active site. In fact, the destabilization caused by introducing four Gly residues into the core structure of DF3 dimer was compensated by the increased stability of the interhelical loop. Thus, the loop had a profound influence on the solubility, stability, and ultimately, on the possibility to introduce destabilizing mutations into the active site.

11.6 CONCLUSIONS AND PERSPECTIVES

The increasing structural data available nowadays on metalloproteins allow to gain insights into the structural and dynamic features of metal binding sites and to correlate these features with the functions they achieve in metalloproteins. The examples discussed in this chapter, mainly regarding iron-containing metalloproteins, demonstrate that, beside the coordinate bond, a plethora of interactions are responsible of finely tuning the metal site activity. An important contribution to understanding the role of the protein matrix in modulating the metal site properties comes from metalloprotein models, developed through different design strategies. Knowledge derived from natural metalloproteins and from their models is synergic, in that more accurate information on natural systems allows better designs, which in turn contribute to shed light on the properties of the natural counterparts.

Our examples of artificial metalloproteins, herein reported, together with outstanding results reported by other authors, demonstrate how, starting from structural models, the design has evolved toward the construction of functional models. Metalloprotein design has proven to be a powerful method for addressing questions in metalloprotein stability, folding, and function. The features required for stabilizing secondary structure have been elucidated, and the gross features that direct the folding into a native-like state have been demonstrated. Computational methods for structure and sequence prediction are also having an enormous impact on metalloprotein studies, allowing one to design uniquely packed structures and metal binding sites within proteins. Further, genetic approaches and methods of combinatorial synthesis allow one to more rapidly optimize initial designs in order to include all the interactions supposed to be valuable for structure and function.

ACKNOWLEDGMENTS

The authors are very grateful to Prof. Vincenzo Pavone for numerous stimulating discussions and continuous support. A special thanks to Dr. Marina Faiella and Dr. Claudia Vicari for their help in the elaboration of the drawings, and to Dr. Luigi Di Costanzo for helping with metalloprotein PDB search. The authors also wish to thank all members of the research groups at the University of Napoli and at the University of Pennsylvania who have contributed to the results on mimochrome and DF models described in this chapter.

REFERENCES

1. P. M. Harrison. *Metalloproteins*. Parts I and II. P. Macmillan, ed. Academic Press, New York, 1985.
2. H. B. Gray, E. I. Stiefel, J. S. Valentine, I. Bertini, eds. *Biological Inorganic Chemistry: Structure and Reactivity*. University Science Books, Sausalito, CA, 2007.
3. T. Dudev and C. Lim. *Annu Rev Biophys* **2008**, 37, 97.
4. T.-Y. Yang, T. Dudev, C. Lim. *J Am Chem Soc* **2008**, 130, 3844.
5. F. A. Cotton. *Chemical Applications of Group Theory*, 3rd ed. John Wiley & Sons, London and New York, **1990**.
6. F. A. Cotton and G. Wilkinson. *Advanced Inorganic Chemistry*. John Wiley & Sons, London and New York, 1972.
7. D. F. Shriver, P. W. Atkins, T. I. Overton, I. P. Rourke, M. T. Weller and F. A. Armstrong, *Inorganic Chemistry*, 4th ed. Oxford University Press, Oxford, UK, **2006**.
8. N. N. Greenwood and A. Earnshaw. *Chemistry of Elements*, 1st ed. Pergamon Press, Oxford, 1984.
9. S. F. A. Kettle. *Physical Inorganic Chemistry: A Coordination Chemistry Approach*. Oxford University Press, Oxford, MA, 2000.
10. P. Müller. *Pure Appl Chem* **1994**, 66, 1077.
11. K. Degatyarenko and S. Contrino. *BMC Struct Biol* **2004**, 4, 3.
12. K. Degatyarenko. *Bioinformatics* **2000**, 16, 851.
13. R. H. Holm, P. Kennepohl, E. I. Solomon. *Chem Rev* **1996**, 96, 2239.
14. B. Nagel, H. Dellweg, L. M. Gierasch. *Pure Appl Chem* **1992**, 64, 143.
15. H. A. Bethe. *Ann Phys* **1929**, 3, 135.
16. S. R. Baten, S. M. Neville, D. R. Turner, eds. *Coordination Polymers*. Royal Society of Chemistry, Cambridge, UK, **2009**.
17. J. A. Cowan. *Inorganic Biochemistry: An Introduction*. Wiley-VCH, New York, 1997.
18. J. J. R. Frausto da Silva and R. J. P. Williams. *The Biological Chemistry of the Elements*. Oxford University Press, Oxford, 1994.

19. S. J. Lippard and J. M. Berg. *Principles of Bioinorganic Chemistry*. University Science Books, Mill Valley, CA, 1994.
20. R. P. J. Williams. Metallo-enzymes and metallo-proteins, chemistry of. In *Wiley Encyclopedia of Chemical Biology*. John Wiley & Sons, Inc., Oxford, UK, 2008; pp. 1–8.
21. D. W. Christianson. *Adv Protein Chem* **1991**, 42, 281.
22. J. P. Glusker, A. K. Katz, C. W. Bock. *Rigaku J* **1999**, 16, 8.
23. J. McMaster. *Annu Rep Prog Chem A* **2001**, 97, 567.
24. C. J. Wilson, D. Apiyo, P. Wittung-Stafshede. *Q Rev Biophys* **2004**, 37, 285.
25. R. R. Crichton and J.-L. Pierre. *BioMetals* **2001**, 14, 99.
26. T. Dudev and C. Lim. *Chem Rev* **2003**, 103, 773.
27. J. A. Cowan. *Chem Rev* **1998**, 98, 1067.
28. D. E. Fenton. *Biocoordination Chemistry*. Oxford University Press, Oxford, 1995.
29. Y. Zhang and V. N. Gladyshev. *J Biol Chem* **2010**, 285, 3393.
30. R. J. P. Williams. *Cell Mol Life Sci* **1997**, 53, 816.
31. K. J. Waldron, J. C. Rutherford, D. Ford, N. J. Robinson. *Nature* **2009**, 460, 823.
32. U. Ryde. *Dalton Trans* **2007**, 607.
33. M. M. Harding, M. W. Nowicki, M. D. Walkinshaw. *Crystallogr Rev* **2010**, 16, 4, 247.
34. I. Dokmanić, M. Šikić, S. Tomić. *Acta Cryst* **2008**, D64, 257.
35. J. A. C. Tamames and M. J. Ramos. *J Mol Model* **2010**, DOI: 10.1007/s00894-010-0733-5.
36. M. M. Harding. *Acta Cryst* **2004**, D60, 849.
37. J. M. Castagnetto, S. W. Hennessy, V. A. Roberts, E. D. Getzoff, J. A. Tainer, M. E. Pique. *Nucleic Acids Res* **2002**, 30, 379.
38. A. Golovin, D. Dimitropoulos, T. Oldfield, A. Rachedi, K. Henrick. *Proteins* **2005**, 58, 190.
39. K. Hsin, Y. Sheng, M. M. Harding, P. Taylor, M. D. Walkinshaw. *J Appl Cryst* **2008**, 41, 963.
40. K. Hemavathi, M. Kalaivani, A. Udayakumar, G. Sowmiya, J. Jeyakanthan, K. Sekar. *J Appl Cryst* **2010**, 43, 196.
41. P. Chakrabarti. *Protein Eng* **1990**, 4, 57.
42. J. P. Glusker. *Adv Protein Chem* **1991**, 42, 1.
43. S. Karlin, Z.-Y. Zhu, K. D. Karlin. *Proc Natl Acad Sci U S A* **1997**, 94, 14225.
44. A. Messerschmidt, R. Ladenstein, R. Huber, M. Bolognesi, L. Avigliano, R. Petruzzelli, A. Rossi, A. Finazzi-Agro. *J Mol Biol* **1992**, 224, 179.
45. A. Messerschmidt. *Adv Inorg Chem* **1993**, 40, 121.
46. L. Leinartaite, K. Saraboji, A. Nordlund, D. T. Logan, M. Oliveberg. *J Am Chem Soc* **2010**, 132, 13495.
47. W. R. Rypniewski, S. Mangani, B. Bruni, P. L. Orioli, M. Casati, K. S. Wilson. *J Mol Biol* **1995**, 251, 282.
48. C. J. Carrell, H. L. Carrell, J. Erlebacher, J. P. Glusker. *J Am Chem Soc* **1988**, 110, 8651.
49. K. A. McCall, C.-C. Huang, C. A. Fierke. *J Nutr* **2000**, 130, 1437S.

50. T. Dudev and C. Lim. *Acc Chem Res* **2007**, 40, 85.
51. T. Dudev and C. Lim. *J Phys Chem B* **2004**, 108, 4546.
52. J. J. Falke, S. K. Drake, A. L. Hazard, O. B. Peersen. *Q Rev Biophys* **1994**, 27, 219.
53. R. L. Rardin, W. B. Tolman, S. J. Lippard. *New J Chem* **1991**, 15, 417.
54. R. G. Pearson. *J Am Chem Soc* **1963**, 85, 3533.
55. R. G. Parr and R. G. Pearson. *J Am Chem Soc* **1983**, 105, 7512.
56. H. Sigel and D. McCormick. *Acc Chem Res* **1970**, 3, 201.
57. R. B. Martin. *J Chem Educ* **1987**, 64, 402.
58. H. Irving and R. J. P. Williams. *J Chem Soc* **1953**, 3192.
59. J. W. Torrance, M. W. MacArthur, J. M. Thornton. *Proteins* **2008**, 71, 813.
60. D. S. Auld. *Biometals* **2009**, 22, 141.
61. D. S. Auld. *Handbook on Metalloproteins*, I. Bertini, A. Sigel, H. Sigel, eds. Marcel Dekker, New York, 2001; pp. 881–959.
62. D. S. Auld. *Zinc Enzymes: in Encyclopedia of Inorganic Chemistry*, 2nd ed. Vol. IX, B. King, ed. John Wiley & Sons, Chichester, 2005; pp. 5885–5927.
63. K. Patel, A. Kumar, S. Durani. *Biochim Biophys Acta* **2007**, 1774, 1247.
64. B. L. Vallee and D. S. Auld. *Biochemistry* **1990**, 29, 5647.
65. D. S. Auld. *The Handbook of Metalloproteins*, A. Messerschmidt, W. Bode, M. Cygler, eds. John Wiley & Sons, Chichester, 2004; pp. 403–415.
66. B. L. Vallee and D. S. Auld. *Biochemistry* **1993**, 32, 6493.
67. D. S. Auld. *The Handbook of Metalloproteins*, A. Messerschmidt, W. Bode, M. Cygler, eds. John Wiley & Sons, Chichester, 2004; pp. 416–431.
68. D. S. Auld. *Biometals* **2001**, 14, 271.
69. S. A. Lister, D. R. Wetmore, R. S. Roche, P. W. Codding. *Acta Crystallogr D* **1996**, 52, 543.
70. F. Colonna-Cesari, D. Perahia, M. Karplus, H. Eklund, C. I. Braden, O. Tapia. *J Biol Chem* **1986**, 261, 15273.
71. B. Chevrier, C. Schalk, H. D'Orchymont, J. M. Rondeau, D. Moras, C. Tarnus. *Structure* **1994**, 2, 283.
72. A. C. R. da Silva and F. C. Reinach. *Trends Biochem Sci* **1991**, 16, 53.
73. R. Kuroki, Y. Taniyama, C. Seko, H. Nakamura, M. Kikuchi, M. Ikehara. *Proc Natl Acad Sci U S A* **1989**, 86, 6903.
74. C. A. Bonagura, M. Sundaramoorthy, H. S. Pappa, W. R. Patterson, T. L. Poulos. *Biochemistry* **1996**, 35, 6107.
75. X. Wang, M. Kirberger, F. Qiu, G. Chen, J. J. Yang. *Proteins* **2009**, 75, 787.
76. W. Yang, H. W. Lee, H. Hellinga, J. J. Yang. *Proteins* **2002**, 47, 344.
77. E. Pidcock and G. R. Moore. *J Biol Inorg Chem* **2001**, 6, 479.
78. Y. S. Babu, J. S. Sack, T. J. Greenhough, C. E. Bugg, A. R. Means, W. J. Cook. *Nature* **1985**, 315, 37.
79. M. A. Wilson and A. T. Brunger. *J Mol Biol* **2000**, 301, 1237.
80. J. L. Gifford, M. P. Walsh, H. J. Vogel. *Biochem J* **2007**, 405, 199.
81. W. I. Weiss, K. Drickamer, W. A. Hendrickson. *Nature* **1992**, 360, 127.
82. M. Kuwahara, J. K. Glenn, M. A. Morgan, M. H. Gold. *FEBS Lett* **1984**, 169, 247.

83. H. Wariishi, L. Akileswaran, M. H. Gold. *Biochemistry* **1988**, 27, 5365.
84. V. V. Barynin, M. M. Whittaker, S. V. Antonyuk, V. S. Lamzin, P. M. Harrison, P. J. Artymiuk, J. W. Whittaker. *Structure* **2001**, 9, 725.
85. R. Banerjee and S. W. Ragsdale. *Annu Rev Biochem* **2003**, 72, 209.
86. T. H. Tahirov, H. Oki, T. Tsukihara, K. Ogasahara, K. Yutani, K. Ogata, Y. Izu, S. Tsunasawa, I. Kato. *J Mol Biol* **1998**, 284, 101.
87. R. A. Bradshaw, W. W. Brickey, K. W. Walker. *Trends Biochem Sci* **1998**, 23, 263.
88. L. Rulíšek and J. Vondrášek. *J Inorg Biochem* **1998**, 71, 115.
89. S. W. Ragsdale. *J Biol Chem* **2009**, 284, 18571.
90. N. Sukdeo, E. Daub, J. F. Honek. *Nickel and Its Surprising Impact in Nature*, A. Sigel, H. Sigel, R. K. O. Sigel, eds. John Wiley & Sons, West Sussex, 2007; pp. 445–472.
91. D. P. Barondeau, C. J. Kassmann, C. K. Bruns, J. A. Tainer, E. D. Getzoff. *Biochemistry* **2004**, 43, 8038.
92. E. Jabri, M. B. Carr, R. P. Hausinger, P. A. Karplus. *Science* **1995**, 268, 998.
93. S. Ciurli. *Nickel and Its Surprising Impact in Nature*, A. Sigel, H. Sigel, R. K. O. Sigel, eds. John Wiley & Sons, West Sussex, 2007; pp. 241–278.
94. A. Vila and A. S. Fernandes. *Handbook on Metalloproteins*, I. Bertini, A. Sigel, H. Sigel, eds. Marcel Dekker, New York, 2001; pp. 813–856.
95. M. A. Halcrow, P. F. Knowles, S. E. V. Phillips. *Handbook on Metalloproteins*, I. Bertini, A. Sigel, H. Sigel, eds. Marcel Dekker, New York, 2001; pp. 709–762.
96. A. G. Sykes. *Adv Inorg Chem* **1991**, 36, 377.
97. E. T. Adman and L. H. Jensen. *Isr J Chem* **1981**, 21, 8.
98. M. H. M. Olsson, U. Ryde, B. O. Roos, K. Pierloot. *J Biol Inorg Chem* **1998**, 3, 109.
99. B. L. Vallee and R. J. P. Williams. *Proc Natl Acad Sci U S A* **1968**, 59, 498.
100. R. J. P. Williams. *Eur J Biochem* **1995**, 234, 363.
101. N. Ito, S. E. Phillips, C. Stevens, Z. B. Ogel, M. J. McPherson, J. N. Keen, K. D. Yadav, P. F. Knowles. *Nature* **1991**, 350, 87.
102. T. Klabunde, C. Eicken, J. C. Sacchettini, B. Krebs. *Nat Struct Biol* **1998**, 5, 1084.
103. M. Wolfgang and L. Yuan. *Chem Rev* **2009**, 109, 4682.
104. H. Eklund, B. Nordstrom, E. Zeppezauer, G. Soderlund, I. Ohlsson, T. Boiwe, C. I. Branden. *FEBS Lett* **1974**, 44, 200.
105. R. B. Honzatko, J. L. Crawford, H. L. Monaco, J. E. Ladner, B. F. Ewards, D. R. Evans, S. G. Warren, D. C. Wiley, R. C. Ladner, W. N. Lipscomb. *J Mol Biol* **1982**, 160, 219.
106. S. Krishna, I. Majumdar, N. V. Grishin. *Nucleic Acids Res* **2003**, 31, 532.
107. S. A. Wolfe, R. A. Grant, M. Elrod-Erickson, C. Pabo. *Structure* **2001**, 9, 717.
108. J. H. Laity, B. M. Lee, P. E. Wright. *Curr Opin Struct Biol* **2001**, 11, 39.
109. O. Leon and M. Roth. *Biol Res* **2000**, 33, 21.
110. A. Klug and J. W. Schwabe. *FASEB J* **1995**, 9, 597.
111. S. Iuchi. *Cell Mol Life Sci* **2001**, 58, 625.
112. Y. Lee and C. Lim. *J Mol Biol* **2008**, 379, 545.
113. D. C. Rees, M. Lewis, W. N. Lipscomb. *J Mol Biol* **1983**, 168, 367.

114. M. A. Holmes and B. W. Matthews. *Biochemistry* **1981**, 20, 6912.
115. D. Hewett-Emmett and R. E. Tashian. *Mol Phylogenet Evol* **1996**, 5, 50.
116. A. Liljas, K. K. Kannan, P. C. Bergstén, I. Waara, K. Fridborg, B. Strandberg, U. Carlbom, L. Järup, S. Lövgren, M. Petef. *Nat New Biol* **1972**, 235, 131.
117. K. Håkansson, M. Carlsson, L. A. Svensson, A. Liljas. *J Mol Biol* **1992**, 227, 1192.
118. E. Hough, L. K. Hansen, B. Birknes, K. Jynge, S. Hansen, A. Hordvik, C. Little, E. Dodson, Z. Derewenda. *Nature* **1989**, 38, 357.
119. J. L. Pierre and M. Fontecave. *Biometals* **1999**, 12, 195.
120. R. R. Crichton. *Iron Metabolism: From Molecular Mechanism to Clinical Consequences*. John Wiley & Sons, Chichester, UK, 2009.
121. P. Aisen, C. Enns, M. Wessling-Resnick. *Int J Biochem Cell Biol* **2001**, 33, 940.
122. H. Drakesmith and A. Prentice. *Nat Rev Microbiol* **2008**, 6, 541.
123. L. L. Dunn, Y. S. Rahmanto, D. R. Richardson. *Trends Cell Biol* **2006**, 17, 93.
124. J. Barasch and K. Mori. *Nature* **2004**, 432, 811.
125. P. F. Lindley. *Rep Prog Phys* **1996**, 59, 867.
126. H. Boukhalfa and A. L. Crumbliss. *BioMetals* **2002**, 15, 325.
127. W. R. Harris. *Molecular and Cellular Iron Transport*, D. M. Templeton, ed. Marcel Dekker, New York, 2002.
128. J. M. Gutteridge and B. Halliwell. *Iron Transport and Storage*, P. Ponka, H. M. Schulman, R. C. Woodworth, eds. CRC Press, Boca Raton, FL, 1990.
129. M. W. Hentze, M. U. Muckenthaler, B. Galy, C. Camaschella. *Cell* **2010**, 142, 24.
130. H. M. Baker, B. F. Anderson, E. N. Baker. *Proc Natl Acad Sci U S A* **2003**, 100, 3579.
131. J. B. Neilands. *J Biol Chem* **1995**, 270, 26723.
132. C. J. Reedy and B. R. Gibney. *Chem Rev* **2004**, 104, 617.
133. C. J. Reedy, M. M. Elvekrog, B. R. Gibney. *Nucleic Acids Res* **2008**, 36, D307.
134. M. Hansson and C. von Wachenfeldt. *FEMS Microbiol Lett* **1993**, 107, 121.
135. D. H. Gonzales and W. Neupert. *J Bioenerg Biomembr* **1990**, 22, 753.
136. R. A. Scott, A. G. Mauk, eds. *Cytochrome c: A Multidisciplinary Approach*. University Science Books, Sausalito, CA, 1996.
137. L. J. Smith, A. Kahraman, J. M. Thornton. *Proteins* **2010**, 78, 2349.
138. D. Dolphin, ed. *The Porphyrins*, Vol. 7. Academic Press, New York, 1979.
139. R. Liddington, Z. Derewenda, E. Dodson, R. Hubbard, G. Dodson. *J Mol Biol* **1992**, 228, 551.
140. M. Gajhede, D. J. Schuller, A. Henriksen, A. T. Smith, T. L. Poulos. *Nat Struct Biol* **1997**, 4, 1032.
141. L. Banci, I. Bertini, H. B. Gray, C. Luchinat, T. Reddig, A. Rosato, P. Turano. *Biochemistry* **1997**, 36, 9867.
142. M. Vidakovic, S. G. Sligar, H. Li, T. L. Poulos. *Biochemistry* **1998**, 37, 9211–9219.
143. K. N. Degtyarenko, A. C. North, J. B. Findlay. *Nucleic Acids Res* **1999**, 27, 233.
144. P. Turano and Y. Lu. *Handbook on Metalloproteins*, I. Bertini, H. Sigel, A. Sigel, eds. Marcel Dekker, New York, 2001.
145. T. L. Poulos. *J Biol Inorg Chem* **1996**, 1, 356.
146. D. B. Goodin. *J Biol Inorg Chem* **1996**, 1, 360.

147. L. Banci, A. Rosato, P. Turano. *J Biol Inorg Chem* **1996**, 1, 364.
148. Z. Gross. *J Biol Inorg Chem* **1996**, 1, 368.
149. I. M. C. M. Rietjens, A. M. Osman, C. Veeger, O. Zakharieva, J. Antony, M. Grodzicki, A. X. Trautwein. *J Biol Inorg Chem* **1996**, 1, 372.
150. R. Weiss, D. Mandon, T. Wolter, A. X. Trautwein, M. Muther, E. Bill, A. Gold, K. Jayaraj, J. Terner. *J Biol Inorg Chem* **1996**, 1, 377.
151. P. Rydberg, E. Sigfridsson, U. Ryde. *J Biol Inorg Chem* **2004**, 9, 203.
152. T. Mashiko and D. Dolphin. In *Comprehensive Coordination Chemistry*, Vol. 2. G. Wilkinson, ed. Pergamon, Oxford, 1987; p. 813. Chapter 21.1.
153. J. W. Buchler. *The Porphyrins*, Vol. 1. Academic Press, New York, 1979; p. 389.
154. R. C. E. Durley and F. S. Mathews. *Acta Crystallogr D Biol Crystallogr* **1996**, 52, 65.
155. T. M. Iverson, D. M. Arciero, B. T. Hsu, M. S. P. Logan, A. B. Hooper, D. C. Rees. *Nat Struct Biol* **1998**, 5, 1005.
156. F. A. Walker, B. H. Huynh, W. R. Scheidt, S. R. Osvath. *J Am Chem Soc* **1986**, 108, 5288.
157. F. A. Walker. *Chem Rev* **2004**, 104, 589.
158. D. K. Menyhard and G. M. Keseru. *J Am Chem Soc* **1998**, 120, 7991.
159. F. A. Walker. *Coord Chem Rev* **1999**, 185–186, 471.
160. S. D. Zaric, D. M. Popovic, E. W. Knapp. *Biochemistry* **2001**, 40, 7914.
161. C. Fufezan, J. Zhang, M. R. Gunner. *Proteins* **2008**, 73, 690.
162. W. R. Scheidt and D. M. Chipman. *J Am Chem Soc* **1986**, 108, 1163.
163. K. Hatano, M. K. Safo, F. A. Walker, W. R. Scheidt. *Inorg Chem* **1991**, 30, 1643.
164. O. Q. Munro, J. A. Serth-Guzzo, I. Turkovska-Tyrk, K. Mohanrao, T. K. Shokhireva, F. A. Walker, P. G. Debrunner, W. R. Scheidt. *J Am Chem Soc* **1999**, 121, 11144.
165. M. K. Safo, M. J. M. Nesset, F. A. Walker, P. G. Debrunner, W. R. Scheidt. *J Am Chem Soc* **1997**, 119, 9438.
166. A. S. Galstyan, S. D. Zaric, E. W. Knapp. *J Biol Inorg Chem* **2005**, 10, 343.
167. S. E. Phillips. *J Mol Biol* **1980**, 142, 531.
168. O. V. Kosmachevskaya and A. F. Topunov. *Appl Biochem Microbiol* **2009**, 45, 627.
169. M. F. Perutz, A. J. Wilkinson, M. Paoli, G. G. Dodson. *Annu Rev Biophys Biomol Struct* **1998**, 27, 1.
170. A. J. Lukin and H. Chien. *Chem Rev* **2004**, 104, 1219.
171. S. Y. Park, T. Yokoyama, N. Shibayama, Y. Shiro, J. R. H. Tame. *J Mol Biol* **2006**, 360, 690.
172. E. Antonini and M. Brunori. *Hemoglobin and Myoglobin in Their Reactions with Ligands*. North-Holland, Amsterdam, 1971.
173. J. Rinku and M. K. Chan. *J Biol Inorg Chem* **2003**, 8, 1.
174. G. Fermi, M. F. Perutz, B. Shaanan, R. Fourme. *J Mol Biol* **1984**, 175, 159.
175. T. G. Traylor, M. J. Mitchell, S. Tsuchiya, D. H. Campbell, D. V. Stynes, N. Koga. *J Am Chem Soc* **1981**, 103, 5234.
176. T. G. Traylor, D. K. White, D. H. Campbell, A. P. Berzinis. *J Am Chem Soc* **1981**, 103, 4932.
177. T. G. Traylor, N. Koga, L. A. Deardurff. *J Am Chem Soc* **1985**, 107, 6504.
178. B. A. Springer, S. G. Sligar, J. S. Olson, G. N. Phillips Jr. *Chem Rev* **1994**, 94, 699.

179. J. Vojtechovsky, K. Chu, J. Berendzen, R. M. Sweet, I. Schlichting. *Biophys J* **1999**, 77, 2153.
180. A. Ostermann, R. Waschipky, F. G. Parak, G. U. Nienhaus. *Nature* **2000**, 404, 205.
181. G. N. Phillips. *Handbook of Metalloproteins*, A. Messerschmidt, R. Huber, K. Wieghardt, T. Poulos, eds. John Wiley & Sons, Chichester, UK, 2001.
182. J. Everse, K. E. Everse, M. B. Grisham. *Peroxidases in Chemistry and Biology*, Vol. 1. CRC Press, Boca Raton, FL, 1991.
183. L. Banci. *J Biotechnol* **1997**, 53, 253.
184. K. G. Welinder. *Curr Opin Struct Biol* **1992**, 2, 388.
185. T. L. Poulos, S. T. Freer, R. A. Alden, S. L. Edward, U. Skogland, K. Takio, B. Eriksson, N. Xuong, T. Yonetani, J. Kraut. *J Biol Chem* **1980**, 255, 575.
186. B. C. Finzel, T. L. Poulos, J. Kraut. *J Biol Chem* **1984**, 259, 13027.
187. T. L. Poulos. *Arch Biochem Biophys* **2010**, 500, 3.
188. C. Obinger. *Arch Biochem Biophys* **2010**, 500, 1.
189. M. I. Savenkova, J. M. Kuo, P. R. Ortiz de Montellano. *Biochemistry* **1998**, 37, 10828.
190. J. H. Dawson. *Science* **1988**, 240, 433.
191. G. Battistuzzi, M. Bellei, C. A. Bortolotti, M. Sola. *Arch Biochem Biophys* **2010**, 500, 21.
192. F. S. Mathews. *Prog Biophys Mol Biol* **1985**, 45, 1.
193. P. D. Barker and S. J. Ferguson. *Structure* **1999**, 7, R281.
194. F. Lederer. *Biochimie* **1994**, 76, 674.
195. R. J. Kassner. *J Am Chem Soc* **1973**, 95, 2675.
196. E. Stellwagen. *Nature* **1978**, 275, 73.
197. F. A. Tezcan, J. R. Winkler, H. B. Gray. *J Am Chem Soc* **1998**, 120, 13383.
198. J. J. Mao, K. Hauser, M. R. Gunner. *Biochemistry* **2003**, 42, 9829.
199. A. G. Mauk and G. R. Moore. *J Biol Inorg Chem* **1997**, 2, 119.
200. M. R. Gunner, E. Alexov, E. Torres, S. Lipovaca. *J Biol Inorg Chem* **1997**, 2, 126.
201. G. Naray-Szabo. *J Biol Inorg Chem* **1997**, 2, 135.
202. D. S. Wuttke and H. B. Gray. *Curr Opin Struct Biol* **1993**, 3, 555.
203. H. B. Gray and J. R. Winkler. *Annu Rev Biochem* **1996**, 65, 537.
204. P. R. Ortiz de Montellano, ed. *Cytochrome P450: Structure, Mechanism, and Biochemistry*, 3rd ed. Kluwer Academic/Plenum Publishers, New York, 2005.
205. D. Mansuy. *Ann Pharm Fr* **2011**, 69, 62.
206. W. D. Woggon. *Top Curr Chem* **1997**, 184, 39.
207. M. Flinspach, H. Li, J. Jamal, W. Yang, H. Huang, R. B. Silverman, T. L. Poulos. *Biochemistry* **2004**, 43, 5181.
208. S. Colonna, N. Gaggero, C. Relmi, P. Pasta. *Trends Biotechnol* **1999**, 17, 163.
209. D. R. Nelson. *Hum Genomics* **2009**, 4, 59.
210. T. L. Poulos, B. C. Finzel, A. J. Howard. *Biochemistry* **1986**, 25, 5314.
211. P. R. Ortiz de Montellano. *Chem Rev* **2010**, 110, 932.
212. A. L. Feig and S. J. Lippard. *Chem Rev* **1994**, 94, 759–805.
213. B. J. Wallar and J. D. Lipscomb. *Chem Rev* **1996**, 96, 2625.
214. P. Nordlund, B.-M. Sjöberg, H. Eklund. *Nature* **1990**, 345, 593.

215. S. J. Lange and L. Que. *Curr Opin Chem Biol* **1998**, 2, 159.
216. K. K. Andersson and A. Gräslund. *Adv Inorg Chem* **1995**, 43, 359.
217. P. Nordlund and H. Eklund. *Curr Opin Struct Biol* **1995**, 5, 758.
218. J. D. Lipscomb. *Annu Rev Microbiol* **1994**, 48, 371.
219. S. C. Gallagher, A. George, H. Dalton. *Eur J Biochem* **1998**, 254, 480.
220. F. Bouvier, Y. Keller, A. D'Harlinghe, B. Cammara. *Biochim Biophys Acta* **1998**, 1391, 320.
221. M. Lee, M. Lenman, A. Banas, M. Bafor, S. Singh, M. Schweizer, R. Nilsson, C. Liljenberg, A. Dahlqvist, P. O. Gummeson, S. Sjodahl, A. Green, S. Stymne. *Science* **1998**, 280, 915.
222. J. A. Broadwater, J. Ai, T. M. Loehr, J. Sanders-Loehr, B. G. Fox. *Biochemistry* **1998**, 37, 14664.
223. H. Sigel, ed. *Metal Ions in Biological Systems*. Marcel Dekker, New York, 1998; p. 35.
224. S. Macedo, C. V. Romao, E. Mitchell, P. M. Matias, M. Y. Liu, A. V. Xavier, J. LeGall, M. Teixeira, P. Lindley, M. A. Carrondo. *Nat Struct Biol* **2003**, 10, 285.
225. R. E. Stenkamp. *Chem Rev* **1994**, 94, 715.
226. D. A. Berthold, M. E. Andersson, P. Nordlund. *Biochim Biophys Acta* **2000**, 1460, 241.
227. D. A. Berthold and P. Stenmark. *Annu Rev Plant Biol* **2003**, 54, 497.
228. M. E. Andersson and P. Nordlund. *FEBS Lett* **1999**, 449, 17.
229. E. I. Solomon, T. C. Brunold, M. I. Davis, J. N. Kemsley, S. K. Lee, N. Lehnert, F. Neese, A. J. Skulan, Y. S. Yang, J. Zhou. *Chem Rev* **2000**, 100, 235.
230. Y. Jin and J. D. Lipscomb. *J Biol Inorg Chem* **2001**, 6, 717.
231. J. Stubbe and W. van der Donk. *Chem Rev* **1998**, 98, 705.
232. J. Du Bois, T. J. Mizoguchi, S. J. Lippard. *Coord Chem Rev* **2000**, 200–202, 443.
233. J. Stubbe and W. A. van der Donk. *Chem Biol* **1995**, 2, 793.
234. J. Stubbe. *Proc Natl Acad Sci U S A* **1998**, 95, 2723.
235. U. Uhlin and H. Eklund. *Nature* **1994**, 370, 533.
236. M. Ericksson, U. Uhlin, S. Ramaswamy, M. Ekberg, K. Regnstrom, B. M. Sjoberg, H. Eklund. *Structure* **1997**, 5, 1078.
237. P. Nordlund and H. Eklund. *J Mol Biol* **1993**, 232, 123.
238. D. T. Logan, X. D. Su, A. Åberg, K. Regnstrom, J. Hajdu, H. Eklund, P. Nordlund. *Structure* **1996**, 4, 1053.
239. K. J. Gibson. *Biochim Biophys Acta* **1993**, 1169, 231.
240. M. Merkx, D. A. Kopp, M. H. Sazinsky, J. L. Blazyk, J. Müller, S. J. Lippard. *Angew Chem Int Ed* **2001**, 40, 2782.
241. P. Moenne-Loccoz, J. Baldwin, B. A. Ley, T. M. Loehr, J. M. Jr. Bollinger. *Biochemistry* **1998**, 37, 14659.
242. X. Yang, Y. Chen-Barrett, P. Arosio, N. D. Chasteen. *Biochemistry* **1998**, 37, 9743.
243. L. Shu, J. C. Nesheim, K. Kauffmann, E. Munck, J. D. Lipscomb, L. Jr. Que. *Science* **1997**, 275, 515.
244. M. Torrent, D. G. Musaev, H. Basch, K. Morokuma. *J Comput Chem* **2002**, 23, 59.

245. E. C. Tinberg and S. J. Lippard. *Biochemistry* **2009**, 48, 12145.
246. B. J. Brazeau and J. D. Lipscomb. *Biochemistry* **2000**, 39, 13503.
247. A. C. Rosenzweig, C. A. Frederick, S. J. Lippard, P. Nordlund. *Nature* **1993**, 366, 537.
248. A. C. Rosenzweig, P. Nordlund, P. M. Takahara, C. A. Frederick, S. J. Lippard. *Chem Biol* **1995**, 2, 409.
249. N. Elango, R. Radhakrishnan, W. A. Froland, B. J. Wallar, C. A. Earhart, J. D. Lipscomb, D. H. Ohlendorf. *Protein Sci* **1997**, 6, 556.
250. S.-L. Chang, B. J. Wallar, J. D. Lipscomb, K. H. Mayo. *Biochemistry* **1999**, 38, 5799.
251. K. J. Walters, G. T. Gassner, S. J. Lippard, G. Wagner. *Proc Natl Acad Sci U S A* **1999**, 96, 7877.
252. D. A. Whittington and S. J. Lippard. *J Am Chem Soc* **2001**, 123, 827.
253. A. C. Rosenzweig and S. J. Lippard. *Acc Chem Res* **1994**, 27, 229.
254. A. C. Rosenzweig, H. Brandstetter, D. A. Whittington, P. Nordlund, S. J. Lippard, C. A. Frederick. *Proteins* **1997**, 29, 141.
255. D. A. Whittington, M. H. Sazinsky, S. J. Lippard. *J Am Chem Soc* **2001**, 123, 1794.
256. D. N. Vaz, D. F. McGinnity, M. J. Coon. *Proc Natl Acad Sci U S A* **1998**, 95, 3555.
257. H. Eklund, U. Uhlin, M. Färnegårdh, D. T. Logan, P. Nordlund. *Prog Biophys Mol Biol* **2001**, 77, 177.
258. M. Ormo, K. Regnstrom, Z. Wang, L. Que Jr., M. Sahlin, B. M. Sjöberg. *J Biol Chem* **1995**, 270, 6570.
259. N. Mitić, M. D. Clay, L. Saleh, J. M. Bollinger, E. I. Solomon Jr. *J Am Chem Soc* **2007**, 129, 9049.
260. D. Yun, R. García-Serres, B. M. Chicalese, Y. H. An, B. H. Huynh, J. M. Bollinger Jr. *Biochemistry* **2007**, 46, 1925.
261. M. H. Sazinsky and S. J. Lippard. *Acc Chem Res* **2006**, 39, 558.
262. Y. Lindqvist. *Δ^9-Stearoyl-Acyl Carrier Protein Desaturase*. John Wiley & Sons, Chichester, 2001.
263. M. Moche, J. Shanklin, A. Ghoshal, Y. Lindqvist. *J Biol Chem* **2003**, 278, 25072.
264. Y. Lindqvist, W. Huang, G. Schneider, J. Shanklin. *EMBO J* **1996**, 15, 4081.
265. E. Whittle, E. B. Cahoon, S. Subrahmanyam, J. Shanklin. *J Biol Chem* **2005**, 280, 28169.
266. J. E. Guy, E. Whittle, D. Kumaran, Y. Lindqvist, J. Shanklin. *J Biol Chem* **2007**, 282, 19863.
267. J. Guy, I. A. Abreu, M. Moche, Y. Lindqvist, E. Whittle, J. Shanklin. *Proc Natl Acad Sci U S A* **2006**, 103, 17220.
268. C. P. Mangum, ed. In *Advances in Comparative and Environmental Physiology*, Vol. 13. Springer-Verlag, New York, 1992; p. 173.
269. D. M. Kurtz Jr. *Comprehensive Coordination Chemistry II*, Vol. 8. J. A. McCleverty, T. J. Meyer, eds. Elsevier, Oxford, 2004; p. 229.
270. C. E. French, J. M. L. Bell, F. B. Ward. *FEMS Microbiol Lett* **2008**, 279, 131.
271. C. E. Isaza, R. Silaghi-Dumitrescu, R. B. Iyer, D. M. Kurtz Jr., M. K. Chan. *Biochemistry* **2006**, 45, 9023.

272. W.-C. Kao, V. C.-C. Wang, Y.-C. Huang, S. S.-F. Yu, T.-C. Chang, S. I. Chan. *J Inorg Biochem* **2008**, 102, 1607.
273. R. C. Reem, J. M. McCormick, D. E. Richardson, F. J. Devlin, P. J. Stephens, R. L. Musselman, E. I. Solomon. *J Am Chem Soc* **1989**, 111, 4688.
274. S. Pulver, W. A. Froland, B. G. Fox, J. D. Lipscomb, E. I. Solomon. *J Am Chem Soc* **1993**, 115, 12409.
275. R. E. Stenkamp. *J Mol Biol* **1991**, 218, 583.
276. M. A. Holmes, I. Le Trong, S. Turley, L. C. Sieker, R. E. Stenkamp. *J Mol Biol* **1991**, 218, 583.
277. Y. Lu. *Angew Chem Int Ed* **2006**, 45, 5588.
278. Y. Lu, N. Yeung, N. Sieracki, N. M. Marshall. *Nature* **2009**, 460, 855.
279. S. J. Lippard. *Nature* **2002**, 416, 587.
280. S. J. Lippard. *Nat Chem Biol* **2006**, 2, 504.
281. A. Ciferri. *Chem Eur J* **2010**, 16, 10930.
282. A. Ciferri. *Biopolymers* **2008**, 89, 700.
283. S. Borovik. *Acc Chem Res* **2005**, 38, 54.
284. C. M. Yang, X. Li, W. Wei, Y. Li, Z. Duan, J. Zheng, T. Huang. *Chem Eur J* **2007**, 13, 3120.
285. S. M. Berry, M. H. Baker, N. J. Reardon. *J Inorg Biochem* **2010**, 104, 1071.
286. R. Yerushalmi, A. Brandis, V. Rosenbach-Belkin, K. K. Baldridge, A. Scherz. *J Phys Chem A* **2006**, 110, 412.
287. R. L. Shook and A. S. Borovik. *Inorg Chem* **2010**, 49, 3646.
288. Y. Lu, S. M. Berry, T. D. Pfisterand. *Chem Rev* **2001**, 101, 3047.
289. T. M. Penning and J. M. Jez. *Chem Rev* **2001**, 101, 3027.
290. W. B. Tolman and D. J. E. Spencer. *Curr Opin Chem Biol* **2001**, 5, 188.
291. E. Y. Tshuva and S. J. Lippard. *Chem Rev* **2004**, 104, 987.
292. P. V. Rao and R. H. Holm. *Chem Rev* **2004**, 104, 527.
293. M. Costas, M. P. Mehn, M. P. Jensen, L. Que. *Chem Rev* **2004**, 104, 939.
294. *Ullmann's Encyclopedia of Industrial Chemistry*, 6th ed. Electronic Release, Wiley-VCH, Weinheim, 1998.
295. J. A. Kent ed., *Riegel's Handbook of Industrial Chemistry*, 10th ed. Kluwer Academic/Plenum, New York, 2003.
296. A. Le Goff, V. Artero, B. Jousselme, P. D. Tran, N. Guillet, R. Métayé, A. Fihri, S. Palacin, M. Fontecave. *Science* **2009**, 326, 1384.
297. H. W. Hellinga. *Protein Engineering: Principles and Practice*. Wiley-Lyss, New York, 1996.
298. W. D. Kohn and R. S. Hodge. *Trends Biotechnol* **1998**, 16, 379.
299. W. F. DeGrado, C. M. Summa, V. Pavone, F. Nastri, A. Lombardi. *Annu Rev Biochem* **1999**, 68, 779.
300. L. Baltzer. *Curr Opin Struct Biol* **1998**, 8, 466.
301. L. Baltzer, H. Nilsson, J. Nilsson. *Chem Rev* **2001**, 101, 3153.
302. O. Maglio, F. Nastri, V. Pavone, A. Lombardi, W. F. DeGrado. *Proc Natl Acad Sci U S A* **2003**, 100, 3772.

303. A. Pasternak, J. Kaplan, J. D. Lear, W. F. DeGrado. *Protein Sci* **2001**, 10, 958.
304. Y. Lu and S. M. Berry. *Encyclopedia Life Sci* **2005**. DOI: 10.1038/npg.els.0002983.
305. G. Xing and V. J. DeRose. *Curr Opin Chem Biol* **2001**, 5, 196.
306. A. F. A. Peacock, O. Iranzo, V. L. Pecoraro. *Dalton Trans* **2009**, 2271.
307. P. D. Barker. *Curr Opin Struct Biol* **2003**, 13, 490.
308. F. Rosati and G. Roelfes. *ChemCatChem* **2010**, 2, 916.
309. R. Razeghifard, B. B. Wallace, R. J. Pace, T. Wydrzynski. *Curr Protein Pept Sci* **2007**, 8, 3.
310. M. L. Kennedy and B. R. Gibney. *Curr Opin Struct Biol* **2001**, 11, 485.
311. D. Ghosh and V. L. Pecoraro. *Curr Opin Chem Biol* **2005**, 9, 97.
312. V. Muñoz Robles, E. Ortega-Carrasco, E. Gonzalez Fuentes, A. Lledos, J. Marechal. *Faraday Discuss* **2011**, 148, 137.
313. T. Ueno, S. Abe, N. Yokoi, Y. Watanabe. *Coord Chem Rev* **2007**, 251, 2717.
314. M. A. Case and G. L. McLendon. *Acc Chem Res* **2004**, 37, 754.
315. G. Gilardi, A. Fantuzzi, S. J. Sadeghi. *Curr Opin Struct Biol* **2001**, 11, 491.
316. T. Albrecht, W. Li, J. Ulstrup, W. Haehnel, P. Hildebrandt. *Chemphyschem* **2005**, 6, 961.
317. R. L. Koder and P. L. Dutton. *Dalton Trans* **2006**, 3045.
318. A. Lombardi, D. Marasco, O. Maglio, L. Di Costanzo, F. Nastri, V. Pavone. *Proc Natl Acad Sci U S A* **2000**, 97, 11922.
319. Z. Dauter, L. C. Sieker, K. S. Wilson. *Acta Crystallogr B* **1992**, 48, 42.
320. W. F. DeGrado, Z. R. Wasserman, J. D. Lear. *Science* **1989**, 243, 622.
321. J. W. Bryson, S. F. Betz, H. S. Lu, D. J. Suich, H. X. Zhou, K. T. O'Neil, W. F. DeGrado. *Science* **1995**, 270, 935.
322. K. T. O'Neil and W. F. DeGrado. *Science* **1990**, 250, 646.
323. R. Aurora and G. D. Rose. *Protein Sci* **1998**, 7, 21.
324. M. J. McGregor, S. A. Islam, M. J. Sternberg. *J Mol Biol* **1987**, 198, 295.
325. R. Munoz and L. Serrano. *Proteins* **1994**, 20, 301.
326. R. L. Dunbrack Jr. and M. Karplus. *Nat Struct Biol* **1994**, 1, 334.
327. M. A. Willis, B. Bishop, L. Regan, A. T. Brunger. *Structure* **2000**, 8, 1319.
328. B. Hill, D. P. Raleigh, A. Lombardi, W. F. DeGrado. *Acc Chem Res* **2000**, 33, 745.
329. D. P. Raleigh, S. F. Betz, W. F. DeGrado. *J Am Chem Soc* **1995**, 117, 7558.
330. C. D. Waldburger, J. F. Schildbach, R. T. Sauer. *Nat Struct Biol* **1995**, 2, 1122.
331. P. B. Harbury, T. Zhang, P. S. Kim, T. Alber. *Science* **1993**, 262, 1401.
332. E. K. O'Shea, K. J. Lumb, P. S. Kim. *Curr Biol* **1993**, 3, 658.
333. K. J. Lumb and P. S. Kim. *Science* **1996**, 271, 1137.
334. K. J. Lumb and P. S. Kim. *Science* **1995**, 268, 436.
335. Y. Yu, O. D. Monera, R. S. Hodges, P. L. Privalov. *J Mol Biol* **1996**, 255, 367.
336. P. Lavigne, F. D. Sönnichsen, C. M. Kay, R. S. Hodges. *Science* **1996**, 271, 1136.
337. D. J. Lockhart and P. S. Kim. *Science* **1993**, 260, 198.
338. R. H. Holm and E. I. Solomon. *Chem Rev* **2004**, 104, 347.
339. J. G. Saven. *Curr Opin Colloid Interface Sci* **2010**, 15, 13.

340. G. A. Lazar and T. M. Handel. *Curr Opin Chem Biol* **1998**, 2, 675.
341. G. L. Butterfoss and B. Kuhlman. *Annu Rev Biophys Biomol Struct* **2006**, 35, 49.
342. M. H. Hecht, A. Das, A. Go, L. H. Bradley, Y. Wei. *Protein Sci* **2004**, 13, 1711.
343. L. H. Bradley, P. P. Thumfort, M. H. Hecht. *Methods Mol Biol* **2006**, 340, 53.
344. L. H. Bradley, Y. Wei, P. P. Thumfort, C. Wurth, M. H. Hecht. *Methods Mol Biol* **2007**, 352, 155.
345. J. Minshull and W. P. Stemmer. *Curr Opin Chem Biol* **1999**, 3, 284.
346. R. H. Hoess. *Chem Rev* **2001**, 101, 3205.
347. R. L. Koder, J. L. R. Anderson, L. A. Solomon, K. S. Reddy, C. C. Moser, P. L. Dutton. *Nature* **2009**, 458, 305.
348. N. Yeung, Y.-W. Lin, Y.-G. Gao, X. Zhao, B. S. Russell, L. Lei, K. D. Miner, H. Robinson, Y. Lu. *Nature* **2009**, 426, 1079.
349. A. Lombardi, F. Nastri, V. Pavone. *Chem Rev* **2001**, 101, 3165.
350. F. Nastri, A. Lombardi, L. D. D'Andrea, M. Sanseverino, O. Maglio, V. Pavone. *Biopolymers* **1998**, 47, 5.
351. J. R. Calhoun, F. Nastri, O. Maglio, V. Pavone, A. Lombardi, W. F. DeGrado. *Pept Sci* **2005**, 80, 264.
352. O. Maglio, F. Nastri, R. Torres Martin de Rosales, M. Faiella, V. Pavone, W. F. DeGrado, A. Lombardi. *C R Chimie* **2007**, 10, 703.
353. H. M. Marques. *Dalton Trans* **2007**, 4371.
354. S. Nicolis, L. Casella, R. Roncone, C. Dallacosta, E. Monzani. *C R Chimie* **2007**, 10, 1.
355. M. L. Kennedy, S. Silchenko, N. Houndonougbo, B. R. Gibney, P. L. Dutton, K. R. Rodgers, D. R. Benson. *J Am Chem Soc* **2001**, 123, 4635.
356. D. R. Benson, B. R. Hart, X. Zhu, M. B. Doughty. *J Am Chem Soc* **1995**, 117, 8502.
357. J. M. Cordova, P. L. Noack, S. A. Hilcove, J. D. Lear, G. Ghirlanda. *J Am Chem Soc* **2007**, 129, 512.
358. I. Obataya, T. Kotaki, S. Sakamoto, A. Ueno, H. Mihara. *Bioorg Med Chem Lett* **2000**, 10, 2719.
359. S. Sakamoto, I. Obataya, A. Ueno, H. Mihara. *Chem Commun* **1999**, 1111.
360. C. T. Choma, J. D. Lear, M. J. Nelson, P. L. Dutton, D. E. Robertson, W. F. DeGrado. *J Am Chem Soc* **1994**, 116, 856.
361. D. E. Robertson, R. S. Farid, C. C. Moser, J. L. Urbauer, S. E. Mulholland, R. Pidikiti, J. D. Lear, A. J. Wand, W. F. DeGrado, P. L. Dutton. *Nature* **1994**, 368, 425.
362. J. M. Shifman, C. C. Moser, W. A. Kalsbeck, D. F. Bocian, P. L. Dutton. *Biochemistry* **1998**, 37, 16815.
363. G. Ghirlanda, A. Osyczka, W. Liu, M. Antolovich, K. M. Smith, P. L. Dutton, A. J. Wand, W. F. DeGrado. *J Am Chem Soc* **2004**, 126, 8141.
364. F. Nastri, A. Lombardi, G. Morelli, O. Maglio, G. D'Auria, C. Pedone, V. Pavone. *Chem Eur J* **1997**, 3, 340.
365. G. D'Auria, O. Maglio, F. Nastri, A. Lombardi, M. Mazzeo, G. Morelli, L. Paolillo, C. Pedone, V. Pavone. *Chem Eur J* **1997**, 3, 350.
366. F. Nastri, A. Lombardi, G. Morelli, C. Pedone, V. Pavone, G. Chottard, P. Battioni, D. Mansuy. *J Biol Inorg Chem* **1998**, 3, 671.

367. A. Lombardi, F. Nastri, M. Sanseverino, O. Maglio, C. Pedone, V. Pavone. *Inorg Chim Acta* **1998**, 301, 275.
368. L. Di Costanzo, S. Geremia, L. Randaccio, F. Nastri, O. Maglio, A. Lombardi, V. Pavone. *J Biol Inorg Chem* **2004**, 9, 1017.
369. A. Lombardi, F. Nastri, D. Marasco, O. Maglio, G. De Sanctis, F. Sinibaldi, R. Santucci, M. Coletta, V. Pavone. *Chem Eur J* **2003**, 9, 5643.
370. F. Nastri, L. Lista, P. Ringhieri, R. Vitale, M. Faiella, C. Andreozzi, P. Travascio, O. Maglio, A. Lombardi, V. Pavone. *Chem Eur J* **2011**, 17, 4444.
371. A. Ranieri, S. Monari, M. Sola, M. Borsari, G. Battistuzzi, P. Ringhieri, F. Nastri, V. Pavone, A. Lombardi. *Langmuir* **2010**, 26, 17831.
372. E. Tshuva and S. J. Lippard. *Chem Rev* **2004**, 104, 987–1012.
373. J. U. Rohde, J. H. In, M. H. Lim, W. W. Brennessel, M. R. Bukowski, A. Stubna, E. Munck, W. Nam, L. Que Jr. *Science* **2003**, 299, 1037–1039.
374. C. He and Y. Mishina. *Curr Opin Chem Biol* **2004**, 8, 201–208.
375. M. Jarenmark, H. Carlsson, E. Nordlander. *C R Chimie* **2007**, 10, 433.
376. A. Lombardi, C. Summa, S. Geremia, L. Randaccio, V. Pavone, W. F. DeGrado. *Proc Natl Acad Sci U S A* **2000**, 97, 6298.
377. C. M. Summa, A. Lombardi, M. Lewis, W. F. DeGrado. *Curr Opin Struct Biol* **1999**, 9, 500.
378. L. Di Costanzo, H. Wade, S. Geremia, L. Randaccio, V. Pavone, W. F. DeGrado, A. Lombardi. *J Am Chem Soc* **2001**, 123, 12749.
379. W. F. DeGrado, L. Di Costanzo, S. Geremia, A. Lombardi, V. Pavone, L. Randaccio. *Angew Chem Int Ed Engl* **2003**, 42, 417.
380. S. Geremia, L. Di Costanzo, L. Randaccio, D. E. Engel, A. Lombardi, F. Nastri, W. F. DeGrado. *J Am Chem Soc* **2005**, 127, 17266.
381. S. J. Lahr, D. E. Engel, S. E. Stayrook, O. Maglio, B. North, S. Geremia, A. Lombardi, W. F. DeGrado. *J Mol Biol* **2005**, 346, 1441.
382. O. Maglio, F. Nastri, J. R. Calhoun, S. Lahr, H. Wade, V. Pavone, W. F. DeGrado, A. Lombardi. *J Biol Inorg Chem* **2005**, 10, 539.
383. H. Wade, S. E. Stayrook, W. F. DeGrado. *Angew Chem Int Ed Engl* **2006**, 45, 4951.
384. P. Wei, A. J. Skulan, H. Wade, W. F. DeGrado, E. I. Solomon. *J Am Chem Soc* **2005**, 127, 16098.
385. C. M. Summa, M. M. Rosenblatt, J. K. Hong, J. D. Lear, W. F. DeGrado. *J Mol Biol* **2002**, 321, 923.
386. E. N. G. Marsh and W. F. DeGrado. *Proc Natl Acad Sci U S A* **2002**, 99, 5150.
387. J. Kaplan and W. F. DeGrado. *Proc Natl Acad Sci U S A* **2004**, 101, 11566.
388. M. Faiella, C. Andreozzi, R. Torres Martin de Rosales, V. Pavone, O. Maglio, F. Nastri, W. F. DeGrado, A. Lombardi. *Nat Chem Biol* **2009**, 5, 882.
389. R. Torres Martin de Rosales, M. Faiella, E. Farquhar, L. Que Jr., C. Andreozzi, V. Pavone, O. Maglio, F. Nastri, A. Lombardi. *J Biol Inorg Chem* **2010**, 15, 717.
390. D. W. Banner, M. Kokkinidis, D. Tsernoglou. *J Mol Biol* **1987**, 196, 657.
391. C. Cohen and D. A. D. Parry. *Proteins* **1990**, 7, 1.
392. F. Frolow, A. J. Kalb (Gilboa), J. Yariv. *Nat Struct Biol* **1994**, 1, 453.

CHAPTER 12

CHARGE-INDUCED EFFECTS ON ACID–BASE TITRATION AND CONFORMATIONAL STABILITY OF PROTEINS AND POLYPEPTIDES

JAN HERMANS

12.1 INTRODUCTION

Many biopolymers assume not only the disordered statistical coil state typical of most polymers but also a specific folded conformation; typically, the latter is the native state indentified with biological activity. As these are polar molecules, the stability of the native state of proteins (and nucleic acids) is dependent, in part, on charge–charge interactions. This chapter pursues two distinct but related objectives: It reviews methods by which the dependence of the stability on charge interactions can be modeled, and it discusses how this dependence may be used to quantitate conformational stability in terms of the change in standard free energy without the need to invoke a specific model of the charge interaction. Examples will concern the stability of folded proteins and of α-helical polypeptides. That field is covered by a very sizeable literature; this chapter seeks to highlight principles, as well as focus in more detail on somewhat older work.

12.2 STUDIES OF PROTEINS

12.2.1 Ionizable Groups in Proteins

Proteins contain a number of amino acids with side chains that carry a net charge of +1 or −1 elementary charge (electronic charge, 1.60218×10^{-19} C = 4.8032×10^{-10} e.s.u.) under physiological conditions; nucleic acids contain one

Ionic Interactions in Natural and Synthetic Macromolecules, First Edition. Edited by Alberto Ciferri and Angelo Perico.

negative charge per phosphate group in the backbone. Many of these groups can be made to revert to their neutral form by a pH change, as a result of either binding or giving up a hydrogen ion. Some protein side chains and the side chains of nucleic acids are not charged at pH 7 but can bind or give up a hydrogen ion in the accessible pH range.

The following rule of thumb characterizes folded globular proteins: Charged side chains are typically exposed to solvent; electrically neutral side chains with polar groups are distributed both on the surface and in the interior, while hydrophobic side chains tend to be buried in the interior. The presence of charged, solvated side chains on the surface is thought to counter aggregation and enhance solubility, while as was first deduced by Kauzmann,[1] the packing of hydrophobic side chains is generally felt to be critical for the stability of the folded conformation.

Interactions between polar groups with zero net charge (dipoles) are ubiquitous in compact, native structures of both proteins and nucleic acids, and are thought to make further important contributions to stabilizing these structures. Direct contacts between oppositely charged side chains (so-called salt bridges) are relatively rare in proteins; these will be discussed in more detail in what follows.

The side chains of four types of amino acids in proteins are normally in their charged form at neutral pH (positively charged for arginine and lysine; negatively charged for aspartate and glutamate). The side chains of two amino acids (cystine and tyrosine) convert into the negatively charged ion form at elevated pH, while the side chains of aspartate and glutamate residues convert into the neutral form at low pH and those of lysine residues do so at high pH. The charged (low pH) and neutral (high pH) forms of the side chain of histidine are in equilibrium in equal concentrations near pH 7.

As a reference for the hydrogen ion binding properties of an amino acid side chain (or nucleic acid base), one may take those of small molecule models having a related chemical structure. For example, for the dissociation of the carboxyl group of a glutamic acid side chain in a protein, one might use as models the dissociation of acetic or propionic acid, of glutamic acid itself, and of peptides containing a glutamic acid residue flanked by other residues. Such comparisons have yielded a set of "normal" dissociation constants (here expressed as $pK_a = \log_{10} K_b$; K_a refers to the acid dissociation constant and K_b to the equilibrium binding constant) for this and other side chains, plus the chain terminal, or α-, amino and carboxyl groups; (see Appendix 1). These numbers are not very precise, as there are small variations among the pK_a values of any reasonably chosen set of reference compounds.

It turns out that the side chain of a free amino acid is actually not a good model of a side chain of the same type of residue in a peptide because in the free residue the nearby charged α-amino and α-carboxyl groups perturb the pK_a of the side chain. The best values of normal pK_as of side chains in proteins are based on measurements on series of peptides containing a single residue of an ionizable side chain flanked by alanine and glycine residues, in which the effect of the terminal charges is found to rapidly diminish as the chain becomes longer.

The positively charged form of the histidine side chain has a hydrogen atom bonded to both nitrogen atoms. In the neutral form, there is a hydrogen atom on only one of the nitrogen atoms; in water, the two forms are in equilibrium, with the species with the hydrogen on N^{δ} favored by a factor of approximately 10, while for an uncharged histidine side chain buried in the interior of a protein, the placement of the hydrogen atom is determined by hydrogen bonding. In either case, the NH group acts as hydrogen donor and the "lone pair" nitrogen acts as an acceptor.

12.2.2 Acid–Base Titration of Ribonuclease

The measurements required to obtain the acid–base titration curve of a protein, such as the one for bovine pancreatic ribonuclease A of Figure 12.1,[2] are quite simple. A fairly concentrated solution of protein in salt solution (no buffer) is brought to higher and lower pH by the addition of small amounts of (usually) 1 M HCl and KOH, and one determines the difference in titrant added to a solution of protein and to the same volume of solvent.

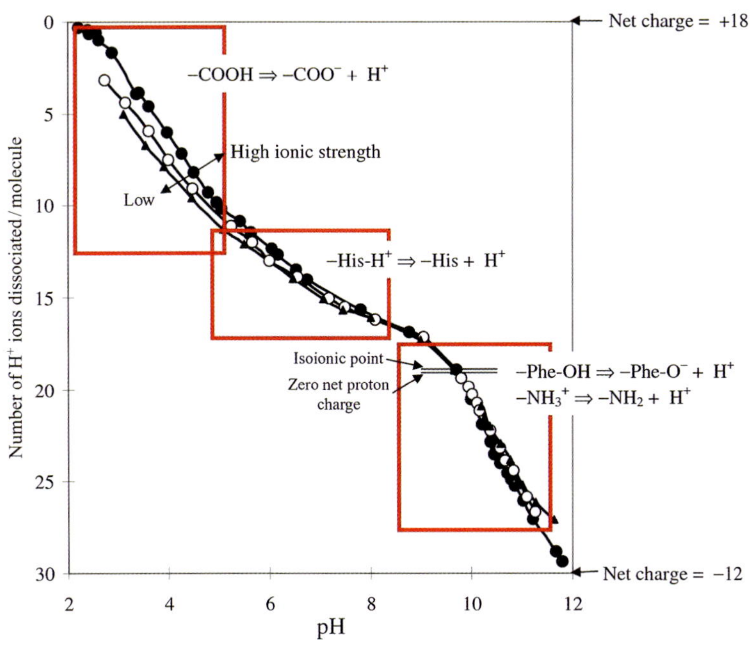

FIGURE 12.1. Titration curves of ribonuclease at three ionic strengths. Data from Tanford and Hauenstein.[2] The boxes indicate titration (from left to right) of side chains of aspartic and glutamic acid, of histidine, and of tyrosine and lysine (see text for additional explanation).

A first approximation to the acid–base titration curve of a protein is a set of dissociation curves for independently ionizing groups. Typically, the titration curve will show several smooth steps, one for each different category of side chains in the protein. These steps are indicated in the figure with boxes. Closer inspection shows that the midpoints of the three major steps do not lie at the "normal" pK_as. The titration is spread out over a larger range of pH than expected for normally titrating side chains, and the effect is larger at low ionic strength. This general effect is attributable to charge–charge interaction—the higher the (positive) charge, the more difficult it is to bind another hydrogen ion—and this results in a lower than normal effective pK_a; conversely, a negative net molecular charge raises pK_a. We shall come back to this effect below.

The titration can be expressed in terms of two scales. One, the number of hydrogen ions dissociated relative to the state at low pH, is indicated with numbers from 0 to 30. The other, the mean net charge of the molecule, covers the same range, as dissociation of one hydrogen ion lowers the net charge by 1, but this scale starts at +18 at low pH, which corresponds to the sum of the numbers of histidine, lysine, and arginine side chains in the molecule. The isoionic or isoelectric point is the pH where the mean net charge is zero and the molecules have zero electrophoretic mobility; for this protein, it is found to coincide within experimental error with the pH at which the net proton charge is zero. Binding of ions other than hydrogen ion from the solution can cause these to be different, but is not observed here.

12.2.3 Lang's Simple Model of the Effect of Charge–Charge Interactions on pK_a

An early model based on the Debye–Hückel theory (see Chapter 3, Section 3.2) was proposed by Linderstrøm-Lang, who sought to account for interactions between charged groups on a protein's surface and how these influence binding of hydrogen ion. This model represents the protein as a sphere in a medium with dielectric constant ε_a, and distributes over the surface of the sphere a charge Ze, where Z is the average net charge (number of positively minus negatively charged groups) and e is the elementary charge. The electrostatic potential at the surface of this model is

$$\Phi_a(R) = \frac{Ze}{\varepsilon_a R}\frac{1}{1+\kappa R}. \tag{12.1}$$

This changes the standard free energy of binding an additional hydrogen ion by an amount

$$\delta G_{el} = \frac{Ze^2}{\varepsilon_a R}\frac{1}{1+\kappa R}. \tag{12.2}$$

This again changes the binding constant from its value in the absence of charge–charge interaction ($Z = 0$)

$$\frac{K_a}{K_{a,0}} = \exp\frac{\delta G_{\text{el}}}{k_{\text{B}}T}, \tag{12.3}$$

and the corresponding difference in pK_a will be given by

$$\Delta \text{p}K_a = \frac{-\delta G_{\text{el}}}{\ln 10\; k_{\text{B}}T}. \tag{12.4}$$

(The factor $\ln 10 = 2.303\ldots$ is needed to convert to 10-based logarithms.) For the simple spherical model with a uniform surface charge density, this then gives

$$\Delta \text{p}K_a \approx \frac{-1}{\ln 10\; k_{\text{B}}T}\frac{Ze^2}{\varepsilon_{\text{a}}R}\frac{1}{1+\kappa R}. \tag{12.5}$$

This model assumes that the side chains are solvated as they would be in a peptide, and thus have normal pK_a, *except* for this repulsion.

According to this model, pK_a is lowered when Z is positive and increased when Z is negative and the effect is greater farther from the isoelectric point, as observed. This model produces reasonable agreement with the observed dependence of the hydrogen ion titration curve on net charge and ionic strength (Fig. 12.1).

12.2.4 Abnormal Ionizable Groups

The behavior of side chains of individual residues, as opposed to that of classes of residues, cannot be inferred from global (all-residue) titration data such as these. Even before global titrations were performed by Tanford's group, the titration of the tyrosine side chains of ribonuclease was studied by following changes in the UV absorption spectrum that accompany ionization of tyrosine side chains in model compounds (including tyrosine itself; similar changes are observed for phenol). The neutral form of the side chain absorbs weakly above 290 nm, while the ionized form has an absorption maximum at 295 nm; calibration with tyrosine and peptides containing tyrosine allows the absorbance of a protein at 295 nm to be used as a measure of the number of tyrosine residues in the ionized (high-pH) form. pH titration of ribonuclease[3,4] shows that of the six tyrosine side chains, three ionize with an (effective) pK_a between 9.5 and 10, while the other three ionize only at much higher pH (Fig. 12.2).

The model for the behavior of these three side chains is that in native ribonuclease these groups are an intrinsic part of the folded structure; this results

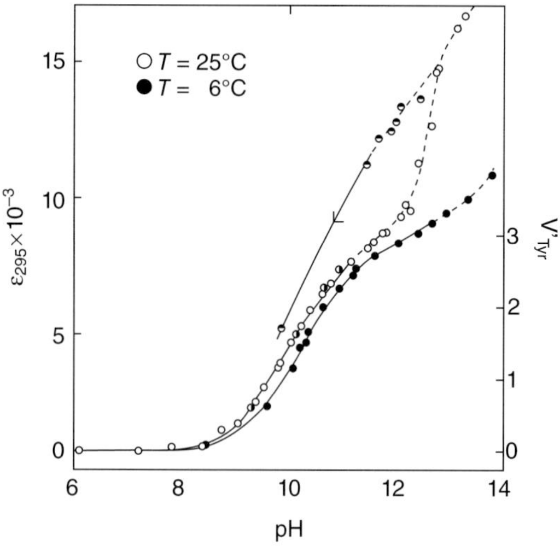

FIGURE 12.2. Titration of the tyrosine side chains of ribonuclease. The half-filled points are observed after exposure to pH 12.5 (upper) or 11.5 (lower).[3]

in their having a very high pK_a, so that their ionization is not observed below the pH where unfolding occurs. The suggestion that ionization of the latter three occurs with concomitant unfolding of the protein was already made by Tanford et al.[4] and was subsequently confirmed by (the author's unpublished) measurements of changes in optical rotatory dispersion at high pH. The ionization of these three groups is then *coupled* to the unfolding equilibrium of ribonuclease. Given these circumstances, the actual pK_as of the three abnormal tyrosine side chains in native ribonuclease have not been determined experimentally. Subsequent experiments[5] revealed the presence in ribonuclease of carboxyl groups with abnormally low pK_a, and binding of hydrogen ion to these groups at low pH was found to be coupled to the unfolding equilibrium; however, the coupling is weak and at temperatures below 20°C the protein remains largely folded even at a low pH of 0.9,[6] where dissociation of all carboxyl groups in ribonuclease is minimal.[2]

The three tyrosine side chains having a high effective pK_a are termed "abnormal." Abnormal ionizable groups are not uncommon. For example, the unfolding of sperm whale myoglobin at low pH is coupled to binding of hydrogen ion to six histidine side chains.

Global titration studies do not distinguish individual side chains; since these experiments were done, it has become possible to determine the pK_a of individual residues by measuring the pH dependence of nuclear magnetic resonances (NMRs) of carbon-bound hydrogens. An early application to the C^{δ} protons of the four histidine side chains of ribonuclease shows that two of these have normal pK_a of 6.6, while the other two (tentatively identified as active-site histidines) have lower pK_a of 5.4 and 5.8, respectively.[7] In later studies, this technique is typically applied to H resonances after these have been assigned to specific residues.[8,9]

12.2.5 Coupling of Ionization and Conformation Change in Myoglobin

A striking case of coupling of ionization and conformation change occurs in sperm whale myoglobin. The protein unfolds over a small range of pH near 4.3 at 25°C as the pH is lowered. The equilibria can be described with a simple model with six abnormally titrating side chains. These side chains remain in the unprotonated state in the native protein at low pH, but each takes up a hydrogen ion only when the protein unfolds; the six groups in question have been identified as side chains of histidine residues, and in the unfolded molecule their pK_a has a value in the normal range of 6.62.[10] (Of the 154 residues in the amino acid sequence of sperm whale myoglobin, 12 are histidines.)

The resulting coupling between binding of hydrogen ion and conformation change allows one to determine the stability of the folded state of myoglobin relative to the unfolded state at neutral pH.[11] To understand this, consider two different titration curves: one, the experimentally observed titration curve of the protein, and the other, the titration curve of unfolded myoglobin. At low pH, the experimental state is identical with the unfolded state, but titration to neutral pH converts this to the folded state. By now applying the relation between chemical potential and proton binding (Eq. A12.6), one finds that integration of the experimental hydrogen ion binding curve gives

$$\mu^{\circ}_{\text{unfolded}}(\text{pH}=0)-\mu^{\circ}_{\text{folded}}(\text{pH}=9)=-k_{\text{B}}T\int_{\text{pH}=9}^{\text{pH}=0}\nu_{\text{H}^+,\text{expt}}\,d\ln a_{\text{H}^+}. \tag{12.6}$$

On the other hand, integration of a hypothetical binding curve for unfolded protein with all histidines titrating with a normal pK_a gives

$$\mu^{\circ}_{\text{unfolded}}(\text{pH}=0)-\mu^{\circ}_{\text{unfolded}}(\text{pH}=9)=-k_{\text{B}}T\int_{\text{pH}=9}^{\text{pH}=0}\nu_{\text{H}^+,\text{unfolded}}\,d\ln a_{\text{H}^+}. \tag{12.7}$$

Subtraction of these two equations then gives the stability at near-neutral pH as

$$\mu^{\circ}_{\text{folded}}(\text{pH}=9)-\mu^{\circ}_{\text{unfolded}}(\text{pH}=9)=-k_{\text{B}}T\int_{\text{pH}=9}^{\text{pH}=0}(\nu_{\text{H}^+,\text{unfolded}}-\nu_{\text{H}^+,\text{expt}})\,d\ln a_{\text{H}^+}. \tag{12.8}$$

The cited paper[11] describes how to compute the hypothetical binding curve for unfolded protein using the known amino acid composition and taking into

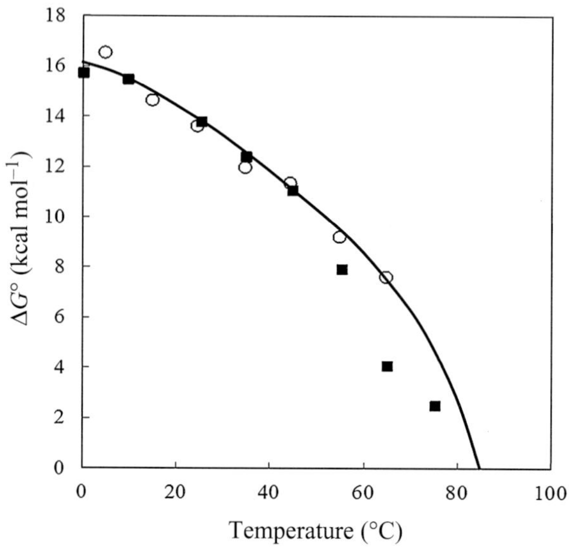

FIGURE 12.3. Temperature dependence of the standard free energy of unfolding sperm whale myoglobin at neutral pH.[11]

account charge–charge repulsion according to Lang's model, with an empirical repulsive parameter. At 25°C, the difference in chemical potentials between folded and unfolded myoglobin at neutral pH was found to be approximately −14 kcal mol^{-1}. (Sperm whale myoglobin has turned out to be one of the more stable small globular proteins.) Additional measurements at other temperatures then allowed determination of the temperature dependence of the standard free energy difference $\Delta\mu^\circ$ (or ΔG°), as shown in Figure 12.3. The data points indicate two series of experiments; points at higher temperatures are systematically low due to partially irreversible unfolding.

The marked curvature of the line drawn in Figure 12.3 corresponds to a large specific heat difference ΔC_P°; at 25°C, the enthalpy difference for forming the folded state from the unfolded state is $\Delta H^\circ = -40$ kcal mol^{-1} and the specific heat difference is $\Delta C_P^\circ = -1.4$ kcal mol^{-1} K^{-1}. These values are typical of unfolding equilibria of globular proteins; the behavior of myoglobin parallels that found by Brandts and coworkers in studies of temperature and pH dependence of chymotrypsin and ribonuclease.[12–14] It should be pointed out that in order to convert the measurements to standard thermodynamic parameters, Brandts et al. chose to rely on the two-state assumption, according to which partial unfolding is the result of an equilibrium between just two states, folded and unfolded, and part-folded, part-unfolded molecules are not observed.[15] On the other hand, this assumption was not made in the study on myoglobin, as it is not required when one estimates the standard free energy difference for folding from measurements of titration with ligand, in this case hydrogen ion, according to Equation 12.8. In fact, the cited study on myoglobin constitutes a proof that the two-state model is applicable to this protein.[11]

12.2.6 Local Interactions Cause pK_as to be Abnormal

The interior of a protein molecule is a less polar medium than water; however, buried polar side chain and backbone groups are nearly always stabilized by hydrogen bonds. Typically, the hydrogen bonds stabilize that particular form of the ionizable group that is present in the native structure at neutral pH; the other form of the side chain will interact more poorly with this environment, and as a result, the pK_a of fully buried side chains will be abnormal (high if the native buried form is the low-pH form, and low if this is the high-pH form). The abnormality will vary, depending on the degree of sequestration.

The uncharged forms of the side chains of Tyr, His, and Cys are frequently found in the core of the protein, with their polar, but neutral, ends forming hydrogen bonds with other parts of the protein. This is observed for the three tyrosine side chains of ribonuclease and the histidine side chains of myoglobin; these residues are usually identifiable on the basis of their interactions with other groups in high-resolution three-dimensional protein structures. As mentioned, the pK_as of these groups in the native protein cannot be determined because the groups titrate only if the protein unfolds. The equilibria for unfolding and binding of hydrogen ions are then said to be coupled.

In less extreme cases, the surrounding structure of the native protein may shift a side chain's pK_a within the accessible pH range. Thus, a shift of one or even two units in the pK_a of a histidine side chain may yet allow this residue to be titrated without the protein unfolding; also, a pK_a shift of several units may move the titration of a carboxyl group (of glutamate or aspartate) into the physiological pH range.

Compared with what is the case for the side chains of histidine and tyrosine, the neutral forms of the side chains of aspartate, glutamate, and lysine are less common in the core of proteins. One can understand this in terms of the energetics of a two-step process: In step 1, the group is converted into its electrically neutral form, which for these side chains is unfavorable at pH 7; and in step 2, the electrically neutral form is transferred to the core (with formation of intramolecular hydrogen bonds). The first step then carries an extra cost in energy that must be overcome when the protein folds.

12.2.7 Internal Charge–Charge Interactions: Ion Pairs or Salt Bridges

The charged forms of aspartate, glutamate, histidine, and lysine side chains and the charged side chains of arginine are predominantly found on the protein surface, and may form hydrogen bonds with nearby charged surface groups or surface groups with dipoles. Hydrogen-bonded ion pairs (such as Lys and Glu; -C^{β}-C^{γ}-C^{δ}-NH_3^+...CO_2^--C^{γ}-C^{β}-) are also called salt bridges. Barlow and Thornton[16] analyzed the available three-dimensional structures of 38 proteins and, using accepted geometric criteria for what constitutes an ion pair, determined the presence of 224 such pairs, for an average of six per protein. They found that one-third of the charged residues in these proteins are involved in ion pairs and that three-fourths of those are concerned with stabilizing the tertiary (rather than the secondary) structure. Only 17% of ion pairs in these proteins were found buried in the interior. (In considering these statistics, it is useful to remember that the analyzed structures are of protein in crystals, where the close proximity of neighboring protein molecules and the use of peculiar crystallization-promoting solvents may favor arrangements in which surface side chains do not protrude from the protein as much as in aqueous solution, something that could favor formation of salt bridges.) Barlow and Thornton, citing also early estimates of the stabilization contributed by such ion pairs,[17,18] concluded that, overall, ion pairs make a significant contribution to the stability of globular proteins. Nevertheless, some proteins have no ion pairs at all, while others contain a great many.

12.2.8 Measuring Stability of Salt Bridges from Double Mutant Cycles

Many studies have investigated the contributions of salt bridges by site-directed mutagenesis; most, but not all, of these studies found that the removal of a surface salt bridge destabilized the folded structure (or that the introduction of a new surface salt bridge stabilized it) to an extent that varied from case to case. The stabilization is context dependent: the stabilization varies from case to case because the local effect of replacing any particular residue with one of another type varies from case to case, and the local effects due to replacing two residues with a pair that might form a salt bridge need to be accounted for before one can conclude that the introduced salt bridge has a stabilizing effect and, if so, of what magnitude.

In order to separate the contribution that arises from the *interaction* between the longer side chains (denoted by X and Y) when these are replaced by shorter, noninteracting side chains A and B, from the context dependence, one uses "double-mutant cycle" analysis. This involves the preparation of four variants, each with a different combination of residues at the two sites XY, XB, AY, and AB, and measuring the stability (as the standard free energy of unfolding ΔG°) of each of these four structures. The overall change in stability due to replacing the AB pair with the XY pair is

$$\Delta\Delta G^{\circ}_{\text{overall}} = \Delta G^{\circ}_{\text{XY}} - \Delta G^{\circ}_{\text{AB}}. \tag{12.9}$$

The context-dependent part is found as

$$\Delta\Delta G^{\circ}_{\text{context}} = \Delta G^{\circ}_{\text{XB}} - \Delta G^{\circ}_{\text{AB}} + \Delta G^{\circ}_{\text{AY}} - \Delta G^{\circ}_{\text{AB}}, \tag{12.10}$$

and the contribution due specifically to interactions between the side chains X and Y is then found by subtracting these two quantities, which gives

$$\Delta G^{\circ}(\text{DMC}) = \Delta G^{\circ}_{\text{XY}} - (\Delta G^{\circ}_{\text{XB}} + \Delta G^{\circ}_{\text{AY}}) + \Delta G^{\circ}_{\text{AB}}. \tag{12.11}$$

Applications to a variety of proteins gave a range of values for the stabilization contributed by the salt bridges, $\Delta G°(\text{DMC})$ between 0 and 6 kJ mol^{-1} (0–1.5 kcal mol^{-1}).[19–28]

In a recent example, Makhatadze et al.[29] applied this to the small globular protein, ubiquitin, which contains a pair of oppositely charged residues, Lys-11 and Glu-34, located properly on the protein surface so as to be able to form a salt bridge, and analyzed the effects of replacing each of these residues with an uncharged residue in two independent three-mutant cycles (mutating K11/E34 to A11/S34, to A11/L34, and to A11/K34; here Ala = A, Ser = S, Leu = L, Glu = E, and Lys = K) and also investigated a double-mutant cycle in which the two charged side chains have changed places (mutating E11/K34 to A11/S34). The values of $\Delta G°(\text{DMC})$ for these four cycles varied little; the salt bridges were found to stabilize the folded protein by an amount that varied from 3.5 to 3.8 kJ mol^{-1} (ca. 0.85 kcal mol^{-1}). These experiments also demonstrated a very marked context dependence, as the K11/E34 and E11/K34 variants are not equally stable, and the K11/E34 variant is slightly less stable than the A11/S34 variant and is considerably less stable than the A11/L34 variant.

Makhatadze et al.[29] analyzed the salt bridges of references[19–29] in terms of the Tanford–Kirkwood electrostatic model[30,31] (a variant of the continuum dielectric model discussed in Section 12.3 of this chapter). Here they also used a double-mutant cycle, in which they compared the Kirkwood–Tanford electrostatic free energies $\Delta G^{\text{o}}_{\text{q-q}}$ of the four variants XY, AX, XB, and AB by evaluating

$$\Delta G^{\text{o}}_{\text{q-q}}(\text{DMC}) = \Delta G^{\text{o}}_{\text{q-q,XY}} - (\Delta G^{\text{o}}_{\text{q-q,AY}} + \Delta G^{\text{o}}_{\text{q-q,XB}}) + \Delta G^{\text{o}}_{\text{q-q,AB}}. \quad (12.12)$$

The observed correlation between $\Delta G^{\text{o}}_{\text{q-q}}(\text{DMC})$ and $\Delta G°(\text{DMC})$ indicates that charge–charge interactions dominate the energetics of salt bridge formation.

12.2.9 Salt Bridges Stabilize Proteins from Thermophilic Organisms

Proteins from thermophilic organisms usually show high intrinsic thermal stability, but their structures are very similar to their mesophilic homologues. With only a few available sequences of highly thermostable proteins, Perutz and Raidt proposed that the greater stability was achieved by the presence of a large proportion of salt bridges,[17] but when additional structures became available, this did not appear to be a general rule.

As an example of recent work, we cite a study by Ge et al.,[32] who used a double-mutant cycle approach in order to estimate the effect of temperature on the contribution that two highly conserved salt bridges make to the stability of the hyperthermophilic protein Ssh10b at two different temperatures. The coupling free energies were found to be 2.4 ± 0.4 kJ mol^{-1} at 298 K and 2.2 ± 0.4 kJ mol^{-1} at 353 K for Glu-54/Arg-57, and 6.0 ± 0.2 kJ/mol at 298 K and 5.9 ± 0.6 kJ/mol at 353 K for Glu-36/Lys-68. One sees that the contribution of these two salt bridges to the stability of this protein changes little with temperature; these salt bridges are resilient to temperature increases and thus are indeed suited for stabilizing this protein at the high temperatures encountered by hyperthermophilic organisms, as had been proposed by Perutz and Raidt.[17]

Szilágyi and Závodszky[33] have compiled and analyzed a nonredundant data set comprising all high-quality structures of thermophilic proteins and their mesophilic homologues from the Protein Data Bank (64 mesophilic and 29 thermophilic protein subunits, representing 25 protein families). From the atomic coordinates, 13 structural parameters were calculated, compared, and evaluated using statistical methods. The authors conclude that different protein families adapt to higher temperatures by different sets of structural devices. The only generally observed rule is an increase in the number of ion pairs with increasing growth temperature. The number of hydrogen bonds and the polarity of buried surfaces exhibit no clear-cut tendency to change with growth temperature.

12.3 THE CONTINUUM DIELECTRIC MODEL

12.3.1 Accounting for Charge–Charge and Charge–Solvent Interactions

We have noted a variety of effects involving charged side chains. These include the mild interaction between charged side chains on the surface of a protein that spreads out the titration curve, the stronger interactions between titratable groups and the protein environment, and the propensity of charged side chains (and other polar groups) to remain solvated (hydrated). In principle, one should seek to understand these effects with a common theory of charge–charge and charge–solvent interactions based on statistical mechanics. A problem with this approach is that one would have to consider a multitude of solvent configurations for *each* conformation of the solute; while these solvent configurations might be accessible *via* dynamics or Monte Carlo simulations, this would require a lengthy process.

Instead, excellent results are obtained with a much easier-to-use method in which the solvent is represented as a continuum with a high dielectric constant. To understand why this is a good model, one should remember that the interaction of a charged group with solvent results in an energetically favorable reorientation and reordering of the nearby solvent water molecules, whose dipole moments tend to point toward negative and away from positive charges. From studies by Debye, it is known that orientation of solvent dipoles makes a major contribution to the dielectric behavior; the dielectric constant of water is around 80 times that of vacuum, thanks to the ability of water molecules to orient themselves in the described manner.

Thus, the continuum dielectric model treats the solvent as a continuum with a high dielectric constant; this solvent continuum is polarized by the charges in the solute molecules, and the energetics are described with equations of classical electrostatics in terms of the electrostatic potential inside the protein and in the surrounding solvent. This model allows one to estimate the electrostatic contribution to the solvation free energy and overall produces a highly successful quantitative representation of solvation of biological macromolecules and its effect on the interaction of charged groups.

12.3.2 Definition of the Model

The continuum dielectric model represents a folded macromolecule in solvent in terms of two volumes having very different dielectric constant, ε, low in the macromolecule's interior and high in the surrounding volume that represents the solvent. The shape of the macromolecule is based on the available information about the molecule's structure, and a surface is assumed that separates the two volumes with different values of ε. Finally, a charge distribution is chosen for the protein (Fig. 12.4). More specifically, this model makes the following assumptions:

1. The volume is divided into clearly separated "inside" and "outside" spaces. The boundary between inside and outside is usually chosen to be the *molecular surface*.[34]
2. Inside the macromolecule, the charge distribution is known (we have a reasonable idea since we have atomic coordinates from X-ray or NMR

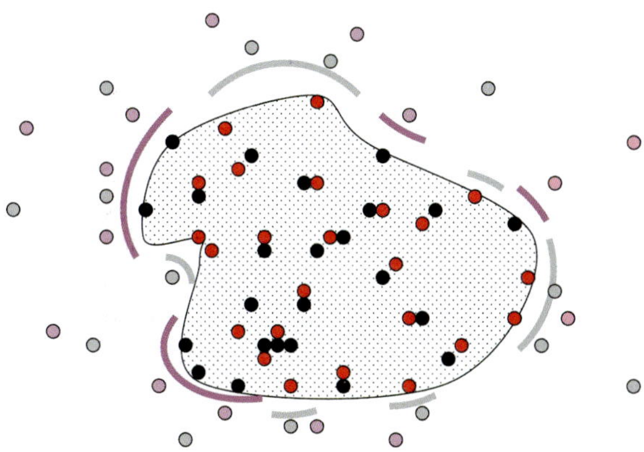

FIGURE 12.4. Schematic representation of the continuum dielectric model. The protein (stippled) is enclosed by a boundary (molecular surface), inside of which the dielectric constant is set to 1 and the atoms are represented with explicit charges (red and black). Outside the protein, the solvent (with the dielectric constant of water) is polarized in response to the charge distribution inside the protein (pink and gray bars). Counterions of a different sign (pink and gray) are in excess near protein charges of an opposite sign.

studies and estimates of charge distribution from models such as molecular mechanics).

3. Inside the macromolecule, the medium is a vacuum, except for the known charges ($\varepsilon = 1$; some prefer to use a higher dielectric constant of 2 or 4 for the interior).
4. Outside the macromolecule, the medium has a very high dielectric constant (ε of water = 78 at 25°C).

This medium is polarized by the charges inside, and the polarization tends to mitigate the direct Coulomb interactions. The charge distribution resulting from polarization and the polarization energy can be found by solving the Poisson equation,[31]

$$\nabla \cdot \varepsilon \nabla \Phi = -4\pi\rho, \tag{12.13}$$

where ε is the dielectric constant of the medium (relative to that of vacuum). This equation relates the electrostatic potential Φ to the (macromolecule's) charge density ρ in a medium of nonuniform dielectric constant. The energy of the entire system is given in terms of the electrostatic potential and the charge distribution by

$$\Delta G_{\text{el}} = \frac{1}{2}\int_V dV \rho(V)\Phi(V) = \frac{1}{2}\sum_i q(r_i)\Phi(r_i), \tag{12.14}$$

where alternate forms are given for a continuous charge distribution and a distribution of point charges, q. The energy can be thought of as consisting of two parts: the energy for the same charge distribution *in vacuo*, and a contribution to the solvation free energy,

$$\Delta G_{\text{el}} = \Delta U_{\text{coul},0} + \Delta G_{\text{sol}} = \frac{1}{2}\sum_i \sum_j q(r_i)q(r_j) + \frac{1}{2}\sum_i q(r_i)\Phi_{\text{pol}}(r_i), \tag{12.15}$$

where the potential Φ_{pol} reflects the polarization of the solvent. Free energy, because polarization of solvent requires an increased ordering of solvent molecules, the lower energy that results is, in part, offset by an entropy loss.

12.3.3 Application to a Charged Spherical Particle[31]

The continuum dielectric model has analytical solutions only in simple cases. We discuss here one such system, a sphere with a uniform charge distribution throughout its volume, with different dielectric constants inside and outside the sphere. This so-called *Born model* represents a solute ion as a spherical particle with radius r and a charge q inside. Solution of the Poisson equation gives the following expression for the electrostatic contribution to the free energy of solvation (transfer of the ion from vacuum to a solvent):

$$\Delta G_{\mathrm{el,sol}} = -\frac{q^2}{2r}\left(1-\frac{1}{\varepsilon}\right). \tag{12.16}$$

This free energy is inversely proportional to the radius of the ion, and for water (which has $\varepsilon = 80$) the magnitude of this free energy is almost equal to the Coulomb energy of a pair of charges in vacuum at a distance equal to twice the radius of the sphere.

Precisely the same electrostatic potential results for an alternative system with uniform dielectric constant equal to 1, in which an additional charge, q_{pol} is distributed over the surface of the sphere, with

$$q_{\mathrm{pol}} = -q\left(1-\frac{1}{\varepsilon_{\mathrm{a}}}\right). \tag{12.17}$$

Solvent polarization is seen to result in an induced negative charge near the surface of a positively charged particle; this was to be expected from a model based on preferential orientation of dipoles. Note that if the result of polarization is formulated in terms of an additional induced charge distribution, then both the original charges and the induced charges contribute to the electrostatic potential, but the energy is found by integrating the product of only the original charge distribution and this electrostatic potential.

12.3.4 Numerical Treatment via Finite Differences

The continuum dielectric model for the general case of a protein molecule surrounded by a medium of high dielectric coefficient can be dealt with only numerically. A much used approach divides the system containing protein and a large (finite) volume of solvent into a large number of discrete volume elements, each with a given value of ε, ρ, and Φ; the values of Φ are iteratively adjusted to obey the Poisson or Poisson–Boltzmann equation. The finite difference method has been implemented in several readily available computer programs, such as DELPHI,[35,36] UHBD, and the more recently developed and faster APBS.[37,38]

As was already mentioned, the dielectric coefficient of the protein, ε_{P} is frequently set to 1, sometimes 2 or 4; for the values of the atomic charges of the protein, q_i, one may use the charges from a molecular mechanics force field. The calculated polarization energy is very sensitive to the atomic radii used to determine the molecular surface, and these radii are parameters that are usually optimized by a fit to experimental data for small molecules, such as free energies of transfer from vacuum to water (see below).

Many approaches to molecular thermodynamics are based on molecular energies, and deal with one conformation at a time. For these, estimating the corresponding *free* energies requires an additional integration over conformation space that can be extremely difficult, and sometimes impossible, to accomplish. The continuum dielectric approach is an exception, as it, in fact, integrates over all solvent configurations, if only for a single conformation of the macromolecule, and gives the *electrostatic contribution to the solvation free energy*. It is also uniquely able to account for the effect of ionic strength. It should, however, be kept in mind that numerical evaluation of the continuum dielectric model for a protein molecule takes a considerable amount of computation, and that with the same effort a large number of different configurations of solvent and macromolecule can be evaluated for a molecular mechanics model, even with explicit representation of solvent.

The Poisson and Poisson–Boltzmann equations assume a *linear polarization*, that is, a polarization of the medium that is proportional to the local electric field (the gradient of the electrostatic potential). In very strong electric fields, polarization is typically less than that expected applying a linear model to the polarization observed in weaker fields; such saturation easily occurs if the dielectric constant has a strong component due to orientation of molecular dipoles, which in strong fields are essentially perfectly oriented, so that further increase of the field may effect only the smaller component due to redistribution of electrons and nuclei that enhance the molecular dipole moment.

The continuum treatment will be least appropriate at short distances from the molecular surface, where the details of the solvent's molecular structure can be expected to be important. A linear treatment will break down in situations where water molecules are strongly ordered, such as water molecules adjacent to small multivalent ions, or multiply hydrogen-bonded water molecules in pockets on the surface of protein molecules. (One may, of course, explicitly represent the bound water molecule, i.e., as a part of the molecular space, rather than the solvent space.)

The reader should be conscious of these limitations when considering the following applications.

12.3.5 Solvation Free Energies Provide Model Parameters

One can simplify the solvation of a (macro)molecule in terms of a three-step process, if one assumes that molecular shape and charge distribution are not affected by the solvation process (Fig. 12.5). In the first step the charges are set to zero, in the second step a cavity is made in the solvent in which to fit the protein and the uncharged protein inserted into this cavity, and in the third step the charges are reintroduced. The free energy for the first step can be estimated by Coulomb's law; the free energy for the second step is, in first approximation, equal to the product of a cavity constant and the area of the protein–water interface (either the solvent-accessible or the molecular surface); the free energy for the third step is the electrostatic free energy and can be estimated using the continuum dielectric model with the Poisson–Boltzmann equation (Eq. 12.15).

Parameter sets consisting of atomic radii, partial charges, and the cavity constant have been developed using, on the one hand, parameters of molecular mechanics force fields and, on the other hand, sets of experimentally determined transfer free energies of small molecules. As an example, we cite a study

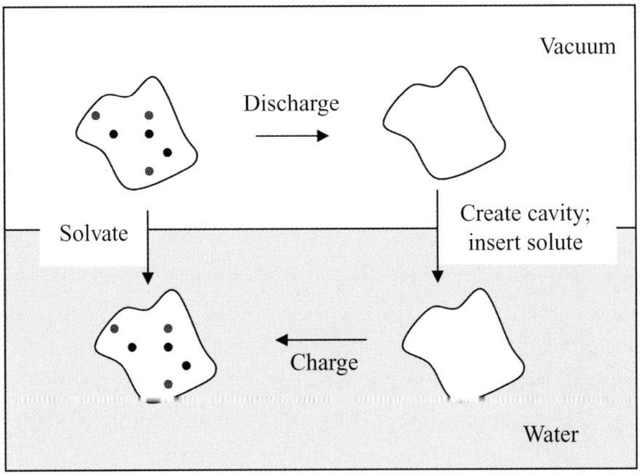

FIGURE 12.5. Decomposition of the solvation process into three steps.[36]

in which such parameters were adjusted to reproduce the solvation free energies of 65 small molecules, ranging from –0.3 to –10 kcal mol^{-1}, with an average error of 0.44 kcal mol^{-1}.[36] For two additional molecules, 1,2-ethanediamine and 1,2-ethanediol, the transfer from vacuum to water was much more favorable for the model than found experimentally (by 3.4 and 4.3 kcal mol^{-1}, respectively).

12.3.6 Macromolecular Charge Distribution and pK_a

By numerically solving the electrostatic potential, Φ, of the continuum dielectric model, this can be used to estimate electrostatic contributions to ion binding, and therefore also the deviations of pK_as from normal values, for any specific model of a protein's shape and internal charge distribution. The procedure is simple; the free energy difference for changing the charge by $+e$ by attaching a hydrogen ion to a side chain at position $\mathbf{r}$ is evaluated as

$$\delta G_{\text{el}} = e\Phi(\mathbf{r}). \tag{12.18}$$

This must be calculated twice: once for the side chain on the protein, and once for the same side chain in a short peptide in a conformation in which the side chain is unencumbered by short-range contacts. The pK_a shift then follows by applying Equation 12.4 to the difference between the two values of δG_{el}.

A complication arises if more than one group can bind a hydrogen ion in the same pH range. In that case, the electrostatic contribution to the solvation free energy of all alternatively charged species (2^n for n such groups) must be evaluated with the Poisson–Boltzmann equation; the extent of saturation with bound ions is then found by considering all these species. Tests of this method performed on a number of small proteins show limited success; the root mean square (RMS) deviation of pK_as of four small proteins from measured values varies between 0.7 and 1.4.[39,40]

There are two reasons why these calculations do not agree better with the experimental results. *The model assumes a single conformation* for the protein in both ionization states, for which one then uses the experimental structure, here the X-ray coordinates. This is a shortcoming not only because the conformation is a dynamic average but also because the average conformations of the two forms, with and without the added hydrogen ion, need not be the same. This will be the case especially if addition or extraction of this hydrogen ion is unusually difficult because the side chain interacts with the rest of the protein structure and if this either requires or excludes the hydrogen ion due to the presence of one or more hydrogen bonds. A second difficulty is related to the presence of *bound water molecules*. An example is a salt bridge between the (protonated) His-31 and Asp-70 side chains in T4-lysozyme, which, in the crystal structure, is connected to four ordered water molecules. If these water molecules are treated as part of the solvent, the pK_a shifts of the side chains are too small (to 8.0 and 3.3, respectively), but if these are explicitly included in the protein structure, then the pK_a shifts are too large (to 9.9 and ca. 0) compared with the experimental values of 9.0 and 1.4.[39,40] In the former case, the four water molecules are underpolarized and contribute too little to the electric field, and in the latter case, they are overpolarized and contribute too much.

The continuum dielectric model has been found to perform well in modeling the electrostatic interactions between side chains and side chains and solvent that affect the titration of ionizable polypeptides and determine the pH dependence of the helix–coil equilibrium.[41]

12.3.7 Molecular Interactions

Charge–charge interaction between enzyme and substrate can have a large effect on the catalytic rate. A favorable interaction enhances the rate of diffusion, which is important when the reaction rate is diffusion limited, and, if this is not the case, raises the concentration of substrate near the enzyme. A well-known simple model treats the diffusion-limited reaction in which enzyme and substrate are represented as featureless spheres. This model can be extended to incorporate an electrostatic force between enzyme and substrate, and slower than diffusion-limited catalysis. However, the spherical symmetry of the model limits its application, and an attractive alternative is to treat both molecules in atomic detail, and represent the electrostatic interactions using the continuum dielectric method, the dynamics with a stochastic model using frictional and random forces according to the Langevin equation (this simulation algorithm is also called Brownian dynamics), and the catalysis with an appropriate model such as the Michaelis–Menten scheme. For details, we refer the reader to the applications cited in the next paragraph.[40,42,43]

This approach has been applied with considerable success in reproducing the rate enhancement of several enzymes with their charged substrates—triose

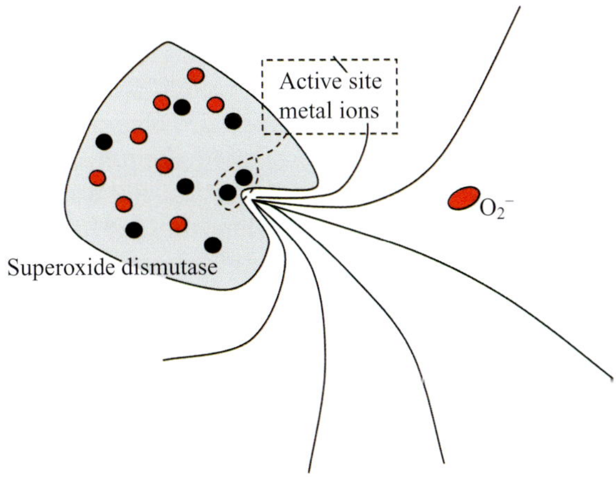

FIGURE 12.6. Electrostatic lines of force "directing" the diffusion of superoxide ion to the active site of superoxide dismutase.

phosphate isomerase with glyceraldehyde phosphate, super oxide dismutase with super oxide, and acetylcholine esterase with acetyl choline—that operate near or at the diffusion-controlled limit.[40,42,43] These studies reproduce the favorable long-range interaction of the charged substrate with the charge distribution of the enzyme, which guides the substrate's diffusive motion toward the site with the largest opposite charge (see Fig. 12.6). The rate enhancement due to the electrostatic force compensates for the fact that the active site is only a quite small fraction of the enzyme surface, and is effective even if the enzyme molecule has zero net charge, but the active site contains metal ions or side chains having charge opposite that of the substrate.

12.4 POLYPEPTIDE HELICES

Studies with synthetic polypeptides offer excellent examples of charge effects on structural stability. We shall discuss, first, measurements of charge-coupled stability in high-molecular-weight homopolymers with ionizable side chains, poly(L-glutamic acid) and poly(L-lysine), and, second, discuss helix-stabilizing interactions in oligopeptides of known sequence.

12.4.1 Polypeptides with Ionizable Side Chains

Early studies of helix formation in synthetic polypeptides were done with water-insoluble homopolymers in organic solvents. The results of these measurements with the polymer of the benzyl ester of L-glutamic acid[44] were critical in that they verified Schellman's theory of the helix–coil equilibrium as elaborated by Zimm and Bragg[45] and Gibbs and DiMarzio[46] (see Appendix

3). Homopolymers of glutamic acid and lysine (poly-Glu and poly-Lys) were the first water-soluble polypeptides to be studied. These were (by measurements of optical rotatory dispersion and, later, circular dichroism) found to be in a random coil state when the side chains were charged (at pH near 7, the charge being negative for poly-Glu and positive for poly-Lys), but to assume the helical state as the molecules' average net charge was lowered by acid–base titration. The helical state is stable when the molecules are electrically neutral, but is destabilized due to mutual repulsion of the charged side chains.

It was soon realized that the coupling between binding of hydrogen ion and conformation change allows one to determine the stability of the helix (in terms of the parameters s and σ) from careful acid–base titrations. We explain the idea with reference to the titration of poly-Glu. Consider a number of different titration curves: as a reference, the titration curve of a carboxylic acid such as acetic acid, the titration curve of helical poly-Glu, the titration curve of randomly coiled poly-Glu, and the experimentally observed titration curve of the helix–coil mixture. The titration of acetic acid follows the Henderson–Hasselbalch equation with a pK_a of around 4.75:

$$\mathrm{p}K_a = \mathrm{pH}\text{-}\log\frac{\alpha}{1-\alpha} = \mathrm{pH}+\log\frac{\alpha'}{1-\alpha'}, \tag{12.19}$$

where α is the degree of ionization and $\alpha' = 1 - \alpha$ is the degree of saturation of the carboxyl group. One can describe the titration of poly-Gly with the same equation, except now with the use of an *apparent* pK_a whose value is not constant but increases with the degree of ionization of the side chains as a result of charge–charge repulsion. The repulsion is less severe for the coil state; ideally, in the limit of low α the apparent pK_a of helix and coil are the same, and the values diverge as α increases. Finally, the experimental data follow the values for helix at low values of α and for coil at high values of α, with a transition somewhere at intermediate values.

Zimm and Rice[47] demonstrate the advantage of plotting the titration curves in the form of apparent pK_a as a function of degree of ionization α. An example taken from a number of experimental studies in the 1960s[47–53] is given in Figure 12.7. Included are linear extrapolations that lead to a common value of apparent pK_a at $\alpha = 0$; studies indicate that these linear extrapolations appear to be adequate at high ionic strength. (The polymer precipitates at low degree of ionization.)

12.4.2 Coupling of Ionization and Helix Formation: Theory and Example

To understand how to exploit the coupling between binding of hydrogen ion to the polymer and its conformation change, consider that by measuring the titration of the polymer with acid, one establishes the change in its chemical

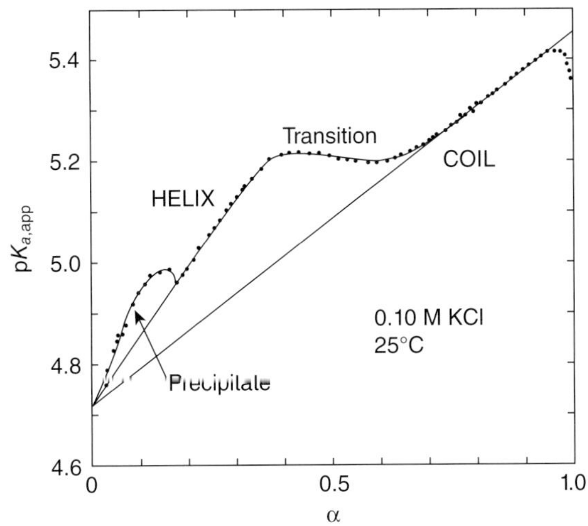

FIGURE 12.7. Acid–base titration of poly-Glu with degree of polymerization $n = 500$.[52]

potential according to Equation A12.6. For the experimental data, this includes a change in conformation, so that we have

$$\frac{1}{n}\left[\mu^{\circ}_{\text{helix}}(\alpha=0)-\mu^{\circ}_{\text{coil}}(\alpha=1)\right]=-k_BT\int_{a_{H^+}=0}^{a_{H^+}=\infty}\alpha'_{\text{expt}}\,d\ln a_{H^+}, \qquad (12.20)$$

where we have replaced the average number ν of hydrogen ions bound to a polymer molecule by the product of the average degree of saturation α' and the number of residues n.

If we perform the same operation on the (in part extrapolated) titration of the coil, we have

$$\frac{1}{n}\left[\mu^{\circ}_{\text{coil}}(\alpha=0)-\mu^{\circ}_{\text{coil}}(\alpha=1)\right]=-k_BT\int_{a_{H^+}=0}^{a_{H^+}=\infty}\alpha'_{\text{coil}}\,d\ln a_{H^+}, \qquad (12.21)$$

and subtraction of the two equations then gives the difference in the per-residue chemical potential between uncharged polymers in helix and coil conformations,

$$\frac{1}{n}\left[\mu^{\circ}_{\text{helix}}(\alpha=0)-\mu^{\circ}_{\text{coil}}(\alpha=0)\right]=-k_BT\int_{a_{H^+}=0}^{a_{H^+}=\infty}\left[\alpha'_{\text{expt}}-\alpha'_{\text{coil}}\right]d\ln a_{H^+}\approx -k_BT\ln s, \qquad (12.22)$$

where Equation A12.8 relates this to the equilibrium constant s for helix elongation.

The integral in Equation 12.22 corresponds to the area between the titration curve (α' vs. ln a_H) for the experimental system and the (in part, extrapolated) titration curve for the coil state. Zimm and Rice[47] have shown that this area equals the area between the curves of $pK_{a,app}$ as a function of α for experimental helix–coil mixture and for pure coil (except for a factor ln 10). Application of Equation 12.22 to the example given in Figure 12.7, gives a per-residue free energy difference $k_B T \ln s = 105$ cal mol^{-1}, for $s = 1.19$ for poly-Gly and $k_B T \ln s = 80$ cal mol^{-1}, for $s = 1.14$ for poly-Lys. Zimm and Rice, using a single set of titration curves, with two different analyses obtained different estimates for the helix initiation parameter, $\sigma = 1.4 \times 10^{-3}$ and $\sigma = 10^{-4}$ and were unable to account for the discrepancy between these two values.

12.4.3 Coupling of Ionization and Helix Formation: Applications

Acid–base titrations of ionizable homopolypeptides have been used effectively to quantitate helix destabilization in urea solutions (a value of $k_B T \ln s =$ 20 cal mol^{-1} is found in 8 M urea) and helix stabilization in water–ethanol mixtures ($k_B T \ln s = 350$ cal mol^{-1} is found in 40% ethanol)[52] and to measure the enthalpy of helix formation at around $\Delta H^{\circ}_{hel} = -1000$ cal mol^{-1} per residue,[53] a value that was confirmed by calorimetric measurements.[54] Effects of salts on the acid–base equilibria of poly-Glu and poly-Lys indicated specific interactions of thiocyanate ion, SCN^- with the former and lithium ion, Li^+ with the latter; these interactions produce a shift of the transition to larger degree of ionization in poly-Gly and larger degree of saturation in poly-Lys, but the thermodynamic effects on helix and coil turn out to cancel, in that no measurable change is found in the value of s in the presence of these perturbing ions.[55,56] A simple explanation is that the ions interact specifically neither with the uncharged coil, nor with the uncharged helix state, but that the interaction requires the presence of the charged forms of the ionizable side chain groups, $-COO^-$ and $-NH_3^+$.

Experiments with copolymers of glutamic acid and leucine gave a remarkably high estimate of $k_B T \ln s = 840$ cal mol^{-1} for leucine, which is, however, not supported by measurements of relative helix propensity in alanine-based oligomers discussed below.[50] Finally, we mention titrations that allowed determination of the stability of the β-structure formed by uncharged poly-Lys at high temperature; the per-residue change in chemical potential for forming the β structure was found to be $k_B T \ln s_\beta = 350$ cal mol^{-1} at 25°C.[57] This shows that the β structure is more stable than the α-helix; however, at room temperature the α-helix is observed because it forms much more rapidly. At higher temperature, the stability of the β-structure increases relative to that of the coil, while that of the α-helix decreases, and then the β-structure forms quite readily.

Problems due to insolubility of uncharged poly-Glu already at room temperature and conversion of poly-Lys to β-structure at elevated temperatures precluded the measurement of temperature-dependent unfolding of the helix, which together with the known value of the per-residue enthalpy of helix formation would otherwise have allowed estimation of the helix initiation parameter σ for these polymers.

12.4.4 Experiments with Oligopeptides

Early experiments used oligopeptide fragments of proteins prepared by specific proteolysis and fractionation of the fragments by column chromatography (e.g., Klee,[58] Epand and Scheraga,[59] and Hermans and Puett[60]). Some peptides were found to form a measurable extent of α-helix in solution and undergo a temperature-dependent transition from helix to coil (with the highest observed helix content at 0°C), while others had little or no tendency to do so. This tendency correlated positively with whether the peptides were in a helical conformation in the folded protein.[60] However, many peptides prepared in this manner were poorly soluble in water, and ultimately, the number of peptides studied was insufficient to establish useful relations between amino acid composition and tendency to form an α-helix, unlike what had been possible for natively folded proteins.[61]

The field languished until chemical methods of peptide synthesis became available in the 1980s, and allowed studies containing single-sequence oligopeptides of medium length (up to 50 residues long). Results of numerous studies in several laboratories have been reviewed by Scholtz and Baldwin,[62] some of which will be summarized here.

Problems of solubility were readily solved by including in the amino acid sequence of the oligomers some amino acids with charged and other hydrophilic side chains. By applying the theory of the helix–coil equilibrium using parameters of s and σ for poly-Glu and poly-Lys obtained from the cited results for polymers containing hundreds of residues, one predicts that such relatively short oligomers of just (uncharged) lysine or glutamic acid show only partial helicity, which turned out to be the case. However, the helicity of these oligomers could be enhanced by a variety of strategies. For example, a combination of equal numbers of residues of Glu and Lys in the same sequence created a soluble molecule with a measurable helix propensity at neutral pH. It was then found that by placing Glu and Lys residues 4 residues apart (in positions i and $i + 4$) in an oligomer containing three such pairs and otherwise only alanine residues, as shown in Figure 12.8, the helicity of a peptide only 17 residues long was as high as 75% at 0°C.[63] However, a similarly designed 16-residue peptide in which the Glu and Lys residues are 3 residues apart has much lower helicity (Fig. 12.9).

12.4.5 Stabilizing Interactions in Oligopeptides: Salt Bridges

Model building indicated that, with a spacing of 4, but not with other spacings, the side chains of each pair of Lys and Glu residues are located favorably in

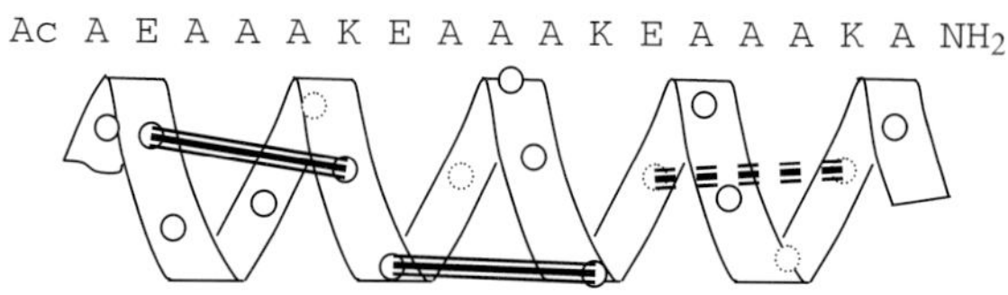

FIGURE 12.8. Amino acid sequence and schematic structure of a 17-residue peptide in which pairs of Glu (E) and Lys (K) residues are separated by 3 Ala (A) residues. (Ac = acetyl.) The bars indicate possible salt bridges between Glu and Lys side chains. Redrawn after Marqusee and Baldwin.[63]

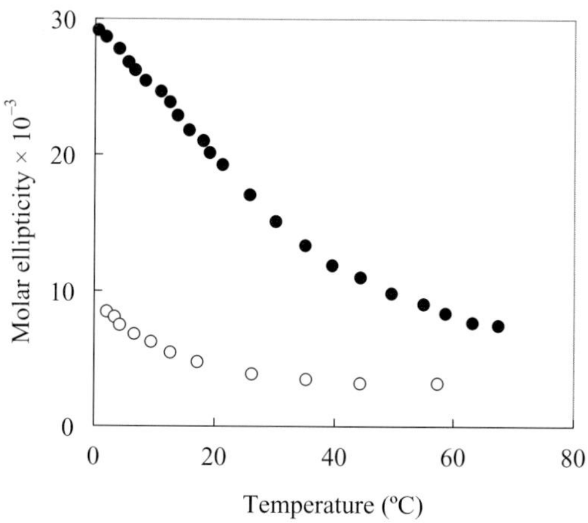

FIGURE 12.9. Thermal unfolding profiles of two peptides each containing 3 Lys and 3 Glu residues, and the remainder Ala. Filled symbols: Peptide described in Figure 12.8, which is capable of forming three salt bridges. Open symbols: Analogous 16-residue peptide incapable of forming salt bridges, as its pairs of Glu and Lys residues are separated by 2 Ala residues. Neutral pH, 0.01 M NaCl. Redrawn from Marqusee and Baldwin.[63]

the helical state for the formation of a salt bridge, as schematically indicated in Figure 12.8. We may account for such a helix-stabilizing local interaction by introducing a factor $\tau\,(\tau > 1)$ in any term in the partition function of Equation A12.7 that corresponds to a helix containing the stabilizing interaction, in this case a Glu–Lys pair separated by 3 Ala residues (in fact, a factor τ for each such a Glu–Lys pair). Values of parameters s and σ were later established for the reference polymers not containing salt bridges (see following paragraph). Using this modified expression for the partition function, it is then relatively easy to make a rough estimate of $\tau = 4$ for the factor that expresses the increase in helix stability by a single salt bridge in these oligomers.

Substitutions of alanine residues with other types in peptides constructed along these lines allowed determination of relative helix propensities of most naturally occurring amino acids; among these, alanine turned out to be the best helix former, whereas glycine and proline were at the bottom of the list.[64–66] Amino acids with two substituents on C^{β} (the so-called β-branched amino acids: valine, isoleucine, and threonine) were found to have lower helix propensity than residues with only a single substituent. Scholtz et al.[67] studied Ala-based peptides up to 50 amino acid residues long, containing repeating sequences (-Ala-Glu-Ala-Ala-Lys-Ala-; in these polymers the helical conformation is *not* stabilized by Lys–Glu salt bridges). In these polymers, helix content increases with chain length; by applying the theory of the helix–coil equilibrium to their results, it was then possible to establish effective values of both parameters of the helix–coil model, $s = 1.35$ and $\sigma = 3 \times 10^{-3}$, at 0°C.

12.4.6 Stabilizing Interactions in Oligopeptides: Interactions with Helix Dipole

The concept of the helix dipole was developed by Hol et al.,[68] in first instance in order to rationalize the high frequency with which phosphate groups are found to bind to proteins near the N-termini of α-helices. Hol et al. estimated that the local dipoles of peptide NH and CO groups (which point approximately along the helix axis in the direction of the N-terminus) together form a global line dipole, called the helix dipole, corresponding to approximately one-half an elementary charge at each helix end, positive charge at the N-terminus, and negative charge at the C-terminus. The former should interact favorably with a negatively charged bound phosphate group.

Experiments with helix-forming oligopeptides[69,70] tested this model by demonstrating considerable stabilization of the helix as a result of changing the charged state of groups near the chain termini. Thus, an oligomer with an electrically neutral acetylated N-terminus (CH_3-CO-NH-) formed helix more readily than the same oligomer with a positively charged free α-amino group (NH_3^+-). Succinylation (giving end group COO^--CH_2-CO-NH-) was found to lead to an even more stable helix, while substitution of the N-terminal residue with lysine, which results in having two positive charges near the N-terminus, gave the lowest tendency to helix formation. Similarly, replacement of the negatively charged free α-carboxyl group (-COO^-) at the C-terminus with an electrically neutral group (-CO-NH_2) increased helix-forming tendency. Some helix stabilization could be observed also by bringing the charge of the α-amino or α-carboxyl group to zero by either lowering or raising the pH, and all these effects were found to depend strongly on the ionic strength.

A very similar effect was noted by Ooi et al.[71] in a study of oligomers containing a block of 20 Ala preceded or followed by a block of 20 Glu or 20 Lys residues. At neutral pH, each side chain of Glu and Lys has a net charge, and the Glu or Lys blocks are in the coil state, while the block of 20 Ala can form α-helix. When the negatively charged Glu block is at the N-terminus of the Ala block (sequence $Glu_{20}Ala_{20}$), the stability of the alanine helix is much greater than when the order is reversed (sequence $Ala_{20}Glu_{20}$), while the reverse is found for a block of Lys residues. The interpretation is that the negative charges interact favorably with the helix dipole if these are located at the N-terminus, while the opposite arrangement corresponds to an unfavorable electrostatic interaction.

12.4.7 Web Server for Estimating Helix Propensity

Muñoz and Serrano[72] used an empirical analysis of these and many other results and from that estimated a set of energy contributions that together can account for the stability of the α-helical conformation of peptides in aqueous solution, at different temperatures, pH, and ionic strength. In the original paper, they analyzed the average helicity of 323 peptides and the helicity per residue derived from NMR. Serrano has established and maintains a Web server that makes this algorithm freely accessible to users.[73]

REFERENCES

1. W. Kauzmann. Some factors in the interpretation of protein denaturation. *Adv Protein Chem* **1959**, 14, 1–63.
2. C. Tanford and J. D. Hauenstein. Hydrogen ion equilibria of ribonuclease. *J Am Chem Soc* **1956**, 78, 5287–5291.
3. D. Shugar. The ultraviolet absorption spectrum of ribonuclease. *Biochem J* **1952**, 52, 142–149.
4. C. Tanford, J. D. Hauenstein, D. G. Rands. Phenolic hydroxyl ionization in proteins. II. *Ribonuclease J Am Chem Soc* **1956**, 77, 6409–6413.
5. J. Hermans and H. A. Scheraga. Structural studies of ribonuclease. 6. Abnormal ionizable groups. *J Am Chem Soc* **1961**, 83, 3293–3300.
6. J. Hermans and H. A. Scheraga. Structural studies of ribonuclease. 5. Reversible change of configuration. *J Am Chem Soc* **1961**, 83, 3283–3292.
7. J. H. Bradbury and H. A. Scheraga. Structural studies of ribonuclease. XXIV. The application of nuclear magnetic resonance spectroscopy to distinguish between the histidine residues of ribonuclease. *J Am Chem Soc* **1966**, 88, 4240–4246.
8. K. Wüthrich, G. Wagner, R. Richarz, S. J. Perkins. Individual assignments of the methyl resonances in the proton nuclear magnetic resonance spectrum of the basic pancreatic trypsin inhibitor. *Biochemistry* **1978**, 17, 2253–2263.
9. R. Richarz and K. Wüthrich. High-field carbon-13 nuclear magnetic resonance studies at 90.5 MHz of the basic pancreatic trypsin inhibitor. *Biochemistry* **1978**, 17, 2263–2269.
10. E. Breslow and F. R. N. Gurd. Reactivity of sperm whale metmyoglobin towards hydrogen ions and p-nitrophenyl acetate. *J Biol Chem* **1962**, 237, 371–381.
11. J. Hermans and G. Acampora. Reversible denaturation of sperm whale myoglobin. II. Thermodynamic analysis. *J Am Chem Soc* **1967**, 89, 1547–1552.
12. J. F. Brandts. The thermodynamics of protein denaturation. 1. The denaturation of chymotrypsinogen. *J Am Chem Soc* **1964**, 86, 4291–4301.

13. J. F. Brandts. The thermodynamics of protein denaturation. 2. A model of reversible denaturation and interpretations regarding the stability of chymotrypsinogen. *J Am Chem Soc* **1964**, 86, 4302–4314.
14. J. F. Brandts and L. Hunt. The thermodynamics of protein denaturation. 3. The denaturation of ribonuclease in water and in aqueous urea and aqueous ethanol mixtures. *J Am Chem Soc* **1967**, 89, 4826–4838.
15. R. Lumry, R. Biltonen, J. F. Brandts. Validity of the "two-state" hypothesis for conformational transitions of proteins. *Biopolymers* **1966**, 4, 917–944.
16. D. J. Barlow and J. M. Thornton. Ion-pairs in proteins. *J Mol Biol* **1983**, 168, 867–885.
17. M. F. Perutz and H. Raidt. Stereochemical basis of heat stability in bacterial ferredoxins and in haemoglobin A2. *Nature* **1975**, 255, 256–259.
18. A. R. Fersht. Conformational equilibria in α- and δ-chymotrypsin: the energetics and importance of the salt bridge. *J Mol Biol* **1972**, 64, 497–509.
19. S. Marqusee and R. T. Sauer. Contributions of a hydrogen bond/salt bridge network to the stability of secondary and tertiary structure in lambda repressor. *Protein Sci* **1994**, 3, 2217–2225.
20. C. A. Blasie and J. M. Berg. Electrostatic interactions across a beta-sheet. *Biochemistry* **1997**, 36, 6218–6222.
21. J. S. Merkel, J. M. Sturtevant, L. Regan. Sidechain interactions in parallel beta sheets: the energetics of cross-strand pairings. *Struct Fold Des* **1999**, 7, 1333–1343.
22. P. Strop and S. L. Mayo. Contribution of surface salt bridges to protein stability. *Biochemistry* **2000**, 39, 1251–1255.
23. K. S. Lassila, D. Datta, S. L. Mayo. Evaluation of the energetic contribution of an ionic network to beta-sheet stability. *Protein Sci* **2002**, 11, 688–690.
24. A. C. Tissot, S. Vuilleumier, A. R. Fersht. Importance of two buried salt bridges in the stability and folding pathway of barnase. *Biochemistry* **1996**, 35, 6786–6794.
25. A. Horovitz, L. Serrano, B. Avron, M. Bycroft, A. R. Fersht. Strength and cooperativity of contributions of surface salt bridges to protein stability. *J Mol Biol* **1990**, 216, 1031–1044.
26. L. Serrano, A. Horovitz, B. Avron, M. Bycroft, A. R. Fersht. Estimating the contribution of engineered surface electrostatic interactions to protein stability by using double-mutant cycles. *Biochemistry* **1990**, 29, 9343–9352.
27. D. Sali, M. Bycroft, A. R. Fersht. Surface electrostatic interactions contribute little of stability of barnase. *J Mol Biol* **1991**, 220, 779–788.
28. H. Nicholson, D. E. Anderson, S. Dao-Pin, B. W. Matthews. Analysis of the interaction between charged side chains and the alpha-helix dipole using designed thermostable mutants of phage T4 lysozyme. *Biochemistry* **1991**, 30, 9816–9828.
29. G. I. Makhatadze, V. V. Loladze, D. N. Ermolenko, X. F. Chen, S. T. Thomas. Contribution of surface salt bridges to protein stability: guidelines for protein engineering. *J Mol Biol* **2003**, 327, 1135–1148.
30. C. Tanford and J. G. Kirkwood. Theory of protein titration curves. I. General equations for impenetrable spheres. *J Am Chem Soc* **1957**, 79, 5333–5339.
31. C. Tanford. *Physical Chemistry of Macromolecules*. Wiley, New York, 1967.
32. M. Ge, X.-Y. Xia, X.-M. Pan. Salt bridges in the hyperthermophilic protein Ssh10b are resilient to temperature increases. *J Biol Chem* **2008**, 283, 31690–31696.

33. A. Szilágyi and P. Závodszky. Structural differences between mesophilic, moderately thermophilic and extremely thermophilic protein subunits: results of a comprehensive survey. *Structure* **2000**, 8, 493–504.

34. M. L. Connolly. Analytical molecular surface calculation. *J Appl Crystallogr* **1983**, 16, 548–558.

35. K. A. Sharp and B. Honig. Calculating total electrostatic energies with the nonlinear Poisson-Boltzmann equation. *J Phys Chem* **1990**, 94, 7684–7692.

36. D. Sitkoff, K. A. Sharp, B. Honig. Accurate calculation of hydration free energies using macroscopic solvent models. *J Phys Chem* **1994**, 98, 1978–1988.

37. N. A. Baker, D. Sept, M. J. Holst, J. A. McCammon. The adaptive multilevel finite element solution of the Poisson–Boltzmann equation on massively parallel computers. *IBM J Res Dev* **2001**, 45, 427–438.

38. B. Luty, M. Davis, J. A. McCammon. Electrostatic energy calculations by a finite-difference method: rapid calculation of charge-solvent interaction energies. *J Comput Chem* **1992**, 13, 768–771.

39. A.-S. Yang, M. R. Gunner, R. Sampogna, K. Sharp, B. Honig. On the calculation of pK_as of proteins. *Proteins* **1993**, 15, 252–265.

40. J. Antosiewicz, J. A. McCammon, M. K. Gilson. The determinants of pK_as in proteins. *Biochemistry* **1996**, 24, 7819–7783.

41. J. A. Vila, D. R. Ripoll, M. E. Villegas, Y. N. Vorobjev, H. A. Scheraga. Role of hydrophobicity and solvent-mediated charge-charge interactions in stabilizing alpha-helices. *Biophys J* **1998**, 75, 2637–2646.

42. J. D. Madura and J. A. McCammon. Brownian dynamics simulation of diffusional encounters between triose phosphate isomerase and D-glyceraldehyde phosphate. *J Phys Chem* **1989**, 93, 7285–7287.

43. J. Sines, S. Allison, A. Wierzbicki, J. A. McCammon. Brownian dynamics simulation of the superoxide-superoxide dismutase reaction: iron and manganese enzymes. *J Phys Chem* **1990**, 94, 959–961.

44. B. H. Zimm, P. Doty, K. Iso. Determination of the parameters for helix formation in poly-γ-benzyl-L-glutamate. *Proc Natl Acad Sci U S A* **1959**, 45, 1601–1607.

45. B. H. Zimm and J. K. Bragg. Theory of the phase transition between helix and random coil in polypeptide chains. *J Chem Phys* **1959**, 31, 526–535.

46. J. H. Gibbs and E. A. DiMarzio. Statistical mechanics of helix-coil transitions in biological macromolecules. *J Chem Phys* **1959**, 30, 271–282.

47. B. H. Zimm and S. A. Rice. The helix-coil transition in charged macromolecules. *Mol Phys* **1960**, 3, 391–407.

48. A. Wada. Helix-coil transformation and titration curve of poly-L-glutamic acid. *Mol Phys* **1960**, 3, 409–416.

49. M. Nagasawa and A. Holtzer. The helix-coil transition in solutions of polyglutamic acid. *J Am Chem Soc* **1964**, 86, 538–543.

50. W. G. Miller and R. E. Nylund. The Stability of the helical conformation of random L-leucine-L-glutamic acid copolymers in aqueous solution. *J Am Chem Soc* **1965**, 87, 3542–3547.

51. P. Appel and J. T. Yang. Helix-coil transition of poly-L-glutamic acid and poly-L-lysine in D_2O. *Biochemistry* **1965**, 4, 1244–1249.

52. J. Hermans. Experimental free energy and enthalpy of formation of the α-helix. *J Phys Chem* **1966**, 70, 510–415.
53. J. Hermans. The effect of protein denaturants on the stability of the alpha helix. *J Am Chem Soc* **1966**, 88, 2418–2422.
54. G. Rialdi and J. Hermans. Calorimetric heat of the helix-coil transition of poly-L-glutamic acid. *J Am Chem Soc* **1966**, 88, 5719–5720.
55. A. Ciferri, D. Puett, L. Rajagh, J. Hermans. Potentiometric titrations and the helix-coil transformation of poly-L-lysine and poly-L-glutamic acid in aqueous solutions. *Biopolymers* **1968**, 6, 1019–1036.
56. D. Puett, A. Ciferri, E. Bianchi, J. Hermans. Helix formation of poly-L-lysine thiocyanate in aqueous solutions. *J Phys Chem* **1967**, 71, 4126–4128.
57. D. Pederson, D. Gabriel, J. Hermans. Potentiometric titration of poly-L-lysine: the coil to beta transition. *Biopolymers* **1971**, 10, 2133–2145.
58. W. A. Klee. Conformation of ribonuclease S-peptide. *Biochemistry* **1968**, 7, 2731–2736.
59. R. M. Epand and H. A. Scheraga. The influence of long-range interactions on the structure of myoglobin. *Biochemistry* **1968**, 7, 2864–2872.
60. J. Hermans and D. Puett. Relative effects of primary and tertiary structure on helix formation in myoglobin and α-lactalbumin. *Biopolymers* **1971**, 10, 895–914.
61. P. Y. Chou and G. D. Fasman. Prediction of protein conformation. *Biochemistry* **1974**, 13, 222–245.
62. J. M. Scholtz and R. L. Baldwin. The mechanism of α-helix formation of peptides. *Annu Rev Biophys Biomol Struct* **1992**, 21, 95–118.
63. S. Marqusee and R. L. Baldwin. Helix stabilization by Glu^- . . . Lys^+ salt bridges in short peptides of de novo design. *Proc Natl Acad Sci U S A* **1987**, 84, 8898–8902.
64. G. Merutka and E. Stellwagen. *Biochemistry* **1989**, 28, 352–357.
65. G. Merutka and E. Stellwagen. *Biochemistry* **1991**, 30, 1591–1594.
66. P. C. Lyu, M. I. Liff, L. A. Marky, N. R. Kallenbach. *Science* **1990**, 250, 669–673.
67. J. M. Scholtz, H. Qian, E. J. York, J. M. Stewart, R. L. Baldwin. Parameters of helix–coil transition theory for alanine-based peptides of varying chain lengths in water. *Biopolymers* **1991**, 31, 1463–1470.
68. W. G. Hol, P. T. van Duijnen, H. J. Berendsen. The alpha-helix dipole and the properties of proteins. *Nature* **1978**, 273, 443–446.
69. K. R. Shoemaker, P. S. Kim, E. J. York, J. M. Stewart, R. L. Baldwin. Tests of the helix dipole model for stabilization of α-helices. *Nature* **1987**, 326, 563–567.
70. R. Fairman, K. R. Shoemaker, E. J. York, J. M. Stewart, R. L. Baldwin. Further studies of the helix dipole model: effects of a free α-$NH3^+$ or α-COO^- group on helix stability. *Proteins* **1989**, 5, 1–7.
71. S. Takahashi, E.-H. Kim, T. Hibino, T. Ooi. Comparison of α-helix stability in peptides having a negatively or positively charged residue block attached either to the N- or C-terminus of an α-helix: the electrostatic contribution and anisotropic stability of the α-helix. *Biopolymers* **1989**, 28, 995–1009.
72. V. Muñoz and L. Serrano. Elucidating the folding problem of helical peptides using empirical parameters. *Nat Struct Biol* **1994**, 1, 399–409.
73. L. Serrano. Agadir web server: an algorithm to predict the helical content of peptides. http://agadir.crg.es, **1994**.

APPENDIX 1

IONIZABLE GROUPS IN PROTEINS

For reference, we list here the types of ionizable groups (side chains and chain termini) of proteins and polypeptides, together with "normal" pK_a values, established by acid–base titration of short oligopeptides each containing one ionizable side chain (see Tanford[31]).

amino acid	*type of group*	*approximate normal* pK_a
Aspartic acid (Asp)	γ-Carboxyl	3.5–4
Glutamic acid (Glu), chain C-terminus	δ-, α-Carboxyl	4.5–5.0
Lysine (Lys)	ε-Amino	10
Chain N-terminus	α-Amino	8
Histidine (His)	Imidazole	6.5–7
Tyrosine (Tyr)	Phenol	10
Cysteine (Cys)	Sulfhydryl or thiol	9
Arginine (Arg)	Guanidyl	14

APPENDIX 2

COUPLING OF BINDING AND CHEMICAL POTENTIAL

Using strictly thermodynamic principles, the interaction between two solutes, M and L, can be expressed in terms of the dependence of the chemical potential μ of each on the amount N of the other (in a constant amount of solvent), and it can be shown that the effect is mutual, that is,

$$\left(\frac{\partial \mu_{\mathrm{M}}}{\partial N_{\mathrm{L}}}\right)_{N_{\mathrm{M}}} = \left(\frac{\partial \mu_{\mathrm{L}}}{\partial N_{\mathrm{M}}}\right)_{N_{\mathrm{L}}} \tag{A12.1}$$

(Zimm and Rice[47] use a parallel analysis based on statistical mechanics and a grand partition function.)

Another important thermodynamic result relates the components' chemical potentials to the binding (or, as the case may be, exclusion) that results from the interaction. The preferential binding of component L to component M, ν_{L}, is expressed as the amount of component L required to maintain a constant chemical potential of component L when the amount of component M is changed; the following equation relates ν_{L} to the dependence of the two chemical potentials,

$$\left(\frac{\partial \mu_{\mathrm{M}}}{\partial \mu_{\mathrm{L}}}\right)_{N_{\mathrm{M}}} = -\left(\frac{\partial N_{\mathrm{L}}}{\partial N_{\mathrm{M}}}\right)_{\mu_{\mathrm{L}}} \equiv -\nu_{\mathrm{L}}. \tag{A12.2}$$

Expressing both μ_L and μ_M in the usual way,

$$\mu = \mu^o + k_B T \ln a, \tag{A12.3}$$

we obtain

$$\frac{1}{k_B T}\left(\frac{\partial \mu_M}{\partial \ln a_L}\right)_{N_M} = -\nu_L. \tag{A12.4}$$

This equation can be integrated to give

$$\mu_M(a_L) = \mu_M(a_L = 0) - k_B T \int_{a_L=0}^{a_L} \nu_L \ln a_L. \tag{A12.5}$$

If the solution of component M is ideal, one may replace activity a_M and concentration c_M in Equation A12.3. This replacement gives rise to the equation

$$\mu_M^o(a_L) = \mu_M^o(a_L = 0) - k_B T \int_{a_L=0}^{a_L} \nu_L \ln a_L, \tag{A12.6}$$

which expresses the standard free energy of component M as a function of ligand activity in terms of two quantities ν_L and a_L that define the titration curve of M with L. In particular, titration of macromolecules with hydrogen ion can be measured quite easily, in which case also the measured pH directly relates directly to the activity of the hydrogen ion, a_H.

APPENDIX 3

THEORY OF THE HELIX–COIL EQUILIBRIUM IN HOMOPOLYPEPTIDES

The theory of the helix-oil transition of polypeptides was formulated by John Schellman and was subsequently elaborated to account for the presence of multiple helical segments in a single polymer.[45,46] We give here a brief summary of the main results.

The theory assigns a state, either helix or coil, to each residue of a polymer molecule and is conveniently expressed in terms of two parameters: One of these, called the helix growth parameter s, is the equilibrium constant for enlarging a helix by one residue; the other, called the helix initiation parameter σ, is the equilibrium constant for starting a helix in an all-coil molecule. The equilibrium constant for forming a single stretch of helix containing i residues is then given by σs^i. The equilibrium constant for converting an entire polypeptide containing n residues from the coil state into the helical conformation is σs^n.

For forming a particular state containing two separated stretches of helix, of lengths i_1 and i_2, the equilibrium constant is $\sigma s^{i_1} \sigma s^{i_2}$ and so on. For mixed helical-coiled molecules, there are many arrangements, $m(i,j,n)$, possible of i helical residues distributed over j helices in a polypeptide containing n residues (especially when n is large), and the partition function, q_{hc}, describing the helix–coil equilibrium is given by

$$\frac{q_{\mathrm{hc}}}{q_{\mathrm{coil}}} = \sum_{i=0}^{n} \sum_{j} m(i,j,n)\sigma^j s^i, \qquad \text{(A12.7)}$$

where q_{coil} is the partition function of molecules without helical residues.

This simple description is a good approximation if, as indeed turns out to be the case for polypeptides, the helix initiation constant σ is small ($\sigma \ll 1$), and consequently, helices that are only very few residues long are not prevalent. Values of helix growth parameters, on the other hand, are near 1.

For a given equilibrium mixture of helix and coil states, the difference in standard chemical potentials is

$$\mu^{\mathrm{o}}_{\mathrm{helix}} - \mu^{\mathrm{o}}_{\mathrm{coil}} = -k_{\mathrm{B}}T \ln(q_{\mathrm{hc}} / q_{\mathrm{coil}}). \qquad \text{(A12.8)}$$

A closed form of the double sum in Equation A12.7 was derived by Zimm and Bragg[45]; in the limit of large n this expression is

$$\ln(q_{\mathrm{hc}} / q_{\mathrm{coil}}) \approx n \ln\left(s + 1 + \sqrt{(s-1)^2 + 4\sigma s}\right) - n \ln 2. \qquad \text{(A12.9)}$$

The average number of helical residues $\langle i \rangle$ is found as

$$\langle i \rangle = \frac{\partial \ln q_{\mathrm{hc}}}{\partial \ln s} \qquad \text{(A12.10)}$$

The theory predicts that the helix content of short polymers increases with polymer length n, but that this reaches a limit for large values of n, as is observed for homopolymers.[44] Also, for large n, the helix content of a homopolymer is predicted to be 50% when $s = 1$. Comparison of predicted helix content with these experimental observations allows one to determine the parameters s and σ. For the cited system,[44] the helix initiation parameter was found to be on the order of $\sigma = 10^{-3}$.

PART III

FUNCTIONS AND APPLICATIONS

CHAPTER 13

IRON TRANSPORT IN LIVING CELLS

ALVIN L. CRUMBLISS and CLAIRE J. PARKER SIBURT

13.1 OVERVIEW

Iron is required to sustain life in almost all organisms, from bacteria to humans. However, excess iron is toxic and can lead to cell death. The tight regulation of iron is therefore a well-tuned process in all living cells. *In vivo* iron transport pathways involve many ionic and mixed interactions within and between biological macromolecules and ions. Dominated by proteins, the pathways for iron uptake are complex delivery systems built on the interactions between these natural macromolecules and small ions including the iron ion itself.

Here we set out to dissect the variety of ionic and mixed interactions involved in iron uptake. We will provide examples from both the human iron transport system and the various ways through which bacteria acquire iron from their respective environments.

As in the previous chapters of this book, we define ionic interactions as an electrostatic interaction between permanent fixed charges. Mixed interactions on the other hand involve partial charges (dipoles), hydrogen bonds, and solvophobic effects (Chapters 6 and 8). In iron uptake systems, the structures and functions of the protein polyampholytes (polymers with multiple mixed charges; Chapters 4, 6, and 8) are determined by the ionic and mixed interactions in which the protein participates. There can be interactions within the polyampholyte, between the polyampholyte and mobile ions, or between multiple polyampholytes.

Ionic and mixed interactions involved in iron transport can be categorized as attractive forces, multivalent associations, or background electrolyte effects.[1] Most ionic and mixed interactions that occur during iron transport fall into the category of attractive forces. The attraction between the small

Ionic Interactions in Natural and Synthetic Macromolecules, First Edition. Edited by Alberto Ciferri and Angelo Perico.

anion phosphate, for example, and a bacterial iron transport protein, ferric-binding protein A (FbpA), preorganizes the protein to be loaded with iron (see Section 13.6). Attractive forces between supramolecular assemblies of macromolecules and mobile ions are also common and in fact necessary in iron transport. We will discuss several examples of attractive forces between polyampholytes and ions throughout this chapter.

Associations created by multivalent mobile ions and multivalent proteins are also exhibited in iron uptake systems. For example, ternary complex formation between two polyampholytes and an iron ion enables the exchange of the iron cargo between two sequential protein carriers in a transport pathway. Protein–protein interactions are one example of multivalent associations that we will discuss at length. Analogous multivalent associations of multiple ions with one protein will also be discussed.

Finally, an underappreciated constellation of ionic and mixed interactions is the possible effects of background electrolytes both *in vivo* and *in vitro*. We will present data from *in vitro* experiments designed to delineate the critical electrolyte effects *in vivo*. These data illustrate the dramatic influence of background electrolytes on experimental findings and point to the possible effects of background/environmental electrolytes on iron transport *in vivo*. We will discuss observations, both published and unpublished, that suggest influences of auxiliary electrolytes (possibly Hofmeister effects) on the systems of interest.

The goal of this chapter is twofold. First, we set out to provide a representative collection of examples, in which ionic and mixed interactions involving natural macromolecules chemically guide the biological transport of the essential nutrient iron. The complexity of these transport systems provides fertile ground for the study of ionic and mixed interactions in biological systems of significant importance. Second, we seek to convince the reader that a multilayered consideration of the many types of ionic and mixed interactions involved in biological iron transport is needed to develop a complete understanding of the transport process. This multilayered approach, integrated by the language and principles of ionic and mixed interactions presented in this book, provides a unique opportunity to further the science necessary to combat pathogenesis and prevent disease.

We will begin with a brief overview of the chemical properties of iron and their relevance in biology (Section 13.2). Then, we will introduce the various iron transport strategies found in biology, which are used to move iron across barriers and through solutions. In Section 13.3, we will define the suite of possible interactions between ions and macromolecules involved in iron transport and describe the complexity that arises when multiple interactions are required for each biological system. The remaining bulk of the chapter (Sections 13.4–13.6) will then be divided into three sections: (1) interactions between multiple proteins; (2) interactions between a protein and an iron–small molecule complex; and (3) interactions between a protein and anions.

13.2 BACKGROUND

13.2.1 Iron: An Overview

The chemical properties of iron make it an attractive element for use in biological systems, but it is *also* required that a delicate balance be struck. This balance is known as the iron paradox. On one hand, iron is the fourth most abundant metal in the earth's crust and is essential for almost all living organisms.[2] On the other hand, however, iron is effectively insoluble in water under physiological conditions and, if uncontrolled, can be toxic. The way in which biological systems maintain control over the reactivity of iron and increase the bioavailability of iron is briefly reviewed in the remainder of this section.

Iron Is Important in Biology The electronic properties of iron make the element useful in biology because iron expands the number and type of chemical reactions that can be accomplished compared with biomacromolecules alone.[3] Iron is found in enzymes, cofactors, energy transduction machinery, small-molecule storage proteins, and redox proteins that all utilize iron to perform their biological functions. For example, iron is integral in the highly conserved ribonucleotide reductase that produces DNA.[4] Given its usefulness, biological systems must acquire large amounts of iron from the environment. For example, 20–25 mg of iron per day is used to produce hemoglobin in the red blood cells of a human.[5]

As introduced previously, the iron paradox is a delicate balance. Imbalance in humans can lead to many well-known diseases such as hemochromatosis or anemia.[6] Bacteria also require iron, and one promising method of killing pathogenic bacteria is to restrict their supply of iron.[7] Given that iron acquisition by all organisms is a necessary function and that even during transport the iron paradox must remain balanced, highly controlled mechanisms of iron acquisition have evolved. Although many examples of ionic and mixed interactions are involved in the biochemistry of iron, including its use and storage in biological systems, the focus of this chapter will be iron transport.

Ionic Properties of Iron Iron commonly exists in biology as a +2, +3, or +4 cation, with Fe^{2+} and Fe^{3+} being the most important.[3] The ionic radius of Fe^{3+} is 64.5 pm.[8] This high-charge-density ion therefore is quite hydrophilic. The hydration enthalpy for Fe^{2+} is -458.9 kcal mol^{-1}, while the hydration enthalpy for Fe^{3+} is -1046 kcal mol^{-1}.[9] The Gibbs free energy of hydration for Fe^{2+} is -456.4 kcal mol^{-1} and for Fe^{3+} is -1035.5 kcal mol^{-1}.[9]

Coordination Chemistry of Iron Iron is a Lewis acid, which normally forms five and six coordinate complexes with electron-donating Lewis base ligands. The bond between the ligand and the positively charged iron ion is called a coordinate covalent bond. Coordinate covalent bonds are not just ionic interactions between a positively charged metal and the electron-donating ligand. Rather, coordinate covalent bonds form when a ligand donates an electron pair to a vacant orbital shared between the metal and the ligand (Chapters 10 and 11). Coordinate covalent bonds are a mixture of ion–dipole interactions, electrostatic interactions, and covalent bond formation.[10] The first coordination shell of iron normally assumes a distorted octahedral geometry.

The hard–soft acid–base (HSAB) principle correlates the thermodynamic strength of a ligand–metal interaction with a match between the polarizability of the metal cation electron pair acceptor and the ligand electron pair donor.[11] HSAB serves to categorize nonpolarizable ligand–metal interactions (hard–hard) and polarizable ligand–metal interactions (soft–soft) as being thermodynamically favored over mixed ligand–metal interactions (soft–hard or hard–soft). The high-charge-density, low-polarizability Fe^{3+} ion is considered as a hard Lewis acid, while Fe^{2+} is a borderline hard/soft Lewis acid. The HSAB principle reliably predicts that Fe^{3+} *in vivo* will normally be found bound to nonpolarizable electronegative donor atoms, such as the O atom in carboxylic acid, phenol, catechol, or alcohol moieties found in protein side chains or small-molecule chelators (e.g., siderophores), while the borderline hard/soft Fe^{2+} ion is normally found bound to donor atoms with polarizable electron clouds, such as S atoms or N atoms in aromatic rings.

Insolubility of Iron In aqueous environments, the "free" iron ion forms a coordination complex with six water molecules through coordinate covalent bonds. The fully aquated form of Fe^{3+} readily undergoes hydrolysis, leading to the formation of insoluble iron hydroxides (Eqs. 13.1–13.3)[12–14]:

$$[Fe(H_2O)_6]^{3+} \rightleftarrows [Fe(H_2O)_5(OH)]^{2+} + H^+ \quad (13.1)$$

$$[Fe(H_2O)_5(OH)]^{2+} \rightleftarrows [Fe(H_2O)_4(OH)_2]^{1+} + H^+ \quad (13.2)$$

$$[Fe(H_2O)_4(OH)_2]^{1+} \rightleftarrows [Fe(OH)_3]_{(s)} + H^+. \quad (13.3)$$

The resulting low solubility of iron(III) hydroxide (or hydrated iron oxide; $K_{sp} = 10^{-36}$) must be overcome in order to make this necessary metal bioavailable. Iron chelators (proteins or small molecules) prevent hydrolysis by replacing the H_2O ligands in the first coordination shell around the iron, resulting in a more water-soluble coordination complex.

Redox Chemistry of Iron An important chemical property of iron is the existence of multiple stable oxidation states. In biology, the most common oxidation states of iron are Fe^{3+} and Fe^{2+}.[14] The ability to cycle between the ferric (Fe^{3+}) and ferrous (Fe^{2+}) ions provides a range of possible reactions in a biological setting. In aqueous solution, the redox potential of the reduction reaction in which the hexaaquoferric iron is reduced to the hexaaquoferrous ion is +0.771 V versus normal hydrogen electrode (NHE) (Eq. 13.4)[14]:

$$[Fe(H_2O)_6]^{3+} + e^- \rightleftarrows [Fe(H_2O)_6]^{2+}. \quad (13.4)$$

Uncontrolled, the spontaneous redox cycling of aqueous iron ions can produce reactive oxygen species through a series of reactions known as the Haber–Weiss cycle (Eqs. 13.5 and 13.6).[15] In this cycle, Fe^{3+} reacts with superoxide to produce Fe^{2+}. Then in the Fenton reaction, Fe^{2+} reacts with hydrogen peroxide restoring the ferric ion, but producing the deleterious and highly toxic hydroxyl radical:

$$Fe^{3+} + O_2^- \rightarrow Fe^{2+} + O_2 \quad (13.5)$$

$$Fe^{2+} + H_2O_2 \rightarrow Fe^{3+} + OH^- + OH. \quad (13.6)$$

In addition to preventing hydrolysis, iron chelators alter the reduction potential of the metal, making it easier or more difficult to reduce the metal.[16] The identities of the specific ligands in the first coordination shell dictate the reduction potential of the metal, while the atoms in the second coordination shell can fine-tune the reduction potential.[12,17,18] The second coordination shell consists of atoms that are not directly bound to the metal, but are close enough to affect the electronic environment of the metal and the effective dielectric constant surrounding the metal.[12,19] As shown in Figure 13.1, the redox potentials of iron complexes vary over 2 V, depending on first coordination shell composition. Iron chelators that stabilize the Fe^{2+} oxidation state relative to the hexaaquo complex raise the reduction potential, while Fe^{3+} stabilizing chelators lower the reduction potential.

Biological Iron Chelators Biological iron chelators, whether they are small molecules or proteins, act as Lewis bases, donating electron density to the iron and forming coordinate covalent bonds. By forming coordination complexes with the iron ion, biological chelators combat both deleterious properties of this necessary iron: insoluble hydroxide formation and radical oxygen production.[18] Some biological chelators, such as small-molecule siderophores and iron transport proteins, solubilize the iron and provide a molecular signature that can be recognized by a cell. These two functions increase the bioavailability of this necessary nutrient.

Biological chelators also control the redox properties of the metal in order to protect the organism and harness the chemical power of the iron ion. Most biological ligands alter the reduction potential of the iron by stabilizing Fe^{3+}. As shown in Figure 13.1, many biological chelators substantially lower the reduction potential of iron, making the iron complex unable to participate in the Haber–Weiss cycle described above (Eqs. 13.5 and 13.6). By lowering the reduction potential of iron, biological chelators protect the system from the production of reactive oxygen species. However, metalloproteins with functions other than iron transport, such as hemoglobin, have evolved to have a more neutral reduction potential. In this case, the redox chemistry in which iron participates is largely controlled by limiting access to the iron center.

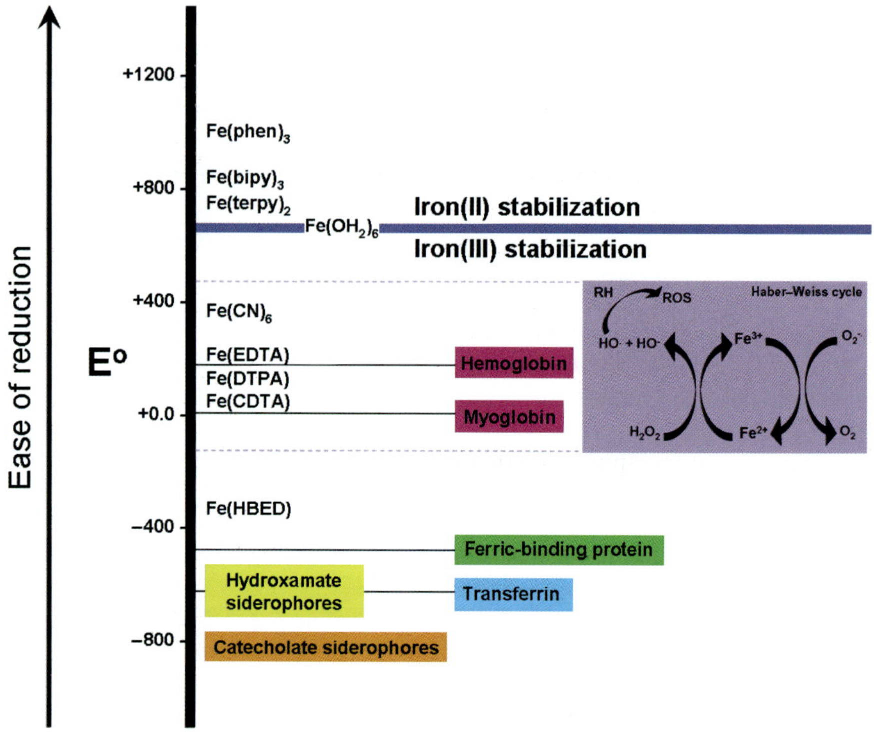

FIGURE 13.1. Scale of iron complexes showing the wide range of reduction potentials, E°, reported in millivolts versus NHE.[12,13,18,19,52,193] Biological chelators, both small molecules and proteins, are highlighted.

In iron transport pathways, iron is either carried by small-molecule chelators or proteins. The human iron transport protein transferrin (Tf) is well studied and will serve as a prototypical example in much of this chapter. Non-Tf-bound iron in the human system is chelated by small biological molecules such as citrate. Bacteria also transport iron in a variety of ways. Both low-molecular-weight chelators, termed siderophores, and iron transport proteins similar to human Tf are utilized. In all organisms, protein receptors also play an important role in iron transport. Whether complexed by a small molecule, a protein carrier, or bound to a protein receptor, intricate systems of ionic and mixed interactions control the acquisition of iron necessary for growth and survival.

13.2.2 Introduction to Iron Transport

Across biology, several strategies are used to transport iron from one compartment to another. In humans, iron must be transported from the gut where it is absorbed, to the bloodstream where it is delivered to various parts of the body, and then inside a cell such as a red blood cell where it is incorporated into its functional form such as hemoglobin. In bacteria, iron must be acquired from the environment, brought into the cell by crossing at least one membrane, and then incorporated or stored as need dictates. The chemical properties of ionic iron and the way that it interacts with biological macromolecules define the transport systems implemented by all organisms, from the most highly ordered species to unicellular pathogens.

In general, iron transport involves moving a charged, insoluble, redox-active ion through the biological system. The iron must be transported across membranes, through aqueous environments, and then deposited into the appropriate site for immediate functional use or into a stable storage site. In order to provide an introduction to the field of iron transport, we will outline the various strategies used to complete each of these tasks. Once the reader is familiar with these general aspects of iron transport, we will examine classic examples of iron transport investigations through the lens of ionic and mixed interactions. *We seek to demonstrate the multiple levels onto which the principles of ionic and mixed interactions of polymers can be applied to iron transport in biology.*

Transport across a Barrier In all cases, from bacteria to humans, one challenge is to move iron, a high-charge-density hydrophilic cation, across a hydrophobic barrier or biological membrane. The three general strategies used to move iron across a biological membrane are (1) transport across a membrane while sequestered within a protein, (2) transport of the unbound or "naked" iron ion through a membrane pore or protein receptor, and (3) transport across a membrane while sequestered by a small-molecule carrier. For the task of transporting iron across a membrane, the ionic interactions must be designed for two functions: (1) molecular recognition on one side of the membrane and (2) moving a hydrophilic iron through a hydrophobic barrier.

Transport of Iron-Containing Proteins into a Cell Transport of an iron–protein complex from outside a cell to inside a cell does not usually involve moving such a large molecule directly across the membrane. Rather, the iron–protein complex is brought into the cell through endocytosis. The human iron transport protein Tf is an example of this receptor-mediated endocytosis strategy.

Iron in the bloodstream is almost entirely bound to the human iron transport protein Tf. Human Tf tightly sequesters ferric ions for three fundamental purposes: (1) to enable iron transport through plasma from storage to utilization site; (2) to act as a bacteriostatic agent by preventing bacterial cells from acquiring the essential nutrient iron; and (3) to prevent iron-catalyzed toxic free radical formation via Fenton reactions.[16,20,21] Tf is an 80-kDa protein that binds two ferric ions.[22] Each active site (the N-lobe and the C-lobe) of Tf is located between two globular domains that are hinged and clamp the iron in place (Fig. 13.2).

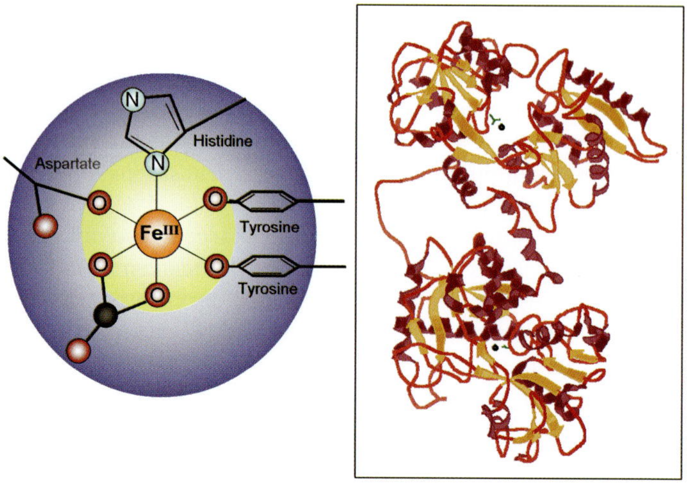

FIGURE 13.2. Ribbon diagram of Fe_2-Tf (shown on the right) and the active site of one lobe of Tf (shown as a cartoon on the left). Tf provides four amino acids and binds iron in the cleft between two globular domains. Aspartate, histidine, and two tyrosines provide four ligands to the iron, while a synergistic anion carbonate completes the first coordination shell. (Kinemage created by Dr. Katherine Weaver.)

Iron chelation by Tf is made possible through two chemically identical iron binding sites containing donor moieties from aspartate, histidine, and two tyrosine amino acid side chains. Therefore, the Tf protein can be viewed as a tetradentate ligand chelating Fe^{3+} with three hard oxygens each with a charge of −1 (Asp, Tyr, and Tyr) and a borderline hard nitrogen (His). The first coordination shell around each iron is completed by a bidentate exogenous carbonate anion providing hard oxygen ligands with a −2 charge. Tf binds ferric iron tightly with a $K_a \sim 10^{20}$ M, a large binding constant that is due to the strength of the coordinate covalent bonds in the first coordination shell and ionic and mixed interactions in the second coordination shell.[23]

The process of Tf-mediated iron uptake by human cells is depicted in Figure 13.3. Fe_2-Tf binds to the human transferrin receptor (TfR) on the cell surface and the Fe_2-Tf/TfR assembly is endocytosed.[24] Using this strategy, the deleterious effects of the redox active and insoluble iron are controlled through constant sequestration within the protein carrier. This protein coating provides a characteristic surface for molecular recognition guiding the receptor to specifically gather iron. Finally, the process of receptor-mediated endocytosis of the iron–protein complex facilitates the movement of the ionic cargo from outside the cell to inside the cell by avoiding moving the charged species directly through the hydrophobic membrane and by avoiding changing the immediate coordination environment of the iron.

After the iron–protein–receptor complex is engulfed, the pH of the fully formed endosome is decreased to ~5.5 and iron is thought to be released from

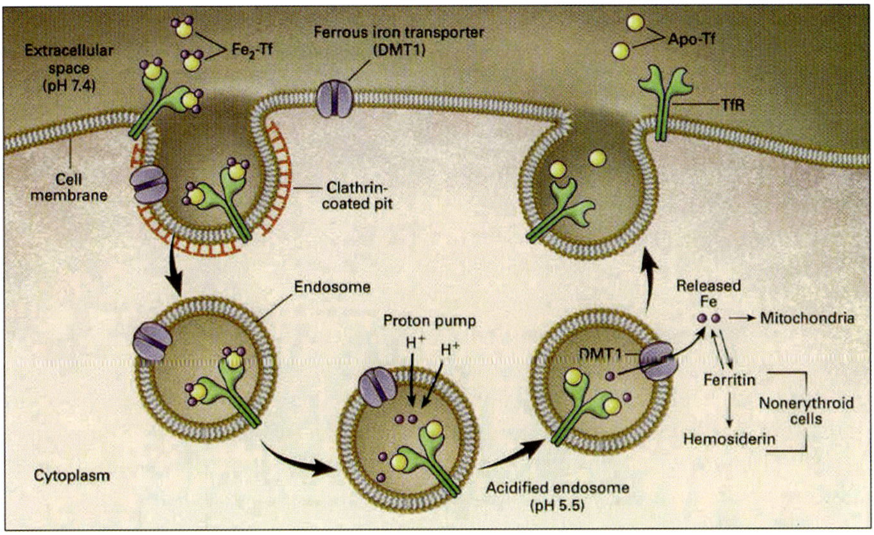

FIGURE 13.3. Schematic representation of cellular iron uptake by endocytosis. Fe_2-Tf binds to the human TfR. The protein–receptor complex is brought into the cell by endocytosis. Inside the endosome (pH 5.5), iron is released from Tf, possibly facilitated by reduction. The protein–receptor complex is recycled back to the surface of the cell and apo-Tf is released into the serum. Fe^{2+} exits the endosome through DMT1 and is incorporated or stored as needed. (Figure used with permission.[20])

Tf via reduction of the ferric iron cargo (Fig. 13.3).[25] The ferrous ion is then transported into the cytosol through the divalent metal transport protein DMT1.[20] The receptor/apo-Tf complex is subsequently recycled to the surface of the cell and apo-Tf is released. Although we will not go into detail here, DMT1 is a classic example of the second transport strategy used to move the iron ion across a membrane: "naked" iron transport.

Transport of "Naked" Iron through a Membrane Protein In order to move "naked iron" across a membrane, the transport pathway must always maintain control of the iron by formation of coordinate covalent bonds. The term "naked" is used to describe iron that is not bound by a protein carrier or chelated by a small molecule during movement across the membrane. However, "naked" iron does not necessarily imply Fe^{3+} or Fe^{2+} is without ligands. In a biological environment, aquated iron or a ligand-substituted iron (e.g., $[Fe(H_2O)_5(X)]^{n+}$, where X is a prevalent anion such as Cl^-) may be the actual transported cargo, and donor side chains from the protein receptor may be involved.

As mentioned previously, movement of Fe^{2+} through DMT1 in the red blood cell is an example of naked iron transport through a pore that spans a membrane. DMT1 is thought to be a glycoprotein with 12 transmembrane helices.[26] Iron transport through DMT1 coincides with proton transport and is highly pH sensitive.[26] No crystal structure of DMT1 is available, but it has been shown by mutagenesis studies that helix 1 and the loop between helices 1 and 2 are important for binding the divalent metal ion and the proton.[5] In addition to iron, DMT1 can also transport other divalent cations such as $Mn,^{2+}$ $Co,^{2+}$ $Zn,^{2+}$ $Cu,^{2+}$ and Pb^{2+}.[6] At this time it is unclear what ionic and mixed interactions allow this transporter to be specific for divalent metals, but non-specific for atomic identity. However, we expect that both the size and charge of the ionic cargo are important.

Another example of this "naked iron" transport strategy that will be discussed in detail in this chapter is the Tf-dependent bacterial iron uptake system used by some gram-negative bacteria such as *Neisseria gonorrhoeae*, the causative agent of the sexually transmitted disease gonorrhea. *N. gonorrhoeae* employ a complex system of proteins to strip iron from the human host protein Tf and transport the iron ion, but not the Tf, across the bacterial outer membrane, through the periplasm, and across the inner membrane (Fig. 13.4).

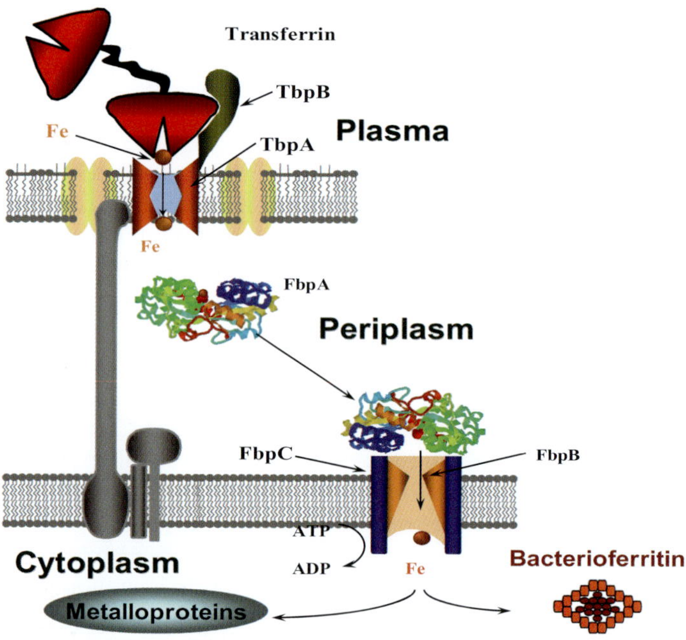

FIGURE 13.4. Schematic representation of iron transport by the human gram-negative pathogen *Neisseria gonorrhoeae*. The outer membrane receptor assembly TbpA/TbpB docks holo-Tf, and "naked" iron is transported through the pore formed by TbpA. Iron is sequestered by FbpA and transported across the periplasm. Iron is subsequently released to the cytoplasm by docking at the FbpB/FbpC inner membrane receptor. (Figure adapted from Bruns et al.[33] and used with permission.)

Transferrin-binding protein A (TbpA) and transferrin-binding protein B (TbpB) form the outer membrane receptor, which binds holo-Tf and hijacks the iron.[27–29] This process requires energy supplied by TonB and involves many ionic and mixed interactions between and within proteins.

Once iron has moved across the outer bacterial membrane, the soluble periplasmic protein FbpA binds iron with a synergistic anion and delivers it to the inner bacterial membrane.[17,30–34] FbpA is part of an ATP-binding cassette or ABC transporter that also includes FbpB and FbpC. FbpB and FbpC are associated with the inner membrane and serve as the channel through the membrane and the ATPase, respectively. Therefore, the *N. gonorrhoeae* uptake pathway involves moving "naked" iron across two membrane barriers. After iron has crossed the inner membrane, it can be utilized or stored as needed.[31,35,36]

Naked iron transport across the outer and inner membranes of this *Neisseria* system involves a variety of ionic and mixed interactions that function to transport iron directly across the hydrophobic membrane while maintaining control of the chemical properties of the iron ion. Control is achieved through protein–protein interactions that have been shown to be necessary to pass the iron cargo from one component of the transport pathway to another.[37] In Section 13.4, we will examine the components of this *N. gonorrhoeae* system in detail, focusing on protein–protein interactions. In Section 13.6, we will examine the protein–ion interactions involved in the transport of naked iron by looking at FbpA, a protein that binds both the iron ion and an auxiliary anion.

Transport of Iron Complexes across a Membrane In contrast to hijacking the naked iron bound by host metalloproteins, bacterial iron acquisition can also be accomplished by transporting iron complexes such as the heme cofactor in hemoglobin[38,39] or by incorporating extracellularly chelated iron in the form of ferric siderophores. The mechanisms of iron transport in gram-negative bacteria have been extensively studied[38,40–48] and the crystal structures of several key proteins have been described.[30,49–51] We have elected to focus on siderophore-mediated systems rather than heme acquisition systems for the purposes of this chapter.

A siderophore is a low-molecular-weight iron-specific chelator that is released into the environment by bacteria, where it sequesters and solubilizes iron, making iron bioavailable for uptake. The formation of several coordinate covalent bonds between a siderophore and Fe^{3+} ($K_a > 10^{30}$ M) ensures the iron does not participate in Fenton chemistry and imparts specificity to the nutrient uptake system.[52]

Acquisition of siderophore-bound iron involves binding to and passage through integral outer membrane TonB-dependent transporters.[53–57] These transporters have similarities to the TbpA component of the outer membrane receptor in *N. gonorrhoeae* discussed in the previous section, in that siderophore receptors exhibit the same global protein structure and are dependent on energy from TonB. However, the ionic and mixed interactions involved in the molecular recognition and transport of an iron–siderophore complex by the outer membrane receptor are different than those of naked iron transport and will be discussed in Section 13.5. At some stage of the transport pathway the iron–siderophore complex must be dissociated, the apo-siderophore degraded or recycled, and the iron sequestered in the cytoplasm. Similar to the FbpA/B/C system utilized in the naked iron uptake pathway of *N. gonorrhoeae*, most siderophore-dependent systems involve ABC transporters to complete the transport task of crossing the inner membrane. This common bacterial machinery will be discussed at length in Section 13.4.3.

Transport through an Aqueous Environment As alluded to in each example discussed in the section "Transport across a Barrier," iron must not only be transported across membranes but must also be moved through aqueous environments in a controlled fashion preventing hydroxide formation and toxic radical production. Tf controls the chemistry of its iron cargo while moving iron throughout the blood serum of the human body. FbpA controls the chemistry of its iron cargo while moving iron throughout the periplasm of the gram-negative *N. gonorrhoeae*. Similarly, siderophores make iron bioavailable in the extracellular environment of the respective bacteria and control the redox chemistry of iron during transport in the aqueous periplasm. Therefore, on either side of the membranes discussed in the previous section, ionic and mixed interactions between iron and a transport carrier dictate the movement of iron in aqueous environments.

Although similarities exist between the variety of transport strategies, movement of an iron–protein complex, naked iron, or an iron–siderophore complex either across a membrane barrier or through aqueous solution each entails unique recipes of ionic and mixed interactions that dictate the specificity and efficiency of the uptake system.

13.3 A COMPLEX SUITE OF IONIC AND MIXED INTERACTIONS DRIVES IRON TRANSPORT

Examining iron transport in biology in the context of ionic and mixed interactions is complex because of the variety of possible interactions. In general, iron transport can involve interactions between

- iron(III/II) and water
- iron(III/II) and small molecules
- iron(III/II) and proteins
- iron(III/II)–small molecule complexes and proteins
- iron(III/II)–protein complexes and other proteins
- iron(III/II)–protein complexes and anions
- proteins and anions

The exact suite of ionic interactions involved in a particular pathway depends on the iron transport strategies (Section 13.2.2) employed to move iron across membranes, through aqueous compartments, and into storage sites. All of the above interactions can include attractive forces between ions and/or dipoles. Interactions between an iron–small molecule complex and a protein, or between multiple proteins, or even between anions and proteins, can also involve multivalent interactions forming associations. Finally, host–guest chemistry and electrolyte effects such as the Hofmeister series are also observed in each iron transport pathway.

Significant work has been published regarding the interactions between iron and small molecules relevant to iron uptake.[52] In light of the polymer focus of this book, we will refrain from further discussing the details of the interactions between iron and water and between iron and siderophores in the absence of proteins. We will focus instead on the interactions between iron and proteins and between iron complexes and proteins, a field which is rapidly evolving and expanding.

Significant progress has been made toward understanding the ionic interactions that govern iron and iron complexes interacting with iron transport proteins. An excellent review by Davies et al. outlines the electrostatic, hydrogen bonding, and hydrophobic interactions that guide the formation of a supramolecular assembly between a metal complex and a protein.[58] This elegant review represents the leading perspective through which iron–protein and iron complex–protein interactions are analyzed. Namely, the protein can serve as the second coordination shell of the metal, but it can also serve as an electron donor in the first coordination shell of the metal.

Here we will examine these prototypical bioinorganic examples of ionic and mixed interactions between ionic species (iron or iron complexes) and natural polymers (proteins), but we will also go further. We will examine the interactions between auxiliary anions present in the biological system and interactions between multiple proteins. These later topics characterizing the protein–protein interactions involved in iron uptake systems and understanding the effects of auxiliary anions on the iron uptake systems are areas of research that will benefit greatly from the polymer and physical chemistry perspective put forth in the preceding chapters of this book.

Taking a metallocentric view or a protein centric view is not sufficient alone to understand an iron uptake system. Rather, a multipronged approach that recognizes the variety of interactions involved in the uptake process is required to fully characterize a particular pathway. The current state of the bioinorganic field is based on a dialogue between inorganic chemists, protein chemists, molecular biologists, and microbiologists. Here we seek to describe how the physical chemistry perspective of ionic and mixed interactions in which polymers participate can complement current research and answer questions that are not yet being addressed.

The remainder of this chapter is divided into three sections: (1) protein–protein interactions, (2) protein/iron–small molecule interactions, and (3) protein–anion interactions. In the third section, we will also highlight the design of *in vitro* environments that are used to mimic *in vivo* conditions and discuss the effects of these environmental conditions on protein behavior. Each of these topics includes many ionic interactions. We will demonstrate that the definitions and principles of ion–polymer interactions described in this book can be applied to the field of iron transport in biological systems in many different and complementary ways.

13.4 PROTEIN–PROTEIN INTERACTIONS

All along the pathway from outside the cell to inside the cell, protein–protein interactions play a large role in the flow of iron for each uptake system. In order to provide examples for the types of ionic and mixed interactions important when two (or more) proteins associate, we will discuss (1) the Tf/TfR complex involved in human iron transport, (2) the Tf/TbpA/TbpB receptor complex involved in bacterial iron transport, and (3) the family of multiprotein ABC transporters commonly found in bacterial iron uptake pathways.

13.4.1 Human TfR and Tf

As introduced in Section 13.2.2, Tf is the human iron transport protein, which is recognized at the surface of the red blood cell by the human TfR and internalized when the protein–receptor complex is endocytosed (Figs. 13.2 and 13.3).[59] The effects of TfR binding on Fe_2-Tf have been the subject of much investigation. A particular focus has been the characterization of the thermodynamic properties of iron binding by Tf when bound to TfR and the role TfR plays in the release of iron from Tf. For example, it remains a matter of debate as to whether reduction of the iron(III) in the endosome occurs prior to dissociation from Tf in the Fe_2-Tf/TfR assembly, or subsequent to dissociation from Tf. A redox-mediated release of iron from Tf is an attractive hypothesis because the binding constant of Tf for Fe(II) is 17 orders of magnitude lower than that for Fe(III), and Fe(II) exchanges its first coordination shell at a much faster rate than Fe(III).[60,61] The very negative reduction potential of Fe(III) in Fe_2-Tf at the conditions of the endosome (–526 mV vs. NHE) suggests that reduction to Fe(II) is energetically unfavorable.[62] However, when bound to the receptor, Fe_2-Tf has a reduction potential of –300 mV and a redox-mediated release mechanism is plausible.[25,26,63–65]

Another plausible mechanism of iron release is the acidification of the endosome. The acidic pH of the endosome is not low enough to remove a significant amount of iron from Fe_2-Tf in solution.[66,67] However, binding of Fe_2-Tf to TfR changes the pH sensitivity of the iron release process compared with proton-induced iron release from free Tf.[68] A conformational change in the global structure of Tf has been suggested as the cause of the altered pH

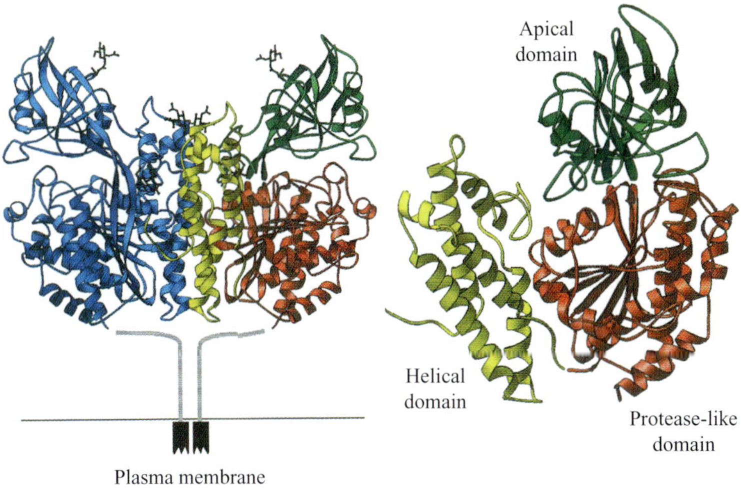

FIGURE 13.5. Ribbon diagram based on the crystal structure of the soluble globular portion of human TfR at 3.2-Å resolution. One monomer is shown on the right and the homodimer is shown on the left. (Figure used with permission.[59])

sensitivity and redox potential of Tf.[25,68] Clearly, the interaction between Tf and TfR has some energetic effect.

Structural characterizations of Tf and TfR have also been a focus of the biophysical community studying iron transport. TfR is a homodimeric receptor that can bind two Fe_2-Tf molecules at pH 7.4 on the surface of the cell. Each monomer of TfR consists of 760 amino acids.[59] Lawrence et al. published the 3.2-Å crystal structure of the soluble globular domain of TfR that binds one Tf molecule (Fig. 13.5).[69] The globular ectodomain of TfR consists of three subdomains: an apical domain, a protease-like domain, and a helical domain. The helical domain has been implicated in the noncovalent dimerization process (there are no disulfide bonds formed) and in the binding of Tf.[59] Together two monomers form a homodimer with a Y-shaped structure anchored by a stalk that crosses the cytoplasmic membrane.[70] TfR is glycosylated with four oligosaccharides, lipidated through a palmitylation site, and can be phosphorylated.[59] For a candid analysis of what is known about the structure of the TfR the reader is referred to a review by Aisen.[59] Here we will discuss a few salient features of ionic and mixed interactions that have been identified as important for interaction between TfR and Tf.

Several approaches to identify the interfacial contacts between Tf and TfR have been utilized. However, the details of how Tf and TfR interact on a molecular level and what ionic and mixed interactions control this transient protein–protein interaction are disputed and under intense investigation. Mutagenesis of TfR has shown that three amino acids in the helical domain

(Arg-Gly-Asp, called the RGD triad) are critical for Fe_2-Tf binding. Further, the significant decrease in binding affinity when L619, Y643, R629, F650, or Q648 are exchanged for alanine suggests that both electrostatic and hydrophobic interactions are involved in Fe_2-Tf binding.[71] Similar mutation experiments probing the surface of TfR on which apo-Tf interacts implicated G647, L619, and R651 of TfR, in apo-Tf recognition, supporting the hypothesis that a mixture of electrostatic, hydrophobic, and van der Waals interactions are important for the binding of apo- or holo-Tf to TfR.[59]

Focusing on the binding surface(s) of Tf, it has been shown that only one domain of the C-lobe of Tf (Tf_{CI}) binds to the helical domain of TfR and that the other domain of the C-lobe of Tf (Tf_{CII}) remains free.[59] This $Tf_{C\text{-lobe}}$/TfR interaction is mediated by an electrostatic binding event between negatively charged areas on Tf_{CI} of apo-Tf (at pH 5.6) or holo-Tf (pH 7.4) and positively charge areas of the helical domain of TfR (at both pH levels).[72] Interestingly, it has been shown that the N-lobe of Tf ($Tf_{N\text{-lobe}}$) makes multiple contacts with TfR, binding to the helical domain, the protease domain, and possibly interacting with the stalk of the TfR.[73–75] Analysis of the global structure of the Tf/TfR complex observed with cryo-electron microscopy suggests the multiple contacts between the $Tf_{N\text{-lobe}}$ and TfR are possible because both domains of the N-lobe of Tf bind to TfR.[73]

As expected for the asymmetric protein–protein interaction of the complex, binding to the TfR affects the iron release properties of each Tf lobe differently, both in terms of kinetics and thermodynamics, and it has been shown that the association between $Tf_{N\text{-lobe}}$ and TfR is much weaker than that of $Tf_{C\text{-lobe}}$ and TfR.[59,76,77] In concert with these findings, atomic force microscopy experiments show that the force required to remove holo-Tf from TfR is stronger than apo-Tf at all conditions encountered during the endocytosis process. Further, holo-Tf removal involves surmounting two energy barriers, whereas apo-Tf involves only one.[78] Taken together, these biophysical techniques have shown that TfR interacts with apo-Tf and holo-Tf differently, and differences between the $Tf_{N\text{-lobe}}$ and $Tf_{C\text{-lobe}}$ are recognized by the receptor. Further, a variety of ionic and mixed interactions work together to guide this complicated, transient, and sensitive protein–protein association.

Now that crystal structures for apo- and holo-Tf are available, significant insight has come from computational docking models between the transport protein and the receptor.[72] *In silico* docking has confirmed that for both apo- and holo-Tf the C-lobe of Tf binds to helical domain of TfR and the N-lobe of Tf is located between the ectodomain of TfR and the membrane.[72] The docking model published by Sakajiri et al. was made by linking the W641/F760 hydrophobic pocket of TfR with the C-lobe of apo-Tf at H349 (Fig. 13.6).[72] This interaction has been shown to occur under acidic conditions and is important for the kinetic differences between iron removal from the $Tf_{C\text{-lobe}}$ and $Tf_{N\text{-lobe}}$.[74] In the model, the imidazole ring of the Tf histidine is sandwiched between the indole and the benzene through predicted π-stacking interactions. Maintaining this hydrophobic interaction, the two protein structures were

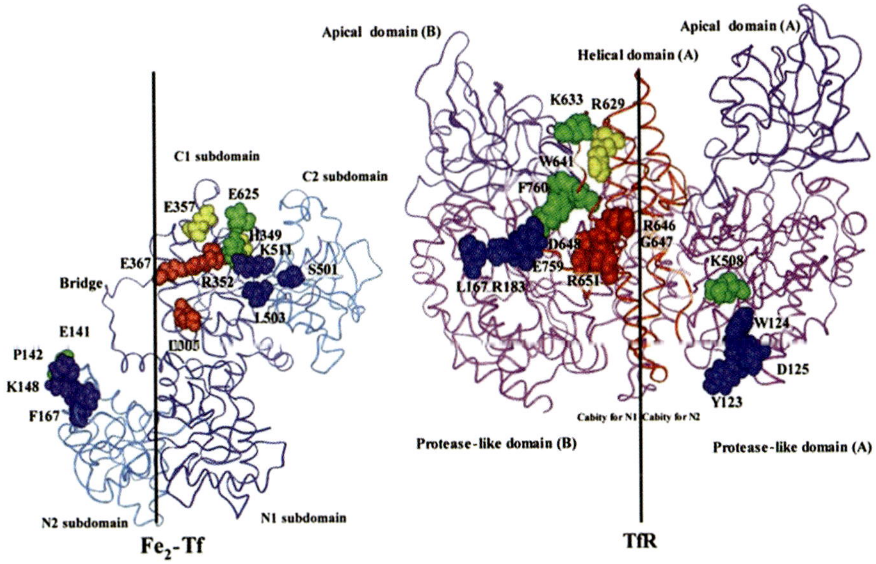

FIGURE 13.6. Ribbon diagrams of Fe_2-Tf and TfR with protein–protein interactions denoted with matching colors. The central protein–protein interaction is shown in red and yellow. Predicted interactions for the Fe_2-Tf/TfR complex are shown in blue, while predicted interactions for the apo-Tf/TfR complex are shown in green. Rotating the Fe_2-Tf structure by 180° around the vertical axis allows superposition on the TfR structure with matching residues interacting. (Figure used with permission.[72])

organized such that the positive charge of R651 of TfR was matched with a negative surface of Tf created by E385, D387, and D393 that are conserved across species and provide the correct geometry to form productive ionic interprotein interactions.[72] In this orientation, R646 of TfR (part of the RGD triad) aligns with E367 of Tf_{CI}, forming an ionic bond. Many more interactions productive for protein–protein assembly formation were also modeled and can be seen in Figure 13.6.

The assembly between holo-Tf and TfR was also modeled and in combination with the apo-Tf/TfR docking model support the hypothesis that ionic and conformational changes in the complex are important for the iron release mechanism.[72] It has also been shown that when Al^{3+} is bound to Tf in the place of Fe^{3+}, structural changes in Tf (toward an intermediate form between apo-Tf and holo-Tf) prevent binding between Tf and TfR due to mismatched ionic and mixed interactions.[79] As more data are collected regarding the effect of pH on these protein–protein interactions, it will be interesting to see how the *in silico* models are refined. Important ongoing work addresses the interactions between the TfR and $Tf_{N\text{-lobe}}$ as both the $Tf_{N\text{-lobe}}$ and the $Tf_{C\text{-lobe}}$ are required for tight TfR binding at neutral pH. Recently, it has been shown that the NII domain of the $Tf_{N\text{-lobe}}$ may serve as a recognition factor between species, allowing TfR to specifically bind its cognate Tf.[80–82]

From the evidence collected thus far, it is obvious that attractive forces between ions of opposite charges and the charge complementarity of proteins within an assembly (Chapter 8) significantly impact the regulation of Tf binding and the recognition of the iron-loading status of Tf by TfR.[1] One question that has not been addressed yet in the literature, but we suspect will be interesting and important, is the role of mobile ions as mediators of the protein association.[1] How do background electrolytes affect the mechanism and strength of Tf binding to TfR?

Another line of thinking that should be addressed in the future is the role of the membrane in this iron uptake process (Chapter 9). We hypothesize that membrane association affects the interaction of Tf with TfR and, possibly, the dynamics of the Tf/TfR complex. Many of the experiments reported in the literature involve solubilized TfR due to the insolubility of the complete receptor and difficulties in isolating the protein.[83,84] Therefore, *in vivo*, the interactions between Tf and TfR may be quite different from the current model based on *in vitro* data. Recently, it has been shown that the Tf/TfR complex colocalizes with MRCKα, a protein involved in the organization of the cytoskeleton, and that the function of MRCKα influences the human iron uptake pathway.[85] These findings indicate that the cytoskeleton and the membrane are key to the endocytotic uptake process and that interactions between multiple biomacromolecules may regulate human iron acquisition.

Another frontier in this field is the possibility of communication between sites on these multivalent polymers.[1] Multivalent interactions can be defined here as (1) Tf binding two iron ions at separate sites, (2) each lobe of Tf binding to TfR in different locations, or (3) each monomer of the receptor binding to a Tf molecule in multiple places while concurrently interacting at the dimer interface of the receptor. From both the structural and energetic (kinetic and thermodynamic) perspectives, communication between multiple binding sites on a particular polymer are likely important. Efforts are under way to address these issues, and the reader is referred to two articles from the Mason group that describe the complexity of teasing out all of the possible interactions and their contributions to the overall iron release process.[77,83] For example, the signal produced by the fluorescent amino acids of Tf is a convoluting factor because these spectroscopic handles report on the solvent accessibility of the residue in addition to the ionic and polymer interactions in which the residue is participating.

A more detailed understanding of the ionic and mixed interactions involved in the protein–protein interactions of human iron uptake is important for combating disease. Another protein, HFE, binds to TfR and acts as an inhibitor of Tf-mediated iron uptake.[59] We will not discuss this particular protein–protein association here, but surely mixed interactions will be critical as hydrophobic interactions between all three proteins (HFE, Tf, and TfR) have been identified.[59] We would be remiss if we did not also mention that, recently, another human Tf receptor, TfR2, has been identified. Little is known about the structure of TfR2 or the way it interacts with Tf, but it is hypothesized that TfR2 serves a supporting role in the complicated human iron uptake pathway.[59]

13.4.2 Bacterial Tf Receptor and Tf

With this strong iron chelation ability, Tf serves a bacteriostatic function gathering essentially all of the free iron in the human blood serum and thus preventing iron acquisition by pathogens living in the human host.[16,20,21,23] The exclusive human pathogen *N. gonorrhoeae* is one of the few types of bacteria that has developed a mechanism to hijack iron from Tf.[27] Here we will focus on the interactions between Tf and the bacterial Tf receptor that have important implications for iron acquisition by *N. gonorrhoeae*.

As introduced in Section 13.2.2, TbpA and TbpB form the TonB-dependent bacterial receptor for Tf that is expressed on the outer membrane of *N. gonorrhoeae* (Fig. 13.4). TbpA, a 100-kDa outer membrane protein, is required for utilization of Tf-bound iron and functions synergistically with TbpB.[27–29] While the surface-exposed lipoprotein TbpB is not absolutely required for *in vitro* growth by *N. gonorrhoeae* on Tf as the sole iron source, TbpB makes iron utilization more efficient.[28] The bacterial receptor binds Tf with an affinity of 5–20 nM, similar to the mammalian TfR.[38]

The receptor proteins in *N. gonorrhoeae* (TbpA/TbpB), which acquire iron from the human metalloprotein Tf, have not yet been crystallized. However, by analogy with the other TonB-dependent transporters, it is accepted that the integral membrane protein TbpA forms a β-barrel, which is periplasmically occluded by a plug domain.[7,27,86] The structure of other homologous receptor proteins (used in siderophore-mediated transport) will be discussed at length in Section 13.5. Recently, the crystal structure of a homologous TbpB from *Actinobacillus pleuropneumoniae* was solved and has shed light on the structure and function of TbpB from *N. gonorrhoeae*.[87] Here we will focus on the outside face of the bacterial outer membrane and the protein–protein interactions between TbpA/B and Tf. We will discuss studies involving the TbpA/B receptor of *N. gonorrhoeae* as well as information that has been gained from similar bacterial receptors.

Specific parts of the structure of TbpA (*N. gonorrhoeae*) have been shown to be surface exposed and/or important for functioning (Fig. 13.7). Hemagglutinin epitope tags were used to identify surface-exposed regions of the proposed 11 loops that tie together the β-strands of the barrel domain.[88] As well, insertions in the proposed plug domain of TbpA revealed partial surface exposure of the plug.[88] Tags inserted into the plug and loop 3 decreased Tf binding to the receptor and retarded bacterial growth. It is unclear whether the insertion into loop 3 disrupted a Tf binding site directly formed by loop 3 or at a distant location.[88] Surprisingly, while mutants with insertions in loops 2, 9, and 11 were still able to bind Tf, these altered forms of TbpA could not transport iron.[88] Loop 2 is expected to be hydrophobic and possibly form an

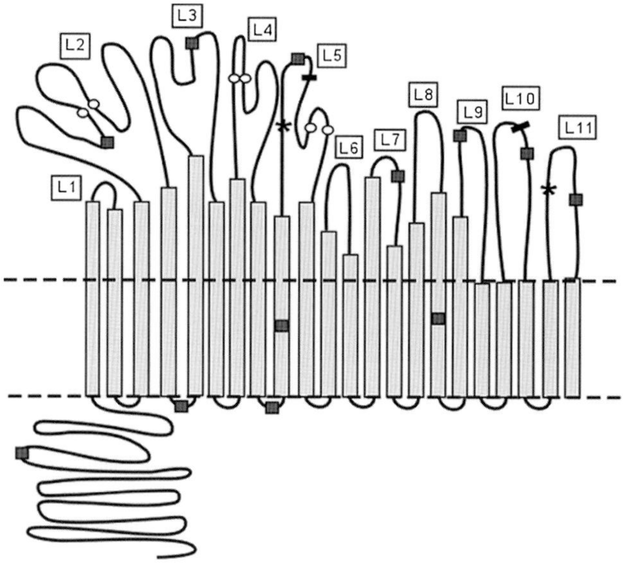

FIGURE 13.7. Hypothetical two-dimensional topological model of TbpA. The horizontal dashed lines show the edges of the outer membrane, and suspected transmembrane β-strands are drawn as vertical boxes. The numbering of the 11 putative extracellular loops is labeled. (Figure used with permission.[88])

internal disulfide bond.[49,51,89] However, it is unclear how this loop interacts with Tf. Significant work has been done to show the effect of certain loops on the binding affinity of Tf to the cell surface, but individual ionic and mixed interactions between the proteins have not been identified yet in *N. gonorrhoeae*.[88] Interestingly, TbpB is able to compensate for some mutations in TbpA, restoring utilization of iron from the Tf source.[88]

Other techniques have implicated loops 4 and 5 of TbpA in Tf binding in *N. gonorrhoeae*. Both TbpA and TbpB have been shown to be immunogenic toward antibodies designed from Tf, even in the absence of the rest of the protein. As well, when loop 5 is deleted Tf binding by TbpA is abolished and similar mutagenesis studies show that loop 4 is also critical for Tf binding.[90] Further, when loop 5 is expressed as an epitope fusion Tf binding was observed.[90,91] These findings suggest that loops 4 and 5 are important for Tf recognition.[90] Interestingly, the primary sequence of loop 5 is quite variable and predicting the ionic and mixed interactions important for Tf binding has been challenging.[90] This begs the following question: What are the important chemical features of such a polymer (the protein loop) that appear to be significant in the function of binding the substrate protein to the receptor?

The iron uptake machinery of *Neisseria* bacteria (*N. gonorrhoeae* and *Neisseria meningitidis*) is unique in that it involves an auxiliary protein in addition to the TonB-dependent transporter. The auxiliary protein, TbpB, varies in size from 64 to 85 kDa depending on the strain and species of bacteria.[28] Although not required in *N. gonorrhoeae*, TbpB makes iron utilization more efficient. Interestingly, TbpB *is* required in *N. meningitidis*.[38] Increased iron uptake efficiency is partially due to the ability of TbpB to discriminate between apo- and holo-Tf.[92,93] In terms of binding constants, TbpB alone can compete with the mammalian Tf receptor for binding the iron transport protein.[29,92] For *N. meningitidis*, TbpA prefers apo-Tf, while TbpB prefers holo-Tf.[45] Various mechanisms of receptor function have been proposed[45,94] and it is thought that the interaction between TbpB and TbpA causes a conformational shift in TbpA that changes its affinity for apo-Tf.[45] As well, it is thought that TbpA has one site for Tf binding, while TbpB has two.[45]

The structure and location of TbpB have also been investigated. TbpB is proposed to consist of two lobes that can each bind Fe_2-Tf.[95] For *N. gonorrhoeae*, the N-terminal lobe of TbpB binds Tf with greater affinity than the C-lobe of TbpB.[45,93] Hemagglutinin epitope tags inserted into TbpB show that residues 293–328 are particularly important for Tf binding by the high-affinity TbpB N-lobe binding site.[95] As well, epitope insertions were used to demonstrate that TbpB is completely surface exposed.[95] It is thought that a lipid moiety anchors TbpB to the surface of the outer membrane.

The TbpA/B receptor of *N. meningitidis* binds to the C-lobe of Tf.[43] It has been shown that TbpA specifically interacts with the C-lobe of Tf and that glycosylation of Tf is not involved in the binding event.[38] Residues 346–588 of Tf are involved in this binding event, but no specific residues have been identified.[38] Recent work has shown that the N-lobe of TbpB interacts exclusively with the C-lobe of Tf.[45,96] Although the N-lobes of various TbpBs have little sequence identity, TbpB protein of various strains and species are thought to have similar tertiary structures. It is of great interest to determine if the overall interaction between TbpB and Tf is conserved and to understand if and how divergent TbpBs bind to the same region of Tf.[96]

A. pleuropneumoniae, a porcine pathogen, requires both TbpA and TbpB for growth on a Fe_2-Tf source. Although the resulting receptor may function slightly differently from that of *N. gonorrhoeae*, important insights can be gained from the 1.98-Å crystal structure of the TbpB protein from *A. pleuropneumoniae*. This first structure of a bacterial Tf receptor protein confirms the general two-lobe structure of the outer membrane protein.[87] The N-lobe and C-lobe of TbpB are similar in terms of fold, but the "cap" of the N-lobe is electropositive, while the "cap" of the C-lobe is electronegative (Fig. 13.8).[87] In *A. pleuropneumoniae*, a phenylalanine residue is essential for Tf binding and in the crystal structure is surface exposed and located in the N-lobe.[87] Therefore, similar to the *Neisseria* receptors, the TbpB N-lobe of *A. pleuropneumoniae* is predicted to the be Tf binding site. However, significant variation in the primary sequence of this region of TbpB between strains of *A. pleuropneumoniae* and between different species have led to the designation of the N-lobe cap as hypervariable.[87] It is thought that although the N-lobe may have a conserved architecture, the hypervariable primary sequence may

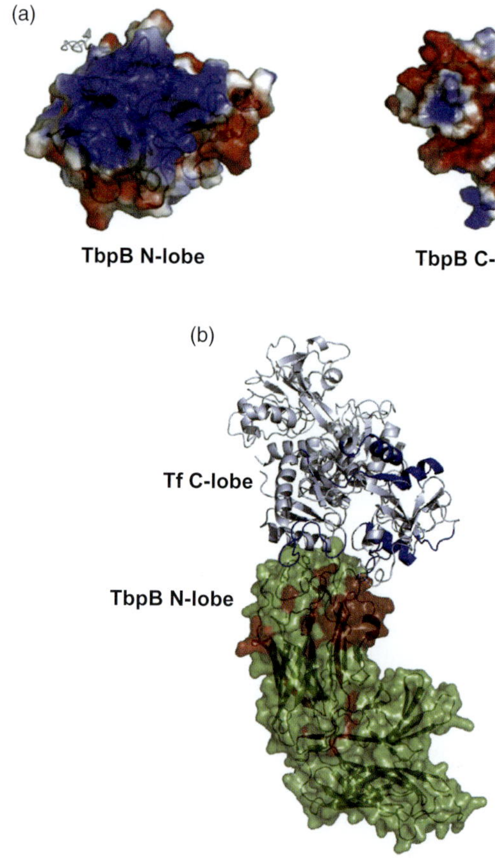

FIGURE 13.8. (a) Electrostatic surface potentials of the N-lobe and the C-lobe of TbpB in *Actinobacillus pleuropneumoniae*. Red indicates negatively charged areas and blue indicates positively charged areas. (b) *In silico* docking of a proposed complex between $pTf_{C\text{-lobe}}$ (ribbon) bound to TbpB (transparent surface). (Figure modified from Moraes et al.[87] and used with permission.)

be important for evading host immunity.[87] Using the crystal structure of *A. pleuropneumoniae* TbpB, Moraes et al. have devised an *in silico* docking model predicting the binding event between pTf and TbpB (Fig. 13.8). Several ion and mixed interactions are hypothesized and some residues have been demonstrated as important for binding Tf, but specific interactions remain to be identified.[87] The hypervariability of the Tf binding site in TbpB is an interesting challenge to determine what ionic and mixed interactions are essential and if there are multiple residues that can provide this binding surface.

As mentioned previously, studies of *N. gonorrhoeae* have shown that TbpB is able to restore the action of Fe_2-Tf utilization in some bacterial mutants that have specific defects in TbpA. For example, mutants with a defect in loop 9 of

TbpA are able to bind Fe_2-Tf, but are not able to acquire iron. TbpB can compensate for this mutation, restoring the utilization of Fe_2-Tf and promoting growth.[95] This compensation function performed by TbpB requires both lobes of TbpB. Further, it has been shown that TbpB can block the surface exposure of some parts of TbpA and the two proteins are closely associated.[88,95] The exact role TbpB plays in this situation in this bacterium is unclear. It is also not known if this function of TbpB is common in other types of bacteria. Given the differences in TbpB sequences and the variability in whether TbpB is essential or not, many questions regarding the structure and function of the TbpA/TbpB receptor remain to be resolved.

Work is ongoing to determine the mechanism of action and the functionally important parts of TbpA and TbpB.[37,43,94,97] Because there is no crystal structure of the intact receptor, the specific interactions between particular residues of TbpA and TbpB with and without Tf have not been identified. Ideally, a cocrystal structure of the complete receptor with Tf will reveal important information regarding the receptor assembly and the way it interacts with its protein substrate.

In addition to crystallization efforts, investigations probing the possible ionic and polymer interactions that might contribute to the Tf/TbpA/TbpB assembly and function are necessary. The physical chemistry perspective expressed in the preceding chapters of this book can shed light on several important questions which need to be addressed: Are the interactions between these polymers (TbpA and TbpB) associations between complementary surfaces of the proteins within the receptor assembly? What interactions drive receptor assembly? What interactions are most important in leading to a binding event between the bacterial receptor and Tf that is strong enough to compete with the human TfR discussed in Section 13.4.1? Are the associations between the receptor and Tf mediated by mobile ions? Do attractions between fixed opposites[1] contribute significantly to the energetics of the binding event (TbpA/B with Tf)? Are the same residues of Tf involved in interacting with both the human TfR and the bacterial receptor?

Another line of investigation for which an ion/polymer perspective can guide our thinking is as follows: How does each interaction that these multivalent proteins participate in affect other interactions? For example, it is not known if binding between TbpB and TbpA causes a conformational change in TbpA that affects its function or vice versa. Additionally, it is unclear how TbpA and TbpB assemble in the membrane and what role the membrane plays in the structure of the functional receptor.

Lastly, another important issue is the effect(s) on the two iron binding sites of Tf when bound by the bacterial receptor. A rich body of literature is available regarding the structure–function relationships important for iron removal from Tf in the context of human iron regulation diseases. However, little is known about communication between the lobes of Tf when the protein is located on the bacterial surface. When bound to the human TfR, changes in the conformation of Tf, induced either by protonation or protein–protein interaction, are thought to be important for the release of iron from Tf.[98,99] Further, Mason et al. have demonstrated that conformational changes in one lobe of Tf are communicated to the other.[100] A reasonable hypothesis follows that conformational changes are imposed on Tf when bound by the bacterial receptor TbpA/TbpB. By extension, communication between the lobes of Tf may facilitate the removal of both iron ions upon binding to the bacterial receptor. Such hypotheses have not yet been investigated, but are important for a complete understanding of the pathogenic iron uptake system. The principles of multivalent interactions involved in the formation of associations between polymers and ions discussed in the earlier chapters of this book are likely at play in the Tf/TbpA/TbpB assembly.

13.4.3 ABC Transporters and Cargo Delivery Proteins

As introduced previously (Fig. 13.4), once iron has been stripped from Tf and transported across the outer membrane of *Neisseria*, the iron is sequestered in ferric-binding protein, FbpA. FbpA carries the iron across the periplasm and delivers it to the ABC transporter FbpB/C. In this section, we will examine the ionic and mixed interactions important in the protein–protein interaction between a periplasmic transporter such as FbpA and an inner membrane receptor such as FbpB/C. First, we will define the ABC transporter family, members of which are ubiquitous in biology. Then, we will discuss in detail the protein–protein interactions of an ABC transporter involved in siderophore-mediated iron uptake in *Escherichia coli*.

Although we will focus on iron transport examples here, ABC transporters are used across biology to transport many different types of cargo such as metal ions, vitamins, sugars, and peptides.[101–105] Members of the ABC transporter family can be found in mammals such as humans and in both gram-positive and gram-negative bacteria.[102] For gram-negative bacteria utilizing iron uptake pathways following the naked iron transport strategy or the iron-complex transport strategy (sections "Transport of 'Naked' Iron through a Membrane Protein" and "Transport of Iron Complexes across a Membrane"), most express an ABC transporter to move the iron cargo from the periplasm, across the inner membrane, to the cytosol.

ABC transporters are *A*TP-*b*inding *c*assettes composed of a transmembrane permease formed by two transmembrane domain proteins and an ATPase formed by two nucleotide-binding domain proteins.[106] In bacteria, a cargo-binding protein delivers the cargo to the permease. Commonly, the set of three components is considered collectively. In gram-negative bacteria, the cargo delivery protein is a soluble periplasmic binding protein (PBP), while in gram-positive bacteria the cargo-binding protein is tethered to the membrane or the permease through a lipid moiety.[102] In general, PBPs are composed of two lobes connected by a hinge and the cargo binding site lies between the two lobes.[102]

On the cytoplasmic side of the ABC transporter, much work has been done to elucidate the structure–function relationships governing the ATPase activity and to determine how energy is translated from the ATPase to the permease. The reader is referred to two excellent reviews by Davidson and Chen and Jones and George for more information regarding the ATPase and its function.[102,106] For the purposes of this chapter, we will limit our discussion to the interactions between the PBP and the permease. Less is known about how these two interact, but two important examples have led to a working model describing the molecular recognition event and the dynamics of this protein–protein interaction. We will begin with an example from a siderophore-mediated uptake system in *E. coli*.

E. coli utilize the iron complex transport strategy by acquiring iron as an iron–siderophore complex such as aerobactin, ferrichrome, or coprogen.[52] The transport system specific for these three siderophores includes the ABC transporter FhuB/FhuC and the PBP FhuD.[107] This system represents one of the first examples of ABC transport complex for which the interactions between the permease and the PBP have been investigated. In general, it is accepted that the binding event between the permease and the PBP activates the ATPase, "turning on" the transporter.[102] Importantly, FhuB has been shown to interact preferentially with holo-FhuD over apo-FhuD.[101,108]

A variety of techniques have been used to probe the interactions between FhuD (with and without cargo) and FhuB. Cross-linking studies and proteinase protection studies showed that both forms of FhuD do interact with FhuB, but suggest that the interaction sites are different depending on the status of the cargo-binding protein.[107] A structural model of FhuB (the permease), based on a hydrophobicity/hydrophilicity analysis of the primary sequence, predicts that transmembrane portions of the protein are α-helices and that loops connecting the transmembrane portions extend from both sides of the inner bacterial membrane (four loops on the periplasmic side and three loops on the cytoplasmic side).[108] Loop 7 on the cytoplasmic side of the protein has a conserved sequence, called the EAA–G site, which is thought to interact directly with the ATPase FhuC.

Loops on the periplasmic side of the integral membrane permease include several positive charges and as will be discussed are predicted to interact electrostatically with FhuD.[108] Competitive peptide mapping techniques have revealed a FhuD interaction site in the transmembrane portion of FhuB, indicating extensive interaction between the two proteins. This finding suggests that FhuD may bind to the permease by embedding itself in a pore of FhuB. It has also been shown that FhuD may interact with a cytoplasmic loop 7 of FhuB.[108] Given that loop 7 of FhuB is thought to be the site of interaction between FhuB and FhuC, it has been proposed that loop 7 may provide a mechanism for signaling to the ATPase the presence of the PBP on the surface of the permease. This signal may, in turn, trigger the nucleotide hydrolysis activity required for transport. Interestingly, FhuD has been shown to not interact directly with G559, the residue of FhuB required for interaction with FhuC. Therefore, it is possible that loop 7 of FhuB could concurrently bind FhuD and FhuC, forming an association between multivalent polymers.[108]

We briefly diverge from the specific protein–protein interactions of the ABC transporter to highlight the importance of the physical properties of ionic and polymer interactions in the application of the competitive peptide mapping technique utilized in the study discussed above. In competitive peptide mapping, one assumes that a small peptide that is not conformationally constrained will bind to surfaces of the target protein for which it has sequence specificity. The target protein acts as a template directing the binding event. Observed binding events are interpreted as positive signals representing "real" binding events. However, conclusions regarding the absence of observed binding events are avoided.[108] Therefore, this technique is touted to be useful for finding some protein–protein interactions, but is not capable of identifying all interactions. A robust analysis and quantification of false positives are required in order to push this technique toward widespread use. The principles of ionic and mixed interactions between polymers and ions will be quite useful in optimizing and assessing the limitations of this technique.

We now return to the protein–protein interactions between the PBP, FhuD, and the ABC transport permease FhuB. FhuB functions as a dimer. Peptide mapping has been used to demonstrate that FhuD binds more strongly with one side of FhuB rather than symmetrically between the two components of the permease dimer. This asymmetric interaction appears to be a common feature of ABC transporters.[42] Separately, the oscillations of FhuD between a closed and a more open conformation with respect to active site accessibility have been shown to occur even in the absence of bound cargo.[109] Taken together, these observations support the hypothesis that binding to FhuB may induce similar motions in FhuD, leading to a mechanism for cargo release from FhuD.[108]

A homologue of the Fhu system is the Btu system, which transports vitamin B12. Although not involved in iron transport, the Btu system has provided the most insight to date regarding the interactions between a PBP and its corresponding ABC transporter. Here we will briefly highlight a few aspects of the Btu system and then return to interactions between FhuD and FhuB.

The crystal structures of both the Btu transporter (BtuC/BtuD) and BtuF (the cargo-binding protein) have been solved.[105,110] Further, it has been demonstrated that the purified proteins used to make each of the crystal structures form a stable complex with the following stoichiometry: one BtuF, two BtuC, and two BtuD.[105] *In silico* docking of the crystal structures, shown in Figure 13.9, revealed a negatively charged region at the tip of each lobe of BtuF that are 46 Å apart from each other. The active site containing vitamin B12 lies directly between these two negatively charged regions that were identified as E72 and E202.[105] On the periplasmic face of the permease BtuC, two positively charged areas that lie 48 Å apart are composed of three Arg residues. All five residues, the acidic residues on the PBP and the basic residues on the perme-

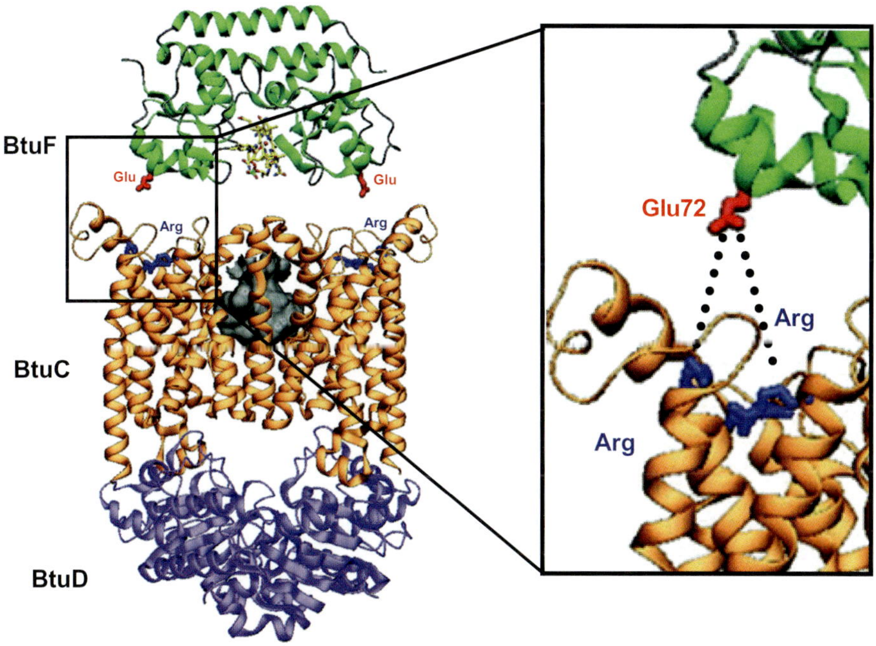

FIGURE 13.9. ABC transporter involved in vitamin B12 uptake. BtuF delivers the cargo to the periplasmic side of the permease BtuC. Ionic interactions between the proteins guide the protein–protein assembly formation. Temporary salt bridges are formed between glutamine residues on BtuF and arginine residues on BtuC (shown as dotted lines in detail). (Figure adapted from Borths et al.[105] and used with permission.)

ase, are conserved in homologues of a variety of iron transport systems in several organisms including the Fhu system for ferrichrome-mediated iron uptake in *E. coli*.

This model of protein–protein docking has recently been extended to the Fhu system used for ferrichrome uptake in *E. coli*.[109] Crystal structures of FhuD (the PBP) with and without gallichrome (cognate siderophore bound to Ga^{3+}, an isoelectric ion of Fe^{3+}) were first used in conjunction with molecular dynamics simulation techniques to determine the possible motions of the protein in solution. As seen in Figure 13.10, it is predicted that the C-lobe of FhuD is quite dynamic compared with the N-lobe.[109] Thicker lines in the figure denote movement between a greater number of possible conformations.

Combining the model described previously for the Btu system[111] and the molecular dynamics simulations of apo-FhuD, Krewulak et al. proposed a mechanism of protein-protein assembly formation and dissociation.[109] These simulated motions are shown as snapshots in Figure 13.11. Similar to the previously described hypothesis for the Btu system, it is thought that an area of

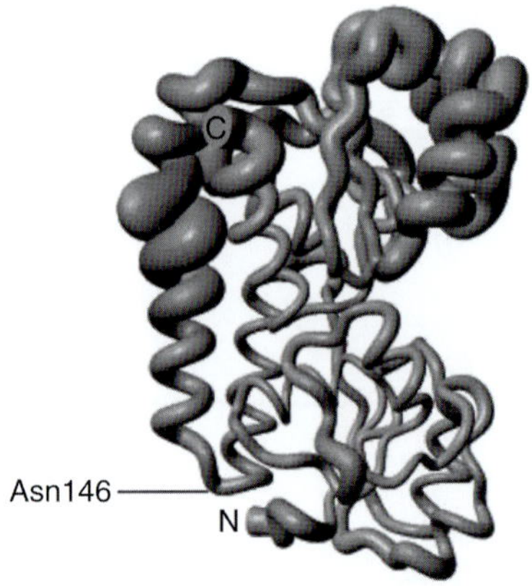

FIGURE 13.10. Sausage diagram of FhuD demonstrating the range of motions observed during *in silico* simulation. Thicker sausage lines denote movement between a greater number of possible conformations. The C-terminal domain is the most mobile, with its motion originating from Asn146. (Figure adapted from Krewulak et al.[109] and used with permission.)

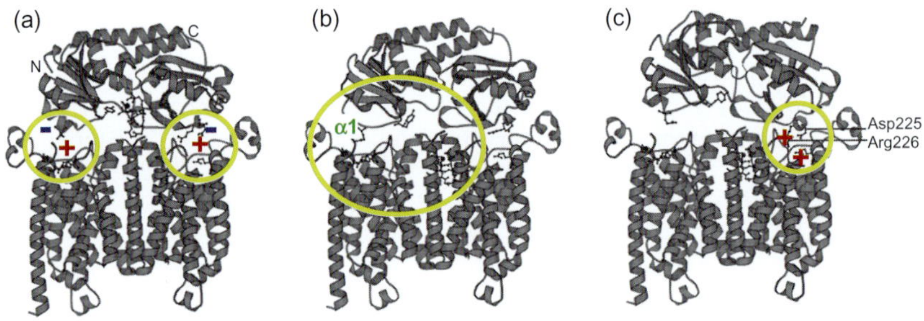

FIGURE 13.11. *In silico* docking suggests possible mechanism for protein–protein assembly between FhuD (top protein) and FhuB (bottom protein). (a) Negatively charged residues of FhuD form salt bridges with the positively charged Arg residues of FhuB (yellow circles). (b) Ionic interaction between FhuD (N-lobe) and FhuB unwinds α-helix 1 and dissociates salt bridge (yellow circle). (c) Motion in the C-lobe of FhuD creates charge repulsions between proteins (yellow circle) leading to protein–protein dissociation and FhuD release to inner membrane permease. (Figure adapted from Krewulak et al.[109] and used with permission.)

negative charge at the tips of FhuD (Asp and Glu residues) is attracted to positively charge areas (Arg) on FhuB.[109] Like the Btu system, this interaction of polymers with complementary charged surfaces guides protein assemblage mediated by ionic forces (Chapter 8). Movement in the C-lobe of FhuD (seen in Fig. 13.10) is then expected to weaken the interaction between FhuD and the ferrichrome cargo and relocates E111, a putative residue in the salt bridge formed with the Args of FhuB (Fig. 13.11). It is also predicted that movement of the C-lobe of FhuD brings Arg 226 (of FhuD) into close contact with the Arg pocket at the apex of FhuB. The resulting charge repulsion is hypothesized to be part of the release mechanism leading to the dissociation of FhuD back to the periplasm and the relaxation of FhuB after the siderophore has been transported through the inner membrane.[109]

Although the protein–protein interactions in other ABC transport systems are predicted to follow a similar global mechanism, the details of each system may be quite diverse. Molecular recognition is a key factor in cargo-specific transport and so recruiting a specific PBP to the proper ABC transporter is likely controlled by the particular ionic and mixed interactions that constitute the complementary charged surfaces of each protein pair.

Going forward, two interesting areas that should be addressed are the roles of mobile ions and of the lipid membrane in protein–protein assembly formation (Chapter 9).[1] Recently, the study of gram-positive bacteria has led to the identification of FhuD2, a lipoprotein version of FhuD.[112] In this gram-positive system, both the cargo-binding protein and the ABC transporter are associated with the membrane. This fact may dramatically affect the way in which these protein–protein assemblies form and dissociate.

Another frontier in understanding the protein–protein interactions involved in iron transport is determining the thermodynamic and kinetic properties of the ionic and mixed interactions between polymers. Now that the key players have been identified and the pathways of iron uptake defined, the next step is to understand how, in concert, the constellation of interactions guides the transport function and how protein dynamics control the transient interactions between proteins.

13.5 PROTEIN/IRON–SMALL MOLECULE INTERACTIONS

In the previous section, we focused on protein–protein interactions involved in (1) iron transport in humans, (2) bacteria hijacking iron from the human host, and (3) iron transport in bacteria. Here we will continue using examples from bacterial iron transport to explore ionic and mixed interactions involved in the interactions between proteins and low-molecular-weight iron complexes. By investigating the structures of specific outer membrane receptor proteins, significant progress has been made in the understanding of bacterial iron acquisition. Here we will explore a few examples of how protein structure dictates the ionic and mixed interactions involved in siderophore-mediated bacterial acquisition of iron.

As discussed in Section 13.2.2, iron acquisition can be accomplished by hijacking the iron bound by host heme proteins or metalloproteins, or by incorporating extracellularly chelated iron in the form of ferric siderophores. The pathways of siderophore-mediated iron transport in gram-negative bacteria have been studied extensively, and several excellent reviews of siderophore-mediated iron uptake have been published.[42,52,53,113,114] As well, reviews focusing on the iron uptake pathways of specific bacterial pathogens such as those causing cholera and whooping cough are also recommended.[40,41]

Acquisition of siderophore-bound iron involves binding to and passage through integral outer membrane TonB-dependent transport proteins. Similar to TbpA discussed in Section 13.4.2, these siderophore transporters are composed of a 22-stranded β-barrel and an amino-terminal plug domain that is sequestered within the lumen of the β-barrel.[42] In this section, we will examine the interactions between natural polymers (protein receptors) and iron–siderophore complexes that can be charged or uncharged. We will focus our discussion on the interactions that occur at the outer or extracellular face of the outer membrane receptor. In general, both the extracellular loops linking the strands of the β-barrel together and the plug domain are partially surface exposed and participate in binding the iron–siderophore complex.

The principles of ionic and mixed interactions presented in this book can be applied to these cargo–receptor interactions in several ways. Iron is sequestered by the siderophore through coordinate covalent bonds and bound to the macromolecule or receptor in a supramolecular assembly. Various electrostatic and mixed interactions are involved in the formation of this supramolecular assembly and thereby define the molecular recognition events that are unique to each receptor–siderophore pair (Chapter 2). Although ferric iron has a charge of 3+, the net charge of iron–siderophore complexes ranges from negative to neutral to positive. Therefore, both attractive forces and charge-independent interactions such as hydrophobic effects are important for the supramolecular binding between the receptor and its cognate iron–siderophore complex (Chapters 7 and 8).

Siderophore-mediated iron acquisition can also be thought of in terms of host–guest chemistry. The charged iron must be transported through the hydrophobic membrane and so the combination of the siderophore and protein receptor must effectively screen the charge of the hydrophilic cation (Chapter 9). Additionally, host–guest complex formation involves prepayment of an entropic price (Chapter 2) that allows for specific and tight binding of the cargo. This is accomplished with a preformed binding site within the receptor poised to receive the cargo. The extent to which a particular receptor is preformed in the ideal conformation that complements its cargo largely determines the specificity of the binding event (the molecular recognition process) and the transport process across the outer membrane. As we will see in the several cases discussed in this section, some bacterial siderophore receptors are quite specific for one siderophore, while others are more promiscuous, binding to and/or transporting a variety of siderophores.

This promiscuity has led in recent years to the use of what is called a Trojan horse approach to targeted drug design, in which an antibiotic is synthetically conjugated to a siderophore.[115] The siderophore provides a recognition surface for the targeted cell receptor and the antibiotic provides the therapeutic agent. Consequently, like the infamous Trojan horse, the cognate siderophore is recognized by its bacterial receptor and the whole conjugate is transported into the cell. Once inside the cell, the antibiotic is activated and kills the bacteria. The interactions between the small-molecule siderophore and the macromolecule receptor, therefore, must remain intact to provide a molecular recognition site and guide cargo binding. However, the ionic and mixed interactions involved in binding must be "flexible" enough to allow the conjugated drug to also reside in the siderophore binding site. Further, the receptor must maintain the role of host, screening the cargo from unwanted interactions with the membrane, but the host–guest interactions must be "flexible" enough to accommodate a larger and effectively different cargo molecule.

Here we will examine several siderophore–receptor pairs involved in iron transport, which illustrate the variety of ionic and mixed interactions involved in the initial stages of siderophore-mediated bacterial iron acquisition.

13.5.1 Siderophore Receptor Binding: Prototypes

Transport of iron across the outer membrane requires a receptor that recognizes the iron–siderophore complex and actively moves the charged iron across the hydrophobic membrane. The crystal structures of three key examples of outer membrane siderophore receptors have been described and have led to a working model of this step in the siderophore-mediated iron transport process.[49–51] In general, it is thought that the iron–siderophore complex binds to the outer membrane receptor and the intact iron–siderophore complex is transported across the membrane. The mechanism and energetics of the transport process, particularly the dynamics of the receptor during transport, have been and continue to be the subject of much investigation.

It should be noted, however, that as more and more systems are examined new models for transport appear to also be utilized by bacteria. For instance, instead of the intact iron–siderophore complex being recognized and transported through the receptor, there is evidence that some receptors continuously have an apo-siderophore inside the receptor and transport is accomplished through exchange between the exterior holo-siderophore and the apo-siderophore in the membrane receptor interior. This exchange model may allow the bacteria to take up iron from a variety of siderophores, utilizing a preformed iron binding site with the siderophore–receptor interactions already formed.[116]

Here we will examine the structures of three prototypical examples as well as a few receptors that have been the subject of intense investigation more recently. In each case, we will focus our discussion on the interactions between the siderophore and the receptor.

The three classical receptors for iron–siderophore uptake expressed by *E. coli* are FepA, FhuA, and FecA, which transport iron-loaded enterobactin (ENT), ferrichrome, and ferric citrate, respectively (Fig. 13.12).[101] Crystal

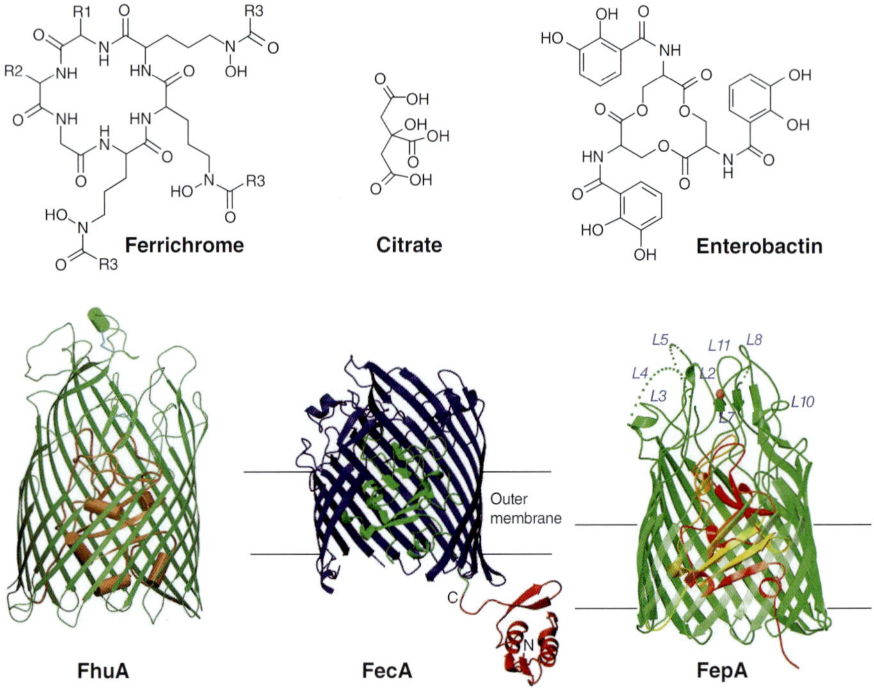

FIGURE 13.12. Prototypical outer-membrane β-barrel receptors each with a plug domain occluding the pore. FepA, FhuA, and FecA transport iron-loaded enterobactin (ENT), ferrichrome, and ferric citrate, respectively. The structure of each siderophore is shown above its respective receptor. (Figure adapted from Krewulak and Vogel,[42] Buchanan et al.,[49] and Locher et al.[51] and used with permission.)

structures for these three receptors are available and have contributed significantly to the accepted understanding of siderophore transport through the outer membrane.[42] Our discussion here is not meant to be a comprehensive review of this work, but instead we seek to highlight the variety of ionic and mixed interactions involved in this aspect of siderophore-mediated iron transport (Fig. 13.12).

We will begin with FepA and ENT. When bound by ENT, iron is chelated through three catechol moieties with a binding constant of 10^{49} M^{-1} and the first coordination shell around the iron is chiral.[117,118] The reader is referred to an excellent review by Raymond et al. for a detailed description of Fe-ENT solution chemistry and the entire Fe-ENT uptake pathway.[118] Fe-ENT is recognized on the surface of the bacteria by the TonB-dependent receptor FepA.

Defining the interactions between Fe-ENT and FepA has been the subject of much investigation from the protein structure perspective. It has been shown that the tris-catechol iron center with amide linkages is required for recognition by FepA, but the trilactone backbone is not.[118] Given that the first coordination shell of the iron molecule is critical for recognition by the receptor, it was thought that the chirality around the iron would be a key in the recognition process and might be a general principle guiding selective iron uptake. Interestingly, it was subsequently shown that both enantiomers of Fe-ENT are taken up by the FepA receptor, but only one enantiomer, the Δ configuration, supports bacterial growth. Therefore, although apparently not important for recognition on the surface of the bacterial membrane in this case, the chirality of the first coordination shell does affect the uptake process further down the pathway.[119,120]

Further biochemical experiments have revealed a more detailed picture of what interactions occur when Fe-ENT is bound to FepA. Inside the extracellular binding pocket of the receptor, positively charged residues (R286 and R316) and aromatic residues (Y260, Y272, and F329) have been shown to be important for Fe-ENT binding.[49,91,118,121] The crystal structure of FepA in the presence of Fe-ENT (although the Fe-ENT molecule could not be modeled unambiguously) demonstrated that additional aromatic residues ring the outer face of the receptor.[49,121] Given that the tris-catechol Fe-ENT has a net charge of 3–, a receptor with positive arginines strategically located in a cavity lined with aromatic side chains leads to the conclusion that both electrostatic and mixed interactions between the ionic siderophore and the biopolymer are important for specific and efficient transport.

By comparing the structures and interactions of multiple siderophore receptors, general trends can be drawn regarding this facet of iron uptake. FhuA is another siderophore receptor expressed by *E. coli*, which recognizes and transports a cyclic hexapeptide siderophore ferrichrome (Fig. 13.12). The reader is referred to an excellent article by Ferguson and Deisenhofer, which compares the structure of FhuA and FepA.[122] Here we will highlight the major findings relevant to our theme.

Crystal structures are available for FhuA with and without bound Fe-ferrichrome.[51] Inside FhuA, the iron binding end of the siderophore (i.e., the iron and its first coordination shell within ferrichrome) are buried in the conical binding site of the receptor. However, the peptide tail of ferrichrome remains solvent exposed. Several hydrogen bonds are formed between the neutral Fe-ferrichrome molecule and R81, Y116, and G99 from the plug domain of FhuA and with Y244 and Y315 from the barrel domain (Fig. 13.13).[51]

Compared with FepA, which has a high specificity for Fe-ENT, FhuA acts as a multifunctional transporter that can recognize and move other cargo molecules in addition to Fe-ferrichrome. Several antibiotics that generally resemble ferrichrome have been crystallized with FhuA.[122] The three residues from the plug that interact with Fe-ferrichrome through hydrogen bonds also interact with these noncognate cargo molecules, indicating the importance

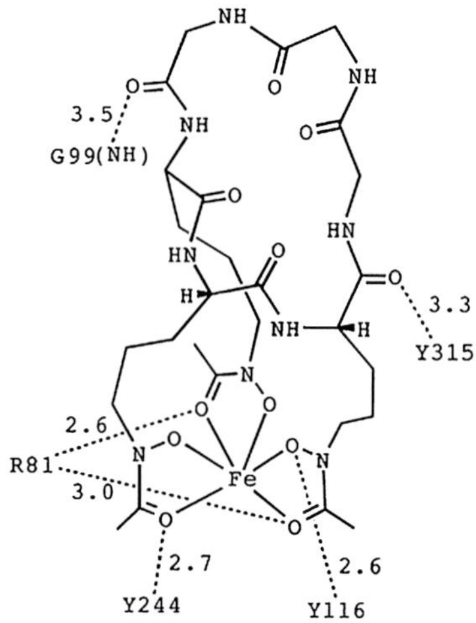

FIGURE 13.13. Interaction of Fe-ferrichrome with residues of FhuA determined from the 2.7-Å crystal structure. Hydrogen bonds are formed with residues R81, G99, and Y116 from the plug and residues Y244 and Y315 from the β-barrel of FhuA. (Figure modified from Locher et al.[51] and used with permission.)

of these residues for receptor function.[122] Given the absence of specific interactions between the peptide tail of ferrichrome and the receptor, FhuA may be a good target for antibiotics designed using the Trojan horse approach.

Similar to FepA in which aromatic residues are important, the binding site of FhuA also includes several aromatic amino acids. In FhuA, F115 and Y116 are provided by the plug domain and Y244, W246, Y313, Y315, and Y391 are located on various loops of the barrel.[51] Although not particularly hydrophobic, the neutral Fe-ferrichrome molecule does interact with aromatics as it can be extracted from environmental mixtures with benzoyl alcohol.[123] It has been proposed that the aromatic residues lining the pocket of FhuA interact through mixed electrostatic interactions between the quadropole moment of the π-electron system of the aromatics and the dipoles of the peptide bonds within Fe-ferrichrome.[123]

By comparing FepA and FhuA, it can be seen that subtle differences in the chemical properties of the residues in a receptor binding pocket dictate the type of siderophore that is recognized and eventually transported by the receptor. FepA has a high specificity for Fe-ENT and involves two positively charged residues that mask the negatively charged cargo as it is passed through the hydrophobic membrane. FhuA, on the other hand, is less specific for its cargo (Fe-ferrichrome) and binds this neutral cargo with hydrogen bonds and other mixed interactions. Both FepA and FhuA have a large number of aromatic residues in their respective binding pockets, but the structural orientation of the residues within the two pockets define unique specificities and functions to each receptor expressed by *E. coli*.

In addition to ferrichrome and ENT, citrate can serve as a siderophore and is recognized and transported by a specific outer membrane receptor. The citrate transporter of *E. coli*, FecA, has been crystallized alone, with diferric dicitrate and with apo-dicitrate.[50,124] In contrast to FepA and FhuA, FecA utilizes several positively charged arginines (R155, R365, R380, and R438) to bind the highly negatively charged apo-citrate cargo.[124] In the first crystal structure published for a receptor bound to the iron-free form of the siderophore, apo-dicitrate forms 13 hydrogen bonds with the positively charged arginines and five water molecules bridging the two apo-citrate molecules. For diferric dicitrate, hydrogen bonds are also formed with R365 and R380 and with additional residues Q176, N721, Q570, T138, and S180, while van der Waals contacts are made with L156 and F333, and the extracellular loops provide additional electrostatic interactions.[50,124] The reader is referred to Yue et al. for an extensive comparison between the various forms of citrate that have been crystallized with FecA.[124] In short, the extensive network of ionic and mixed interactions between FecA and its highly charged cargo are necessary for FecA to move this ionic iron–siderophore through the hydrophobic membrane.

13.5.2 Siderophore Receptor Binding: Newcomers

FecA, FepA, and FhuA have served as the prototypical receptors for siderophore-mediated iron transport through the outer membrane and are the foundation of our understanding of this process. However, recently, receptors that differ structurally in important ways from these prototypes have been investigated. We will briefly describe two additional receptors, maintaining our focus on the interactions between the iron–siderophore complex and the extracellular binding pocket of the receptor.

Campylobacter jejuni, a gram-negative foodborne pathogen, produces an outer membrane receptor, CfrA, which consists of the classic β-barrel and plug shape and binds Fe-ENT. However, alignment studies predict that different residues are involved in binding Fe-ENT compared with those of FepA.[125] Particularly intriguing is that CfrA lacks the aromatic residues so integral for FepA binding to the same siderophore (Fe-ENT) or in FhuA, the ferrichrome receptor. CfrA also lacks residues corresponding to Y260 and Y272 in the binding pocket of FepA.[125] Compared with FepA, FhuA, and FecA, CfrA contains only a few positive residues. There is, however, a lysine residue in CfrA near the binding site of the Fe-ENT tris-catechol-iron head group, which matches with R286 of FepA.[125] This conserved basic residue suggests the importance of a positive charge dictating the location of the iron within the receptor.

It is thought that CfrA is the only siderophore receptor expressed by *C. jejuni*. Given that CfrA lacks many of the key residues making FepA selective for Fe-ENT in *E. coli*, it has been proposed that CfrA may be able to take advantage of multiple siderophores allowing *C. jejuni* to be more opportunistic.[125] This difference in Fe-ENT interaction between FepA and CfrA may also make CfrA a good target for therapeutics that selectively kill *C. jejuni* and not *E. coli*. Finally, the possible promiscuity of CfrA and its role as a nodal point in iron uptake by *C. jejuni* make this receptor a good target for Trojan horse therapeutics.

Another outer membrane transporter for which crystal structures have recently been published is FpvA.[126] FpvA is expressed by *Pseudomonas aeruginosa*, an opportunistic pathogen best known for causing hospital-acquired infections. Fluorescent pseudomonads (one of five homology groups within the genus) produce and secrete a large class of siderophores called pyoverdines (Pvd).[126] There are over 100 known Pvds with a wide variety of structures. However, all Pvds consist of a fluorescent chromophore, 2,3-diamino-6,7-dihydroxyquinoline, and a peptide backbone that can be linear or cyclic (Fig. 13.14).[126] The reader is referred to an excellent review of the Pvd field by Schalk.[126]

P. aeruginosa (strain ATCC 15692) utilize PvdI, which binds ferric iron through a catechol functional group and two hydroxamic acids.[127] PvdI is recognized on the cell surface by FpvAI, a TonB-dependent β-barrel receptor with a plug and signaling domain. Given that the structure of FpvA consists of several tryptophan amino acids in the vicinity of the siderophore binding site, interesting insight has been gained by monitoring the FRET between the fluorescent siderophore and receptor. The mechanism of siderophore binding and uptake is different for FpvA compared with FepA, FhuA, and FecA in that FpvA is loaded with apo-Pvd at all times. The apo-Pvd siderophore must be released from the receptor prior to binding of holo-Pvd, adding additional steps to the transport mechanism. It has also been shown that Pvd is recycled to the extracellular environment after iron has been removed from the siderophore in the periplasm.[128–131] The function of apo-Pvd binding to the receptor in this case is still unclear. However, these interesting features of FpvA are expanding the model of outer membrane transport and the proposed roles of β-barrel receptors. To date, FpvA is the only Pvd transporter for which crystal structures have been published.

Crystal structures are available for FpvA alone, bound to apo-Pvd, and with either holo-Pvd or Ga-Pvd.[55,132,133] The receptor binding pocket consists of six tyrosines, three tryptophans, and three glutamines from both the plug and the barrel (Fig. 13.14).[55,132,133] With a large number of aromatic residues lining the binding pocket, FpvA is similar in structure to FepA and FhuA. In FpvA, the peptide moiety of Pvd interacts mostly with the barrel, while the chromophore of Pvd interacts directly with the plug.[126] Interestingly, fluorescence studies have shown that the solvent exposure of the chromophore is different if the siderophore and receptor are copurified or incubated together *in vitro*.[134] Nuclear magnetic resonance studies have also recently shown that Pvd is able to access multiple conformations in solution, changing both chirality at the iron center and the global structure of the cyclic peptide.[135]

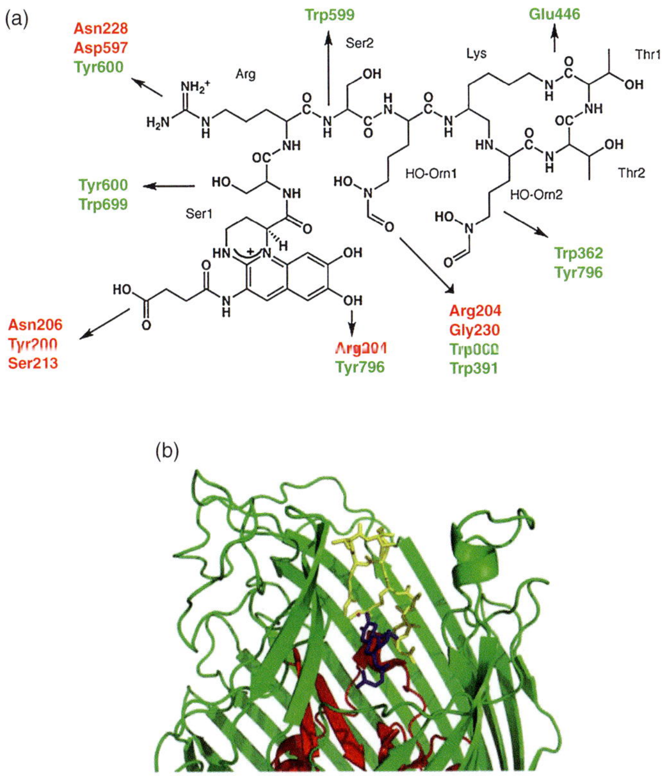

FIGURE 13.14. (a) Structure of Pvd. Interactions with specific residues of FpvA are shown and color indicates in which part of the protein the residue is located (red for plug and green from barrel). (b) Structure of FpvA-Pvd-Fe focusing on the binding site. The siderophore is shown in yellow and blue (peptide and fluorophore respectively). As in part (a), the barrel is green and the plug is red. (Figure adapted from Schalk et al. [137] and used with permission.)

Conformational changes of Pvd have been hypothesized to aid in the entrance of Pvd into the FpvA binding pocket following an induced fit model, but as of yet complexes of FpvA with various conformations of Pvd have not been visualized.[135]

The succinyl tail extending from the chromophore and the first amino acid, a serine, of the Pvd structure can be modified with little impact on binding (Fig. 13.14). However, arginine, the second amino acid beginning the peptide backbone of Pvd, has been shown to be particularly important in the binding of PvdI to FpvA.[136] The crystal structure of FpvA bound to Fe-Pvd shows that this arginine of Pvd points toward the FpvA plug and electrostatically interacts with D597 of loop 7 of the barrel that folds into the binding pocket.[137]

It is thought that the peptide backbone of the siderophore provides the structural differences between possible cargo molecules that lead to receptor specificity.[126] Noncognate Pvds are able to bind to FpvA and have been crystallized with the receptor. However, noncognate Pvds are not transported into the cell. The crystal structures of these nonproductive receptor/cargo complexes show that the features that are common to all Pvds (the chromophore, the iron, and the first amino acids that bind to iron) are important for binding to the receptor.[138] The chirality of the octahedral first coordination shell of iron is the same for all Fe-Pvds, so this may also be an important factor in receptor/cargo binding. For those siderophores that bind to FpvA with high affinity, W599 and Y600 on loop 7 of the receptor move closer to the peptide bonds of the first few residues of Pvd. There is also an electrostatic attraction between the positively charged second amino acid (arginine) of Pvd and D597 on loop 7 of FpvA, although the two attracted atoms are not close enough to form a hydrogen bond.[138] Finally, a hydrogen bond or dipole–dipole interaction (its not clear from the crystal structure) between the carbonyl of the Pvd arginine and N228 of FpvA stabilizes a hydrogen bond between N228 and G230 within the receptor.[138] It is clear that several types of interactions are involved in cargo binding and additional interactions fine-tune the stability of the cargo–receptor complex.

In contrast to FecA, FepA, and FhuA, for FpvA the interactions that define cargo binding may be significantly different from those that define transport. This difference or shift in balance between the two functions of recognition and transport may provide *P. aeruginosa* with a set of evolutionary advantages such as quorum sensing, heightened environmental scavenging, and metal promiscuity.

Interestingly, the interactions seen for apo-Pvd with FpvA are similar to those seen for holo-Pvd.[129] It is thought that although the binding constant for apo-Pvd and holo-Pvd to FpvA are similar, the differences between the two complexes (one leads to transport and one does not) lie with the kinetic properties of these interactions.[128] The biological repercussions of this have been studied extensively in terms of binding mechanism and uptake kinetics.[126] W326 from loop 3 of FpvA is particularly important for recognition of Fe-Pvd and interacts with hydroxyl ornithines in the peptide portion of Pvd.[137] However, both apo-Pvd and Fe-Pvd interact with R204 of FpvA.[55] It is likely that different residues will be involved in determining the rate constants for each form of the siderophore associating and dissociating from the receptor.[126] Further, it has been shown that release of apo-Pvd is regulated by TonB, while TonB is involved in the uptake mechanism of Fe-Pvd.[126]

Another interesting property of FpvA is that it has broad metal specificity.[139] This finding supports the idea that the residues affecting cargo specificity interact with the peptide backbone of Pvd and that the interactions closer to the metal site are more "flexible," as they are able to accommodate metals of varying size. The broad metal specificity of FpvA also indicates that the specificity of the entire uptake pathway for iron must be regulated by factors other than cargo recognition at and transport through the outer membrane.[139]

This series of five β-barrel TonB-dependent outer membrane receptors that initiate siderophore-mediated iron uptake pathways in a variety of gram-negative bacteria demonstrate the commonalities and important differences leading to efficient iron acquisition. The constellation of ionic and mixed interactions between the macromolecule and the metal siderophore complex are unique to each pair (siderophore and receptor). From a structural perspective, many electrostatic, hydrogen bonding, and dipole interactions have been identified. As well, the importance of the three-dimensional placement of interacting atoms and the possible motions of the proteins has dramatically enhanced our ability to visualize these ionic and polymeric host–guest complexes. In addition to deepening our understanding of these biological interactions, the findings outlined above have opened the opportunity to design antibiotics such as inhibitors or Trojan horse conjugates that will be increasingly important in the arms race against antibiotic-resistant bacteria.

13.5.3 Additional Structural Elements and Future Directions

We would be remiss not to mention a few other structural elements of these β-barrel receptors that have been the focus of many important studies. This outline serves as a list of some other important ionic and mixed interactions in which β-barrel receptors participate. As more detailed information is gathered on each of these topics, an understanding of how biological polymers, iron, and ions interact will become more and more useful. First, FhuA has now been crystallized with a liposaccharide that helps to orient the receptor in the membrane.[101] As the study of this outer membrane receptor moves forward, it will be interesting to see how the interactions between the receptor and the membrane are defined and how liposaccharide posttranslational modifications of the receptors guide their orientation and function in the membrane.

Second, the loops of the β-barrel receptors extend to varying lengths into the extracellular space and may serve a variety of functions. For example, the loops of FepA extend 30–40 Å out from the membrane and are expected to be involved in recruiting Fe-ENT to the receptor.[42,118] The motions of these loops are also important, as it has been shown that they close over the mouth of the receptor in FecA, locking its cargo (iron citrate) in place. However, it is unclear if loop closure is a common feature of β-barrel receptors or if FecA is an anomaly. It will be interesting to see the variety of roles the loops play as more systems are elucidated.[140] It will also be interesting to investigate the strength of the possible interactions between a siderophore and these flexible polymers (the loops of the respective receptor) in order to understand the energy required to recruit a molecule to the receptor and the specificity imparted at this recruitment stage.

Third, some of the β-barrel receptors are also involved in cellular signaling. FecA and FpvA include an additional domain called the signaling domain that serves both a signaling function and is involved in gene regulation.[42,50,55,134,141] As more is learned about the functions of the signaling domain, it will be very interesting to see how information is transmitted from one part of the β-barrel receptor to another and how the signal is passed on to other proteins.

Finally, significant work has been done to elucidate the mechanism of cargo transport *through* the receptor from one side of the membrane to the other. The reader is referred to several reviews and key papers discussing topics such as (1) the role of TonB in providing the energy required for transport, (2) the possible movements of the plug domain that allow cargo to pass through the receptor, (3) the role of the signaling domain as TonB interacts with the receptor, and (4) the role of the plug in binding the cargo during transport *through* the membrane.[48,50,55,56,113,132,133,142–152] In each case, the interactions between proteins, between the iron cargo and the receptor, and between domains within the receptor involve a transient and unique recipe of ionic and mixed interactions that guide the process of transport. We regret that we are not able to review those studies here, but we predict that the finely tuned interactions of each system and the chemical diversity utilized by each system will fascinate researchers in the field for years to come.

Going forward, the impact of environmental conditions on these cargo–receptor complexes will also lead to a greater understanding of ionic and mixed interactions between ions and macromolecules. For example, some work has been done on changes in Pvd/FpvA binding due to changes in pH and temperature.[134] As well, a sulfate anion has been modeled as binding to several side chains of FpvA in both the crystal structures of apo-Pvd bound to FpvA and of holo-Pvd bound to the receptor.[133] However, it is unclear what role these exogenous anions play *in vivo* or *in vitro*. We predict, based on the data and theories presented in this book, that environmental ions do affect the structural properties of the protein receptors (polymeric macromolecules) and may mediate a variety of yet unappreciated interactions between iron–siderophore complexes and bacterial receptors.

In the transport of an iron complex across a hydrophobic membrane, many ionic and mixed interactions guide the recognition process and host–guest assembly formation. The specific combination of types of interactions and spatial orientation of the protein residues provide a wide range of promiscuity and specificity depending on the required function of a given receptor. Attractive forces are the dominant feature of these ion–polymer interactions. In the next section, we will address more explicitly the effects of background electrolytes and multivalent communication between polymers and ions.

13.6 PROTEIN–ANION INTERACTIONS

Thus far, we have discussed interactions between multiple proteins and between proteins and small iron complexes that are involved in iron transport.

Interactions between proteins and small exogenous ions are also common in biology and important for iron transport. Protein–ion interactions can serve many roles in biology, such as (1) participating in the chemical activity of metalloenzymes, (2) defining the structure of a protein subdomain (Chapter 11), (3) transporting the ion through a membrane, and (4) interacting nonspecifically with the protein (background electrolyte effects). As discussed in Chapter 12, charge-induced effects on the conformation of a biopolymer and the formation of salt bridges are essential chemical interactions for both protein function *in vivo* and for synthetic protein design. Generally, the interactions between proteins and ions can be thought of as attractive forces, such as ionic bonds.[1] However, often, ions are multivalent in terms of participating in multiple interactions within a protein and can form extensive networks of hydrogen bonds and van der Waals contacts. Given the enormous possibilities for nonspecific interactions between polyampholytic proteins in aqueous electrolyte solutions, ionic strength can dramatically affect the chemical properties of a protein and is therefore important to consider.

The effects of ion–protein interactions, particularly anions, are ubiquitous and important in iron transport in biology. Both Tf (the human iron transport protein; section "Transport of Iron-Containing Protein into a Cell") and FbpA (a naked iron transporter produced by some pathogenic bacteria; section "Transport of 'Naked' Iron through a Membrane Protein") require an exogenous anion, called the synergistic anion, in order to effectively sequester iron. In FbpA, the identity of the synergistic anion modulates both the thermodynamic and kinetic properties of the transport protein and directly affects its function. Although iron is also an ion that interacts with these proteins, here we will focus on the synergistic anion–protein interactions rather than the interactions between the protein and iron. Additionally, the behavior of both FbpA and Tf are affected by the background electrolyte anions in the environment around the protein. Given that aqueous environments in which these proteins exist *in vivo* (the bacterial periplasm for FbpA and the human serum for Tf) are dynamic, an understanding of the electrolyte effects on protein function is important. To illustrate the elaborate and important interactions between anions and proteins, we will use FbpA as our primary example and where appropriate compare the findings with Tf. Rather than survey the literature, we will focus mainly on examples from our laboratory.

13.6.1 Introduction to FbpA

FbpA, the 34-kDa PBP of the ABC transporter in *N. gonorrhoeae* introduced in Section 13.4.3, is required for growth and acquires iron after the metal is stripped from Tf and crosses the outer membrane. A homologue with 71% amino acid sequence fidelity is produced by *Haemophilus influenzae* and called hFbpA (formally HitA).[31,101,153–156] In the respective bacteria, FbpA serves as a nodal point in iron uptake because transport by FbpA is the only way to move "naked iron" across the periplasm.[31]

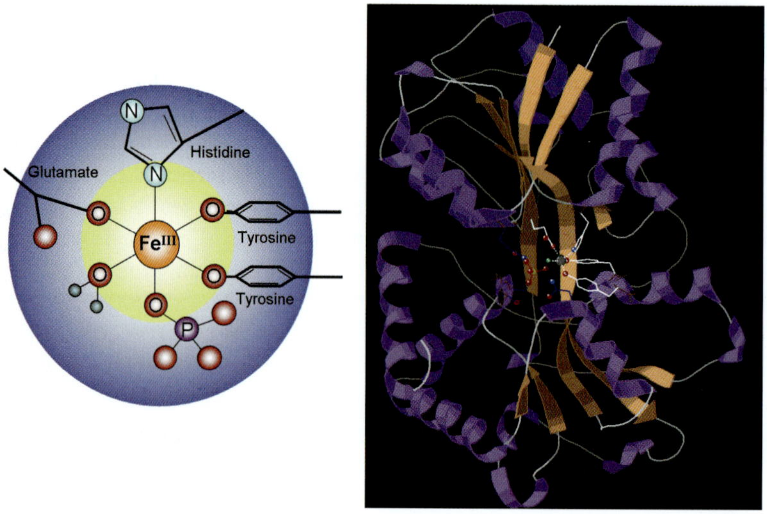

FIGURE 13.15. Ribbon diagram for FbpA, which shows Fe^{3+} at the cleft between two lobes of the monomeric protein (shown on the right). Four amino acids bind the iron ion, while the first coordination shell is completed by a labile phosphate anion and a water molecule (shown on the left). (Kinemage created by Dr. Katherine Weaver.)

The structure of FbpA is depicted in Figure 13.15. On the right is a ribbon diagram of the crystal structure of holo-FbpA (PDB code ID9Y) and on the left is a cartoon diagram of the iron binding site of the protein.[157] FbpA binds iron with four amino acids (glutamate, histidine, and two tyrosines) and a synergistic anion.[30] This first coordination shell is similar to that of human Tf except that Tf utilizes aspartate instead of glutamate (Fig. 13.2, in the section "Transport of Iron-Containing Protein into a Cell"). Here again the protein can be seen as a tetradentate ligand binding iron with three hard oxygens each with a −1 charge (Glu, Tyr, and Tyr) and a borderline hard nitrogen (His). In both Tf and FbpA, an exogenous anion is required for tight iron binding. In Tf, the synergistic anion is the bidentate carbonate at physiological conditions, while in FbpA the monodentate phosphate anion is the synergistic anion and a water molecule completes the first coordination shell.[30,32] Although phosphate is thought to be the synergistic anion of the FeFbpA-X assembly *in vivo*, a variety of anions are capable of serving this role *in vitro*.[17,34,158–160] As will be discussed further, the variety of exogenous anions that can serve as the synergistic anion exhibit a range of charges under physiological conditions. The FeFbpA-PO_4 assembly has a large Fe^{3+} binding constant, ~10^{18} M^{-1}, due to the strength of the coordinate covalent bonds in the first coordination shell and important ionic and mixed interactions in the second coordination shell.[158]

FbpA is a member of a family of PBPs, which is characterized as containing bilobal proteins whose substrate binding motion resembles a Pac-Man or a

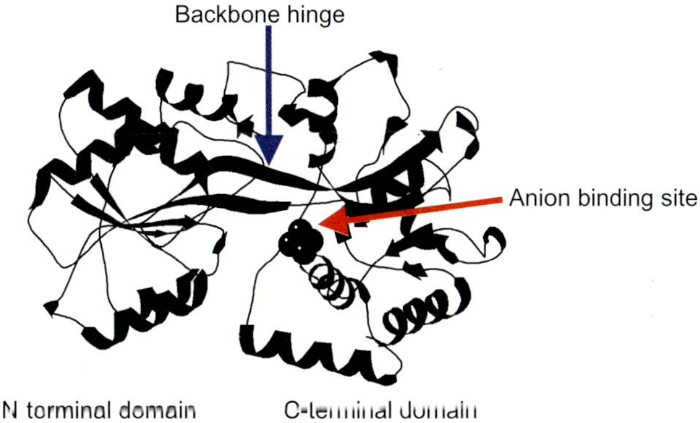

FIGURE 13.16. Apo-FbpA with phosphate bound at the anion binding site located in the C-domain (PDB code 1D9V). The backbone hinge of this periplasmic binding protein consists of two β-strands, which allows the two domains to move closer to each other in a "Pac-Man" or clamping motion. (Kinemage created by Dr. Katherine Weaver.[33,101])

Venus flytrap.[101] Sulfate-binding protein, phosphate-binding protein, and maltose-binding protein are classic examples of PBPs. FhuD, discussed in Section 13.4.3, is another example of this extensive family. FbpA binds iron in the cleft between two globular lobes each made of α-helices and β-strands. The lobes are connected by a hinge which, in the case of FbpA, is composed of a β-sheet (Fig. 13.16).[30,33] The members of the PBP family are categorized by the type of hinge connecting the two lobes.[101,161] Therefore, FbpA and FhuD, which has a single α-helix hinge (Fig. 13.10), are placed in two different categories even though both are periplasmic iron transport proteins.

When iron is bound to FbpA, the protein is in the closed conformation, binding the cargo like a clamp (Pac-Man or Venus flytrap) with a 21° rotation around the protein hinge axis. Apo-FbpA, on the other hand, has a more open conformation. The conformational change between the apo and holo forms is also observed for the two iron binding lobes of Tf discussed in the section "Transport of Iron-Containing Protein into a Cell." The similarities between FbpA and Tf are as follows: (1) both transport naked or unchelated iron; (2) both require a synergistic anion; (3) both provide four similar iron donor groups; (4) both sequester iron through a clamping motion; and (5) both function to transport iron between membrane receptors. These structural and functional similarities have led to FbpA being called "bacterial Tf."

As introduced above, the iron binding site located between the two globular domains of FbpA consists of electronegative or polar groups (mostly hard Lewis bases) which form coordinate covalent bonds with the +3 charged metal ion. Conversely, adjacent to the iron binding site, the anion binding site includes positively charged residues that stabilize the negative anion. As shown in Figure 13.16, the anion binding site is located at the N-terminus of an α-helix in the C-lobe. Binding at this site, the exogenous anion preforms the iron binding site in apo-FbpA so that holo-FbpA formation involving rotation of the C and N domains about the hinge axis is poised for facile protein side chain–iron bond formation.

In vitro experiments have demonstrated that the identity of the synergistic anion can modulate both the thermodynamic and kinetic properties of the FeFbpA-X assembly, where X is the anion.[17,158,162] Within the context of this chapter, the FeFbpA-X can be defined as a ternary complex of three multivalent components: the protein, the synergistic anion, and the metal ion. Many ionic and mixed interactions are involved in the chemical and biological properties of this association. In this section, we will focus on the interactions between the protein and anions, both the synergistic anion and the anions present in the surrounding aqueous environment (the periplasm *in vivo* and the experimental background electrolyte *in vitro*). However, it is impossible to exclude the interactions of the metal cation with the anion and the protein as all are interrelated.

13.6.2 The Periplasm

The periplasmic space between the inner and outer membranes of gram-negative bacteria has a dynamic chemical composition and serves to buffer the cytoplasm from the external environment. The periplasm comprises about 30% of the bacterial total volume.[163] Low-molecular-weight (<600 Da) molecules and ions enter the periplasm by passive diffusion through outer membrane porins.[163,164] Many nutrients are bound in and transported through the periplasm by anion binding proteins such as phosphate-binding protein and sulfate-binding protein.[165] Due to its anion-binding ability, FbpA is thought to have evolved from these anion-binding proteins.[30]

The chemical composition of the periplasm is directly influenced by the environment in which the bacteria live and grow. Therefore, for human pathogens, such as *Neisseria*, that colonize the biological fluids, anions such as phosphate and sulfate are expected to be prevalent in the bacterial periplasm. The phosphate and sulfate concentrations within human serum are between 0.1 and 1.0 mM.[163,166–168] Importantly, the pH of the periplasm is expected to be lower than the surrounding environment and *in vitro* is mimicked using solutions buffered at pH 6.5.[17,169–173]

13.6.3 Synergistic Anion Modulates FeFbpA-X Thermodynamic Properties

As mentioned previously, the identity of the synergistic anion and its interactions with the protein modulates both the thermodynamic and kinetic properties of the FeFbpA-X assembly. By changing which synergistic anion is bound, the affinity of the protein for iron(III) is adjusted, and in concert the Fe^{3+}/Fe^{2+} reduction potential of the FeFbpA-X complex is modulated.[17] In this section, we will briefly review the effects of the synergistic anion in FeFbpA-X and propose a functional role for the anion. Throughout this discussion, we will focus on the specific ionic and mixed interactions between the synergistic anion and the protein.

Analogous to the reductive release hypothesis for iron removal from Tf (Section 13.4.1), Fe^{3+} reduction has been proposed as the method of iron removal from FbpA at the inner membrane receptor (Eq. 13.7)[17,158]:

$$Fe^{3+}\text{-FbpA-X} + e^- \rightleftarrows Fe^{2+}\text{-FbpA-X} \qquad E_{1/2}. \tag{13.7}$$

Upon reduction to Fe^{2+}, the affinity of FbpA for iron is decreased by >10 orders of magnitude. The redox potentials of FeFbpA-X ($X^{n-} = SO_4^{2-}$, PO_4^{3-}, AsO_4^{3-}, citrate^{3-}, NTA^{3-}, $P_2O_7^{4-}$, oxalate^{2-}) have been shown to be in the range where reduction may be driven by NADH/NADPH or similar biological reducing agents. Depending on the identity of the synergistic anion, the redox potential of the ternary complex FeFbpA-X is modulated by ~170 mV or ~14 kJ.[17,158] Varying the synergistic anion also modulates the iron binding affinity (K'_{eff}) of FbpA by ~14 kJ (Eq. 13.8):

$$Fe^{3+} + \text{apo-FbpA-X} \rightleftarrows \text{FeFbpA-X} \quad K'_{eff}. \tag{13.8}$$

Figure 13.17 illustrates the trend in $E_{1/2}$ plotted as a function of log K'_{eff} for the ternary assembly FeFbpA-X with a variety of anions. This correlation suggests that the variation in $E_{1/2}$ of the FeFbpA-X largely represents the degree of ground state stabilization of the Fe^{3+}oxidation state by the synergistic anion X^{n-}.[160] While the functional role of the synergistic anion *in vivo* is debated,[174] the influence of anion identity on the reduction potential suggests that anion exchange could serve as a regulator of a reductive iron release mechanism.

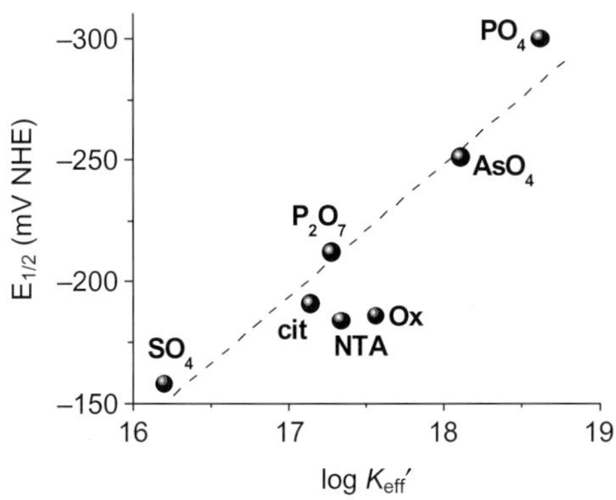

FIGURE 13.17. Measured reduction potentials, $E_{1/2}$, for FeFbpA-X plotted as a function of log K'_{eff}, the effective Fe^{3+} affinity constant for the FeFbpA-X assembly. (Figure adapted from Heymann et al.[160] and used with permission.)

The factors that influence the ability of various anions to tune the stability, K'_{eff}, of the FeFbpA-X ternary assembly have been investigated in detail. The properties of the anion and its interaction with the protein are of utmost importance and will be outlined below. However, additional factors must also be considered for the ternary complex stability such as the oxidation state of the iron and the amino acid residues responsible for the binding of both metallic and anionic substrates. The reader is referred to an extensive analysis of the chemical properties and ionic and mixed interactions involved in determining the characteristics of the protein–anion–metal association.[160] Here we will highlight a few of the findings. We will start with a discussion of the FbpA protein residues involved in binding the synergistic anion.

The structures for holo-FbpA from *N. gonorrhoeae* (1D9Y) and apo-FbpA (apo-HitA) from *H. influenzae* (1D9V) are available.[33,157] The structure for the apo-form of Neisserial FbpA has not been deposited into the data bank so the apo-form of a homologous protein, HitA, was used to analyze the iron-free protein interactions.[160] In both crystal structures, phosphate is bound within the active site as the synergistic anion. As mentioned previously, the anion is bound at the N-terminus of an α-helix that is adjacent to the iron binding site. This binding event occurs through a nitrogen of the backbone amide of Ala141 and the hydroxyl group of S139. Both residues are from the C-terminal domain. As seen in Figure 13.18, the synergistic phosphate anion also forms a hydrogen bond with Q58 from the N-terminal domain of the protein and coordinates to the iron in the holo-protein. In contrast, phosphate interacts with residues from only the C-terminal domain in the apo-FbpA crystal structure (Fig. 13.18). The ionic and mixed interactions of the phosphate binding site in both the holo- and apo- forms of FbpA is similar to that of the anion binding sites in sulfate-binding protein and in phosphate-binding protein.[165,175] Interestingly, both of these classical anion-binding proteins are specific for sulfate and phosphate, respectively, but FbpA is capable of binding a variety of tetrahedral anions.[17,160]

The stability of the anion interaction with the apo-protein depends on the identity of the anion (Eq. 13.9):

$$\text{apo-FbpA} + X^{n-} \rightleftarrows \text{apo-FbpA-X} \quad \Delta G_{\text{FbpAX}}. \tag{13.9}$$

One factor influencing the interactions between the anion and protein is the shape of the anion.[160] Of the anions that have been shown to be capable of serving as the synergistic anion, phosphate, arsenate, and sulfate are tetrahedral, while pyrophosphate, oxalate, citrate, EDTA, and NTA are larger and less spherically symmetrical.[17,160,176,177] Pyrophosphate (K_d = 18 μM) and citrate (K_d = 60 μM) bind with the highest affinity to apo-FbpA (Eq. 13.9), while the tetrahedral anions (sulfate, phosphate, and arsenate) bind with weaker affinity (K_d ~ 1 mM).[17,160] However, this trend does not explain the

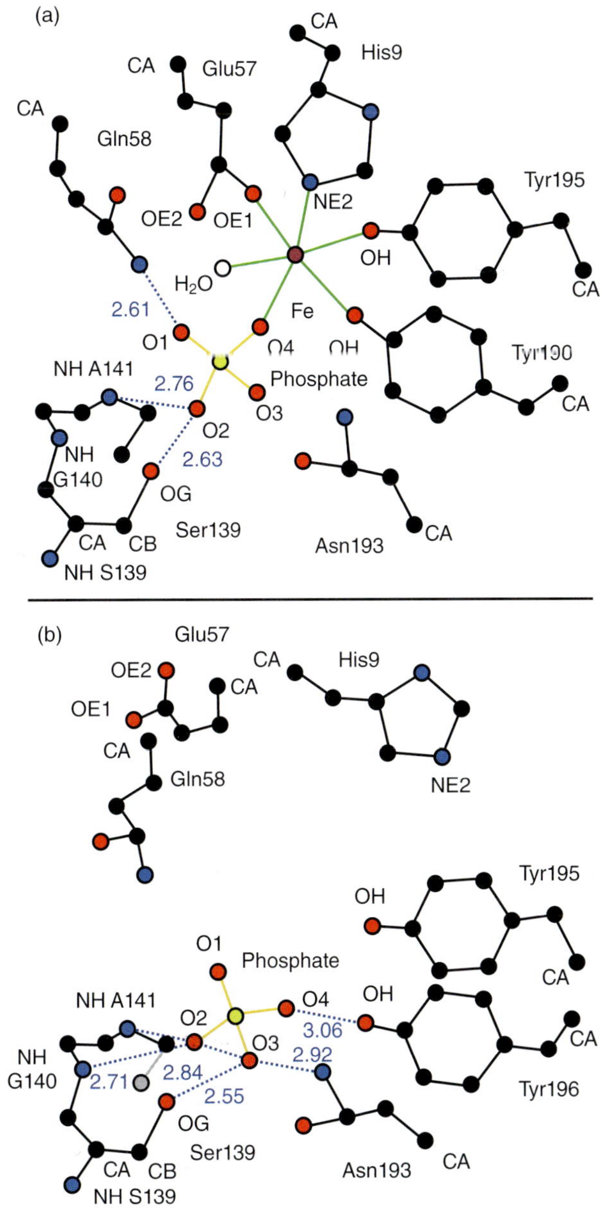

FIGURE 13.18. (a) Iron and synergistic anion binding site for holo-FbpA from *Neisseria gonorrhoeae*. Phosphate interacts with Ser139, Ala141, and Gln58. (b) Iron and synergistic anion binding site for apo-FbpA (HitA) from *Haemophilus influenzae*. Phosphate interacts with Ser139, Gly140, Ala141, Tyr196, and Asn193. Blue dashed lines represent hydrogen bonds with bond length (Å) printed in the same color. (Figure used with permission.[160])

observation that FeFbpA-PO_4 is the most stable iron-loaded form of the protein (Eq. 13.8).

The protonation state of the anion is also a factor in ternary complex association (Eq. 13.8), as other anion-binding proteins are selective for particular protonation states.[160] Sulfate-binding protein prefers fully ionized tetrahedral oxydianions and phosphate-binding protein prefers partially protonated (dibasic and monobasic) species.[160,165] In apo-FbpA, the hydroxyl group of S139 at the end of the anion-binding helix can rotate away from the anion, altering its ability to donate or accept hydrogen bonds from the anion and changing the affinity of the protein for the anion.[178] The flexibility of the anion-binding helix in FbpA is thought to reduce this selection factor for a particular protonation state, making FbpA more promiscuous than sulfate-binding protein or phosphate-binding protein.[165,179,180] The proximity of E57 to the anion in the anion binding site of FbpA has also been suggested to influence the decrease in the binding affinity between FbpA and sulfate compared with phosphate as corresponding residues in the classic anion binding proteins function in this way. However, in FbpA, the effect of E57 is muted somewhat by the presence of Fe^{3+} and an intervening water molecule.[160]

In addition to shape and protonation state, a third factor influencing the anion–protein interaction in FbpA is hydration enthalpy.[160] For example, sulfate, which has a more negative hydration enthalpy due to its high charge density, has a very stable primary hydration shell. This stable hydration shell suggests that the sulfate ion is more likely to remain solvated in the bulk solvent rather than binding to the protein, an event that would require stripping of the primary hydration shell.[160] This analysis is consistent with the observation that the effective equilibrium constant (K'_{eff}; Eq. 13.8) for FeFbpA-SO_4 formation is the weakest of all the anions studied so far.[160]

Thermodynamic analysis of each of these anion properties as well as others, such as (1) the interaction between the metal and the anion and (2) the average molecular charge or electrostatic properties of the anions, has led to the conclusion that the anion exerts primary control over the variations in stability of the FeFbpA-X assembly.[160] By analyzing a linear free energy relationship (LFER) between the anion/apo-protein interaction, $\Delta(\Delta G_{FbpAX})$, and the overall stability of the ternary assembly, $\Delta(\Delta G_{FeFbpAX})$, two groups of anions have been identified (Fig. 13.19). The parameters $\Delta(\Delta G_{FeFbpAX})$ and $\Delta(\Delta G_{FbpAX})$ are defined in Figure 13.19 and are related to Equations 13.8 and 13.9 respectively. Group 1 in Figure 13.19 consists of the tetrahedral anions PO_4^{3-}, AsO_4^{3-}, and SO_4^{2-}. Each of these geometrically similar tetraoxyanions binds as a monodentate ligand to iron. Although each has a similar apo-protein/anion affinity, $\Delta(\Delta G_{FbpAX})$, the affinity of the respective FeFbpA-X complex for Fe^{3+}, $\Delta(\Delta G_{FeFbpAX})$, is quite different. Group 2 in Figure 13.19 includes NTA^{3-}, oxalate^{2-}, citrate^{3-}, and pyrophosphate^{4-}, which are all potentially polydentate ligands. Here the stabilities of the respective ternary complexes FeFbpA-X are similar, but the thermodynamics of the apo-protein/anion association varies. From these data, we conclude that electrostatic, hydrogen

$$\text{apo-FbpA} + X^{n-} \leftrightarrows \text{apo-FbpA-X} \qquad \Delta G_{FbpAX}$$

$$\Delta(\Delta G_{FbpAX}) = \Delta G_{FbpAX} - \Delta G_{FbpASO4}$$

$$Fe^{3+} + \text{apo-FbpA-X} \leftrightarrows \text{FeFbpA-X} \qquad \Delta G_{FeFbpAX}$$

$$\Delta(\Delta G_{FeFbpAX}) = \Delta G_{FeFbpAX} - \Delta G_{FeFbpASO4}$$

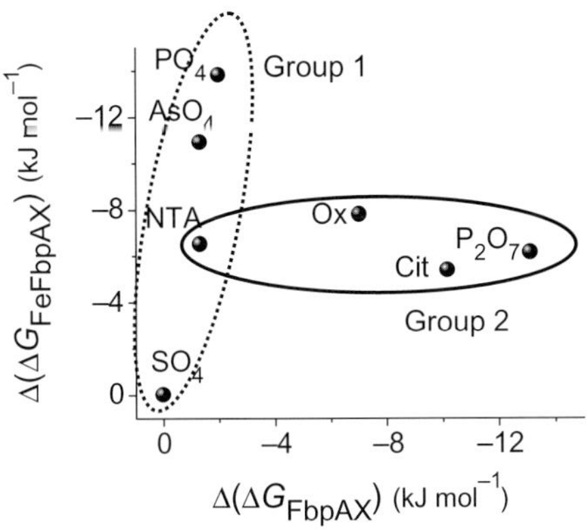

FIGURE 13.19. LFER for $\Delta(\Delta G_{FeFbpAX})$, the free energy ($\Delta G_{FeFbpAX}$) of FeFbpA-X complex formation (Eq. 13.8) relative to that of FeFbpA-SO_4, plotted as a function of ($\Delta(\Delta G_{FbpAX})$), the free energy of the apo-protein/anion interaction (Eq. 13.9; ΔG_{FbpAX}), relative to that of FeFbpA-SO_4. Anion groups denoted with a dotted circle for Group 1 and a solid circle for Group 2.[181]

bonding, and van der Waals interactions between the protein and the anion are all important.[181]

However, when we factor out the electrostatic interactions between the apo-protein and the anion, we observe a LFER demonstrating that van der Waals and hydrogen bonding interactions between the anion and the apo-protein directly modulate the stability of the FeFbpA-X assembly.[160] As shown in Figure 13.20, a LFER is apparent when $\Delta(\Delta G_{FbpAX})$ is plotted as a function of $\Delta(\Delta G_{FeFbpAX})$ corrected for the electrostatic properties of the anion (Z_{ave}). These data suggest that intermolecular forces such as van der Waals interactions and hydrogen bonding, rather than ionic interactions, are responsible for the *variability* in the stability of the FeFbpA-X assembly with changes in X.[160]

It remains to be determined how the various properties of each anion work together to control the stability of a given protein–anion–iron ternary assembly, FeFbpA-X. However, it is clear that a variety of ionic and mixed interactions between the anion and the protein are deeply involved in the overall stability of this ion polymer assembly.

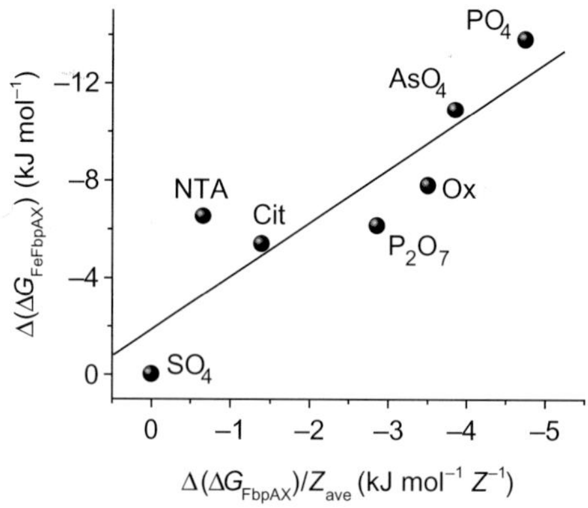

FIGURE 13.20. LFER for $\Delta(\Delta G_{FeFbpAX})$, the free energy of FeFbpA-X complex formation (Eq. 13.8) relative to that of FeFbpA-SO_4, plotted as a function of $(\Delta(\Delta G_{FbpAX})/Z_{ave})$, the free energy of the apo-protein/anion interaction (Eq. 13.9) relative to that of FeFbpA-SO_4 corrected for the electrostatic properties of the anion. (Figure adapted from Heymann et al.[160] and used with permission.)

13.6.4 Anion Influence on FeFbpA-X Kinetic Properties

The kinetics and mechanism of iron loading and unloading also depend on the synergistic anion. In the iron loading process, the synergistic anion serves two kinetic functions.[182–184] First, the synergistic anion enters the active site and preorganizes the iron binding residues prior to iron binding (Eq. 13.10)[33]:

$$Fe^{3+} + \text{apo-FbpA-X} \rightarrow \text{FeFbpA-X}. \qquad (13.10)$$

The ability to bind the anion even in the absence of iron further supports the hypothesis that FbpA evolved from anion-binding proteins. Specifically, the anion binds in the anion binding site adjacent to the iron binding site and influences the orientation of the two iron-binding tyrosines (Y195 and Y196) through hydrogen bonding interactions.[33,183] Second, the anion assists in assembly formation when iron is inserted as an Fe-X complex (Eq. 13.11)[184]:

$$\text{Fe-X} + \text{apo-FbpA-X} \rightarrow \text{FeFbpA-X} + X^{n-}. \qquad (13.11)$$

In the iron release process, the synergistic anion also serves two functions. In a kinetic role, the synergistic anion controls the release of Fe(III) by acting as the gate keeper; that is, the synergistic anion must first be displaced before the Fe(III) can be released.[162,177] Second, as discussed previously, the anion controls the release of the reduced form of iron, Fe(II), by modulating the reduction potential.[17,160]

The ability of the synergistic anion to alter both the kinetic and thermodynamic properties of the FeFbpA-X ternary assembly may provide a regulatory mechanism to control iron delivery and temporary iron storage in the periplasm. By varying the identity of the synergistic anion, which depends directly on the ionic composition of the periplasmic milieu at any given time, we propose that the bacteria may be able to "sense" fluctuations in its environment and regulate iron transport accordingly.

13.6.5 FeFbpA-X Anion Exchange and Environmental Effects

The ability to exchange one anion for another while maintaining the interaction between iron and the protein (Eq. 13.12) has also been investigated extensively for both the Group 1 anions and the Group 2 anions defined in Figure 13.19.[17,160,173] In short, synergistic anion exchange is possible and depends on (1) the identity of both the entering and leaving anion, and (2) the ability of the entering ligand to serve as a competitive chelator[17,173,181]:

$$\text{FeFbpA-X} + \text{Y}^{m-} \rightleftarrows \text{FeFbpA-Y} + \text{X}^{n-}. \tag{13.12}$$

In order to understand the ability of the anion to influence the thermodynamics and kinetics of the ternary complex, FeFbpA-X, structural issues beyond the active site (the anion binding site and the iron binding site) must also be considered.[173] In FbpA, a channel leads from the bulk solvent (the aqueous periplasm) to the active site of the protein, located between the two globular lobes. The channel is partially solvent exposed and the extent of the exposure depends on the iron-loaded status of the protein due to the conformational change between the apo- and holo- forms of the protein. As shown in Figure 13.21, the entrance to the channel from the bulk solvent is surrounded by five positive residues.[33,160,173]

A positively charged entrance to the channel leading to the active site is thought to provide two functions for the protein.[33,160,173] First, once bound, iron is effectively trapped in the active site due to electrostatic repulsion at the only exit. This repulsion is not a significant problem prior to iron loading because in the apo- form of the protein positively charged lysines are rotated away from each other as the protein hinges open. A second function of this positive cluster may be to recruit anions to the channel.[173] In this way, the effective concentration of anions in the bulk solvent is increased at the entrance of the channel, possibly facilitating synergistic anion exchange and or dissociation of Fe^{3+}. In this section, we will examine the kinetic properties of a particular anion exchange mechanism (Eq. 13.12). Further, the *in vitro* composition of the environment around the protein has been shown to influence this exchange reaction; consequently, we will discuss the ability of nonspecific anion–protein interactions to influence protein function.

As discussed in Chapter 6, polymers and in particular proteins are greatly affected by the ions in the surrounding environment. Here we will discuss how ions in the environment can affect the behavior of a protein, focusing on

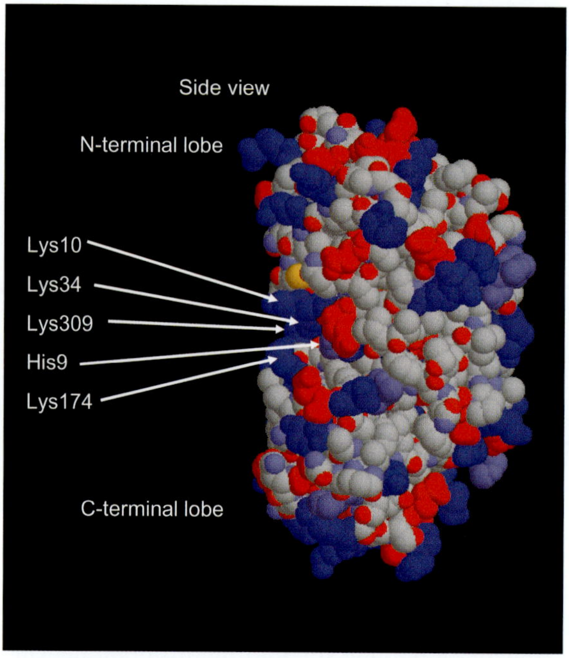

FIGURE 13.21. Space-filled model of holo-FbpA depicting the cluster of positive residues at the entrance of the active site channel. Color coding: neutral residues (gray); negative residues (red); histidine residues (lilac); and positive residues (blue). (Kinemage created by Dr. Katherine Weaver.[181])

in vitro background electrolyte studies. Specifically, we will address the influence of the environmental anions (background electrolytes) on a particular FbpA reaction (exchange of the synergistic anion). *In vitro*, the identity of the background electrolyte has been shown to greatly influence the behavior of FbpA in the synergistic anion exchange reaction, in which, for example, the synergistic anion sulfate is replaced by phosphate (Eq. 13.13; charge omitted due to degree of protonation variability with pH)[181]:

$$\text{FeFbpA-SO}_4 + \text{PO}_4 \rightleftarrows \text{FeFbpA-PO}_4 + \text{SO}_4 \qquad (13.13)$$
$$(\text{Background electrolyte: Cl}^- \text{ or ClO}_4^-).$$

We will demonstrate how the iron transport process is sensitive to the aqueous environment surrounding the protein, a finding that can be extended to the dynamic ion composition of the periplasm *in vivo*.

The kinetic properties of the anion exchange reaction between FeFbpA-SO_4 and PO_4^{3-} (Eq. 13.13) have been observed *in vitro* with both chloride and perchlorate as the background electrolyte.[181] To compare the effect of the identity of the background electrolyte anion, the same ionic strength was used for each system. We will briefly outline the findings of these *in vitro* kinetic experiments and then discuss these results using the principles of the Hofmeister series introduced in Chapter 6.

A preassociation mechanism has been assigned to the synergistic anion exchange reaction (Eq. 13.13) in both perchlorate and chloride background electrolyte solutions.[173,181] In a preassociation mechanism, the entering phosphate anion is expected to bind to the protein in a rapidly established equilibrium (Eq. 13.14). Then in an exchange step the phosphate replaces sulfate in the first coordination shell of the iron (Eq. 13.15; charge omitted due to degree of protonation variability with pH). We propose that this initial preassociative complex is formed at an initial anion binding site that may be located at the opening of active site channel seen in Figure 13.21:

$$\text{FeFbpA-SO}_4 + \text{PO}_4 \rightleftarrows \text{FeFbpA-SO}_4(\text{PO}_4) \qquad (13.14)$$
$$(\text{Background electrolyte: Cl}^- \text{ or ClO}_4^-)$$

$$\text{FeFbpA-SO}_4(\text{PO}_4) \rightarrow \text{FeFbpA-PO}_4 + \text{SO}_4 \qquad (13.15)$$
$$(\text{Background electrolyte: Cl}^- \text{ or ClO}_4^-).$$

By varying the concentration of the background electrolyte for the chloride and perchlorate systems, respectively, it was shown that neither electrolyte participates directly in the anion exchange mechanism shown in Equations 13.14 and 13.15. However, the rate of the reaction (Eq. 13.13) is faster in the presence of chloride.[181] Interestingly, kinetic studies in the presence of mixtures of background electrolytes with chloride and perchlorate in varying ratios showed that the presence of perchlorate dictates the value of the rate constant even when there are equal parts chloride and perchlorate. It is not until very high ratios of chloride that an increase in the rate constant is observed.[181] The exchange reaction in the absence of either chloride or perchlorate was also examined. In this case, excess sulfate served as the background electrolyte in order to maintain protein solubility. These experiments (no chloride and no perchlorate) showed that high concentrations of excess sulfate slowed the reaction.[181]

Taken together, these kinetic observations indicate that the identity of the background electrolyte affects the rate of the anion exchange reaction (Eq. 13.13). These observations can be explained by examining the folding stability of the protein in each environment. Here we invoke the idea of the Hofmeister series, discussed at length in Chapter 6, as one possible explanation the findings reported above.[1,181] The Hofmeister series is an ordering of ions corresponding to their ability to affect the physical properties of macromolecules after charge-dependent effects have been maximized (Chapter 6).[185] As shown below, anions are ordered based on their ability to affect the solubility and folding structure of proteins. Anions that promote protein stabilization are referred to as "kosmotropes," and "anions" that destabilize folded proteins are called "chaotropes"[185]:

Kosmotropes...Chaotropes

$$\mathbf{SO_4^{2-}} > \mathbf{H_2PO_4^-} > OH^-, F^- > \mathbf{Cl^-} > Br^- > NO_3^- > \mathbf{ClO_4^-} > I^- > SCN^-$$

The physical explanation of the Hofmeister series is a matter of ongoing debate.[185–187] However, the observed phenomenon that the stability of a protein in solution is affected by background electrolytes in predictable ways is repeatable and well accepted.

In the case of chloride and perchlorate, one background electrolyte anion stabilizes protein folding, while one anion destabilizes protein folding. It was observed that the presence of perchlorate as the background electrolyte slows the rate of the FbpA synergistic anion exchange reaction (Eq. 13.13).[181] Given that perchlorate is a chaotrope, perchlorate destabilizes the protein fold and may disrupt the preassociative anion binding site located at the opening of active site channel. This explanation is consistent with the conformational changes described previously, in which the cluster of positively charge residues at the entrance to the active site channel (Fig. 13.21) move away from each other in the more open apo- form of the protein.[181] Although the true apo- form of the protein is not achieved during the anion exchange reaction because iron remains in the active site, there is evidence to suggest that the extent of the conformation change from apo- to holo- is fluid.[159]

The opposite is true for the chloride system. Chloride is classified as a kosmotrope, enhancing protein folding stability. In the presence of chloride, the rate of the anion exchange reaction (Eq. 13.13) is increased and may indicate that the pre-equilibrium (Eq. 13.14) lies further toward assembly formation ($FeFbpA\text{-}SO_4(PO_4)$). Stabilization of the more folded form of the protein would bring the positive residues of the active site channel entrance in proximity to one another, stabilizing the preassociation anion binding site (Fig. 13.21).[181]

This interpretation of kinetic data using the Hofmeister series can also be applied to the anion exchange reaction in the absence of chloride and perchlorate, for which excess sulfate served as the background electrolyte and the initial synergistic anion.[181] Both sulfate and phosphate are kosmotropic and enhance folding stability. At the highest concentration of entering synergistic anion phosphate, the excess sulfate (background electrolyte) can be viewed as similar to chloride: Protein folding stability is promoted and anion exchange is rapid. Indeed the highest rate constant for anion exchange was observed under these conditions compared with all others tested. However, sulfate can also compete with phosphate for the role of synergistic anion. Therefore, increasing sulfate concentration compared with phosphate concentration slows the anion exchange reaction even though it is a kosmotrope promoting protein folding and the formation of an anion binding preassociation site.[181]

In summary, exchange (Eq. 13.13) of the synergistic anion in FeFbpA-X occurs through a preassociation mechanism (Eqs. 13.14 and 13.15), in which both the leaving and entering synergistic anions can affect the process.[17,160,181,184] We hypothesize that the site of this preassociative anion binding may be located at the entrance to the active site channel that consists of several positive residues. Finally, the ionic composition of the aqueous environment surrounding the protein greatly influences the behavior of the protein and may be another example of the Hofmeister series effect.

Given that all three of the components of the ternary assembly FeFbpA-X are multivalent (i.e., each interact with one another and influence all of the ionic and mixed interactions of the system), FbpA is an excellent example of how complicated teasing out the exact interactions that in concert determine the relationship between two or more binding partners. In the FbpA example, anion–protein interactions dramatically influence both the kinetic and thermodynamic properties of the FeFbpA-X assembly and regulate the iron release process. This complex suite of interactions between the protein, the anion, and the metal cation provides an elegant and highly tunable nodal point in the transport of iron through the periplasm of these gram-negative bacteria.

We would be remiss if we did not also mention that similar to FbpA, the chemical properties of Tf are also affected by anion–protein interactions. The synergistic anion of Tf and background electrolytes have been shown to affect (1) the thermodynamic properties of the FeTf-X assembly and (2) the cooperativity between the two lobes of Fe_2-Tf in the iron release process.[188–191] As expected, the rate and mechanism of iron release also change as a function of increased salt concentration.[77] While similar to FbpA, the active site of Tf is less solvent exposed as can be seen in part by the greater stability of the Fe_2-Tf complex. Even though they are more secluded in the protein, nonbinding residues (histidines) in the iron binding pocket have been shown to affect the rate and mechanism of iron release particularly as pH is varied. It can be inferred that auxiliary anions might also interact with these histidines and modulate the kinetic properties of Tf.[192]

Implied but not stated explicitly until now is the relationship between *in vitro* experimental observations and *in vivo* chemical behavior. As illustrated by the FbpA system, protein–anion interactions can directly and indirectly affect the chemical properties of an iron transport protein *in vitro*. Given the complicated nature of the multiple, various, and dynamic ionic and mixed interactions involved in these multivalent polyampholyte/ion assemblies, *in vitro* conditions may not sufficiently mimic the *in vivo* environment in which a particular protein functions. Caution is required when interpreting *in vitro* data regarding the role and influence of protons and salts, whose concentrations may be dynamic or poorly defined *in vivo*.

13.7 CONCLUSION

Protein transport systems that carry the essential nutrient iron function to move this insoluble and redox active metal cation across hydrophobic membranes and through aqueous solutions. Each transport system involves unique recipes of ionic and mixed interactions that dictate the specificity and efficiency of the uptake system. Protein–protein interactions, protein–iron complex interactions, and protein–anion interactions can all be viewed through the lens of the definitions and principles of ion–polymer interactions described in this book. Each of these complex and interrelated interactions has been studied in detail individually. However, going forward a multilayered consideration of the many types of ionic and mixed interactions involved in iron transport is needed to develop a complete understanding of each transport process. A multilayered approach, integrated by the language and principles of ionic and mixed interactions presented in this book, can complement current research and answer questions that have yet to be addressed. The challenge will be to understand the role synergy plays between the many and varied types of interactions discussed in this chapter and to organize the complexity biology has presented.

ACKNOWLEDGMENTS

Drs. Jared J. Heymann and Katherine D. Weaver are acknowledged for many helpful conversations and for lending both illustrations and unpublished data. We thank the National Science Foundation (grant no. CHE0809466) for financial support.

GLOSSARY

Apo-	without metal present
BtuF/C/D	vitamin B12 transport system
CfrA	ferric enterobactin outer membrane receptor in *Campylobacter jejuni*
DMT1	divalent metal transporter 1
EDTA	ethylenediamminetetraacetate
Endocytosis	process through which a receptor on the outer surface of a membrane is invaginate in a vesicle and brought into the cytosol of a cell
ENT	enterobactin; common bacterial siderophore
FbpA	ferric binding protein A
FbpB/C	ABC transporter that recognizes FbpA and moves naked iron across the inner membrane of *Neisseria gonorrhoeae*

FecA	ferric citrate outer membrane receptor
FepA	ferric enterobactin outer membrane receptor
FhuA	ferrichrome outer membrane receptor
FhuD	ferrichrome periplasmic binding protein
FhuB/C	ABC transporter that recognizes FhuD at the inner membrane of *Escherichia coli*
FpvA	ferric pyoverdine outer membrane receptor in *Pseudomonas aeruginosa*
Gram-negative	classification of bacteria that have an outer and an inner membrane; bacteria are identified using a staining process developed by Gram
Holo-	metal bound
Lipoprotein	protein with a lipid covalently attached as a posttranslational modification
LFER	linear free energy relationship
NHE	normal hydrogen electrode
NTA	nitrilotriacetate
PBP	periplasmic binding protein
Periplasm	aqueous space between the outer and inner membranes of gram-negative bacteria
Pvd	pyoverdine; siderophore produces by psuedomonads
Siderophore	small-molecular-weight iron chelator
TbpA	transferrin-binding protein A; a Neisserial outer membrane receptor protein that binds Tf
TbpA-plug	plug domain of transferrin-binding protein A
TbpB	transferrin-binding protein B; a Neisserial outer membrane receptor protein that binds Tf
Tf	transferrin; human plasma iron transport protein
TfR	human transferrin receptor
TonB	protein that provides mechanical energy to outer membrane receptors by harnessing the chemical energy created by the proton gradient across the inner membrane in gram-negative bacteria

REFERENCES

1. A. Ciferri. *Chem Eur J* **2010**, 16, 10930–10945.
2. R. R. Crichton and M. Charloteaux-Wauters. *Eur J Biochem* **1987**, 164, 485–506.
3. J. J. R. Frausto da Silva and R. J. P. Williams. *The Biological Chemistry of the Elements: The Inorganic Chemistry of Life*. Oxford University Press, New York, 2005.

4. B. C. Chu, A. Garcia-Herrero, T. H. Johanson, K. D. Krewulak, C. K. Lau, R. S. Peacock, Z. Slavinskaya, H. J. Vogel. *BioMetals* **2010**, 23, 601–611.
5. M. R. Bleackley, A. Y. K. Wong, D. M. Hudson, C. H.-Y. Wu, R. T. A. MacGillivary. *Transfus Med Rev* **2009**, 23, 103–123.
6. N. C. Andrews. *Blood* **2008**, 112, 219–230.
7. C. N. Cornelissen and P. F. Sparling. *Mol Microbiol* **1994**, 14, 843–850.
8. A. L. Crumbliss and J. M. Garrison. *Comments Inorg Chem* **1988**, 8, 1–26.
9. J. Burgess. *Metal Ions in Solution*. Ellis Horwood Ltd., Sussex, 1978.
10. D. F. Shriver and P. Atkins. *Inorganic Chemistry*, 4th ed. Oxford University Press, Oxford, 2006.
11. R. G. Pearson. *J Am Chem Soc* **1963**, 85, 3533–3539.
12. H. Boukhalfa and A. L. Crumbliss. *BioMetals* **2002**, 15, 325–339.
13. S. Dhungana and A. L. Crumbliss. *Geomicrobiol J* **2005**, 22, 87–98.
14. F. A. Cotton, G. Wilkinson, C. A. Murillo, M. Bouchmann. *Advances in Inorganic Chemistry*, 6th ed. Wiley-Interscience, New York, 1999.
15. F. Haber and J. Weiss. *P R Soc London A* **1934**, 147, 332–337.
16. R. Crichton. *Iron Metabolism: From Molecular Mechanism to Clinical Consequences*, 3rd ed. John Wiley & Sons, West Sussex, 2009.
17. S. Dhungana, C. H. Taboy, D. S. Anderson, K. G. Vaughan, P. Aisen, T. A. Mietzner, A. L. Crumbliss. *Proc Natl Acad Sci U S A* **2003**, 100, 3659–3664.
18. K. A. Mies, J. I. Wirgau, A. L. Crumbliss. *BioMetals* **2006**, 19, 115–126.
19. C. H. Taboy, C. Bonaventura, A. L. Crumbliss. *Bioelectrochem Bioenerg* **1999**, 48, 79–86.
20. N. C. Andrews. *N Engl J Med* **1999**, 341, 1986–1995.
21. M. D. Garrick and L. M. Garrick. *Biochim Biophys Acta* **2009**, 1790, 309–325.
22. R. T. A. MacGillivary, S. A. Moore, J. Chen, B. Anderson, H. Baker, Y. Luo, M. Bewley, C. A. Smith, M. E. P. Murphy, Y. Wang, A. B. Mason, R. C. Woodworth, G. D. Brayer, E. N. Baker. *Biochemistry* **1998**, 37, 7919–7928.
23. P. Aisen, A. Leibman, J. Zweier. *J Biol Chem* **1978**, 253, 1930–1937.
24. P. Aisen, M. Wessling-Resnick, E. A. Leibold. *Curr Opin Chem Biol* **1999**, 3, 200–206.
25. S. Dhungana, C. H. Taboy, O. Zak, M. Larvie, A. L. Crumbliss, P. Aisen. *Biochemistry* **2004**, 43, 205–209.
26. R. M. Graham, A. C. G. Chua, C. E. Herbison, J. K. Olynyk, D. Trinder. *World Gastroenterol* **2007**, 13, 4725–4736.
27. C. N. Cornelissen, G. D. Biswas, J. Tsai, D. K. Paruchuri, S. A. Thompson, P. F. Sparling. *J Bacteriol* **1992**, 174, 5788–5797.
28. J. E. Anderson, P. F. Sparling, C. N. Cornelissen. *J Bacteriol* **1994**, 176, 3126–3170.
29. C. N. Cornelissen and P. F. Sparling. *J Bacteriol* **1996**, 178, 1437–1444.
30. C. M. Bruns, A. J. Nowalk, A. S. Arvai, M. A. McTigue, K. G. Vaughan, T. A. Meitzner, D. E. McRee. *Nat Struct Biol* **1997**, 4, 919–924.
31. T. A. Mietzner, S. B. Tencza, P. Adhikari, K. G. Vaughan, A. J. Nowalk. *Current Topics in Microbiology and Immunology, Vol. 225 Bacterial Infection: Close*

Encounters at the Host Pathogen Interface, P. K. Vogt and M. J. Maham, eds. Springer-Verlag, Berlin and Heidelberg, Germany, 1998; pp. 113–135.

32. P. L. Roulhac, K. D. Weaver, P. Adhikari, D. S. Anderson, P. D. DeArmond, T. A. Mietzner, A. L. Crumbliss, M. C. Fitzgerald. *Biochemistry* **2008**, 47, 4298–4305.
33. C. M. Bruns, D. S. Anderson, K. G. Vaughan, P. A. Williams, A. J. Nowalk, D. E. McRee, T. A. Mietzner. *Biochemistry* **2001**, 40, 15631–15637.
34. M. Guo, I. Harvey, W. Yang, L. Coghill, D. J. Campopiano, J. A. Parkinson, R. T. A. MacGillivary, W. R. Harris, P. J. Sadler. *J Biol Chem* **2003**, 278, 2490–2502.
35. T. A. Mietzner, G. Bolan, G. K. Schoolnik, S. A. Morse. *J Exp Med* **1987**, 165, 1041–1057.
36. C.-Y. Chen, S. A. Berish, S. A. Morse, T. A. Mietzner. *Mol Microbiol* **1993**, 10, 311–318.
37. C. J. Parker Siburt, P. L. Roulhac, K. D. Weaver, J. L. M. Noto, T. A. Mietzner, C. N. Cornelissen, M. C. Fitzgerald, A. L. Crumbliss. *Metallomics* **2009**, 1, 249–255.
38. D. Perkins-Balding, M. Ratliff-Griffin, I. Stojiljkovic. *Microbiol Mol Biol Rev* **2004**, 68, 154–171.
39. E. P. Skaar and O. Schneewind. *Microbes Infect* **2004**, 6, 390–397.
40. T. J. Brickman, M. T. Anderson, S. K. Armstrong. *BioMetals* **2007**, 20, 303–320.
41. E. E. Wyckoff, A. R. Mey, S. M. Payne. *BioMetals* **2007**, 20, 405–416.
42. K. D. Krewulak and H. J. Vogel. *Biochim Biophys Acta* **2008**, 1778, 1781–1804.
43. I. C. Boulton, A. R. Gorringe, B. Gorinsky, M. D. Retzner, A. B. Schryvers, C. L. Joannou, R. W. Evans. *Biochem J* **1999**, 339, 143–149.
44. A. G. Khan, S. R. Shouldice, S. D. Kirby, R.-H. Yu, L. W. Tari, A. B. Schryvers. *Biochem J* **2007**, 404, 217–225.
45. T. Krell, G. Renauld-Mongenie, M.-C. Nicolai, S. Fraysse, M. Chevalier, Y. Berard, J. S. Oakhill, R. W. Evans, A. Gorringe, L. Lissolo. *J Biol Chem* **2003**, 278, 14712–14722.
46. J. S. Oakhill, C. L. Joannou, S. K. Buchanan, A. Gorringe, R. W. Evans. *Biochem J* **2002**, 364, 613–616.
47. A. Ekins, A. G. Khan, S. R. Shouldice, A. B. Schryvers. *BioMetals* **2004**, 17, 235–243.
48. S. Letoffe, K. Wecker, M. Delepelaire, P. Delepelaire, C. Wandersman. *J Bacteriol* **2005**, 187, 4637–4645.
49. S. K. Buchanan, B. S. Smith, L. Venkatramani, D. Xia, L. Esser, M. Palnitka, R. Chakraborty, D. van der Helm, J. Deisenhofer. *Nat Struct Biol* **1999**, 6, 56–63.
50. A. D. Ferguson, R. Charkraborty, B. S. Smith, L. Esser, D. van der Helm, J. Deisenhofer. *Science* **2002**, 295, 1715–1719.
51. K. P. Locher, B. Rees, R. Koebnik, A. Mitschler, L. Moulinier, J. P. Rosenbusch, D. Moras. *Cell* **1998**, 95, 771–778.
52. J. M. Harrington and A. L. Crumbliss. *Adv Inorg Chem* **2009**, 61, 179–250.
53. C. Wandersman and P. Delepelaire. *Annu Rev Microbiol* **2004**, 58, 611–647.
54. G. S. Moeck and J. W. Coulton. *Mol Microbiol* **1998**, 28, 675–681.
55. K. Brillet, L. Journett, H. Celia, L. Paulus, A. Stahl, F. Pattus, D. Cobessi. *Structure* **2007**, 15, 1383–1391.

56. S. Devanathan and K. Postle. *Mol Microbiol* **2007**, 65, 441–453.
57. I. J. Schalk, P. Kyslik, D. Prome, A. van Dorsselaer, K. Poole, M. A. Abdallah, F. Pattus. *Biochemistry* **1999**, 38, 9357–9365.
58. C. L. Davies, E. L. Dux, A.-K. Duhme-Klair. *Dalton Trans* **2009**, 10141–10154.
59. P. Aisen. *Int J Biochem Cell Biol* **2004**, 36, 2137–2143.
60. W. R. Harris. *J Inorg Biochem* **1986**, 27, 41–52.
61. R. G. Wilkins. *Kinetics and Mechanism of Reactions of Transition Metal Complexes*, 2nd ed. VCH Publishers, New York, 1991.
62. D. C. Kraiter, O. Zak, P. Aisen, A. L. Crumbliss. *Inorg Chem* **1998**, 37, 964–968.
63. R. S. Ohgami, D. R. Campagna, E. L. Greer, B. Antiochos, A. McDonald, J. Chen, J. J. Sharp, Y. Fujiwara, J. E. Barker, M. D. Fleming. *Nat Genet* **2005**, 37, 1264–1269.
64. A. T. McKie. *Nat Genet* **2005**, 37, 1159–1160.
65. M. Scarpellini, A. Casellato, A. Bortoluzzi, I. Vencato, A. S. Mangrich, A. Neves, S. P. Machado. *J Braz Chem Soc* **2006**, 17, 1617–1626.
66. J. V. Princiotto and E. J. Zapolski. *Nature* **1975**, 255, 87–88.
67. D. A. Baldwin, D. M. R. de Sousa, R. M. A. Von Wandruszka. *Biochim Biophys Acta* **1982**, 719, 140–146.
68. D. M. Sipe and R. F. Murphy. *J Biol Chem* **1991**, 266, 8002–8007.
69. C. M. Lawrence, S. Ray, M. Babyonyshev, R. Galluser, D. W. Borhani, S. C. Harrison. *Science* **1999**, 286, 779–782.
70. H. Fuchs, U. Lucken, R. Tauber, A. Engel, R. Gessner. *Structure* **1998**, 6, 1235–1243.
71. V. Dubljevic, A. Sali, J. W. Goding. *Biochem J* **1999**, 341, 11–14.
72. T. Sakajiri, T. Yamamura, T. Kikuchi, H. Yajima. *Protein J* **2009**, 28, 407–414.
73. Y. Cheng, O. Zak, P. Aisen, S. C. Harrison, T. Walz. *J Struct Biol* **2005**, 152, 204–210.
74. A. M. Giannetti, P. M. Snow, O. Zak, P. J. Bjorkman. *PLoS Biol* **2003**, 1, 341–350.
75. A. M. Giannetti, P. J. Halbrooks, A. B. Mason, T. M. Vogt, C. A. Enns, P. J. Bjorkman. *Structure* **2005**, 13, 1613–1623.
76. P. K. Bali and P. Aisen. *Biochemistry* **1991**, 30, 9947–9952.
77. S. L. Byrne, N. D. Chasteen, A. N. Steere, A. B. Mason. *J Mol Biol* **2010**, 396, 130–140.
78. A. Yersin, T. Osada, A. Ikai. *Biophys J* **2008**, 94, 230–240.
79. T. Sakajiri, T. Yamamura, T. Kikuchi, K. Ichimura, T. Sawada, H. Yajima. *Biol Trace Elem Res* **2010**, 136, 279–286.
80. J. Wally, P. J. Halbrooks, C. Vonrhein, M. A. Rould, S. J. Everse, A. B. Mason, S. K. Buchanan. *J Biol Chem* **2006**, 281, 24934–24944.
81. Y. Cheng, O. Zak, P. Aisen, S. C. Harrison, T. Walz. *Cell* **2004**, 116, 565–576.
82. A. B. Mason, S. L. Byrne, S. J. Everse, S. E. Roberts, N. D. Chasteen, V. C. Smith, R. T. A. MacGillivary, B. Kandemir, F. Bou-Abdallah. *J Mol Recognit* **2009**, 22, 521–529.
83. N. G. James, S. L. Byrne, A. N. Steere, V. C. Smith, R. T. A. MacGillivary, A. B. Mason. *Biochemistry* **2009**, 48, 2858–2867.

84. P. K. Bali, O. Zak, P. Aisen. *Biochemistry* **1991**, 30, 324–328.

85. R. Cmejla, P. Ptackova, J. Petrak, F. Sacculidi, J. Cerny, O. Sebesta, D. Vyoral. *Biochem Biophys Res Commun* **2010**, 395, 163–167.

86. I. C. Boulton, M. K. Yost, J. E. Anderson, C. N. Cornelissen. *Infect Immun* **2000**, 68, 6988–6996.

87. T. F. Moraes, R.-H. Yu, N. C. J. Strynadka, A. B. Schryvers. *Mol Cell* **2009**, 35, 523–533.

88. M. K. Yost-Daljev and C. N. Cornelissen. *Infect Immun* **2004**, 72, 1774–1785.

89. C. N. Cornelissen, J. E. Anderson, I. C. Boulton, P. F. Sparling. *Infect Immun* **2000**, 68, 4725–4735.

90. H. P. Masri and C. N. Cornelissen. *Infect Immun* **2002**, 70, 732–740.

91. S. M. C. Newton, J. S. Allen, Z. Cao, Z. Qi, X. Jiang, C. Sprencel, J. D. Igo, S. B. Foster, M. A. Payne, P. E. Klebba. *Proc Natl Acad Sci U S A* **1997**, 94, 4560–4565.

92. M. D. Retzer, R.-H. Yu, Y. Zhang, G. C. Gonzalez, A. B. Schryvers. *Microb Pathog* **1998**, 25, 175–180.

93. G. Renauld-Mongenie, M. Latour, D. Poncet, S. Naville, M.-J. Quentin-Millet. *FEMS Microbiol Lett* **1998**, 169, 171–177.

94. I. C. Boulton, A. R. Gorringe, J. K. Shergill, C. L. Joannou, R. W. Evans. *J Theor Biol* **1999**, 198, 497–505.

95. A. J. DeRocco and C. N. Cornelissen. *Infect Immun* **2007**, 75, 3220–3232.

96. J. M. L. Ling, C. H. Shima, D. C. Scheriemer, A. B. Schryvers. *Mol Microbiol* **2010**, 77, 1301–1314.

97. I. C. Boulton, A. R. Gorringe, R. J. G. Carr, B. Gorinsky, C. L. Joannou, R. W. Evans. *FEBS Lett* **1997**, 414, 409–413.

98. S. A. Kretchmar Nguyen, A. Craig, K. N. Raymond. *J Am Chem Soc* **1993**, 115, 6758–6764.

99. P. J. Halbrooks, Q.-Y. He, S. K. Briggs, S. J. Everse, V. C. Smith, R. T. A. MacGillivary, A. B. Mason. *Biochemistry* **2003**, 42, 3701–3707.

100. A. B. Mason, Q.-Y. He, P. J. Halbrooks, S. J. Everse, D. R. Gumerov, I. A. Kaltashov, V. C. Smith, J. Hewitt, R. T. A. MacGillivary. *Biochemistry* **2002**, 41, 9448–9454.

101. T. E. Clarke, L. W. Tari, H. J. Vogel. *Curr Top Med Chem* **2001**, 1, 7–30.

102. A. L. Davidson and J. Chen. *Annu Rev Biochem* **2004**, 73, 241–268.

103. E. Loisel, L. Jacquamet, L. Serre, C. Bauvois, J. L. Ferrer, T. Vernet, A. M. Di Guilmi, C. Durmort. *J Mol Biol* **2008**, 381, 594–606.

104. D. C. Desrosiers, Y. C. Sun, A. A. Zaidi, C. H. Eggers, D. L. Cox, J. D. Radolf. *Mol Microbiol* **2007**, 65, 137–152.

105. E. L. Borths, K. P. Locher, A. T. Lee, D. C. Rees. *Proc Natl Acad Sci U S A* **2002**, 99, 16642–16647.

106. P. M. Jones and A. M. George. *Cell Mol Life Sci* **2004**, 61, 682–699.

107. M. R. Rohrbach, V. Braun, W. Koster. *J Bacteriol* **1995**, 177, 7186–7193.

108. A. Mademidis, H. Killmann, W. Kraas, I. Flechsler, G. Jung, V. Braun. *Mol Microbiol* **1997**, 26, 1109–1123.

109. K. D. Krewulak, C. M. Shepherd, H. J. Vogel. *BioMetals* **2005**, 18, 375–386.

110. K. P. Locher, A. T. Lee, D. C. Rees. *Science* **2002**, 296, 1091–1098.
111. J. H. Crosa, A. R. Mey, S. M. Payne. *Iron Transport in Bacteria*. ASM Press, Washington, DC, 2004.
112. M. T. Sebulsky, B. H. Shilton, C. D. Speziali, D. E. Heinrichs. *J Biol Chem* **2003**, 278, 49890–49900.
113. J. D. Faraldo-Gomez and M. S. P. Sansom. *Nat Rev Mol Cell Biol* **2003**, 4, 105–116.
114. M. Miethke and M. A. Marahiel. *Microbiol Mol Biol Rev* **2007**, 71, 413–451.
115. J. M. Roosenberg, Y. M. Lin, Y. Lu, M. J. Miller. *Curr Med Chem* **2000**, 7, 159–197.
116. A. Stintzi, C. Barnes, J. Xu, K. N. Raymond. *Proc Natl Acad Sci U S A* **2000**, 97, 10691–10696.
117. F. Peuckert, M. Miethke, A. G. Albrecht, L.-O. Essen, M. A. Marahiel. *Angew Chem Int Ed Engl* **2009**, 48, 7924–7927.
118. K. N. Raymond, E. A. Dertz, S. S. Kim. *Proc Natl Acad Sci U S A* **2003**, 100, 3584–3588.
119. J. B. Neilands, T. J. Erickson, W. H. Rastetter. *J Biol Chem* **1981**, 26, 3831–3832.
120. P. Thulasiraman, S. M. C. Newton, J. D. Xu, K. N. Raymond, C. Mai, A. Hall, M. A. Montague, P. E. Klebba. *J Bacteriol* **1998**, 180, 6689–6696.
121. Z. H. Cao, Z. B. Qi, C. Sprencel, S. M. C. Newton, P. E. Klebba. *Mol Microbiol* **2000**, 37, 1306–1317.
122. A. D. Ferguson and J. Deisenhofer. *Biochim Biophys Acta* **2002**, 1565, 318–332.
123. A. D. Ferguson, E. Hofman, J. W. Coulton, K. Diederichs, W. Welte. *Science* **1998**, 282, 2215–2220.
124. W. W. Yue, S. Grizot, S. K. Buchanan. *J Mol Biol* **2003**, 332, 353–368.
125. C. L. Carswell, M. D. Rigden, J. E. Baenziger. *J Bacteriol* **2008**, 190, 5650–5662.
126. I. J. Schalk. *J Inorg Biochem* **2008**, 102, 1159–1169.
127. A.-M. Albrecht-Gary, S. Blanc, N. Rochel, A. Z. Ocaktan, M. A. Abdallah. *Inorg Chem* **1994**, 33, 6391–6402.
128. E. Clement, P. J. Mesini, F. Pattus, I. J. Schalk. *Biochemistry* **2004**, 43, 7954–7965.
129. I. J. Schalk, M. A. Abdallah, F. Pattus. *Biochem Soc Trans* **2002**, 30, 702–705.
130. I. J. Schalk, M. A. Abdallah, F. Pattus. *Biochemistry* **2002**, 41, 1663–1671.
131. I. J. Schalk, C. Hennard, C. Dugave, K. Poole, M. A. Abdallah, F. Pattus. *Mol Microbiol* **2001**, 39, 351–360.
132. D. Cobessi, H. Celia, N. Folschweiller, I. J. Schalk, M. A. Abdallah, F. Pattus. *J Mol Biol* **2005**, 347, 121–134.
133. C. Wirth, W. Meyer-Klaucke, F. Pattus, D. Cobessi. *J Mol Biol* **2007**, 368, 398–408.
134. N. Folschweiller, J. Gallay, M. Vincent, M. A. Abdallah, F. Pattus, I. J. Schalk. *Biochemistry* **2002**, 41, 14591–14601.
135. E. Wasielewski, D.-L. Tzou, B. Dillmann, J. Czaplicki, M. A. Abdallah, R. A. Atkinson, B. Kieffer. *Biochemistry* **2008**, 47, 3397–3406.
136. V. Schons, R. A. Atkinson, C. Dugave, R. Graff, G. L. A. Mislin, L. Rochet, C. Hennard, B. Kieffer, M. A. Abdallah, I. J. Schalk. *Biochemistry* **2005**, 44, 14069–11079.

137. I. J. Schalk, I. L. Lamont, D. Cobessi. *BioMetals* **2009**, 22, 671–678.

138. J. Greenwald, M. Nader, H. Celia, C. Gruffaz, V. A. Geoffroy, J.-M. Meyer, I. J. Schalk, F. Pattus. *Mol Microbiol* **2009**, 72, 1246–1259.

139. A. Braud, F. Hoegy, K. Jezequel, T. Lebeau, I. J. Schalk. *Environ Microbiol* **2009**, 11, 1079–1091.

140. K. J. James, M. A. Hancock, V. Moreau, F. Molina, J. W. Coulton. *Protein Sci* **2008**, 17, 1679–1688.

141. V. Braun. *Arch Microbiol* **1997**, 167, 325–331.

142. M. Oke, R. Sarra, R. Ghirlando, S. Farnaud, A. Gorringe, R. W. Evans, S. K. Buchanan. *FEBS Lett* **2004**, 564, 294–300.

143. G. Ramakrishnan, A. Meeker, B. Dragulev. *J Bacteriol* **2008**, 190, 5353–5361

144. H. Adams, G. Zeder-Lutz, I. J. Schalk, F. Pattus, H. Celia. *J Bacteriol* **2006**, 188, 5752–5761.

145. M. Nader, W. Dobbelaere, M. Vincent, L. Journett, H. Adams, D. Cobessi, J. Gallay, I. J. Schalk. *Biochemistry* **2007**, 46, 11707–11717.

146. D. D. Shultis, M. D. Purdy, C. N. Banchs, M. C. Wiener. *Science* **2006**, 312, 1396–1399.

147. J. Gumbart, M. C. Wiener, E. Tajkhorshid. *Biophys J* **2007**, 93, 496–504.

148. K. C. Usher, E. Ozkan, K. H. Gardner, J. Deisenhofer. *Proc Natl Acad Sci U S A* **2001**, 98, 10676–10681.

149. W. Kaserer, X. Jiang, X. Qiaboin, D. C. Scott, M. Bauler, D. Copeland, S. M. C. Newton, P. E. Klebba. *J Bacteriol* **2008**, 190, 4001–4016.

150. E. Udho, K. S. Jakes, S. K. Buchanan, K. J. James, X. Jiang, P. E. Klebba, A. Finkelstein. *Proc Natl Acad Sci U S A* **2009**, 106, 21990–21995.

151. K. J. James, M. A. Hancock, J.-N. Gagnon, J. W. Coulton. *Biochemistry* **2009**, 48, 9212–9220.

152. S. M. Newton, V. Trinh, H. Pi, P. E. Klebba. *J Biol Chem* **2010**, 285, 17488–17497.

153. N. Chin, J. Frey, C.-F. Chang, Y.-F. Chang. *FEMS Microbiol Lett* **1996**, 143, 1–6.

154. E. Saken, A. Rakin, J. Heesemann. *Int J Med Microbiol* **2000**, 290, 51–60.

155. S. Gong, S. W. Bearden, V. A. Geoffroy, J. D. Fetherston, R. D. Perry. *Infect Immun* **2001**, 69, 2829–2837.

156. P. Adhikari, S. A. Berish, A. J. Nowalk, K. L. Veraldi, S. A. Morse, T. A. Mietzner. *J Bacteriol* **1996**, 178, 2145–2149.

157. D. E. McRee, C. M. Bruns, P. A. Williams, T. A. Mietzner, R. Nunn. 1999, Cambridge PDB code: ID9Y.

158. C. H. Taboy, K. G. Vaughan, T. A. Mietzner, P. Aisen, A. L. Crumbliss. *J Biol Chem* **2001**, 276, 2719–2724.

159. P. L. Roulhac, K. D. Powell, S. Dhungana, K. D. Weaver, T. A. Mietzner, A. L. Crumbliss, M. C. Fitzgerald. *Biochemistry* **2004**, 43, 15767–15774.

160. J. J. Heymann, K. D. Weaver, T. A. Mietzner, A. L. Crumbliss. *J Am Chem Soc* **2007**, 129, 9704–9712.

161. W. W. Ho, H. Li, S. Eakanumkul, Y. Tong, A. Wilks, M. Guo, T. L. Poulos. *J Biol Chem* **2007**, 282, 35796–35802.

162. H. Boukhalfa, D. S. Anderson, T. A. Mietzner, A. L. Crumbliss. *J Biol Inorg Chem* **2003**, 8, 881–892.

163. S. J. Ferguson. *Prokaryotic Structure and Function: A New Perspective*, S. Mohan, C. Dow, J. A. Coles, eds. Cambridge University Press, Cambridge, 1991.
164. A. D. Ferguson and J. Deisenhofer. *Cell* **2004**, 116, 15–24.
165. F. A. Quiocho. *Curr Opin Struct Biol* **1991**, 1, 922–933.
166. C. M. Blinn, E. R. Dibbs, L. J. J. Hronowski, P. S. Vokonas, J. E. Silbert. *Arthritis Rheum* **2005**, 52, 2808–2813.
167. L. Politi, R. Chiaraluce, V. Consalvi, N. Cerulli, R. Scandurra. *Clin Chim Acta* **1989**, 184, 155–165.
168. L. A. Patterson and P. M. C. DeBlieux. Hyperphosphatemia. In emedicine.com. http://emedicine.medscape.com/article/767010-overview. 2007.
169. P. Bruscella, L. Cassagnaud, J. Ratouchniak, G. Brasseur, E. Lojou, R. Amilis, V. Bonnefoy. *Microbiology* **2005**, 151, 1421–1431.
170. J. C. Wilks and J. L. Slonczewski. *J Bacteriol* **2007**, 189, 5601–5607.
171. D. O. O'Keefe and R. J. Collier. *Proc Natl Acad Sci U S A* **1989**, 86, 343–346.
172. J. B. Stock, B. Rauch, S. Rosaman. *J Biol Chem* **1977**, 252, 7850–7861.
173. J. J. Heymann, M. Gabricevic, T. A. Mietzner, A. L. Crumbliss. *J Biol Inorg Chem* **2010**, 15, 237–248.
174. A. Khan, S. R. Shouldice, L. W. Tari, A. B. Schryvers. *Biochem J* **2006**, 403, 43–48.
175. J. W. Pflugrath and F. A. Quiocho. *Nature* **1985**, 314, 257–260.
176. S. R. Shouldice, D. R. Dougan, R. J. Skene, L. W. Tari, D. E. McRee, R.-H. Yu, A. B. Schryvers. *J Biol Chem* **2003**, 278, 11513–11519.
177. S. Dhungana, D. S. Anderson, T. A. Mietzner, A. L. Crumbliss. *Biochemistry* **2005**, 44, 9606–9618.
178. H. Luecke and F. A. Quiocho. *Nature* **1990**, 347, 402–406.
179. P. Adhikari, K. D. Weaver, T. A. Mietzner. **2010**, personal communication.
180. J. J. He and F. A. Quiocho. *Protein Sci* **1993**, 2, 1643–1647.
181. J. J. Heymann and A. L. Crumbliss, unpublished results.
182. M. Gabricevic and A. L. Crumbliss. *Inorg Chem* **2003**, 42, 4098–4101.
183. M. Gabricevic, D. S. Anderson, T. A. Mietzner, A. L. Crumbliss. *Biochemistry* **2004**, 43, 5811–5819.
184. K. D. Weaver, M. Gabricevic, D. S. Anderson, P. Adhikari, T. A. Mietzner, A. L. Crumbliss. *Biochemistry* **2010**, 49, 6021–6032.
185. Y. Zhang and P. S. Cremer. *Curr Opin Chem Biol* **2006**, 10, 658–663.
186. A. W. Omta, M. F. Kropman, S. Woutersen, H. J. Bakker. *Science* **2003**, 301, 347–349.
187. W. Kunz, P. L. Nostro, B. W. Ninham. *Curr Opin Colloid Interface Sci* **2004**, 9, 1–37.
188. H. Sun, M. C. Cox, H. Li, P. J. Sadler. *Metal Sites in Proteins and Models*, H. A. O. Hill, P. J. Sadler, A. J. Thomsom, eds. Springer-Verlag, Berlin and Heidelberg, Germany, 1999; pp. 71–102.
189. R. Kumar and A. G. Mauk. *J Phys Chem B* **2009**, 113, 12400–12409.
190. D. H. Hamilton, I. Turcot, A. Stintzi, K. N. Raymond. *J Biol Inorg Chem* **2004**, 9, 936–944.

191. S. L. Byrne, A. N. Steere, N. D. Chasteen, A. B. Mason. *Biochemistry* **2010**, 49, 4200–4207.

192. C. P. Thompson, J. K. Grady, N. D. Chasteen. *J Biol Chem* **1986**, 261, 13128–13134.

193. J. M. Harrington and A. L. Crumbliss. *BioMetals* **2009**, 22, 679–689.

CHAPTER 14

DNA–LIPID AMPHIPHILES FOR DRUG AND GENE THERAPY

PEGGY P.Y. CHAN and LISHAN WANG

14.1 INTRODUCTION

Gene therapy is defined as the introduction of exogenous genetic materials including both DNA and RNA into a target tissue with the aim of providing therapeutic benefit. Gene therapy holds great promise in treating diseases ranging from inherited disorders, to cancer, to acquired conditions. The genetic material being introduced has encoded genetic sequence functions to replace defective genes. Gene expression occurs by synthesizing functional gene products using the newly introduced genetic material. The introduced genetic material can also function to substitute missing genes, silence unwanted gene expression, or introduce new cellular biofunctions.[1,2] Gene therapy is often hampered by the need to overcome both the extracellular and the intracellular barriers. Before the gene can reach the target cells, it is subject to removal from opsonins, phagocytes, and degradative enzymes. Once it has penetrated through extracellular matrices and reaches the target cells, it is often limited by poor penetration into the cell membrane, lack of recognition characteristics necessary for intracellular transport, degradation within lysosomal compartments, and inability to release from transport vesicles. An efficient delivery carrier is therefore required to protect the genetic materials and facilitate cellular uptake.[2] Gene delivery carriers can be classified as viral and nonviral. In this chapter, the use of viral and nonviral gene delivery carriers will be discussed, with a focus on nonviral cationic liposome gene carriers. This chapter will also discuss on cationic liposomes' therapeutic applications, their structural–functional relationship, preparation methods, characterization methods, toxicity effects, challenges in delivery, their cellular pathways, and performance evaluation.

Ionic Interactions in Natural and Synthetic Macromolecules, First Edition. Edited by Alberto Ciferri and Angelo Perico.

14.2 VIRAL AND NONVIRAL GENE CARRIERS

Viruses are known for their ability to penetrate cells to deliver their genome. Viruses have therefore been considered as a gene delivery carrier to deliver foreign genes into cells. Adenoviral vectors are the most commonly employed viral carrier, as they can infect a wide range of mammalian cells with high transfection efficiency. The measure of a gene carrier's performance in terms of transfection is referred to as transfection efficiency. Viral carriers can carry DNA with size up to 7.5 kb and transduce these transgenes into nonproliferating cells.[3] These carriers can be manipulated using recombinant DNA techniques that are well-established protocols. However, the drawback of using viral carriers is that they have low packaging capacity, they are difficult and expensive to manufacture, they raise safety issues such as provoking carcinogenesis, and they can induce an immune response that abolishes transgenic expression.[4,5]

Numerous studies have attempted to develop nonviral, synthetic carriers to mimic the efficient gene delivery capacity of viral carriers. Cationic polymers, cationic lipids, cationic peptides, and recombinant protein are the major types of nonviral carriers. Although nonviral carriers yield lower gene expression compared with viral carriers, the use of nonviral carriers are attractive due to their low toxicity and low immunogenicity, lack of pathogenicity, and ease of pharmacological production.[6] Cationic polymers can be classified into two groups: natural polymers and synthetic polymers. Cationic polymers are composed of protonatable amine residues that enable them to spontaneously associate with negatively charged nucleic acid via electrostatic interactions.[7] Among various gene delivery approaches, cationic liposomal gene delivery was found to be rather effective and safe[8]; the following chapter will therefore discuss the use of cationic lipids for gene therapy.

14.2.1 Cationic Liposomes for DNA Delivery

Negatively charged polyelectrolytes form supramolecular (ion pair) associations with low-molecular-weight cationic surfactants, or lipids, that have single or double chain aliphatic tails. The association between a single anionic site on the polyelectrolyte and a single cationic group of the amphiphile occurs when the latter is molecularly dispersed or when it forms intramolecular suprastructures. Chapter 8 describes the association of DNA types of varying molecular length and rigidity with surfactants, either molecularly dispersed or forming micellar-like structures. Perico and Ciferri[9] described DNA as a rigid, rodlike molecule for molecular lengths corresponding to about 150 base pairs, and as a semirigid molecule for larger contour lengths. It was suggested that the rigidity of short DNA rods interferes with the formation of spherical micelles. However, the rodlike DNA molecules, decorated by the complementary charged surfactants, undergo self-assembly by interdigitation of the hydrophobic tails into a cylindrical symmetry stabilized by hydrophobic interactions. The resulting structure is a hexagonal distribution of DNA rods imbibed in a lattice of less ordered chains (Chapter 8).

In addition to micelles, single- and double-chained amphiphiles are also known to form liposomes, characterized by closed, interdigitated double layers with a distribution of cationic head groups in the external as well as in the internal rim. At variance with spherical micelles, the association of a negatively charged polyelectrolyte can therefore occur also in the internal, water-filled cavity of the vesicle. Cationic liposomes have been extensively used as DNA delivery carriers due to their advantages, including simplicity in preparation, safety in human use, and versatility in compositions.[5] The complexes formed by DNA and cationic liposomes are known as lipoplexes. The introduction of nucleic acids (DNA or RNA) into cells by a nonviral method is referred to as a transfection. If said introduction is performed using a liposome carrier, it is referred as a lipofection, first described by Felgner et al.[10]

Theoretically, there is no limitation on the length of DNA that can be complexed with cationic liposomes.[11] The final size of the lipoplex is determined by the size of the supercoiled DNA plasmid. It is reported that lipoplexes have the potential to transfer DNA with size up to 1 million base pairs.[12] The plasmid DNA can be surrounded by three to five amphiphilic bilayers to form lipoplexes.[13] It is reported that the membrane charge density, σ_M, a measure of average charge per unit area of membrane, can be a key parameter that governs the transfection efficiency of lipoplexes. One of the major problems with cationic liposomal carriers appears to be the tight association of DNA and its inability to dissociate intracellularly.[12] Manipulation the σ_M of lipoplexes can therefore be an important approach to tackle this problem.

14.2.2 Cationic Liposomes for Drug and Small Interfering RNA (siRNA) Delivery

Using the same electrostatic interaction principle, cationic liposomes are also employed as carrier for other negatively charged molecules such as messenger RNA (mRNA), siRNA, synthetic oligonucleotides (ODNs), negatively charged proteins, and peptides.[14] siRNA is a short double-stranded RNA (typically 21-base pair RNA) that can mediate sequence-specific inhibition of gene expression known as gene silencing or RNA interference (RNAi). These siRNA can assemble into RNA-induced silencing complexes (RISCs), which result in cleaving of complementary mRNA. RNAi technology is an emerging field, as RNAi is an important tool for gene function analysis, and has also been evaluated as a therapeutic agent to cure diseases such as cancer, infection, respiratory disease, neuronal disease, and autoimmune disease.[15] Similar to naked DNA, naked siRNA is highly negatively charged and cannot pass through the cytoplasmic membrane on its own; siRNA have a half-time of less than an hour in human plasma.[15] Mammalian cells appear to lack the effective dsRNA uptake mechanism. In addition, the uptake of siRNA by fluid phase endocytosis does not necessary result in the endosomal release of siRNA into

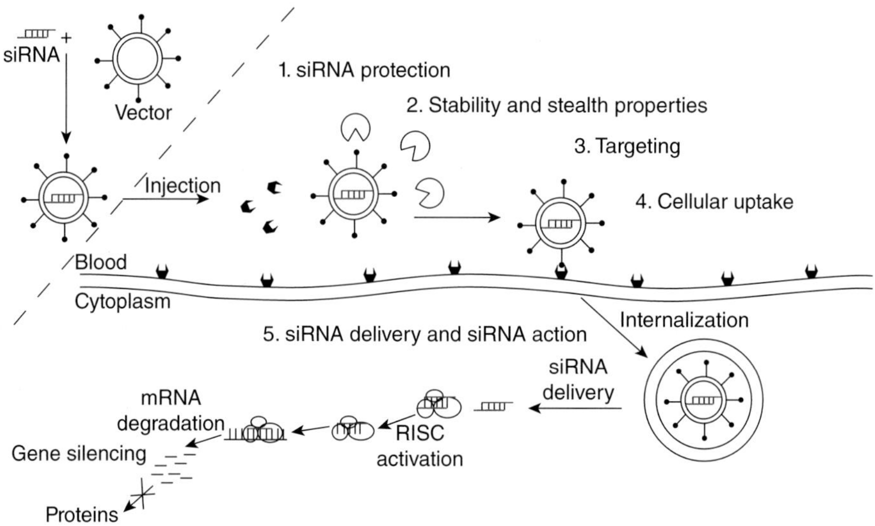

FIGURE 14.1. Illustration of siRNA delivery using a nonviral carrier, which protects siRNA from degradation, facilitates cell uptake and internalization, and releases siRNA into the cytosol. Once the siRNA is free in the cytoplasm, it can assemble into RNA-induced silencing complexes, leading to cleavage of complementary mRNA and knockdown protein expression. Reproduced with permission from David et al.[22] (Copyright 2010 Elsevier)

the cytoplasm. Due to these reasons, the addition of naked siRNA to mammalian cells does not result in effective gene silencing.[16] Therefore, various cationic liposomes have been employed as a carrier for siRNA delivery.[17–20] Issues involving the use of liposomes for siRNA and plasmid DNA delivery are similar, with the exception that the targeted delivery site for siRNA is the cytoplasm, where the targeted site for plasmid DNA is the nucleus.[21] Figure 14.1 summarizes how a nonviral carrier protects siRNA in the extracellular matrix, follows by delivery, and facilitates siRNA internalization into target cells, after which the siRNA is released into the cytosol and activates the gene silencing pathway.[22]

Apart from siRNA, biologically active ODNs are short nucleic acids that are commonly used as gene function examining tools, and have also found therapeutic applications as polynucleotide drugs. Problems relating to administration, delivery, and cellular uptake of ODNs are similar to those of siRNA. The use of polycation conjugates with amphiphilic copolymers for ODN delivery has been attempted to increase dispersion stability of polymer/DNA and reduce serum inhibition. Representative examples include conjugates of hydrophilic nonionic polymers poly(ethylene oxide) (PEO), amphiphilic block copolymers of ethylene oxide (EO) and propylene oxide (PO), copolymer of lipophilic poly(propylene oxide) (PPO), and hydrophilic PEO. These polycation conjugates consist of lipophilic segments and hydrophobic segments that can form micelle-like structures with ODNs in a similar fashion as lipoplex. Cationic copolymers of hydrophilic polyethylenimine (PEI) and amphiphilic Pluronic (BASF, Mount Olive, NJ) have been reported to form stable suspensions of small micelle-like particles with ODNs, and have shown increased cellular uptake and increased sequence-specific activity of antisense ODNs.[23]

Liposomes have been extensively studied as drug delivery systems, and macromolecules including protein and anticancer agents (e.g., doxorubicin, donaumycin) can be encapsulated in liposomes by hydrophobic interaction. Wang et al.[24] demonstrated that gene expression can be enhanced by the synergistic effect of drug and gene delivery both *in vitro* and *in vivo*, where attempts have been made to codeliver drug and DNA using a biodegradable amphiphilic copolymer containing cholesterol side chains to form liposome-like polymeric nanoparticles.

14.3 CHEMICAL STRUCTURE OF CATIONIC AMPHIPHILES

A prototypical cationic lipid amphiphile molecule contains three basic domains: a positively charged polar head and a hydrophobic part connect through a linker. Figure 14.2 shows a typical structure of an amphiphile DOTAP representation.[25] The hydrophobic moiety of DOTAP consists of two unsaturated diacyl side chains, the linker consists of an ester group, and the positively charged group consists of a propylammonium group.[25]

14.3.1 Hydrophobic Chain

The hydrophobic part of cationic lipid amphiphiles generally consists of either aliphatic chains or cholesterol-based derivatives.[26] Traditionally, it is believed that a cationic lipid with a single aliphatic chain has lower transfection

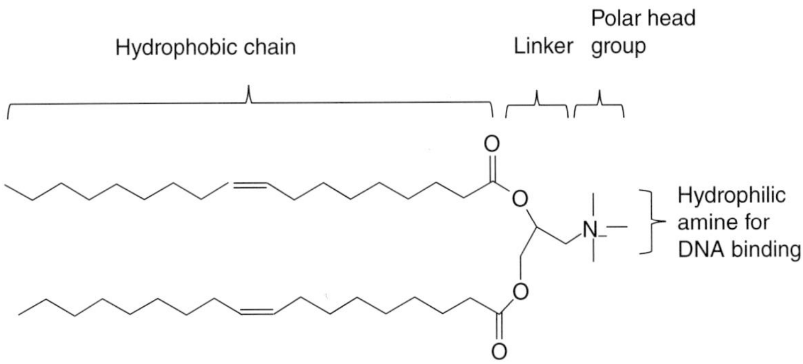

FIGURE 14.2. Schematic representation of a cationic amphiphile DOTAP. A typical cationic amphiphile, DOTAP consists of a hydrophobic part and a polar head group anchored by a linker. Adapted from El-Aneed.[25]

efficiency compared with a cationic lipid with double chains. In additional, single-chain cationic lipids such as CTAB are less efficient and induce higher cytotoxicity compared with their double-chain analogs, such as DOTMA[15]. However, some studies demonstrated that single-chain LHON is more efficient and less cytotoxic compared with double-chain DOTAP. There are two major types of double chains: unsaturated and saturated chains. Representative examples of unsaturated double chains are DOTMA, DOSPA, DOTAP, and DOSPER. For saturated double chains, representative examples are DOGS, DDAB, TM-TPS, and DMRIE. Both single and double aliphatic chains can form stable liposome by itself, but the incorporation of helper lipids in the formulation is not uncommon.[27] It was shown by Felgner et al.[28] that by decreasing the carbon number of the acyl chain from 18 to 14, the phase transition temperature (T_m) of lipids can be lowered. A cationic lipid that exhibits T_m below 37°C can result in higher transfection.[28–30] The presence of the double bond has a similar effect as acyl chain shortening and gives rise to a less rigid bilayer.[31] Obika and coworkers[32,33] have attempted the synthesis and utilization of cationic triglycerides for lipofection, and it was found that the transfection activity of symmetrical cationic triglycerides is comparable with asymmetrical cationic triglycerides.

The cholesterol-derived cationic lipids generally cannot form stable bilayers on their own. To overcome this, these lipids can be formulated with helper lipids such as neutral lipids.[27] Neutral lipids such as DOPE, cholesterol, and DOPC are the three most commonly used helper lipids. It is believed that the incorporation of helper lipids facilitates lipid membrane destabilization and allows higher transfection activity. For instance, DOPE is known to destabilize the endosome membrane and facilitate endosomal escape of lipoplexes.[8] It is reported that antisense ODNs carried by the DC-CHOL/DOPE formulation were preferentially transfected into target cells' nuclei.[8] Despite the need for helper lipids, the use of cholesterol-based hydrophobic moiety is still attractive due to its role in membrane fusion during lipofection. The effect of lipid hydrophobic moiety on cytotoxicity has not been fully understood, though it is postulated that their cytotoxicity effect is dependent on the physiochemical features of the positively charged head group and linker.[26]

14.3.2 Positively Charged Group

The positively charged group of cationic lipid amphiphiles is responsible for the ionic interaction with negatively charged molecules such as peptide and phosphate groups in nucleic acid (including DNA, mRNA, siRNA, and ODNs).[14] Typically, amine group substitution on the lipid chain is carried out via methylation.[31] These positively charged groups generally consist of primary, secondary, and tertiary amines, or quaternary ammonium salts with varying degrees of substitution. Polyvalent cationic lipid amphiphiles have also been prepared using polyamine to improve DNA binding and transfection efficiency; some well-known examples are DOGS and DOSPA[8]. Guanidine- and imidazole-containing groups have also been used as the positively charged group on cationic lipid amphiphiles. The cytotoxicity of cationic lipid amphiphiles is mainly caused by the positively charged head group; it is also known that tertiary amine is less toxic compared with quaternary amine. The cytotoxicity can be reduced significantly by substituting the linear amine head group by a heterocyclic ring such as pyridinium and guanidine, which spreads the positive charge of the cationic group.[26]

14.3.3 Linker

The linker of cationic lipid amphiphiles can be composed of amide, ester, carbamate, ether, or a phosphate bond. Besides functioning as an anchor for both the hydrophobic group and the positively charged group, the linker group also has some influence on stability, biodegradability, and transfection activity. It is known that the substitution of amide or urea linkages for the carbamate group can yield lower transfection activity due to the reduced chemical stability.[31] Cationic lipid amphiphiles with an ether linker tend to give better transfection activity compared with those with an ester or amide bond.[26] The use of ester linkage enables fast hydrolysis after transfection, which reduces cytotoxicity of cationic lipid amphiphiles.[32] Generally, cationic lipid amphiphiles that are composed of ether linkages are more toxic compared with those that are composed of biodegradable linkages, such as ester, amide, and carbamoyl bonds. The use of a linker segment with a length of 3–6 atoms can optimize transfection activity[34] as well as lower cytotoxicity, given that the linkage is degradable.[26] Figure 14.3 shows some other commonly used cationic amphiphiles for nucleic acid delivery.

14.3.4 Helper Lipids

Neutral lipids such as DOPE, DOPC, and cholesterol are often used in synergy with cationic lipid amphiphiles as helper lipids. Figure 14.4 shows some commonly used neutral lipids. Cationic liposome formulated from equimolar amounts of neutral lipid and cationic lipid has been reported to give high gene transfection. The incorporation of neutral lipids can also decrease toxicity of the liposome formulation, which contributes to higher transfection levels.[26] The choice of helper lipids has a strong influence on the activity of liposomes. For example, cholesterol as a helper lipid can enable the formation of more stable liposomes with cationic lipids but results in less efficient transfection compared with a DOPE-containing formulation.[35] The improved gene transfection may partly result from reducing the DNase I susceptibility of liposomes in the presence of helper lipids. Neutral lipids such as DOPE can facilitate membrane fusion and aid the destabilization of endosomes, while DOPC does not destabilize lipid bilayers.[26] Some studies have reported that DOPE can also facilitate the disassembling of lipopolexes to enable DNA escape from endocytotic vesicles, which contributes to higher transfection levels.[35]

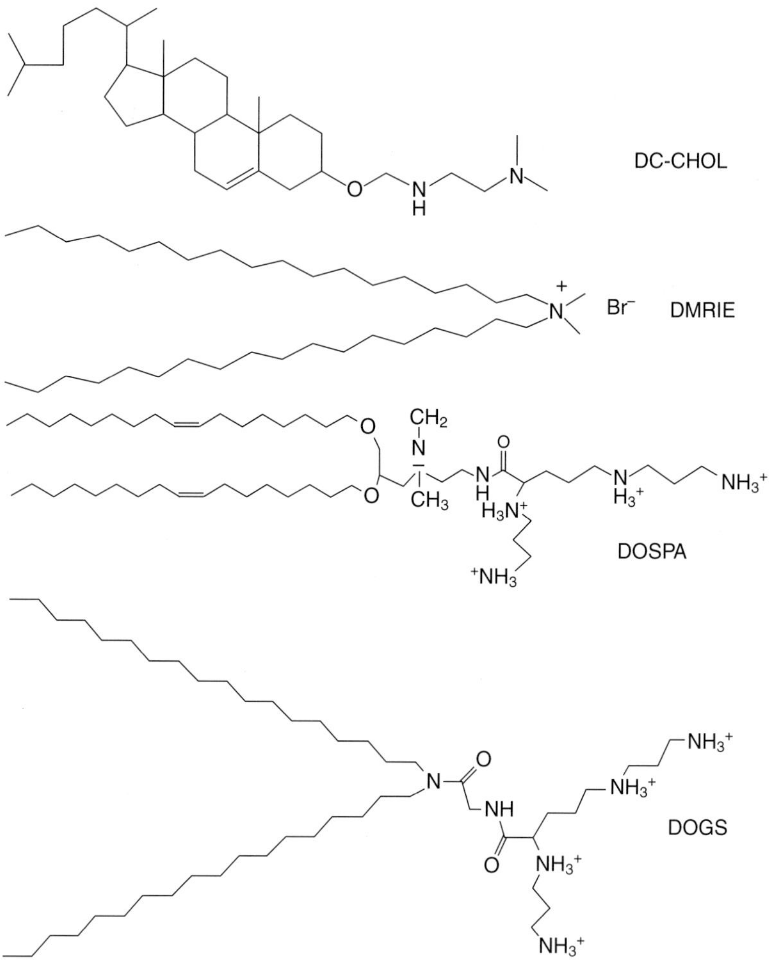

FIGURE 14.3. Structures of some other commonly used cationic amphiphiles for nucleic acid delivery: DC-CHOL, DMRIE, DOSPA, and DOGS.[12,27,34]

14.4 COMPLEXATION

In gene therapy, plasmid DNA is the most commonly employed nucleic acid due to its advantages compared with retroviral vectors. Plasmid DNA is an autonomous, extrachromosomal, circular DNA that exhibits two topological forms: supercoiled and nicked relaxed (open circular). Plasmid preparation is simple, quick, safe, and inexpensive; plasmid DNA can be modified to contain a promoter and the gene coding for the protein of interest. Plasmids can be expressed in the transfected cell nucleus in a transient or stable (in rare events) manner.[36]

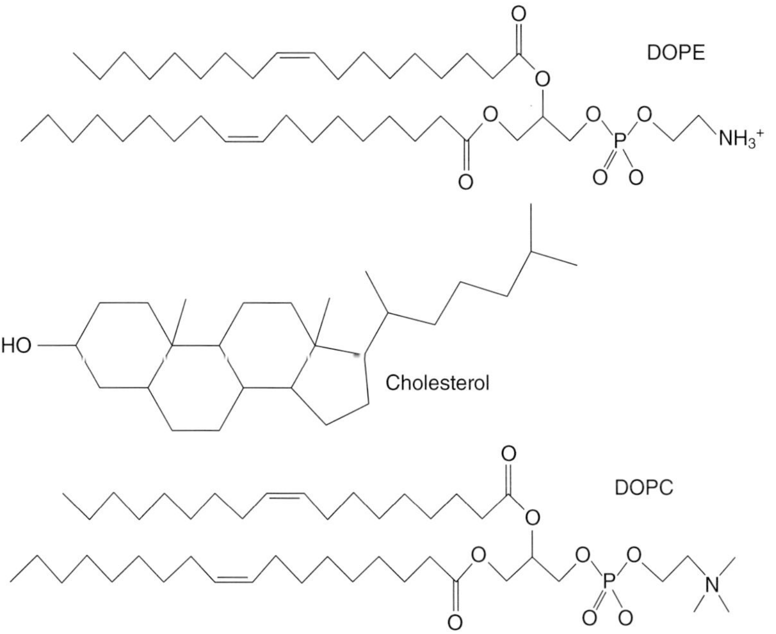

FIGURE 14.4. Structures of some commonly used neutral lipids: DOPE, cholesterol, and DOPC.[12,96]

When the solutions of cationic lipid amphiphiles and DNA are combined, they can spontaneously self-assemble to form complexes via electrostatic interaction. These complexes can be bilayered lipoposomal structures, depending on the lipid compositions; some lipoplexes form multilayered structures consisting of plasmids sandwiched between cationic lipid.[7] The driving force of this process comes from the release of counterions that were tightly bound to the cationic lipid and negatively charged DNA[37]. The complex formation involves a three-step mechanism. The DNA plasmid first interacts with the lipid amphiphile monolayer, then unwraps itself into lamellae with lipid amphiphile molecules. At the end, the wrapped DNA assemble with lipid amphiphiles to form a complex.[11] The resulting complex known as lipoplex can be anionic, neutral, or cationic depending on the charge ratio (ρ_{chg}) of cationic lipid to DNA. For lipofection, a positively charged ($\rho_{chg} > 1$) is generally preferred, as the lipoplex can be attached on the negatively charged cell surface proteoglycans by electrostatic attractions.[37]

The mixing method[18,38–41] and lipid film hydration method[42–45] are the most commonly used methods for the preparation of lipoplex for nucleic acid delivery. Figure 14.5 illustrates the two lipoplex preparation methods using siRNA as the representative nucleic acid.[46]

The mixing method involves simply mixing solutions of plasmid DNA and cationic liposomes under serum-free conditions. The lipid film hydration

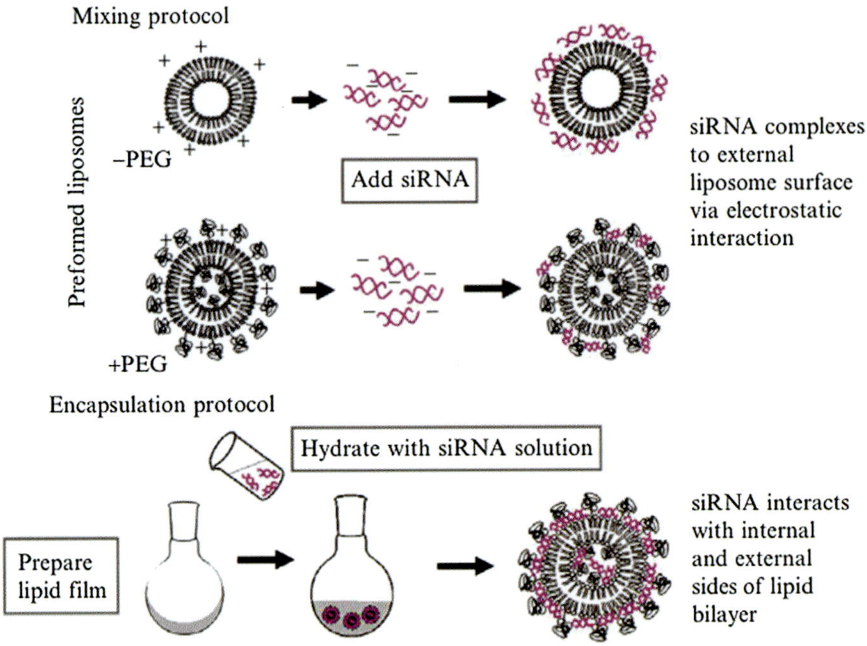

FIGURE 14.5. Preparation of lipoplex for nucleic acid (e.g., siRNA) delivery using (a) the mixing method and (b) the lipid film hydration method. Reproduced with permission from Podesta and Kostarelos.[46] (Copyright 2009 Elsevier)

method involves dissolving the lipid in a volatile solvent such as chloroform, and evaporating the solvent to obtain a thin film, which will then be hydrated to obtain crude liposomes. The resulting liposomes are heterogeneous in size and can be mechanically fragmentated or extruded to obtain liposomes of a more uniform size. Next, a plasmid DNA solution will be added dropwise to the aqueous liposome to form a complex.[47] Both of the methods are amenable to cationic lipid and/or neutral helper lipid substitutions. Some studies have simplified the lipid film hydration procedure; for example, Podesta and Kostarelos[46] reported a preparation method where the lipid film is hydrated directly with an siRNA solution, followed by extrusion. Regardless of the method being employed, it should be noted that the complexation between liposomes and nucleic acids may result in localized precipitation. To overcome this, the nucleic acid should be added to the liposomes with rapid mixing, not vice versa.

The following protocols for preparation of lipoplexes are adapted from Whitemore et al.[47] and Girao da Cruz et al.[38] The lipoplexes should be prepared using sterile solutions under aseptic conditions; these complexes should be used while fresh to avoid aggregation. All solutions and glassware used should be DNase-free or RNase-free, whichever is applicable. DNase-free water can be prepared by autoclaving deionized water. RNase-free dH_2O can be prepared by treating the water with 0.1% (v/v) diethylpyrocarbonate (DEPC) overnight, followed by autoclaving the solution at 121°C for a minimum of 20 minutes.[46] The protocol is amenable for liposome/siRNA complexation.

Reagents and Solutions

1. *5× Dextrose Stock Solution:* Dissolve 2.6 g dextrose in 10.0 mL sterile water; sterile filtrate the solution through a 0.2-μm membrane.
2. *Cholesterol Stock Solution 20.0 mg mL^{-1} (51.7 mM):* Dissolve 100.0 mg cholesterol into 5.0 mL chloroform solvent.
3. *DOTAP Stock Solution (20 mg mL^{-1}):* Dissolve 20 mg DOTAP in 1 mL chloroform solvent.
4. *Dextrose 5.2% Buffer Solution:* Dissolve 2.6 g dextrose in 50 mL sterile water; sterile filtrate the solution through a 0.2-μm membrane.

Preparation of DOTAP/DNA Complexes Using the Mixing Method

1. Prepare DOTAP liposome stock solution by adding 1.0 mL DOTAP stock solution to 0.55 mL cholesterol stock solution; follow by swirl mixing.
2. Prepare liposome/polycation solution by adding 30.0 μL of 5× dextrose, 43.0 μL DOTAP liposomes, 3.0 μL protamine sulphate, and 74.0 μL sterile water into a 50-mL sterile conical tube; mix the solution by swirling.
3. Prepare DNA solution by adding 50.0 μL plasmid DNA, 30.0 μL of 5× dextrose, and 70.0 μL sterile water to a sterile microcentrifuge tube; follow by tapping the tube to mix.
4. Add DNA solution dropwise to a gently swirling liposome/polycation solution.
5. Incubate the complexes at room temperature for 10–15 minutes. Use the complexes immediately.

Preparation of Cationic Lipid/DNA Complexes Using the Lipid Film Hydration Method

1. Rinse a 30.0-mL Corex glass centrifuge tube three times with chloroform.
2. Mix 1.0 mL DOTAP stock solution to 0.55 mL cholesterol stock solution into the Corex tube; follow by swirl mixing.
3. Dry the chloroform from DOTAP/cholesterol in a chloroform mixture under a vacuum using a rotator evaporator. Alternatively, blow-dry the chloroform using N_2 gas to form a thin lipid film on the glass while rotating the tube by hand.
4. Dry the lipid film completely in vacuum desiccators for another 2–3 hours.

5. Hydrate the lipid film to suspension by adding 2.0 mL of 5.2% dextrose buffer solution to the film and vortex several times.
6. Incubate the suspension for 2–3 hours at room temperature.
7. To disperse lipid aggregates, the resulting lipid suspension can be bath sonicated, vortexed, or extruded through a 100-nm polycarbonate filter hereafter, until lipids are entirely resuspended.
8. The resulting liposomes can be diluted with distilled water, and the phospholipid concentration can be determined by a phosphate assay.
9. Add predetermined volume of DNA solution to the liposomes to obtain desired lipid/DNA charge ratio.
10. Incubate the complexes for 15 minutes at room temperature. Use lipolexes immediately after being prepared.

14.5 CHARACTERIZATION OF CATIONIC LIPID CARRIER

14.5.1 Agarose Gel Electrophoresis

The ability of a cationic lipid amphiphile to bind DNA is usually evaluated using agarose gel electrophoresis, followed by staining with nucleic acid stain. Agarose gel electrophoresis is a method used to separate DNA or RNA molecules by size. The agarose gel electrophoresis protocol can be divided into three steps: (1) prepare agarose gel with concentration appropriate for the size of DNA to be separated; (2) load the lipoplexes and controls (DNA molecular ladder, naked DNA) into the sample wells and run the gel at a voltage and time period to obtain optimal separation; and (3) stain the gel by a nucleic acid stain and follow by visualization using UV light illumination. Standard protocols for running agarose gel electrophoresis can be found in Voytas[48] and Armstrong and Schulz.[49] The commonly used nucleic acid stains are ethidium bromide, or fluorescent cyanine DNA dyes such as SYBR Green I and SYBR Gold. The use of SYBR Green I dye is gaining popularity due to several reasons: Its concentration-dependent fluorescence intensity is identical for the nicked-relaxed form and supercoiled form DNA, it has higher sensitivity for both double-strand and single-strand DNA compared with ethidium bromide,[50] and because ethidium bromide is more mutagenic.[51,52]

Using the agarose gel electrophoresis technique, the mobility of the DNA in the presence of various formulations is monitored and compared with the control (naked DNA).[53] This assay usually cannot provide quantitative information, but qualitative conclusions can usually be drawn from the results. Figure 14.6 illustrates the separation of DNA on an agarose gel. The separation is achieved by using an electric field to move negatively charged nucleic acid through the agarose matrix. Lane 1 represents a DNA ladder that is a molecular weight size marker that provides a scale for size estimation. The molecular weight of DNA is inversely proportional to its migration rate, as smaller molecules can move faster than longer molecules. Lane 2 represents naked plasmid

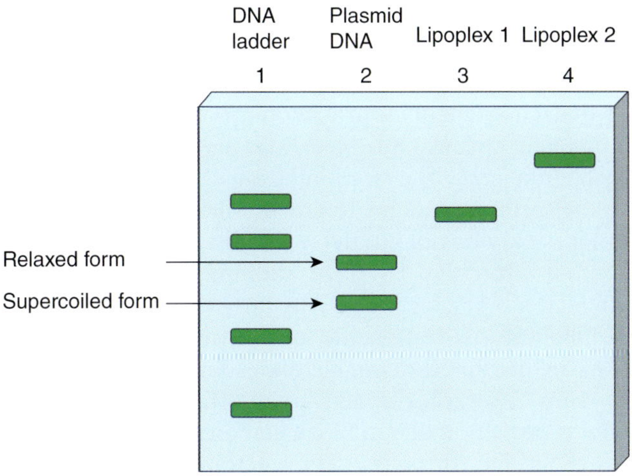

FIGURE 14.6. Illustration of DNA separation on agarose gel.

DNA control; two DNA bands can usually be found, one corresponds to the relaxed form and the other corresponds to the supercoiled form. The supercoiled form can migrate further through the gel due to its complex structure compared with the relaxed form. Lane 3 represents a formulation of lipoplex; the change in mobility relative to the control is an indication of complexation. Lane 4 represents a different formulation of lipoplex; the DNA band is retarded at the origin, indicating that the corresponding cationic lipid amphiphiles can form stable complexes with DNA.

Lipoplex stability generally increases with the number of charges in the positively charged amine group. Cationic lipid amphiphiles composed of more hydrophobic bile acid skeletons can retard DNA mobility more efficiently due to their ability to form more stable complexes compared with their hydrophilic counterparts.[53] Even-Chen and Barenholz[50] reported that cationic liposomes preferentially bind to plasmid DNA in nicked-relaxed form over the supercoiled form, as the binding of supercoiled DNA is weaker and easier to dissociate from the lipoplex. The negative charges of the nicked-relaxed DNA are more readily available and can induce changes in positive charge distribution during complexation.

14.5.2 Size and Surface Charge of Lipoplex

The size of a lipoplex is an important parameter that affects transfection efficiency, as it determines the cell internalization pathway via endocytosis, which will be discussed further in a later section. The type of liposomes employed, the charge ratio, incubation time in polyanion-containing media, and the presence of serum can all influence the size of lipoplex, and thus the transfection efficiency. The lipoplex size and heterogeneity usually increase with increasing lipid-to-DNA ratio. Lipofection of lipoplexes with size less than 250 nm is known to be inhibited by the presence of serum. Lipoplexes of size larger than 700 nm are known to induce efficient transfection in the presence or absence of serum. For this reason, many transfection protocols in the literature employ a serum-free reagent such as Opti-MEM (GIBCO, Invitrogen, Carlsbad, CA) for gene transfection.[18,54–57] The size inhibition effect of serum is suggested to contribute to the serum resistance to ethanol-dried lipid-DNA (EDL). Large cationic lipoplexes may be able to delay the dissociation of DNA from lipids, and can therefore overcome EDL. In addition, it is postulated that larger liposomes can increase phagocytic activity due to their maximum contact with cells. Upon cellular internalization, the formation of large intracellular vesicles can be more easily disrupted to enable endosomal escape of DNA into cytoplasm. The *in vitro* criteria for effective gene transfection may not be applicable to *in vivo* transfection due to bioavailability problems; lipoplexes of size 200–400 nm are optimal for *in vitro* transfection, while smaller lipoplexes of size 40–90 nm are better for *in vivo* transfection.[58]

Another important parameter that determines the transfection efficiency is the surface charge of the lipoplex.[59] The surface charge of lipoplex increases as the ratio of cationic lipid to DNA increases, and higher charge ratios are generally more toxic to a variety of cells. Some nonspecific efficacious issues are due to the toxicity of lipoplex and should be taken into consideration during lipoplex design.[60] Lipoplexes that carry different surface charges display different structures: positively charged lipoplex exhibits aggregated multilamellar structure where DNA is encapsulated within the lipid coat, while negatively charged lipoplex exhibits free plasmids or protruding DNA-strings structure where DNA is bound to the exterior of the complex.[58]

The surface charge often determines the biodistribution of lipoplex *in vivo*. Chonn et al.[61] reported that surface charge is the key determinant of complement-activating liposomes. Complement activation by negatively charged liposomes containing phosphatidylglycerol, phosphatidic acid, cardiolipin, phosphatidylinositol or phosphatidylserine occurred via the Ca^{2+}-dependent classical pathway. Neutral liposomes such as PC:cholesterol (CHOL) and PC:CHOL:DPPE are less likely to trigger complement activation. Complement activation triggered by positively charged liposomes containing stearylamine occurred via an alternative pathway.

Experimentally, the particle diameter (z-average diameter) of lipoplexes and the surface charge (ζ potential) of lipoplexes can be measured using photon correlation spectroscopy (PCS)/dynamic light scattering and laser Doppler electrophoresis (LDE), respectively. PCS is a technique that measures the time-dependent fluctuation in light scattering intensity caused by particles in suspension that undergo Brownian motion. LDE is a technique that measures the frequency shift caused by electrophoretic mobility of particles in suspension under the influence of an electric field.[62] A polydispersity index can be calculated at the same time; this parameter gives an indication of the homogeneity of the liposome population.[46]

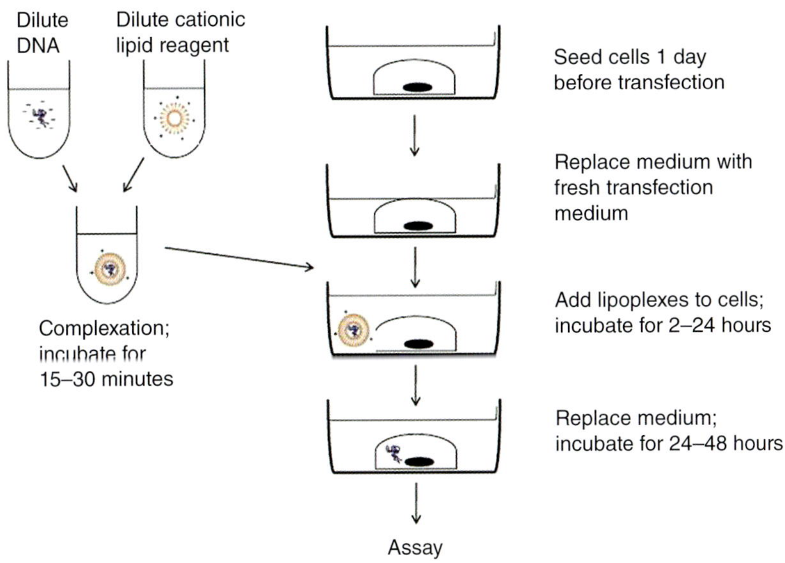

FIGURE 14.7. Diagram demonstrating lipoplex *in vitro* transfection protocol adapted from Hawley-Nelson et al.[66]

14.5.3 Transfection of Mammalian Cells *In Vitro*

Many studies performed *in vitro* transfection using cultured mammalian cells in order to study the performance and optimize the lipoplex formulation prior to *in vivo* transfection using animals.[63–65] This section describes the basic protocol of how to transfect adherent mammalian cells (primary cells or cell lines) with lipoplexes (Fig. 14.7).

The following protocol for the preparation of lipoplexes is an adaptation from Hawley-Nelson et al.[66] It should be noted that the preparation and transfection experiments should be carried out using sterile solutions under aseptic conditions with the aid of a biosafety cabinet. All mammalian cell cultures should be incubated at 37°C in a humidified 5% carbon dioxide incubator. Table 14.1 shows the recommended working volumes for lipoplex transfection in culture well plate of different sizes. The protocol is amenable for liposome/siRNA transfection.

1. Trypsinize and count adherent cells 1 day prior to transfection. Seed cells into a sterile well plate with complete cell culture medium (with serum) to obtain 50–80% confluent on the transfection day.
2. On the day of transfection, dilute DNA into dilution medium in a polystyrene or polypropylene tube. Dilute cationic lipid reagent into dilution medium. Mix diluted DNA and diluted cationic lipid reagent, and allow mixture to incubate for 15–30 minutes at room temperature.

TABLE 14.1. Suggested Working Volumes for Lipoplex Transfection for 48–72 Hours Study (Adapted from Hawley-Nelson et al.[66])

Well Plate	Seeding Density (cells per well)	DNA (μg)	Dilution Medium (μL)	Cationic Lipid Reagent (μL)	Transfection Medium (μL)
96 wells	$0.5–1 \times 10^4$	0.05–0.4	10–25	0.075–1.5	0.08–0.1
48 wells	$1.5–2 \times 10^4$	0.1–0.8	12–37	0.25–5	0.1–0.2
24 wells	$3–4 \times 10^4$	0.2–1.6	25–50	0.5–10	0.2–0.5
12 wells	$6–8 \times 10^4$	0.4–3.2	50–100	1–20	0.4–1
6 wells	$15–16 \times 10^4$	1–8	100–250	2.5–50	0.8–2.5

3. While complexation is taking place, carefully replace culture medium with fresh transfection medium without scratching adherent cells. Avoid the use of serum and antibiotics during transfection.
4. Add lipoplex into each well. Incubate for 2–4 hours at 37°C in 5% CO_2 incubator.
5. Remove medium, add fresh cell culture medium (with serum), and allow cells to cultivate for another 24–48 hours before assay.

The cell morphology during transfection interval should be monitored, especially when cells are maintained in a serum-free medium during lipoplex treatment. Some cells may lose viability under these conditions; in such cases, the lipoplex/serum-free medium transfection time should be reduced. The transfected cells should generally be harvested 24–72 hours post transfection for gene expression or protein expression analysis. To optimize transfection efficiency, one can vary the charge ratio of cationic lipid to nucleic acid; the amount of DNA or RNA added to each well can also be varied.

14.6 EXTRACELLULAR BARRIERS

The use of cationic liposomes to deliver nucleic acid can increase drug availability due to the prevention of nucleic acid degradation. However, these nucleic acid carriers are limited by extracellular barriers such as the opsonization process, in which opsonin proteins present in the blood serum often bind quickly to conventional nanoparticles; these opsonin-binded nanoparticles can be recognized by macrophages.[67,68] For this reason, the administration route should be carefully selected; for instance, these carriers have limited applicability for intravenous administration routes,[69] as they are subjected to clearance by the reticuloendothelial system (RES), a complex system that consists of macrophages, dendritic reticular cells, lymphocytes, and plasma cells that contribute to pathogenic microorganism resistance.[70] About 90% of any immunological recognized particles will be taken up by the liver macrophage within 5 minutes, 1–5% of these particulates will be taken up by spleen macrophage, and the rest may be taken up by macrophage population in lung and bone marrow. Larger particles are taken up by the RES faster than small particles. Particles with higher charge are taken up by the RES faster than lower or noncharged particles. Particles that display higher surface hydrophobicity tend to enhance phagocytosis due to their hydrophobic interactions with phagocytes. In addition, the surface hydrophobicity also affects the amount and composition of apolipoproteins adsorbed to the particles and leads to binding with apolipoprotein receptors.[71]

The most common approach to minimizing RES uptake is to modify the lipid by polyethylene glycol (PEG), also known as pegylation, to reduce the opsonization of liposomes. Pegylation has also been employed in various nanoparticle designs to improve circulation half-time and enhance drug delivery.[72] Nonpegylated liposomes often encounter phagocyte system uptake and undergo rapid clearance by the RES compared with pegylated liposomes. Uptake by the RES results in irreversible sequestering of liposomes and can induce toxicity or acute impairment of the phagocyte system in some cases.[73]

The effectiveness of a drug or gene carrier can be assessed by measuring the *in vivo* biodistribution of the liposomes. Usually, it can be carried out by intravenously injecting animals using a lipid and/or DNA marker. The animals are sacrificed and blood samples are withdrawn from the heart at different time intervals post injection. Organs including the liver, spleen, lung, kidney, and heart will be excised and the amount of marker can be quantified according to a calibration curve of the administered particles, after subtracting the background measurement of untreated animals.[74]

14.7 INTRACELLULAR BARRIER: CELL ADHESION, INTERNALIZATION, AND INTRACELLULAR TRAFFICKING

14.7.1 Cell Adhesion

Whether a lipoplex can be transfected into cells is determined by several critical factors in addition to the nucleic acid's own innate biological activity. Once overcoming extracellular barriers, the lipoplex is required to overcome a few intracellular barriers before the encapsulated gene can be expressed. The first intracellular barrier is the cell membrane (Fig. 14.8; see also Chapter 13). The first step of successful lipofection involves the binding of lipoplex to negatively charged membrane-associated proteins such as sialyated glycoproteins or proteoglycans, mainly via electrostatic interactions. It is therefore possible to enhance cell adhesion by manipulating the charge ratio of cationic lipid to nucleic acid. Following cell adhesion, the internalization of lipoplex occurs through the fusion of liposome and plasma membrane.[31] Although the mechanisms involved are still under debate, it is generally accepted that endocytosis and endocytosis-like mechanisms are the main internalization pathways of lipoplex; these mechanisms will be discussed in the following sections.

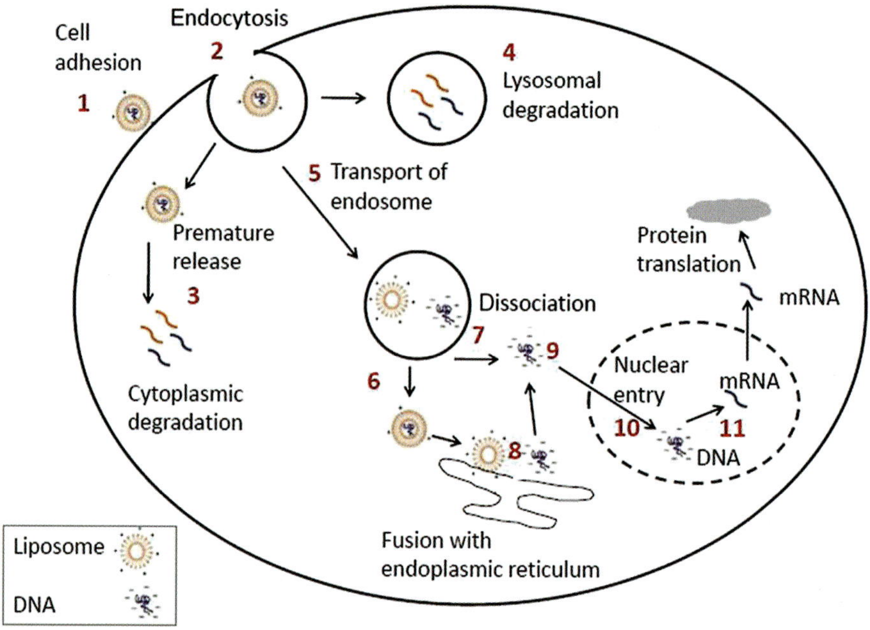

FIGURE 14.8. Summary of intracellular barriers involved in lipofection: (1) the lipoplex needs to bind to the membrane surface for cell adhesion; (2) the lipoplex may internalize inside the cell by endocytosis; (3) if the lipoplex can be degraded by cytoplasmic nucleases before reaching the nucleus if it is released too early from endosomes; (4) the lipoplex may be degraded in lysosomes if it is unable to escape from endosomes; (5) cytoplasmic transport of endosomes can bring the lipoplex near the perinuclear region; (6) the lipoplex can escape from endosomes if osmotic rupture of endosomes occur; (7) the DNA may dissociate from liposomes during endosomal escape; (8) the DNA may dissociate from liposomes in the cytoplasm; (9) the plasmid may diffuse through the cytoplasm; (10) the DNA may translocate to the nucleus if nuclear entry is successful; and (11) successful gene expression will lead to DNA transcription to mRNA followed by protein translation. Diagram adapted from Barron and Szoka[31] and Elouahabi and Ruysschaert.[77]

14.7.2 Cell Internalization

Once the lipoplex has successfully adhered to the cell membrane, the lipoplex has to overcome the next intracellular barrier, where it will have to be internalized into the cell cytosol in order for it to reach the cell nucleus. Phagocytosis and pinocytosis are two of the major endocytosis mechanisms that have been described as possible pathways of internalization of lipoplex. Phagocytosis is a specific cell surface receptor and signal cascade-mediated process that is highly regulated. Phagocytosis occurs only in specialized mammalian cells including macrophages, monocytes, and neutrophils. The major function of phagocytosis is to remove large pathogens or debris. The phagocytosis is activated by antibodies bound to surface antigens on pathogen cells such as bacteria; next, a signal cascade triggers actin assembly and the formation of cell surface extensions that enclose and engulf antibody bond pathogens.[75]

Pinocytosis, also known as fluid-phase uptake, can occur in many cell types, and there are five morphologically distinct pinocytosis mechanisms: macropinocytosis, clathrin-mediated endocytosis, caveolae-mediated endocytosis, lipid raft-mediated endocytosis, and clathrin- and caveolae-independent endocytosis. The cargo molecule and its receptor determine which pinocytic pathway they will use to internalize into the cell.

Macropinocytosis is a membrane ruffling process that is generally considered to be a nonreceptor-mediated process, where cells internalize a large volume of extracellular fluids. In this process, an actin-driven formation of membrane protrusions is triggered by a signal cascade stimulated by growth factors or other signals. Unlike phagocytosis, these protrusions collapse onto ligand-coated particles and fuse with the plasma membrane to generate endocytic vesicles known as macropinosomes. Macropinosomes do not fuse into lysosomes, so particles that are internalized via this pathway can therefore avoid lysosomal degradation, and macropinosome escape occurs due to their leaky nature.[75,76]

Clathrin-mediated endocytosis occurs in all mammalian cells and functions to continuously uptake essential nutrients, antigens, growth factors, and pathogens. Clathrin-mediated endocytosis was previously known as receptor-mediated endocytosis, and plays a crucial role in intercellular communication during tissue development. Clathrin-coated vesicles with size ranging from 100 to 150 nm in diameter can carry concentrated receptor–ligand complexes into cells. Clathrin-mediated endocytosis is also involved in regulating the internalization of membrane pumps and membrane transport of ions and small molecules, which play an important role in cell and serum homeostasis. Molecules internalized through clathrin-mediated endocytosis experience a pH change from neutral to 5.9 or 6.0 in the lumen of early endosomes, followed by transition to pH 5 from late endosomes to lysosomes.[76]

Caveolae are flask-shaped invaginations of the plasma membrane that are present on many cells and constitute 10–20% of the cell surface. After activation, caveolae are slowly internalized to form small vesicles 50–60 nm in diameter. Caveolin, a dimeric protein that binds cholesterol in caveolae, can self-associate on the surface of membrane invagination to form a striated caveolin coat. It is postulated that caveolae may be involved in intracellular cholesterol trafficking and intracellular cholesterol homeostasis. Caveolae-mediated endocytosis is nonacidic and nondigestive, and is receptor-mediated. It could be advantageous if it is utilized as a drug delivery route; however, this internalization route has a low capacity uptake, which can only internalize the fluid phase with small volume.[75]

Lipid rafts are small, cholesterol-free and sphingolipid-rich planar microdomains of cell surfaces that lack caveolin and caveolae. These 40–60-nm-diameter lipid rafts are found in lymphocytes, many human cancer cells, and rodent

TABLE 14.2. Multiple Endocytic Pathways with Regard to Size of the Endocytic Vesicle (Adapted from Prokop and Davidson[76])

Triggering Mechanism	Vehicle Size	Entry Ports
Receptor mediated, signal cascade mediated	1–10 μm	Phagocytosis
Growth factors, other signals	1–5 μm	Macropinocytosis
Receptor mediated	120–150 nm	Clathrin-mediated endocytosis
Receptor mediated	<60 nm	Caveolae-mediated endocytosis
Hydrophobic interaction, scavenger receptor mediated	40–60 nm	Lipid rafts
Unclear	20–300 nm	Clathrin- and caveolin-independent endocytosis

macrophages. Anionic, neutral, and hydrophobic particles can be internalized into cells via lipid raft-mediated endocytosis. Lipid raft-mediated endocytosis includes hydrophobic uptake via lipid membrane fusion, and uptake of lipophilic and anionic groups via scavenger receptor-mediated membrane fusion. Unlike caveolae-mediated endocytosis, lipid raft-mediated endocytosis lacks an energy requirement and the uptake process bypasses the lysosomal compartment, so it is therefore a potential route for gene delivery.[76]

Apart from caveolae, there are other cholesterol-rich microdomains 40–50 nm in diameter, referred to as rafts, that diffuse freely on the cell surface. Clathrin- and caveolae-independent endocytosis mechanisms are poorly understood. It is only known that these raft structures have unique lipid compositions that are specific for membrane protein and glycolipid sorting.[75] Table 14.2 summarizes the different mechanisms for entry of nanovehicular drugs/carriers.

14.7.3 Intracellular Trafficking

Once lipoplexes internalize into cells, they need to overcome another intracellular barrier, where they need to escape from endocytic vesicles and diffuse through the cytoplasm before they can import into the nucleus. If the endosome release of lipoplex occurs too early, the nucleic acid will be degraded by cytoplasmic nucleases before reaching the nucleus. Cytoplasmic transport of endosomes can determine the success of lipoplex; endosomal release near the perinuclear region provides a better chance for the nucleic acid to enter the nucleus.[77]

Endosomal escape of lipoplex can occur via three major mechanisms: (1) endosomal membrane destabilization; (2) cationic lipid and anionic phospholipid exchange that occurs mostly on the cytoplasm-facing monolayer of cell membrane; and (3) endosomolysis induced by osmotic swelling.[78]

Dissociation of nucleic acid may occur during endosomal escape or after fusion of released lipoplexes with the endoplasmic reticulum (ER), where nucleic acid will separate from the carrier. Lipoplexes that fail to escape from endosomes will end up being degraded by the lysosome. The DNA may translocate into the nucleus during mitosis depending on the cell division activity and the plasmid lifetime. The nuclear translocation efficiency of plasmid DNA, determined as the nuclear fraction divided by the fraction of plasmid DNA that escaped from the endosome,[79] is estimated to be 1/1000. If the DNA is encoded with specific sequences that mediate its interaction with the transcription factor, the DNA–protein complex can be translocated into the nucleus via a nuclear localization signal-mediated transport mechanism. It is also postulated that the lipoplex-enclosed endosomes may fuse with the nuclear membrane and facilitate the direct entry of DNA into the nucleus.[77] The cell binding, internalization processes, and intracellular barrier presence are summarized in Figure 14.8.[31,77]

14.8 CYTOTOXICITY OF LIPOSOMES

Cationic lipids can cause *in vivo* and *in vitro* toxic effects; for example, they may induce cell shrinking, reduced number of mitoses, and vacuolization. Their toxicity effect is often dose-dependent.[26] In order for gene therapy to be successful, a cationic lipid amphiphile with low cytotoxicity is required. The cytotoxicity of cationic lipid or liposomes can be evaluated by measuring the cell viability. Traditionally, cell viability is measured using thiazolyl blue formazan (MTT) assay.[80,81] More recently, other viability assays such as tetrazolium compound [3-(4,5-dimethylthiazol-2-yl)-5-(3-carboxymethoxyphenyl)-2-(4-sulfophenyl)-2H-tetrazolium] reduction assay (MTS) and Alamar Blue™ (Trek Diagnostic Systems, Cleveland, OH) have gained popularity due to their simplicity of use.[38,82–87] Unlike MTT, Alamar Blue assay is nontoxic and is less likely to interfere with normal metabolic activity of cells, and can therefore give more reliable measurements.[88] Alamar Blue assay can be reduced by $FMNH_2$, $FADH_2$, NADH, DADPH, and cytochromes, whereas MTT assay cannot be reduced by cytochromes. The reduction in Alamar Blue causes a color change that is detectable by both UV absorbance spectrometer and fluorometer. The fluorescence is measured at 530–560 nm excitation wavelength and 590 nm emission wavelength,[89,90] whereas the absorbance is determined at 570 and 600 nm. The following section describes the general procedure for determining the viability of transfected mammalian cells using the Alamar Blue assay, adapted from Chan et al.[82]

1. Seed cells into a sterile 96-well plate with complete cell culture medium (with serum) to obtain 50–80% confluent on the transfection day; perform lipofection using the transfection protocol described above.
2. Incubate cells at 37°C in 5% CO_2 incubator for 3 days.
3. Remove spent medium. Add 250 μL of 10% Alamar Blue (in complete medium) to each well.
4. Return well plate to incubator for 3–4 hours; measure the absorbance using a microplate reader.

The cell viability can be calculated as a percentage of Alamar Blue reduction using nontransfected cells as the control group, according to the equation,[62]:

$$\% \text{ Cell viability} = \left(\frac{A_{570} - A_{600}}{A_{570,\text{control}} - A_{600,\text{control}}} \right) \times 100,$$

14.9 DETERMINATION OF TRANSFECTION EFFICIENCY

Various experiments can be performed to assess the transfection efficiency of cationic liposome carriers. The transfection efficiency is a measure of target gene or protein expression compared with the gene or protein expression result from that of a control. Usually, naked DNA/siRNA transfected cells are employed as negative control, and cells that are transfected by DNA/siRNA complex with a commercial carrier are employed as positive control. The commercial carrier Lipofectamine™2000 (mixture of DOSPA : DOPE) (Invitrogen, Carlsbad, CA) is the most commonly used positive control for both DNA and siRNA delivery.[91] A cationic liposome carrier is considered to be efficient if its resulting gene expression or protein expression is higher than that given by the controls.

14.9.1 RNA Expression

Post transfection, the targeted gene products may require isolation of RNA or protein prior to analysis. RNA expression level can be quantified by assaying mRNA using reverse transcription polymerase chain reaction (RT-PCR). For quantification by RT-PCR, the total RNA is first extracted and isolated from cell lysates. Next, a reverse transcriptase will be employed to synthesize complementary DNA (cDNA) from the isolated total RNA at a temperature range of 42–60°C. RNA extraction and isolation and cDNA conversion can be performed with the aid of commercial kits such as the RNeasy Mini Kit (Qiagen, Valencia, CA) and SuperScript™ III First-Strand Synthesis System for RT-PCR (Invitrogen). The cDNA will be amplified by PCR and quantified using a thermocycler. The target gene expression level will be normalized to the amount of total RNA by comparing with an endogenous reference gene expression level.[62]

14.9.2 Protein Expression Quantified by Western Blot

Protein expression can be quantified using enzymatic assays or immunoassays such as Western blotting. For Western blot, protein extracts are obtained from transfected cells, and a sodium dodecyl sulfate–polyacrylamide gel electrophoresis (SDS-PAGE) will then be performed. After electrophoresis, the separated protein bands will be transferred to a polyvinylidene fluoride (PVDF) membrane, and the protein of interest will be detected using the immunodetection method. The target protein's band intensity is normalized against reference protein expression to correct for variation in total protein content, where β-actin is the most commonly used reference protein. The transfection efficiency is usually expressed as percentage of protein intensity with respect to nontransfected controls.[62]

14.9.3 Flow Cytometry and Confocal Microscopy Studies

For analysis of lipoplex internalization or intracellular distribution, it is often useful to perform the transfection using fluorescently labeled DNA or siRNA, or fluorescently labeled liposomes. Post transfection, fluorescent cells are first washed with phosphate buffered saline (PBS), trypsinized, and fixed by a reagent such as paraformaldehyde. The intracellular trafficking of the lipoplexes can then be visualized by laser confocal scanning microscopy (LCSM).[92] Figure 14.9 shows an example of intracellular trafficking of lipoplexes being visualized using LCSM; the plasma membrane of target cell was stained with PKH67 (green fluorescent dye), and the lipoplexes were labeled with red fluorescent dye, so the cellular uptake and nuclear entry of lipoplex can be observed.[93]

For DNA delivery, it is useful to employ a plasmid that encodes reporter genes, such as green fluorescence protein (GFP) or luciferase, which allows the transfection efficiency to be measured using simple assays. For siRNA delivery, it may be useful to employ cell lines that stably express reporter gene GFP and luciferase, so that the transfection efficiency and the degree of gene knockdown can be measured easily.[62] Post transfection, the GFP expressing

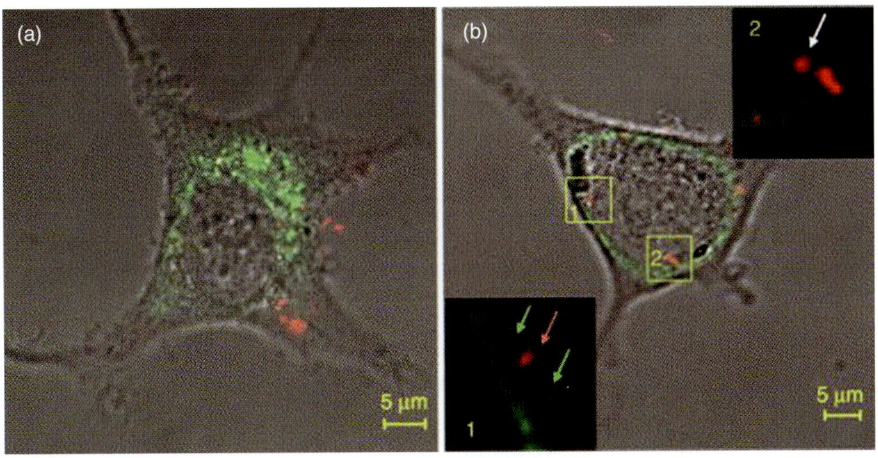

FIGURE 14.9. Imaging of (a) cellular uptake and (b) intracellular distribution of lipoplexes by confocal laser scanning microscopy. Plasma membranes were stained with PKH67 (green fluorescent); the labeled lipoplexes are detected as red clusters The white and red arrowheads (in inset 1 and 2) indicate intracellular uptake of lipoplex; the green arrowhead represents plasma membrane signals. Reproduced with permission from Iwasa et al.[93] (Copyright 2006 Elsevier)

cells will be washed with PBS, trypsinized, resuspended in PBS, and can be quantified by flow cytometry.[92] Flow cytometry is an important technique that simultaneously measures and analyzes multiple cell characteristics including size, quantity, granularity, and relative fluorescence intensity as cells flow through the laser beam detector. The principle of this method is to determine the different characteristics of cells by measuring how the cell scatters incident laser light and emits fluorescence.[62]

14.9.4 Luciferase Gene Expression

Genetic reporters are commonly used to study transfection efficiency, wherein a reporter gene-encoded plasmid is transfected into mammalian cells using the delivery vehicle, and the reporter protein expression is evaluated. Firefly luciferase is widely used as a reporter as its activity is available immediately upon gene translation, where protein posttranslational processing is not required. Luciferin gene expression can be measured by a luminescence assay. The procedure involves harvest and lyses transfected cells, followed by mixing cell lysate with a luciferin-based reagent. The firefly luciferase expressed by cells will catalyze luciferin oxidation through an electron transition using ATP. With Mg^{2+} as a cosubstrate, light is produced from the formation of product oxyluciferin.[94,95] The luminescence light emission can be quantified using a luminometer, and the intensity is normalized by the amount of total protein content to correct for variation in cell number. The total protein content can be measured using bicinchoninic acid protein assay (BCA).[92]

ABBREVIATIONS

CHOL	cholesterol
CTAB	acetyl trimethylammonium bromide
DC-CHOL	3β-[*N*-(*N*′,*N*′-dimethylaminoethane)-carbamoyl]cholesterol hydrochloride
DDAB	dimethyldioctadecylammonium bromide
DMRIE	2,3-di(myristyloxy)propyl(2-hydroxyethyl)dimethylammonium bromide
DOGS	dioctadecylamidoglycylspermine
DOPC	dioleoylphosphatidylcholine
DOPE	1,2-dioleoyl-*sn*-glycerophosphatidylethanolamine
DOSPA	1,3-dioleoyloxy-*N*-[2-(sperminecarboxamido)ethyl]-*N*,*N*-dimethyl-1-propanaminium trifluoroacetate
DOSPER	1,3-di-oleoyloxy-2-(6-carboxy-spermyl)-propylamid
DOTAP	1,2-dioleoyl-3-trimethylammonium propane
DOTMA	*N*-[1-(2,3-dioleyloxy) propyl]-*N*,*N*,*N*-trimethylammoniumchloride

DPPE	dipalmitoylphosphatidylethanolamine
LHON	6-lauroxyhexyl ornithinate
MTS	[3-(4,5-dimethylthiazol-2-yl)-5-(3-carboxymethoxyphenyl)-2-(4-sulfophenyl)-2H-tctrazolium]
MTT	1-(4,5-dimethylthiazol-2-yl)-3,5-diphenylformazan, or thiazolyl blue formazan
PC	phosphatidylcholine
TM-TPS	*N*,*N*1,*N*2,*N*3 tetramethyltetrapalmitylspermine

REFERENCES

1. M. Raponi and G. Symonds. Gene delivery technology: nonviral and viral vector systems. In *Ex Vivo Cell Therapy*, K. Schindhelm and R. Nordon, eds. Academic Press, San Diego, CA, 1999.
2. Z. Liu, Z. Zhang, C. Zhou, Y. Jiao. Hydrophobic modifications of cationic polymers for gene delivery. *Prog Polym Sci* **2010**, 35, 1144–1162.
3. R. S. Slack and F. D. Miller. Viral vectors for modulating gene expression in neurons. *Curr Opin Neurobiol* **1996**, 6, 576–583.
4. C. H. Evans, E. Gouze, J. N. Gouze, P. D. Robbins, S. C. Ghivizzani. Gene therapeutic approaches—transfer *in vivo*. *Adv Drug Deliv Rev* **2006**, 58, 243–258.
5. W. C. Tseng and L. Huang. Liposome-based gene therapy. *Pharm Sci Technol Today* **1998**, 1 (5), 206–213.
6. L. K. Medina-Kauwe, J. Xie, S. Hamm-Alvarez. Intracellular trafficking of nonviral vectors. *Gene Ther* **2005**, 12, 1734–1751.
7. L. De Laporte, J. C. Rea, L. D. Shea. Design of modular non-viral gene therapy vectors. *Biomaterials* **2006**, 27, 947–954.
8. S. Zhang, Y. Xu, B. Wang, W. Qiao, D. Liu, Z. Li. Cationic compounds used in lipoplexes and polyplexes for gene delivery. *J Control Release* **2004**, 100, 165–180.
9. A. Perico and A. Ciferri. The supramolecular association of polyelectrolytes to complementary charged surfactants and protein assemblies. *Chem Eur J* **2009**, 15, 6312–6320.
10. P. L. Felgner, T. R. Gadek, M. Holm, R. Roman, H. W. Chan, M. Wenz, J. P. Northrop, G. M. Ringold, M. Danielsen. Lipofection: a highly efficient, lipid mediated DNA-transfection procedure. *Proc Natl Acad Sci U S A* **1987**, 84, 7413–7417.
11. A. Cao, D. Briane, R. Coudert. Cationic liposomes as transmembrane carriers of nucleic acids. In *Advances in Planar Lipid Bilayers and Liposomes*, A. L. Liu, ed. Academic Press, California, **2006**; pp. 135–190.
12. K. Ewert, N. L. Slack, A. Ahmad, H. M. Evans, A. J. Lin, C. E. Samuel, C. R. Safinya. Cationic lipid-DNA complexes for gene therapy: understanding the relationship between complex structure and gene delivery pathways at the molecular level. *Curr Med Chem* **2004**, 11, 133–149.
13. V. Oberle, U. Bakowsky, I. S. Zuhorn, D. Hoekstra. Lipoplex formation under equilibrium condition reveals a three-step mechanism. *Biophys J* **2000**, 79, 1447–1454.

14. C. Lonez, M. Vandenbranden, J. M. Ruysschaert. Cationic liposomal lipids: from gene carriers to cell signalling. *Prog Lipid Res* **2008**, 47, 340–347.
15. S. Zhang, B. Zhao, H. Jiang, B. Wang, B. Ma. Cationic lipids and polymers mediated vectors for delivery of siRNA. *J Control Release* **2007**, 123, 1–10.
16. D. B. Rozema and D. L. Lewis. siRNA delivery technologies for mammalian systems. *Targets* **2003**, 2 (6), 253–260.
17. H. K. Kim, E. Davaa, C. S. Myung, J. S. Park. Enhanced siRNA delivery using cationic liposomes with new polyarginine-conjugated PEG-lipid. *Int J Pharm* **2010**, 392, 141–147.
18. M. Mével, N. Kamaly, S. Carmona, M. H. Oliver, M. R. Jorgensen, C. Crowther, F. H. Salazar, P. L. Marion, M. Fujino, Y. Natori, M. Thanou, P. Arbuthnot, J. J. Yaouanc, P. A. Jaffrès, A. D. Mille. DODAG; a versatile new cationic lipid that mediates efficient delivery of pDNA and siRNA. *J Control Release* **2010**, 143, 222–232.
19. S. E. Han, H. Kang, G. Y. Shim, M. S. Suh, S. J. Kim, J. S. Kim, Y. K. Oh. Novel cationic cholesterol derivative-based liposomes for serum-enhanced delivery of siRNA. *Int J Pharm* **2008**, 353 (1–2), 260–269.
20. B. Geusens, J. Lambert, S. C. De Smedt, K. Buyens, N. N. Sanders, M. Van Gele. Ultradeformable cationic liposomes for delivery of small interfering RNA (siRNA) into human primary melanocytes. *J Control Release* **2009**, 133 (3), 214–220.
21. N. M. Rao. Cationic lipid-mediated nucleic acid delivery: beyond cationic. *Chem Phys Lipids* **2010**, 163, 245–252.
22. S. David, B. Pitard, J. P. Benoit, C. Passirani. Non-viral nanosystems for systemic siRNA delivery. *Pharmacol Res* **2010**, 62, 100–114.
23. S. V. Vinogradov, E. V. Batrakova, S. Li, A. V. Kabanov. Mixed polymer micelles of amphiphilic and cationic copolymers for delivery of antisense oligonucleotides. *J Drug Target* **2004**, 12 (8), 517–526.
24. Y. Wang, S. Gao, W. H. Ye, H. S. Yoon, Y. Y. Yang. Co-delivery of drugs and DNA from cationic core-shell nanoparticles self-assembled from a biodegradable copolymer. *Nat Mater* **2006**, 5, 791–796.
25. A. El-Aneed. An overview of current delivery systems in cancer gene therapy. *J Control Release* **2004**, 94, 1–14.
26. H. Lv, S. Zhang, B. Wang, C. Shaohui, J. Yan. Toxicity of cationic lipids and cationic polymers in gene delivery. *J Control Release* **2006**, 114, 100–109.
27. S. Bhattacharya and L. Huang. Cationic liposome-DNA complexes in gene therapy. In *Medical Applications of Liposomes*, D. D. Lasic and D. Papahadjopoulos, eds. Elsevier Science, Netherlands, 1998; pp. 371–394.
28. J. H. Felgner, R. Kumar, C. N. Sridhar, C. J. Wheeler, Y. J. Tsai, R. Border, P. Ramsey, M. Martin, P. L. Felgner. Enhanced gene delivery and mechanisms studies with a novel series of cationic lipid formulations. *J Biol Chem* **1994**, 269, 2550–2561.
29. E. R. Lee, J. Marshall, C. S. Siegel, C. Jiang, N. S. Yew, M. R. Nichols, J. B. Nietupski, R. J. Ziegler, M. B. Lane, K. X. Wang, N. C. Wan, R. K. Scheule, D. J. Harris, A. E. Smith, S. H. Cheng. Detailed analysis of structures and formulations of cationic lipids for efficient gene transfer to lung. *Hum Gene Ther* **1996**, 7, 1701–1717.
30. T. Akao, T. Osaki, J. Mitoma, A. Ito, T. Kunitake. Correlation between physiochemical characteristics of synthetic cationic amphiphiles and their DNA transfection ability. *Bull Chem Soc Jpn* **1991**, 64, 3677–3681.

31. L. G. Barron and F. C. Szoka. The perplexing delivery mechanism of lipoplexes. In *Nonviral Vectors for Gene Delivery*, L. Huang, M. C. Hung, E. Wagner, eds. Elsevier Academic Press, London, 1999.
32. S. Obika, W. Yu, A. Shimoyama, T. Uneda, K. Miyashita, T. Doi, T. Imanishi. A symmetrical and biodegradable cationic lipid synthesis and application for efficient gene transfection. *Bioorg Med Chem Lett* **1997**, 7 (14), 1817–1820.
33. S. Obika, W. Yu, A. Shimoyama, T. Uneda, K. Miyashita, T. Doi, T. Imanishi. symmetrical cationic triglycerides: an efficient synthesis and application to gene transfer. *Bioorg Med Chem* **2001**, 9, 245–254.
34. S. Li, X. Gao, K. Son, F. Sorgi, H. Hofland, L. Huang. DC-Chol lipid system in gene transfer. *J Control Release* **1996**, 39, 373–381.
35. M. C. Pedroso de Lima, S. Simoes, P. Pires, H. Faneca, N. Duzgunes. Cationic lipid-DNA complexes in gene delivery: from biophysics to biological applications. *Adv Drug Deliv Rev* **2001**, 47, 277–294.
36. A. R. Thierry, V. Norris, F. Molina, M. Schmutz. Lipoplex nanostructures reveal a general self-organization of nucleic acids. *Biochim Biophys Acta* **2009**, 1790, 385–394.
37. K. Ewert, H. M. Evans, A. Ahmad, N. L. Slack, A. J. Lin, A. Martin-Herranz, C. R. Safinya. Lipoplex structures and their distinct cellular pathways. *Adv Genet* **2005**, 53, 119–155.
38. T. G. Girao da Cruz, S. Simoes, M. C. Pedroso de Lima. Improving lipoplex-mediated gene transfer into C6 glioma cells and primary neurons. *Exp Neurol* **2004**, 187, 65–75.
39. R. Koynova, Y. S. Tarahovsky, L. Wang, R. C. MacDonald. Lipoplex formulation of superior efficacy exhibits high surface activity and fusogenicity, and readily releases DNA. *Biochim Biophys Acta* **2007**, 1768, 375–386.
40. C. Lavigne and A. R. Thierry. Specific subcellular localization of siRNAs delivered by lipoplex in MCF-7 breast cancer cells. *Biochimie* **2007**, 89, 1245–1251.
41. Y. Zhang, W. Garzon-Rodriguez, M. C. Manning, T. J. Anchordoquy. The use of fluorescence resonance energy transfer to monitor dynamic changes of lipid–DNA interactions during lipoplex formation. *Biochim Biophys Acta* **2003**, 1614, 182–192.
42. M. D. Kearns, Y. N. Patel, M. Savva. Physicochemical characteristics associated with transfection of cationic cholesterol-based gene delivery vectors in the presence of DOPE. *Chem Phys Lipids* **2010**, 163, 755–764.
43. C. N. C. Sobral, M. A. Soto, A. M. Carmona-Ribeiro. Characterization of DODAB/DPPC vesicles. *Chem Phys Lipids* **2008**, 152, 38–45.
44. E. Pupo, A. Padron, E. Santana, J. Sotolongo, D. Quintana, S. Duenas, C. Duarte, M. C. de la Rosa, E. Hardy. Preparation of plasmid DNA-containing liposomes using a high-pressure homogenization–extrusion technique. *J Control Release* **2005**, 104, 379–396.
45. E. Pozharski and R. C. MacDonald. Thermodynamics of cationic lipid-DNA complex formation as studied by isothermal titration calorimetry. *Biophys J* **2002**, 83, 556–565.
46. J. E. Podesta and K. Kostarelos. Engineering cationic liposome: siRNA complexes for *in vitro* and *in vivo* delivery. *Methods Enzymol* **2009**, 464, 343–354.

47. M. Whitmore, S. Li, L. Huang. Liposome vectors for *in vivo* gene delivery. *Curr Protoc Hum Genet* **1999**, Suppl. 20, 12.8.1–12.9.9.

48. D. Voytas. Agarose gel electrophoresis. *Curr Protoc Immunol* **2000**, Suppl. 51, 2.5A.1–2.5A.9.

49. J. A. Armstrong and J. R. Schulz. Agarose gel electrophoresis. *Curr Protoc Essent Lab Tech* **2008**, 7.2.1–7.2.20, Core Publication.

50. S. Even-Chen and Y. Barenholz. DOTAP cationic liposomes prefer relaxed over supercoiled plasmid. *Biochim Biophys Acta* **2000**, 1509, 176–188.

51. V. L. Singer, T. E. Lawlor, S. Yue. Comparison of SYBR green I nucleic acid gel stain mutagenicity and ethidium bromide mutagenicity in the Salmonella/mammalian microsome reverse mutation assay (Ames test). *Mutat Res* **1999**, 439, 37–47.

52. K. I. Kirsanov, E. A. Lesovaya, M. G. Yakubovskaya, G. A. Belitsky. SYBR gold and SYBR green II are not mutagenic in the Ames test. *Mutat Res* **2010**, 699, 1–4.

53. S. Walker, M. J. Sofia, H. R. Axelrod. Chemistry and cellular aspects of cationic facial amphiphiles. *Adv Drug Deliv Rev* **1998**, 30 (1–3), 61–71.

54. S. Resina, S. Abes, J. J. Turner, P. Prevot, A. Travo, P. Clair, M. J. Gait, A. R. Thierry, B. Lebleu. Lipoplex and peptide-based strategies for the delivery of steric-block oligonucleotides. *Int J Pharm* **2007**, 344, 96–102.

55. Y. Maitani, S. Igarashi, M. Sato, Y. Hattori. Cationic liposome (DC-Chol/DOPE = 1:2) and a modified ethanol injection method to prepare liposomes, increased gene expression. *Int J Pharm* **2007**, 342, 33–39.

56. S. Zou, K. Scarfo, M. H. Nantz, J. G. Hecker. Lipid-mediated delivery of RNA is more efficient than delivery of DNA in non-dividing cells. *Int J Pharm* **2010**, 389, 232–243.

57. M. A. Ilies, B. H. Johnson, F. Makori, A. Miller, W. A. Seitz, E. B. Thompson, A. T. Balaban. Pyridinium cationic lipids in gene delivery: an *in vitro* and *in vivo* comparison of transfection efficiency versus a tetraalkylammonium congener. *Arch Biochem Biophys* **2005**, 435, 217–226.

58. B. Ma, S. Zhang, H. Jiang, B. Zhao, H. Lv. Lipoplex morphologies and their influences on transfection efficiency in gene delivery. *J Control Release* **2007**, 123, 184–194.

59. L. Wasungu and D. Hoekstra. Cationic lipids, lipoplexes and intracellular delivery of gene. *J Control Release* **2006**, 116, 255–264.

60. C. R. Dass. Lipoplex-mediated delivery of nucleic acids: factors affecting *in vivo* transfection. *J Mol Med* **2004**, 82, 579–591.

61. A. Chonn, P. R. Cullis, D. V. Devine. The role of surface charge in the activation of the classical and alternative pathways of complement by liposomes. *J Immunol* **1991**, 146 (12), 4234–4241.

62. A. Cardoso, S. Trabulo, J. N. Moreira, N. Duzgunes, M. C. Pedroso de Lima. Targeted lipoplexes for siRNA delivery. *Methods Enzymol* **2009**, 465, 267–287.

63. H. S. Kim, I. H. Song, J. C. Kim, E. J. Kim, D. O. Jang, Y. S. Park. *In vitro* and *in vivo* gene-transferring characteristics of novel cationic lipids, DMKD (*O*,*O*′-dimyristyl-*N*-lysyl aspartate) and DMKE (*O*,*O*’dimyristyl-*N*-lysyl glutamate). *J Control Release* **2006**, 115, 234–241.

64. V. Floch, P. Delepine, C. Guillaume, S. Loisel, S. Chasse, G. Le Bolch, E. Gobin, J. P. Leroy, C. Ferec. Systemic administration of cationic phosphonolipids/DNA complexes and the relationship between formulation and lung transfection efficiency. *Biochim Biophys Acta* **2000**, 1464, 95–103.

65. C. Tros de Ilarduya, M. A. Arangoa, M. J. Moreno-Aliaga, N. Duzgunes. Enhanced gene delivery *in vitro* and *in vivo* by improved transferring-lipoplexes. *Biochim Biophys Acta* **2002**, 1561, 209–221.

66. P. Hawley-Nelson, V. Ciccarone, M. L. Moore. Transfection of cultured eukaryotic cells using cationic lipid reagents. *Curr Protoc Mol Biol* **2008**, Suppl. 81, 9.4.1–9.4.17.

67. D. E. Owens and N. A. Peppas. Opsonization, biodistribution, and pharmacokinetics of polymeric nanoparticles. *Int J Pharm* **2006**, 307, 93–102.

68. D. Liu, Q. Hu, Y. K. Song. Liposome clearance from blood: different animal species have different mechanisms. *Biochim Biophys Acta* **1995**, 1240, 277–284.

69. M. D. Brown, A. G. Schatzlein, I. F. Uchegbu. Gene delivery with synthetic (non viral) carriers. *Int J Pharm* **2001**, 229, 1–21.

70. G. Lazar, E. Husztik, G. Lazar. The role of the reticuloendothelial system in natural immunity. In *Neuroimmune Biology: Nature Immunity*, L. Bertok and D. A. Chow, eds. Elsevier, The Netherlands, 2005.

71. H. Carstensen, R. H. Muller, B. W. Muller. Particle size, surface hydrophobicity and interaction with serum of parenteral fat emulsions and model drug carriers as parameters related to RES uptake. *Clin Nutr* **1992**, 11 (5), 289–297.

72. S. D. Li and L. Huang. Nanoparticles evading the reticuloendothelial system: role of the supported bilayer. *Biochim Biophys Acta Biomembr* **2009**, 1788 (10), 2259–2266.

73. W. C. Zamboni. Liposomal, nanoparticle, and conjugated formulations of anticancer agent. *Clin Cancer Res* **2000**, 11, 8230–8234.

74. H. Eliyahu, A. Joseph, J. P. Schillemans, T. Azzam, A. J. Domb, Y. Barenholz. Characterization and *in vivo* performancc of dextran-spermine polyplexes and DOTAP/cholesterol lipoplexes administered locally and systemically. *Biomaterials* **2007**, 28, 2339–2349.

75. S. D. Conner and S. L. Schmid. Regulated portals of entry into the cell. *Nature* **2003**, 422, 37–44.

76. A. Prokop and J. M. Davidson. Gene delivery into cells and tissues. In *Principles of Tissue Engineering*, R. Lanza, R. Langer, J. Vacanti, eds. Elsevier Academic Press, China, 2007.

77. A. Elouahabi and J. M. Ruysschaert. Formation and intracellular trafficking of lipoplexes and polyplexes. *Mol Ther* **2005**, 11 (3), 336–347.

78. J. Rejman, A. Bragonzi, M. Conese. Role of clathrin and caveolae-mediated endocytosis in gene transfer mediated by lipo- and polyplexes. *Mol Ther* **2005**, 12 (3), 468–474.

79. S. Hama, H. Akita, R. Ito, H. Mizuguchi, T. Hayakawa, H. Harashima. Quantitative comparison of intracellular trafficking and nuclear transcription between adenoviral and lipoplex systems. *Mol Ther* **2006**, 13 (4), 786–794.

80. P. Pires, S. Simoes, S. Nir, R. Gaspar, N. Duzgunes, M. C. Pedroso de Lima. Interaction of cationic liposomes and their DNA complexes with monocytic leukemia cells. *Biochim Biophys Acta* **1999**, 1418, 71–84.

81. F. Sakurai, R. Inoue, Y. Nishino, A. Okuda, O. Matsumoto, T. Taga, F. Yamashita, Y. Takakura, M. Hashida. Effect of DNA/liposome mixing ratio on the physiochemical characteristics, cellular uptake and intracellular trafficking of plasmid DNA/ cationic liposome complexes and subsequent gene expression. *J Control Release* **2000**, 66, 255–269.

82. P. Chan, M. Kurisawa, J. E. Chung, Y. Y. Yang. Synthesis and characterization of chitosan-g-poly(ethylene glycol)-folate as a non-viral carrier for tumor-targeted gene delivery. *Biomaterials* **2007**, 28, 540–549.

83. O. Veiseh, F. M. Kievit, C. Fang, N. Mu, S. Jana, M. C. Leung, H. Mok, R. G. Ellenbogen, J. O. Park, M. Zhang. Chlorotoxin bound magnetic nanovector tailor for cancer cell targeting, imaging, and siRNA delivery. *Biomaterials* **2010**, 31, 8032–8042.

84. L. Garcia, M. Bunuales, N. Duzgunes, C. Tros de Ilarduya. Serum-resistant lipopolyplexes for gene delivery to liver tumour cells. *Eur J Pharm Biopharm* **2007**, 67, 58–66.

85. K. Romoren, B. J. Thu, N. C. Bols, O. Evensen. Transfection efficiency and cytotoxicity of cationic liposomes in salmonid cell lines of hepatocyte and macrophage origin. *Biochim Biophys Acta* **2004**, 1663, 127–134.

86. J. E. Adrian, H. W. M. Morselt, R. Suss, S. Barnert, J. W. Kok, S. A. Asgeirsdottir, M. H. J. Ruiters, G. Molema, J. A. A. M. Kamps. Targeted SAINT-O-Somes for improved intracellular delivery of siRNA and cytotoxic drugs into endothelial cells. *J Control Release* **2010**, 144, 341–349.

87. G. Kramer-Marek, C. Serpa, A. Szurko, M. Widel, A. Sochanik, M. Snietura, P. Kus, R. M. D. Nunes, L. G. Arnaut, A. Ratuszna. Spectroscopic properties and photodynamic effects of new lipophilic porphyrin derivatives: efficacy, localisation and cell death pathways. *J Photochem Photobiol B* **2006**, 84, 1–14.

88. G. Kumar Das, P. P. Y. Chan, A. Teo, J. S. C. Loo, J. M. Anderson, T. Y. Tan. *In vitro* cytotoxicity evaluation of biomedical nanoparticles and their extracts. *J Biomed Mater Res A* **2010**, 93A, 337–346.

89. R. Hamid, Y. Rotshteyn, L. Rabadi, R. Parikh, P. Bullock. Comparison of alamar blue and MTT assays for high through-put screening. *Toxicol In Vitro* **2004**, 18, 703–710.

90. S. A. Back, R. Khan, X. Gan, P. A. Rosenberg, J. J. Volpe. A new alamar blue viability assay to rapidly quantify oligodendrocyte death. *J Neurosci Methods* **1999**, 91, 47–54.

91. W. T. Godbey and A. G. Mikos. Recent progress in gene delivery using non-viral transfer complexes. *J Control Release* **2001**, 72, 115–125.

92. I. Kopatz, J. S. Remy, J. P. Behr. A model for non-viral gene delivery: through syndecan adhesion molecules and powered by actin. *J Gene Med* **2004**, 6, 769–776.

93. A. Iwasa, H. Akita, I. Khalil, K. Kogure, S. Futaki, H. Harashima. Cellular uptake and subsequent intracellular trafficking of R8-liposomes introduced at low temperature. *Biochim Biophys Acta* **2006**, 1758, 713–720.

94. A. J. Palomares, M. A. DeLuca, D. R. Helinski. Firefly luciferase as a reporter enzyme for measuring gene expression in vegetative and symbiotics *Rhizobium meliloti* and other gram-negative bacteria. *Gene* **1989**, 81, 55–64.

95. T. M. Williams, J. E. Burlein, S. Ogden, L. J. Kricka, J. A. Kant. Advantages of firefly luciferase as a reporter gene: application to the interleukin-2 gene promoter. *Anal Biochem* **1989**, 176, 28–32.

96. L. Scheffer, I. Solomonov, M. J. Weygand, K. Kjaer, L. Leiserowitz, L. Addadi. Structure of cholesterol/ceramide monolayer mixtures: implications to the molecular organization of lipid rafts. *Biophys J* **2005**, 88, 3381–3391.

CHAPTER 15

POLYELECTROLYTE INTELLIGENT GELS: DESIGN AND APPLICATIONS

PIERO CHIARELLI and DANILO DE ROSSI

15.1 INTRODUCTION AND TECHNICAL OVERVIEW

Synthetic, biological, and hybrid hydrogels have found a large variety of uses, mostly related to biomedical applications.[1] Stimulus-responsive polyelectrolyte gels represent a niche area in which interest is growing; these environmentally responsive gels are often referred to as "smart" or "intelligent" gels.[2]

Reversible swelling of partially ionized polyelectrolyte gels mediated by external stimuli has attracted considerable interest from scientists and engineers in the last 60 years.[3]

The field essentially started with the independent works of J.W. Braitenbach,[4] W. Kuhn,[5] and A. Katchalsky,[6] all published in 1949, reporting how reversible titration of weakly ionized polyelectrolyte gels provides conspicuous swelling and deswelling of the polymer network. These works inaugurated the field of "polymer mechanochemistry" or, as redefined later, "polymer chemomechanics."[7]

In more recent years, theoretical work[8] and experimental observations of discrete phase transitions in ionic gels under appropriate conditions[9] were reported.

These observations sparked active research in the field. Many other external stimuli beyond pH changes proved to be effective in generating macroscopic volume changes in polyelectrolyte gels. These changes are also accompanied by large variations in physicochemical properties of the gels. Temperature, ion or solvent exchange, electric and magnetic fields, radiant energy, and also biochemical reactions or immobilized ligand interactions have all been demonstrated to be effective external stimuli on specifically tailored gels.[10] Several areas of application are thought to greatly benefit smart

Ionic Interactions in Natural and Synthetic Macromolecules, First Edition. Edited by Alberto Ciferri and Angelo Perico.

properties of polymer gels and a large amount of research work is now devoted to the development of materials and devices that could exploit stimulus-responsive behavior, particularly in the biomedical field.[11–14]

Early work on stimulus-responsive gels was largely focused on the implementation of macroscopic muscle-like actuators driven by chemicals,[15] by electric fields,[16] or by electrochemical reactions.[17] More recent work has addressed the development of drug delivery systems,[18] microfluidic components and circuits,[19] sensors,[20] biosensors,[21] optical components,[22] active membranes for separation,[23] smart surfaces,[24] and scaffolds for tissue engineering.[25]

It is now evident that research efforts in the field of macromolecular engineering have led to substantial advances to the endeavor of mimicking through synthetic pathways some functional characteristics of biological engines.

While applications in tissue engineering, active surfaces, and drug delivery systems typically do not require gels to be mechanically strong, other applications do, in particular, muscle-like actuators need tough materials, fast swelling-deswelling response, and the capability to undergo many cycles without damage and with minimum hysteresis. Stimulus-responsive gels endowed with these optimized properties are not yet available and devices that have been proposed are just proof of the difficulties involved.

Strong hydrogels have been recently obtained.[26] These gels might open new avenues in more demanding applications; however, the intrinsic biphasic (liquid-solid) nature of gels makes it necessary for most devices to be encapsulated, adding additional complexity to a design that engineers could only accept in view of unique performances.

In this chapter, polyelectrolyte intelligent gels are examined along three broad lines. The effects of different physical, chemical, and biological stimuli on gels' responses are analyzed and mechanisms of response are outlined. The broad range of biomedical applications of smart gels is reviewed and limits and perspectives of the proposed techniques and devices are critically discussed. Finally, continuous modeling of gel electromechanochemistry is described, providing quantitative tools to assess swelling equilibrium conditions and coupled kinetics.

15.2 MODELS OF SWELLING EQUILIBRIUM AND KINETICS

Modeling the response of polyelectrolyte gels to stimuli of different nature to quantify their swelling equilibrium and kinetic behavior is a complex task and it can be attempted at different scales and very different levels of approximation.[27] Section 15.2.1 summarizes the thermodynamic approach to the equilibrium behavior of gels. Modeling with the intent of designing devices and predicting their responses is best accomplished by referring to a macroscopic, phenomenological continuum or discrete description of the solvent–network system. Continuum models of gel electromechanochemistry are now available in the literature[28] as much as analytical solutions for simple geometries and strong approximations and numerical solution of more rich formulations and descriptions. We summarize in the Appendices the basic concepts and derivations that attempt to grasp the essentials of the modeling efforts that have been undertaken.

15.2.1 Gel Biphasic Systems: Free Energy and Phase Diagrams

A gel consists of a solid polymeric network and a liquid phase that fills its interstices. The polymeric matrix may have ionizable groups on its backbone with small ions of opposite charge (counterions) dissolved in the interstitial fluid to maintain electroneutrality.

The free energy streaming out by the solid–fluid interactions prevents the polymer network from collapsing, conferring to polyelectrolyte gels their characteristic shape, volume, elasticity, and porosity. All these equilibrium properties depend on physical and chemical parameters such as temperature, pH, ionic strength of the solution, and type of solvent.

The studies of Flory and Huggins[29] have put into evidence that a gel can be schematized as a physical system governed by osmotic forces that define its volume, shape, and elasticity. These osmotic forces Π_i are defined by how free energy $\Delta G_{tot} = \sum_i \Delta G_i$ changes in respect to variation of the gel's extensive variables, such as its volume, following the rule,

$$\Pi_i = -\partial(\Delta G_i)/\partial V. \tag{15.1}$$

The most important forces that generally operate in gels originate from rubber elasticity and polymer–solvent affinity, while those deriving from the counterion osmotic pressure and electrostatic repulsion (or attraction) on the polymer chains are present in charged gels.

The polymer–solvent affinity quantifies how the polymer–solvent contacts are energetically preferred with respect to the polymer–polymer ones. The greater the polymer solvent-affinity, the higher the osmotic pressure that sucks the fluid into the matrix network. This part of the free energy ΔG_m depends on the polar character of the solvent molecules and the hydrophobic-hydrophilic character of the groups linked to the polymer backbone following the Flory–Huggins formula,[29]

$$\Delta G_m = RT[n_1 \ln v_1 + n_2 \ln v_2 + n_1 v_2 \chi_{12}], \tag{15.2}$$

where n_1 and n_2 are the moles of solvent and polymer, respectively, and v_1 and v_2 are their respective volume fractions; χ_{12} is a "dilution" parameter that takes into account the increase of energy due to the polymer–solvent contacts. From Equation 15.2, the osmotic pressure of polymer–solvent mixing reads

$$\Delta\Pi_m = -(V_1)^{-1}\partial(\Delta G_m)/\partial n_1 = (V_1)^{-1} RT[\ln(1-v_2) + v_2 + v_2\chi_{12}], \tag{15.3}$$

where the identity $v_1 + v_2 = 1$ has been used, and V_1 is the molar volume of solvent.

The rubber elasticity of the polymer network comes from the tendency of each statistical coil to maintain its end-to-end distance between two network cross-links. Rigid units that are free to rotate with respect to the axis of the preceding one generate the statistical coil of the polymeric chain. Thermal fluctuations make these units continuously change their reciprocal angles, letting the end-to-end length of the coil fluctuate around a defined mean distance. Changing the intensity of the thermal fluctuations, the mean end-to-end distance of the polymeric coil will vary. Moreover, the higher the temperature, the stronger the tendency of the polymeric coil to maintain its mean undeformed end-to-end distance, leading to an increase of the macroscopic network elastic moduli. This behavior is described by the free energy of the polymer network that reads[29]

$$\Delta G_r \cong \tfrac{1}{2} n_v\, RT\left\{\alpha_x^2 + \alpha_y^2 + \alpha_z^2 - 3 - \ln\left[\alpha_x \alpha_y \alpha_z\right]\right\}, \tag{15.4}$$

where n_v is the number of moles of polymer chains, $\alpha_i = r_i/r_{0i}$, $r = (r_x, r_y, r_z)$ is the end-to-end distance, and r_0 is the reference (isotropic, undeformed) length. In the affine deformation approximation, the chain deformation parameters correspond to the macroscopic strain of the network.

For isotropic swelling (i.e., $(\alpha_x\alpha_y\alpha_z) = V/V_0,\ \alpha_x^2 + \alpha_y^2 + \alpha_z^2 = 3(V/V_0)^{2/3}$), Equation 15.4 leads to the osmotic pressure of the rubber elasticity that reads

$$\begin{aligned}\Delta\Pi_r &= -(V_1)^{-1}\,\partial(\Delta G_r)/\partial n_1 = -\tfrac{1}{2} n_v\, RT\,\partial\left\{3(V/V_0)^{2/3} - \ln\left[V/V_0\right]\right\}/\partial V \\ &= -n_v\, RT\left\{(V_0/V)^{1/3} - \tfrac{1}{2}(V_0/V)\right\}/V_0^3,\end{aligned} \tag{15.5}$$

where V is the gel volume and $V_0 = r_{0x}\, r_{0y}\, r_{0z}$. For polyelectrolytes that undergo large swelling (i.e., $V \gg V_0$ and $(V/V_0)^{1/3} \ll V/V_0$), it holds

$$\Delta\Pi_r \cong -n_v\, RT/V^{1/3}V_0^{2/3}. \tag{15.6}$$

The electrostatic repulsion of the charges fixed to the polymer coil leads to positive gel swelling pressure. This term can be calculated by means of the Debye–Hückel theory applied to the polyelectrolyte chains. Even if the majority of the counterions are captured by the ionic atmosphere around the charged polyelectrolyte chains, two contributions to the gel's free energy remain: the electrostatic interaction between the fixed charges and the osmotic pressure of the free counterions that cannot leave the gel network because of the Donnan potential. The first contribution[30] (which is influenced by the dielectric constant of the interstitial solvent, its ionic strength, and pH) is smaller than the other ones and is usually disregarded, particularly when a salt is added to the swelling solvent.

At equilibrium, the diffusional (chemical) force of the free counterions to leave the gel is balanced by the electrostatic force so that the total force as well as the electrochemical gradient at the gel's external bath is null. The Donnan electrical potential $\Delta\Psi$ equals the chemical one following the formula

$$\Delta\Psi = \pm RT \ln\left[C_i^{\pm} / C_i^{*\pm}\right] / F, \tag{15.7}$$

where $C_i^{\pm}$ and $C_i^{*\pm}$ are the anionic and cationic concentrations inside the gel and in the external bath, respectively. The relation (Eq. 15.7) fixes the ratio of each ionic species inside and outside the gel so that the mass conservation defines the ionic concentrations of the gel system. The Donnan equilibrium in the presence of an added salt requires a disproportion of individual mobile ions within the gel and in the equilibrium excess solution. For instance, for a cationic polymer of fixed charge Z, there will be more mobile anions in the solution within the gel (m_-) than in the excess solution (m'_-). The disproportion decreases with the overall salt concentration, and the disproportion coefficients ($m_- / m'_- = m'_+ / m_+$) attain the value of 1 above the electrolyte concentration of 0.2 M, when screening of the polymer charge is almost complete (see Chapters 3 and 6).

The osmotic force of the free counterions, which are confined inside the gel, exerts a positive pressure that inflates it like a gas confined inside a container. This contribution increases with temperature and with the dissociation of the ionizable groups along the polymer chain. Dissociation, in turn, is influenced by the pH and ionic strength of the interstitial solution. Given the Donnan potential (Eq. 15.7), the free energy and osmotic pressure read

$$\Delta G_{os} = \alpha_i n_i F \Delta\Psi = \sum_i \alpha_i^2 \left\{ n_i RT \ln[n_i / V] + n_i^* RT \ln[n_i^* / V^*] \right\} \tag{15.8}$$

$$\Delta\Pi_{os} = -(V_1)^{-1} \partial(\Delta G_{os}) / \partial n_1 = \sum_i n_i RT / V - \sum_i n_i^* RT / V^*, \tag{15.9}$$

where $\alpha_i = q_i/|q_i|$, q_i is the charge of the ith ionic species, n_i^* is the ith ionic concentration, and $V^* = V_{tot} - V$ is the volume of the external bath in equilibrium with the gel. Equation 15.9 is quite complicated since the disproportion between n_i and n_i^* as well as the effective charge on polymer chains are functions of the gel volume.

Nevertheless, it is interesting to derive the osmotic pressure in a simple case. If we suppose that the external bath is infinitely larger than the gel (i.e., $V^* \gg V$, and hence $V_{tot} \gg V$) and that the gel is electroneutral (we disregard the infinitesimal amount of ions needed to generate the Donnan potential), $C_i^{*\pm}$ as well as the effective fixed polymer charge at different gel volumes can be considered practically constant. Then, in the absence of support electrolyte so that the moles of counterions n_c inside the gel are constant, Equation 15.9 leads to the expressions

$$\Delta\Pi_{os} \cong n_c RT / V + \Pi_{ext} \tag{15.10}$$

$$\Pi_{ext} = -n_c^* RT/(V_{tot} - V) \cong -C_c^* RT = \text{Constant.} \tag{15.11}$$

A compact expression for the degree of swelling of a polyelectrolyte network as a function of the degree of cross-linking (elastic contribution), the interaction parameter (mixing contribution), the fixed charge, and ionic strength (ionic contribution) was earlier given by Flory.[29]

This plurality of energetic contributions, having no equal in other nonbiological material systems, is able to confer to gels a complex behavior sensible to many physicochemical inputs.

The counterions' osmotic pressure is usually positive and tends to inflate the gel, while the contribution of rubber elasticity is typically negative and it counterbalances the small ions' osmotic pressure, giving to the gel volume a stable condition. The other terms can be either positive or negative, displacing the equilibrium point.

Generally speaking, the isothermal curves in the gel pressure–volume phase diagram have a hyperbolic, decreasing shape. If we try to increase the gel volume (at zero total pressure), the positive osmotic pressure of the small ions decreases faster than the negative rubber elasticity, so that the total gel pressure needed to maintain this new state is lower and of negative value.

Since the rate of change of each osmotic pressure term is not linear, we may have the appearance of a flex in the phase diagram with a critical point as well as unstable states with the typical bell-shaped area.

In this domain of states, for instance, the rate of decrease of the small ions' pressure may be slower than the rate of increase of the polymer–solvent affinity, and therefore, a volume increase leads to an increase of total gel pressure, generating a further expansion. As is well known, this kind of instability is associated with phase transitions. In this case, the gel makes an abrupt, large-step expansion to a new equilibrium volume.[31]

A continuum kinetic model for the network–solvent readjustment dynamics, induced, for example, by a phase transition, is outlined in the Appendix.

15.3 PHYSICALLY RESPONSIVE GELS

Behind similar kinetics of the gel network readjustment, there may be very different phenomena on the molecular scale activated by the external stimulus.

In this section, we are going to examine the variety of actuation mechanisms in gel systems grouped by the typology of the stimulus.

15.3.1 Thermally Sensitive Gels

Temperature-sensitive gels are based on the principle that a polymer coil may undergo conformational changes with temperature, eventually resulting in gel volume changes. This process is induced by the change of the polymer–polymer contact energy with respect to the polymer–solvent one when the temperature is changed.

Poly-*N*-isopropylacrylamide (NIPAM) gels constitute one of the most investigated systems. These supermolecular aggregates transduce on a macroscopic scale the conformational changes that happen on a molecular one where polymer shrinks as temperature increases.

In a solution, the polymer solute separates into a polymer-rich and diluted phase as the lower critical solution temperature (LCST) is reached.[32] When the polymer coils are cross-linked, the resulting gel undergoes volume deswelling at a temperature that is close to that of the polymeric solution.

A swollen NIPAM gel undergoes shrinking at 40°C[33]. Homopolymer gels have a sharp volume collapse that can be recognized as true phase transitions, while functionalized copolymers with, for instance, ionizable acrylic acid groups[34] have a more smooth gel deswelling at the critical temperature.

The gel readjustment kinetics and the thermal diffusion in these gels are coupled with each other. The simplest behavior happens in macroporous gels wherein the thin gel pore wall mechanically responds very quickly to the temperature change.

In macroporous gels, therefore, the limiting rate phenomenon is the heat diffusion into the macropores. In the gel swelling process (temperature decrease below the LCST), the solvent uptake by the gel enhances, by convection, heat transfer to the inner part of the network. In the inverse process (gel deswelling), the heat transfer is hindered by the outflowing of the solvent from the inside of the polymer matrix, thus generating an asymmetry (hysteresis) in the contraction–elongation cycle.

In homogeneous gels, the heat transfer inside the gel can only occur by thermal diffusion since the water uptake or release, due to the mechanical readjustment, is very slow. Moreover, given that the mechanical parameters depend also on the state of the gel, in NIPAM gel with a large gel deswelling, the solvent diffusion through the collapsed domains can be lowered practically to zero.

When the temperature change is induced by gel immersion in an external bath, the collapsed domains may constitute an external skin around the swollen gel and can prevent the outflow of fluid from inside the gel and the consequent mechanical readjustment.

Other gels that also show thermal response are the polyvinyl methyl ether (PVME),[35] which progressively undergoes swelling below a temperature of about 37°C; poly-*N*-vynilcaprolactam/ethylene glycol and dimethacrylate/divinyl sulfone,[36] which have a glass-like collapsed state above the LCST; and hydroxypropylcellulose,[37,38] which has the peculiarity of decreasing its elastic modulus in the deswelled state.

Even if the thermal response is highly reversible, the major obstacle to the use of these kinds of gel actuators (e.g., in the field of drug delivery) is the very low cooling rate obtainable by spontaneous heat diffusion.

15.3.2 Electromagneto-Sensitive Gels

The responsitivity of polyelectrolyte gels in solution to electric fields is governed by different phenomena whose importance depends on several factors such as potential difference at the electrodes and geometry.

Even from the very first experimental observation of gels' responsitivity to electrical stimuli, diverse mechanisms have been postulated to operate, although many, if not all of them, are concomitantly active.[39]

Hamlen et al.[17] and subsequently Fragala et al.[40] exploited electrolytic reactions or proton electrodialysis to swell and shrink pH-sensitive polyelectrolyte gels, and were the first to exploit electrical stimuli.

Tanaka et al.[41] reported progressive collapse of a weakly ionized polycationic gel in contact with a metal anode, immersed in acetone–water (50:50) mixture under an electrode potential difference ranging from 1.25 to 2.15 V. Gel deswelling was ascribed to the electrophoretic attraction of the charged gel network to the electrode, resulting in the generation of a mechanical stress gradient (orthogonal to the electrode) squeezing the gel.

Osada and Hasebe[42] reported poly(2acrylamido-2-methylpropane sulfonic) (pAMPS) and other gels shrinking and water exudation when in contact with carbon electrodes under direct current (DC) excitation.

The phenomena were tentatively ascribed to concomitant electrophoretic attraction of the gel to the electrode and electrostatically induced gel dehydration. De Rossi et al.,[43] working with polyvinylalcohol (PVA)–polyacrylic acid (PAA) gels in solution with slightly higher potential difference at the electrodes, observed gel swelling or shrinking governed by water electrolysis at interfaces, resulting in local pH changes.

Gel bending under the action of an electric field was first reported by Shiga and Karauchi[44] under appropriate conditions and geometry.

Alternate bending of gels leading to a wormlike motility was reported by Osada et al.[45] by exploiting selective charged surfactant binding driven by an electric field and causing osmotic pressure changes at the gel surface, resulting in mechanical actions.

The electrochemical response of gels to copper oxidation to Cu^{2+} has also been described.[46] In this case, the double interaction of each ion of Cu^{2+} with the anionic ionizable groups on the polymer network generates additional cross-links, increasing its rubber elasticity and leading to deswelling as the Cu^{2+} concentration increases.

Gels have been prepared containing magnetic particles in their soft matrices that react to magnetic fields[47] as well as to electrical ones[48] by changing their shape and physical properties, similar to magnetorheological and electrorheological fluids.

Gels of silicone rubber or oil, and polyurethanes with dispersed micron-sized particles of iron, have been synthesized.[49] Such gels respond to the magnetic field by increasing their stiffness, so that an increase of stress can be obtained by a prestrained sample.

A carrageenan hydrogel with barium ferrite microparticles showed a decrease in elasticity when subjected to a magnetic field.[50]

15.3.3 Light-Sensitive Gels

Light can also be used to swell and deswell gels. One possibility relies on the indirect stimulation of thermally sensitive gels by absorption of radiant energy to increase their temperature.

One example is given by Suzuki of a synthesized NIPAM gel containing copper chlorophyllin that swells and contracts as light is turned on and off.[51]

A similar response was also obtained in NIPAM gels with dispersed nanoparticles of gold[52] that respond to near-infrared (IR) electromagnetic radiation.[53]

A different mechanism leading to light-sensitive gels is constituted by the use of light-induced conformational changes such as the photoisomerization of azobenzene between the *cis* and *trans* forms[54] that changes the rigidity of the polymer network under UV irradiation. Ionization of leucocyanide groups under UV that can lead to osmotic swelling of the hosting gels[55] and nitrocinnamate groups that reversibly undergoes cross-linking and cleavage under photoirradiation have also been reported.[56] The *keto*-to-*enol* tautomerization was also shown to be able to induce gel shape change under UV irradiation.[57]

PAA gels cross-linked by means of copper ions and containing titanium dioxide have been shown to undergo swelling due to the pH change induced by UV light.[58] Other similar gel systems have been synthesized by making use of silver-coated titanium dioxide microparticles.

15.4 CHEMICALLY RESPONSIVE GELS

15.4.1 pH- and Salt-Sensitive Gels

Since the pioneering works of Kunh and Katchalski in the 1950s, gels have been investigated as chemomechanical systems stimulated by means of monovalent and bivalent cations. The muscle-like contraction–elongation behavior has attracted the attention of many researchers recently, even if the biological analog is much more complex and it exploits a very different contractile mechanism.

The most investigated hydrogel is cross-linked PAA. The solubility of the polymer in water as a function of pH almost defines the state of the gel network in the macromolecular aggregation. At low pH, the ionizable groups are not charged, so that the electrostatic repulsion is almost absent as well as the counterions responsible for the positive osmotic pressure. In this state, the gel is in the shrunken state.

At high pH (>5), much above the carboxylic groups' pK_a, the functional groups are ionized with counterions that warrant electroneutrality. In this salt form, the gels adsorb water and remain in the swollen state.

As shown in the Appendix, the rate of the gel volume change depends on its geometrical form. A polymer that can be shaped in the form of very thin fibers can undergo a rapid contraction–elongation cycle (1–10 seconds). Moreover, the anisotropic orientation of the network chains can increase the elastic modulus of the gel and its stress generation along a preferred direction in response to pH stimuli. This is the case of polyacrylonitrile (PAN) fibers, which, under thermal treatment in basic conditions, can be converted into ionized gel fibers.[59]

Alternatively, the time response can be shortened by synthesizing a gel with a macroporous structure by means of freezing–thawing[60] or by freezing–drying cycles.[61]

Other polyacid gels that react to pH and salt concentration variations have been synthesized,[62,63] such as those based on polymethacrylic acid.[7] If the gel network owns strong acid groups, such as the sulfonated ones, the swelling is translated to lower pH as well the pK_a value.

Other gels, such as phosphated ones, can undergo multiple ionization processes wherein the functional groups progressively reach the dissociated form.

Symmetrically to polyacids, polybasic gels shrink in basic conditions and swell in acid ones. In this case, the functional groups, such as that of the aminomethylmethacrylate monomer, lead to polybasic gels. Another type of polybase is given by the epoxy-amino gels.[64]

A polymeric network owning both acidic and basic side groups has anphoteric properties, showing a more complex swelling behavior.[65,66] These gels also react to changes of salt concentration in the interstitial solution as a consequence of the shielding of the charges present on the matrix by the salt ions.

15.4.2 Chemical Reaction-Sensitive Gels

The number of gels responding to chemical or biochemical reactions and nonlinear covalent cooperative binding[67] is rapidly increasing due to the perspective of innovative applications in the biosensors and drug delivery fields.

In this section, we first consider those gels that are sensible to simple chemical reactions, leaving the others to the next sections.

An interesting class of gels concerns those ones that swell and contract cyclically in response to oscillating chemical reactions.

A copolymer of NIPAM gel has been shown[68–70] to cyclically contract and swell following the oscillating Belousov–Zhabotinsky reaction.

The mechanical oscillation is generated by the oxidation and reduction of the ruthenium(II) tris(2,2′-bipyridine) group, covalently bonded to the polymer chain, which autonomously periodically switches between the two states in a closed solution without any external action.

Also, pH oscillations around acidic block copolymers have been shown to induce cyclic swelling and shrinking in gels.[71,72]

Even if there exists a formal analogy with the biological contraction, the energy densities involved in those systems are much smaller than those of natural muscles.

Gel systems that respond to a chemical input via covalent bond formation to the analyte have been recently developed for the detection of amines, alcohols, aldehydes, carbon dioxide, saccharides, thiols, hydrogen sulfite, hydrogen disulfide, cyanide, and amino acids.[73]

The glucose-sensitive gels via boronic acids[74] are the most investigated since enzyme-based hydrogels raised concerns about their stability, toxicity, and undesirable immunogenic responses.

A gel has been synthesized via the free radical polymerization of *N*-vinyl-2-pyrrlidone and m-acrylamidophenylboronic acid. A polymeric complex was then formed via boronate ester formation to the diol units of PVA. Viscosity was shown to markedly increase upon complexation. Upon the addition of glucose as a competitive binding agent, the viscosity decreased significantly, while only minor changes were observed upon the addition of other similar chemical substances.

Choi et al. published an analogous boronate-containing gel for insulin delivery based on competitive displacement between glucose and boronic acid binding sites.[75]

A similar material, containing both phenylboronic acid and tertiary amine moieties complexed to PVA, was used as an electrochemical sensor for glucose with a membrane-coated platinum electrode.[76] The polymer complex exhibited enhanced swelling effects proportional to glucose concentration at physiological pH, as glucose displaced the PVA from the boronate groups. Swelling of the cast gel membrane upon glucose addition resulted in enhanced diffusion of ions and a concomitant increase in current.

15.5 BIORESPONSIVE GELS

The bioresponsiveness of gels can be achieved by different mechanisms that can be classified as follows: (1) change of electrostatic and small ion osmotic pressure; (2) change of network elasticity by variation of the number of its cross-links; and (3) conformational change of polymer chains between cross-links.

The most common examples are given by the glucose-responsive hydrogels investigated for the treatment of diabetes. For instance, Hoffman et al.[74] have synthesized a network of poly-glucosyloxy-ethyl methacrylate (PGEMA) interacting with concavalin A (ConcA) present in the interstitial solution. In the presence of competitive free glucose molecules, the number of the PGEMA–ConcA binding sites decreases, inducing gel swelling.

Another type of biosensitive gel makes use of natural proteins incorporated into its matrix undergoing conformational transition. An example is given by Mrksich et al.[77] who prepared a gel by covalently linking the calmodulin

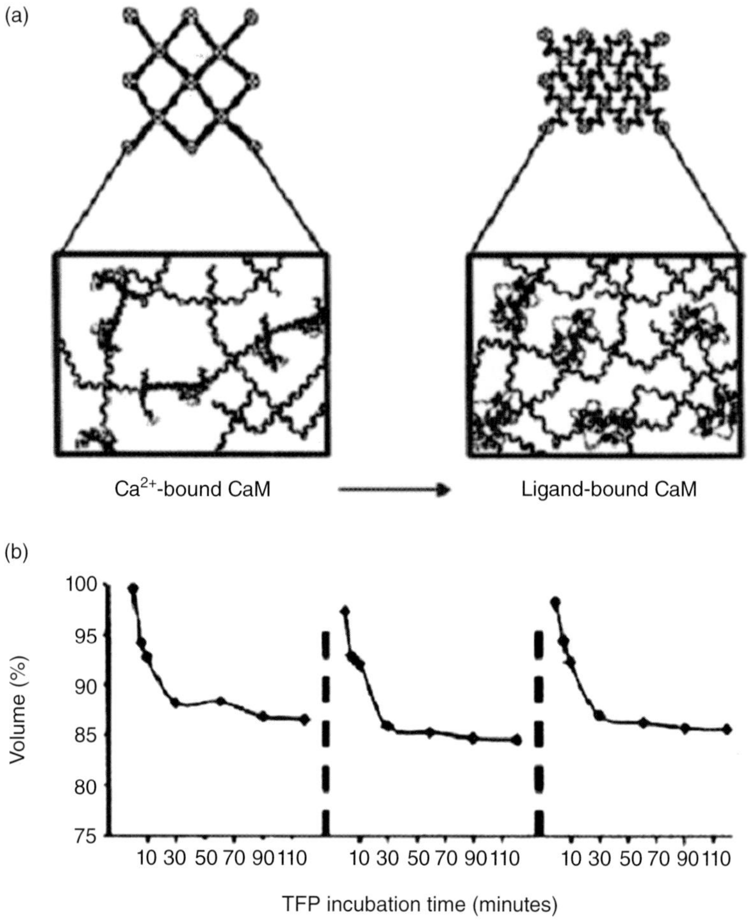

FIGURE 15.1. (a) The extended (left) and collapsed (right) states of CaM in the hydrogel network; (b) the volume of hydrogel exposed to TFP as a function of time and after being washed repeatedly in a calcium containing bath to restore the extended CaM configuration. (Reprinted with permission from Z. M. Yang et al. Copyright 2006 American Chemical Society.)

(CaM) protein to a polymer network by means of polyethylene glycol (PEG) molecules (Fig. 15.1). The gel undergoes shrinking when a ligand such as trifluoperazine (TFP) diffuses into the gel and CaM collapses from the dumbbell shape to a random coil. Since TFP is a calcium-binding ligand, when the calcium is depleted from the interstitial solution the gel swells quasireversibly to the initial volume.

Even if several bioactuated gel systems have been realized[78,79] and many others can be designed, the deficiency of specificity to the target substance may limit their use. To overcome this difficulty, enzyme- as well as antigen-responsive gels have been synthesized and investigated.

15.5.1 Enzyme-Responsive Gels

A glucose-induced gel response can be achieved by immobilizing glucose oxidase together with the catalase protein. In the gel matrix, the first enzyme converts glucose into gluconic acid and hydrogen peroxide in presence of oxygen, while the latter hydrogen peroxide into water and oxygen. In this case, a polymer gel matrix containing amine groups[80] undergoes swelling as a consequence of the pH lowering generated by gluconic acid.

Another hydrogel system sensible to protease has been synthesized by Ulijn et al.[81] The hydrogel has amino acid chains with specific protease-cleavable peptide sites. When the protease is introduced into the gel, the anionic peptide is detached and it freely diffuses out of the gel, leaving the cationic peptide attached to the polymer matrix. The decrease of cross-links of the gel network and the generation of the fixed charged groups on it leads to a noticeable swelling.

A more complex hydrogel system, conceptually similar to the previous one, has been synthesized to detect elastase.[82] The irreversibility of these kinds of actuation mechanism does not make them useful for repeated release.

15.5.2 Antigen- and Ligand-Responsive Gels

Antigen-sensitive gels have been recently synthesized[21,83] by incorporating in the gel network both the antigen as well as the antibody (Fig. 15.2). The antigen–antibody binding increases the cross-link density of the polymer matrix and induces gel deswelling. When the gel is exposed to a solution with the free antigen that competitively binds itself to the antibody, the number of cross-links decreases and the gel undergoes swelling and it raises its porosity.

More recently, a tumor marker-responsive gel that exhibited volume changes in response to the tumor-specific marker glycoprotein (α-fetoprotein [AFP]) was prepared by using lectin and antibody molecules as ligands.[84,85]

15.6 BIOMEDICAL APPLICATIONS

15.6.1 Gel Actuators

The analogy between natural muscle and a contractile gel resides just in the outward performance where the actuator element is a material that increases its elastic modulus and shortens its equilibrium rest length to engender force and displacement. Beyond this, the similarity ends. On a microscopic scale, the mechanism that generates movements and forces in natural muscle is very different from that of gels.

The biological muscles almost base their unattainable performance on the actin–myosin highly organized system with a maximum stress generation of 300 kPa, contractile strain of about 25%, and 50 W kg^{-1} of aerobic power generation that can reach 200 W kg^{-1} in a peak supply.

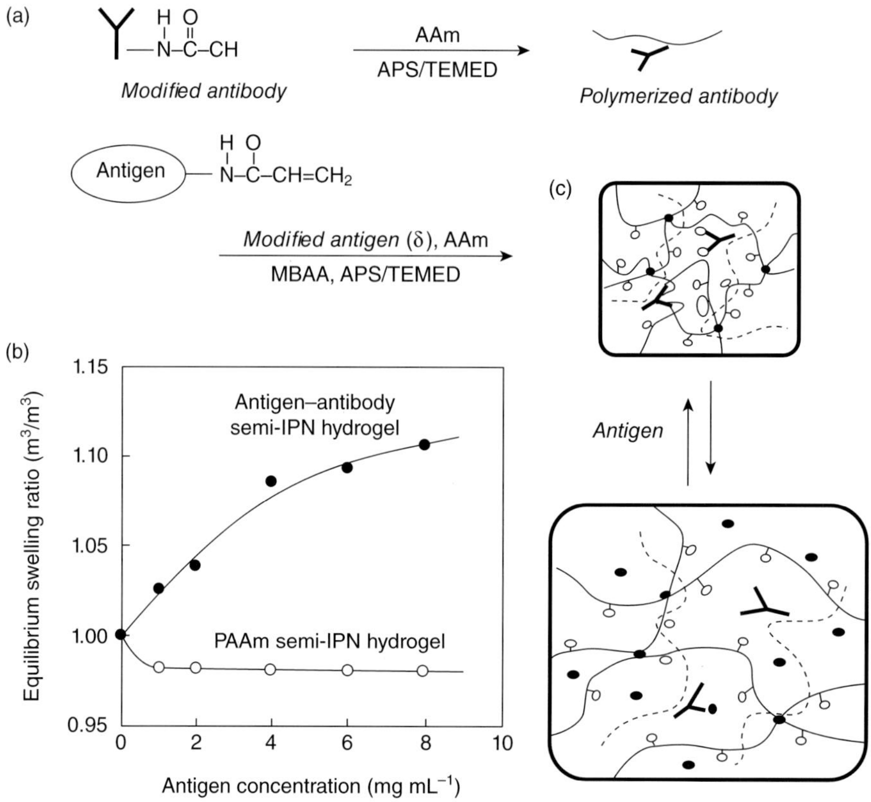

FIGURE 15.2. Antigen-responsive hydrogel. (a) Synthesis of the antigen-bounded network; (b) hydrogel swelling as a function of the free antigen concentration; (c) mechanism of free antigen competitive binding. (Reprinted with permission from T. Miyata et al. Copyright 1999 Nature Publishing Group.)

Even if gels are not able to mimic muscle, gel contractility is useful and nature utilizes it in some cases. For instance, sea cucumbers, starfish, and other echinoderms that embody natural hydrogels made by proteoglycans, elastin, collagen, and muscle fibers can quickly switch their elastic properties from soft to hard.[86–89] The elastic change is due to the release of proteins that temporarily bond to the collagen fibers of the matrix.

Even if the gel actuator is not yet a reliable solution as a muscle-like engine, laboratory prototypes that attain the muscle performance with a contraction time of 1 second have been realized by using macroporous NIPAM gel[90,91] and PAN fibers.[92,93]

As shown by macroporous NIPAM actuators, the thermal stimulation is highly reversible but the heat dispersion, when used in the human body, is very slow due to the small thermal gradient that can be established with the surrounding living tissues.

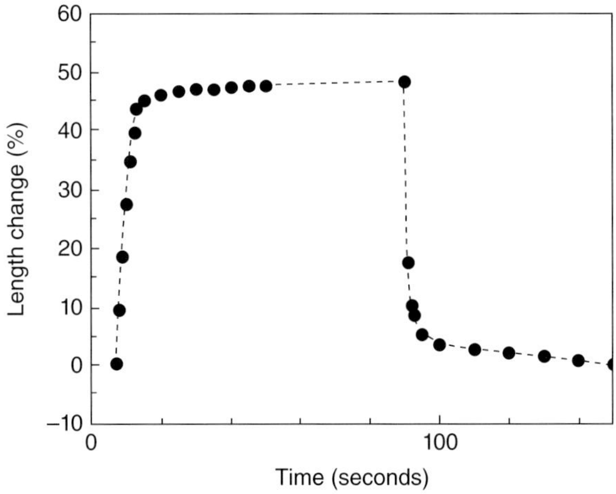

FIGURE 15.3. The isotonic fractional length change of PAN gel fibers as a function of time when the external bath pH is suddenly changed from 1 to 13. Electrochemical potentials are actually observed.[200,201]

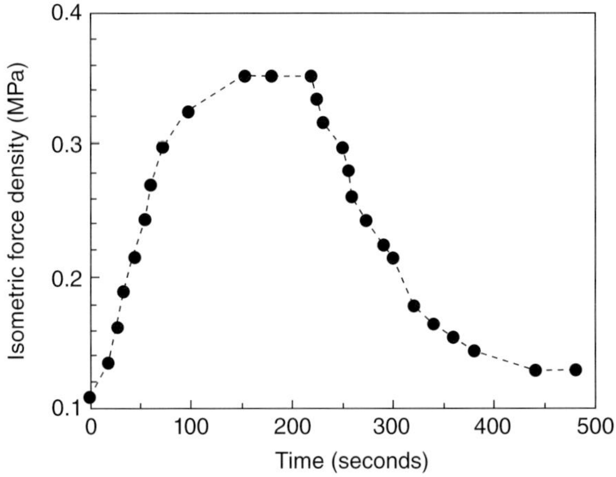

FIGURE 15.4. The isometric force density generated by the PAN gel fibers as a function of time when the carbon fiber electrodes are excited by steps of electric potential between +10 and −10 V.

The chemical actuation of PAN fibers gives reproducible force–elongation cycles (Fig. 15.3), but the delivery of HCl and NaOH solutions requires a complex system of pumping and piping with a relevant lost of chemical energy due to the unused reagents wetting the surface of the gel fibers.

The electrical activation (Fig. 15.4) has engineering problems too. The generation of acid fronts in 1 g of acrylic gel needs about 10 mM of cations and a current of about 15 A for 1 minute. If irreversible reactions with production

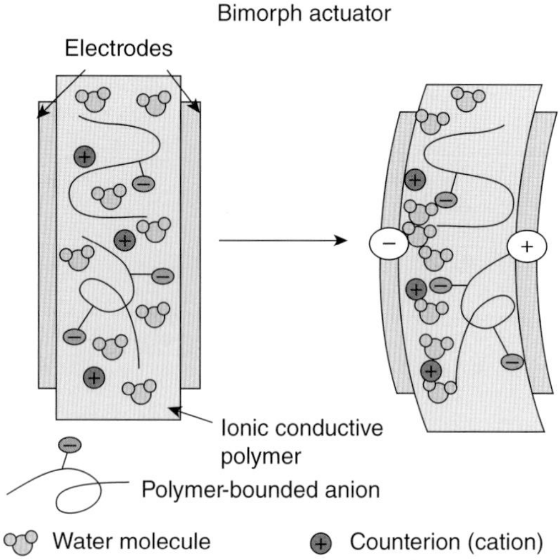

FIGURE 15.5. Schematic view of an ionic polymer–metal composite (IPMC) bimorph gel actuator.

of gas occur at the electrodes, the life of the actuator is very limited. Moreover, degradation of the gel (PAN fibers) due to these problems has received a first solution by the "bimorph" gel actuators where a thin layer of an ionic polymer gel is enclosed between two thin metal electrodes.[94] This configuration has the advantage of also utilizing the electrochemical reaction of the counterelectrode. The ionic fronts of opposite characteristics (e.g., acid and the basic ones) of the electrodes produce the bending of the device due to the gel swelling, on one side, and its deswelling on the opposite one (see Fig. 15.5).

The thin actuator shape leads to a fast response, relatively low currents, and energy dissipation. The most promising application for this type of actuator is the realization steerable microcatheter,[95] but also Braille display and tactile stimulator are proposed. Recently, the "bimorph" actuator has been improved by using electrodes made by a network of carbon nanotubes.[96] It has also been shown that such actuators respond to a mechanical deformation with the generation of an electric potential allowing, in principle, an inverse sensing function.[97]

On the same principle of the "bimorph" actuator, an improved layered actuator consisting of multiple thin elements, each one composed of two films of gels of opposite chemical characteristic (e.g., a polyacid–polybase gel), has been proposed.[98] The two gel films are sandwiched between two electrodes of conducting polymer (CP) containing doping ions that are released (and/or adsorbed) during the electrical inputs (see Fig. 15.6). This actuator has the benefit of avoiding the gas generation at the electrodes since the electrochemical stimulation is brought by the release of doping ions. Moreover, given that

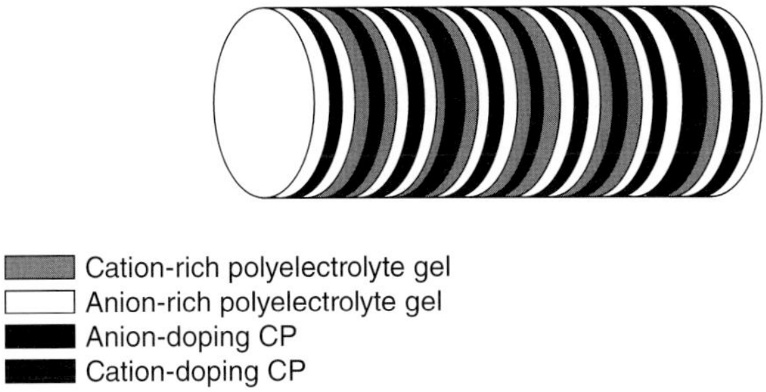

FIGURE 15.6. Schematic view of a layered electrically driven gel actuator.

the gel attached to the electrode cannot freely contract in the radial direction, due to the high rigidity of the electrodes, the gel contraction in the axial direction is magnified. This thin-layered assembling leads to a fast response as to the bimorph actuator. The practical limit is constituted by a large number of contractile elements, about 200 per centimeter, needed by a macroscopic actuator.

Given that chemomechanical gel dynamics are mostly diffusive (the electric field is damped by the high concentration of ions), the layers' thickness needed to have a contraction of 1 second is of the order of 10 μm for gel, with shear and bulk elastic moduli of about 1 and 5 MPa, respectively, and a friction coefficient of 10^{16} Ns m^{-4}. Of thc same order of magnitude (10 μm) is the characteristic length of the structure of the natural muscle whose limiting dynamics are given by the diffusion of calcium ions.

Even if an electromechanochemical gel actuator might be shortly realized, the amount of the energy needed to power it would constitute a great limitation to its use. The diffusive dynamics of the gel contraction result in a large dissipation into the viscous flow of the gel interstitial fluid.

Arndt and collaborators[99] have shown that a PVA-PAA gel actuator utilizing 175 J kg^{-1} of energy (more than the double that of natural muscle) during its contraction gives an output of 2 W kg^{-1} compared with the 200 W kg^{-1} of the biological analog (Fig. 15.7). The efficiency of the gel-like contraction is less than one hundredth of the natural one.

In order to increase the gel actuator efficiency (delivered power), the shortening of its response time becomes very important since a fixed amount of energy is dissipated during each cycle of contraction. Given that the gel contraction scales with the square of its physical length, it is clear that the realization of an efficient gel actuator is based on the development of a micron-structured system.

In the frame of the present state of the art, the mature applications for gel actuators are almost addressed to "low-energy" tasks such as valves, light modulators, drug delivery systems, and sensing functions described in the next sections.

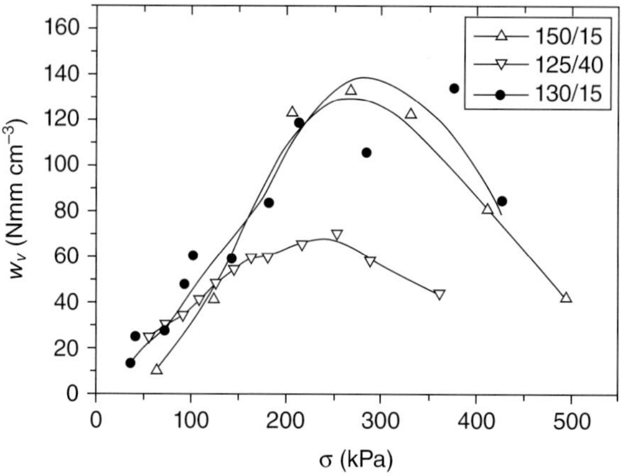

FIGURE 15.7. Work energy generated by PVA-PAA hydrogel under different temperatures and durations of cross-linking. (Reprinted with permission from Arndt et al.[99] Copyright 1999 Wiley-VCH Verlag GmbH.)

15.6.2 Gel Sensors

The intrinsic energy transduction properties of polyelectrolyte gels and the versatility of functionalized intelligent gels have been widely exploited to design soft sensors for a large variety of measurands.

Sensors for physical, chemical, and biochemical measures either as standalone devices or fully embedded into closed-loop systems have been reported.

Possibly the least investigated gel sensors are those intended to detect physical quantities; this fact may originate from the scarce involvement of engineers in the use of soft, wet materials. Besides the above uses, polyelectrolyte gel sensors have been described to sense dynamic contact forces, mimicking mechanoelectrical (streaming potentials) properties of the human dermis.[100] Other tactile sensors that exploit gel ionization caused by mechanical deformation have also been reported.[101,102] The softness and skinlike mechanical properties of water-swollen gels were claimed to be important whenever gentle object grasp is important, such as in prosthetics and humanoid robotics.

Temperature sensors using intelligent gels have also been described, based on temperature-modulated fluorescence in a PVA/borax hydrogel system containing 2-napthhol.[103]

Tunable Bragg et al.[104] reflections in photonic gels have also been exploited to sense different parameters including salt concentration, pressure, and humidity.[105]

Substantial work has been reported on the use of ionic conduction polymers and polyelectrolyte gels.[106]

Work in this area is still intense, trying to overcome problems such as poor stability, hysteresis, and slow response.

The use of intelligent gels to implement chemical sensors is essentially related to the detection of the amount of swelling or shrinking of a gel-based system using different methods in response to the presence of chemical analytes in its aqueous phase. The most investigated gel chemical sensors are those measuring pH[107] where reversible swelling and shrinking is quantified by a piezoresistive element sensitive to gel swelling pressure changes.[108] A CO_2 sensor was also developed as a medical indwelling probe based on a pH-sensitive hydrogel whose response is governed by the CO_2/bicarbonate equilibrium in water.[109]

Colloidal crystal hydrogel films have also been developed and used to sense different chemical analytes, based again on Bragg diffraction peak shifts generated by reversible swelling and shrinking.[110]

The use of stimulus-responsive gels as a sensor element has found in sensing biochemical analytes its largest interest.[111] A very broad range of biomolecules of medical interest has been investigated, the most relevant being glucose due to the strong need of a sensor to close the loop in insulin delivery systems.[112]

Since early studies[113] the use of intelligent gels sensitive to pH changes generated by enzymatic conversion of glucose to gluconic acid has attracted large research interest. Subsequently, boronic acid derivatives were used to sense glucose to generate reversible gel expansion and contraction because of the higher stability of the chemical ligand in comparison with glucose oxidase.[114]

Holographic glucose sensors for body fluid measurements[115] and photonic gel sensors[116] for noninvasive monitoring of glucose in tears have been recently reported, both based on boronic acid chemistry.

The literature in this field is broad and the reader may refer to a specialized review to get more insight.[67] Another important biochemical analyte whose detection has been accomplished through sensitive polyelectrolyte gels is glutathione, a polypeptide having an important role in several cellular processes whose controlled delivery has therapeutic relevance. A dual responsive and delivery system has been recently proposed that uses both pH- and glutathione-sensitive gels to properly tune the release of trapped oligodeoxynucleotides.[117]

Antigen sensing through antibody and antigen grafted onto the chains of an intelligent gel has been accomplished through competitive binding, resulting in volume changes caused by reversible, noncovalent cross-link breaking.[21] Other protein–ligand recognition systems have been disclosed using stimulus-responsive gels and more work is expected in view of the strong interest in biosensors in biotechnology-related areas.[118]

15.6.3 Gel Microfluidic Circuits

Sensors, channels, valves, and pumps are the basic components of microfluidic circuits which are needed for the development of lab-on-a-chip systems for biochemical analysis,[119,120] genomics and proteonomics,[121] and cell studies.[122]

In the early 1980s, the exploitation of silicon technology and the use of silicon as a mechanical material[123] provided powerful tools for the fabrication of micro-electromechanical systems (MEMS) and bio-micro-electromechanical systems (BioMEMS) for miniaturized analytical systems on a chip.[124,125] The manipulation of fluid at the microscale to implement fully operative separation and analytical systems has been since then very intensively studied.[126] The main drive for lab-on-chip development still is the need for cheap, reliable, simple, and even disposable analytical systems at the point of care.

Despite tremendous R&D efforts, however, several technical obstacles still impede the full exploitation of BioMEMS devices. Possibly, the major difficulties rely on the fact that the superior mechanical properties of silicon and the powerful microfabrication technique nowadays available are largely vanified by the absence or inefficiency of silicon transduction properties. These properties are absolutely necessary to realize active components such as pumps, valves, and sensors.

The advent of hybrid technologies that rely on the integration of silicon with other solid-state materials endowed with piezoelectric, photoemissive, and other transduction properties is complex and expensive.

Stimulus-responsive gels have been proposed and intensively studied as possible alternative materials that can be easily microfabricated and possess all the needed properties of sensing, actuation, and even self-regulation.[127]

Besides the easy patterning and microfabrication techniques that can be adopted to build gel active microcircuits, the main advantages of stimulus-responsive gels in this field of application is the possibility to use chemicals dissolved in the liquid phase as stimuli to trigger functions without the need of external action and control[128] or to use convenient photoirradiation with time and space selective control.[129]

Thermally[130] and pH[131]-activated gel valves for microfluidic chips have been developed as much as micromixers, micropumps,[132,133] and adjustable focus microlenses integrated into gel microfluidic systems for optical sensing.[134–136]

Three-dimensional (3D) patterning and direct writing of intelligent gel microstructures have also been reported,[137,138] opening up further avenues toward the achievement of the difficult goal of fully integrated 3D active gel microfluidic channels.

15.6.4 Gel Drug Delivery Systems

The “drug delivery system” is a very wide concept that embraces all means that can lead to a release of drugs to the desired target site, at the right time, and in the right dose. This has the advantage of almost avoiding side effects, of making drugs more efficient, and of possibly substituting medical interventions by pharmaceutical ones.

Most of the affirmed drug delivery systems are solid polymers to sustain a prolonged release of active substance since hydrogels would release the small molecules very quickly for most of the applications. Recently, the use of drugs made of larger molecules such as proteins and peptides, which are effective at

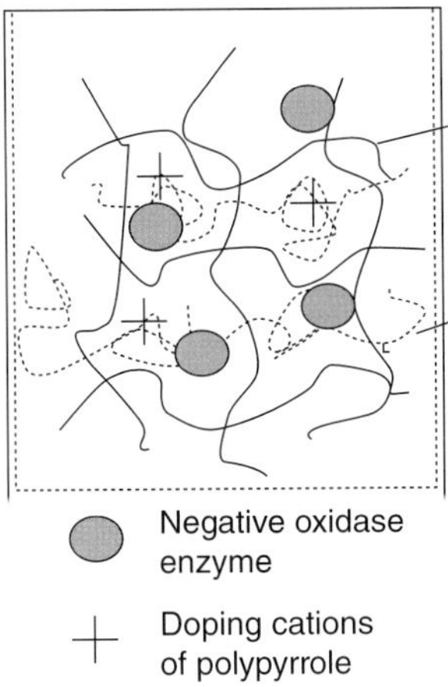

FIGURE 15.8. Cross-linked poly(hydroxyethyl methacrylate) (pHEMA) and interpenetrating polypyrrole with entrapped glucose oxidase (GOx) coating a platinum electrode for amperometric glucose sensing.

a very low dose, has brought the attention on gels as tools for efficient drug delivery. The more recent applications use hydrogels for protection of protein drugs (e.g., by the digestive system), for the adhesion of patches releasing drugs, and for the protection of nanoparticles of drugs by the immune system[139].

Moreover, the use of the electrical activation to enhance drug outflow by patches or internal gel systems has been proposed by many authors.[140] Drug eluting stents have also been realized by using thermal-sensitive gels.[141] A more sophisticated electrically controlled system made with gels and conducting polymers for the release of a drug has been realized by Brahim and coworkers[142] (Fig. 15.8). Moreover, the ability of gels (described in the previous sections) to react to the presence of glucose leads to the design of a variety of insulin delivery systems at a controlled rate.[74–76]

Finally, it is worth mentioning the great interest in smart gels for encapsulating microparticles of drugs that are absorbed at the target biological site or organ.[143–150]

15.6.5 Gels for Chromatographic and Membrane Separation

Since the early observations of large changes in solute permeability of biological[151] and synthetic[152] polyelectrolyte membranes upon changes in their ionization state, substantial research activity has been devoted to better comprehending and potentially exploiting these properties for solute separation.[153] Different stimuli have been used to alter and control polyelectrolyte gel membranes, the most common being pH or ionic strength,[154] electric field,[23] and temperature[155] changes. The field has been recently reviewed[156] with particular attention to separation by chromatographic techniques through stimulus-responsive gels.

Most of the work related to solute separation by intelligent polyelectrolyte gels has been addressed to chromatographic separation through ionic or hydrophobic interactions,[157–159] temperature-responsive stationary phases for size exclusion,[160] and affinity-based separation.[118]

Temperature control of gels mostly affects the hydrophilic–hydrophobic balance of the network and average pore size. On the other hand, pH, salts, and electric fields mostly affect ionic interactions and Donnan partition.

Conjugating temperature-responsive gels with ligands offers a very interesting way of modulating noncovalent interactions between stationary and mobile phases without resorting to changes in buffer composition or to other cumbersome techniques. A large variety of biomolecules has been selectively separated through affinity modulation by smart gels,[161] potentially providing cost-effective and simple separation of biologically active compounds.

15.6.6 Gel Tissue Analogs

The ubiquity of gels in natural systems and biological tissues has been a great source of interest for these materials. The biphasic architecture of gels that displays both a solid-like rheology (undergoing to morphological changes) and hosts electroconvective and chemodiffusional processes, typical of fluids, elicits a series of complex biological functionalities.

The comprehension of gel systems has lead to a better understanding of many biological phenomena.[162–164] Obviously, the knowledge about gels comes through an approximated model that is a synthetic and incomplete representation of real systems.

Between the theoretical "biphasic" material and the real system of a biological tissue, there are two main differences. The "biphasic" model, shown in the preceding sections, was developed by Biot for the study of impregnated rooks in the geological field. Taken as it is, in principle, it does not make any differences with gels.

If the low-frequency model gets a satisfying experimental validation and gives interesting insights about biological phenomena such as the streaming potentials in bone,[165] the sensing ability of skin,[162] and the compressional behavior of cartilage, the response of gels to ultrasounds deviates from that of the standard poroelastic theory.[166] As shown recently, the response of gels to ultrasounds is sensibly influenced by the presence of bounded water.[167]

Moreover, even if a sophisticated biphasic model can describe the gel behavior, the tissue usually hosts an amount of cells that own a membrane with a jelly internal structure whose poroelastic properties can differ much from the extracellular matrix.

Therefore, the model of a soft biological tissue must introduce a syncytium of dispersed cells into the gel matrix.[168] In this case, it is possible to define an overall poroelasticity of the composed mean that depends on the extracellular matrix characteristics, the cell membrane permeability, the cell internal poroelasticity, and its fractional volume. Since all these variables depend on the physiological state of the cells, in principle, the ultrasound tissue response can be correlated to its health state. The possible applications in this field are the development of noninvasive methodologies for health tissue evaluation such as elastography for cirrhosis[169] and biphasic imaging.[170]

An accurate model of gel response allows also the realization of phantoms by which to replicate and to investigate biological phenomena in a laboratory in a repetitive manner. The main applications in this field are ultrasound thermal therapy for tumors[171,172] and focused ultrasound surgery.[173] Rovai and collaborators[174] have studied the effect of cavitation ultrasound microbubbles on thrombi in an experimental cell by using a gel tissue phantom. Additionally, pulmonary comets[175] can be replicated by means of a gel phantom mimicking the lung structure leading to quantitative analysis of the phenomenon and its correlation with the water present in the tissue[176] and its origin.

15.6.7 Gels for Cell Culture and Tissue Engineering

In vitro cell and tissue culture has experienced tremendous advances in the last 20 years, mostly due to the discovery and exploitation of new techniques in cell pattering and printing[177] and tissue engineering.[178] Cell encapsulation, cell sheet adhesion modulation, and two-dimensional (2D) and 3D polymer scaffolding for cell support have all benefited from the use of hydrogels[165] and responsive gels.[179] Temperature- and biomolecule-responsive hydrogels are currently used to endow the extracellular artificial cell matrix with specific functional properties that allow controlled surface modification or the release of bioactive compounds to interact with the cellular component. The study of temperature-sensitive gels and their application to cellular engineering has been pioneered by T. Okano. Dramatic changes in the wettability of poly(N-isopropylacrylamide) (PIPAm) grafted onto cell culture substrates[170] after temperature variation from 32°C (PIPAm LCST) and 20°C have been exploited to detach cell cultures from their substrate without using digestive enzymes or chelating agents. Cells adhere to and proliferate onto PIPAm-treated culture dishes at 37°C when the surface is hydrophobic and they detach when temperature is lowered to 20°C, when the surface becomes hydrophilic.[180]

Based on this technique, an alternative approach to 3D scaffolding for tissue engineering has been proposed with the name of "cell sheet engineering,"[181] which promises to broaden the capabilities and applicability of tissue engineering in the clinical setting.[182,183]

The process of biomineralization has also been shown to be affected by temperature-sensitive hydrogels when PIPAm was grafted onto poly(L-lactic) acid and bioglass.[184] Apatite was found to form on the substrate when the material was kept at the temperature of 37°C, above LCST of PIPAm, while no precipitation occurred below 32°C. Poly(L-lactic) acid and bioglass substrates were modified by grafting chitosan through plasma treatment, and pH chitosan responsively was exploited to trigger apatite deposition.[185]

APPENDIX 1

NETWORK READJUSTMENT KINETICS

When the equilibrium conditions change, the gel starts to readjust itself to a new stationary state.

Given the multivariable nature of the gel's free energy, the readjustment kinetics may concern a large number of thermodynamic variables such as temperature, volume, pressure, length, mechanical stress, concentration of chemical species of the interstitial solution, and electrical potential.

When such a system changes its state, forces and related fluxes of all sets of extensive-intensive variables appear. Generally speaking, the kinetics are coupled with each other and the overall system of equations is not manageable.

In certain cases, these kinetics have very different characteristic times so that the related thermodynamical coordinates can be considered quasiconstant or at equilibrium during the process allowing simplified descriptive models.[186]

For instance, if the diffusion of the chemical species into the gel network is much slower than its mechanical readjustment, the gel volume can be considered to be at equilibrium with the local chemical conditions.

In other cases, under controlled laboratory conditions, some of the gel variables can be held constant, allowing the investigation of one type of readjustment kinetic at a time. This is the case of free swelling experiments in which it is possible to observe the mechanical readjustment of the gel network while the gel temperature and its chemical environment are held constant.

By means of laboratory-controlled experiments it is possible to determine the chemical and elastic constants of the gel present in the theoretical model. Moreover, when the gel deformations occur on a scale much larger than the molecular one, the continuum approach can be assumed.

Even with some limitations,[166] continuum poroelastic models have been shown to satisfactorily describe the readjustment of a gel system.[187,188]

In the low-frequency limit, the Biot continuum poroelastic model[189] is a simple and elegant theory that also describes the diffusion kinetics of gels.

In gel with diluted polymer (solid) content and with incompressible fluid and solid constituents, the mechanical readjustment of the gel network is satisfactorily described by the stationary solvent approximation. In this case, the Biot model disembogues in the Tanaka, Hocker, and Benedek (THB) model[190] frictional equation

$$f\,\partial U_i/\partial t = \partial \sigma_{ij}/\partial x_j, \tag{A15.1}$$

where U_i is the displacement vector of a gel element, f is the gel friction coefficient (the inverse of the gel hydraulic permeability[191]), σ_{ij} is the gel stress tensor that, in the linear approximation, reads,

$$\sigma_{ij} = k\,\varepsilon_{\alpha\alpha}\delta_{ij} + 2\mu(\varepsilon_{ij} - \varepsilon_{\alpha\alpha}\delta_{ij}/3) + \alpha\,\delta_{ij}, \tag{A15.2}$$

where k and μ are, respectively, the bulk and the shear elastic moduli of the gel, α is the chemically or thermally induced stress (at zero strain) for isotropic materials, δ_{ij} is the Kroneker delta and $\varepsilon_{\alpha\alpha}$ is the gel dilatation given by the trace of the strain matrix ε_{ij}:

$$\varepsilon_{ij} = \left(\partial U_i / \partial x_j + \partial U_j / \partial x_i\right)/2 \tag{A15.3}$$

Generally speaking, the material parameters in Equations A15.1 and A15.2 are functions of the physical variables of the material (e.g., temperature [see Figs. A15.1 and A15.2] and strain) as well as of the chemical ones (e.g., pH, ionic strength, and type of solvent).

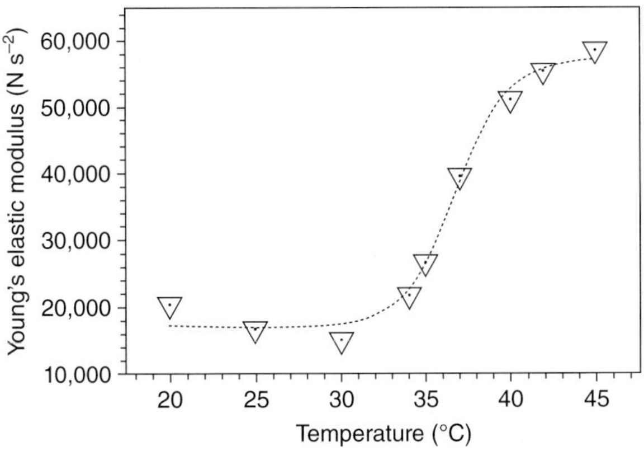

FIGURE A15.1. The Young's elastic modulus $E = 9\mu k/(3k + \mu)$ of the PVME macroporous gel as a function of temperature.

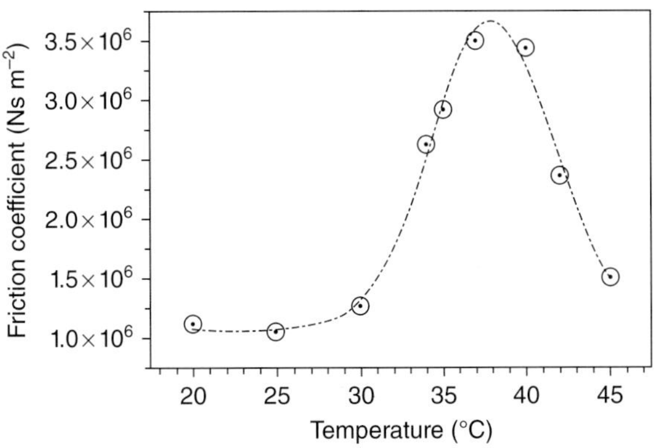

FIGURE A15.2. The fluid-matrix friction coefficient f of the PVME macroporous gel as a function of temperature.

Because the dependence of the material parameters on the mechanical deformation is usually weak, μ, k, and f can be assumed constants if the gel state is far away from a volume phase transition.

By introducing Equation A15.2 in Equation A15.1, by taking the divergence of both members and inserting the incompressibility of solid and liquid constituents, the equation of motion finally reads[192]

$$f\,\partial\varepsilon_{\alpha\alpha}/\partial t = (k + 4\mu/3)\partial^2\varepsilon_{\alpha\alpha}/\partial x_i\,\partial x_i. \tag{A15.4}$$

Actually, the poroelastic model is a semiempirical model, wherein the link between the material parameters and the physicochemical structure of the gel is not explicit.

This fact allows the model to hold for a large typology of gel systems, where the dependence of the material parameters on the environmental conditions is usually experimentally measured.

As a first example, we consider the free swelling experiments, a well-known technique that allows characterization of the poroelastic constants of a gel system. In such tests, an initially compressed gel network is allowed to freely expand in a bath held under constant physicochemical conditions.

In the case of gel shapes with particular symmetries such as the form of a sphere, a thin cylinder, or a thin planar film, Equation A15.4 can be simplified and leads to analytical solutions for the time evolution of the local gel strain.[193,194] Here we report the solution concerning the case of gel in the form of a thin planar film or disk. When the gel sample has the shape of a thin quasiplanar layer, assuming the z-axis perpendicular to the gel layer plane, Equations A15.2 and A15.4 can be simplified to read[195]

$$\partial\varepsilon_{xx}/\partial t = D\,\partial^2\varepsilon_{xx}/\partial z^2 \tag{A15.5a}$$

$$\partial\varepsilon_{yy}/\partial t = D\,\partial^2\varepsilon_{yy}/\partial z^2 \tag{A15.5b}$$

$$\partial\varepsilon_{\alpha\alpha}/\partial t = D_b\,\partial^2\varepsilon_{\alpha\alpha}/\partial z^2, \tag{A15.5c}$$

where $D = \mu/f$ and $D_b = (k + 4\mu/3)/f$.

The spatiotemporal solutions for the strains ε_{xx} (Fig. A15.3) and ε_{zz} (Fig. A15.4) are[193]

$$\varepsilon_{xx} = \frac{4\varepsilon_0}{\pi}\sum_{n=1}^{\infty}\left(\frac{(-1)^n}{2n+1}\right)\exp\left[\frac{(2n+1)^2\,t}{\tau}\right]\cos\left(\frac{(2n+1)z\pi}{a}\right) \tag{A15.6a}$$

$$\begin{aligned}\varepsilon_{zz} &= \varepsilon_{\alpha\alpha} - 2\varepsilon_{xx}\\ &= \frac{4\varepsilon_0}{\pi}\sum_{n=1}^{\infty}\left(\frac{(-1)^n}{2n+1}\right)\cos\left(\frac{(2n+1)z\pi}{a}\right)\left\{3\exp\left[\frac{(2n+1)^2\,t}{\tau_b}\right] - 2\exp\left[\frac{(2n+1)^2\,t}{\tau}\right]\right\},\end{aligned} \tag{A15.6b}$$

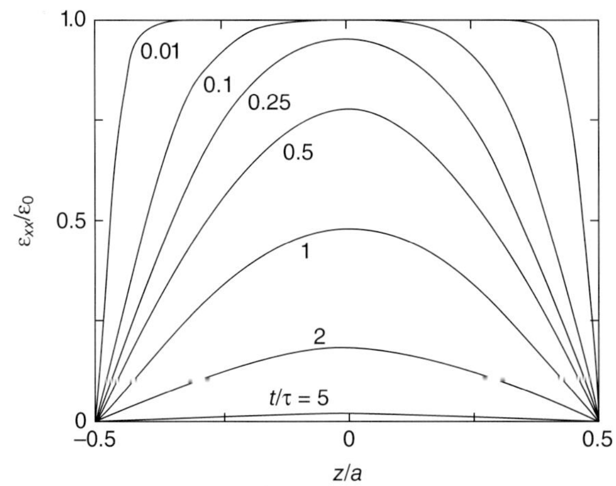

FIGURE A15.3. Normalized strain $\varepsilon_{xx}/\varepsilon_0$ in a partially dried hydrogel strip undergoing free swelling as a function of the reduced variable z/a at various scaled times t/τ given by Equation A15.6a.

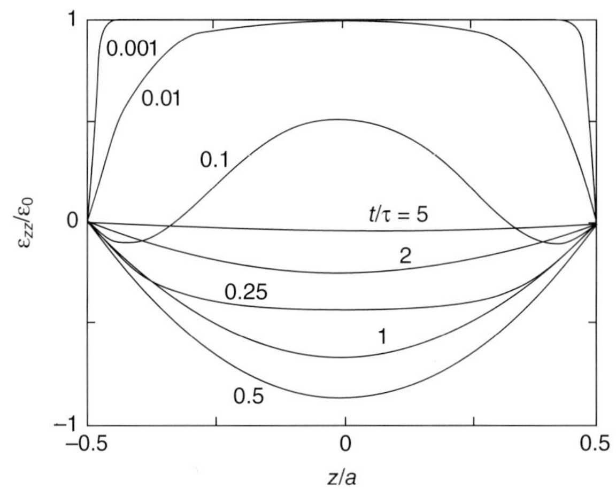

FIGURE A15.4. Normalized dilatation $\varepsilon_{zz}/\varepsilon_0$ in a partially dried hydrogel strip undergoing free swelling as a function of the reduced variable z/a at various scaled times t/τ given by Equation A15.6b.

where $\tau = a^2/\pi^2 D$ and $\tau_b = a^2/\pi^2 D_b$ are the characteristic time constants for the "shear" and "bulk" gel readjustment where $\tau_b < \tau$, since $D_b > D$; ε_0 is the initial uniform strain of the gel sample with respect to the final one (at $t = \infty$) assumed as reference ($\varepsilon_\infty = 0$), and a is the gel layer thickness at $t = \infty$.

From Equation A15.6a,b, the thickness $a_{(t)}$ (Fig. A15.5) and the length $L_{(t)}$ (Fig. A15.6) of the gel are obtained as a function of time, respectively, to read[193]

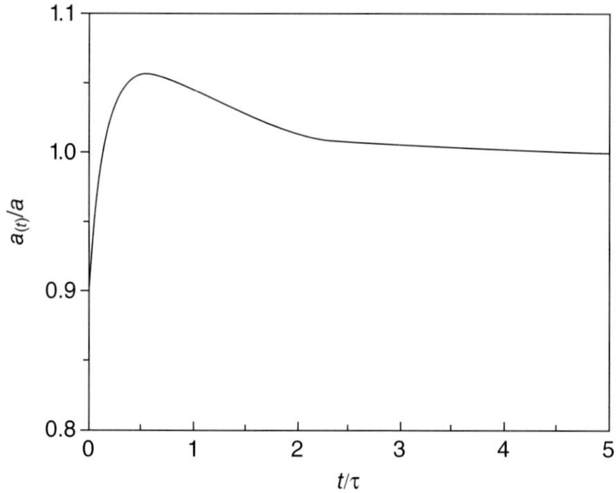

FIGURE A15.5. Normalized thickness of a thin hydrogel strip undergoing free swelling as a function of the reduced time t/τ given by Equation A15.7a.

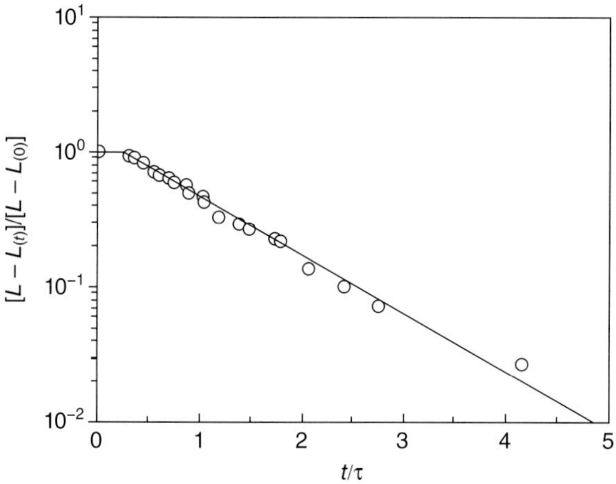

FIGURE A15.6. Normalized length variation of a thin hydrogel strip undergoing free swelling as a function of the reduced time t/τ given by Equation A15.8.

$$a_{(t)} = a\left\{1 + 2\int_0^{a/2} \varepsilon_{zz}\right\}$$

$$= a\left\{1 + \frac{8\varepsilon_0}{\pi^2}\sum_{n=1}^{\infty}\left(\frac{(-1)^n}{2n+1}\right)\left\{3\exp\left[\frac{(2n+1)^2 t}{\tau_b}\right] - 2\exp\left[\frac{(2n+1)^2 t}{\tau}\right]\right\}\right\} \tag{A15.7a}$$

$$L_{(t)} = L_\infty\{1 + \varepsilon_{xx}(z=0)\} = L_\infty\left\{1 + \frac{4\varepsilon_0}{\pi}\sum_{n=1}^{\infty}\left(\frac{(-1)^n}{2n+1}\right)\exp\left[\frac{(2n+1)^2 t}{\tau}\right]\right\}. \tag{A15.7b}$$

For $t > \tau/9$ in Equation A15.7a, the slower exponential relaxation prevails so that the gel length reads

$$L_{(t)} \cong L_{\infty}\left\{1-\frac{4\varepsilon_0}{\pi}\exp\left[-\frac{t}{\tau}\right]\right\} = L_{\infty}\left\{1-\frac{4\varepsilon_0}{\pi}\exp\left[-\frac{\pi^2 D}{a^2}t\right]\right\} \qquad t > 9\tau. \qquad \text{(A15.8)}$$

By fitting the exponential length relaxation of the gel, it is possible to obtain the characteristic time, τ, and the gel diffusion coefficient, D.

APPENDIX 2

ELECTRODIFFUSION–REACTION KINETICS

In the case where the concentrations of the chemical species may change in the gel bath, the presence of gradients will generate chemical currents in the interstitial solution.

Given that the poroelastic parameters μ, k, f, and α depend on the chemical concentrations C_i of the interstitial solution (see, e.g., Fig. A15.7), their redistribution may induce mechanical swelling or deswelling of the gel system.

In this case, the mechanical Equations A15.1–A15.3 are coupled to the Nernst–Planck electroconvective equations for the motion of each charged chemical species (together with the electrical charge conservation and the Gauss equation for the definition of the electric field).[196]

The electroconvective kinetics introduces a mathematical complexity that can usually be circumvented when they are very fast with respect to the other

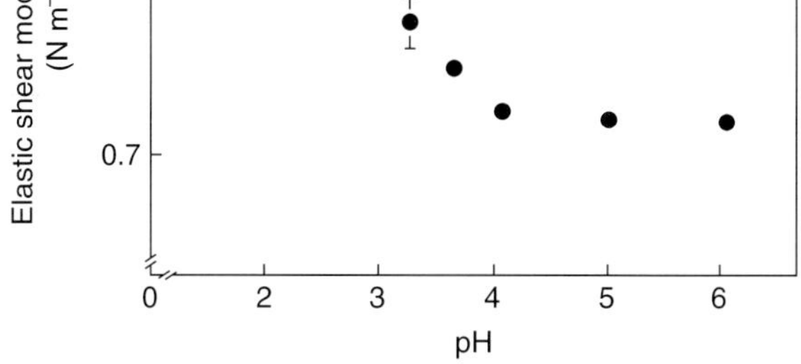

FIGURE A15.7. The shear elastic modulus of polyvinylalcohol–polyacrylic acid gel as a function of external bath pH.

ones, so that the electrical equilibrium (null electrical currents) or the stationary electrical conditions can be imposed.

Assuming that the convective and migration currents can be disregarded, the full system of equations reads

$$f_{(C_i)}\,\partial U_i/\partial t = \partial\sigma_{ij}/\partial x_j \tag{A15.9}$$

$$\sigma_{ij} = k_{(C_i)}\varepsilon_{\alpha\alpha}\delta_{ij} + 2\mu_{(C_i)}(\varepsilon_{ij} - \varepsilon_{\alpha\alpha}\delta_{ij}/3) + \alpha_{(C_i)}\delta_{ij} \tag{A15.10}$$

$$\varepsilon_{ij} = (\partial U_i/\partial x_j + \partial U_j/\partial x_i)/2 \tag{A15.11}$$

$$\partial C_i/\partial t = -D_{(C_i)}\partial^2 C_i/\partial x_j \partial x_j + \partial C_{i\,\mathrm{chem}}/\partial t, \tag{A15.12}$$

where $C_{i\mathrm{chem}}$ is the concentration in moles per liters of the ith species that has reacted.

In the above equations, we have considered that the temperature is constant through all the processes.

Actually, temperature can vary as a consequence of chemical reactions as well as of external inputs (as happens in thermally activated gel systems). In this case, the equation of thermal diffusion must be added to the system of equations.

The resulting overall system of equations is complex and usually cannot be treated analytically.

Nevertheless, some insight can come by investigating the simplest case of diffusion of hydrogen ions in a gel matrix with ionizable groups (RAH) undergoing dissociation

$$\mathrm{RAH} \Leftrightarrow \mathrm{RA}^- + \mathrm{H}^+. \tag{A15.13}$$

In this case, the system of equations reads

$$f_{(\mathrm{H}^+)}\,\partial U_i/\partial t = \partial\sigma_{ij}/\partial x_j \tag{A15.14}$$

$$\sigma_{ij} = k_{(\mathrm{H}^+)}\varepsilon_{\alpha\alpha}\delta_{ij} + 2\mu_{(\mathrm{H}^+)}(\varepsilon_{ij} - \varepsilon_{\alpha\alpha}\delta_{ij}/3) + \alpha_{(\mathrm{H}^+)}\delta_{ij} \tag{A15.15}$$

$$\varepsilon_{ij} = (\partial U_i/\partial x_j + \partial U_j/\partial x_i)/2 \tag{A15.16}$$

$$\partial[\mathrm{H}^+]/\partial t = -D_{(\mathrm{H}^+)}\partial^2[\mathrm{H}^+]/\partial x_j \partial x_j - \partial[\mathrm{RAH}]/\partial t. \tag{A15.17}$$

Usually, the chemical reaction is faster than the gel mechanical readjustment as well as the chemical diffusion, so the reaction equilibrium condition can be applied to read

$$K_A^* = [\mathrm{RA}^-][\mathrm{H}^+]/[\mathrm{RAH}], \tag{A15.18}$$

where K_A^* is not a simple reaction constant since, in the gel matrix, the ionization of each functional group is influenced by the state of the groups in its neighborhood.[197]

Therefore, K_A^* depends on the hydrogen ion concentration and is related to the free acid equilibrium constant $K_A = [A^-][H^+]/[AH]$ by the relation

$$K_A^* = [H^+] + K_A. \tag{A15.19}$$

Moreover, given that the variation of the chemical stress $\alpha_{(H)}$ as a function of the proton concentration is much bigger than those of $f_{(H^+)}$, $k_{(H^+)}$, and $\mu_{(H^+)}$ at the zero order of approximation, we can consider the latter ones as constants to end with the motion equations,

$$f\,\partial U_i / \partial t = \partial\sigma_{ij} / \partial x_j \tag{A15.20}$$

$$\sigma_{ij} = k\varepsilon_{\alpha\alpha}\delta_{ij} + 2\mu\left(\varepsilon_{ij} - \varepsilon_{\alpha\alpha}\delta_{ij}/3\right) + \alpha_{(H^+)}\delta_{ij} \tag{A15.21}$$

$$\varepsilon_{ij} = \left(\partial U_i/\partial x_j + \partial U_j/\partial x_i\right)2 \tag{A15.22}$$

$$\partial[H^+]/\partial t = -D_{\text{eff}}\partial^2[H^+]/\partial x_j \partial x_j \tag{A15.23}$$

$$D_{\text{eff}} = D_{(H^+)} / \left(1 + [RA^-]K_A / \left([H^+] + K_A\right)^2\right) \tag{A15.24}$$

$$[RA^-]_{(x,t)} = [RA^-]_0\, \varepsilon_{\alpha\alpha(x,t)}, \tag{A15.25}$$

where $[RA^-]_0$ is the ionizable group concentration at the initial gel volume.

From the motion equations (Eqs. A15.20–A15.25), we can see that the chemical problem is coupled to the mechanical one through the gel dilatation evolution $\varepsilon_{\alpha\alpha(x,t)}$ and that the mechanical dynamics are coupled to the chemical through the chemical stress $\alpha_{(H^+(x,t))}$.

If we start from a basic condition (swollen polyacid gel with most of the ionizable groups dissociated), as far as $[H^+] \ll K_A \sim 10^{-4\div5}$, we can use the approximation[198]

$$D_{\text{eff}} \approx D_{(H^+)} / \left(1 + [RA^-]/K_A\right). \tag{A15.26}$$

From the above formula, when the ionizable group density $[RA^-]$ on the gel network is much larger than the acid dissociation constant $K_A \sim 10^{-4\div5}$, it follows that the effective diffusion coefficient may be much smaller than the free proton diffusion one. Initially, in the acidification process (inflow of H^+), there is an excess of binding sites available in the gel so that almost all the incoming hydrogen ions are immediately bound to these sites and thus cannot freely diffuse through the gel.

Since the mechanical readjustment, as well as the chemical process, follow diffusive kinetics, their characteristic times, which scale by the square of the same characteristic length of the system, depend on the elastic constant $k_{(H^+)}$, $\mu_{(H^+)}$, on the friction coefficient $f_{(H^+)}$ of the gel network, on the hydrogen diffusion coefficient $D_{(H^+)}$, and on the ratio between the ionizable group density $[RA^-]$ of the gel network and the acid constant K_A, respectively. Therefore, in a soft, weakly charged gel ($[RA^-]/K_A \ll 1$) with a highly viscous solvent–network interaction, it may result

$$D_{\text{eff}} \approx D_{(\text{H}^+)} \gg D_b = (k + 4\mu/3)/f. \tag{A15.27}$$

In this case, the network motion is much slower that the chemical kinetics so that they decouple themselves: The gel volume change can be described to happen as a consequence of a step change of the chemical conditions to the final ones.

The chemomechanical decoupling can also happen in a strong elastic gel (μ and k are very high) with heavily charged network ($[\text{RA}^-]/K_A \gg 1$) where it may result

$$D_{\text{eff}} \approx D_{(\text{H}^+)} K_A / [\text{RA}^-] \ll D = \mu / f < D_b. \tag{A15.28}$$

In this case, the gel kinetics is simplified since its mechanical state is always at equilibrium with the local chemical conditions.

The proton diffusion is a simple example of chemomechanoelectrical kinetics in gels. More recently, the development of bioresponsive hydrogel for biomedical application has been brought to the synthesis of many complex gel systems where the diffusion of a specific chemical species or biological compound elicits a process leading the gel volume and poroelasticity to change. Even much more complex, this kinetics follows the scheme of a diffusion of a compound in the gel matrix coupled to a reaction that leads to the readjustment of the network as well as of the ionic concentrations in the interstitial solution. This is the case of enzyme-loaded gels that, in presence of the target substance, may change the pH of the interstitial solution, or the case of gels using the antigen–antibody binding reaction to change the degree of cross-linking of its network.

APPENDIX 3

A NONEQUILIBRIUM THERMODYNAMICS VIEW OF ELECTROMECHANICAL PHENOMENA

When the gel has a charged polymer matrix, or ions are dissolved in the interstitial solution, in addition to the polymer network motion we have fluxes of ionic charges.

In this case, the hydraulic flux governed by the Darcy law is coupled to the electric one (Fig. A15.8) that at first order can be described by the Onsager relations that, in the interstitial stationary fluid approximation, read

$$J_i^{\text{netw}} = \partial\left(U_i - u_i^{\text{fluid}}\right)/\partial t \cong \partial U_i/\partial t = K_{11}\partial P/\partial x_i + K_{12}\partial\Psi/\partial x_i \tag{A15.29}$$

$$J_i^{\text{charge}} = K_{12}\partial P/\partial x_i + K_{22}\partial\Psi/\partial x_i, \tag{A15.30}$$

where K_{ij} are the electro-osmotic Onsager coefficients,[199] P is the hydraulic gel pressure, Φ is the electric potential, J_i^{netw} is the current of the polymer with respect to the fluid, and J_i^{charge} is the ionic charge current of the ith species.

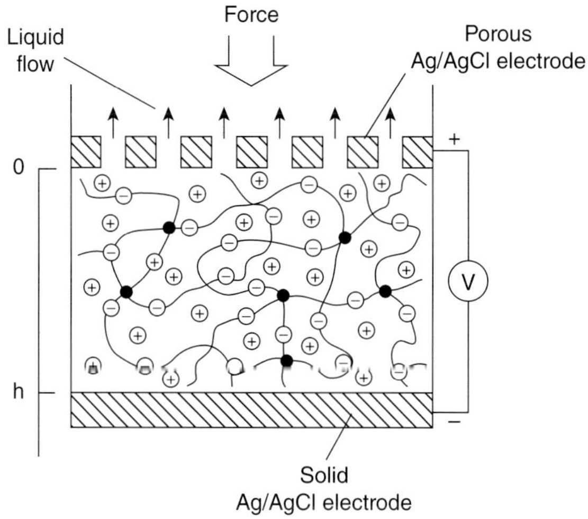

FIGURE A15.8. Schematic representation of the streaming potential operating in a sample of a negatively charged polymer network with positive mobile counterions. When the sample is compressed, the generated water flow through the porous electrode induces the mobile charge displacement that originates an electric potential.

The introduction of the electric field increases the number of variables as well as the number of equations. As already noted in the preceding paragraph, the Gauss equation for the electric field as well as the charge conservation law must be introduced in order to know how the electric potential builds itself up when the electric charges move.

Even if no external electric field is applied, there exists the one generated by the motion of the charges due to the hydraulic pressure gradients in the coupled equations (Eqs. A15.29–A15.30).

The introduction of the Gauss equation makes it more difficult to have a tractable solution of the electromechanical problem. Nevertheless, in gels when the ionic conductivity is so high that the electrical equilibrium is much faster than the network diffusive readjustment, we can assume that the electric charges are always at the stationary state to read

$$J_i^{\text{charge}} = K_{21}\partial P/\partial x_i + K_{22}\partial \Psi/\partial x_i = 0. \qquad \text{(A15.31)}$$

This simplifies very much the problem leading to the expression for the electric potential that reads

$$\partial \Psi/\partial x_i = -(K_{21}/K_{22})\partial P/\partial x_i, \qquad \text{(A15.32)}$$

from which we obtain,

$$\partial U_i/\partial t = \{K_{11} - K_{12}(K_{21}/K_{22})\}\partial P/\partial x_i. \qquad \text{(A15.33)}$$

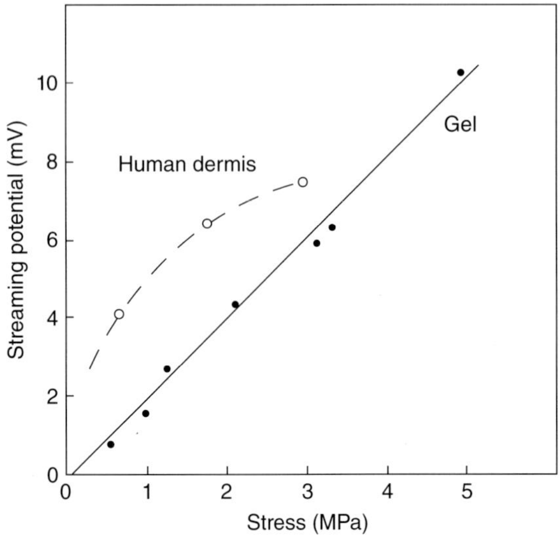

FIGURE A15.9. Stress-generated potentials versus applied load for 400-μm-thick polyvinylalcohol–polyacrylic acid gels and human skin samples in water.

Moreover, even if the solvent is stationary, in the low-frequency limit a non-null force acting on the polymer network exists, which is at equilibrium with the hydraulic pressure gradient to read

$$\partial P / \partial x_i \cong \partial \sigma_{ij} / \partial x_j, \tag{A15.34}$$

so that the friction coefficient by Equation A15.33 reads

$$f = (K_{11} - K_{12} K_{21} / K_{22})^{-1}. \tag{A15.35}$$

It is straightforward to see that f reduces to the inverse of hydraulic permeability K_{11} when no charges are present into the gel system and $K_{12} = K_{21} = 0$.

This electromechanical coupling in gels is responsible for the generation of electrical potentials in the presence of gel readjustment induced both by mechanical inputs such as compression (Fig. A15.9) and deformation and by physicochemical inputs such as temperature, pH, salt concentration, and solvent affinity. Inversely, it is responsible for the gel mechanical response to the electrochemical currents and potentials.

REFERENCES

1. K. Park, T. Okano, R. M. Ottembrite, N. A. Peppas, eds. *Biomedical Applications of Hydrogels Handbook*. Springer-Verlag, New York, 2010.
2. A. S. Hoffman. *Artif Organs* **1995**, 19, 5.

3. J. F. Mano. *Adv Eng Mater* **2008**, 10 (6), 515.
4. J. W. Breitenbach and H. Karlinger. *Monatsh Chem* **1949**, 80, 2.
5. W. Kuhn. *Experientia* **1949**, 5, 8.
6. A. Katchalsky. *Experientia* **1949**, 5, 8.
7. Y. Osada. *Adv Polym Sci—Polym Phys* **1987**, 82, 1.
8. J. Hasa, M. Ilavsky, K. Dusek. *J Polym Sci—Polym Phys* **1975**, 13, 253.
9. T. Tanaka. *Phys Rev Lett* **1978**, 40, 12.
10. S. Ahn, R. M. Kasi, S. Kim, N. Sharma, Y. Zhon. *Soft Matter* **2008**, 4, 1151.
11. Y. Osada, J. P. Gong, Y. Tanaka. *J Macromol Sci—Polym Rev* **2004**, C44, 1.
12. D. De Rossi, K. Kajiwara, Y. Osada, A. Yamauchi, eds. *Polymer Gels: Fundamentals and Biomedical Applications*. Plenum Press, New York, 1991.
13. S. Chaterji, I. K. Kwon, K. Park. *Prog Polym Sci* **2007**, 32 (8–9), 1083.
14. N. A. Peppas, J. Z. Hilt, A. Khademhosseini, R. Langer. *Adv Mater* **2006**, 18, 1345.
15. I. Z. Steinberg, A. Oplatka, A. Katchalsky. *Nature* **1966**, 210, 568–571.
16. A. J. Grodzinsky and N. A. Shoenfeld. *Polymer* **1977**, 18, 5.
17. R. P. Hamlen, C. E. Kent, S. N. Shafer. *Nature* **1965**, 206, 1149.
18. T. Miyata, T. Uragami, K. Nakamae. *Adv Drug Deliv Rev* **2002**, 54, 1.
19. J. Atencia and D. J. Beebe. *Nature* **2005**, 437, 648–655.
20. G. Gerlach and K. F. Arndt, eds. *Hydrogel Sensor and Actuators*. Springer, Berlin and Heidelberg, 2009.
21. T. Miyata, N. Asami, T. Uragami. *Nature* **1999**, 399, 766–769.
22. L. Dong, A. K. Agarwal, D. J. Beebe, H. Jiang. *Nature* **2006**, 442, 3.
23. S. R. Eisenberg and A. J. Grodzinsky. *J Membr Sci* **1984**, 19, 2.
24. T. P. Russell. *Science* **2002**, 297, 5583.
25. N. E. Fedorovich, J. Alblas, J. R. De Wijn, W. E. Hennink, A. J. Verbout, W. J. A. Dhert. *Tissue Eng* **2007**, 13, 8.
26. P. Calvert. *Adv Mater* **2009**, 21, 743–756.
27. T. Wallmersperger. Modelling and simulation of the chemo-electro-mechanical behaviour. In *Hydrogel Sensors and Actuators*, G. Gerlach, K. F. Arndt, eds. Springer, Berlin and Heidelberg, 2009.
28. Y. Osada, M. Sato, *Polymer* **1980**, 21, 1057–1061.
29. P. J. Flory. *J Chem Phys* **1941**, 9 (8), 660; P. J. Flory. *Principles of Polymer Chemistry*. Cornel Univ. Press, Ithaca, NY, 1967; M. L. Huggins. *J Chem Phys* **1941**, 9 (5), 440.
30. I. Hasa, M. Ilavsky, K. Dusek. *J Polym Sci* **1975**, 13, 253.
31. T. Tanaka, et al. *Phys Rev Lett* **1980**, 45 (20), 1636.
32. M. Heskins and J. E. Guillet. *J Macromol Sci Chem A* **1986**, 28, 923.
33. Y. H. Bae, T. Okano, S. W. Kim. *J Polym Sci B Polym Phys* **1990**, 28, 223.
34. G. Chen and A. S. Hoffman. *Nature* **1995**, 375, 49.
35. O. Hirasa, Y. Morishita, R. Onomura. *Kobunshi Ronhbunshu* **1989**, 46, 661.
36. J. A. Hinkley, L. D. Morgret, S. H. Gehrke. *Polymer* **2004**, 45, 8837.
37. A. Chilkoti, et al. *Biocon Chem* **1994**, 5, 504.
38. S. Murdan. *J Controll Release* **2003**, 92, 1.

39. T. Shiga. *Adv Polym Sci* **1997**, 134, 131.
40. A. Fragala, et al. *Electrochim Acta* **1972**, 17, 8.
41. T. Tanaka, et al. *Science* **1982**, 18 (4571), 467.
42. Y. Osada and M. Hasebe. *Chem Lett* **1985**, 14, 1285–1288.
43. D. De Rossi, et al. *Trans Am Soc Artif Intern Organs* **1986**, 32, 157.
44. T. Shiga and T. Karauchi. *J Appl Polym Sci* **1990**, 39, 2305.
45. Y. Osada, et al. *Nature* **1992**, 355, 242.
46. K. Takada, N. Tanaka, T. Tatsuma. *J Electroanal Chem* **2005**, 585, 120.
47. G. Filipcsei, I. Csetneki, A. Szilagyi, M. Zrinyi. *Adv Polym Sci* **2007**, 206, 137.
48. C. Bellan and G. Bossis. *Int J Mod Phys B* **2002**, 16, 2447.
49. Y. An, B. Liu, M. T. Shaw. *Int J Mod Phys B* **2002**, 16, 2440.
50. T. Mitsumata, K. Sakai, J. I. Takimoto. *J Phys Chem B* **2006**, 23, 2217.
51. A. Suzuki. *J Intell Mater Syst Struct* **1994**, 5, 112.
52. A. Shiotani, T. Mori, T. Niidome, et al. *Langmuir* **2007**, 23, 4012.
53. N. Hosono, H. Furukawa, Y. Masubuchi, et al. *Colloid Surface B* **2007**, 56, 285.
54. M. Irie. *Macromolecules* **1986**, 19, 2890.
55. M. Mycic, Y. Zheng, V. Moy, et al. *Colloid Surface B* **2002**, 27, 147.
56. T. Watanabe, M. Akiyama, K. Totani, et al. *Adv Funct Mater* **2002**, 12, 611.
57. K. Takada, T. Miyazaki, N. Tanaka, T. Tatsuma. *Chem Commun* **2006**, 2024–2026.
58. T. Tatsuma, K. Takada, T. Miyazaki. *Adv Mater* **2007**, 19, 1249.
59. H. B. Schreyer, N. Gebhart, K. J. Kim, M. Shahinpoor. *Macromolecules* **2000**, 1 (4), 642.
60. M. Suzuki and O. Hirasa. *Adv Polym Sci* **1993**, 110, 241.
61. X. Zang and R. Zhuo. *Eur Polym J* **2000**, 36, 2301.
62. E. Rodriguez and I. Katime. *J Polym Sci* **2003**, 90, 530.
63. E. Rodriguez and I. Katime. *J App Polym Sci* **2003**, 3, 612–619.
64. Y. Yoshioka and P. Calvert. *Exp Mech* **2002**, 42, 404.
65. H. Tamagawa and M. Taya. *Mater Sci Eng A* **2000**, 285–314.
66. Z. Zang, T. Chao, S. Jiang. *J Phys Chem B* **2008**, 112, 5327.
67. H. J. Schneider and R. M. Strongin. *Acc Chem Res* **2009**, 42 (10), 1489.
68. R. Yoshida, T. Takahashi, T. Yamaguchi, H. Ichijo. *J Am Chem Soc* **1996**, 118, 5134.
69. V. V. Yashin and A. C. Balazs. *J Chem Phys* **2007**, 126, 124707 (online).
70. P. D. Topham, J. R. Howse, C. J. Crook, et al. *Macromolecules* **2007**, 40, 4393.
71. J. R. Howse, P. D. Topham, C. J. Crook, et al. *Nano Lett* **2006**, 6, 73.
72. G. J. Mohr. *Sens Actuators B* **2005**, 107, 2.
73. G. J. Mohr. *Anal Bioanal Chem* **2006**, 386, 1201.
74. Y. Kitano, K. Koyama, O. Kataoka, T. Kazunori, et al. *J Controll Release* **1992**, 19, 161.
75. Y. K. Choi, S. Y. Jeong, Y. H. Kim. *Int J Pharm* **1992**, 80, 9.
76. T. Miyata, A. Jikihara, K. Nakamae, A. S. Hoffman. *Macromol Chem Phys* **1996**, 197, 1135.

77. W. L. D. Murphy, W. S. Dillmore, J. Modica, M. Mrksich. *Angew Chem Int Edn* **2007**, 46, 3066.
78. Z. Sui, W. King, W. L. Murphy. *Adv Mater* **2007**, 19, 3377.
79. J. S. Kim, N. Singh, L. A. Lion. *Chem Mater* **2007**, 19, 2527.
80. T. Traitel, Y. Cohen, J. Kost. *Biomaterials* **2000**, 21, 1679.
81. P. D. M. Thornton, R. J. Mart, R. V. Ulijn. *Adv Mater* **2007**, 19, 1252.
82. A. G. Patrick and R. V. Ulijn. *Mater Res Soc Proc* **2007**, 1063, 06–05.
83. R. Zhang, A. Bowyer, R. Eisenthal, J. Hubble. *Biotechnol Bioeng* **2007**, 97, 976.
84. Y. Aoyagi, Y. Suzuki, M. Isemura, et al. *Cancer* **1988**, 61, 769–774.
85. T. Miyata, M. Jige, T. Nakaminami, T. Uragami. *Proc Natl Acad Sci U S A* **2006**, 103 (5), 1190–1193
86. W. M. Megil, J. M. Gosline, R. W. Blake. *J Exp Biol* **2005**, 208, 3819–3834.
87. J. Gosline. *J Exp Biol* **1971**, 55, 763–774.
88. T. J. Koob, M. M. Koob-Emunds, J. A. Trotter. *J Exp Biol* **1999**, 202, 2291–2301.
89. J. Gosline. *J Exp Biol* **1971**, 55, 775–785.
90. X. Zhang, F. Wang, C. Chu. *J Mater Sci Mater Med* **2003**, 14, 451–455.
91. X. Zhang and R. Zhuo. *Eur Polym J* **2000**, 36, 2301–2303.
92. D. Y. Lee, Y. Kim, S. J. Lee, et al. *Mater Sci Eng C Biomim Mater Sens Syst* **2007**, 28, 220–226.
93. K. Choe and K. J. Kim. *Sens Actuators A* **2006**, 126, 165–172.
94. K. J. Kim and M. Shahinpoor. *Smart Mater Struct* **2002**, 11, 1–5.
95. S. Guo, T. Fukada, K. Kosuge, et al. Proceedings of the 5th International Symposium on Micromachine and Human Science, Nagoya, **1994**, pp.191–197.
96. T. Fukushima, K. Asaka, et al. *Angew Chem Int Ed* **2005**, 44 (16), 2410–2413.
97. M. Mojarrad and M. Shahinpoor. Proceedings of the SPIE International Symposium on Smart Structures and Materials, Smart Sensing, Processing, and Instrumentation, **1997**, 3042, pp. 52–60.
98. P. Chiarelli and D. De Rossi. *Polym Gels Netw* **1996**, 4, 499–508.
99. K. Arndt, A. Richter, S. Ludwig, et al. *Acta Polimer* **1999**, 50, 383–390.
100. D. De Rossi, et al. *IEEE Trans Biomed Eng* **1988**, 35 (2), 83.
101. D. De Rossi, et al. *Sens Actuators A Phys* **1989**, 17 (1–2), 107.
102. K. Sawahata, J. P. Gong, Y. Osada. *Macromol Rapid Commun* **1995**, 16 (10), 713.
103. S. M. Lee, et al. *Polym Mater Sci Eng* **2004**, 90, 469.
104. Y. J. Kang, et al. *Nat Mater* **2007**, 6, 957.
105. Y. Sakai, Y. Sadaoka, M. Matsuguchi, et al. *IEEE International Conference on Solid State Sensors and Actuators* **1991**, 562–565.
106. Y. Osada and D. De Rossi, eds. *Polymer Sensors and Actuators*. Springer-Verlag, Berlin, Heidelberg, 2000.
107. G. Gerlach, et al. *Sens Actuators B* **2005**, 111–112, 555.
108. A. Richter, et al. *Sens Actuators B* **2004**, 99 (2–3), 579.
109. S. Herber, et al. Proceedings on Transducers ’05, **2005**, 1146.
110. J. H. Holtz and S. A. Asher. *Nature* **1997**, 389, 829.

111. Y. Qiu and K. Park. *Adv Drug Deliv Rev* **2001**, 53 (3), 321.
112. A. Guiseppi-Elie, S. I. Brahim, D. Narinesingh. *Adv Mater* **2002**, 14 (10), 743.
113. R. A. Siegel and B. A. Firestone. *J Controll Release* **1990**, 11 (1–3), 181.
114. I. Hisamitsu, et al. *Pharm Res* **1997**, 14, 289.
115. S. Kabilan, et al. *Biosens Bioelectron* **2005**, 20 (8), 1602.
116. V. L. Alexeev, et al. *Clin Chem* **2004**, 50, 2353.
117. M. E. H. El-Sayed, A. S. Hoffman, P. S. Stayton. *J Controll Release* **2005**, 101, 47.
118. P. S. Stayton, et al. *Nature* **1995**, 378, 472.
119. G. M. Whitesides. *Nature* **2006**, 442, 368.
120. D. Janasek, J. Franzke, A. Manz. *Nature* **2006**, 442, 374.
121. G. H. W. Sanders and A. Manz. *Trends Anal Chem* **2000**, 19, 364.
122. J. El-Ali, P. K. Sorger, K. F. Jensen. *Nature* **2006**, 442, 403.
123. K. E. Peterson. Proceedings of the IEEE 70, **1982**, 420.
124. D. J. Harrison, et al. *Science* **1993**, 261, 895.
125. A. Manz, N. Graber, H. M. Widmer. *Sens Actuators B* **1990**, 1, 244.
126. J. Atencia and D. J. Beebe. *Nature* **2005**, 437, 648.
127. D. J. Beebe, et al. *Nature* **2000**, 404, 588.
128. B. Ziaie, et al. *Adv Drug Deliv Rev* **2004**, 56 (2), 145.
129. S. Sugiura, et al. *Lab Chip* **2009**, 9, 196.
130. Q. Luo, et al. *Electrophoresis* **2003**, 24 (21), 3694.
131. R. H. Liu, et al. *J Microelectromech Syst* **2001**, 11, 45.
132. R. H. Liu, et al. *J Microelectromech Syst* **2000**, 9, 190.
133. D. Kim and D. J. Beebe. *Lab Chip* **2007**, 7, 193.
134. L. Dong, et al. *Nature* **2006**, 442, 551.
135. L. Dong and H. Jiang. *Appl Phys Lett* **2006**, 89, 211120.
136. X. Zeng and H. Jiang. *Appl Phys Lett* **2008**, 93, 151101.
137. G. M. Gratson, M. Xu, J. A. Lewis. *Nature* **2004**, 428, 386.
138. A. Sidorenko, et al. *Science* **2007**, 315, 487.
139. C.-C. Lin and A. T. Matters. *Adv Drug Deliv Rev* **2006**, 58, 1379–1408.
140. N. A. Peppas, P. Bures, W. Leobandung, H. Ichikawa. *Eur J Pharm Biopharm* **2000**, 50, 27–46.
141. C. A. Kavanagh, Y. A. Rochev, W. M. Gallagher, et al. *Pharmacol Ther* **2004**, 102, 1–15.
142. S. Brahim, D. Narinesingh, A. Guiseppi-Elie. *Macromol Symp* **2002**, 186, 63–73.
143. M. Yokoyama, M. Miyauchi, N. Yamada, et al. *J Control Release* **1990**, 11, 269–278.
144. M. Yokoyama, S. Inoue, K. Kataoka, et al. *Makromol Chem Rapid Commun* **1987**, 8, 431–435.
145. A. Harada and K. Kataoka. *Prog Polym Sci* **2006**, 31, 949–982.
146. S. Commas-Marion, T. Okano, K. Kataoka. *Colloid Surface B* **1999**, 16, 207–215.
147. A. Chourcair, P. L. Soo, A. Eisenberg. *Langmuir* **2005**, 21, 9308–9313.
148. K. Kataoka, A. Harada, Y. Nagasaki. *Adv Drug Deliv Rev* **2001**, 47, 113–131.

149. M. L. Adams, A. Lavasanifar, G. S. Kwon. *J Pharm Sci* **2003**, 92, 1343–1355.
150. A. V. Kabanov, E. V. Batrakova, V. Y. Alakhov. *J Control Release* **2002**, 82, 189–212.
151. A. Gliozzi, V. Vittoria, A. Cifferi. *J Membr Biol* **1972**, 8, 149.
152. J. Kopecek, J. Vacik, D. Lim. *J Polym Sci A-1* **1971**, 9 (10), 2801.
153. Y. Osada, J. P. Gong, Y. Tanaka. *J Macromol Sci C* **2004**, C44 (1), 87.
154. P. E. Grimshaw and A. J. Grodzinsky. *Chem Eng Sci* **1989**, 44 (4), 827.
155. H. Kanazawa, K. Yamamoto, Y. Matsushima. *Anal Chem* **1996**, 68 (1), 100.
156. P. Maharjan, et al. *Innovat Food Sci Emerg Technol* **2008**, 9, 232.
157. J. Kobayashi, et al. *Anal Chem* **2003**, 75 (13), 3244.
158. H. Kanazawa, T. Sunamoto, Y. Matsushima. *Anal Chem* **2000**, 24, 5961.
159. T. Peng and Y. L. Cheng. *J Appl Polym Sci* **1998**, 70, 11.
160. K. Hosoya, et al. *Macromolecules* **1994**, 27 (24), 3973.
161. A. S. Hoffman and P. S. Stayton. *Prog Polym Sci* **2007**, 32 (8–9), 922.
162. D. De Rossi, A. Nannini, C. Domenici. *IEEE Trans Biomed Eng* **1988**, 35, 83–90.
163. S. R. Eisenberg and A. J. Grodzinsky. *J Orthop Res* **1985**, 3, 148–259.
164. R. A. Saltzstein, S. R. Pollack, et al. *J Biomech* **1987**, 20, 261–270.
165. N. E. Fedorovich, et al. *Tissue Eng* **2007**, 13 (8), 1905.
166. D. L. Johnson. *J Chem Phys* **1982**, 77, 1531.
167. P. Chiarelli, A. Lanatà, et al. *J Acoust Soc Am* **2010**, 127, 1197–1207.
168. V. Gismondi. Graduate thesis, Faculty of Biomedical Engineering, University of Pisa, **2008**.
169. M. Ziol, A. Handra-Luca, et al. *Hepatology* **2005**, 41, 48–54.
170. G. P. Berry, J. C. Bamber, et al. *Ultrasound Med Biol* **2006**, 32, 547–567.
171. S. Lockhead, D. Bradwell, et al. IEEE Ultrasonic Symposium, Canada, **2004**, 2, 1481–1483.
172. G. Divkovic, M. Liebler, et al. *Ultrasound Med Biol* **2007**, 33, 981–986.
173. I. M. Hallaj, R. O. Cleveland, K. Hynynen. *J Acoust Soc Am* **2001**, 109 (5), 2245.
174. G. Basta, C. Lupi, G. Lazzerini, et al. *Thromb Haemost* **2004**, 91 (6), 1078–1083.
175. L. Avruch and P. L. Cooperberg. *J Ultrasound Med* **1985**, 4, 21–28.
176. L. Buono. Graduate thesis, Faculty of Biomedical Engineering, University of Pisa, 2009.
177. P. Calvert. *Science*, **2007**, 318 (5848), 208.
178. R. Langer and J. P. Vacanti. *Science* **1993**, 260, 920.
179. R. V. Ulijn, et al. *Mater Today* **2007**, 10 (4), 40.
180. T. Okano, et al. *Biomaterials* **1995**, 16, 297.
181. M. Yamato and T. Okano. *Mater Today* **2004**, 7, 42.
182. K. Nishida, et al. *N Engl J Med* **2004**, 351, 1187.
183. M. Yamato, et al. *Prog Polym Sci* **2007**, 32, 1123.
184. J. Shi, N. M. Alves, J. F. Mano. *Adv Funct Mater* **2007**, 17 (16), 3312.
185. C. I. Dias, J. F. Mano, N. M. Alves. *J Mater Chem* **2008**, 18, 2493.

186. P. E. Grimshaw, J. H. Nussbaum, A. J. Grodzinsky, M. L. Yarmush. *J Chem Phys* **1990**, 93, 6.
187. P. G. De Gennes. *Macromolecules* **1976**, 8, 587.
188. J. C. Bacri and R. Rajaonarison. *J Phys Lett* **1979**, L-5, 40.
189. M. A. Biot. *J Acoust Soc Am* **1956**, 28, 168.
190. T. Tanaka, L. O. Hocker, G. B. Benedek. *J Chem Phys* **1973**, 59, 5151.
191. A. Biot and D. G. Willis. *J Appl Mech* **1957**, 24, 594.
192. P. Chiarelli and D. De Rossi. *J Intell Mater Syst Struct* **1992**, 3, 396.
193. P. Chiarelli, C. Domenici, G. Genuini. *J Mater Sci Mater Med* **1993**, 4, 5.
194. P. Chiarelli. *Mater Sci Eng* **2004**, C24, 463.
195. P. Chiarelli, D. De Rossi, P. J. Basser, S. Goldstein. *Biorheology* **1992**, 29, 383.
196. S. M. Josè, J. A. Manzanares, A. E. English, T. Tanaka. *Phys Rev Lett* **1997**, 79 (16), 3086.
197. J. H. Nussbaum and A. J. Grodzinsky. *J Membr Sci* **1981**, 8, 193.
198. A. J. Grodzinsky, H. Lipshitz, M. J. Glimcher. *Nature* **1978**, 275, 18.
199. A. Katchalsky and P. F. Curran. *Non Equilibrium Thermodynamics in Biophysics*. Harvard University Press, Cambridge, MA, 1960.
200. G. Casalino, P. Chiarelli, D. De Rossi, et al. *Nato ASi Serie*, P. Dario, et al., eds. Springer Verlag, Berlin, 1991; pp. 495–505.
201. H. Schreyer, N. Gebhart, K. J. Kim, M. Shahinpoor. *Biomacromolecules* **2000**, 126, 165–172.

CHAPTER 16

IONIC POLYMER–METAL COMPOSITES FOR SENSORS AND ARTIFICIAL MUSCLES: MECHANOELECTRIC PERSPECTIVES

RASHI TIWARI and KWANG J. KIM

16.1 INTRODUCTION

Polyelectrolyte gels, considered in the previous chapter (Chapter 15), when plated with metal electrodes, demonstrate bending upon application of electric field. One such composite system fabricated from ionomeric membranes are ionic polymer–metal composites (IPMCs), which are a subclass of electroactive polymers (EAPs).[1–11] Ionomers are a special type of polyelectrolyte polymer that contains relatively few ionic repeating units and nonionic repeating units. The polar ionic groups have the tendency to aggregate forming clusters (see Chapter 9). The bulk properties of ionomers are dependent on the ionic interaction in specific regions in the material. The ionomer in IPMCs is semipermeable and allows only certain ions to pass through while blocking the others. These semipermeable membranes made with ionomer are called ion-selective membranes. Chemical reduction of metal ions on the membrane surface is used for plating electrodes (electroless deposition) on IPMCs.

The ion exchange membrane acts as a pathway for selective ions to move. IPMCs (see Fig. 16.1) not only demonstrate bending in response to applied field (electromechanical transduction) but also are capable of generating current on mechanical deformation (mechanoelectric transduction). IPMC transduction depends on the ionomer, the counterions, and the state of hydration of the membrane. Oguro et al.[12] initially demonstrated IPMC bending upon application of low voltage (1–2 V). Similar observations were also reported by Shahinpoor.[13] The application of an electric field causes cations,

Ionic Interactions in Natural and Synthetic Macromolecules, First Edition. Edited by Alberto Ciferri and Angelo Perico.

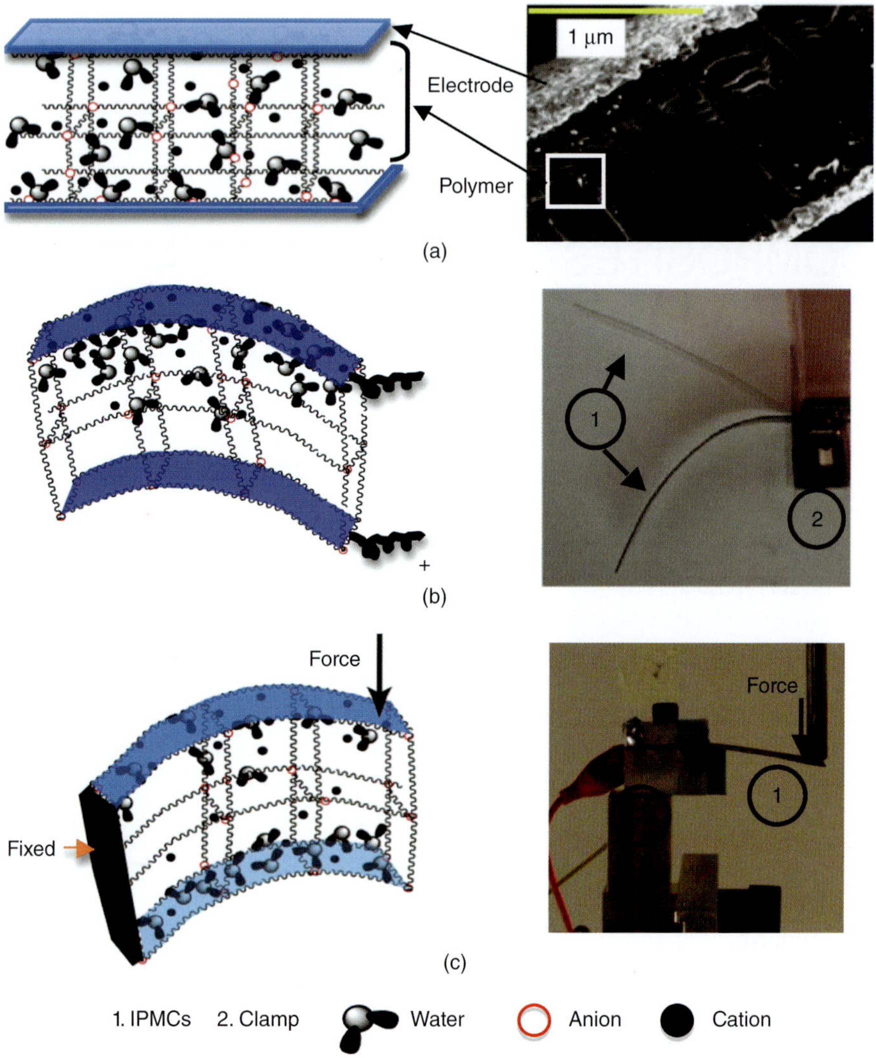

FIGURE 16.1. Illustration showing ions, water, and polymer chain inside IPMCs, along with the scanning electron microscopy (SEM) image of the cross section when no external input is exerted (a); electromechanical effect on IPMC causing IPMC to bend, with a photograph showing IPMC bending under electric field (b); and mechanoelectrical effect due to charge motion on mechanical bending (c).[75]

along with water molecules, to move toward the cathode, forming a cation-rich layer (Fig. 16.1a,b). Presence of excess water molecules causes the IPMCs to swell near the cathode. The strain in the cation-rich layer induces stresses in the polymer, resulting in a fast bending motion toward the anode (cation-poor layer).[14] The pressure from the strained polymer matrix causes water to diffuse out of the cation-rich areas, resulting in a slow relaxation toward the cathode.

On the other hand, mechanical deformation of the IPMC membrane forms localized cation-rich regions. A charge is developed across the membrane due to the motion of cations from its higher density region to its lower density region upon mechanical deformation such as bending (Fig. 16.1c).[15,16]

Relatively fast response (several hundred Hertz), flexibility, operation in aqueous and dry environments, and chemical stability are some of the advantages offered by IPMCs. Lifetime of up to 250,000 cycles has been recorded for IPMCs operating in air.[17] IPMC transduction has been exploited in various applications in biomedical, space exploration, robotics, and naval applications. Artificial muscle,[18] robotic fish that can swim in water,[19] aircraft wings,[20] catheter tips,[21,22] microrobotics,[23] automobiles and spacecraft,[24] or stand-alone sensor actuators for biomedical devices[25]—IPMCs are proving their capability everywhere. Use of mechanoelectric transduction for the development of a pressure sensor has been reported by Sadeghipour et al.[1] Due to the growing need for self-powered sensors and systems, the application of IPMCs for energy harvesting has also been considered.[26,27]

Despite of various advantages, application of IPMCs for sensor and actuator development has been limited to the research environment. In order for researchers to tap into the full potential of this material, some fundamental questions need to be answered. The first key issue is to understand the basic ion transport in electromechanical and mechanoelectrical transduction. Although a large amount of effort has been put into the electromechanical aspect, the mechanoelectric side still remains underexplored and less understood. Second, the effect of factors such as the type of polymer used, counterion in the polymer, the polymer conductivity, the degree of hydration, the electrode quality, and the electrode thickness on IPMC behavior needs to be understood. Lastly, IPMCs suffer from low output charge, nonlinear response, low force output, and limited bandwidth of operation. These limitations must be overcome for the successful transition of this technology to market. Many research works are being performed on exploring various manufacturing techniques, understanding IPMC layers (electrode, polymer, and intermediate), modeling, and application.

In this chapter, the electroless deposition technique for IPMC fabrication is discussed. The interaction of various IPMC layers that are ramifications of the typical fabrication technique is described through a phenomenological model. The model aims to provide insight into the fundamental physics of IPMCs. Since most of the previous research have focused on electromechanical transduction, the focus of this chapter is on understanding the mechanoelectric transduction of IPMCs. Potential energy harvesting using mechanoelectric transduction is also discussed.

16.2 MANUFACTURING

Typically, IPMCs consist of a thin ionomeric polyelectrolyte membrane sandwiched between metal electrodes deposited chemically.[1,28] Two ionomers are commonly available: Nafion® (perfluorosulfonate) by DuPont (Wilmington, DE; http://www2.dupont.com) or Flemion® (perfluorocarboxylate) by Asahi Glass (Tokyo, Japan; http://www.agc.co.jp). Nafion contains a poly(tetrafluoroethylene) backbone and perfluoroether side chains that terminate in a sulfonate group. Nafion as a base polymer offers several advantages such as chemical resistance, a higher working temperature (up to 120°C) as compared with other polymers, and a high ion conductivity of around 0.1 S cm^{-1} in 1 M H_2SO_4 at 20°C measured using a direct current (DC) pulse.[29] Platinum and gold are commonly used for electroding due to their corrosion resistance and high conductivity. Other metals such as copper[30] and nickel[31] could also be used as electrode materials. Electrodes could be deposited using painting or the physical vapor deposition (PVD) process.[32,33] Electrode deposition on dry and solvated membranes, using the direct assembly process, have also been considered.[34] However, electroless/chemical deposition[35] is the most commonly used manufacturing process because this process results in mechanically and chemically stable electrodes.

During the electroless deposition process, Nafion membrane is hydrated in deionized (DI) water for 24 hours and cleaned in water bath. The cleaned sample is immersed in a platinum salt solution for 2 hours. The reduction process using ammonia and sodium borohydride added every half hour follows, with the bath temperature increasing from 40 to 70°C. The reduction process was repeated until a surface resistivity of less than 10 Ω sq^{-1} is achieved. This deposition method typically results in IPMCs with thickness between 0.17 and 0.20 mm. Figure 16.2 summarizes the steps of electroless deposition.

A layered structure results due to the electroless deposition technique as shown in Figure 16.3.[76] The incipient metal particles with a diameter less than 10 nm coagulate to 50–100 nm during reduction[36] to form the outermost *electrode layer*. The electrode thickness is around a few micrometers (1–5 μm) each. The granular electrode morphology with peak and valley depth of approximately 50 nm has been measured using atomic force microscopy (AFM).[37] In addition, average particle size has been measured to be around 47 nm with the aid of transmission electron microscopy (TEM).[37] The *intermediate layer*, or the composite layer consisting of scarcely dispersed metal particles inside the polymer matrix, follows the electrode layer and gradually blends into the *polymer layer*. Electrode and intermediate layers inherit capacitive behavior as shown in the electrochemical impedance spectroscopy (EIS) graphs in Figure 16.3a (inset). The capacitance was observed to exponentially decrease with increase in frequency, as capacitive impedance is inversely proportional to frequency. IPMC impedance is affected by surface roughness, electrode morphology, as well as the intermediate layer thickness. Figure 16.3b shows the X-ray diffraction (XRD) of an electrode layer made of platinum metal particles. It is evident that the platinum particles were mostly crystalline with a (111) orientation, indicating a face-centered cubic (FCC) crystal structure. The crystallite size was approximately calculated using the XRD pattern and the Scherrer formula. The resulting crystallite size for platinum electrode

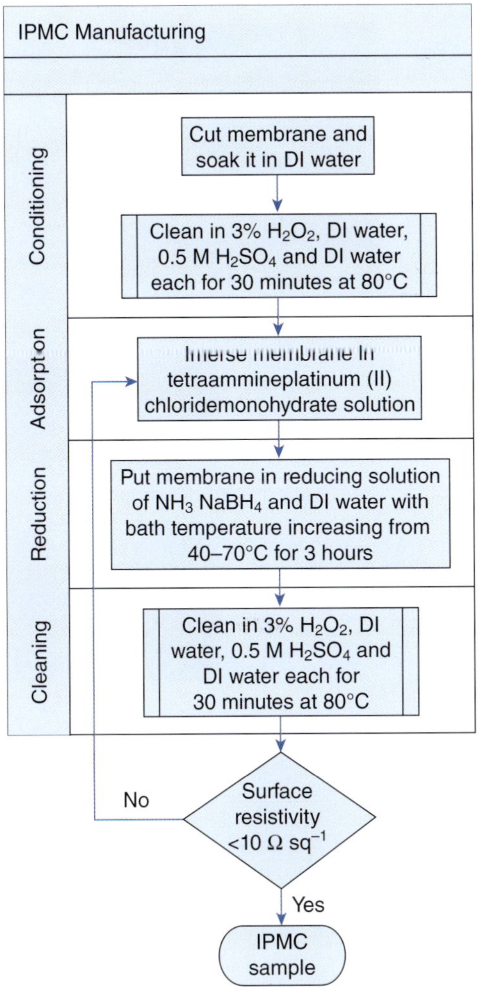

FIGURE 16.2. IPMC fabrication using chemical deposition technique. The process results in uniform electrode layer with the surface resistivity lower than 10 Ω sq^{-1}.[75]

at (111) reflection was measured to be 13 nm. An energy dispersive spectroscopy (EDS) of the IPMC with platinum electrode shows the platinum particle distribution across the thickness of the sample (Fig. 16.3c).

Nafion, which constitutes the major part of the IPMC, includes a hydrophilic side chain[38] and a hydrophobic backbone (forming a semicrystalline matrix). The polymer matrix reorients to form discrete clusters on hydration. The diameter of the cluster is estimated to be around 4 nm.[39,40] The chemical structure and the cluster dimension is illustrated in Figure 16.4 (additional details on Nafion structures are discussed in Chapter 9). Electrostatic dipole interaction is believed to be the dominant force for clustering.[41] The number

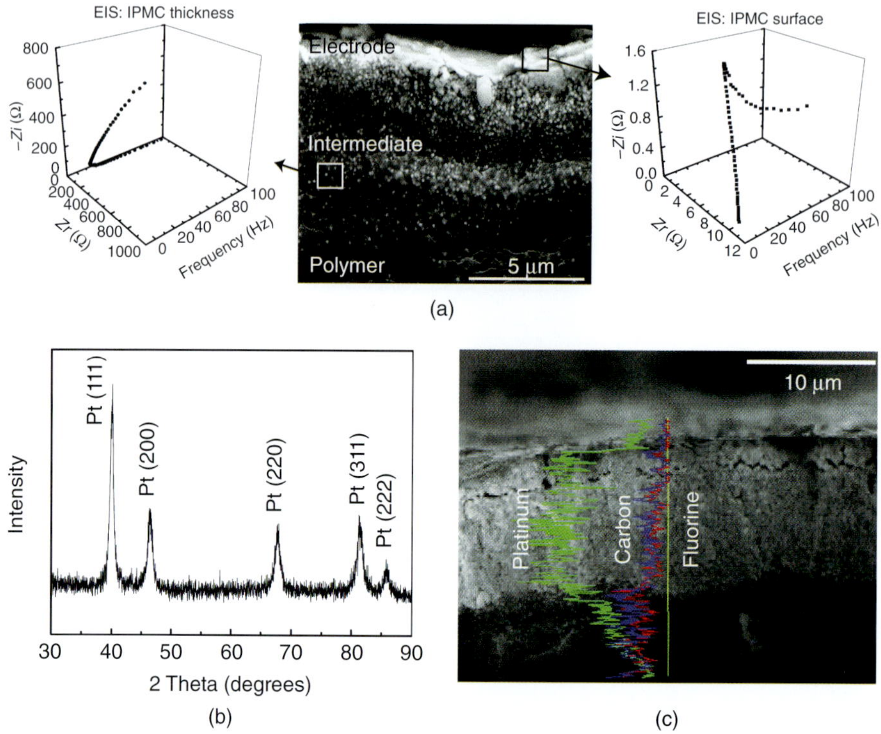

FIGURE 16.3. (a) Scanning electron microscopy (SEM) image of the cross section of IPMC manufactured using the electroless deposition method. The distinct electrode layer on the surface, the intermediate layer having highly dispersed metal particles in polymer matrix, and the polymer layer are shown. Electrochemical impedance spectroscopy (EIS) shows the capacitive-resistive behavior of IPMC, (b) X-ray diffraction (XRD) pattern of platinum electrode IPMC, and (c) energy dispersive spectroscopy (EDS) showing platinum particle distribution across IPMC thickness.[57]

and shape of the cluster, in a sample, is dependent on its equivalent weight, EW (EW is the weight in grams of dry polymer per mole of ion exchange sites), and hydration of the membrane, respectively. This is because the EW governs the number of SO_3^- sites, which in turn governs the cluster formation in Nafion. Water molecules, on the other hand, are believed to penetrate inside the cluster and hence dictate the shape of the cluster. The distance between the clusters increases with an increase in the EW of Nafion.[39]

Extensive investigation has been dedicated to understand the polymer membrane structure.[42] Also, research has suggested that the spherical shape of the cluster is an oversimplification.[43] In another study, two different techniques, AFM and small-angle X-ray scattering (SAXS), were employed to demonstrate the spheroidal cluster geometry.[44] Datye et al.[45] studied cluster formation and ionic transport in an ionomer membrane. A number of works

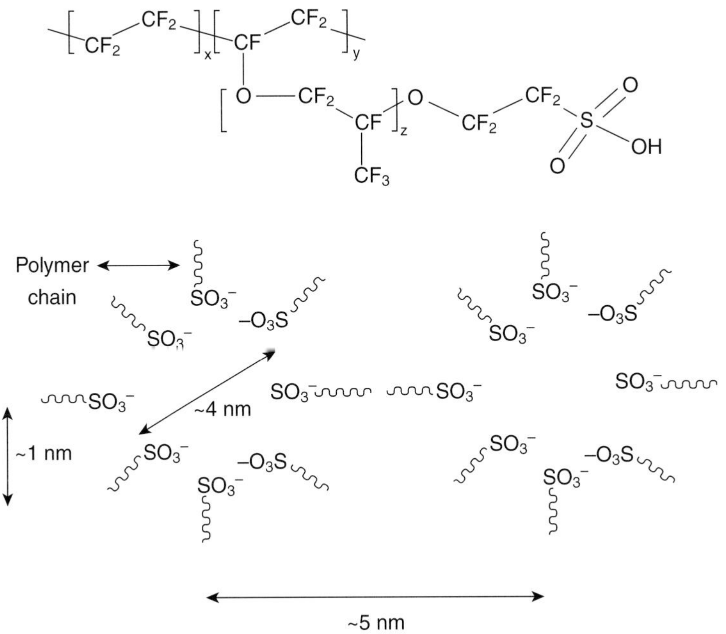

FIGURE 16.4. (a) Chemical structure and (b) pendant structure of Nafion in IPMCs.[39]

have been conducted to study the various aspects of water fraction in the ion exchange membrane such as modeling the effect of hydration on ionic conductivity,[46] microconfiguration of pore structure,[47] and acid forms of Nafion at different humidity levels.[48]

16.3 IPMC MECHANOELECTRIC MODEL

As mentioned before, clusters are formed upon hydration due to the reorientation of hydrophilic side chains and hydrophobic backbone in Nafion. The cluster can be described as two concentric circles with anions distributed on the outer circle while cations are distributed on the inner circle perimeter. While anions are fixed, cations are free to move in response to applied field or mechanical deformation. Models have been developed to explain two widely accepted mechanoelectric transduction mechanisms: charge transport and electrostatic interactions that result from the cluster structure of Nafion.

Charge transport models take into account forces exerted on ions as they migrate through the polymer network. Internal charge dynamics has been studied previously using one-dimensional linear partial differential equations.[49] Charge sensing in IPMCs can be related to piezoelectric materials.[50] To capture both the electromechanical and mechanoelectric properties of IPMCs, a static model based on irreversible thermodynamics has been developed by Newbury and Leo.[11] A dynamic model incorporating the effect of distributed surface resistance has also been formulated[51] for studying the frequency response of IPMCs. Electric field produced due to effective dipole in the cluster has been considered in electrostatic models. Nemat-Nasser and Li[9] studied the cation redistribution in the cluster along with the effect of hydration to explain the electromechanical response of IPMCs. Later, Weiland and Leo[52] studied the effect of polarization on mechanoelectric transduction. They incorporated pendant chain stiffness, cluster surface energy, and external loading into the model. Gray-box models that do not account for the internal charge dynamics or electrostatic forces in IPMCs have been developed for simplicity and applicability.[50,53] These models are formulated by considering IPMC response to certain input. Table 16.1 lists some of the key models that have been employed for understanding different aspects of mechanoelectric transduction of IPMCs. Although these models provide a tool to understand mechanoelectric transduction, the role of a specific layer and its effect on IPMC response has not been studied. Besides, the intermediate layer has been mostly ignored for simplicity. Since IPMC layers are tightly coupled, it is difficult to experimentally understand their individual role during transduction. Thus, there is a need to develop a model that can provide an insight into the effect of individual layers on IPMC response.

A phenomenological model to assess the effect of *surface* and *intermediate layer impedance* on IPMC response is herein described. The model consists of fundamental physical parameters, geometric dimensions, and impedance of IPMC layers. Although the works by Farinholt and Leo[49]and Chen et al.[51] form the basis of our research, our approach differs in the following ways. First of all, material response to dynamic input is studied. Second, both surface and intermediate layer impedances were incorporated into the ionic model in order to address the effect of surface and polymer nonlinearity. Lastly, the model is suitable for a study of electrode properties (such as electrode thickness, morphology, and surface area) on the mechanoelectric response of IPMCs. These electrode properties are incorporated in the model in terms of their impedance.

When an IPMC is exposed to a stress, it produces charge. This can be partially due to the internal motion of ions.[54] The charge density on the polymer composite surface is assumed to be linearly proportional to the stress on the beam.[9,51,55] The characteristic equation describing the spatial and temporal distribution of the charge on the IPMC surface is represented as[51,56]:

$$s\rho(x,y,s) - d\nabla^2\rho(x,y,s) + \frac{dF_c^2C^-}{RTk_e}\left(1 - C^-\Delta V\right)\rho(x,y,s) = 0, \qquad (16.1)$$

where ρ is charge density, s is Laplace operator, F_c is Faraday's constant, d is the ionic diffusivity, T is the absolute temperature, R is the gas constant, C^- is the time invariant anion concentration in the polymer matrix, k_e is the dielectric permittivity, and ΔV is the volumetric change due to water

TABLE 16.1. Comparison of Some Representative Models of the Mechanoelectric Principle in IPMC

Model	Reference	Description
Electrostatic model	Nemat-Nasser and Li[9]	Electrostatic forces are believed to play a dominant role. The mechanoelectric behavior is projected to be due to the differential displacement of the effective charge centers of the anions and cations of each cluster.
	Nemat-Nasser[14]	The model focuses on the electromechanical response based on cluster morphology and cation redistribution upon application of electric field. Hydration level and water migration are also incorporated. The model was used to study the electromechanical response of IPMCs.
	Weiland and Leo[52]	The model depicts the dominance of polarization in the mechanoelectric stage of IPMCs. Forces on dipoles, pendant chain stiffness, cluster surface energy, and external loading are accounted for.
Charge transport model	Farinholt and Leo[49]	Internal charge dynamics is studied using one-dimensional linear partial differential equations.
	Chen et al.[51]	The model is based on the charge transport inside IPMCs. The effect of distributed surface resistance due to porous electrodes is included.
Gray-box model	Newbury and Leo[50]	A transformer-based equivalent circuit is developed to derive both electromechanical and mechanoelectric responses of IPMC in cantilever beam configuration. The coupling coefficients are derived using a pair of linearly coupled equations.
	Kanno et al.[53]	The model consisted of three key domains: electrical, stress generation, and mechanical. The formulation is based on experimental results.
Irreversible thermodynamic model	De Gennes et al.[7]	Linear irreversible thermodynamics-based model to formulate the mechanoelectrical and electromechanical behavior of IPMCs.
Surface resistance model	Punning[73]	The model illustrates the dependence of electrical parameters such as surface resistance on bending. The model is based on electrochemical analysis.
	Kim et al.[74]	Porous surface electrode is modeled to estimate resistance and capacitance. Parameters such as electrode thickness and porosity of electrode were found to affect resistivity, whereas interlayer electrode and particle gap in that layer affect capacitance.

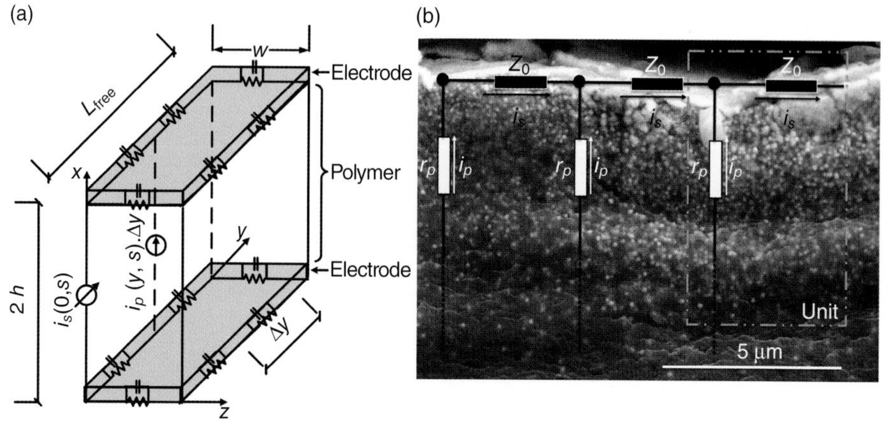

FIGURE 16.5. IPMC (a) surface (Z_0) and (b) intermediate layer (r_p) impedances. Both the surface and intermediate layer impedances result due to metal particle dispersion in the polymer matrix. The density of metal particles on the surface is higher than in the intermediate layer.[57]

absorption. Figure 16.5 describes some of the model parameters along with the coordinate system. Neutral axis is defined at $x = 0$ for a polymer of thickness $2h$ and modulus Y. L_{free} and L_t are the free and total length of the sample, respectively. The surface impedance r_s and intermediate layer impedance r_p, per unit length, represented in Figure 16.5 are defined as

$$r_s = \frac{Z_0}{w}; \quad r_p = \frac{r_p'}{w}, \tag{16.2}$$

where Z_0 and r_p' are the surface and intermediate layer impedances per unit length and width, w. For every section of length, Δy, a current, $i_p \Delta y$, is generated inside the polymer. This current adds to the surface current, i_s.

The characteristic equation was solved to predict the surface current for a given tip displacement, $u(t)$ for three respective cases: without impedance; with only surface impedance; and with both surface and intermediate layer impedances. The surface current, $i(s)$, derived for these cases is summarized in Table 16.2.[57]

The mechanoelectric model without impedance was employed to study the step response as shown in Figure 16.6. The voltage due to the ionic effect shows an exponential decay with time. This could be because ions move back to their original position to achieve equilibrium (back diffusion). As shown in the graph, the charge density is zero inside the polymer. This value increases toward surface electrodes. The plots show the same trend as reported by Farinholt and Leo.[49] The mechanoelectric model with surface and intermediate layer impedances can be simplified to a model considering only the surface by considering $r_p' = 0$. The intermediate layer impedance could be considered

TABLE 16.2. Comparing Mechanoelectric Models "with" and "without" Impedance

Without impedance	$i(s)=\frac{3sYw(h\tau.\coth(h\tau)-1)}{2\varphi L_{\text{free}}\tau^2}u(s)$ $\tau=\frac{1}{d}\left(s+\frac{dF_c^2C^-}{RTk_e}(1-C^-\Delta V)\right)$	
With impedance	$i(s)=\frac{j_2}{\cosh(\sqrt{j_1}y)}\left(\frac{1-\cosh(\sqrt{j_1}y)}{j_1}+\frac{L\sinh(\sqrt{j_1}y)}{\sqrt{j_1}}\right)$	
	Without intermediate layer impedance	With intermediate layer impedance
	$j_1=\frac{sZ_0k_e}{h}$	$j_1=\frac{sZ_0k_e}{h+sk_er_p'}$
	$j_2=\frac{3sYw(h\tau.\coth(h\tau)-1)}{\varphi L_{\text{free}}^3\tau^2}u(s)$	$j_2=\frac{3sYw(h\tau.\coth(h\tau)-1)}{\varphi L_{\text{free}}^3\tau^2(h+sk_er_p')}u(s)$

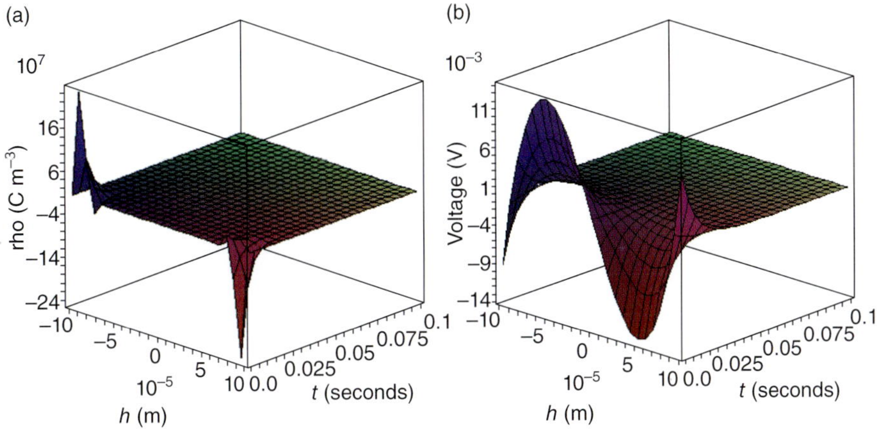

FIGURE 16.6. (a) Simulated charge density and (b) simulated surface potential on a 10 mm × 50 mm × 0.17 mm IPMC under 0.5-in. tip deflection.

purely resistive or capacitive for convenience. The value of r_p' if assumed to be purely resistive is in the order of 10^{-5} Ω·m.

The model can also be used for determining some material properties such as diffusion, anion concentration, coupling coefficient, and dielectric constant by data fitting as listed in Table 16.3. The estimated values were close to those reported in the literature.[51] Figure 16.7 compares the predicted output from the model with the measured current. The model with impedance shows lower current output. This was because there are more losses associated with the impedance on the surface.

TABLE 16.3. Values of the Parameters Determined by Data Fitting

Model	d ($m^2 s^{-1}$)	C^- ($mol\ m^{-3}$)	k_e ($F\ m^{-1}$)	ψ (JC^{-1})
No impedance	6.11×10^{-11}	1891.10	0.0062	308.82
With impedance	6.91×10^{-11}	2308.90	0.0018	328.98
Literature[51]	3.32×10^{-11}	1091.00	0.0018	104–550

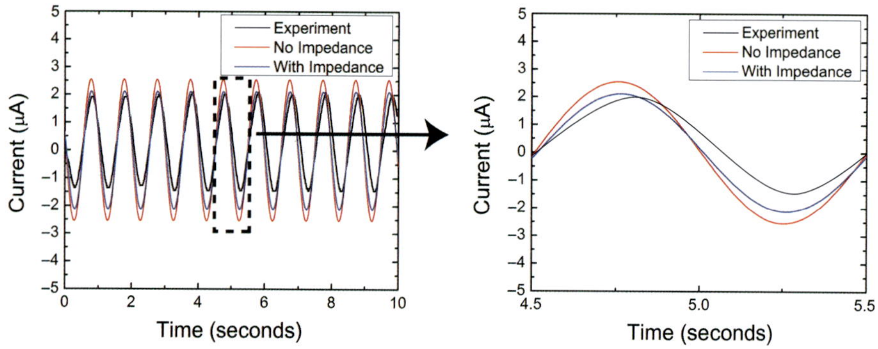

FIGURE 16.7. Graph comparing predicted current output with measured current upon application of 0.5-in., 1 Hz displacement showing 10 cycles and one cycle.[57]

Figure 16.8a,b shows the frequency response of IPMCs for samples with thickness 230 and 290 μm, respectively. The length and width of the samples were 50 and 10 mm, respectively. It is evident from the figures that both the models, with and without impedance, predict the current magnitude fairly well. However, the model incorporating the effect of surface impedance predicted the phase of current produced by IPMCs more accurately. The phase was observed to vary from 90° at 1 Hz frequency to around 40° at 25 Hz for both the samples.

16.4 ENERGY HARVESTING APPLICATIONS

Due to the advancements in portable electronics and wireless technology, there has been a growing need to develop energy sources for the uninterrupted operation of these devices. Vibration energy harvesters mainly comprise a transducer with mechanoelectric capability, a conditioning circuit to convert generated signal into useful form, and storage device such as batteries or supercapacitors. Although a large amount of research has been concentrating on the signal conditioning circuit, for optimal power output from the transducer[58–60] and storage devices,[61] the IPMC as transducer is the focus of this research.

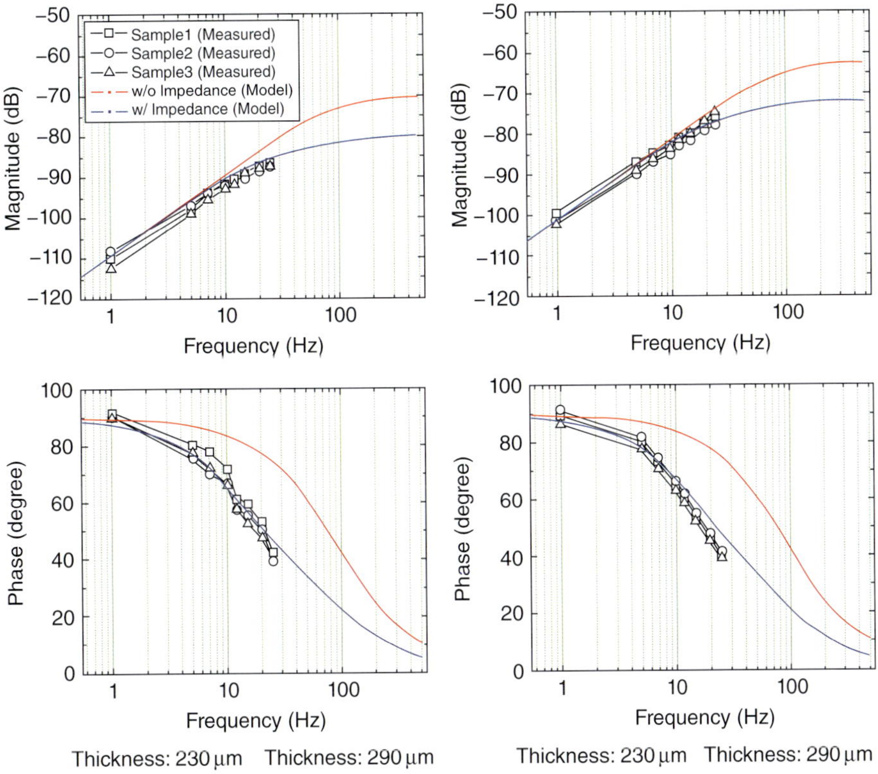

FIGURE 16.8. Comparing frequency response of IPMCs using the mechanoelectric model with and without surface impedance with experimental results. Response of samples with thickness (a) 230 μm and (b) 290 μm vibrated at frequency ranging from 1–25 Hz with vibration amplitude at 0.5 in. (pk-pk) are plotted.[57]

Piezoelectric materials have been commonly used as vibration energy harvesters[62] in vibration frequencies above 100 Hz.[63] However, most of the ambient sources of vibration fall in the lower frequency region (1–100 Hz). Hence, there is an opportunity to develop low-frequency-vibration energy harvesters.

Due to the inherent flexibility, mechanoelectric coupling, ability to operate in both water and air,[64] and long operating life, IPMCs are great candidates for harvesting energy in low-frequency regime. Numerous sources of mechanical vibration might be taken into consideration such as automobiles, ships, process instrumentations, and aircrafts. IPMCs are commonly employed in the bender configuration.[65–67] The energy harvesting capability of IPMCs has been considered in the past. Brufau-Penella et al.[68] developed a gray-box model to determine that the optimal dimension of IPMCs (length 12 cm and width 6.2 cm) generate 3 nW power when vibrated at 7.09 rad s^{-1} in air. Tiwari

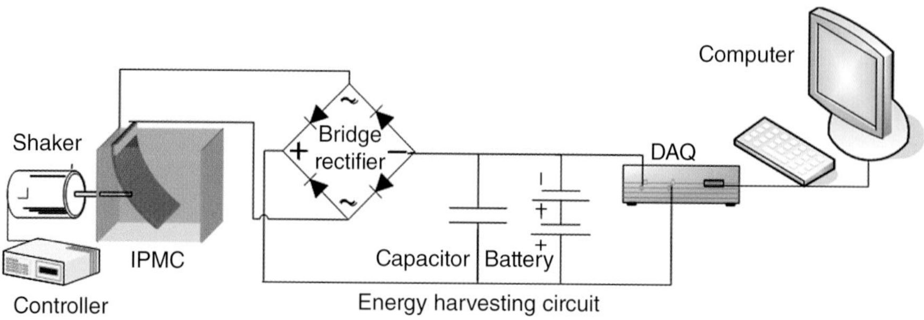

FIGURE 16.9. Energy harvesting setup consisting of an IPMC as transducer, full bridge rectifier circuit, and NiMH as the storage device.

et al.[26] demonstrated the harvesting capability of IPMCs not only in bending but also in shear and tension. Energy harvesting from base excitation in fluid environment has also been reported.[69] The maximum power harvested was 1 nW for a base excitation of 1 mm. Recently, energy harvesting from a three-dimensional IPMC disk has also been reported.[27] Most of the reported work on energy harvesting using IPMCs have focused on analyzing the energy harvesting capability.[70] In addition, previous research has shown that the power generated from IPMCs suffers from low power density of around 1.219 μW kg^{-1}.

For energy harvesting experiments, a shaker assembly (TIRA Vibration System, TV52110; TIRA Gmbh, Schalkau, Germany) is used as the source of mechanical input. The shaker is programmed using a VR8500 controller to produce a sinusoidal mechanical vibration at a frequency of 10 Hz, 0.5-in. (pk-pk) displacement. The sample is vibrated in clamp-free configuration (Fig. 16.9). Under dynamic excitation, an IPMC generates a sinusoidal output that can be stored on a battery after rectification using a full bridge rectifier. The rectified signal is used for charging a capacitor (typically >1000 μF) followed by a 40-mAh rechargeable NiMH battery (VARTA, White Plains, NY). Battery voltage is recorded using an IOtech Personal Daq 56 data acquisition system (IOtech, Cleveland, OH). Due to fewer number of electronics in the circuit, the transmission losses are reduced.

The inherent flexibility of IPMCs allows for not only bending but also extension and shear motions. The experimental setup for the bending, extension, and shear modes is shown in Figure 16.10. Platinum electroded IPMC with dimensions 63 mm × 8.6 mm × 0.20 mm is used. Shear load is applied on the beam acting perpendicular to its longitudinal axis. Figure 16.10 shows recorded battery voltage for 1 hour.

It is important to determine the power generated by IPMCs to identify its feasibility for real-world applications. Experiments are conducted to determine the optimal resistive load for platinum IPMC samples in bending, extension, and shear configurations (Fig. 16.11). The experiment is performed by

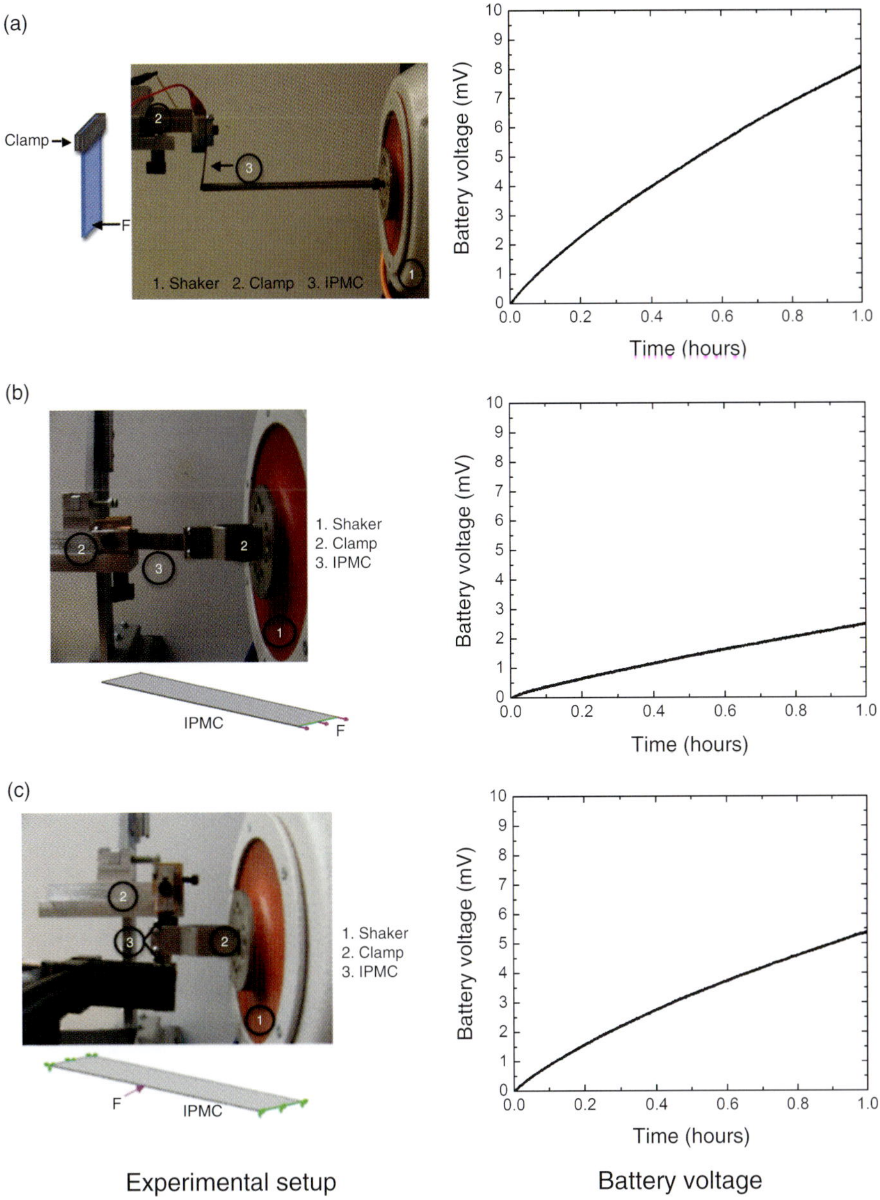

FIGURE 16.10. Energy harvesting experimental setup and battery voltage after 1 hour charging in (a) bending, (b) extension, and (c) shear modes.

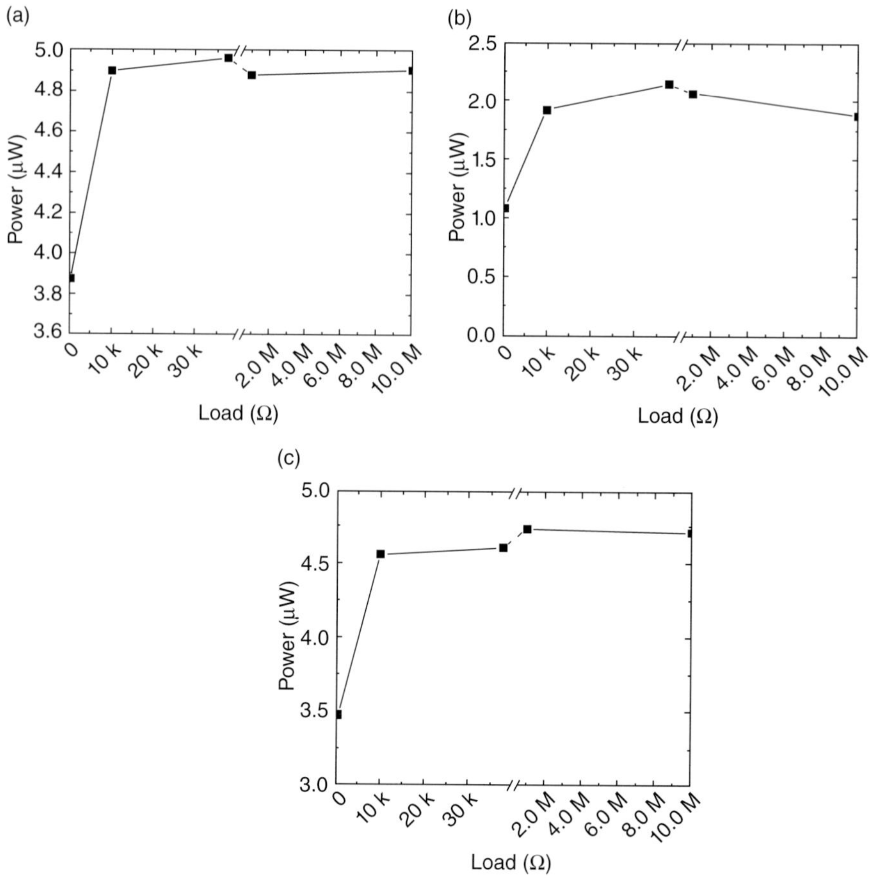

FIGURE 16.11. Output power as a function of resistive load measured experimentally in (a) bending, (b) extension, and (c) shear modes.

measuring the voltage drop across the resistor connected in series with the output.

It is evident from the graph that the IPMC produced slightly more power in bending mode than in shear and extension modes. Optimal resistive load for maximum power dissipation is experimentally evaluated to be 38 KΩ for the bending mode. Maximum power dissipated at the optimal load is 2.2 μW for a platinum IPMC sample in shear mode. Optimal resistive load for maximum power dissipation is around 1 MΩ in shear mode.

Other applications such as velocity sensing, impact sensing, vibration control, or structural health monitoring can exploit the mechanoelectric transduction in IPMCs as well.

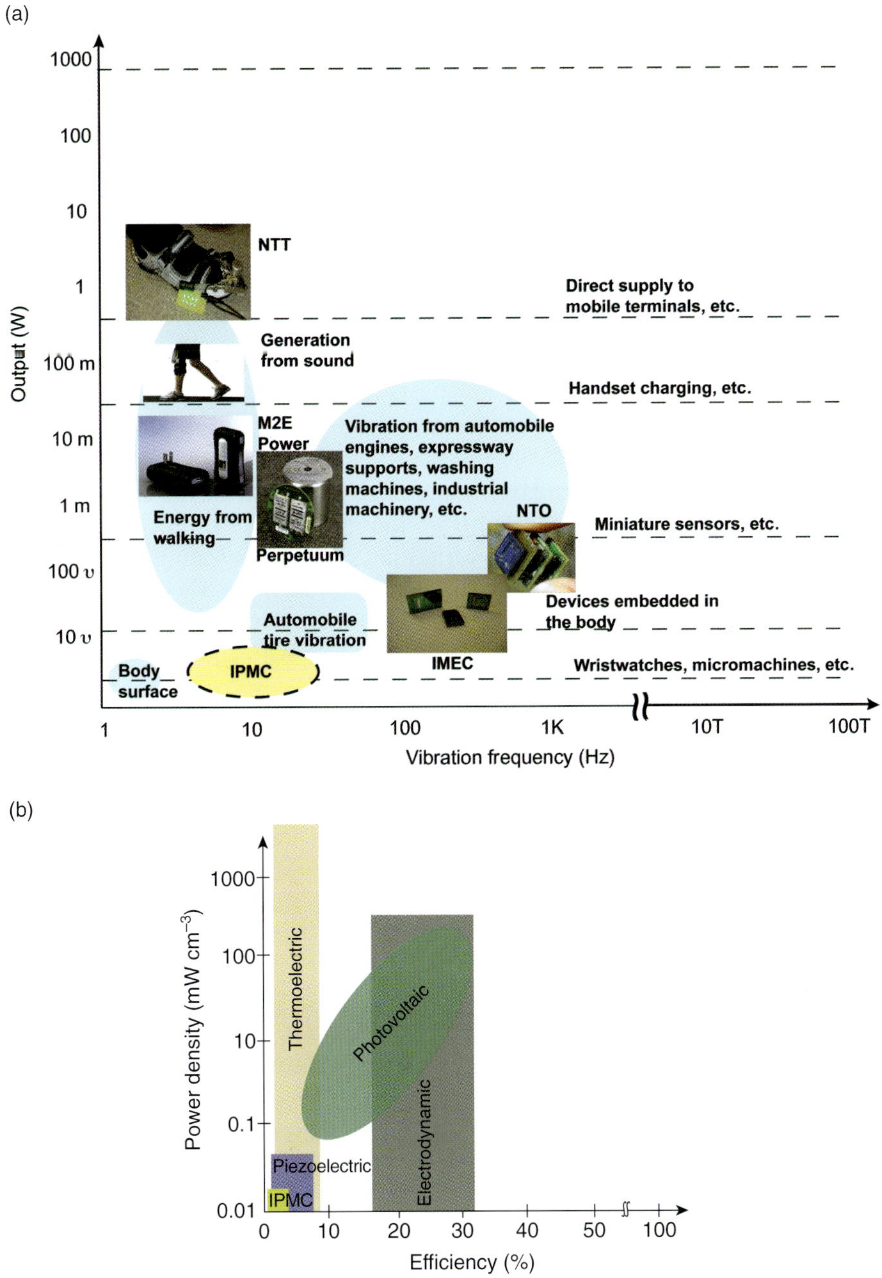

FIGURE 16.12. (a) Power output along with possible operating frequency from existing technologies as vibration-based energy harvesters manufactured by NTT (Toyosu, Tokyo, Japan), M2E (Boise, ID), Omron (Kyoto, Japan), Perpetuum (Southampton, UK), and IMEC (Leuven, Belgium)[71]; and (b) power density comparison.[72]

16.5 DISCUSSION AND CONCLUSION

The mechanoelectric model that comprises not only the widely accepted ionic transport but also the surface and intermediate layer impedances is developed. Understanding the effect of various parameters that can impact mechanoelectric property can be useful during manufacturing of IPMCs. Effect of the mechanical and electrical components of IPMCs on the mechanoelectric property of IPMCs was studied. It was observed that the material's mechanoelectric output could be tailored by changing the thickness of the material. The other solution to tailor the mechanoelectric output was by reducing surface or intermediate layer impedance. This could be achieved by increasing the capacitance. The intermediate layer impedance could be reduced either by increasing the thickness of the electrode deposition or by dispersing more metal particles inside the polymer as more particles lead to an improvement in polymer conductivity.

Figure 16.12a lists the available technologies along with the frequency of vibration and the output generated from them[71]. Figure 16.12b compares the power density produced by IPMCs with other energy harvesting technologies including piezoelectric and solar.[72] As can be seen from the figure, IPMCs produced less power density as compared with other technologies such as solar and thermoelectric, with an efficiency comparable with that of piezoelectric materials.

The low value of electromechanical coupling coefficient of IPMC imposes a limitation on its efficiency. In addition, the coefficient "d_c" for IPMC is dependent on sample dimensions and is not represented as an intrinsic property. Further research is required in order to intrinsically measure the overall coupling coefficient.

REFERENCES

1. K. Sadeghipour, R. Salomon, S. Neogi. *Smart Mater Struct* **1992**, 1 (2), 172.
2. K. Oguro, K. Asaka, H. Takenaka. U.S. Patent 5,268,082, 1993.
3. K. Asaka, K. Oguro, Y. Nishimura, M. Mizuhata, H. Takenaka. *Polym J* **1995**, 27 (4), 436.
4. M. Mojarrad and M. Shahinpoor. Proc. SPIE Smart Struct. Mater. Symp., EAPAD Conf. 1997, 294.
5. Y. Bar-Cohen, M. Shahinpoor, J. O. Harrison, J. Smith. Proc. SPIE Smart Struct. Mater. Symp., EAPAD Conf. 1999, 51.
6. K. Asaka and K. Oguro. *J Electroanal Chem* **2000**, 480, 186.
7. P. G. De Gennes, K. Okumura, M. Shahinpoor, K. J. Kim. *Europhys Lett* **2000**, 50 (4), 513.
8. E. T. Enikov and B. J. Bradley. Proc. SPIE Smart Struct. Mater. Symp., EAPAD Conf. 2000, 129.
9. S. Nemat-Nasser and J. Li. *J Appl Phys* **2000**, 87, 3321.

10. S. Tadakoro, S. Yamagami, T. Kimura, T. Takamori, K. Oguro. *J Rob Mechatron* **1999**, 11 (4), 244.
11. K. M. Newbury and D. J. Leo. Proc. ASME Int. Mech. Eng. Congr. Exposition. 2001, AD-23705.
12. K. Oguro, Y. Kawami, H. Takenaka. *J Micromachine Soc* **1992**, 5, 27.
13. M. Shahinpoor. *Smart Mater Struct* **1992**, 1 (1), 91.
14. S. Nemat-Nasser. *J Appl Phys* **2002**, 92 (5), 2899.
15. C. S. Kothera. MS thesis, 2002, Blacksburg ,VA.
16. C. Bonomo, C. D. Negro, L. Fortuna, S. Graziani. Proc. Int. Symp. Circuits Syst. 2003, 4, IV-816.
17. B. J. Akle and D. J. Leo. *J Intell Mater Syst Struct* **2008**, 19, 905.
18. J. D. W. Madden, N. A. Vandesteeg, P. A. Anquetil, P. G. A. Madden, A. Takshi, R. Z. Pytel, S. R. Lafontaine, P. A. Wieringa, I. W. Hunter. *IEEE J Oceanic Eng* **2004**, 29 (3), 706.
19. EAMEX Corporation. http://www.eamex.co.jp/video/fish.WMV.
20. A. Colozza. *IEEE Spectr* **2007**, 44 (5), 38.
21. S. Sewa, K. Onishi, K. Asaka, N. Fujiwara, K. Oguro. Proc. IEEE Workshop Microelectromech Syst: Invest. Micro Struct. Sens. Actuators Mach. Syst. 1998, 148.
22. K. Oguro, N. Fujiwara, K. Asaka, K. Onishi, S. Sewa. Proc. SPIE Smart Struct. Mater. Symp., EAPAD Conf. 1999, 3669, 64.
23. J. Paquette and K. J. Kim. *IEEE J Oceanic Eng* **2004**, 29 (3), 729.
24. Y. Bar-Cohen, S. Leary, A. Yavrouian, K. Oguro, S. Tadokoro, J. Harrison, J. Smith, J. Su. Proc. SPIE Smart Struct. Mater. Symp., EAPAD Conf. 2000, 3987, 140.
25. E. Smela. *Adv Mater* **2003**, 15 (6), 481.
26. R. Tiwari, K. J. Kim, S.-M. Kim. *Smart Struct Syst* **2008**, 4 (5), 549.
27. R. Tiwari and K. J. Kim. *Smart Mater Struct* **2010**, 19, 065016.
28. K. Oguro. Preparations procedure: ionic polymer metal composites (IPMC) membranes. http://ndeaa.jpl.nasa.gov/nasa-nde/lommas/eap/IPMCPrepProcedure.htm. 2001.
29. M. Verbrugge and R. Hill. *J Electrochem Soc* **1990**, 137, 3770.
30. U. Johnson, U. Maeorg, V. Sammelselg, D. Brandell, A. Punning, M. Kruusmaa, A. Aabloo. *Sens Actuators, B* **2008**, 131 (1), 340.
31. I.-S. Park and K. J. Kim. *Sens Actuators, A* **2007**, 135 (1), 220.
32. M. Siripong, S. Fredholm, Q. A. Nguyen, J. Itescu, B. Shih, J. Stolk. Proc. MRS Fall Meeting. 2006, 889, W04-03.1.
33. S.-G. Lee, H.-C. Park, S. D. Pandita, Y. Yoo. *Int J Control Autom Syst* **2006**, 4 (6), 748–755.
34. B. J. Akle. PhD dissertation. Virginia Polytechnic Institute and State University, Blacksburg, VA, 2005.
35. H. Takenaka, E. Torikai, Y. Kawami, N. Wakabayashi. *Int J Hydrogen Energy* **1982**, 7, 397.
36. M. Shahinpoor and K. J. Kim. *Smart Mater Struct* **2001**, 10, 819.
37. M. Shahinpoor and K. J. Kim. *Smart Mater Struct* **2000**, 9, 543.
38. C. Heitner-Wirguin. *J Membr Sci* **1996**, 120, 1.

39. T. D. Gierke, G. E. Munn, F. C. Wilson. *J Polym Sci Polym Phys Ed* **1981**, 19, 1687.
40. W. Y. Hsu and T. D. Gierke. *Macromolecules* **1982**, 15 (1), 101.
41. A. Eisenberg. *Macromolecules* **1970**, 3 (2), 147.
42. Y. Wang and C.-Y. Wang. *Electrochim Acta* **2004**, 50, 1307.
43. S. J. Paddison. *Annu Rev Mater Res* **2003**, 33, 289.
44. P. J. James, J. A. Elliott, T. J. McMaster, J. M. Newton, A. M. S. Elliott, S. Hanna, M. J. Miles. *J Mater Sci* **2000**, 35, 5111.
45. V. K. Datye, P. L. Taylor, A. J. Hopfinger. *Macromolecules* **1984**, 17 (9), 1704.
46. Y. Zhen, P. Xiaofeng, W. Buxuan, L. Duujong, D. Yuanyuan. *Front Energy Power Eng China* **2007**, 1 (1), 58.
47. Y. L. Jiang and N. Sia. *J Mech Mater* **2000**, 32, 303.
48. J. J. Fontanella, M. C. Wintersgill, R. S. Chen. *Electrochim Acta* **1995**, 40, 2321.
49. K. Farinholt and D. J. Leo. *Mech Mater* **2004**, 36, 421.
50. K. Newbury and D. J. Leo. *J Intell Mater Syst Struct* **2002**, 13, 51.
51. Z. Chen, X. Tan, A. Will, C. Ziel. *Smart Mater Struct* **2007**, 16, 1477.
52. L. M. Weiland and D. J. Leo. *Smart Mater Struct* **2004**, 13, 323.
53. R. Kanno, S. Tadokoro, T. Takamori, M. Hattori, K. Oguro. Proc. IEEE Int. Conf. Rob. Autom. 1996, 1, 219.
54. S. Nemat-Nasser. *Handb. Exp. Solid Mech*, W. Sharpe, eds. Springer-Verlag Publishers, New York, 2008.
55. F. Fahy. *Foundations of Engineering Acoustics*. Academic Press, San Diego, CA, 2001.
56. Z. Chen and X. Tan. *IEEE/ASME Trans Mechatron* **2008**, 13 (5), 519.
57. R. Tiwari and K. J. Kim. *J Appl Phys* **2010**, 97, 244104.
58. A. Kasyap, J. Lim, D. Johnson, S. Horowitz, T. Nishida, K. Ngo, M. Sheplak, L. Cattafesta. Proc. Int. Cong. Sound Vib. 2002, 271.
59. G. K. Ottman, H. F. Hofmann, C. A. Bhatt, G. A. Lesieutre. *IEEE Trans Power Electron* **2002**, 17 (5), 1–8.
60. H. Hofmann, G. K. Ottman, G. A. Lesieutre. *IEEE Trans Power Electron* **2002**, 18 (2), 696.
61. H. A. Sodano, D. J. Inman, G. Park. *Shock Vib Digest* **2004**, 36 (3), 197.
62. S. R. Anton and H. A. Sodano. *Smart Mater Struct* **2007**, 16 (3), R1.
63. E. Lefeuvre, A. Badel, C. Richard, D. Guyomar. *J Electroceram* **2007**, 19, 349.
64. M. D. Bennett, D. J. Leo, G. L. Wilkes, F. L. Beyer, T. W. Pechar. *Polymer* **2006**, 47 (19), 6782.
65. C. Bonomo, L. Fortuna, P. Giannone, S. Graziani. *Proc IEEE Sens* **2004**, 1, 489.
66. C. Bonomo, P. Brunetto, L. Fortuna, P. Giannone, S. Graziani, S. Strazzeri. *IEEE Sens J* **2008**, 8 (8), 1486.
67. J. J. Pak, J. Kim, S. W. Oh, J. H. Son, S. H. Cho, S.-K. Lee, J.-Y. Park, B. Kim. Proc. SPIE Smart Struct. Mater. Symp., EAPAD Conf. 2004, 5385, 272.
68. J. Brufau-Penella, M. Puig-Vidal, S. Graziani, S. Strazzeri. *Smart Mater Struct* **2008**, 17, 015009.
69. M. Aureli, W. Lin, M. Porfiri. *J Appl Phys* **2009**, 105, 104911.

70. K. M. Farinholt, N. A. Pedrazas, D. M. Schluneker, D. W. Burt, C. R. Farrar. *J Intell Mater Syst Struct* **2009**, 20 (5), 633.
71. Energy harvesting from heat and vibration. Nikkei Electronics. **2009**. http://techon.nikkeibp.co.jp/article/HONSHI/20090127/164653/.
72. P. Harrop. IDTechEx. http://www.greenspacetoday.com/green-guide/paybacks-energy-harvesting. 2009.
73. A. Punning. PhD dissertation, University of Tartu, Estonia, 2007.
74. S. J. Kim, S.-M. Kim, K. J. Kim, Y. H. Kim. *Smart Mater Struct* **2007**, 16, 2286.
75. R. Tiwari. PhD dissertation, University of Nevada, Reno, NV, 2009.

CHAPTER 17

FUNCTIONAL LAYER-BY-LAYER POLYELECTROLYTES: ASSEMBLY STRATEGIES, CHARACTERIZATION, AND SELECTED APPLICATIONS

NICEL ESTILLORE, WOLFGANG KNOLL, and RIGOBERTO ADVINCULA

17.1 INTRODUCTION

Numerous reviews on the layer-by-layer (LbL) technique involving polyelectrolytes and amphiphiles have been made available since its reintroduction in the early 1990s by Decher[1] The LbL technique has shown to be versatile and simple in terms of the deposition process. A number of studies have been reported investigating the deposition of charged molecules on various surfaces.[1–4] The strong ionic interactions between polyelectrolytes hold the multilayered films together and the sequential deposition of oppositely charged molecules is made possible by the charge reversal on the surface. Scheme 17.1 displays some of the frequently used commercially available polymers and polyelectrolytes. In addition to this list, we have synthesized a wide array of functional, water-soluble (conducting) polymers also shown in this scheme.[5,6] Because of this wide availability of materials, the composition of the multilayered films can easily be tuned for specific applications. In general, the LbL multilayer assembly can be carried out using flat and spherical substrates and because of this wide spectrum of versatility, multilayered LbL films have found usefulness in a range of applications including optical devices,[7–9] biomedical,[10–12] functional membranes[13] (cf. Chapter 18), and even for the preparation of free-standing films.[14–16]

Besides the conventional manual dipping of the substrates, there has been a surge for deposition methods in order to shorten the lengthy deposition time. For example, Hammond and coworkers have employed an automated and

Ionic Interactions in Natural and Synthetic Macromolecules, First Edition. Edited by Alberto Ciferri and Angelo Perico.

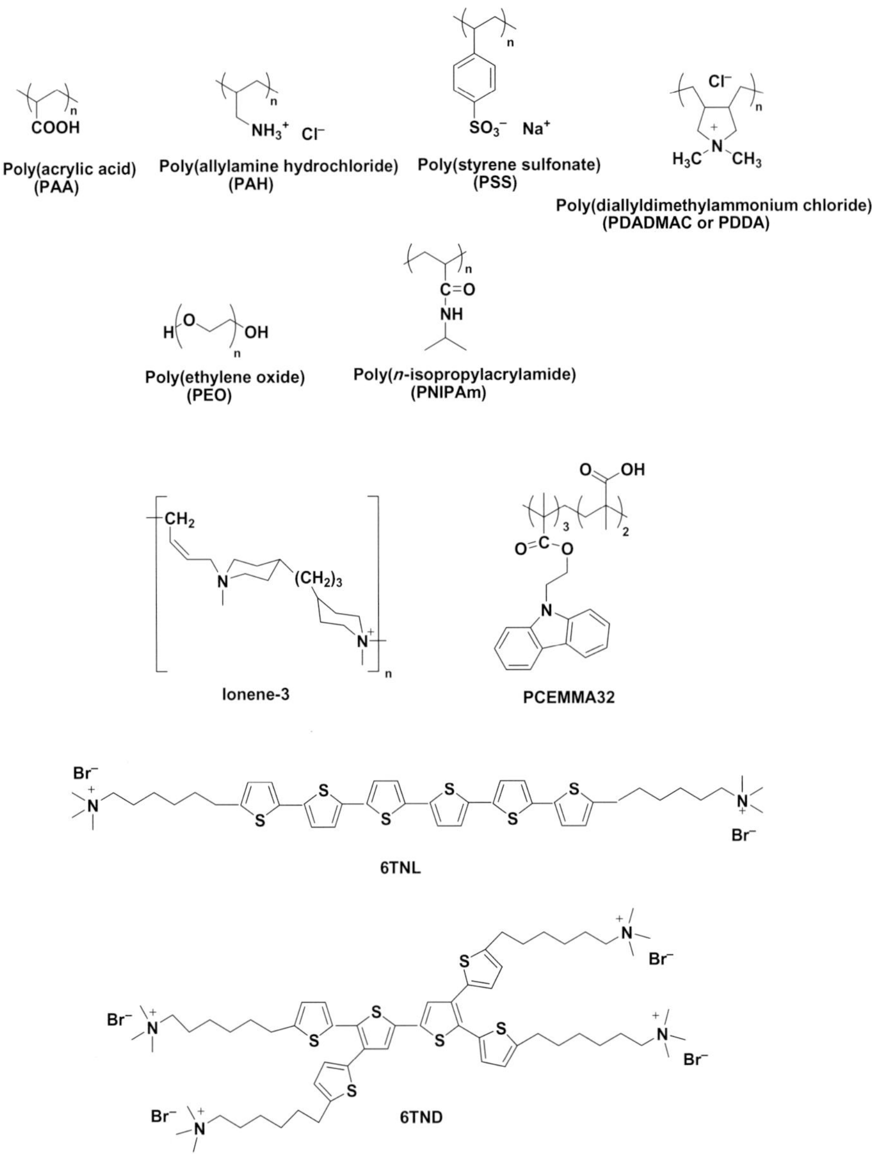

SCHEME 17.1. Chemical structures of commonly used and uniquely designed polymers, polyelectrolytes, and amphiphiles for the LbL assembly.

programmable dipping system allowing for highly reproducible LbL films.[17,18] Decher and coworkers have demonstrated the spraying method in which the polyelectrolyte solutions are contained in aerosol cans.[15,19] This method has significantly cut the time of deposition from minutes to seconds while simultaneously attaining the homogeneity and smoothness of the LbL films over large areas. Several other groups have used the spin-assisted LbL (SA-LbL) assembly, which has shown to shorten the deposition time from minutes to seconds as well.[20,21] The deposition of the polyelectrolyte material is based on the rapid spin coating and evaporation of the solutions and even the rinsing step is also carried out through spinning.

Typically, the multilayer assembly is based on the primary electrostatic interaction. However, there are polymers of interest such as polyethylene oxide (PEO) and poly(*n*-isopropylacrylamide) (PNIPAm) that have potential in the biomedical field but cannot be deposited through the sequential polycation/polyanion approach. Therefore, the use of secondary hydrogen bonding interactions has been extensively studied.[17,18,22] Furthermore, click reactions, most notably the 1,3-dipolar cycloaddition of azide and terminal acetylenes generating the 1,2,3-triazole moiety, result in the covalent attachment of each deposited layer, rendering the LbL assembly more robust and stable from possible delamination issues due to harsh conditions.[23] The preparation of a variety of click capsules has attracted interest in the biomedical field as potential drug carriers.[24–26]

Although there have been many other characterization methods for monitoring the multilayer growth of the LbL films, the surface plasmon resonance (SPR) technique has emerged as a leading candidate for optical determination of ultrathin films on metallic substrates.[27] When paired with other techniques such as electrochemistry (EC-SPR), the SPR technique becomes more useful for determining not only optical but also electrochemical properties of conducting films.[28–30] For instance, the mechanism into the electrochemical process, that is, doping–dedoping events, occurring in conducting LbL films can be investigated.[28] SPR, along with several other analytical methods including quartz crystal microbalance (QCM), can provide more meaningful elucidation of film properties.[29,31,32]

One relatively emerging application of the LbL technique is its applicability in a variety of polymer brush coatings. Recently, we introduced the concept of the combined LbL and surface-initiated polymerization (SIP) techniques.[33,34] Similar to the use of polyelectrolytes in the LbL assembly, a pair of charged macroinitiators is employed to grow dense polymer brushes. Of particular interest is the use of living radical polymerization technique such as atom transfer radical polymerization (ATRP) for the controlled growth of stimuli-responsive brushes.[35]

This chapter is organized into three parts. The first part includes discussions on the different substrates and interactions employed for the multilayer growth of the LbL films. This is then followed by an emphasis on the use of the SPR technique and its applications. Finally, a relatively emerging field in the SIP technique is also briefly reviewed.

17.2 USEFUL SUBSTRATES

A variety of suitable substrates, both flat and colloidal supports, have been employed for the multilayer assembly of LbL films (Fig. 17.1).[4] The electrostatic

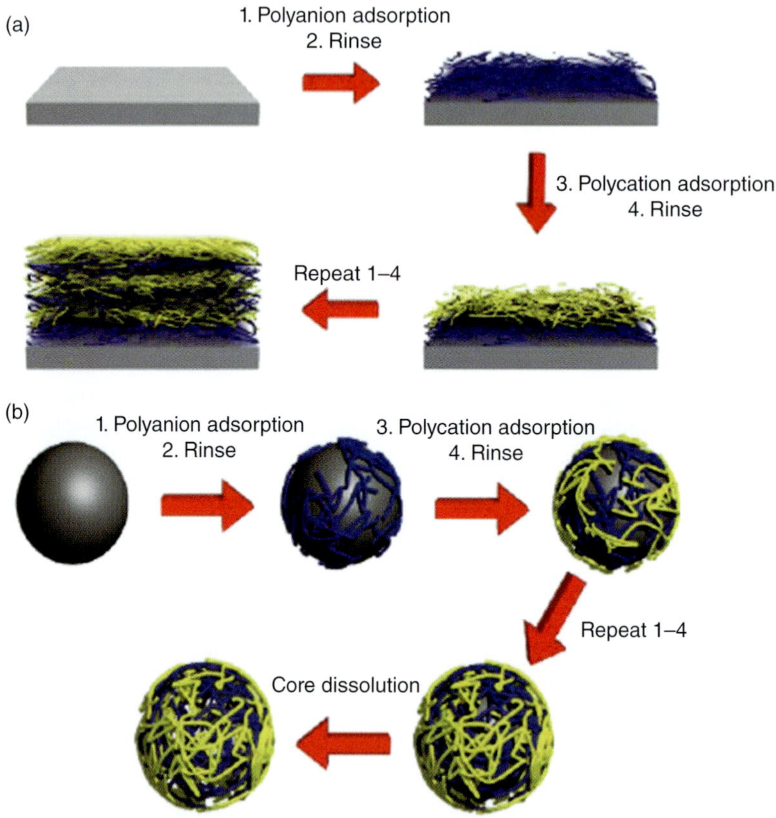

FIGURE 17.1. General schemes for the electrostatic self-assembly onto (a) flat and (b) colloidal substrates. (Reprinted from Quinn et al.[4]; with permission.)

self-assembly of polyelectrolyte pairs relies on a primed substrate. Therefore, in order for a substrate to be useful, it must carry an inherent minimal charge or a charge created following a surface treatment. Depending on the charge of the primed substrate, the substrate is dipped into an oppositely charged polyion solution for a short length of time followed by a water rinse to get rid of any loosely bound molecules. The sequential deposition to the other polyion solution is then performed under the same conditions. The charge reversal on the substrate surface allows for the sequential deposition to occur.

17.2.1 Multilayer Assembly Grown from Flat Substrates

Numerous studies based on the electrostatic self-assembly of LbL films have been done on flat substrates.[1,2] The choice of substrates greatly depends on the application as well as the analytical methods to characterize the films. Typically, silicon (Si) wafers, glass, indium titanium oxide (ITO), quartz, mica, and gold-coated substrates are frequently utilized. The Si wafer substrates are generally used for tracking the multilayer growth by ellipsometry but also for surface characterizations such as Fourier transform infrared (FTIR), atomic force microscopy (AFM), X-ray photoelectron spectroscopy (XPS), X-ray reflectivity (XRR), and water contact angle measurements. On the contrary, gold-coated substrates are used for electrochemical, QCM,[29,31,32] and SPR measurements.[36,37] When a relatively flat and smooth surface is required for AFM measurements, mica is typically used. For optical measurements including UV-vis and fluorescence spectroscopy, glass and quartz substrates are employed. Although most commercially available substrates have no net surface charge, a simple surface functionalization is required. For example, cleaving the surface of mica renders it negatively charged in water. The silanol groups of Si wafer, glass, and quartz substrates can easily be deprotonated in a basic solution rendering them negatively charged. Likewise, the surfaces of Si wafer and gold-coated substrates can be functionalized with silanes and thiols, respectively. The use of flat substrates has been extensively studied as model substrates for a variety of applications.

17.2.2 Multilayer Assembly Grown from Colloidal Substrates

The use of colloidal and spherical substrates has also been extensively studied. Similar to the LbL assembly onto flat substrates, colloidal core shell multilayer particles are prepared from a colloidal solution exposed to an excess amount solution of the polyelectrolyte pairs (Fig. 17.2).[33,38–45] The most commonly studied systems are prepared from silica nanoparticles, sulfonated polystyrene latex spheres, and melanine formaldehyde particles.[33,38–45] The properties of the particles can be finely tuned by the choice of the polyelectrolyte pair, number

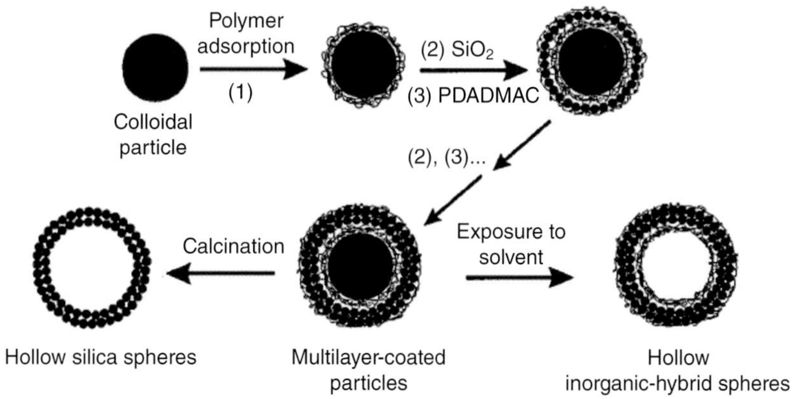

FIGURE 17.2. Multilayer LbL assembly onto colloidal substrates followed by the decomposition of the core. The core can be decomposed either by dissolving in an organic solvent or by a calcination reaction. (Reprinted from Caruso et al.[38]; with permission.)

of deposition cycles, and response to external stimuli. A series of centrifugation and rinsing steps are conducted for the sequential deposition of the polyelectrolytes. In order to monitor the multilayer buildup, electrophoretic mobility measurements are often performed. The alternating charges on the colloidal particles can be tracked with zeta (ζ) potential where the alternating signs indicate the stepwise growth of the multilayer assembly.[33,42–45] In addition, the size and shape of the modified particles are analyzed by scanning electron microscopy (SEM) and transmission electron microscopy (TEM).[33,38–45] More interestingly, the decomposition (by hydrofluoric acid [HF] solution, dissolution in organic solvents, calcinations, and acidic solution) of the core has led to the fabrication of hollow core-shell particles or capsules that have found applications in the biomedical field.[41–45] Various biomedical applications ranging from biomimetics to tissue engineering employing the LbL assembly on both flat and spherical substrates have been recently reviewed.[10–12]

17.3 INTERACTIONS IN MULTILAYERED LbL FILMS

The LbL assembly is not limited to electrostatic interactions. Although the electrostatic is the primary interaction in an LbL assembly, assemblies based on hydrogen bonding, charge transfer (CT), biological recognition, hydrophobic, and covalent bonding interactions are also possible. In this section, a brief overview of the various interactions is presented as well as the recent developments in click chemistry as the leading covalent attachment method for the LbL assembly.

17.3.1 Electrostatic Interactions

Various noncovalent interactions drive the multilayer assembly in an LbL film. For example, uncharged species such as PEO and PNIPAm have tremendous implications in the biomedical field.[4] PEO is biocompatible and highly resistant to protein, while PNIPAm exhibits a sharp volume transition at its lower critical solution temperature (LCST) at 32°C. The multilayer deposition of these two polymers has been made possible through the secondary hydrogen bonding interaction.[22,46]

Ionic The most widely studied multilayer assembly is based on the sequential deposition of charged species. The strong ionic interaction between the positive and negative polyelectrolytes is the driving force for the multilayer assembly.[1,47] There are a variety of complementary charged polyelectrolyte pairs. For instance, the commonly studied polyelectrolyte pair is that of weak polyelectrolytes, poly(acrylic acid) (PAA) and poly(allylamine hydrochloride) (PAH).[48,49] Throughout this chapter, various other polyelectrolyte pairs are discussed in detail. Relevant theoretical considerations are presented in Chapter 5.

Hydrogen Bonding One of the most studied nonelectrostatic interactions is based on hydrogen bonding where a complementary hydrogen bond donor and acceptor pair is employed. The earliest studies of hydrogen bonding interaction were reported by the groups of Stockton and Rubner and Zhang.[22,46] Stockton and Rubner studied the alternate deposition of polyaniline (PANI) with various polymers including poly(vinylpyrrolidone) (PVPON), poly(vinyl alcohol) (PVA), polyacrylamide (PAAm) and PEO.[22] FTIR spectroscopy was used to confirm that hydrogen bonding interaction between the polymers is the driving force for the multilayer assembly. Of particular interest is the N-H peak around ~3300 cm^{-1}; a nonhydrogen-bonded N-H would appear at 3380 cm^{-1}, while a hydrogen-bonded N-H would appear at 3310 cm^{-1}. In this case, it was determined that a hydrogen-bonded N-H for PVPON/PANI and PEO/PANI appeared at 3304 and 3300 cm^{-1}, respectively, indicating that indeed the polymer pairs were bound to the surface through hydrogen bonding.

An LbL film assembled via hydrogen bonding interaction has shown applicability as stimuli-responsive nanomaterials. Granick and Sukhishvili have studied the construction and deconstruction of hydrogen-bonded PAA or poly(methacrylic acid) (PMAA) with PEO or PVPON.[50,51] The suppressed ionization in the hydrogen-bonded LbL film is what keeps the multilayer assembly together. Therefore, a change in pH would lead to the disruption in the hydrogen bonding of the LbL assembly. At a critical pH, the PAA or PMAA becomes ionized and since PAA and PMAA are hydrogen bond donors and PVPON and PEO are acceptors, this leads to the breaking of the hydrogen bond interaction of the multilayer film. Although this presents possibility of pH-tunable multilayer disassembly, the instability of hydrogen-bonded films at high pH values is a concern. Yang and Rubner have reported the cross-linking of hydrogen-bonded films rendering them stable at a high pH.[52] In this case, the temperature was elevated to 175°C during the multilayer assembly of PAA and PAAm, causing the thermal imidization reaction between the amide groups of PAAm and the carboxylic acid groups of PAA.

Hydrogen-bonded multilayers have also found usefulness in release applications.[51] It was shown that rhodamine 6G can be easily loaded during the multilayer assembly and that rinsing with a buffer solution at pH 4 did not release rhodamine. However, at a higher pH 5.5, rhodamine 6G was released. Hammond and coworkers have also extensively studied hydrogen-bonded LbL films.[17,18] They have demonstrated the possible applications of these hydrogen-bonded films for capture and release of drugs and fabrication of free-standing films.

CT Yamamoto and coworkers reported the use of nonionic polymers, poly[2-(9-carbazolyl)ethyl methacrylate] and poly[2-((3,5-dinitrobenzoyl)oxy)ethyl methacrylate], which have electron-donating and electron-withdrawing groups as the side chains, respectively.[53] The multilayer assembly on gold and quartz substrates was made possible based on the CT between the donating and withdrawing polymers (Fig. 17.3). UV-vis spectroscopy and SPR techniques

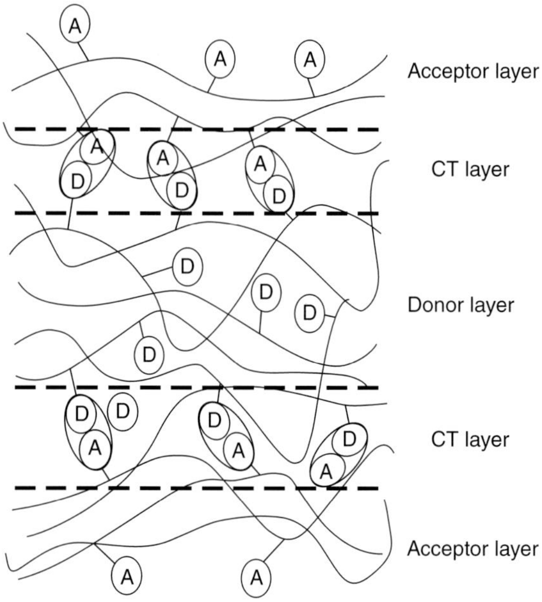

FIGURE 17.3. The sequential deposition of electron-donating and electron-withdrawing polymers based on the charge transfer interaction. (Reprinted from Shimazaki et al.[53]; with permission.)

were used to monitor the multilayer growth and indicated a linear growth, but after the 10th layer a stagnant growth was observed. Since a nonaqueous solvent was used for the dipping solutions, there is potential for a variety of organic-based nanostructured materials.

Biological Recognition There are just a few reports on the multilayer assembly based on biological recognition.[54–56] Decher and coworkers demonstrated the specific binding of streptavidin–biotin and complementary base pairs of nucleotides as the driving force for the LbL assembly.[57] In another example, Hoshi and coworkers studied the multilayer formation using avidin and biotin-labeled polymers.[54] When the unmodified polymers were used, there was no multilayer growth as evident by the QCM data, which further exemplifies that the strong binding of avidin and biotin is responsible for the multilayer assembly.

17.3.2 Covalent Assembly of Multilayered Films

The majority of work regarding multilayered films has mainly focused on the assembly of polyelectrolytes or polymers via ionic[1] and hydrogen bonding[17,18,22] interactions. However, under harsh environmental conditions, these noncovalently assembled thin films are prone to disassembly or delamination. For example, weak polyelectrolytes such as PAA and PAH are susceptible to dis-

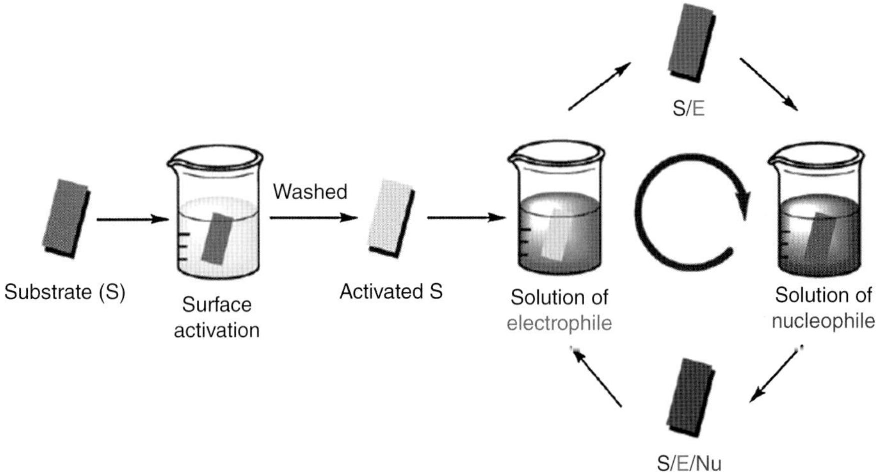

SCHEME 17.2. General preparation of substrates for the covalent LbL assembly. S, substrate; E, electrophile; Nu, nucleophile. (Reprinted from Bergbreiter and Liao[58]; with permission.)

sociate under alterations in the medium pH or salt.[50,51] Therefore, the covalent attachment of the multilayered assembly has emerged in the recent decade, proving useful in rendering films more chemically and mechanically stable. There are various useful reactions for chemically attaching polymer molecules via covalent linkages, and the recent development in the field of click chemistry has made this technique particularly very versatile in preparing a variety of functional multilayered films. The covalent LbL assembly is not limited to planar substrates, and as will be discussed below, spherical substrates have also been employed. Scheme 17.2 depicts a generalized fabrication of a covalently LbL assembled film.[58]

Various Reactions In order for the covalent interaction to take place, appropriate reactant pairs are necessary for a particular reaction. These pairs must possess complementary functional groups that can form stable covalent linkages *in situ*. There have been different reactions employed for the covalent LbL assembly of multilayered films. Some of the earliest examples involved the condensation reactions of nucleophiles and carbonyl derivatives.[59–61]

Blanchard and Kohli demonstrated the covalent LbL multilayer assembly through the use of diphenylmethane derivatives.[59,60] As shown in Scheme 17.3, the interlayer linkage is through a urea moiety that is made possible by the reaction between isocyanate and amine monomers. The alternating reaction led to the controlled LbL growth on the surface, yielding a thickness of 625 Å after seven layers. However, a reaction time of 4 hours under an argon atmosphere was required for each layer to be formed. Bergbreiter and Crooks' groups have described using commercially available poly(maleic anhydride)-*c*-poly(methyl vinyl ether) copolymer as an electrophile in the

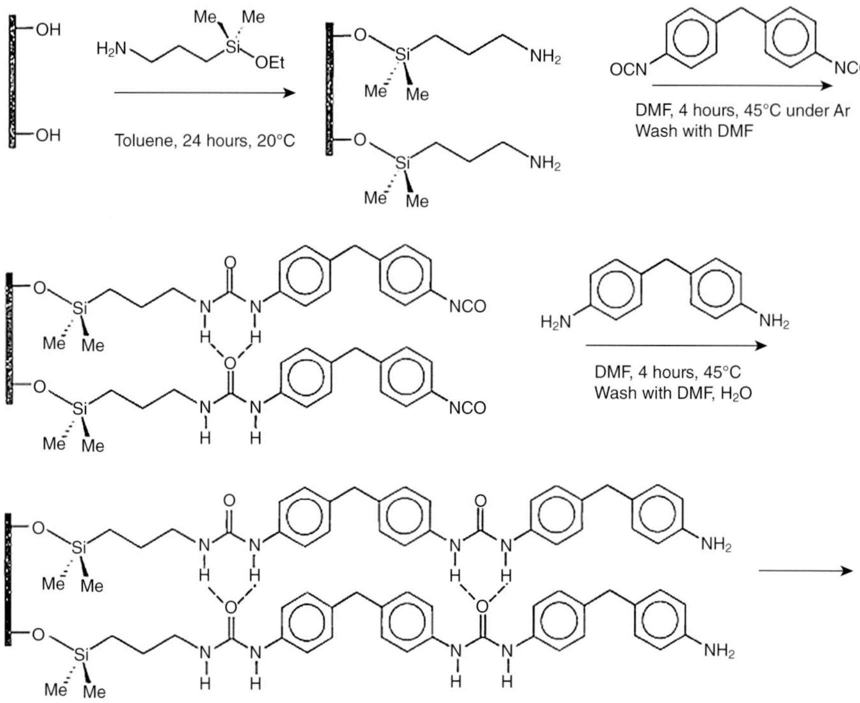

SCHEME 17.3. LbL multilayer assembly using diphenylmethane derivatives. DMF, dimethyformamide. (Reprinted from Kohli and Blanchard[59]; with permission.)

reaction of nucleophilic amine or hydroxyl-terminated dendrimers, poly(amidoamine), or poly(iminopropane-1,3-diyl).[62–64]

The use of conjugated polymers such as poly(*p*-phenylenevinylenes) (PPVs) has also been investigated.[65] The first step in the assembly was to functionalize a gold substrate to possess carboxylic acid end groups. The PPVs were modified with amino and pentafluorophenyl ester groups as side chains where the driving force for the covalent LbL assembly is based on the condensation reaction between amines and activated esters, but some hydrogen bonding interactions also contribute to the assembly. The multilayer assembly was monitored by UV-vis spectroscopy and ellipsometry where a linear growth was observed. The covalent LbL assembly was also performed on patterned self-assembled monolayers (SAMs) on gold where the patterned regions contained the carboxylic acid groups and the selective reaction of the modified PPVs occurred only on these patterned regions.

The covalent LbL assembly based on aromatic substitution reactions assembled on quartz or on sulfonated polystyrene latex has been reported.[66,67] Cao and coworkers demonstrated the use of electrophilic aromatic substitution by a polymeric aryl diazonium salt on electron-rich polymers.[66,67] The assembly proved stable in organic solvents and because an azo dye group is simultaneously introduced after each coupling reaction on the surface, UV-vis was used in monitoring the multilayer growth.

Recently, Char and coworkers have demonstrated the covalent LbL assembly based on activated esters.[68] The deposited layers were formed through amide bond formation by the reaction between activated esters, poly(pentafluorophenyl-4-vinylbenzoate) or poly(pentafluoroacrylate), with poly(allyl amine) (PAAm). The activated esters and PAAm solutions were made from tetrahydrofuran (THF) and ethanol, respectively, with reaction times usually between 10 and 20 minutes. Because of only ionic interaction between the poly(ethyleneimine) (PEI) primed surface and the first layer of PAAm, detachment of the film was possible under mild acidic conditions, generating stable free-standing films.

Click Chemistry Click chemistry is a term describing a number of reactions that can be easily performed and worked up, high yielding, and tolerant to air and moisture.[69] Although click chemistry was initially intended for bioconjugation, it now provides an efficient method for the construction of covalently bonded multilayered films. The most documented click reaction involves the Cu(I)-catalyzed reaction of Huisgen 1,3-dipolar cycloaddition of azides and terminal acetylenes (or alkynes) forming the 1,2,3-triazole moiety.[23] The combination of click chemistry and LbL assembly offers several advantages. First, the reaction is fast and quantitative with high yields under mild aqueous conditions. Second, the resulting triazole linkage is extremely stable to hydrolysis, oxidation, or reduction.[23] Lastly, a wide array of functional materials ranging from polymers, to proteins, to nanoparticles are compatible with this reaction. Because excess unreacted alkyne and azide moieties reside after the LbL-click assembly, further modification with various materials is realized. In general, clickable groups have to be synthesized, and various groups have reported the use of living radical polymerizations.[70] A typical LbL assembly involves the exposure of the substrate (both planar and spherical) to solution of an alkyne- and azide-terminated polymer in the presence of copper sulfate and sodium ascorbate for 20 minutes, followed by a water rinsing step.[23] Because there is no production of by-products, further purification or workup is not necessary. The LbL-click assembly can be repeated until the desired number of layers is achieved.

For instance, Caruso and coworkers reported the first LbL-click assembly using a single polyelectrolyte, PAA, by first synthesizing the alkyne-terminated PAA (PAA-Alk) and azide-terminated PAA (PAA-Az).[23] Figure 17.4 illustrates the LbL-click chemistry assembly. The LbL-click assembly of PAA-Alk/PAA-Az onto a planar substrate was monitored with UV-vis spectroscopy, and it was determined that a linear growth of the click groups was formed. In addition, the formation of the triazole groups was also monitored by XPS. The functionalization of multiwalled carbon nanotubes (MWNTs) with clickable groups has also been demonstrated.[70] In order to modify the surfaces of the MWNTs, alkyne groups were first introduced through an acylation reaction,

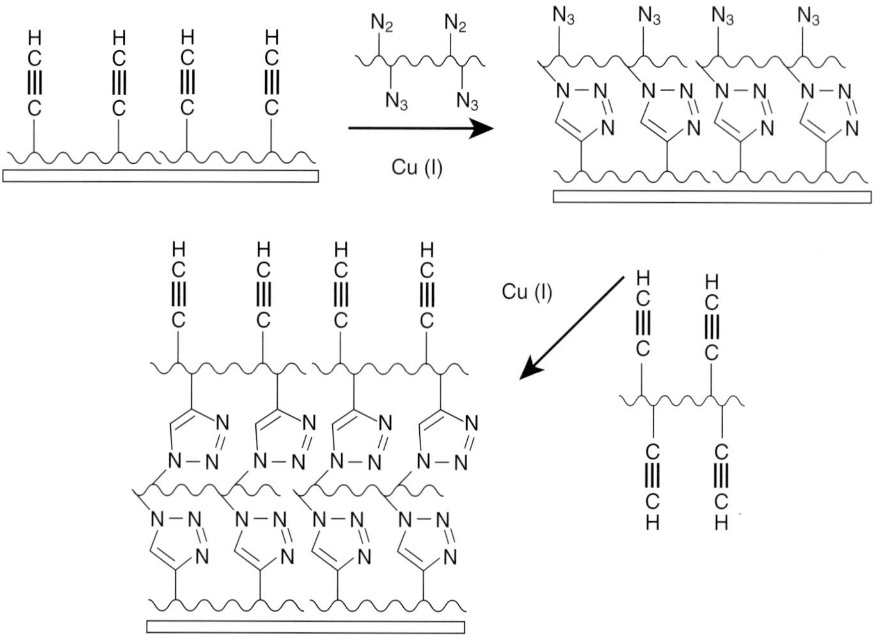

FIGURE 17.4. Schematic of the LbL assembly of the substrate using click chemistry. (Reprinted from Such et al.[23]; with permission.)

yielding the alkyne-terminated MWNTs (MWNT-Alk). The clickable groups, poly(2-azidoethyl methacrylate) (poly(AzEMA)) and poly(propargyl methacrylate) (poly(PgMA)), which were synthesized via living free radical polymerization techniques were then sequentially deposited. Surface analysis with thermogravimetric analysis (TGA), FTIR, Raman, XPS, SEM, and TEM all confirmed the successful LbL assembly of the MWNTs. With the last layer containing the azido groups, further functionalization with an alkyne-terminated rhodamine B and a monoalkyne-terminated poly(styrene) was demonstrated. Figure 17.5 displays fluorescence images of the resulting MWNT-RhB. Various other functionalities including porphyrins and perylene diimides have shown more complex multilayer structures achieved by the LbL-click assembly.[71] Hawker and coworkers have fabricated dendritic thin films on silicon wafer substrates through the use of clickable dendrimers.[72]

Several groups have reported the LbL-click assembly of various functionalities on spherical substrates.[24–26] These so-called click capsules have demonstrated responsive properties, making them widely applicable for biomedical and drug delivery applications.[24–26] For instance, the reversible shrinking and swelling of click capsules composed of PAA-Az/PAA-Alk were illustrated.[24] The presence of free alkyne click moieties in these capsules was verified by further reacting them with an azido-modified rhodamine dye.[24] Removal of the core was conducted and the resulting hollow capsules underwent reversible size changes under basic and acidic conditions, which were illustrated

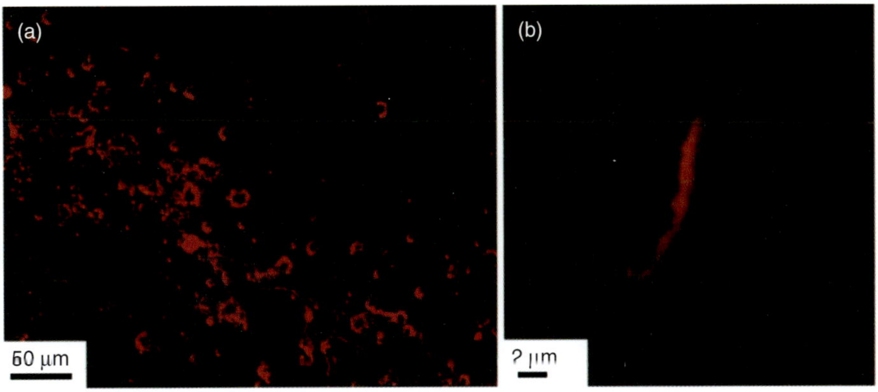

FIGURE 17.5. (a) Fluorescence microscopy image of MWNT-RhB (λ_{max} = 520–550 nm). (b) Confocal fluorescence image of a singular tube of MWNT-RhB (λ_{max} = 543 nm). (Reprinted from Palomaki and Dinolfo[71]; with permission.)

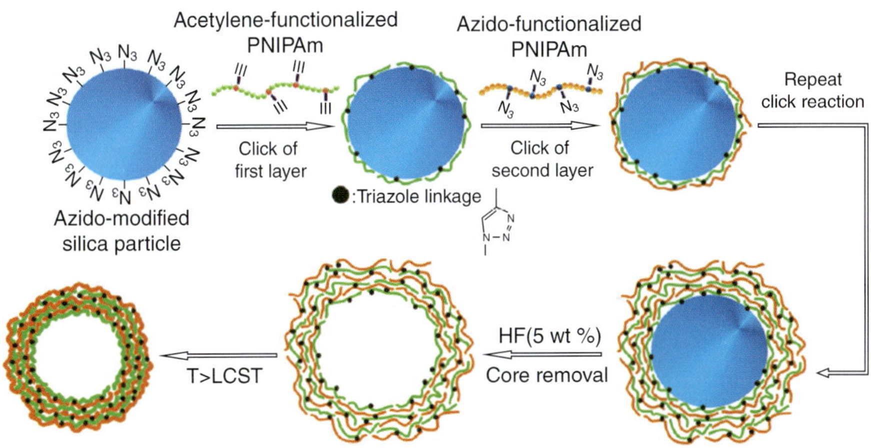

FIGURE 17.6. Schematic illustration of the functionalization of azido-modified silica particles and the subsequent click reaction with alkyne-terminated PNIPAm. After the repeated LbL-click assembly, the silica core was removed through exposure to HF solution, producing hollow core-shell particles. (Reprinted from Huang and Chang[26]; with permission.)

by TEM. Various other low-fouling and biodegradable click capsules have also been demonstrated.[25] For example, biodegradable poly(L-lysine) (PLL) and poly(L-glutamic acid) (PGA) polymers were modified to possess clickable moieties. The amine groups of PLL were used for postfunctionalization with a heterobifunctional poly(ethylene glycol) (PEG), which provided low-fouling properties and enhanced specific protein binding.[25] Temperature-sensitive microcapsules have also been reported.[26] Figure 17.6 shows the functionalization of silica particles with clickable PNIPAm copolymers with various compositions. The swelling and collapsing of the microcapsules as a function of temperature was demonstrated and tracked with TEM.

Although the copper-catalyzed reaction between an azide and terminal alkyne is the most popular click reaction, other click-type reactions including poly(2-alkenyl azlactone)s and thiolene have also been demonstrated for the covalent LbL assembly of multilayered films.[73,74]

17.4 SPR SPECTROSCOPY

SPR technique is a valuable tool for characterizing the optical properties of ultrathin films at the nanometer thickness scale.[27] This technique is highly sensitive to changes in the film thickness and dielectric constant. Upon total reflection of light at an interface, an evanescent wave is created in the medium above the surface. During adsorption or desorption of organic molecules, the evanescent wave and propagation of the surface plasmons are altered. As a consequence, the resonance angle is shifted. The magnitude, direction, and shape of the resonance angle shift are dependent on a specific polymer combination pair. In terms of multilayered nanostructured films, both *ex situ* and *in situ* SPR measurements have been conducted. Typically, SPR reflectivity-angular scans are performed to monitor the stepwise and uniform growth of the LbL assembly in which the shift of the angular curve to higher incident angles is indicative of more materials being deposited on the surface. Furthermore, the SPR technique allows for the *in situ* kinetic studies of the adsorption of the alternating deposition of the polyelectrolytes. In such cases, the change in reflectivity as a function of time is measured. For instance, LbL hybrid films composed of gold nanoparticles (AuNPs) and various polyelectrolytes (PAH and poly(styrene sulfonate) [PSS]) under different pH conditions have shown an SPR signal enhancement.[75] SPR, in this case, was utilized not only to monitor the multilayer growth but also to investigate the effect of pH switching on the SPR signal. It was concluded that the SPR response was greatly affected by the pH and distance of the AuNPs from the metal film.[75]

More interestingly, the combined EC-SPR has provided a way for the simultaneous measurements of the optical and electrochemical properties of a given film. This allows for the *in situ* elucidation of the mechanism of the electrochemical events occurring, that is, doping–dedoping, of conjugated ultrathin polymer films. For example, repeated cyclic potential scans were employed to grow PANI onto a gold electrode surface.[28] The shift of the SPR reflectivity-angular curve to higher incident angles correlated well with increasing cyclic scans.

17.4.1 *In Situ* Adsorption of Polyelectrolytes by SPR

We have reported the *in situ* adsorption of cationic ionenes and anionic polyelectrolyte pairs using the SPR technique.[37,76] Figure 17.7a depicts the setup of the *in situ* SPR experiment based on the Kretschman/attenuated total reflec-

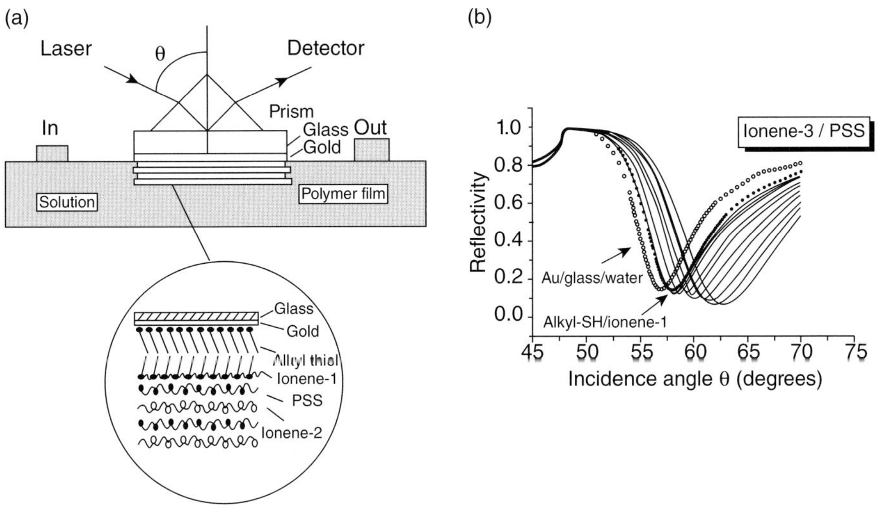

FIGURE 17.7. (a) *In situ* SPR setup in the Kretschman configuration and attenuated total reflection (ATR) setup. (b) SPR reflectivity-angular plot for the ionene-3/PSS polyelectrolyte pair. A shift to higher incident angles signified the stepwise deposition of the LbL film. (Reprinted from Advincula et al.[76]; with permission.)

tion (ATR) configuration. Briefly, a Teflon holder is utilized as the sample cell where the gold substrate is mounted and the polyelectrolyte solutions are injected (in) and removed (out) by means of a syringe.[37,76,77] Prior to the adsorption of the polyelectrolyte solutions, the gold substrate is first functionalized with the ionene-1 polymer using the Langmuir–Blodgett–Kuhn (LBK) technique, rendering the film with a uniform positively charged surface.[76] It is clearly evident that the shift in the angular curve and thickness obtained are specific for the polyelectrolyte combination pair. For instance, the ionene-3/PSS polymer pair displayed the biggest shift in the angular curve, giving an average thickness of 3.0 and 1.62 nm for ionene-3 and PSS, respectively (Fig. 17.7b). On the contrary, both the PES (poly-[propyl-1,3-phenyldicarboxylate-5-sulfonate] ester sodium salt)/ionene-2 and PSS/ionene-2 polyelectrolyte pairs showed the smallest shift in the angular curve, with minimal changes in the thicknesses.[76] Therefore, the shift in the angular curve is unique for each polyelectrolyte pair. The *in situ* SPR measurement also affects the conformation of the polyelectrolyte pairs on the surface. In this case, the film is exposed to a continuous aqueous solution, yielding much swelled layers as compared with film measurements performed in the dry state.[76,78] Moreover, the adsorption of the polyelectrolyte solutions as a function of time showed that a 90% adsorption is taking place in the first 15 minutes and no significant change in thickness was observed for the next 3 hours. This result indicates that an excess surface charge is created on the film during the first 15 minutes and repulsion between the like charges is occurring afterward.

The same groups also investigated the LBK deposition of ionenes and several ionic nonlinear optically (NLO) active amphiphiles.[37] *Ex situ* and *in situ* SPR measurements were conducted on the LBK deposited films and showed a shift of the angular curve to higher incident angles corresponding directly to the number of deposited layers. The thicknesses for each deposited layer were also elucidated from the SPR measurements using the Fresnel equation, and the increase in thickness correlated well with increasing layers deposited.

17.4.2 Combined EC-SPR Measurement of Conducting Polymers

We have also extensively studied the optical and electrochemical properties of multilayered conducting films through the combined electrochemistry and SPR (EC-SPR) techniques.[79–87] The combined EC-SPR technique allows for the simultaneous measurement of the electrochemical and optical properties of a conducting ultrathin film. Consequently, this technique permits the *in situ* determination of the complex dielectric constant and thickness of the conducting ultrathin film. In EC-SPR (Fig. 17.8), the gold film is utilized as both the surface plasmon medium and the working electrode for electrochemistry. Since the surface plasmons are located on the surface of the electrode, electrochemical events, including doping–dedoping, affect how the plasmons are excited; thus, the EC-SPR technique is useful in investigating the processes occurring on the surface of the electrode. For example, the optical properties and electropolymerization of poly(3,4-ethylenedioxythiophene) (PEDOT) have been investigated in detail using the EC-SPR technique.[80] The changes in the SPR reflectivity-angular curve correlate well with the doping state of PEDOT, which is then translated into changes in the dielectric constant and thickness of the electropolymerized PEDOT.

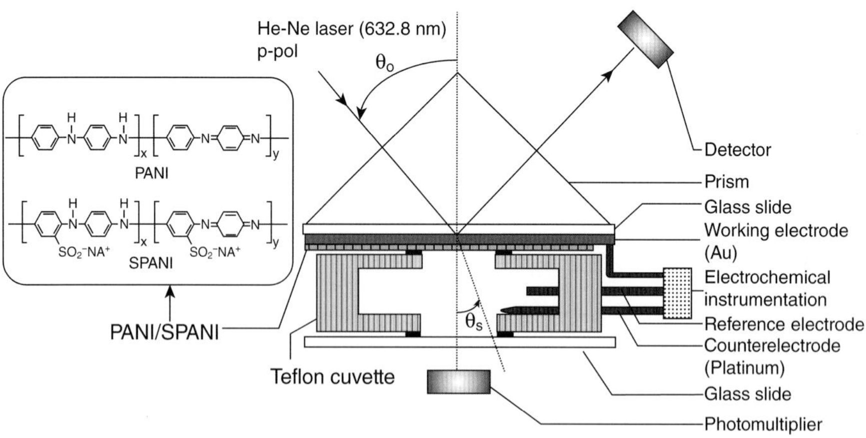

FIGURE 17.8. Schematic diagram of the EC-SPR setup in the Kretschman and ATR configurations. (Reprinted from Baba et al.[81]; with permission.)

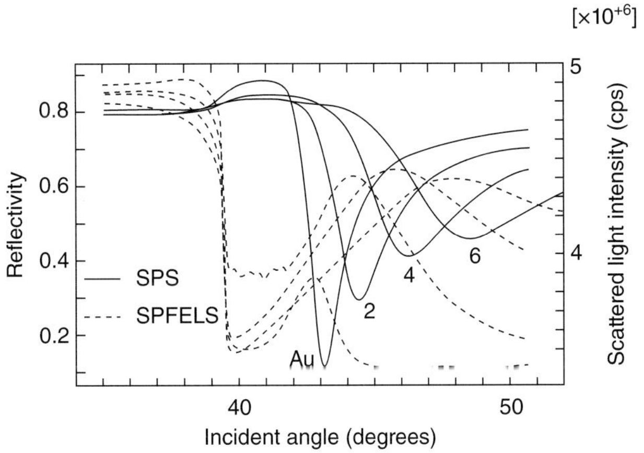

FIGURE 17.9. SPR reflectivity-angular curves and SPFELS curves measured after the bilayer deposition of each polyelectrolyte. (Reprinted from Baba et al.[81]; with permission.)

EC-SPR Measurement of PANI The optical and electrochemical properties of multilayers prepared from PANI and sulfonated polyaniline (SPANI) by simultaneous EC-SPR and surface plasmon field enhanced light scattering (SPFELS) have been reported.[81,82] Figure 17.9 shows the SPR reflectivity-angular curves and the SPFELS curves after deposition of each bilayer of PANI/SPANI.[81] A satisfactory linear relationship is found between the deposition cycle and shift of the angular curve to higher incident angles. This means that more material is being deposited on the surface, which in turn is accompanied by a change in the dielectric property of the film. In addition, the incident light scattered off the conducting polymer/electrode interface enabled the characterization of the film surface roughness by SPFELS. The peak intensities of SPFELS also exhibited a strong correlation with the number of layers deposited, indicating that an increase in surface roughness and inhomogeneity are occurring with more layers of PANI/SPANI deposited.[81,88] Moreover, a strong correlation between the change in the dielectric constant and thickness of the PANI/SPANI with the doping–dedoping or protonation–deprotonation processes is observed. Specifically, SPR reflectivity decreases and increases during the doping and dedoping of the anion, respectively, as shown in Figure 17.10. However, these results are unlike previous reports by Orata and Buttry where they observed an increase and decrease in the film mass (probed by E-QCM) during the doping and dedoping of the ion, respectively.[89] In the present case, the dielectric constant of PANI/SPANI, and not the film thickness change, imposed a significant effect on the SPR reflectivity behavior.[81,83] This was further confirmed with the time differential SPR data. Repeated potential cycling measurements suggest that the thickness and dielectric constant of PANI/SPANI remained unchanged. The cycling experiments also led to an increase in the film surface roughness.[80–82,90]

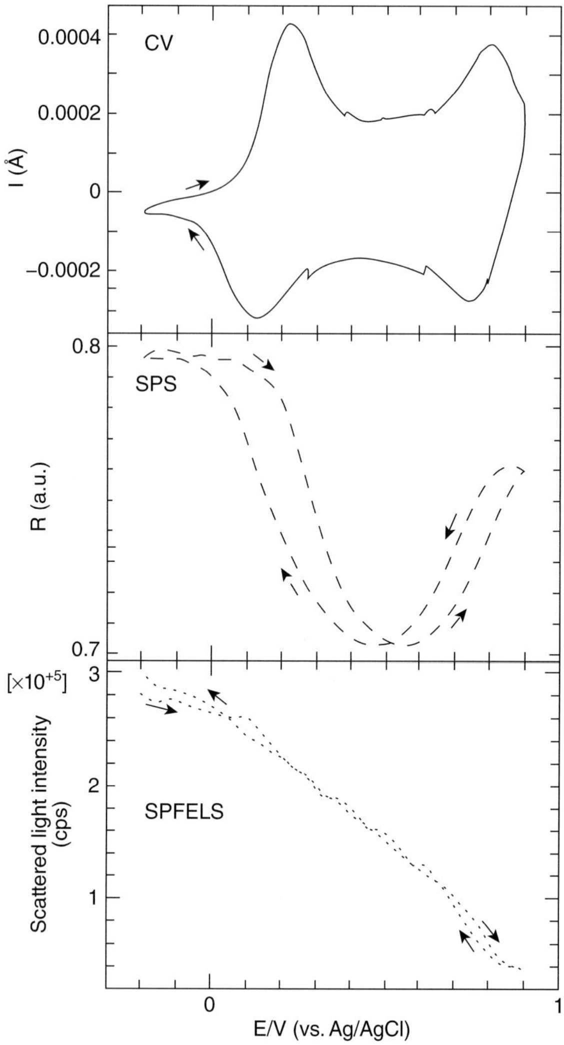

FIGURE 17.10. Simultaneous measurements of the cyclic voltammogram (CV) (top), SPR curve (middle), and SPFELS curve (bottom) for four bilayers of PANI/SPANI. Potential cycling was carried out in 0.5 M H_2SO_4 solution at the scan rate of 100 mV s^{-1}. (Reprinted from Baba et al.[81]; with permission.)

EC-SPR Measurement of Polythiophene and Polycarbazole Derivatives

Conducting films composed of polythiophene and polycarbazole derivatives have been investigated.[84–86]. Specifically, water-soluble sexithiophene bolaform amphiphiles (6TNL, 6TND) and polycarbazole (PCEMMA) derivatives have displayed interesting optical and electrochemical properties. The SPR kinetic curve (reflectivity vs. time) displayed a stepwise and controlled adsorption of

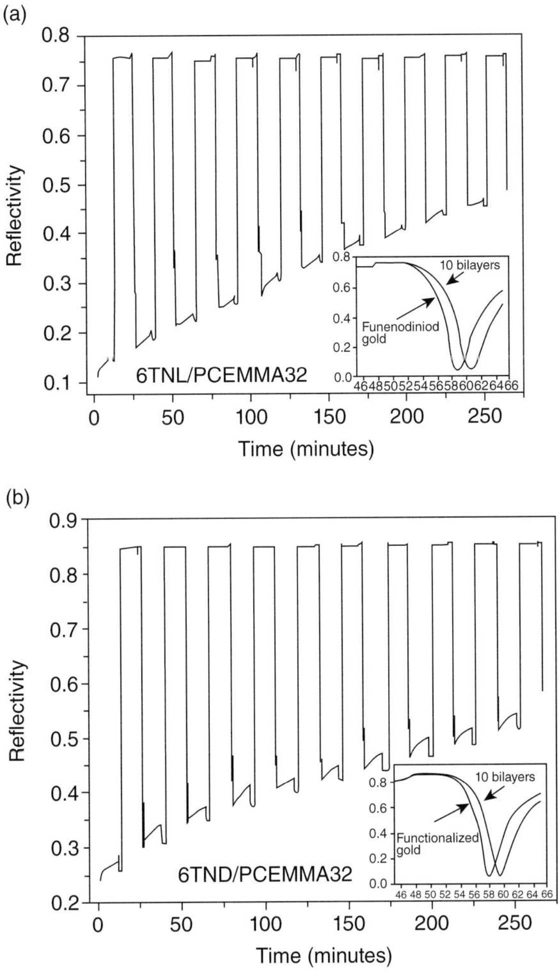

FIGURE 17.11. SPR kinetic plots (reflectivity vs. time) for 10 bilayers of (a) 6TNL/PCEMMA32 and (b) 6TND/PCEMMA32. The insets show the reflectivity-angular plots after deposition. (Reprinted from Sriwichai et al.[84]; with permission.)

10 bilayers of 6TNL/PCEMMA32 and 6TND/PCEMMA32.[84] Figure 17.11 shows that the SPR reflectivity increases with each deposition of 6TNL/PCEMMA32 and 6TND/PCEMMA32 and decreases during the rinsing steps. The insets in this figure display the reflectivity-angular measurements of the functionalized gold substrate and after deposition of 10 bilayers. The shift of the SPR curve to higher incident angles indicates the increase in the overall film thickness as more layers are deposited.[5,84] Total film thicknesses (after 10 bilayers deposited) of 177 and 180 Å were obtained for the 6TNL and 6TND systems, respectively. The multilayer assembly for both 6TNL and 6TND was also further confirmed by UV-vis data.

From the cyclic voltammograms, the doping–dedoping currents for both 6TNL and 6TND systems increased with each cycle. The 6TNL showed an oxidation onset at 1.37 V but shifted to 1.4 V with the succeeding cycles. On the other hand, the 6TND showed a lower oxidation onset at 1.32 V and shifts to 1.4 V with the succeeding cycles. Based on the doping current, it is concluded that 6TND is more electrochemically active than 6TNL. The doping–dedoping process was verified by UV-vis data where the carbazole peaks at 330 and 345 nm disappeared and the appearance of the new peak at 400 nm was observed. The red shift also is an indication of the formation of more conjugated PCEMMA32 after electropolymerization.[84,85,91]

Figure 17.12 shows the combined EC-SPR measurements for which the doping–dedoping process is correlated to the changes in the dielectric constant and thickness of the LbL film.[80,84] The potential cycling measurements led to a stepwise increase in the reflectivity for both 6TNL and 6TND systems. The changes in reflectivity (which are correlated to changes in the dielectric constant) are attributed to the oxidation–doping and reduction–dedoping

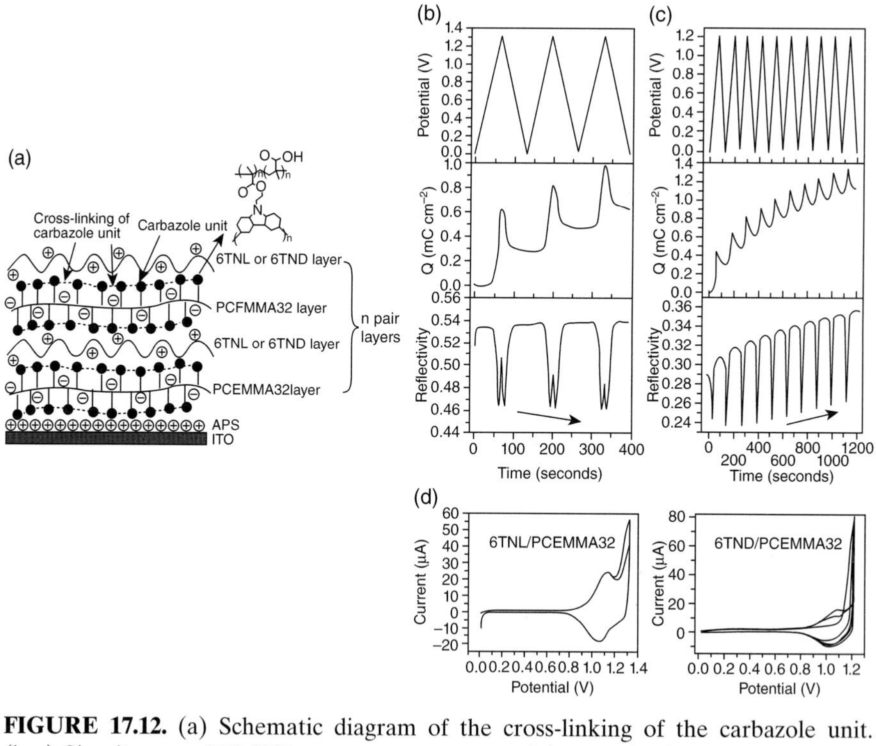

FIGURE 17.12. (a) Schematic diagram of the cross-linking of the carbazole unit. (b, c) Simultaneous EC-SPR measurements: potential ramp (top), amount of charge (middle), and kinetic plot (bottom) during the cross-linking of the carbazole unit of 6TNL and 6TND. (d) Corresponding cyclic voltammetry (CV) traces. (Reprinted from Sriwichai et al.[84]; with permission.)

processes. The higher current/charge for the 6TNL/PCEMMA32 suggests that there is a higher degree of cross-linking or doping effect of the PCEMMA. The thicker layers deposited for the 6TND/PCEMMA32 resulted in a lesser electron tunneling for the PCEMMA. Overall, the EC-SPR displayed a reversible oxidation–reduction process for both systems.

The SPR reflectivity-angular measurements after the electropolymerization of 6TNL and 6TND are shown in Figure 17.13. The 6TNL/PCEMMA32 showed a shift of the angular curve to a lower incident angle, while the angular curve for the 6TND/PCEMMA32 shifted to a higher incident angle. These results are attributed to the changes in the thickness and dielectric constant

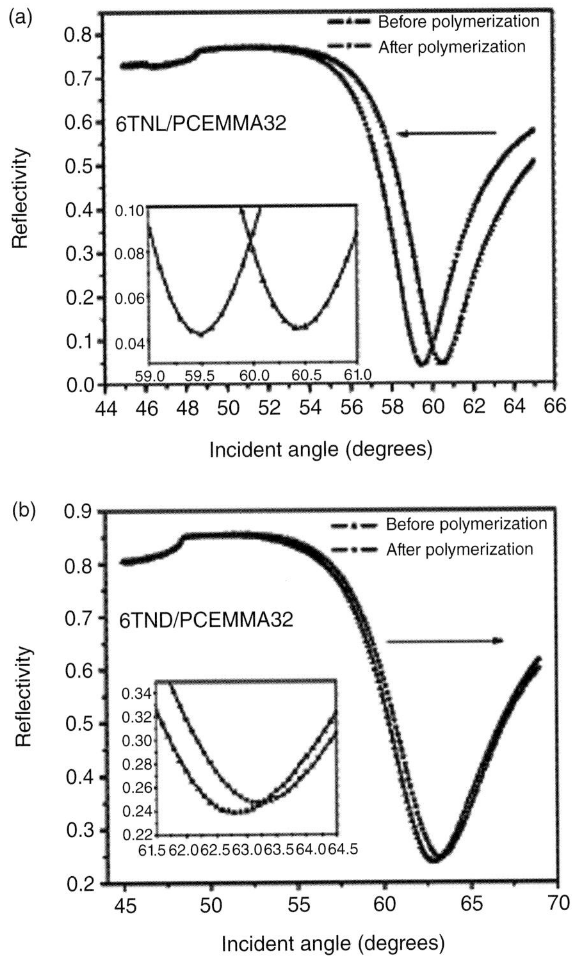

FIGURE 17.13. SPR reflectivity-angular scans of the electropolymerized films (a) 6TNL and (b) 6TND. (Reprinted from Sriwichai et al.[84]; with permission.)

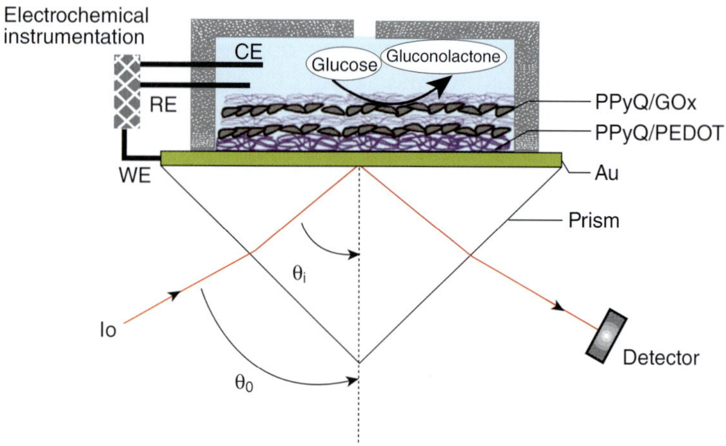

FIGURE 17.14. Schematic diagram of EC-SPR glucose biosensor in the Kretschman and ATR configurations with a three-electrode cell setup. CE, counter electrode; RE, reference electrode; WE, working electrode. (Reprinted from Baba et al.[92]; with permission.)

for these films, which are affected by the nanostructure ordering of 6TNL and 6TND as well as the electropolymerizability of PCEMMA 32.[84]

17.4.3 Applications of EC-SPR

Glucose Biosensors Recently, we published an EC-SPR/waveguide glucose biosensor prepared from multilayers of polypyrrole/glucose oxidase (PPy-Q/GO_x).[92] The detection of enzymatic reactions can easily be done using electrically conducting polymers due to their high conductivity and specific electrochemical redox properties.[92,93] The EC-SPR technique allows for the direct electrochemical oxidation of glucose where enhancements in both the optical and electrochemical detection signals are obtained. In addition, the incorporation of a conducting polymer such as polypyrrole allows for a better electron transfer from the enzyme to the electrode surface.[92,94] Figure 17.14 depicts the schematic diagram for the glucose biosensor. In this case, the fabrication of the conducting polymer/enzyme electrode started out with the initial deposition of PPy-Q and PEDOT-PSS to provide an enhancement of electron transfer between glucose and the gold electrode followed by the alternating deposition of PPy-Q and GO_x in a phosphate buffered saline (PBS) buffer solution. The SPR kinetic curve displayed a stepwise deposition of PPyPEDOT-PPyGO_x and is evident that some desorption of GO_x is occurring during the adsorption of PPy-Q. The detection of glucose was conducted in a neutral PBS buffer upon the addition of 10 mM glucose every 5 minutes, and during the sensing experiment, the potential was set to 0.3 V.[92,95] The change in reflectivity is highly affected by the doping state of the film. Specifically, the dedoped state

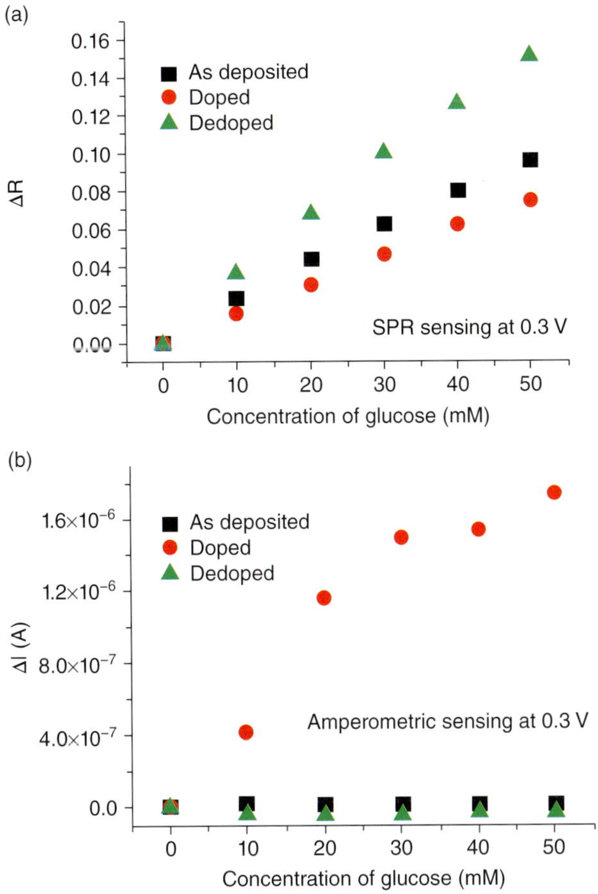

FIGURE 17.15. Plot of amperometric versus SPR sensing with increasing glucose concentration. (Reprinted from Baba et al.[92]; with permission.)

showed the largest change in reflectivity as compared with the as-deposited and doped states. Moreover, as the concentration of glucose increased, the change in reflectivity was highest for the dedoped state, while the amperometric measurement showed the largest current change for the doped state (0.3 μA/10 mM) as displayed in Figure 17.15.

Furthermore, the EC-SPR/waveguide setup was employed for the glucose sensing in order to distinguish the enzymatic activity from changes in the thickness of the film and the interface dielectric constant (buffer solution).[92,96] Specifically, when more negative potentials were applied, the SPR dip was shown to move toward lower angles, while the waveguide mode remained constant (Fig. 17.16a). In addition, the reflectivity-angular scan for 25 bilayers of PPy-PEDOT/PPy-GO_x before and after glucose (10 mM) sensing (at 0.3 V)

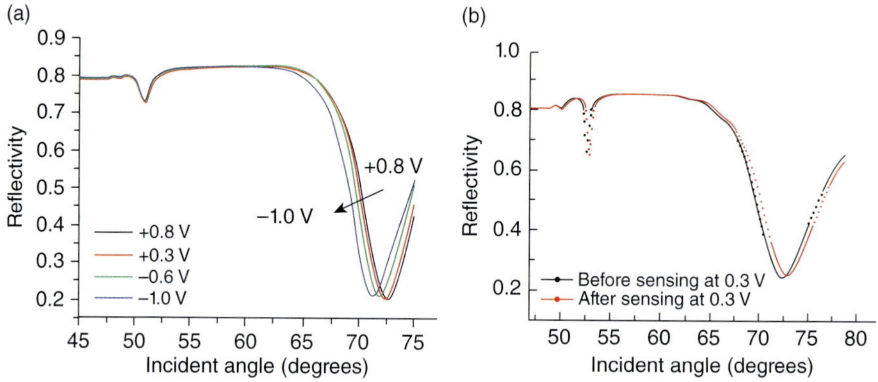

FIGURE 17.16. (a) SPR reflectivity-angular scan of PPyPEDOT-PPyGO$_x$ on an SPR/waveguide combined mode at various potentials. (b) SPR reflectivity-angular curves before and after glucose sensing at 0.3 V and 10 mM concentration. (Reprinted from Baba et al.[92]; with permission.)

showed a slight shift to a higher incident angle for both the SPR and waveguide modes (Fig. 17.16b). These results explicitly show the high activity of the enzyme after the glucose addition and that the shift in the SPR mode is attributed to the change in the electrochemical behavior of the PPy matrix. In this case, the dielectric constant of the PPy-Q/GO$_x$ layer is altered due to the oxidation reaction of this layer. By analyzing the SPR ($m = 0$) and waveguide ($m = 1$) modes, specific changes in the enzymatic reaction during the sensing of glucose can be distinguished or separated from the dielectric constant and thickness parameters.

Electrochemical Nanopatterning of Conducting LbL Films The electrochemical nanopatterning of conducting or semiconducting polymer films is one of the most attractive applications for high-density information storage or memory devices. Advincula and coworkers has reported several studies dealing with the patterning of semiconducting LbL films.[80,97–100] For instance, the fabrication of a write-read-erasable memory device based on the LbL deposition of PEDOT-PSS and poly(diallyldimethylammonium chloride) (PDDA or PDADMAC) was demonstrated using current-sensing atomic force microscopy (CS-AFM).[97] As shown in Figure 17.17a, the nanopatterns are created by applying a bias voltage between the conductive tip and the polymer film deposited on a gold substrate. The multilayer growth of PDDA/PEDOT-PSS layers was monitored by SPR. Figure 17.18a displays a time-resolved SPR measurement for 10 bilayers of PDDA/PEDOT-PSS, which reveals a stepwise deposition of the LbL film. In addition, average thicknesses of 5.0, 19.6, and 35.4 nm were obtained from the reflectivity-angular curve for 10, 40, 70 bilayers of PDDA/PEDOT-PSS films, respectively (Fig. 17.18b). Figure 17.17 illustrates that when 8 (10 bilayers) and 10 (40 bilayers) V bias voltages were

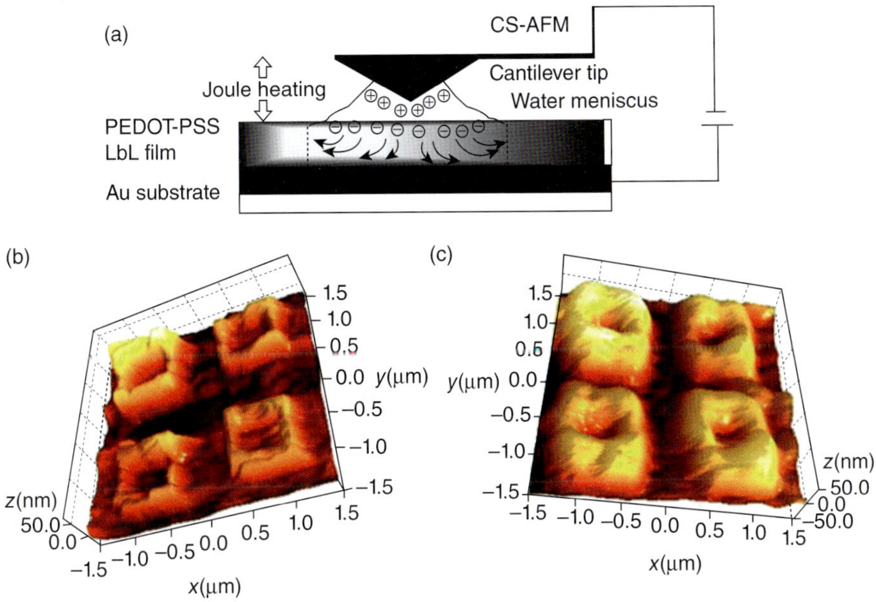

FIGURE 17.17. (a) Schematic diagram illustrating the fabrication of the write-read-erasable memory device. (b) Three-dimensional topographic images of four squares by applying 8 V on 10 bilayers. (c) Three-dimensional topographic images of four squares by applying 10 V on 70 bilayers. (Reprinted from Jiang et al.[97]; with permission.)

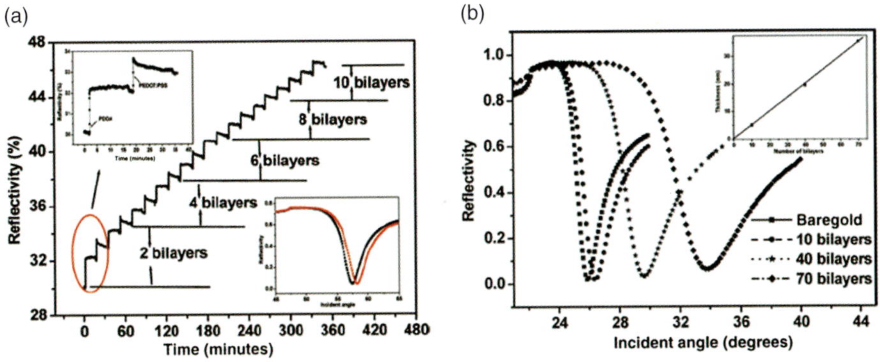

FIGURE 17.18. (a) SPR kinetic curve for 10 bilayers of PDDA/PEDOT-PSS. Upper left inset shows the first bilayer adsorption with a slight decomplexation of PEDOT. Lower right inset shows the reflectivity-angular scan after deposition of 10 bilayers. (b) SPR reflectivity-angular curves of PDDA/PEDOT-PSS LbL films. The inset displays the film thickness as a function of number of bilayers. (Reprinted from Jiang et al.[97]; with permission.)

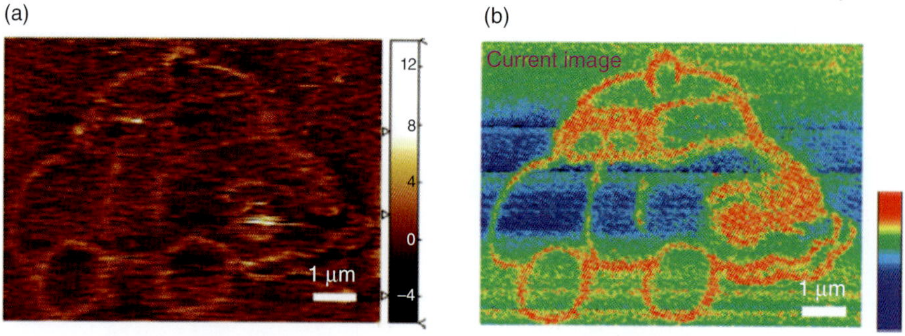

FIGURE 17.19. CS-AFM nanopatterning of a "nanocar" on a 10-bilayer P4VPCBZ/PAA film at 10 V with a writing speed 0.8 μm s^{-1}: (a) topographic image and (b) current image. Color bar range is 0–14.7 pA. (Reprinted from Jiang et al.[97]; with permission.)

applied, the square patterns were obtained. The write-read-erase feature of the LbL film was demonstrated on the 10- and 40-bilayer films by changing the doping state of the film succeeding the writing step. In this case, a bias voltage of 0 V was applied with an erasing speed of 8 μm s^{-1}. Several cycles of the writing and erasing were performed and for the 10 bilayer film, it showed complete erasing of the line patterns as evident by the AFM height and thickness change. However, the erasing property of the 40-bilayer film was reduced due to the thicker film deposited. It was determined that the nanopatterns created and the erasing capability of the semiconducting LbL films depended on the thickness of the film, applied bias, and writing and erasing speeds.

Another example is the electrochemical nanopatterning of semiconducting LbL films composed of the sequential deposition of pendant carbazole-modified polyelectrolytes (P4VPCBZ/PAA, P4VPCBZ/PCEMMA32, and P4VPCBZ [poly(-4-vinylpyridine carbazole)]/PCEMMA41).[6] The *ex situ* SPR data revealed a direct correlation in the shift of the angular curve to higher incident angles with increasing deposition cycles. Again, CS-AFM was used to create a nanocar pattern as displayed in Figure 17.19.

17.5 POLYMER BRUSHES AND SIP TECHNIQUES

In the last decade or so, the emergence of polymer brushes as the premiere material for surface coatings has garnered them the forefront for a variety of advanced applications.[101,102] High-density tethered polymer brushes are particularly attractive for smart surface and coating applications due to the fact that they can undergo reversible conformational changes (i.e., swelling and collapsing) as a result of alterations in the environment. "Smart" polymer brushes that respond to external stimuli including pH, light, solvent, temperature, electric field, and so on, have all been reported.[102–104] Advincula and coworkers have published numerous reports on a variety of intelligent polymer brush surfaces both through radical and nonradical polymerization techniques.[105,106] Moreover, they were the first to introduce the concept of utilizing a combined approach based on the LbL and SIP techniques for the synthesis of highly dense polymer brushes.[33,34]

17.5.1 Grafting of Polymer Brushes

Polymer brushes by definition are nothing more than polymer chains that are grown from surface-bound initiators in such a way that with a sufficiently high grafting density, these tethered chains are forced to stretch or grow in a direction perpendicular to the substrate surface.[101] These strongly stretched polymer chains are arranging in a more favorable conformation in order to minimize their interaction with the adjacent propagating chains, thus attaining the brush morphology. However, in addition to the brush morphology, other polymer architectures such as mushroom and pancake are also possible depending on the choice of solvent as well as the affinities of the polymer chains toward the substrate surface.[101] Figure 17.20 illustrates the different architectures possible with surface-tethered polymer chains.

While the physisorption method has been employed for the attachment of polymer brushes to surfaces,[107] the chemisorption methods, namely the "grafting to" and "grafting from," have been more widely explored for the synthesis of polymer brushes[101] (Fig. 17.21). Although the "grafting to" is a suitable method for the covalent attachment of polymer brushes to reactive groups inherent to a surface, this method is self-limiting in that as soon as the preformed polymer attaches to the surface, steric hindrance starts to accumulate, thereby preventing the diffusion of the incoming polymer molecules to the available anchor groups on the surface.[101] In comparison with the "grafting from" or more commonly known as the SIP technique, a relative small monomer grows directly onto an initiator-functionalized surface.[101] This route is then more preferred over the "grafting to" for the synthesis of highly dense polymer brushes. However, in order to ensure that the polymer chains grow

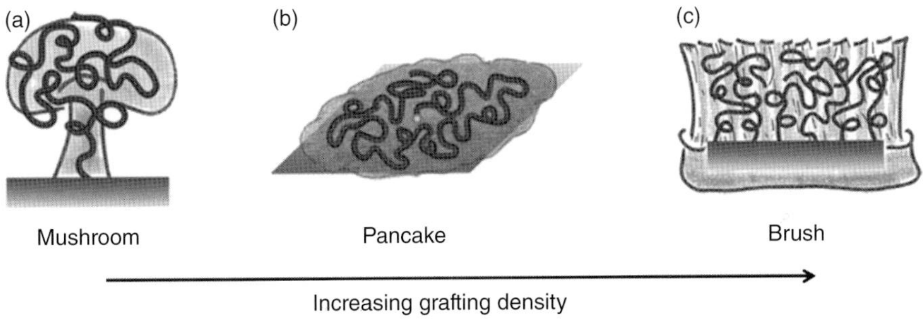

FIGURE 17.20. Depiction of surface-tethered polymer chains: (a) "mushroom," (b) "pancake," and (c) "brush." (Reprinted from Advincula et al.[101]; with permission.)

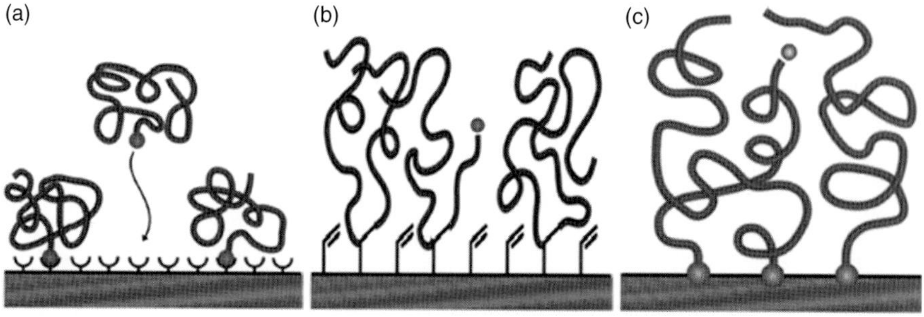

FIGURE 17.21. Grafting strategies of surface-tethered polymer chains: (a) "grafting to," (b) grafting through, and (c) "grafting from" or SIP. (Reprinted from Advincula et al.[101]; with permission.)

in the brush regime, careful manipulation regarding the control of the initiator density must be taken into consideration.

17.5.2 Control of Initiator Density

Polymer brushes are grown on an initiator-functionalized surface taking that there is a sufficiently high grafting or initiator density. Therefore, controlling the initiator density is crucial in establishing a proper brush conformation. Scrupulous control of the initiator density requires one to account for the number of available initiating sites on a surface as well as the length of time the substrate is in contact with the initiator solution. Some of the more common techniques in controlling the initiator density include varying the initiator concentration,[108] photodecomposition,[109] gradient approach,[110] and the Langmuir–Blodgett[111] and Langmuir–Schaeffer techniques.[112]

Recently, the use of water-soluble macroinitiators has offered an advantage over the typical synthesis of silane-based initiators.[113] More specifically, these macroinitiators allow for their deposition onto various substrates using water as the solvent medium, which minimizes the damage to certain substrates caused by organic solvents. However, only a single macroinitiator layer was deposited onto a prefunctionalized substrate.[114]

Advincula and coworkers introduced the concept of employing both the LbL and SIP techniques for control of the underlying layers and the subsequent growth of polymer brushes, respectively.[33,34] Scheme 17.4 displays a typical surface modification for the combined LbL and SIP techniques. This combined approach offers a unique advantage in that the LbL underlayers can be prepared to respond to various external stimuli while at the same time the outerlayer is covered with polymer brushes. Succeeding publications by Armes and coworkers have employed the LbL technique for the sole deposition of polyelectrolyte macroinitiators, which provided control of the initiator density.[115]

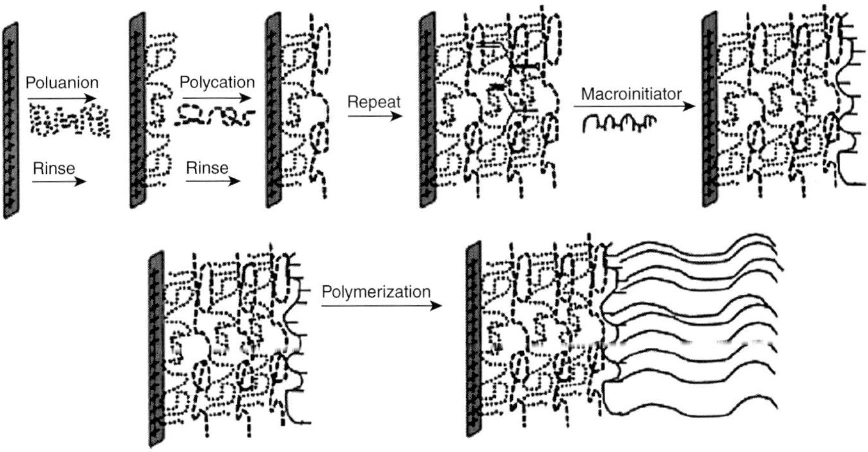

SCHEME 17.4. Surface modification based on the combined LbL and SIP techniques. (Reprinted from Fulghum et al.[34]; with permission.)

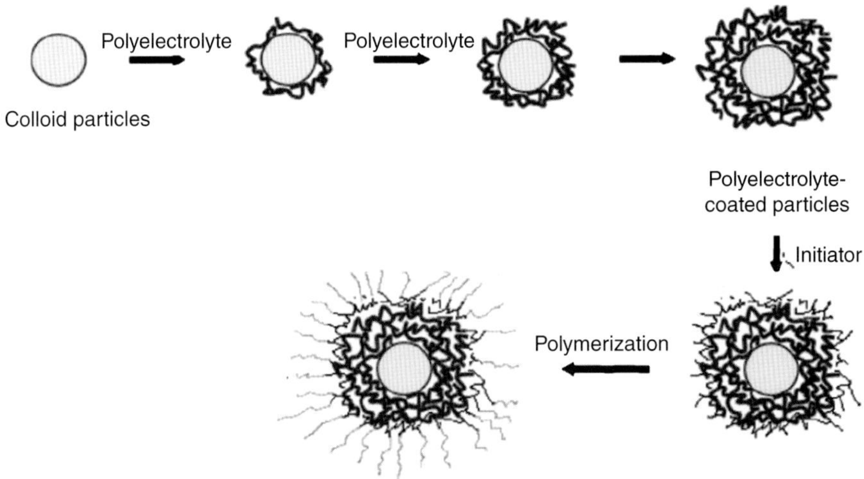

FIGURE 17.22. Surface modification of PS particles using the LbL PAH/PAA layer pair and the single deposition of the ATRP macroinitiator. (Reprinted from Fulghum et al.[33]; with permission.)

17.5.3 LbL Multilayer Deposition and Subsequent SIP of Functional Polymer Brush Surfaces

In 2006, Advincula and coworkers demonstrated the combined LbL-SIP approach, which involved the synthesis of fuzzy particles based on the surface-initiated atom transfer radical polymerization (SI-ATRP) onto the LbL polyelectrolyte modified macroinitiator PS particles (Fig. 17.22).[33]

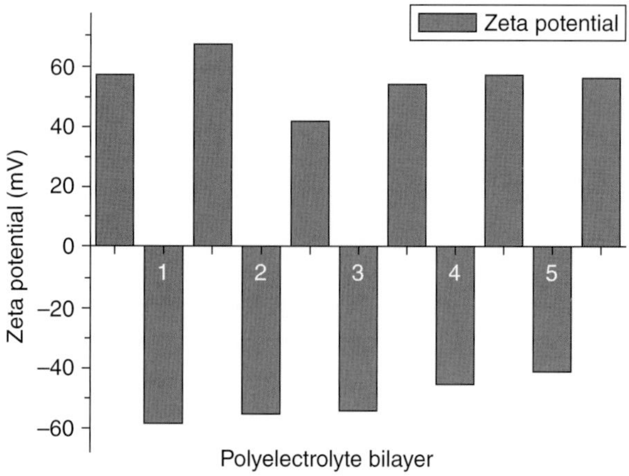

FIGURE 17.23. The LbL buildup of the PS core-shell using ζ potential measurements. The surface charges alternated from positive to negative, indicating the successful deposition of the polyelectrolyte pair. (Reprinted from Fulghum et al.[33]; with permission.)

The LbL multilayer assembly of the PDADMAC/PAA core shell was tracked using the ζ potential. Figure 17.23 shows alternating surface charges after the deposition of PDADMAC and PAA with variations in the potential ranging from ±40 to 45 mV.[33] In order to incorporate the ATRP initiator, a simple esterification reaction of the PAA with bromoisobutyryl ATRP initiator afforded the esterified PAA macroinitiator. The PAA was only 30% functionalized with the bromoisobutyryl initiator so as to allow it to adsorb onto the prior PDADMAC layer. By having the outermost layer to be that of the ATRP macroinitiator, the polymerization of methyl methacrylate (MMA) was conducted successfully. TGA measurements were performed on the PS modified polyelectrolyte/polymethylmethacrylate (pMMA) brush particles. The thermograms clearly showed the decomposition of the PS core at 413°C, while at 192°C revealed a 7% mass in the particles after the deposition of 12 layers. Moreover, the peak maximum at 387°C was attributed to the decomposition of the pMMA brush. TEM was also utilized to observe changes in the particle shape and size before and after the depositions and brush growth. Although there was no change in the shape of the particles after the LbL depositions, the TEM images (Fig. 17.24) confirmed the deformation of the particles after the SIP growth of the pMMA brushes. Furthermore, there was a 50 nm increase in diameter size of the PS particles after the deposition of six bilayers of PDADMAC/PAA. More interestingly, the size of the particle dramatically changed to between 1100 and 1500 nm after the grafting of the pMMA brushes.

In a similar manner, Bruening and coworkers reported the LbL deposition of polyelectrolytes and a polycationic ATRP macroinitiator layer onto Au and porous PES membrane substrates.[116] They prepared two films of different compositions with alternating depositions of (1) (polycationic macroinitiator/

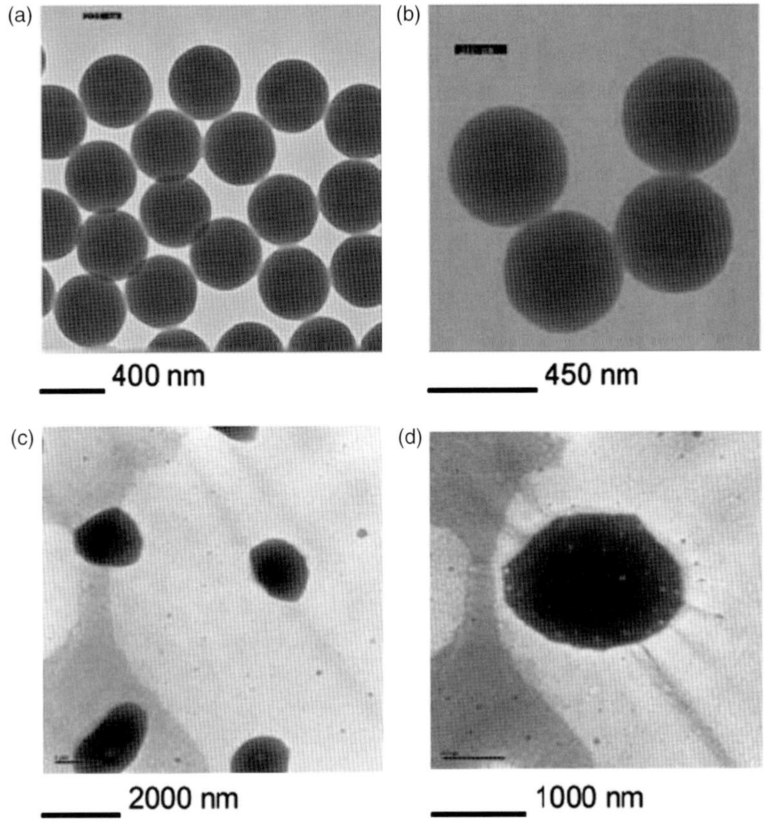

FIGURE 17.24. TEM images of the PS particles: (a) pristine PS particles (400 nm); (b) particles with six bilayers of PAA/PAH deposited (450 nm) with a roughened surface; (c, d) particles after the polymerization of pMMA (diameter increased between 1100 and 1500 nm). (Reprinted from Fulghum et al.[33]; with permission.)

$PSS)_n$ and (2) $(PDADMAC/PSS)_n$ and a single layer of the polycationic macroinitiator. With the alternate deposition of the PSS and polycationic macroinitiator, they concluded that each layer with the macroinitiator throughout the film could act as an initiation site for the polymer brush growth. Film thicknesses of 160 nm were obtained after the ATRP growth of pHEMA brushes in a relatively short time of 2 hours. Kinetic studies were also performed on the single deposition of the polycationic macroinitiator on top of the $(PDADMAC/PSS)_n$ layers.

Armes and coworkers also observed an enhanced initiator density for the sole deposition of both positive and negative ATRP macroinitiators.[115] Scheme 17.5 illustrates the multilayer growth of the ATRP macroinitiators followed by the growth of polymer brushes. When each layer within the film acts as an initiation site, the initiator density increases, and consequently, thick and

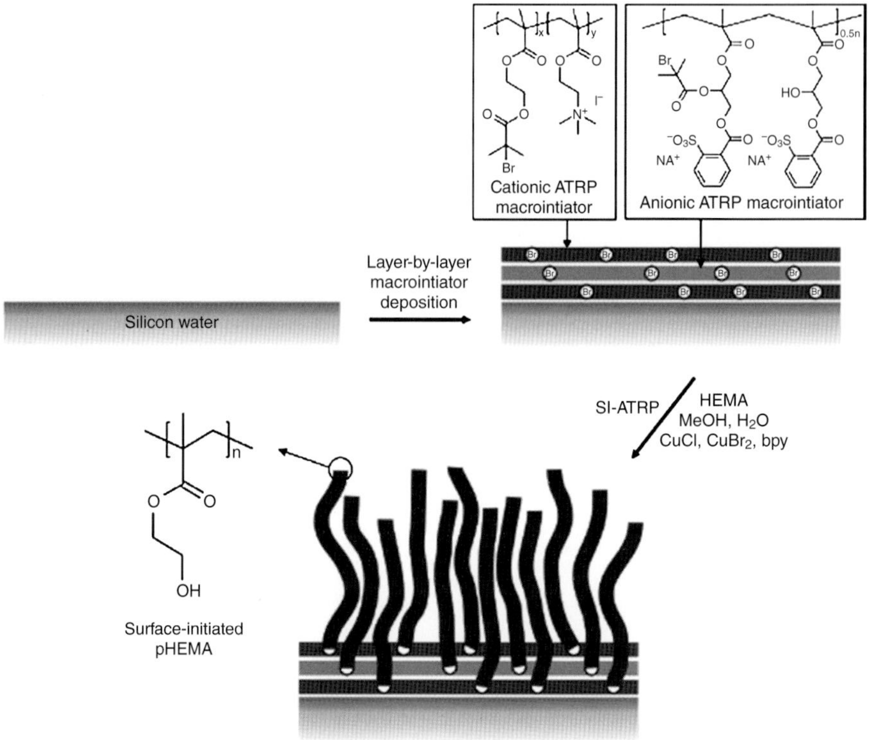

SCHEME 17.5. Electrostatic adsorption of ATRP macroinitiators and the subsequent growth of pHEMA brushes. (Reprinted from Edmonson et al.[115]; with permission.)

dense polymer brushes are produced. Ellipsometry, dual polarization interferometry (DPI), and XPS were all employed for monitoring the buildup of the macroinitiator layers. Figure 17.25a shows a linear relationship between the layers deposited and film thickness with an average thickness of 2.3 Å per macroinitiator layer deposited, indicating that these macroinitiators are strong polyelectrolytes and exhibited lower charge densities as compared with weak polyelectrolytes. Furthermore, DPI measurements estimated an initiator density of $4.9 \pm 0.2\ nm^{-2}$ for 17 layers of macroinitiators deposited. Films (LbL 1, 3, 5, 11, and 21) with varying amounts of the macroinitiators deposited were fabricated. XPS is used to probe the bromine (Br 3d) signal, which is the elemental marker for the ATRP macroinitiators. The Br signal increased proportionally with LbL film, indicating that 21 layers of macroinitiators deposited had the highest initiator density as shown in Figure 17.25b. Unlike the bromine signal, the silicon (Si 2p) signal decreased with increasing LbL film, suggesting an increase in the overall surface coverage of the substrate. The subsequent growth of the pHEMA brushes yielded increasing brush thicknesses with LbL film. For a single layer of macroinitiator deposited, a brush

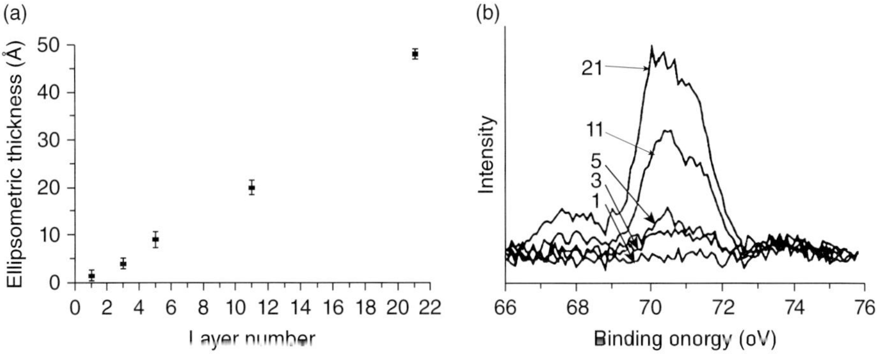

FIGURE 17.25. (a) Ellipsometry film thickness versus number of macroinitiator layers deposited. (b) High-resolution XPS of the bromine (Br 3d) signal for the different LbL films prepared. (Reprinted from Edmonson et al.[115]; with permission.)

thickness of only 9 nm was obtained, while a thickness of over 100 nm was obtained for the 21 layers. By having each layer capable of participating during the polymerization, this produces extremely crowded polymer brushes.

17.5.4 Applications of the Combined LbL and SIP Techniques

The LbL technique has provided a way of manipulating the initiator density, and consequently, varying brush thicknesses are obtainable. However, what is more novel with the combined use of the LbL and SIP techniques is that one can prepare a dual intelligent surface where the polyelectrolytes and polymer brushes behave differently upon inducing a change in the environment.

We demonstrated a dual stimuli-responsive surface based on the LbL deposition of weak polyelectrolytes (PAA/PAH) and the subsequent growth of a temperature-sensitive, PNIPAm, brush.[34] In particular, the induced pH swelling behavior of the PAA/PAH polyelectrolyte layers and the temperature response of the PNIPAm brushes were investigated. Weak polyelectrolytes such as PAA and PAH respond accordingly to variations in the pH where they undergo conformational changes (i.e., swelling and collapsing) as a result of the extent of ionization. In a similar manner, PNIPAm is a well-known temperature-sensitive polymer exhibiting an LCST of 32°C where the polymer undergoes a coil-to-globule transition above and below the LCST.[34] This dual response was demonstrated by performing permeability studies using $Fe(CN)_6^{3-}$ as the probe molecule. When the temperature is at room temperature (RT), below the LCST, the PNIPAm brushes are in a hydrated state where they are at an extended conformation; therefore, the determining factor in the diffusion of the probe molecule will be that of the pH value. Cyclic voltammetry shown in Figure 17.26 confirmed that the anionic probe molecule had the highest mobility at pH 3, while there was a decrease of 20% in the current density at pH 7. Furthermore, at pH 9, the anionic probe molecule displayed

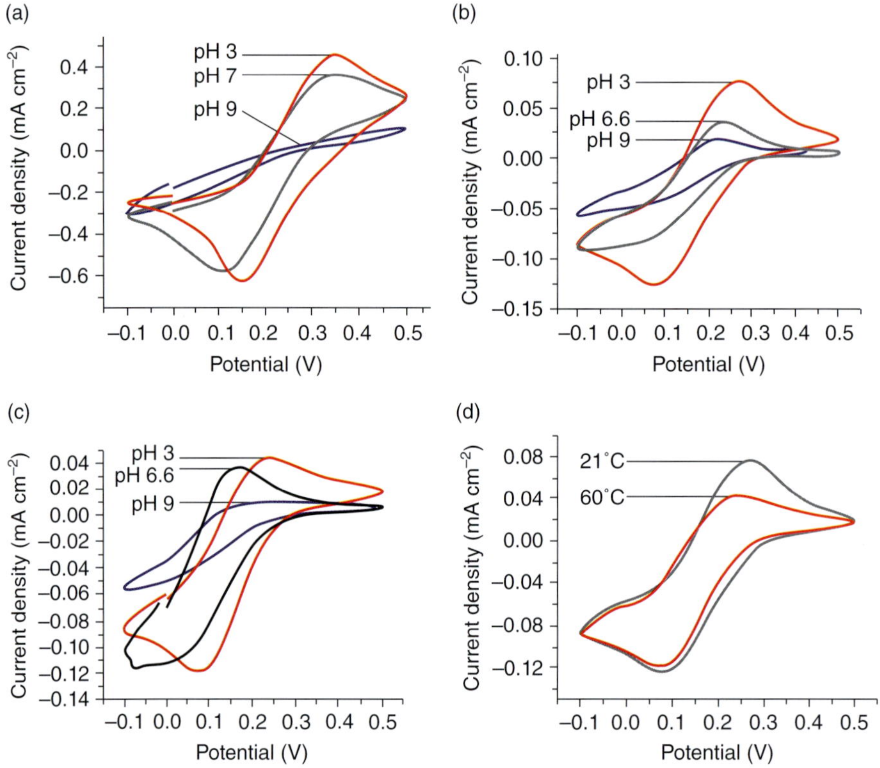

FIGURE 17.26. Cyclic voltammetry plots using $Fe(CN)_6^{3-}$ as the anionic probe molecule: (a) pH-dependent scans for the LbL PAA/PAH film at RT; (b) pH-dependent scans for the LbL PAA/PAH/PNIPAm brush film at RT; (c) pH-dependent scans for the LbL PAA/PAH/PNIPAm brush film at 60°C; (d) comparative scans of the LbL PAA/PAH/PNIPAm brush films at pH 3 at RT and 60°C. (Reprinted from Fulghum et al.[34]; with permission.)

zero mobility. On the other hand, when the temperature was elevated to 60°C, above the LCST, the collapsed conformation of the PNIPAm brushes contributes to the permeability of the anionic probe in addition to the pH value. Again, at pH 3, the anionic probe displayed the highest mobility as compared at pH 7 and 9. A 45% decrease in the maximum current density is obtained at 60°C when compared with the cyclic voltammetry at RT (pH 3).

Advincula and coworkers have shown applicability of the LbL-SIP techniques toward the preparation of free-standing nanosheets.[117] In this case, they have employed the SA-LbL of polysaccharides (chitosan and sodium alginate), which have amino and hydroxyl groups. These functional groups can then react with various bromoester ATRP initiators in the presence of a base. As a consequence, each layer of the polysaccharide bears an ATRP initiator, which can initiate the polymerization of NIPAM. For 10.5 polysaccharide layer

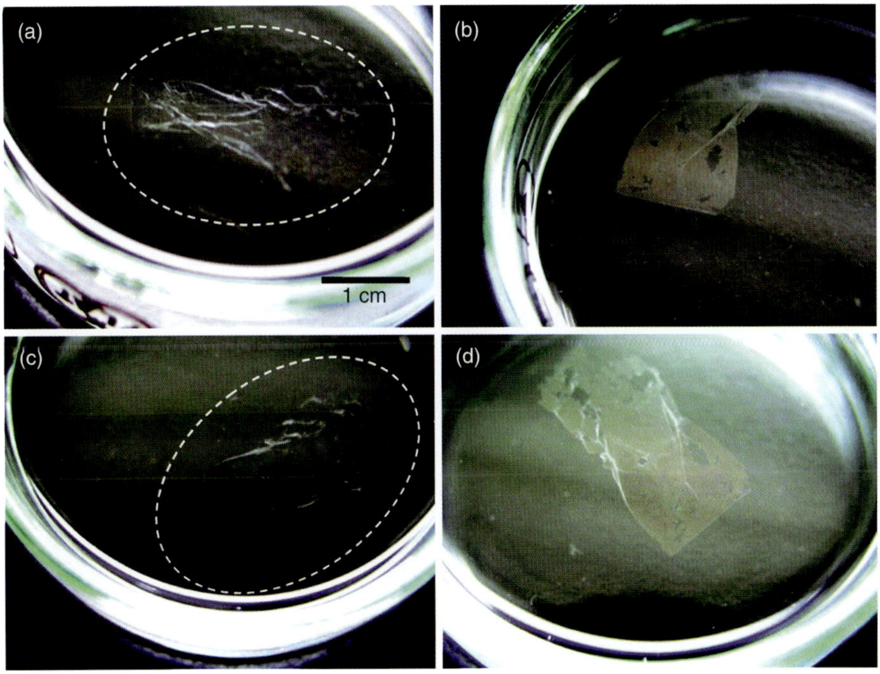

FIGURE 17.27. Photographic images of the free-standing PNIPAm nanosheets as a function of temperature: (a) 25°C, (b) 40°C, (c) 25°C, and (d) 40°C. (Reprinted from Fujie et al.[117]; with permission.)

pairs, an ellipsometric thickness of 41 nm was obtained. The successful incorporation of the ATRP bromo initiator was confirmed by the appearance of the bromine signal in the XPS, which was absent prior to the reaction of the initiator. A concentration of 0.5 M and 1-hour polymerization time yielded a PNIPAm brush thickness of 83 nm (overall film thickness 124 nm). Although a thick film was obtained, the resulting free-standing PNIPAm nanosheet exhibited flexibility by undergoing reversible conformational changes above and below the LCST as shown in Figure 17.27. Furthermore, the structural coloration is attributed to the changing film thickness.

Laschewsky and coworkers have demonstrated their LbL-SIP system toward cell adhesion applications.[118] Although only a single ATRP macroinitiator was deposited on top of the interlayer polyelectrolytes, they have illustrated how a nonlinear growth of the interlayer polyelectrolyte affects the thickness of the deposited macroinitiator. The composition of the interlayer polyelectrolyte (i.e., different combinations of weak and strong polyelectrolytes) can be used to fine-tune the succeeding deposition of an ATRP macroinitiator and the subsequent growth of the polymer brushes. Different thicknesses of the thermoresponsive brushes were shown to affect

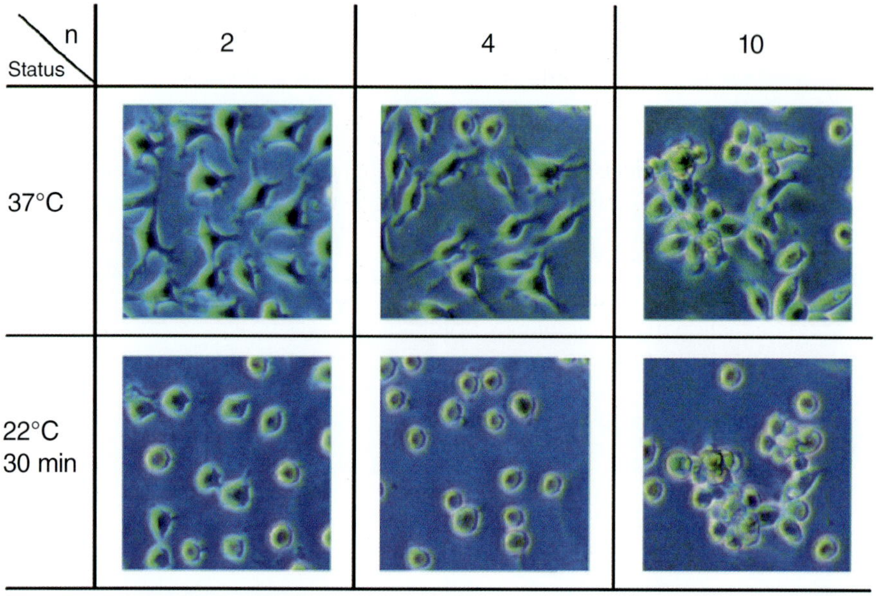

FIGURE 17.28. Cell adhesion experiments on the layer sequence {PEI/PSS/(F-PDADMAC/PSS)$_n$/F-PDADMAC/MA01}. The attachment of the cell is evident at 37°C for $n = 2, 4$, and 10. However, upon decreasing the temperature to 22°C, only the $n = 4$ displayed a well-rounded shape of the fibroblast cell. (Reprinted from Wischerhoff et al. [118]; with permission.)

cell adhesion upon switching the temperature. For the layer sequence {PEI/PSS/(F-PDADMAC/PSS)$_n$/F-PDADMAC/MA01} with $n = 2$, 4, and 10, only $n = 4$ showed the optimum conditions for the cell adhesion studies. Figure 17.28 illustrates images of L929 mouse fibroblasts on the different surfaces. All the films ($n = 2$, 4, and 10) show the successful fibroblasts attachment at 37°C. However, upon decreasing the temperature to 22°C, only the $n = 4$ film display a good rounding of the cell.

The versatility of the combined LbL and SIP techniques has so far been beneficial in the tuning of a wide range of surface coatings. Different polyelectrolyte combination pairs can be used to fine-tune the overall film thickness and internal composition, and at the same time, the single or multiple deposition of polyelectrolyte macroinitiators can be used to graft "smart" or responsive polymer brushes.

17.6 CONCLUSIONS

To summarize, the LbL technique has led to the preparation of various nanostructured materials with well-defined compositions and controllable thicknesses from the angstrom to the nanometer scale. The attractiveness of the LbL technique comes from its versatility and simplicity. The LbL assembly can be carried out utilizing flat and spherical substrates, and there exists a variety of commercially available polymers or polyelectrolytes. Moreover, the SPR technique is proving very valuable in assessing the multilayer composition and has widened its applicability to include hyphenated techniques such as electrochemistry. The EC-SPR technique has permitted for the investigation of both optical and electrochemical properties of conducting LbL films. These conducting LbL films find many applications in glucose biosensors and memory devices. Lastly, the combined LbL-SIP approach is becoming a versatile method in achieving very dense polymer brush surfaces. Consequently, a variety of intelligent polymer brush surfaces are obtainable.

REFERENCES

1. G. Decher. *Science* **1997**, 277, 1232.
2. P. Bertrand, A. Jonas, A. Laschewsky, R. Legras. *Macromol Rapid Commun* **2000**, 21, 319.
3. G. Decher and J. Schlenoff. *Multilayer Thin Films*. Wiley-VCH, Weinheim, Germany, 2003.
4. J. F. Quinn, A. P. R. Johnston, G. K. Such, A. N. Zelikin, F. Caruso. *Chem Soc Rev* **2007**, 36, 707.
5. J. Locklin, J. H. Youk, C. Xia, M.-K. Park, X. Fan, R. C. Advincula. *Langmuir* **2002**, 18, 877.
6. C. Huang, G. Jiang, R. Advincula. *Macromolecules* **2008**, 41, 4661.
7. R. C. Advincula, E. Fells, M.-K. Park. *Chem Mater* **2001**, 13, 2870.
8. R. Advincula, M.-K. Park, A. Baba, F. Kaneko. *Langmuir* **2003**, 19, 654.
9. S. Dante, R. Advincula, C. W. Frank, P. Stroeve. *Langmuir* **1999**, 15, 193.
10. Z. Tang, Y. Wang, P. Podsiadlo, N. A. Kotov. *Adv Mater* **2006**, 18, 3203.
11. T. Boudou, T. Crouzier, K. Ren, G. Blin, C. Picart. *Adv Mater* **2010**, 22, 441.
12. J. A. Lichter, K. J. Van Vliet, M. F. Rubner. *Macromolecules* **2009**, 42, 8573.
13. M. L. Bruening, D. M. Dotzauer, P. Jain, L. Ouyang, G. L. Baker. *Langmuir* **2008**, 24, 7663.
14. C. Jiang, S. Markutsya, Y. Pikus, V. V. Tsukruk. *Nat Mater* **2004**, 3, 721.
15. S. S. Ono and G. Decher. *Nano Lett* **2006**, 6, 592.
16. Y. Ma, J. Sun, J. Shen. *Chem Mater* **2007**, 19, 5058.
17. B. S. Kim, S. W. Park, P. T. Hammond. *ACS Nano* **2008**, 2, 386.
18. J. L. Lutkenhaus, K. D. Hrabak, K. McEnnis, P. T. Hammond. *J Am Chem Soc* **2005**, 127, 17228.
19. J. B. Schlenoff, S. T. Dubas, T. Farhat. *Langmuir* **2000**, 16, 9968.
20. P. A. Chiarelli, M. S. Johal, J. L. Casson, J. B. Roberts, J. M. Robinson, H.-L. Wang. *Adv Mater* **2001**, 13, 1167.
21. J. Cho, K. Char, J.-D. Hong, K.-B. Lee. *Adv Mater* **2001**, 13, 1076.

22. W. B. Stockton and M. F. Rubner. *Macromolecules* **1997**, 30, 2717.
23. G. K. Such, J. F. Quinn, A. Quinn, E. Tjipto, F. Caruso. *J Am Chem Soc* **2006**, 128, 9318.
24. G. K. Such, E. Tjipto, A. Postma, A. P. R. Jonhston, F. Caruso. *Nano Lett* **2007**, 7, 1706.
25. C. J. Ochs, G. K. Such, B. Stadler, F. Caruso. *Biomacromolecules* **2008**, 9, 3389.
26. C.-J. Huang and F.-C. Chang. *Macromolecules* **2009**, 42, 5155.
27. W. Knoll. *Annu Rev Phys Chem* **1998**, 49, 569.
28. X. Kang, Y. Jin, G. Cheng, S. Dong. *Langmuir* **2002**, 18, 1713.
29. S. Tian, A. Baba, J. Liu, Z. Wang, W. Knoll, M.-K. Park, R. Advincula. *Adv Funct Mater* **2003**, 13, 473.
30. O. A. Raitman, E. Katz, I. Willner, V. I. Chegel, G. V. Popova. *Angew Chem Int Ed* **2001**, 40, 3649.
31. A. Baba, F. Kaneko, R. C. Advincula. *Colloids Surf A* **2000**, 173, 39.
32. A. F. Frau, N. C. Estillore, T. M. Fulghum, R. C. Advincula. *ACS Appl Mater Interfaces* **2010**, 2 (12), 3726–3737.
33. T. M. Fulghum, D. L. Patton, R. C. Advincula. *Langmuir* **2006**, 22, 8397.
34. T. M. Fulghum, N. C. Estillore, C. D. Vo, S. P. Armes, R. C. Advincula. *Macromolecules* **2008**, 41, 429.
35. K. Matyjaszewski and J. Xia. *Chem Rev* **2001**, 101, 2921.
36. R. C. Advincula, A. Baba, F. Kaneko. *Polym Mat Sci Eng* **1999**, 81, 95.
37. R. Advincula and W. Knoll. *Colloids Surf A* **1997**, 123, 443.
38. F. Caruso, R. A. Caruso, H. Mohwald. *Science* **1998**, 282, 1111.
39. F. Caruso, H. Lichtenfeld, M. Giersig, H. Mohwald. *J Am Chem Soc* **1998**, 120, 8523.
40. E. Donath, G. B. Sukhorukov, F. Caruso, S. A. Davis, H. Mohwald. *Angew Chem Int Ed* **1998**, 37, 2202.
41. G. B. Sukhorukov, A. A. Antipov, A. Voigt, E. Donath, H. Mohwald. *Macromol Rapid Commun* **2001**, 22, 44.
42. S. Y. Yang, D. Lee, R. E. Cohen, M. F. Rubner. *Langmuir* **2004**, 20, 5978.
43. M.-Y. Park, C. Xia, R. C. Advincula. *Langmuir* **2001**, 17, 7670.
44. M.-Y. Park, K. Onishi, J. Locklin, F. Caruso, R. C. Advincula. *Langmuir* **2003**, 19, 8550.
45. M.-Y. Park, S. Deng, R. C. Advincula. *Langmuir* **2005**, 21, 5272.
46. L. Wang, Z. Wang, X. Zhang, J. Shen. *Macromol Rapid Commun* **1997**, 18, 509.
47. Y. Lvov, G. Decher, H. Mohwald. *Langmuir* **1993**, 9, 481.
48. S. S. Shiratori and M. F. Rubner. *Macromolecules* **2000**, 33, 4213.
49. J. Choi and M. F. Rubner. *Macromolecules* **2005**, 38, 116.
50. S. A. Sukhishvili and S. Granick. *J Am Chem Soc* **2000**, 122, 9550.
51. S. A. Sukhishvili and S. Granick. *Macromolecules* **2002**, 35, 301.
52. S. Y. Yang and M. F. Rubner. *J Am Chem Soc* **2002**, 124, 2100.
53. Y. Shimazaki, M. Mitsuishi, S. Ito, M. Yamamoto. *Langmuir* **1997**, 13, 1385.
54. J. Anzai, Y. Kobayashi, N. Nakamura, M. Nishimura, T. Hoshi. *Langmuir* **1999**, 15, 221.

55. K. Fujita, S. Kimura, Y. Imanishi. *J Am Chem Soc* **1994**, 116, 5479.
56. K. Ijiro, H. Ringsdorf, E. Birch-Hirschfeld, S. Hoffmann, U. Schilken, M. Strube. *Langmuir* **1998**, 14, 2796.
57. G. Decher, B. Lehr, K. Lowack, Y. Lvov, J. Schmitt. *Biosens Bioelectron* **1994**, 9, 677.
58. D. E. Bergbreiter and K.-S. Liao. *Soft Matter* **2009**, 5, 23.
59. P. Kohli and G. J. Blanchard. *Langmuir* **2000**, 16, 4655.
60. P. Kohli and G. J. Blanchard. *Langmuir* **2000**, 16, 8518.
61. T. Serizawa, K. Nanameki, K. Yamamoto, M. Akashi. *Macromolecules* **2002**, 35, 2184.
62. Y. Liu, M. L. Bruening, D. E. Bergbreiter, R. M. Crooks. *Angew Chem Int Ed Engl* **1997**, 36, 2114.
63. Y. Liu, M. Zhao, D. E. Bergbreiter, R. M. Crooks. *J Am Chem Soc* **1997**, 119, 8720.
64. M. Zhao, Y. Liu, R. M. Crooks, D. E. Bergbreiter. *J Am Chem Soc* **1999**, 121, 923.
65. Z. Liang and Q. Wang. *Langmuir* **2004**, 20, 9600.
66. J. Chen, G. Luo, W. Cao. *Macromol Rapid Commun* **2001**, 22, 311.
67. Y. Zhang, S. Yang, Y. Guan, W. Cao, J. Xu. *Macromolecules* **2003**, 36, 4238.
68. J. Seo, P. Schattling, T. Lang, F. Jochum, K. Nilles, P. Theato, K. Char. *Langmuir* **2010**, 26, 1830.
69. R. A. Evans. *Aust J Chem* **2007**, 60, 384.
70. Y. Zhang, H. He, C. Gao, J. Wu. *Langmuir* **2009**, 25, 5814.
71. P. K. B. Palomaki and P. H. Dinolfo. *Langmuir* **2010**, 26, 9677.
72. R. Vestberg, M. Malkoch, M. Kade, P. Wu, V. V. Fokin, K. B. Sharpless, E. Drocken-Muller, C. J. Hawker. *J Polym Sci [A1]* **2007**, 45, 2835.
73. M. E. Buck, J. Zhang, D. M. Lynn. *Adv Mater* **2007**, 19, 3951.
74. L. A. Connal, C. R. Kinnane, A. N. Zelikin, F. Caruso. *Chem Mater* **2009**, 21, 576.
75. G. Jiang, A. Baba, H. Ikarashi, R. Xu, J. Locklin, K. R. Kashif, K. Shinbo, K. Kato, F. Kaneko, R. Advincula. *J Phys Chem C* **2007**, 111, 18687.
76. R. Advincula, E. Aust, W. Meyer, W. Knoll. *Langmuir* **1996**, 12, 3536.
77. W. Knoll. *MRS Bull* **1991**, 16, 29.
78. G. Decher, J. Hong, J. Schmitt. *Thin Solid Films* **1992**, 210, 831.
79. A. Badia, S. Arnold, V. Scheumann, M. Zizlsperger, J. Mack, G. Jung, W. Knoll. *Sens Actuators B* **1999**, 54, 145.
80. C. Xia and R. C. Advincula. *Langmuir* **2002**, 18, 3555.
81. A. Baba, M.-K. Park, R. C. Advincula, W. Knoll. *Langmuir* **2002**, 18, 4648.
82. A. Baba, R. C. Advincula, W. Knoll. *J Phys Chem B* **2002**, 106, 1581.
83. A. Baba, S. Tian, F. Stefani, C. Xia, Z. Wang, R. C. Advincula, D. Johannsmann, W. Knoll. *J Electroanal Chem* **2004**, 562, 95.
84. S. Sriwichai, A. Baba, S. Deng, C. Huang, S. Phanichphant, R. C. Advincula. *Langmuir* **2008**, 24, 9017.
85. P. Taranekar, T. Fulghum, A. Baba, D. Patton, R. Advincula. *Langmuir* **2007**, 23, 908.
86. A. Baba, K. Onishi, W. Knoll, R. C. Advincula. *J Phys Chem B* **2004**, 108, 18949.
87. A. Baba and W. Knoll. *Adv Mater* **2003**, 15, 1015.

88. R. C. Advincula, W. Knoll, C. W. Frank, D. Roitman, R. Moon, J. Sheats. *Mater Res Soc Symp Proc* **1998**, 488, 115.
89. D. Orata and D. A. Buttry. *J Am Chem Soc* **1987**, 109, 3574.
90. N. Sarkar, M. K. Ram, R. Narizzano, S. Paddeu, C. Nicolini. *Nanotechnology* **2000**, 11, 30.
91. P. Waenkaew, P. Taranekar, S. Phanichphant, R. C. Advincula. *Macromol Rapid Commun* **2007**, 28, 1522.
92. A. Baba, P. Taranekar, R. R. Ponnapati, W. Knoll, R. C. Advincula. *ACS Appl Mater Interfaces* **2010**, 2, 2347.
93. K. Peter, R. Nilsson, O. Inganas. *Nat Mater* **2003**, 2, 419.
94. A. Kros, S. Hovell, N. Sommerdijk, R. Nolte. *Adv Mater* **2002**, 13, 1555.
95. M. Umana and J. Waller. *Anal Chem* **1986**, 58, 2979.
96. A. Aulasevich, R. Roskamp, U. Jonas, B. Menges, K. Dostalek, W. Knoll. *Macromol Rapid Commun* **2009**, 30, 872.
97. G. Jiang, A. Baba, R. Advincula. *Langmuir* **2007**, 23, 817.
98. A. Baba, J. Locklin, R. Xu, R. C. Advincula. *J Phys Chem B* **2006**, 110, 42.
99. A. Baba and W. Knoll. *J Phys Chem B* **2003**, 107, 7733.
100. A. Baba, J. Lubben, K. Tamada, W. Knoll. *Langmuir* **2003**, 19, 9058.
101. R. C. Advincula, W. J. Brittain, K. C. Caster, J. Ruhe, eds. *Polymer Brushes: Synthesis, Characterization, Application*. Wiley-VCH, Weinheim, Germany, 2004.
102. N. Ayres. *Polym Chem* **2010**, 1, 769.
103. L. Ionov. *J Mater Chem* **2010**, 20, 3382.
104. S. Edmonson, V. L. Osborne, W. T. S. Huck. *Chem Soc Rev* **2004**, 33, 14.
105. G. Sakellariou, M. Park, R. Advincula, J. W. Mays, N. Hadjichristidis. *J Polym Sci [A1]* **2006**, 44, 769.
106. R. Advincula. *Adv Polym Sci* **2006**, 197, 107.
107. G. F. Belder, G. T. Brinke, G. Hadziioannou. *Langmuir* **1997**, 13, 4102.
108. D. M. Jones, A. A. Brown, W. T. S. Huck. *Langmuir* **2002**, 18, 1265.
109. S. Yamamoto, M. Ejaz, Y. Tsujii, T. Fukuda. *Macromolecules* **2000**, 33, 5608.
110. T. Wu, K. Efimenko, J. Genzer. *J Am Chem Soc* **2002**, 124, 9394.
111. M. Ejaz, S. Yamamoto, K. Ohno, Y. Tsujii, T. Fukuda. *Macromolecules* **1998**, 31, 5934.
112. N. C. Estillore, J. Y. Park, R. C. Advincula. *Macromolecules* **2010**, 43, 6588.
113. S. Edmonson and S. P. Armes. *Polym Int* **2009**, 58, 307.
114. S. Edmonson, C. D. Vo, S. P. Armes, G. F. Unali. *Macromolecules* **2007**, 40, 5271.
115. S. Edmonson, C. D. Vo, S. P. Armes, G. F. Unali, M. P. Weir. *Langmuir* **2008**, 24, 7208.
116. D. Jain, J. Dai, S. Grajales, S. Saha, G. L. Baker, M. L. Bruening. *Langmuir* **2007**, 23, 11360.
117. T. Fujie, J. Y. Park, A. Murata, N. C. Estillore, M. C. R. Tria, S. Takeoka, R. C. Advincula. *ACS Appl Mater Interfaces* **2009**, 1, 1404.
118. E. Wischerhoff, S. Glatzel, K. Uhlig, A. Lankenau, J. F. Lutz, A. Laschewsky. *Langmuir* **2009**, 25, 5949.

CHAPTER 18

POLYELECTROLYTES AT INTERFACES: APPLICATIONS AND TRANSPORT PROPERTIES OF POLYELECTROLYTE MULTILAYERS IN MEMBRANES

BASTIEN SEANTIER and ANDRÉ DERATANI

18.1 INTRODUCTION

The development of nanotechnology in all fields of science comes with a tremendous interest in techniques enabling the formation of nanostructured architectures. Molecular self-assembly is one of the most popular routes for nanoengineering surface and interface properties of materials offering potentially novel developments for advanced applications. The layer-by-layer (LbL) assembly of polyelectrolytes (PEs) belonging to this approach allows thin films to be designed with a fine nanometer length scale control over thickness and spatial charge distribution. This chapter follows a comprehensive overview on layered PE assemblies published in 2005, including more than 900 references.[1] The LbL technique has been extensively investigated due to its simplicity and efficiency in building up nanostructured thin films with controlled properties and in providing nano- and microscale objects.[2] Therefore, the number of papers published on this topic is steadily increasing with a broad range of applications and has become so large as to make impossible an exhaustive review in a limited space. This chapter and the following one focus on specific applications using polyelectrolyte multilayer (PEM) properties in transport of molecular and ionic species and in construction of photonic devices, respectively.

Modification of interface properties by PE adsorption on oppositely charged surfaces is a very active domain for over more than 40 years[3,4] because of the

Ionic Interactions in Natural and Synthetic Macromolecules, First Edition. Edited by Alberto Ciferri and Angelo Perico.

large number of applications such as stabilizing/destabilizing colloids,[5] proteins and biomolecules adsorption in biomaterials,[6] and separation processes.[7,8] The electrostatic interaction between the opposite charges of PE and the surface is the driven force of PE adsorption. For instance, the adsorption of the weak polycation poly(vinylimidazole) onto negatively charged interfaces has been shown to be strongly dependent on the ionic strength and pH9 that are the external parameters governing the surface charge and the PE ionization degree.[10] The entropic gain related to the release of small counterions and water molecules linked to the surface also plays an important role on the stability of the PE layer at the interface.[11,12] The balance of these forces is of primary importance on the amount and conformation of the adsorbed polymer.[13,14] Subsequent formation of secondary interactions like H-bonding and the hydrophobic effect may take place between the adsorbed PE and interface depending on their respective chemical natures.[15] An interesting feature of PE adsorption is that the amount of adsorbed PE on an oppositely charged surface can exceed simple neutralization and lead to a charge reversal of the interface. This overcompensation phenomenon is the key point underlining the successful construction of PEM films. So by choosing appropriate operating conditions, each adsorption step yields the charge inversion of the surface, and alternated deposition of oppositely charged species (PE or particles), the so-called LbL technique, can produce stable thin films of multilayered assemblies held together by strong electrostatic as well as nonelectrostatic interactions. For more information, see Chapter 5, which is devoted to the theoretical aspects of PE adsorption and the modeling of PEM building up using numerical simulations.

It is generally admitted that the development of LbL PE self-assembly results from the pioneering work of Decher and coworkers.[16,17] It should be mentioned that some earlier papers have explored the technique of electrostatic self-assembly. For instance, the multilayer fabrication using oppositely charged colloidal particles has been described by Iler in 1966.[18] On the other hand, the LbL technique is not limited to electrostatic interactions and any function that has affinity for another one can be used. A large variety of driving forces has been reported to give rise to nanostructured thin films, including the H-bonding between nonionic polymers bearing phenol and pyridine groups[19]; base pairing using the specific recognition of the four DNA bases[20]; charge transfer between donor and acceptor group-containing polymers[21]; host–guest accommodation[22,23]; and also by formation of covalent binding, as for instance using click chemistry[24]; and metal coordination complexation.[25] More information can be found in Chapter 17. Although these recent developments extend the spectrum of potential applications, the vast majority of reports found in literature is based on PEMs.

In fact, the LbL PE self-assembly has attracted considerable attention because it possesses remarkable advantages. First of all, the process of fabrication based on a stepwise exposure to species containing complementary charged groups (Fig. 18.1) is quite simple and inexpensive to carry out. It is

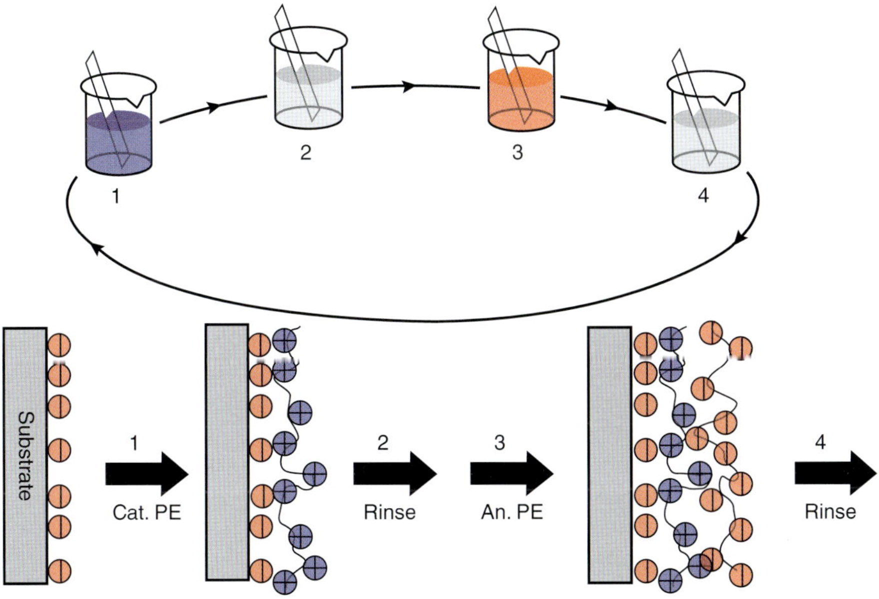

FIGURE 18.1. Schematic representation of LbL assembly using the dipping method. (1) The negatively charged substrate is dipped into a solution of a cationic (Cat.) PE, resulting in the formation of the first PE layer; (2) rinsing step into pure water; (3) a second monolayer is deposited by dipping in the oppositely charged (anionic, An.) PE; (4) rinsing step. The same procedure is repeated until obtaining the desired number of bilayers. (Redrawn from Decher[17]).

also worth mentioning that the LbL assembly is robust and environmentally friendly since PEs are soluble in water. In addition, the technique is versatile and flexible as there is virtually no limitation in the choice of substrates and chemical structures of PE that can be used to build up PEM assemblies provided that each adsorption step results in a stable adsorbed layer.[26] Various charged substances, including synthetic PEs[27] and biopolymers as proteins,[28,29] polypeptides,[30,31] and polysaccharides,[32,33] have been successfully assembled into thin films by the LbL technique. On the other hand, inorganic particles[34] and biological materials[35,36] can serve as supports or be inserted in PEMs. Some examples of component pairs are presented in Table 18.1. Because of their availability, poly(allylamine hydrochloride) (PAH), poly(diallyldimethylammonium chloride) (PDADMA), poly(L-lysine) (PLL), and chitosan (CHI) as positively charged PEs, and poly(sodium styrene sulfonate) (PSS), poly(acrylic acid) (PAA), poly(*L*-glutamic acid) (PGA), and hyaluronic acid (HA) as the negatively charged PE are among the most commonly studied PEs for creating multilayered films.

The most common approach is based on the alternated deposition of a single pair of oppositely charged PEs. However, it has been demonstrated that

TABLE 18.1. Chemical Structures of Commonly Studied PEs for the LbL Assembly

Cationic PE	Chemical Structure	Anionic PE	Chemical Structure
Poly(allylamine hydrochloride) (PAH)	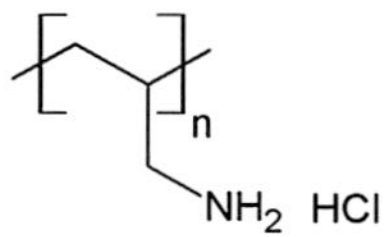	Poly(sodium styrene sulfonate) (PSS)	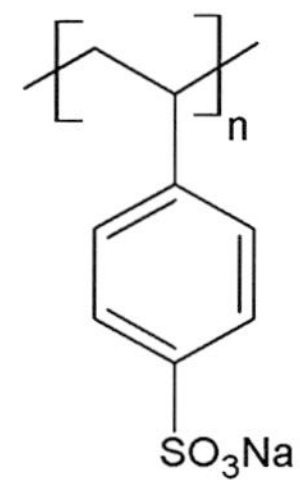
Poly(diallyl dimethyl ammonium chloride) (PDADMA)	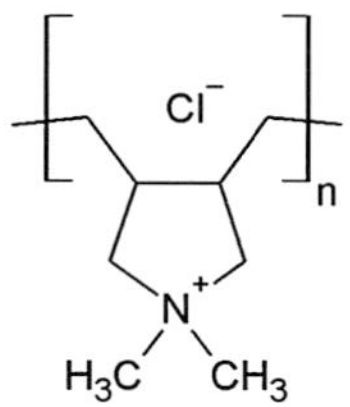	Poly(acrylic acid) (PAA)	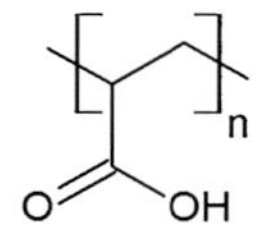

Poly(L-lysine) (PLL)	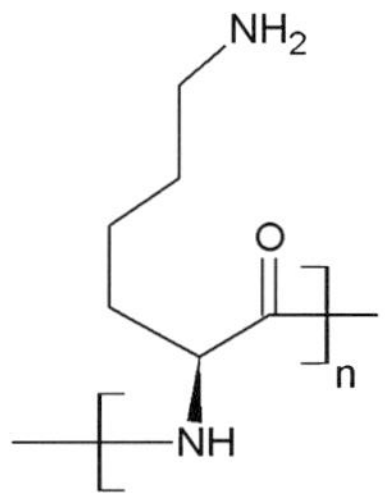	Poly(*L*-glutamic acid) (PGA)	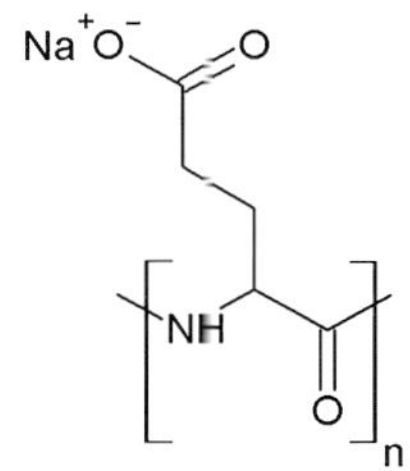
Chitosan (CHI)	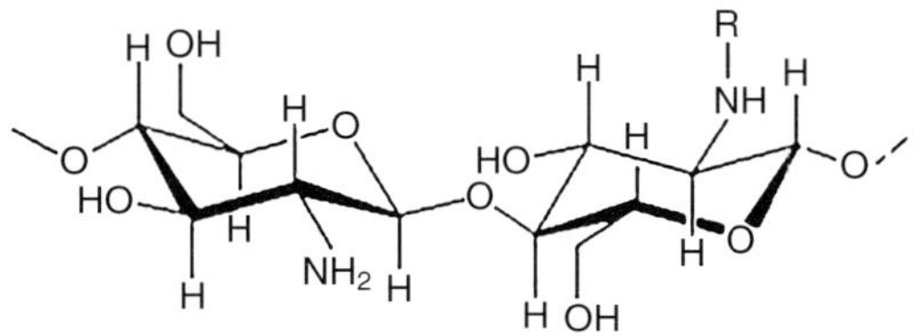 R = H or $COCH_3$	Hyaluronic acid (HA)	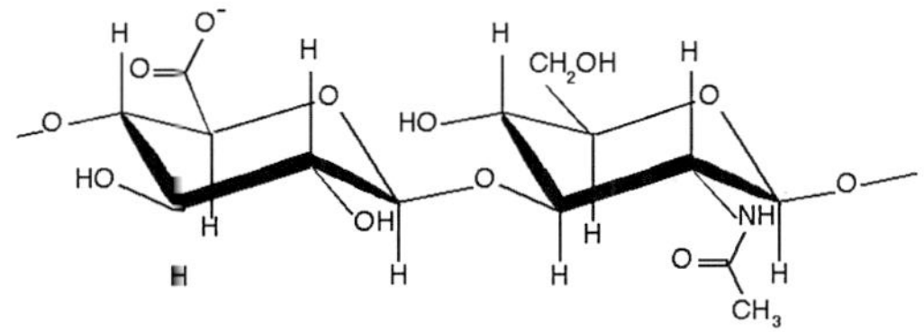

multilayer assembly from a blend PE solution is a good alternative with such advantages as the dilution of expensive and highly functional materials and the tailoring of PEM films for specific applications. This approach has been recently reviewed.[37] Two routes have been identified from the literature data. In the first one, a polyelectrolyte complex (PEC) composed of two oppositely charged PEs with an overall net positive (negative) charge is alternatively assembled with a negatively (positively) charged PEC (Fig. 18.2a). Low-molecular-weight molecules and enzymes can then be incorporated in PEM by precomplexing with higher molecular weight PEs. The second route consists of the alternated deposition of a blend solution containing similarly charged PEs with an oppositely charged PE (Fig. 18.2b). This approach is assumed to

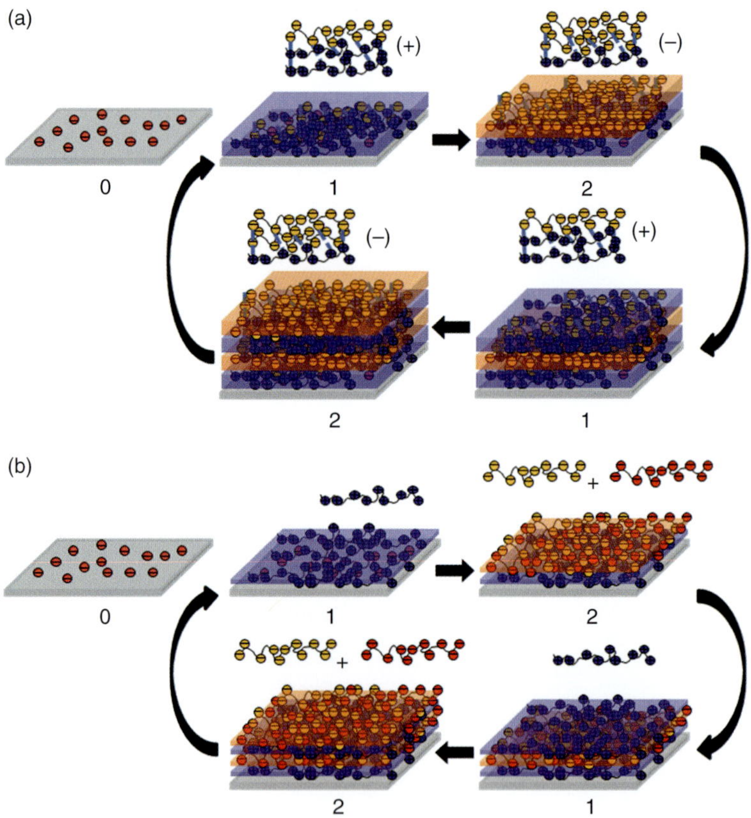

FIGURE 18.2. Schematic representation of LbL assembly from blend PE solutions on a negatively charged substrate. (a) A polyelectrolyte complex (PEC) composed of two oppositely charged PEs with an overall net positive charge is alternatively assembled with a negatively charged PEC; (b) a single positive PE is alternatively assembled with a blend solution containing two negatively charged PEs.

be more versatile as many combinations can be envisioned as double blends[38] or blends of three or four components, hence it is more commonly studied. The use of blended solutions in LbL assembly compared to the conventional route can afford a precise engineering of the PEM film properties and be used, for instance, to tune film thickness and composition, as well as interfacial properties and protein adsorption characteristics.[39]

Several reviews on the potential applications of PEMs have been published in recent years. Biopolymers and biocompatible materials can be easily self-assembled using the LbL technique, making PEM structures especially attractive for biomedical and biological applications.[40] Interfacial and mechanical properties that are key parameters for controlling protein and cell adhesion can be matched with respect to the targeted application.[41] Coating of biocompatible and implantable materials with PEMs has proved to be useful for providing anticoagulation, anti-inflammatory, antibacterial, and localized gene therapy.[35,42,43] LbL films can serve as a biocompatible platform for cell attachment, and growth and tissue engineering is now emerging as a promising application.[35,43] A large number of works have been devoted to the use of PEM films containing enzymes and proteins in biosensor and immunosensor applications.[35] Taking advantage of the mechanical properties of PEMs[44] and their permeability to molecules, design of nano-sized delivery systems is another field of active research.[27,35,45–47] The core–shell structures enables encapsulation of active molecules and their controlled release upon various external stimuli.[48–50] On the other hand, nanoengineering of the LbL shells is assumed to allow the controlled delivery by remote guiding and activation using, for instance, infrared or ultrasound radiation.[46] PEM films comprising functional PEs with optical, electro-optical, and electronic responsive groups have found useful applications in the design of optical and optoelectronic devices (see Chapter 19).[51] A perspective on the future of PEM films has been recently published that summarizes some of the most promising ideas and applications.[52]

The control of molecules and ions transport through thin PEM films is one of the main applications, including nanocapsules for drug delivery, sensors, nanoreactors, and membranes. Here, we focus our attention on the transport properties of PEM films in membrane filtration and separation with a particular emphasis on the most recent advances. In that case, the LbL assembly acts as a permselective barrier to the transport of molecules and ionic species. Because the properties of PEM films are deeply related to their morphology, the LbL buildup of PEMs and structure characterization are briefly summarized in a first part. LbL assembly is also useful for modifying membrane interface as a passive coating owing to its biocompatibility, and as an active coating to combine transport and reaction. The second part includes a survey of applications in membrane science and reviews the selectivity achieved using PEM membranes. Finally, the transport mechanism is discussed in relation with the morphology and composition of PEMs.

18.2 PE MULTILAYER BUILD-UP AND MORPHOLOGY

The LbL assembly technique consists of alternating deposition of oppositely charged PEs on a substrate. Because surfaces usually have a net negative charge due to the presence of ionized oxygenated groups, the first layer generally comprises a positively charged PE. In many studies, a primer layer ensures the anchorage of the LbL film to the underlying substrate. Poly(ethylene imine) (branched Poly(ethylene imine), b-PEI), a cationic PE, is commonly used as a primer with negatively charged substrates owing to a high-charge density over a large pH range due to its branched chemical structure containing primary, secondary, and tertiary amine functions[53] and the possibility to develop secondary hydrogen bonding interactions with the substrate. Subsequent adsorption of a negatively charged PE leads to the formation of the first bilayer. The procedure can then be repeated as many times as needed with another more appropriate cationic PE to obtain the desired number of layers.

In the case of neutral and hydrophobic substrates, PE modified with suitable groups can afford a good anchoring of the LbL assembly. For instance, hydrophobically modified PEs[54] and mussel adhesive-inspired PEs containing catechol and amine functional groups[55] have been proven to be efficient primers to build-up PEM films, the latter approach being probably the more universal. Surface modification such as plasma treatment techniques has been also developed to functionalize and anchor PEM films to poly(tetrafluoroethylene) substrate[56] and poly(acrylonitrile) PAN/poly(ethylene terephtalate) (PET) membranes.[57]

This section describes the different techniques for creating PEMs. The structure that can be obtained is also presented since the transport properties of LbL films depend to a large extend on their internal organization. During the process, PEs self-assemble and self-organize at the support surface to yield the PEM film. The final morphology of the obtained PEMs depends on a large number of parameters,[15] including the number of bilayers, the operating parameters (pH, ionic strength, and build-up technique), the chemical nature, and charge density of PEs. Some of them will be summarized in the following paragraphs.

18.2.1 Techniques of PE Multilayer Build-Up

Dip Coating The technique initially developed by Decher in the early 1990s was based on alternatively dipping a substrate in oppositely charged PE solutions (Fig. 18.1).[16] The dip coating method for depositing PE and building up PEM films is still largely employed and certainly the more popular owing to its simplicity. A large choice of substrates can be used, such as glass, silicon wafers, mineral particles, or any kind of modified surfaces that exhibit a net charge surface density. Due to electrostatic forces, the charged macromolecules diffuse to the surface and adsorb onto it. The net charge of interface is then reversed because of the overcompensation phenomenon when the amount of adsorbed PE reaches the plateau of the adsorption isotherm.[58] The saturation step usually takes about 15 to more than 60 minutes to be completed depending on the PE molecular weight and molar mass distribution. At that point it is necessary to rinse the surface in pure solvent to remove the excess of loosely adsorbed polymers and to avoid flocculation coming from the formation of PEC by contacting with the solution of the oppositely charged macromolecules. Repeating the adsorption step with the oppositely charged PE solution results in the buildup of a PE bilayer and so on.

Because the dip coating method is mainly a diffusion-controlled process, the procedure is time consuming due to the formation of a depletion zone near the interface. Typically about 1 hour is needed to deposit a PE bilayer with good properties in spite of the development of automatic dipping systems. On the other hand, mastering the deposit of PEMs onto large surfaces is not an easy task. These reasons explain why the dipping method can hardly fulfill the requirements for industrial applications. In order to overcome those drawbacks, much effort has been done to develop alternative methods for shortening the contact time and maximizing the treated surface area.

Spin Coating It has been demonstrated from the late 1990s that spinning the samples after each dipping step is efficient for drying, spreading, and flattening the deposited layers.[59] Later, it has been proposed to directly spin PE solutions onto a surface.[60,61] A few droplets of PE solution are deposited on the substrate and rotated at a speed between 3000 and 4000 rpm for 2–20 seconds (Fig. 18.3). The combination of centrifugal and air shear force thus permit one to obtain thin and homogeneous PEM films in a very short process time.

Most of the studies have reported that more ordered PEMs were obtained by the spinning procedure when compared to the conventional dipping technique. It is assumed that the spinning process induces drainage of the excess solution and the fast evaporation of water, increasing the PE concentration at interface. Two parameters, namely the spinning speed and the initial concentration of PE solutions, significantly impact the film buildup; for example, a low rotation rate or a high concentration of PE induces an increased amount of adsorbed materials. The PEM film thickness per bilayer can then be easily controlled in a certain range by adjusting these two parameters.[62] The process of build-up LbL PAH/PSS films has been modeled using a scaling law as a function of the spinning rate (2000–6000 rpm) and the PE concentration (1–10 mmol L^{-1}) by fitting the experimental data:

$$A(\propto H) \sim \Omega^{-\alpha} \cdot C_i^{\beta} \quad \text{with} \quad \alpha \approx 0.34 \quad \text{and} \quad \beta \approx 0.74, \qquad (18.1)$$

where A is the absorbance at 225 nm per bilayer, H is the bilayer thickness in nanometers, Ω the spinning speed in revolutions per minute, C_i the initial PE concentration in millimolar, and α and β the fitting parameters.

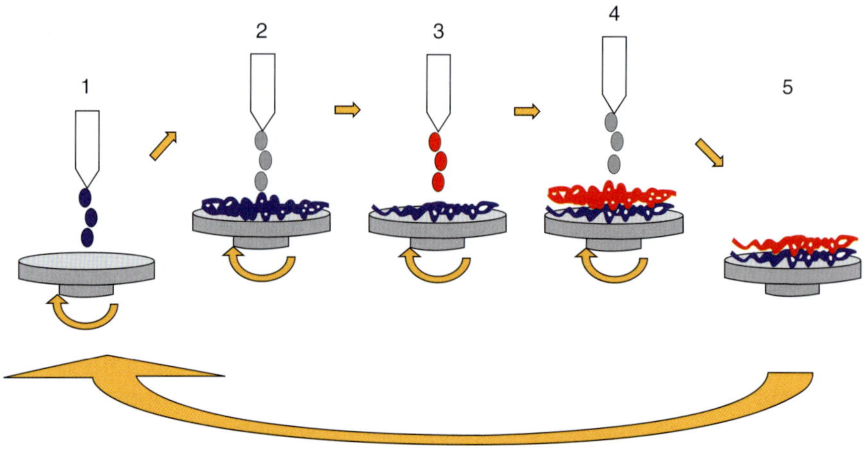

FIGURE 18.3. Schematic representation of LbL assembly using the spinning method. (1) The negatively charged substrate is spun during the deposition of a certain amount of a cationic PE solution; (2) washing with the rinsing solution; (3) a second monolayer is deposited by spinning the oppositely charged PE solution; (4) rinsing step; (5) the first bilayer is formed. The same procedure is repeated to obtain the predetermined number of bilayers.

More recently, it has been observed by neutron reflectivity that highly stratified internal structures are formed for LbL assemblies (PAH/PSS) prepared by spinning while more intermixed internal structures were found for films prepared by the dipping technique.[63]

Dewetting A new method of PEM formation has been more recently developed based on the dewetting of PE solutions from the substrate.[64] It consists of dipping a surface into a DMF/water solution of PE for 10 seconds (Fig. 18.4). A dewetting step is subsequently carried out by self-cleansing the surface for 30 seconds, leaving the first adsorbed layer onto the surface. In fact, adding DMF to the PE solution drastically reduces the stability of the solution wetting state as showed by contact angle measurements, enabling quick dewetting of the PE solution from the surface. Without any rinsing step, it is then possible to directly proceed to the deposition of the second layer by dipping the substrate in the oppositely charged PE solution. The dewetting procedure avoids the rinsing step, thus shortening the deposition time of one PE layer. The applicability of the dewetting concept to the LbL buildup of PE systems has been proved for the PDADMA/PSS system. It has to be noted that this approach can be used to prepare composite PEM films with the SnO_2/PDADMA/PAA system that do not form LbL assembly with the conventional dipping method.

The spin coating and dewetting techniques appear to be interesting alternatives to the dipping method as they allow much faster buildup of LbL assembly. Nevertheless, these two techniques are hardly applicable to the treatment of

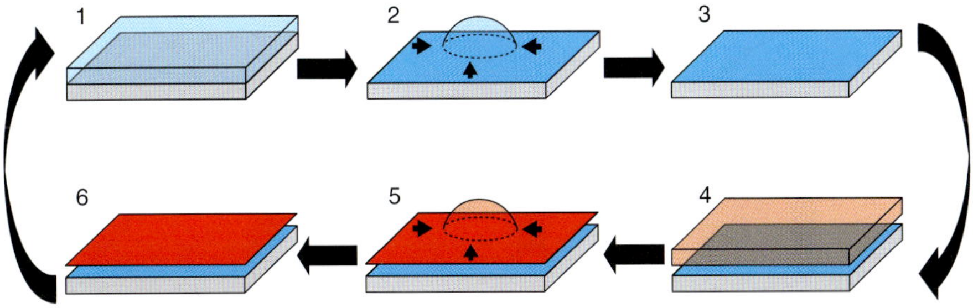

FIGURE 18.4. Schematic representation of LbL assembly using the dewetting method. (1) Wetting the surface by dipping into a PE solution; (2) dewetting step; (3) the first monolayer is deposited; (4) wetting the surface by dipping into the oppositely charged PE solution; (5) dewetting step (6) yielding the second monolayer. The same procedure is repeated to obtain the predetermined number of bilayers.

substrates with large surface area, as needed in membrane technology in contrast to the two following techniques.

Spray Coating Spraying first reported by Schlenoff et al.[65] and further developed by Schaaf et al.[66] is another technique which has proven to be very effective for LbL buildup. The method is based on the spray depositing of PE solutions onto a substrate positioned vertically and perpendicularly to the spray axis (Fig. 18.5). This setup allows the drainage by gravity of the solution in excess from the surface and only a thin film of liquid remains on the surface. It is assumed that the PE concentration is always identical even in the close vicinity of the receiving surface as fresh PE solution is continuously arriving at the surface, replacing the liquid draining off. The typical deposition time for a PE bilayer can then be shortened by about two orders of magnitude compared to the dipping method. The sequence starts with a spraying time of a few seconds followed by a waiting period (20–30 seconds), enabling PE interaction and drainage of solution in excess. A rinsing step may be done but it has been shown that it is not compulsory.[66] Moreover, the method was extended showing that simultaneous and continuous spraying of the oppositely charged PE solutions can afford PEM growth, reducing even more the film building time.[67]

The different parameters entering into the spraying process, namely the spraying rate and the spraying time,[68] pH and ionic strength,[69] and the concentration of PE solutions[70] were extensively investigated in relation to the obtained PEM structure. In particular, it was shown that the spraying technique minimizes the interlayer penetration, leading to the presence of more counterions trapped within LbL coating.[71] The comparison between the morphology and internal structure of PEM films obtained by using the different techniques of LbL buildup is reviewed in the next section.

One of the major advantages compared to the previously described techniques is that the spraying process can be easily scaled up for treating large

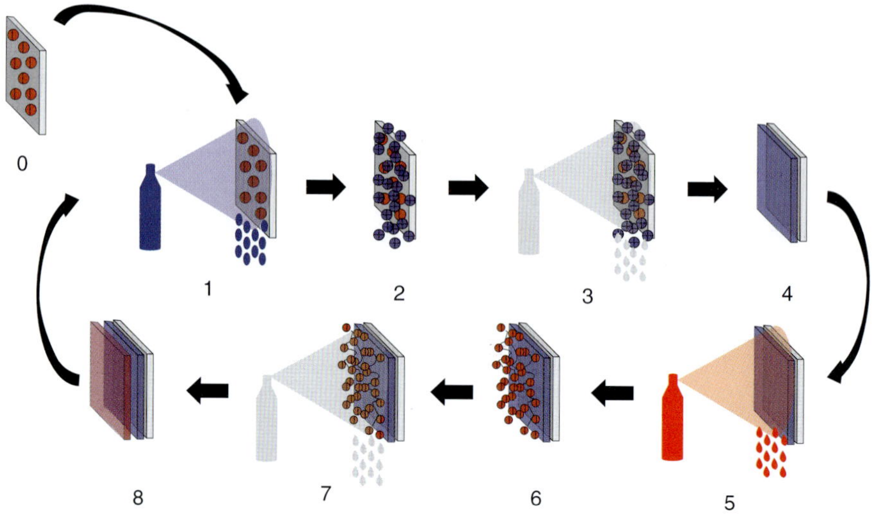

FIGURE 18.5. Schematic representation of LbL assembly using the spraying method. A negatively charged surface (0) is sprayed (stage 1) with a positively charged PE; the obtained surface (2) is subsequently washed with the rinsing solution (stage 3) to yield the first PE adsorbed layer (4); repeating the procedure by spraying a solution of oppositely charged solution (stage 5) and rinsing the obtained surface (6) (stage 7) results in the formation of a firmly adsorbed PE bilayer (8) ready for a new deposition cycle.

surface area as, for example, rolls of flat membrane. Therefore, automated spraying systems were developed and their ability to form uniform LbL films has been demonstrated on a large range of materials using strong and weak PEs, hydrogen-bonded interactions, dendritic compounds, and charged nanoparticles.[72] The technique was applied to the LbL coating of textile fibers.[73] Depending on the operating conditions, a conformal coverage of individual fibers (electrospun nylon 6,6) by LbL spray coating of PDADMA/TiO_2 or a coating bridging the fibers (poly(amidoamine) (PAMAM)/PAA, PEI/PAA, PDADMA/PSS) can be obtained (Fig. 18.6).

The technique developed in this study was applied to the engineering of a selective reactive membrane with tunable mass-transfer properties. The conformal film of PDADMA/TiO_2 (Fig. 18.6b) serves as a photocalytic film of high active surface area (ca. 20–50 $m^2\ g^{-1}$). The fiber bridging by PE spray deposition (Fig. 18.6c) allows one to dramatically increase the catalytic activity of the membrane due to the contribution of the solution–diffusion transport mechanism.

Dynamic Coating The deposition of PE layer *in situ* by alternatively circulating the oppositely charged PE solution is an LbL coating technique well suited for functionalizing the inner surface area of devices. The dynamic LbL process was used to coat with PEM various substrates, as for example fused

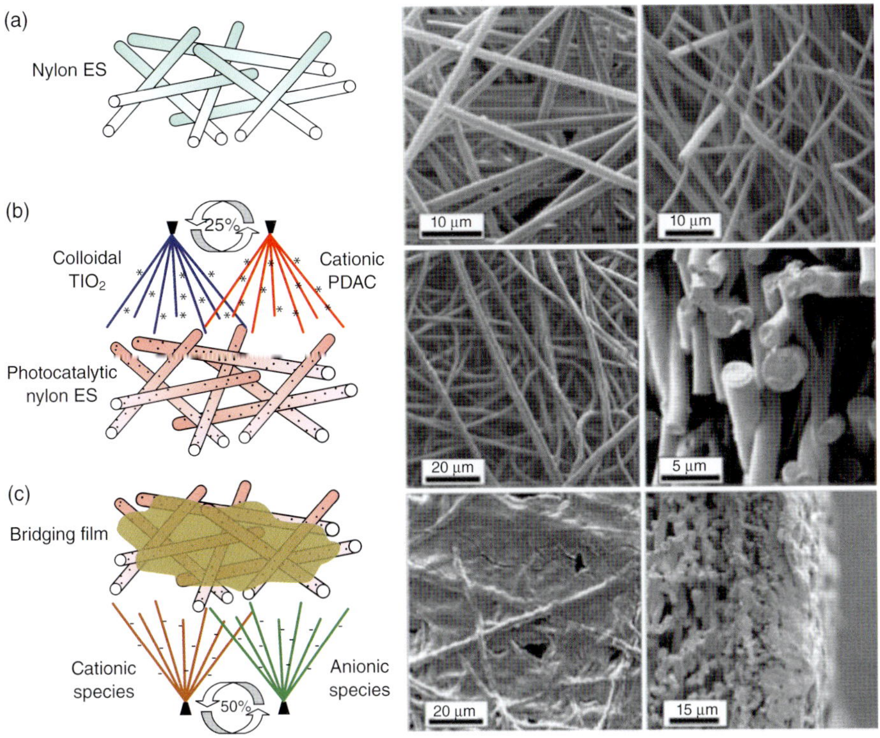

FIGURE 18.6. Conformal versus bridging effects in sprayed LbL deposition: (a) Starting nylon electrospun (ES) fibers; (b) conformal coating (PDADMA/TiO_2) by vacuum-assisted spray deposition; and (c) bridged fibers by coating using spray deposition of PDADMA/PSS. Schemes of the system are represented in the left column, surface observation by scanning electron microscopy in the center column, and cross-sectional observation in the right column. (Taken from Krogman et al.,[73] with permission).

silica capillaries for electrophoretic separation[74] and poly(dimethylsiloxane) separation channels in capillary electrophoresis microchips.[75] Although only slightly faster than the dipping technique, LbL adsorption using the dynamic coating process has been proven to be efficient at preparing PEM membranes in flat sheet[76] and hollow fiber[77] geometry as it enables denser layering to form and, consequently, the number of bilayers required to reach the same range of performance becomes lower.

In conclusion, the method used to prepare the PE LbL assembly has a dramatic impact on the morphology of the formed PEM film and, hence, on its properties. For instance, spin coating could be used to prepare very ordered layers[63] and dewetting to produce unique adsorption topographies, including fractal networks and aligned fibers.[64] It has become obvious that more fundamental studies are still needed to clarify this point as many other parameters can influence the structure of PEM films (see Sections 18.2.2 and 18.2.3).

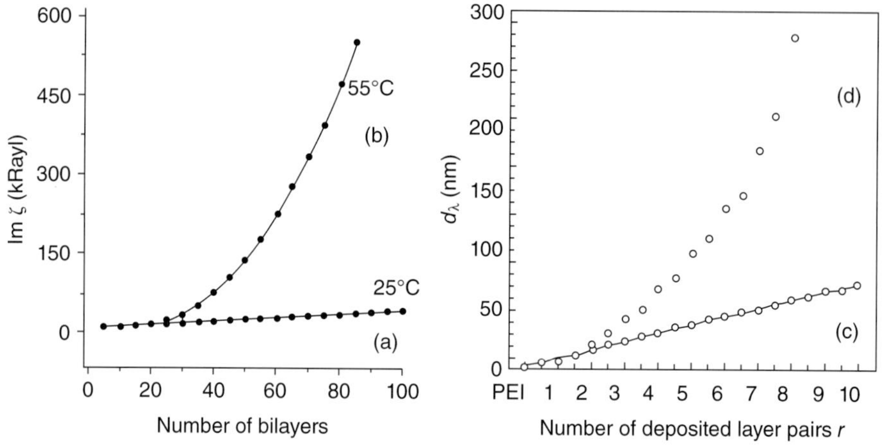

FIGURE 18.7. Linear versus exponential growth. (a) Mass increment as a function of deposited bilayer number (dipping) determined for the PDADMA/PSS system in 0.1 mol L^{-1} NaF at 25°C and (b) 55°C (Taken from Salomäki et al.,[78] with permission). (c) Film thickness variation as a function of deposited bilayer number (dipping) for the PSS/PAH and (d) PGA/PLL systems. PEI on the abscissa highlights the fact that cationic PEI is the first deposited layer. (Reprinted from Lavalle et al.,[80] with permission).

18.2.2 Morphology and Growth Regime of PE Multilayer

Two kinds of growth regimes have been identified for LbL assembly of PEs. The first one termed as "linear growth" is used to describe a linear mass uptake (Fig. 18.7a) and film thickness increment (Fig. 18.7c) with the number of deposited layers. Later, a second type of growth regime termed as "exponential growth" was observed, showing an upward curvature indicating that the increment per deposition cycle increases with the number of deposited layers (Fig. 18.7b,d). Both regimes of growth can be encountered for the same pair of PEs depending on the LbL operating conditions as in the case of the PSS/PDADMA system for which a linear growth is observed at room temperature while an exponential increase occurs at higher temperature (Fig. 18.7a,b).[78] This point is detailed in the next section.

The linear growth is featured by low mass uptake and thickness increment (typically a few nm per bilayer) so that thin PEM films are generally obtained with that regime (Fig. 18.7c). A stratified structure with poor interpenetration between successive layers is believed to occur in these films.

By contrast, the exponential growth yields film thickness that can reach the micron scale after a limited number of bilayers (Fig. 18.7d). It was observed that surface morphology of the thicker films is also quite different from that of PEMs showing a linear growth. Surface AFM observation generally reveals island and vermiculate structures and high interface roughness.[79] The roughness is more or less constant after a certain number of layers and this parameter cannot be related to the growing mechanism. It was then stated that the driving force for exponential growth is the diffusion through the multilayer of

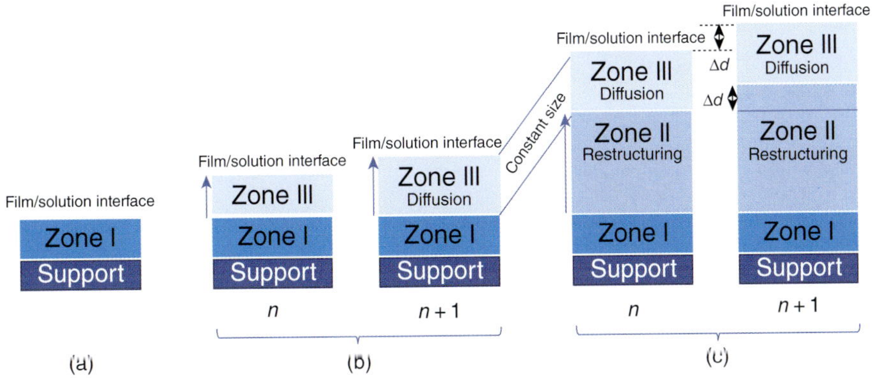

FIGURE 18.8. Schematic representation for the three-zone build-up mechanism model of an exponentially growing PEM film. (a) The structure of the first few bilayers (I) is highly influenced by the substrate; (b) after the deposition of a certain number of bilayers, a diffusion zone (III) is growing, inducing an exponential growth of the film thickness; (c) during the film buildup, a restructuring occurs in the lower layers of the film, denoted as restructuring zone (II). This process induces changes in the film properties, stopping the diffusion of PEs. Therefore, the diffusion zone reaches maximal thickness and the film grows linearly. (Redrawn from Porcel et al.[70]).

at least one PE.[80] This mechanism was proved for the PLL/HA system by showing the diffusion of labeled PLL molecules into the film over its whole thickness, whereas HA did not.[81] Because these cationic chains do not interact strongly with polyanions, they are called "free polycations" and can also diffuse out of the film during the subsequent steps of rinsing and contacting with the polyanion.

Recently, a new model for a multilayer buildup mechanism has been proposed based on a three-zone growing film (Fig. 18.8)[70] that can account for the crossover to a linear growth regime generally observed after a certain number of exponentially growing deposited bilayers as shown in Figure 18.9 for the PLL/HA system. The mechanism is assumed to take place in three steps defined by three kinds of growing zones:

- Zone (I) comprises the first bilayers with a structure that is largely influenced by the substrate and by the nature of the adsorbed polymers as well.
- Zone (III), called the diffusion zone, is the region located at the interface with the solution where PE molecules can diffuse in and out of the film. It is exponentially growing and keeps a constant thickness after a certain number of layers.
- Zone (II), called the restructuring zone, is a region located in the depth of the building-up film where the interdiffusion of polymer is hindered and the PE molecules can no longer diffuse out of the film. It occurs when the PEM film reaches a certain thickness. The zone (II) grows linearly with the number of deposited layers.

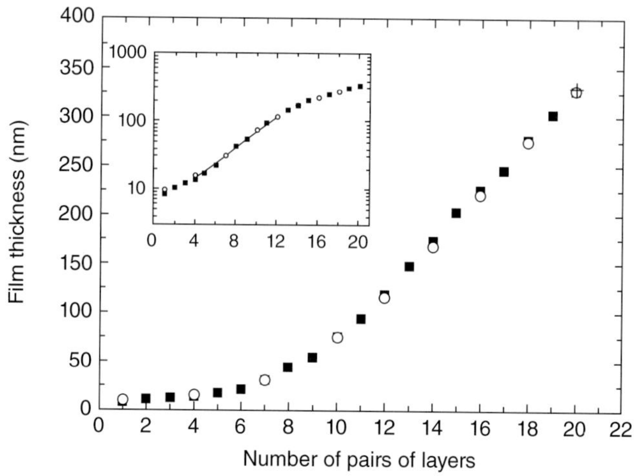

FIGURE 18.9. Exponential growth crossover to linear growth. Film thickness variation determined by ellipsometry as a function of the number of bilayers (spraying) for the PEI-HA/PLL system. The inset represents the same data with log-scaled axes and shows the exponential growth as a linear variation, in this case between the 4th and 12th pair of layers (plain lines). (Taken from Porcel et al.,[68] with permission).

The three-zone mechanism of PEM growing was confirmed by experimental measurement of lateral diffusion coefficient that showed the coexistence of three PLL populations: two within the film, one being almost immobile ($D < 0.001 \times 10^{-8}$ cm^2 s^{-1}) and one mobile ($D \sim 0.1 \times 10^{-8}$ cm^2 s^{-1}), and at the interface, another population which is highly mobile ($D \sim 1 \times 10^{-8}$ cm^2 s^{-1}).[82]

It should be noted that the diffusion coefficient in linearly growing films (as PAH/PSS system) was found to be very low, confirming that the PE diffusion within the LbL assembly is limited for this morphology.[83] The latter mechanism of growing is generally observed with PEs having a high charge density along the polymer chain like PSS. Interestingly, some studies have taken advantage of this property to hinder the interlayer diffusion of diffusing PE by incorporating blocking layers as PAH/PSS.[84] Discrete and compartmentalized films made of alternate PEMs of exponential and linear growth can then be created[85] and could be interesting in the design of new separation membranes. The molar mass average and distribution are two other parameters to consider in the sense that they control the diffusivity of PEs. For instance, a transition from linear to exponential growth in the system PAH/PAA was observed by using well-defined, low-molecular-weight PAA instead of high-molecular-weight PAA.[86]

18.2.3 Influence of Preparation Parameters on PEM Morphology

It is obvious that the structure and morphology of PEM films is of prime importance for the functional properties and many works can be found in literature dealing with the influence of external parameters on the control and engineering of film thickness and morphology. For recent reviews, see Lichter et al.[41]; Boudou et al.[43]; and Tokarev and Minkov[49] This section aims at summarizing some significant and new insights concerning the preparation parameters to be adjusted in order to direct the growing regime and obtain the expected properties of a multilayer film in view of PEM design optimization for membrane separation.

As indicated before, the morphology of PE LbL self-assemblies results from a subtle balance between electrostatic and hydrophobic interactions, gain of entropy, and polymer diffusion. Several parameters such as ionic strength, pH, and temperature can affect this balance. An increase in mobility of PE chains within the LbL film may give rise to the transition from linear to exponential growth. This effect can be achieved by an increase of the chain flexibility and a decrease of connecting points between the oppositely charged PEs constituting the LbL film.[87]

Ionic Strength Since the earlier development of the LbL technique, the salt concentration has been recognized as a key parameter to control the deposition of PEs. An increase of the ionic strength leads to the screening of PE charges which causes the chain to adopt a more coiled and flexible conformation. As a result, thicker deposited layers and high film roughness are obtained (PDADMA/PSS system).[12] In fact, transition from linear to exponential growth regime for the PDADMA/PSS system was demonstrated to occur for NaCl concentrations higher than 0.1 mol L^{-1}.[79] On the other hand, the ion pairing between PEs is also screened and should yield a dramatic decrease of the adsorbed amount at very high ionic strength. However, this was not observed (at NaCl = 1 mol L^{-1}) for this system probably due to other forces like the hydrophobic interaction and the gain of entropy owing to the release of small counterions.

Once the LbL film formed, an increase of the salt concentration may induce large-scale rearrangements due to a change of the interactions between the PEs. It was reported that linearly growing PEM films swell on exposure to solutions containing salt.[88] The surface film roughness was also observed to decrease significantly upon annealing in salt solutions, indicating polymer interdiffusion within the film leading to smoothing of the surface. During the linearly growing PEM formation, the ion pairing that forms between oppositely charged PEs is called "intrinsic" charge compensation. In this case, only a very low concentration of counterions is entrapped within the film. The swelling observed when these films are brought into contact with solutions of high ionic strength results from exchange of the polymer charge compensation ("intrinsic") by small counterions charge compensation called "extrinsic" charge compensation.[89] It was then assumed that the PE chains can diffuse into oppositely charged layers through the whole film thickness by exchange of "extrinsic charges." This assumption was confirmed in the case of the exponentially growing film prepared with the PLL/HA system, in which about half of the PLL molecules were found to be unpaired and compensated with chloride anions as against only approximately 10% for HA molecules compensated by sodium cations.[90] This result is in complete agreement with the diffusion observed for labeled PLL (see Section 18.2.2).

It can then be concluded that the rate of "extrinsic" charge compensation is one of the most important parameters governing the PEM film morphology. This point must carefully be considered for the application of PEM membrane for desalination and ion separation.

Polarizability of anions may also play a role in the morphology of PEM films. The thickness and surface roughness of PDADMA/PSS LbL films prepared in 0.1 mol L^{-1} NaX were found to increase with the anion polarizability.[91] This effect is comparable to that observed with an increase of the ionic strength. Because polarizable anions are known to strongly interact with polycation, a decrease of "intrinsic" in favor of "extrinsic" charge compensation along the polymer chain probably takes place in the LbL film, leading to a more coiled conformation of PEs. Because cations have much lower difference in polarizability, anions have much larger effect on PEM film morphology than cations.[92]

pH of the Deposition Solution The degree of ionization of weak PEs strongly depends on local pH. This parameter is therefore another means of manipulating the morphology and properties of PEM films.[41] Weak PEs can be LbL self-assembled over a large range of pH values to create films with variable layer thickness and compositions.

It should be kept in mind that a shift of the effective pKa is often observed when the polymer is assembled into an LbL assembly. For instance, PAA becomes more acidic when part of a PEM with PAH so that its degree of ionization for a given pH can be significantly different from that in solution.[93]

Temperature of LbL Buildup PEM films fabricated at elevated temperatures are significantly thicker than similar films deposited at room temperature.[94,95] As seen in Figure 18.7a, the mass increment as a function of the number of layers deposited for the PDAPDMA/PSS system increases faster at elevated temperature of deposition.[78] At room temperature, the PDAPDMA/PSS pair shows a linearly growing LbL buildup, whereas it follows an exponential regime at high temperature. This effect is accounted for an increase of the diffusion coefficient of at least one PE that becomes high enough to carry the polymer within the entire thickness of the film.

18.2.4 Concluding Remarks

The packing density of PE chains in PEM films depends on a large extent to the number of layers deposited and on the balance between "intrinsic" and "extrinsic" charge compensation.

The number of bilayers is a limiting factor from an economical point of view as the buildup of PEMs is quite time consuming. However, only a few bilayers are often required to successfully modify the interfacial properties, such as biofouling resistance and biocompatibility. When a small number of bilayers is deposited, it must be taken into account that the structure of the film is strongly directed by the surface properties the PEM is growing on.

In the case of a larger number of deposited bilayers, the LbL self-assembly of PEs can grow according to two build-up mechanisms (linear and exponential) depending on parameters related to the chemical nature of the PE pair, the solution properties, and the preparation conditions. A linear build-up regime leads to a thinner and PE close packing than the exponential regime. The packing density can then be tuned to a more or less loose morphology by choosing appropriate conditions of ionic strength, pH, and temperature. Beyond a certain thickness, a more compact region may arise from the inner restructuration as predicted by the three-zone model. Thus a gradient of packing density happens from a loose structure in the outer layers at the film/solution interface to a close packing near the film/support interface. The versatility of PEM construction enables a variety of film morphology to be engineered with permeation properties tailored to a given membrane application.

PEM films, once prepared, are stimuli-responsive materials since the structure can self-rearrange upon various external stimuli like ionic strength, pH, and temperature. In most of membrane applications, PEM films are in contact with aqueous phases containing ionic or neutral solutes that can affect the initial morphology obtained during preparation. Chemical and photo cross-linking can be used if required to lock and maintain the integrity of the film morphology.[43]

18.3 PEM-BASED MEMBRANES

A membrane can be defined as a thin and selective barrier which enables the transport or the retention of compounds between two media. The separation layer must be as thin as possible to limit the membrane resistance to the permeate flow rate. The LbL deposition is then a particularly attractive route to design semipermeable membranes with improved permeation properties as the thickness and morphology of the resulting PE films can be easily controlled at the nanoscale range. This reason probably accounts for the large number of works dedicated to the development of PEM-based membranes since the first papers were published about 10 years ago.[96,97] LbL build-up is usually carried out at the top surface and within the pores of a supporting porous membrane to overcome the poor mechanical strength related to a nanometer thickness (Fig. 18.10). Water and ion transport across PEM films and interfacial compatibility with biomolecules are the main properties that are looked for in PEM-based membranes.

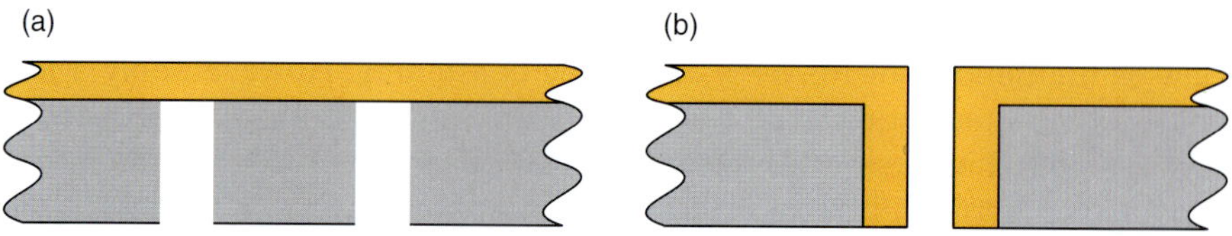

FIGURE 18.10. Schematic representation of PEM membranes (a) at the top surface: the solutes must cross the PEM film; and (b) within the pores of a porous support: the solutes only interact with the PEM surface.

Potential applications have been found in various fields including water desalination, artificial organs, fuel cells, chemistry, and the biotechnology industry. Tables 18.2–18.5 summarize most of the studies carried out during the recent years on the use of PEM in membrane separation techniques.

18.3.1 Permeability of PEM Film

Membranes are typically classified into porous and nonporous materials because the mechanism of separation and transport in these two systems is fundamentally different. Porous refers to the presence of permanent pores in the membrane structure.

The transport mechanism across nonporous membranes, called the solution–diffusion mechanism, implies that permeants dissolve in the polymer matrix and then diffuse though the whole membrane thickness down a concentration gradient.[98] The separation performance is based on the difference in permeant solubility into the polymer matrix and its diffusivity through the membrane (Fig. 18.11a). The solution–diffusion mechanism applies to reverse osmosis (RO), pervaporation (PV), and gas separation (GS). The flux of permeant is proportional to the transmembrane pressure and inversely related to the thickness of the membrane. Furthermore, the transport can be selectively facilitated by the presence in the membrane of specific recognition sites to one permeant (called the facilitated transport mechanism, Fig. 18.11c).

PEM can serve as a dense separation barrier supported on a porous structure as for example in PV applications presented in Table 18.2. A homogeneous and compact layer is a prerequisite condition to obtain an efficient separation. If the thickness of the PEM film is too large, the flux of permeant is very low. The challenge in that case is to prepare a stable, ultra-thin dense separation layer.

The separation through porous membranes is mainly achieved by steric exclusion of solutes and by electrostatic interactions between solutes and pores. Microfiltration (MF), ultrafiltration (UF) and nanofiltration (NF) are pressure-driven membrane processes that belong to this class. For neutral solutes, the key factor is the ratio between the solute and the pore sizes (sieving effect). The hindered transport of solutes through the membrane is described by both diffusion and convective terms, which come from the transmembrane concentration and pressure gradient, respectively (Fig. 18.11b).[99]

TABLE 18.2. Recent Works Published on Dehydration of Hydroxylated Solvents by PV Using PEM-Based Membranes

Solvent	Support	PE Pair (Deposition Method)	Bilayer Number	Separation Factor	Flux ($kg\ m^{-2}\ h^{-1}$)	Temperature (°C)	References
IPA	Polyamide RO	PDADMA/PSS (vibration)	4.5	156 132	2.32 2.38	50	Yin et al.[115]
	Polyamide RO	PDADMA/CMCNa = PEC^+ PDMC/CMCNa = PEC^- (dipping)	24	1013	1.36	50	Zhao et al.[112]
	Polyamide RO	b-PEI/PAA (electric-field-assisted dipping)	4.5	1062	4.05	70	Zhang et al.[116]
	Polyamide RO	$(PDADMA/PSS)_4$/PDADMA $(PDADMA/PAA)_4$/PDADMA $(b\text{-}PEI/PAA)_4$/b-PEI (electric-field-assisted dipping)	4.5	180 285 1289	2.50 2.20 1.71	50 50 50	Zhang et al.[117]
	Hydrolyzed PAN UF (MWCO 20,000)	b-PEI/PAA CHI/PAA (dipping)	4	495	1.30	70	Zhu et al.[111]
	Hydrolyzed PAN UF (MWCO 20,000)	b-PEI/PAA (dipping)	6	2900	0.50	50	Zhu et al.[107]

(Continued)

TABLE 18.2. (*Continued*)

Solvent	Support	PE Pair (Deposition Method)	Bilayer Number	Separation Factor	Flux (kg m^{-2} h^{-1})	Temperature (°C)	References
Ethanol	Hydrolyzed PAN UF (MWCO 60,000)	b-PEI/PAA (dipping)	1–4.5	—	—	—	Wang et al.[171]
	Hydrolyzed PAN UF (MWCO 20,000)	b-PEI/PAA (dynamic)	2.5	604	0.30	70	Zhang et al.[172]
	Hydrolyzed PAN hollow fibers	b-PEI/PAA (dynamic)	3.5	750	1.40	50	Zhang et al.[118]
	Poly(ethersulfone) UF (MWCO 20,000)	b-PEI/PAA (dipping and dynamic)	10	1207	0.14	40	Zhang et al.[76]
	ZrO_2/Al_2O_3 tubular MF (0.2 μm)	b-PEI-PVS/b-PEI b-PEI- PSS/b-PEI b-PEI-PVS/PAH b-PEI-PSS/PAH b-PEI-PSS/PDADMA (dipping)	60	8.0	18.4	65	Chen et al.[173]
	Hydrolyzed PAN UF hollow fibers	b-PEI/PAA (dynamic)	6.5	1338	0.12	50	Zhang et al.[174]
Ethylene glycol	Polyamide RO	b-PEI/PAA (dipping)	10	340	0.4	40	Zhang et al.[77]
Alcohols	Hydrolyzed PAN UF (MWCO 20,000)	l-PEI/PAA (dynamic)	2.5	—	—	—	Ji et al.[175]

TABLE 18.3. Recent Works Published on Selective Ion Separation by NF Using PEM-Based Membranes

Application	Support	PE Pair (Deposition Method)	Bilayer Number	Permeability ($kg\ m^{-2}\ h^{-1}\ bar^{-1}$)	*R* (%)	References
Water softening	PES MWCO 50,000	PSS/PAH (dipping)	4.5	7.5	95 (Na_2SO_4) –8 (NaCl)	Malaisamy et al.[128]
	Porous alumina UF (20 nm)	PSS or PAA/PDADMA or PAH	3.5	23.5	96/-17 SO_4^{2-}/Cl^- –5/96/95 $Na^+/Ca^{2+}/Mg^{2+}$	Hong et al.[130]; Ouyang et al.[131]
	PAN/PET plasma-treated UF	Aza6/PVS (dipping)	30 60	0.04 0.03	75/42 99/56 SO_4^{2-}/Cl^-	El-Hashani et al.[129]
	Hydrolyzed PAN UF MWCO 50,000	$(PAH/PSS)_n/PSSMA$ (dipping, dynamic)	1.5	14	91 (Na_2SO_4) 66 ($MgCl_2$) 33 (NaCl)	Deng et al.[141]
	Cellulose acetate NF	CHI/SA (dipping)	15–30	2	98 (Na_2SO_4) 92 ($CaCl_2$) 60 (NaCl)	Lajimi et al.[124]
	PES MWCO 50,000	PSS and/or PAA/PDADMA or PAH (dipping)	3–4.5	Fouling	~100 ($MgSO_4$)	Shan et al.[125]
	PSf MWCO 50,000	$QCMC^+/CMCNa^-$ PEC^+, PEC^- (casting)	1	1.4	87 (K_2SO_4) 23 (NaCl) 99 (XO)	Ji et al.[123]
Desalination (NaCl)	Hydrolyzed PAN MWCO 20,000	b-PEI/SPEEK (dynamic)	5	0.82	89 (2 $g\ L^{-1}$)	Wang et al.[122]
	Cellulose acetate electrospun fibers MF	CHI/SA CHI/PSS (dipping)	15–25	—	6–15 (2 $g\ L^{-1}$)	Ritcharoen et al.[176]

(*Continued*)

TABLE 18.3. (*Continued*)

Application	Support	PE Pair (Deposition Method)	Bilayer Number	Permeability ($kg\ m^{-2}\ h^{-1}\ bar^{-1}$)	R (%)	References
F^-/Cl^-	Porous alumina UF (20 nm)	PSS/PDADMA (dipping)	4.5	30	73 (F^-) 9 (Cl^-)	Hong et al.[139]
Phosphate recovery	Hydrolyzed PAN	PDADMA/SPEEK (dipping)	10^a	37 33 29	9 (Cl^-) 30 (SO_4^2) 80 (HPO_4^2)	Ahmadiannamini et al.[132]
	Porous alumina UF (20 nm)	PSS/PDADMA (dipping)	4.5	21	21 (Cl^-) 98 (HPO_4^{2-})	Hong et al.[133]
Fe^{3+}/Fe^{2+}	PAN $^+$ PAN^- UF	PAA/CTAC CTAC/PAA (dipping)	3 2.5	– –	80/45 90/51 Fe^{3+}/Fe^{2+}	Fu and Kobayashi[134]
Dyes, sugars, amino acid	Porous alumina UF (20 nm)	PSS/PAH (dipping)	4.5	15	>99.9 (dye) 99.4 $(SUC)^d$	Hong et al.[136]
			7	11	34.4 $(GLY)^d$ 98.7 $(GLN)^d$	Hong and Bruening[135]
SRNF	Hydrolyzed PAN	PDADMA/SPEEK (dipping)	4^b 15^b 15^c	0.06 0.98 0.85	99 $(RB)^d$ 85 $(RB)^d$ 95 $(RB)^d$	Li et al.[143,144]

[a] NaCl = 0.1 mol L^{-1} + HPO_4^{2-}.
[b] NaCl = 0 mol L^{-1}.
[c] NaCl = 0.5 mol L^{-1}.
[d] SUC = Sucrose; GLY = Glycine; GLN = Glutamine; RB = Rose Bengale.

TABLE 18.4. Applications of PEM-Based Membranes in Fuel Cells and Batteries

Application	Support	PE Pair (Deposition Method)	Bilayer Number	Conductivity (mS cm^{-1})	Performance	References
H_2/air FC	PC 0.1 μm	PEO/PAA	40	0.24 (25°C)	17 mW cm^{-2}	Farhat and Hammond[150]
		PDADMA/PAMPS		0.036 (25°C)	12 mW cm^{-2}	
		l-PEI/PAA (dipping)		0.033 (25°C)	6 mW cm^{-2}	
		Nafion 112	0	1.5 (25°C)	43 mW cm^{-2}	
DMFC	Nafion 1135	PDADMA/PSS (dipping)	4	107 (30°C)	37 mW cm^{-2}	Jiang et al.[151]
		Nafion 1135	0	124 (30°C)	26 mW cm^{-2}	
	Nafion 117	PAH/PSS (dipping)	5	88 (22°C)	4.5×10^{-7} cm^2 s^{-1a}	Yilmaztürk et al.[152]
		Nafion 1135	0	38 (22°C)	6.0×10^{-7} cm^2 s^{-1a}	
	Nafion 112	PAH/PSS (dipping)	20	152 (30°C)	5.6×10^{-8} cm^2 s^{-1a}	Deligöz et al.[153]
		Nafion 112	0	73 (30°C)	9.6×10^{-8} cm^2 s^{-1a}	
	SPAEK-C	PANI/PWA (dipping)	5	240 (80°C)	4.6×10^{-8} cm^2 s^{-1a}	Zhao et al.[154]
		Nafion 117	0	146 (80°C)	10×10^{-7} cm^2 s^{-1a}	
VRB	Nafion 117	PDADMA/PSS (dipping)	5	50 (25°C)	CE = 97.6%	Xi et al.[155]
		Nafion 117	0	59 (25°C)	CE = 92.1%	

[a] Methanol permeability.

H_2/air FC, hydrogen/air fuel cell; DMFC, direct methanol fuel cell; VRB, Vanadium redox flow battery; CE, coulombic efficiency at 80 mA cm^{-2}.

TABLE 18.5. Applications of PEM as Porous Membrane Coating

Application	Support	PE Pair (Deposition Method)	Bilayer Number	Comments	References
Tunable pore size	PES 0.45 μm MF	CHI/PSS (dipping)	7–11	Rejection of proteins: bovine serum albumin, ovalbumin	Mathew et al.[156]
	PES 0.45 μm MF	CHI/PSS (dipping)	5–12	Rejection of proteins: bovine serum albumin, ovalbumin, lysozyme.	Aravind et al.[158]
	PES hollow fiber UF	b-PEI/PSS (dipping)	1–3	Permeability and MWCO by rejection of dextrans	Kochan et al.[157]
	PES 0.45 μm MF	CHI/PSS (dipping)	5.5–20.5	Removal of color and COD[a] from industrial effluents	Aravind et al.[159]
Fouling resistance	b-PEI grafted Poly(etherimide) NF	PAA PVS PVA (dynamic)	1	Foulants: bovine serum albumin, humic acid, sodium alginate	Ba et al.[160]
Enzyme catalytic membrane	PC 0.4 and 0.2 μm MF	Peroxidase-PSS/PAH (dipping, sonication)	1–12	Sandwich peroxidase immobilization; activity as a function of the bilayer number and the pore size	Yu et al.[162]
	Nylon 0.45 μm MF	PGA/PAH/PSS PLL/PSS/PAH (dynamic)	1	Glucoxidase immobilization as the outermost layer: reusability and regeneration	Smuleac et al.[164]
	Nylon 0.45 μm PVDF 0.45 μm MF	PLL/(PSS/PAH)$_n$ grafted PAA/PAH (dynamic)	1–3 1	Glucoxidase immobilization as the outermost layer: activity under convective flow	Tristan et al.[166]
	PAN-b-PEI UF	SA/trypsin(/SA)$_n$ Urease(/b-PEI)$_n$ (dipping)	0–1	Enzyme immobilization as a sandwiched or outermost layer: activity and stability	Guedidi et al.[163]
Ion-exchange	Nylon, PES, PVDF 5 μm Nylon 1.2 μm RC 1 μm MF	PSS/b-PEI PSS/PDADMA PSS/PAH (dynamic)	1–3 3.5	Binding capacity: 38 mg mL^{-1} negatively charged Au colloids (b-PEI outermost layer), 16 mg mL^{-1} lysozyme (PSS outermost layer)	Liu et al.[167]
Hemodialysis	PTAT PU	DSS/CHI (dipping)	5.5	Resistance to platelet adhesion and fibrinogen adsorption	Yu et al.[168,169]
	PLA	DSS-Ag/CHI (dipping)	5.5	Human endothelial cell attachment antibacterial activity	Yu et al.[170]

[a] COD = chemical oxygen demand.

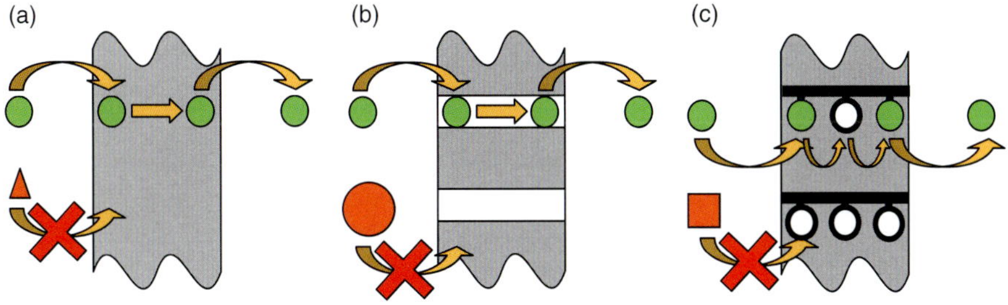

FIGURE 18.11. Schematic representation of the different membrane transport mechanisms. (a) Solution–diffusion mechanism across dense membranes; (b) hindered transport mechanism across porous membranes; (c) facilitated transport across membranes containing specific recognition sites.

The permeate flux through a pore is proportional to the transmembrane pressure and inversely proportional to the length of the pore. For charged species, electrostatic interactions become preponderant when the pore size is in the same order of magnitude as the Debye length (NF and tight UF) or when the solute size is close to the pore size.

Porosity of PEM originates from the average length between two cross-linking points (ξ), that is, two points of ion pairing occurring between the PE chains. The pore size in a PSS/PAH assembly composed of seven bilayers prepared from solutions containing 0.5 mol L^{-1} salt was estimated by measuring the rejection of neutral solutes in NF experiments.[100] Simulation of the filtration data indicates that the film contains pores with radii of 0.4–0.5 nm. The pore size can then be controlled through the balance of "intrinsic" and "extrinsic" charge compensation adjusted under specific pH and ionic strength conditions. It makes it possible to tune the permeability to molecules or ions that can permeate across the membrane. Phase separation of PEM obtained by further treatment of preassembled films by changing ionic strength and pH often results in the formation of a large distribution of pore size, typically in the range 20–200 nm. Some examples can be found in Tokarev and Minkov.[49]

An alternative approach for inducing more well-defined porosity in PEM is based on the sacrificial templates incorporated in the LbL assembly.[101] As an example, hybrid multilayers were constructed by alternatively depositing an anionic blend of monodisperse SiO_2 nanoparticles and PAA, and a cationic PE, PAH.[102] After chemical cross-linking of the PAA/PAH multilayer film, the nanoparticles were dissolved, resulting in a nanoporous film. Absorption experiments of bovine serum albumin provide some evidence for the interconnectivity of the pores in the PEM.

18.3.2 Interfacial Properties of PEM Coating

LbL deposition of PEs is also useful for modifying the top surface and the inner pore surface of UF and MF membranes. Dramatic change of the interfacial properties can then be obtained.

Adsorption of PE onto an interface is known to increase its hydrophilic character. Most of the commercially available membranes are fabricated from hydrophobic polymers which promote the surface adhesion of hydrophobic molecules and of many biomolecules, such as proteins causing membrane fouling. (Bio)fouling is still the main unsolved problem for the membrane applications as it often leads to a permanent flux decline. One approach consists of carrying out a coating with a highly hydrated polymer enabling the membrane surface to change from hydrophobic to *hydrophilic*.[103] Protein adsorption-resistant interface can be created by surface modification of membranes with PEM.[104]

Moreover, coating of a thin PEM film has been proved to be useful for preparing hemocompatible biomaterials.[43] LbL assemblies incorporating natural PEs exhibit strong anticoagulant properties when the outermost layer of the film comprises anionic polysaccharides including HA, dextran sulfate (DSS), and heparin. This property was put to use in the design of new membranes for hemodialysis application (see the next section).

Active coating can be prepared by incorporating a PE containing a recognition site or a reactive site during the LbL buildup. It was shown that LbL films are a good platform for immobilizing enzymes, providing a good environment probably because of the high degree of hydration of PE assembly.[105] Enzymes can then be immobilized in PEM without loss of their bioactivity. These systems were studied for applications in biosensors and bioreactors.[35] Applications in the domain of biocatalytic membranes are also described below.

18.3.3 Applications of PEM-Based Membranes

Applications of PEM membranes are classified according to the mechanism of transport involved in the considered process. The two first sections deal with the pressure-driven membrane processes in which the LbL assembly is the active layer, that is, dense PEM films governed by the solution—diffusion mechanism (GS, PV, and RO) and porous PEM films governed by the hindered transport mechanism (NF and tight UF). Ion exchange is addressed in a separate section as the newest application of ion exchange PEM membranes in fuel cells. Finally, coating of membranes with PEM is examined for applications in low fouling membranes and membranes for artificial organs.

Applications of Dense PEM Films GS, PV, and RO require dense materials to bring about a high selectivity process. As mentioned earlier, the permeation rate is inversely related to the thickness of the active layer. Composite membranes made of a PE LbL assembly deposited onto a highly permeable porous support are particularly attractive because of the nanoscale control over the thickness and the morphology of PEM films. However, a large number of deposited layers are often needed to achieve high membrane selectivity. As an illustration, the first studies carried out to create PEM films for GS[106] and PV[96,107] found that more than 60 bilayers deposited on an MF membrane are needed to obtain selectivity. Therefore, the general motivation for the recent works has been essentially guided by a reduction of the bilayer number required for fabricating dense and defect-free PEM film all while maintaining high selectivity and flux.

Most of the works of GS using LbL films showed little selectivity.[108] To our knowledge, no recent paper on the application of LbL films for GS has been published. In contrast, PEM-based membranes have been recognized as promising materials for dehydration[109] of solvents by PV because of their high hydrophilicity.[87,105] Each ionic group in LbL assemblies regardless of sign was found to be associated to approximately eight water molecules. Table 18.2 gathers most of the works recently published on PV dehydration of alcohols using PEM membranes.

PV is a membrane technique that permits one to break azeotropes, such as those commonly found in alcohol/water systems. In PV a liquid feed is circulated over the membrane surface and one component is able to be preferentially transported across the membrane.[110] The vapor produced on the permeate side has a different composition from that obtained by distillation. The parameters that characterize the membrane performance are the flux of permeate J (kilogram per square meter per hour) and the separation factor α between two permeants A and B defined as follows:

$$\alpha = \frac{Y_A / Y_B}{X_A / X_B}, \tag{18.2}$$

where X_A and Y_A are the weight fractions of the solute A in the feed and in the permeate, respectively. High temperature of the feed solution generally increases the flux of permeate and decreases the separation factor.

From Table 18.2, it can be seen that highly efficient PEM membranes for dehydration of alcohols ($\alpha > 1000$ and $J > 1$ kg m^{-2} h^{-1}) were successfully prepared. A significant improvement was obtained by using membrane support with lower pore size since the PE molecules should be sufficiently larger than the pore size of the support to reduce significantly the number of layers required to form a defect-free PEM film.[111] This is exemplified in Table 18.2 with the use of UF and RO membranes instead of MF membranes.

As expected for a dense packing density, most of the studies reported in Table 18.2 confirm a previous finding that high charge density PEs[101] and the linear growth regime were favorable parameters to obtain good performance. The PE pair leading to the best PV performance in dehydration of solvents is the PEI/PAA system. A novel approach was recently developed based on the alternative deposition of PEC^+ and PEC^- nanoparticles.[112] The principle of the method is described in Figure 18.2a. For a number of PEC^+/PEC^- bilayers higher than 7, membranes exhibited good PV performance for the dehydration of isopropanol with a maximum separation factor for 24 bilayers. It should be noted that high permeate fluxes were reported in spite of a film thickness of 500–3000 nm.

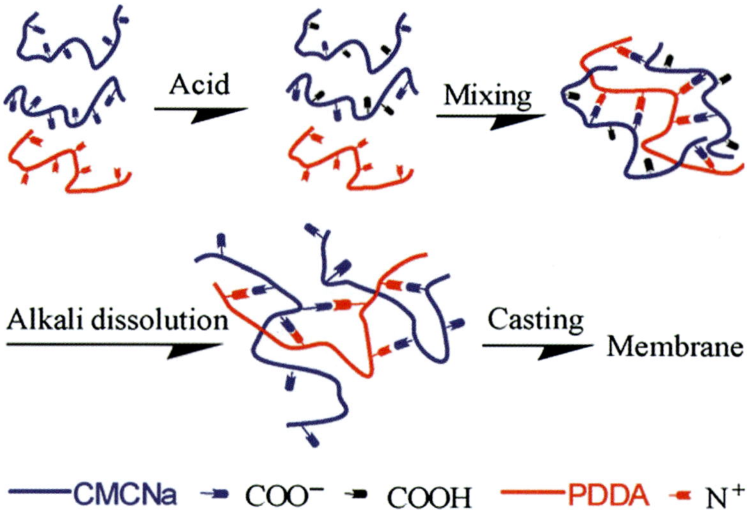

FIGURE 18.12. Schematic representation of membrane fabrication from preformed PEC composed of sodium carboxymethyl cellulose (CMCNa) and poly(diallyldimethylammonium chloride) (PDADMA). (Taken from Zhao et al.,[113] with permission).

An alternative way that allows one to limit the number of deposited layer consists of casting soluble PEC preformed in solution (Fig. 18.12).[113,114] This approach is more convenient than the successive layer deposition and can be easily adapted to a large-scale fabrication process of membranes. Defect-free skin layer was obtained with separation properties that can be tuned by varying the nature of PEs, the mixing, and the cross-linking ratios. These membranes exhibited high PV performance for dehydration of isopropanol ($J = 3$ kg m^{-2} h^{-1}, $\alpha = 930$ at 75°C) and ethanol ($J = 0.9$ kg m^{-2} h^{-1}, $\alpha = 1420$ at 70°C). It has to be noted that the morphology of these membranes is probably different from those obtained by LbL deposition.

On the other hand, the PV selectivity was found to decline when the LbL assembly was prepared in the presence of salt[115] or after exposure to salt-containing solutions.[107] This effect is consistent with the transition from a close packing to a loose structure related to the increase of "extrinsic" charge compensation (see Section 18.2.3).

The influence of the self-assembly method on the PV performance of membranes has been investigated by several authors. LbL assembly was built up by dipping under vibration condition.[115] The obtained PEM films are denser with fewer defects so that the same level of PV performance in isopropanol dehydration can be achieved by membranes with fewer bilayers. The LbL deposition using the dipping method was proven to be strongly enhanced by applying an electric field.[116,117] This technique results in a PV performance higher than the vibration-assisted dipping one for PEM films composed of the

PDADMA/PSS pair.[117] Two electrodes were inserted in parallel into a cell and a voltage was imposed on the two electrodes during the assembly process. It was assumed that the technique enables the PE chains to closely pack in the LbL assembly. The optimal voltage was found to be 5 V.[116] Beyond this value the deposition speed of PE chains is probably too fast for them to relax and form the best structure on the surface. However, the scaling up of the previously described methods seems to be difficult compared to the dynamic method. PEM membranes made of four PEI/PAA bilayers prepared using dynamic LbL deposition (under a pressure of 0.1 MPa) on a UF membrane (molecular weight cutoff [MWCO] = 20,000) showed a separation factor in ethanol/water PV approximately 67-fold higher than the same PEM membrane prepared by the conventional dipping method.[76] The same approach was successfully applied to the preparation of PEM-based hollow fibers by using a negative pressure (−0.04 MPa).[77,118]

Membranes prepared upon alternating LbL deposition of polyvinylamine (PVA) and polyvinylsulfate (PVS) onto a UF membrane were studied for desalination under "RO conditions."[119,120] In the RO (and NF) process, the desalination efficiency is defined by the ion or salt rejection R^i, defined as:

$$R^i(\%) = \left(1 - \frac{C_p^i}{C_r^i}\right) \times 100, \tag{18.3}$$

where C_p^i and C_r^i denote the concentration in the permeate and in the retentate, respectively. The permeability was one order of magnitude lower (approximately 0.1 L m^{-2} h^{-1} bar^{-1}) than that usually obtained with commercially available thin-film composite RO membranes, probably because of the high number of deposited bilayers (60). In dilute solute solutions, rejection of divalent ions was almost quantitative and that of monovalent ions was approximately 90%. The salt rejection increases in the series $1^+,1^- < 1^+,2^- < 2^+,1^- \leq 2^+,2^-$. Na_2SO_4 was less rejected than $MgCl_2$, probably because of the strong affinity of Mg^{2+} for PVS. Indeed, these results indicate that the prepared membranes should be considered as tight NF rather than RO membranes since the ion rejection is selective and not quantitative. Interestingly, seawater filtration at 40 bar gave only a small decrease of the monovalent ion rejection (approximately 75%). Some interesting questions arise from this work. No pressure offset due to osmotic pressure can be seen in the flux measurements and the permeability shows only 20% decrease between pure water and seawater. A partial destructuring of the PEM film resulting in a loose structure should originate from the high salt content in seawater, which might explain the absence of pressure offset. Applying pressures higher than 20 bar should lead to a redensification of the PEM layer and hence a partial recovery of the monovalent rejection.

Furthermore, it was shown that replacement of one PE by charged macrocyclic compounds leads to an improvement of membrane performance with a permeate flux about three times higher than the previous $(PVA/PVS)_{60}$ membranes with very similar rejection rates for divalent and monovalent ions.[121]

Applications of Porous PEM Films Over the past few years, NF has been recognized as another potential application for PEM membranes owing the high charge density of these systems and the nanometric mesh of the PE network formed. Furthermore, the high hydrophilicity of these films ensures high permeation flux and low fouling properties for aqueous filtration.[122–125]

Contrary to RO, selectivity according to the solute size and charge can be achieved by NF due to the hindered transport mechanism and electrostatic interactions. For instance, multivalent ions are more rejected by a NF membrane than monovalent ions due to the Donnan exclusion from the pores. Therefore, the performance of NF membranes is evaluated in terms of permeate flux J and solute i rejection R^i (%). The selectivity $\alpha(i/j)$ of solute i over solute j can be expressed as:

$$\alpha(i/j) = \frac{100 - R^j}{100 - R^i}. \tag{18.4}$$

The main applications of NF can also be found for PEM-based membranes including water softening, partial desalination, and selective removal and separation of metal ions and charged species (Table 18.3).

As indicated above for PV applications, the dramatic reduction of the number of bilayers deposited on porous supports from several tens[126] to just a few ones[127,128] has been one of the major improvements for the preparation of membranes of economical interest. It was also demonstrated in the case of the PSS/PAH pair that the porous support should have an MWCO lower than 100,000 to obtain a full coverage of the surface with as low as 4.5 bilayers and to lead to efficient NF membranes.[128] The permeate flux was found to be dramatically increased when compared to thicker membranes[124,129] and was generally higher than conventional commercial NF membranes. NF is a technique that permits one to alleviate water hardness by rejecting divalent ions such as Ca^{2+}, Mg^{2+}, and SO_4^{2-}. As removal of monovalent ions is often not necessary, the selectivity in rejection of multivalent ions is an important criterion of performance for NF membranes. Selectivities as high as 27 and 22 were reported in the case of PEM membranes for SO_4^{2-}/Cl^-[130] and Ca^{2+}or Mg^{2+}/Na^+,[131] respectively. Based on the same principle, high selectivity was found in phosphate recovery[132,133] and Fe^{3+}/Fe^{2+} separation.[134]

Neutral and charged species can also be selectively separated using PEM membranes according to their size and charge.[135,136]

Rejection of halide anions by NF membranes can be strongly affected by their size and polarizability ($R^{F-} > R^{Cl-} > R^{Br-} > R^{I-}$) due to dielectric effects.[137] It was shown that a PEM membrane made of 4.5 PSS/PDADMA bilayers on porous alumina has higher F^-/Cl^- selectivity[138] and higher permeate flux than commercial membranes.[139]

As for PV applications, a dense packing is preferable to obtain high salt rejection. Therefore, the most studied polymers are PEs with high charge density, including PSS and SPEEK as polyanions and PAH, PDADMA, and b-PEI as polycations. The performance and selectivity can be tuned by changing the inter distance ξ between two ion pairings. For instance, PSS/PDADMA films give higher flux but lower selectivity than PSS/PAH films.[130,131] However, the capping of PSS/PDADMA films with a bilayer PSS/PAH enables one to reach a selectivity similar to pure PSS/PAH films with higher flux for Na^+/Mg^{2+} separation.[131] Solution ionic strength and pH during the deposition steps are also important parameters for optimizing the membrane performance. For instance, the surface charge of the PEM films can be enhanced by depositing the outermost layer (PSS) from a high ionic strength solution, hence increasing the rejection of multivalent ions.

It is known that a decrease of the solvent quality by increasing the weight fraction of ethanol in the PE deposition solution results in the formation of thicker and probably denser PEM films.[12,140] Thus, tight NF membranes were produced by LbL deposition of five bilayers of b-PEI dissolved in water and SPEEK dissolved in ethanol.[122] This PEM membranes exhibited 89% NaCl rejection from a 2 g L^{-1} saline solution. However, the good rejection performance came with a low permeability (approximately 0.9 L h^{-1} m^2 bar^{-1}).

Various compositions of NF PEM membranes, including copolymers, PECs, composite with inorganic colloids, alternated PE, and hydrophobic layers, were studied, offering alternative routes for nanoengineering NF membranes with improved performance.

Dynamic LbL deposition of copolymers containing both weakly and strongly ionized poly(4-styrenesulfonic acid-*co*-maleic acid) sodium salt (PSSMA), with PAH and PSS, was used to prepare membranes with tailored permeation properties by changing the ratio between the two comonomers and the deposition pH.[141] Membranes composed of only two bilayers prepared at pH 2.5 showed higher rejection and flux performances (Table 18.3).

NF membranes comprising a thin PEC top layer were prepared according to the procedure described in Figure 18.12, including a chemical cross-linking of the PE pair.[121] The obtained membranes showed a nearly quantitative rejection of xylenol orange (XO) and selective rejections of sulfate salts with α (K^+/Na^+) and α (K^+/Mg^{2+}) of 2.3 and 5.6, respectively. On the other hand, the anion selectivity (SO_4^{2-}/Cl^-) is only of 2.6.

In a study on the fouling and regeneration of NF PEM membranes, it was found that membranes fouled by colloidal SiO_2 nanoparticles showed higher $MgSO_4$ rejection.[125] This observation might open the way to a new design for PEM membranes.

Another interesting structure consists of the LbL assembly of PE with a cationic surfactant, cetyl trimethyl ammonium chloride (CTAC), creating alternative PE and lipid bilayers. The arrangement with CTAC as the outermost layer exhibited the higher Fe^{3+}/Fe^{2+}selectivity (4.9).[134]

Chemical grafting of functional groups onto PE chains has opened a new research area in designing stimuli responsible PEM films, enabling remote and reversible control for application in nanocapsules and membranes.[49] Photoresponsive ion gating membrane was successfully achieved by using the alternative deposition (spin coating) of an azobenzene-modified ammonium polyacrylamide and PSS on a nanoporous alumina support.[142] Activation by ultraviolet (UV) light of a PEM membrane made of 10 bilayers enhanced the SO_4^{2-} ion transport by 1.6 while the Cl^- ion transport is only increased by 1.2.

PEM-based membranes have been successfully applied to NFs of nonaqueous solutions, so-called solvent-resistant nanofiltration (SRNF).[143,144] SRNF requires the preparation of nanoporous membranes stable under conditions of organic solvent-caused swelling that is hardly achievable with the most common polymer architectures. PEM-based membranes are promising candidates for this novel application as the multilayer assembly is stabilized by strong and multiple ion pairing that can be regarded as hardly influenced by low dielectric constant solvents. Membranes were prepared by LbL PDADMA/SPEEK assembly deposited onto hydrolyzed PAN. Performances of the resulting membranes were tuned by varying the ionic strength of the PE solution. As expected, increasing the ionic strength induced the formation of looser PEMs, resulting in a decrease of solute rejection and an increase of permeate flux.

The transport of neutral and charged solutes was also investigated under dialysis conditions.[145,146] The mean pore size of the PEM membrane was tailored by varying the charge density of PEs in the following order: PDADMA/PSS > PAH/PSS > PVA/PVS, with pore diameters estimated from the permeate rates of neutral aromatic compounds in ethanol of 0.82, 0.67, and 0.54 nm, respectively.[147] These results can be considered consistent with the pore size of 0.8–1.0 nm determined by NF of neutral solutes in aqueous solutions for the PAH/PSS multilayer membrane,[100] the discrepancy probably coming from the difference in preparation conditions and in solvent permeation (water instead of ethanol).

The LbL buildup of PEM incorporating charged macrocyclic compounds including azacrown, calixarene, and cyclodextrin derivatives was proven to result in highly selective ion and neutral solute transport. These findings were accounted for by evoking specific interaction occurring between the solutes and the cyclic ring as those expected from a facilitated transport mechanism.[148]

Application in Fuel Cells and Battery Fuel cells that consume hydrogen (H_2/air FC) and methanol (DMFC) rather than crude oil to generate electricity are expected to become alternative power sources in the coming years for many applications. The key component in these systems is the polymer electrolyte membrane. Nafion® (Dupont, Wilmington, DE), so far the benchmark membrane owing to its excellent properties of conductivity and chemical resistance, has still some limitations in terms of cost and permselectivity for gas crossover so that extensive research efforts have been undertaken to improve its performances and to find alternative polymers. LbL assemblies might be good candidates since they possess a good conductivity and a high resistance to the transport of alcohols as mentioned above (see the section dealing with PV applications).[149]

As an illustration, composite membranes were created by depositing PEM films on both sides of MF track-etched polycarbonate supports: water fills the pores, enabling a high proton conductance to take place while the PEM films block the gas crossover.[150] The hydrogen-bonded assembly made up of 40 PEO/PAA bilayers gave the best performances compared to conventional PE LbL assemblies (Table 18.4). The resulting H_2/air FC, operating at 55% relative humidity, yielded a power density that is about five times less than Nafion 112 for a total thickness of 10 μm and a much lower cost. The difference in power performance of the studied PEM systems was explained by three internal structure factors, that is, hydrophilicity, weak ion pairing, and high segmental motion, which enhance the proton exchange in PEM films.

PE LbL-modified Nafion membranes were demonstrated to block the methanol crossover in DMFC while maintaining a high proton conductance. Thus, PDADMA/PSS film built up on Nafion 1135 entailed a significant decrease of the methanol permeability with the increase of the number of deposited layers.[151] Although the proton conductivity decreased at the same time, a 42% enhancement of the power performance occurred at four PDADMA/PSS bilayers. On the other hand, by using the PAH/PSS pair, both high proton conductivity and methanol blocking permeability were achieved (see Table 18.4).[152,153] The increase of proton conductivity was obtained by deposition in 0.01 M NaCl PE solution and a subsequent inversion of the PEM assembly into H^+ form. A higher number of bilayers led to mechanically stronger membranes.

PEM-modified sulfonated poly(arylene ether ketone)-bearing carboxyl groups (SPAEK-C) was proposed as an alternative support to Nafion.[154] The LbL deposition of positively charged polyaniline (PANI) and phosphotungstic acid (PWA) onto SPAEK-C gave membranes having a methanol permeability two orders of magnitude lower than Nafion 117 and with a higher proton conductivity at 80°C.

Application of PEM was also described for the design of batteries. For example, the crossover of vanadium ions through proton exchange membranes is the main reason for the low efficiency and high self-discharge of vanadium redox flow batteries (VRB). As indicated in the section dealing with NF, PEM films enable the rejection of multicharged ions. Composite membranes composed of five PDADMA/PSS bilayers deposited onto Nafion strongly decrease the vanadium ion transport (Table 18.3), yielding VRB having high coulombic efficiency and energy efficiency together with a slow discharge rate.[155]

Application as Porous Membrane Coatings In this section, the LbL coating serves to modify the interfacial properties and may provide additional functionality such as hemocompatibility, biocatalysis by immobilized enzyme, and ion exchange. PEM coating has been used for tuning membrane pore size by partially blocking large pores of MF and UF membranes (Table 18.5).[156,157] Membranes with selective transport of proteins[158] and good performance for rejection of industrial effluents[159] were then obtained. It was observed that these membranes had low protein binding properties.

Surface modification is a popular route for controlling the interactions between solutes and membranes, and consequently membrane fouling. Thus, it is generally accepted that fouling decreases with an increase in surface hydrophilicity and with the presence of repulsive charges. Many works dealt with the surface modification by PE adsorption and grafting to improve fouling resistance to nonspecific protein adsorption.[103] The nonfouling behavior of PEM construction has been reported many times over the past decade[104] and some examples have been mentioned before.[122–125] A recent example of PEM coating to prevent fouling of NF membranes is given in Table 18.5.[160]

Intriguingly, LbL buildup has been also recognized as an efficient system for immobilizing enzymes[161] either sandwiched between two PE layers[162,163] or as the outermost layer[164,165] to prepare biocatalytic membranes (Table 18.5). This approach, which was proven to be simple and versatile, provides stable and reusable enzymatic membrane systems. The apparent inconsistency between protein resistance behavior and immobilization of enzymes was recently investigated.[166] The adsorption of glucose oxidase (pI = 4.2) on PEI/(PAA/PAH)$_4$ film occurred under buffered conditions (pH 7.4 and ionic strength 0.15 mM) as expected from a negatively charged protein contacting a positively charged surface. Surprisingly, the adsorption was nearly zero under unbuffered conditions (pH 5.8 and ionic strength 0.10 mM). In contrast, glucose oxidase was readily absorbed when strong and hydrophobic PE PSS was used instead of weak PE PAA. On the other hand, lysozyme, a positively charged protein, did not absorb in both cases due to repulsive electrostatic interactions. The fouling-resistance property, observed when oppositely charged proteins and surfaces are contacted, is assumed to originate from the low surface charge and the large film hydration as occurs at low pH with weak PEM. This probably indicates that the entropy gain by release of water molecules and small ions is not favorable. In the other cases, the protein adsorption is promoted and PEM containing immobilized enzyme can be built up.

As seen before, active PEM coating of a membrane surface can be used for affinity-based separations and biocatalysis. LbL PE deposition within membrane pores also provides a simple way to prepare ion exchange membranes.[167] The binding capacity of negatively charged Au colloids was shown to increase with the number of bilayers. PEM coatings having an outermost b-PEI layer extracted more Au particles than those with PAH and PDADMA. In contrast, the binding of lysozyme on the outermost PSS layer is independent of the pair composition and increases with the ionic strength (results in agreement with Tristan et al.[166]).

Hydrophilicity and hemocompatibility of hemodialysis membrane materials generally composed of hydrophobic polymers need to be strongly improved to avoid severe membrane fouling and device clogging. PEM coatings using biopolymers such as polysaccharides have been successfully proven to meet these requirements. For instance, five CHI/DSS bilayers with DSS as the outermost layer exhibited resistance to the platelet adhesion and to the adsorption of human fibrinogen.[168,169] Incorporation of silver nanoparticles embedded into the CHI/DSS PEM led to additional antibacterial activity.[170]

18.4 CONCLUSION

This chapter reviews the fabrication of LbL PE assemblies and their potential applications in the domain of membrane filtration. The control over thickness and morphology at the nanoscale of the fabricated PEM enables the membrane surface to be engineered as very thin permselective films. Thus, advanced membrane materials can be designed with high selectivity and high flux. As presented above, many processes of significant importance—including dehydration of alcohols, desalination, power sources, and artificial organs—can be improved and novel functionalities can be introduced by using PEM-modified membranes.

One of the main limitations lies in the number of bilayers required to reach the desired performances because the technique is time consuming if one goes beyond the first layers. Dynamic LbL buildup is probably the most appropriate method in the case of membrane applications and it was shown that only a few layers are most often necessary to obtain improved performances. Spray coating, though still little used, may be an alternative method. It was also exemplified that the transport properties across the PEM films along with the interfacial properties are dramatically affected by the internal structure and morphology of LbL assemblies.

To summarize, the development of PEM-based membranes appears to be still in its infancy, as only a small number of PE pairs have been used so far and no in-depth investigations have been undertaken to elucidate the dependence of the deposition operating parameter on the membrane performance. Probably, future studies will be more concerned with the interplay between the transport properties and the morphology, which is governed by the PE nature and the deposition conditions.

ABBREVIATIONS AND SYMBOLS

General

DMF	Dimethylformamide
DMFC	Direct methanol fuel cell
FC	Fuel cell
LbL	Layer-by-Layer
PE	Polyelectrolyte
PEC	Polyelectrolyte complex
PEM	Polyelectrolyte multilayer
VRB	Vanadium redox flow batteries

Polymers

Aza6	1,4,7,10,13,16-hexaazacyclooctadeca
ALG	Alginate
CHI	Chitosan
CMCNa	Sodium carboxymethyl cellulose
CTAC	Cetyl trimethyl ammonium chloride
DSS	Dextran sodium sulfate
HA	Hyaluronic acid
PAA	Poly(acrylic acid)
PAH	Poly(allylamine hydrochloride)
PAMAM	Poly(amidoamine)
PAMPS	Poly(2-acryloylamido-2-methylpropanesulfonic acid)
PAN	Poly(acrylonitrile)
PANI	Poly(aniline)
PC	Polycarbonate
PDADMA	Poly(diallyl dimethyl ammonium chloride)
b-PEI	Branched poly(ethylene imine)
l-PEI	Linear poly(ethylene imine)
PEO	Poly(ethylene oxide)
PES	Poly(ether sulfone)
PET	Poly(ethylene terephtalate)
PGA	Poly(*L*-glutamic acid)
PLA	Poly(L-lactic acid)
PLL	Poly(L-lysine)
PSS	Poly(sodium styrene sulfonate)
PSSMA	Poly(4-styrenesulfonic acid-*co*-maleic acid)
PTAT	Poly(tetramethylene adipate-co-terephtalate)
PU	Polyurethane
PVS	Poly(vinylsulfate)
PVA	Poly(vinylamine)
PVDF	Poly(vinylidene difluoride)
PWA	Phosphotungstic acid
QCMC	Quaternary ammonium cellulose ether
RC	Regenerated cellulose
SA	Sodium alginate
SPAEK-C	Sulfonated poly(arylene ether ketone)-bearing carboxyl groups
sPEEK or SPEEK	Sulfonated poly (ether ether ketone)

Physical Values

A	Absorbance
C_i	Initial concentration of polyelectrolyte
C_p^i	Concentration of i in permeate
C_r^i	Concentration of i in retentate
D	Diffusion coefficient
H	Bilayer thickness
J	Flux of permeate
MWCO	Molecular weight cutoff
R^i	Ion or salt rejection
X_i	Weight fraction of solute i in the feed
Y_i	Weight fraction of solute i in the permeate
α	Separation factor
ξ	Average length between two cross-linking points
Ω	Spinning speed

Techniques

GS	Gas separation
MF	Microfiltration
NF	Nanofiltration
PV	Pervaporation
RO	Reverse osmosis
SRNF	Solvent-resistant nanofiltration
UF	Ultrafiltration

REFERENCES

1. X. Arys, A. M. Jonas, A. Laschewsky, R. Legras, F. Mallwitz. *Supramolecular Polymers*, 2nd ed. A. Ciferri, ed. CRC Press, Taylor & Francis Group, Boca Raton, FL, 2005; pp. 651–710.
2. P. T. Hammond. *Adv Mater* **2004**, 16, 1271.
3. M. A. Cohen Stuart, G. J. Fleer, J. Lyklema, W. Norde, J. M. H. M. Scheutjens. *Adv Colloid Interface Sci* **1991**, 34, 477.
4. G. J. Fleer, M. A. Cohen Stuart, J. M. H. M. Scheutjens, T. Cosgrove, B. Vincent. *Polymers at Interfaces*. Chapman and Hall, London, 1993.
5. P. Molyneux. *Water-Soluble Synthetic Polymers: Properties and Behavior*. CRC Press, Boca Raton, FL, 1983.
6. H. Elwing. *Biomaterials* **1999**, 20, 1621.
7. J. Ethève, P. Déjardin, M. Boissière. *Colloids Surf B* **2003**, 28, 285.

8. N. Thuaud, G. Lelièvre, A. Deratani, B. Sébille. *Eur Polym J* **1997**, 33, 1015.
9. B. Popping, A. Deratani, B. Sebille, N. Desbois, J.-M. Lamarche, A. Foissy. *Colloids Surf* **1992**, 64, 125.
10. B. Cabot, A. Deratani, A. Foissy. *Colloids Surf A* **1998**, 139, 287.
11. T. Roques-Carmes, F. Membrey, A. Deratani, M. Boehmer, A. Foissy. *J Colloid Interface Sci* **2002**, 256, 273.
12. S. T. Dubas and J. B. Schlenoff. *Macromolecules* **1999**, 32, 8153.
13. M. R. Bohmer, W. H. A. Heesterbeek, A. Deratani, E. Renard. *Colloids Surf A* **1995**, 99, 53.
14. M. R. Bohmer, O. A. Evers, J. M. H. M. Scheutjens. *Macromolecules* **1990**, 23, 2288.
15. P. T. Hammond. *Curr Opin Colloid Interface Sci* **2000**, 4, 430.
16. G. Decher and J. D. Hong. *Ber Bunsenges Phys Chem* **1991**, 46, 321.
17. G. Decher. *Science* **1997**, 277, 1232.
18. R. K. Iler. *J Colloid Interface Sci* **1966**, 21, 569.
19. H. Zhang, Z. Wang, Y. Zhang, X. Zhang. *Langmuir* **2004**, 20, 9366.
20. A. P. R. Johnston, H. Mitomo, E. S. Read, F. Caruso. *Langmuir* **2006**, 22, 3251.
21. Y. Shimazaki, M. Mitsuishi, S. Ito, M. Yamamoto. *Langmuir* **1997**, 13, 1385.
22. A. Van der Heyden, M. Wilczewski, P. Labbé, R. Auzély. *Chem Commun* **2006**, 30, 3220.
23. T. Kida, T. Minabe, S. Nakano, M. Akashi. *Langmuir* **2008**, 24, 9227.
24. G. K. Such, J. F. Quinn, A. Quinn, E. Tjipto, F. Caruso. *J Am Chem Soc* **2006**, 128, 9318.
25. H. Krass, G. Papastavrou, D. G. Kurth. *Chem Mater* **2003**, 15, 196.
26. S. A. Sukhishvili, E. Kharlampieva, V. Izumrudov. *Macromolecules* **2006**, 39, 8873.
27. C. S. Peyratout and L. Dähne. *Angew Chem Int Ed* **2004**, 43, 3762.
28. F. Caruso and H. Möhwald. *J Am Chem Soc* **1999**, 121, 6039.
29. G. Ladam, P. Schaaf, F. J. G. Cuisinier, G. Decher, J.-C. Voegel. *Langmuir* **2001**, 17, 878.
30. D. T. Haynie, L. Zhang, J. S. Rudra, W. Zhao, Y. Zhong, N. Palath. *Biomacromlecules* **2005**, 6, 2895.
31. D. T. Haynie. *J Biomed Mater Res B* **2006**, 78B, 243.
32. X. Qiu, S. Leporatti, E. Donath, H. Möhwald. *Langmuir* **2001**, 17, 5375.
33. F. J. Pavinatto, L. Caseli, O. N. Oliveira Jr. *Biomacromolecules* **2010**, 11, 1897.
34. H. Zou, S. Wu, J. Shen. *Chem Rev* **2008**, 108, 3893.
35. Z. Tang, Y. Wang, P. Podsiadlo, N. A. Kotov. *Adv Mater* **2006**, 18, 3203.
36. H. Kerdjoudj, C. Boura, V. Moby, K. Montagne, P. Schaaf, J.-C. Voegel, J.-F. Stoltz, P. Menu. *Adv Funct Mater* **2007**, 17, 2667.
37. A. Quinn, G. K. Such, J. F. Quinn, F. Caruso. *Adv Funct Mater* **2008**, 18, 17.
38. N. Benkarine-Jessel, P. Lavalle, E. Hubsch, V. Holl, B. Senger, Y. Haikel, J.-C. Voegel, J. Ogier, P. Schaaf. *Adv Funct Mater* **2005**, 15, 648.
39. A. Quinn, E. Tjipto, A. Yu, T. R. Gengenbach, F. Caruso. *Langmuir* **2007**, 23, 494.
40. J.-C. Voegel. *ACS Nano* **2009**, 3, 3343.
41. J. A. Lichter, K. J. Van Vliet, M. F. Rubner. *Macromolecules* **2009**, 42, 8573.

42. J. A. Jaber and J. B. Schlenoff. *Curr Opin Colloid Interface Sci* **2006**, 11, 324.
43. T. Boudou, T. Crouzier, K. Ren, G. Blin, C. Picart. *Adv Mater* **2010**, 22, 441.
44. A. Fery and R. Weinkamer. *Polymer* **2007**, 48, 7221.
45. A. P. R. Johnston, C. Cortez, A. S. Angelatos, F. Caruso. *Curr Opin Colloid Interface Sci* **2006**, 11, 203.
46. G. B. Sukhorukov and H. Möhwald. *Trends Biotechnol* **2007**, 25, 93.
47. S. F. M. van Dongen, H.-P. M. de Hoog, R. J. R. W. Peters, M. Nallani, R. J. M. Nolte, J. C. M. van Hest. *Chem Rev* **2009**, 109, 6212.
48. F. Liu and M. W. Urban. *Prog Polym Sci* **2010**, 35, 3.
49. I. Tokarev and S. Minkov. *Adv Mater* **2010**, 22, 3446.
50. J. F. Wong and W. Richtering. *Curr Opin Colloid Interface Sci* **2008**, 13, 403.
51. K. S. Schanze and A. H. Shelton. *Langmuir* **2009**, 25, 13698.
52. J. B. Schlenoff. *Langmuir* **2009**, 25, 14007.
53. D. Horn. *Polyamines and Polyquarternary Ammonium Salts*. E. J. Goethals, ed. Pergamon Press, Oxford, 1980; pp. 333–355.
54. R. H. Lajimi, A. B. Abdallah, E. Ferjani, M. S. Roudesli, A. Deratani. *Desalination* **2004**, 163, 193.
55. H. Lee, Y. Lee, A. R. Statz, J. Rho, T. G. Park, P. B. Messersmith. *Adv Mater* **2008**, 20, 1619.
56. M. C. Hsieh, R. J. Farris, T. J. McCarthy. *Macromolecules* **1997**, 30, 8453.
57. L. Krasemann and B. Tieke. *J Membr Sci* **1998**, 150, 23.
58. J. B. Schlenoff and S. T. Dubas. *Macromolecules* **2001**, 34, 592.
59. H. Hong, R. Steitz, S. Kirstein, D. Davidov. *Adv Mater* **1998**, 10, 1104.
60. J. Cho, K. Char, J.-D. Hong, K.-B. Lee. *Adv Mater* **2001**, 13, 1076.
61. P. A. Chiarelli, M. S. Johal, J. L. Casson, J. B. Roberts, J. M. Robinson, H.-L. Wang. *Adv Mater* **2001**, 13, 1167.
62. J. Cho, S.-H. Lee, H. Kang, K. Char, J. Koo, B. H. Seung, K.-B. Lee. *Polymer* **2003**, 44, 5455.
63. E. Kharlampieva, V. Kozlovskaya, J. Chan, J. F. Ankner, V. V. Tsukruk. *Langmuir* **2009**, 25, 14017.
64. B. S. Shim, P. Podsiadlo, D. G. Lilly, A. Agarwal, J. Lee, Z. Tang, S. Ho, P. Ingle, D. Paterson, W. Lu, N. A. Kotov. *Nano Lett* **2007**, 7, 3266.
65. J. B. Schlenoff, S. T. Dubas, T. Farhat. *Langmuir* **2000**, 16, 9968.
66. A. Izquierdo, S. S. Ono, J.-C. Voegel, P. Schaaf, G. Decher. *Langmuir* **2005**, 21, 7558.
67. C. H. Porcel, A. Izquierdo, V. Ball, G. Decher, J.-C. Voegel, P. Schaaf. *Langmuir* **2005**, 21, 800.
68. C. Porcel, P. Lavalle, V. Ball, G. Decher, B. Senger, J.-C. Voegel, P. Schaaf. *Langmuir* **2006**, 22, 4376.
69. M. Kolasinska, R. Krastev, T. Gutberlet, P. Warszynski. *Langmuir* **2009**, 25, 1224.
70. C. Porcel, P. Lavalle, G. Decher, B. Senger, J.-C. Voegel, P. Schaaf. *Langmuir* **2007**, 23, 1898.
71. K. C. Krogman, K. F. Lyon, P. T. Hammond. *J Phys Chem B* **2008**, 112, 14453.
72. K. C. Krogman, N. S. Zacharia, S. Schroeder, P. T. Hammond. *Langmuir* **2007**, 23, 3137.

73. K. C. Krogman, J. L. Lowery, N. S. Zacharia, G. C. Rutledge, P. T. Hammond. *Nat Mater* **2009**, 8, 512.
74. T. W. Graul and J. B. Schlenoff. *Anal Chem* **1999**, 71, 4007.
75. Y. Liu, J. C. Fanguy, J. M. Bledsoe, C. S. Henry. *Anal Chem* **2000**, 72, 5939.
76. G. Zhang, W. Gu, S. Ji, Z. Liu, Y. Peng, Z. Wang. *J Membr Sci* **2006**, 280, 727.
77. G. Zhang, X. Song, S. Ji, N. Wang, Z. Liu. *J Membr Sci* **2008**, 325, 109.
78. M. Salomäki, I. A. Vinokurov, J. Kankare. *Langmuir* **2005**, 21, 11232.
79. R. A. McAloney, M. Sinyor, V. Dudnik, M. C. Goh. *Langmuir* **2001**, 17, 6655.
80. P. Lavalle, C. Gergely, F. J. G. Cuisinier, G. Decher, P. Schaaf, J.-C. Voegle, C. Picart. *Macromolecules* **2002**, 35, 4458.
81. C. Picart, J. Mutterer, L. Richert, Y. Luo, G. D. Prestwich, P. Schaaf, J.-C. Voegel, P. Lavalle. *Proc Natl Acad Sci U S A* **2002**, 99, 12531.
82. L. Jourdainne, S. Lecuyer, Y. Arntz, C. Picart, P. Schaaf, B. Senger, J.-C. Voegel, P. Lavalle, T. Charitat. *Langmuir* **2008**, 24, 7842.
83. P. Nazaran, V. Bosio, W. Jaeger, D. F. Anghel, R. von Klitzing. *J Phys Chem B* **2007**, 111, 8572.
84. J. M. Garza, P. Schaaf, S. Muller, V. Ball, J.-F. Stoltz, J.-C. Voegel, P. Lavalle. *Langmuir* **2004**, 20, 7298.
85. N. S. Zacharia, D. M. DeLongchamp, M. Modestino, P. T. Hammond. *Macromolecules* **2007**, 40, 1598.
86. B. Sun, C. M. Jewell, N. J. Fredin, D. M. Lynn. *Langmuir* **2007**, 23, 8452.
87. R. von Klitzing. *Phys Chem Chem Phys* **2006**, 8, 5012.
88. S. T. Dubas and J. B. Schlenoff. *Langmuir* **2001**, 17, 7725.
89. H. W. Jomaa and J. B. Schlenoff. *Macromolecules* **2005**, 38, 8473.
90. T. Crouzier and C. Picart. *Biomacromolecules* **2009**, 10, 433.
91. M. Salomäke, P. Tervasmäki, S. Areva, J. Kankare. *Langmuir* **2004**, 20, 3679.
92. J. E. Wong, H. Zastrow, W. Jaeger, R. von Klitzing. *Langmuir* **2009**, 25, 14061.
93. J. Choi and M. F. Rubner. *Macromolecules* **2005**, 38, 116.
94. K. Büscher, K. Graf, H. Ahrens, C. A. Helm. *Langmuir* **2002**, 18, 3585.
95. H. L. Tan, M. J. McMurdo, G. Pan, P. G. Van Patten. *Langmuir* **2003**, 19, 9311.
96. F. van Ackern, L. Krasemann, B. Tieke. *Thin Solid Films* **1998**, 327-329, 762.
97. J. J. Harris, J. L. Stair, M. L. Bruening. *Chem Mater* **2000**, 12, 1941.
98. M. Mulder. *Basic Principles of Membrane Technology*. Kluwer Academic Publishers, Dordrecht, the Netherlands, 1996.
99. M. G. Davidson and W. M. Deen. *J Membr Sci* **1988**, 35, 167.
100. X. Liu and M. L. Bruening. *Chem Mater* **2004**, 16, 351.
101. Y. Wang, A. S. Angelatos, F. Caruso. *Chem Mater* **2008**, 20, 848.
102. Q. Li, J. F. Quinn, Y. Wang, F. Caruso. *Chem Mater* **2006**, 18, 5480.
103. D. Rana and T. Matsuura. *Chem Rev* **2010**, 110, 2448.
104. J. Meier-Haack, T. Rieser, W. Lenk, D. Lehmann, S. Berwald, S. Schwarz. *Chem Eng Technol* **2000**, 23, 114.
105. M. Schönhoff, V. Ball, A. R. Bausch, C. Dejugnat, N. Delorme, K. Glinel, R. von Klitzing, R. Steitz. *Colloids Surf A* **2007**, 303, 14.

106. P. Stroeve, V. Vasquez, M. A. N. Coelho, J. F. Rabolt. *Thin Solid Films* **1996**, 284-285, 708.
107. Z. Zhu, X. Feng, A. Penlidis. *Mater Sci Eng C* **2006**, 26, 1.
108. M. L. Bruening, D. M. Dotzauer, P. Jain, L. Ouyang, G. L. Baker. *Langmuir* **2008**, 24, 7663.
109. L. Krasemann, A. Toutianoush, B. Tieke. *J Membr Sci* **2001**, 181, 221.
110. P. D. Chapman, T. Oliveira, A. G. Livingston, K. Li. *J Membr Sci* **2008**, 318, 221.
111. Z. Zhu, X. Feng, A. Penlidis. *Mater Sci Eng C* **2007**, 27, 612.
112. Q. Zhao, J. Qian, Q. An, Z. Sun. *J Membr Sci* **2010**, 346, 335.
113. Q. Zhao, J.-W. Qian, Q.-F. An, Q. Yang, P. Zhang. *J Membr Sci* **2008**, 320, 8.
114. H. Jin, Q.-F. An, Q. Zhao, J.-W. Qian, M. Zhu. *J Membr Sci* **2010**, 347, 183.
115. M. Yin, J. Qian, Q. An, Q. Zhao, Z. Gui, J. Li. *J Membr Sci* **2010**, 358, 43.
116. P. Zhang, J. Qian, Q. An, X. Lui, Q. Zhao, H. Jin. *J Membr Sci* **2009**, 328, 141.
117. P. Zhang, J. Qian, Y. Yang, Q. An, X. Lui, Z. Gui. *J Membr Sci* **2008**, 320, 73.
118. G. Zhang, N. Wang, X. Song, S. Ji, Z. Liu. *J Membr Sci* **2009**, 338, 43.
119. W. Jin, A. Toutianoush, B. Tieke. *Langmuir* **2003**, 19, 2550.
120. A. Toutianoush, W. Jin, H. Deligöz, B. Tieke. *Appl Surf Sci* **2005**, 246, 2550.
121. K. Hoffmann and B. Tieke. *J Membr Sci* **2009**, 341, 287.
122. J. Wang, Y. Yao, Z. Yue, J. Economy. *J Membr Sci* **2009**, 337, 200.
123. Y. Ji, Q. An, Q. Zhao, H. Chen, J. Qian, C. Gao. *J Membr Sci* **2010**, 357, 80.
124. R. H. Lajimi, E. Ferjani, M. S. Roudesli, A. Deratani. *Desalination* **2011**, 266, 78.
125. W. Shan, P. Bacchin, P. Aimar, M. L. Bruening, V. V. Tarabara. *J Membr Sci* **2010**, 349, 268.
126. L. Krasemann and B. Tieke. *Langmuir* **2000**, 16, 261.
127. M. D. Miller and M. L. Bruening. *Langmuir* **2004**, 20, 11545.
128. R. Malaisamy and M. L. Bruening. *Langmuir* **2005**, 21, 10587.
129. A. El-Hashani, A. Toutianoush, B. Tieke. *J Membr Sci* **2008**, 318, 65.
130. S. U. Hong, R. Malaisamy, M. L. Bruening. *J Membr Sci* **2006**, 283, 366.
131. L. Ouyang, R. Malaisamy, M. L. Bruening. *J Membr Sci* **2008**, 310, 76.
132. P. Ahmadiannamini, X. Li, W. Goyens, B. Meesschaert, I. F. J. Vankelecom. *J Membr Sci* **2010**, 360, 250.
133. S. U. Hong, L. Ouyang, M. L. Bruening. *J Membr Sci* **2009**, 327, 2.
134. H. Fu and T. Kobayashi. *Desalination* **2010**, 264, 115.
135. S. U. Hong and M. L. Bruening. *J Membr Sci* **2006**, 280, 1.
136. S. U. Hong, M. D. Miller, M. L. Bruening. *Ind Eng Chem Res* **2006**, 45, 6284.
137. P. Jungwirth and D. J. Tobias. *Chem Rev* **2006**, 106, 1259.
138. A. Lhassani, M. Rumeau, D. Benjelloun, M. Pontie. *Water Res* **2001**, 35, 3260.
139. S. U. Hong, R. Malaisamy, M. L. Bruening. *Langmuir* **2007**, 23, 1716.
140. E. Poptoshev, B. Schoeler, F. Caruso. *Langmuir* **2004**, 20, 829.
141. H.-Y. Deng, Y.-Y. Xu, B.-K. Zhu, X.-Z. Wei, F. Liu, Z.-Y. Cui. *J Membr Sci* **2008**, 323, 125.
142. S. K. Kumar and J.-D. Hong. *Langmuir* **2008**, 24, 4190.

143. X. Li, S. De Feyter, D. Chen, S. Aldea, P. Vandezande, P. Du Prez, I. F. J. Vankelecom. *Chem Mater* **2008**, 20, 3876.
144. X. Li, W. Goyens, P. Ahmadiannamini, W. Vanderlinden, S. De Feyter, I. Vankelecom. *J Membr Sci* **2010**, 358, 150.
145. W. Jin, A. Toutianoush, B. Tieke. *Appl Surf Sci* **2005**, 246, 444.
146. A. Toutianoush, J. Schnepf, A. El Hashani, B. Tieke. *Adv Funct Mater* **2005**, 15, 700.
147. B. Tieke, A. Toutianoush, W. Jin. *Adv Colloid Interface Sci* **2005**, 116, 121.
148. B. Tieke, A. El Hashani, A. Toutianoush, A. Fendt. *Thin Solid Films* **2008**, 516, 8814.
149. J. L. Lutkenhaus and P. T. Hammond. *Soft Matter* **2007**, 3, 804.
150. T. R. Farhat and P. T. Hammond. *Adv Funct Mater* **2005**, 15, 945.
151. S. P. Jiang, Z. Liu, Z. Q. Tian. *Adv Mater* **2006**, 18, 1068.
152. S. Yilmaztürk, H. Deligöz, M. Yilmazoglu, H. Damyan, F. Öksüzömer, S. Naci Koç, A. Durmus, M. Ali Gürkaynak. *J Membr Sci* **2009**, 343, 137.
153. H. Deligöz, S. Yilmaztürk, M. Yilmazoglu, H. Damyan. *J Membr Sci* **2010**, 351, 131.
154. C. Zhao, H. Lin, Q. Zhang, H. Na. *Int J Hydrogen Energy* **2010**, 35, 10482.
155. J. Xi, Z. Wu, X. Teng, Y. Zhao, L. Chen, X. Qiu. *J Mater Chem* **2008**, 18, 1232.
156. J. Mathew, C. T. Aravindakumar, U. K. Aravind. *J Membr Sci* **2008**, 325, 625.
157. J. Kochan, T. Wintgens, J. E. Wong, T. Melin. *Desalination* **2010**, 250, 1008.
158. U. K. Aravind, J. Mathew, C. T. Aravindakumar. *J Membr Sci* **2007**, 299, 146.
159. U. K. Aravind, B. George, M. S. Baburaj, S. Thomas, A. P. Thomas, C. T. Aravindakumar. *Desalination* **2010**, 252, 27.
160. C. Ba, D. A. Ladner, J. Economy. *J Membr Sci* **2010**, 347, 250.
161. B. Krajewska. *J Mol Catal , B Enzym* **2009**, 59, 22.
162. A. Yu, Z. Liang, F. Caruso. *Chem Mater* **2005**, 17, 171.
163. S. Guedidi, Y. Yurekli, A. Deratani, P. Déjardin, C. Innocent, S. Alsoy Altinkaya, S. Roudesli, A. Yemenicioglu. *J Membr Sci* **2010**, 365, 59.
164. V. Smuleac, D. A. Butterfield, D. Bhattacharyya. *Langmuir* **2006**, 22, 10118.
165. S. Datta, C. Cecil, D. Bhattacharyya. *Ind Eng Chem Res* **2008**, 47, 4586.
166. F. Tristan, G. Palestino, J.-L. Menchaca, E. Perez, H. Atmani, F. Cuisinier, G. Ladam. *Biomacromolecules* **2009**, 10, 2275.
167. G. Liu, D. M. Dotzauer, M. L. Bruening. *J Membr Sci* **2010**, 354, 198.
168. D.-G. Yu, C.-H. Jou, W.-C. Lin, M.-C. Yang. *Colloids Surf B* **2007**, 54, 222.
169. D.-G. Yu, W.-C. Lin, C.-H. Lin, Y.-H. Yeh, M.-C. Yang. *J Biomed Mater Res B Appl Biomater* **2007**, 83B, 105.
170. D.-G. Yu, W.-C. Lin, M.-C. Yang. *Bioconjug Chem* **2007**, 18, 1521.
171. N. Wang, G. Zhang, S. Ji, Z. Qin, Z. Liu. *J Membr Sci* **2010**, 354, 14.
172. G. Zhang, H. Yan, S. Ji, Z. Liu. *J Membr Sci* **2007**, 292, 1.
173. Y. Chen, F. Xiangli, W. Jin, N. Xu. *J Membr Sci* **2007**, 302, 78.
174. G. Zhang, X. Song, S. Ji, N. Wang, Z. Liu. *J Membr Sci* **2008**, 325, 109.
175. S. Ji, G. Zhang, Z. Liu, Y. Peng, Z. Wang. *Desalination* **2008**, 234, 300.
176. W. Ritcharoen, P. Supaphol, P. Pavasant. *Eur Polym J* **2008**, 44, 3963.

CHAPTER 19

SELF-ASSEMBLY OF POLYELECTROLYTES FOR PHOTONIC CRYSTAL APPLICATIONS

DARIO CAVALLO and DAVIDE COMORETTO

The use of polymers in microelectronics and photonics is continuously developing along two main paths: the preparation of polymer structures spanning from nano- to mesoscale, and the improvement of semiconducting/conducting functions into the macromolecule, which makes their electronic and optical properties suitable to specific goals (charge transport or generation, light emission, optical nonlinearities, etc.).[1–4] In order to proceed with these topics, the exploitation of bottom-up self-organization properties of polymers, which work through quasi-equilibrium steps, enabling the self-repulsion of defects, is a powerful approach. Indeed, suitable structures ranging from the nano- to the mesoscale might provide the required property or functionality from either the optical or the electronic point of view.[2,5] Moreover, the use of self-organizing processes is often a need for soft matter since the usual techniques used in the top-down approach, such as UV or X-ray lithography, could be expensive, energy consuming, and, in particular, potentially destructive for organic materials.

The length scales of ordered structures obtained by the self-organization process separates their possible application into either optoelectronics or to photonic. Polymer structures of few nanometers might have two different functions: to be a template or to impart a specific electronic/optical role. Block copolymers (BCPs) have a phase behavior of considerable scientific and technological interest due to the formation of different ordered structures.[6] When the composing blocks do not have electronic functionalities, that is, are insulating polymers, the nanostructures can be used as an efficient and low-cost mask for the semiconductor industry. Indeed, by using suitable molecular

Ionic Interactions in Natural and Synthetic Macromolecules, First Edition. Edited by Alberto Ciferri and Angelo Perico.

masses, critical dimensions below current lithographic resolution limits can be achieved.[2,7]

When the polymer or the copolymer is semiconducting, aggregation phenomena between chains or blocks can improve the charge carrier transport, which is usually limited by disorder.[8] For instance, regio-regular poly(alkylthiophenes) are known to give rise to a lamellar structure. Within each lamella, the backbone is extended and the twisting angle between the thiophene rings is close to zero, thus allowing an efficient interaction of π-electrons with those of the chains in the adjacent lamella, which is only 3 Å apart. In this way, charge transport is very efficient and this material is the prototype of conjugated polymers for field effect transistors.[1,4,9] Conjugated BCPs possessing both electron donating and withdrawing properties upon photoexcitation are another example of macromolecules, which might induce a nanostructure due to block phase segregation, thus improving the performances of devices. Recently, the Hadziioannou group reported several papers on the use of donor–acceptor BCPs, which are able to improve light harvesting, photoinduced charge transfer, and charge transport in organic photovoltaic cells.[10,11] Aggregation phenomena among ordered assemblies of molecules might also give rise to J- or H-aggregates, a sort of supramolecular polymer wherein the molecular dipole transition moments are suitably aligned.[1,12] In this way, collective optical excitations might occur, giving rise to enhanced nonlinear optical phenomena.[12,13] In order to provide the correct assembly of molecules, polyelectrolyte templates can be used.[12,14] Details on the layer-by-layer (LbL) self-assembly and multilayer modeling have been provided in Chapters 5 and 18 of this book.

When changing the length scale of the structures to the mesoscopic scales, that is, a length scale ranging from 100 nm to 1 μm, two polymer systems assume a relevant importance: colloids and, again, BCPs having very large molecular mass. In particular, in this chapter we will deal with highly ordered periodic structures where materials possessing different refractive indices are assembled in a periodic structure extending over large areas in one (1D), two (2D), or three dimensions (3D). We call these periodic dielectric lattices photonic crystals (PhCs).

In the following sections, we will introduce the definition and main properties of PhC and then we will review the use of polyelectrolytes for producing both colloidal (3D) and BCP PhC (mainly 1D). Additional information related to items discussed in this chapter can be found in Chapters 16, 17, and 22.

19.1 PHOTONIC CRYSTAL PROPERTIES

PhCs are composite materials made of media possessing different dielectric constants (or refractive indices) and arranged in a well-ordered periodical lattice (1D, 2D, or 3D) having a periodicity comparable to the wavelength of visible light (Fig. 19.1).

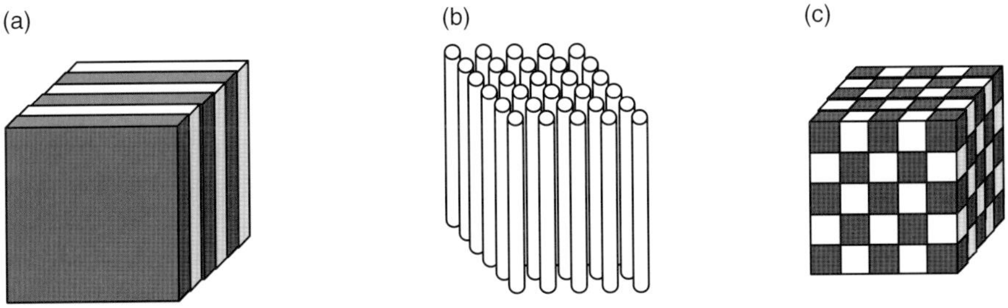

FIGURE 19.1. Scheme of 1D (a), 2D (b), or 3D (c) PhCs.

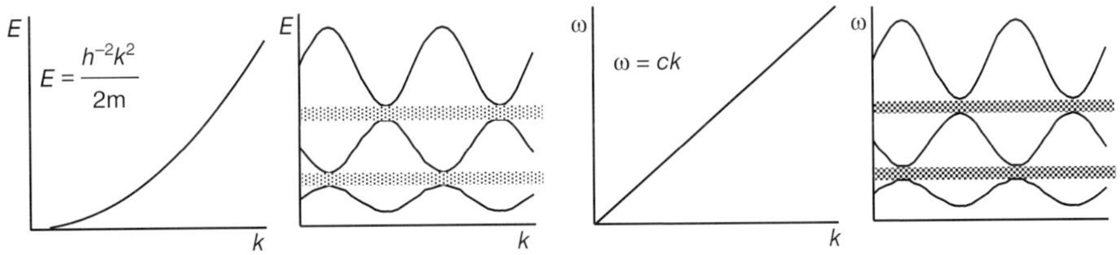

FIGURE 19.2. Comparison of dispersion properties of electron and photons in the free space and inside a crystal (left) or a PhC (right).

PhCs were discovered about 20 years ago by Yablonovitch[15] and John[16] with two seminal papers tackling fundamental issues in physics such as the inhibition of spontaneous emission and light localization. Nowadays, PhCs find application in different fields spanning from photonics to photovoltaics and are gaining a great interest for sensing. The fabrication of novel PhC structures with unconventional methods, as well as their chemical functionalization in order to tailor their responsive properties, is currently of great interest.[2]

Typical dimensions of composing media in a PhC are much larger than the Bohr's radius (0.5 Å), therefore we can use in their description the concept of dielectric constant (or refractive index), which is a macroscopic property. A simple comparison with standard crystals allows one to better highlight the properties of PhC, which can be considered dielectric lattices. A free electron possesses a well-known parabolic dispersion relation. When this electron is inserted inside a crystal, it feels a periodic potential and it is scattered at the edges of the Brillouin zone. Back and forth scattered electrons interfere with each other, giving rise to allowed and forbidden bands separated by an energy gap (Fig. 19.2).[17]

An analogous situation occurs in PhC where the interaction between free photons and the dielectric lattice modifies their linear dispersion law, creating allowed and forbidden bands for photon propagation. These two bands are separated by a photonic band gap (PBG): there, photons cannot propagate inside the PhC (Fig. 19.2). Again, the formation of the PBG can be viewed as

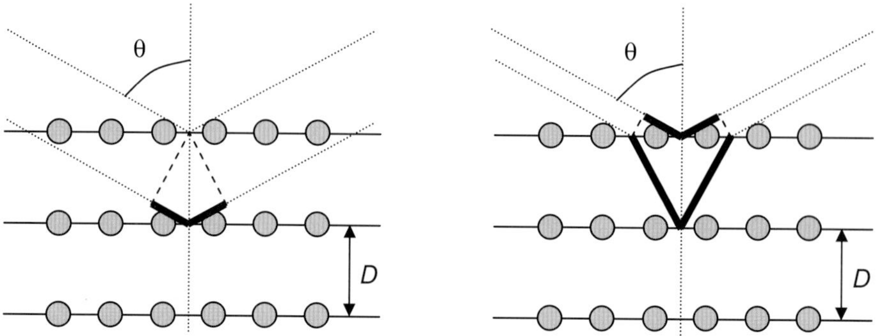

FIGURE 19.3. Diffraction scheme in a crystal and in a PhC.

a complicated combination of diffraction and interference phenomena occurring within the dielectric lattice.

The PBG in a PhC has several important consequences. Among them, two are particularly interesting. First of all, photons possessing energy within the band gap cannot propagate through the PhC, thus being backward diffracted, giving rise to strong chromatic effects and iridescence. Indeed, the fingerprint of the PBG is a typical peak in the reflectance spectrum of the sample as well as one deep in the transmission one occurring at the same energy. Moreover, their spectral position is strongly dependent on sample orientation since the photonic band properties depend on the wave vector (the momentum of the photon), that is, on the incidence angle of the light with respect to the normal to the surface. In particular, the reflectance peak (transmission minimum) is expected to shift toward higher photon energies upon increasing the incidence angle.

More meaningful in showing the strength of the PBG concept is the fate of a fluorescent molecule embedded in the PhC possessing an emission spectrum tuned within the band gap. In the free space, an excited molecule undergoes spontaneous radiative emission (fluorescence). However, being inside the PhC, it cannot radiatively relax to the ground state since the spontaneous emission is inhibited due to the matching of the emission spectrum with the photonic gap. As a consequence, the radiative lifetime increases and the molecule survives in the excited states. This is the so-called Purcell's effect, which can be exploited both for the production of innovative low threshold lasers or for photoluminescence (PL) directional enhancement, a phenomenon which can be relevant for sensing.[18–21]

Let us now work out a simple diffraction model to discuss the main properties of PhCs. Again, we compare the case of a standard crystal with that of a PhC. The diffraction of X-ray in crystal is described by the well-known Bragg law (Fig. 19.3):

$$m\lambda = 2D\cos(\theta) = 2D\sqrt{1-\sin^2(\theta)}, \qquad (19.1)$$

where θ is the incidence angle (measured from the normal to the crystal surface) and D is the interplanar spacing. In a PhC the same picture holds on a different scale since the interplanar spacing, due to the dielectric lattice, is of the order of hundredths of nanometers. Then, diffraction occurs in the visible–near infrared (IR) spectral region. This fact has an important consequence on photon propagation at the interfaces between the two media composing the PhC. Indeed, light is also refracted as a consequence of the Snell's law (Fig. 19.3). By combining the Bragg and the Snell laws we obtain the Bragg–Snell law:

$$m\lambda = 2D\sqrt{n_{eff}^2 - \sin^2(\theta)}, \tag{19.2}$$

which describes the main spectroscopic effects related to PhCs. The effective refractive index n_{eff} (or effective dielectric constant $n_{eff}^2 = \varepsilon_{eff}$) plays a major role. The effective refractive index is an average index obtained from the refractive indices of composing media (a and b) weighted over the correspondent volume fractions (f_a and f_b):

$$n_{eff}^2 = \varepsilon_{eff} = f_a\varepsilon_a + f_b\varepsilon_b. \tag{19.3}$$

More sophisticated models for the determination of n_{eff} can also be used.[22,23] By comparing Equation 19.2 with Equation 19.1, we notice that the refractive index of a standard crystal is assumed to be 1, like the void, since at those wavelengths (few Ångströms) the crystal is described as an ordered array of small atoms dispersed into a wide void. Improvements to this picture for crystals have been proposed (from dynamical X-ray scattering measurements[24]) but the correction for the refractive index is almost negligible if compared to that occurring at optical wavelengths.

It is worth adding some further comments about the role of the refractive index contrast of composing media on the PhC properties. The full width half maximum (ΔE) of the reflectance peak at the PBG ($E = hc/\lambda$, where λ is provided in Eq. 19.2) is a function of the dielectric contrast: $n_a - n_b$:

$$\Delta E = \frac{4}{\pi} E \frac{|n_a - n_b|}{n_a + n_b}. \tag{19.4}$$

By assuming that the disorder does not play a major role in the PhC, an increase of the dielectric contrast gives rise to a wider PBG.[22,23]

We would like to stress at this point that the properties of PhCs only depend on their structure and on the dielectric contrast between the composing materials. Often, one of them is the air, due to the cavities available inside the ordered nanostructured medium. Filling these cavities with suitable materials, characterized by particularly high dielectric constant, photophysical and nonlinear optical properties, or chemical responsivity opens intriguing perspectives.[25–27] The discussion of properties and applications of all kinds of

PhCs is beyond the aim of this chapter, which will be limited to the opportunities offered by polymer materials and polyelectrolytes.

19.2 PhC PREPARATION AND APPLICATIONS

Two methods are commonly used to obtain PhCs: a "top-down" approach, which consists of obtaining structures with the required dimensions by lithographic or etching techniques applied to bulk materials (usually inorganic semiconductors and insulators), and a "bottom-up" one. The latter is based on the preparation of fundamental units which self-assemble into the designed structure. For instance, the fabrication of 2D PhCs (selective etching process on a mask) is a standard top-down method, while the creation of 3D PhCs is more complex, requiring expensive lithographic techniques using electron beams or UV photons, and cannot be easily applied to materials different from the ones commonly used in microelectronics. To overcome these difficulties, alternative techniques have to be used and it is evident that supramolecular chemistry plays a fundamental role here.[1] At present, techniques relying on supramolecular chemistry used in the preparation of organic PhCs are photopolymerization induced by two-photon absorption,[28] BCP assemblies,[29,30] liquid crystals,[31,32] self-assembly of building blocks like moist air bubbles,[33] or nanospheres in the case of opals.[34–41] The last process seems particularly interesting since it is based on the spontaneous growth of stable, well-defined structures, starting from elements like molecules or mesophases that interact through noncovalent bonds.

It is not the purpose of this chapter to review all methods used to prepare organic PhCs. We will thus restrict to methods of fabrication and properties of organic PhCs where electrostatic interactions between complementary charged polyelectrolytes are adopted to impart suitable properties to colloids and BCPs.

19.2.1 Polyelectrolytes for Colloidal PhCs

A widely investigated colloidal PhC is the opal, which is formed by an ordered face-centered cubic lattice of monodisperse silica or polymer microspheres. Opals are a very simple and cheap playground in which to investigate optical effects of PhC.[15,16] Despite the fact that opals do not show a complete PBG but only a pseudogap (stop band, i.e., a gap only along one crystallographic direction), their versatility allows researchers to obtain interesting structures. Artificial opals are grown by different methods based on the spontaneous or forced assembly of microspheres into stable and well-defined structures.[5,34,36,38,39,42–45] From the PhC point of view, the main drawback of opals is their structure, which, as mentioned above, does not allow the formation of a complete PBG, that is, a gap for photons having suitable energy for all polarizations and directions. At variance with opals, inverse opals with the proper dielectric contrast allow the opening of a complete PBG.[45,46] Inverse opals are obtained by infiltrating the opal template with different functional materials and subsequent sphere removal.

In order to control or modulate opal properties, we must remember that the main features of PhCs depend on their structure and dielectric contrast.[46] Being that the opal's structure is well defined (face-centered cubic) with a unit cell length ($L = a\surd 2 = D\surd 3$) depending only on sphere diameter (a), the possibilities of tuning their properties are provided either by the modulation of refractive index of composing materials or by the use of spheres having different diameter. This latter approach, even though useful to prepare opals suitable to work in selected spectral regions, is not compatible for operating photonic devices.[47] For these reasons, the problem of optical response modulation is approached by refractive index tuning thorough photoactive materials.[21]

Several different methods based on self-assembly have been used for opal preparation, including gravity sedimentation, flow-through channels or pores, meniscus and vertical deposition, shear-induced crystallization, convective assembly, vortical methods, spin coating, and electrophoretic deposition.[5,37,38,42,48–63] Electrophoretic deposition exploits the presence of charged groups on the surface of the microspheric colloid. The application of an electric field in a suspension with suitable conditions allows migration of microspheres to oppositely charged electrodes where the deposition of high-quality opals occurs.[58] The same technique can be applied to an infiltrating (polyelectrolytic) material by using the opal as a template.[63] In this case, the infiltration process of the polyelectrolyte is favored and, after template removal, a well-ordered inverse opal structure can be obtained. Depending on the nature of the polyelectrolyte it is possible to obtain polymer, metal, or nanocrystal (NC) inverse opals. As previously described, inverse opals are usually obtained by using an opal as a template and then by infiltrating it with different materials[37]; however, alternative methods can be used. For instance, it is well known that breathing figures rassembling the inverse opal structure can be created in a polymer film during solvent evaporation in a wet environment. In this case, water microdroplets dispersed in the air float on the surface of the film or sink into the solution and then leave their shape during solvent evaporation and consequent film formation. A detailed discussion of this process and others that can be brought to similar structures can be found in References 1, 30, and 33, and the papers cited therein. In spite of the simplicity of this method to obtain inverse opal structures, some drawbacks affect this approach. First of all, the order of the structure is relatively poor for photonic applications since the periodicity can be achieved only on a scale of few tens of microns and the polydispersity of the cavity radius is too wide. Moreover, it is very difficult to reduce the cavity diameter below a few microns, thus making it impossible to use these structures in the visible–near IR spectral range. For these reasons, different approaches have been investigated. We briefly review here those related to the use of polyelectrolytes in order to obtain 3D PhC. In particular,

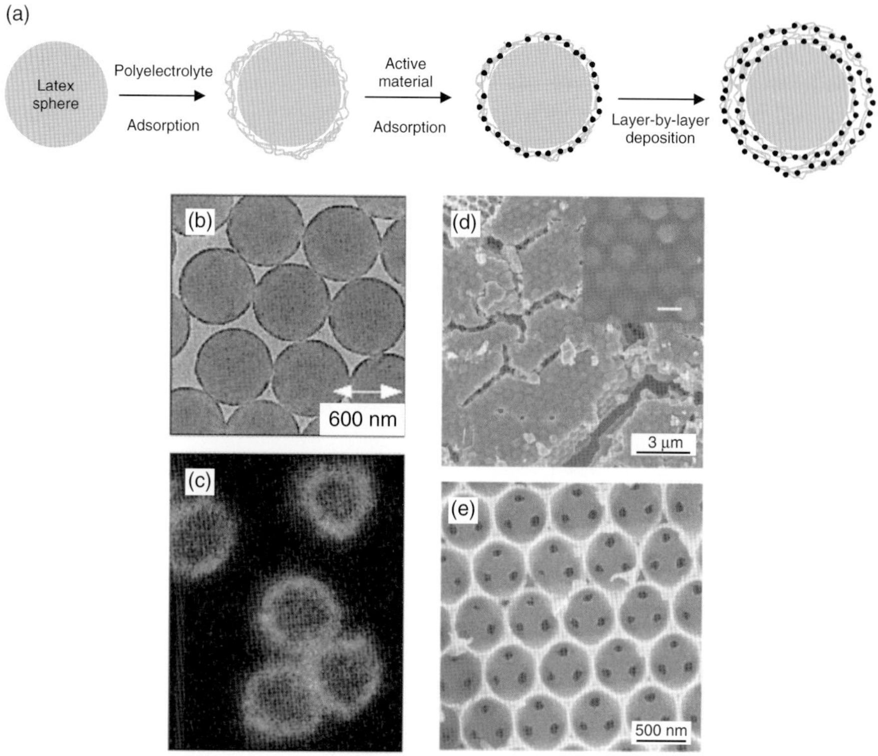

FIGURE 19.4. (a) Scheme of the LbL deposition on latex microsphere substrates. Images of polystyrene microspheres coated by the LbL process with poly(allyl-amine hydrochloride)/CdTe(S). NC as probed by TEM (b) and confocal fluorescence microscopy (c). SEM micrographs of TiO_2/HgTe inverse opals before (d) and after (e) polystyrene sphere template calcination. Adapted with permission from Rogach et al.[64] and Wang et al.[65]

polyelectrolytes are used for coating microspheres, thus improving the possibility of functionalizing the PhC structure.

Figure 19.4a shows the typical function of polyelectrolytes applied to colloids. The polyelectrolyte is adsorbed on the surface of the latex microsphere and then suitable semiconductor NCs or conjugated polymers can be coated on it. The process can be repeated several times[58,64–67] as in standard LbL deposition as described in Chapter 18 of this book. In the case of semiconducting NCs, the quality of the sphere coating can be probed not only by transmission electron microscopy (TEM) imaging (Fig. 19.4b), but also with confocal fluorescence microscopy (Fig. 19.4c), due to the high fluorescence quantum yield of CdTe(S) NC.[64] The apparent different thickness of the NC coating is due to the lower spatial resolution of confocal microscopy with respect to TEM. The coating quality is guaranteed by the chemical functionalization of the NC, which are surface capped with 1-mercapto-2,3-propandiol and 1,2-dimercapto-3-propanol, thus ensuring incorporation in the polyelectrolyte (poly(allyl-amine hydrochloride)) by the strong electrostatic interaction between the negatively charged –OH and –SH groups on the NC surface and the positively charged $-NH_3^+$ groups of the polymer[64] (general principles concerning these interactions can be found in Chapters 1 and 8 of this book). Coated microspheres can then be used to prepare opals by sedimentation,[64] crystallization,[65] or electrophoretic deposition.[58]

An example of such an opal is shown in Figure 19.4d. The use of precoated spheres allows researchers to slightly modify the diameter of the sphere as well as its refractive index, thus finely tuning the spectral position of the photonic stop band even though sometimes cracks and other defects can be observed (Fig. 19.4d). Once the opal is formed, calcination in a nitrogen atmosphere allows one to remove the latex sphere template by leaving the inverse opal structure made of NCs (Fig. 19.4e).

It is worthwhile to notice that such relatively easy techniques for the preparation of high quality PhCs allows one to separately engineer the electron and photon density of states within the same mesoscopic structure, which can then be used to investigate PL enhancement and lasing effects. Indeed, the electronic structure is selected by choosing the suitable NC, according to its composition and structure. On the other hand, the photonic structure is spectrally tuned by growing opals composed of microspheres whose diameter and refractive index can be engineered by the LbL coating process (see Chapter 17 of this book).

The versatility of the LbL deposition method applied to spherical colloids can be applied not only to inorganic systems but also to organic semiconductors. For instance, silica spheres have been coated with a water-soluble polyanionic derivative of poly(phenylene ethynylene), a well-known conjugated polymer possessing very high fluorescence quantum yield, and a polycationic polymer (poly(dimethyldiallylammonium chloride)) (Fig. 19.5).[66] The advantage of using the LbL technique instead of the standard procedure consisting of the infiltration with solutions after opal deposition is the possibility to avoid the disruption of the opal during the infiltration process as often observed.[21,43] Moreover, by controlling the number of LbL depositions, it is possible to tune the sphere diameter as well as the refractive index. Figure 19.5a shows the absorption spectrum of coated spheres by increasing the number of bilayers. The absorption is dominated by the spectral features of the conjugated polymers in the aggregated form at about 440 nm.[66] It is interesting to notice that the absorption maximum linearly increases upon increasing the number of bilayer coatings[66] (Fig. 19.5a, inset), thus demonstrating the high precision of the LbL method (see Chapter 17). By increasing the bilayer number, a red shift of the absorption is also observed, indicating a more efficient aggregation effect.[68,69] In Figure 19.5b the effect of the number of coated bilayers on the PL properties is shown. Two emission species have been identified: the singlet exciton of the isolated molecule at about 460 nm and the excimer emission at

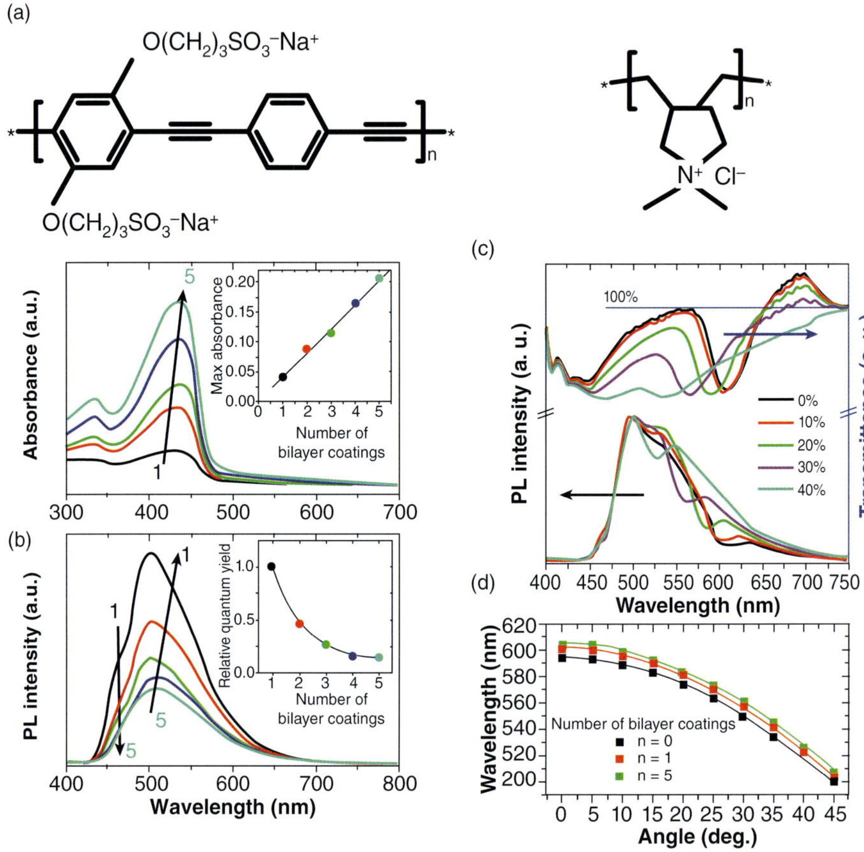

FIGURE 19.5. Chemical structure of the polyelectrolytes (sulfonated poly(phenylene ethynylene) and poly(dimethyldiallylammonium chloride)) used for LbL sphere coating. (a) Absorption and (b) PL spectra of colloidal suspension in ethanol with increasing bilayers concentration. (c) Incidence angle-dependent transmission and photoluminescence of an opal made with spheres covered by five bilayers. (d) Bragg–Snell fit of the photonic stop band as a function of the incidence angle for different bilayer coating opals. All spectra are adapted with permission from Kim et al.[66]

about 500 nm. Upon increasing the number of layers, isolated molecule emission decreases, while the aggregate one increases. Moreover, the PL lifetime is reduced upon increasing the coating thickness.[66] This scenario is consistent with the increased role of aggregation upon increasing the number of bilayers, which is known to reduce the emission intensity.[69] A stronger support of this interpretation comes by analyzing the relative intensity of the singlet to the aggregate emission as shown in the inset of the same figure. The singlet emission intensity decreases much faster than the aggregated one upon increasing the number of bilayers.

In spite of the interest into the photophysics of poly(phenylene ethynylene) film upon LbL deposition, it is worthwhile to notice the effects of coated spheres on the optical properties of opals as reported in Figure 19.5c. A minimum in the transmission spectra is clearly observed at about 600 nm as the signature of the photonic stop band. This is further confirmed by the dispersion of this feature upon increasing the incidence angle toward higher energies as predicted by Equation 19.2. Since these opals have been engineered in order to overlap the photonic stop band with the emission spectrum of the conjugated polymer, a strong modification of its shape is observed. As a matter of fact, by comparing the PL spectra of opals (Fig. 19.5c) with those of dispersed coated spheres (Fig. 19.5b), minima on the PL spectra of the polymer are observed.[66] This is due to the presence of the stop band, which does not allow the emission of light from the polymer to come out from the PhC. From the physical point of view, at the photonic stop band frequencies, the density of photonic states is reduced, thus reducing the spontaneous emission rate and then the fluorescence intensity.[19,21] Photonic states removed from the stop band are squeezed out at the edges, thus generating at that wavelength a directional enhancement of the emission.[18–21,70] Even though this effect seems only suggested (and more detailed investigations are required) it allows us to envisage a new generation of fluorescence-based sensors. In this sense, the development of conjugated polymers/polyelectrolyte spheres is very promising, in particular when joined to the possibility of chemically engineering polymer properties.

An additional important characteristic of the use of the LbL technique to coat the microspheres, in particular when joined to the use of highly emissive source embedded in it, is the fine-tuning of the stop band position. In a photonic scenario where several PhC elements are integrated to perform various signal-processing functions, fine adjustment and control of the individual gaps and their resonance is essential to the operation of these advanced generation optoelectronic chips.[71] Tuning of a PBG can be pursued by acting either on the PhC lattice periodicity or on the refractive index of the material. Liquid crystals,[72] ferroelectrics,[73] electrochemically controlled materials,[74] and gold nanoparticles[71] have been explored with this aim by applying external voltages, temperature, or laser pulses to modify their susceptibility. In the case of LbL deposition of conjugated polymers (or NCs) on the spheres, all meaningful parameters can be properly engineered, since the sphere diameter and the refractive index depend on the deposition conditions, while conjugated polymers (or NCs) allow one to also use the photoexcitation tools for further tuning. Indeed, the effect of the number of bilayers on the PhC stop band and its dispersion is reported in Figure 19.5d. By using the Bragg–Snell law (Eq. 19.2) and the effective refractive index model (Eq. 19.3), it was shown that coating the spheres with five bilayers provided an increase of both the sphere diameter (from 279 to 283 nm) and the refractive index of the interstices among the spheres (from 1.1 to 1.125), thus confirming the interest of this system for photonics.

The role of polyelectrolytes in the infiltration of opals and, eventually, in the preparation of inverse opals so far described, is limited to their ability to coat the spheres and to embed photoactive materials such as NCs or conjugated polymers. However, an active role for the polyelectrolyte itself for tuning the photonic properties can also be found. For instance, opal films can be infiltrated by electrostatic self-assembly with polyferrocenylsilanes (Fig. 19.6a) without any loss in their optical quality.[75] By controlling the number of deposi-

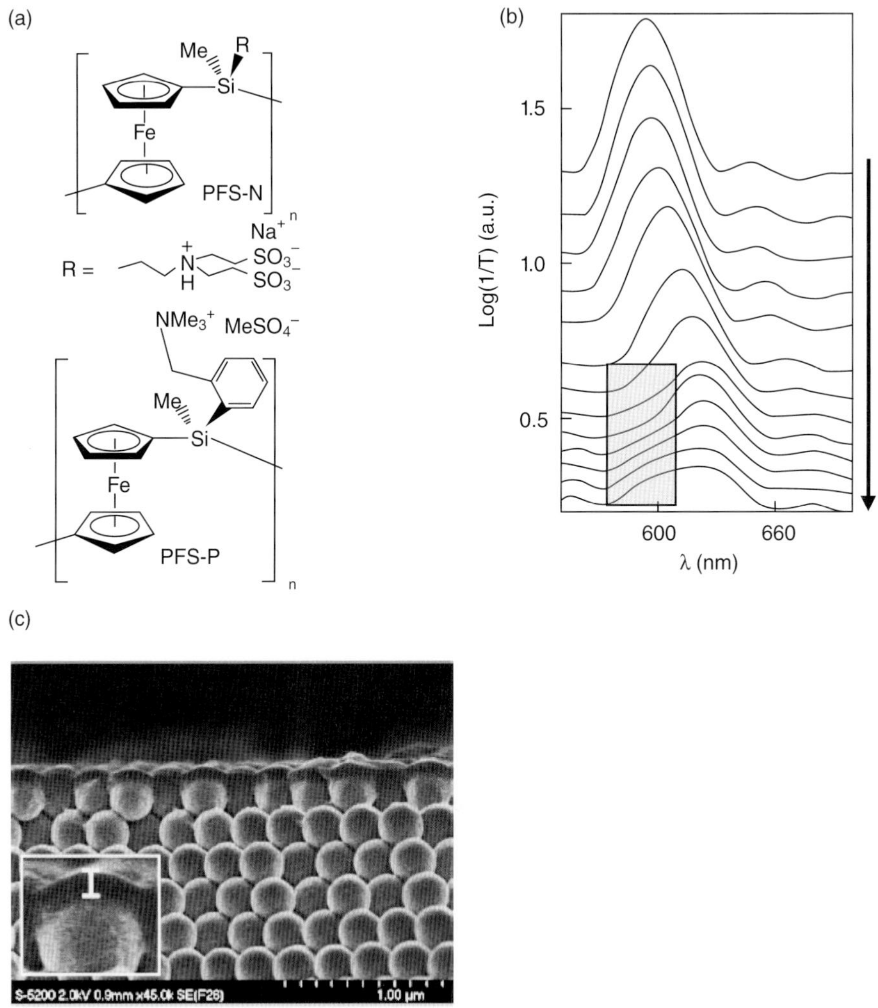

FIGURE 19.6. (a) Chemical structure of polyelectrolytes used for opal infiltration. (b) Experimental optical spectra of silica opal infiltrated as a function of the number of deposition cycles indicated by the arrow. (c) Scanning electron microscope image of a polyelectrolyte film formed on the opal surface. Adapted with permission from Arsenault et al.[75]

tion cycles by alternately dipping in solutions containing polymer electrolytes, it is again possible to control the position of the opal stop band (Fig. 19.6b). Indeed, upon increasing the number of dipping cycles the position of the stop band bathochromically shifts due to the progressive infiltration of the polyelectrolyte within the interstices in agreement with the description provided by Equation 19.2. However, as soon as the windows interconnecting the octahedral and tetrahedral interstices are closed, the polymers cannot be further infiltrated and a polyelectrolyte layer is formed on the opal surface as observed in Figure 19.6c. This layer affects the optical response of the photonic structure generating a satellite peak blue shifted with respect to the stop band (as evidenced in the dashed region of Fig. 19.6b).[75]

The importance of this result goes beyond the academic demonstration and description of tailoring the PhC properties by using polyelectrolyte multilayers and it is strongly connected to applications in the field of color displays for e-readers.

Several technologies for cheap and portable flat displays are currently under development. Among them, organic light-emitting displays are more than a promise since they are cheap, light, and rollable,[1,76] and they could be a superior technology for displays with respect to those based on inorganic semiconductors. However, the market opportunities for e-readers open new perspectives for alternative technologies. In this case, emissive or back-illuminated displays, even of high quality, might create reading difficulties when strong environment illumination occurs. In facts, in such displays the image is created by light emission, thus tiring the eye during long reading times. In this sense, e-readers based on the electronic ink technology (E-Ink), where the images are created by an electronic ink and can be observed only by reflection of the environmental light, provide a more comfortable way of reading. Usually, such readers have only black and white displays and a great improvement to the market could be the development of color technologies. PhCs can provide important opportunities to this end.[77,78]

The basic idea is to create an electrochemical cell made of an opal PhC infiltrated and coated with the suitable polyelectrolyte system (Fig. 19.7a), called Photonic-Ink (P-Ink).[77] Upon applying a voltage to the cell, the redox properties of the polyelectrolytes are changed (Fig. 19.7b). Oxidation of the P-Ink composite causes drawing of electrons from iron atoms in the polyferrocenylsilanes backbone while anions from the electrolyte neutralize its positive charge. This exchange of charges might induce a change of the refractive index of the composing materials of the photonic structures. In addition, a relevant flux of ions and solvent occurs inside the PhC where the polymer has been infiltrated, thus causing its swelling. As a consequence, a modification of the distance between the sphere layers is induced. The final effect of the redox reaction inside the structure is a change of both the effective refractive index and the periodicity. As described in Equation 19.1, these two parameters modify the diffracted wavelength, that is, the color of the system. Since the oxidation state can be continuously tuned, a range of colors can be obtained

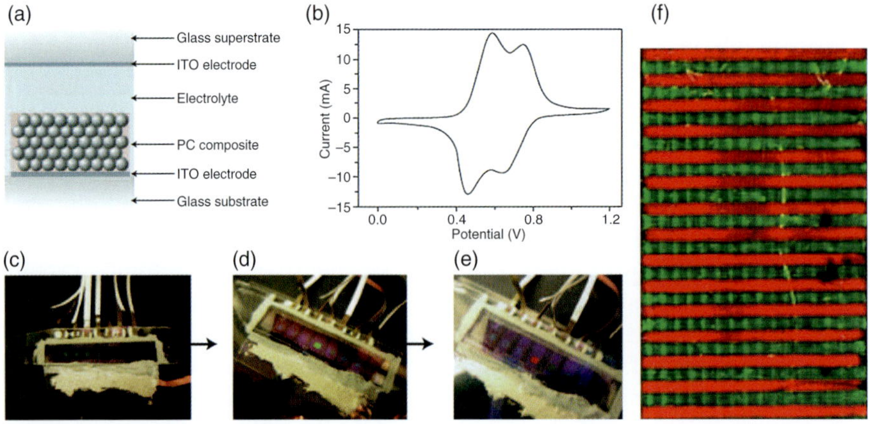

FIGURE 19.7. (a) Schematic of electrochemical PhC cell; (b) cyclic voltammogram of the cell; (c–e) image of the cell with applied potential 0, 1.5, and 2.0 V; (f) pixelation of the PhC electrochemical display. Adapted with permission from figures 2 and 4 of Arsenault et al.[77]

(Fig. 19.7c–e). The process so far described can then be reproduced for several different small adjacent cells with a very simple and cheap technology able to produce 25 micron large pixels. Since each single pixel can be individually driven by an electrical signal, it is possible to guarantee the suitable resolution for a screen (Fig. 19.7f).[77]

We have so far discussed different methods by which to achieve the modulation of the PhC properties by changing the dielectric contrast or the PhC cell length. Now we will show an alternative strategy. Optical and photonic properties of PhC can be tuned by engineering structural defects of the dielectric lattice.[79] Indeed, defects act as photonic dopants and, depending on the structure, can trap the light inside the material. Point defects represent optical cavities where the light is highly localized within the structure.[80] Linear defects work as waveguides in which the light is effectively confined within the structure by the PBG.[80–82] Moreover, the presence of a structural defect inside the dielectric lattice allows one to test light localization effects.[79,80,83,84] Inside such localized states, the electric field associated to the electromagnetic wave is spatially concentrated. Localization effects, combined with a high quality factor of the optical cavity, can provide an optical feedback in order to achieve laser or amplified spontaneous emission when an emitting material is embedded in the planar defect.[85–93] Without going into much detail on the spectroscopical properties of structural defects in PhC, we would like to stress here that their main effect is to introduce inside the PBG an allowed optical state. At that specific energy, photons are transmitted within the structure (instead of being backward diffracted), thus providing a characteristic sharp on the top of the broader reflectance peak as shown in Figure 19.8 for an opal having a defect with thickness $d = 0.3\ L$ (where L is the lattice cell length, $L = D\sqrt{3}$).

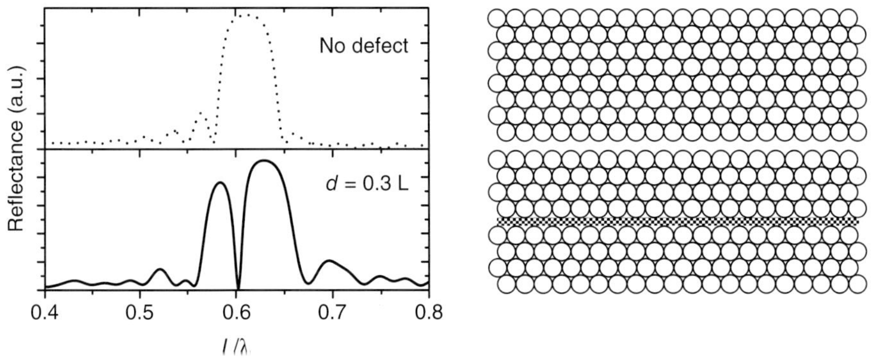

FIGURE 19.8. Calculated reflectance spectra for an opal (top) and an opal with a structural defect (bottom).

Several methods have been used to prepare defects inside opals' and inverse opals' PhCs.[59,80,82,85,87,88,90,94–105] Here, we will briefly review those involving the use of polyelectrolytes.[96,100]

A polyelectrolyte layer, which can be eventually functionalized with an active material,[100] is directly coated onto the surface of the colloidal crystal when both the wettability of the surface and that of the bulk are properly engineered. Alternatively, the polyelectrolyte layer is cast on a flat piece of poly(dimethyl-siloxane) and then transfer printed on the opal surface (Fig. 19.9a).[96] After deposition of the polyelectrolyte layer, a second opal is grown on top of the surface layer, thus embedding a structural defect (Fig. 19.9b).[96]

The spectral position of the defect state is strongly dependent on the dielectric properties of the environment. In Figure 19.9c,d, the reflectance spectra of an opal containing a defect exposed to the vapor of water and acetone is reported as a function of the vapor pressure.[96] Water vapor infiltration causes two effects: swelling of the polyelectrolyte layer and condensation inside the interstices on the sphere surfaces. Swelling is more effective for water vapor and gives rise to a logarithmic dependence of the defect mode wavelength on the partial pressure; condensation is favored for acetone, providing a change in the dielectric contrast of the system, causing an exponential dependence on the partial pressure. The different effects of the two phenomena on the optical spectra of the defect are highlighted in Figure 19.9d and can be exploited as a proof of concept for a novel generation of gas sensors.[96]

A different approach to the preparation of PhC by using polyelectrolytes is provided by inkjet printing.[106–108] Inkjet printing allows a rapid and flexible route for the fabrication of 3D PhCs at micrometric scale. Even though at present the technique does not allow shrinking dimensions to submicrometric wavelength, it retains the advantage of using flexible polymer materials which can be then replicated by proper chemical vapor deposition techniques with silicon. In particular, inkjet printing has been used to prepare self-supporting woodpile structures of polyelectrolytes (Fig. 19.10a).[106,108] Details on the use

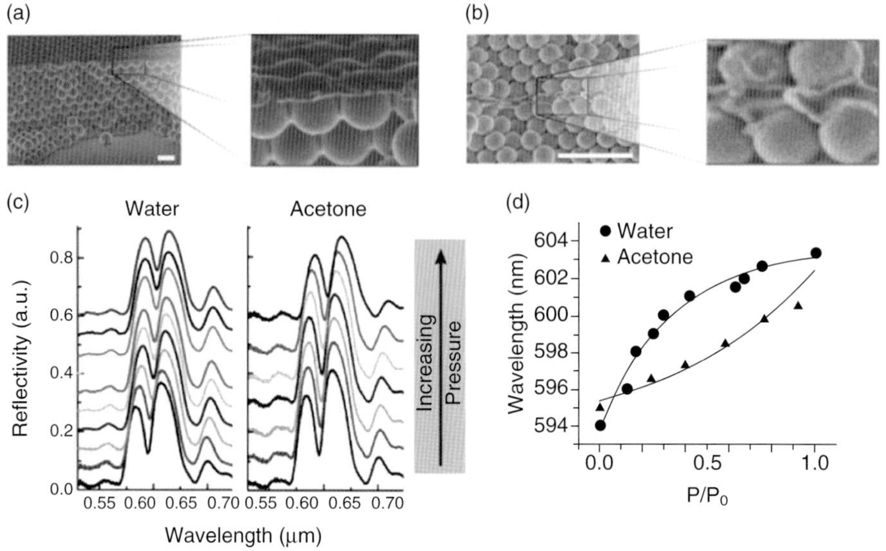

FIGURE 19.9. SEM images of the polyelectrolyte layer cast over the opal (a) and embedded within the PhC, thus forming the structural defect (b). (c) Reflectance spectra of opals with defects after sorption of water and acetone at different partial pressures. (d) Defect state wavelength versus water and acetone partial pressures. Adapted with permission from Tetreault et al.[96]

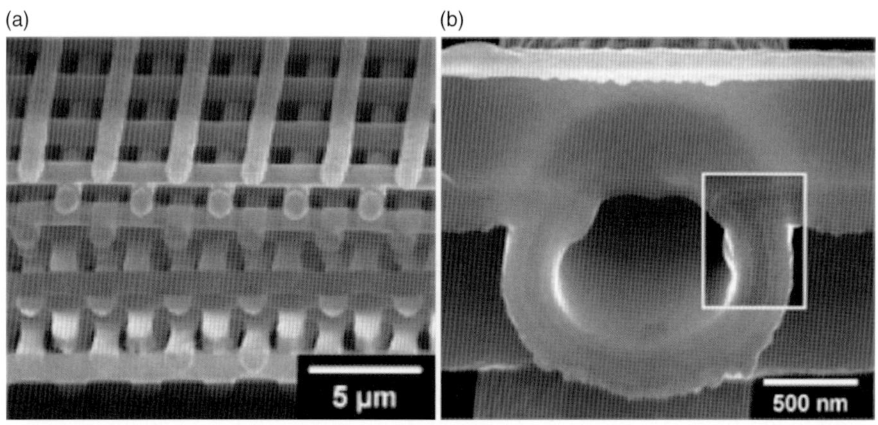

FIGURE 19.10. (a) Polyelectrolyte woodpile structure created by inkjet printing. (b) Replica of (a) with silicon after template removing. Adapted with permission from Gratson et al.[106]

of the inkjet printing technique applied to organic electronics and other technologies can be found in more specific literature.[1,109,110] Here, we would like to stress the possibility of using this technique for the preparation of PhC structures. Indeed, the woodpile structure reported in Figure 19.10a is interesting since it can be directly modified in order to create engineered defects in an easier way than for the inverse opal structure.[106] The ink is composed of a mixture in water of poly(acrylic acid) and poly(ethylenimine) with suitable viscosity (which can be tuned in the 5–150 Pa s range) for deposition through microcapillary nozzles (diameter 0.5–5 μm).[106] When the printer head scans the surface, it leaves the polyelectrolyte ink, which rapidly solidifies, thus yielding cylindrical filaments with a diameter comparable to the nozzle size. After deposition, filaments retain their cylindrical shape in the woodpile structure (Fig. 19.10a),[106–108] which can be used as a template for coating with a very thin silicon layer by a low temperature process that preserves the polymer mesostructure. After removal of the polymer template, the hollow woodpile structure can be further enriched with silicon by standard high temperature chemical vapor deposition techniques (Fig. 19.10c).[108] The new structure has a much stronger dielectric contrast with respect to the polyelectrolyte template, thus enhancing its photonic properties.

19.2.2 Self-Assembled BCPs PhCs

We have so far described the use of polyelectrolytes for different functions ranging from the formulation of an ink to the ability to favor interactions with a suitable active medium. However, we have not really exploited their intrinsic polymer properties to develop a PhC structure. To this end, we will report in the following sections results obtained with BCPs.

Fundamental theoretical and morphological aspects of BCPs' phase behavior have been extensively discussed in specific publications (References 6, 111, and 112). Herein we would like to recall some basic concepts that will be useful to the current PhC topic.

BCPs are macromolecules composed of two or more chemically distinct parts (blocks) in the same chain. The simplest case is that of linear AB di-BCPs, in which each chain is composed of an A and a B segment joined at their endpoint. If we consider an analogous system of A and B homopolymers, that is, a binary blend, phase separation on a micrometer scale is commonly observed. In fact, provided that no special interaction (such as hydrogen bonding) between A and B monomers occurs, homogeneous polymer mixtures cannot be formed due to the positive free energy mixing contribution. The tendency to segregate is quantified by the value of χN, χ being the segment–segment Flory–Huggins interactions parameter and N the total degree of polymerization.[111] Also, in AB BCP systems, the two different blocks have a tendency to phase separate but, being covalently linked at one end, a macroscopic phase separation is inhibited. Therefore, the segregation happens on the length scale of the blocks themselves, leading to nanoscopic domains. These domains are characterized by uniform sizes and undergo a self-assembling process driven by a competition between the requirement of minimal interfacial area and maximum chain conformational entropy.[113] The final morphology of sufficiently segregated BCP is dictated by the relative volume fraction of the two blocks and is described in peculiar equilibrium phase diagrams.[111,112] Different periodic structures are encountered upon increasing the volume fraction of one of the blocks: spheres arranged in a body center cubic lattice, hexagonally packed cylinders, bicontinuous double gyroid cubic, and, finally, layered lamellae under symmetric composition conditions. A schematic summary of the equilibrium phases, along with some typical TEM images for the various morphologies, is presented in Figure 19.11.

BCP self-assembly, being tens of nanometers its common length scale with the usual molecular weights ($M_n \approx 10^4$ g mol^{-1}), has been widely exploited for nanotechnology applications such as the preparation of nanostructured membranes or templates and high-density storage media.[7,114] By considering the structures sketched in Figure 19.11a, an obvious possible application is the fabrication of PhCs, since the self-assembled structures closely resemble the periodic dielectric lattices of Figure 19.1. In fact, 1D ,2D, and 3D PhCs have been obtained from lamellar,[115] cylindrical,[116] and bicontinuous gyroid[29] morphologies, respectively. In the section "BCP Periodical Mesostructures," the main issues involved in the use of BCP as PBG materials will be introduced

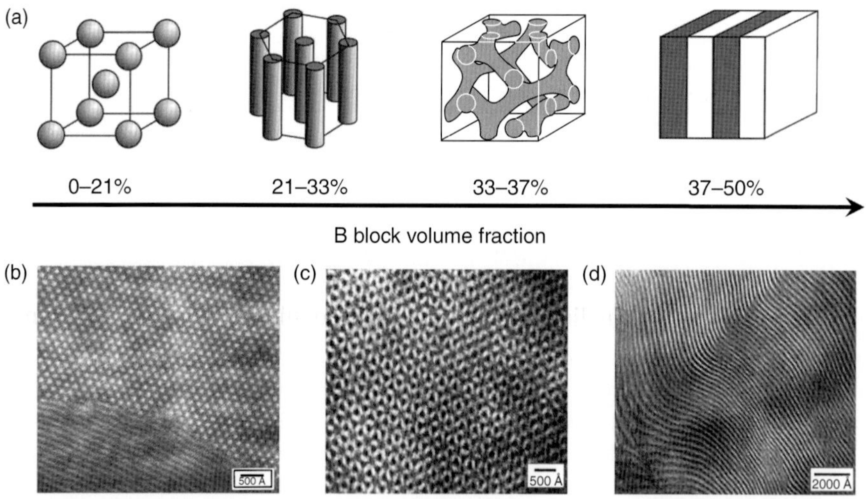

FIGURE 19.11. (a) Schematic representation of the morphologies obtained with increasing B content in a linear AB BCP. TEM images of RuO_4-stained PS-b-PI BCPs showing (b) HCP cylinders, (c) cubic bicontinuous gyroid, and (d) lamellae. PS, polystyrene; PI, polyisoprene. Adapted with permission from Khandpur et al.[148] and Schulz et al.[149]

and briefly discussed. Finally, in the section "Stimuli-Responsive BCP PhCs," the main results concerning tunable or stimuli responsive BCP PhC, which mostly rely on the properties of polyelectrolytes, will be reviewed.

BCP Periodical Mesostructures As already highlighted in previous parts of this chapter, in order to develop a PBG, fundamental requirements for the material have to be fulfilled. These requirements are a suitable refractive index contrast between composing media and a periodicity of the dielectric lattice on the scale of hundredths of nanometers extending at long range in order to obtain macroscopic ordered samples. The fulfillment of these needs is actually the main challenge one has to face in order to use BCP as PhCs.[117–119] Indeed, in common BCP systems, the domain spacing is usually of the order of a few tens of nanometers only, and the dielectric contrast between the two polymer blocks is intrinsically low. Moreover, self-assembled BCP structures usually present a polygranular texture, where the order is maintained in the single grain only, typically on a length scale of 1–2 μm, which is too small for any photonic application. Some elegant solutions have been proposed to circumvent all these problems.[115,116,120–122]

For what concerns the required domain spacing, a rough estimate of its value is possible by considering that a symmetric di-BCP with lamellar morphology is equivalent to a quarter-wave stack. In this case, the domain periodicity at which the maximum of reflectivity occurs is linked to the wavelength λ and the average refractive index n through the approximate relation $d \approx \lambda/4n$.[123] Thus, for a PhC in the visible range ($\lambda \approx 500$ nm), given the typical value for the refractive index of a polymer material ($n = 1.5$), it turns out that BCP lamellar spacing needs to be around 100 nm. The simplest way to access this range of periodicities is by varying the BCP molecular weight. In fact, the lamellar period scales with two-thirds the power of the molecular weight,[124,125] and a total molecular weight close to 1000 kg mol^{-1} (instead of the usual 50–100 kg mol^{-1}) would give the domain spacing required for photonic applications. Indeed, employing such high molecular weight BCPs, 1D, 2D, and 3D PhCs have actually been obtained.[29,116,119] Nevertheless, ultrahigh molecular weight polymers are extremely viscous in the melt, and their highly entangled nature hinders the self-assembly process. As a consequence they must be processed in solution, thus limiting the range of possible applications. Recently, lamellar spacing suitable to obtain PhC in the visible range has been obtained in low molecular weight BCPs, thanks to a novel synthetic route which produces polydisperse block lengths.[126] A more convenient way to reach the required domain size without reducing processability is to use a somewhat lower molecular weight BCP in a ternary blend with the two respective homopolymers.[117,127] Indeed, theoretical[128,129] and experimental[125,130] studies demonstrated that when a mixture of BCPs with homopolymers is produced, the homopolymer will be preferentially confined in the domains of the like block. This results in a swelling of the microdomain, the extent of which is controlled by homopolymer molecular weight and content.[124,125,130]

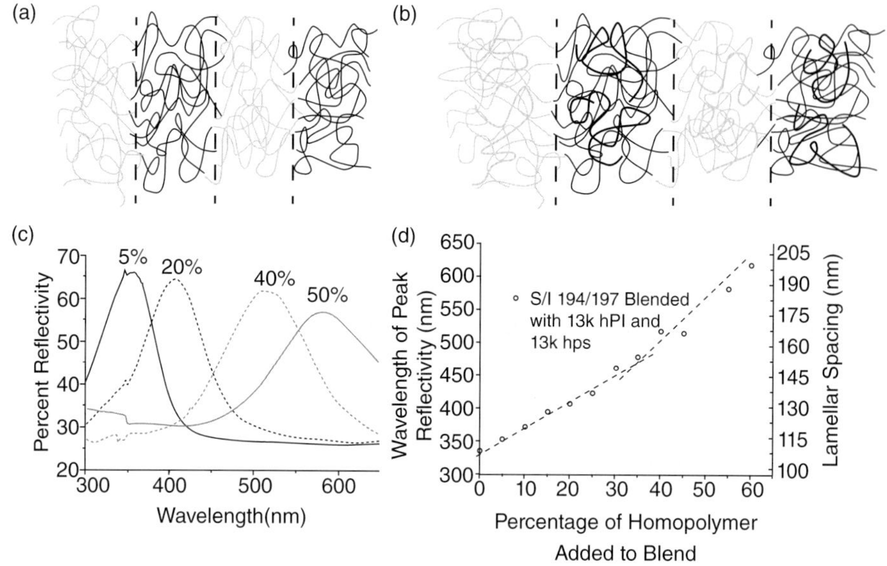

FIGURE 19.12. (a) Scheme of the pure BCP 1D PhC. (b) Same as (a) after homopolymer addition. Homopolymer is represented by thicker lines. (c) Reflectivity of sample containing 5, 20, 40, and 50% of total added homopolymer (b) Peak reflectivity wavelength as a function of homopolymer weight fraction. The right axis reports the associated lamellar domain spacing. Lines are guide for eyes. Adapted with permission from Urbas.[127]

The potential photonic application of a BCP is evident when the domain periodicity can be tailored by controlled swelling of the blocks with different amounts of additives, according to the schematic mechanism shown in Figure 19.12a,b. As a matter of fact, a 1D reflector, whose reflectance peak was tunable in a wide range of visible wavelengths, was realized with polystyrene–polyisoprene (PS-b-PI) BCPs blended with different percentages of PS and PI homopolymers.[127] Figure 19.12c shows the reflectivity of such ternary blends having different compositions. It can be noticed that there is both a red shift of the peak reflectivity wavelength and a ΔE increase (Eq. 19.4) upon increasing the homopolymers content. These effects can be explained considering both the increase in lamellar spacing (Eq. 19.2) and the increase of the effective refractive index. In particular, the broadening of the reflectivity peak, as previously discussed, is indicative of an increased dielectric contrast (Eq. 19.4). In Figure 19.12d the shift in the wavelength of the reflectivity maximum (and the correspondent estimated lamellar spacing) is reported as a function of the homopolymer content in the blend. The observed trend, which is linear up to about 25% and then deviates from linearity at larger contents, results from the abovementioned evolution of the effective refractive index and from a complex microdomain swelling behavior, which results in a power law dependence of the domain spacing on the volume fraction of added homopolymer.[128]

While the fine-tuning of blocks' periodicity can be achieved with relatively easy methods, the achievement of a high dielectric contrast is a far more problematic task. Considering the abovementioned PS-b-PI BCP, which can be regarded as a general example, the dielectric contrast is poor (1.05), since the refractive index of polystyrene and polyisoprene blocks is 1.59 and 1.51, respectively. Such values are typical for polymer systems and the possibility of chemically varying the refractive index, for instance by means of copolymerization,[119] is rather limited, as can be deduced from Figure 19.13a, where the dispersion properties of the refractive index of selected polymers are reported. Even the use of semicrystalline–amorphous BCPs, in which the semicrystalline part of the chain has a higher density and thus a higher refractive index,[126] does not improve the contrast significantly. Instead, a successful strategy that has been used is to selectively embed inorganic NCs[117] or metallic nanoparticles[120,131] in one of the domains. Obviously, the size of the nanoparticles should be much lower than the visible light wavelength in order to avoid detrimental light-scattering effects. The spatial distribution of an inorganic

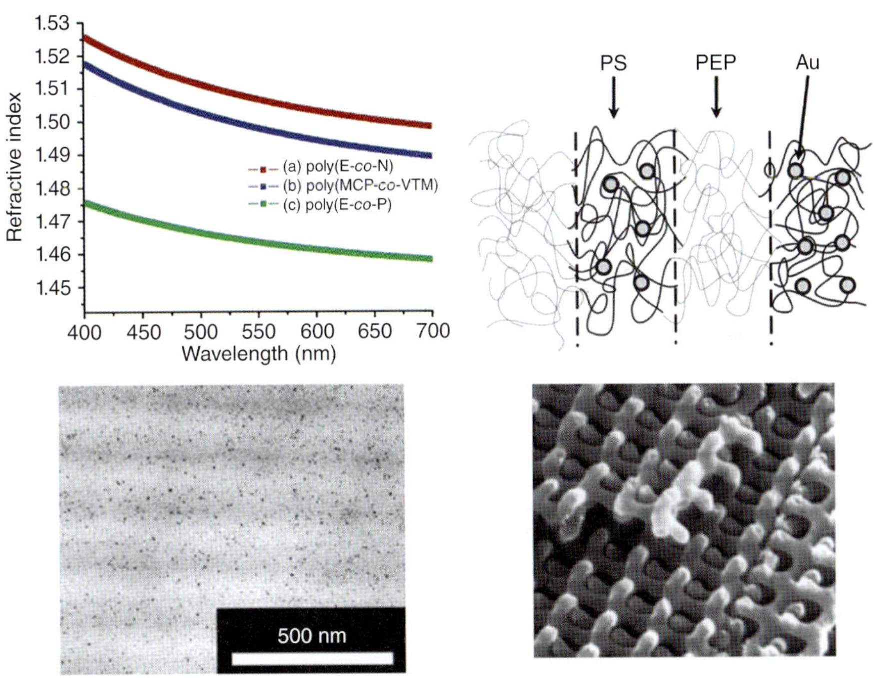

FIGURE 19.13. (a) Refractive index as a function of wavelength for polyolefin random copolymers. (b) Schematic representation of polystyrene-poly(propylene-ethylene) BCP with a dispersion of surface-modified Au nanospheres. (c) TEM image of gold nanoparticles dispersed in a PS-b-PI system. (d) PS-air double gyroid network obtained by selective etching of PI phase in a PS-b-PI BCP. Adapted with permission from Urbas et al.,[29] Yoon et al.,[119] and Bockstaller et al.[133]

nanofiller in microphase-separated BCPs depends on factors such as the size of the additive and the specific surface grafting.[131–133] By using small polystyrene-grafted gold NCs, a selective and homogeneous dispersion of the metallic particles in the PS domain of a PS-b-PI BCP was achieved, leading to improved dielectric contrast and enhanced photonic properties.[120,121] A schematic representation and a TEM image of this system are shown in Figure 19.13b and Figure 19.13c, respectively, indicating that indeed the nanoparticles are preferentially dispersed in a single phase.

Another method to obtain a somewhat higher dielectric contrast is based on the selective etching of one of the microphases. Of course this approach requires the availability of an open bicontinuous structure such as that of double gyroid morphology. Figure 19.13d shows a scanning electron microscope (SEM) image of an ultrahigh molecular weight PS-b-PI BCP possessing a cubic double gyroid structure, which has been treated with UV and ozone to selectively degrade the polyisoprene phase, leaving a polystyrene–air double gyroid network with increased refractive index contrast (about 1.6).[29] Using a silicon containing BCP, such as poly(dimetylsiloxane), a freestanding silicon oxycarbide network is produced upon etching and oxidation, leading to ceramic 3D PhCs from a self-assembled polymer precursor.[122]

A suitable domain spacing and a proper dielectric contrast alone are often not sufficient to obtain a high-quality PhC. In fact, a high number of lattice periods, that is, layers for 1D PhC, and the absence of symmetry-breaking defects on a long range scale, are also required to reach optimal reflectivity. Self-assembled BCP structures are well known to be locally very precise, while the order on the macroscopic scale is usually poor due to the occurrence of several defects such as grain boundaries and dislocations.[134] The control of BCP microdomain orientation and regularity over a large dimensional scale has been extensively investigated, due to its paramount importance for optical and nanotechnological applications. The common strategy to produce well-ordered BCP nanostructures involves the application of a biasing force field during the microphase separation. Various external forces, such as mechanical, electrical, and surface interaction, have been successfully used.

In the roll casting technique,[135] BCP solutions are placed in a gap between two counter-rotating rollers and the solvent is allowed to evaporate while the rollers are moving. The counter-rotating elements generate complex flow fields involving both shear and elongation deformation, which eventually produce globally oriented BCP microdomains with near-single crystal features. Since this method is applied to solutions, it allows the processing of higher molecular weight BCPs, interesting for photonic applications, well below their degradation temperature. Indeed, roll casting was employed to produce a 2D PhC from a 5% w/v solution of ultrahigh molecular weight PS-b-PI BCP ($M_n = 10^6$ g mol^{-1}).[116] The atomic force microscope (AFM) phase image of a cross section is reported in Figure 19.14a, showing well-ordered PS cylinders in a PI matrix, with an intercylinder spacing of about 100 nm.

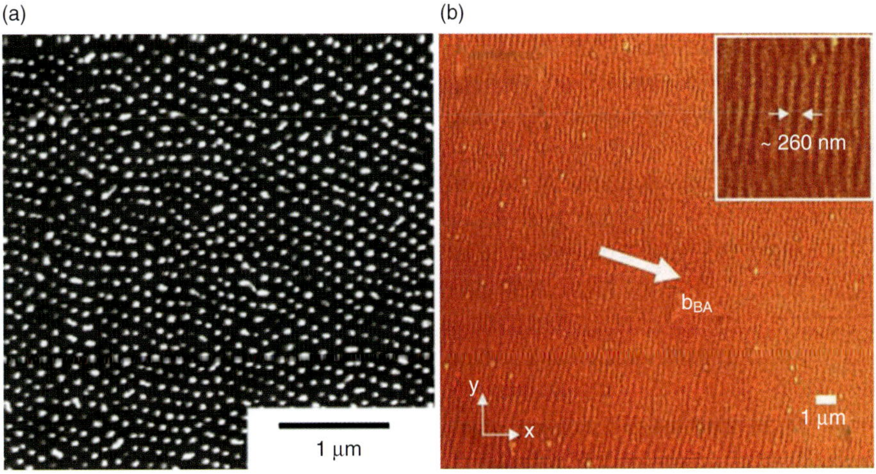

FIGURE 19.14. (a) Cross-section AFM phase image of the polystyrene microdomains viewed parallel to the cylinder axis. Polystyrene cylinders have higher modulus than polyisoprene matrix and are shown as white spots in the image (b) Laser scanning confocal microscopy image of PS-b-PI BCPs showing highly oriented cylindrical microdomains obtained from directional solidification of benzoic acid. The inset is a high magnification image of the cylinders for which the inter-cylinder spacing is shown. Adapted with permission from Deng et al.[116] and Park et al.[137]

Another very efficient process used to control microdomain orientation over large areas is based on the directional solidification of a solvent in a solution containing the BCPs.[136] The key aspect of this method is the choice of an organic solvent able to dissolve the BCP at a high temperature and with the tendency to crystallize directionally into large, plate-like crystals. Examples of such solvents are benzoic acid and anthracene.[136] The hot BCP solution is placed in a temperature gradient where the solvent can crystallize along its fast growth direction. As soon as the crystallization proceeds, the BCP undergoes microphase separation because of the rapid decrease of the solvent concentration. The microdomains are then nucleated from the solvent crystals and their orientation is dictated by the directional growth of the organic diluents. This process was successfully applied to both normal and ultrahigh molecular weight PS-b-PI BCPs.[136,137] In the latter case, long-range ordered lamellae and cylinders, with spacings relevant to the production of PhC, were obtained within few seconds.[137] An example of these highly regular nanostructures is shown in Figure 19.14b.

Stimuli-Responsive BCP PhCs The possibility of obtaining PhC with simple and low-cost processes based on BCP self-assembly, which has just been illustrated, is one of the many advantages compared to top-down methods. In fact, BCP self-assembly can be combined with interesting properties of other systems, such as liquid crystals or supramolecular aggregates, to produce easily tunable or stimuli-responsive PhCs, in particular when they include a polyelectrolyte segment. The optical characteristics of PhCs can be finely tuned over the chosen wavelength range by tailoring the dielectric contrast and/or the characteristic size. As previously discussed for opal-based PhC, several polymeric materials and a variety of external stimuli (e.g., electrical fields, strain fields, and chemical compounds)[138] have been used to prepare tunable or switchable PBG systems. Tunability of BCP-based PhCs in response to changes in the temperature, chemical environment, or electric field has recently been investigated.[25,139–145] In the following, the main results of this research will be reviewed, focusing on the (supra)molecular mechanism responsible for the modification of the optical properties.

Temperature-dependent PBG materials based on supramolecular assemblies were the first examples reported.[139–141] Osuji et al.[139] incorporated guest liquid crystalline (LC) molecules into one domain of a host BCP, exploiting hydrogen bonding interactions. A high molecular weight polystyrene-*block*-poly(methacrylic acid) (PS-b-MAA) and an imidazole-terminated mesogen (LC) were used. Solvent casting of PS-b-MAA/LC leads to a 1D lamellar microdomain structure that exhibits an optical stop band in the green spectral region (560 nm). The liquid crystal in the MAA domains is initially organized in a smectic structure and undergoes a smectic-to-isotropic transition when heated to about 70°C (see the scheme in Fig. 19.15a). The occurrence of this transition implies a change in the average refractive index of the MMA/LC layers, which in turn, according to Equation 19.2, produces a shift of the maximum reflectivity of the film toward the red (600 nm). Since the peculiar morphology of the solvent cast system cannot be obtained again upon cooling, this shift in the reflectivity peak is irreversible and the supramolecular assembly acts as a temperature-sensitive PhC sensor. A fully reversible PBG switch was instead realized by using a supramolecular complex between BCP and amphiphilic molecules, giving rise to a lamellar structure.[140,141] Polystyrene-*block*-poly(4-vinylpyridine) (PS-b-P4VP) is able to form comb-shaped supramolecular assemblies with long chain surfactant molecules,[140] which can bond as side chains to the pyridine by protonation or hydrogen bonding. Thus, P4VP chains assume a stretched conformation to accommodate the "combs," resulting in a particularly large lamellar periodicity consistent with PhC behavior. The thermal stability and the temperature-dependent solubility of the guest molecules in the different polymer blocks were exploited to obtain a temperature-switchable PhC. Valkama et al.[141] used a complex between PS-b-P4VP and 3-*n*-pentadecylphenol (PDP) for that purpose. At room temperature, the abovementioned comb-shaped architecture forms a dielectric reflector and the sample appears green. On heating, hydrogen bonds are broken and PDP additionally becomes soluble in the polystyrene block, leading to the collapse of the lamellar periodicity and thus to a sharp transition, which is reversible on cooling, to uncolored material. This mechanism is illustrated in

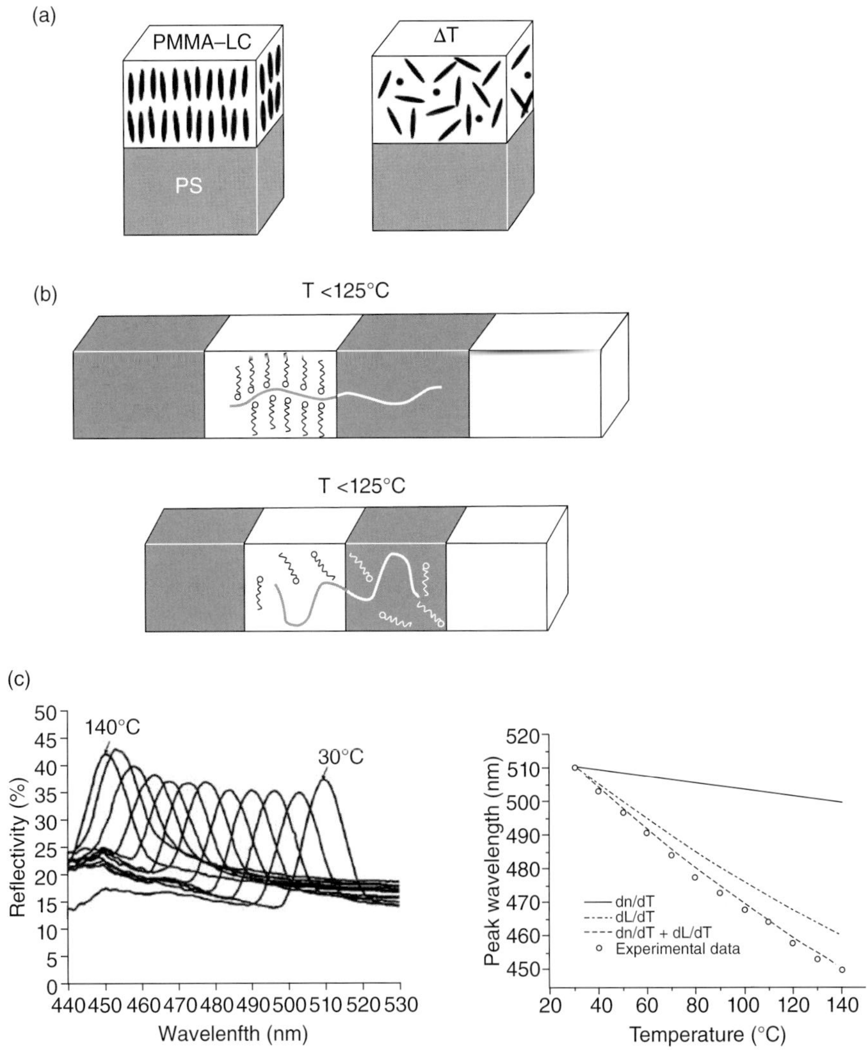

FIGURE 19.15. (a) Schematic explanation of LC randomization upon heating. (b) Schematic view of complex disgregation and domain period collapse above 125°C. (c) Reflectivity spectra of the PS-b-PI photonic gel between 30 and 140°C in 10°C increments (left); experimental and calculated reflectivity peak position as a function of temperature (T) (right). Adapted with permission from Yoon et al.[142]

Figure 19.15b. Also in these cases, the conformational change induces a modification of both the refractive index and periodicity of the 1D PhC structure.

More recently, a thermochromic BCP photonic gel was obtained[142] by sealing a high molecular weight PS-b-PI BCP swollen with a neutral solvent between two glass plates. The measured reflectivity spectra as a function of temperature are reported in Figure 19.15c. It can be noticed that the reflectivity peak moves toward shorter wavelengths by more than 50 nm on increasing temperature of 110°C. This spectral shift cannot be accounted for by the usual temperature dependence of the refractive index ($dn/dT \approx 10^{-4}$ K^{-1}),[146] which according to Equation 19.2 would provide a negligible change of the spectral position of the PBG in the explored temperature range.[147] Indeed, the most important contribution to the observed thermochromic effect comes from the variation of the lamellar domain spacing, as a consequence of a varying degree of segregation between the blocks. In fact, upon increasing temperature, the Flory–Huggins interaction parameter χ, the segregation strength between the blocks becomes smaller, that is, blocks are "less incompatible," resulting in a relaxation of the stretched chains and therefore a decrease of the domain thickness. By also considering this effect, the peak position as a function of temperature can be properly modeled as shown in Figure 19.15c.

BCP-based photonic gels have also been revealed to be promising materials in view of their responsiveness to the chemical environment. By coupling the high swelling capability of polyelectrolytes with the self-assembling properties of BCPs, chemically tunable photonic gels can be easily obtained.[25,143] Kang et al. used a polystyrene-*b*-quaternized poly (2-vinyl pyridine) (PS-b-QP2VP) to obtain a photonic gel responsive to salt concentration in aqueous solution.[25]Lamellar films were prepared by spin casting a BCP solution and subsequently annealing it in chloroform vapor. The pyridine groups were then quaternized with 1-bromoethane and partially cross-linked to various extents using 1,4-dibromobutane. The self-assembled BCP microdomains present a periodicity in the dry state of about 100 nm. With the given effective refractive index and according to the Bragg–Snell law, the PBG of this structure is expected to occur in the UV spectral range (Fig. 19.16a). By varying the relative humidity, the polyelectrolyte layers selectively swell, while the glassy polystyrene domains remain unaltered, forcing the expansion of the hydrophilic layer along the normal to the film. At 100% relative humidity, extremely large domain periodicity is attained, enabling a broad tunability of the PhC over the visible and near IR spectral range. Moreover, thanks to the presence of pores and defects, the uptake of fluids into and out of the lamellar structure is particularly rapid. Since polyelectrolyte gels undergo osmotic deswelling when placed in a salt solution (see Chapters 5, 6, and 16 of this book), optical properties of PS-b-QP2VP photonic gel are expected to change if it is in contact with a fluid reservoir with different salt concentrations due to a variation of the refractive index and domain periodicity as a consequence of different swelling conditions. In fact, this can be actually observed in the UV–Vis–near IR absorbance spectra of a BCP photonic gel exposed at various NH_4Cl concentration reported in Figure 19.16a. As the salt concentration decreases from 2.5 M to pure water, the position of the primary stop band shifts from 364 to 1627 nm, allowing tunability across the UV–Vis and near IR regions. It is important to notice that the fully swollen BCP shows up to the fifth-order PBG as a consequence of the highly ordered lamellar structure,

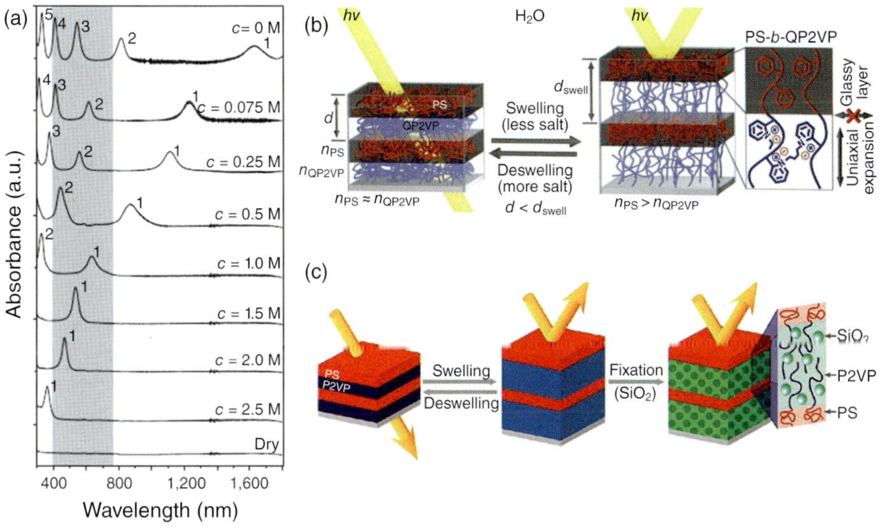

FIGURE 19.16. UV–Vis–near IR absorbance spectra of PS-b-QP2VP photonic gels swollen by contact with NH_4Cl solutions at different concentrations. The gray region represents the visible range. (b) Schematic diagram of the structure and tuning mechanism of the photonic gel. (c) Illustration of the swelling–freezing method. Adapted with permission from Kang et al.[25,143]

thus indicating the high optical quality of these multilayers (1D PhC). The swelling/deswelling mechanism leading to tunability of the photonic properties is summarized in the scheme of Figure 19.16b. Obviously, the optical characteristics of the swollen photonic gels disappear as the gels dry, thus limiting their application in solvent-free environments. In a recent paper, BCP photonic gels were used as a template to produce organic/inorganic hybrid 1D PhC by incorporating a component which turned the soft gel to a stable solid.[143] The so-called swelling–freezing method applied to analogous PS-b-QP2VP BCP is illustrated in Figure 19.16c. P2VP domains are first selectively swollen with the solvent, that is, methanol, and then infiltrated with SiO_2 by sol-gel processing of a silica precursor. By aging in basic conditions, the colloidal particles of SiO_2 aggregate in a continuous network, and after drying the fixed photonic gel shows persistent colored reflections. By varying the reaction time, the amount of silica infiltrated increases, along with the P2VP/SiO_2 domain size and the overall periodicity. The wavelength of the reflectivity peak shifts accordingly to the above morphological changes, thus providing an easy way to tune the stop band by choosing the preparation conditions.

In view of their potential application in modern display technology, due to their inherent passive color which spares light-absorbing filters and backlights, the electrical tunability of BCP photonic gels has recently been investigated.[144,145] The working principle of such systems is again the controlled swelling of the BCP polyelectrolyte layer, as a result of a variation of the chemical environment in an electrochemical cell. In fact, by applying a given voltage an electrochemical oxidation or reduction reaction at the electrode/electrolyte interface occurs, thus giving rise to a change of the local pH. By properly choosing the BCP and the electrolyte, electrically tunable BCP pixels can be produced. The first example involves the usual PS-b-P2VP BCP, which is enclosed between two electrodes of indium tin oxide (ITO)-coated glass and swollen with 2,2,2-trifluoroethanol (TFE) as electrolytic fluid.[144] The mechanism underlying the electrochemical color change is sketched in Figure 19.17a. When the polymer is initially swollen with TFE, which has a lower refractive index than both PS and P2VP, the combination of layer thickness and refractive index contrast gives reflectivity peaks in the "red" region (657 nm). By

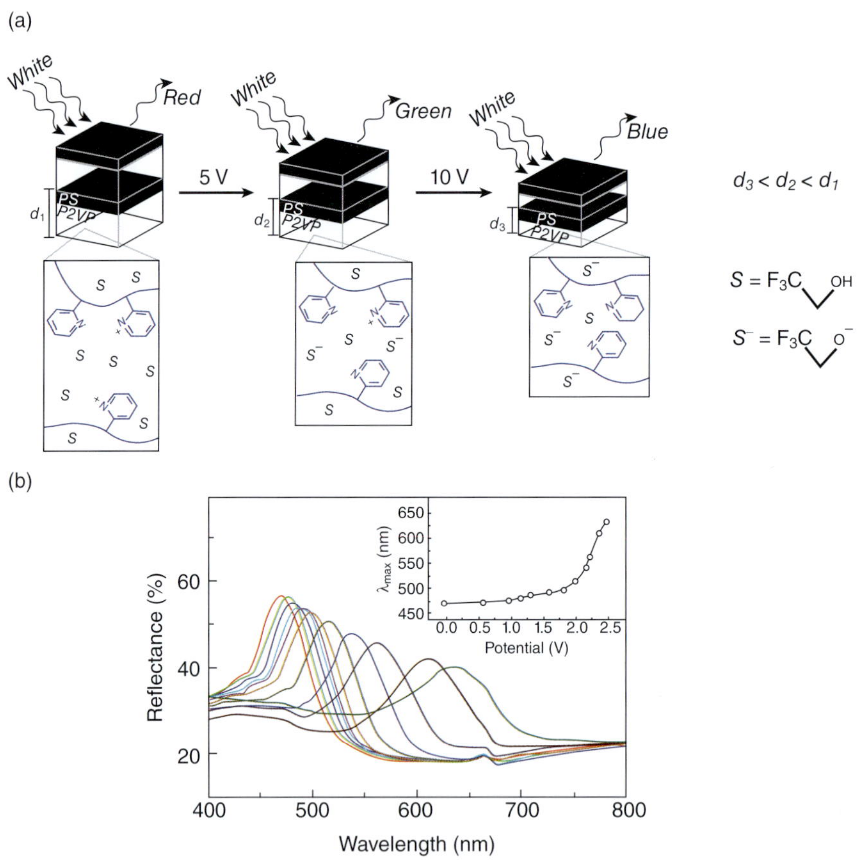

FIGURE 19.17. Mechanism responsible of the electrochemical color change in PS-b-P2VP swollen with TFE as electrolytic fluid. (b) Reflectivity spectra of PS-b-P2VP photonic gel for different electrical potentials. The inset shows the voltage dependence of the reflectivity peak position. Adapted with permission from Walish et al.[144] and Lu et al.[145]

applying a certain voltage to the film, TFE is converted into trifluoroethoxide (TFX), which is a base and thus partially deprotonates P2VP. The increase of P2VP/P2VP$^+$ ratio as a consequence of the electrochemical reaction causes the shrinkage of P2VP layer because of a reduced solubility in the solvent. Consequently, the reflected color shifts from red to green and even to blue by further pushing the TFX$^-$ production, that is, providing a larger voltage to the electrochemical cell. Analogous results were obtained by swelling the same polymer in a water/ethanol mixture.[145] In this case the low voltages applied do not induce an electrochemical reaction, but lead the positively charged P2VP groups to move away from the anode. In this way layer expansion is produced, enabling electrical tunability of the PhC in the entire visible range as shown in Figure 19.17b.

19.3 CONCLUSIONS

We have shown the opportunities offered by polymer science to obtain highly ordered periodical structures having lattice constant at the mesoscale. Self-assembled PhCs, made of colloids or macromolecules, have improved optical quality and functionality when polyelectrolytes are used. Indeed, these ionic polymers allow not only to facilitate the PhC growth, but also to impart the system-specific tunability. Polyelectrolytes inside PhCs offer the possibility to control their optical and chromatic response, becoming stimulus responsive systems and thus allowing us to envisage their use in different fields such as photonics, sensing, and display technologies (see also Chapters 16, 17, and 22).

REFERENCES

1. D. Comoretto. *Supramolecular Polymers*, 2nd ed, A. Ciferri, ed. CRC Press Taylor & Francis Group, Boca Raton, FL, 2005.
2. G. A. Ozin and A. C. Arsenault. *Nanochemistry—A Chemical Approach to Nanomaterials*. Royal Society of Chemistry (Great Britain), Cambridge, 2006. CB4 0WF.
3. *Photophysics of Molecular Materials: From Single Molecules to Single Crystals*, G. Lanzani, ed. Wiley-VCH Verlag GmbH & Co. KGaA, 2006.
4. J. Clark and G. Lanzani. *Nat Photon* **2010**, 4, 438.
5. Y. Xia, B. Gates, Z. Y. Li. *Adv Mater* **2001**, 13, 409.
6. *Supramolecular Polymers*, 2nd ed, A. Ciferri, ed. CRC Press Taylor & Francis Group, Boca Raton, FL, 2005.
7. I. A. Zucchi, E. Poliani, M. Perego. *Nanotechnology* **2010**, 21, 185304.
8. J. Kim. *Pure Appl Chem* **2002**, 74, 2031.
9. H. Sirringhaus, T. Kawase, R. H. Friend, T. Shimoda, M. Inbasekaran, W. Wu, E. P. Woo. *Science* **2000**, 290, 2123.
10. L. Biniek, C. L. Chochos, G. Hadziioannou, N. Leclerc, P. Léveque, T. Heiser. *Macromol Rapid Commun* **2010**, 31, 651–656.

11. N. Sary, F. Richard, C. Brochon, N. Leclerc, P. Léveque, J.-N. Audinot, S. Berson, T. Heiser, G. Hadziioannou, R. Mezzenga. *Adv Mater* **2010**, 22, 763–768.
12. T. Kobayashi. *J-Aggregates*. World Scientific Publishing Co Pte Ltd, Singapore, 1996.
13. *Supramolecular Photosensitive and Electroactive Materials*, H. S. Nalwa, ed. Academic Press, San Diego, CA, 2001.
14. H. Fukumoto and Y. Yonezawa. *Thin Solid Films* **1998**, 327–329, 748.
15. E. Yablonovitch. *Phys Rev Lett* **1987**, 58, 2059.
16. S. John. *Phys Rev Lett* **1987**, 58, 2486.
17. L. Solymar and D. Walsh. *Lectures on the Electrical Properties of Materials*. Oxford University Press, Oxford, 1993.
18. P. Lodhal, A. F. vanDriel, I. S. Nikolaev, A. Irman, K. Overgaag, D. Vanmaekelbergh, W. L. Vos. *Nature* **2004**, 430, 654.
19. M. Barth, A. Gruber, F. Cichos. *Phys Rev B* **2005**, 72, 085129.
20. L. Bechger, P. Lodahl, W. L. Vos. *J Phys Chem B* **2005**, 109, 9980.
21. L. Berti, M. Cucini, F. Di Stasio, D. Comoretto, M. Galli, F. Marabelli, N. Manfredi, C. Marinzi, A. Abbotto. *J Phys Chem C* **2010**, 114, 2403.
22. V. Morandi, F. Marabelli, V. Amendola, M. Meneghetti, D. Comoretto. *Adv Funct Mater* **2007**, 17, 2779.
23. V. Morandi, F. Marabelli, V. Amendola, M. Meneghetti, D. Comoretto. *J Phys Chem C* **2008**, 112, 6293.
24. W. H. Zachariasen. *Theory of X-Ray Diffraction in Crystals*. Dover, New York, 1967.
25. Y. Kang, J. J. Walish, T. Gorishnyy, E. L. Thomas. *Nat Mater* **2007**, 6, 957.
26. B. V. Lotsch, C. B. Knobbe, G. A. Ozin. *Small* **2009**, 5, 1498.
27. S. A. Asher, K. W. Kimble, J. P. Walker. *Chem Mater* **2008**, 20, 7501.
28. B. H. Cumpston, S. P. Ananthavel, S. Barlow, D. L. Dyer, J. E. Ehrlich, L. L. Erskine, A. A. Heikal, S. M. Kuebler, I. Y. S. Lee, D. McCord-Maughon, J. Qin, H. Rockel, M. Rumi, X.-L. Wu, S. R. Marder, J. W. Perry. *Nature* **1999**, 398, 51.
29. A. M. Urbas, M. Maldovan, P. DeRege, E. L. Thomas. *Adv Mater* **2002**, 14, 1850.
30. Z. Nie and E. Kumacheva. *Nat Mater* **2008**, 7, 277.
31. H. Finkelmann, S. T. Kim, A. Muaeoz, P. Palffy-Muhoray, B. Taheri. *Adv Mater* **2001**, 13, 1069.
32. M. Ozaki, M. Kasano, T. Kitasho, D. Ganzke, W. Haase, K. Yoshino. *Adv Mater* **2003**, 15, 974.
33. A. Bolognesi, C. Mercogliano, S. Yunus, M. Civardi, D. Comoretto, A. Turturro. *Langmuir* **2005**, 21, 3480.
34. A. A. Zakhidov, R. H. Baughman, Z. Iqbal, C. Cui, I. Khayrullin, S. O. Dantas, J. Marti, V. G. Ralchenko. *Science* **1998**, 282, 897.
35. P. Jiang, K. S. Hwang, D. M. Mittleman, J. F. Bertone, V. L. Colvin. *J Am Chem Soc* **1999**, 121, 11631.
36. P. Jiang, J. F. Bertone, K. S. Hwang, V. L. Colvin. *Chem Mater* **1999**, 11, 2132.
37. C. Lopez. *Adv Mater* **2003**, 15, 1679.
38. P. Jiang, J. F. Bertone, V. L. Colvin. *Science* **2001**, 291, 453.

39. J. E. G. J. Wijnhoven and W. L. Vos. *Science* **1998**, 281, 802.
40. J. E. G. Wijnhoven, L. Bechger, W. L. Vos. *Chem Mater* **2001**, 13, 4486.
41. E. Pavarini, L. C. Andreani, C. Soci, M. Galli, F. Marabelli, D. Comoretto. *Phys Rev B* **2005**, 72, 045102.
42. D. J. Norris, E. G. Arlinghaus, L. Meng, R. Heiny, L. E. Scriven. *Adv Mater* **2004**, 16, 1393.
43. D. J. Norris and Y. A. Vlasov. *Adv Mater* **2001**, 13, 371.
44. S. G. Romanov, T. Maka, C. M. S. Torres, M. Muller, R. Zentel. *Synth Met* **2001**, 124, 131.
45. A. Blanco, E. Chomski, S. Grabtchak, M. Ibisate, S. John, S. W. Leonard, C. Lopez, F. Meseguer, H. Miguez, J. P. Mondia, G. A. Ozin, O. Toader, H. M. V. Driel. *Nature* **2000**, 405, 437.
46. J. D. Joannopulos, R. D. Meade, J. N. Win. *Photonic Crystals: Molding the Flow of the Light*. Princeton University Press, Princeton, NJ, 1995.
47. F. Garcia-Santamaria, H. Minguez, M. Ibisate, F. Meseguer, C. Lopez. *Langmuir* **2002**, 18, 1942.
48. S. H. Park and Y. Xia. *Langmuir* **1999**, 15, 266.
49. O. D. Velev, T. A. Jede, R. F. Lobo, A. M. Lenhoff. *Chem Mater* **1998**, 10, 3597.
50. A. van Blaaderen, R. Ruel, P. Wiltzius. *Nature* **1997**, 385, 321.
51. N. D. Denkov, O. D. Velev, P. A. Kralchevsky, I. B. Ivanov, H. Yoshimura, K. Nagayama. *Nature* **1993**, 361, 26.
52. S. Wong, V. Kitaev, G. A. Ozin. *J Am Chem Soc* **2003**, 125, 15589.
53. T. Solomon and M. J. Solomon. *J Chem Phys* **2006**, 124, 134905.
54. F. Pan, J. Zhang, C. Cai, T. Wang. *Langmuir* **2006**, 22, 7101.
55. J. Ye, R. Zentel, S. Arpiainen, J. Ahopelto, F. Jonsson, S. G. Romanov, C. M. S. Torres. *Langmuir* **2006**, 22, 7378.
56. J. Xia, Y. Ying, S. H. Foulger. *Adv Mater* **2005**, 17, 2463.
57. A. S. Dimitrov and K. Nagayama. *Langmuir* **1996**, 12, 1303.
58. A. L. Rogach, N. A. Kotov, D. S. Koktysh, J. W. Ostrander, G. A. Ragoisha. *Chem Mater* **2000**, 12, 2721.
59. R. Pozas, A. Mihi, M. Ocaña, H. Míguez. *Adv Mater* **2006**, 18, 1183.
60. B. Hatton, L. Mishchenko, S. Davis, K. H. Sandhage, J. Aizenberg. *PNAS* **2010**, 107, 10354.
61. A. Mihi, M. Ocaña, H. Míguez. *Adv Mater* **2006**, 18, 2244.
62. J. A. Lee, L. Meng, D. J. Norris, L. E. Scriven, M. Tsapatsis. *Langmuir* **2006**, 22, 5217.
63. P. V. Braun and P. Wiltzius. *Curr Opin Colloid Interface Sci* **2002**, 7, 116.
64. A. Rogach, A. Susha, F. Caruso, G. Sukhorukov, A. Kornowski, S. Kershaw, H. Mohwald, A. Eychmuller, H. Weller. *Adv Mater* **2000**, 12, 333.
65. D. Wang, A. L. Rogach, F. Caruso. *Chem Mater* **2003**, 15, 2724.
66. K. Kim, S. Webster, N. Levi, D. L. Carroll, M. R. Pinto, K. S. Schanze. *Langmuir* **2005**, 21, 5207.
67. Y. Jin, Y. H. Zhu, X. L. Yang, H. B. Jiang, C. Z. Li. *J Colloid Interface Sci* **2006**, 301, 130.

68. M. Pope and C. E. Swenberg. *Electronic Processes in Organic Crystals and Polymers*. Oxford University Press, New York, 1999.
69. D. Comoretto, V. Morandi, M. Galli, F. Marabelli, C. Soci. *Functional Supramolecular Architectures*, P. Samori and F. Cacialli, eds. WILEY-VCH Verlag GmbH & Co. KGaA, Weinheim, 2010.
70. S. Kubo, A. Fujishima, O. Sato, H. Segawa. *J Phys Chem C* **2009**, 113, 11704.
71. A. Pasquazi, S. Stivala, G. Assanto, V. Amendola, M. Meneghetti, M. Cucini, D. Comoretto. *Appl Phys Lett* **2008**, 93, 091111.
72. D. P. Gaillot, E. Graugnard, J. S. King, C. J. Summers. *J Opt Soc Am B* **2007**, 24, 990.
73. T. V. Murzina, F. Y. Sychev, I. A. Kolmychek, O. A. Aktsipetrov. *Appl Phys Lett* **2007**, 90, 161120.
74. S.-L. Kuai, G. Bader, P. V. Ashrit. *Appl Phys Lett* **2005**, 86, 221110.
75. A. C. Arsenault, J. Halfyard, Z. Wang, V. Kitaev, G. A. Ozin, I. Manners, A. Mihi, H. Miguez. *Langmuir* **2005**, 21, 499.
76. *Organic Light Emitting Devices: Synthesis, Properties and Applications*, K. Müllen, ed. Wiley-VCH, 2006.
77. A. C. Arsenault, D. P. Puzzo, I. Manners, G. A. Ozin. *Nat Photon* **2007**, 1, 468.
78. http://www.opalux.com/.
79. J. D. Joannopoulos, P. R. Villeneuve, S. Fan. *Nature* **1997**, 386, 143.
80. P. V. Braun, S. A. Rinne, F. Garcia-Santamaria. *Adv Mater* **2006**, 18, 2665.
81. Y. H. Ye, T. S. Mayer, I. C. Khoo, I. B. Divliansky, N. Abrams, T. E. Mallouk. *J Mater Chem* **2002**, 12, 3637.
82. Q. Yan, Z. Zhou, X. S. Zhao, S. J. Chua. *Adv Mater* **2005**, 17, 1917.
83. J. F. Galisteo-Lopez, M. Galli, L. C. Andreani, A. Mihi, R. Pozas, M. Ocana, H. Miguez. *Appl Phys Lett* **2007**, 90, 101113.
84. D. L. C. Chan, E. Lidorikis, J. D. Joannopoulos. *Phys Rev E* **2005**, 71, 056602.
85. E. Palacios-Lidon, J. F. Galisteo-Lopez, B. H. Juarez, C. Lopez. *Adv Mater* **2004**, 16, 341.
86. Q. Yan, Z. Zhou, X. S. Zhao. *Chem Mater* **2005**, 17, 3069.
87. S. Furumi, H. Fudouzi, H. T. Miyazaki, Y. Sakka. *Adv Mater* **2007**, 19, 2067.
88. N. Tetreault, A. Mihi, H. Miguez, I. Rodriguez, G. A. Ozin, F. Meseguer, V. Kitaev. *Adv Mater* **2004**, 16, 346.
89. J. R. Lawrence, Y. Ying, P. Jiang, S. H. Foulger. *Adv Mater* **2006**, 18, 300.
90. L. Wang, Q. Yan, X. S. Zhao. *Langmuir* **2006**, 22, 3481.
91. S. Noda, A. Chutian, M. Imada. *Nature* **2000**, 407, 608.
92. O. Painter, R. K. Lee, A. Scherer, A. Yariv, J. D. O'Brien, P. D. Dapkus, I. Kim. *Science* **1999**, 284, 1819.
93. F. DiStasio, L. Berti, M. Burger, F. Marabelli, S. Gardin, T. Dainese, R. Signorini, R. Bozio, D. Comoretto. *Phys Chem Chem Phys* **2009**, 11, 11515.
94. A. Arsenault, F. Fleischhaker, G. V. Freymann, V. Kitaev, H. Miguez, A. Mihi, N. Tetreault, E. Vekris, I. Manners, S. Aitchison, D. Perovic, G. A. Ozin. *Adv Mater* **2006**, 18, 2779.
95. Q. Yan, L. K. Teh, Q. Shao, C. C. Wong, Y. M. Chiang. *Langmuir* **2008**, 24, 1796.

96. N. Tetreault, A. C. Arsenault, A. Mihi, S. Wong, V. Kitaev, I. Manners, H. Miguez, G. A. Ozin. *Adv Mater* **2005**, 17, 1912.
97. F. Fleischhaker, A. C. Arsenault, J. Schmidtke, R. Zentel, G. A. Ozin. *Chem Mater* **2006**, 18, 5640.
98. N. Tetreault, H. Miguez, S. M. Yang, V. Kitaev, G. A. Ozin. *Adv Mater* **2003**, 15, 1167.
99. Y. Jun, C. A. Leatherdale, D. J. Norris. *Adv Mater* **2005**, 17, 1908.
100. F. Fleischhaker, A. C. Arsenault, V. Kitaev, F. C. Peiris, G. von Freymann, I. Manners, R. Zentel, G. A. Ozin. *J Am Chem Soc* **2005**, 127, 9318.
101. Q. Yan, A. Chen, S. J. Chua, X. S. Zhao. *Adv Mater* **2005**, 17, 2849.
102. P. Massé, S. Reculusa, K. Clays, S. Ravaine. *Chem Phys Lett* **2006**, 422, 251.
103. Q. Yan, L. Wang, X. S. Zhao. *Adv Funct Mater* **2007**, 17, 3695.
104. T. A. Taton and D. J. Norris. *Nature* **2002**, 416, 685.
105. W. Lee, S. A. Pruzinsky, P. V. Braun. *Adv Mater* **2002**, 14, 271.
106. G. M. Gratson, M. Xu, J. A. Lewis. *Nature* **2004**, 428, 386.
107. G. M. Gratson and J. A. Lewis. *Langmuir* **2005**, 21, 457.
108. G. M. Gratson, F. Garcia-Santamaria, V. Lousse, M. Xu, S. Fan, J. A. Lewis, P. V. Braun. *Adv Mater* **2006**, 18, 461.
109. T. Sekitani, Y. Noguchi, U. Zschieschang, H. Klauk, T. Someya. *PNAS* **2008**, 105, 4976.
110. M. Singh, H. M. Haverinen, P. Dhagat, G. E. Jabbour. *Adv Mater* **2010**, 22, 673.
111. *The Physics of Block Copolymers*. I. W. Hamley, ed. Oxford University Press, Oxford, 1998.
112. *Developments in Block Copolymers Science and Technology*, I. W. Hamley, eds. Wiley, New York, 2004.
113. E. L. Thomas, D. M. Anderson, C. S. Henkee, D. Hoffman. *Nature* **1988**, 334, 598.
114. C. Park, J. Yoon, E. L. Thomas. *Polymer* **2003**, 44, 6725.
115. A. Urbas, Y. Fink, E. L. Thomas. *Macromolecules* **1999**, 32, 4748.
116. T. Deng, C. Chen, C. Honeker, E. L. Thomas. *Polymer* **2003**, 44, 6549.
117. Y. Fink, A. M. Urbas, M. G. Bawendi, J. D. Joannopoulos, E. L. Thomas. *J Light Tech* **1999**, 17, 1963.
118. A. C. Edrington, A. M. Urbas, P. DeRege, C. X. Chen, T. M. Swager. *Adv Mater* **2001**, 13, 421.
119. J. Yoon, R. T. Mathers, G. W. Coates, E. L. Thomas. *Macromolecules* **2006**, 39, 1913.
120. M. Bockstaller, R. Kolb, E. L. Thomas. *Adv Mater* **2001**, 13, 1783.
121. M. R. Bockstaller and E. L. Thomas. *J Phys Chem B* **2003**, 107, 10017.
122. V. Z. H. Chan, J. Hoffman, V. Y. Lee, H. Iatrou, A. Avgeropoulos, N. Hadjichristidis, R. D. Miller, E. L. Thomas. *Science* **1999**, 286, 1716.
123. M. Scalora, J. P. Dowling, C. M. Bowden, M. J. Blomer. *Phys Rev Lett* **1994**, 73, 1368.
124. T. Hashimoto, M. Shibayama, H. Kawai. *Macromolecules* **1980**, 13, 1237.
125. T. Hashimoto, H. Tanaka, H. Hasegawa. *Macromolecules* **1990**, 23, 4378.
126. P. D. Hustad, G. R. Marchand, E. I. Garcia-Meitin, P. L. Roberts, J. D. Weinhold. *Macromolecules* **2009**, 42, 3788.

127. A. Urbas. *Adv Mater* **2000**, 12, 812.
128. A. N. Semenov. *Macromolecules* **1993**, 26, 2273.
129. J. Huh and W. H. Jo. *J Chem Phys* **2002**, 117, 9920.
130. N. Torikai, N. Takabayashi, I. Noda, S. Koizumi, Y. Morii, Y. Matsushita. *Macromolecules* **1997**, 30, 5698.
131. M. R. Bockstaller, Y. Lapetnikov, S. Margel, E. L. Thomas. *J Am Chem Soc* **2003**, 125, 5276.
132. M. R. Bockstaller, R. A. Mickiewicz, E. L. Thomas. *Adv Mater* **2005**, 17, 1331.
133. M. R. Bockstaller and E. L. Thomas. *Phys Rev Lett* **2004**, 93, 166106.
134. S. P. Gido, J. Gunther, E. L. Thomas, D. Hoffman. *Macromolecules* **1993**, 26, 4506.
135. R. J. Albalak and E. L. Thomas. *J Polym Sci B Polym Phys* **1993**, 31, 37.
136. J. Yoon, W. Lee, E. Thomas. *Adv Mater* **2006**, 18, 2691.
137. C. Park, C. D. Rosa, E. L. Thomas. *Macromolecules* **2001**, 34, 2602.
138. J. Hu and S. Liu. *Macromolecules* **2010**, 43, 8315.
139. C. Osuji. *Adv Funct Mater* **2002**, 12, 753.
140. H. Kosonen, S. Valkama, J. Ruokolainen, M. Torkkeli, R. Serimaa, G. ten Brinke, O. Ikkala. *Eur Phys J E Soft Mat Biol Phys* **2003**, 10, 69.
141. S. Valkama, H. Kosonen, J. Ruokolainen, T. Haatainen, M. Torkkeli, R. Serimaa, G. Brinke, O. Ikkala. *Nature Mater* **2004**, 3, 872.
142. J. Yoon, W. Lee, E. L. Thomas. *Macromolecules* **2008**, 41, 4582.
143. C. Kang, E. Kim, H. Baek, K. Hwang, D. Kwak, Y. Kang, E. L. Thomas. *J Am Chem Soc* **2009**, 131, 7538.
144. J. J. Walish, Y. Kang, R. A. Mickiewicz, E. L. Thomas. *Adv Mater* **2009**, 21, 3078.
145. Y. Lu, H. Xia, G. Zhang, C. Wu. *J Mater Chem* **2009**, 19, 5952.
146. *Polymer Handbook*, 4th ed, J. I. Brandrup, H. Edmund, E. A. Grulke, A. Abe, D. R. Bloch, eds. John Wiley & Sons, 2005.
147. M. Cucini, M. Alloisio, A. Demartini, D. Comoretto. *Biomimetic and Supramolecular Systems Research*. Nova Science Publishers Inc., Hauppauge, NY, 2008.
148. A. K. Khandpur, S. Foerster, F. S. Bates, I. W. Hamley, A. J. Ryan, W. Bras, K. Almdal, K. Mortensen. *Macromolecules* **1995**, 28, 8796.
149. M. F. Schulz, A. K. Khandpur, F. S. Bates, K. Almdal, K. Mortensen, D. A. Hajduk, S. M. Gruner. *Macromolecules* **1996**, 29, 2857.

CHAPTER 20

APPLICATIONS OF CHARGED MEMBRANES IN SEPARATION, FUEL CELLS, AND EMERGING PROCESSES

GÉRALD POURCELLY, VICTOR V. NIKONENKO,
NATALIA D. PISMENSKAYA, and ANDREY B. YAROSLAVTSEV

20.1 INTRODUCTION

A great advantage of membrane separation processes is their high selectivity and very low need for chemical additives. It makes them a very effective, environmentally friendly tool for resolving numerous industrial problems related to liquid and gas separation; in particular, the treatment of certain industrial effluents containing toxic components, which should not be released into the environment, or valuable materials, which can be recovered and reused. Membrane processes seem indispensable in realization of zero liquid discharge (ZLD) systems, meaning such industrial engineering where only solid wastes may leave the boundary of the plant.[1]

Techniques involving charged membranes occupy an important place among membrane technologies in general. The structure and properties of charged membranes were considered in Chapter 9. Their main structural feature is a network of nanometer aqueous domains serving as conducting channels and embedded into a hydrophobic polymer matrix. The ions fixed on the charged pore walls are the reason for the channel hydrophilicity and permselectivity. The most extensively used charged membranes are ion exchange, reverse osmosis (RO), and nanofiltration (NF) membranes. Historically, ion exchange membranes (IEMs)[2,3] and their practical applications[4–7] have appeared earlier than the pressure-driven ones.[8,9] The RO process is definitively suitable in water and wastewater desalination where the pressure-driven processes have demonstrated greater growth then that of

Ionic Interactions in Natural and Synthetic Macromolecules, First Edition. Edited by Alberto Ciferri and Angelo Perico.

electrodialysis (ED) and electrodeionization (EDI). Desalination systems produce a huge amount of freshwater; *Global Water Intelligence*[10] evaluates worldwide desalination as 40 million m^3 day^{-1} in 2006 with a growth up to 64 million m^3 day^{-1} in 2010 and up to 97.5 million m^3 day^{-1} in 2015. This large market is dominated by RO (about 50%) and thermal (40%) technologies while ED and EDI occupy only about 5%.[1] However, the interest in IEMs and electrotransfer phenomena in membrane systems increases rather rapidly due to novel or renovated applications. The opinion that for water treatment applications hybrid systems (including RO together with ED and/or EDI units) are the most effective solution is becoming generally recognized. Some researchers call the use of ED for concentrating RO brines a "new niche" for ED[11] (see also Section 20.2). Second, electromembrane processes are much demanded in other domains that emerged within the last decade: clean energy systems (fuel cells and other devices for the production and accumulation of electric power), micropumps and other micro- and nanofluidic devices (microelectromechanical-systems), analytical sensors, and others. Finally, there are traditional applications, such as chlor-alkali production, electrochemical synthesis, and the wine, food, and whey industries, where the use of IEM is rapidly growing.

Strathmann[12] proposed a quite useful classification of electromembrane processes. He separated them in developed (conventional ED, diffusion, and Donnan dialysis) and developing (bipolar membrane electrodialysis [BMED], continuous electrodeionization [CEDI], capacitive deionization [CDI]) categories, and marked out the processes emerging from the ongoing research (catalytic membrane reactors, reverse electrodialysis (RED), overlimiting current density operation, and others). In this chapter, we give a survey of traditional and emerging applications of charged membranes, including overlimiting and pulse current ED, micro- and nanofluidic desalination devices, and others.

This chapter is conceived to describe the main ideas of how membranes can be used to solve different practical problems.

20.2 DESALINATION AND DEIONIZATION

The predominance of charge membrane processes in desalination and deionization is attributed to their remarkable ion removal efficiency, particularly in natural and wastewater treatment. These technologies include mainly RO and ED, as was mentioned in the introduction. Both of these processes use membranes to separate the feed stream into two streams differing in concentration: a dilute and a concentrated stream.

The main product of a desalination process is freshwater (potable water as well as water for agriculture and industry). This product is critical to human habitation on this planet. Actually, most of the 98% of our waters are either sea or brackish waters. Typical concentration of dissolved salts in seawater and brackish water is 35,000 and 1000–10,000 mg L^{-1}, respectively. Besides RO and ED with its derivatives (electrodialysis reversal [EDR], EDI), novel techniques involving charged membranes have emerged: CDI[13] and micro-/nanofluidic desalination.[14]At the same time, there is a stable trend of applying hybrid membrane systems accumulating the advantages of individual membrane processes. In this section we consider all of the abovementioned techniques.

20.2.1 Reverse Osmosis

Consider a system wherein a charged membrane separates an electrolyte solution and water. An osmotic pressure difference causing water to move from the dilute to the concentrated compartments occurs in the system. The phenomenon is known as osmosis. When a pressure higher than the osmotic pressure difference is applied in the concentrated compartment, the direction of water transfer changes and water passes through the membrane into the diluate compartment. This is known as RO. The main difference of RO from other baromembrane processes is that the solvent molecules and dissolved substances, which are separating, are of comparable size.

The scheme of RO desalination mechanism is shown in Figure 20.1. The main role is played by charge sites fixed on the pore walls (see Chapter 9 for the structure of charged membranes). Under the action of pressure gradient, water molecules pass freely through a pore with a charged surface; however, the charged particles are retained. In the case where the fixed ions are charged negatively (Fig. 20.1), the anions from the external solution cannot pass because they are excluded from the pore as co-ions (Donnan exclusion, Chapter 9); the cations are retained by the electric field formed by themselves: a small amount of these ions transferred with the forced convection flow creates an excess of positive charge at the outlet of the pore, while an excess of negative charge appears at the inlet. As a result, an electric potential difference (pd) (named streaming potential)[15,16] is created. This electric pd produces an electric force opposite to the mechanic force created by the pressure gradient. In steady state, both forces are equal and directed in opposite sides, so that the resultant driving force is zero.

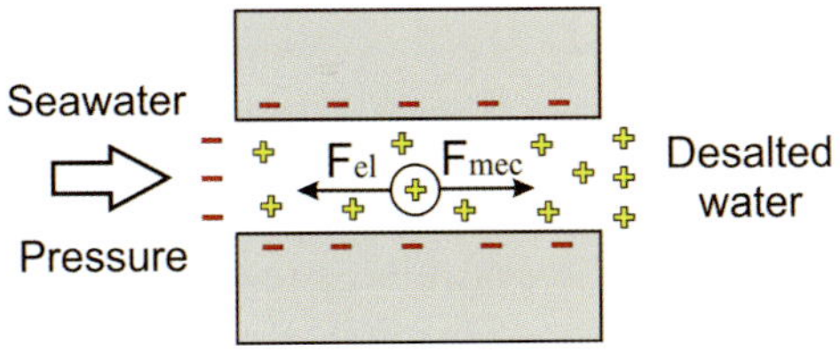

FIGURE 20.1. Scheme of seawater desalination by a RO membrane; a nanopore with negatively charged walls, and the forces applied to a cation within the pore are shown. Adapted from Nikonenko et al.[29]

Besides RO, another pressure-driven process essentially related to the charge of pore walls of the membrane is widely applied: NF. The main difference between two kinds of membranes is in the pore radius, r. NF membranes contain mesopores with a radius in the range 2–4 nm, while in RO membranes there are micropores whose radius is about two times lower.[17] Larger pores of NF membranes produce a lower Donnan exclusion, so that monovalent ions pass through NF membranes relatively easily, while multivalent ones do not. As a result, RO membranes effectively reject all ions and only water (or another solvent) may pass through. NF is used where mono- and multivalent ions should be separated, for example, in the water softening process.[18]

Cationic NF membranes have negative groups attached to the polymer backbone. Their negative charges selectively repel negative, particularly multivalent, anions such as SO_4^{2-}, while attracting positive cations, particularly divalent cations such as Ca^{2+}. The result is an order of salt rejection[19]:

$$Na_2SO_4 > NaCl > CaCl_2.$$

In the case of anionic NF membranes, their positive fixed groups repel cations, particularly divalent ones, and attract anions. The order of salt rejection is then[19]:

$$CaCl_2 > NaCl > Na_2SO_4.$$

In order to enhance the rejection performance, NF membranes frequently combine both size and Donnan exclusion effects. This kind of pressure-driven membranes, called low pressure RO membranes, have very high salt rejections and low hydraulic resistance at low salt concentrations, but lose their selectivity at salt concentrations above 1000 or 2000 ppm in feedwater since the thickness of the electric double layer (EDL) at the pore wall becomes small compared to the pore diameter. The membranes are therefore used to remove salt from already relatively clean water. The membranes are usually operated at a very low pressures (3–15 bars).[7]

Mathematical description of RO and NF processes is carried out by applying capillary space charge models.[20–22] These models allow quantitative description for the increase of the capacity and separation ability of charged membranes with the charge of the walls, and why multivalent ions are better rejected.

The well-known applications of RO and NF stem from the seminal discovery by Loeb and Sourirajan.[23] They invented bilayer RO membranes that contained a thin (about 1 μm) selective surface layer and a much thicker (about 100 μm), more permeable macroporous support providing the mechanical strength. The water permeability of the first RO membrane prepared by Loeb and Sourirajan was 10 times larger than that of any membrane then available.

Nowadays, about 20 million m^3 day^{-1} of water are desalted by RO.[10] According to Baker,[19] about half of this capacity is mainly used to produce ultrapure industrial water in the United States, Europe, and Japan; the other half is used to produce municipal drinking water in the Middle East and other desert regions from seawater or brackish groundwater.

There is a wide variety of designs of RO/NF membrane modules.[17] The most common are spiral-wound modules and those with hollow fiber membranes. The outer diameter of these fibers can be less than 100 μm, wall thickness is about 20 μm, and specific surface area can be larger than 1000 $m^2 m^{-3}$.[17,19] Modern composite membranes provide more then 99% salt rejection at 1.2 $m^3 m^{-2}$ day at 55 bars in the seawater version and 15 bars in the brackish water version.[19]

In addition to desalination, RO is also increasingly being used in other applications[24,25] due to the high and stable quality of the water produced and the relatively low cost. Some of these applications are described in Section 20.6. The economically justifiable range of RO application is from 2–5 mg L^{-1} of dissolved salts (the resistivity of 0.2 Mohm cm) in the permeate (desalted product) to 50–70 g L^{-1} in the retentate (brine), which are obtained in processing seawater and other electrolyte solutions. In the production of more dilute solutions, the energy costs are larger, since it is necessary to separate water from the solution. Production of more concentrated brines is limited by the need to overcome the osmotic pressure, the magnitude of which in the case of seawater (salt concentration about 30 g L^{-1}) is close to 25 bars.

The main problems in the operation of RO/NF modules are fouling of the membranes and relatively large liquid waste emission.[7] Moreover, RO membranes are sensitive to chlorine and other oxidants often present in feedwater, which provokes their degradation. Heavy metal ions such as iron are able to catalyze chlorine degradation.[26,27] As a result, RO operation needs pretreatment, entailing high investment costs.

20.2.2 Conventional ED

ED is an electromembrane process for the separation of ions across charged membranes from one solution to another under the influence of an electrical potential difference used as the driving force. Nonionic and macromolecular species are retained as they cannot permeate through IEMs. The cation exchange membranes (CEMs) and anion exchange membranes (AEMs) are normally separated by a spacer gasket and form individual cells. As a rule, an electrolyte solution is pumped through these cells. An electrical potential difference applied between the electrodes forces the cations (C^+) to migrate toward the cathode. They pass through the CEMs but are retained by the anion exchange ones. Likewise, the anions (A^-) migrate toward the anode and pass through the AEMs but are retained by the cation exchange ones. The overall result is that the feed electrolyte solution is separated into a concentrated solution, found in concentrate compartments (CC), and a dilute solution, in diluate compartments (DC). CC and DC alternate in an ED stack (Fig. 20.2).

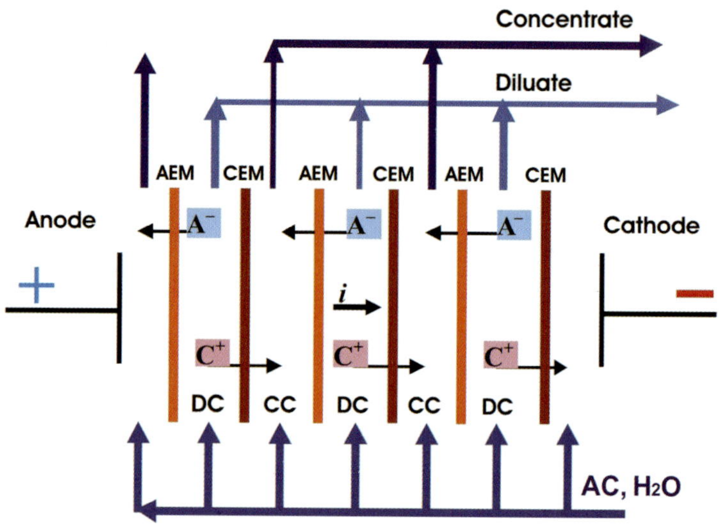

FIGURE 20.2. Scheme of anions (A^-) and cations (C^+) transfer through anion (AEM) and cation exchange (CEM) membranes forming alternate diluate (DC) and concentrate (CC) compartments in a conventional ED stack.

In an industrial-sized ED stack, 100–1000 cell pairs are arranged between the electrodes. Various stack constructions such as the plate-and-frame (another name for sheet flow) or the spiral-wound stack design are used in practical applications.[12,28]

The flowing of current through an IEM results in concentration polarization, which is the formation of concentration gradients at the membrane/solution interface.[7,29,30] The cause of this phenomenon is that the membrane has the ability to transport some species (counterions) more readily than the other(s) (co-ions) (see also Chapter 9, Section 9.3). The selective transport of counterions through an IEM leads to a decrease in salt concentration at the depleted interface facing a DC, and to an increase at the other interface facing a CC. Concentration polarization is an inherent but undesirable phenomenon in ED process. The salt concentration at the depleted interface decreases with increasing current density. At a certain value called the limiting current density, i_{lim}, the interfacial concentration becomes much lower than the bulk concentration. The equation relating i_{lim} to the thickness of diffusion boundary layer at an IEM, δ, and phenomena occurring when the current density is close or higher than i_{lim} are considered in Chapter 9. As the interfacial concentration approaches zero, an increase in the cell resistance and potential difference occurs. Depletion of the interface in salt ions also leads to a water splitting process producing H^+ and OH^- ions. These ions carry a part of the current, thus lowering the ED current efficiency. Moreover, this process gives rise to variation in pH of the solutions in the DC and CC. Normally, the rate of H^+ and OH^- ion generation is higher at the AEM in comparison with the CEM forming a cell pair. In this case, the DC solution becomes acid, and the CC one, alkaline. High pH value at the enriched interface of an IEM provokes here the deposition of salts (scaling), mainly $CaCO_3$.[31,32] When treating solutions containing macromolecules or colloidal substances, their deposition is also possible. Feed solutions often contain negatively charged colloid particles or large organic anions. Under the action of applied electric field these components migrate to the AEM and can be deposited on its surface to form a so-called fouling layer.[12] The formation of a fouling (or biofouling, if biological substances such as bacteria or algae are present) layer is favored by a pH variation, and an increase in temperature (T) at membrane surface provides polymerization of organic matter.

Membrane scaling and fouling lead to a significant increase in the resistance of IEMs, which can eventually reduce limiting current density and raise the operating costs. However, it should be remarked that ED membranes show better resistance to fouling and scaling in comparison with the RO ones.[12] This resistance can be improved by application of ED reversal and pulsed current (see Section 20.2.3). As a consequence, RO modules demand a more careful pretreatment than ED. Other advantages of ED compared to RO are high water recovery rates, long useful life of membranes up to 5 years or higher, and operation at elevated T up to 50°C.[31,33] Among the disadvantages of ED compared to RO, Strathmann[12] remarks that neutral toxic components such as viruses or bacteria are not removed from a feed stream. Note that overlimiting operation of ED results in killing bacteria and allows production of apyrogenic water.[34,35]

Conventionally, desalination ED process is carried out in underlimiting current modes in order to prevent scaling and fouling of IEMs as well as elevated energy costs. Since the limiting current density is approximately proportional to the feed solution concentration, the rate of ED decreases with decreasing concentration. As a result, conventional ED becomes inefficient when the salinity of the feed solution becomes lower than about 400 mg L^{-1}. The upper limit for the application of this process is approximately 5000 mg L^{-1} of total dissolved salts (TDS) in the feed solution, as higher salt concentration causes large energy costs.[12,36] In the range of salinity of feed solution between 400 and 5000 mg L^{-1}, other membrane processes are more effective: EDI in the low concentration range, and RO in the high one.

As in RO, the macrocomponent, water, is moved away from the feed solution, the energy consumption only slightly depends on the feedwater salinity and constitutes about 5 kWh m^{-3}.[37] In ED, the energy consumption is approximately proportional to the amount of the extracted salt. It is evaluated as approximately 300 kWh per 1 ton of extracted salt in the case of brackish water treatment,[38] and 150 kWh t^{-1} in the case of seawater.[39] (Note that the energy consumption depends on the current density applied, type of membranes, and other parameters.[38,39]) To desalinate seawater, ~7–15 kWh m^{-3} are needed[40]; for brackish water with 10 g L^{-1} salinity, the energy consumption is close to 3 kWh m^{-3}.[38]

ED has an economic advantage over other desalination processes in cases of particular feedwater salt composition. This chemical-free technology competes directly with RO especially when it is necessary to remove nitrates[41] or fluorides[42] from brackish waters.

Conventional ED, which is now frequently displaced by EDR, is widely used in desalination of brackish river waters. This application was developed in the United States[43,44] and recently in Spain where 4.5 million inhabitants of the Barcelona metropolitan area are mainly supplied with surface water treated by EDR.[45] More often, ED is used in small- to medium-sized plants with capacities of less than a few 100 $m^3\ d^{-1}$ to more than 20,000 $m^3\ d^{-1}$.[12,46] Another large application of this process is predemineralization of industrial solutions. For example, this process is applied in boiler feedwater treatment as a step before the use of ion exchange (or EDI).[12] Such systems allow significant savings in recovery chemicals needed for the ion exchange step. The savings increase with increasing feedwater salinity. In ultrapure water production needed for microelectronics and power plants, ED is used for predemineralization in water recycling systems.[35,47] A typical application of industrial water reuse is the recycling of cooling tower blowdown water. ED is particularly suited for this purpose since high recovery rates up to 95% and high brine concentrations can be achieved, resulting in saving of feedwater costs and in a reduction of wastewater discharge. Furthermore, in a majority of cases, IEMs can be operated at T of up to 50°C, which is in the range of most cooling systems.

The characteristics of IEMs important in ED desalination process are as follows: (1) mechanical strength, thermal stability, and durability in aggressive environments, as well as (2) high permselectivity and ionic conductivity in combination with low diffusion permeability.[7,48]

Mechanical strength is needed since rather large IEM sheets are used, up to 1 m and even greater. Often the T of solutions under treatment is elevated, and pH changes in a wide range; normally IEMs elaborated for ED are stable between 1 and 11 of the pH range.[33] The properties denoted by (2) should provide a high current efficiency (high permselectivity and low diffusion permeability) as well as a low electricity consumption. The salt counterion transport number characterizing the permselectivity of commercial IEM in NaCl solution (up to 0.1 M) for homogeneous membranes (such as CM2, Neosepta®, Tokuyama Corporation, Shibuya, Japan) is higher then 0.99. For heterogeneous membranes (MK-40, SchekinoAzote, Schekino, Russia), this parameter is higher than 0.97.[49] The salt diffusion permeability is about 0.1×10^{-11} and $1 \times 10^{-11}\ m^2\ s^{-1}$ for these membranes, respectively.[49] The surface membrane resistance is about 2 ohm cm^2 for CM2 and 6 ohm cm^{-2} for MK-40 in 1 M NaCl.

20.2.3 EDR and Pulsed Current ED

EDR is a variation of the ED process, wherein electrode polarity reversal is used to automatically clean membrane surfaces.[50] EDR works in the same way as the conventional ED, except that the polarity of the DC power is reversed at specific time intervals ranging from a few minutes to several hours. When the polarity is reversed, the DC and CC are also reversed and so are the chemical reactions at the electrodes. If water splitting takes place at IEM surface, these reactions will also be reversed. As was explained in the previous section, water splitting reaction can produce a shift in pH value in the diluate and the CC: an increase in pH together with increasing salt concentration leads to the scale formation, mainly calcium and magnesium carbonates. The electrode polarity reversal also results in reversal of pH shift. The acidification of the earlier alkaline solution tends to dissolve any calcium carbonate present on membrane surface.[43,51] Similar effects occur with the organic matter fouling the membrane surface. Negatively charged particles or large organic anions present in feed solution may form a deposit on the AEM in DC under the action of a direct current (Fig. 20.3a). When the polarity is reversed, the large organic anions will migrate from the AEM surface back into the core stream and the membrane properties are restored (Fig. 20.3b).

A similar effect takes place when a pulsed current is applied without reversing the solution streams. Imposing of current pulses contributes to the formation of loose, easy-to-wash with the flowing solution, fouling layers.[52–55] Moreover, the local variation in T and pH near the membrane interface are essentially lower than in conventional ED that mitigates the fouling of IEMs.[52,54] However, this promising technique needs to be more comprehensively studied.

We have also mentioned the rapidly rising application of EDR in the desalination of brackish waters. The recently started potable water production in the Barcelona area with capacity up to 4 $m^3 s^{-1}$ (about 350,000 m^3 day^{-1})[45] serves as a telling example.

Another application of EDR is the treatment of water obtained in the oil and gas industries. In recent years, a systematic study of various desalination

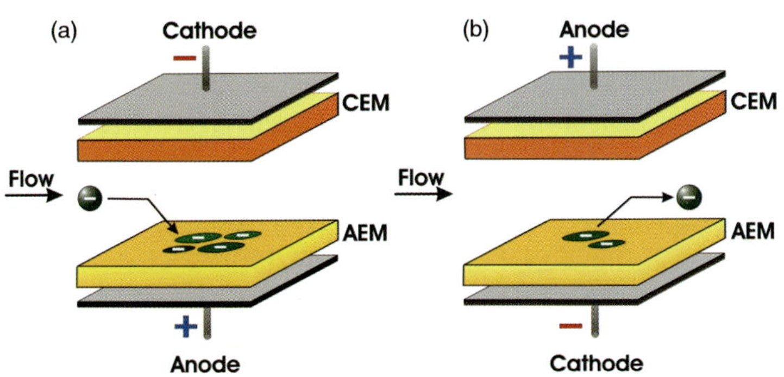

FIGURE 20.3. Schematic drawing illustrating the removal of deposited negatively charged colloidal components from the surface of an AEM by reversing the electric field. (a) Direct current; (b) reverse current. Adapted from Strathmann.[12]

methods for polymer flooding-produced water has been made in Canada, the United States, and China.[56,57] It was found that RO and EDR are reliable and economically feasible techniques. After membrane treatment, polymer flooding-produced water may have two beneficial uses. First, the diluate obtained from this water can be used for preparing polymer solutions. Second, the concentrate can be used as the injecting water in the water-flooding process for high permeability layers.[57]

The use of EDR allows a significant decrease in the requirements to the quality of the feed solution as for the presence of substances critical for membrane scaling and fouling. Consequently, very little feed pretreatment is required, making EDR attractive in treating natural and wastewaters. As compared to RO, IEMs are more stable in aggressive and oxidizing solutions: they can operate on waters with up to 0.5 mg L^{-1} chlorine, and can also be shock chlorinated up to 30 mg L^{-1}.[45]

However, to be effective, the reversal of the polarity of a stack has to be accompanied with a reversal of the flow streams. This always leads to some loss of product and requires a more sophisticated flow control. Besides, some pretreatment of the feedwater is required. In particular, the iron and manganese ion concentrations must be kept below 0.3 and 0.05 mg L^{-1}.[12] We have also mentioned the fact that neutral toxic components such as viruses or bacteria are not removed from the feed stream in ED and EDR. The use of pulsed current mode may be quite beneficial as it makes it possible to avoid the loss of the product.

Even if IEMs are less prone to scaling and fouling in EDR stacks, pH variation in the stack compartments should be avoided. Besides, high roughness of membrane surface leads to the consolidation of the deposit on the membrane. Therefore, in order to minimize the surface fouling, it is preferable to use membranes with smooth surface, which does not generate H^+ and OH^- ions and has minimal adhesion to the solid precipitations from the feed solutions. The latter may be obtained, for instance, by hydrophilization of the membrane surface.

20.2.4 Deionization

CEDI To obtain high purity water, electrodialyzers with a special stack design are applied, the process is named CEDI.[7,58] These apparatuses are able to produce deionized water, the resistivity of which is close to that of pure water (18 Mohm cm at 25°C). The process design and the different hardware components needed in CEDI are very similar to those used in conventional ED. The main difference is that in a CEDI stack the diluate cells and sometimes also the concentrate cells are filled with an ion exchange filler: a bed of ion exchange resin granules,[6,7,59–61] ion exchange textile,[35,62,63] conducting spacer,[59,64–67] or organic porous ion exchange material.[68] The stacks with profiled (relief/undulated) IEMs[12,60,69–71] occupy an intermediate position between CEDI with ion exchange bed and conventional ED with smooth membranes and nonconductive spacer. The range of feed solutions with TDS from 5 to

500 mg L^{-1} (from 10^{-4} to 10^{-2} M NaCl) is the most effective field of profiled membranes application, as it can be seen in Figure 20.4.

The ion exchange resin increases the conductivity in the diluate cells; as a result, the resistance of CEDI stack is significantly lowered. However, the main difference from conventional ED lies in the mechanism of deionization: the electromigration transfer of salt ions in CEDI is combined with ion exchange: the salt ions are exchanged with the H^+ and OH^- ions present in the ion exchange material.[59,70,72] The latter enter ion exchange resins or membranes due to the water-splitting reaction[73] occurring at $\overrightarrow{A/C}$ contacts where A is an anion and C is a cation exchange material; the arrow shows the direction of the current flow (Fig. 20.5). The active mass exchange surface in a CEDI stack is much higher than that in an ED stack due to the presence of an ion exchange bed. The ion exchange can occur on the all ion exchange particle surface, including the part normal to the current flow direction, thus enhancing the rate and the depth of deionization. Besides, ion exchange particles or profiles on the membrane surface contribute to improving hydrodynamics and reducing the diffusion layer thickness.

The source of H^+ and OH^- ions in CEDI can be not only the $\overrightarrow{A/C}$ contacts between ion exchange fillers, but also BPMs[58,60] or electrochemical reactions occurring at the electrodes.[74]

Two different concepts of CEDI stacks are shown in Figure 20.6.

Most often, in the production of ultrapure water, EDI/CEDI devices are combined with RO, which provides predemineralized water after pretreatment (normally including water softening) and sterilization of feedwater.[12,75–77] There are different solutions to the layout of such hybrid schemes, differing in pretreatment methods and the direction of diluate and concentrate flows.[76] CEDI devices replace in these systems the mixed-bed ion exchanger. The integration of CEDI in the ultrapure water production systems results in a simpler process which does not need regeneration chemicals. Besides, the raw water consumption is lower and the costs are essentially reduced; in systems with separated beds of ion exchange resins and BPMs, the removal of weakly dissociated acids is more efficient.[12]

Among the limitations of EDI devices, note the higher requirements on the quality of feedwater in comparison to conventional ED units.[35] Normally, these requirements can be met by applying single- or double-pass RO.[35] When the quality of produced water approaches high purity one (the resistance higher than 15 Mohm cm or the conductivity less than 0.07 mS cm^{-1}), high permselectivity of AEM becomes crucial.[12]

There are now several thousand continuous EDI systems in commercial operation for the production of high purity water at capacities ranging from less than 0.1 to more than 1500 m^3h^{-1}.[35] This includes a CEDI system that has been in continuous operation for over 18 years, producing deionized water for a university laboratory.[35]

CEDI technology of high purity, industrial-process water production is used in microelectronic/semiconductor, chemical, and pharmaceutical manufacturing; steam generation at power plants[12,35,78,79]; academic and clinical

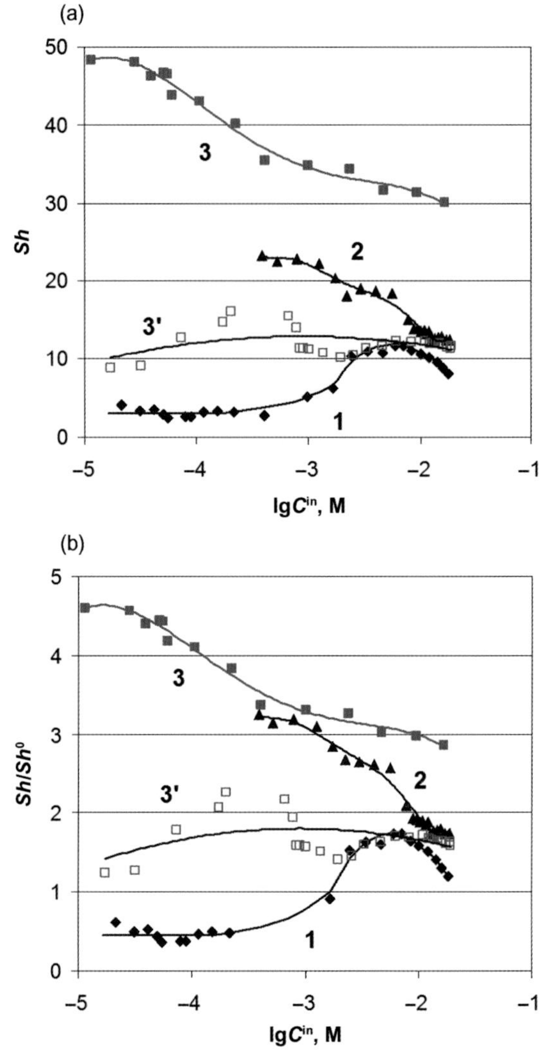

FIGURE 20.4. The Sherwood number, *Sh*, (a) and the ratio Sh/Sh^0 (b) as functions of the inlet feed NaCl solution concentration for different ED and CEDI stacks (of length 2 dm) formed by MK-40 and MA-40 membranes and containing: a nonconducting network spacer (**1**); an ion exchange monolayer bed KU-2/AV-17 (**2**); or a profiled membrane MA-40P (instead of MA-40) (**3**, **3′**). **1**, **2**, **3**: the solution flow rate through one desalination compartment normalized to a width of 1 dm, *w*, is close to 3 $dm^3\ h^{-1}\ dm^{-1}$; **3′**: $w = 7.5\ dm^3\ h^{-1}\ dm^{-1}$. The potential difference per cell pair in all cases was 2.5 V. Sh^0 is calculated for a reference channel formed by flat membranes without spacer and the same intermembrane spacing as the channel under study; Sh/Sh^0 is equal to the ratio of the effective diffusion boundary layer thickness in the reference channel to that in the channel under study. Adapted from Larchet et al.[70]

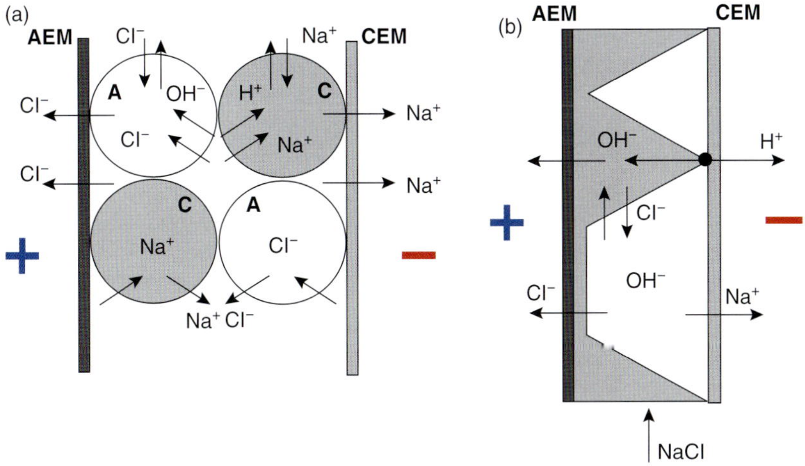

FIGURE 20.5. Scheme of ion fluxes in an EDI diluate compartment with an ion exchange bed (a) and with a profiled membrane (b). **A**, An anion exchange resin bead; **C**, a cation exchange resin bead. Arrows show the ion fluxes flowing through the interfaces of ion exchange materials. Adapted from Larchet et al.[70]

laboratories; the food and beverage industry; and other such processes.[80] Additionally, this process is used for removal of traces of heavy metal and hardness ions[74,77] or radioactive salts.[81–83]

In ED of dilute solutions under intensive currents, at least when using profiled membranes, the electrical conductivity and diffusion permeability of IEMs no longer play a decisive role. The resistivity of the solution in the diluate compartment is much higher than that of the membranes. The membrane permselectivity increases with diluting solution, hence, the electrolyte back diffusion is insignificant (if only it is not the last step of high purity water production).[12] Properties which grow in significance are those which allow enhancing the mass transfer and the control of water splitting. Besides, prevention of the surface degradation in intensive electric fields represents another important problem.

Coating the surface of a heterogeneous membrane with a conducting homogeneous film seems to be a promising way to improve the membrane properties pertinent to ED of diluted solutions. The use in the coating film of functional groups not catalytically active toward the water splitting allows decreasing generation of H^+ and OH^- ions. Even if the chemical nature of the functional groups in the film is the same as in the original membrane, this aim is achieved because the distribution of current lines becomes more homogeneous. Intensification of electroconvective mixing is obtained due to two effects. First, the use of a film with a high surface hydrophobicity leads to enhancement of fluid slipping at the surface and facilitates the generation of electroconvective vertices. Second, the use of coating allows optimization of the current line distribution. More details are given in Chapter 9 and in Nikonenko et al.[29]

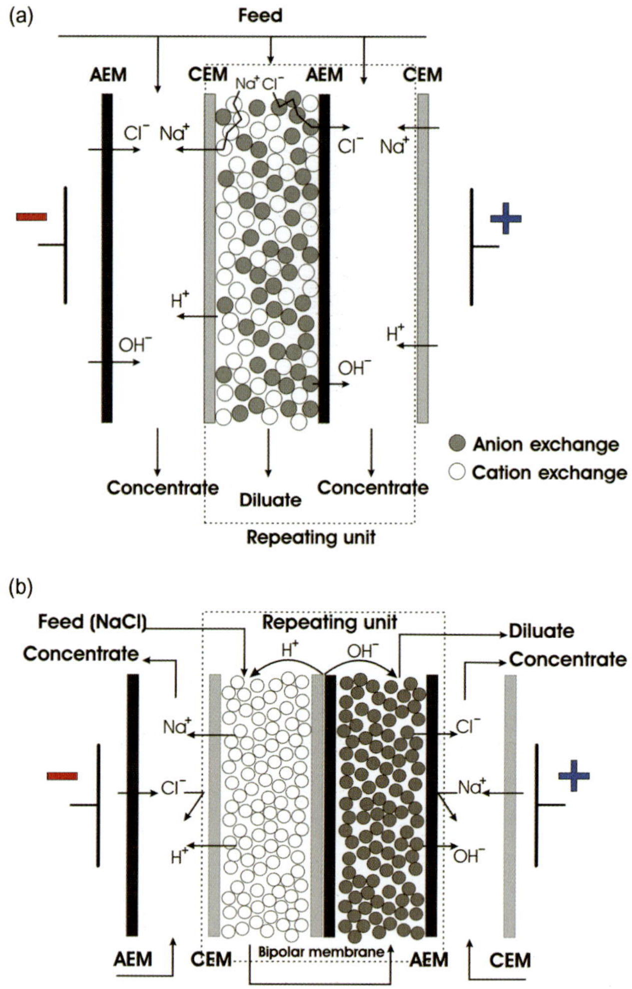

FIGURE 20.6. Schemes illustrating different CEDI stack concepts. (a) Conventional stack with diluate cells filled with a mixed-bed ion exchange resin; (b) stack with cation exchange and anion exchange resins in different diluate cells and regeneration of the ion exchange resins by H^+ and OH^- ions produced in a bipolar membrane. Adapted from Strathmann.[12]

CDI with IEMs CDI is an electrosorption process that can be used to remove ions from an aqueous solution by charge separation.[12,13,84,85] CDI operates by adsorbing ions in the double layer formed at the electrodes by the application of an electric potential difference. The principles of the process are traced to the studies of Helmholtz and to the modeling of the electrical double layer by Guoy and Chapman.[13]

A CDI cell unit consists of two electrodes made out of activated carbon and separated by a spacer. A solution to be treated flows in the channel between the electrodes. When one electrode is charged positively, and the other negatively, anions from the solution are adsorbed on the first, and cations on the second. The solution outgoing from the channel is found partially desalted. At the next (ion release) step, the polarization of the electrodes is reduced to zero or inversed, and the retained ions are desorbed. A small product stream concentrated in salt is obtained. The potential difference applied in CDI is not high, in the range of 0.8–1.5 V, in order to avoid electrochemical reactions (usually water splitting) at the electrodes.

Conventional CDI is known to be energy inefficient because ion adsorption and desorption steps are not completely separated. When an electric potential is applied, counterions in the pore are adsorbed onto the electrode, but simultaneously co-ions are expelled from the pore volume of this electrode, seriously reducing the degree of desalination.[86] To solve this problem, Andelman[87] suggested a charge barrier (IEMs) placed adjacent to an electrode of a flow-through capacitor.

The membrane capacitive deionization (MCDI) is a two-step process. In the first step, application of an electrical potential between two electrodes results in the migration of cations through a CEM, and anions through an AEM, separating the cathode and the anode, respectively, from the feed solution (Fig. 20.7a). The amount of the ions passed through the membranes will be determined by the electric capacity of the anode and cathode. During this step, a partially deionized water is obtained. In the second step, the polarity of the electrodes is reversed and the ions are desorbed at the electrodes and

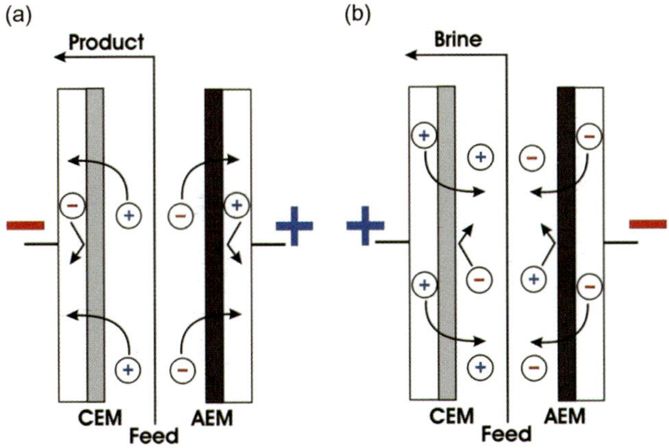

FIGURE 20.7. Schematic description of the MCDI process: an IEM is placed between the feed solution and the porous carbon electrodes. In the deionization step, the ions from the feed solution are adsorbed onto the electrodes (a); in the ion release step, desorption of ions takes place due to a reverse of electrode polarity. Adapted from Strathmann.[12]

then released into the feed solution, thus producing the brine. The advantage of the use of IEMs is that the efficiency of both steps becomes much higher in comparison to that of the conventional CDI.[85,88,89] During the deionization step, the CEM prevents the transfer of anions from the cathode pore volume, and the AEM the transfer of cations from the anode pore volume, into the feed solution, as shown in Figure 20.7b. Hence, the process of ion adsorption is not deteriorated by the ion desorption. Similarly, the use of IEM avoids ion adsorption at the electrodes during the regeneration step.

Besides IEMs, the key component in this process is the carbon electrode.[13] The amount of ions adsorbed at the electrodes is directly proportional to the available surface area. The role of the electrode nature and properties in MCDI processes has been intensively investigated. Many studies utilize carbon aerogels,[90] while others use carbon cloths,[91] carbon sheets,[92] carbon nanotubes,[93] or carbon nanofibers[94] coated with a thin-film nanoporous inorganic[95] or organic membranes.[84,86] The electrode internal area for ion adsorption is in the order of 1000 $m^2\ g^{-1}$.

Note that MCDI technology is young and needs testing. The benefits of MCDI systems over those of RO and CEDI desalination are in the low pressure applied in the process and, therefore, as expected at least for brackish waters, lower energy consumption (1/3 less energy).[13]

The disadvantage of CDI is the large surface area of the electrode which is necessary when feed solutions with high salt concentration are deionized. MCDI systems may also suffer from fouling similar to that observed in ED operation. However, it is found that the fouling may be drastically reduced by switching potentials of electrodes.[96] Lastly, it is conceivable to apply a pulsed field to the electrodes to reduce both inorganic and organic fouling.[97]

Micro- and Nanofluidics Desalination Device There are nontraditional, exciting applications resulting from research in the very novel and pioneering field of micro- and nanofluidics.[14,29,85,98,99] A promising micro-/nanofluidic desalination device was proposed by J. Han's group[14] (Fig. 20.8).

Seawater is pumped through a microchannel (with a width of 500 μm and depth of 100 μm) and bifurcated into two other microchannels (each with a width of 250 μm and depth of 100 μm), one of which is used for collecting desalted and the other for concentrated streams. The desalination is due to an ion-selective (cation-selective in Fig. 20.8) membrane, which serves as a nanojunction between the region of bifurcation and an additional microchannel. An electric current through the membrane is generated with help of three anodes and one (or two) grounded (GND) electrode(s). A high voltage (of the order of 70 $V\ cm^{-1}$) applied across the membrane gives rise to a space charge region (SCR) repulsion zone just before the bifurcation of the seawater microchannel. This zone plays the same role as the charged solution inside an RO membrane with fixed negatively charged sites (Fig. 20.1). Water molecules pass through this zone under the action of pressure gradient, while the charged particles are retained. The anions do not enter this zone because these ions

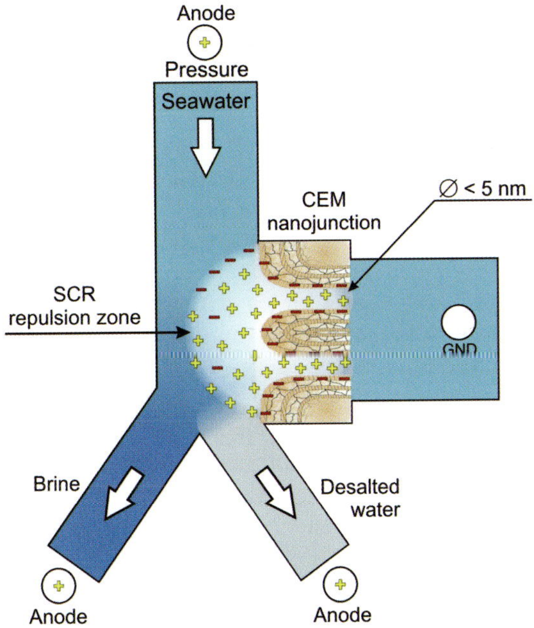

FIGURE 20.8. A micro-/nanofluidic desalination device (description in the text). Adapted from Kim et al.[14]

are expelled by the fixed ions and the applied electric force. The cations are retained by the electric field induced by the separation of charges. A small amount of cations transferred with the convection flow creates an excess of positive charge at the outlet of the SCR, while an excess of negative charge appears at the inlet. In steady state, the mechanical force (due to the pressure gradient) and the convection-induced electric force applied to a cation are equal and directed to opposite sides, so that the resultant driving force is zero.

Preliminary tests have shown[14] that both salts and larger particles (cells, viruses, and microorganisms) are expulsed from the SCR near the membrane, which significantly reduces the risk of membrane fouling. The tested device has shown ~99% salt rejection in the case of seawater desalination at 50% recovery rate and at a power consumption of less than 3.5 W h L^{-1}, which is comparable to current state-of-the-art systems. The authors[14] believe that this method could be used to produce small- or medium-scale systems (for families or small villages), with the possibility of battery-powered operation, instead of competing with larger desalination plants.

Overlimiting Current ED As was shown in Chapter 9, the limiting current density, which is characterized by a sharp increase in the electric resistance of ED cells due to depletion of electrolyte concentration at the membrane surface, is attained at about 1 V per cell pair. In most cases of practical use of ED, this or a slightly lower voltage is steadily applied. The increase in voltage leads to some undesirable effects: growing electric consumption, lowering current efficiency, and increasing scaling and fouling. However, the use of overlimiting current modes attracts a number of researchers and engineers.[12,29,30,71,78,79,100,101] The interest is in the fact that the overlimiting mass transfer rate at reasonable voltages close to 2 V may be up to three times higher than i_{lim} (see, e.g., Fig. 9.24 in Chapter 9). This means that even if the costs related to electric consumption increase, the total costs per 1 m^3 of treated water may be lower owing to decreasing capital costs since a lesser amount of membranes and membrane devices is needed. Besides, smaller sizes of the membrane installation may give other advantages, especially in production facilities of limited dimensions.

The mass transfer rate in the overlimiting current range strongly depends on the morphology and chemical composition of the surface membrane layer that promotes the need for the optimizing surface (see Chapter 9 and Nikonenko et al.,[29] Balster et al.[71,100]). The main mechanism of mass transfer enhancement at overlimiting currents is electroconvection, which arises due to the action of the electric field on the electric space charge in the depleted solution near the membrane surface produced by the same electric field (current-induced electroosmosis or electroosmosis of the second kind).[29,30,101,102] The electroconvection provides additional mixing of depleted solution, in addition to the forced convection. By applying a more hydrophobic surface with tailored heterogeneity, it is possible to essentially enhance the overlimiting mass transfer (Chapter 9, Section 9.5.1).

To reduce the risk of scaling and fouling, several methods may be used. The first one is in pretreatment of water prior to ED. The elimination of salts causing hardness from water eliminates the cause of salt deposition on the membrane surface. Another method is the application of membrane pairs where the water-splitting rate is higher at the CEM surface than at the AEM one. In this case, the desalinating stream becomes alkalized while the concentrating stream becomes acid.[79] The latter prevents the deposition of carbonates and oxides on the concentrating membrane surface. The alkalinization of the desalinating stream increases the degree of weak acid anion ionization that improves the removal of weak acids (such as silicic and boric) from feedwater. The third way is the use of pulsed current mode. This mode not only allows mitigation of scaling and fouling, but also results in an increasing rate of ED desalination. Mishchuk[30] supposes that the gain in applying pulsed currents may be due to electroconvection mixing, which continues during the pause owing to inertial properties of liquid movement. This residual convection could partially restore the electrolyte concentration near the membrane.

The application of overlimiting currents in ED is in accordance with the rather general trend of process intensification in chemical engineering. However, as was rightly noted by Strathmann,[12] this emerging mode of ED process is under development and needs further study.

20.3 ED CONCENTRATION

In conventional ED (Fig. 20.2), the final product can be both a demineralized solution and a concentrate. The degree of concentration that can be achieved in passing the feed solution through an ED stack is a function of the feed solution concentration, the applied current density, and the residence time of the solution in the stack concentrate. The latter can be increased by reducing the flow rate of the concentrate stream or by using concentrate stream recirculation. In ED, a much higher brine concentration can be achieved, in comparison with RO, since there are no osmotic pressure limitations. To obtain extremely concentrated solutions (close to the saturated ones), CC without entrance for feed solution are used[4,39,103–105] (Fig. 9.15 in Chapter 9). Ionic species are transferred from the feed solution circulating through the diluate compartments into the CC under the action of the applied electric potential difference. A small amount of water is transported there by electroosmosis, mainly in the hydration shell of ions. The produced concentrate is removed from the compartment through a capillary.

The fundamental basis (briefly presented in Chapter 9) and the main concepts of ED concentration were developed in the 1970s and 1980s.[33,104–106] Large plants with a capacity of 20,000 to more than 200,000 tons of salt per year are in operation in Japan. Of very special interest is the concentration of NaCl from seawater prior to evaporation for the production of table salt in Japan, which has no native salt deposit. The use of ED as a preconcentration step prior to evaporation leads to substantial savings in energy costs. Similar applications of ED and EDR are described by Turek.[40,107] A coal-mine brine containing 32.8 g L^{-1} Cl^- was desalinated and concentrated by an ED–EDR system[107]; the obtained concentrate was then treated by evaporation and crystallization in order to produce a solid salt and lye. The comparison of two systems of solid state production, without and with ED–EDR units, has shown that the overall energy consumption decreased from 970 kWh per 1 ton of salt to 500 kWh t^{-1} in the case where ED–EDR were used to the brine preconcentration.[107]

The main advantage of the use of ED–EDR in electrolyte solution concentration is the possibility of obtaining highly concentrated brines (up to saturated solutions) at relatively low energy consumption. The degree of water recovery when treating brackish waters with TDS 1–2 g L^{-1} may be higher than 99%.

Furthermore, ED is used for concentration of organic acids (citric, formic, and others) in the food industry,[108,109] for concentration of liquid radioactive waste,[81,110,111] and other applications.[111] It is possible also to recover valuable components of processing solutions and close the loop on the solvent.[111] As an example, consider the recycling of dimethylacetamide used as a solvent and extraction and concentration (up to 3.5–4 M) of lithium chloride from wastewater in chemical industry.[112] Quite successful is the use of ED for the recovery

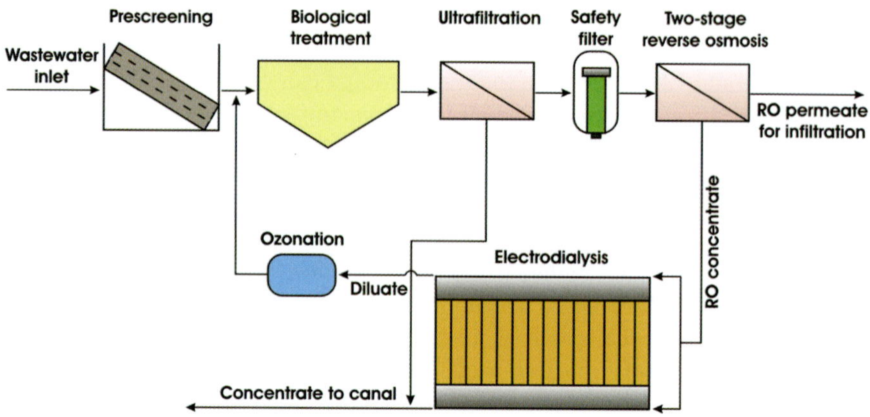

FIGURE 20.9. Schematic diagram of a hybrid system for wastewater treatment. Adapted from Zhang et al.[116]

of the condensate of juice vapor in the manufacture of ammonium nitrate fertilizer.[104] The condensate of juice vapor contains NH_4NO_3 with a concentration from 0.3 to 3 g L^{-1}. ED allows production of pure water (containing less than 3 mg L^{-1} of NH_4NO_3), which is then used to generate steam and a concentrate of NH_4NO_3 (up to 200 g L^{-1}).

The use of ED–EDR for concentrating RO brine is of a great interest. Traditionally, the RO concentrate is discharged into the natural water body (with or without dilution, depending on the local discharge regulations) or treated by evaporation. The former method is not environmentally friendly and the latter process is very costly.[113] A number of studies[114,115] have been carried out in order to reduce the liquid discharge and to increase the water recovery in RO–ED–EDR systems. The use of bipolar ED for producing acids and alkalis from the concentrate allows for the generation of agents which can be used for restoring pressure-driven units from fouling.[114,115]

Figure 20.9 shows a schematic diagram of a hybrid system developed by Zhang et al.[116] to treat (mainly domestic) wastewater to produce infiltration water for groundwater recharge. This recharge is needed in coastal areas where overconsumption of groundwater takes place, and, as a consequence, seawater intrusion and soil salinization occur. The wastewater first passes a mechanical screen, biological, and then ultrafiltration (UF) treatment prior to being treated by a two-stage RO. The RO permeate is used as the infiltration water; the concentrate is treated by ED. The ED diluate is reinserted into the biological treatment process, and the concentrate is mixed with the UF concentrate to meet regulations and then discharged to the canal.

Korngold et al.[117] have realized a similar process in which RO concentrates were treated by the EDR and the produced diluate was mixed together with the RO permeate. Thus, the overall water recovery of the RO unit was improved (to 97–98%) and the volume of brine was essentially reduced.

As the main mechanism of water permeation through IEMs contacting concentrated solutions is electroosmosis,[118] the water transport number in the membranes used for ED concentration should be minimized. Besides, the electrolyte diffusion permeability should be decreased as well. For these aims, modifications of perfluorosulfonated membranes MF-4SK were carried out with tetraethoxysilane[118] and polyaniline.[119,120] It was shown that these modifications result in high reduction of free water electroosmotic transfer, therefore, water molecules are transported only within hydrated shells.[118] After modifications, diffusion permeability drops by an order of magnitude, and water transport numbers are reduced by 50–70%. In the process of sodium chloride concentration by ED, the salt content of the concentrate increased by 50–70%.[120]

20.4 BMED

A BPM is a bilayer film containing a cation and an anion exchange layer (Fig. 20.10).

If an electric field is applied in a way that mobile salt cations and anions (Na^+ and Cl^- in Fig. 20.10) move away from the bipolar junction, the junction becomes depleted of salt ions. The charge transfer is then provided by new carriers, the ions H^+ and OH^-, which are generated by water dissociation reaction. The reaction occurs in a thin boundary layer of a thickness of 1–10 nm, where it is catalyzed by functional fixed groups (or a specially introduced catalyst) and strong electric field in the EDL.[73,121–124] The water splitting rate also depends on the topology/roughness of the bipolar junction.[122,124–126]

Electromembrane methods for producing acids, alkalis, and bases with the use of BPMs from the respective salts and water are reduced to the realization of the reaction:

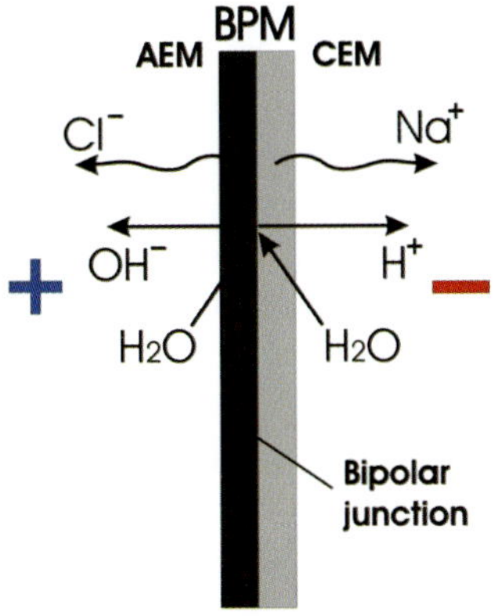

FIGURE 20.10. Schematic drawing of the water splitting function of a BPM.

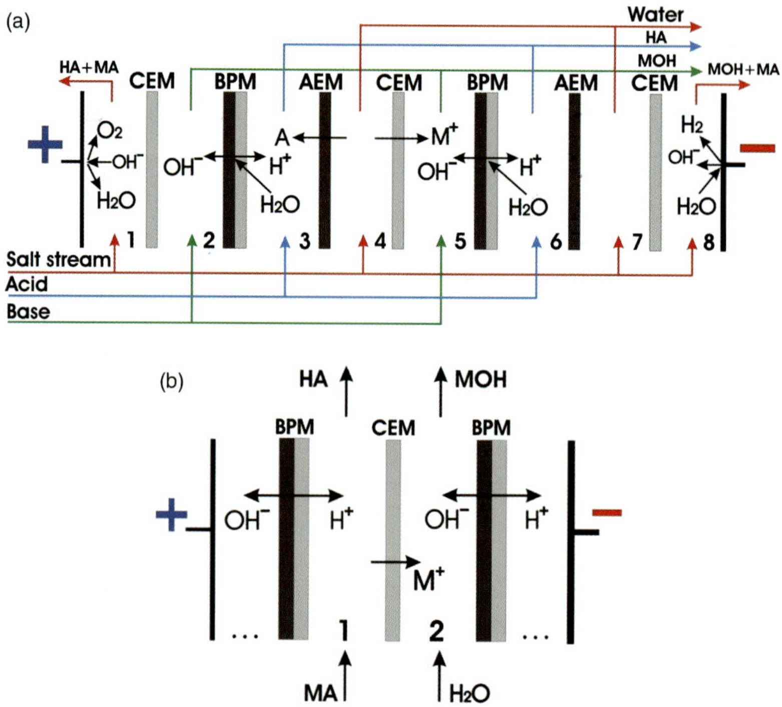

FIGURE 20.11. Schematic diagram of BMED stack for producing strong acids and alkalis (a) and weak bases (b) from their salts.

$$MA + H_2O \rightarrow MOH + HA, \tag{20.1}$$

where M^+ is a (metal) cation, and A^- is an anion.

In the case where strong acids and alkalis are to be obtained from their salts, generally a BMED stack configuration is used with two monopolar membranes (shown in Fig. 20.11a).[7,12,127,128] The salt anions migrating under the action of applied electric field from compartment 4 through an AEM, and hydrogen ions generated in a BPM, produce an acid in compartment 3. A similar process, but with the participation of hydroxyl ions and salt cations, leads to the formation of an alkali in compartments 2 and 5. A salt solution feeds compartments 1, 4, 7, and 8. Compartments 2, 3, 5, and 6 are fed with distillate or a diluted solution of acid/alkali (Fig. 20.11a). Generation of H^+ and OH^- ions in BMED allows obtaining solutions with controlled pH value up to concentrated (2–3 M) solutions of acids and alkalis.

To obtain weak acids or bases, simpler configurations with a two-compartment repeating cell are mainly used. In the case where a base is produced, the cell with alternate BPM and CEMs is applied (Fig. 20.11b); in the case where the product is an acid, AEMs are used instead of the CEMs. When operating such a cell, hydrogen ions generated into compartment 1 are linked in weak acid molecules with anion A^- and do not penetrate into compartment 2. The current efficiency and the concentration of the weak acid in this case are higher than these parameters in the case of strong acid, obtained in the cell shown in Figure 20.11a.

An alternative method of producing acids and alkalis is electrolysis. The main advantage of BMED in comparison with electrolysis is that in the ED stack there are no metal electrodes, hence, the solutions are not contaminated by the products of their destruction. Emission of gas, which is especially dangerous in the case of generation of oxidants such as chlorine and oxygen at the anode, is also absent. The elementary cell, including compartments 3–5 (Fig. 20.11a), can be repeated in an ED stack an unlimited number of times. As the stack contains only one cathode and one anode, the requirements for the cost of the anode, often manufactured from platinum or platinized titanium, are essentially reduced. Moreover, the use of buffer compartments 1 and 8 allows one to abandon platinum and to use anodes of base metals.[129]

Typical examples of the use of BMED are the production of acids and bases from corresponding salts,[12,128,130] the acidification/deacidification of product streams,[12,128,131,132] and the production of organic acids.[133–135] The case where BPMs are successfully used in an EDI stack was considered in the section "CEDI" (Fig. 20.6b). Very intensive is the use of BMED in the food, dairy, and pharmaceutical industries.[132,134,136] This method is effective in the treatment of amino acid-containing solutions, in particular, in the separation of amino acids and proteins on the basis of their isoelectric points.[128,137,138]

Particularly promising is the production with BMED of malic, succinic, and other organic acids used as precursors for the synthesis of biodegradable polymers and plastic materials.[139,140] In up-to-date membrane technologies, the process of organic acid recovery from fermentation broth is preceded by the step of concentration of these components (e.g., glyceric acid salts) using conventional ED.[134,141]

Further examples can be found in books and reviews concerning BMED applications.[128–130,133,136]

BMED allows for the conversion of salts into acids without yielding byproducts. This advantageously distinguishes it from traditional chemical processes, which require significant amounts of sulfuric acid and are accompanied by the formation of large quantities of gypsum (2 kg of gypsum are needed for production of 1 kg of citric acid).

The main problems in the case of production of strong acids and alkalis by BMED are the pollution of the products with the feed salt ions, and the relatively high specific energy consumption of the electromembrane process. The performance is improved by increasing the hydrogen transport number in the cationic layer and the hydroxide transport number in the anionic layer (and, respectively, decreasing co-ion transport number in both layers), and by enhancing the kinetic energy of the water splitting reaction in the bipolar junction. The latter is generally achieved by introducing between the cationic and anionic layers a thin interface layer, containing a water dissociation catalyst, and by increasing interface layer area, for example by increasing the surface roughness.[125] Good catalytic activities have been found for weak acids such as amino groups, pyridines, carboxylic acids, and phosphoric acid groups.[121,124,125,130] Besides, good catalysts are metal oxides and hydroxides[142] or heavy metal ion complexes.[124,143]

The introduction of a catalytic interface layer gives more freedom in choice of cationic and anionic layers. Use of an anion exchange layer with quaternary ammonium groups results in BPMs with low electrical resistance.[124]

Note also that the water permeability of both layers should be sufficiently high in order to avoid water lack in the bipolar junction for generation of H^+ and OH^- ions. High concentrations of ionic groups in the bipolar junction not only increase the rate of the water dissociation reaction, but give rise to the hydrophilicity of the interface layer. As a result, water activity in the bipolar junction increases, as well as the water flux from the ion-exchange layers to the interface region.[124]

Another drawback of BPMs is their high value that limits the range of applications, making BMED economically feasible only in the case of expensive products. Some authors[121,130] investigate the method of preparing low cost BPMs by using heterogeneous cation and/or anion exchange layers and a catalytic interface layer. An effective way to minimize the costs and the impact on the environment is the use of hybrid systems. Wang et al.[134] report that the integration of conventional ED in the production of gluconic acid by BMED results in the decrease of the process costs from \$0.39 kg^{-1} to \$0.31 kg^{-1}. Mondor et al.[138] describe the advantages of combining BMED with UF/diafiltration to produce a soy protein isolate.

20.5 FRACTIONATION AND SEPARATION PROCESSES WITH CHARGED MEMBRANES

Membrane fractionation and separation processes are perhaps the most developed and commonly used. They are extensively applied in the electrochemical industry, water treatment, wastewater treatment, and the food industry. A very large area of their application is medicine. Following Baker,[19] the total membrane area produced for medical applications almost matches all industrial membrane applications together. In the United States, the medical membrane market approaches \$1.5 billion per year and grows steadily.[144] The biggest part of this market involves membranes in drug delivery, hemodialysis (kidney–blood purification), other artificial organs (oxygenators, pancreas, liver, etc.), and tissue engineering. Historically, in medical applications, porous membranes separating the components via a sieve mechanism were used. However, in recent years the share of charged membranes in this area is continuously increasing, especially in combination with porous membranes. Some of these applications are considered in this section.

20.5.1 Electro-Electrodialysis

Electro-electrodialysis (EED) or membrane electrolysis[133] combines electrolysis and ED. The term "membrane electrolysis" is sometimes reserved for two-compartment cells with one IEM. In this kind of ED technique, electrode reactions play a critical role in feed treatment. A large number of different EED processes using CEMs and/or AEMs have been proposed so far.[48,133] The most important example of this technique is chlor-alkali production. The chlor-alkali industry is currently one of the largest industries producing annually more than 48 million tons of sodium hydroxide (NaOH) and more than 42 million tons of chlorine (Cl_2).[145] The raw material is sodium chloride (NaCl) and the reaction proceeds according to:

$$2NaCl + 2H_2O = 2NaOH + Cl_2 + H_2. \tag{20.2}$$

To obtain the products of this reaction with a sufficiently high yield, it is necessary to separate chlorine from hydrogen (because they react explosively) and from NaOH (because chlorine dissolves in contact with NaOH to form a hypochlorite solution). As can be seen in Figure 20.12, a two-compartment EED cell with a CEM simultaneously allows the realization of reaction (20.2) and the separation of the reaction products.

A saturated NaCl brine is fed to the anolyte compartment where Cl_2 gas is produced at the anode. Under the action of applied electric field, Na^+ ions migrate through the CEM into the catholyte compartment where they form

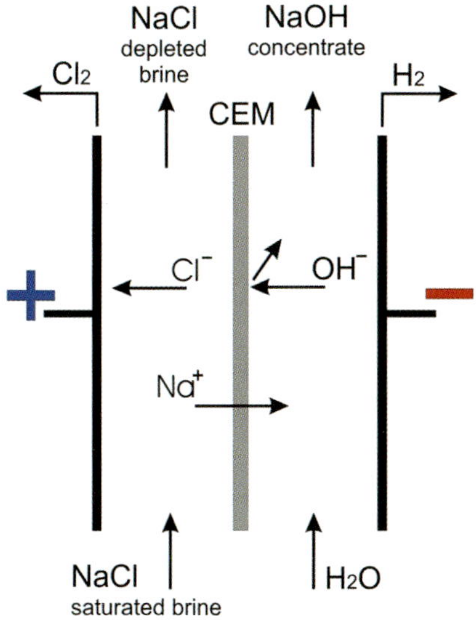

FIGURE 20.12. Schematic diagram illustrating the chlorine-alkaline production process by EED. Adapted from Nagarale et al.[48]

sodium hydroxide with OH^- generated at the cathode.[48] The process utilizes perfluorinated CEMs such as Nafion® (DuPont, Wilmington, DE), and is operated at an elevated T (80–90°C). The produced caustic soda is concentrated up to 35%. In this medium, even extremely corrosion-resistant Nafion® membranes suffer from degradation. Moreover, back migration of OH^- at high NaOH concentration results in significant loss in current efficiency. The situation can be improved by using composite CEM containing carboxylic groups.[48] These membranes are more stable; they have a lower water content and are less permeable to OH^- ions. Their lifetime in a chlor-alkali plant is evaluated as more than 5 years.[48]

Another example of application EED, where the role of the electrode reactions is essential, is the chromic acid recovery from plating bath. A schematic diagram illustrating the process in the cases of two-compartment and three-compartment cells is shown in Figure 20.13.

In the case of two-compartment EED, exhausted plating solution containing chromic acid and metallic impurities, such as copper, iron, zinc, aluminum, nickel, and trivalent chromium flows through the anolyte compartment where trivalent chromium is oxidized (Fig. 20.13a). The other cations (impurities) migrate through a CEM (normally, a Nafion® membrane) into the cathode compartment where they are collected. Some amount of Cr(III) and Cr(VI) may be lost due to the migration into the cathode compartment.[146]

In the case of three-compartment EED, the rinse water from the plating process containing chromic acid and metallic impurities passes through the central compartment (Fig. 20.13b). Cationic impurities migrate toward the cathode through the CEM. The cathode compartment is fed with diluted sulfuric acid to form dissolved metal sulfates. The acidic conditions prevent precipitation of metal hydroxides. Hexavalent chromium anions (chromates) migrate toward the anode through the AEM. There, the chromate and dichromate anions form chromic acid together with the protons generated at the anode. The treated rinse water is reused in the rinsing process and the pure chromic acid is returned to the plating bath. This process provides purification of chromium plating solutions and the treatment of the rinse water in a single step without need of either water or chemicals. The main process limitations are the poor stability of the AEM against the oxidative chromic acid solution and the increase in the AEM resistance due to the formation of polychromates in the membrane.[147]

When ED is applied to organic acid production, the role of electrode reactions is similar to that of BPMs (see Section 20.4). Both electrodes and BPMs provide H^+ and/or OH^- for the acidification or ionization of organic anions. Generally, a monomembrane EED process (sometimes coupled with extraction)[133] is used to recover organic acids (butyric, valeric, adipic, caproic, and oxalic acids).[148]

In hydrogen energy, one of the most suitable process of H_2 production is the iodine–sulfur (IS) cycle.[149,150] Iodhydric acid (IH) is an intermediate product which generates H_2 during its decomposition. It is difficult to concentrate HI

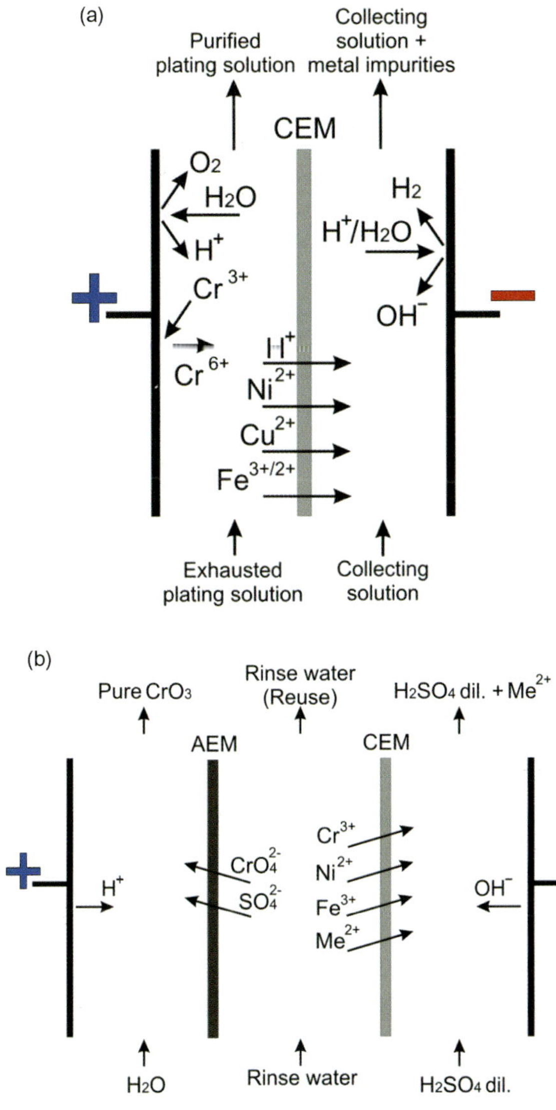

FIGURE 20.13. Principles of electro-electrodialysis (EED) for chromic acid recovery by applying two-compartment (a) and three-compartment (b) cells. Adapted from Huang et al.[146]

solution by conventional distillation, since HI forms with water an azeotropic solution $HI/H_2O = 1{:}5$ (molar ratio). However, HI acid can be effectively concentrated through EED with CEM, and its concentration could exceed the azeotropic composition.[151]

EED can be used in other separation process, such as phosphoric and sulfuric acid separation[152] and concentration of inorganic acids over its azeotropic composition.[151] There are designs with more than three compartments in an EED cell. Cheng et al.[153] have proposed a system for bovine serum albumin and hemoglobin separation with six compartments. The electrode compartments served as sources of H^+ (the anolyte one) and OH^- (the catholyte one) ions. The H^+ ions migrated from the anolyte compartment through a CEM into the neighboring compartment in order to maintain there a stable pH close to 4. In turn, the OH^- ions migrated through an AEM into the neighboring compartment in order to establish there a pH equal to 9.8. There are also two central compartments wherein pH is equal to 7 and 8.5, the first one being closer to the anode, and the second one to the cathode. Quasi-stable pH values in the four central compartments allow separation of proteins due to the fact that they can be ionized under various pH. A protein can spontaneously move toward the location where the medium pH is equal to its pI (isoelectric point) value and be stationary at that location. Thus the protein molecules can migrate across membranes and be relocated into different compartments according to their pI values.

20.5.2 Dialysis (Diffusion Dialysis, Donnan Dialysis)

When charged membranes are used in diffusion separation processes, referred to as diffusion dialysis (DD) and Donnan dialysis, the contribution of electric force is essentially due to the electroneutrality requirement, which must be held on a macroscopic scale.

The principle of DD is presented in Figure 20.14, which describes the process in the case where a mixture of a salt and an acid is separated. The dialysis cell contains AEMs only. A feed solution containing a mixture of NaCl and HCl flows through alternating cells while water is fed in countercurrent flow through the other cells of the stack. Cl^- (or SO_4^{2-}, NO_3^-, PO_4^{3-}) anions pass through the AEMs into neighboring compartments, while the salt cations are rejected from AEMs due to the Donnan exclusion effect. The transfer of anions leads to the formation of an excessive positive charge in the feed compartments and a negative charge in the neighboring ones. The separation of charges results in the creation of an electric force (due to diffusion potential difference). Under the influence of the electric and diffusion forces acting in the same direction, the H^+ ions pass through AEMs. Although positively charged, the H^+ ions are very mobile, and hence have higher competition in diffusion than metal ions.[154] Therefore they can diffuse through AEMs along with salt anions to meet the requirement of electrical neutrality.[155] The net result is the removal of acids from a mixture with salts. Similarly, bases can be removed from salt solutions using a stack with CEMs only.[12,48]

In the case of Donnan dialysis, there is ion exchange across an IEM. For example, in softening of hard water, divalent ions, such as Ca^{2+} and Mg^{2+}or SO_4^{2-} are exchanged for monovalent ions such as Na^+ or Cl^-.[48,156,157] The principle of this process is as follows. Feedwater containing NaCl and ions of hardness (Ca^{2+}) is separated by a CEM from an auxiliary NaCl solution of a

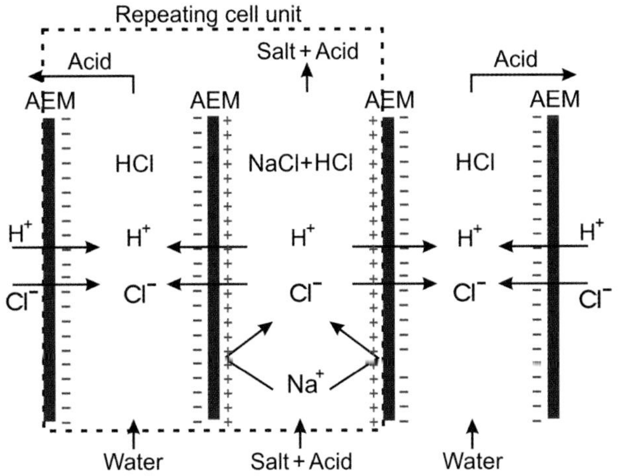

FIGURE 20.14. Schematic drawing illustrating the principle of DD used to recover an acid from a mixture with salt in a stack containing AEMs only. Adapted from Strathmann.[12]

sufficiently high concentration. Since the Na^+ concentration in the auxiliary solution is significantly higher than the concentration in the feed solution, there will be diffusion of this ion into the feed solution. The transfer of Na^+ ions gives rise to an excessive positive electric charge in the feed solution. These electric forces produced by the separation of the charges will assist the transport of Ca^{2+} ions from the feedwater into the auxiliary solution.

The advantages of both methods of dialysis are in low energy consumption and low installation and operating costs. The methods are stable, reliable, and easy for operation, as well as being less susceptible to membrane fouling.[155,158] They are effective in the recovery of acids and alkalis from mixtures with salts in wastewaters from steel production, metal refining, electroplating, cation exchange resin regeneration, nonferrous metal smelting, aluminum etching, and tungsten ore smelting.[155,159]

Besides its uses for water softening mentioned above, Donnan dialysis is successfully applied for boron,[160] nitrate,[158] and bromate[161] removal from aqueous solutions and the recovery and concentration of transition and rare metals.[162,163] An example of the anion Donnan dialysis is the sweetening of citrus juice. In this process, hydroxide ions furnished by a caustic solution replace the citrate ions in the juice.[155]

The dialysis methods are well developed and find many applications in the United States and Japan as well as in China and North African countries.[155]

In comparison with pressure-driven and electro-driven processes, dialysis has a relatively low processing capability and efficiency: a relatively high amount of membranes is needed to provide a given quantity of product. Another weakness is the low stability of AEMs in acidic solution and their high H^+ permeability. The dialysis membranes should have a high selectivity and a high diffusion permeability that is difficult to achieve simultaneously. As a result, the variety of membranes suitable for dialysis is limited, and only few of them satisfactorily meet the requirements.[155,161]

20.5.3 Electrofiltration (ED with UF Membranes)

The electrofiltration process occurs under the action of pressure and electrical potential gradients as driving forces. This process is used for the separation of polyamino acids, proteins or bioactive peptides, and other organic compounds.[144,164,165] An important example is the treatment of β-lactoglobulin, which is one of the major whey components.[166] β-Lactoglobulin can release, by enzymatic hydrolysis, different bioactive peptides according to the enzyme used. Figure 20.15 shows an ED cell in which UF and IEMs are used for fractionation of β-lactoglobulin hydrolysate, to obtain peptides in a more purified form. Under the action of applied electric field, negatively charged peptides (P^-) migrate through a UF membrane from the central feed compartment to the neighboring compartment fed with a KCl solution. Cl^- ions leave this compartment through an AEM toward the anolyte one. As a result, they are substituted by P^-. Similar processes occur in the right-hand side of the cell: in the compartment between a UF membrane and a CEM, K^+ ions are substituted by positively charged peptides, P^+. The neutral fraction of peptides remains in the central compartment.

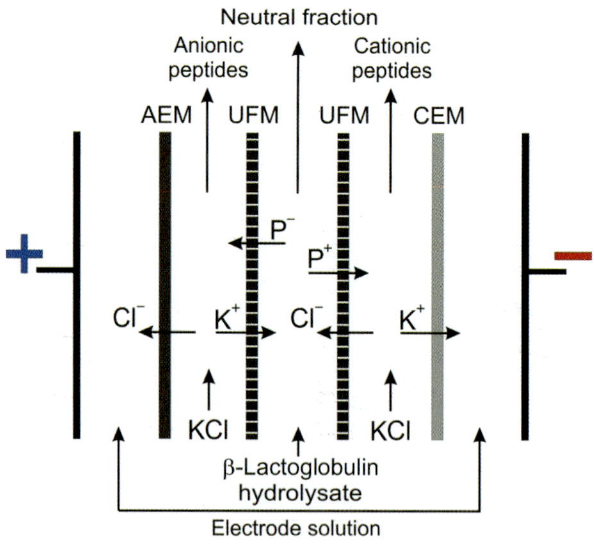

FIGURE 20.15. Configuration of the ED module using two UF membranes for the simultaneous electroseparation of cationic (P^+) and anionic (P^-) peptides from β-lactoglobulin hydrolysate. Adapted from Poulin et al.[166]

ED with UF membranes appeared to be a very selective method for separating peptides from protein hydrolysates. The system presented in Figure 20.15, or similar configurations, were used by Vanhoute et al.[164] for the fractionation of a bovine hemoglobin peptic hydrolysate mixture in order to increase the value of slaughterhouse by-products such as blood protein. Firdaous et al.[165] also examined a similar process for alfalfa white protein hydrolysate.

20.6 MEMBRANE-BASED HYBRID ZLD TECHNOLOGIES

Nowadays, ZLD or near-ZLD systems have become more and more often a goal of water and waste treatment industrial engineering for recovery of clean water and minimization of brine streams. Mandatory elements of such systems are thermal methods such as brine concentrators, crystallizers, thermal evaporators, and spray driers, which reduce concentrate to a slurry or solid product that can be disposed of in landfills. These processes are capable of recovering high purity water (95–99% recovery from waste streams) and revenue-generating mineral salts.[167] Although these processes have been proven effective for volume minimization, the capital and operating costs often exceed the cost of the desalting facility.[168]

For economic reasons, ZLD has been regarded for many years as an overly expensive solution and therefore was employed in limited cases. However, the increasing rate of population growth, scarcity of water in many places around the globe, and the growing awareness about the need for environment protection has brought this issue back to focus. Accordingly, new ZLD technologies are today developed and new, more cost-effective options are now available. The ZLD approach is becoming an important strategy in managing water resources toward maximizing water savings, reducing desalination and water treatment costs, and protecting water resources and the environment.[169]

Nowadays, the situation of water regulation and new technical solutions has prompted those in the field to reconsider the economics of ZLD. First, the requirements to the quality of potable water become stricter and its price rises. And, perhaps more importantly, regulations on the discharge of waste fluids into open waterways are more stringent. When making assessments for water treatment systems, the cost of ZLD should be compared to the cost of freshwater and the savings on sludge disposal. Hence, regulation represents the biggest incentive to ZLD implementation so far.[1]

Another element improving the economics of ZLD is novel separation techniques. More effective evaporation processes, such as mechanical vapor recompression (MVR), are available. Use of membrane distillation (MD) and membrane crystallization (MC) systems could essentially decrease the power consumption in ZLD systems. These are thermally driven processes, in which a relatively low feed T (<90°C) is used. This allows the utilization of waste heat, for example, the condensate from turbines or heat exchangers.[170]

ED concentration may be a key technique to essentially reduce energy consumption in ZLD systems. By applying ED concentration, it is possible to diminish the volume of liquid, which should be evaporated.

Turek[171] has realized several ZLD systems for treatment of coal mine brackish waters. According to Turek,[171] water of approximately 10 g L^{-1} salinity was desalted/concentrated by four ED units arranged in cascade. In another case, a part of coal mine water containing about 35.45 g L^{-1} Cl^- was softened by NF, and the other part of water from another underground level of about 2.25 g L^{-1} Cl^- content was desalinated and simultaneously concentrated by ED with univalent permselective membranes.[172] Partially evaporated, pretreated by NF water was mixed with ED brine containing about 55 g L^{-1} of Cl^- and underwent further evaporation and then crystallization. The value of produced salt was found to be close to the total cost of the system utilization. The benefit was due to the decrease of cost for chloride and sulfate discharge.

Another system using RO and EDR was applied to treat brackish water with high scaling potential from the Sahel region in Tunisia.[173]

The combination of RO and ED concentration seems a very effective way to obtain concentrated brine from feedwater. A schematic diagram of wastewater treatment including an ED unit concentrating RO retentate developed by Zhang et al.[116] is shown in Figure 20.9.

Mavrov et al.[174] have applied BMED to produce an acid and an alkaline stream used for regeneration of cation and anion exchange resins, respectively, in a hybrid system including RO and conventional ED. The ion exchange units were installed to produce deionized water from RO permeate.

More recently, Oren et al.[169,175] proposed an RO–EDR process with a side loop crystallizer to improve water recovery and produce mineral byproducts from brackish groundwater in the Negev Highland, Israel. Inland brackish water desalination is becoming a significant source for water particularly in areas that are remote from the sea. An important element of the process is a wind-aided intensified evaporation (WAIV) unit, which benefits the climate peculiarities of dry desert to produce a dry or semidry solid salt from a slurry obtained in a crystallizer treating the brine produced in an EDR unit (Fig. 20.16). The RO retentate is concentrated in EDR to over 10% TDS concentrate while producing the diluate that was recycled to the RO permeate. The risk of scaling in the CC of the EDR unit was prevented, besides of operation in a reversal mode, by acidification, and the use of a side loop crystallizer. Settlers, inline microfiltration, and side loop UF kept suspended solids from returning to the EDR unit.

In the above cited publications[169,173,174] the authors have applied underlimiting current modes when using conventional or reversal ED. This allowed them to minimize the risk of scaling as well as the energy consumption. However, the use of low current density results in high membrane surface area needed to provide necessary production capacity. Moreover, the alternation of diluate and concentrate streams leads to losses of product. To avoid the precipitation of poorly soluble salts when applying intensive current densities, some special measures can be undertaken: pH correction (by using special AEMs with

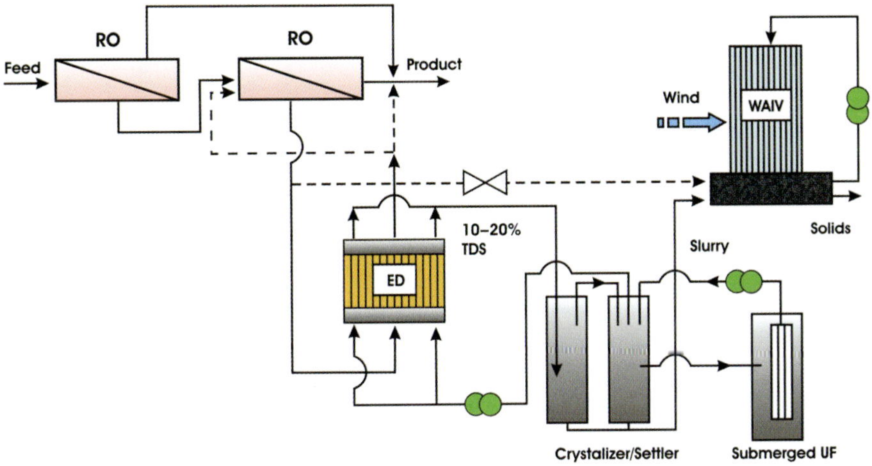

FIGURE 20.16. Schematic presentation of a ZLD system for treatment of brackish water in desert areas using wind-aided intensified evaporation (WAIV). Adapted from Oren et al.[169]

lower H^+/OH^- ion generation functions[29,79] or by adding acid in the concentrate compartment),[169] the use of pulsed currents,[53,54] and/or reversal ED mode. Some authors[78,116,117] propose an even more radical solution: removing the scaling components from the feedwater prior to its treatment in membrane modules. The ions causing hardness can be removed by ion exchange techniques. For example, Ca^{2+} and Mg^{2+} can be removed by exchange with Na^+ initially associated to a cation exchange resin. The regeneration of the resin may be carried out by using a part of the ED brine.[176] Another possibility is given by the application of BMED,[78,174] which produces an acid and an alkaline solution for regeneration of, respectively, cation and anion exchange resins. Tskhay[177] considers different schemes providing sedimentation of salts causing hardness by changing pH of the feedwater and application of BMED for producing acid and alkaline streams. Zhang et al.[116] studied an RO–ED system where RO retentate, prior to being treated by ED, was decarbonated via an acidification and an aeration. They found that after this decarbonation, it was possible to operate the ED unit under a steady state for a long-term experiment with a high overall water recovery (95%).

A very important advantage of removing the scaling components from the feedwater is that all membrane units, the NF and/or RO, as well as ED, may be operated at high fluxes; in particular, ED may be operated at overlimiting current density. Moreover, the maintenance of the membrane stacks might be realized much more easily, and the lifetime of membranes significantly increases.

In brief, the idea of realization of ZLD systems with removing scaling components from the feedwater is in the precipitation of hardness salts before the membrane modules and that of well-dissolved salts after.[29]

Among the areas where ZLD systems seem to be the most suitable are:

- *Power Plants*: steam production, regeneration of liquid wastes from flue gas scrubbers;
- *Oil/Gas Industry (Refineries, Petrochemical)*: steam production for heavy oil and bitumen recovery;
- *Mining*: treatment of drainage wastewater to reuse the water and create saleable salt product (e.g., for deicing of roads);
- *Coal Production, Shale Gas*: reuse of water for fracturing the rocks;
- *Chemical Industry*: ethanol production (production of 1 m^3 of ethanol needs 3.5 m^3 of water), fertilizer production, and other industries (electronics, food and beverages, pulp and paper, etc) which use industrially desalinated water as process water, boiler feed water, make-up water, and ultra-pure water.

20.7 MEMBRANE-BASED ENERGY CONVERSION TECHNIQUES

20.7.1 Fuel Cells

Fuel cells are electrochemical devices with high energy conversion efficiency, minimized pollutant emission, and other advanced features. Proton exchange membrane fuel cells (PEMFC; see Fig. 20.17 for general knowledge) are considered a key issue against oil rarefaction and greenhouse gas emissions.[178] Although some companies have produced vehicles fuelled by PEMFCs, they have to face problems of water management, carbon monoxide poisoning of

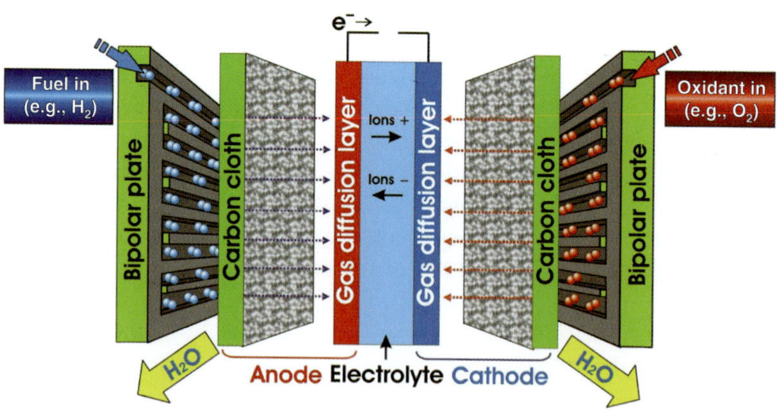

FIGURE 20.17. Scheme diagram illustrating the structure of PEMFC and the principle of operation. In the case of H_2 and O_2, the key reactions are: $H_2 – 2e^- \rightarrow 2H^+$; $0.5O_2 + 2H^+ + 2e^- \rightarrow H_2O$. The electrons ceded by the hydrogen at the anode pass via the external load circuit to be accepted by the oxygen at the cathode side; the arriving protons form water molecules.

fuel cell catalysts, membrane behavior, and costs. High T PEMFCs have been proposed to solve problems of catalyst poisoning by CO and fuel cell electrode flooding, as well as to improve fuel cell efficiency, reduce the amount of noble metal catalyst, and avoid reactant humidification.[179]

Most PEMFC research efforts are aimed at increasing performances (yield efficiency, power density, reduction of catalyst content, durability); improving mechanical, thermal, and electrochemical stabilities; and decreasing mass, volume, and costs. Operating at an increased T (120–150°C) causes greater challenges for PEMFC.[180] Novel materials that can give high performance and high durability under such conditions are prerequisites for high T PEMFC, among which alternative electrolyte membranes that can work at high T (120–150°C) and low relative humidity (RH = 25–50%) are of the greatest relevance. Many current research efforts are therefore devoted to the development of alternative electrolyte membranes,[180] including nonfluorinated hydrocarbon polymers,[181] inorganic polymer composites,[182] and anhydrous proton-conducting polymers[183] such as PBI/H_3PO_4 [184,185] (phosphoric acid doped polybenzimidazole polymer) or Nafion®/H_3PO_4[186] (phosphoric acid doped perfluorosulfonic acid-*tetrafluoroethylene* copolymer).

In this section we discuss the remaining bottlenecks for low T PEMFC and focus on the material challenges for high T PEMFC membranes. A great number of publications are devoted to fuel cells.

We will consider new and powerful methods allowing the study of the dynamics of solvent and molecules concentration profiles, which are developed during diffusion and current flow through the membrane.

Voltage–Current Response of a Membrane Electrode Assembly (MEA)

Voltage–current response (VCR, Fig. 20.18) is perhaps the most important dynamic characteristics of an MEA.

The VCR curve reveals three regions. The first one is governed by the electrode catalysis (reduction of the charge transfer overvoltage); the second one reflects the membrane behavior (reduction of the ohmic losses); and the third one reflects the cell design (improvement of fluid management). The development of an optimized MEA entails work in all these aspects.

In most PEMFCs, the proton exchange membranes are currently based on perfluorosulfonic acid (PFSA) polymers, such as Nafion®. This membrane material exhibits high conductivity and excellent chemical stability, mechanical strength, and flexibility. However, its operation requires a highly hydrated state, and therefore is limited to T of up to around 90°C under ambient pressure. Moreover, several challenges for the PEMFC power technology are associated with low operating T.[187] Fuel processors such as hydrogen storage tanks and hydrocarbon or alcohol reformers with subsequent carbon monoxide (CO) removers are voluminous, heavy, and costly. Water management involves appropriate humidification of fuel and oxidant, airflow rate, and power load regulation. Temperature cooling is more critical for larger stacks, and the heat produced is of low value. PEMFC operating at higher T

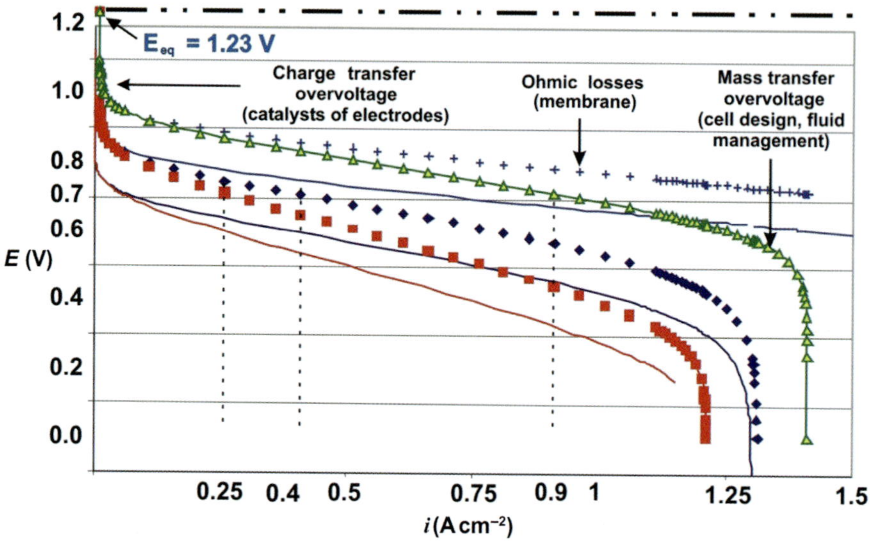

FIGURE 20.18. Voltage–current response of an MEA.

(120–150°C) has been recognized as a promising solution to meet these challenges. Hence, starting in 2001, researchers have aimed to increase operating T (120–150°C) with a 25–50% RH.

Increase of the Operating Temperature of PEMFC: The New Challenges

The theoretical analysis and experimental investigations have shown that working at high T (120–150°C) can provide the following advantages for PEMFC[188–190]:

- *Electrode Reaction Kinetics*: the overall electrochemical kinetics of PEMFC are mainly determined by the slow oxygen reduction reaction (ORR)[180] accounting for the major overvoltage loss of PEMFC.[180] The reaction kinetics of hydrogen oxidation and ORR are enhanced at high T.[191]
- *CO Tolerance*: trace CO in hydrogen feed gas drastically depresses the performance of PEMFC due to the strong irreversible adsorption of CO on Pt electrocatalysts.[192] The adsorption of CO on Pt is weakened at high T and CO tolerance is enhanced.
- *Heat Management*: a PEMFC operating at 80°C with an efficiency of 40–50% produces a large amount of heat that has to be removed in order to maintain the working T. The rate of heat transport is proportional to the T difference between PEMFC and environment. For a PEMFC working at low T (≤80°C), the heat rejection rate of the conventional automotive radiators is not sufficient to reject continuous full power waste heat. Increase of PEMFC working T to >120°C will allow the use of the cooling system of present internal combustion engines.

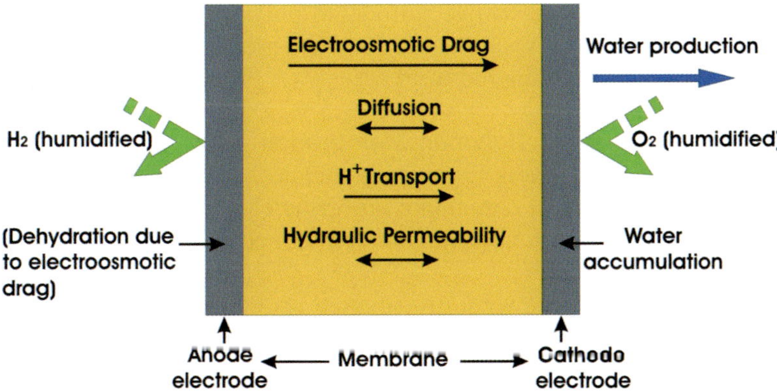

FIGURE 20.19. Modes of water transport in an operating H_2/O_2 PEMFC; adapted from Zhang et al.[180]

- *Water Management*: PEMFC working at T ≤80°C under atmospheric pressure involves a dual-phase system (liquid and vapor water). When the humidification is too high, water condenses and the electrodes are flooded (mainly the cathode), which makes water management difficult.[193] If a PEMFC is running at atmospheric pressure and T > 100°C, only a single phase exists,[194] and therefore, the transport of water in the membrane, catalyst layers, and diffusion layers is easier to balance. The water balance in a PEMFC involves the following mechanisms: (1) water supply from the fuel and oxidant (humidification), (2) water production at the cathode (current density), (3) water drag from the anode to the cathode (current density, humidity, T), and (4) back-diffusion of water from the cathode to the anode (gradient concentration, capillary forces, etc.) (Fig. 20.19).
- *Nonplatinum Catalysts*: at higher T, the electrode reaction kinetics are strongly enhanced, making possible the use of nonplatinum catalysts, reducing PEMFC cost.

The advantages of high T PEMFC are therefore very attractive and many researchers are presently devoted to this challenge.

High Temperature Proton Exchange Membranes Great efforts are being devoted to synthesize proton-conducting membranes and other materials operating at T above 100°C.[180,181,183,188,194,195] Membranes under development can be classified as follows: (1) modified PFSA membranes,[38,189,190,194,196,197] (2) alternative membranes based on partially fluorinated and aromatic hydrocarbon polymers,[198,199] (3) inorganic–organic composites,[182,200,201] and (4) acid–base polymer membranes,[202] typically a basic polymer doped with a nonvolatile inorganic acid or blended with a polymeric acid.

Note only that the major drawback of PFSA membranes is their low conductivity and their poor performance under low humidification at elevated T (>90°C) due to the water loss. Here we will consider some new results in the development of alternative sulfonated polymer membranes and their composites. Among others, sulfonated polymer membranes have the advantage of being low cost compared to PFSA.

Alternative Sulfonated Polymer Membranes and Their Composites Two main groups of polymers having high chemical and thermal stability have been investigated, the first containing inorganic elements such as fluorine in fluoropolymers and silicon in polysiloxanes, the second is aromatic polymers with phenylene backbones.

Sulfonated polystyrenes were investigated in the 1960s and were the first generation of polymer electrolytes for fuel cells.[203] However, in this type of polymer membrane, the tertiary C-H bonds in the styrene chains are sensitive to oxidation by oxygen and hydrogen peroxide. The bond strength for C-F is about 485 kJ mol^{-1} and that of C-H bonds in the range of 350–435 kJ mol^{-1}. Polymers containing C-F bonds therefore have high chemical and thermal stability. Partially fluorinated membranes have also been investigated, on a base of poly(tetrafluoroethylene-hexafluoropropylene) (FEP) films, by Scherer's group,[204] and polyvinylidene fluoride (PVDF), by Sundholm's group.[205] The combination of the PVDF properties with the conductive properties of sulfonated polystyrene gives both high water uptake and conductivity, but low thermal stability.[205]

Another type of temperature-resistant polymers of interest includes the Si-O chemical bond (445 kJ mol^{-1}). Si-O networks are formed at high T (ceramics) but can also be developed at low T in organic or aqueous solutions. Organic groups can be attached to the silica matrix to give organic-modified silicates (ORMOSIL), organic-modified ceramics (ORMOCER), or organic-modified silicate electrolyte (ORMOLYTE). Attempts have also been made to develop proton-conducting membranes for PEMFC by using arylsulfonic anions or alkylsulfonic anions grafted to the benzyl groups.[206] These structures exhibit a proton conductivity of 0.01 S cm^{-1} at room temperature and thermal stability up to 120°C.

A large group of low cost and commercially available polymers is based on aromatic hydrocarbons. Polymers consisting entirely of linked benzene rings such as poly-p-phenylene are resistant to oxidation. Polyphenylene sulfide and polyphenylene oxide have high melting points with good thermal and oxidative stability above 200°C. Aromatic polymers containing ether links have also been widely investigated, such as polyether(ether)ketones (PEEK).

To create proton conductivity, charged groups must be attached to the polymer chains. This can be done by chemical modification of the polymers (postfunctionalization) through the introduction of an anion, typically sulfonate ($-SO_3^-$). This sulfonation can be performed (1) by direct sulfonation in concentrated sulfuric acid or chlorosulfonic acid, sulfur trioxide, or its complex with tri-ethyl-phosphate[207]; (2) by lithiation–sulfonation–oxidation[208]; (3) by chemically grafting a group containing a sulfonic acid onto a polymer[208]; (4) by graft copolymerization using high radiation followed by sulfonation of the aromatic components[204]; or (5) by synthesis from monomers bearing sulfonic groups.[209]

Generally, PFSA membranes and sulfonated aromatic polymers have different microstructures. The water-filled channels in sulfonated polyaryls (sulfonated PEEK) are narrow compared to those in hydrated perfluorosulfonic polymers (Nafion®). They are less separated and more branched with more dead-end spaces.[210] For PFSA membranes, the water content within the membrane is balanced by the extreme hydrophilicity of the sulfonic ion exchange groups. In the presence of water, the hydrophilic domains of the nanostructure are hydrated to maintain the proton conductivity, while the hydrophobic domains provide the mechanical strength. The water uptake of PFSA membranes is high but their conductivity is very sensitive to RH. In the case of sulfonated hydrocarbon polymers, the hydrocarbon backbones are less hydrophobic and the sulfonic acid exchange groups are less acidic and polar. Water molecules are therefore distributed within the nanostructure.[211] Sulfonation of polyphenylene sulfide results in the increase of proton conductivity up to 0.01 S cm^{-1} in the range 30–180°C.

Similar to PFSA, sulfonated hydrocarbon polymers can be modified. In particular, they can be used as a host matrix for preparing inorganic/organic composites for high operating T.

Another effective approach to the development of proton-conducting membranes is acid–base complexation. Three ways may be used there: (1) basic polymers can be doped with an amphoteric acid acting as a donor and with an acceptor in proton transfer, allowing proton migration; (2) H_3PO_4-doped PBI polymer in which proton hopping from one N-H site to phosphoric acid anions contributes significantly to the conductivity; and (3) organic acid–base blends developed by Kerres.[202,212]

New Routes in PEMFC Membrane Development

Cross-Linked Terpolymers via a Sol-Gel Strategy To overcome the issues of polymer mechanical stability and swelling, a new class of polymer electrolyte membrane based on a continuous thermostable, nonconductive, organic polymer matrix mixed with a proton-conductive, sulfonated mesostructured silica network has been recently investigated.[213,214] This approach aims to reproduce the behavior of PFSA membranes where hydrophobic and hydrophilic regions coexist. The hydrophilic regions contain the ionic groups and are supported by the functionalized mesostructured silica network while the hydrophobic regions contain the fluorobackbone of the polymer. The volume fraction of the membrane supporting the proton conductivity represents only 40% of the total volume of the hybrid organic/inorganic membrane. Thus *in situ* sol-gel growth of an acid-functionalized inorganic network in a nonporogenic organic matrix has recently been presented (Fig. 20.20).[201]

The improvement in the water uptake is related to the surface hydroxyl groups and/or the $-SO_3H$ groups in the hybrid interfacial region, which can

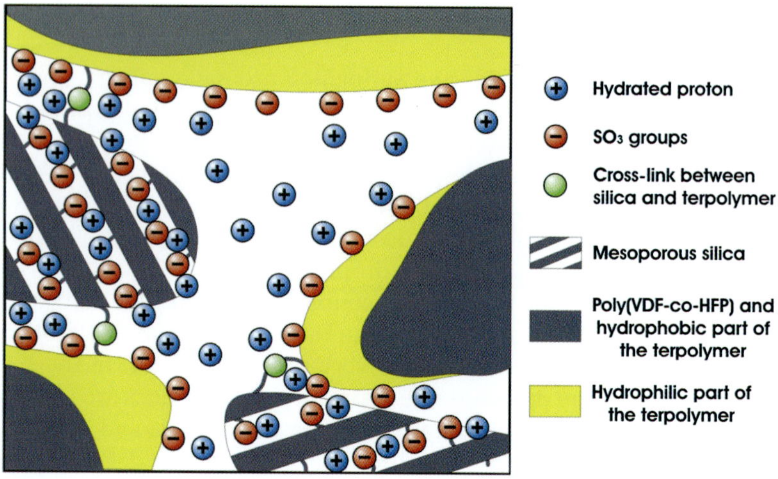

FIGURE 20.20. Schematic illustration of the hybrid SiO_2–SO_3H/terpolymer/poly(VDF-co-HFP) copolymer membrane. The silica domains exhibit a lamellar mesostructure with characteristic size 10 nm, based on SAXS studies. Adapted from Sel et al.[201]

strongly attract water molecules through hydrogen bonding. These organic/inorganic membranes show proton-conductivity values of 0.043 S cm^{-1} at 65°C under 100% RH and a conductivity value of 0.012 S cm^{-1} at 120°C.

Inert Polymer Matrix: Proton-Conducting Hybrid Inorganic Particles New composite membranes have been prepared by inserting polystyrene sulfonic acid-grafted silica particles into an inert polymer matrix of poly(vinylidene fluoride-co-hexafluoropropylene), PVDF-HFP.[215] Power density of 1 W cm^{-2} was recorded at 70°C using nonhydrated gas feeds, suggesting that this composite membrane is capable of self-humidification. Figure 20.21 shows single-cell performances of the hybrid membranes, exhibiting higher values than for Nafion® 112 membranes.

New Characterization* In Situ *Methods As indicated above, a shortcoming of water-mediated ion conductors is that their ion conduction varies widely with water content, the conductivity becoming too low at low water content. The most powerful method to quantify the membrane water content in a running fuel cell is small-angle neutron scattering (SANS). SANS spectra are usually recorded without any driving force applied to the membrane. Since several dynamic operations influence the water management in a PEMFC, a special cell has been designed to record SANS spectra during PEMFC functioning.[216] The data analysis leads to the determination of water concentration profiles across the membrane that could be used to validate mass transfer models and predict the best operating conditions of the PEMFC (Fig. 20.22).

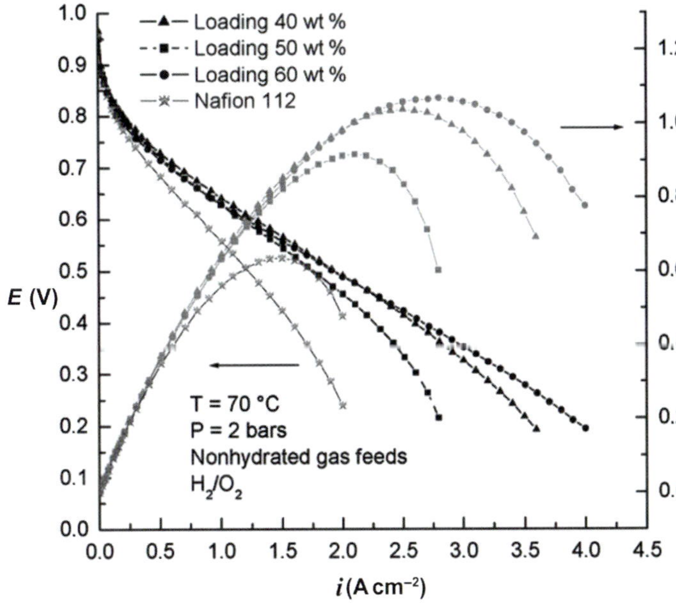

FIGURE 20.21. Single-cell PEMFC performances of hybrid membranes as function of filler loading and a Nafion® 112 membrane at 70°C with nonhydrated gas feeds (H_2/O_2; 2 bars). Adapted from Niepceron et al.[215]

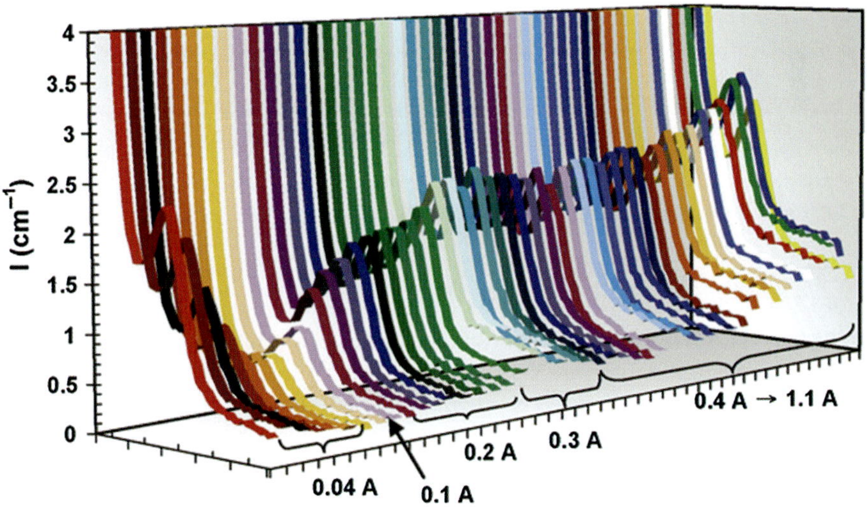

FIGURE 20.22. Series of SANS spectra obtained from a Nafion® 117 membrane using highly porous gas distribution. The membrane is first dried and then the current density is increased step by step from 0 to 1.1 A. Adapted from Gebel et al.[216]

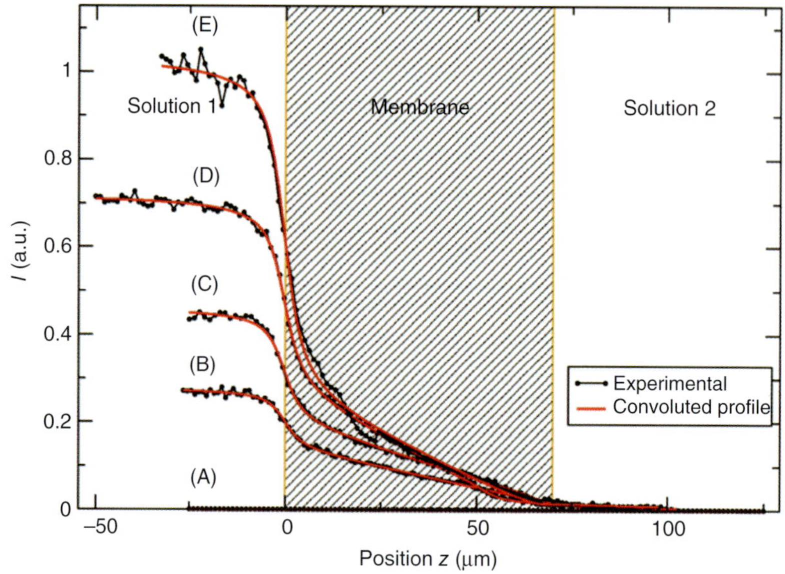

FIGURE 20.23. Experimental and convoluted Raman spectra intensity (in arbitrary units, a.u.) profiles of methanol diffusion through a Nafion® 112, obtained at different methanol (mol percent) content in the water–methanol feed compartment (pure water is in the other compartment). (A) 0%; (B) 25%; (C) 50%; (D) 75%; (E) 100%. Adapted from Deabate et al.[217]

Another tool for measuring the local water (or/and methanol) concentration/gradients within a proton conducting membrane are the *in situ* confocal Raman measurements.[217] A specially developed microfluidic cell allows the study of concentration profiles (within and near the surface of IEMs) through the acquisition of Raman spectra during dynamic transport. The results of such measurements are presented in Figure 20.23.

20.7.2 Salinity Gradient Power Generation

Pressure-Retarded Osmosis (PRO) A huge potential for clean energy is based on the mixing of water streams with different salt concentrations: sea water and river water. Salinity gradient power (SGP) is a relatively recent kind of renewable form of energy production, such as wind power, hydropower, or solar power. The global energy output produced by SGP is estimated to be between 2.6 [218] and 2.8 TW,[219] while the average world energy consumption in 2008 was about 15 TW.[222] Two membrane-based energy conversion techniques were proposed several years ago: pressure-retarded osmosis[220] and RED.[221] A considerable number of papers have since been published.[12,219,222–226] A brief review is presented below.

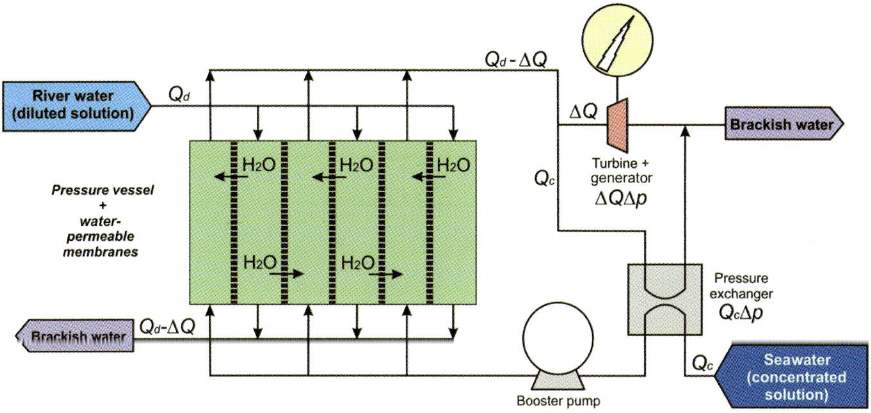

FIGURE 20.24. Schematic representation of an energy conversion scheme using pressure-retarded osmosis; Q is the volumetric solution flow rate (cubic meters per second); ΔQ the volume flux of water through the membranes (cubic meters per second) from the dilute to concentrated solution; and Δp the hydrostatic pressure difference between both solutions (pascals), whereas the power generated by means of a turbine and generator is $\Delta Q \Delta p$ (watts). Adapted from Post et al.[222]

In a pressure-retarded osmosis system, two solutions of different salinity are brought into contact by a semipermeable membrane (Fig. 20.24). The membrane allows the solvent (i.e., water) to permeate and retains the solute (i.e., dissolved salts). Under the action of the chemical potential difference between the solutions, water passes through the membrane from the diluted salt solution to the more concentrated one. This transport of water results in an increasing volumetric flow rate (from Q to $Q + \Delta Q$) and hydrostatic pressure (from p to $p + \Delta p$) of the concentrated solution. This pressurized, transported water can be used to generate electrical power in a turbine equal to $\Delta Q \Delta p$.[222]

Theoretical evaluations made by Post et al.[222] showed that in the case of sea and river waters, the maximum generated power density is 1.2 W m^{-2}. In real systems, the amount of generated power is between 0.11 and 1.22 W m^{-2}.[222] The higher value was obtained using a brine with osmotic pressure 39 bars (larger than the osmotic pressure of seawater, which is in the range 20–25 bars). The loss in power is mainly due to the fact that the driving force, which is determined by the osmotic pressure difference over the semipermeable active skin, is essentially lower than the osmotic pressure difference between the bulk solutions because of concentration polarization in the porous support.

Reverse ED with Monopolar Membranes In a RED system, CEMs and AEMs are stacked in an alternating pattern between a cathode and an anode, similarly to conventional ED (Fig. 20.25). The compartments between the membranes are alternately fed with a concentrated salt solution (e.g.,

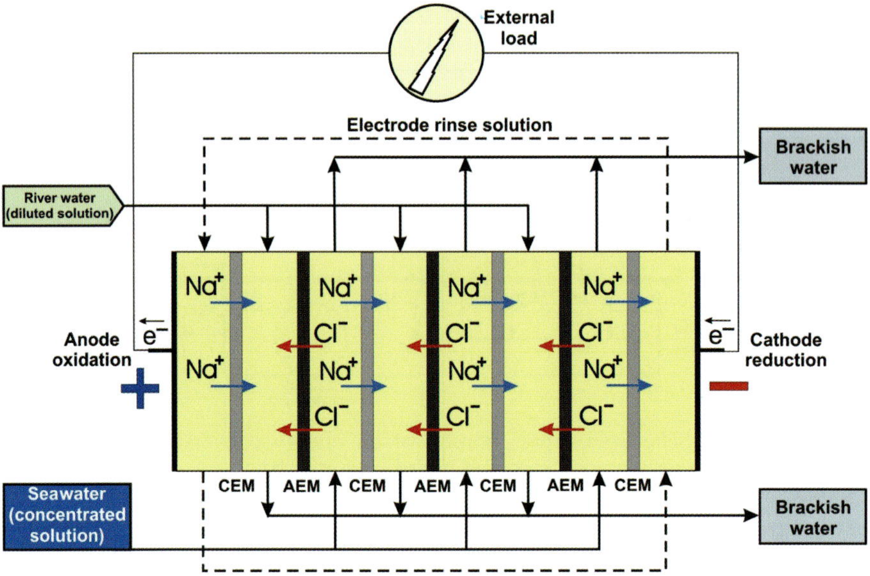

FIGURE 20.25. Schematic representation of an energy conversion system using RED. Adapted from Post et al.[222]

seawater) and a diluted salt solution (e.g., river water). The salinity gradient results in the formation of a potential difference (the so-called membrane potential) equal to about 80 mV for seawater and river water for each membrane. The potential difference between the outer compartments of the membrane stack is the sum of the potential differences for each membrane.

The salinity gradient produces ion transport through the membranes from the concentrated to the diluted solution. The cations pass through the CEMs toward the cathode, and the anions pass through the AEMs toward the anode. The electroneutrality condition causes equal fluxes of charges transported by cations and anions. In the electrode compartments, this condition is maintained via electrochemical reactions at the electrode surface with generation of electrons. The latter, transferred from the anode to the cathode in the external circuit, provide the electric current.

The power generated in the system is equal to the product of the current, I, and the electric potential difference between the electrodes. The latter is approximately proportional to the number of cell pairs and the pd over one cell pair: $P = I(N\Delta\varphi)$. The value of $\Delta\varphi$ is lower than the membrane potential for a pair of a CEM and an AEM due to ohmic losses and concentration polarization, reducing the concentration drop over the membrane.

Theoretical evaluations made for the case of seawater and river water show that RED has larger potential maximum power density than pressure-retarded osmosis. For RED this value is 2–4 W m^{-2} according to Post et al.[222] or higher than 6 W m^{-2} according to Długołecki et al.[223] The losses in power density are mainly determined by the resistance of the diluate compartment; they depend weakly on membrane properties.[223] Thinner conductive spacers could essentially improve the characteristics of a RED system. Currently available ED membranes in a RED application on concentrated brines and freshwater yield a maximum power density of 1.2 W m^{-2} [224] or 1.6 W m^{-2}.[219] It is important that the obtained power density decreases with increasing size of membranes: the values found for a small stack of 10 × 10 cm^2 are essentially higher than those found for a stack of 24 × 75 cm^2.[219]

Analysis made by Długołecki et al.[225,226] shows that in order to improve SGP processes, the development of pressure-retarded osmosis must focus on membrane characteristics, that is, increasing water permeability of the membrane and optimization of the porous support. On the other hand, the development of RED must focus on system characteristics, that is, optimization of the internal resistance, which is mainly determined by the thickness and conductivity of the spacers. Moreover, the sensitivity for fouling is an important characteristic highly influencing the performance of SGP devices.

The main drawback of membrane-based conversion techniques are the high price of membranes.[227] However, the decreasing prices of membranes for desalination and water reuse applications, as well as the increasing prices of fossil fuels, should make salinity gradient power attractive in near future.

RED with BPMs A minimum potential of 0.828 V could start water dissociation by a BPM unit at 25°C of 1 mol acid and 1 mol base.[125,228] A similar potential drop will be produced by reacting 1 mol acid and 1 mol base. Thus, electrical energy can be produced in a system involving BPMs and two feed (acid and alkaline) solutions. Figure 20.26 shows the electricity generation process of a BPM battery, which is constructed with a BPM and a couple of electrodes. When acid and alkaline solutions are fed, respectively, from the cationic and the anionic sides of the BPM, H^+ ions diffuse from the acid solution, and OH^- ions diffuse from the alkaline side toward the middle of the BPM and react at the junction. As a result, a current density passes through the BPM, which can be used in the external circuit through the electrodes.

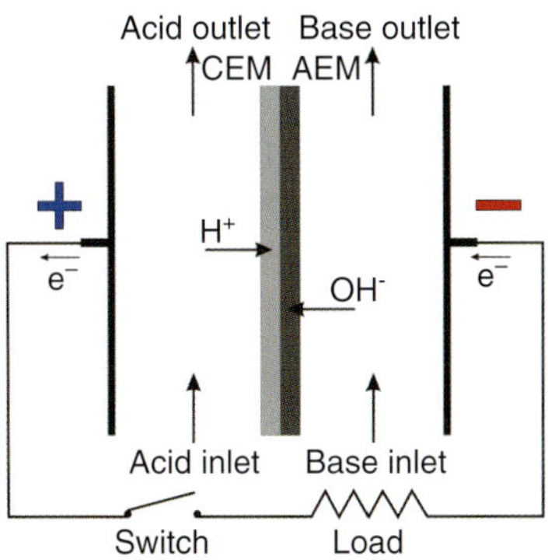

FIGURE 20.26. Schematic representation of an energy conversion system using a bipolar membrane cell.

A BPM with 100 cell units (consisting of a BPM, a CEM, and an AEM) may produce 40–80 V with output power 30–60 kW m^{-2}.[229] One of the advantages of such devices is that they can provide energy from acid and base wastes.

ACKNOWLEDGMENTS

The chapter has been prepared in the framework of collaboration within French–Russian laboratory "Ion-Exchange Membranes and Related Processes." We are grateful to CNRS, France, and RFBR, Russia, for financial support of this work.

LIST OF SYMBOLS

Abbreviations

AEM	anion exchange membrane
BMED	bipolar membrane electrodialysis
BPM	bipolar membrane
CC	concentrate compartments
CDI	capacitive deionization
CEDI	continuous electrodeionization
CEM	cation exchange membrane
DBL	diffusion boundary layer
DC	diluate compartment
DD	diffusion dialysis
ED	electrodialysis
EDI	electrodeionization
EDL	electrical double layer
EDR	electrodialysis reversal
EED	electro-electrodialysis
IEM	ion exchange membrane
MCDI	membrane capacitive deionization
MC	membrane crystallization
MD	membrane distillation
MEA	membrane electrode assembly
MVR	mechanical vapor recompression
NF	nanofiltration
PBI	polybenzimidazole
pd	potential difference
PEEK	polyether(ether)ketones

PEMFC	proton exchange membrane fuel cells
PFSA	perfluorosulfonic acid polymers
PRO	pressure-retarded osmosis
PVDF	polyvinylidene fluoride
PVDF-HFP	poly(vinylidene fluoride-co-hexafluoropropylene)
RED	reverse electrodialysis
RH	relative humidity
RO	reverse osmosis
SANS	small-angle neutron scattering
SCR	space charge region
SGP	salinity gradient power
T	temperature
TDS	total dissolved salts
UF	ultrafiltration
VCR	voltage–current response
WAIV	wind-aided intensified evaporation
ZLD	zero liquid discharge (system)

Symbols

C^{in}	electrolyte concentration at inlet of DC
i	current density
$i_{\lim}$	limiting current density
p	hydrostatic pressure
P	power
Sh	Sherwood number
w	solution flow rate

Greek Symbols

δ Nernst's diffusion layer thickness

REFERENCES

1. Market profile: zero liquid discharge systems. From zero to hero—the rise of ZLD. *Glob Water Intell* **2009**, 10 (12), 37.
2. M. Juda and W. A. McRac. *J Am Chem Soc* **1950**, 72, 1044.
3. T. R. E. Kressman. *Nature* **1950**, 165, 568.
4. W. E. Katz. In *Industrial Processing with Membranes*, R. E. Lacey, S. Loeb, eds. Wiley-Interscience, New York, 1972; p. 39.

5. V. A. Shaposhnik and K. Kesore. *J Membr Sci* **1997**, 136, 35.
6. V. D. Grebenyuk and O. V. Grebenyuk. *Russ J Electrochem* **2002**, 38 (8), 806.
7. H. Strathmann. *Ion-Exchange Membrane Separation Processes. Membrane Science and Technology Ser. 9*. Elsevier, Amsterdam, 2004.
8. S. Loeb and S. Sourirajan. *Adv Chem Ser* **1962**, 38, 117.
9. J. Glater. *Desalination* **1998**, 117, 297.
10. C. Gasson and P. Allison. *Desalination Markets 2007. A Global Forecast, Global Water Intelligence*. Media Analytics Ltd., Oxford, 2007.
11. D. Lawler, Y. Kim, W. Walker. Proceedings of ICOM'2008, Honolulu **2008**.
12. H. Strathmann. *Desalination* **2010**, 264, 268.
13. M. Anderson and A. L. Cudero. *J Palma Electrochim Acta* **2010**, 55, 3845.
14. S.-J. Kim, S.-H. Ko, K. H. Kang, J. Han. *Nature Nanotech* **2010**, 5, 297.
15. J. Lyklema. *Fundamentals of Interface and Colloid Science*. Academic Press, London, 1995.
16. V. S. Bogotsky. *Fundamentals of Electrochemistry*. Wiley-Interscience, Hoboken, NJ, 2006.
17. J. Mulder. *Basic Principles of Membrane Technology*. Kluwer Academic Publisher, Boston, 1996.
18. C. O. Anne, D. Trébouet, P. Jaouen, F. Quéméneur. *Desalination* **2001**, 140 (1), 67.
19. R. W. Baker. *Membrane Technology and Applications*. Wiley, Chichester, 2004.
20. X.-L. Wang, T. Tsuru, S.-I. Nakao, S. Kimura. *J Membr Sci* **1995**, 103 (1–2), 117.
21. A. E. Yaroshchuk. *Adv Colloid Interface Sci* **1995**, 60 (1–2), 1.
22. A. Yaroshchuk, Y. Boiko, A. Makovetskiy. *Langmuir* **2009**, 25 (16), 9605.
23. S. Loeb and S. Sourirajan. US Patent No 3,133,132. **1964**.
24. E. Drioli and E. Fontananova. *Chem Eng Res Des* **2004**, 82, 1557.
25. G. Pearce. *Filtr Sep* **2007**, 44, 28.
26. G.-D. Kang, C.-J. Gao, W.-D. Chen, X.-M. Jie, Y.-M. Cao, Q. Yuan. *J Membr Sci* **2007**, 300 (1–2), 165.
27. I. C. Tessaro, J. B. A. Silva, K. Wada. *Desalination* **2005**, 181 (1–3), 275.
28. J. R. Wilson. Design and operation of electrodialysis plants. In *Demineralization by Electrodialysis*. J. R. Wilson, ed. Butterworth, London, 1960.
29. V. V. Nikonenko, N. D. Pismenskaya, E. I. Belova, P. H. Sistat, P. Huguet, G. Pourcelly, C. H. Larchet. *Adv Colloid Interface Sci* **2010**, 160, 101.
30. N. A. Mishchuk. *Adv Colloid Interface Sci* **2010**, 160 (1–2), 16–39.
31. Y. Tanaka. *Ion Exchange Membranes: Fundamentals and Applications. Membrane Science and Technology Ser. 12*. Elsevier, Amsterdam, 2007.
32. C. Casademont, G. Pourcelly, L. Bazinet. *Langmuir* **2010**, 26 (2), 854.
33. T. Sata. *Ion Exchange Membranes. Preparation, Modification, Characterization and Application*. The Royal Society of Chemistry, Cambridge, 2004.
34. T. Sato, T. Tanaka, H. Ohya. *Bioelectrochem Bioenerg* **1989**, 21, 47.
35. J. Wood, J. Gifford, J. Arba, M. Shaw. *Desalination* **2010**, 250, 973.
36. M. B. S. Ali, A. Mnif, B. Hamrouni, M. Dhahbi. *Ionics* **2010**, 16, 621.
37. L. F. Greenlee, D. F. Lawler, B. D. Freeman, B. Marrot, P. Moulin. *Water Res* **2009**, 43, 2317.

38. M. Turek and P. Dydo. *Desalination* **2008**, 221, 455.
39. Y. Tanaka. *J Membr Sci* **1999**, 163, 227.
40. M. Turek. *Desalination* **2003**, 153 (1–3), 371.
41. A. Elmidaoui, F. Elhannouni, M. A. Menkouchi Sahli, L. Chay, H. Elabbassi, M. Hafsi, D. Largeteau. *Desalination* **2001**, 136, 325.
42. Z. Amor, B. Bariou, N. Mameri, M. Taky, S. Nicolas, A. Elmidaoui. *Desalination* **2001**, 133, 215.
43. R. P. Alison. *Desalination* **1995**, 103, 11.
44. American Water Works Association (AWWA) manual M38. *Electrodialysis and electrodialysis reversal (M38)*. Denver, CO, **2007**.
45. F. Valero and R. Arbós. *Desalination* **2010**, 253 (1–3), 170.
46. C. Larchet, G. Eigenberger, A. Tskhay, K. Tastanov, V. Nikonenko. *Desalination* **2002**, 149, 383.
47. B. P. Hemon, R. H. Zanapalidou, L. Zhang, K. J. Sims, L. R. Siwak. *Ultrapure Water* **1994**, 11, 33.
48. R. K. Nagarale, G. S. Gohil, V. K. Shahi. *Adv Colloid Interface Sci* **2006**, 119 (2–3), 97.
49. C. Larchet, L. Dammak, B. Auclair, S. Parchikov, V. Nikonenko. *New J Chem* **2004**, 28 (10), 1260.
50. W. E. Katz. *Desalination* **1979**, 28, 31.
51. Y. M. Chao and T. M. Liang. *Desalination* **2008**, 221, 433.
52. J.-S. Park, H.-J. Lee, S.-H. Moon. *Sep Pur Technol* **2003**, 30 (2), 101.
53. C. H. Casademont, P. H. Sistat, B. Ruiz, G. Pourcelly, L. Bazinet. *J Membr Sci* **2009**, 328 (1–2), 238.
54. B. Ruiz, P. H. Sistat, P. Huguet, G. Pourcelly, M. Araya-Farias. *J Membr Sci* **2007**, 287 (1), 41.
55. B. Ruiz, P. H. Sistat, P. Huguet, G. Pourcelly, M. Araya-Farias, L. Bazinet. *Desalination* **2006**, 200 (1–3), 208.
56. J. Guolin, W. Xiaoyu, H. Chunjie. *Desalination* **2008**, 220, 386.
57. J. Guolin, X. Lijie, L. Yang, D. Wenting, H. Chunjie. *Desalination* **2010**, 264 (3), 214.
58. A. Grabowski, G. Zhang, H. Strathmann, G. Eigenberger. *Sep Pur Technol* **2008**, 60 (1), 86.
59. N. P. Gnusin and V. D. Grebenyuk. *Elektrokhimiya Granulirovannykh Ionitov (Electrochemistry of Granulated Ionites)*. Naukova Dumka, Kiev, Ukraine, 1972.
60. G. Eigenberger, H. Strathmann, A. Grabowski. Germany Patent WO 2005/009596 A1, B01D 61/44. 2005.
61. V. A. Shaposhnik, N. N. Zubets, B. E. Mill, I. P. Strigina. *Desalination* **2001**, 133, 211.
62. E. Dejean, J. Sandeaux, R. Sandeaux, C. Gavach. *Separ Sci Technol* **1998**, 33, 801.
63. C. E. M. Ben, M. B. Amor, G. Pourcelly. *Sep Purif Technol* **2006**, 51, 285.
64. O. Kedem. *Desalination* **1975**, 16, 105.
65. R. L. Messalem, Y. Mirsky, N. Daltrophe, G. Saveliev, O. Kedem. *J Membr Sci* **1998**, 138, 171.
66. V. A. Shaposhnik, O. V. Grigorchuk, E. N. Korzhov, V. I. Vasil'eva, V. Ya. Klimov. *J Membr Sci* **1998**, 139, 85.

67. K. H. Yeon and S. H. Moon. *Sep Sci Technol* **2003**, 38 (10), 2347.
68. Y. Koji, I. Hiroshi, T. Naoyuki. US Patent No. 7,201,832, B01D61/48; B01D61/44. 2007.
69. V. I. Zabolotski, V. V. Nikonenko, N. D. Pismenskaya, V. F. Pismenski, E. V. Lactionov. Russ Patent No 2033850 B 01 D 13/02. 1995.
70. C. Larchet, V. I. Zabolotsky, N. Pismenskaya, V. V. Nikonenko, A. Tskhay, K. Tastanov, G. Pourcelly. *Desalination* **2008**, 222, 489.
71. J. Balster, D. F. Stamatialis, M. Wessling. *J Membr Sci* **2010**, 360, 185.
72. M. Henley. *Ultrapure Water* **1997**, 9, 15.
73. R. Simons. *Electrochim Acta* **1984**, 29, 151.
74. B. P. Spoor, L. Grabovska, L. Koene, L. J. J. Janssen, W. R. Ter Veen. *Chem Eng J* **2002**, 89, 193.
75. G. Ganzi and P. Parise. *J Parenteral Sci Technol* **1990**, 44 (4), 231.
76. D. Avijit, P. Vista, T. G. Hugh, H. Towers. US Patent No 6,929,748 B01D 61/00 B01D 59/44. 2005.
77. L. Fu, J. Wang, Y. Su. *Sep Pur Technol* **2009**, 68, 390.
78. V. I. Zabolotsky, V. V. Nikonenko, N. D. Pismenskaya, A. G. Istoshin. *Desalination* **1996**, 108, 179.
79. N. D. Pismenskaya, E. I. Belova, V. V. Nikonenko, V. I. Zabolotsky, G. Yu. Lopatkova, Yu. N. Karzhavin, C. Larchet. *Desalinat Water Treat* **2010**, 21, 109.
80. K. H. Yeon, J. H. Song, S. H. Moon. *Water Res* **2004**, 38, 1911.
81. I. Demkin, Y. A. Tubashov, V. I. Panteleev, Y. V. Karlin. *Desalination* **1987**, 64, 367.
82. M. I. Ojovan and W. E. Lee. In *An Introduction to Nuclear Waste Immobilisation*, Elsevier, Amsterdam, 2005; pp. 149–178.
83. J. H. Song, M. C. Song, K. H. Yeon, J. B. Kim, K. J. Lee, S. H. Moon. *J Radioanal Nucl Chem* **2004**, 262 (3), 725.
84. J.-S. Kim and J.-H. Choi. *J Membr Sci* **2010**, 355, 85.
85. P. M. Biesheuvel and A. van der Wal. *J Membr Sci* **2010**, 346, 256.
86. Y.-J. Kim, J. H. Wisup Bae, J.-H. Choi. *Desalination* **2010**, 253, 119.
87. M. D. Andelman. CA Patent No 2444390. **2002**.
88. Y. Oren. *Desalination* **2008**, 228, 10.
89. Y. J. Kim and J. H. Choi. *Sep Purif Technol* **2010**, 71, 70.
90. H. H. Jung, S. W. Hwang, S. H. Hyun, K. H. Lee, G. T. Kim. *Desalination* **2007**, 216, 377.
91. A. Afkhami and B. E. Conway. *J Colloid Interface Sci* **2002**, 251, 248.
92. K. K. Park, J. B. Lee, P. Y. Park, S. W. Yoon, J. S. Moon, H. M. Eum, C. W. Lee. *Desalination* **2007**, 206, 86.
93. Y. Gao, L. Pan, H. Li, Y. Zhang, Z. Zhang, Y. Chen, Z. Sun. *Thin Solid Films* **2009**, 517, 1616.
94. D. Zhang, L. Shi, J. Fang, K. Dai. *J Mater Sci* **2007**, 242, 2471.
95. K. C. Leonard, J. R. Genthe, J. L. Sanfilippo, W. A. Zeltner, M. A. Anderson. *Electrochim Acta* **2009**, 54, 5286.
96. M. I. Kerwick, S. M. Reddy, A. H. L. Chamberlain, D. M. Holt. *Electrochim Acta* **2005**, 50, 5270.

97. R. E. Perez-Roa, D. T. Tompkins, M. Paulose, C. A. Grimes, M. A. Anderson, D. R. Noguera. *Biofouling* **2006**, 22, 383.
98. M. Z. Bazant, M. S. Kilic, B. D. Storey, A. Ajdari. *Adv Colloid Interface Sci* **2009**, 152, 48.
99. J. H. Lee and Y.-A. Song. *J Lab Chip* **2008**, 8, 596.
100. J. Balster, M. H. Yildirim, D. F. Stamatialis, R. Ibanez, R. G. H. Lammertink, V. Jordan, M. Wessling. *J Phys Chem* **2007**, 111, 2152.
101. I. Rubinstein and B. Zaltzman. *Adv Colloid Interface Sci* **2010**, 159 (2), 117.
102. S. S. Dukhin. *Adv Colloid Interface Sci* **1991**, 35, 173.
103. V. K. Egorov. USSR Patent No 587960 B 01D 13/02. **1978**.
104. V. F. Pismenskiy, V. I. Zabolotsky, N. P. Gnusin, A. A. Shudrenko. *J Appl Chem (Russ)* **1986**, 59 (4), 773.
105. V. D. Grebeniuk, B. K. Veysov, R. D. Chebotareva, K. P. , G. Z. Nefedova. *J Appl Chem (Russ)* **1986**, 59 (4), 916.
106. T. Nishiwaki. Concentration of electrolytes prior to evaporation with an electromembrane process. In *Industrial Process with Membranes*, R. F. Lacey and S. Loeb, eds. Wiley-Interscience, New York, 1972.
107. M. Turek, P. Dydo, R. Klimek. *Desalination* **2005**, 84 (1–3), 439.
108. L. P. Ling, H. F. Leow, M. R. Sarmidi. *J Membr Sci* **2002**, 199, 59.
109. G. S. Luo, S. Pan, J. G. Liu. *Desalination* **2002**, 150, 227.
110. I. A. Sobolev, S. A. Dmitriev, V. I. Demkin, E. M. Timofeev, V. I. Panteleev, D. V. Adamovich. Russ. Patent No 2160473. **1999**.
111. K. Scott and R. Hughes. *Industrial Membrane Separation Technology*, Blackie Academic & Professional, Glasgow, 1996.
112. A. V. Demin and V. I. Zabolotsky. *Russ J Electrochem* **2008**, 44 (9), 1058.
113. B. Van der Bruggen, L. Lejon, C. Vandecasteele. *Environ Sci Technol* **2003**, 37, 3733.
114. V. Mavrov, H. Chmiel, B. Heitele, F. Rögener. *Desalination* **1999**, 124, 205.
115. M. Badruzzaman, J. Oppenheimer, S. Adham, M. Kumar. *J Membr Sci* **2009**, 326, 392.
116. Y. Zhang, K. Ghyselbrecht, B. Meesschaert, L. Pinoy, B. Van der Bruggen. *J Membr Sci* **2010**, 378, 101.
117. E. Korngold, L. Aronov, N. Daltrophe. *Desalination* **2009**, 242, 215.
118. V. Zabolotsky, K. Protasov, M. Sharafan. *Russ J Electrochem* **2010**, 46 (9), 979.
119. N. P. Berezina, N. A. Kononenko, A. A.-R. Sytcheva, N. V. Loza, S. A. Shkirskaya, N. Hegman, A. Pungor. *Electrochim Acta* **2009**, 54 (8), 2342.
120. K. Protasov, S. Shkirskaya, N. Berezina, V. Zabolotsky. *Russ J Electrochem* **2010**, 46, 1131.
121. V. I. Zabolotsky, N. V. Shel'deshov, N. P. Gnusin, N. D. Pis'menskaya. *Sov Electrochem* **1985**, 21 (8), 993.
122. V. I. Zabolotsky, N. V. Shel'deshov, N. P. Gnusin. *Russ Chem Rev* **1988**, 57, 801.
123. S. Mafé, P. Ramirez, A. Alcaraz, V. Aguilella. *Handbook on Bipolar Membrane Technology*, A. J. B. Kemperman, ed. Twente University Press, Twente, the Netherlands, 2000.
124. J. Balster, S. Srinkantharajah, R. Sumbharaju, I. Pünt, R. G. H. Lammertink, D. F. Stamatialis, M. Wessling. *J Membr Sci* **2010**, 365, 389.

125. F. G. Wilhelm, N. F. Vegt van der, M. Wessling, H. Strathmann. Bipolar membrane preparation. In *Handbook on Bipolar Membrane Technology*, A. J. B. Kemperman, ed. Twente University Press, Enschede, the Netherlands, 2000.
126. M.-S. Kang, Y.-J. Choi, S.-H. Kim, S.-H. Moon. *J Membr Sci* **2004**, 229, 137.
127. G. Pourcelly. *Russ J Electrochem* **2002**, 38, 919.
128. X. Tongwen and Y. Weihua. *J Membr Sci* **2002**, 203, 145.
129. G. Pourcelly and C. Gavach. Electrodialysis water splitting—applications of electrodialysis with bipolar membranes. In *Handbook on Bipolar Membrane Technology*, A. J. B. Kemperman, ed. Twente University Press, Enschede, the Netherlands, 2000.
130. M. Kumar and V. K. Shahi. *J Membr Sci* **2010**, 349 (1–2), 130.
131. J. Balster, I. Pünt, D. F. Stamatialis, H. Lammers, A. B. Verver, M. Wessling. *J Membr Sci* **2007**, 303, 213.
132. E. Vera, J. Sandeaux, F. Persin, G. Pourcelly, M. Dornier, J. Ruales. *J Food Eng* **2007**, 78, 1427.
133. Ch. Huang, T. Xu, Y. Zhang, Y. Xue, G. Chen. *J Membr Sci* **2007**, 288, 1.
134. Y. Wang, X. Zhang, T. Xu. *J Membr Sci* **2010**, 365 (1–2), 294.
135. K. Zhang, M. Wang, D. Wang, C. Gao. *J Membr Sci* **2009**, 341 (1–2), 246.
136. G. Pourcelly and L. Bazinet. Developments of bipolar membrane technology in food and bio-industry. In *Handbook of Membrane Separation; Chemical, Pharmaceutical, Food, and Biotechnological Applications*, A. K. Pabby, S. S. H. Rizvi, A. M. Sastre, eds. CRC Press, Taylor & Francis Group, New York, 2009.
137. T. V. Eliseeva, A. Yu. Tekuchev, V. A. Shaposhnik, I. G. Lushchik. *Russ J Electrochem* **2001**, 37 (4), 423.
138. M. Mondor, F. Ali, D. Ippersiel, F. Lamarche. *Innovat Food Sci Emerg Tech* **2010**, 11 (3), 491.
139. K. Bechthold, S. Bretz, R. Kabasci, A. Kopitzky. *Chem Eng Technol* **2008**, 31 (5), 647.
140. H. G. Joglekar, R. Imran, B. Suresh, B. D. Kulkarni, J. Ajit. *Sep Pur Technol* **2006**, 52, 1.
141. H. Habe, Y. Shimada, T. Fukuoka, D. Kitamoto, M. Itagaki, K. Watanabe, H. Yanagishita, K. Sakaki. *J Biosci Bioeng* **2010**, 110 (6), 690.
142. M.-S. Kang, Y.-J. Choi, H.-J. Lee, S.-H. Moon. *J Colloid Interface Sci* **2004**, 273, 523.
143. M.-S. Kang, Y.-J. Choi, S.-H. Moon. *J Colloid Interface Sci* **2004**, 273, 533.
144. F. D. Stamatialis, B. J. Papenburg, M. Giron′es, S. Saiful, S. N. M. Bettahalli, S. Schmitmeier, M. Wessling. *J Membr Sci* **2008**, 308, 1.
145. J. Balster, D. F. Stamatialis, M. Wessling. *Chem Eng Process* **2004**, 43, 1115.
146. K.-L. Huang, T. M. Holsen, T.-C. Chou, J. R. Selman. *Environ Sci Technol* **2003**, 37 (9), 1992.
147. I. Frenzel, H. Holdik, D. F. Stamatialis, G. Pourcelly, M. Wessling. *J Membr Sci* **2005**, 261, 49.
148. Z. X. Wang, Y. B. Luo, P. Yu. *J Membr Sci* **2006**, 280, 134.
149. A. Terada, J. Iwatsuki, S. Ishikura, H. Noguchi, S. Kubo, H. Okuda, S. Kasahara, N. Tanaka, H. Ota, K. Onuki, R. Hino. *J Nucl Sci Technol* **2007**, 44 (3), 477.
150. P. Zhang, S. Z. Chen, L. J. Wang, J. M. Xu. *Int J Hydrogen Energy* **2010**, 35, 2883.

151. K. Onuki, G.-J. Hwang, Arifal S. Shimizu. *J Membr Sci* **2001**, 192, 193.
152. S. Koter and M. Kultys. *Sep Pur Technol* **2010**, 73, 219.
153. J.-H. Cheng, Y. Xiao, C. Wu, T.-S. Chung. *Chem Eng J* **2010**, 160, 340.
154. Y. Kobuchi, H. Motomura, Y. Noma, F. Hanada. *J Membr Sci* **1987**, 27, 173.
155. J. Luo, C. Wu, T. Xu, Y. Wu. *J Membr Sci* **2011**, 366 (1–2), 1.
156. Q. Wang, J. J. Lenhart, H. W. Walker. *J Membr Sci* **2010**, 360, 469.
157. J. Wisniewski and A. Rozanska. *Desalination* **2007**, 212, 251.
158. O. Altintas, A. Tor, Y. Cengeloglu, M. Ersoz. *Desalination* **2009**, 239, 276.
159. Z. Palatý and A. Zakova. *J Appl Polym Sci* **2006**, 101, 1391.
160. E. Kir, B. Gurler, A. Gulec. *Desalination* **2011**, 267 (1), 114.
161. J. A. Wisniewski and M. Kabsch-Korbutowicz. *Desalination* **2010**, 261, 197.
162. D. E. Akretche and H. Kerdjoudj. *Talanta* **2000**, 51, 281.
163. T. Sardohan, E. Kir, A. Gulec, Y. Cengeloglu. *Sep Pur Technol* **2010**, 74, 14.
164. M. Vanhoute, L. Firdaous, L. Bazinet, R. Froidevaux, D. Lecouturier, D. Guillochon, P. Dhulster. *J Membr Sci* **2010**, 365, 16.
165. L. Firdaous, P. Dhulster, J. Amiot, A. Gaudreau, D. Lecouturier, R. Kapel, F. Lutin, L. P. Vezina, L. Bazinet. *J Membr Sci* **2009**, 329, 60.
166. J. F. Poulin, J. Amiot, L. Bazinet. *J Biotechnol* **2006**, 123, 314.
167. C. R. Martinetti, A. E. Childress, T. Y. Cath. *J Membr Sci* **2009**, 331, 31.
168. B. Durham and M. Mierzejewski. *Water Sci Technol Water Supply* **2003**, 3, 97.
169. Y. Oren, E. Korngold, N. Daltrophe, R. Messalem, Y. Volkman, L. Aronov, M. Weismann, N. Bouriakov, P. Glueckstern, J. Gilron. *Desalination* **2010**, 261, 321.
170. F. Macedonio and E. Drioli. *Membr Water Treat* **2010**, 1 (1), 75.
171. M. Turek. *Chem Pap* **2003**, 1, 50.
172. M. Turek, P. Dydo, A. Surma. *Desalination* **2005**, 185 (1–3), 275.
173. M. Turek, J. Was, P. Dydo. *Desalinat Water Treat* **2009**, 7, 263.
174. V. Mavrov, H. Chmiel, B. Heitele, F. Rögener. *Desalination* **1996**, 108, 159.
175. L. Katzir, Y. Volkmann, N. Daltrophe, E. Korngold, R. Messalem, Y. Oren, J. Gilron. *Desalinat Water Treat* **2010**, 13 (1–3), 63.
176. R. Passino and G. Boari. U.S. Patent No 3901781, **1973**.
177. A. Tskhay. *Abstracts of International Conference Ion Transport in Organic and Inorganic Membranes*. Kuban State University, Tuapse, Russia, 2009.
178. F. Barbir. Fuel cells for clean power generation: status and perspectives. In *Assessment of Hydrogen Energy for Sustainable Development*, J. W. Sheffield, C. Sheffield, eds. Springer, Dordrecht, 2007; pp. 113–121.
179. S. Reichman, A. Ulus, E. Peled. *J Electrochem Soc* **2007**, 154, 327.
180. J. L. Zhang, Z. Xie, J. Zhang, Y. Tang, C. Song, T. Navessin, Z. Shi, D. Song, H. Wang, D.-P. Wilkinson, Z.-S. Liu, S. Holdcroft. *J Power Sources* **2006**, 160, 872.
181. J. Rozière and D. Jones. *Ann Rev Mater Res* **2003**, 33, 503.
182. A.-M. Herring. *Polym Rev* **2006**, 46, 245.
183. M. F.-H. Schuster and W. H. Meyer. *Ann Rev Mater Res* **2003**, 33, 233.
184. J. Weber, K. D. Kreuer, J. Maier, A. Thomas. *Adv Mater* **2008**, 20, 2595.
185. Y. L. Ma, J. S. Wainright, M. H. Litt, R. F. Savinell. *J Electrochem Soc* **2004**, 151, 8.

186. S. Wasmus, A. Valeriu, G. D. Mateescu, D. A. Tryk, R. F. Savinell. *Solid State Ionics* **1995**, 80, 87.
187. P. Costamagna and S. Srinivasan. *J Power Sources* **2001**, 102, 242.
188. C. Yang, P. Costamagna, S. Srinivasan, J. Benziger, A. B. Bocarsly. *J Power Sources* **2001**, 103, 1.
189. S. J. Paddison. *Ann Rev Mater Res* **2003**, 33, 289.
190. Y. Shao, G. Yin, Z. Wang, Y. Gao. *J Power Sources* **2007**, 167, 235.
191. A. Parthasarathy, S. Srinivasan, A. J. Appleby, C. R. Martin. *J Electrochem Soc* **1992**, 139, 530.
192. A. K. Santra and D. W. Goodmann. *Electrochim Acta* **2002**, 47, 3595.
193. W. Dai, H. Wang, X.-Z. Yuan, J.-J. Martin, D. Yang, J. Qiao, J. Ma. *Intern J Hydrogen Energy* **2009**, 34, 9461.
194. Q. Li, R. He, J.-O. Jensen, N.-J. Bjerrum. *Chem Mater* **2003**, 15, 4896.
195. S. Li, J. O. Jensen, R. F. Savinell, N. J. Bjerrum. *Progr Polymer Sci* **2009**, 34, 449.
196. S. Thayumanasundaram, M. Piga, S. Lavina, E. Negro, M. Jeyapandian, L. Ghassemzadeh, K. Muller, V. Di. *Noto Electrochim Acta* **2010**, 55, 1355.
197. A. B. Yaroslavtsev. *Russ J General Chem* **2010**, 80 (3), 675.
198. D. J. Jones and J. Rozière. *J Membrane Sci* **2001**, 185, 41.
199. W. L. Harrison, M. A. Hickner, Y. S. Kim, J. E. McGrath. *Fuel Cells* **2005**, 2, 201.
200. D. J. Jones and J. Rozière. Inorganic/organic composite membranes. In *V.3 Handbook of Fuel Cells: Fundamentals, Technology and Applications*, W. Vielstichm, A. Lamm, H. A. Gasteiger, eds. Wiley, New York, 2003; pp. 447–455.
201. O. Sel, A. Soules, B. Ameduri, B. Boutevin, C. Laberty-Robert, G. Gebel, C. Sanchez. *Adv Funct Mater* **2010**, 20, 1090.
202. J. A. Kerres. *Fuel Cells* **2005**, 5, 230.
203. R. H. Wiley and T. K. Venkatachalam. *J Polym Sci A* **1966**, 4, 182.
204. B. Gupta, F. N. Buechi, G. G. Scherer. *J Polym Sci A Polym Chem* **1994**, 32, 1931.
205. S. Hietala, S. L. Maunu, F. Sundholm. *J Polym Sci B* **2000**, 38, 3277.
206. L. Depre, J. Kappel, M. Popall. *Electrochim Acta* **1998**, 43, 1301.
207. M. I. Litter and C. S. Marvel. *J Polymer Sci Polym* **1985**, 23, 2205.
208. D. Kerres, W. Cui, S. Reiche. *J Polym Sci A* **1996**, 34, 2421.
209. C. Genies, R. Mercier, B. Sillon, N. Cornet, G. Gebel, M. Pineri. *Polymer* **2001**, 42, 359.
210. K. D. Kreuer. *J Membr Sci* **2001**, 185, 26.
211. M. Rikukawa and K. Sanui. *Prog Polym Sci* **2000**, 25, 1463.
212. J. A. Kerres. *J Membr Sci* **2001**, 185, 3.
213. K. Valle, P. Belleville, F. Pereira, C. Sanchez. *Nat Mater* **2006**, 6, 107.
214. F. Pereira, K. Valle, P. Belleville, A. Morin, S. Lambert, C. Sanchez. *Chem Mater* **2008**, 20, 1710.
215. F. Niepceron, B. Lafitte, H. Galiano, J. Bigarré, E. Nicol, J. F. Tassin. *J Membr Sci* **2009**, 338, 100.
216. G. Gebel, O. Diat, S. Escribano, R. Mosdale. *J Power Sources* **2008**, 179, 132.
217. S. Deabate, R. Fatnassi, P. Sistat, P. Huguet. *J Power Sources* **2008**, 176, 39.

218. G. L. Wick and W. R. Schmitt. *Mar Technol Soc J* **1977**, 11 (5–6), 16.

219. J. Veerman, M. Saakes, S. J. Metz, G. J. Harmsen. *Chem Eng J* **2011**, 166 (1), 256.

220. S. Loeb. *J Membr Sci* **1976**, 1 (1), 49.

221. R. E. Pattle. *Nature* **1954**, 174 (4431), 660.

222. J. W. Post, J. Veerman, H. V. M. Hamelers, G. J. W. Euverink, S. J. Metz, K. Nymeijer, C. J. N. Buisman. *J Memb Sci* **2007**, 288 (1–2), 218.

223. P. Długołecki, K. Nymeijer, S. Metz, M. Wessling. *J Memb Sci* **2008**, 319, 214.

224. J. Jagur-Grodzinski and R. Kramer. *Ind Eng Chem Process Des Dev* **1986**, 25 (2), 443.

225. P. Długołecki, A. Gambier, K. Nijmeijer, M. Wessling. *Environ Sci Technol* **2009**, 43, 6888.

226. P. Długołecki, J. Dabrowska, K. Nijmeijer, M. Wessling. *J Memb Sci* **2010**, 347, 101.

227. M. Turek and B. Bandura. *Desalination* **2007**, 205, 67.

228. H. Strathmann, J. J. Krol, H. J. Rapp, G. Eigenberger. *J Memb Sci* **1997**, 125, 123.

229. T. Xu. *Resour Conserv Recycl* **2002**, 37, 1.

CHAPTER 21

POLYMER GEL ELECTROLYTES: CONDUCTION MECHANISM AND BATTERY APPLICATIONS

IAN M. WARD and HUGH V. ST. A. HUBBARD

21.1 INTRODUCTION

Polymer gel electrolytes (PGEs) are essentially membranes that possess ionic transport properties comparable with those of common liquid ionic solutions. They are principally composed of polymer, organic solvent, and salt. They find applications not only in lithium batteries but also in other electrochemical devices such as supercapacitors, electrochromic devices, and so on.[1,2] PGEs have several advantages over their liquid counterparts; reducing leakage, they can act as a separator or can bind electrodes into a mechanically stabilizing laminate.

In principle, a PGE battery can be formed by sandwiching the PGE between a composite carbon (or lithium metal or silica) anode and a composite cathode containing a suitable lithiated metal oxide. PGEs for use in lithium batteries should aim for (1) high ionic conductivity at ambient and subambient temperatures, (2) good mechanical strength, (3) appreciable lithium transference number, (4) thermal and electrochemical stabilities, and (5) compatibility with electrodes. The PGE should allow for good cycle life at low temperatures and must withstand any internal pressure build up during the battery operations.[1,3–6]

PGEs are capable of incorporating a diverse range of materials, including mixed solvents, mixed salts, ceramic fillers, and so on. Several polymer hosts have been developed as promising PGE materials for industrial applications: poly(vinylidene fluoride) (PVDF), poly(vinylidene fluoride–hexafluoro propylene) (PVDF-HFP), poly(ethylene oxide) (PEO), poly(propylene oxide)

Ionic Interactions in Natural and Synthetic Macromolecules, First Edition. Edited by Alberto Ciferri and Angelo Perico.

(PPO), poly(acrylonitrile) (PAN), poly(methyl methacrylate) (PMMA), poly(vinyl chloride) (PVC), and polyacrylamide.[1,7–11] Table 21.1 lists the various polymers, solvents, and salts used to prepare the PGEs discussed in this chapter.

The ionic conductivity and electrochemical properties of the PGEs depend not only on the polymer host, but also on the host polymer membrane properties such as pore size, porosity, and pore distribution, which are strongly dependent on the processing method.[12] Different methods such as solvent casting,[13] plasticizer extraction,[14] thermoreversible gel crystallization,[7,8] foaming,[15] phase inversion (PI),[16,17] and electrospinning (ES)[12,18] have been adopted for preparing such porous membranes. With the PI method (in which the polymer is dissolved in solvent which is then washed out with a nonsolvent for the polymer and the polymer dried), microporous membranes of pore sizes between 0.1 and 1 μm with porosities greater than 70% can generally be obtained.[19] ES is a simple and efficient method to prepare polymer membranes with fully interconnected pore structure[12,18] and with morphologies quite different from those prepared by other methods. Electrospun materials can show marked improvements in electrical properties such as electrolyte uptake, ionic conductivity, anodic stability, interfacial resistance, cell capacity, and cyclability, but generally poorer mechanical properties and robustness.

Some PGEs are crystallizable (e.g., PVDF), some are noncrystallizable (e.g., PMMA), and both can act as separator or purely as containment to inhibit liquid flow which would otherwise lead to leakage. Both crystallizable and noncrystallizable polymers form gels through entanglements; crystallizable gels also form strong crystallite cross-links. For polymers, which form crystalline phases on cooling from the melt, thermoreversible gels can be made, which provide enhanced processability and a firm robust gel phase below the melt temperature. Thermoreversible gels formed from a range of crystallizable polymers including polyvinylidene fluoride (PVDF), polyethylene terephthalate (PET), polyhydroxybutyric acid (PHBA), polyoxymethylene (POM), poly(m-xylene adipamide), Nylon 6,6,[7,8] and polyacrylonitrile (PAN)[20–22] have been investigated. It has been found that PVDF gels have shown superior performance in terms of mechanical and conductive properties compared with gels made with other polymers and much research has focused on the PVDF gels.[7,8,23–27] This chapter will therefore focus principally on PVDF PGEs.

21.2 PGEs

21.2.1 Thermoreversible PVDF Gel Electrolytes

PVDF has many advantages as a polymer host for lithium battery applications. PVDF-based polymer electrolytes are highly anodically stable due to the presence of strong electron-withdrawing functional groups (–C–F) and the polymer itself has a moderate dielectric constant ($\varepsilon = 8.4$), which helps for greater dissolution of lithium salts and thus can support a high concentration

TABLE 21.1. The Various PGE Materials Investigated

Polymer		Solvent		Salt
Poly(vinylidene fluoride)	PVDF	Propylene carbonate	PC	$LiBF_4$
Poly(vinylidene fluoride–hexafluoro propylene)	PVDF-HFP	Ethylene carbonate	EC	$LiPF_6$
Poly(methyl methacrylate)	PMMA	Diethyl carbonate	DEC	$LiClO_4$
Poly(acrylonitrile)	PAN	Diethoxyethane	DEE	$LiCF_3SO_3$
Poly(vinyl chloride)	PVC	N,N-dimethylformamide	DMF	LiBOB
Poly(ethylene oxide)	PEO	Tetraglyme		$LiAsF_6$
Poly(propylene oxide)	PPO	γ-Butyrolactone	GBL	$LiN(CF_3SO_2)_2$
Poly(acrylamide)		N-methyl pyrrolidone	NMP	$LiN(SO_2C_2F_5)_2$
Poly(ethylene terephthalate)	PET			$LiC(CF_3SO_2)_3$
Poly(hydroxybutyric acid)	PHBA			
Poly(oxymethylene)	POM			
Poly(m-xylene adipamide)				
Nylon 6,6				

Tetraethylene glycol dimethyl ether = (tetraglyme) [$CH_3(OCH_2CH_2)_4OCH_3$]
Lithium bis(oxalate)borate = (LiBOB)

of charge carriers.[1] Homopolymer PVDF forms crystalline phases on cooling from the melt. Thermoreversible PGEs based on PVDF have been extensively investigated.[7,8,23–27]

Thermoreversible PVDF PGEs are prepared from a mixture of polymer, solvent, and lithium salt, which is then heated and stirred until a homogeneous melt is formed; upon cooling, a thermoreversible gel is formed. Thermoreversible PVDF gels can be formed from a range of high boiling point solvents, for example, propylene carbonate (PC), ethylene carbonate (EC), N,N-dimethylformamide (DMF), tetraethylene glycol dimethyl ether (tetraglyme), γ-butyrolactone (GBL), and N-methyl pyrrolidone (NMP), either singly or in combination; more volatile solvents can also be incorporated, for example, DEE (diethoxyethane). A range of lithium salts have also been investigated: $LiClO_4$, $LiBF_4$, $LiCF_3SO_3$, $LiPF_6$, and lithium bis(oxalate)borate (LiBOB).[7,8,24–27]

Figure 21.1 shows the ionic conductivity behavior over a range of temperatures and compares the PVDF PGEs with the corresponding liquid electrolyte and some typical polymer electrolytes. Room temperature conductivities are above 10^{-3} S cm^{-1} for all the gel systems, and for gels made with polar solvents, such as DMF or PC, conductivities are greater than 10^{-2} S cm^{-1} at −20°C.

The PVDF–$LiCF_3SO_3$ with DMF or tetraglyme gels were thoroughly researched through thermal, mechanical, rheometric, ionic conductivity, vis-

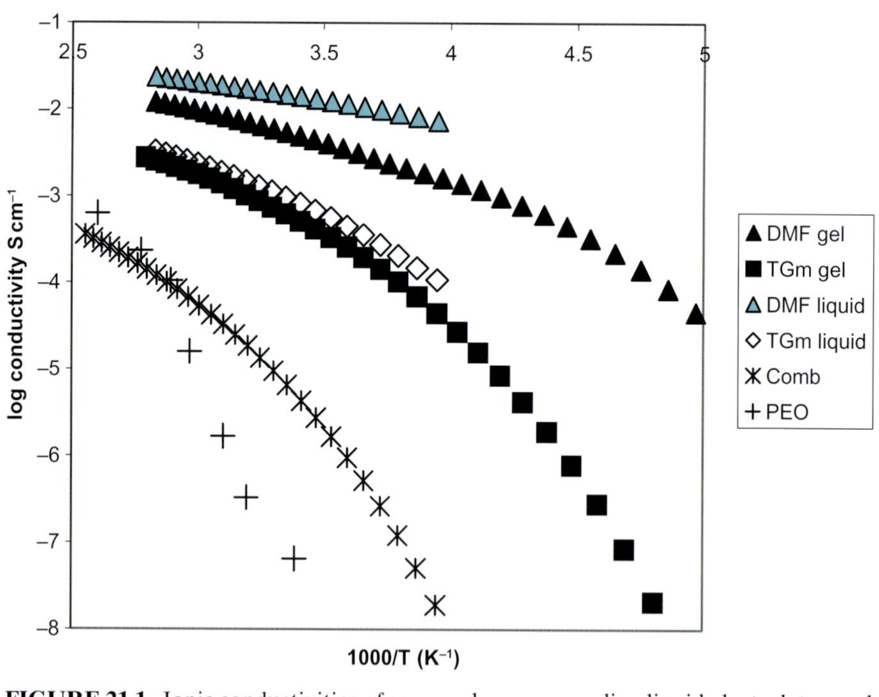

FIGURE 21.1. Ionic conductivities of some gels, corresponding liquid electrolytes, and polymer electrolyte complexes with $LiCF_3SO_3$ compared. T, temperature; TGm, tetraglyme; comb, poly(methoxy polyethylene glycol monomethacrylate).

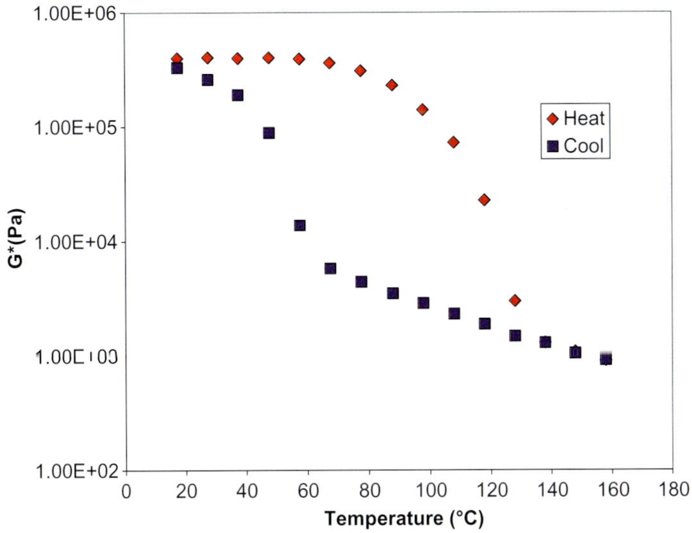

FIGURE 21.2. Complex modulus G* for 30% PVDF-tetraglyme-$LiCF_3SO_3$ sample (heating rate 2°C/min) showing the difference between melting and gelation temperatures.

cosity, and nuclear magnetic resonance (NMR) investigations.[7,8,24,25,27] Two solvents, DMF and tetraglyme, were chosen to represent limits of solvents that are suitable for lithium ion conduction; DMF is a highly polar solvent and tetraglyme is a poor solvent for Li ions and model analog for the PEO polymer electrolytes; the properties of solvents readily suitable for PGEs generally lie between those of these two solvents.

Figure 21.2 shows the complex modulus G* of the thermoreversible gels as a function of temperature. The figure clearly shows that the temperature at which the gel melts is very much higher than that at which the gel is formed on cooling. The gelation behavior of these thermoreversible gels is a function both of temperature and time. This unique property of these gels is exploited in the fabrication of cells by extrusion lamination of the PGE.

Gels with polymer concentrations between 15 and 40% (by weight) maintain their mechanical rigidity up to temperatures well above 100°C. However once melted, the gel structure only reforms at much lower temperatures. The variation of ionic conductivity of salted gels with temperature shows no such hysteresis, and it is concluded that the ionic conductivity is independent of the mechanical state of the gel.[7]

Differential scanning calorimetry (DSC) and dynamical mechanical analysis have been used to show that the majority of the network junctions of the gel are crystalline in nature.[8] Wide angle X-ray diffraction has revealed that when no salt is included in the gel, these crystal junctions are almost an order of magnitude larger in their lateral dimensions than when salt is present. The salt is thought to nucleate crystallization. The modulus is reduced by the inclusion of salt while the apparent crystallinity is only slightly reduced by the presence of salt; the discrepancy is attributed to the formation of small crystalline particles that are not incorporated in the network junctions.

Thermal, mechanical, rheometric, ionic conductivity, viscosity, and NMR investigations including T_2 and $T_{1\rho}$ measurements (as discussed in detail below)[7,8,24–27] indicate a structure for these gels containing up to four regions:

1. Crystalline PVDF also acting as junction cross-links.
2. Interlamellar amorphous PVDF.
3. Solvated chains of PVDF in the form of liquid electrolyte solvation sheaths around the amorphous PVDF.
4. A pure liquid electrolyte component.

21.2.2 Other PVDF-Based Gel Electrolytes

Various other methods of forming PVDF gels have also been explored, including plasticizing, electrospun PVDF nanofiber webs, foaming, and PI.

PGEs formed by plasticizing PVDF with a low molecular weight organic solvent and an ionic salt have been studied.[28–30] The structure of these materials is different from the thermoreversible gels with lower ionic conductivities in the region of 10^{-5} S cm^{-1} at ambient temperature.

Porous PGEs based on PVDF have been prepared via a foaming technology using salicylic acid as a foaming agent.[15] The pores were evenly distributed with an average diameter of 400 nm and room temperature conductivities up to 5×10^{-3} S cm^{-1}.

PVDF gels in the form of nanofiber webs with ionic conductivities above 10^{-3} S cm^{-1} can also be prepared by ES where the large surface-area-to-volume ratio and high porosity are crucial to increase the ionic conductivity.[18,31] However, these gels have poor mechanical properties. These can be improved by thermal treatment which increases tensile strength, elongation at break, tensile modulus, and level of crystallinity and which also improves the dimensional stability.

PVDF polymer electrolyte membranes were prepared by a PI method using deionized water as the nonsolvent.[16] The PVDF was dissolved in NMP and cast as a thin film on a Teflon sheet. PI of the film was carried out by washing with deionized water and further drying under vacuum at 80°C for 24 hours. The PGE electrolyte was formed by the infusion of tetra(ethylene glycol) dimethyl ether/1,3-dioxolane with LiCF3SO3 and evaluated for discharge capacity and cycle performance of Li/S cells at room temperature.

21.2.3 PVDF-HFP-Based Gel Electrolytes

PGEs based on a noncrystallizable (or weakly crystallizable) copolymer of PVDF-HFP have been extensively studied,[1,32] particularly due to their use in the Bellcore process.[33] This process has been pioneering in the development of PGE batteries.

Poly(VDF-co-HFP) gel polymer electrolytes have typically been prepared by dissolving polymer, lithium salt, and low molecular weight organic plasticizers in a low boiling point solvent,[34] and the resulting slurry is cast as a film. This process, however, requires a moisture-free environment because of the hydroscopic nature of the lithium salt. The Bellcore technology[33] attempts to circumvent these problems by using a liquid activation and extraction process to prepare the polymeric electrolyte material for plastic lithium ion batteries. The Bellcore process is a two-step method. PVDF-HFP copolymer is added to acetone together with dibutyl phthalate (DBP). The mixture is stirred vigorously and the solution cast onto a glass plate, covered by a Petri dish and left for slow evaporation. The resulting membranes are immersed in diethyl ether and left overnight under stirring to remove the DBP. The membranes are then dried in a vacuum dryer at 60°C for 24 hours, transferred to a dry box, and immersed in the solution of the liquid electrolyte (typically 1M $LiPF_6$ in ethylene carbonate/diethyl carbonate [EC/DEC]), for 24 hours to ensure saturation of the pores with the liquid electrolyte.

PGEs using PVDF-HFP have been extensively studied with a wide range of electrolyte solutions commonly used in the battery industry: particularly EC, PC, and DEC with the salts $LiN(CF_3SO_2)_2$, $LiCF_3SO_3$, $LiBF_4$, and $LiClO_4$.[1,35,36] As expected, the ionic conductivities and both diffusion coefficients of lithium (D^+_{Li}) and fluoride species (D^-_F) decrease with increase of polymer content. Comparison[37] of PGE made with films formed from PVDF-HFP with those from PVDF show that those from the PVDF-HFP host exhibited higher conductivity and transport number than those from PVDF. The higher conductivity of PVDF-HFP films was attributed to the higher amorphous content as there are two randomly mixed monomers providing greater mobility for the Li^+ ions. On the other hand, PVDF has higher degree of crystallinity, which results in lower conductivity but often better mechanical properties.[8]

Hybrid organic–inorganic PGEs for lithium batteries based on a PVDF-HFP polymeric matrix have been studied extensively.[38,39] These employ ceramic fillers typically nanometer-sized silicas up to 10 weight percent (wt %).[40] The addition of even small quantities of inorganic nanofiller causes an increase in ionic conductivity, ion transference number, and also improves the lithium electrode–electrolyte interfacial stability. This leads to less capacity fade and better high rate performance and is ascribed to the formation of a protective layer to cover the active material in the electrode during cycling, reducing electrolyte decomposition.[41]

Nanocomposite polymer electrolytes comprising room temperature ionic liquids (RTILs) with nano-sized ceramic fillers hosted in electrospun poly(VDF-co-HFP) membranes have also been prepared.[12] These exhibit higher ionic conductivity at room temperature, good interfacial and oxidation stability at 5.5–6 V, and enhanced charge–discharge performance.[40] RTILs have been used as electrolyte components in lithium rechargeable cells, electric double-layer capacitors, fuel cells, and solar cells. However as with electrospun PVDF PGEs, those prepared with electrospun PVDF-HFP show improved electrical properties, but are mechanically quite poor.[12]

It is also found[42] that the addition of small quantities of polymerizable organic additives—for example, thiophene, 3,4,5-ethylenedioxythiophene (EDOT), and biphenyl—to the GPE can reduce the interfacial resistance in the cell during cycling, resulting in reduced capacity fade and better high rate performance.

21.2.4 PMMA-Based Gel Electrolytes

PGEs based on PMMA have been widely studied since the early 1990s for potential use in lithium ion cells and for electrochromic windows.[43–47]

These totally amorphous noncrystalline gels have the specific advantage of excellent transparency, elasticity, and good adhesion,[48,49] especially to smooth glass surfaces without syneresis, which make them excellent candidates for large area electrochromic devices. They have also been studied for use in lithium batteries.

The use of PMMA as a gelatinized agent in $LiC1O_4$–PC electrolytes was first reported by Iijima et al.[50] The gels are most usually formed by dissolving the PMMA in PC/EC with a suitable alkali salt and cast as a sticky film.

The rheological and electrochemical properties of PMMA–$LiClO_4$–PC membranes were obtained by Bohnke et al.[43–45] The addition of PMMA greatly increases the viscosity of the macromolecular solution by up to nine orders of magnitude from 0 to 50 wt %. However the conductivity was reduced upon the addition of PMMA; at ambient temperature the ionic conductivity decreased from 5×10^{-3} to 5×10^{-5} S cm^{-1} over the same weight percent range.

The experimental results show that some change occurs in the conduction path of the gel around 30–35 wt % of polymer. For low concentrations of PMMA, the gelatinized electrolyte can be considered as a liquid electrolyte encaged in a polymer matrix. The conduction paths are preserved through the polymer matrix and thus the conductivity is very close to the liquid electrolyte, although the macroscopic viscosity of the gel is increased because of the presence of a polymer matrix. However, for high concentrations the decrease in ionic conductivity and an increase in activation energy at high concentration are attributed to some interactions created between the polymer chain and the ionic species. Lithium ions can be solvated either by PC or by the carbonyl group of the PMMA, with these interactions becoming very strong at 45 wt %. Cross-linking between two or several polymer chains can occur through the cation solvation, leading to a decrease of the ionic mobility and to an increase of the activation energy of conduction. However, NMR investigations suggest that although the immediate environments of the cations and anions differ in the gel and in the liquid, there is no direct evidence that the ions may be partially solvated by the PMMA.[46] The glass transition temperature Tg increases because of a decrease in the polymer motion and the macroviscosity of the gel material increases as salt concentration increases. These gels can then be considered as elastomeric materials where long-range displacements are sometimes difficult to occur as in solid electrolytes.

DSC, NMR, and electrical conductivity studies have been made by Stallworth et al.[46,51] for the gel electrolytes synthesized from PMMA, with EC, PC, and various lithium salts ($LiClO_4$, $LiAsF_6$, $LiN(CF_3SO_2)_2$). DSC and ^7Li, ^{19}F, and ^{75}As NMR spectroscopy reveal that these gel electrolytes exhibit single-phase behavior, as evidenced by their single glass transitions and by dramatic changes in NMR linewidth, which occur in the vicinity of the glass transition. Although the ionic conductivities of the gels approach those of liquid electrolytes above room temperature, the NMR results indicate that the immediate environments of both the cations and anions differ significantly in the gel and in the liquid. Thus the presence of microscopic regions of pure liquid electrolyte in the gel can be ruled out.

This observation emphasizes the fundamental difference between these amorphous PMMA gels and crystallizable PVDF gels for which there is clear evidence in the latter for the presence of microscopic regions of pure liquid electrolyte in the gel.[25,27] Thus PDVF gels for the same polymer content tend to exhibit slightly higher ionic conductivities. Moreover, in PMMA there is a long-term tendency to flow because the gelation is due to only entanglements and a consequent increase in viscosity, whereas PVDF gels also contain crystallized polymer cross-links that can only be broken through melting.

Studies of the electrochemical stability of PMMA gels with a lithium metal electrode show that although better stability is obtained with PMMA gels when compared with PAN, the voltammetric results and the efficiency tests revealed that a consistent fraction of lithium is lost upon cycling and thus a large excess of lithium would be required eventually to assure an acceptable life to the battery.[52] One useful way of improving the conductive and electrical properties of gels is through the addition of ceramic fillers, for example, nano-sized Al_2O_3[47] or TiO_2[53] particles up to 8 wt %, which enhances the diffusion of the cations, resulting in an increase in ionic conductivity.

21.2.5 PAN-Based Gel Electrolytes

As with amorphous PMMA gel electrolytes, PAN gel electrolytes offer homogeneous, hybrid electrolyte films in which the salt and the plasticizer are molecularly dispersed.[20–22] Watanabe et al.[20,21] used PAN plasticized with EC:PC complexed with $LiClO_4$ and concluded that the PAN host is inactive in the ionic transport mechanism but acts as a matrix for structural stability. Fully amorphous gels of PAN–EC–$LiClO_4$ show room-temperature conductivities of the order of 10^{-3} S cm^{-1} at ambient temperature. The PAN-based gels were found to have lithium ion transference numbers above 0.5 due to the absence of oxygen atoms in the PAN matrix. Also the transference number was enhanced to 0.7 when lithium salts such as $LiAsF_6$ or $LiN(CF_3SO_2)_2$ were used.[52]

Cyclic voltammetry studies of PAN–EC:PC with various Li–salt solutions of $LiN(CF_3SO_2)_2$, $LiAsF_6$, $LiCF_3SO_3$, and $LiPF_6$ revealed that these gel electrolytes have an inherent oxidation stability window exceeding 5 V versus Li_+/Li.[54] Despite their poor compatibility with lithium metal anodes, the several advantages of PAN-based electrolytes—for example, high ionic conductivity of the order of 10^{-3} S cm^{-1} at room temperature, good electrochemical stability to 5 V, and a lithium transference number around 0.6—make them suitable candidates for use in lithium-ion battery applications.

21.3 ROUTES TO A MOLECULAR UNDERSTANDING OF IONIC CONDUCTIVITY: NMR MEASUREMENTS

NMR has proved to be a useful tool to gain an understanding of ionic conductivity in electrolyte solutions and PGEs. It is appropriate to discuss the several key techniques in turn to include pulse field gradient spin echo (PFG) measurements of self-diffusion, electrophoresis measurements of ionic mobility, and T_1 and $T_{1\rho}$ measurements.

21.3.1 PFG Measurements of Self–Diffusion

In a PFG NMR experiment, a spin echo is formed at time 2τ by the application of a $(\pi/_2)(+x)$ radiofrequency (rf) pulse at time zero, followed by a $(\pi/2)(+y)$ rf pulse at time τ. The amplitude of the spin echo is attenuated by adding two identical, square-shaped intense magnetic field gradient pulses of duration δ and magnitude +G to the spin-echo pulse sequence. The first gradient pulse, applied between the rf pulses, rapidly dephases the spins, while the second gradient pulse, applied following the $\pi(+y)$ rf pulse at a time Δ after the first, undoes the dephasing effect of the first gradient pulse.

The echo amplitude is attenuated by an amount determined by how much the position of the spin has changed by self-diffusion in the time interval Δ. A very small field gradient pulse G_{min} can be applied to obtain stable "unattenuated spin echoes." The attenuation of the spin echo is then:

$$Ln\left\{\frac{M(G)}{M(G_{\min})}\right\} = -\gamma^2\delta^2 D(G^2 - G_{\min}^2)\left(\Delta - \frac{\delta}{3}\right), \tag{21.1}$$

where D is the self-diffusion coefficient, $M(G)$ is the magnitude of the spin echo in the presence of gradient pulses of magnitude $+G$, and $M(G_{\min})$ is the magnitude of the "unattenuated" spin echo.

The measurement of diffusion coefficients allows the calculation of the mean displacement of the diffusing species on different experimental timescales. Using a random walk argument,[55] it can be shown that the mean displacement d of a diffusing particle in a time t can be approximated to:

$$d = \sqrt{\overline{\Delta d^2}} = \sqrt{6Dt}, \qquad (21.2)$$

where $\sqrt{\overline{\Delta d^2}}$ is the mean square displacement.

The timescale for the diffusion experiments is the diffusion time of the PFG experiments, given by $t = \Delta - \delta/3$, which is of the order of 40 ms. For electrolyte solutions based on $LiCF_3SO_3$ in either tetraglyme or DMF, displacements in the range 3–30 μm were calculated[56] for the tetraglyme and DMF molecules, with slightly smaller displacements for the Li^+ and CF_3SO_3 ions, but still of the order of a few micrometers.

A key application of the diffusion measurements is to determine the degree of ionic association, by comparing the measured ionic conductivity σ with that calculated from the Nernst–Einstein equation. This gives:

$$\sigma = \frac{nq^2}{kT}[D(\text{cation}) + D(\text{anion})], \qquad (21.3)$$

where n is the number of anions and/or cations per unit volume, q is the charge on each ion, and D(cation) and D(anion) are the diffusion coefficients of the cation and anion, respectively.

Typically the ionic conductivity σ_{calc} calculated from Equation 21.3 will be greater than that measured σ_{meas} because the NMR measurements do not distinguish between a free ion and an ion pair or correlated motion of anions or cations in neutral pairs or clusters.

Defining the measure of ionic association by a parameter Γ, we have:

$$\sigma_{\text{meas}} = \sigma_{\text{calc}}(1 - \Gamma),$$

where $\Gamma = 0$ for full ion dissociation and $\Gamma = 1$ for ion association in neutral pairs or clusters.

Studies of the conductivities of lithium perchlorate $LiClO_4$ and lithium tetrafluoroborate $LiBF_4$ in tetraglyme were undertaken by Hall et al.[57] as a model for ionic conductivity in polymers such as polyethylene oxide (PEO). The results suggested that the predominant species in solution were ion pairs and that the major charged species may be ion triplets.

This result was confirmed by NMR PFG spin-echo measurements by Clericuzio et al.[58] for $LiBF_4$/polyether(PE), tetraglyme, or PC polymer electrolytes.

The Nernst–Einstein equation showed conductivities calculated from D_{Li} and D_{F}, the diffusion coefficients for the lithium ion and fluorinated anion, respectively, to be greater than those observed by direct measurements of ionic conductivity. This was attributed to short-lived correlated motions of cations and anions (ion–ion associations). The deviation parameter Γ was greater for tetraglyme than for PC, and decreased with increasing temperature, very markedly for tetraglyme. D_F was twice D_{Li} at room temperature but at 60°C, $D_F \approx D_{Li}$, perhaps due to the decrease in solvation sheath size with increasing temperature.

In parallel with this, Boden et al.[59] undertook NMR PFG measurements of lithium triflate $LiCF_3SO_3$ in polyethylene glycol (PEG) with molecular weight 60 as a model for PEO. Comparison of NMR data and ionic conductivity showed that at low salt concentrations, ion pairs, triplets, and so on occurred, but at high salt concentrations, the behavior was akin to a molten salt, with the ions so close that they cannot behave as ion pairs, and so on, but the system can be regarded as a polymer dissolved in a molten salt.

In a later investigation, Williamson and coworkers undertook PFG measurements on $LiCF_3SO_3$ in tetraglyme or DMF for O:Li concentrations of from 100:1 to 12:1 for tetraglyme and from 100:1 to 3:1 for DMF.[56]

It was found that:

1. D_F was always greater than D_{Li} due to some degree of association of the lithium ions with the solvent.
2. D_F and D_{Li} became equal at high salt concentrations either due to ionic association or correlated motions of the ions. In both the tetraglyme and DMF systems, there was increasing ionic association with increasing temperature.

Two possible explanations were offered for this. First, by considering the free energy arguments that $\Delta G° = \Delta H° - T\Delta S°$, a negative $\Delta H°$ could give rise to a decrease in free ions. Secondly, a reduction in the dielectric constants with increasing temperature would favor the formation of ionic pairs. With regard to the effect of salt concentration, Γ increased with increasing salt concentration for DMF as might be expected, but fell for tetraglyme, consistent with the conclusion of Boden et al. that the electrolyte is behaving more like an ionic melt.[59]

Contemporaneous with the PFG measurements reviewed above, PFG measurements were undertaken by Hayimiza et al. and Aihara et al.[60–63] on gel electrolytes composed of lithium salts (predominantly $LiBF_4$ but also several others) and cross-linked acrylated PEO with molecular weight 4000 and GBL or PC as solvent. Application of the Nernst–Einstein equation to a comparison between the PFG measurements and ionic conductivities showed high degrees of ion association (Γ values from 0.72 to 0.45). D_{Li} was always less than D_F, similar to the results observed above. Lithium transport numbers were ~0.4 for solutions and ~0.2 for gels. Further papers describe results for a very wide range of salts with GBL as solvent ($LiPF_6$, $LiBF_4$, $LiSO_3CF_3$, $LiN(CF_3SO_2)_2$, and LiBOB). A general conclusion was that the Li ion diffusion was governed by solvent diffusion, hence the bulk viscosity, but was insensitive to the composition of the anions. The degree of association always increased with increasing salt concentration. Furthermore, the degree of association in gels was lower than in the liquid electrolytes. It was suggested that the polymer matrix interacts with the Li ions, and the differences in association suggest that the PEO matrix behaves like a solvent in the gel electrolyte, as previously proposed by Boden et al.[59] for their studies of tetraglyme.

Finally, another early paper on self-diffusion is that by Johansson et al.,[64] who used the static fringe field gradient method[65] to study low molecular weight $LiCF_3SO_3$ PEG_{10} and $LiN(CF_3SO_2)_2$ PEG_{10} with 10 ether oxygens/Li^+ ion. Their conclusions were that the Li cation transport correlated with polymer chain diffusion, and that the anion diffusion was faster than that of the Li cations even though the anions are the larger species. Application of the Nernst–Einstein equation showed similar differences between calculated and measured ionic conductivities for $LiCF_3SO_3$ PEG_{10} and confirmed their agreement with the proposal of Boden et al. that this is due to correlated motions of the cations and anions. Interestingly, no discrepancies were observed for $LiN(CF_3SO_2)_2$ PEG_{10} where no ion pairs exist. These workers also undertook T_1 measurements (to be discussed in this chapter later) and proposed that these showed that the polymer chains undergo local conformational transformations in addition to translational diffusion.

Further research, including PEG measurements of ionic diffusion, combined with conductivity measurements and Raman spectroscopy, have been reported by Adebahr et al.[53] The addition of nano-sized TiO_2 particles to PMMA-EC:PC-$LiClO_4$ gels was shown to give a small but significant enhancement of Li ion mobility. Because the PFG measurements were only undertaken in the Li ions, it was assumed that $D_{Li} = D_{anion}$. The Nernst–Einstein equation showed a large discrepancy between calculated and measured conductivities and the Raman measurements showed that ion-pair formation is constant and small. It was suggested that the filler particles interact with the anions. A later paper[66] used PFG measurements to study Li ion diffusion and solvent diffusivity in polyelectrolyte gels based on N,N-dimethyl acrylamide with added TiO_2 nanoparticles and a zwitterionic compound. Surprisingly, it was found that there was an increase in lithium ion mobility with a decrease in solvent mobility, suggesting a different pathway for the transport mechanism.

Watanabe and coworkers[67] have reported ion transport properties of lithium ionic liquids and their ion gels prepared by incorporating poly(ethylene oxide-co-propylene oxide) triacrylate macromonomer and lithium ionic liquids based on lithium trifluoroacetic acid-n, where n is an oligo-ethylene glycol monoethyl ether (n = 3, 4,and 7.2). The Nernst–Einstein equation based on PFG measurements and conductivities showed very high degrees of ion association and the self-diffusion coefficients were low (10^{-8} cm^2 s^{-1}) due to the high viscosity of the ionic liquids. However, room temperature conductivities as high as 10^{-4} S cm^{-1} were reported.

Stephan and Saito[68] undertook PFG measurements on PVDF-HFP polymer electrolytes prepared by the PI technique developed by Bellcore. The gel formation process was considered to occur in several distinct steps. First, the electrolyte solution enters the cavities present in the host solvent. Next, the trapped solutions in the cavities reach saturation. Finally, the fully swollen polymer gel reaches equilibrium and provides maximum conductivity. Stephan and Saito reported two values for the ion diffusion: slow diffusion attributed to trapped solution in the cavities, and fast diffusion attributed to the solution in the swollen polymer gel.

21.3.2 Correlations between Ion Diffusion and Viscosity Measurements: Use of Stokes–Einstein Equation and Walden Product

Another useful route to a molecular understanding of the ionic conductivity of the electrolytes is to explore the possible relationship between the conductivity and the macroscopic viscosity.

The Stokes–Einstein equation relates the diffusion coefficient D of the mobile species (i.e., ions and solvent molecules) to the viscosity η, giving:

$$D = \frac{kT}{6\pi a\eta}, \qquad (21.4)$$

where a is the effective radius of the diffusion species.

The Walden product is the product of the molal conductivity $\Lambda = \sigma/\mathrm{m}$, where m is the molality in mole per kilogram, and the viscosity η.[69] The combination of Stokes–Einstein and Nernst–Einstein equations (Eqs. 21.4 and 21.3) leads to the relationship:

$$\Lambda\eta \propto \frac{\delta e^2}{6\pi a}, \qquad (21.5)$$

where δ is the fractional salt dissociation ($\delta \propto n/m$). Alternatively, the conductivity σ is given by:

$$\sigma = \left(\frac{e^2}{6\pi a}\right) n\varphi, \qquad (21.6)$$

where $\varphi = \eta^{-1}$ is the fluidity and n is the number density of charge carriers, which is determined by the concentration of salt in the electrolyte.

Southall et al.[23] determined the correlations between ionic conductivity and viscosity in liquid electrolytes based on $LiCF_3SO_3$/DMF or tetraglyme.

The DMF electrolytes showed a monotonic fall in Λ with increasing salt concentration, which is attributed to two factors: increasing ionic interaction and increased viscosity. The behavior of the tetraglyme electrolyte is more complex but was observed previously in solvents of low dielectric constant[70,71] and in tetraglyme complexed with other salts.[57,72]

At very low level salt concentrations Λ falls with increasing salt concentration before passing through a minimum and rising again (Fig. 21.3). At higher salt concentrations the molal conductivity passes through a maximum before

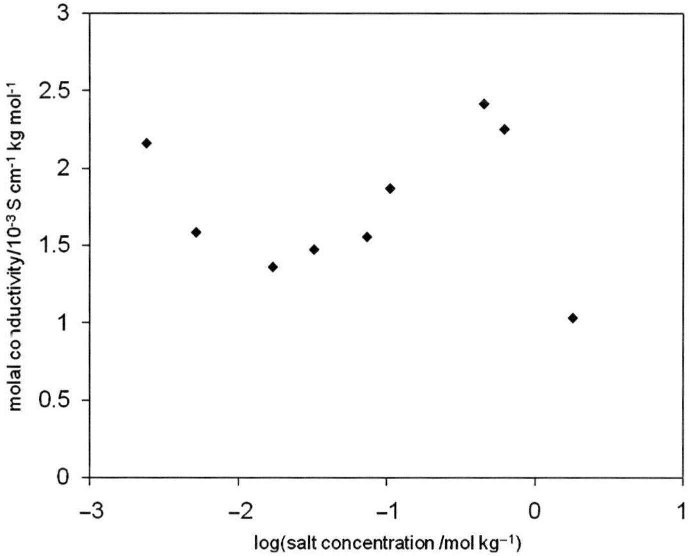

FIGURE 21.3. Molal conductivity versus log salt concentration for tetraglyme with $LiClO_4$.

finally falling again. The initial fall in Λ is attributed to the formation of ion pairs, whereas the final fall at high salt concentrations is attributed to the increase in solution viscosity. Southall et al.[23] explored the effect of temperature as well as salt concentration on the Walden product. In both tetraglyme and DMF electrolytes, $\Lambda\eta$ falls as temperature rises, that is, ionic association increases at higher temperatures in agreement with previous studies.[73,74] The change in $\Lambda\eta$ with salt concentration is however different for these two systems. In the case of DMF, $\Lambda\eta$ falls with increasing concentration, which implies that the ionic association increases, whereas in the tetraglyme electrolyte $\Lambda\eta$ rises monotonically. This suggests that the ionic association decreases. Although this result is surprising, it is in agreement with previous research by Watanabe et al.,[75] McLin and Angell,[76] and Mendolia and Farrington.[77] This result is consistent with the view that tetraglyme electrolytes at high levels of salt concentration are better considered as ionic melts. A point that has been raised in this regard is that the results of infrared and Raman spectroscopic studies of ion association suggest increasing numbers of ion pairs with increasing salt concentration. These results may be misleading with respect to ionic conductivity and the NMR data which measure ion diffusion over distances of the order of micrometers, whereas the spectroscopic data are sensitive to the relative proximity of the anions and cations.

Finally, it is relevant to summarize the research of Ward and coworkers on PFG measurements of PVDF gel electrolytes incorporating either tetraglyme or DMF, which enables direct comparison with the liquid electrolyte systems discussed above.[24,25,56,78] Consistent with measurements of the transverse relaxation time T_2, these gel electrolytes can be considered to contain regions of liquid electrolyte so that the results obtained for the latter with regard to the ionic association are relevant. The major effect of the PVDF gels is to reduce the ionic conductivities and diffusion coefficients of the protons, fluorine, and lithium to about half that of the corresponding liquid electrolytes. Similarities in behavior of the liquid and gel systems are greater than any differences: proton diffusion is greater than fluorine and lithium in both liquids and gels, the temperature dependence is Arrhenius with similar activation energies, and the degree of ionic association Γ and its variation with salt concentration are similar. However, there are some detailed differences in the activation energies and the degree of ionic association. A major difference is observed in the temperature dependence where there is a decrease in ionic association with temperature in the PVDF gel electrolytes, whereas the liquid electrolytes showed an increase in ionic association. It has been proposed that these results show that the PVDF does interact with the solvent electrolyte in the gels. This issue is discussed below when recent $T_{1\rho}$ measurements on the PVDF gels are reported.

21.3.3 Electrophoresis Measurements

Although much valuable information has been obtained from the PFG measurements, as extensively discussed above, it has been emphasized that these measurements measure the mobilities of all the ions including those which are associated, hence the value of the Nernst–Einstein equation in determining the degree of association. It is, however, very desirable to determine the mobility in an electric field, that is, undertake electrophoresis measurements.

Dai and Zawodzinski[79] undertook electrophoresis measurements by measuring the phase shift Φ of the spin-echo for lithium ions

$$\Phi = \gamma \Delta G \delta t_{Li} \frac{1}{CFA} \tag{21.7}$$

where γ is the nuclear gyromagnetic ratio, δ is the magnetic gradient pulse width, Δ is the time between the two gradient pulses, G is the magnetic field strength, I is the current, A the sample cross sectional area, C is the salt concentration, F is the Faraday constant and t_{Li} is the lithium transference number. The gel electrolyte was based on PVDF-HFP copolymer/PC with a solution to polymer weight ratio of 10:3 and a range of lithium salts including $LiCF_3SO_3$, $LiN(CF_3SO_2)_2$, $LiN(SO_2C_2F_5)_2$ and $LiC(CF_3SO_2)_3$.

For $LiCF_3SO_3$, t_{Li} was 0.30–0.35 for a temperature range of 30–60°C. Somewhat larger transfer numbers were found for the other salts, up to a value of 0.45 for $LiC(CF_3SO_2)_3$. It was concluded that larger anions resulted in higher values of t_{Li}. For $LiCF_3SO_3$, t_{Li} decreased with salt concentration.

In similar electrophoresis measurements by Kataoka et al.[80] on PVDF-HFP/EC:PC gels containing 1 mol $LiN(CF_3SO_2)_2$, the transport number of the lithium ions was determined as 0.34. In both Dai and Zawodzinski[79] and Kataoka et al.,[80] the transport numbers were significantly different from those measured by straight PFG measurements.

These lithium transport numbers for gels are quite similar to those obtained for DMF solutions containing $LiCF_3SO_3$ by Hubbard et al. (unpublished).

21.3.4 Studies of Gel Structures by $T_{1\rho}$ Measurements

Although it is recognized that the gel electrolyte structures are more complex than a two-phase gel-sol system, there is comparatively little molecular understanding of this issue. This led Hubbard et al.[27] to undertake $T_{1\rho}$ measurements on PGEs of composition 30 wt % PVDF-PC, 30 wt %PVDF-DMF, and 30 wt %PVDF-DMF-$LiCF_3SO_3$(1M), and on the pure solvents PC, DMF, and DMF-$LiCF_3SO_3$(1M) as functions of temperature and frequency.

Four components could be recognized in the decay of magnetization curve:

1. Crystalline PVDF—a very slow short-time component.
2. Interlamellar amorphous PVDF—intermediate but still a short-time component.
3. Solvated chains of PVDF—intermediate time.
4. A long-time solvent component.

The following conclusions could be made:

1. As expected, the short $T_{1\rho}$ value for the crystalline phase is relatively constant in different gels.
2. The interlamellar amorphous PVDF phase contains a significant amount of solvent, confirmed by examining dried gels.
3. There is a solvated PVDF component, probably in the form of a solvent sheath around the PVDF chain.
4. There is a solvent-rich phase which is not pure solvent.
5. There is little difference between salted and unsalted gels except for the increased viscosity of the former.

These results confirm the observations that the ionic conductivity of the gels mirror those of the electrolyte solutions, but lower levels of conductivity are observed in the gels. They also offer a molecular explanation for the increased conductivity of the gels with increasing temperature, compared with the liquid electrolyte where the conductivity falls. It does seem likely that this can be correctly attributed to ions in the solvent sheath around the PVDF, which can be released at higher temperatures.

21.3.5 T_1 and T_2 Measurements

Johansson et al.[64] undertook spin-lattice relaxation measurements on the Li cation, ^{16}F in the anion, and on the polymer chain protons in $LiCF_3SO_3$ PEG_{10} and $LiN(CF_3SO_3)_2$ PEG_{10}. In all cases the relaxation could be described by a simple exponential. The relaxation rate $1/T_1$ was fitted to the Bloembergen, Purcell, and Pound (BPP)[81] model with correlation times τ_c fitted to the Arrhenius equation:

$$\tau_c = \tau_0 \exp\left(\frac{E_A}{RT}\right). \tag{21.8}$$

Similar values for E_A in the range 14–26 kJ mol^{-1} were obtained for the different species. These were about one-half those obtained for the diffusion coefficients from PFG experiments. It can be concluded that there are activated processes occurring other than those relating to translational diffusion.

In a related study, Williamson et al.[82] undertook T_1 measurements of ^{7}Li, and ^{19}F and protons in tetraglyme/$LiCF_3SO_3$ and DMF/$LiCF_3SO_3$ liquid electrolytes.

From comparisons with values of diffusion coefficients from PFG measurements and solution viscosity data, it was concluded that in general there are contributions to T_1 from both transitional and rotational motions. More specifically, the fluorine T_1 values for DMF/$LiCF_3SO_3$ and tetraglyme/$LiCF_3SO_3$ liquid electrolytes could be quite well modeled by rotational motion, following BPP theory for identical spin 1/2 nuclei separated by a distance r,where:

$$\left(\frac{1}{T_1}\right)_{\text{rot}} = \frac{3}{2}\frac{\gamma^4\hbar^2}{r^6}\tau_c, \tag{21.9}$$

and the correlation time τ_c is given by:

$$\tau_c = \frac{4\pi\eta a^2}{3kT}, \tag{21.10}$$

for spheres of radius a in a liquid of viscosity η. It was even possible to calculate fluorine T_1 values on the basis of Equations 21.9 and 21.10 above using measured viscosity data, effective spherical radii of the triflate ions from the Stokes–Einstein equation, and an interfluorine distance of 2.2 Å. Calculated values were of the order of a few seconds, comparable with those determined experimentally by inversion-recovery measurements.

Abragam theory[83] for the calculation of T_1 due to translational motions leads to:

$$\left(\frac{1}{T_1}\right)_{\text{trans}} = \frac{\pi}{5}\frac{N\gamma^4\hbar^2}{aD} \tag{21.11}$$

where N is the number of spins per unit volume. Substitution of the experimentally determined diffusion coefficients and effective radii calculated from the Stokes–Einstein equation into Equation 21.3 again produced values for the proton T_1 of the order of seconds, comparable with those observed experimentally.

In a rather different publication, Greenbaum et al.[84] undertook T_1 measurements in conjunction with PFG measurements of diffusion on PVDF-co-HFP gel electrolytes containing layered nanoparticles such as fluorohectorite. ^{19}F and ^{1}H spectra both showed two peaks and there was no T_1 minimum for ^{7}Li, suggesting that the presence of the filler leads to a heterogeneous cation environment and hence a broad distribution of correlation times.

21.4 COMMERCIAL ISSUES FOR PGEs

From a commercial point of view, it can be argued that the key important factors determining the performance of a battery, in descending order of importance, are:

1. Safety
2. Cost
3. Performance

It is therefore appropriate to review the potential advantages of using a PGE as the separator under these three headings.

21.4.1 Safety

The most obvious advantage of a PGE is that it can largely eliminate the presence of a volatile liquid in both the production process[11] and the final product. In the production process, liquid electrolyte can be fed into a twin-screw extruder to be mixed with powdered polymer (e.g., PVDF) to produce a slurry that can then be introduced into a single screw extruder to produce a melt extruded film. The film is then passed between the electrodes so the final solid gel electrolyte is obtained. The advantages of a polymer electrolyte battery include less critical packaging in terms of being able to use plastic for packaging rather than a metal container. In terms of subsequent handling, the polymer battery is clearly able to withstand puncture and mechanical deformation without the release of dangerous liquid electrolyte.

21.4.2 Cost

There are two advantages of a gel electrolyte in terms of cost. First, the gel acts as both a separator and as an ion conductor so that the cost of a separator is eliminated. Second, it is possible to follow the route of continuous production, which has been satisfactorily demonstrated for prototype rechargeable lithium batteries. This will be a major advantage for the production of large batteries, where the present conventional rate of assembling batteries by final introduction of the liquid electrolyte is both tedious and potentially involves high safety requirements.

21.4.3 Performance

The energy density of a rechargeable lithium battery is essentially determined by the total number of lithium ions that can be shuttled between the cathode and anode when the battery is discharged. The lithium ion content is then quoted per unit volume or more often per unit total mass of the battery. It is important that all the lithium ions are active, that is, ions are not lost by being absorbed on "passivation layers" on the surface of the anode. Also ideally, the ionic charge is transported by movement of the lithium ions, that is, the transference number is as close to unity as possible. This is not usually achieved, as we have seen from the discussion above.

Another important factor is the rate of charge and discharge. so that the battery can be fully charged in a short time and the current can be drawn in a short time. Figure 21.4 shows typical results for a prototype cell made with a PGE separator, for two discharge rates C/t, where t is the time in hours to fully discharge the cell.[26] Furthermore, this cycling performance should be repeatable for many cycles with minimum reduction in performance (i.e., battery capacity).

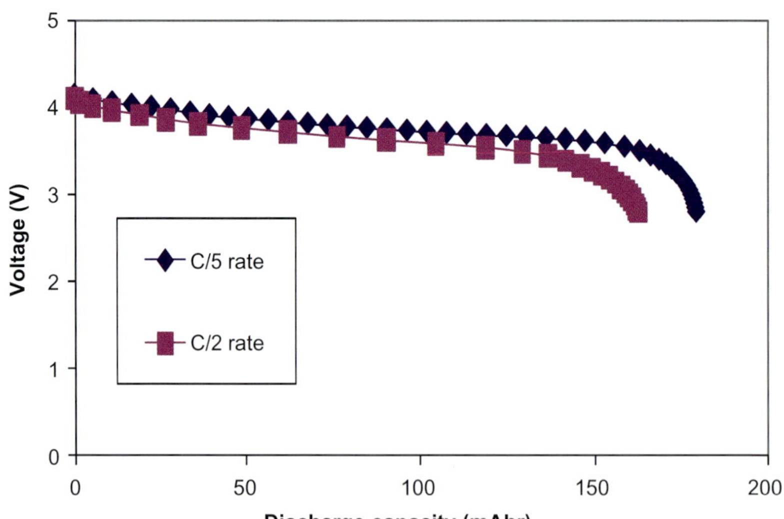

FIGURE 21.4. Discharge capacity (mAhr) versus voltage (V) for two discharge rates, C/5 and C/2 for a prototype cell with a thermoreversible PVDF PGE separator.

21.5 SUMMARY

PGEs are reviewed with particular reference to their application as conducting separators for rechargeable lithium batteries. Gel electrolytes based on polyvinylene fluoride (PVDF) have been studied extensively but other polymers, including PMMA and PAN, are also reviewed.

NMR measurements are shown to be a vital tool for obtaining a molecular understanding of the ionic conductivity, with pulsed field spin echo measurements being especially useful. The gel electrolyte conductivities can be understood in terms of the conductivities of electrolyte solutions and the gel structure based on NMR $T_{1\rho}$ measurements. The combination of Nernst–Einstein and Stokes–Einstein equations is extremely valuable.

Finally, the essential ingredients required for battery performance, that is, safety, cost and performance, are discussed.

REFERENCES

1. A. M. Stephan. *Eur Polym J* **2006**, 42, 21.
2. P. Raghavan, J. W. Choi, J. H. Ahn, G. Cheruvally, G. S. Chauhan, H. J. Ahn, C. Nah. *J Power Sources* **2008**, 184, 437.
3. B. Scrosati. *Applications of Electroactive Polymers*. Chapman Hall, London, 1993.
4. F. M. Gray. *Polymer Electrolytes. RSC Materials Monographs*. The Royal Society of Chemistry, Cambridge, 1997.
5. J. R. MacCallum and C. A. Vincent. *Polymer Electrolytes Reviews-I*. Elsevier, London, 1987.
6. J. R. MacCallum and C. A. Vincent. *Polymer Electrolytes Reviews-II*. Elsevier, London, 1987.
7. A. M. Voice, J. P. Southall, V. Rogers, K. H. Matthews, G. R. Davies, J. E. McIntyre, I. M. Ward. *Polymer* **1994**, 35, 3363.
8. A. M. Voice, G. R. Davies, I. M. Ward. *Polymer Gels Networks* **1997**, 5, 123.
9. S. A. Dobrowski, G. R. Davies, J. E. McIntyre, I. M. Ward. *Polymer* **1991**, 32, 2887.
10. J. E. McIntyre, I. M. Ward, H. V. St. A. Hubbard, V. Rogers. Patent PCT/GB92/01781, 1992.
11. I. M. Ward, H. V. St. A. Hubbard, P. L. Carr, F. Tyldesley. Patent PCT/GB01/03937, 2001.
12. P. Raghavan, X. Zhao, J. Manuel, C. Shin, M.-Y. Heo, J.-H. Ahn, H.-S. Ryu, H.-J. Ahn, J.-P. Noh, G.-B. Cho. *Mater Res Bull* **2010**, 45, 362.
13. Z. Wang and Z. Tang. *Mater Chem Phys* **2003**, 82, 16.
14. J. W. Choi, J. H. Kim, G. Cheruvally, J. H. Ahn, K. W. Kim, H. J. Ahn, J. U. Kim, J. Ind. *Eng Chem* **2006**, 12, 939.
15. H. P. Zhang, P. Zhang, G. C. Li, Y. P. Wu, D. L. Sun. *J Power Sources* **2009**, 189, 594.
16. J. W. Choi, J. K. Kim, G. Cheruvally, J. H. Ahn, H. J. Ahn, K. W. Kim. *Electrochim Acta* **2007**, 52, 2075.

17. V. Arcella, A. Sanguineti, E. Quartarone, P. Mustarelli. *J Power Sources* **1999**, 81–82, 790.
18. S. S. Choi, Y. S. Lee, C. W. Joo, S. G. Lee, J. K. Park, K. S. Han. *Electrochim Acta* **2004**, 50, 339.
19. H. Wang, H. Huang, S. L. Wunder. *J Electrochem Soc* **2000**, 147, 2853.
20. M. Watanabe, M. Kanba, K. Nagaoka, I. Shinohara. *J Appl Polym Sci* **1982**, 27, 4191.
21. M. Watanabe, M. Kanba, K. Nagaoka, I. Shinohara. *J Polym Sci Polym Phys Edn* **1983**, 21, 939.
22. K. M. Abraham and M. Alamgir. *J Electrochem Soc* **1990**, 137, 1657.
23. J. P. Southall, H. V. St. A. Hubbard, S. F. Johnston, V. Rogers, G. R. Davies, J. E. McIntyre, I. M. Ward. *Solid State Ionics* **1996**, 85, 51.
24. I. M. Ward, M. J. Williamson, H. V. St. A. Hubbard, J. P. Southall, G. R. Davies. *J Power Sources* **1999**, 81–82, 700.
25. M. J. Williamson, H. V. St. A. Hubbard, I. M. Ward. *Polymer* **1999**, 40, 7177.
26. I. M. Ward, H. V. St. A. Hubbard, S. C. Wellings, G. P. Thompson, J. Kaschmitter, H. P. Wang. *J Power Sources* **2006**, 162, 818.
27. H. V. St. A. Hubbard and I. M. Ward. *Polym Preprints* **2008**, 49, 709.
28. E. Tsuchida, H. Ohno, K. Tsunemi. *Electrochimica Acta* **1983**, 28, 591.
29. M. Watanabe and M. Kanba. *Makromol Chem Rapid Comm* **1981**, 2, 741.
30. K. Tsunemi, H. Ohno, E. Tsuchida. *Electrochim Acta* **1983**, 28, 833.
31. C. Yang, Z. Jia, Z. Guan, L. Wang. *J Power Sources* **2009**, 189, 716.
32. H. S. Kim and S. I. Moon. *J Power Sources* **2005**, 141, 293.
33. A. S. Gozdz, C. W. Schmutz, J. M. Tarascon, P. C. Warren. US Patent 5540741, **1997**.
34. A. M. Stephan and D. Teeters. *Electrochim Acta* **2003**, 48, 2143.
35. C. Capiglia, Y. Saito, H. Kataoka, T. Kodama, E. Quartarone, P. Mustarelli. *Solid State Ionics* **2001**, 131, 291.
36. A. M. Stephan, S. G. Kumar, N. G. Renganathan, M. A. Kulandainathan. *Eur Polym J* **2005**, 41, 15.
37. D. Saika and A. Kumar. *Electrochim Acta* **2004**, 49, 2581.
38. A. Zalewska, M. Walkowiak, L. Niedzicki, T. Jesionowski, N. Langwald. *Electrochim Acta* **2010**, 55, 1308.
39. M. Stolarska, L. Niedzicki, R. Borkowska, A. Zalewska, W. Wieczorek. *Electrochim Acta* **2007**, 53, 1512.
40. J. K. Kim, G. Cheruvally, X. Li, J. H. Ahn, K. W. Kim, H. J. Ahn. *J Power Sources* **2008**, 178, 815.
41. S. M. Eo, E. Cha, D. W. Kim. *J Power Sources* **2009**, 189, 766.
42. J. A. Choi, S. M. Eo, D. R. MacFarlane, M. Forsyth, E. Cha, D. W. Kim. *J Power Sources* **2008**, 178, 832.
43. O. Bohnke, C. Rousselot, P. A. Gillet, C. Truche. *J Electrochem Soc* **1992**, 139, 1862.
44. O. Bohnke, G. Frand, M. Rezrazi, C. Rousselot, C. Truche. *Solid State Ionics* **1993**, 66, 97.
45. O. Bohnke, G. Frand, M. Rezrazi, C. Rousselot, C. Truche. *Solid State Ionics* **1993**, 66, 105.

46. P. E. Stallworth, S. G. Greenbaum, F. Croce, S. Slane, M. Salomon. *Electrochim Acta* **1995**, 40, 2137.
47. O. Krejza, J. Velicka, M. Sedlarıkova, J. Vondrak. *J Power Sources* **2008**, 178, 774.
48. J. Vondrak, M. Sedlarikova, J. Velicka, B. Klapste, V. Novak, J. Reiter. *Electrochim Acta* **2001**, 46, 2047.
49. J. Vondrak, M. Sedlarikova, J. Reitera, T. Hodalb. *Electrochim Acta* **1999**, 44, 3067.
50. T. Iijima, Y. Toyoguchi, N. Eda. *Denki Kagaku* **1985**, 53, 619. (in Japanese).
51. F. Croce, S. D. Brown, S. G. Greenbaum, S. M. Slane, M. Salomon. *Chem Mater* **1993**, 5, 1268.
52. G. B. Appetecchi, F. Croce, B. Scrosati. *Electrochim Acta* **1995**, 40, 991.
53. J. Adebahr, N. Byrne, M. Forsyth, D. R. MacFarlane, P. Jacobsson. *Electrochim Acta* **2003**, 48, 2099.
54. H. S. Choe, B. G. Carroll, D. M. Pasquarillo, K. M. Abraham. *Chem Mater* **1997**, 9, 369.
55. W. Jost. *Diffusion*. Academic Press, New York, 1960.
56. M. J. Williamson, J. P. Southall, H. V. St. A. Hubbard, G. R. Davies, I. M. Ward. *Polymer* **1999**, 40, 3945.
57. P. G. Hall, G. R. Davies, I. M. Ward, J. E. McIntyre. *Polymer Comm* **1986**, 198, 100.
58. M. Clericuzio, W. O. Parker, Jr, M. Soprani, M. Andrei. *Solid State Ionics* **1995**, 82, 179.
59. N. Boden, S. A. Leng, I. M. Ward. *Solid State Ionics* **1991**, 40, 261.
60. K. Hayamiza, Y. Aihara, S. Arai, W. S. Price. *Solid State Ionics* **1998**, 107, 1.
61. K. Hayamiza, Y. Aihara, S. Arai, W. S. Price. *Electrochim Acta* **2000**, 45, 1313.
62. Y. Aihara, S. Arai, K. Hayamiza. *Electrochim Acta* **2000**, 45, 1321.
63. Y. Aihara, T. Bardo, H. Nakagawa, H. Yoshida, K. Hayamiza, E. Akiba, W. S. Price. *J Electrochemical Soc* **2004**, 151 (1), A119.
64. A. Johansson, A. Gogoll, J. Tegenfeld. *Polymer* **1996**, 37, 1387.
65. D. E. Demer, A. Johansson, J. Tegenfeld. *J Magn Reson* **1994**, 110, 183.
66. N. Byrne, D. R. MacFarlane, M. Forsythe. *Electrochimica Acta* **2005**, 50, 3917.
67. H. Shobukawa, H. Tokuda, M. A. Hasan Susan, M. Watanabe. *Electrochim Acta* **2005**, 50, 3872.
68. A. M. Stephan and Y. Saito. *Solid State Ionics* **2002**, 148, 475.
69. J. O. M. Bockris and A. K. N. Reddy. *Modern Electrochemisty*. 1.Plenum Press, New York, 1970.
70. O. Popovych and R. P. T. Tomkin. *Non-Aqueous Solution Chemistry*. Wiley Interscience, New York, 1981.
71. A. Cisak and L. Werblan. *High Energy Non-Aqueous Batteries*. Ellis Horwood, New York, 1993.
72. G. G. Cameron, M. D. Ingram, S. A. Sorrie. *J Chem Soc Faraday Trans* **1987**, 183, 3345.
73. A. Wobbe. *J Elecrochem Soc* **1991**, 138, 2586.
74. L. M. Torrell, P. Jacobsson, G. Peterson. *Polym Adv Technol* **1993**, 4, 152.
75. M. Watanabe, J. Ikeda, I. Shinohara. *Polymer J* **1983**, 15, 175.

76. M. G. McLin and C. A. Angell. *J Phys Chem* **1991**, 95, 9464.
77. M. S. Mendolia and G. C. Farrington. *Chem Mat* **1993**, 5, 174.
78. M. J. Williamson, J. P. Southall, H. V. St. A. Hubbard, S. F. Johnston, G. R. Davies, I. M. Ward. *Electrochem Acta* **1998**, 43, 1415.
79. H. Dai and T. A. Zawodzinski. *J Electroanal Chem* **1998**, 459, 111.
80. H. Kataoka, Y. Saito, Y. Miyazaki, S. Deki. *Solid State Ionics* **2002**, 152–153, 175.
81. N. Bloembergen, E. M. Purcell, R. V. Pound. *Phys Rev* **1948**, 73, 679.
82. M. J. Williamson, J. P. Southall, I. M. Ward. *J Chem Phys* **1998**, 109, 7893.
83. A. Abragam. *Principles of Nuclear Magnetism*. Oxford University Press, New York, 1961; p. 302.
84. S. Abbrent, S. H. Chung, S. G. Greenbaum, J. Muthu, E. P. Giannelis. *Electrochem Acta* **2003**, 48, 2113.

SYMBOLS

a	radius of a cylindrical polyelectrolyte
b	bond length, segment length, separation charge distance on a polyelectrolyte
c_s	salt concentration
e	absolute electronic charge
E	electric field strength
F	Helmoltz free energy
G	Gibbs free energy
H	enthalpy
$\langle h^2 \rangle$	mean square end-to-end distance of a polymer chain
I	ionic strength
k_B	Boltzmann constant
L	chain extended length
l_B	Bjerrum length
l_p	persistence length
M	polymer molecular weight
N	number of ("effective") monomers or sites on a polymer
N_A	Avogadro number
Q	total charge of a macroion
R	molar gas constant
R	radius of a spherical macroion
T	absolute temperature
V	volume
Z	partition function

Ionic Interactions in Natural and Synthetic Macromolecules, First Edition. Edited by Alberto Ciferri and Angelo Perico.

Z	counterion valence
ε	relative dielectric constant
ζ	Manning linear charge density
θ	condensed counterions of valence Z per site
κ	Debye screening constant
ρ	charge density
σ	surface charge density
ψ	electrostatic potential

INDEX

Ionic Interactions in Natural and Synthetic Macromolecules, First Edition. Edited by Alberto Ciferri and Angelo Perico.